DISCRETE EVENT SYSTEMS 2004
(WODES'04)

A Proceedings volume from the 7th IFAC Workshop,
Reims, France, 22 – 24 September 2004

Edited by

J. ZAYTOON
CReSTIC, University of Reims Champagne Ardenne,
Reims, France

V. CARRÉ-MÉNÉTRIER
CReSTIC, University of Reims Champagne Ardenne,
Reims, France

C. CASSANDRAS
CISE, Boston University,
Boston, MA, USA

and

X. CAO
EEE, Hong Kong University of Science and Technology,
Hong Kong

Published for the

INTERNATIONAL FEDERATION OF AUTOMATIC CONTROL

by

ELSEVIER LTD

ELSEVIER Ltd
The Boulevard, Langford Lane
Kidlington, Oxford OX5 1GB, UK

Elsevier Internet Homepage
http://www.elsevier.com

Consult the Elsevier Homepage for full catalogue information on all books, journals and electronic products and services.

IFAC Publications Internet Homepage
http://www.elsevier.com/locate/ifac

Consult the IFAC Publications Homepage for full details on the preparation of IFAC meeting papers, published/forthcoming IFAC books, and information about the IFAC Journals and affiliated journals.

Library of Congress Cataloging in Publication Data

A catalogue record for this book is available from the Library of Congress

British Library Cataloguing in Publication Data

A catalogue record for this book is available from the British Library

ISBN 0-08-0441688
ISSN 1474-6670

To Contact the Publisher

Elsevier welcomes enquiries concerning publishing proposals: books, journal special issues, conference proceedings, etc. All formats and media can be considered. Should you have a publishing proposal you wish to discuss, please contact, without obligation, the publisher responsible for Elsevier's industrial and control engineering publishing programme:

Christopher Greenwell
Senior Publishing Editor
Elsevier Ltd
The Boulevard, Langford Lane
Kidlington, Oxford
OX5 1GB, UK

Phone: +44 1865 843230
Fax: +44 1865 843920
E.mail: c.greenwell@elsevier.com

General enquiries, including placing orders, should be directed to Elsevier's Regional Sales Offices – please access the Elsevier homepage for full contact details (homepage details at the top of this page).

Transferred to Digital Printing 2010

7th IFAC WORKSHOP ON DISCRETE EVENT SYSTEMS 2004

Sponsored by

International Federation of Automatic Control (IFAC)
Technical Committee on Discrete Event and Hybrid Systems (TC 1.3)

Co-Sponsored by

Technical Committee on Control Design (TC 2.1)
Technical Committee on Manufacturing Plant Control (TC 5.1)
GDR MACS of CNRS
Centre National de la Recherche Scientifique (CNRS)
French Ministry of Education, Research & Technology
University of Reims Champagne Ardenne
Region Champagne Ardenne
City of Reims
Department of Marne

Organized by

SEE and the *Centre de Recherche en STIC* (CReSTIC) of Reims University

International Program Committee (IPC)

Chairman: J. Zaytoon (F)
Co-Chairmen: X. Cao (RC-HK) – C. Cassandras (US)

H. Alla (Fr)
P. Antsaklis (US)
Z. Banaszak (Pl)
R. Boel (Be)
J.L. Boimond (Fr)
P. Borne (Fr)
L. Boullart (Be)
V. Carré-Ménétrier (Fr)
J. Cury (Br)
I. Demongodin (Fr)
B. De Schutter (Nl)
A. Di Febbraro (It)
A. Dolgui (Fr)
M. P. Fanti (It)
J.M. Faure (Fr)
L. Ferrarini (It)
J.L. Ferrier (Fr)
B. Gaujal (Fr)
A. Giua (It)
B. Grabot (Fr)
H. Guégen (Fr)
S. Haddad (Fr)
Y.C. Ho (US)
L. Holloway (US)
M. Jafari (US)
G. Juanole (Fr)
C. Jard (Fr)
N. Koussoulas (Gr)
V. Krebs (De)
R. Kumar (US)
A. Kusiak (US)
S. Lafortune (US)
K. Larsen (Dk)
B. Lennartson (Se)
J.J. Lesage (Fr)
J.J. Loiseau (Fr)
P.B. Luh (US)
J. Lunze (De)
T. Moor (De)
P. Moreaux (Fr)
G. Morel (Fr)
E. Niel (Fr)
M. Nourelfath (Ca)
G. Olsder (Nl)
A. Perkusich (Br)
J. Raisch (De)
N. Rakoto (Fr)
A. Ramírez (Mx)
L. Recalde (Es)
S. Reveliotis (US)
B. Rozoy (Fr)
K. Rudie (Ca)
E. Rutten (Fr)
C. Seatzu (It)
M. Silva (Es)
J. R. Silva (Br)
M. P. Spathopoulos (UK)
R.S. Sreenivas (US)
O. Stursberg (De)
J. Thistle (Ca)
D. Tilbury (US)
T. Ushio (Jp)
R. Valette (Fr)
P. Valigi (It)
J. van Schuppen (Nl)
F. Vazquez-Abad (Ca)
G. Villermain-Lecolier (F)
Y. Wardi (US)
X. Xie (Fr)
P. Yim (Fr)
Q. Zhao (RC)

Other Referees

L. Aguirre
E. Almeida
G. Alpan
K. Altisen
K. Barkaoui
G. Barrett
M. Boccadoro
L. Capra
A. Carrilho
S. Chafik
J.-M. Colom
O. Contant
D. Corona
B. Cottenceau
M. Dotoli
E. Dumitrescu
A. Duret-Lutz
C. Dutheillet
C. Dutilleul Simon
M. Fabian
P. Falkman
L. Fang
H. Flordal
S. Genc
D. Giglio
D. Gromov
T. E. Hodrus
I. Ihnatsenka
J. Julvez
O. Kamach
S. I. Kamau
M. Kluwe
S. Lahaye
A. B. Leal
S. Lee
M. Lhommeau
H. Lin
E. López-Mellado
H. Marchand
F. Martinelli
E. Mayer
L. Piétrac
J. Reger
P.-O. Ribet
B. Riera
E.E. Roubtsova
J.-M. Roussel
K. Schmidt
G. Schullerus
D. Simon
J. Stanczyk
P. Supavatanakul
A. Vahidi
V. Verdult
L. Xia
X. Xu

PREFACE

The interdisciplinary field of Discrete Event Systems combines different formalisms, methodologies and tools from control, computer science and operations research. The research activity in this field is driven by the needs of many different applications domains: manufacturing, process control, supervisory systems, transportation, software engineering, and so on. Discrete Event Systems is now a mature field and many interesting applications have been developed.

This volume contains the proceedings of IFAC Workshop on Discrete Event Systems, WODES'04, which is one of the most distinguished international events entirely devoted to discrete event systems. This 7th edition of the WODES follows earlier successful meetings in Prague (1992), Antibes (1994), Edinburgh (1996), Cagliari (1998), Ghent (2000) and Zaragoza (2002). It aims to provide researchers from different fields (control theoreticians and control engineers, software engineers and computer scientists, operations research specialists) with an opportunity to exchange information and new ideas, and to discuss new developments in the field of Discrete Event Systems.

Further to the review process, 77 papers were accepted from the 130 contributions that were received. Besides the three plenary lectures, this volume contains the 75 papers that were presented by their authors at the Workshop. The published papers are coming from France (20), USA (10), Italy (10), Germany (7), Canada (5), Netherlands (3.5), Brazil (3), Japan (2.5), Sweden (2.5), Tunisia (1.5), Belgium, Mexico, New Zealand, Poland, Slovenia, Spain, Turkey and UK (1), Romania, Czech Republic, Cyprus and Thailand (0.5).

The papers are organized in two tracks of sessions as well as a poster session, as follows:

- 3 invited lectures given by distinguished speakers: Oded Maler, whose talk is titled "On optimal and sub-optimal control in the presence of adversaries"; Stefan Kowalewski, whose talk is titled "Discrete control systems for cars – quality is more than correct function"; and Alessandro Giua whose talk, co-authored with Carla Seatzu, is titled "A systems theory view of Petri nets".
- 2 invited sessions on theory & application of supervisory control (organised by T. Ushio and N. Rakoto) and another invited session on theory & applications of colored Petri nets (organised by C. Seatzu and M.-P. Fanti).
- 9 contributed sessions covering the following topics related to Discrete Event Systems: modelling formalisms, verification, control synthesis, diagnosis, scheduling, performance evaluation and optimisation. 2 other sessions are related to the verification, analysis and control of hybrid systems.
- 1 poster session including 21 papers.

The editors are grateful to the members of the IPC and the outside reviewers for their valuable contribution to the selection process and for their help in keeping high the standards of the WODES. We also wish to express our sincere thanks to the members of the Organising Committee for their precious help. In particular, we thank the authors and all those who have assisted us by acting as scientific or financial sponsor or cosponsor.

We hope that this volume will be a valuable source of information on the state of the art in Discrete Event Systems not only for the workshop participants but also for the DES research community in general.

The editors

J. Zaytoon, V. Carré-Ménétrier, C. Cassandras and X. Cao

CONTENTS

KEYNOTE PAPERS

CONTROL SYNTHESIS PRACTICE

PETRI NETS

SUPERVISORY CONTROL: THEORY AND APPLICATIONS I

DIOIDS ALGEBRA (MIN,+), (MAX,+)

CONTROL OF HYBRID SYSTEMS

DIAGNOSIS OF DES I

SUPERVISORY CONTROL: THEORY AND APPLICATIONS II

SCHEDULING AND OPTIMIZATION

THEORY AND APPLICATIONS OF COLORED PETRI NETS

PERFORMANCE EVALUATION AND OPTIMIZATION

CONTROL SYNTHESIS

ELSEVIER
IFAC
PUBLICATIONS
www.elsevier.com/locate/ifac

ON OPTIMAL AND SUB-OPTIMAL CONTROL IN THE PRESENCE OF ADVERSARIES [1]

Oded Maler *

* CNRS-VERIMAG, *2, av. de Vignate, 38610, Gières, France.*
`Oded.Maler@imag.fr`
`www-verimag.imag.fr/~maler/`

Abstract: This paper constitutes a sketch of a unified framework for posing and solving problems of optimal control in the presence of uncontrolled disturbances. After laying down the general framework we look closely at a concrete instance where the controller is a scheduler and the disturbances are related to uncertainties in task durations.

1. INTRODUCTION

I would like to use this opportunity to present a paper free of any "new original results", not pretending to put something new under the sun. I will describe what I consider to be the essence of many activities concerned with the design of systems based on mathematical models, without, I believe, saying anything that is not known in some of the many disciplines and communities that occupy themselves with these issues. Yet, I think that putting all this together in the present form has some "synergetic" value that goes beyond a collection of informal definitions, results and algorithms.

The paper has two main parts. In the first I lay down a kind of a "special theory of everything" where system design is viewed as synthesizing an optimal strategy in some dynamic game. We start in Section 2 with a general discussion on how to *define* the performance of a system which is subject to external disturbances. In section 3 we introduce the generic model that we use, a dynamic "multi-stage" game between a controller and its environment. The restriction of the problem to behaviors of bounded length is the topic of Section 4 where it is reduced (for discrete time systems) to standard finite-dimensional constrained optimization. The classical solution technique known as dynamic programming is described in Section 5 followed, in section 6, by the alternative method of heuristic forward search.

In the second part of the paper I focus on one concrete instance of this scheme, the problem of scheduling under bounded uncertainty which is modeled as a discrete game on continuous time using the timed automaton model. Section 7 attempts to convince the reader that scheduling falls into the class of problems described in the first part. The job shop problem is described in Section 8 along with its traditional solution scheme based on non-convex "combinatorial" optimization. In Section 9 we show how this problem can be solved using shortest path algorithms on timed automata. Section 10 extends the problem by considering bounded uncertainty in task durations, while Section 11 sketches a dynamic programming algorithm that can find adaptive strategies that are better than worst-case strategies.

The topics discussed in this paper are treated by different disciplines and under diverse titles such as System Verification, Controller Synthesis, Sequential Decision Making, Game Theory, Markov Decision Processes, AI Planning, Optimal and Model Predictive Control, Shortest Path Algorithms, Dynamic Programming, Optimization, Differential Games and more, each with its own terminology. For example, what is called in one

[1] This research was supported in part by the European Community projects IST-2001-33520 CC (Control and Computation) and IST-2001-35304 AMETIST (Advanced Methods for Timed Systems).

context a strategy, can be called elsewhere a policy, a feed-back law, a controller or a dispatching rule. Even worse, the same term can mean different things to different audiences. I did my best to pick each time the term (or terms) that I felt are the most appropriate for the discussion and tried not to switch between terms too often. In any case I apologize for the potential inconvenience for those who are accustomed to their own specific terminology. I also tried not to bias the description toward my own automata-theoretic background.

2. STATIC OPTIMIZATION

Throughout this paper we will be concerned with situations that resemble two player games. One player, henceforth the *controller*, represents the system that we want to design while the other player, the *environment*, represents external disturbances beyond our control. The controller chooses actions $u \in U$, the environment picks $v \in V$ and these choices determine the outcome of the game. The controller wants the outcome to be as good as possible, according to some predefined criterion, while the environment, unless one is paranoid, is indifferent to the results. To illustrate the problematics of optimizing something that depends not only on one's own actions we start with a simple one-shot game of the type introduced in the seminal work of von Neumann and Morgenstern.

Let $U = \{u_1, u_2\}$, $V = \{v_1, v_2\}$ and let the outcome be defined as a function $c : U \times V \to \mathbb{R}$ which can be given as a table

c	v_1	v_2
u_1	c_{11}	c_{12}
u_2	c_{21}	c_{22}

We want to choose among u_1 and u_2 the one that minimizes c but since different choices of v may lead to different values we need to specify how to take these values into account. There are basically three generic approaches for evaluating our decisions:

- *Worst-case*: each action of the controller is evaluated according to the worst outcome that may result from taking the action:

$$u = argmin\ \max\{c(u, v_1), c(u, v_2)\}.$$

- *Average case*: the environment is modeled as a stochastic agent acting randomly according to a probability function $p : V \to [0, 1]$ and the controller actions are evaluated according to the expected value of c:

$$u = argmin\ p(v_1) \cdot c(u, v_1) + p(v_2) \cdot c(u, v_2).$$

- *Typical case*: the evaluation is done with respect to a *fixed* element of V, say v_1, which represents the most "typical" behavior of the adversary. This amounts to denying the existence of uncontrolled disturbances and the problem is reduced to ordinary optimization:

$$u = argmin\ c(u, v_1).$$

Before moving to dynamic games let us note what happens when c is a continuous function over continuous U and V. The average case analysis stays within the standard framework of continuous optimization, that is, optimization of a real-valued function, while the worst-case min-max analysis does not. This may partially explain why in domains such as continuous control stochastic disturbances are much more popular.

3. DYNAMIC GAMES

A dynamic game is a game where the players are engaged in an *ongoing interaction* extended over time. In the computer science context, the term *reactive systems*, coined by Harel and Pnueli, is used to denote such objects. A game is characterized by a state-space X and a dynamic rule of the form

$$x' = f(x, u, v)$$

stating that at each time instant the "next" value of x is a function of its current value and of the actions of both players. In this part of the paper we focus on *discrete time "synchronous" games* where such a dynamics is often written as a recurrence equation of the form

$$x_i = f(x_{i-1}, u_i, v_i)$$

but we keep in mind the existence of other models such as differential games on dense time defined via

$$\dot{x} = f(x, u, v)$$

or games with a more "asynchronous" flavor where actions may occur on a non-periodic time set (event-triggered rather then time-triggered) and where actions of both players need not occur simultaneously. Such asynchronous games will be used later to model scheduling problems.

We assume all games to start from an initial state x_0 and use the notation $\bar{x}$ for a state-sequence $x[0], x[1], \ldots, x[k]$. Likewise, we will use $\bar{u}$ and $\bar{v}$ for sequences of players actions. The predicate (constraint) $B(\bar{x}, \bar{u}, \bar{v})$ denotes the fact that $\bar{x}$ is the behavior of the system when the two players apply the action sequences $\bar{u}$ and $\bar{v}$, respectively:

$$B(\bar{x}, \bar{u}, \bar{v})\ \textit{iff}\ x[0] = x_0$$
$$x[t] = f(x[t-1], u[t], v[t])\ \forall t$$

It is sometimes useful to view the game as a *labeled directed graph* whose nodes are the elements of X and its edges are all the pairs (x, x') such that $f(x, u, v) = x'$ for some u and v. The reachable part of this transition graph is the subgraph obtained by restriction to elements of X for which a path from x_0 exists. An alternative useful notation for $B(\bar{x}, \bar{u}, \bar{v})$ is:

$$x[0] \xrightarrow{u[1],v[1]} x[1] \cdots \xrightarrow{u[k],v[k]} x[k].$$

There are many ways to assign performance measures to such behaviors in order to compare them and find the optimal one. For example, any cost function $c : X \to \mathbb{R}$ on states can be lifted to a cost function on sequences by letting

$$c(\bar{x}) = \sum_{t=1}^{k} c(x[t])$$

or

$$c(\bar{x}) = \max\{c(x[t]) : t \in 1..k\}.$$

Another useful measure is the minimal time to reach a goal set $F \subseteq X$:

$$c(\bar{x}) = \min\{t : x[t] \in F)\}.$$

In the more general case the cost function may also take into account the costs associated with the controller actions $\bar{u}$. The nature of cost functions depends on the application domain. In discrete verification $c(x)$ is typically a $\{0,1\}$-valued function indicating bad states, and checking invariance properties (whether the system always avoids those bad states) reduces to checking whether $c(\bar{x}) = 0$ for the max-extension of c to sequences. In continuous domains some quadratic functions on $\bar{x}$ or other "norms" are often used to indicate the distance of the sequence (trajectory) from a reference. Sometimes the choice of the cost function is influenced less by its adequacy for the problem and more by the existence of a corresponding optimization method, especially when the optimum is to be computed analytically.

4. BOUNDED HORIZON PROBLEMS

We will now restrict ourselves to situations where we compare only behaviors of a fixed finite length. There are several reasons to focus on bounded decision horizons. The first is that there are certain problems of the "control to target" or "shortest path" type, where all reasonable behaviors converge to a goal state in a bounded number of steps (but via paths of different costs). Another reason is the common sense intuition that as we look further into the future, our models becomes less reliable, and hence it is better to plan for a shorter horizon and revise the plan during execution (this is the basis of model-predictive control). Finally, bounded horizon problems in discrete time can be reduced to *finite dimensional* optimization problems.

We first illustrate the formulation of the problem for adversary-free situations with dynamics of the form $x' = f(x,u)$. In this case we look for a sequence $\bar{u} = u[1], \ldots, u[k]$ which is the solution of the constrained optimization problem

$$\min_{\bar{u}} c(\bar{x}) \text{ subject to } B(\bar{x}, \bar{u}).$$

Here we used a cost function based only on $\bar{x}$ while the fact that $\bar{x}$ is the result of following the dynamics f under control $\bar{u}$ is part of the *constraints*. For linear dynamics, specified by $x' = Ax + Bu$, and a linear cost function, the problem reduces to standard linear programming. In discrete verification where the dynamics and cost are defined logically, the problem reduces to Boolean satisfiability (this is the essence of bounded model checking).

If we dispose of a constrained optimization procedure for the domain in question, we can compute the desired $\bar{u}$. Note that in the absence of external disturbances $\bar{u}$ completely determines $\bar{x}$ and no feed-back from x is needed. The control "strategy" reduces to an open-loop "plan": at each time instant t apply the element $u[t]$ of $\bar{u}$. We could have rephrased it as a feed-back function (strategy) s defined over all $x[t]$ in $\bar{x}$ as $s(x[t]) = u[t+1]$ but this would be an overkill.

Let us re-introduce the adversary and use, without loss of generality, the worst-case criterion. We now need to find $\bar{u}$ which is the solution of

$$\min_{\bar{u}} \max_{\bar{v}} c(\bar{x}) \text{ subject to } B(\bar{x}, \bar{u}, \bar{v}).$$

Consider the case where $U = \{u_1, u_2\}$ and $V = \{v_1, v_2\}$ which is captured, for horizon 2, by the game tree of Figure 1. We assume, for simplicity, that the cost of each behavior is determined by its terminal state. In this case we can enumerate all the 4 possible control sequences and compute the cost they induce as:

$$\begin{aligned}
u_1u_1 &: \max\{c(x_5), c(x_6), c(x_9), c(x_{10})\}\\
u_1u_2 &: \max\{c(x_7), c(x_8), c(x_{11}), c(x_{12})\}\\
u_2u_1 &: \max\{c(x_{13}), c(x_{14}), c(x_{17}), c(x_{18})\}\\
u_2u_2 &: \max\{c(x_{15}), c(x_{16}), c(x_{19}), c(x_{20})\}
\end{aligned}$$

The sequence which minimizes these values is the optimal open-loop control that can be achieved. Using feed-back, however, one can do better. While the choice of $u[1]$ is done without any knowledge of the adversary's action, the choice of $u[2]$ is done *after* the effect of $v[1]$, that is, the value of $x[1]$, is known. Consider the case where $u[1] = u_1$ and we need to choose $u[2]$. If, for example, $\max\{c(x_5), c(x_6)\} < \max\{c(x_7), c(x_8)\}$ but $\max\{c(x_9), c(x_{10})\} > \max\{c(x_{11}), c(x_{12})\}$ then the optimal thing to do is to apply u_1 when $x[1] = x_1$ and u_2 when $x[1] = x_2$.

A *control strategy* is thus a function $s : X \to U$ telling the controller what to do at any reachable state of the game. The following predicate indicates the fact that $\bar{x}$ is the behavior of the

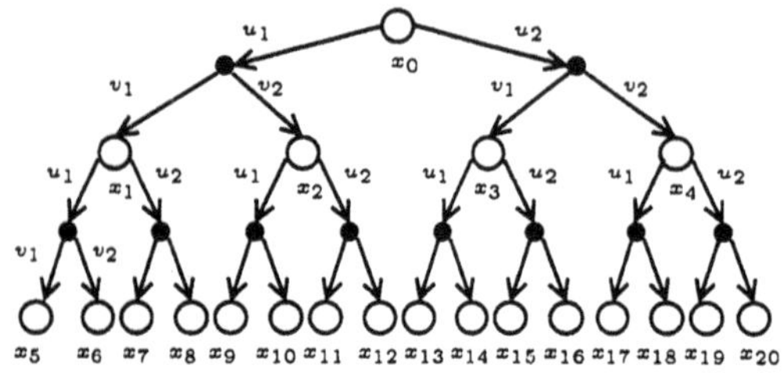

Fig. 1. A game tree of depth 2.

system in the presence of disturbance $\bar{v}$ when the controller employs strategy s:

$$B(\bar{x}, s, \bar{v}) \text{ iff } x[0] = x_0$$
$$u[t] = s(x[t-1]) \ \forall t$$
$$x[t] = f(x[t-1], u[t], v[t]) \ \forall t$$

Finding the best (worst-case) strategy s becomes the following second-order optimization problem:

$$\min_{s} \max_{\bar{v}} c(\bar{x}) \text{ subject to } B(\bar{x}, s, \bar{v}).$$

Finding an optimal strategy is usually much more difficult than finding an optimal sequence. Discrete finite-state systems admit $|U|^{|X|}$ potential strategies and each of them induces $|V|^k$ behaviors of length k. In continuous domains (on continuous time), where the enumeration of all strategies is not an option, such a strategy is the solution of a partial differential equation known as the Hamilton-Jacobi-Bellman-Isaacs equation. Note that a strategy need not be defined all over X, only for *elements reachable from* x_0 when the controller employs that strategy.

5. DYNAMIC PROGRAMMING

Dynamic programming, or backward value iteration, is a technique, advocated by Bellman, for computing optimal strategies in an incremental way. For discrete systems the algorithm is polynomial in the size of the transition graph, which is better than the exponential enumeration of strategies. However, as we will see later, this is not of much comfort in many situations where the transition graph itself is *exponential* in the number of system variables.

We will illustrate dynamic programming on the following shortest path problem. A subset F of X is designated as a target set, and a cost $c(x, u, v)$ is associated with each transition. The cost of a path

$$x[0] \xrightarrow{u[1],v[1]} x[1] \cdots \xrightarrow{u[k],v[k]} x[k]$$

from the initial state to a target state is

$$c(\bar{x}, \bar{u}, \bar{v}) = \sum_{t=1}^{k} c(x[t-1], u[t], v[t]),$$

and our goal is to find the strategy that minimizes the worst-case.

Dynamic programming uses an auxiliary function (value function, cost-to-go) $\overleftarrow{\mathcal{V}}$: $X \to \mathbb{R}$ such that $\overleftarrow{\mathcal{V}}(x)$ is the performance of the optimal strategy for the sub-game starting from x. For "leveled" acyclic transition graphs (where all paths that reach a state x from x_0 have the same number of transitions), $\overleftarrow{\mathcal{V}}$ admits the following backward recursive definition:

$$\overleftarrow{\mathcal{V}}(x) = 0 \qquad \text{when } x \in F$$

$$\overleftarrow{\mathcal{V}}(x) = \min_{u} \max_{v} (c(x, u, v) + \overleftarrow{\mathcal{V}}(f(x, u, v))).$$

In more general settings, including cycles, the value function is the fixed-point of the following iteration:

$$\overleftarrow{\mathcal{V}}_0(x) = \begin{cases} 0 & \text{when } x \in F \\ \infty & \text{when } x \notin F \end{cases}$$

$$\overleftarrow{\mathcal{V}}_{i+1}(x) = \min \left\{ \begin{array}{l} \overleftarrow{\mathcal{V}}_i(x), \\ \min_{u} \max_{v} (c(x, u, v) + \overleftarrow{\mathcal{V}}(f(x, u, v))) \end{array} \right\}$$

We recall that the choice of max and summation in this "local" operator is specific to this particular performance criterion. When max is replaced by weighted sum, we obtain the solution procedure for Markov decision processes, leading to a strategy with optimal expected value. When summation is replaced by max, you obtain essentially the backward synthesis algorithm for discrete event systems (automata). In this case, the value of $\overleftarrow{\mathcal{V}}_i$ stands for the characteristic function of the set of states from which the controller cannot postpone reaching a forbidden state for more than i steps.

This elegant procedure is guaranteed (if it converges) to find the optimal value $\overleftarrow{\mathcal{V}}(x_0)$ of the game, as well as the strategy that attains this optimum: just take for each x the u that achieves the local optimum. For finite-state systems with positive transition costs, finite convergence is guaranteed. The only deficiency of dynamic programming is the need to compute the function for too many states, sometimes states that are not reachable from x_0 at all, and sometimes states that are not reachable by any reasonable strategy. This prevents the straightforward application of the algorithm to systems having a huge state-space (Bellman's "curse of dimensionality").

6. FORWARD SEARCH

Shortest path problems (without an adversary) admit a dual *forward* procedure, due to Dijkstra. Here we use a backward value function $\overleftarrow{\mathcal{V}}$ such that $\overleftarrow{\mathcal{V}}(x)$ indicates the minimal cost for reaching x from x_0, and we want to compute $\overleftarrow{\mathcal{V}}(x)$ for $x \in F$. This is done iteratively using:

$$\overleftarrow{\mathcal{V}}(x_0) = 0$$

$$\bar{\mathcal{V}}(x) = \min_u (c(x', u) + \bar{\mathcal{V}}(f(x', u)))$$

where x' ranges over all the immediate predecessors of x. This algorithm is polynomial as well but, as noted before, this is not of much help for exponential transition graphs. The advantage of this procedure is the ability to apply intelligent search and sometimes find the optimum without exploring all the reachable states. If we do not insist on optimality, we can find reasonable solutions while exploring a small fraction of the paths.

To demonstrate this idea in a way that can be extended later for problems with adversaries, we view each incomplete path as a *partial strategy* defined only on states encountered along the path. We will store triples of the from $(s, x, \bar{\mathcal{V}})$ where s is a partial strategy, x stands for the last node in the path and $\bar{\mathcal{V}}$ is the cost for reaching x along the path. The first version of the algorithm explores all the paths (and all the partial strategies). We use a waiting list W in which we store partial paths that need to be explored further. The algorithm is given below:

$W := \{(\emptyset, x_0, 0)\}$
repeat
  Pick a non-terminal node $(s, x, \bar{\mathcal{V}}) \in W$
  for every $u \in U$ **do**
    $(s', x', \bar{\mathcal{V}}') :=$
      $(s \cup \{x \mapsto u\}, f(x, u), \bar{\mathcal{V}} + c(x, u))$
    Insert $(s', x', \bar{\mathcal{V}}')$ into W
  end
  Remove $(s, x, \bar{\mathcal{V}})$ from W
until W contains only terminal nodes

If new nodes are inserted at the end of W, the graph is explored in a breadth-first manner. The nodes explored by this procedure on the example of Figure 2 appear in the table below. The four entries for the terminal nodes are compared and the path/strategy with minimal cost is selected.

s	x	$\bar{\mathcal{V}}$
$\emptyset$	x_0	0
$\{x_0 \mapsto u_1\}$	x_1	$c(x_0, u_1)$
$\{x_0 \mapsto u_2\}$	x_2	$c(x_0, u_2)$
$\{x_0 \mapsto u_1, x_1 \mapsto u_1,\}$	x_3	$c(x_0, u_1) + c(x_1, u_1)$
$\{x_0 \mapsto u_1, x_1 \mapsto u_2,\}$	x_3	$c(x_0, u_1) + c(x_1, u_2)$
$\{x_0 \mapsto u_2, x_2 \mapsto u_1,\}$	x_3	$c(x_0, u_2) + c(x_2, u_1)$
$\{x_0 \mapsto u_2, x_2 \mapsto u_2,\}$	x_3	$c(x_0, u_2) + c(x_2, u_2)$

This algorithm can be modified in order to find an optimal strategy without exhaustive exploration. The idea is to associate with every partial strategy an *estimation function* which gives a lower-bound on the cost of any extension of the strategy. This function is defined for every $(s, x, \bar{\mathcal{V}})$ in W as $\mathcal{E}(s, x, \bar{\mathcal{V}}) = \bar{\mathcal{V}} + \underline{\mathcal{V}}(x)$ where $\underline{\mathcal{V}}$ is an under-approximation of the cost-to-go function $\bar{\mathcal{V}}$. This function can be derived from domain-

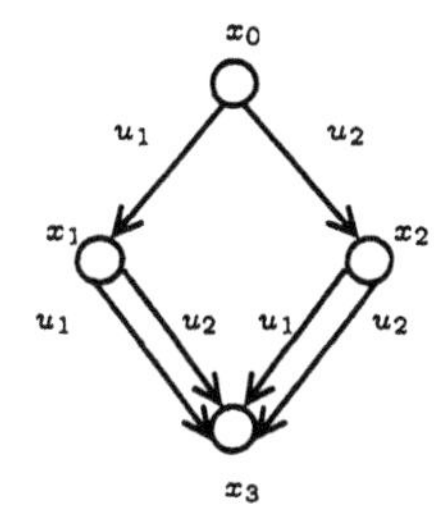

Fig. 2. A shortest path problem.

specific knowledge and it provides an optimistic estimation of the remaining cost to reach the target by any strategy that extends s. Note that as x gets deeper, the past component in $\mathcal{E}(x)$ becomes more dominant and the estimation — more realistic.

A *best-first* version of the algorithm maintains W ordered according to $\mathcal{E}$ and explores the more promising nodes first. Moreover, the algorithm can stop exploring when the value of $\mathcal{E}$ for the first element in W is larger than solutions that were already found. To increase the efficiency of the algorithm we can first perform depth-first random search to obtain some solutions at early stages of the search (branch and bound). Best-first search is guaranteed to find the optimum, and if we relax the optimality requirements, we can explore even less states, for example, by exploring only a subset of the successors of each node, or by stopping the algorithm when a solution smaller than some pre-specified value is found.

The adaptation of forward search to game graphs situations is done as follows. First let us redefine the value function as:

$$\bar{\mathcal{V}}(x) = \min_u \max_v (c(x', u, v) + \bar{\mathcal{V}}(f(x', u, v))).$$

Here, because of the adversary, every partial strategy results in a *set of states*. To avoid heavy notations we omit the cost from the reachable sets of states and the reader should keep in mind that every reachable x is, in fact, $(x, \bar{\mathcal{V}})$ where $\bar{\mathcal{V}}$ is the accumulated cost to reach x via the path in question. The set of successors of a state x via a controller action u is

$$f(x, u) = \{f(x, u, v) : v \in V\}.$$

The u-successor of (s, x) with s being a partial strategy is

$$\sigma((s, x), u)) = (s \cup \{x \mapsto u\}, f(x, u)).$$

Consider now a node of the form (s, L) where s is a partial strategy and L is the set of states reachable while following s. The successors of (s, L), that is, the partial strategies the extend s and their respective sets of reachable states, are all combinations of all possible choices of u for all $x \in L$:

$$\sigma(s, L) = \bigotimes_{x \in L} \{(\sigma(s, x), u) : u \in U)\},$$

where

$$L_1 \otimes L_2 = \{(s_1 \cup s_2, m_1 \cup m_2) : (s_1, m_1) \in L_1, (s_2, m_2) \in L_2\}.$$

Let is illustrate this on the game tree of Figure 1. The root node, $(\emptyset, \{x_0\})$, has two successors:

$$\sigma(\emptyset, \{x_0\}) = \{\sigma((\emptyset, x_0), u_1), \sigma((\emptyset, x_0), u_2)\}$$

$$= \left\{ \begin{array}{l} (\{x_0 \mapsto u_1\}, \{x_1, x_2\}) \\ (\{x_0 \mapsto u_2\}, \{x_3, x_4\}) \end{array} \right\}$$

Let us compute the successors of the first:

$$\sigma(\{x_0 \mapsto u_1\}, \{(x_1, x_2\}) =$$

$$\left\{ \begin{array}{l} (\{x_0 \mapsto u_1, x_1 \mapsto u_1\}, \{x_5, x_6\}), \\ (\{x_0 \mapsto u_1, x_1 \mapsto u_2\}, \{x_7, x_8\}) \end{array} \right\} \otimes$$

$$\left\{ \begin{array}{l} (\{x_0 \mapsto u_1, x_2 \mapsto u_1\}, \{x_9, x_{10}\}), \\ (\{x_0 \mapsto u_1, x_2 \mapsto u_2\}, \{x_{11}, x_{12}\}) \end{array} \right\}$$

which gives the following four nodes

$$\left(\begin{array}{l} \{x_0 \mapsto u_1, x_1 \mapsto u_1, x_2 \mapsto u_1\}, \\ \{x_5, x_6, x_9, x_{10}\} \end{array} \right),$$

$$\left(\begin{array}{l} \{x_0 \mapsto u_1, x_1 \mapsto u_1, x_2 \mapsto u_2\}, \\ \{x_5, x_6, x_{11}, x_{12}\} \end{array} \right),$$

$$\left(\begin{array}{l} \{x_0 \mapsto u_1, x_1 \mapsto u_2, x_2 \mapsto u_1\}, \\ \{x_7, x_8, x_9, x_{10}\} \end{array} \right)$$

and

$$\left(\begin{array}{l} \{x_0 \mapsto u_1, x_1 \mapsto u_2, x_2 \mapsto u_2\}, \\ \{x_7, x_8, x_{11}, x_{12}\} \end{array} \right)$$

An exhaustive version of the forward search algorithm for game graphs is exponential in the size of the graph (unlike the backward procedure) but with the help of an estimation function on sets of nodes, one can hope to prune many branches of the search tree using best-first search. Note that the procedure can be adapted easily to the average-case criterion by replacing sets of states by probabilities on states.

The alert reader might have traced some inaccuracies in the description which was specialized to game trees, rather to the more general case of game graphs. In the latter case, a set of reachable states my include several copies of the same state, and all but the one with the worst cost can be removed. We also assumed that a strategy cannot reach the same state twice along the same path. A strategy that does that has an infinite value and can be discarded. For clarity of exposition we assumed that both players can apply any element of U and V at any time, while in reality some actions are possible only at certain states.

This concludes the first part of the paper in which we presented three approaches to solve optimal control problems in the presence of an adversary. The first approach was based on bounded horizon and finite dimensional optimization. The two other approaches were based on propagation of costs along paths, either backward (dynamic programming) or forward. These approaches were described using discrete U and V and their adaptation to continuous domains is not straightforward, unless they are discretized. Discretization of U may lead to loss of optimality while discretization of V — to optimistic values of the chosen strategy. In the following sections we demonstrate this approach on an interesting type of a game played with discrete values over continuous time, namely, scheduling under uncertainty in task durations.

7. SCHEDULING AS A GAME

Scheduling problems appear in diverse situations where the use of bounded resources over time has to be regulated. A scheduler is a mechanism that decides at each time instant whether or not to allocate a resource to one of the tasks that needs it. Unfortunately, scheduling research is spread over many application domains, and in many of them problems are often solved using domain specific methods, without leading to a more general theory (except for, perhaps, operations research where scheduling is treated as a static optimization problem, similar to the the approach described in Section 4). In this section we will reformulate scheduling in our terminology of dynamic two player games.

On one side of the problem we have the *resources*, a set $M = \{m_1, \ldots, m_k\}$ of "machines" that we assume to be fixed. On the other side we have *tasks*, units of work that require the allocation of certain machines for certain durations in order to be accomplished. In a world of unbounded resources scheduling is not a problem: each task picks resources as soon as it needs them and terminates at its earliest convenience. When this is not the case, two tasks may need the same resource at the same time and the scheduler has to resolve the conflict and decide to whom to give the resource first. The tasks may be related to each other by various inter-dependence conditions, the most typical among them is *precedence*: a task can start only after some other tasks (its predecessors) have terminated. In this paper we assume the set of tasks to be fixed and known in advance.

To model such situations as dynamic games we need first to fix the state-space. For our purposes we take the state of the system at any given instant to include the states of the tasks (waiting, active, finished), the time already elapsed (for active tasks) and the corresponding states of the machines (idle, or busy when it is used by an active task). The actions of the scheduler are of two types, the first being actions of the form *start(p)* which means allocating a machine m to task p so that it can execute. The effect of such an action on a state where p is enabled (all its predecessors have terminated) and m is idle, is to make p active and m occupied. Let us denote this set of actions by S. The other "action" of the scheduler is to do nothing, denoted by $\perp$. In this case the active tasks continue to execute, the waiting tasks keep on waiting and time elapses. The actions of the environment consist of similar waiting and a set of actions of the form *end(p)* whose effect, when the task spent enough time in an active state, is to move the task to a terminal state and release the machine. We assume that the environment is deterministic, that is, every *end(p)* transition occurs exactly d time after the *start(p)* where d is the pre-specified duration of the task (later, we will relax this assumption). In this case the strategy can be viewed as a single schedule, a function $s : \mathbb{R}_+ \rightarrow S \cup \{\perp\}$. For all but a finite number of time instances we have $s(t) = \perp$ and the schedule is determined by a finite number of start times for each task.

8. DETERMINISTIC JOB SHOP SCHEDULING

A job shop problem consists of a finite set $J = \{J^1, \ldots, J^n\}$ of jobs to be processed on a finite set M of machines. Each job J^i consists of a finite sequence of tasks to be executed one after the other, where each task is characterized by a pair of the form (m, d) with $m \in M$ and $d \in \mathbb{N}$, indicating the required utilization of machine m for a fixed time duration d. Each machine can process at most one task at a time and, due to precedence constraints, at most one task of each job can be active at any time. Tasks cannot be preempted once started. We want to determine the starting times for each task so that the total execution time of all jobs (the time the last task terminates) is minimal.

As an example consider the problem

$$J^1 : (m_1, 4), (m_2, 5) \quad J^2 : (m_1, 3)$$

which exhibits a conflict on m_1. This conflict can be resolved in two ways, either by giving priority to J^1 or to J^2 (schedules s_1 and s_2 of Figure 3). The length induced by s_1 is 9 and it is the optimal schedule for this example. The hardness of the problem stems from the fact that sometimes an optimal schedule is achieved by *not* executing a task as soon as it is ready in order to keep the machine free for another task that will need it in the future.

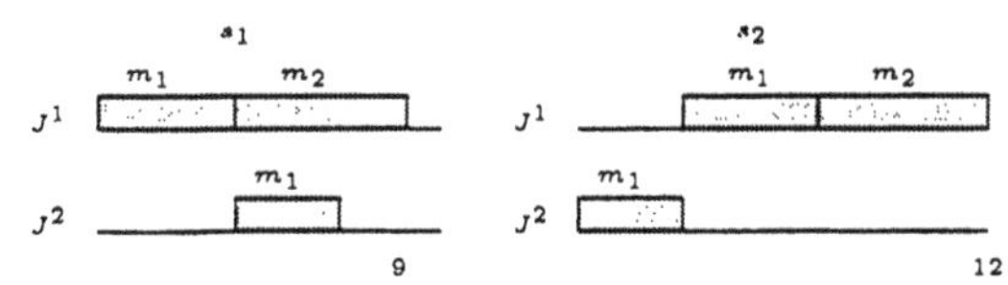

Fig. 3. Two schedule s_1 and s_2 for the example.

The traditional way to solve this problem is to assign variables, z_1, z_2 and z_3 for the start times of the three tasks, a variable z_4 for the total length of the schedule and solve a constrained optimization problem. The precedence constraints for J^1 are expressed by $z_2 \geq z_1 + 4$ (it cannot use m_2 before finishing using m_1). The fact that only one task can use m_1 at a given time is expressed by the condition

$$[z_1, z_1 + 4] \cap [z_2, z_2 + 3] = \emptyset$$

stating that the utilization periods of m_1 by both jobs should not coincide. The whole problem is thus formulated as:

$$\begin{aligned} &\min(z_4) \quad \text{subject to} \\ &z_2 - z_1 \geq 4 \\ &z_4 - z_2 \geq 5 \\ &z_4 - z_3 \geq 3 \\ &(z_2 - z_1 \geq 4 \vee z_1 - z_2 \geq 3) \end{aligned}$$

The format of this problem is both simpler and more complex than general linear programming. On one hand the constraints are always of the form $z_i - z_j \geq d$ rather than arbitrary linear inequalities. On the other, the last disjunctive constraint, which expresses a discrete choice, makes the set of feasible solutions *non-convex*. As the problem gets larger, the set of feasible solutions gets more and more fragmented into a disjoint union of convex polyhedra whose number is exponential in the number of conflicts. Like many other combinatorial optimization problems, job shop scheduling is NP-hard and this suggests that any algorithm might, in some cases, end up enumerating all possible solutions.

It is worth mentioning that people accustomed to continuous optimization tend to transform the problem into mixed integer-linear program by introducing auxiliary integer variables with which it is possible to encode disjunctions as arithmetical constraints. The problem is then transformed via relaxation (assuming temporarily that these variables are real-valued) into a convex linear program which can be solved efficiently. Then it rests to transform the obtained "solution" to a feasible solution with integer values for the relaxed variables. While this approach has been reported to work well for some classes of problems, I have doubts concerning its usefulness for scheduling, a problem dominated by discrete choices that *have no numerical interpretation.*

9. SCHEDULING WITH TIMED AUTOMATA

In this section I sketch in more detail the modeling of scheduling situations as a dynamical system on which optimal paths and optimal strategies can be computed using the forward search algorithm described in Section 6. We use the timed automaton model which has established itself as the formalism of choice for describing discrete time-dependent behaviors. Timed automata are automata operating in the dense time domain. Their state-space is a product of a finite set of discrete states (locations) and the clock-space $\mathbb{R}_+^m$, the set of possible valuations of clock variables. The behavior of the automaton consists of an alternation of time-passage periods where the automaton stays in the same location and the clock values grow uniformly, and of instantaneous transitions that can be taken when clock values satisfy certain conditions and which may reset some clocks to zero. The interaction between clock values and discrete transitions is specified by conditions on the clock-space which determine what future evolution, either passage of time or one or more transitions, is possible at a state.

When timed automata model scheduling problems, the discrete states record the qualitative state of the scheduling problem (who is executing, who has terminated) and the clocks provide the quantitative component of the state, namely the times that each active task has already spent executing. We assume here that there is a single machine of each type and hence the states of the machines are implied by the states of the tasks. We will spare from the reader the exact formal definition of timed automata and illustrate our modeling approach via an example.

We start by modeling each job as a simple automaton with one clock. The automata for our example, depicted in Figure 4, have a straightforward structure. Automaton $\mathcal{A}_1$ starts with state $\overline{m}_1$ where it waits for machine m_1. It stays at this state until a transition to active state m_1 is taken. This "start" transition is issued by the scheduler and it resets clock c_1 to zero. The automaton stays at that state until the clock reaches 4 and then moves to state $\overline{m}_2$, waiting for the next task and so on until it reaches a final state. The "end" transitions outgoing from active states are made by the environment and are considered as actions uncontrolled by the scheduler. Clocks are considered "inactive" at waiting states as they are reset to zero before they are tested.

These automata describe the possible behaviors of each job in isolation. Their joint behavior under resource constraints is captured by their product shown in Figure 5. This is essentially a Cartesian product of the job automata, where

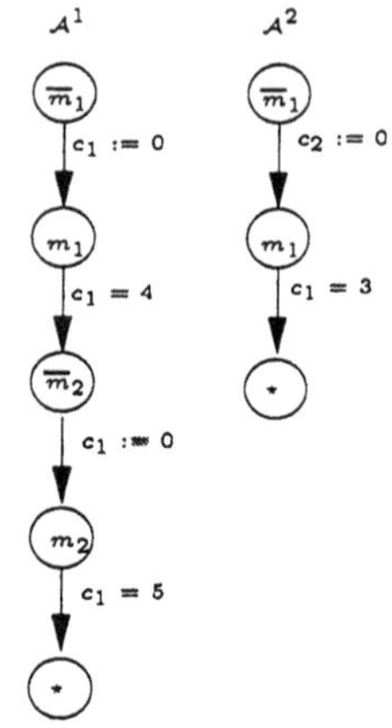

Fig. 4. Automata for the two jobs.

resource constraints are expressed by removing states such as (m_1, m_1) where more than one job uses a machine. This results in a "hole" in the automaton and the scheduler has to decide how to bypass this hole, either by giving the machine first to J^1 or to J^2. The two schedules of Figure 3 correspond to the following two behaviors (runs) of the automaton (we use notation $\perp$ to indicate inactive clocks, and $\xrightarrow{0}$ for discrete actions such as starting or ending a task):

s_1 :
$$(\overline{m}_1, \overline{m}_1, \perp, \perp) \xrightarrow{0} (m_1, \overline{m}_1, 0, \perp) \xrightarrow{4} (m_1, \overline{m}_1, 4, \perp) \xrightarrow{0}$$
$$(\overline{m}_2, \overline{m}_1, \perp, \perp) \xrightarrow{0} (\overline{m}_2, m_1, 0, \perp) \xrightarrow{0} (m_2, m_1, 0, 0) \xrightarrow{3}$$
$$(m_2, m_1, 3, 3) \xrightarrow{0} (m_2, \star, 3, \perp) \xrightarrow{2} (m_2, \star, 5, \perp) \xrightarrow{0}$$
$$(\star, \star, \perp, \perp)$$

s_2 :
$$(\overline{m}_1, \overline{m}_1, \perp, \perp) \xrightarrow{0} (\overline{m}_1, m_1, \perp, 0) \xrightarrow{3} (\overline{m}_1, m_1, \perp, 3) \xrightarrow{0}$$
$$(\overline{m}_1, \star, \perp, \perp) \xrightarrow{0} (m_1, \star, 0, \perp) \xrightarrow{4} (m_1, \star, 4, \perp) \xrightarrow{0}$$
$$(\overline{m}_2, \star, \perp, \perp) \xrightarrow{0} (m_2, \star, 0, \perp) \xrightarrow{5} (m_2, \star, 5, \perp) \xrightarrow{0}$$
$$(\star, \star, \perp, \perp)$$

It is not hard to see the correspondence between the set of possible behaviors of the automaton that reach the final state and the set of all feasible schedules. Hence the problem of optimal scheduling reduces to finding the shortest run in a timed automaton, where the length of the run is the total elapsed time. The number of such runs is uncountable (each automaton may stay any amount of time in a waiting state) however we have shown that the optimum is found among a *finite* number of runs and each node in the search tree has a finite number of successors worth exploring. The number of such paths is still exponential and an exhaustive search is infeasible. Our implementation of a best-first search algorithm on this model could find optimal schedules for problems with 6 jobs, 6 machines and 36 tasks. Beyond that we had to apply a heuristic that could find solutions with 5% from the known optimum for problems with up to 15 jobs, 15 machines and 225 tasks.

The reader probably noticed that the dynamic model used here does not fit exactly into the discrete time synchronous framework previously

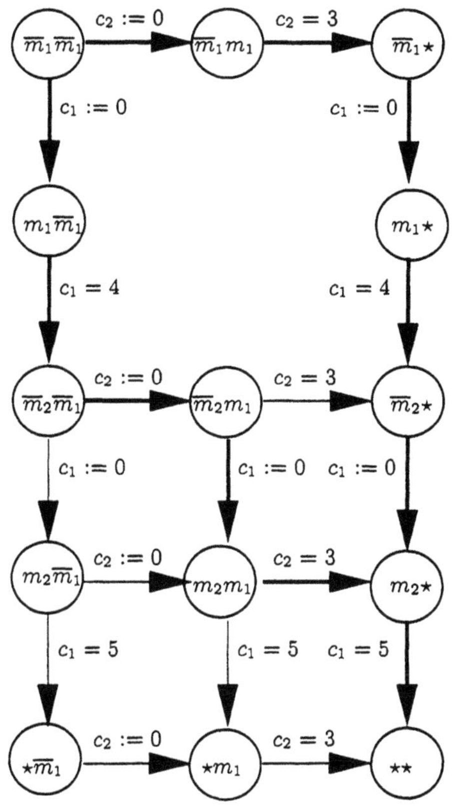

Fig. 5. The global timed automaton for the two jobs. The paths that correspond to the two schedules are indicated by thicker arrows.

described. By using a "sampled" approach and restricting events to occur and to be observed only at multiples of some constant δ, we can approximate any timed automaton by a discrete time system. However, when events occur sparsely over time, the continuous time asynchronous approach is computationally more efficient as it allows to "accelerate" the evolution of the system by letting time advance until the next event.

10. SCHEDULING UNDER UNCERTAINTY

Although the approach just described is elegant, one may argue that the world of scheduling could live without yet another technique for solving the job shop problem. The advantage of using state-based dynamic models is manifested when we move to the more complex problems of scheduling under uncertainty. Academic scheduling research has often been criticized from a practical point of view for making unrealistic assumptions and it was noted that real schedules are rarely executed as planned. During execution it may happen that tasks terminate sooner or later then expected, new tasks may appear, machines may break down, etc. In such situations what we need is a scheduling policy, a strategy which adapts to the evolution of the plant and modifies its decisions accordingly. In this section we augment the job shop problem with one type of uncertainty, namely bounded uncertainty in task durations. This means that a task description gets the form $(m, [l, h])$ indicating that the actual duration of the task is some $d \in [l, h]$. Each actual *instance* of the job shop problem consists of picking such a d for each interval and we need to evaluate a strategy according to its performance on all such instances. Consider the problem

$$J^1 : (m_1, 10), (m_3, [2, 4]), (m_4, 5) \; J^2 : (m_2, [2, 8]), (m_3, 7)$$

where the only resource under conflict is m_3 and the order of its utilization is the only decision of the scheduler. The uncertainties concern the durations of the first task of J^2 and the second task in J^1. Hence an instance is a pair $d = (d_1, d_2) \in [2, 4] \times [2, 8]$. Figure 6-(a) depicts the optimal schedules for the instances $(8, 4)$, $(8, 2)$ and $(4, 4)$ that could have been found by a non-causal *clairvoyant* scheduler who knows the whole instance in advance. But instances reveal themselves *progressively* during execution — the value of d_1, for example, is known *only after the termination* of the second task of J^1.

It turns out that for this particular type of uncertainty, optimization with respect to the worst-case criterion is somewhat trivial. There is always a maximal (critical) instance, $(8, 4)$ in this example, having two important properties: 1) The optimal schedule for this instance is valid also for all other smaller instances (just ignore earlier termination of certain tasks and keep the machine busy until h time elapses); 2) No strategy can perform better on this instance. Figure 6-(b) shows the behavior of a static worst-case strategy based on instance $(8, 4)$ and one can see that is is rather wasteful for other instances. We want a smarter adaptive scheduler which takes the actual duration of m_2 into consideration.

One of the simplest ways to be adaptive is the following. First we choose a *nominal instance* d and find a schedule s which is optimal for that instance. Rather than taking s "literally" as an assignment of absolute start times to tasks, we extract from it only the *qualitative information*, the order in which conflicting tasks utilize each resource. In our example the optimal schedule for instance $(8, 4)$ is associated with giving priority to J^1 on m_3. Then, during execution, we start every task as soon as its predecessors have terminated, provided that the ordering is not violated. As Figure 6-(c) shows, such a strategy is better than the static schedule for instances such as $(8, 2)$ where it takes advantage of the earlier termination of the second task of J^1 and "shifts forward" the start times of the two tasks that follow.

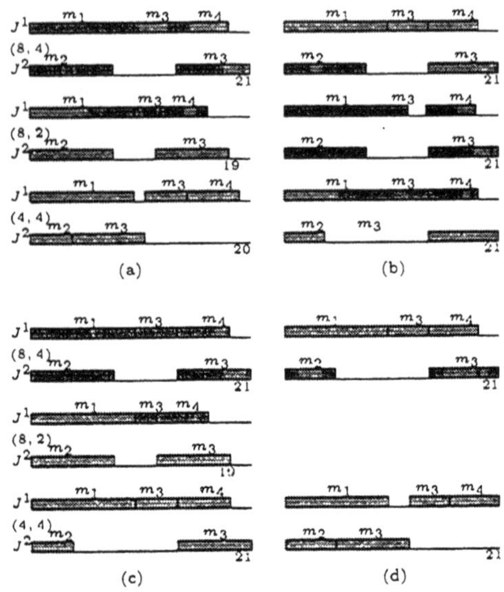

Fig. 6. (a) Optimal schedules for three instances; (b) A static schedule based on the worst instance $(8,4)$; (c) The behavior of a hole filling strategy based on instance $(8,4)$; (d) The equal performance of the two strategies on instance $(5,4)$.

Note that this "hole filling" strategy is not restricted to the worst-case. One can use any nominal instance and then shift tasks forward or backward in time as needed while maintaining the order. On the other hand, a static schedule can only be based on the worst-case — a schedule based on another nominal instance may assume a resource available at some time point, while in reality it will be occupied.

The hole filling strategy is optimal for all instances whose optimal schedule has the same ordering as that for the nominal instance. It is not good, however for instances such as $(4,4)$ which cannot benefit from the early termination of m_2 because shifting m_3 of J^2 forward will violate the priority on m_3. For such cases a more refined form of adaptiveness is required. Looking at the optimal schedules for $(8,4)$ and $(4,4)$ in Figure 6-(a), we observe that in both of them the decision whether or not to give m_3 to J^2 is taken at the same qualitative state where m_1 is executing and m_2 has terminated. The only difference is in the elapsed execution time of m_1 at the decision point. Hence an adaptive scheduler should base its decisions also on *quantitative* information encoded by clock values.

Consider the following approach: initially we find an optimal schedule for some nominal instance. During execution, whenever a task terminates we reschedule the "residual" problem, assuming nominal duration for tasks that have not yet terminated. In our example, we first build an optimal schedule for $(8,4)$ and start executing it. If task m_2 in J^2 terminated after 4 time units we obtain the residual problem

$$J_1' : (\mathbf{m_1}, \mathbf{6}), (m_3, 4), (m_4, 5) \qquad J_2' : (m_3, 7)$$

where the boldface letters indicate that m_1 must be scheduled immediately (it is already executing and we assume no preemption). For this problem the optimal solution will be to give m_3 to J^2. Likewise, if m_2 terminates at 8 we have

$$J_1' : (\mathbf{m_1}, \mathbf{2}), (m_3, 4), (m_4, 5) \qquad J_2' : (m_3, 7)$$

and the optimal schedule consists of waiting for the termination of m_1 in order to give m_3 to J^1. The property of the schedules obtained this way is that at any state reachable during execution they are optimal with respect to the nominal assumption concerning the *future*. We call such strategies *d-future optimal*.

This is the principle underlying model-predictive control where at each step, actions at the current "real" state are re-optimized while assuming some nominal prediction for a bounded horizon future. A major drawback of this approach is that it involves a lot of *online* computation, solving a new scheduling problem each time a task terminates. This restricts its applicability to "slow" processes. In the next section we present an alternative approach where an equivalent strategy is synthesized *offline* using a symbolic variant of dynamic programming adapted for timed automata.

11. DYNAMIC PROGRAMMING ON TIMED AUTOMATA

The state-space of a timed automaton consists of pairs of the form (q, c) where $q = (q_1, \ldots, q_n)$ is a discrete state, indicating the local states of all jobs, and $c = (c_1, \ldots, c_n)$ is a vector of clock valuations ranging over a bounded subset of the non-negative reals. On these we define a value function $\overleftarrow{\mathcal{V}}$ such that $\overleftarrow{\mathcal{V}}(q, c)$ denotes the minimal time to reach the final state from (q, c), assuming nominal values for tasks that have not terminated. Before giving the formal definition let us give an intuitive explanation. Being at (q, c), all the local choices of the scheduler can be brought into the following form: *let some t time pass and then execute one transition that is enabled by the clock values.* This definition covers also the possibility of an immediate action ($t = 0$), as well as the possibility of waiting until an uncontrolled transition is taken by the environment. The value induced by this choice is the sum of the waiting time t and the value of the state reached after the transition. This is captured by the following recursive definition:

$$\overleftarrow{\mathcal{V}}(\star, c) = 0$$
$$\overleftarrow{\mathcal{V}}(q, c) = \min\{t + \overleftarrow{\mathcal{V}}(q', c') : (q, c) \xrightarrow{t} (q, c + t1) \xrightarrow{0} (q', c')\}$$

To illustrate the computation of $\vec{\mathcal{V}}$ we consider a simplified version of the example from the previous section with only one uncertain duration:

$$J^1 : (m_1, 10), (m_3, 4), (m_4, 5) \quad J^2 : (m_2, [2, 8]), (m_3, 7).$$

Figure 7 shows the final part of the global automaton corresponding to the problem, which includes state $(m_1, \overline{m}_3)$ where a decision of the scheduler has to be taken. The computation starts with $\vec{\mathcal{V}}\ (\star, \star, \bot, \bot) = 0$. The value of $(m_4, \star, c_1, \bot)$ is the time it takes to satisfy the condition $c_1 = 5$, which is $5 \dot{-} c_1$. Likewise $\vec{\mathcal{V}}\ (\star, m_3, \bot, c_2) = 7 \dot{-} c_2$. In state (m_4, m_3) the two jobs are active and the transition to be taken depends on which of them will "win the race" and terminate before:

$$\vec{\mathcal{V}}\ (m_4, m_3, c_1, c_2)$$

$$= \min \left\{ \begin{array}{l} 7 \dot{-}\ c_2 + \vec{\mathcal{V}}\ (m_4, \star, c_1 + (7 \dot{-}\ c_2), \bot), \\ 5 \dot{-}\ c_1 + \vec{\mathcal{V}}\ (\star, m_3, \bot, c_2 + (5 \dot{-}\ c_1)) \end{array} \right\}$$

$$= \min\{5 \dot{-}\ c_1, 7 \dot{-}\ c_2\}$$

$$= \left\{ \begin{array}{l} 5 \dot{-}\ c_1 \text{ if } c_2 \dot{-}\ c_1 \geq 2 \\ 7 \dot{-}\ c_2 \text{ if } c_2 \dot{-}\ c_1 \leq 2 \end{array} \right.$$

Note that both transitions are uncontrolled *end* transitions and no decision of the scheduler is required in this state. The computation proceeds backwards, computing $\vec{\mathcal{V}}$ for all states. In particular, for state $(m_1, \overline{m}_3)$ where we need to choose between giving m_3 immediately to J^2 or waiting for the termination of m_1 to give m_3 to J^1, we obtain:

$$\vec{\mathcal{V}}\ (m_1, \overline{m}_3, c_1, \bot) = \min\{16, 21 \dot{-}\ c_1\}$$

$$= \left\{ \begin{array}{ll} 16 & \text{if } c_1 \leq 5 \\ 21 \dot{-}\ c_1 & \text{if } c_1 \geq 5 \end{array} \right.$$

Hence, if m_1 terminates after less than 5 time units it is better to give m_3 to J^2, otherwise it is worth waiting and giving it to J^1. Figure 6-(d) shows that, indeed, the two strategies coincide in performance when $c_1 = 5$.

The reader should not be misled by the success of our strategy to match the performance of a clairvoyant scheduler for this small example. In slightly more complex problems with several uncertainties it is impossible to compete with knowing the future and being d-future optimal is good enough.

The actual computation of the value function is implemented using standard reachability techniques for timed automata which are outside the scope of the present paper. We have tested our implementation on a problem with 4 jobs, 6 machines and 24 tasks, 8 of which having uncertain durations. we fixed two instances, one "optimistic" where each task duration is set to l, and one "pessimistic" with h durations. We applied our algorithm to find two d-future optimal strategies and two hole filling strategies based on these instances. We have generated 100 random instances with durations drawn uniformly from each $[l, h]$ interval, and compared the results of the abovementioned strategies with an optimal clairvoyant scheduler that knows each d in advance, and with the static worst-case scheduler. It turns out that the static schedule is, on the average, longer than the optimum by 12.54%. The hole filling strategy deviates from the optimum by 4.90% (for optimistic prediction) and 4.44% (for pessimistic prediction). Our strategy produces schedules that are longer than the optimum by 1.40% and 1.14%, respectively.

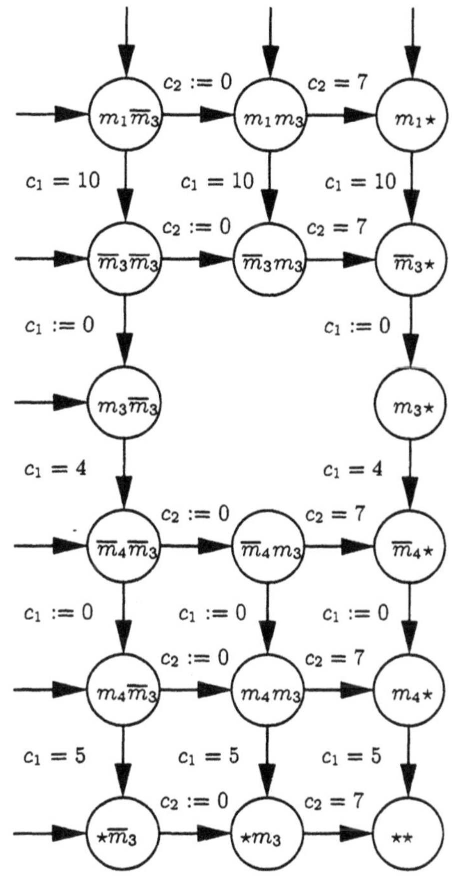

Fig. 7. Part of the global automaton.

The good news is that our strategy is much better than static scheduling, and can be considered as a useful tool for systems with "soft" real-time performance criteria. The bad news is that it is much more costly than the hole filling strategy. The latter solves an adversary-free problem and can use intelligent forward search while the computation of our strategy has to explore the whole state-space. The adaptation of forward game tree search to this problem is not straightforward, due to the density of the set of adversary actions, and it is subject to ongoing research along with the adaptation of this approach to other types of uncertainty such as imprecise arrival times or discrete uncertainty associated with conditional dependencies between tasks.

12. DISCUSSION

People who are experts in their domain are often skeptical toward proposals for unified theories. Indeed, compared to successes of domain specific research, various holistic trends such as "general systems theory" proved in the past to be rather sterile. Saying that "everything is systems" and that many things that look so different are, at a certain level of abstraction, similar, does not necessarily solve problems. I hope that the framework presented in this paper will have a better fate. It is less ambitious than some of its predecessors in the sense of not trying to predict the unpredictable and pretend to give optimal recipes for complex socio-economic or biological phenomena for which we do not even know the appropriate modeling vocabulary. Rather it is restricted to situations where useful dynamic models and performance criteria do exist, models which are already used, implicitly or explicitly, for simulation, verification or optimization. This framework is geared toward a *concrete* goal: developing a tool for defining and solving optimal control problems for systems with diverse types of dynamics.

Some principles underlying such a framework (some of which already exist in respective domains) are mentioned below. First, I believe that systems should be defined with a clear semantics from which it is easy to see who are the players, what are the variables they can observe and influence, what constitutes a behavior of the system, what is assumed about the environment and what are the natural performance criteria. At this level, the description should be *separated* from the specific computational techniques that are used to reason about the model. This is in contrast with the domain-specific approaches where problems are often phrased in terms biased toward particular and, sometimes, accidental solution techniques which are common in the domain.

After an ideal optimal controller has been mathematically defined, computational issues should be addressed. Here the difference between classes of system dynamics is manifested by the type of constrained optimization problem to be solved, discrete (logical), continuous (numerical) or hybrid. In most cases the global optimality of the solution is a ceremonial matter. No one really intends to be optimal and models are imprecise anyway. In some cases, proving some relation between approximate solutions and the optimum is a good measure for the quality of a technique, but this is neither a necessary nor a sufficient condition for its usefulness.

Since some space is left, let me add some controversial remarks. It seems to me that in many domains relevant to this paper, there is a tension between the mathematical (theoretical) and engineering (hacking) approaches. The (real) practitioner cannot choose the problems he has to solve and also does not have time to develop nice theories. In many cases he will adapt solutions provided by mathematicians of previous generations to get the job done. The theoretician is supposed to be more open-minded and explore new classes of models for new phenomena but the structure of academe does not always encourage him to do so. Members of scientific communities often impose upon themselves some intrinsic evaluation criteria that deviate over time from the raison d'être of the domain. There is nothing wrong with (good) mathematics for its own sake, but one should not confuse it with solving real engineering problems or even with laying the foundations for future solutions. What is really needed is a middle road between mathematics and engineering, which allows us to see the generic mathematical objects behind the engineering instances, together with a strong sense of criticism toward the traditions of the respective academic fields, which are often by-products of the sociology of scientific communities, rather than the result of a genuine attempt to be relevant.

Acknowledgments: Any attempt to give a fair survey of related work, spanning over more than 50 years and numerous scientific communities, will either be pretentious or require an effort that goes beyond the scope of this paper. Consequently, I provide no references to the first part of the paper. The results concerning scheduling were obtained in collaboration with Y. Abdeddaïm and E. Asarin. Readers interested in more details might want to look at the thesis [A02] or the paper [AAM04].

This work was inspired partly by an interaction with control theorists on various occasions related to hybrid (discrete/continuous) systems, including my partners in the CC project, whom I would like to thank (the opinions are, of course, my own). The work on scheduling enjoyed similar relations with partners in the AMETIST project. This manuscript benefited from feed-back provided by Anil Nerode, Anders Ravn, Tariq Samad, Michel Sintzoff and Pravin Varaiya, and from numerous past discussions with Eugene Asarin and Bruce Krogh.

References

[A02] Y. Abdeddaïm, *Scheduling with Timed Automata.* PhD thesis, Institut National Polytechnique de Grenoble, Laboratoire Verimag, 2002.

[AAM04] Y. Abdeddaïm, E. Asarin and O. Maler, Scheduling with Timed Automata, *Theoretical Computer Science*, 2004.

ELSEVIER
IFAC
PUBLICATIONS
www.elsevier.com/locate/ifac

DISCRETE CONTROL SYSTEMS FOR CARS – QUALITY IS MORE THAN CORRECT FUNCTION

Stefan Kowalewski *

* *RWTH Aachen University, Computer Science Chair XI – Embedded Software Laboratory, Ahornstr. 55, 52074 Aachen, Germany*

Abstract: Based on the situation and trends in the automotive industry, the paper presents current challenges for the development of automotive electronic control systems, in particular discrete control functions. It is argued that the current situation in the automotive industry should motivate the discrete event and hybrid systems research community to extend their attention from formally defined, single-application focussed control design problems to so-called non-functional view-points or qualitative requirements like, e.g., maintainability or inter-operability of control software components. *Copyright © 2004 IFAC*

Keywords: Automotive control systems, quality requirements, non-functional properties, system architecture

1. INTRODUCTION

Modern cars contain numerous discrete control systems. For example, in the body and interior domain, most control systems are dominantly discrete, take for instance the keyless entry and immobilizer logic, the sequential motion control of windows, sun-roofs etc., or the airbag firing logic. But also in domains traditionally being dominated by continuous control, like drive-train or chassis, discrete control functions become more important. One reason for this is that single functions are integrated to increasingly complex systems, and this usually makes it necessary to introduce a discrete mode management level into the control software. One instance of this development is the integration of ABS, ESP, suspension and traction control to integrated vehicle dynamics management systems. Modern engine control systems with complex relationships between the air-fuel ratio, spray characteristics and exhaust quality represent a further example. An additional source for discrete control logic are the emerging domain-crossing services like car-wide diagnostics or power management.

This trend clearly provides an opportunity for the discrete event (and hybrid) systems research community to transfer more of the existing theory, methodology, and models to the automotive industry than it was possible in the past. Design support methods like formal specification, verification or even synthesis should gain increasing interest, in particular with respect to safety-critical applications. This trend will be fostered by the expected adoption of the IEC 61508 standard (International Electrotechnical Commission, 1998) by the automotive industry which explicitly recommends formal methods for systems with safety integrity level SIL 3 or higher.

However, it should be observed that the most pressing problems in the design of automotive electronic control systems, including discrete and continuous control functions, will not be solved by formal verification or synthesis procedures. Problems of the kind, "given a design and some desired properties, show that the properties hold for the design", are of minor interest and can be solved satisfactorily in most cases by extensive testing. More attention is currently put on questions like how the appropriate requirements are captured, analyzed, and stabilized in the first place, or how requirements, designs, and implemented functions can be maintained and re-used for future projects. These questions relate to the quality or so-called non-functional properties of systems and less to their correct function in one particular application. Still, the discrete event and hybrid systems research has the potential to contribute solutions to these problems. How this can be done will be sketched in the sequel.

The paper is organized as follows. In the next two section, we sketch the current situation and trends in the automotive industry with respect to the development of control software. Section 4 points to current initiatives and approaches either in the automotive industry or being investigated by the industry. The paper concludes with a list of challenges for research.

2. SITUATION

In less than fifty years, cars have changed from dominantly mechanical products with only little electric or electronic components – mainly lighting and engine ignition – to moving computer networks. Modern high end cars include 50 to 80 electronic control units (ECUs), connected via three different bus systems, and with one million lines or 50 MB of control software code. For a Mercedes S, the development lead from 1 MB code in 1990 over 100 MB in 1998 to expected more than 500 MB in the 2005 model (Bortolazzi, 2004). Much of this increase, of course, is caused by the infotainment sector. However, also for basic control tasks we find an enormous raise of complexity. In the case of door controllers, a domain with dominantly discrete control functions, the processors moved from 8 bit and 32 KB ROM in 1997 to 32 bit and 128 KB Flash memory today in only 7 years (Bortolazzi, 2004).

It is expected that the software size per ECU will double every 2–3 years. Already today, 50–70% of the development cost for an ECU is for the software. An often cited prognosis is that 90% of all future innovations in cars will be realized by electronics and software.

The downside of this development are the prominent reliability problems caused by electronics and software. In Germany, according to the statistics of the road assistance services, the software and electronics related breakdowns increased by more than 20% while the breakdowns with other causes remained nearly constant. The number of recalls increased from 52 in 1993 to 127 in 2003. However, these figures have to be related for example to an enormous gain of safety in the last 30 years. While the number of cars doubled during this time in Germany, the number of injuries and deaths in car accidents decreased by more than 70%.

3. TRENDS

With respect to the latest trends, the automotive industry is currently subject to a technology shift. Cars and automotive control systems are becoming software-intensive products. Software emerged from an added-on technology to the most dominant factor for product quality. This change in technology fosters a change of the relationship between the car manufacturers (OEMs for short, for Original Equipment Manufacturer) and suppliers: The traditional development of electronic control systems in cars has been based on a bilateral partnership between OEM and supplier for each subsystem: The supplier develops a complete control subsystem consisting of sensors, actuators, a dedicated ECU and the control software.

These premises are currently changing: New functions and services in modern and future cars require the connection of distributed components in a car-wide network, cost considerations result in the need to share resources and to integrate control software from different sources on common hardware platforms. As a consequence, the perspective on the development of automotive control systems is changing: It will become insufficient to focus on a single control system and develop an optimal solution. Instead, issues like the integration of the control software into an only partly known environment must be considered from the outset. The OEMs need to specify properly and check early that all the elements from the different sources will interact properly not only with respect to the control requirements but also regarding non-functional properties like reliability, performance, memory and power consumption, maintainability, system cost, etc.

Concrete trends are listed in the sequel.

Resource sharing. With the increasing number of interrelated functions and services, it becomes more and more necessary to break up the one function – one subsystem paradigm and to share sensors, hardware and actuators between different functions. One example is the field of driver assistance systems based on car periphery supervision. It would be impossible for mounting space and economical reasons to offer cars with a set of functions like parking assistance, blind spot supervision, adaptive cruise control and pre-crash detection when all of these functions would have their own sensors, control units etc. Instead, we need platforms in which they will share the sensors and most of the signal processing.

The consequences of this trend are that the scheduling becomes crucial but also more complicated. The real-time properties of the whole system – including bus communication – must be analyzed thoroughly.

Domain crossing functionality. The traditional mapping of automotive control systems to exactly one of the so-called automotive domains, i.e., drive train, chassis, body and infotainment, is beginning to erode. More and more functions and services are domain-crossing. Examples are the energy management, car-wide diagnosis with central access, or profiling services.

The consequences are manifold: The communication between subsystems becomes more complex and has to be analyzed. The traditional failure containment regions macerate and, maybe most important, the traditional company organizations and system architectures which were tailored according to the domain boundaries do not fit anymore.

Changing OEM/supplier relation. Due to the mentioned problems, the OEMs are currently trying to gain back expertise in electronics and software from the suppliers. This is necessary for them to be able to do the specification, contracting and quality control competently. They also want more independence from the suppliers. One aim is to be able to buy functions, i.e. software, from different sources than the hardware and to deploy it on arbitrary ECUs. As they already do, OEMs will increasingly want to include own, competition relevant software on ECUs from suppliers.

Here the consequences are that the integration will become the major quality determining development phase in the future. And it will be much more complex than in the past. For example, problems of intellectual property (IP) protection are arising. Methodical support is badly needed.

New quality requirements arise. Due to the rising connectivity and complexity, new quality properties move into focus which were not in the center of interest for embedded systems in the past. One example is security. IP prtection during distributed development and integration was already mentioned. Another source of this requirement is the possibility to download software to ECUs in the field. What is a convenient way for OEMs to update the electronic systems during the regular maintenance intervals, becomes a threat to safety and reliability if it is done using third party software of doubtful origin.

Here it is necessary for the both OEMs and suppliers to be able to prove that such manipulations took place.

4. APPROACHES

The automotive industry has reacted to the situation described above by investigating methods which shall support the future development of automotive software by meeting the new quality requirements. The most prominent initiative is the project Automotive Open Systems Architecture (AUTOSAR) (AUTOSAR-consortium, 2004) which was started in 2002 and has nearly all relevant OEMs and suppliers in the consortium. The main objectives are:

- standardization of non-competition-relevant components
- definition of standard abstraction layers and interfaces to support integrability and transferability of functions
- improvement of maintainability over the whole life cycle

The basic approach is to define a layered architecture and a static middleware layer which provides the interfaces for cross-ECU communication between software components. Critics say that AUTOSAR is focussing too much on functionality issues and is neglecting the quality properties like modifiability, reliability or resource consumption and costs. Being able to distribute software components arbitrarily over the ECU network, for example, makes it necessary to evaluate the distribution with respect to these properties. Given the ambitious schedule of AUTOSAR, it is often doubted that the project will achieve methodological support for these challenges.

A further initiative which the automotive industry is observing thoroughly is the Model Driven Architecture (MDA) approach (Object Management Group, 2004). Here, concrete instantiations and tools suitable for automotive software are still missing or only subject to research projects. Already applied is the Product Line Approach (PLA) to cope with software and systems which have to be developed and maintained in large varieties (Software Engineering Institute, 2004). The experience shows that it is necessary not only to include but to put the main emphasis on non-technical issues, like market analysis or organization structuring, to run a successful product line.

5. CHALLENGES FOR RESEARCH

The paper concludes with a collections of challenges for research. Given the current situation

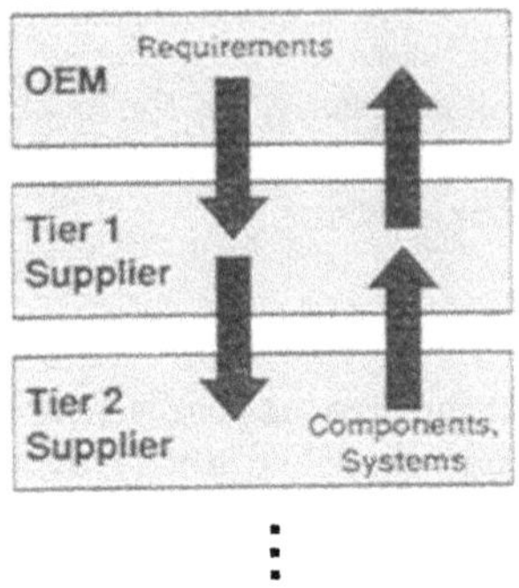

Fig. 1. Layered design chain

and the future challenges in automotive control systems, we need to move from a model-based function development to a model-based system engineering. This means, system properties must be considered from the outset and throughout the life cycle.

To make this more concrete we consider the interaction between the OEMs and the suppliers. It is often characterized as a layered design chain in which requirements are passed down from the OEM to the tier 1 suppliers and so on, whereas components and systems are made available from the different supplier levels bottom-up to the OEM (Fig. 1). The relation between the layers gives rise to the following challenges.

Iteration cycles. It is obvious, that the number and depth of iteration cycles between the layers should be as low as possible. It is therefore necessary for the higher layers to represent their requirements and specifications in an unambiguous way. Moreover, executable models are needed which make it possible to evaluate the interaction between components from different sources before they are actually available (virtual integration).

IP protection. It must be possible for an upper layer to buy components from a lower layer and integrate it into a system, first, without having to uncover confidential information about the rest of the system and, second, without needing to know confidential details about the design and implementation of the component by the lower level supplier. To achieve this, models are needed in which we can specify *what* a component does and how it has to be integrated in varying environments without uncovering *how* the component realizes its function.

Re-use. The models being used for specification and early simulation must be easily adaptable to different, sometimes unexpectedly varying environments in which they have to be integrated. Thus, libraries of re-usable components must offer more than functional interfaces. For example, consequences of deploying control functions differently over an ECU network must be analyzable from the components' description.

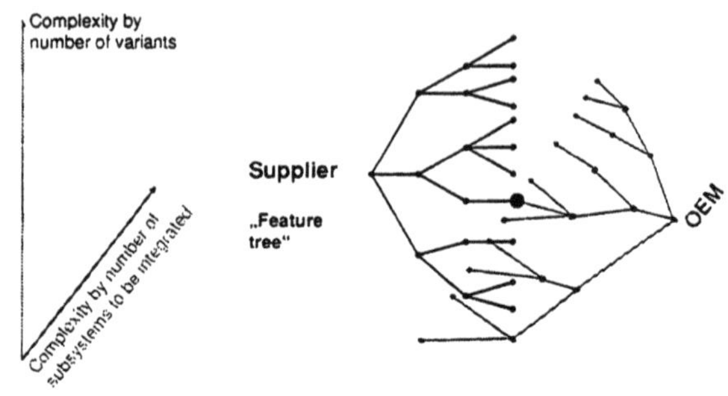

Fig. 2. Different sources for complexity

The main aim is to be able to deal with the complexity of the modern development of automotive software. If we develop methods and models for that, we should be aware that the complexity has different sources for OEMs and suppliers. This is illustrated in Fig. 2. For the OEM, the main problem is the huge number and interconnection of components to be integrated. Whereas for the supplier, complexity arises from the huge number of variants. Models of re-usable specifications and components have to support both issues.

REFERENCES

AUTOSAR-consortium (2004). www.autosar.org. AUTOSAR web site, last access: 7 October 2004.

Bortolazzi, J. (2004). Prozesse und Methoden fur die Entwicklung von Kfz–Software bei der Daimler-Chrysler AG. In: *P. Dadam, M. Reichert (Hrsg.): Informatik 2004*. Vol. P-51 of *Lecture Notes in Informatics – GI-Edition*. Springer. p. 2. Talk at the annual conference of the German Society for Computer Science, in German.

International Electrotechnical Commission (1998). International standard 61508: Functional safety of electrical, electronic, and programmable electronic systems. Geneva. First edition.

Object Management Group (2004). www.omg.org/mda. MDA web site, last access: 7 October 2004.

Software Engineering Institute (2004). www.sei.cmu.edu/productlines/. PLA web site, Carnegie Mellon University, last access: 7 October 2004.

ELSEVIER
IFAC
PUBLICATIONS
www.elsevier.com/locate/ifac

A SYSTEMS THEORY VIEW OF PETRI NETS

Alessandro Giua, Carla Seatzu

DIEE, Università di Cagliari, Piazza d'Armi,
09123 Cagliari, Italy. {giua,seatzu}@diee.unica.it

Abstract: The objective of this talk is that of surveying the literature of the last 15 years dealing with Petri nets and Discrete Event Systems theory. Given time and space constraints, we will only focus on logical models and theories. We attempt neither exhaustive detail nor analytical profundity, but rather a pastiche of scattered parts, which, so it hoped, will coalesce into a focused picture. We believe that this survey, albeit incomplete, contains clear evidence of the fact that Petri net research is a cornerstone of systems theory.

Keywords: Petri nets, discrete event systems, automatic control.

The field of Discrete Event Systems (DES) is a relatively new research area of systems theory that combines different formalisms, methodologies and tools from automatic control, computer science and operations research. The research activity in this field is driven by the needs of many different applications domains: manufacturing, process control, supervisory control and data acquisition systems, failure diagnosis, software engineering, transportation, and so on.

The field of DES can be partitioned into two domains.

Logical models and theories are used to describe qualitative properties and aim to control the sequences of events generated by a DES. To achieve this goal, the timing of event occurrence can be ignored and it is only required to keep track of the order between events. Typical problems that are considered within this framework are: avoiding forbidden sequences of events or forbidden states, deadlock avoidance, liveness enforcing, and so on.

Performance models and theories are used to describe quantitative properties and aim to control the timed behavior of a DES. In this case it is necessary to keep track of the occurrence time of each event. Typical problems that are considered within this framework are: satisfaction of timing constraints, optimization of process time and routing, scheduling, and so on.

A Petri net (PN) is an algebraic and graphical formalism that generalizes a state machine and has primitives to explicitly represent the notion of concurrency among events. Properly speaking, in fact, a PN can be seen as a particular model of concurrent systems: however, the basic research on PN has assumed such a relevance and has found application in so many domains, that it is common to consider PN a full-fledged research field.

Petri nets represent a research area that is extremely important for DES for several reasons.

- *Family of models.* There exist several types of Petri nets that are effective for modeling both logical and performance DES (place/transition nets, timed nets, time nets, generalized stochastic PN, etc.)
- *States and actions on equal basis.* Petri nets use two simple primitives, the *marking of places* and the *transitions*, to represent states and actions on equal footing. This is a major advantage over state-based approaches —

that give no easy way to express concurrent systems because of the *dimensionality curse* — and action-based approaches — that have no easy way to represent specifications such as counters or semaphores.

- *Locality of states and actions.* The overall state of a PN is described in terms of "local" states, i.e., the marking of each place. The overall behavior of a PN is described in terms of "local" actions, i.e., the firing of a transition that updates the marking of only a subset of places. This feature offers a structured model of DES dynamics that often allows the designer to derive more computationally efficient algorithms for controller synthesis.
- *States and actions as vectors.* The states and events of a DES are often described using non-numerical labels. When a PN is used as discrete event model, however, a state corresponds to a non-negative integer vector $x \in \mathbb{N}^m$. Similarly, an event can be described by an integer vector $y \in \mathbb{Z}^m$ that from a state x yields a new state $x' = x + y$. This important property allows one to use linear algebraic techniques for the analysis and control of these models.

The overly ambitious objective of this talk is that of reviewing the literature of the last 15 years dealing with Petri nets and DES theory. Given time and space constraints, we will only focus on logical models and theories. We attempt neither exhaustive detail nor analytical profundity, but rather a pastiche of scattered parts, which, so it hoped, will coalesce into a focused picture.

In particular, there will be three main topics addressed in this talk.

(1) *Net models and computational issues.* We will discuss several logical models including the basic place/transition net model (Murata, 1989), continuous nets (Alla and David, 1998; Silva and Recalde, 2003), hybrid nets (Di Febbraro *et al.*, 2002). We will also consider labeled Petri nets, seen as language generators and review the main properties of Petri nets languages (Peterson, 1981; Gaubert and Giua, 1999). Finally, we will consider extended Petri net models such as nets with inhibitory arcs that are equivalent to Turing machines (Sreenivas and Krogh, 1992).

(2) *Analysis techniques.* There exists a large spectrum of analysis techniques that have successfully been applied to solve many theoretical problems. Among these we will discuss *path-based methods* (Holloway and Krogh, 1990; Ghaffari *et al.*, 2003*b*), *structural analysis* based on integer, linear or quadratic programming (Silva *et al.*, 1998; Giua *et al.*, 2004), *unfolding* (McMillan, 1995; Esparza *et al.*, 2002) and the *theory of regions* (Darondeau, 2000; Ghaffari *et al.*, 2003*a*).

(3) *Control theory issues.* Many properties and problems that are the object of classical control theory for time-driven systems, can also be framed in the field of DES, and solved using Petri net models. We will discuss reachability (Reutenauer, 1990; Balduzzi *et al.*, 2001; Júlvez *et al.*, 2003) and controllability (Murata, 1977), state feedback control (Giua *et al.*, 1992; Li and Wonham, 1993; Li and Wonham, 1994; Yamalidou *et al.*, 1996; Holloway *et al.*, 1997; Moody and Antsaklis, 2000; Stremersch and Boel, 2002), state estimation and observer design (Giua and Seatzu, 2002), identification (Hiraishi, 1992; Meda *et al.*, 1998), diagnosis (Benveniste *et al.*, 2003), model reduction (Sreenivas, 2002).

As we have already mentioned, many interesting and important PN approaches for logical systems will not be discussed in this talk. Examples of these approaches are: event-based supervisory control of PN, top-down and bottom-up synthesis and validation of PN models by trasformation, deadlock avoidance and prevention, etc.

However, we believe that this survey, albeit incomplete, contains a clear evidence of the fact that Petri net research is a cornerstone of systems theory.

REFERENCES

Alla, H. and R. David (1998). A modeling and analysis tool for discrete event systems – continuous Petri net. *Performance Evaluation* **33**(3), 175–199.

Balduzzi, F., A. Di Febbraro, A. Giua and C. Seatzu (2001). Decidability results in first-order hybrid petri nets. *Discrete Event Dynamic Systems: Theory and Applications,* **11**(1 & 2) pp. 41–57.

Benveniste, A., E. Fabre, S. Haar and C. Jard (2003). Diagnosis of asynchronous discrete event systems, a net unfolding approach. *IEEE Trans. Automatic Control* **48**(5), 714–727.

Darondeau, Ph. (2000). Region based synthesis of P/T-nets and its potential applications. In: *Lecture Notes in Computer Science: 21st International Conference on Application and Theory of Petri Nets (Aarhus, Denmark).* Vol. 1825. Springer-Verlag. pp. 16–23.

Di Febbraro, A., A. Giua and G. Menga (Eds.) (2002). Special issue on hybrid Petri nets. *Discrete Event Dynamic Systems: Theory and Applications,* **11**(1 & 2).

Esparza, J., S. Römer and W. Vogler (2002). An improvement of McMillan's unfolding algorithm. *Formal Methods in System Design* **20**, 285–310.

Gaubert, S. and A. Giua (1999). Petri net languages and infinite subsets of N^m. *Journal of Computer and System Sciences* **59**(3), 373–391.

Ghaffari, A., N. Rezg and X. Xie (2003*a*). Design of a live and maximally permissive Petri net controller using the theory of regions. *IEEE Trans. Robotics & Automation* **19**(1), 137–141.

Ghaffari, A., N. Rezg and X. Xie (2003*b*). Feedback control logic for forbidden-state problems of marked graphs: application to a real manufacturing system. *IEEE Trans. Automatic Control* **48**(1), 18–29.

Giua, A. and C. Seatzu (2002). Observability of place/transition nets. *IEEE Trans. Automatic Control* **47**(9), 1424–1437.

Giua, A., C. Seatzu and F. Basile (2004). Observer based state-feedback control of timed Petri nets with deadlock recovery. *IEEE Trans. Automatic Control* **49**(1), 17–29.

Giua, A., F. DiCesare and M. Silva (1992). Generalized mutual exclusion constraints on nets with uncontrollable transitions. In: *Proc. 1992 IEEE Int. Conf. on Systems, Man, and Cybernetics*. pp. 974–979.

Hiraishi, K. (1992). Construction of a class of safe petri nets by presenting firing sequences. In: *Lecture Notes in Computer Science: Proc. of the 13th Int. Conf. Application and Theory of Petri Nets (Sheffield, UK)*. Vol. 616. Springer-Verlag. pp. 244–262.

Holloway, L.E. and B.H. Krogh (1990). Synthesis of feedback control logic for a class of controlled Petri nets. *IEEE Trans. Automatic Control* **35**(5), 514–523.

Holloway, L.E., B.H. Krogh and A. Giua (1997). A survey of Petri net methods for controlled discrete event systems. *Discrete Event Dynamic Systems* **7**, 151–190.

Júlvez, J., L. Recalde and M. Silva (2003). On reachability in autonomous continuous Petri net systems. In: *Lecture Notes in Computer Science: Proc. of the 24th Int. Conf. Application and Theory of Petri Nets (Eindhoven, The Netherlands)*. Vol. 2679. Springer-Verlag. pp. 221–240.

Li, Y. and W.M. Wonham (1993). Control of vector discrete-event systems - part I: The base model. *IEEE Trans. Automatic Control* **38**(8), 1215–1227.

Li, Y. and W.M. Wonham (1994). Control of vector discrete-event systems - part II: Controller synthesis. *IEEE Trans. Automatic Control* **39**(3), 512–531.

McMillan, K.L. (1995). A technique of state space search based unfolding. *Formal Methods in System Design* **6**(1), 45–65.

Meda, M.E., A. Ramírez and A. Malo (1998). Identification in discrete event systems. In: *Proc. 1998 IEEE Int. Conf. on Systems, Man, and Cybernetics*. pp. 740–745.

Moody, J.O. and P.J. Antsaklis (2000). Petri net supervisors for des with uncontrollable and unobservable transitions. *IEEE Trans. Automatic Control* **45**(3), 462–476.

Murata, T. (1977). State equation, controllability, and maximal matchings of Petri nets. *IEEE Trans. Automatic Control* **22**(3), 412–416.

Murata, T. (1989). Petri nets: properties, analysis and applications. *Proceedings of the IEEE* **77**(4), 541–580.

Peterson, J.L. (1981). *Petri net theory and the modeling of systems*. Prentice-Hall.

Reutenauer, C. (1990). *The Mathematics of Petri Nets*. Prentice-Hall.

Silva, M. and L. Recalde (2003). On fluidification of Petri nets: from discrete to hybrid and continuous model. In: *Proc. IFAC Conf. on Analysis and Design of Hybrid Systems*. pp. 9–20.

Silva, M., E. Teruel and J.M. Colom (1998). Linear algebraic and linear programming techniques for the analysis of place/transition net systems. *Lecture Notes in Computer Science: Lectures on Petri Nets I: Basic Models* **1491**, 309–373.

Sreenivas, R.S. (2002). On minimal representations of Petri net languages. In: *Proc. 6th Int. Work. on Discrete Event Systems (Zaragoza, Spain)*. pp. 237–244.

Sreenivas, R.S. and B.H. Krogh (1992). On Petri net models of infinite state supervisors. *IEEE Trans. Automatic Control* **37**(2), 274–277.

Stremersch, G. and R.K. Boel (2002). Structuring acyclic Petri nets for reachability analysis and control. *Discrete Event Dynamic Systems, Volume 12, Issue 1* pp. 7–41.

Yamalidou, E., J.O. Moody, P.J. Antsaklis and M.D. Lemmon (1996). Feedback control of Petri nets based on place invariants. *Automatica* **32**(1), 15–28.

ELSEVIER
IFAC
PUBLICATIONS
www.elsevier.com/locate/ifac

CONTROL SYNTHESIS FOR PRODUCT-DRIVEN AUTOMATION

David GOUYON, Jean-François PETIN, Gérard MOREL

Nancy Research Centre for Automatic Control (CRAN)
Henry Poincaré University (UHP)
{gouyon, petin, morel}@cran.uhp-nancy.fr

Abstract: We study the problem of controlling a flexible manufacturing cell by means of automatically computed supervisors. Such a flexible control can be seen as a modular combination of product-driven supervisors and components-driven ones, mainly for control integration in a Manufacturing Execution System. Our method associates a top-down synthesis approach based on customized product specification and a bottom-up synthesis approach based on standardised components on the shelves. Practical result is a routings supervisor devoted to each product occurrence which is consistent with the manufacturing stations supervisors. *Copyright © 2004 IFAC*

Keywords: Control synthesis, Flexible Automation, Discrete-Event Systems, Intelligent Manufacturing Systems.

1. INTRODUCTION

The design of control systems (Cassandras and Lafortune 1999) consists in defining the (unknown) control rules of the (known) dynamics of a physical system that satisfy some (known) behavioural goals according to the following predicate (Fusaoka *et al.*, 1983):

$$\textit{Dynamics} \wedge \textit{Unknown Control Rules} \supset \textit{Goal} \quad (1)$$

In this way, the Supervisory Control Theory (SCT) (Ramadge and Wonham, 1987) defines a formal framework aiming to analyse Discrete Event Systems (DES) and provides algorithms (Wonham and Ramadge, 1987)(Kumar and Garg 1995) that enable an automatic synthesis of supervisory controllers in such a way that the controlled plant behaves according to some given goals.

This paper studies the application of such a formal approach for the design of the control of a Flexible Manufacturing System (FMS), mainly for control integration in a Manufacturing Execution System. Control flexibility is seen as the combination of two apparently opposite properties: re-configurability of the control system to manage the variability and customisation of the products, interoperability of the control devices that promotes an automation based on standardised and reusable components on the shelves. This problem has been partially addressed in the area of FMS control synthesis. Indeed, some approaches highlight the needs to introduce generic models of the workstations in order to synthesise modular and component-based controllers according to a given set of manufacturing routes (Lauzon *et al.*, 1997) while other ones integrate product features within the synthesis method in order to generate automatic routings and allocation of resources (i.e. workstations) according to their given controlled behaviour (Lennartson *et al.*, 1998).

Our objective is to automatically synthesise FMS controller that combines product-driven supervisors that accommodate alternate manufacturing routes when needed and component-based supervisors that control the workstations behaviour according to the product requests.

This paper presents an attempt toward such a product-driven control synthesis through the study of the flexible assembly cell presented in section 2. An overview of our method is given in section 3. It combines a top-down approach for the synthesis of supervisors associated to each product occurrences, with a bottom-up synthesis that progressively aggregate field devices supervisors to give rise to the assembly stations supervisors. These approaches are illustrated in sections 4 and 5. In section 6, some conclusions and prospects are drawn.

2. CASE STUDY: FLEXIBLE ASSEMBLY CELL

Our case study is a *Flexible Assembly Cell* of the training centre AIP-PRIMECA[1] Lorraine. This application enables flexibility in the routing of products assembly through six workstations: one for loading pallets circulating on a conveyor, four similar

[1] http://www.aipl.uhp-nancy.fr/lorraine/

workstations to built products by parts assemblage, and one to unload products on pallets (Fig. 1). Six different product families can be assembled (Fig. 2). The manipulation of parts and products is ensured by a vacuum generator and a manipulator made of three air cylinders (Fig. 3).

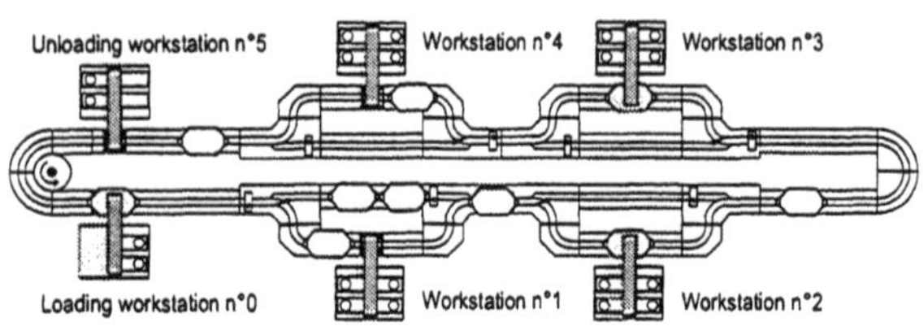

Fig. 1. Flexible Assembly Cell

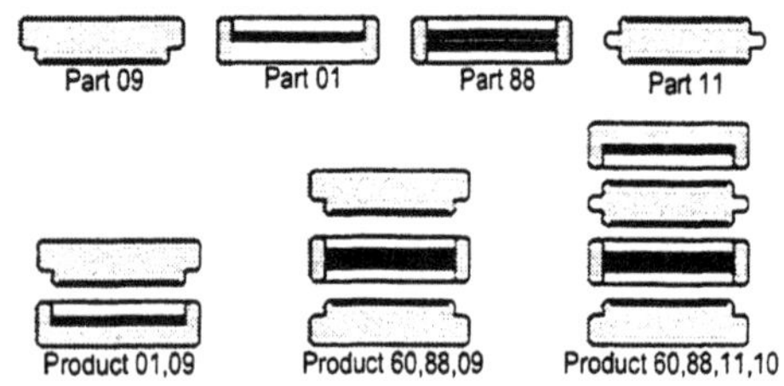

Fig. 2. Examples of products assemblies

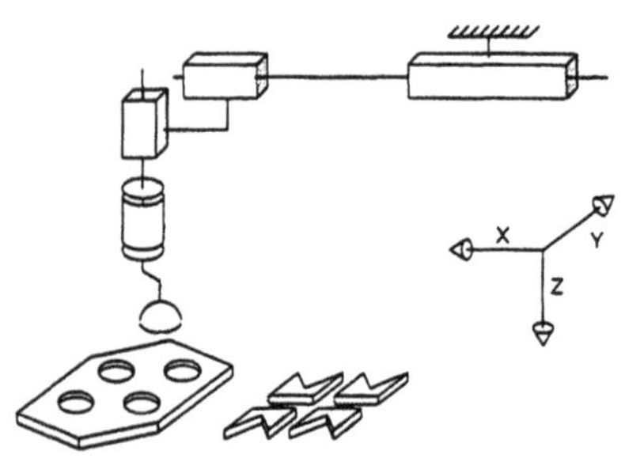

Fig. 3. Cinematic schema of workstation manipulators

Control flexibility of this assembly cell is ensured by applying a product-driven way of automation (McFarlane *et al.*, Valckenaers *et al.*, in Morel and Grabot 2003). It means that the product itself is assumed to embed abilities for information processing, decision making and communication in order to have an active role in production control and scheduling. From a technological point of view, this approach is supported by recent evolutions in the area of micro-electronics and auto-identification. In this way, pallets circulating on the *Flexible Assembly Cell* are equipped with RFID tags that enable automatic identification of products and the storage of the eligible assembly routes. Control of the product routes is not directly embedded into the product tag but is implemented in intelligent remote I/O that support the tag read/write procedures.

Consequently, structure of the cell control is divided in two main parts: the product controllers and the assembly stations controllers. The first ones have to manage the product routing within the cell. Associated to each product occurrence, these controllers involve a routing map that define the whole alternate routes according to stations capabilities and a route controller that dynamically chooses the accurate way within the cell according to simple criteria such as the stations availability or more complex ones such as product priority or stations current and future status.

The controllers of the assembly stations are called by the product controllers in order to perform one or more services required by the product. Main functions of these controllers are the management of the parts feeding and the coordination of the manipulator moves in order to assemble a given part on the product.

The following sections study the problem of the design of such a product-driven control by using the SCT framework and synthesis techniques in order to satisfy flexibility, interoperability and modularity properties.

3. OVERWIEW OF THE APPROACH

Our approach is based on two consistent interpretations of the predicate (1).

3.1. Synthesis of product supervisors

The first one aims to synthesise the product supervisors that manage the product routings according to product specification, i.e. the product manufacturing plan given in terms of operations that must be applied on the product, and according to the various resources capabilities of the manufacturing processes, i.e. the workstations services given in terms of supplied operations. Predicate (1) becomes:

Manufacturing processes behaviour
$\wedge$ *Unknown product-driven control rules*
$\supset$ *Product manufacturing planning* (2)

Product manufacturing plans are elaborated by the product designer. From a control point of view, this ordered set of manufacturing operations the product has to undergo can be described by a finite-state automaton (Fig. 4), which alphabet is composed of uncontrollable events. These events correspond to reports of tasks achievement.

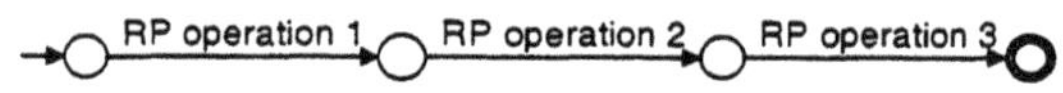

Fig. 4. Example of operational product planning

The model of the manufacturing processes involves both a description of the workstations behaviour and of the cell topology. According to (Wright and Bourne, 1988) (Lauzon *et al.*, 1997) and our previous works in manufacturing automation, any device can be characterised by three main states: *working*, *idle* and *down*. We extend this generic model by splitting the *working* state into several states associated to the realisation of the different supplied operations and we restrict it by cancelling the *down* state with regard to our modelling objective aiming to consider first the normal behaviour (Fig. 5).

Operations requests (noted RQ) and achievement reports (noted RP) are the events seen at the product level. They respectively correspond to the events that launch a manufacturing operation (e.g. request for drilling or tapping) and indicate that the operation is well achieved (e.g. end of drilling or tapping report).

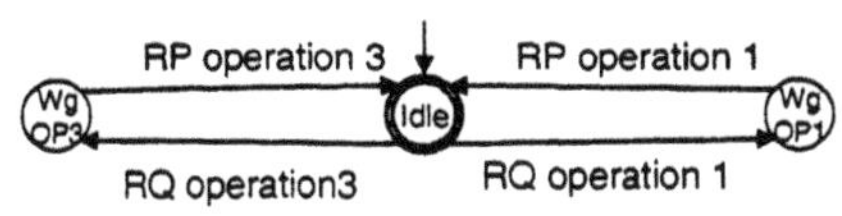

Fig. 5. Model of a workstation: an example

Model of the manufacturing processes is given by a certain composition of all workstations models according to the cell topology. This composition does not result from a synchronous product of the generic workstations models. Indeed, it has to take into account the cell topology in terms of spatial distribution of the workstations and of the transport resources between them. At that time, this analysis is supposed to be done by hand. An example of such a composition is given by (Fig. 6). States 1 and 2 are known as *idle state* of respectively the workstations 1 and 2 while states 3 and 4 represent the use of transport resource between these two workstations.

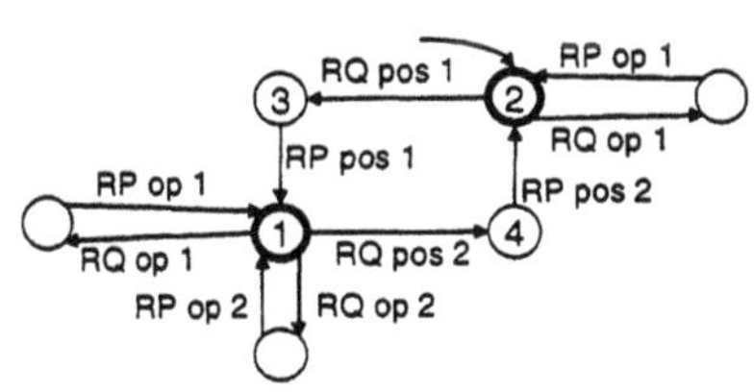

Fig. 6. Example of process model: 2 workstations.

Classical synthesis techniques can then be used to obtain a supervisor that controls the alternate routes of a product within the cell. This supervisor is expected to be generated in-line and implemented into tags of product occurrences.

3.2. *Synthesis of the workstations supervisors*

A second interpretation of predicate (1) is used in order to provide a modular control for the manufacturing workstations according to:

Workstation processes dynamics
$\wedge$ *Unknown workstation control rules*
$\supset$ *Workstation behavioural goal* (3)

Our approach is built on a component-based way of modelling that aggregate, in a bottom-up approach, simple behaviour to give rise to more complex ones. In this way, automation object-oriented methods provide guidelines for a structured design of the control systems starting from generic models of the field device behaviour. Note that these models may be elaborated by the designer, defined by a standard (functions blocks of the IEC 61499 for example) or given by the supplier of intelligent devices that embed their own local controllers (Component On The Shelves). Resulting control is modular and hierarchically coordinated. Complexity of models and combinatory explosion are decreased thanks to this modular decomposition.

Synthesis algorithms are applied to automatically generate the supervisor associated to each module of the control architecture. For a given module, the supervisor will be computed from specifications describing the mission allocated to this module and from a process model (*generator*) that is given by the projection and the synchronous product of the lower level supervisors associated to its resource modules. Note that projection of the lower level supervisors is done in order to only keep the alphabet understandable by the coordination module. This approach is iteratively applied for each hierarchical level of the control architecture (Fig. 7).

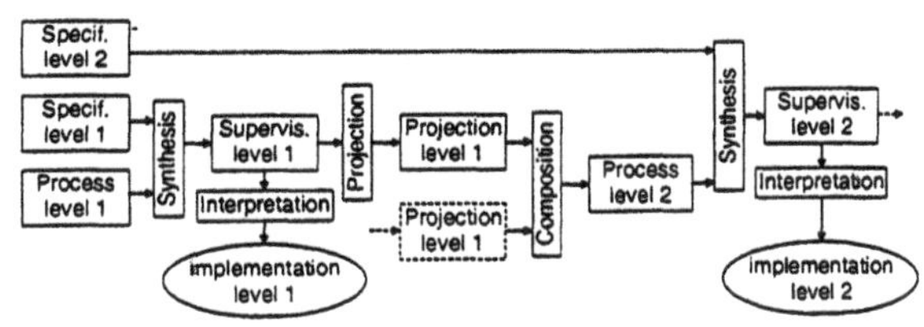

Fig. 7. Modular synthesis principle

Further details about this incremental approach of control synthesis and implementation are given in (Gouyon *et al.* 2004).

3.3. *Consistency between product and workstations supervisors*

Product and workstations supervisors resulting from the above top-down and bottom-up approach are assumed to work together: product supervisors invoke operations to be realised under the control of workstations supervisors which provide in return the product supervisors with information allowing product traceability. In order to be consistent, these two families of supervisors have to share a common alphabet.

This constraint is satisfied by considering that the workstation models involved in the manufacturing processes model, which is used as "*generator*" in the predicate (2), are used again, but as "*goal specification*" basis in the predicate (3). These models can indeed be seen as the desired behaviour of each workstation from the product point of view. A complete model of the goal specification used in predicate (3) is then obtained by enriching the workstation models with the internal functioning rules of the workstation. Local rules to be satisfied by the management of part feeding or manipulator moves are examples of these added properties.

In other words, consistency proof between the top-down and bottom-up approaches is given by demonstrating that the model of the workstation behaviour as used in predicate (2) is in fact a projection of the model of the workstations goal involved in the predicate (3). Such a projection keeps the only events that are shared by the alphabets used in the models of predicate (2) and (3), i.e. events related to operation report and request.

In order to illustrate the contents of this section, our structured approach is applied for the synthesis of product and workstation supervisors of the *Flexible Assembly Cell* presented in section 2.

4. PRODUCT-DRIVEN CONTROL SYNTHESIS

First step deal with the synthesis of product supervisors from a description of product manufacturing plans and available assembly stations. Let's take the example of supervisor synthesis for a product 01-09 (represented in Fig. 2), i.e. a product obtained by assembling part 01 with part 09. The alphabet used for this application is given in Table 1.

Label	Significance
rq n	Request for assembling of a part of type n
rp n	Report of assembling of a part of type n
rq pos j	Request of move to workstation j
rp pos j	Report of move to workstation j achievement

Table 1: alphabet for synthesis of product supervisors

The product designer gives product manufacturing plans. In their simplest form, they can be represented as a finite state automaton where the states represent the different assembly step from raw to final product and where transitions are validated by the achievement report of the needed operations (Fig. 8).

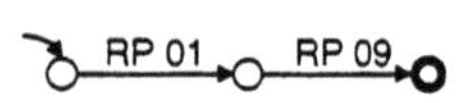

Fig. 8. Manufacturing plan for product 01-09

Workstation models presented in (Fig. 9) are designed by adapting the generic model of the section 3.1. (Fig. 5) to the following capabilities: workstation 0 stores parts 60 and 01, workstation 1 stores parts 10, workstation 2 stores parts 09, workstation 3 is unused, workstation 4 stores parts 09 and 10

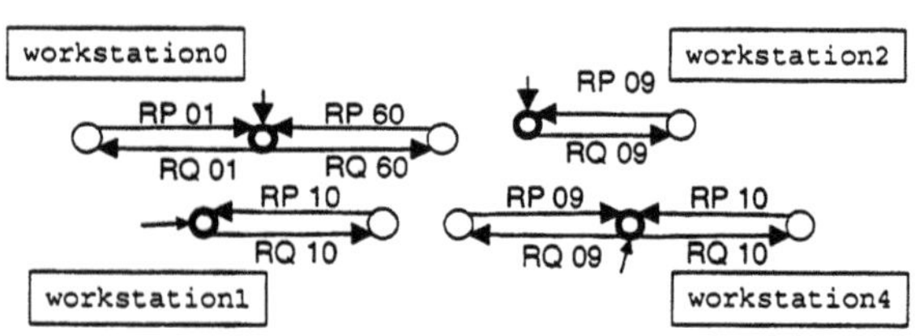

Fig. 9. Workstations models

The model of Manufacturing Processes as used in predicate (2) is then obtained by a particular composition of the workstation models which takes into account the cell topology: pallets can only transport one part, workstation 0 is the loading workstation for all products, workstation 5 unloads products, a product is assumed to be built in an only one cycle within the cell.

According to these restrictive constraints, but representative enough for our purpose, the model of the manufacturing processes is given by (Fig. 10). Initial state 1 means that the product has to begin its cycle within the cell by the loading workstation 0. States 2, 3, 4 and 8 represent the reservation of transport resource toward respectively workstations 1, 2, 4 and 5. Cycle ends at state 9 by the unloading operation performed by workstation 5. Links between transport states 2,3,4,8 and the workstations states 1,5,6,7 strictly depends on the cell topology (mandatory workstations, parallel circuits, etc).

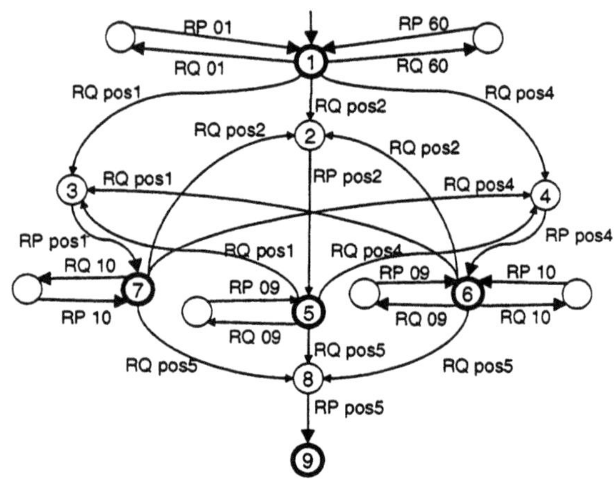

Fig. 10. Manufacturing processes model

A product supervisor can then be computed using TCT[2] tool for the synthesis procedures (Fig. 11). This supervisor limits the behaviour of the manufacturing processes to the only one that are compliant with the product specification. In other words, it represents the maximal set of alternate routes for a given product within the cell.

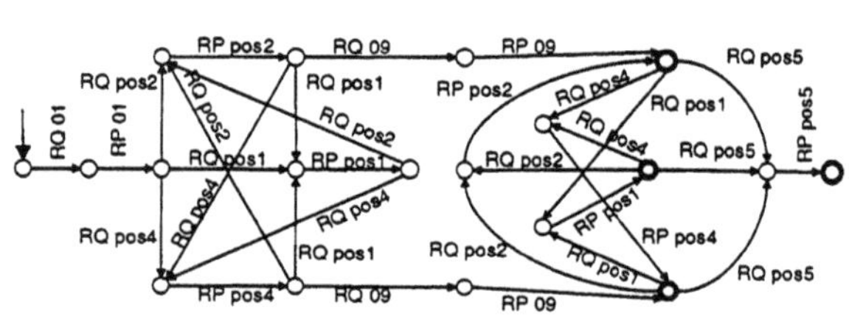

Fig. 11. Alternate routes for product 01,09

This supervisor can be used as the product controller when considering request as controllable events and report as uncontrollable ones that can be respectively emitted/received to/from the workstation controllers. Note that this controller is quite limited with regards to complex decision making we can wait for in the context of a product-driven automation. That means that specification of Fig. 8 should be enriched by negotiation and anticipation rules that could be based on product priority within the cell and workstation availability in a near future.

[2] Operations upon automata (product, projection, synthesis) are made using TCT software tool from the University of Toronto (http://odin.control.toronto.edu/DES/)

This kind of rules is not easily captured through finite state automata and can not consequently be integrated within synthesis procedures. It rather refers to Distributed Artificial Intelligence techniques applied in agile automation, which we consider out of the scope of control synthesis.

5. WORKSTATIONS CONTROL SYNTHESIS

Second step aims at synthesising supervisors for the workstations. Our bottom-up approach starts with the supervisors of the field devices (step 1: actuators, transmitters, etc). These supervisors can be provided by the devices suppliers or by international standard, such as the IEC 61499 which defines some behavioural functions blocks for the distributed applications. If no standardised components on the shelves exist, generic supervisor for a given class of field device can be synthesised. Let us consider this last case. Generic field devices of the Flexible Assembly Cell are the actuators involved in a workstation – mainly air cylinders and vaccum generator – and their associated magnetic sensors.

Table 2 gives the alphabet used in this application.

Label	Significance
rqR	Cylinder Retracting request
rqP	Cylinder Pushing request
rpR	Cylinder Retracted position report
rpP	Cylinder Pushed position report
Rord	Cylinder Retracting order
Pord	Cylinder Pushing order
Rpos	Cylinder in retracted position (given by detectors)
Ppos	Cylinder in pushed position (given by detectors)
rqS	Sucking request to vacuum cups
rqF	Freeing request for vacuum cups
rpS	Sucked Part report
rpF	Free Part report
Sord	Sucking order for vacuum cups
Ford	Freeing order for vacuum cups
rqT	Rise request (go to top)
rqB	Go down request (go to bottom)
rpT	"At the Top" report
rpB	"At the Bottom" report
rqPK	Pick request
rqPL	Place request
rpPK	Part picked
rpPL	Part placed

Table 2: alphabet for workstations supervisor synthesis

5.1. Level 1: generic supervisors synthesis for devices

The workstations manipulators double-acting air cylinders with their control valve are represented (Fig. 12) by two marked states, drawn by a double circle, in which the observable devices behaviour is steady (cylinder in a pushed or retracted position) and by two moving states which represent the evolution from a steady state to another one.

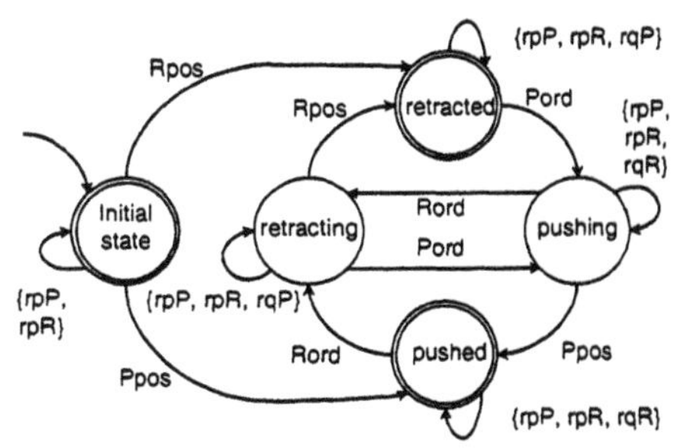

Fig. 12. Model of a double acting air cylinder with a 5/2 bi-stable valve

The specification of the devices behaviour control is done, for the air cylinders, by four automata (Fig. 13): two describing the rules that filter some pushing and retracting requests according to the device current state, and two other automata describing the observations filtering.

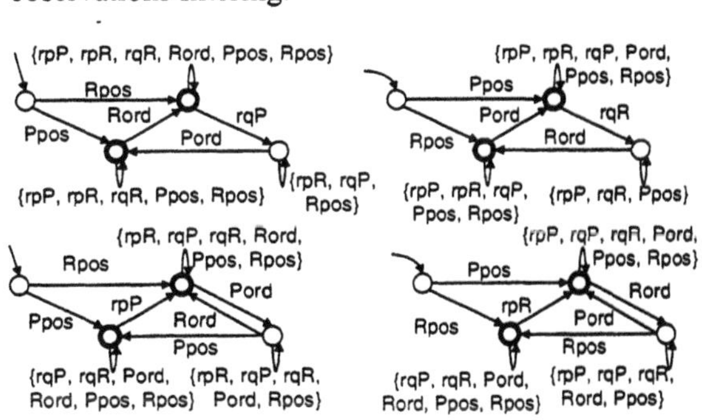

Fig. 13. Specifications for the 5/2 bistable air-cylinder

The complete specification of the cylinder control module results from the synchronous product of these four automata (21 states, 68 transitions). The same approach is used for the specification of the vacuum system (8 states, 25 transitions), and all other devices.

TCT[2] tool is then used to generate the supremal controllable sublanguage that defines the most permissive supervisor (Fig. 14) as usually done in a SCT synthesis framework.

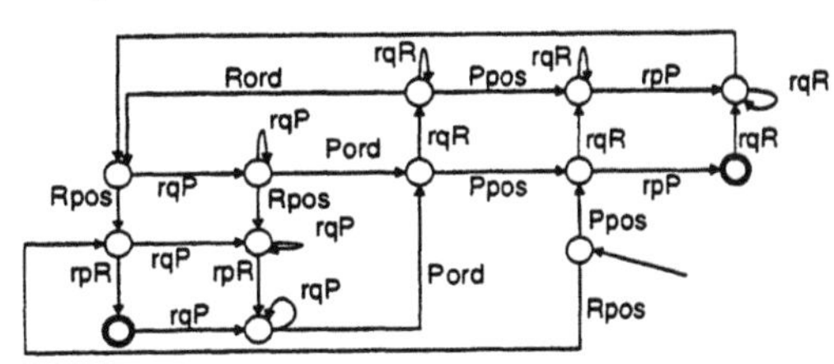

Fig. 14. Supervisor of a 5/2 bistable air- cylinder

5.2. Level 2: coordination modules synthesis

At a higher level, manipulator supervisor can then be seen as a coordinating module that synchronise the field devices in such a way that a requested assembly operation is performed. Process model (*generator* of SCT framework) used for the synthesis of this coordinating supervisor is automatically built from the local supervisors of the field devices, by:

- 1. projection of the local supervisors by preserving the only events interacting with the superior level,
- 2. controllability status modification (*requests* are now seen as controllable while *reports* are now seen as uncontrollable),
- 3. instantiation of the generic supervisors to each of the field devices; it is done by renaming the projected events (for example by adding an "h" for the horizontal cylinder) to create a specific alphabet for each instances (Fig. 15),
- 4. synchronous product of automata so obtained.

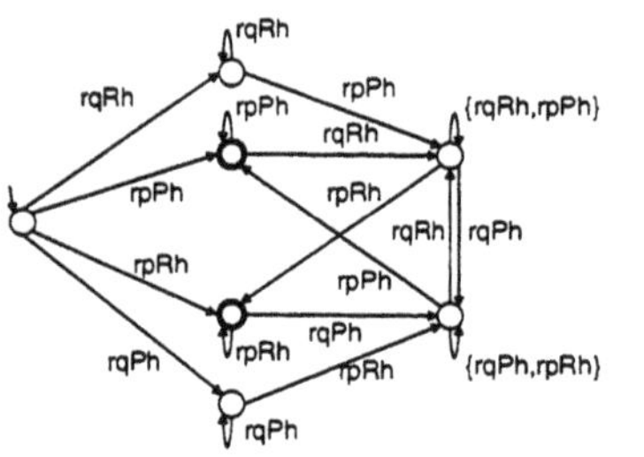

Fig. 15. Projection and instantiation of the air-cylinder generic supervisor

The specification models describe coordinating rules of the field devices in order to realize coordinated actions. An example (Fig. 16) is a "pick and place" function, realized by two field devices (a cylinder and its 5/2 bi-stable valve, and a holding system with a vacuum generator).

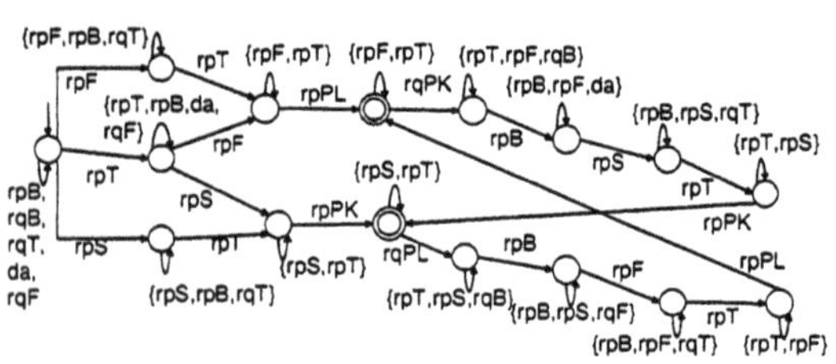

Fig. 16. Coordinating supervisory specification

TCT[2] tool is then used to generate the supervisor for the coordinating module (Fig. 17).

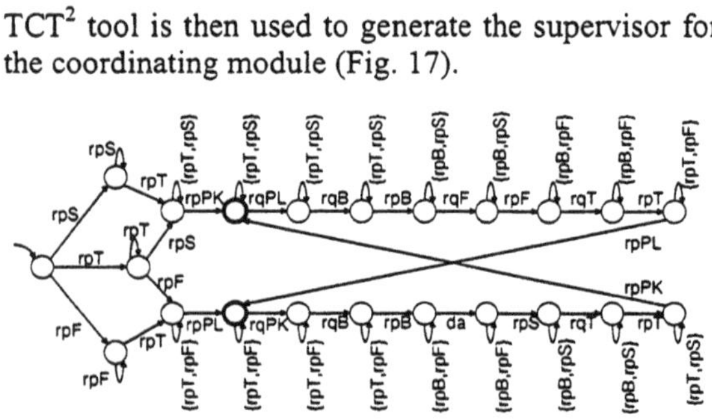

Fig. 17. Coordinating supervisor

This iterative way of modelling and synthesising can be applied again to lead to the workstation supervisor that synchronise manipulator moves (supervisor of Fig. 17) and parts feeding. Its alphabet is shared with the one of product supervisors. Interoperability of product and workstations supervisors is ensured by sharing the same workstation models as *generator* for the first ones and as *specification* for the second ones.

5.3. Level 3: workstation control synthesis

Using the same approach, with a specification model including the alphabet of operations control (Fig. 9), workstations controllers are synthesized.

CONCLUSION

This work is a part of a research project on product-driven automation for product traceability purposes (Morel *et al.* 2003). This paper presents a product-driven automation method based on SCT in order to automatically generate a modular control that satisfies flexibility and interoperability properties. First practical result consists in the synthesis for product-instances supervisors accommodating alternate manufacturing routes. Second practical result consists in the synthesis of consistent supervisors controlling workstations behaviours when executing a requested operation. Works in progress should validate this approach by implementing these synthesized supervisors into IEC 61499 function blocks through their transformation into controllers as addressed in (Gouyon *et al.*, 2004). That implies to support this automation method by the interoperation of two tools, TCT for the synthesis phase and Control Build[3] for the implementation phase.

REFERENCES

Cassandras C.G., Lafortune S., (1999). *Introduction to discrete event systems*. Kluwer Academic. ISBN 0-7923-8609-4.

Fusaoka A., Seki H., Takahashi K., (1983). A description and reasoning of plant controllers in temporal logic. *Proceedings of the 8th International Joint Conference on Artificial Intelligence*, Karlsruhe, Germany.

Gouyon D., Pétin J.-F., Gouin A., (2004). Pragmatic approach for modular control synthesis and implementation. *International Journal of Production Research*, to appear.

Kumar R., Garg V., (1995). *Modeling and control of logical discrete event systems*, Kluwer Academic. ISBN 0-7923-9538-7.

Lauzon S.C., Mills J.K., Benhabib B., (1997). An implementation methodology for the supervisory control of flexible manufacturing workcells. *SME, Journal of Manufacturing Systems*, **vol. 16, n°1.**

Lennartson B., Tittus M., Fabian M., Hellgren A., (1998). Specification structures for supervisory control. *Proceedings of the 4th WODES*, Cagliari, Italy.

Morel G., Grabot B., (Eds.), (2003). Intelligent Manufacturing. *Special issue of Engineering Applications of Artificial Intelligence*, **vol. 16 (4).**

Morel G., Panetto H., Zaremba M., Mayer F. (2003) Manufacturing *Enterprise Control and Management System Engineering: Rationales and Open Issues*. Annual Reviews in Control. **vol. 27, n°2.**

Ramadge P.J., Wonham W.M., (1987). Supervisory control of a class of discrete event processes. *SIAM Journal of Control and Optimization*, **vol. 25, n°1.**

Wonham W.M., Ramadge P.J., (1987). On the supremal controllable of a given language. *SIAM Journal of Control and Optimization*, **vol. 25, n°3.**

Wright P.K., Bourne D.A., (1988). *Manufacturing intelligence*. Addison-Wesley, ISBN 0-201-13576-0.

[3] http://www.controlbuild.com/

Copyright © IFAC Discrete Event Systems,
Reims, France, 2004

ELSEVIER
IFAC
PUBLICATIONS
www.elsevier.com/locate/ifac

IMPLEMENTATION OF LOCAL MODULAR SUPERVISORY CONTROL FOR A PNEUMATIC SYSTEM USING PLC

İ. Tolga Hasdemir **Salman Kurtulan** **Leyla Gören**

Department of Electrical Engineering, Istanbul Technical University
34390, *Maslak, Istanbul, Turkey*
{hasdemir, kurtulan, goren}@elk.itu.edu.tr

Abstract: In this study, we implement a controller, based on Supervisory Control Theory on a pneumatic assembly system. We use local modular approach, which exploits modular structure of the plant and the specifications. An implementation methodology using a hierarchical control structure where PLCs are utilized is introduced. Local modular approach, together with the implementation methodology presented in this study provides an effective way for synthesizing and realizing supervisors for DES control problems. The resulting PLC program is also modular in structure, making it easy to modify and detect errors. *Copyright © 2004 IFAC*

Keywords: Supervisory control; discrete-event systems; modular control; programmable logic controllers

1. INTRODUCTION

An important portion of today's manufacturing systems can be classified as event driven systems or discrete event systems (DESs). Although it is possible to analyse and control this class of industrial systems formally, general methods applied in industry are intuitive rather than being formal. For small sized DES control problems, intuitive methods may yield practical solutions, but as the controlled system gets larger and more complex, formal methods need to be applied.

The supervisory control theory (SCT) introduced by Ramadge and Wonham (1987), provides a powerful framework for control of discrete event systems. The theory enables synthesis of closed loop control systems for DESs. Although the SCT has received a wide acceptance in the academy and some applications of SCT have been reported in the literature (Balemi, *et al.*, 1993; Brandin, 1996; Leduc, 1996; Queiroz and Cury, 2002) it has not been accepted in the industry yet. This is mainly due to the difficulties arising in physical implementation of SCT (Leduc, 1996; Fabian and Hellgren, 1998).

Programmable Logic Controllers (PLCs) have been used in industrial applications for more than 25 years, and today most of the automated manufacturing systems use PLCs as control units. Therefore the potential physical platform to realize a supervisor in industry is the PLC.

Finite state automata are used for representing the plant model and the supervisors in SCT. Implementation of SCT necessitates an appropriate method for developing a PLC program corresponding to the automaton that represents the theoretical supervisor. Therefore implementing the SCT is a matter of developing an appropriate PLC program, which will make the PLC behave as an automaton. The methods for developing PLC codes for this purpose and the problems that may arise in doing so are discussed in (Leduc, 1996; Fabian and Hellgren, 1998; Queiroz and Cury, 2002). In particular, some implementation concepts and a control method using Grafcet is proposed in (Zaytoon and Carré-Ménétrier, 2001).

One of the main difficulties arising in SCT implementations is state-space explosion. System models are generally built as combination of subsystems and the number of states of the global system grows exponentially with the number of subsystems. Computational complexity caused by large number of states (may be more than 10^{20}) can make it impossible to design and implement a supervisor for a given DES control problem. Therefore some formal and informal techniques are used to reduce computational complexity either by reducing sizes of the automata or using different control structures comprising more than one supervisor. Informal techniques generally depend on the designer's experience and preferences such as ignoring some of the details (i.e. some of the states of the physical system) that are not necessary to appear for the given control problem (Leduc, 1996). Modular control, as a formal method, allows designing multiple supervisors whose combining control action is equivalent to a monolithic supervisor for a given system (Wonham and Ramadge, 1988). In this way, memory used for storing states of the supervisor can be reduced. Another formal method, called Local Modular Supervisory Control, extending the results of (Wonham and Ramadge, 1988) reduces computational complexity and provides significant memory savings by using multiple supervisors and local models of the plant (Queiroz and Cury, 2000a; Queiroz and Cury, 2000b). A necessary and sufficient condition called local modularity is given in this method.

In this study, implementation of local modular supervisory control for an educational pneumatic assembly system is presented and a hierarchical control structure for modular supervisory control is given. It is shown that modularity condition holds for the problem at hand. As the last step we introduce an implementation methodology realized by a PLC. This implementation methodology allows considering controllable events that drive the system as in the SCT.

The organization of the paper is as follows. A brief introduction to DESs, SCT and Local Modular Approach is given in Section 2. The pneumatic system is introduced in Section 3. Section 4 discusses modelling of the system and design of the supervisors. Finally, Section 5 introduces control system structure and explains implementation of the supervisors.

2. PRELIMINARIES

Discrete event systems evolve on spontaneously occurring events. Let Σ be finite set of the events that drive the system. The set of all finite concatenations of events in Σ is denoted as Σ^*. An element in this set is called a string. The number of events gives the length of the string. The string with zero length is denoted as ε. A subset $L \subseteq \Sigma^*$ is called a language over Σ. For a string $s \in \Sigma^*$, $\overline{s}$ denotes the prefixes of s and is defined as $\overline{s} = \{s_p \in \Sigma^* : \exists t \in \Sigma^* (s_p t = s)\}$. Extending this definition to languages we get prefix closure of a language L denoted as $\overline{L}$. A language L satisfying the condition $L = \overline{L}$ is said to be prefix closed (Cassandras and Lafortune, 1999).

An automaton, denoted by G, is a six tuple $G = (Q, \Sigma, f, \Gamma, q_0, Q_m)$ where Q is the set of states, Σ is the finite set of events. f: $Q \times \Sigma \rightarrow Q$ is the partial transition function on its domain. Γ: $Q \rightarrow 2^{\Sigma}$ is the active event function. $\Gamma(q)$ is the set defined for every state of G and represents the feasible events of q. q_0 is the initial state and $Q_m \subseteq Q$ is the set of marked states representing a completion of a given task or operation. The language generated by G is denoted by $L(G)$ and defined as $L(G)=\{s \in \Sigma^* : f(q_0,s) \text{ is defined}\}$. The language marked by G is denoted by $L_m(G)$ and defined as $L_m(G)=\{s \in \Sigma^* : f(q_0,s) \in Q_m\}$. A DES G is said to be nonblocking if $L(G) = \overline{L_m(G)}$. Product and parallel composition operations are defined for expressing two forms of joint behaviour of multiple automata that operate concurrently. Product of two automata, say G_1 and G_2 is denoted as $G_1 X G_2$ and represents the synchronous behaviour of the two automata. Thus, an event occurs in the resulting automata if and only if it occurs in both automata. The generated and marked languages of the resulting automaton will be $L(G_1 \text{ X } G_2) = L(G_1) \cap L(G_2)$ and $L_m(G_1 \text{ X } G_2) = L_m(G1) \cap L_m(G2)$, respectively. Parallel composition of automata G1 and G2 is denoted as G1||G2. In the resulting automaton common events occur synchronously, while the other events occur asynchronously. Therefore, if the event sets Σ_1 and Σ_2 of the automata are equal then parallel composition of G1 and G2 will be equivalent to the product of G1 and G2 (Cassandras and Lafortune, 1999).

The supervisory control theory (SCT) uses formal languages to model the uncontrolled behaviour of discrete event systems and specifications for the controlled behaviour. The objective is to restrict the behaviour of the system to a desired behaviour, which is represented by the specifications. This is done by disabling some events to prevent some of undesired strings from occurring in the system. The decision of which event to be disabled is made by another simultaneously executing system, called the supervisor. The supervisor is also represented by an automaton. The active event set associated with a state of the supervisor gives the events that are allowed to occur in the corresponding state of the controlled system.

In SCT events are divided into two classes as controllable events and uncontrollable events. The set of controllable events is denoted by Σ_c, while Σ_{uc} represents the uncontrollable event set. The supervisor has no effect on uncontrollable events, which means that the supervisor cannot prevent an event of Σ_{uc} from happening. The existence of a supervisor is guaranteed if the desired language satisfies a condition. This condition is called as the controllability condition and defined as $\overline{K}\Sigma_{uc} \cap M \subseteq \overline{K}$ where $\overline{K}$ is the language that will be generated under the control of the supervisor and M is the language generated by the uncontrolled system.

The behaviour of a DES G, under the control of supervisor S is denoted by *L(S/G)*. A language K is called $L_m(G)$-closed when the language satisfies the condition $K = \overline{K} \cap L_m(G)$. If the desired behaviour $K \subseteq L_m(G)$ is controllable and $L_m(G)$-closed, then $\overline{K} = L(S/G) = \overline{L_m(S/G)}$, and this means that the controlled behaviour is nonblocking.

Modular Supervisory Control guarantees nonblocking of DES controlled by multiple supervisors if the modularity condition $\overline{\bigcap_i L_m(S_i/G)} = \bigcap_i \overline{L_m(S_i/G)}$, holds. Local Modular Control approach extends this result to exploit the modularity of the controlled system G. When a DES G is given as a composition of $G_{loc,j}$, j=1,...,m, local modularity condition which is given as $\overline{\|_{j=1}^{m} L_m(S_{loc,j}/G_{loc,j})} = \|_{j=1}^{m} \overline{L_m(S_{loc,j}/G_{loc,j})}$ is equivalent to global modularity (Queiroz and Cury, 2000a; Queiroz and Cury, 2000b). While modelling a system as a combination of subsystems, local modular approach uses product system representation (PSR) (Wonham and Ramadge, 1988; Queiroz and Cury, 2000b) of the global system. In the PSR, subsystems are modelled by asynchronous automata.

3. THE PNEUMATIC SYSTEM

The pneumatic system (SMC-MAP 205) shown in Figure 1 is a flexible automation minicell built for educational purposes comprising pneumatic manipulators and a PLC (Siemens S7-200). It demonstrates the assembly of a rotary mechanism composed of a base, a bearing, a shaft and a cover. Assembly of the rotary mechanism is carried out by means of three manipulators and four cylinders. We will call the manipulators as Manipulator E, Manipulator H and Manipulator K. Manipulator E is used for picking up the bearing and inserting it into the base. Manipulator H carries out the process necessary for the assembly of the shaft by picking up and placing it inside the previously inserted bearing. The task of Manipulator K is to pick up and place the cover onto the base after the shaft is inserted into the bearing. The four cylinders are called Cylinder A, B, C and D.

Figure 1. The Pneumatic System (MAP 205)

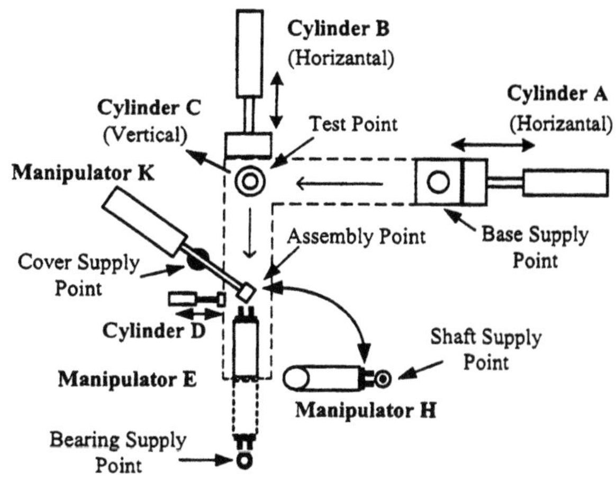

Figure 2. Schematic representation of the system

Cylinder A is used for feeding the base to the test point. Cylinder C tests a possible incorrect position of the base. The task of Cylinder B is to move the tested base to the assembly point. Finally, Cylinder D is used for ejecting the assembled mechanism. If the base fails the test carried out by Cylinder C, then Cylinder D ejects the base before the assembly process begins.

Figure 2 gives a schematic representation of the system. The supply mechanisms of the components are beyond the scope of this study and we assume that the components are supplied by a mechanism not modelled here.

The order of assembly is as follows. First a base arrives at the assembly point. Then a bearing is inserted into the base. Following insertion of the bearing a shaft is inserted into the bearing. And finally, a cover is placed onto the base.

The original control algorithm provided by the vendor allows only one rotary mechanism in an assembly process. Also, picking up and insertion operations of the components are done sequentially causing additional decrease in efficiency. The purpose of this study is to minimize the average time spent during the manufacturing of an assembled mechanism by an efficient operation. There are two steps to increase efficiency: One is to allow the system to work on more than one product simultaneously, and the other is picking up a supplied component immediately, therefore making the component ready for assembly before insertion/placement operation of the component begins. It is also necessary to prevent a possible collision of the product components, i.e. bases.

4. MODELLING AND CONTROL

The model of the uncontrolled system can be built in different levels depending on the control purpose. A low level model gives detailed information of the system such as sensor readings, actuator status and deals with operational procedures (Hasdemir *et al.*, 2003; Queiroz and Cury, 2002). However, this kind of modelling is not suitable for large-scale control problems, since the model may become extremely large. Therefore, in this study modelling is done at higher level, which defines start and end operations of subsystems. Lower level control algorithms are also developed using formal methods (Hasdemir *et al.*, 2003)

In the modelling stage, decomposition of the global system into subsystems exploits physical decomposition of the system. Therefore, there are seven models corresponding to the four cylinders and the three manipulators. Also there are four more models, which describe existence and absence of the supplied components.

Figure 3 shows the cylinder models. For $x \in \{A,B,C,D\}$, the controllable event α_X represents the start operation, while the uncontrollable event β_X represents the end operation of the corresponding subsystem. The uncontrollable events β_{BO} and β_{BNO} appearing in the model of cylinder C represent the test results of the base. If the base position is correct then event β_{BO} is generated, if not event β_{BNO} is generated.

Manipulator models are given in Figure 4. There are two different start and end operations defined for each manipulator. The first start operation is denoted by α_{Y1}, $Y \in \{E,H,K\}$, and this matches to start of picking up operation for the corresponding assembly component. The second start operation is denoted by α_{Y2} and represents start of insertion/placement operation of the component. β_{Y1} and β_{Y2} are used for end of picking up and insertion/placement operations, respectively. Figure 5 shows the automaton models of component existence and absence. The uncontrollable events br, sh, bs and cv correspond to the existence of bearing, shaft, base and cover respectively.

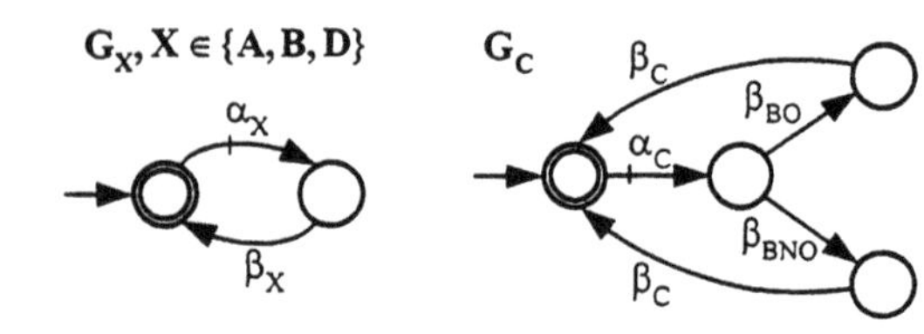

Figure 3. Cylinder Models

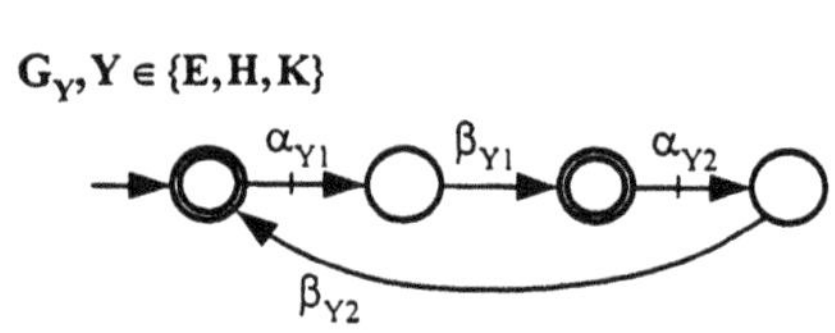

Figure 4. Manipulator Models

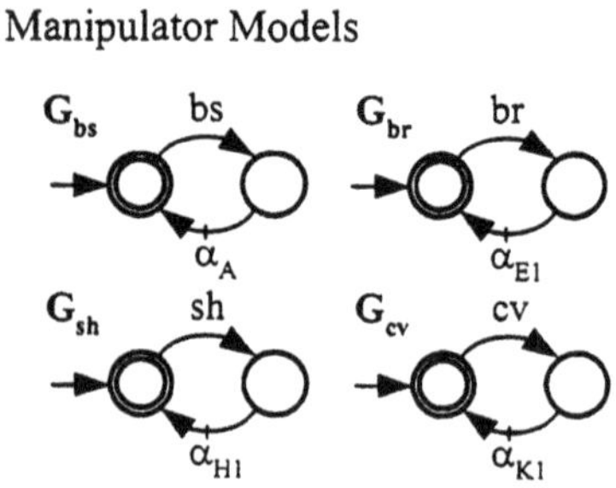

Figure 5. Automata modelling component existence

Cylinder Models and Manipulator Models together with component existence/absence models represent the *Composed System Representation* (Queiroz and Cury, 2000b) of the global system. In the Product System Representation, the following subsystem models should be obtained by parallel compositions.

$$G_{Abs} = G_A \,||\, G_{bs}\,,\; G_{Ebr} = G_E \,||\, G_{br}$$

$$G_{Hsh} = G_H \,||\, G_{sh},\; G_{Kcv} = G_K \,||\, G_{cv}$$

There are three groups of generic specifications corresponding to the desired behaviour of the system. The first group of specifications, shown in Figure 6, are called ordering specifications (OS) and manages the operation of subsystems in desired order. Depending on the test result obtained by Cylinder C, Manipulator E or Cylinder D may be started following the end operation of Cylinder B. This choice is done by OS2 and OS5.

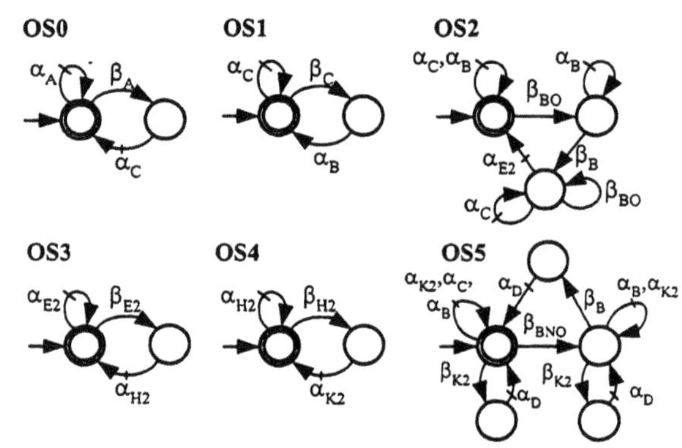

Figure 6. Ordering Specifications

The second group of specifications forces the manipulators to pick up the supplied component immediately and make the components ready for the next mode of operation, which is called insertion/placement operation. These specifications are called preparation specifications (PS). Note that although Cylinder A operates in one mode only, the specification regarding the operation of Cylinder A, named PS0, is included in this group of specifications as given in Figure 7.

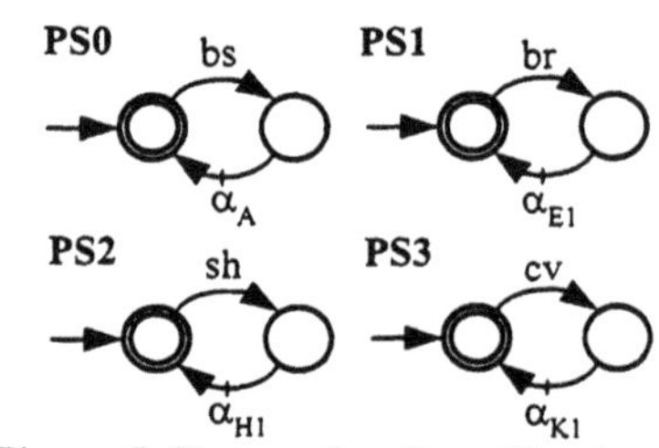

Figure 7. Preparation Specifications

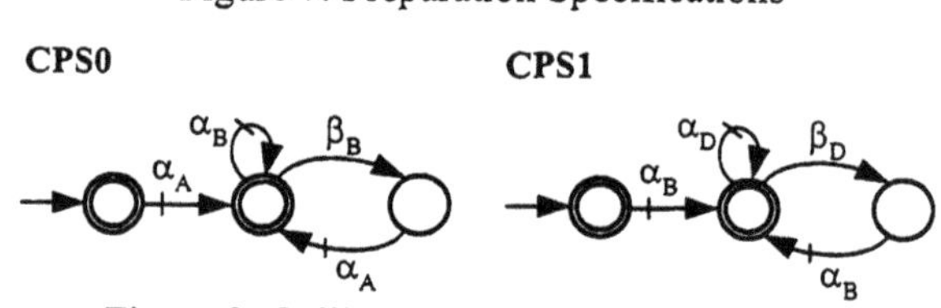

Figure 8. Collision Protection Specifications

The last group of specifications prevents possible collision of two bases, which may occur in the test point or assembly point. These specifications are called collision protection specifications (CPS), and are shown in Figure 8. The specification CPS0 does not allow Cylinder A to push the cover to the test point unless the test point is empty. Similarly, CPS1 disables event α_B until the assembly point is emptied by the occurrence of the event β_D.

As a next step in the local modular control approach, local models and the automata that mark the local specifications of the system are built as in Table 1. The computations necessary for Table 1 and verification of controllability and modularity have been performed by UMDES-LIB on a personal computer using Celeron 1700 processor.

It can be shown that each local specification is controllable with respect to its corresponding local system. Therefore, supervisors $S_{loc,x}$, x={a0,a1,a2,a3,a4,a5,a6,b0,b1,b2,c0,c1}, can be computed directly from their respective local specifications.

Table 1. Local Systems and Local Specifications.

Local System Model	Local Specification
$G_{loc,a0} = G_{Abs}\|\|G_C$	$H_{loc,a0} = OS0\|\|G_{loc,a0}$
$G_{loc,a1} = G_B\|\|G_C$	$H_{loc,a1} = OS1\|\|G_{loc,a1}$
$G_{loc,a2} = G_B\|\|G_C\|\|G_{Ebr}$	$H_{loc,a2} = OS2\|\|G_{loc,a2}$
$G_{loc,a3} = G_{Ebr}\|\|G_{Hsh}$	$H_{loc,a3} = OS3\|\|G_{loc,a3}$
$G_{loc,a4} = G_{Hsh}\|\|G_{Kcv}$	$H_{loc,a4} = OS4\|\|G_{loc,a4}$
$G_{loc,a5} = G_B\|\|G_C\|\|G_D\|\|G_{Kcv}$	$H_{loc,a5} = OS5\|\|G_{loc,a5}$
$G_{loc,b0} = G_{Abs}$	$H_{loc,b0} = PS0\|\|G_{loc,b0}$
$G_{loc,b1} = G_{Ebr}$	$H_{loc,b1} = PS1\|\|G_{loc,b1}$
$G_{loc,b2} = G_{Hsh}$	$H_{loc,b2} = PS2\|\|G_{loc,b2}$
$G_{loc,b3} = G_{Kcv}$	$H_{loc,b3} = PS3\|\|G_{loc,b3}$
$G_{loc,c0} = G_{Abs}\|\|G_B$	$H_{loc,c0} = CPS0\|\|G_{loc,c0}$
$G_{loc,c1} = G_B\|\|G_D$	$H_{loc,c1} = CPS1\|\|G_{loc,c1}$

Local specifications are obtained by parallel composition of generic specifications and local system models. Therefore, in the realization stage, when generic specifications are considered as local supervisors and are operated in parallel with the corresponding local systems, it is guaranteed that the resulting behaviour will be equivalent to the behaviour determined by local specifications. Figure 9 shows 3 of 13 local supervisors as an example. The labels appearing on some of the states of the supervisors correspond to enabling signals of the controllable events that drive the system as defined in section 5.

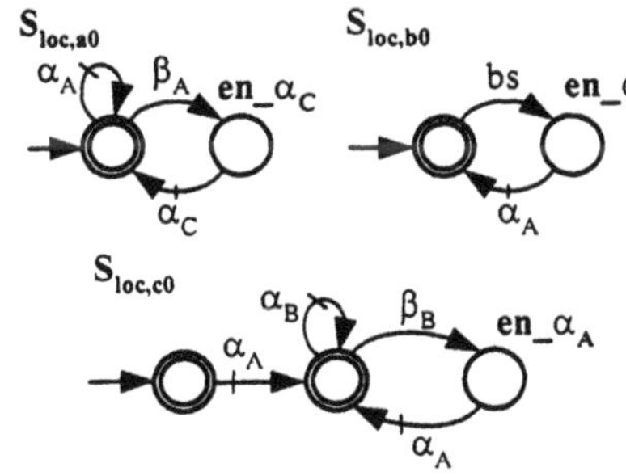

Figure 9. Three of local Supervisors.

By using a computational tool it can also be shown that the necessary and sufficient condition for local modularity, $\overline{\|_x L_m(S_{loc,x}/G_{loc,x})} = \|_x \overline{L_m(S_{loc,x}/G_{loc,x})}$, (Queiroz and Cury, 2000b) holds for the system at hand.

5. IMPLEMENTATION

In the implementation stage, we use the structure shown in Figure 10. In this structure, modular supervisors constitute the highest level of hierarchy and they interact with the low-level control algorithms, which define the operational procedures of the real system. The task of low-level interface is to convert responses obtained as sensor signals into events and to convert low-level enabling signals into outputs that drive the real system.

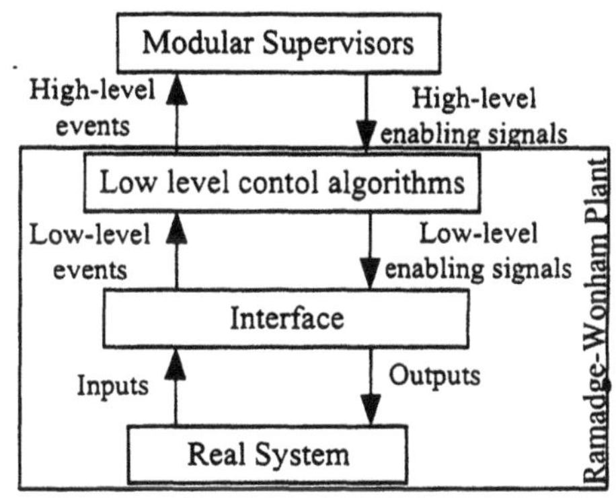

Figure 10. Control System Structure

Modular supervisors update their states by the occurrence of high-level events. When modular supervisors update their states, they also update the enabling signal information sent to low-level control algorithms. Low-level control algorithms send high-level events corresponding to these enabling signals back to modular supervisors as soon as the enabling signals are processed by the plant. This feed back structure for high-level events ensures that modular supervisors update their states only if the plant processes the enabled signals. This is also compatible with SCT, which assumes that events are generated by the plant.

In the implementation stage we try to force the PLC behave as the supervisors that we have designed. The PLC code should fulfil two tasks. One of the tasks is to provide synchronization of the supervisors with the plant and the other is generation of the enabling signals shown on some of the states of the automaton representations of the supervisors as in Figure 9. However, some difficulties may arise in programming PLCs to operate them as the finite state machines that correspond to the designed supervisors (Fabian and Hellgren, 1998).

We will explain the realization methodology on supervisor $S_{loc,c0}$ shown in Figure 9 using Ladder Diagram (LD) programming technique. We will use memory bits to represent states and events. Falling and rising edges of sensor signals will be used to meet the assumption of momentary events in DES theory. At every PLC cycle, the status of the states for the next cycle is calculated according to current status of the states. Therefore, we use two memory bits for every state, one for current PLC scan cycle, and the other for the next cycle. These two bits corresponding to one state are synchronized at the end of each PLC cycle. Using two bits for one state solves the problem named as "avalanche effect" defined in (Fabian and Hellgren, 1998). Although this technique seems to double the memory size used for the states, it is possible to use the technique only for the states that cause avalanche effect in an effective and economic manner.

Figure 10 shows the PLC program parts related with the supervisor $S_{loc,c0}$. In Figure 10a, state transition of $S_{loc,c0}$ is programmed. Figure 10b shows the program part which generates $S_{loc,c0}$'s enabling signal for α_A. The program part shown in Figure 10c is designed to enable α_A if both of $S_{loc,b0}$ and $S_{loc,c0}$ enables α_A. Figure 10d shows the state transition program code of the low-level control algorithm for Cylinder A. As can be seen from the code, when transition from A_0 to A_1 occurs by the enabling signal en_A, the controllable event α_A is generated. Note that the uncontrollable events A_f and A_b correspond to forward and backward positions of cylinder A and are obtained at the interface by rising edges of sensor signals.

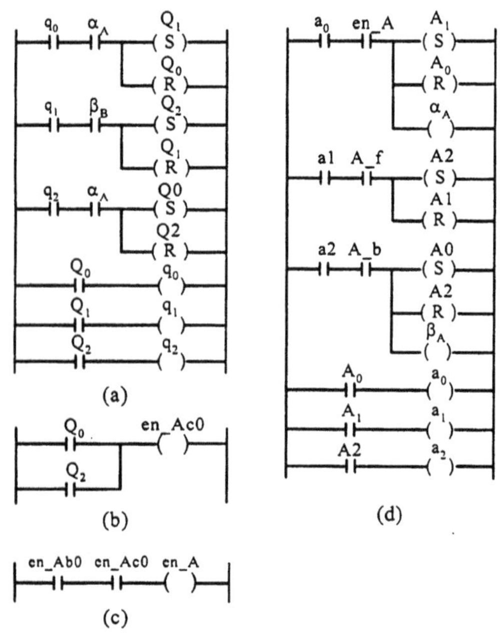

Figure 11. PLC programs for modular supervisor $S_{loc,c0}$ and low-level control logic for Cylinder A.

6. CONCLUSION

In this paper, we have presented an application of supervisory control theory to a pneumatic assembly system. We have used the local modular approach, which makes it possible to design and analyse supervisors by exploiting modular structure of the specification and the system. Generic specifications have been used as supervisors since they correspond to controllable local specifications. We have also shown that modularity condition holds for the defined problem. A hierarchical control structure and a PLC implementation methodology is also introduced. The implementation methodology solves the problem of "avalanche effect" and yields a modular PLC program structure. The used control strategy increased efficiency of the system by allowing multiple products to be assembled simultaneously and by shortening the preparation period of the supplied components.

ACKNOWLEDGMENTS

This work was partially supported by SMC-ENTEK company. The authors also thank Neslihan S. Şengör for her comments on the manuscript.

REFERENCES

Balemi, S., Hoffman, G.J., and others (1993). Supervisory control of a rapid thermal multiprocessor. *IEEE Transactions on Automatic Control*, Vol. 38, No. 7, 1040-1059.

Brandin, B. A. (1996). The real-time supervisory control of an experimental manufacturing cell. *IEEE Transactions on Robotics and Automation*, Vol. 12, No. 1,1-14.

Cassandras, G.C., Lafortune, S. (1999). *Introduction to discrete event systems*. Kluwer Acedemic Publishers, Massachusetts, USA.

Fabian M., Hellgren A. (1998). PLC-based implementation of supervisory control for discrete event systems. In: *Proceedings of the 37th IEEE conf. on Decision & Control, Tampa, Florida, USA.*

Hasdemir İ.T., Kurtulan, S., Gören L. (2003). Supervisory control of a pneumatic system using PLC. In: *Proceedings of the Third International Conference on Electrical and Electronics Engineering, ELECO, Bursa, Turkey.*

Leduc, R. J. (1996). PLC implementation of a DES supervisor for a manufacturing testbed: An implementation Perspective. Master's thesis, Department of Computer and Electrical Engineering, Univ. of Toronto, Toronto, Canada.

Queiroz, M.H. de, Cury, J.E.R. (2000a). Modular control of composed systems. In: *Proceedings of the ACC. Chicago, USA.*

Queiroz, M.H. de, Cury, J.E.R. (2000b). Modular supervisory control of large scale discrete-event systems. In: *Proc. of WODES 2000,Ghent, Belgium.*

Queiroz, M.H. de, Cury, J.E.R. (2002). Synthesis and implementation of local modular supervisory control for a manufacturing cell. In: *Proceedings of Wodes 2002, Zaragoza, Spain.*

Ramadge, P.J., Wonham, W. M. (1987). Supervisory control of a class of discrete event processes. *SIAM Journal of Control and Optimization*, Vol. 25, No 1,206-230.

Wonham, W. M., Ramadge, P. J. (1988). Modular supervisory control of DESs. *Mathematics of control of discrete event systems*, 1(1):13:30

Zaytoon, J., Carré-Ménétrier, V. (2001). Synthesis of control implementation for discrete manufacturing systems. International Journal of Production Research, Vol. 39, No.2,329-345.

ELSEVIER
IFAC
PUBLICATIONS
www.elsevier.com/locate/ifac

AUTOMATIC LOGIC GENERATION FOR RECONFIGURABLE CELL-BASED MANUFACTURING SYSTEMS

Emanuel Almeida [1] **Dawn Tilbury** [2]

Engineering Research Center
The University of Michigan
Department of Mechanical Engineering
Ann Arbor, MI 48109-2125

Abstract: This paper presents a method to automatically generate the control logic for manufacturing cell controllers, using Modular Finite State Machines (MFSMs). Using the proposed method models can be automatically created upon choice of specifications, and used to run as controllers. They are based on a specified behavior between the cell controller and its supervisor, and are verifiably correct from their construction. The method has been implemented on a small manufacturing test bed at the University of Michigan. *Copyright © 2004 IFAC*

Keywords: Logic Controllers, Finite State Machines, Cellular Logic

1. INTRODUCTION

Reconfigurable Manufacturing Systems (RMS) (Koren *et al.*, 1999) are systems that can adapt to rapidly changing manufacturing environments through a modular construction of components. Reconfiguration should involve changing the operations performed by the system and the addition, removal or replacement of physical components. The logic controls will need to be modified when the system is reconfigured. This paper looks at how this modification can be automated for the class of manufacturing systems that can be considered as cells.

In the eighties much work was done in cell control of Flexible Manufacturing Systems (FMS) (Wright and Bourne, 1988), however FMS have not had great acceptance due to their complexity and cost. The controls generated by the method proposed in this paper are designed for the RMS concept, where machine structures are adjustable (vs fixed), systems are focused on part families (vs machines), and flexibility is customized (vs general flexibility).

Typically, automatic logic generation is associated with the translation of some model into implementable logic, and much work has been done to this end, for example (Thieme and Hanisch, 2002; Feldmann *et al.*, 1999). In this paper, automatic logic generation is meant to imply that the model can be created instantly from choice of certain specifications, and it can then be used to run the cell controller.

2. PROBLEM DEFINITION

This section briefly introduces TR FSMs, which will define the logic in the manufacturing cell. Definitions for manufacturing cell and part family

[1] Emanuel Almeida would like to acknowledge support from the Fundação para a Ciência e Tecnologia and the Fundo So... Europeu under the III Quadro Comunitário de Apoio
[2] This re... was supported in part by the NSF under grant nu... EEC95-92125 and CMS98-76039

types are given, and the behavior of the manufacturing cell is formalized.

2.1 TR FSMs

MFSMs are a type of Discrete Event System (DES). They react to and generate events. The logic of how a particular MFSM behaves is defined by Trigger / Response Finite State Machines (TR FSMs). A TR FSM is always in one of a finite number of defined states. When a TR FSM witnesses an event, it will react by changing state and/or generating other events as determined by its current state. The events that the TR FSM reacts to are called triggers and the events that it generates are called responses.

Trigger / Response Transition A trigger / response transition determines when and how a change of state occurs in a trigger / response FSM. The transition τ consists of a trigger t_τ, a source state x_{s_τ}, a destination state x_{d_τ} , and a sequence of responses r_τ.

Trigger / Response FSM A trigger / response FSM observes trigger events and produces response events. The FSM F consists of a set of states X_F, a set of triggers T_F, a set of responses R_F, a set of transitions $\mathcal{T}_F$, an identification of the initial state x_{0_F}, an identification of the dump state x_{dp_F} and a set of marked states X_{M_F}.

The dump state indicates that the TR FSM is in a terminal state, and that it is no longer able to process events, therefore there are no transitions out of the dump state.

Socket A socket bundles events of a module together for connection to another module. Every socket will consist of a set of triggers and a set of responses.

Module A module encapsulates a TR FSM for use within a system. It consists of a trigger / response FSM (as defined above) and a set of sockets which can be connected to other modules.

2.2 Definition of a Manufacturing Cell

This section defines the types of manufacturing systems under consideration, in terms of the format of the cells. The *class* of cells that shall be considered can be defined in the following way:

1-Manufacturing subsystems that contain a single material handling device and m machines.
2-Manufacturing subsystems branched off one or more main arteries (see Figure 1).
3-There is a well defined physical location where parts enter the cell and possibly another where they exit the cell. This characteristic is extended to every artery with which parts are exchanged.
4-The entry and exit locations and the material handling device can hold only 1 part at a time.

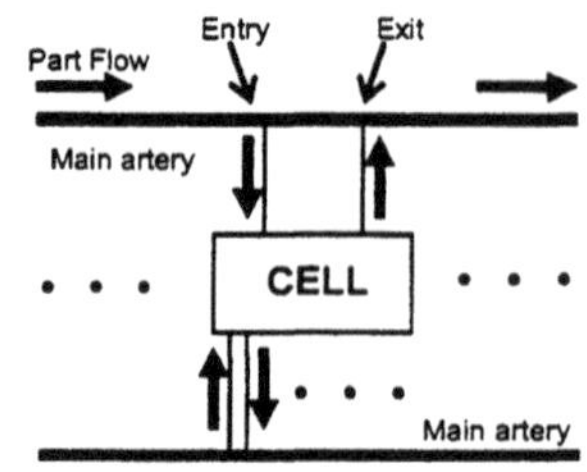

Fig. 1. Cell Abstraction

According to this broad classification, the following examples fall into this class of manufacturing cells:

Example 1 (Figure 2a) depicts a cell that uses a gantry as a material handling device to feed/remove parts to/from the machines. There is a single location for parts to enter and exit the cell.

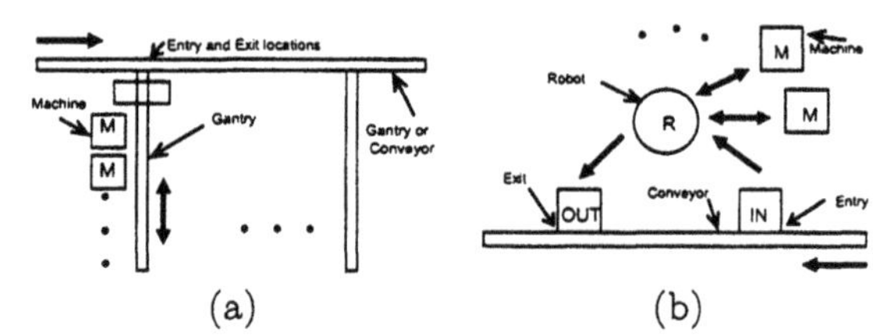

Fig. 2. a) Cell is a linear arrangement of machines connected by an overhead gantry; b) Cell consists of a set of machines loaded and unloaded by a robot

Example 2 (Figure 2b) shows a cell that uses a robot to feed/remove parts to/from the machines. Parts enter and exit the cell through distinct physical locations.

Definition 1. (Resource). A resource is a physical element of the cell that is under the control of the cell controller. Included in this set are machines, clamps and material handlers.

Definition 2. (Operation). An operation can be any of the following three actions: loading machine, running machine program or unloading machine. An operation is typically constituted by one or more programs, for example, to load a part into a machine, first the respective clamp must be open and then a material handler program is run to pick up and place the part in the machine.

2.3 Part Families

A part family is constituted by parts that are not exactly alike, but are physically and geometrically similar, so that the system may consider them to be the same part (there can be significant geometrical differences as long as these do not affect any of the arguments following). This means that, for example, the same robot program may be used for all parts within a part family, because the gripper can handle the geometric differences, and the placement of the parts in machines is not compromised because of these physical perturbations. Within a part family there can be many part type groups. All of the parts in the same group will undergo the same processes in the cell. This way, all of the parts within the same part type group can be considered by the controller to be the same.

2.4 Definition of Cell Control Requirements

Section 2.1 presented the TR FSMs which are to be used to model the cell controller. In order to model, it is necessary to understand the cell's behavior with its environment, namely its interaction with the system level controller (SLC), with whom it will exchange events. The formalization of this behavior is a key requirement to obtain the generation of the model automatically. The events issued by the SLC to the cell will be called Cell Triggers and the events issued by the cell to the SLC will be called Cell Responses (Figure 3). The events in these sets are:

Cell Triggers = {**start_i**, **unload_part**, **cycle_interrupt**, **resume**, **remove_i_j**}

Cell Responses = {*done_i*, *reject_part_i*, *unload*, *cell_resumed*, *status*}

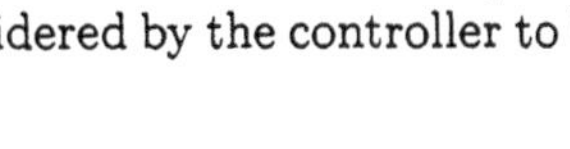

Fig. 3. System-Cell Communication

Following are descriptions of the ways the SLC and the cell controller exchange these events. It shows when the events can be sent and what reaction the other controller has to those events. A distinction is made between the cell working in normal auto mode and during an interruption.

2.4.1. Normal auto mode behavior When the cell is not processing any part it is idle and waiting for jobs. Processing is as follows:

⇛ **start_i** - The SLC informs the cell controller a part_i has arrived at the cell to be processed. At the moment there is no handshaking between the SLC and the cell controller. The cell will then check the availability of the machines.
⇛ *reject_part_i* - The cell may reject the part if it cannot process it at that time. It is assumed the SLC would take appropriate action on receiving this response. If the cell can process the part it will start the sequence of operations to do so, which include a loading of the part into a machine and running a machine program. The cell does not inform the SLC that it has begun operating. It is assumed that the SLC recognizes that a lack of part rejection implies part acceptance by the cell.
⇛ *unload* - When the machine program finishes, the cell controller will issue an unload request to the SLC. It's assumed the SLC will allow unloading, at some future time.
⇛ **unload_part** - Having received the unload request, the SLC knows there is a processed part ready to exit the cell, and so will issue this command whenever the part may be removed.
⇛ *done_i* - The cell controller will issue this command at the end of the unloading operation.

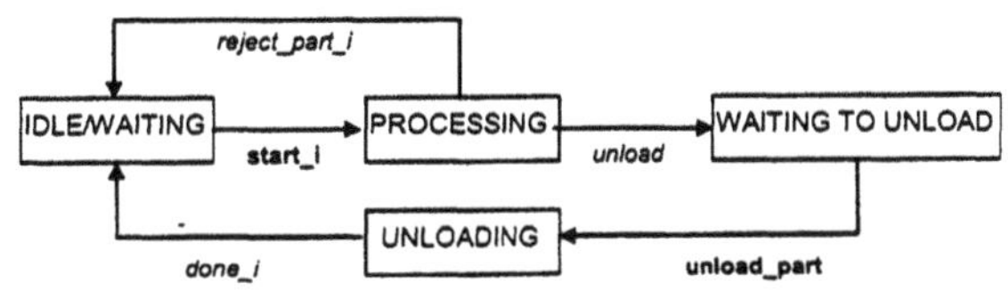

Fig. 4. System-Cell Normal Behavior

It should be noted that the sequence shown in Figure 4 need not run alone, rather there can be concurrent sequences being executed. As the cell has a shared resource (material handler) concurrency can lead to unavailability of the resource. This is dealt with by recourse to a queue, which queues operations in a FIFO manner.

2.4.2. Interruption behavior An interruption can occur at any moment. When an interruption occurs, operations being executed in the cell are finished and no new operation is started:

⇛ **cycle_interrupt** - Interruption command issued by the SLC. The cell will make a self check as to what parts are in the system and where.
⇛ *status* - The cell controller shall inform the SLC what parts are in the cell, and in which machines they're in.
⇛ **remove_i_j** - Once the SLC knows what parts are where in the cell, it can order their individual removal using this command. This command will be used successively $\forall\{i,j\}$ that are to be removed.
⇛ **resume** - This command issued by the SLC will resume normal auto mode operations in the cell.
⇛ *cell_resumed* - Once the cell has been ordered to resume normal operations and has effectively done so it will inform the SLC by issuing this command.

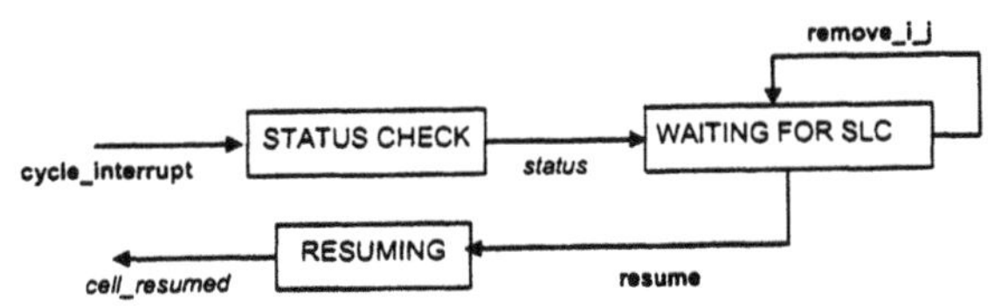

Fig. 5. System-Cell Interruption Behavior

2.4.3. Communication with Resources Independent of the connection between cell controller and resources, the controller has the following actions over the resources:

-Clamps: Cell controller opens and closes clamps;
-Machines: Cell controller starts machine program, and machine informs controller when its finished;
-Material Handler: Cell controller must run several different material handler programs, and material handler should notify the cell controller when it finishes its programs.

3. CONTROL GENERATION

Using the MFSM framework, models are built using communicating TR FSMs. These special kind of FSMs are encapsulated in what will be called modules. Each module contains logic that can represent a control function or a mechanical element of the cell. The controller will thus be made up of many connected modules.

3.1 Automatic Generation of Logic Model

Figure 6 shows the architecture of the model for a *generalized* cell controller. The dots indicate that the number of modules will vary according to different system specifications. Upon choice of these, the particular model and cell controller can be defined. Two important facts must be emphasized about the results. First, the models created respect all the requirements described in Section 2.4. Secondly, the created model is free of any transitions leading to the dump state, as explained in Section 2.1. This is an important feature because if a controller executes a transition to the dump state, it will halt and cease reacting to new input.

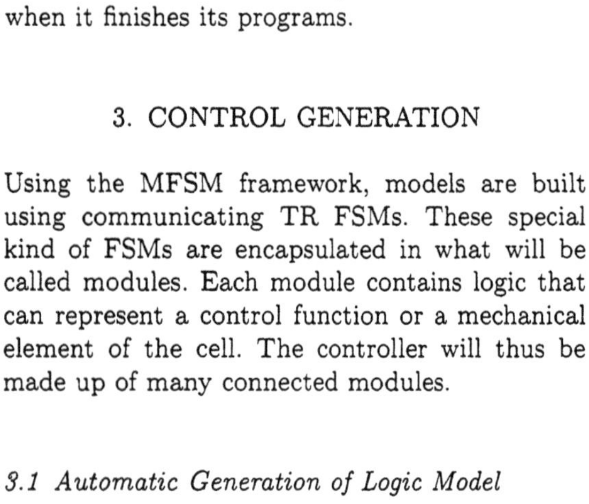

Fig. 6. Overall Modular Architecture of Generalized Cell Controller

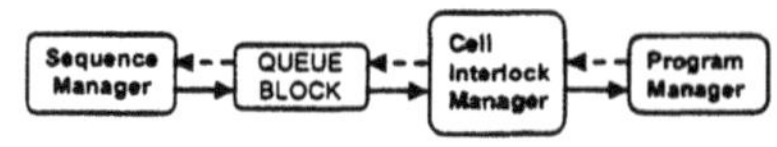

Fig. 7. Model initialization

Construction of Model:

-Let m be the number of machines at the cell;
-Let n be the maximum number of part types being processed;
-$\forall i = 1, .., n$ let $M_j^i \subset M_1, ..., M_m$ be the machines that can process part i;
-Determine if cycle interruptions are to be included in the logic.

Figure 6 shows how the architecture of the model can change with the alteration of these specifications. By knowing what the changes are and what causes them, we can create a method to automatically generate the logic model. Following is an algorithm that could be used for this intent.

Algorithm 1 Automatic Generation of Logic Model

1-Initialize the model, as shown in Figure 7.
2-Connect n modules, S_1,...,S_n, to the SM module, one for each part type.
3-Connect m Part Sequence modules to each module created in step 2, and 1 Unloading module to the SM module.
4-Connect *m-1* Store Cmd modules and 1 Store Done Cmd module to the Queue Block module. The maximum possible number of queued operations can be deduced to be equal to *m-1*.
5-Connect m Machine Interlock modules, m Clamp Interlock modules and 1 Material Handler Interlock module to the CIM module.
6-Connect m Program Modules to the PM module. The number of Program Modules needed is dictated by the maximum number of concurrent operations that can be processed at the cell, which in turn can be deduced to be m.
7-Create m machine blocks as shown in Figure 6. Each one of these blocks corresponds to the connection and control of each pair of machine and clamp.
8-If interruptions are to be included, then a Memory Stop module and a Removal Operations module are created and connected to the SM.

3.1.1. Code Verification The logic output by the model created using Algorithm 1 will respect the requirements of Section 2.4. This method was implemented at the University of Michigan for the case where m=n=*2*. The model was verified by checking transitions to the dump state, and the logic was found to be correct.

4. RECONFIGURATION

Reconfigurability of the cell means that a new part type and new machines can be easily and quickly introduced into the controls. The routing of the parts in the cell can also be easily altered. One can imagine a machine breaking down and the parts being routed to other machines in the cell, through a quick reconfiguration of the control logic. In fact, due to the structure of the model, these processes lend themselves to being automated, as was described in Section 3.

4.1 Adding machines

This section explains how the control code can be automatically reconfigured when new machines (and corresponding clamps) are added to the cell. The part types that may be processed in this new machine must also be defined. Let P_{new} be the set of part types that can be processed in the new machine and let $S_{new} \subseteq S_n$ be the set of modules (as shown in Fig 6) associated to the parts in P_{new}. Let the number of elements in P_{new} be k.

Assuming that there is a cell that holds m machines and processes n part types, then to every new machine added, there would be the following changes in the logic model:

1-Create k Part Sequence modules. To every module in S_{new} connect a Part Sequence module. These Part Sequence modules will run the operations to process any part belonging to P_{new} on this new machine.
2-Create 1 Store Cmd module and connect it to the Queue Block module. One more machine in the cell implies that there may be one more operation queued.
3-Create 1 Clamp Interlock and 1 Machine Interlock module. These modules are by default identical to the existing clamp and machine interlocks for the existing machine, but refer to the new clamp and machine.
4-Create 1 block as shown in Figure 6. This block will be the connection of the controller body to the physical machine.
5-Create 1 new Program Module, as there can now be one more program running concurrently.

All these changes mean that there are new transitions and sockets that have to be created; for example for the Program Manager module, add 3 new sockets and respective transitions. However, all of these changes are mechanical and can be easily automated.

4.2 Adding parts

Using the same cell with m machines and n part types, the addition of a new part type would encounter the following modifications.

Let $M_{new} \subseteq M$ be the set of machines that can process the new part. Let the number of elements in M_{new} be q.

1-Create a module S_{n+1}.
2-To the module created in Step 1 connect q new Part Sequence modules. Each of these Part Sequence modules will run the operations of the new part in a machine belonging to M_{new}.

Once again, there are new transitions and sockets that can be automatically created.

5. IMPLEMENTATION EXAMPLE - RFT

As an implementation example, a cell controller developed for the Reconfigurable Factory Test Bed (RFT), which is present at the University of Michigan's Engineering Research Center, is presented (Moyne *et al.*, 2004). The RFT holds a Serial Parallel Line with 2 manufacturing cells, and its general layout is depicted in Figure 8.

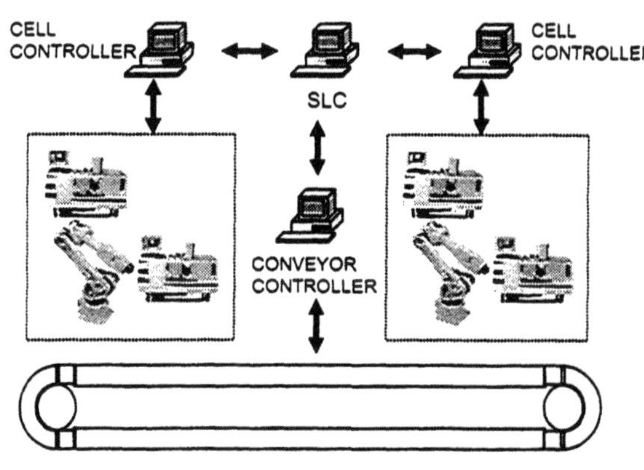

Fig. 8. Reconfigurable Factory Test Bed Simplified

Parts travel on a conveyor and are processed in both cells. Each cell is composed of a robot and two milling machines. The robot is used to transfer parts from conveyor to machines and vice-versa. The part pick-up and drop-off locations are the same spatial location. In this application, the parts travel on pallets that are carried by the moving conveyor. There is a SLC that manages the whole RFT manufacturing process. The communication between the SLC and the cell controller is done using OPC across an Ethernet Network. The connection from the cell controller to the physical machines is done across a DeviceNet Network.

To tailor the generalized model described in Figure 6, the specifications listed in Section 3.1 may be listed as:

(1) 2 machines in cell.
(2) 2 part types.
(3) Both part types can be processed in any of the machines at the cell.
(4) Include cycle interruptions.

From here, the particular model (and consequent logic) can be automatically generated to be as shown in Figure 9. Whereas Figure 6 shows a generalized architecture for the model, Figure 9 shows how that generalized architecture is particularized by introduction of the system specifications (Section 3.1). This logic model was created and verified using FSMTools, a software tool developed at the University of Michigan. FSMTools allows the user to build the modules and connect them as shown in Figure 9. The created system is then verified by building its parallel composition and using filters in the connection with the machines, clamps, material handler and SLC. Filters are TR FSMs that have the function of enforcing the desired behavior to the system.

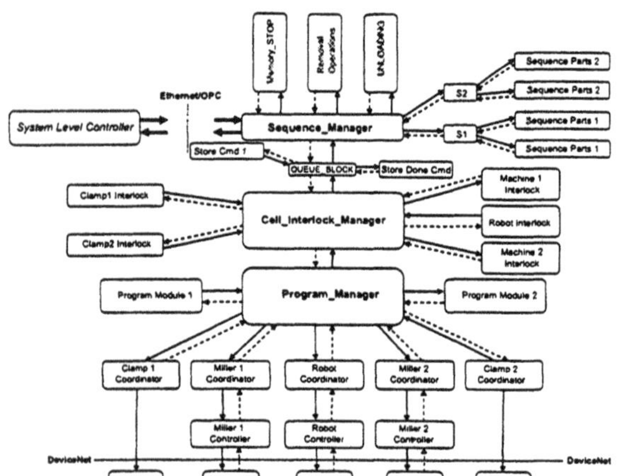

Fig. 9. Overall Architecture of the RFT Cell

The model as shown in Figure 9 has over 100,000 states. If the state space of the parallel composition is too large to build, then the model may be broken down, until the parallel composition of the sub-models may be built and thus verified (Endsley, 2004).

Once the logic was verified, it has been used to run the cell control of the RFT, also using FSMTools, which can run similar to a PLC.

6. CONCLUSIONS AND FUTURE WORK

This paper has described a method for automatic control generation, using MFSMs. This method uses as inputs system specifications and a pre-defined behavioral model for the system (a key requirement in this process), and creates as output a logic that is verifiably correct, making verification unnecessary. This implies that a cell controller for a system (possibly too big to ever be composed), could be built and implemented on the fly, with the confidence that the logic is correct.

Using this method a reconfigurable cell controller has been built and implemented at the University of Michigan. The cell that was studied had 2 machines and processed 2 different part types, and respected all requirements as to fit the class of cells described.

Current work involves encapsulating MFSMs into function blocks for implementation with the IEC 61499 control program. Addition of exception handling and diagnostic capabilities are also sought. Future research might include application of concept to systems whose structure is changing, which would mean the online creation and destruction of logic elements.

REFERENCES

Endsley, E. W. (2004). "Modular Finite State Machines for Logic Control: Theory, Verification and Applications to Reconfigurable Manufacturing Systems". PhD thesis. University of Michigan.

Feldmann, K., A. W. Colombo, C. Schnur and T. Stockel (1999). "Specification, Design, and Implementation of Logic Controllers based on Colored Petri Net Models and the Standard IEC 1131". *IEEE Transactions on Control Systems Technology.*

Koren, Y., U. Heisel, F. Jovane, T. Moriwaki, G. Pritschow, G. Ulsoy and H. Van Brussel (1999). "Reconfigurable Manufacturing Systems". In: *Annals of the CIRP.* Vol. 48. pp. 527–540.

Moyne, J., J. Korsakas and D. M. Tilbury (2004). "Reconfigurable Factory Testbed (RFT): A Distributed Testbed for Reconfigurable Manufacturing Systems". In: *Japan-USA Symposium on Flexible Automation.*

Thieme, J. and H. M. Hanisch (2002). "Model-Based Generation of Modular PLC Code Using IEC 61131 Function Blocks". In: *Proceedings of the IEEE International Symposium on Industrial Electronics.* Vol. 1. pp. 199–204.

Wright, P. K. and D. A. Bourne (1988). *Manufacturing Intelligence.* Addison-Wesley Publishing Company, Inc.

ELSEVIER
IFAC
PUBLICATIONS
www.elsevier.com/locate/ifac

SUPERVISORY CONTROL FOR THE CONFORMANCE OF REAL-TIME DISCRETE EVENT SYSTEMS [1,2]

Ahmed Khoumsi

Université de Sherbrooke, Dep. GEGI, Sherbrooke, CANADA

Abstract: We study the problem of controlling a plant described as a *real-time* discrete event system by means of a supervisor, in order to ensure a conformance relation denoted **tioco** between the plant and the formal specification of the system.

Keywords: Supervisory control, real-time discrete event systems, conformance relation

1. INTRODUCTION

(Supervisory) control consists of using a supervisor *Sup* that *forces* a plant PL to conform to a specification SPEC (Ramadge and Wonham, 1989). In (Jéron *et al.*, 2003), the control problem is solved by using the conformance relation **ioco** (Tretmans, 1996). We extend (Jéron *et al.*, 2003) to the real-time case where PL and SPEC are described by *order* and *timing* constraints of events. We use a variant of Timed Automata (TA) (Alur and Dill, 1994) to describe SPEC and PL, and define **tioco**, a timed adaptation of **ioco**. We express the control problem into a non-real-time form by using a transformation of a TA into a finite state automaton denoted SEA, and then we adapt a non-real-time control method. Unlike previous control methods (Brandin and Wonham, 1994; Gouin, 1999; D'Souza and Madhusudan, 2002), in practice the complexity of our method does not increase significantly with the magnitudes of the constants used in timing constraints. This is also the case for some algorithms (Daws and Yovine, 1995; Larsen, 1995), but the latter are not applicable for control.

Sect. 2 presents the ingredients to formalize the problem to be solved. In Sect. 3, we show how to transform the control problem into a non-real-time form and how supervisor acts upon the plant. Sect. 4 presents a solution to the problem. In Sect. 5, we conclude.

[1] Work partially realized at IRISA, Rennes, France
[2] A longer version of this article is available at http://www.gel.usherb.ca/khoumsi/Research/Public/WODES04.ps

2. PRELIMINARIES AND CONCEPTS

2.1 Model and notations

Definition 1. A *clock* is a real-valued variable that can be reset (to 0) at the occurrence of an event and such that, between two resets, its derivative (w.r.t. time) is equal to 1. $\mathcal{C} = \{c_1, \cdots, c_{N_c}\}$ is a set of clocks.

Definition 2. A *Clock Condition (CC)* is a conjunction of formula(s) in the form "$c_i \sim k$", where c_i is a clock, $\sim \in \{<, >, \leq, \geq, =\}$ and $k \geq 0$.

A CC can be the constant *True* (empty conjunction) or a single formula. We denote by $\Phi_\mathcal{C}$ the set of CCs defined upon $\mathcal{C}$.

Definition 3. A *Timed Automata (TA)* is syntactically defined by $(\mathcal{L}, \Sigma, \mathcal{C}, \mathcal{T}, l_0)$ where: $\mathcal{L}$ is a finite set of locations, l_0 is the initial location, $\mathcal{C}$ is a finite set of clocks, and $\Sigma = \Sigma_? \cup \Sigma_!$ is a finite set of events. An event can be: the reception of an *input* i (written $?i \in \Sigma_?$) or the sending of an *output* o (written $!o \in \Sigma_!$). $\mathcal{T}$ is a set of transitions, and a transition is defined by Tr = $\langle q; \sigma; r; CC; Z; x\rangle$, where : q and r are origin and destination locations, $\sigma \in \Sigma$, CC is a clock condition, $Z \subseteq \mathcal{C}$ is a set of clocks to be reset to zero, $x \in \{f, uf\}$. Tr is said *forced* (resp. *unforced*) if $x = f$ (resp. $x = uf$).

Figure 1 represents an example of TA. Locations are represented by nodes, a transition Tr = $\langle q;\sigma;r;CC;\{c_i,\cdots\};x\rangle$ is represented by an arrow linking q to r and labelled by $(\sigma;CC;\{c_i,\cdots\})$, and the CC *True* and the empty clock reset are indicated by "-".

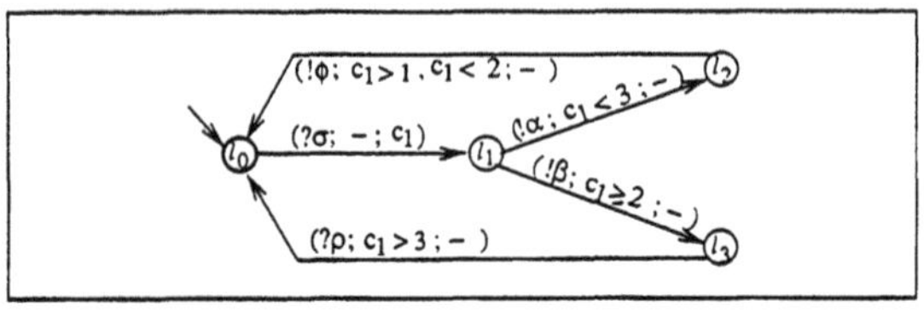

Fig. 1. Example of TA

At time 0, the TA $A = (\mathcal{L},\Sigma,\mathcal{C},\mathcal{T},l_0)$ A is at l_0 with all clocks equal to 0. The transition Tr=$\langle q;\sigma;r;CC;Z;x\rangle$ is *enabled* when q is the current location and CC evaluates to *true*; otherwise, Tr is *disabled*. From q, σ is executed only when Tr is enabled; and after the execution of σ, r is reached and the clocks in Z are reset. The event of a *forced* Tr is necessarily executed if Tr is enabled and on the point of becoming disabled with the passing of time.

A *timed trace* of a TA is a "$(e_1,\tau_1)\cdots(e_i,\tau_i)\cdots$", where $e_i \in \Sigma$, τ_i is the time when e_i occurs, and $0<\tau_1<\cdots<\tau_i<\cdots$. Now given a TA A:

- The *empty* timed trace λ_0 is *accepted* by A *iff* $\forall$ forced transition Tr, $\forall u>0$ s.t. Tr is enabled at time u after no execution: Tr is enabled at every time $\tau \geq u$ after no execution.
- A *finite* timed trace $\lambda_n = (e_1,\tau_1)\cdots(e_n,\tau_n)$ is *accepted* by A *iff* there exists a sequence of consecutive transitions $Tr_1\cdots Tr_n$ of A that starts in l_0 and s.t.:
  (1) $\forall i \in \{1,\cdots,n\}$: the event of Tr_i is e_i; after the execution of λ_{i-1}, Tr_i is enabled at time τ_i and no forced transition is enabled at a time $\tau \in]\tau_{i-1},\tau_i[$ and disabled at τ_i.
  (2) $\forall$ forced transition Tr, $\forall u>\tau_n$ s.t. Tr is enabled at time u after the execution of λ_n: Tr is enabled at every time $\tau \geq u$ after the execution of λ_n.
- *Acceptance* of an *infinite* timed trace λ_∞ is defined from acceptance of finite timed trace by: removing Item 2 and replacing n by ∞.

Definition 4. Timed language of a TA A $(TL_\infty(A))$, is the set of (finite and infinite) timed traces accepted by A. *Timed quiescence language of A* $(TL(A))$ is the set of *finite* timed traces accepted by A.

Coming back to the example of Fig. 1, the TA is initially in l_0; it reaches l_1 at the reception of σ. From l_1, the TA reaches l_2 (resp. l_3) when it sends α (resp. β). From l_2 (resp. l_3), the TA reaches l_0 at the sending of ϕ (resp. reception of ρ). Let $\delta_{u,v}$ be the delay between u and v. We have: $\delta_{?\sigma,!\alpha}<3$, $\delta_{?\sigma,!\beta}\geq 2$, $1<\delta_{?\sigma,!\phi}<2$, $\delta_{?\sigma,?\rho}>3$. When l_1 is reached, then α: either 1) is sent before $c_1=3$, or 2) is never sent; Case 2 is impossible if α is in a forced transition and β is not sent before $c_1=3$. When l_2 is reached, then ϕ: either 1) is sent before $c_1=2$, or 2) is never sent; Case 2 is impossible if ϕ is in a forced transition.

Hypothesis 1. Each TA is deterministic. That is, two transitions executing the same event from the same location and that can be enabled at the same time, lead to the same location and reset the same clock(s).

Hypothesis 2. Input transitions are unforced.

2.2 Conformance relation between TA

Notation 2.1. If X is a set of sequences, $\overline{X}$ denotes the set of ***finite*** prefixes of X.

ioco conformance relation (Tretmans, 1996) defines what it means for PL to be correct w.r.t. a SPEC. We assume that PL and SPEC can be modelled as TA (resp. P and S) over an alphabet Σ. The following conformance relation **tioco** is an adaptation of **ioco** to the real-time case.

Definition 5. P **tioco** $S \Leftrightarrow \forall\lambda \in \overline{TL_\infty(S)}$, $\forall o \in \Sigma_!$, $\forall\tau>0$:

$$(\lambda\cdot(o,\tau) \in \overline{TL_\infty(P)}) \Rightarrow (\lambda\cdot(o,\tau) \in \overline{TL_\infty(S)}) \text{ and}$$
$$(\lambda \in TL(P)) \Rightarrow (\lambda \in TL(S)).$$

The intuition of "P **tioco** S" is that after an execution of PL accepted by S:

(1) PL can generate an output o at time τ only if S accepts o at time τ, and
(2) PL can remain quiescent forever only if S accepts to be quiescent forever.

Definition 6. Let $A = (\mathcal{L},\Sigma,\mathcal{C},\mathcal{T},l_0)$ be a TA. *IComp*(A) is a TA obtained from A as follows:

(1) a trap TL is added to $\mathcal{L}$, by *trap* we mean that $\forall a \in \Sigma$, TL has a self-loop labelled $(a;-;-)$;
(2) $\forall l \in \mathcal{L}$ and $\forall i \in \Sigma_?$, an (unforced) non-specified transition labelled $(i;CC;-)$ is added from l to TL; by *non-specified* we mean that CC defines the domain in which i is *not* enabled in l of A.

A is said *input-complete iff* $A = IComp(A)$. Intuitively, such A accepts every input at any time.

Lemma 2.1. (P **tioco** S) $\Leftrightarrow$ (P **tioco** *IComp*(S))

Lemma 2.2. If S is input-complete then,

$$P \textbf{ tioco } S \Leftrightarrow TL_\infty(P) \subseteq TL_\infty(S)$$

Lemmas 2.1 and 2.2 are inspired from their non-real-time versions (Jéron *et al.*, 2003). Based on those two lemmas, we can assume that SPEC is modelled by an input-complete S.

2.3 The control of a TA

Given PL and SPEC modelled by TAs P and S, resp., the aim of the control is to design a supervisor *Sup* that forces PL to conform to S. The task of *Sup* is to disable and force events in order to achieve this goal.

2.3.1. Controllability and forcibility The alphabet Σ of P is partitioned into Σ_c and Σ_{uc}, the sets of *controllable events* and *uncontrollable events*, resp., meaning that a supervisor can disable the controllable events (Ramadge and Wonham, 1989). The alphabet Σ is also partitioned into Σ_f and Σ_{uf}, the sets of *forcible events* and *unforcible events*, resp., meaning that a supervisor can force the forcible events whenever they are enabled (Brandin and Wonham, 1994).

Remark 2.1. Let us clarify the difference between "forced transition" and "forcible event": In P, the event of a forced Tr *is* forced by the plant when Tr is enabled and on the point of becoming disabled with the passing of time. *A contrario*, in P a forcible event e *can be* forced by the supervisor at any time when e is enabled. Finally, let e be a forcible event of a transition Tr_P of P and of a forced transition Tr_S of S. If at the same time instant, Tr_P and Tr_S are enabled and Tr_S is on the point of becoming disabled by the passing of time, then e must forced by the supervisor.

We make the choice of controlling PL. Thus, *Inputs* of Σ are *uncontrollable* and *unforcible* because under the control of the environment. Moreover, the *forced transitions* of P are *uncontrollable outputs*: "outputs" because inputs are produced by the environment and thus cannot be forced by PL, and "uncontrollable" because outputs cannot be, at the same time, forced by PL and forbidden by the supervisor.

A supervisor can be seen as a function *Sup* : $\overline{TL_\infty(P)} \rightarrow 2^{\Sigma_c} \times 2^{\Sigma_f}$, delivering the actions that are disabled and forced in P by the control after a trajectory in $\overline{TL_\infty(P)}$. Write *Sup*/P for PL modelled by P and controlled by *Sup*.

Definition 7. A TA S is *controllable* w.r.t. P, Σ_{uc}, Σ_{uf} *iff* S can be obtained from P without forbidding events of Σ_{uc} nor forcing events of Σ_{uf}.

We denote by $Cont_P^{TA}$ the set of controllable TAs w.r.t. P, Σ_{uc}, Σ_{uf}.

2.3.2. Formalization of the control problem

Objective 1. Given two TAs P and S, we have to synthesize a supervisor *Sup* s.t. *Sup*/P is described by a TA (also denoted *Sup*/P) which satisfies:

$$Sup/P \ \textbf{tioco}\ S \quad \text{and} \quad Sup/P \in Cont_P^{TA}$$

3. TRANSFORMATION OF THE CONTROL PROBLEM INTO A NON-REAL-TIME FORM

3.1 From a Timed Automaton to a Set-Exp-Automaton

Our control problem will be solved in Sect. 4 by using a transformation "*SetExp*: TA$\mapsto$SEA". *SetExp* transforms a TA into a finite state automaton by adding to the structure of the TA two types of events: *Set* and *Exp* that capture the temporal aspect of the TA.

An event $Set(c_i, k)$ means: clock c_i is set to zero and will expire when its value is equal to k.

An event $Exp(c_i, k)$ means: clock c_i expires and its current value is k.

Thus, k is the delay between $Set(c_i, k)$ and $Exp(c_i, k)$. For a formal description of *SetExp*, see (Khoumsi and Ouedraogo, 2004). Let us explain the principle of *SetExp* in an example using a single clock. We consider the following timing constraint : $1 \leq \delta_{a,b} < 3$, where $\delta_{a,b}$ is the delay between events a and b. In a TA, such a constraint is expressed by: 1) using two transitions Tr_i and Tr_j that represent the occurrences of a and b, resp., 2) resetting a clock c at the occurrence of Tr_i, and 3) associating to Tr_j the CC: $((c \geq 1) \wedge (c < 3))$. This constraint can also be expressed as follows: i) the reset "$c := 0$" of Tr_i is replaced by a $Set(c, 1, 3)$, and ii) the CC "$((c \geq 1) \wedge (c < 3))$" of Tr_j becomes "Tr_j occurs after or simultaneously to $Exp(c, 1)$ and before $Exp(c, 3)$". This timing constraint is represented in a SEA by the following two sequences: "$\langle a, Set(c, 1, 3)\rangle \cdot Exp(c, 1) \cdot b \cdot Exp(c, 3)$" and "$\langle a, Set(c, 1, 3)\rangle \cdot \langle Exp(c, 1), b\rangle \cdot Exp(c, 3)$".

Here are the three types of transitions of a SEA $B = SetExp(A)$, where σ is an event of the alphabet of A, and S (resp. $\mathcal{E}$) is a set of *Set* (resp. *Exp*) events :

Type 1 : a transition labelled $\mathcal{E}$ represents the simultaneous occurrences of the *Exp* events of $\mathcal{E}$.

Type 2 : a transition labelled σ represents the occurrence of σ; a transition labelled (σ, S) represents the simultaneous occurrences of σ and S.

Type 3 : a transition labelled $(\mathcal{E}, \sigma)$ represents the simultaneous occurrences of $\mathcal{E}$ and σ; and a transition labelled $(\mathcal{E}, \sigma, S)$ represents the simultaneous occurrences of $\mathcal{E}$, σ and S.

Note that each transition Tr of type $\neq 1$ of a SEA $B = SetExp(A)$, contains parameter x of the corresponding transition of the TA A.

Let *Exp-Trans* be any transition of type 1 or 3.

Definition 8. "An Exp-Trans *Tr is preempted*" means that Tr is never executed because another transition with the same origin state as Tr is executed before Tr. For a state q of SEA, the set $PreemptExp(q)$ contains the *Exp* event(s) forbidden in q due to the preemption of *Exp-Trans*(s).

For the TA of Fig. 1, we obtain the SEA of Fig. 2 if all transitions are unforced. In the presence of forced transitions, we apply a procedure *ForcTr* after $SetExp$. The idea of *ForcTr* is that forced transitions can preempt *Exp-Trans*s. For example, if in Fig. 1 α is in a forced transition, then in Fig. 2 the two outgoing *Exp-Trans*s of state N are cut because preempted by α, and we have $PreemptExp(N) = \{Exp(c_1, 3)\}$. If ϕ is in a forced transition, then the outgoing $Exp(c_1, 2)$ of state M is cut because preempted by ϕ, and we have $PreemptExp(M) = \{Exp(c_1, 2)\}$.

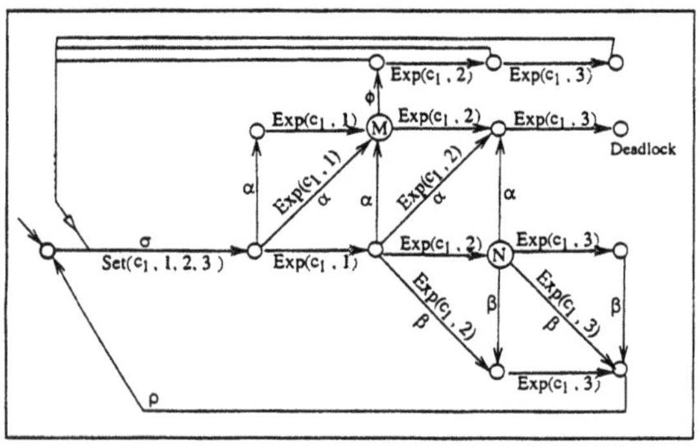

Fig. 2. SEA obtained from the TA of Fig. 1

A SEA B is syntactically defined by (Q, Γ, T, Π, q^0), where Q is a set of states, Γ is the alphabet, $T \subseteq Q \times \Gamma \times Q$ is a set of transitions, Π is a function defined by $\Pi(q) = PreemptExp(q)$, and q^0 is the initial state.

Definition 9. Given a SEA B and a state q of B, q is said to be a *quiescent state*, whenever it has neither outgoing *Exp-Trans* nor preempted *Exp-Trans*.

Intuitively, when B reaches a quiescent state q, it can remain in q indefinitely without executing any event.

A trace is a (finite or infinite) sequence "$E_1 \cdots E_i \cdots$", where $E_i \in \Gamma$. Such a sequence is said "sequence of B" whenever there exist states $q_0, \cdots, q_i, \cdots \in Q$, such that $\forall i > 0$, $(q_{i-1}, E_i, q_i) \in T$.

Definition 10. The *language of a SEA B* ($L_\infty(B)$) contains the finite traces of B terminating in a quiescent state, and the infinite traces of B. The *quiescent language of B* ($L(B)$) contains the *finite* traces of $L_\infty(B)$.

One can remark that $L_\infty(B)$ and $L(B)$ implicitly respect the following condition, called the *Consistency condition*: every $Set(c, k)$ and its corresponding $Exp(c, k)$ are separated by time k.

3.2 Control architecture of a TA based on its SEA

The supervisor *Sup* that controls the plant, is not directly computed on the TAs P and S, but rather on a SEA computed from the two TAs. Hence, in order to make the link between *Sup* and the plant, so that it can control it, we use a particular control architecture that we now present. It is represented in Fig. 3:

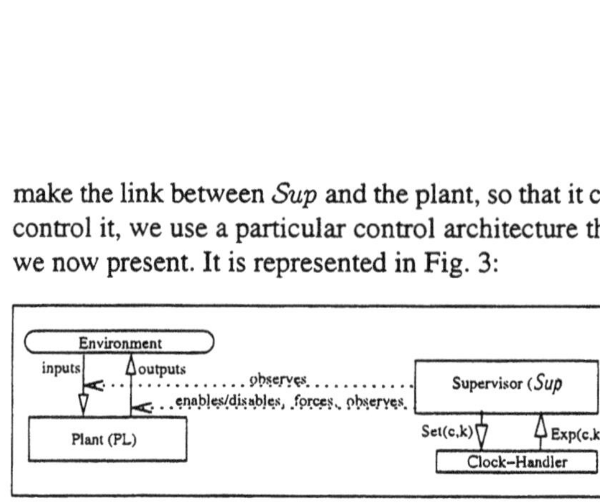

Fig. 3. Control architecture

Environment sends inputs to PL and receives outputs from it.

Clock-Handler is a Timer module that receives *Set* events from the supervisor and sends Exp events to it. It respects the consistency condition.

Supervisor *Sup* is derived from an SEA (See Sect. 4). It is tagged with the Set/Exp events coming from this SEA. It observes inputs and outputs of PL, and disables controllable outputs and forces forcible outputs of PL when necessary. The timing constraints that the supervisor has to deal with are performed via the interaction with the Clock-Handler.

We say that PL is *under* the control of the supervisor if the latter disables/forces events of PL. A contrario, PL is *without* the control of the supervisor if the latter disables/forces *no* event of PL.

Let SYST be described by the interactions: between PL and Environment *without* the control of *Sup*, and between *Sup* and Clock-Handler. That is, SYST consists of PL, Clock-Handler and the supervisor when the latter forces and disables *no* event of PL.

SYST can be seen as an *uncontrolled system*. When SYST can be modelled by a SEA *PP*, then we write *Sup/PP* for the system consisting of SYST *controlled* by the supervisor *Sup*. That is, *Sup/PP* denotes the system described by the interactions : between PL and Environment *under* the control of *Sup*, and between *Sup* and Clock-Handler.

We can now make the link with the control of a TA of Sect. 2.3 by the following lemma which states that if there exists a SEA that model the controlled SYST, then there exists a TA that models the controlled PL:

Lemma 3.1. If *Sup/PP* can be modelled by a SEA, then *Sup/P* can be modelled by a TA.

3.3 Non real-time Conformance relation between SEAs

Definition 11. Let P' and S' be two SEAs over the same alphabet : $P' \;\mathbf{conf}\; S' \equiv L_\infty(P') \subseteq L_\infty(S')$.

Proposition 3.1. Let S be an input-complete TA. If *Sup/P* and *Sup/PP* are described by a TA and a SEA, resp. (also called *Sup/P* and *Sup/PP*, resp.) then: *Sup/PP* $\mathbf{conf}$ $SetExp(S) \Rightarrow$ *Sup/P* $\mathbf{tioco}$ S.

4. RESOLUTION OF THE CONTROL PROBLEM

Let $P = (\mathcal{L}^P, \Sigma, \mathcal{C}^P, \mathcal{T}^P, l_0^P)$ be a TA modeling PL and $S = (\mathcal{L}^S, \Sigma, \mathcal{C}^S, \mathcal{T}^S, l_0^S)$ modeling SPEC with $\Sigma = \Sigma_c \cup \Sigma_{uc} = \Sigma_f \cup \Sigma_{uf}$.

4.1 Control method

In order to ensure the conformity between PL and SPEC, we transform the problem into a control problem over some SEAs derived from P and S. Our method consists of four steps outlined in Fig. 4.

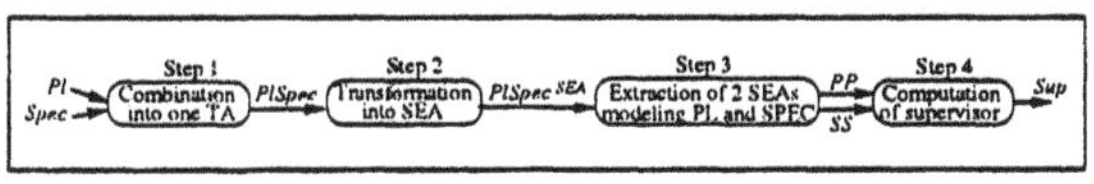

Fig. 4. Steps of the control method

4.2 Step 1 : Combine P and S into a TA PS

We construct a TA *PS* that captures the behaviors of PL and of SPEC plus some error locations. This "combination" of PL and SPEC is necessary for modelling (in Step 3) PL and SPEC by two SEAs over the *same* alphabet.

Definition 12. The *completion* of a TA A can be defined by adapting Def. 6: we add *all* non-specified transitions, instead of only non-specified *input* transitions. The result of completion is denoted $Comp(A)$.

Notation 4.1. Let $A^i = (\mathcal{L}^i, \Sigma, \mathcal{C}^i, \mathcal{T}^i, l_0^i)$, for $i = 1, 2$, be 2 TAs s.t. $\mathcal{C}^1 \cap \mathcal{C}^1 = \emptyset$. The *synchronized product of* A^1 *and* A^2 is a TA denoted $A^1 \otimes A^2$, that models the joint behavior of A^1 and A^2 when synchronized on the intersection of their alphabets.[3]

The TA *PS* is computed as follows:

$$PS = Comp(P) \otimes Comp(S).$$

4.3 Step 2: Transform the TA PS into a SEA PS^{SEA}

We abstract away the real-time aspect, by transforming the obtained TA *PS* into an SEA $PS^{SEA} = SetExp(PS)$. All transitions are assumed *unforced*; the "forced" aspect is taken into account in Step 3.

4.4 Step 3 : Extract PP and SS from PS^{SEA}

We extract from PS^{SEA} two SEAs *PP* and *SS* that model SYST and a specification of its desired behavior, resp. The "forced" aspect, which has been ignored so far, will be taken into account by the use of *ForcTr*.

[3] For lack of space, we do not define the synchronized product formally.

PP is computed from PS^{SEA} by: 1) removing error-states E_P and $E_{P,S}$; and 2) applying *ForcTr* to take into account forced transitions of P.

SS is computed from PS^{SEA} by: a) removing error-states E_S and $E_{P,S}$; b) applying *ForcTr* to take into account forced transitions of S; and c) removing states not in *PP*.

Lemma 4.1. We can consider that: the SEA *PP* models SYST, and the SEA *SS* models the most permissive desired behaviour of SYST.

4.5 Step 4 : Compute Sup from PP and SS

Based on the two SEAs *PP* and *SS*, we compute a supervisor *Sup* ensuring *Sup/PP* **conf** *SS*. Proposition 3.1 ensures that the resulting controlled plant is in conformance with SPEC, i.e., *Sup/P* **tioco** S.

Notations :
- Σ and Γ are the alphabets of P and *PP*, resp.
- for $M \subseteq \Sigma$: M^2 contains every $\lambda \in \Gamma$ of the form σ or (σ, S), and M^3 contains every $\lambda \in \Gamma$ of the form $(\mathcal{E}, \sigma)$ or $(\mathcal{E}, \sigma, S)$, where $\sigma \in M$.
- for $M \subseteq \Gamma$, $[M]_{Exp}$ is obtained from M by keeping only *Exp* events.
- For a SEA B over Γ and $\mu \in \overline{L_\infty(B)}$, $Acc_B(\mu)$ is the set of labels accepted in B after the execution of μ. Formally : $Acc_B(\mu) = \{\sigma \in \Gamma | \mu\sigma \in \overline{L_\infty(B)}\}$.

Controllability of SEA: A SEA B over Γ and s.t. $L_\infty(B) \subseteq L_\infty(PP)$ is said *controllable* w.r.t. *PP* if $\forall \mu \in \overline{L_\infty(B)}$:

$$\text{If } Acc_B(\mu) \cap \Sigma_f^2 \neq \emptyset : Acc_B(\mu) \supseteq Acc_{PP}(\mu) \cap \Sigma_{uc}^2 \quad (1)$$

$$\text{If } Acc_B(\mu) \cap \Sigma_f^2 = \emptyset : Acc_B(\mu) \supseteq$$
$$Acc_{PP}(\mu) \cap (\Sigma_{uc}^2 \cup \Sigma_{uc}^3 \cup [\Sigma_f^3 \setminus Acc_B(\mu)]_{Exp}) \quad (2)$$
$$[(Acc_{PP}(\mu) \cap \Sigma_{uc}^3) \setminus Acc_B(\mu)]_{Exp} \cap [Acc_B(\mu)]_{Exp} = \emptyset \quad (3)$$

Intuitively, B controllable w.r.t. *PP* means that after an execution accepted by *PP* and B: (in the following, $\sigma_{uc} \in \Sigma_{uc}$ and $\sigma_f \in \Sigma_f$)

(1) If B accepts a σ_f or (σ_f, S) then : B accepts every σ_{uc} or (σ_{uc}, S) accepted by *PP*.

(2) If B accepts no σ_f or (σ_f, S) then :
- B accepts, not only every σ_{uc} and (σ_{uc}, S) accepted by *PP*, but also every $(\mathcal{E}, \sigma_{uc})$ and $(\mathcal{E}, \sigma_{uc}, S)$ accepted by *PP*. This is so, because B can preempt a $(\mathcal{E}, \sigma_{uc})$ or $(\mathcal{E}, \sigma_{uc}, S)$ only by the execution of a (σ_f) or (σ_f, S).
- If B does not accept a $L = \mathcal{E}$ accepted by *PP*, then B accepts a (L, σ_f) or (L, σ_f, S) accepted by *PP*. This is so, because B can preempt a $L = \mathcal{E}$ only by the execution of (σ_f) or (σ_f, S) or (L, σ_f) or (L, σ_f, S).

(3) If B forbids a $(\mathcal{E}, \sigma_{uc})$ or $(\mathcal{E}, \sigma_{uc}, S)$ accepted by *PP*, then B forbids every transition with the same expiration(s). This is so, because $(\mathcal{E}, \sigma_{uc})$ or $(\mathcal{E}, \sigma_{uc}, S)$ can be forbidden only by preemption of the corresponding expiration(s).

Let $Cont_{PP}^{SEA}$ be the set of controllable SEAs w.r.t. *PP*, $Cont_{PP}^{SEA}(SS) = \{B \in Cont_{PP}^{SEA} | L_\infty(B) \subseteq L_\infty(SS)\}$, and $SupL_{PP}^{SEA}(SS) = \bigcup_{B \in Cont_{PP}^{SEA}(SS)} L_\infty(B)$.

Lemma 4.2. Let $B_1, B_2 \in Cont_{PP}^{SEA}(SS)$, B be a SEA s.t. $L_\infty(B) = L_\infty(B_1) \cup L_\infty(B_2)$, and B' be a SEA s.t. $L_\infty(B') = SupL_{PP}^{SEA}(SS)$. We have : $B, B' \in Cont_{PP}^{SEA}(SS)$.

From Lemma 4.2, we assert the existence of a unique minimal SEA B s.t. $L_\infty(B) = SupL_{PP}^{SEA}(SS)$. Such SEA is denoted $SupC_{PP}^{SEA}(SS)$ and called "supremal controllable SEA w.r.t. *PP* and *SS*".

Lemma 4.3. $SupC_{PP}^{SEA}(SS)$ **conf** SS.

Proposition 4.1. Objective 1 is reached if *Sup/PP* is described by $SupC_{PP}^{SEA}(SS)$.

Prop. 4.1 allows to replace Objective 1 by:

Objective 2. To synthesize a supervisor s.t. *Sup/PP* is described by $SupC_{PP}^{SEA}(SS)$.

That is, the control problem that was expressed in Obj. 1 by the use of TA, has been transformed into a non-real-time form in Obj. 2 by the use of SEA.

$SupC_{PP}^{SEA}(SS)$ can be computed from *PP* and *SS* by using a fixpoint method [4]. The intuition of the latter is to start with *SS* and to proceed iteratively as follows :

- We remove each state where Eq. (1) or (2) is not satisfied.
- For each state where Eq. (3) is not satisfied, we remove the *Exp-Trans*s that must be removed in order to satisfy Eq. (3).
- We update the sets of preempted expirations of each state.

Note that states non-satisfying Eq. (1) or (2) are processed as usual by removing them. As for Eq. (3), when it is not satisfied in a state q, it can be made satisfied by removing some outgoing *Exp-Trans* of q.

Theorem 1. The supervisor *Sup* computed by the above algorithm guarantees Objective 2 (and thus Objective 1 from Proposition 4.1).

5. CONTRIBUTION AND FUTURE WORK

We propose a method for controlling real-time discrete event systems. The plant and the specification are modelled by a variant of timed automata (TA) that uses the notion of *forced transitions*. Our control method is based on a transformation, called $SetExp$, of TA into a so-called Set-Exp-Automaton (SEA) which uses two additional types of events: Set and Exp. We use a conformance relation between the plant and the specification, called **tioco**, which takes into account the notion of *quiescence*.

In comparison to other control methods (Brandin and Wonham, 1994; Gouin, 1999; D'Souza and Madhusudan, 2002) [5], an advantage of our method is that in practice, its complexity does not increase significantly with the magnitudes of the constants used in timing constraints. This is due to the use of $SetExp$ (Khoumsi and Ouedraogo, 2004).

As future work, we intend to quantify more accurately the gain of our approach in comparison with other methods.

ACKNOWLEDGEMENTS

I thank Hervé Marchand and Thierry Jéron, from IRISA, for their contribution towards the improvement of this manuscript.

REFERENCES

Alur, R. and D. Dill (1994). A theory of timed automata. *Theoret. Comp. Sc.* **126**, 183–235.

Brandin, B. and W.M. Wonham (1994). Supervisory control of timed discrete-event systems. *IEEE Trans. on Aut. Cont.* **39**(2), 329–342.

Daws, C. and S. Yovine (1995). Two examples of verification of multirate timed automata with KRONOS. In: *Proc. 16th IEEE RTSS*. Italy.

D'Souza, D. and P. Madhusudan (2002). Timed control synthesis for external specifications. In: *Proc. 14th STACS, LNCS 2285*. Springer-Verlag.

Gouin, A. (1999). Contribution à la commande de systèmes à événements discrets temporisés : synthèse de superviseur dans le cadre de modèle automate. PhD thesis. Lab. d'Ing. des Syst. Autom., University of Angers. France.

Jéron, T., H. Marchand, V. Rusu and V. Tschaen (2003). Ensuring the conformance of reactive discrete-event systems using supervisory control. In: *42nd CDC*. Hawaii, USA.

Khoumsi, A. and L. Ouedraogo (2004). A new transformation of timed automata into finite state automata. Subm. to: *FORMATS/FTRTFT 2004*.

Larsen, K.G. (1995). Compositional and symbolic model-checking of real-time systems. In: *Proc. 16th IEEE Real-Time Systems Sympos. (RTSS)*. Italy.

Ramadge, P.J. and W.M. Wonham (1989). The control of discrete event systems. In: *Proc. IEEE*. Vol. 77. pp. 81–98.

Tretmans, J. (1996). Test generation with inputs, outputs and repetitive quiescence. *Software-Concepts and Tools*.

[4] For lack of space, we do not present the method formally.

[5] For lack of space, many other references are not cited.

ELSEVIER
IFAC
PUBLICATIONS
www.elsevier.com/locate/ifac

A HPN-BASED CONTROL STRUCTURE FOR PRIVILEGED VEHICLE PERFORMANCE OPTIMISATION

Angela Di Febbraro * **Davide Giglio** ** **Nicola Sacco** *

* *DAUIN - Politecnico di Torino*
Corso Duca degli Abruzzi, 24
10129 - Torino, Italy

** *DIST - Università di Genova*
Via Opera Pia, 13
16145 - Genova, Italy

Abstract:
Urban networks of signalised intersections can be suitably modelled as hybrid systems, where the vehicle flow behaviour is described by means of a time-driven model, and the traffic light dynamics is represented by a discrete event model. In this paper, a Hybrid-Petri-Net based model is used to evaluate the effects of an optimisation procedure intended to improve the performances of public and emergency vehicles by coordinating several traffic lights. To this aim, some experimental results, relevant to the Italian city of Turin, are reported and discussed. *Copyright © 2004 IFAC*

Keywords: Transportation, Traffic control, Petri nets

1. INTRODUCTION

Urban traffic networks can be considered to be complex large-scale systems, in which different, seldom cooperative components interact. The growing necessity of optimising, or at least improving, the performances of public transportation services to make them attractive to users has led many researchers to focus on modelling, analysing, and controlling urban traffic systems. In this framework, the introduction of *Intelligent Transportation Systems* (ITS) has provoked a significant enhancement of the workability of research results.

An urban traffic network is suitable to be modelled via hybrid Petri nets (HPN) (Alla and David, 2001), which allow to represent systems characterised by both continuous (vehicle flows) and discrete-event (traffic lights) dynamics, i.e., hybrid systems. Classical Petri nets have been applied to traffic control in several works (among the first (Di Cesare *et al.*, 1994) and (Gallego *et al.*, 1996)). This paper presents the recent advancements of a research stream that the authors have been following since some years. In particular, based on the HPN-based model described and validated in (Di Febbraro and Sacco, 2003), the problem of coordinating several traffic lights in a signalised urban traffic network, in order to optimise the performance of special public services, is here addressed. That extends the problem presented by the authors in (Di Febbraro *et al.*, 2003), which does not rely on the use of HPN. In fact, the adopted model is suitable to represent at once a signalised urban traffic network and the relevant time-varying regulation strategies.

2. THE TRAFFIC SYSTEM MODEL

The adopted model of a traffic system consists of *signalised intersections* and *roads*. Let I_h, $h = 1, \ldots, H$, and R_k, $k = 1, \ldots, K$, be the generic signalised intersection and road, respectively. A signalised intersection is characterised by the sets of indices of incoming and outgoing roads, namely IN_h and OUT_h, respectively. In the proposed model, the flow f_j, $j \in IN_h$, represents the (input) vehicle flow approaching the intersection from the adjacent road R_j; the flow $f_j^{\rm in}$, $j \in IN_h$, represents the flow crossing the intersection, entering from the incoming direction j when enabled by the traffic light; the flow $f_i^{\rm out}$, $i \in OUT_h$, represents the flow leaving the intersection towards direction i (as an example, consider the intersection in Fig. 1, consisting of 3 incoming and 3 outgoing roads). Moreover, q_j, $j \in IN_h$, denotes the queue of vehicles waiting for the green signal at each incoming direction j (the black circles in Fig. 1 indicate the positions of the inductor loops placed to measure the real vehicle flows).

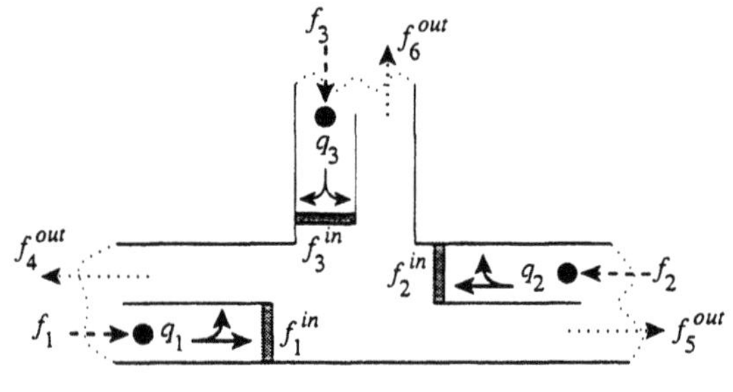

Fig. 1. A signalised intersection.

At each signalised intersection, a multi-phase traffic lights rule the vehicle flows. Each phase is characterised by the flows which are enabled within the phase itself. Let ϑ_φ^g and ϑ_φ^a, $\varphi = 1, \ldots, \Phi$, be the lengths of the green and amber parts, respectively, of the generic phase φ (only the green and the amber signals are taken into account, since all the incoming flows which do not find a green or a amber signal are supposed to find the red one).

As regards the traffic flow dynamics within I_h, each input flow $f_j^{\rm in}$, $j \in IN_h$ – enabled by the traffic lights – is given, at each time τ, by

$$f_j^{\rm in}(\tau) = \begin{cases} f_{{\rm cr}_j} & \text{if } q_j(\tau) > 0 \\ \min\{f_j(\tau), f_{{\rm cr}_j}\} & \text{if } q_j(\tau) = 0 \end{cases} \quad (1)$$

where $f_{{\rm cr}_j}$ is a constant parameter experimentally determined, which represents the number of vehicles leaving the queue at direction j at each time τ. It is obvious that $f_j^{\rm in} = 0$ when $f_j^{\rm in}$ is not enabled by the traffic lights. The queue dynamics is ruled by

$$\dot{q}_j(\tau) = f_j(\tau) - f_j^{\rm in}(\tau) \qquad j \in IN_h \quad (2)$$

Finally, the output flows can be calculated by associating with I_h an *origin-destination* matrix $A_h(\tau)$, whose generic element $\alpha_{i,j}^h(\tau)$ indicates the percentage of vehicles coming from direction j, $j \in IN_h$, and going towards direction i, $i \in OUT_h$, at time τ. Thus, at any time τ, the output flows are given by

$$f_i^{\rm out}(\tau) = \sum_{j \in IN_h} \alpha_{i,j}^h(\tau) \cdot f_j^{\rm in}(\tau) \qquad i \in OUT_h \quad (3)$$

The model of an urban road mainly consists of a road section R_k, of length Δ_k. In such a way,

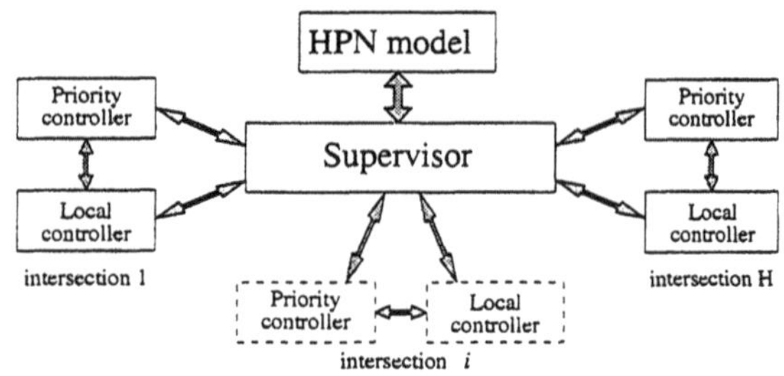

Fig. 2. The proposed control structure.

the traffic behaviour can be described using a well-known second order macroscopic model, first introduced in (Payne, 1971), and where the traffic behaviour is described by means of the vehicle density $\rho_k(\tau)$ in section R_k, and the volume of vehicles $f_k(\tau)$ exiting section R_k towards the downstream intersection, at time τ. The relevant equations are:

$$\dot{\rho}_k(\tau) = \frac{f_{h,k}^{\rm out}(\tau) - f_k(\tau)}{\Delta_k}$$
$$f_k(\tau) = \mu \cdot F\big[\rho_k(\tau)\big] + (1-\mu) \cdot F\big[\rho_{q_{k,l}}(\tau)\big]$$

where $f_{h,k}^{\rm out}(\tau)$ is the flow entering section R_k from the upstream intersection I_h, $F\big[\rho_k(\tau)\big]$ is the *fundamental traffic diagram* (see (Papageorgiou, 1983) for details), μ is a parameter dependent on the road stretch topology, and $\rho_{q_{k,l}}(\tau)$ is the vehicle density at the incoming direction k of the downstream intersection I_l.

2.1 The control structure

The control structure here presented is conceived to perform traffic optimisation by changing the timing of traffic lights, under the hypotheses of both regular and special conditions, i.e., when a special public or emergency vehicle asks for a privilege. To this aim, consider the control architecture represented in Fig. 2. Each signalised intersection is equipped with both a *local controller*, which rules the intersection in regular conditions, by applying an optimal response traffic plan, and a *priority controller*, which can force a modified traffic light timing when a particular event occurs. In turn, a supervisor coordinates the local and priority controllers and solves all the problems involving several intersections.

3. THE HYBRID PETRI NET MODEL

The urban traffic network model is implemented into the supervisor by means of a *hybrid Petri net* (HPN). This choice is motivated by the fact that it is possible to determine the HPN by means of quite simple procedures (for example, by means of a simple *modular* modelling procedure, here partially reported). The formal definition (Alla and David, 2001) of the adopted HPN model is the following.

Definition 1. A *hybrid Petri net* (HPN) is the 6-tuple $HPN = \{P, T, Pre, Post, h, M_0\}$ where P is a finite non-empty set of n; T is a finite non-empty set of m ($P \cap T = \emptyset$, i.e., the sets of places and transitions are disjoint); P (resp., T) is split into two subsets P^{D} and P^{C} (resp., T^{D} and T^{C}) gathering, respectively, the discrete and the continuous places (resp., transitions); Pre is the preset function $Pre_{i,j} : (P \times T) \rightarrow \{\mathbb{R}_0^+ \, \forall p_i \in P^{\mathrm{C}}, \mathbb{N}_0^+ \, \forall p_i \in P^{\mathrm{D}}\}$ which assigns a weight to any arc between a transition t_i and its input place p_j; $Post$ is the postset function $Post_{i,j} : (P \times T) \rightarrow \{\mathbb{R}_0^+ \, \forall p_i \in P^{\mathrm{C}}, \mathbb{N}_0^+ \, \forall p_i \in P^{\mathrm{D}}\}$ which assigns a weight to any arc between a transition t_j and its input place p_i; $h : P \cup T \rightarrow \{\mathrm{D}, \mathrm{C}\}$ is the hybrid function which indicates whether a node (place or transition) is a discrete (D) or a continuous (C) one; M_0 is the initial marking vector, being the marking the function $M : P \rightarrow \{\mathbb{R}_0^+ \, \forall p_i \in P^{\mathrm{C}}, \mathbb{N}_0^+ \, \forall p_i \in P^{\mathrm{D}}\}$ which assigns a non-negative, integer number (i.e., tokens) to each discrete place and a non-negative real number to each continuous place.

The hybrid Petri net model of the intersection in Fig. 1 is depicted in Fig. 3. A detailed description of places and transitions is reported in Tab. 1. Here, it is just remarked that the marking of a continuous place represents the number of vehicles in the incoming/outgoing direction associated with that place.

Table 1. Places and transitions in Fig. 3.

P	Meaning	
p_j^{in}	queue at incoming direction j	
p_i^{out}	vehicles are in the outgoing direction i	
p^{φ}	the φ-th phase is enabled	
p_g^{φ}	green period of φ-th phase	
p_a^{φ}	amber period of φ-th phase	
$p_{d'}^{i}$	vehicles cannot flow through road i	
$p_{e'}^{i}$	vehicles can flow through road i	
T	**Meaning**	**Speed or Delay**
t_j^{cr}	vehicles leave the queue at incoming direction j	f_{cr_j}
t_j	vehicles enter the queue at incoming direction j, $j \in IN_h$	f_j
t_i	vehicles leave the intersection towards direction i, $i \in OUT_h$	f_i
$t_{\varphi > \varepsilon}$	φ-th phase ends and ε-th one begins	ϑ_a^{φ}
$t_{g>a}^{\varphi}$	φ-th green phase ends and φ-th amber phase begins	ϑ_g^{φ}
t_d^i	the road stretch R_i becomes blocked	-
t_e^i	the road stretch R_i becomes unblocked	-
t_s^i	vehicles enter the road stretch R_i from a secondary road	f_i^s
t_w^i	vehicles exit from road stretch R_i towards a secondary road	f_i^w

On the other hand, the marking evolution of the discrete part allows, by alternately putting a token within implicit places p^1, p^2, and p^3, the incoming directions to find green light in turn, and thus cross the intersection. The continuous part of the net evolves following the time-driven relation (3). Such an equivalence (see (Di Febbraro and Sacco, 2003) for a detailed description) comes from the definition of the HPN weights $\beta_j(\tau) = f_j^{\mathrm{in}}(\tau)/\mathrm{f}$, $j \in IN_h$, $i \in OUT_h$, and $\gamma_{i,j}(\tau) = \alpha_{i,j}(\tau) \cdot \beta_j(\tau)$, $j \in IN_h$, $i \in OUT_h$, where $\mathrm{f} = 1\, veh/s$ is the flow rate unity.

4. CONTROL AND OPTIMISATION PROBLEMS

In this section, the problem of controlling traffic in an urban transportation network so as to minimise the travelling time from a specific origin to a specific destination of special *privileged vehicles* is addressed, based on the above described model. In this framework, the categories of vehicles taken into account are those performing public services, or those performing emergency services.

Although the basic traffic regulation problem is the same for both the considered classes of vehicles, a higher priority might be guaranteed to emergency vehicles, so as to make their travelling times as short as possible. In addition, public vehicles follow a-priori determined routes, whereas emergency vehicles can choose the best route among all the possible routes. This consideration implies that the problem of controlling public vehicles can be considered to be a sub-problem of the more general problem of setting an *optimal route* for emergency vehicles. It is assumed that the traffic behaviour is optimised via an optimal responsive plan, which minimises the queue lengths at the incoming directions of intersections.

4.1 Local priority control problem

In this section, the basic problem of managing a priority request is solved for an isolated intersection, by proposing a simple control algorithm.

When a privileged vehicle, at $\bar{\tau}$, asks the priority controller of the near intersection for a phase modification, it communicates to the controller its position. The controller, which knows the traffic conditions around it, computes the time instant $\tilde{\tau}$ when the privileged vehicle is expected to arrive. Then, if at time $\tilde{\tau}$, according to the current traffic light timing, the signal for the direction the vehicle comes from is expected to be red, the priority controller has to decide whether:

(a) extending the precedent green signal till the privileged vehicle crosses the intersection;
(b) anticipating the next green signal, shortening (as much as possible) the red one accordingly.

Note that in the optimisation problems which find the responsive traffic plan there is a constraint which sets the minimum length of each green

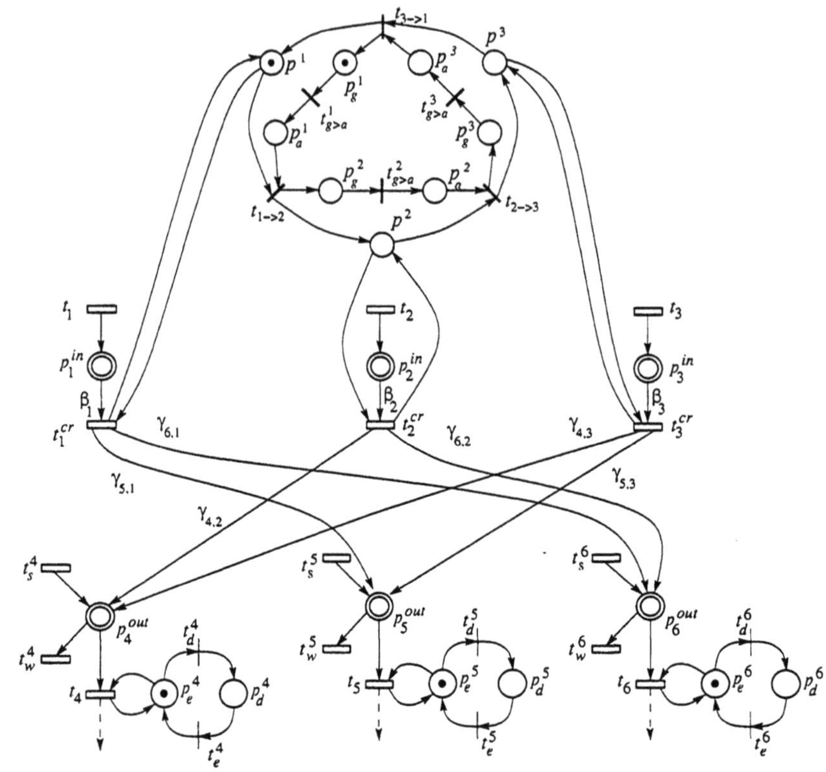

Fig. 3. The hybrid Petri net model of the intersection of Fig. 1.

phase. Then, the priority controller has to solve the almost trivial integer programming problem

$$\min_{y_a, y_b} \ [(1 - y_a)(1 - y_b)\, C + y_a\, C_a + y_b\, C_b]$$
$$y_a + y_b \leq 1 \qquad (4)$$
$$y_a,\, y_b \in \{0, 1\}$$

$$C = C_0 + \sum_{j \in IN_h} \frac{\varsigma_j}{\tau_c^2 - \bar{\tau}} \int_{\bar{\tau}}^{\tau_c^2} [\alpha \cdot q_j(\tau)]^2 d\tau \quad (5)$$

$$C_a = \sum_{j \in IN_h} \frac{\xi_j}{\tau_c^2 - \bar{\tau}} \int_{\bar{\tau}}^{\tau_c^2} [q_j^a(\tau) - q_j(\tau)]^2 d\tau \quad (6)$$

$$C_b = \sum_{j \in IN_h} \frac{\psi_j}{\tau_c^2 - \bar{\tau}} \int_{\bar{\tau}}^{\tau_c^2} [q_j^b(\tau_c) - q_j(\tau_c)]^2 d\tau \quad (7)$$

determining whether it is more convenient to concede the privilege or not. The control variable y_a (resp., y_b) assumes the value 1 if the privilege request is satisfied by extending the current green signal (resp., shortening the current red signal).

C, C_a, and C_b are due to the fact that the modifications of the traffic light plan introduced by the privileged vehicle represent suboptimal solutions to the responsive traffic light plan. Note that it is reasonable to suppose that the perturbations become negligible after two or three traffic light cycles. Thus, such costs are calculated over a time period starting when the privilege is requested and finishing when the second traffic light cycle after the modification ends in τ_c^2. In particular, the cost C represents the cost of non conceding any privilege; In this cost, the term C_0 is a constant cost due to the fact that the privileged vehicle has to wait, and α is the maximum admissible displacement for a privileged vehicle. Then, the cost C_a and C_b represent the mean square deviations of the queue lengths $q_j^a(\tau)$ and $q_j^b(\tau)$, $\forall j \in IN_h$, from the ideal queue length $q_j(\tau)$ in the perturbed cases (a) and (b), respectively. Then, the weights ς_j, ξ_j, and ψ_j, represent the importance of each incoming direction j, $j \in IN_h$. It is worth noting that the above introduced costs can be easily calculated since $q_j^a(\tau)$, $q_j^b(\tau)$, and $q_j(\tau)$ can be determined by means of the traffic model proposed in Sec. 2 by knowing the nominal and the perturbed traffic light timings.

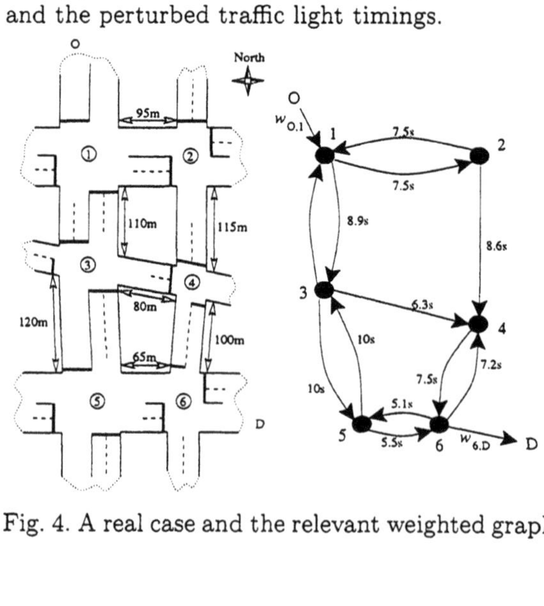

Fig. 4. A real case and the relevant weighted graph

Step	Instructions
1.	The supervisor computes $\mathcal{SP}(\bar{\tau}) = \{\mathcal{P}_i\}_i^N$, i.e., the set of the shortest paths according to the traffic conditions at $\bar{\tau}$; set $i = 0$.
2.	$\forall I_h \in \mathcal{P}_i$, each priority controller solve the problem (4), knowing the time instant $\bar{\tau}_{i,h}$ and the direction the vehicle comes from.
3.	The supervisor computes the real travel time T_i on the route $\mathcal{P}_i$ by means of the traffic model of Sec. 2, based on the traffic conditions and the traffic light timings determined in *3*; then, if $i < N$, set $i = i + 1$ and go to *2*, else go to *4*.
4.	Among all the possible routes $\mathcal{P}_i$, the supervisor chooses the route $\mathcal{P}_i$ such that $i = \arg\min\{T_i\}$; then it communicates the route to the privileged vehicle and to the involved priority controllers.

Table 2. Routing algorithm.

4.2 Emergency vehicle performance optimisation

In this section, a control algorithm which finds the optimal route trying to change the traffic light phases conveniently is described. Such a problem differs from the problem of finding the shortest path in a net, since each intersection of all the possible routes is asked to modify its traffic light timing with the aim of shortening the transit time. The basic idea is that an urban traffic network can be seen as a weighted graph, such as the one in Fig. 4, where each node represents a signalised intersection whereas any road linking two intersections is represented by means of an arc whose weights represent the time vehicles need to reach the downstream intersection starting from the upstream intersection.

This control problem can be solved via the algorithm for privileged vehicle routing reported in Tab.2. Such an algorithm is based on the hypothesis that the traffic behaviour has a slow dynamics which implies that the traffic plan can be considered to keep constant during some traffic light cycles after the privilege request.

4.3 Case study

In this section, an example relevant to the emergency vehicles priority control is presented, with the aim of proving the efficacy of the proposed control procedures. First, consider the traffic net in Fig. 4, which represents a portion of the traffic network of the Italian city of Turin with six signalised intersections and the relevant graph. The flows enabled during each traffic light phase, and the relevant timings, are reported in Fig. 5. Moreover, since the responsive traffic plan privileges the direction North-South, the traffic light cycles of the intersections 1, 2, 4, and 6 begin at the same time τ_0, whereas, on the contrary, the traffic light cycles of the intersections 3 and 5 begin after a offset $8\,s$ and $16.5\,s$, respectively.

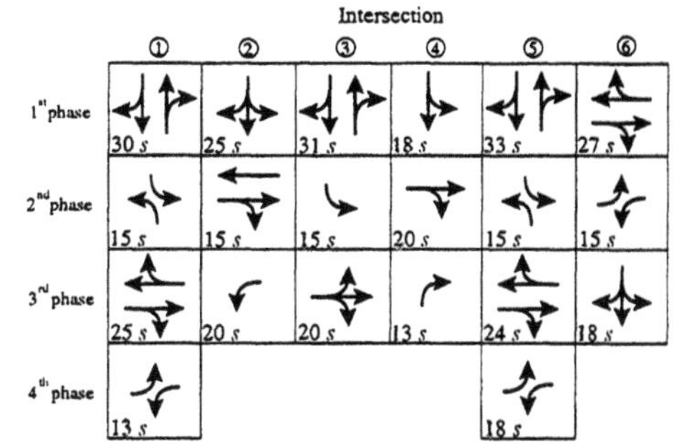

Fig. 5. Enabled flows at each intersections of the net in Fig. 4 during each phase and phase timings.

Then, suppose that an emergency vehicle arrives at the north incoming direction of I_1 at $\tau_1^1 = \tau_0 + 25\,s$, where τ_0 is the initial time instant, and asks the supervisor for a *privileged route* to reach its destination D from the origin O depicted in Fig. 4. It is easy to note that, in this case, the set of the admissible paths is $\mathcal{SP} = \{\mathcal{P}_1, \mathcal{P}_2, \mathcal{P}_3\} = \{(1,3,5,6), (1,2,4,6), (1,3,4,6)\}$. Moreover, the length of the routes and the travelling time, only due to the traffic conditions on the road stretches are $\mathcal{L} = \{295\,m,\, 310\,m,\, 290\,m\}$ and $\mathcal{W} = \{24.3\,s,\, 23.6\,s,\, 23\,s\}$ respectively.

Thus, from both these points of view, the shortest route results to be the third one, but the two performance parameters do not take into account the waiting times at the intersections. On the other hand, when the coordinated control problem is solved, the supervisor has to decide among the following routes:

$\mathcal{P}_1$: if the vehicle follows the first route, it crosses I_1 towards I_3 without stopping. Then, due to the traffic behavior, it reaches I_3 at time $\tau_2^1 = \tau_1^1 + 9.2\,s = 34.2\,s$, and crosses it without stopping towards I_5, where it arrives at $\tau_2^1 = \tau_2^1 + 10\,s = 44.2\,s$, finding the first phase whereas it needs the second one. Thus, a local priority control problem is solved, yielding the costs reported in the second column of the second row of Tab. 3. As it is clear, the best choice for the controller is to shorten the first phase. With this change, the privileged vehicle can cross the intersection towards I_6, where it arrives at $\tau_3^1 = \tau_2^1 + 5.1\,s = 39.3\,s$, finding the third phase instead of the first one, that it needs. Thus, a second control problem has to be solved. It is worth noting that the best choice is to do nothing, as it is clear in the third column of the third row of Tab. 3. Thus, the vehicle has to wait until a new cycle begins in $\tau_4^1 = 60\,s$. At this time, it can proceed. The travelling time on this route is $T_1 = 35\,s$, whereas the total cost is $C_1^{tot} = 31.13$;

Table 3. Costs of the problem of the example in Sec. 4.3.

Route	1st Problem	2nd Problem	T_i	C_i^{tot}
$\mathcal{P}_1$	$C_0 = 14.57$, $C_1 = 14.92$, $C_2 = \mathbf{13.02}$	$C_0 = \mathbf{18.11}$, $C_1 = 20.37$, $C_2 = 19.09$	35 s	31.13
$\mathcal{P}_2$	$C_0 = 15.12$, $C_1 = 24$, $C_2 = \mathbf{14.54}$	$C_0 = 13.39$, $C_1 = 29.11$, $C_2 = \mathbf{12.08}$	30.5 s	**26.62**
$\mathcal{P}_3$	$C_0 = 15$, $C_1 = 25.07$, $C_2 = \mathbf{14.38}$	$C_0 = 15.24$, $C_1 = \mathbf{15.03}$, $C_3 = 17.12$	**26.8** s	29.4

$\mathcal{P}_2$: if the vehicle follows the second route, it finds the first phase of the traffic light cycle instead of the second one, as it needs. Thus, a local priority control problem has to be solved. Being the results those reported in the second column of the third row of Tab. 3, the best choice is to shorten the first phase but, however, the vehicle should have to wait for $2\,s$, due to the constraint on the minimum length of the first phase. Then, it can proceed towards I_2, where it arrives at $\tau_1^2 = 25\,s + 3\,s + 7.5\,s = 34.5\,s$, and it can proceed, without stopping, towards I_4, where it arrives at $\tau_2^2 = \tau_1^2 + 8.6\,s = 43.1\,s$. At this time, it finds the second phase, whereas it needs the third one. Thus, another local priority control problem has to be solved, giving the costs reported in the third column of the third row of Tab. 3. As it is clear in such a table, the best choice for the controller is to shorten the present second phase, but, as for the previous problem, it should wait for $4.9\,s$, due to the constraint on the minimum length of the first phase. After this waiting time, the privileged vehicle can cross the intersection towards I_6, where it arrives at $\tau_3^2 = \tau_2^2 + 4.9\,s + 7.5\,s = 55.5\,s$, finding the third phase, which gives it the right to turn left towards its destination. The travelling time on this route is $T_2 = 30.05\,s$, whereas the total cost is $C_2^{tot} = 26.62$;

$\mathcal{P}_3$: if the vehicle follows the third route, it can cross the intersection towards I_3 without stopping. Due to the traffic behavior, it arrives at I_3 at $\tau_1^3 = \tau_1^1 + 9.2\,s = 34.2\,s$, as in $\mathcal{P}_1$. However, this time it finds the first phase, whereas it needs the second one. Thus, a local priority control problem has to be solved by means of the costs reported in the second column of the fourth row of Tab. 3. As it is clear in such a table, the best choice is to shorten the first phase, but the vehicle has to wait for $1.8\,s$, because the first phase has a minimum length constraint. Then, with this change the privileged vehicle can cross the intersection towards I_4, where it arrives at $\tau_2^3 = \tau_1^3 + 1.8\,s + 6.3\,s = 42.3\,s$, finding the third phase instead of the second one it needs. In such a scenery, a second control problem has to be solved and, in this case, the best choice is to extend the present second phase (with the cost reported in the third column of the fourth row of Tab. 3). Finally, the vehicle can cross the intersection towards I_6, where it arrives at $\tau_3^3 = \tau_2^3 + 7.5\,s = 49.8\,s$ finding the third phase, which is the one it needs to reach its destination D. The travelling time on this route is $T_3 = 24.8\,s$, whereas the total cost is $C_1^{tot} = 29.4$.

Then, by observing the results summarised in Tab. 3, one can conclude that the choice of path $\mathcal{P}_3$ minimises the travelling time, whereas the choice of path $\mathcal{P}_2$ introduces the minimum perturbation in the traffic net in terms of "abnormal" length queue with respect to the nominal case. Thus, the supervisor decides whether to choose the path $\mathcal{P}_2$ or the path $\mathcal{P}_3$, depending on the selected criterium, i.e., the minimum travelling time or the minimum perturbation cost.

5. CONCLUSIONS

In this paper, the performances of a proposed control structure, based on a HPN model of urban networks of signalised intersections, have been evaluated. The adopted representation allows to represent at once a signalised urban traffic network and the relevant time-varying regulation strategies.

Some experimental results, relevant a real case study, have been reported and discussed to show the efficacy of the presented approach.

REFERENCES

Alla, H. and R. David (2001). On Hybrid Petri Nets. *Journal of Discrete Event Dynamic Systems* **8**(1–2), 9–40.

Di Cesare, F., P. Kulp, K. Gile and G. List (1994). *The Applications of Petri Nets to the Modeling Analysis and Control of Intelligent Urban Traffic Networks.* Vol. 815 of *Lecture Notes in Computer Science.* Springer–Verlag.

Di Febbraro, A. and N. Sacco (2003). Experimental validation of a hybrid Petri-net based model of urban transportation networks. *Proc. IFAC Int. Conf. on Analysis and Design of Hybrid Systems, ADHS 2003* pp. 111–115.

Di Febbraro, A., D. Giglio and N. Sacco (2003). On controlling privileged vehicles by means of coordinated traffic lights. *Proc. 2003 IEEE Int. Conf. on Intelligent Transportation Systems* pp. 1318–1323.

Gallego, J. L., J. L. Farges and J. J. Henry (1996). Design by Petri nets of an intersection signal controller. *Transportation Research Part C* **4**, 231–248.

Papageorgiou, M. (1983). *Applications of Automatic Control Concepts to Traffic Flow Modeling and Control.* Springer-Verlag.

Payne, H. J. (1971). Models of Freeway traffic and control. *Simulation Council Proc.* **1**, 51–61.

ELSEVIER
IFAC
PUBLICATIONS
www.elsevier.com/locate/ifac

SUPERVISORY CONTROLLER OF PETRI NETS UNDER PARTIAL OBSERVATION

Zied Achour, Nidhal Rezg and Xiaolan Xie

INRIA/MACSI & LGIPM
Ile du Saulcy, 57045 Metz Cedex, FRANCE
Email : {achour, nrezg, xie}@loria.fr

Abstract: We propose in this paper a control synthesis policy for the forbidden state problems of bounded Petri nets (PN) with liveness requirement and uncontrollable / unobservable transitions. The proposed approach computes a maximally permissive PN controller, whenever such a controller exists. The first step, based on a Ramadge-Wonham theory, determines the refined controlled observer reachability graph of the controlled PN. In the second step, the theory of regions is used to generate a set of control places to be added to the plant PN model. These control places guarantee the desired behavior. The proposed approach is illustrated on a manufacturing example. *Copyright © 2004 IFAC*

Keywords: Supervisory Control, Petri Nets, Unobservable Transitions, Forbidden State Problems.

1. INTRODUCTION

In this paper, bounded Petri nets are used to model Discrete Event Systems (DES) also called plants to control. We are interested in designing a supervisor which guarantees that the marking of the Petri net cannot leave a given set of legal marking. In a general DES, some events are uncontrollable, i.e. cannot be prevented from occurring, and some events are unobservable, i.e. cannot be observed by the supervisor. The type of supervisor, which we intend to design, is a compiled supervisor, whose control policy is represented as a Petri net structure. This paper addresses the synthesis of a Petri net supervisor for the forbidden state problem of a partially observable plant with unobservable / uncontrollable transitions.

We present a synthesis method based on the theory of regions which is a technique of Petri net synthesis from automata models. Plant models under consideration are generalized bounded Petri nets. The Petri net controller is optimal in the sense of maximum permissiveness under the restriction of the liveness of the controlled system.Supervisory control of partially observable Petri net plants has been addressed in (Moody and Antsaklis, 1999; Stremersch, 2000; Moody, *et al.*, 1996) using linear algebra approaches which are extensions of the optimal place-invariant approach proposed in (Yamalidou, *et al.*, 1996) for enforcing linear constraints for totally controllable and observable Petri net plants. The most serious drawback of these approaches is that the optimality of the control policy cannot be guaranteed. An exception is the linear algebra approach proposed in (Darondeau and Xie, 2003) for enforcing linear constraints of firing count vector and/or markings of partially controllable and partially observable marked graphs. Note that the liveness of the controlled system cannot be enforced with these approaches.We follow, in this paper, the approach of (Ghaffari, *et al.*, 2003) to address the forbidden state problems of generalized Petri net model with uncontrollable and unobservable transitions.

Contribution of this paper can be summarized as follows. Supervisory control of partially controllable and partially unobservable Petri nets has been addressed in (Moody and Antsaklis, 1999; Stremersch, 2000; Moody, *et al.*, 1996) using algebra approaches. Compare with these algebra approaches, our paper proposes an optimal control policy while the algebra approaches of (Moody and Antsaklis, 1999; Stremersch, 2000; Moody, *et al.*, 1996) are complicated and do not guarantee the optimality. Another significant improvement of the present paper is the enforcement of livens requirement of the controlled system that cannot be taken into account in the algebra approaches. Finally, as in the algebra approaches, our approach also represents the controller by means of control places. Further, necessary and sufficient conditions are given in this paper for the existence of such an optimal PN controller.

Compared with (Giua and Seatzu, 2002) which considered the case with unknown initial marking but without unobservable transitions, the problem addressed in this paper is different and concerns the supervisory control of Petri nets with known initial marking and unobservable transitions.

More specifically, contributions and motivations of this paper with regard to (Ghaffari, *et al.*, 2003) are as follows. (i) It proves that the supervision problems in the presence of unobservable events can be decomposed into two sub-problems: observability problem and control problem. (ii) It significantly extends of the synthesis method proposed in (Ghaffari, *et al.*, 2003) to a general Petri net with unobservable transitions for a forbidden state specifications by using the theory of regions.

This paper is organized as follows. Section 2 presents some definitions and notations. Section 3 is composed of tree subsections. In the first one, we introduce the observation problem. Observer Reachability Graph and Refined Observer Reachability Graph are then defined in Section 3.2 and Section 3.3. Section 4 deals with the proposed synthesis approach. Section 5 applies results of this paper to the supervisory control problem of manufacturing system.

2. PRELIMINARIES

This section presents notations and some basic definitions and properties that will be used through the paper.

Let the set of transitions $T = T^c \cup T^u$ be partitioned into *controllable* transitions (T^c) and *uncontrollable* transitions (T^u). The set $T^u = T^f \cup T^{uo}$ is again partitioned into *free* transitions (T^f) and *unobservable* transitions (T^{uo}). Note that $T^{uo} \subseteq T^u$ and $T^o = T^c \cup T^f$ is the set of *observable* transitions.

The notation $x[t>$ ($x[t>x'$) means that the transition t is firable at the state x (leading to x'). This is also extended to sequence σ of transitions with $x[\sigma>$ and $x[\sigma>x'$.

For the natural projection P from T to T^o, we adopt the same definition used in the literature (Cassandras and Lafortune, 1999). Namely, we have the function $P : T^* \rightarrow (T^o)^*$ with: $P(\varepsilon) := \varepsilon$,

$$P(t) := \begin{cases} t & if\ t \in T^o \\ \varepsilon & if\ t \in T^{uo} \end{cases},$$

$$P(\sigma t) := P(\sigma)P(t) \quad for\ \sigma \in T^*, t \in T$$

3. OBSERVABILITY

3.1 Observation Problem

Let us first show through an example that the state of the system represented by the current marking is not fully known when some transitions are unobservable.

Example 1: Consider the Petri net of figure 1. From the initial marking $(1, 0, 0, 0)^t$ and after the firing of the transition t_1, the real state of the plant becomes unknown. In fact t_2 and t_3 are unobservable and from the supervisor point of view, the token can be in p_2 or in p_3 or in p_4.

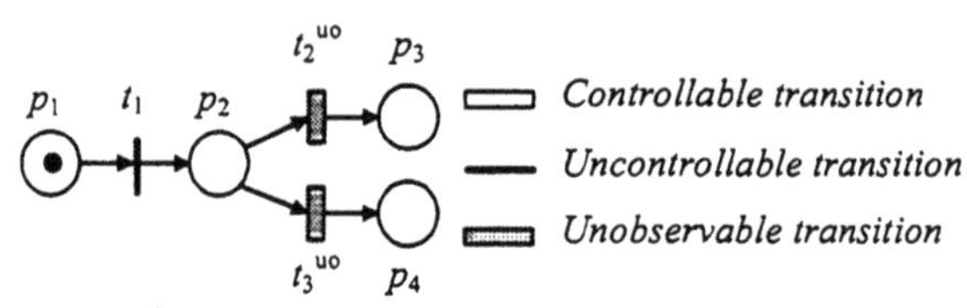

Fig. 1. Petri net model with unobservable transitions

Definition 1. A sequence σ of transitions is said consistent with respect to a sequence of observable transitions σ^o if $M_0[\sigma\rangle$ and $P(\sigma)=\sigma^o$. Let $\Sigma(\sigma^o) = \{\sigma \in T^* \mid M_0[\sigma >, P(\sigma) = \sigma^o\}$ be the set of all consistent sequences.

Definition 2. Let $\mathcal{M}(\sigma^o) = \{M \mid \exists \sigma \in \Sigma(\sigma^o),\ M_0[\sigma\rangle M\}$ be the set of all possible current markings reachable from M_0 by a sequence of transitions that is consistent with respect to σ^o.

For example 1, let $\sigma_1 = t_1$, $\sigma_2 = t_1 t_2$ and $\sigma_3 = t_1 t_3$. $P(\sigma_1)=P(\sigma_2)=P(\sigma_3)=\sigma^o = t_1$. In this case, we have $\mathcal{M}(\sigma^o = t_1) = \{M_1 = (0, 1, 0, 0)^t,\ M_2 = (0, 0, 1, 0)^t,\ M_3 = (0, 0, 0, 1)^t\}$

3.2 The Observer Reachability Graph

The behavior of the Petri Net model (N, M_0) can be represented by a reachability graph $RG(N, M_0)$. With unobservable transitions, the plant state marking is only partially observable. We can always transform a Reachability Graph $RG(N, M_0)$, into an equivalent observer reachability graph, $RG^O(N, M_0)$ using the following algorithm (*algorithm 1*) based on the framework presented in (Cassandras and Lafortune, 1999). While each node of the reachability graph represents a state marking, each node of the observer reachability graph represents a set of possible state markings. This means that if the plant model (N, M_0) is bounded, the reachability graph is finite and the observer reachability graph is finite too. The construction of the observer reachability graph $RG^O(N, M_0)$ is based on the assumption that the initial marking M_0 is known and on the following defintions.

Definition 3: $R^{uo}(x)$ *denotes the set of indistinguishable states with respect to a Petri net marking x, and represents the set of reachable states from x by firing only unobservable transitions. Formally,* $R^{uo}(x) = \{ x' \in RG(N, M_0) \mid \exists\ \sigma \in (T^{uo})^*$ *and* $x[\sigma\rangle x' \}$.

For a set S, $R^{uo}(S) = \bigcup_{x \in S} R^{uo}(x)$. In the rest of the paper, x is called *a micro-state* and $y = R^{uo}(x)$ is called a *macro-state* which is a set of micro-states.

In other terms, the firing of an unobservable transition t^{uo} leading to micro-state x_j from a micro-state x_i does not change the macro-state y_k. On the other hand, the firing of an observable transition t^o leads to another macro-state y' from a macro-state y.

For the reachability graph of example 1 (figure 2), the macro-states can be determined as follows: $y_0 = R^{uo}(x_0) = \{x_0\}$, $y_1 = R^{uo}(x_1) = \{x_1, x_2, x_3\}$, $y_2 = R^{uo}(x_2) = \{x_2\}$ and $y_3 = R^{uo}(x_3) = \{x_3\}$.

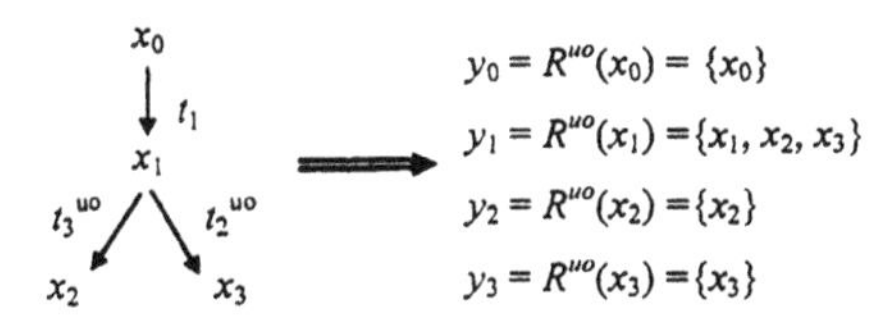

Fig. 2. Reachability graph of the Petri net example of figure 1

Definition 4: $R^o(y)$ *denote the set of reachable macro-states from a macro-state y by firing only observable transitions. Formally,* $R^o(y) = \{ y' = R^{uo}(x') \mid \exists\, x \in y$ *and* $\exists\, t^o \in T^o$ *such* $x[t^o > x' \}$.

Remark 1 : since $R^{uo}(S) = \bigcup_{x \in S} R^{uo}(x)$, Hence, the number of states in $RG^O(N, M_0)$ may be up to 2^n, where n is the number of states in $RG(N, M_0)$.

Remark 2: If $x[t^o > x'$ with $x \in y$ and $y' = R^{uo}(x')$, then we denote $y[t^o > y'$.

For the example 1, the $R^o(y_0) = R^{uo}(x_1) = y_1$ since $x_0\,[t_1 > x_1$. Using the remark 2, we have then $y_0\,[t_1 > y_1$. The set of reachable macro-states of y_1 is $R^o(y_1) = \varnothing$. In this case the macro-states y_2 and y_3 are not accessible. The observer reachability graph $RG^O(N, M_0)$ of the example 1 is presented in figure 3.

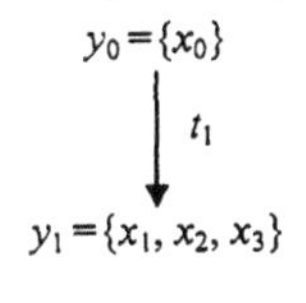

Fig. 3. Observer reachability graph of the Petri net example 1

The construction of the observer reachability graph $RG^O(N, M_0)$ of $RG(N, M_0)$ is summarized in the following algorithm:

Algorithm 1:

1. *Generate the set of macro-states* $\boldsymbol{Y} = \{y_i\}$.
   For each $x \in RG(N, M_0)$ *define* $y = R^{uo}(x)$.
2. *Generate the accessible macro-state from* y_0.
   For each accessible macro-state y_i, *determine* $y_j = R^o(y_i)$.

Example 2: Consider a system of two jobs sharing two resources. Resources R1 and R2 are needed for performing the job A and job B. The PN plant model is given in Figure 4a. The set of controllable transitions is $T^c = \{t_1, t_4\}$. The supervisor controls only the incoming and the outgoing pieces but it cannot observes the operations behaviors of the job A and job B. The set of unobservable transitions is $T^{uo} = \{t_2, t_5\}$. The reachability graph $RG(N, M_0)$ is given in figure 4b.

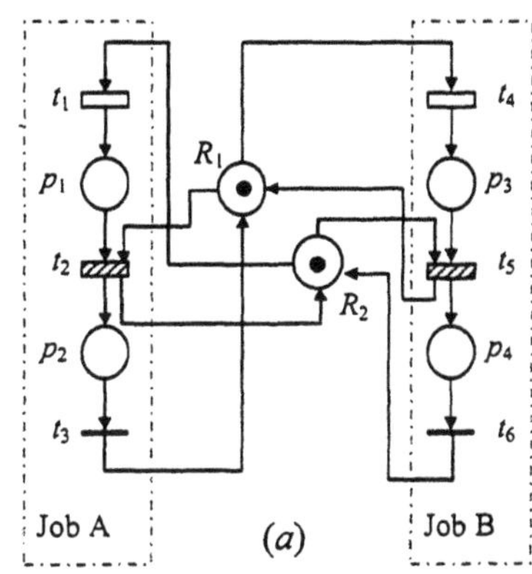

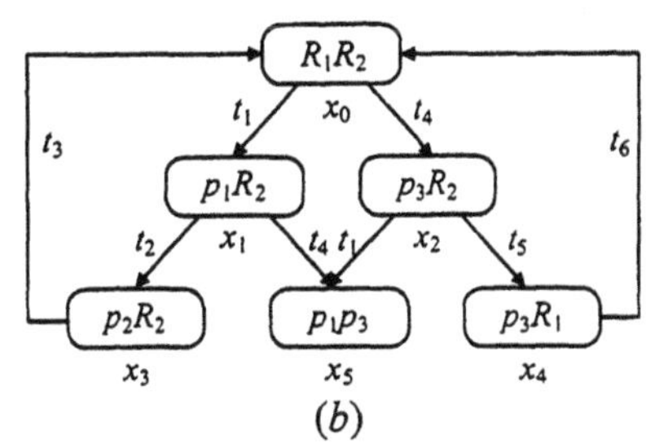

Fig. 4. Production system
(*a*) Petri Net model (N, M_0)
(*b*) Reachability graph $RG(N, M_0)$

For this example, six macro-states are obtained: $y_0 = R^{uo}(x_0) = \{x_0\}$, $y_1 = R^{uo}(x_1) = \{x_1, x_3\}$, $y_2 = R^{uo}(x_2) = \{x_2, x_4\}$, $y_3 = R^{uo}(x_3) = \{x_3\}$, $y_4 = R^{uo}(x_4) = \{x_4\}$ and $y_5 = R^{uo}(x_5) = \{x_5\}$.

The accessible macro-states are as follows:
$R^o(y_0) = \{R^{uo}(x_1) \cup R^{uo}(x_2)\}$ then $y_0[t_1 > y_1$ and $y_0[t_4 > y_2$
$R^o(y_1) = \{R^{uo}(x_0) \cup R^{uo}(x_5)\}$ then $y_1[t_3 > y_0$ and $y_1[t_4 > y_5$
$R^o(y_2) = \{R^{uo}(x_5) \cup R^{uo}(x_0)\}$ then $y_2[t_1 > y_5$ and $y_2[t_6 > y_0$
$R^o(y_5) = \varnothing$ (x_5 is a blocking state).

The building process of $RG^O(N, M_0)$ is completed since all accessible macro-states have been examined. The observer reachability graph $RG^O(N, M_0)$ is presented in figure 5.

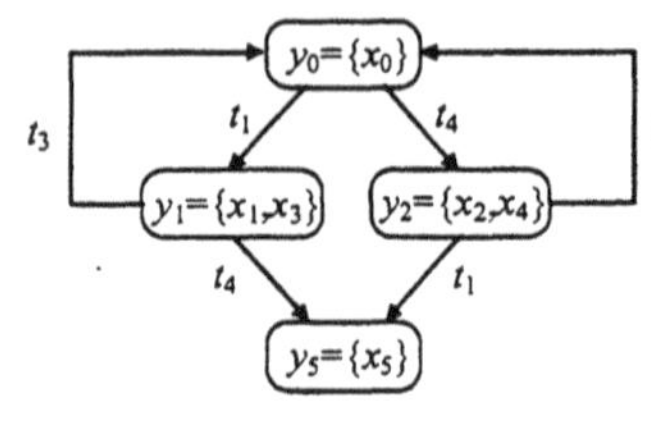

Fig. 5. Observer reachability graph $RG^O(N, M_0)$ of example 2

3.3 The Refined Observer Reachability Graph

We now introduce the refined observer reachability graph $RG^{RO}(N, M_0)$ which is a state graph obtained by refinement of the observer reachability graph $RG^O(N, M_0)$ to introduce unobservable transitions between micro-states. More precisely, each node representing a macro-state y is represented by several nodes each representing a micro-state of y and unobservable transitions between these micro-states are represented. This graph is equivalent to the parallel composition of the $RG^O(N, M_0)$ and $RG(N, M_0)$. The refined observer reachability graph $RG^{RO}(N, M_0)$ related to the example 2 is given in figure 6.

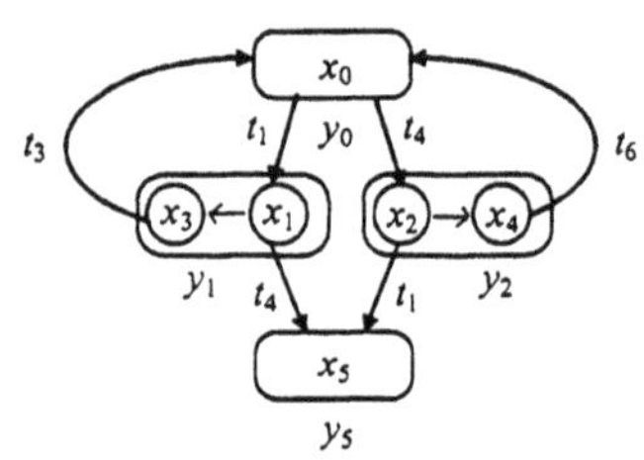

Fig. 6. Refined observer reachability graph $RG^{RO}(N, M_0)$

4. SUPERVISORY CONTROL APPROACH

In this section, we are concerned with the forbidden state problem and liveness requirement. The control specifications are given as a set of forbidden states. The liveness requirement considered in this paper is equivalent to the Ramadge and Wonham's approach. That is, it is required that the so-called marked markings can always be reached by the controlled model. However, we limit ourselves to a particular case with only one marked marking that is the initial marking. In this case, the liveness requirement is equivalent to impose reversibility.

4.1 Methodology of synthesis

Following the Ramadge-Wonham (1998) approach by taking into account forbidden states, liveness conditions and unobservable/uncontrollable transitions, the objective is to determine the desired controlled behaviour starting from the $RG^{RO}(N, M_0)$ with represent the refined observed behaviour and the specifications in terms of forbidden states. We use the theory of regions (Badouel and Daroundeau, 1998) to obtain the optimal PN supervisor for DES, if it exists.

Definition 5: A macro-state is said forbidden if at least one of its micro-states is forbidden. Let Y_F be the set of forbidden macro-states.

Definition 6: A macro-state is said dangerous if it leads to forbidden macro-state by firing uncontrollable transitions. Let Y_D be the set of dangerous macro-states. These states can be determined by the refined observer reachability graph.

Definition 7: A macro-states is said blocking if it contains at least one micro-state that uncontrollably leads outside the Live Sub-Graph (LSG) of the refined observer reachability graph. LSG is a sub-graph of $RG^{RO}(N, M_0)$ containing all macro-states y such that all micro-states x in these macro-states can be led to a micro-state representing the initial marking. Let Y_B be the set of blocking macro-states. In other terms, if a macro-state contains a blocking micro-state then the macro-state if blocking too.

Computation of Live Sub-Graph LSG of controlled system is given by the following algorithm 2:

Algorithm 2: The controlled behavior computation

Step 1. *Generate the reachability graph $RG(N, M_0)$ of the plant Petri net model (N, M_0) to be controlled.*
Step 2. *Generate the observer reachability graph $RG^O(N, M_0)$.*
Step 3. *Build the refined observer reachability graph $RG^{RO}(N, M_0)$.*
Step 4. *Define the set of forbidden macro-states Y_F.*
Step 5. *Identify the set Y_D of dangerous macro-states by exploring $RG^{RO}(N, M_0)$.*
Step 6. *Determine the refined controlled observer reachability graph $RG^{RCO}(N, M_0)$ derived from $RG^{RO}(N, M_0)$ by removing Y_D.*
Step 7. *Verify that $RG^{RCO}(N, M_0)$ is a LSG to enforce the liveness of the controlled system.*
Step 7.1. *If $RG^{RCO}(N, M_0)$ is a LSG then go to step 10.*
Step 7.2. *If $RG^{RCO}(N, M_0)$ is not a LSG then compute the LSG.*
Step 8. *Replace $RG^{RO}(N, M_0)$ by the LSG determined in 7.2 and replace Y_F by the set of macro-states outside LSG.*
Step 9. *Go to step 5.*
Step 10. *$RG^{RCO}(N, M_0)$ is the legal behavior.*

Algorithm 2 gives the maximally permissive controller, since only the dangerous and blocking macro-states are eliminated. Let Y_L be the set of all macro-states of $RG^{RCO}(N, M_0)$ and Ω the set of all macro-state-transitions leading outside $RG^{RCO}(N, M_0)$. Formaly, the set of state-transitions the controller has to disable is $\Omega = \{(y, t) \mid y[t\rangle y' \wedge y \in Y_L \wedge y' \notin Y_L\}$.

4.2 Theory of regions

Theory of regions was proposed for synthesis of pure Petri nets from given finite transition systems

(Cassandras and Lafortune, 1999). Given the plant model of the system to control (N, M_0), the refined observer reachability graph $RG^{RCO}(N, M_0)$, and the set of forbidden states, we use the theory of regions to design the control places $\{p_c\}$ to add (Ghaffari, *et al.*, 2003).Theory of regions is summarized in the following three conditions:

- *Reachability condition.* The reachability of any marking M in G implies that:

$$M_0(p)+C(p,.)\bar{\Gamma}_M \geq 0 \qquad (1)$$

where $\bar{\Gamma}_M$ is the counting vector of the non-oriented path Γ_M from the initial state M_0 to M.

- *Cycle equation.* Consider now any non-oriented cycle γ of the reachability graph.

$$\sum_{t\in T} C(p,t).\bar{\gamma}[t]=0, \quad \forall \gamma \in S \qquad (2)$$

where $\bar{\gamma}[t]$ denotes the algebraic sum of all occurrences of t in γ and S is the set of non-oriented cycles of the graph.

- *Separation conditions.* In order to obtain exactly the desired graph G, for each pair (M, t) such that M is a reachable marking of G and t is a transition not firable at M, t should be disabled by some place p. Since the net is pure, t is disabled at M by a place p if and only if:

$$M_0(p)+C(p,.)\bar{\Gamma}_M + C(p,t) < 0 \qquad (3)$$

The set of all possible pairs (M, t) where M is a reachable marking and t is not firable at M will be called the set of event separation instances.

4.3 Control synthesis policy

We proceed now to determine the controller which guaranteed the $RG^{RCO}(N, M_0)$. This controller is composed of a set of control places to add to the Petri net model plant. The set of control place must guarantee the accessibility of each legal state $y \in Y_L$ which implies that p_c has to satisfy reachability conditions (1), i.e.

$$M_0(p_c)+C(p_c,.)\bar{\Gamma}_y \geq 0 \qquad (4)$$

where $\bar{\Gamma}_y$ is the counting vector of the non-oriented path Γ_y from the initial macro-state y_0 to y. Similarly, each place p_c should satisfy cycle equations (2) for cycles in $RG^{RCO}(N, M_0)$, i.e.

$$\sum_{t\in T} C(p_c,t).\bar{\gamma}[t]=0, \quad \forall \gamma \in S_c \qquad (5)$$

where S_c is the set cycles of the refined controlled observer reachability graph $RG^{RCO}(N, M_0)$.
Therefore each element in Ω will be solved by the addition of a control place p_c. This means that the number of new places to add is at most equal to, in practice much smaller than, the number of state-transitions to inhibit.

Hence, each place p_c to add solves at least one event separation instance (y, t) in Ω, i.e.:

$$M_0(p_c)+C(p_c,.)\bar{\Gamma}_y + C(p_c,t) < 0 \qquad (6)$$

The relations (4)-(6) determine the control place p_c.

Theorem 1: The supervisory control problem can be optimally solved by adding a set of control places to the plant model, if and only if there exists a solution $(C(p_c, .), M_0(p_c))$ satisfying equations (4), (5) and (6) for each event separation instance (y, t) in Ω.

Proof: Theorems 2 of (Ghaffari, *et al.*, 2003).

Remark 3: Different event separation instances may have common solutions. As a result, the number of places needed for solving all event separation instances is generally much smaller than the number of event separation instances.

4.3 Algorithm of the synthesis policy

Adding a set of control places to the plant model can optimally solve the supervisory control problem (if a solution of control exists). The aim of the supervision control is to disable the firing of a transition leading to undesired macro-states. The algorithm of the synthesis method as follows:

Algorithm 3:

Given a refined controlled observer reachability graph $RG^{RCO}(N, M_0)$ and a set Ω of event separation instances (y, t) with $y \in RG^{RCO}(N, M_0)$.

Step 1. *Generate the reachability condition (4) for each macro-state y in* $RG^{RCO}(N, M_0)$.
Step 2. *Generate the cycle equations (5) related to the cycles in* $RG^{RCO}(N, M_0)$.
Step 3. *While* $\Omega \neq \emptyset$ *do:*
Step 3.1. *Generate the event separation condition (6) for any element* (y, t) *of* Ω.
Step 3.2. *Solve the set of relations (4)-(6). Let* $(M_0(p_{(y,t)}),C(p_{(y,t)}))$ be the solution if it exists. Otherwise, exit, as adding control places to the plant Petri net model cannot enforce the maximum permissive controlled behavior.
Step 3.3. *Eliminate from* Ω *all the separation instances that can be solved by* $(M_0(p_{(y,t)}),C(p_{(y,t)}))$.
Step 4. *Remove redundant control places to obtain the controlled net by comparing the set of separation instances solved by each control places.*

Remark 4: The control places determined by algorithm obviously respect the requirement of unobservable constraints as the equations (4)-(6) and the refined controlled observer reachability graph concern only observable transitions.

5. EXAMPLE

Consider the plant Petri net model (N, M_0) of figure 4.a and its refined observer reachability graph RG^{RO} (N, M_0) is given in figure 6. This net is not live since in the macro-state y_3 no transition can be fire. In this case, $y_3=\{x_5=(1,0,1,0,0,0)\}$ is a blocking macro-state. Applying algorithm 2, we obtain the refined controlled observer reachability graph RG^{RCO} (N, M_0) figure 7. where $Y_L=\{y_0, y_1, y_2\}$ and $\Omega=\{(y_1, t_4), (y_2, t_1)\}$ the set of all state-transitions leading to y_3.

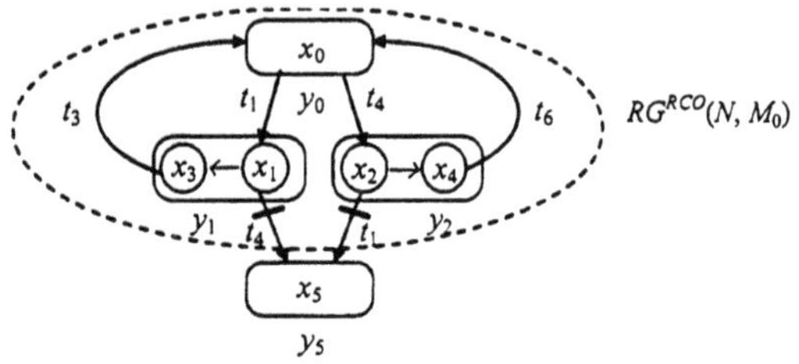

Fig. 7. Refined controlled observer reachability graph RG^{RCO} (N, M_0)

Let us use now algorithm 3 to synthesize the set of control places that will avoid the reachability of y_3. Only one separation condition related to the two separation instances needs to be considered.

$$M_0(p_c)+C(p_c,t_1)+C(p_c,t_4)<0$$

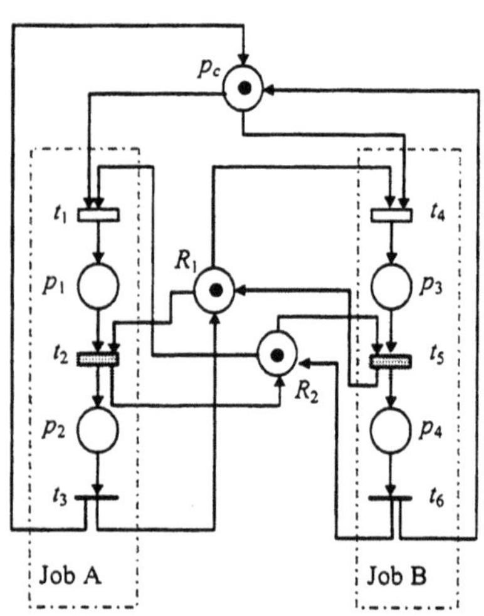

Fig. 8. The controlled model of the system of fig.2a

The refined controlled observer reachability graph RG^{RCO} (N, M_0) contains two different cycles and then two cycle equations

$$C(p_c,t_4)+C(p_c,t_6)=0$$
$$C(p_c,t_1)+C(p_c,t_3)=0$$

whereas the reachability conditions can be expressed as follows:

$$y_0 :\rightarrow M_0(p_c)\geq 0$$
$$y_1 :\rightarrow M_0(p_c)+C(p_c,t_1)\geq 0$$
$$y_2 :\rightarrow M_0(p_c)+C(p_c,t_4)\geq 0$$

The above linear system has a solution $C(p_c, .) = (-1, 0, 1, -1, 0, 1)$ and $M_0(p_c) = 1$. The corresponding control place p_c is an input place of t_1 and t_4 and an output place of t_3 and t_6. The control place p_c contains initially one token.

6. CONCLUSION

In this paper, we have proposed an optimal design methodology of Petri net controllers for the forbidden state problem with both liveness requirement and unobservable transition constraints. The proposed approach uses Ramadge-Wonham's control approach and the theory of regions. It uses a Ramadge-Wonham approach to determine the refined observer reachability graph of the controlled system and then it uses the theory of regions to design a compiled Petri net controller that is a set of control places to add to the initial plant model.

REFERENCES

Yamalidou K., Moody J., Lemmon M. and Antsaklis P., (1996). Feedback control of Petri nets based on place invariants, *Automatica*, **Vol. 32**, n°1, pp. 15-28.

Moody J. O. and Antsaklis P. J., (1999). Petri Net Supervisors for DES with Uncontrollable and Unobservable Transitions, research report isis-99-004.

Stremersch G., (2000). Linear Algebraic Design of Supervisors for Partially Observed Petri Nets, *Proc. IFAC Conf. on Control Systems Design (CSD'2000)*, Bratislava, Slovak Republic, pp 281-286.

Darondeau Ph. and Xie X.L., (2003). Linear Control of Live Marked Graphs, *Automatica*, **Vol. 39**, n°3, pp 429-440.

Ghaffari A., Rezg N. and Xie X., (2003). Design of Live and Maximally Permissive Petri Net Controller Using Theory of Regions, *IEEE Trans. on Robotics and Automation*, **Vol. 19**, n. 1, pp.137-142.

Giua A. and Seatzu C., (2002). Observability of Place/Transition Nets, *IEEE Trans. on Automatic Control*, **Vol. 47**, n°9, pp. 1424-1437.

Moody J. O., Lemmon M. and Antsaklis P. J., (1996). Supervisory control of Petri nets with uncontrollable/unobservable transitions, *35th IEEE CDC*, Kobe, Japan.

Badouel E. and Daroundeau P., (1998). Teory of Regions, Third Advance Course on Petri Nets: Dagstuhl Castle, Vol. 1491 of Springer-Verlag Lecture Notes in Computer Science, pp. 529-586.

Cassandras C.G. and Lafortune S., (1999). *Introduction to Discrete Event Systems*. Kluwer Academic Publishers.

Ramadge P.J. and Wonham W.M., (1989). The Control of Discrete Event Systems, *Proceedings of IEEE*, **Vol. 77**, pp. 81-98.

ELSEVIER
IFAC
PUBLICATIONS
www.elsevier.com/locate/ifac

AN EQUIVALENT LTL/KRIPKE STRUCTURE FOR THE CONDITION SEQUENCE/CONDITION SYSTEM MODEL

Jeff Ashley [*,2] **L. E. Holloway** [**,1]

* *Division of Computer and Technical Sciences*
Kentucky State University
Frankfort, KY, USA 40601
** *Center for Manufacturing and*
Dept. of Electrical and Computer Engineering,
University of Kentucky
Lexington, KY, USA 40506

Abstract: A condition system is a form of Petri net that interacts with other condition systems and the environment via input and output signals called conditions. The Condition Language framework has been used in previous papers to characterize the input/output behavior of such interacting systems, as well as to specify desired control behavior. In this paper, we show that condition sequences (the specification) and condition systems (the model of the system) have an equivalent structure in the temporal logic framework. In particular, we show that there exists a linear-time temporal logic (LTL) specification and a Kripke structure that represents our specification and model. *Copyright © 2004 IFAC*

Keywords: Linear Temporal Logic, Petri Nets, Condition Systems, Languages

A condition system is a model framework based on Petri nets but with input and output signals called *conditions*. The framework is a subclass of the condition/event framework developed in (Sreenivas and Krogh, 1991). The input and output conditions allow causal interactions between systems and their environment.

In (Holloway and Ashley, 1998*a*), *condition languages* were introduced to characterize the input and output behavior of condition systems. A condition language is a set of C-sequences, which is a sequential ordering of condition sets indicating the order in which input and output conditions were true over time. C-sequences can also be used as a specification of desired behavior: By giving incomplete sets of true conditions as they evolve over time, such a C-sequence can represent a high-level specification of desired activity. Control synthesis is then responsible for determining the actuation and sensing details (also represented as conditions) necessary to ensure that the closed-loop system behaviors will satisfy the high-level specification. The *descriptive order* is a partial order for comparing one C-sequence (such as the specification) with another C-sequence (such as a system behavior) with more detail regarding the conditions over time. The descriptive ordering can be used to characterize whether the system behaviors in fact satisfy the specification.

[1] Email: holloway@engr.uky.edu. Supported in part by Rockwell International, NSF grant ECS-9807106 and ECS-0115694, the Office of Naval Research under the grant N000140110621, and the Center for Manufacturing at the University of Kentucky.
[2] Email: jashley@gwmail.kysu.edu.

In this paper, we compare C-sequences with linear-time temporal logic (LTL). Linear-time temporal logic (LTL) is regularly used to specify qualitative properties (desired or otherwise) that can be used in conjunction with a discrete model of some system to evaluate whether the properties will hold for the system as it changes over time (Manna and Pnuelli, 1991; Huth and Ryan, 2000). In this logic system, quantitative timing is ignored which allows for very general statements about the functioning of such systems. This is the formal verification problem that is prevalent in software and systems research. A Kripke structure (a labeled finite state machine) specifies some process, and temporal logic statements can be used to specify desired properties of this process.

In this paper, we show how the condition system and condition sequence map into the LTL framework. In the next two sections we introduce condition systems and the LTL notation used in this paper. Then we introduce a one way translation from condition sequence to an LTL formulation and a one way translation from condition system to fair Kripke Structure. In the last section we show that the systems are equivalent in terms of determining whether a string belongs to the observed language of the system or not. In particular, we show that a C-Sequence is within the language generated by a condition system if and only if the LTL formula generated by the C-Sequence is satisfied by the Kripke Structure generated from the Condition System.

1. CONDITION SYSTEM MODELS

In this section, we define a condition system as a form of Petri Net that requires conditions for enabling of transitions, and that outputs conditions (establishes the truth of certain conditions) according to its marking. Further discussion of the condition systems we consider can be found in (Holloway *et al.*, 2000; Holloway and Ashley, 1998*a*).

A condition system is composed of distinct subsystems(each is represented by a model). The systems that we consider interact with each other and with their outside environment through *conditions.* A condition is a signal that either has value "true", or "false". Let *AllC* be the universe of all conditions, such that for each condition c in *AllC*, there also exists a negated condition denoted $\neg c$, where $\neg(\neg c) = c$. Given a subset of conditions $C \subseteq AllC$, C is said to have a *contradiction* if for some $c \in C$, the negation $\neg c$ is also in C.

Formally, a condition system $\mathcal{G}$ consists of a Petri net with a set of places $\mathcal{P}_\mathcal{G}$, a set of transitions $\mathcal{T}_\mathcal{G}$, a set of directed arcs $\mathcal{A}_\mathcal{G}$ between places and transitions, and a condition mapping function $\Phi_\mathcal{G}(\cdot)$, where $(\forall p \in \mathcal{P}_\mathcal{G})\Phi_\mathcal{G}(p) \subseteq AllC$ maps output conditions to each place, and $(\forall t \in \mathcal{T}_\mathcal{G})\Phi_\mathcal{G}(t) \subseteq AllC$ maps *enabling conditions* to each transition. We define the output condition set for a system $\mathcal{G}$ as $\Phi_{out}(\mathcal{G}) = \{c \in \Phi_\mathcal{G}(p) \mid p \in \mathcal{P}_G\}$. Similarly, define $\Phi_{in}(\mathcal{G}) = \{c \in \Phi_\mathcal{G}(t) \mid t \in \mathcal{T}_\mathcal{G}\}$.

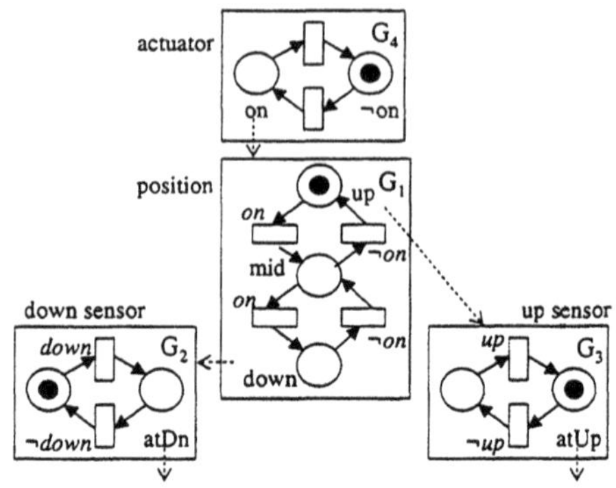

Fig. 1. An example Condition System model

The state of the system is represented by its marking, where a marking m is a function associating each place with a non-negative integer. Define $M_\mathcal{G}$ as the set of markings considered in this paper.

Given a marking m, define
$g_\mathcal{G}(m) := \{c \in \Phi_\mathcal{G}(p) \subseteq \Phi_{out}(\mathcal{G}) \text{ s.t. } m(p) \geq 1\}$
Under m, each condition in $g_\mathcal{G}(m)$ has value true. (Note conditions not in Φ_{out} can also have value true by other systems or external factors.)

Given any $m \in M_\mathcal{G}$ and any set $TrueC \subseteq AllC$ of conditions with value "true" such that $g_\mathcal{G}(m) \subseteq TrueC$, a set of transitions T is enabled if and only if it is *state-enabled* and *condition-enabled,* defined as follows:

(1) T is *state-enabled* if $(\forall p \in \mathcal{P}_\mathcal{G})$
$m(p) \geq |\{t \in T \mid p \text{ is input to } t\}|$
(2) T is *condition-enabled* if
$(\forall t \in T)\ \Phi_\mathcal{G}(t) \subseteq TrueC$

A set of enabled transitions T can *fire*, leading to a next marking m' satisfying $(\forall p \in \mathcal{P}_\mathcal{G})$
$m'(p) = m(p) - |\{t \in T \mid p \text{ is input to } t\}| + |\{t \in T \mid p \text{ is output of } t\}|$

Denote the set of next possible markings from a marking m under a set of true conditions $TrueC$ as $f_\mathcal{G}(m, TrueC)$. Note that since the empty transition set $T = \emptyset$ is trivially marking enabled and transition enabled and does not change the marking upon "firing", then for any marking m and any condition set $TrueC$, then trivially $m \in f_\mathcal{G}(m, TrueC)$.

Note that a condition system can be subdivided into subsystems, where each subsystem is a condition system over a set of connected places and transitions which are disconnected from all other places and transitions. In figure 1, there are four such subsystems. Note that in the figure, the labels indicate conditions associated with places and transitions. For this paper, we use the notation $\mathcal{G}$

to indicate the complete system, and the notation $\{G_1, \ldots G_n\}$ to indicate the set of subsystems in $\mathcal{G}$.

We make the following assumptions for the condition systems considered in this paper:

(1) $M_\mathcal{G}$ is a finite set.
(2) $M_\mathcal{G}$ is closed under $f_\mathcal{G}$: $\forall m \in M_\mathcal{G}, \forall TrueC \subseteq AllC$, $f_\mathcal{G}(m, TrueC) \subseteq M_\mathcal{G}$.
(3) No marking in $M_\mathcal{G}$ results in a contradiction in output conditions: $\forall m \in M_\mathcal{G}$, $\forall c \in AllC$, if $c \in g_\mathcal{G}(m)$, then $\neg c \notin g_\mathcal{G}(m)$.
(4) For any subsystem $G_i \in \mathcal{G}$ and any condition $c \in C_{out}(G_i)$, c is not an output of any other subsystem $\forall j \neq i$, $c \notin C_{out}(G_j)$.

2. CONDITION SYSTEM LANGUAGES AND THE DESCRIPTIVE ORDERING

The behavior of a condition system can be described by sequences of condition sets. A condition set sequence, called a *C-sequence*, is a finite length sequence of condition sets. Each condition set sequence is of the form $(C_0 C_1 \ldots C_k)$ for some integer k and sets $C_i \subseteq AllC$ for all $0 \leq i \leq k$. A set of C-sequences is called a language, and the language consisting of all C-sequences is denoted $\mathcal{L}$.

A C-sequence can be viewed as a sequence of condition sets that must be true over consecutive intervals of the real line, where each interval represents some time period. Given a C-sequence $s = (C_0 C_1 \ldots C_n)$ and some $0 \leq i \leq n$, C_i represents a subset of conditions (or negated conditions) that are true for some (possibly non-unique) period of time immediately preceding the period represented by C_{i+1} and immediately following the period represented by C_{i-1}. Conditions omitted in a condition set C_i can be true or false, so lack of specification in a condition set is analogous to a "don't care" or "don't know" condition on its value.

Comparison of C-sequences is done through a *descriptive ordering*, a partial ordering formally defined in (Holloway and Ashley, 1998*b*; Holloway *et al.*, 2000). The main idea is that a C-sequence s is said to be more descriptive than s' if it contains more detail, i.e. more information on condition truth values over the time than s'. This is useful in using C-sequences as a way of specifying desired behavior by omitting details of sensing and control which can then be determined in implementing (or synthesizing) a control. Similarly, some signals can be specified to be true during periods where they are important, but left unspecified ("don't care" / "don't know") during periods where they are not relevant.

A system's input-output behavior is represented as a set of c-sequences which map to marking sequences of the system. This is formally defined as follows:

Definition 1. : Given a condition system $\mathcal{G}$ and a marking m_0, define the language $L(\mathcal{G}, m_0) \subseteq \mathcal{L}$ to be the set of condition set sequences such that $(C_0 C_1 C_2 \ldots C_n) \in L(\mathcal{G}, m_0)$ if there exists some marking sequence $m_0 m_1 \ldots m_k$ and index mapping function $j(i)$ with $j(0) = 0$, $j(k) = n$ such that

(1) Markings evolve according to conditions: $m_{i+1} \in f_\mathcal{G}(m_i, C_{j(i)})$ *for* $0 \leq i \leq n-1$.
(2) Output conditions result only from the marking:
$C_{j(i)} \cap C_{out}(\mathcal{G}) = g_\mathcal{G}(m_i)$,
(3) Sequencing is maintained: *for any* $0 \leq i < n$, $j(i+1) = j(i)$ *or* $j(i+1) = j(i) + 1$;

The above definition deserves some explanation. The notation $C_{j(i)}$ indicates the condition set associated with the ith marking. By statement 2, the output conditions in set $C_{j(i)}$ correspond to the marking m_i, and by statement 1, marking m_i evolves to marking m_{i+1} only if enabled under condition set $C_{j(i)}$. The marking sequence and condition set sequence have different indices because the mapping between the sequences is not necessarily one-to-one, but do maintain ordering, as per statement 3. Furthermore, consecutive condition sets or consecutive markings in the sequence are not necessarily distinct. Finally, we point out that $L(\mathcal{G}, m_0)$ is obviously prefix-closed (excluding the empty prefix).

3. LINEAR-TIME TEMPORAL LOGIC.

In this section we formally define the linear-time temporal logic syntax that we will use throughout the remainder of this chapter. We adopt the notation of (Huth and Ryan, 2000) for the linear-time temporal logic and a Kripke structure. While subsequent research in temporal logic has extended the logic and temporal operators defined, the set of temporal and propositional operators presented(which is missing $\vee$, $\rightarrow$,$\leftrightarrow$, and $Before$) is sufficient for the translation from C-sequences to LTL formulas.

In definition 2 we define a Kripke structure, and in definitions 3 and 4 we define the LTL syntax and an evaluation respectively. Let AP be a set of atomic propositions(conditions) which have a value of *true* or *false*.

Definition 2. : A Kripke structure is discrete model defined as $\mathcal{M} := \{S, E, Label, n_0\}$, where S is a set of state nodes, $E \subseteq S \times S$ is a transition relation, $Label : S \rightarrow 2^{AP}$ is a labelling function that maps states of the system to atomic *propositions, and $n_0 \in S$. Furthermore, every $n_i \in S$ has some $n_j \in S$ such that $(n_i, n_j) \in E$(liveness property).*

The temporal operators used in LTL allow for the timed specification of properties where these operators have the following intuitive meaning:

(1) U : The operator U represents the until operation between two propositions. Informally, pUq means that proposition p is true until proposition q becomes true.
(2) G : The G operator represents "globally" or "always", in that Gp represents the specification that proposition p is true for all time.
(3) F : The F operator represents "eventually" or "in the future", in that Fp represents the specification that proposition p will hold sometime in the future.
(4) X : The next state operator has the meaning for Xp that p will hold in the next state.

Given these temporal operators, the following formally defines the linear-time temporal logic we will use for this chapter. We use the notation from (Huth and Ryan, 2000).

Definition 3. : Linear-time temporal logic (LTL) has the following syntax given in Backus Naur form:
$\phi := p \mid (\neg\phi) \mid (\phi \wedge \phi) \mid (\phi U \phi) \mid (G\phi) \mid (F\phi) \mid (X\phi)$,
where p is any atomic proposition.

An LTL formula is evaluated over a *path* or a set of paths(also called a *state trace* in the Kripke structure), where a path is $\pi := n_0, n_1, \cdots, n_k$ for some integer $k \geq 0$ where each n_i for $0 \leq i \leq k$ is a element of S. Define $\pi(i) := n_i, \cdots, n_k$ for some integer $0 \leq i \leq k$. These paths represent a qualitative time line of propositional valuations which as we shall see is directly comparable to our C-sequence. With a slight abuse of notation, we will represent some path π that exists in $\mathcal{M}$ as $\pi \in \mathcal{M}$.

The satisfaction relation as defined below shows how a Kripke structure and a LTL formulation are related in the verification process. In the definition below, and in the remainder of the paper, the notation $\top$ indicates a formula that is always satisfied (always true).

Definition 4. : A path $\pi := (n_0, n_1, \cdots, n_k) \in \mathcal{M}$ satisfies *an LTL formula via the satisfaction relation $\models$ for LTL formulas as follows:*

(1) $\pi \models \top$.
(2) $\pi \models p$ iff $p \in Label(s_0)$.
(3) $\pi \models \neg\phi$ iff $\pi \not\models \phi$.
(4) $\pi \models \phi_1 \wedge \phi_2$ iff $\pi \models \phi_1$ *and* $\pi \models \phi_2$.
(5) $\pi \models \phi_1 U \phi_2$ holds iff *there is some $i \geq 0$ such that $\pi(i) \models \phi_2$ and for all $j = 0, \ldots, i-1$ we have $\pi(i) \models \phi_1$.*
(6) $\pi \models G\phi$ *holds* iff *, for all $i \geq 0, \pi(i) \models \phi$.*
(7) $\pi \models F\phi$ *holds* iff *, for some $i \geq 0, \pi(i) \models \phi$.*
(8) $\pi \models X\phi$ iff $\pi(1) \models \phi$.

4. A TRANSLATION FROM THE C-SEQUENCE TO THE LTL FORMULA FRAMEWORK

In this section, we present definitions that define a transform from our model and specification (the condition system and C-sequence respectively) into a Kripke structure and a LTL formula. One problem lies in the fact that a condition system model of a plant is an input output model where inputs to the system(i.e.- the control decisions) are considered part of $L(\mathcal{G}, m)$, and a Kripke structure is a labeled state machine with no equivalent provision for system input. So we will first modify our condition system model of the plant by adding models that represent the inputs to the system which will convert the system into an output model where inputs are unconstrained. We will then show that these models are equivalent in terms of language analysis.

Define $C_{in} := (\Phi_{in}(\mathcal{G})/\Phi_{out}(\mathcal{G}))$ as the set of all observable external input conditions to the system. For the remainder of this chapter we will assume that the values of input conditions to a system $\mathcal{G}$ are known at time 0 of the plant(i.e. at the time of the initial marking m_0). Thus define $C_{in}^0 \subseteq C_{in}$ as the set of input conditions to the system that are *true* at time 0.

The following defines our modified condition system model where each input condition to the system is represented in a new model that is appended to the existing system model $\mathcal{G}$.

Definition 5. : Given $\mathcal{G}$, m_0, and a set of conditions C_{in}^0 without contradictions, define $\mathcal{G}_{MOD}$ as a condition system composed of $\mathcal{G}$ and for each condition $c \in C_{in}$, subsystems $G_{c,\mathcal{M}OD}$. Each of these subsystems have the following properties.

(1) There are two places(p and p′) such that $C_{G_{c,\mathcal{M}OD}}(p) = \{c\}$ and $C_{G_{c,\mathcal{M}OD}}(p') = \{\neg c\}$.
(2) There are two transitions(t and t′) such that $C_{G_{c,\mathcal{M}OD}}(t) = C_{G_{c,\mathcal{M}OD}}(t') = \emptyset$.
(3) These components are connected with edges to create a cycle.
(4) Assign one token to one of these places using the following rule: if $c \in C_{in}^0$ then $m(p) = 1$ else $m(p') = 1$.

Denote the marking space of $\mathcal{G}_{\mathcal{M}OD}$ by $\mathcal{M}_{\mathcal{M}OD}$. We note that the state space of $\mathcal{G}_{\mathcal{M}OD}$ will be $\mathcal{M}_{\mathcal{M}OD} = 2^{|C_{in}|}|\mathcal{M}_{\mathcal{G}}|$.

In (Ashley, 2004) Theorem 3.1 we show that given a condition system $\mathcal{G} = \{G_1 \ldots G_n\}$ then $L(\mathcal{G}, m_0) = \bigcap_{1 \leq i \leq n} L(G_i, m_{0,i})$. This result is needed for the rest of the paper and it is omitted because of space limitations.

The following lemma shows that in terms of language analysis, that $\mathcal{G}$ and $G_{c,\mathcal{MOD}}$ are equivalent. First, however, we must define the language of all possible C-sequences for which C_{in}^0 is true:

$L^0(C_{in}^0) := \{(C_0 C_1 \ldots C_n)$ s.t. $\forall c \in C_{in}$, if $c \in C_{in}^0$, then $c \in C_0$ and otherwise $\neg c \in C_0$.

Lemma 1. : Given $\mathcal{G}$, m_0, C_{in}^0 and a corresponding $\mathcal{G}_{\mathcal{MOD}}$ and $m_{0,\mathcal{MOD}}$ generated from application of definition 5, then $s \in L(\mathcal{G}, m_0) \cap L^0(C_{in}^0)$ if and only if $s \in L(\mathcal{G}_{\mathcal{MOD}}, m_{0,\mathcal{MOD}})$.

Proof : Let $\mathcal{G}'$ represent the set of subsystems created in Definition 5 for each $c \in C_{in}^0$, and m' its corresponding marking for the set of subsystems. Then $L(\mathcal{G}', m') = L^0(c_{in}^0)$. By (Ashley, 2004) Theorem 3.1, since $\mathcal{G}_{mod} = \mathcal{G} \cup \mathcal{G}'$, then $L(\mathcal{G}_{\mathcal{MOD}}, m_{0,\mathcal{MOD}}) = L(\mathcal{G}, m_0) \cap L(\mathcal{G}', m'_0)$, which thus equals $L(\mathcal{G}, m_0) \cap L^0(C_{in}^0)$. ◇◇◇

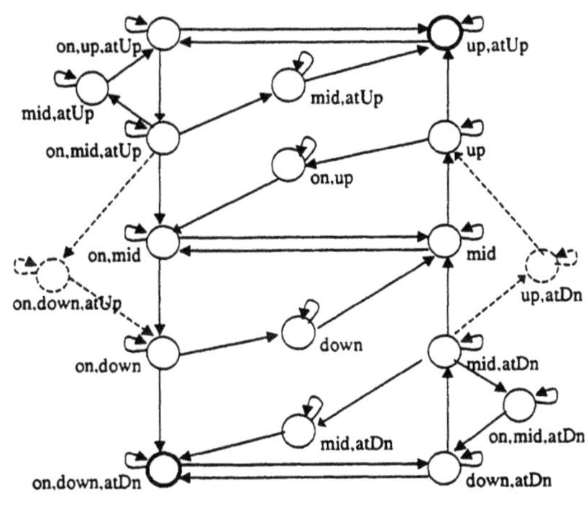

Fig. 2. A Kripke structure for the system of figure 1.

Definition 6. : Given some condition system and initial marking $\mathcal{G}_{\mathcal{MOD}}$ and $m_{0,\mathcal{MOD}}$, define $\mathcal{K}_{\mathcal{G}}$ as a Kripke structure that is created by the following:

(1) Let $AP = AllC$.
(2) For each $m_i \in \mathcal{M}_{\mathcal{MOD}}$, create a new node, n_i in the Kripke structure, with $Label(n_i) = g_{\mathcal{G}_{\mathcal{MOD}}}(m_i)$.
(3) For each $(m_i, m_j) \in \mathcal{M}_{\mathcal{MOD}}$ such that $m_j \in f_{\mathcal{G}_{\mathcal{MOD}}}(m_i, C)$ for some $C \subseteq AllC$, add an edge from n_i to n_j in the corresponding Kripke structure, $(n_i, n_j) \in E$. Assign the initial marking n_0 to the node corresponding to $m_{0,\mathcal{MOD}}$. Remove all nodes from the structure that are unreachable from n_0

Example 1. : Consider the system of figure 1, but excluding G_4. Let $\mathcal{G} = G_1 \cup G_2 \cup G_3$. Then, $C_{in} = \{on, \neg on\}$. Define $C_{in}^0 = \{on\} \subseteq C_{in}$ indicating on is initially true. Then, G_4 is $G_{on,\mathcal{MOD}}$, and $\mathcal{G}_{\mathcal{MOD}} = G_1 \cup G_2 \cup G_3 \cup G_4$. This condition system and its initial marking was used to generate the Kripke structure shown in figure 2 (we list only the non-negated conditions for each node).

Definition 7. : Given some C-Sequence , $s = (C_0 C_1 \ldots C_n)$, define the LTL formula f_s given the following construction rules:

(1) If $C_i \neq \emptyset$ let compound proposition ϕ_i be equal to the conjunction of each condition $c \in C_i$. Thus, for example, if $C_i = \{c_1, \neg c_2, c_4\}$ then $\phi_i = (c_1 \wedge \neg c_2 \wedge c_4))$.
(2) If $C_i = \emptyset$ then $\phi_i = \top$.
(3) The formula is constructed from these compound propositions in the following manner, $f_s = \phi_1 U(\phi_2 U(\ldots U\phi_n)\ldots)$.

Our condition sequence represents a timeline ordering that is directly related to the until operator, U. We note that the parentheses are due to the binding precedence of this operator.

Example 2. : As a continuation of previous examples, consider the C-Sequence
$s_{obs} = (\{up, atUp\}, \emptyset, \{on, down, atDn\}$.

The equivalent LTL formulation is $\{up, atUp\} U(\top U(\{on, down, atDn\}))$. Which we see is achievable via inspection of figure 2.

In the following theorem we show that the LTL/Kripke system is equivalent to its corresponding C-Sequence/condition system in terms of language analysis.

Lemma 2. : Given a $\mathcal{G}_{\mathcal{MOD}}$, $m_{0,\mathcal{MOD}}$, a corresponding Kripke structure $\mathcal{K}_{\mathcal{G}}$ created by application of defn. 6, and a C-sequence s and its corresponding LTL formula f_s generated by defn. 7 then the following is true:
$s \in L(\mathcal{G}_{\mathcal{MOD}}, m_{0,\mathcal{MOD}})$ iff $\exists \pi \in \mathcal{K}_{\mathcal{G}}$ s.t. $\pi \models f_s$

Proof: (Only If:) Let $s \in L(\mathcal{G}_{\mathcal{MOD}}, m_{0,\mathcal{MOD}})$, we need to show that there exists a $\pi \in \mathcal{K}_{\mathcal{G}}$ such that $\pi \models f_s$. Given that $s = (C_0, C_1, \ldots, C_n)$ is in the language of the system then there exists some marking sequence $(m_{0,\mathcal{MOD}}, m_1, \ldots, m_k)$ for some $k \geq n$ in the system that generates s. Note that there may be more than one marking sequence that generates s. We also note that each marking m_i is immediately reachable from m_{i-1} for $i > 0$ by definition and as a direct result of definition 6 we see there exists some

$\pi = (n_0, n_1, \ldots, n_k) \in \mathcal{K}_{\mathcal{G}}$. Now consider f_s, we need to show that $\pi \models f_s$. The path π in the Kripke structure must output $(C_0, C_1, \ldots, C_n)$ which is a direct result of definition 6. f_s, which was generated by application of definition 7 given s, contains each condition set from s joined in order using the Until operator. We see then that the path π satisfies f_s from defn. 4.

(If:) It suffices to show that given some $s \notin L(\mathcal{G}_{\mathcal{MOD}}, m_{0,\mathcal{MOD}})$ that there does not exist a $\pi \in \mathcal{K}_{\mathcal{G}}$ such that $\pi \models f_s$. Our logic follows the first part of this proof. First, since $s \notin L(\mathcal{G}_{\mathcal{MOD}}, m_{0,\mathcal{MOD}})$, then there can't exist a marking sequence in $\mathcal{G}_{\mathcal{MOD}}$ given $m_{0,\mathcal{MOD}}$. Since no such marking sequence exists then there does not exist a path $\pi \in \mathcal{K}_{\mathcal{G}}$ that outputs $(C_0, C_1, \ldots, C_n)$ by definition 6. Hence there exists no path $\pi \in \mathcal{K}_{\mathcal{G}}$ that can satisfy f_s. Thus the lemma is proved. $\diamond\diamond\diamond$

The theorem below then follows directly from lemma 2 using lemma 1.

Theorem 1. : Given a $\mathcal{G}$, m_0, and C^0_{in}, then there exists a corresponding Kripke structure $\mathcal{K}_{\mathcal{G}}$ created by application of definitions 5 and 6, and a C-sequence s and its corresponding LTL formula f_s generated by definition 7 where the following is true:

$s \in L(\mathcal{G}, m_0) \cap L^0(C^0_{in})$ *iff* $\exists \pi \in \mathcal{K}_{\mathcal{G}}$ *s.t.* $\pi \models f_s$

5. DISCUSSION

In this paper, we have shown that the condition sequence representation of system behaviors and specifications can be equivalently represented as a LTL formula. A method is given to convert a C-sequence s representation into its equivalent LTL formula f_s. To show the equivalence of the c-sequence and formula, a Kripke structure was created. This was done by first modifying the condition system such that all input signals are incorporated into the model by appending a generator of those input signals. The Kripke structure is then determined from the reachable marking space of the expanded system, with condition labels on each state. A C-sequence is shown to be included within the language of the condition system if and only if the constructed Kripke structure satisfies the corresponding LTL formula.

The control synthesis methods of (Holloway *et al.*, 2000) begin with C-sequences as high-level specifications of desired behavior, where the synthesis method then determines the necessary low-level actuation and sensing signals necessary to achieve the specification. The control synthesis is thus a finitely recursive expansion of the specification with details until a plant behavior (with appropriate control actions) is determined. Thus, an initial single C-sequence specification may correspond (under the descriptive ordering) to a potentially large family of permissible plant behaviors. Showing that such C-sequence specifications have an LTL formula may make such methods more accessible to a broader community, and allow the use of LTL verification tools in analysis. This paper is a first step towards that goal.

It would be useful to go also the other direction: given an LTL formula, is there a C-sequence or finite set of C-sequences that can represent the formula. This would allow the current control synthesis method to take input as an LTL formula, which would then be converted into a C-sequence for use by the current synthesis methods. Creating such a C-sequence representation from an arbitrary LTL formula is a topic of current research. However, early investigations indicate that this will not be possible for an arbitrary LTL formula, primarily due to the ability of LTL formulas to have negated formulas within them. This is because C-sequences as specifications are used in the existing control synthesis method as minimal (under the descriptive ordering) elements of the allowed system behavior, and the negation operation destroys this structure. Research is continuing in this area.

REFERENCES

Ashley, Jeffrey (2004). Diagnosis of Condition Systems. PhD thesis. University of Kentucky.

Holloway, L. and J. Ashley (1998*a*). Condition languages and condition systems for modeling ambiguous control specifications. In: *IEE International Workshop on Discrete Event Systems (WODES98)*. Cagliari.

Holloway, L. E. and J. Ashley (1998*b*). Elaborative orderings of condition languages. In: *Proceedings of 1998 IEEE Conference on Decision and Control*. Tampa.

Holloway, L. E., X. Guan, R. Sundaravadivelu and J. Ashley (2000). Automated synthesis and composition of taskblocks for control or manufacturing systems. *IEEE Transactions on Systems, Man and Cybernetics: Part B* **30**(5), 696–712.

Huth, M. and M. Ryan (2000). *Logic in Computer Science: Modelling and reasoning about systems*. Cambridge University Press.

Manna, Z. and A. Pnuelli (1991). *The Temporal Logic of Reactive and Concurrent Systems*. Springer-Verlag.

Sreenivas, R.S. and B.H. Krogh (1991). On condition/event systems with discrete state realizations. *Discrete Event Dynamic Theory and Applications*.

ELSEVIER
IFAC
PUBLICATIONS
www.elsevier.com/locate/ifac

CONTROL OF SAFE ORDINARY PETRI NETS WITH MARKING SPECIFICATIONS USING UNFOLDING

Alessandro Giua* Xiaolan Xie**

* *DIEE, Università di Cagliari, Piazza d'Armi, 09123 Cagliari, Italy.* giua@diee.unica.it
** *INRIA/MACSI Team, ISGMP - Bat A, Ile du Saulcy, 57045 Metz, France.* xie@loria.fr

Abstract: In this paper we deal with the problem of controlling a safe place/transition nets so as to avoid a set of forbidden markings $\mathcal{F}$. We say that a given set of markings has property REACH if it is closed under the reachability operator. We assume that all transitions of the net are controllable and that the set of forbidden markings $\mathcal{F}$ has the property REACH. Under these assumptions we show that using the technique of unfolding is possible to efficiently design a maximally permissive supervisor to solve this control problem. The supervisor takes the form of a set of control places to be added to the unfolding.

Keywords: Petri nets, unfolding, control of discrete event systems.

1. INTRODUCTION

The use of *partial order methods* for the efficient verification of *concurrent systems* is a technique that has been used by several authors in the last 10-15 years. In particular, a Petri net is a natural model for this approach because it has primitives to explicitly capture the notion of precedence and independence between events.

The interleaving of concurrent sequences often leads to the well-known problem of *state space explosion* that hinders the applicability of all those Petri net analysis techniques, e.g. the reachability graph, that are based on the exhaustive search over the set of reachable markings. However, the sets of states introduced by concurrency are for the most part intermediate markings that are irrelevant to determine the properties of the system: what matters is the unique marking reached by the firing of all these concurrent sequences.

As Valmari (1994) has lucidly explained, this fact has motivated research along at least two different lines.

(a) A first approach is to let one (or at least as few as possible) interleaving represent all its equivalent interleavings: the notion of *stubborn set* (Valmari, 1991) and *persistent set* (Godefroid, 1996) is inspired by this idea.

(b) A second approach consists in replacing the reachability graph by a net structure which captures the concurrent executions, and does not explicitly show individual interleavings. This technique is based on the *unfolding* of a (bounded) Petri net into an *occurrence net*. A finite prefix of the unfolding can be used to characterize the set of all reachable markings without having to enumerate them (McMillan, 1995; Esparza *et al.*, 2002). Recently, this approach has also been extended to unbounded nets (Neumair, 2002). Note that the occurrence net is much simpler that the original Petri net and can usually be validated using structural analysis.

Although these two type of techniques have proved to be a powerful instrument in the *verification* of concurrent systems, the application of these techniques to the *control* of discrete event systems has not received a lot of attention. We recall here some contributions in this area.

Hellgren *et al.* (1999) have used persistent sets to design supervisors for deadlock avoidance.

Observability and diagnosis are closely related to control: Aghasaryan *et al.*, (1988) were the first to use unfolding for fault detection and diagnosis in distributed systems. This approach has also been extended in two subsequent papers by Benveniste *et. al* (2003a and 2003b).

Recently, in a series of papers He and Lemmon (2000a, 2000b, 2002) have presented an original approach based on unfolding for liveness verification and enforcing. However we have shown (Xie and Giua, 2004) that some key results of these papers need to be refined. As a result, although we still strongly believe that unfolding is an interesting and potentially fruitful technique for Petri net control, the applicability of unfolding for Petri net supervision is still an open issue.

In the paper we consider discrete event systems modeled by safe place/transition nets. The control problem we consider can be framed within the theory of Supervisory Control (Ramadge and Wonham, 1989). In particular, we consider a control specification that requires avoiding a set of forbidden marking $\mathcal{F}$. In the current state of investigation, we assume that all transitions are controllable, i.e., they can be disabled by a controlling agent called supervisor that must enforce the specification.

We use a set of finite prefixes of the unfolding, that we call order 1, to characterize the reachability set of the original net.

We restrict our attention to a special class of forbidden marking specification that have a property we call REACH: once a forbidden marking is reached, all markings reachable from it will also be forbidden. This has a nice advantage over the unfolding structure: if a configuration (i.e., a set of transition firings) is forbidden, any larger configuration should also be forbidden. We show that in this case a simple control structure - that consists in a set of places to be added to the order 1 prefix - can be used to implement a maximally permissive control policy that enforces the specification.

The approach we present in the paper requires an exhaustive enumeration of the set of forbidden markings. It has however the advantage of allowing one to construct a maximally permissive supervisor in the form of a "controlled" occurence net (i.e., an occurence net with the addition of control places) using a procedure where where the set of markings of the plant needs not be exhaustively enumerated. The closed loop system in this approach can also be represented by this controlled occurence net.

2. BACKGROUND ON PETRI NETS

In this section we recall the formalism used in the paper. A more detailed introduction to Petri nets can be found in (Murata, 1989).

The Petri net model considered in this paper is an *ordinary Place/Transition net* (P/T net) denoted $N = (P, T, F)$, where P is a set of m places; T is a set of n transitions; $F \subseteq (P \times T) \cup (T \times P)$ is the flow function that specifies the arcs from places to transitions and from transitions to places.

The *preset* and *postset* of a node $x \in P \cup T$ are denoted ${}^\bullet x \triangleq \{x' \,|(x', x) \in F\}$ and $x^\bullet \triangleq \{x' \,|(x, x') \in F\}$ while ${}^\bullet x^\bullet = {}^\bullet x \cup x^\bullet$. Node x is a *source* (resp., *sink*) if ${}^\bullet x = \emptyset$ (resp.; $x^\bullet = \emptyset$).

Given two nodes $x, x' \in P \cup T$ we define the following relations.

- Node x *precedes* x' (denoted $x \preccurlyeq x'$) if there exists a directed path from x to x' . If we require that the path has length greater than zero we write $x \prec x'$.
- Nodes x and x' are in *conflict* (denoted $x\#x'$) if there exist two different transitions $t, t' \in T$ such that: $t \preccurlyeq x$, $t' \preccurlyeq x'$, ${}^\bullet t \cap {}^\bullet t' \neq \emptyset$. In this case, in fact transitions t and t' are in conflict because they have a common input place, and the conflict propagates to all nodes following them. A node x is in *self-conflict* if $x\#x$ holds.
- Nodes x and x' are *concurrent* (denoted $x \approx x'$) if neither $x \preccurlyeq x'$, nor $x' \preccurlyeq x$, nor $x\#x'$ hold.

Note that given two nodes x and x' it may hold that: ($x \prec x'$ and $x' \prec x$ and $x\#x'$).

A *marking* $M : P \to \mathbb{N}$ that assigns to each place of a P/T net a non–negative integer number of tokens, represented by black dots. A *P/T system* or *net system* $\langle N, M_0 \rangle$ is a net N with an initial marking M_0.

A transition t is *enabled* at M iff $M(p) > 0$ for all $p \in {}^\bullet t$. If t is enabled, it may *fire* yielding the marking $M' = M + C(\cdot, t)$. We write $M\ |\sigma\rangle$ to denote that the sequence of transitions $\sigma = t_{j_1} \cdots t_{j_k}$ is enabled at M, and we write $M\ |\sigma\rangle\ M'$ to denote that the firing of σ yields M'. We can associate to a sequence σ a *firing vector* $X : T \to \mathbb{N}$ such that $X(t) = k$ if the transition t is contained k times in σ.

A marking M is *reachable* in $\langle N, M_0 \rangle$ iff there exists a firing sequence σ such that $M_0\ |\sigma\rangle\ M$. The set of all markings reachable from M_0 defines the *reachability set* of $\langle N, M_0 \rangle$ and is denoted $R(N, M_0)$.

The *incidence matrix* of a net is an $m \times n$ matrix C where; $C(p,t) = 1$ if $(t,p) \in F$ and $(p,t) \notin F$, $C(p,t) = -1$ if $(p,t) \in F$ and $(t,p) \notin F$, else $C(p,t) = 0$.

We denote $PR(N, M_0)$ the *potentially reachable set*, i.e., the set of all markings $M \in \mathbb{N}^m$ for which there exists a vector $X \in \mathbb{N}^n$ that satisfies the *state equation* $M = M_0 + C \cdot X$, i.e., $PR(N, M_0) \triangleq \{M \in \mathbb{N}^m \mid \exists X \in \mathbb{N}^n : M = M_0 + C \cdot X\}$. It holds that $R(N, M_0) \subseteq PR(N, M_0)$.

A place p is *k-bounded* if for all $M \in R(N, M_0)$ it holds $M(p) \leqslant k$. A place 1-*bounded* is called *safe*. A net system $\langle N, M_0 \rangle$ is said k-bounded (resp., safe) if all its places are k-bounded (resp., safe). A marking M of a safe net system is a binary vector and can also be seen as a set of places $M = \{p \in P \,|M(p) = 1\}$.

In the rest of the paper for sake of simplicity we will consider only safe net systems but the results presented in this paper can easily be extended to arbitrary bounded nets.

3. UNFOLDING

In this section we informally recall how it is possible, given a safe net system $\langle N, M_0 \rangle$, to *unfold* it constructing a *labelled occurrence net* $\tilde{N}(M_0)$. This occurrence net, that is also commonly called the *unfolding of* $\langle N, M_0 \rangle$, has a structure that depends both on N and on M_0. A formal description of the unfolding procedure requires a long and tedious series of definitions: we prefer to present the key concepts here. Any of the references (McMillan, 1995; Esparza *et al.*, 2002; He and Lemmon, 2002; Benveniste *et al.*, 2003a) contains a more comprehensive and accurate discussion.

An *occurrence net* is an ordinary P/T net with a special structure: (a) starting from any node, all backward paths are finite, i.e., eventually they reach a source node; (b) each place has at most one input arc; (c) no node is in self-conflict. It is easy to show that in an occurrence net if x and x' are two distinct nodes, one and only one of the following conditions holds: $x \prec x'$, or $x' \prec x$, or $x \# x'$, or $x \approx x'$.

To the unfolding $\tilde{N}(M_0) = (\tilde{P}, \tilde{T}, \tilde{F})$ a *labelling function* $\ell : (\tilde{P} \to P) \cup (\tilde{T} \to T)$ is also associated: it maps each node of the unfolding into a node of the original net N. Note that usually a node p or t of N may correspond to more than one node of the unfolding, i.e., $\ell^{-1}(p) \subset \tilde{P}$ and $\ell^{-1}(t) \subset \tilde{T}$.

The labelling function can also map set of nodes into set of nodes. In particular, in the following procedure given a set of places $P' \subseteq P$ of the original net, we write $P' = \hat{\ell}(\tilde{P}')$ to denote that the set of places $\tilde{P}'$ of the unfolding has the same cardinality of P' and $P' = \left\{ p \in P \,\middle|\, \tilde{p} \in \tilde{P}', p = \ell(\tilde{p}) \right\}$, hence each place of $\tilde{P}'$ maps into a place of P' but no two places in $\tilde{P}'$ map into the same place of P'.

Procedure 1. **(Unfolding of a safe net system $\langle N, M_0 \rangle$ into an occurrence net $\tilde{N}(M_0)$)**

(1) Add to the unfolding a set of source places $\tilde{P}_0$ with $\hat{\ell}(\tilde{P}_0) = \{p \in P \,|\, M_0(p) = 1\}$.
(2) Let $i := 0$.
(3) Let $\tilde{P}_{\text{exp}} := \emptyset$
(4) If $\tilde{P}_i \neq \emptyset$ then STOP.
(5) Let $i := i + 1$.
(6) Let $\tilde{P}_i := \emptyset$.
(7) For all transitions $t \in T$
    For all sets of places $\tilde{P}' \subseteq \tilde{P}_{\text{exp}}$ such that all following 3 conditions are verified:
    - $\hat{\ell}(\tilde{P}') = {}^\bullet t$,
    - all places in $\tilde{P}'$ are concurrent,
    - $\tilde{P}' \cap \tilde{P}_{i-1} \neq \emptyset$,
    (a) Add to the unfolding a new transition $\tilde{t}$ with $\hat{\ell}(\tilde{t}) = t$.
    (b) Add to the unfolding a set of new places $\tilde{P}''$ with $\hat{\ell}(\tilde{P}'') = t^\bullet$.
    (c) Add an arc from each place in $\tilde{P}'$ to $\tilde{t}$ and from $\tilde{t}$ to each place in $\tilde{P}''$.
    (d) Let $\tilde{P}_i := \tilde{P}_i \cup \tilde{P}''$.
    (e) Let $\tilde{P}_{\text{exp}} := \tilde{P}_{\text{exp}} \cup \tilde{P}''$.
(8) Goto 3.

In the procedure at step 1 we add to the unfolding a copy of each place of the original net marked by the initial marking: all places in this set $\tilde{P}_0$ are ranged on the tier 0 and represent the source nodes of the occurrence net. The index i initialised at step 2 denotes the tier on which the places of each set $\tilde{P}_i$ are ranged.

The set $\tilde{P}_{\text{exp}}$ initialised at step 3 keeps track of the places that can be used to expand the unfolding. Strictly speaking, in this version of the procedure this set needs not be defined because it always coincides with the set of places in the unfolding. However we will need it when the procedure is modified to construct a finite prefix (that we call order 1 unfolding) as explained in the following.

Each time a new tier is added we check for all transitions t of the original net if there exists in the unfolding a set $\tilde{P}'$ with the following properties:

- it is a copy of the set of input places of t, hence a marking that marks in the unfolding all places in $\tilde{P}'$ corresponds to a marking of the original net that enables t;
- all places in $\tilde{P}'$ are concurrent, hence they can simultaneously be marked (note that the safeness of the original net ensures that two places on the unfolding with the same label cannot be concurrent);
- at least one place in $\tilde{P}'$ belongs to the lastly added tier, so that the marking of the unfolding that marks all these places has not been considered in the previous steps.

For all such sets $\tilde{P}'$ we add to the net a new copy of transition t, a new copy of all its output places $\tilde{P}''$ and the relative arcs.

The procedure given above is not an algorithm because it is not guaranteed to halt in a finite number the steps. In fact the unfolding of a net that admits repetitive sequences is infinite.

Note that we can consider an unfolding both as a net and as a marked net where the initial marking assigns to each source place in $\tilde{P}_0$ a token, so we need not specify its initial marking and simply write $R(\tilde{N}(M_0))$ to denote its reachability set.

The unfolding is a safe net so we can represent a marking with the set of non-empty place: we write $\tilde{M}_0 = \tilde{P}_0$ and in general $\tilde{M} = \left\{ \tilde{p} \in \tilde{P} \,\middle|\, \tilde{M}(\tilde{p}) = 1 \right\}$. It is also possible to apply the mapping $\hat{\ell}$ to markings.

Definition 2. To each marking $\tilde{M}$ of the unfolding corresponds a marking of the original net $M = \hat{\ell}(\tilde{M}) \triangleq \left\{ p \in P \,|\, p = \ell(\tilde{p}) \,, \tilde{p} \in \tilde{M} \right\}$. This leads to an equivalence relation among markings in $R(\tilde{N}(M_0))$ and if $\hat{\ell}(\tilde{M}) = \hat{\ell}(\tilde{M}')$ we write $\tilde{M} =_P \tilde{M}'$.

A firing vector $\tilde{X}$ of the unfolding is a binary vector that can also be seen as a set of transitions $\tilde{X} = \left\{ \tilde{t} \in \tilde{T} \,\middle|\, \tilde{X}(\tilde{t}) = 1 \right\}$.

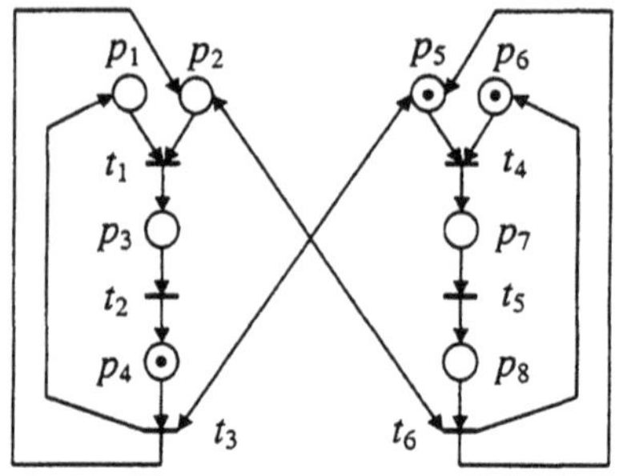

Fig. 1. A safe Petri net.

Definition 3. Given a transition $\tilde{t} \in \tilde{T}$, the minimal firing vector of the unfolding that contains it is called a local configuration; it can be show that this vector is unique and we denote it $[\tilde{t}]$. The marking reached firing configuration $\tilde{X}$ (resp., $[\tilde{t}]$) will be denoted $\tilde{M}(\tilde{X})$ (resp., $\tilde{M}([\tilde{t}])$).

It is also clear that each marking $\tilde{M}$ reachable in $\tilde{N}(M_0)$ corresponds to a unique configuration in $\tilde{N}(M_0)$ (the unfolding net is acyclic) that we sometimes denote Conf($\tilde{M}$).
Given a net system $\langle N, M_0 \rangle$, McMillan (1995) presented a technique to construct a finite prefix of its unfolding. Following Lemmon and He (2002), we consider a slightly different construction of the finite prefix.

Definition 4. (Order 1 unfolding). The order 1 unfolding, denoted $\tilde{N}_1(M_0)$, is a finite prefix of the unfolding obtained by Procedure 1 stopping the construction of the unfolding when we reach a *cut-off transition* $\tilde{t}$, i.e., a transition such that:

- EITHER the firing of the local configuration of $\tilde{t}$ brings back to the initial marking, i.e., $\tilde{M}([\tilde{t}]) =_P \tilde{M}_0$;
- OR there exists another transition $\tilde{t}'$ with the following properties:
(a) $\tilde{t}'$ has a smaller configuration than $\tilde{t}$: $[\tilde{t}'] \subset [\tilde{t}]$;
(b) the markings reached firing the two configurations are equivalent, i.e., $\tilde{M}([\tilde{t}']) =_P \tilde{M}([\tilde{t}])$.

In the following we call $\tilde{t}'$ the mirror transition of $\tilde{t}$ in $\tilde{N}_1(M_0)$.

It should be noted that what we call order 1 unfolding is a net slightly larger than McMillan finite prefix, because our condition (a) is stronger: McMillan requires only that card($[\tilde{t}']$) < card($[\tilde{t}]$).

Algorithm 5. The order 1 unfolding can be constructed using a modified version of Procedure 1 where the instruction 7.(e) is changed to
7.(e') If t is not a cut-off transition, then let $\tilde{P}_{\rm exp} := \tilde{P}_{\rm exp} \cup \tilde{P}''$.

In this case when a cut-off transitions is added to the unfolding, a copy of its output places is also added but they will not be used to expand the unfolding any further. With this small change, Procedure 1 always stops in a finite number of steps if the original system is safe.

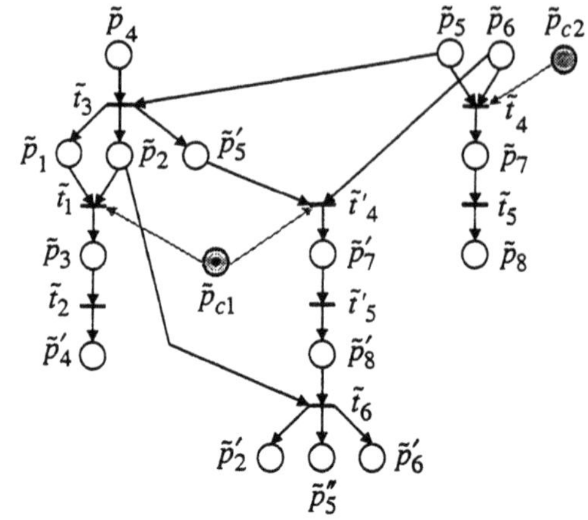

Fig. 2. The order 1 unfolding of the net in Fig. 1 with control places.

Example 6. Consider the net shown in Fig. 1. Its order 1 unfolding is shown in Fig. 2 (ignore the cyan filled places denoted p_{ci}). Note that we have also added to the unfolding the cut-off transitions (transition t_2 on tier 3 and transition t_6 on tier 4, marked in red) and their output places. Transition t_5 on tier 2 is not not a cut-off transition: after its firing the unfolding cannot proceed because a deadlock is reached. ■

The following result follows from an original result presented by McMillan (1995).

Lemma 7. The image through the labelling function of the reachability set of the order 1 unfolding $\tilde{N}_1(M_0)$ is the reachability set of the original system, i.e.,

$$R(N, M_0) = \ell(R(\tilde{N}_1(M_0))) \triangleq \left\{ M \in \mathbb{N}^m \,\middle|\, M = \ell(\tilde{M}) \,, \tilde{M} \in R(\tilde{N}_1(M_0)) \right\}.$$

Proof. McMillan showed this result holds for the total unfolding and also for the finite prefix. Since $\tilde{N}_1(M_0)$ is larger than McMillan finite prefix, the result follows immediately. □

4. A SPECIAL CLASS OF FORBIDDEN MARKINGS PROBLEM

We consider a control problem where the set of forbidden marking $\mathcal{F}$ has a special structure.

Definition 8. A set $\mathcal{F} \subseteq R(N, M_0)$ has property REACH wrt a net system (N, M_0) if

$$M \in \mathcal{F} \text{ and } M' \in R(N, M) \Rightarrow M' \in \mathcal{F}.$$

Thus property REACH implies that the set is closed under the reachability operator.
Meaningful examples of sets that have property REACH are the following: (a) the set of deadlock

markings; (b) the set of markings from which there exists no firing sequence containing a given transition; (c) the set of markings that are not co-reachable, i.e., from which it is not possible to reach a given final marking; (d) the set of markings from which the initial marking is not reachable, i.e., from which no control law can ensure reversibility; (e) the set of markings from which there exists no firing sequence containing all transitions, i.e., from which no control law can ensure liveness.

In the following we focus on the optimal control of $\mathcal{F}$. Property REACH will allow us to use unfolding to design optimal controllers, as we show in the following section.

Theorem 9. Given a set $\mathcal{F}$ with property REACH and a marking $\tilde{M}$ such that $\hat{\ell}(\tilde{M}) \in \mathcal{F}$, if $\tilde{M}$ is reachable with configuration $\tilde{X}$, then any larger configuration $\tilde{X}' \geqslant \tilde{X}$ leads to a marking $\tilde{M}'$ such that $\hat{\ell}(\tilde{M}') \in \mathcal{F}$.

Proof. If $\tilde{M}$ is reachable with configuration $\tilde{X}$, and $\tilde{M}'$ is reachable with configuration $\tilde{X}'$ then:

$$\tilde{M} = \tilde{M}_0 + \tilde{C}\tilde{X}, \quad \textit{and} \quad \tilde{M}' = \tilde{M}_0 + \tilde{C}\tilde{X}'.$$

This implies $\tilde{M}' = \tilde{M} + \tilde{C}(\tilde{X}' - \tilde{X})$ with $\tilde{X}' - \tilde{X} \in \mathbb{N}^{\tilde{n}}$ hence $\tilde{M}'$ is reachable by $\tilde{M}$ (the unfolding is an acyclic net). Thus $\hat{\ell}(\tilde{M}')$ is reachable from $\hat{\ell}(\tilde{M})$ and (by REACH) it belongs to $\mathcal{F}$. □

Based on this property, given a forbidden marking set $\mathcal{F}$ with property REACH we now present a maximally permissive control policy ensuring that no marking in $\mathcal{F}$ is reached. This control policy will be "implemented" in the unfolding net by places with output arcs and no input arcs.

5. CONTROL POLICY FOR $\mathcal{F}$

For marking $\tilde{M} \in \mathrm{R}(\tilde{N}(M_0))$ such that $\hat{\ell}(\tilde{M}) \in \mathcal{F}$ let $\tilde{X}$ be the unique configuration that yields it.

Definition 10. The set of control transitions of $\tilde{M}$ is $\tilde{X}_c = \left\{\tilde{t} \in \tilde{X} \,\middle|\, \nexists \tilde{t}' \in \tilde{X}, \tilde{t} \in [\tilde{t}']\right\}$.

In plain words, these are all transitions inputting into the places that belong to $\tilde{M}$ and that do not precede any other such transition. It is easy to prove that all these transitions are concurrent. In fact, since $\tilde{X}_c \subseteq \tilde{X}$ and $\tilde{X}$ is a firable sequence, no two transitions can be in conflict. Furthermore, all transitions preceding another one in the set $\tilde{X}$ are removed by construction, thus we are left with only concurrent transitions in $\tilde{X}_c$.

We will use the following control structure to prevent reaching $\tilde{M}$.

Definition 11. Given a marking $\tilde{M}$ with set of control transitions $\tilde{X}_c$, the control place $\tilde{p}_c$ for $\tilde{M}$ is a new place initially marked with $\left|\tilde{X}_c\right| - 1$ tokens and with an arc going to each transition in $\tilde{X}_c$. The incidence matrix of the control place is $\tilde{C}(\tilde{p}_c, \tilde{t}) = -1$ if $\tilde{t} \in \tilde{X}_c$, else $\tilde{C}(\tilde{p}_c, \tilde{t}) = 0$.

Theorem 12. The control strategy corresponding to control places for all $\hat{\ell}(\tilde{M}) \in \mathcal{F}$ is maximally permissive, i.e., it does not prevents the unfolding to reach a marking $\tilde{M}'$ with $\hat{\ell}(\tilde{M}') \notin \mathcal{F}$, if the set $\mathcal{F}$ has property reach.

Proof. By construction, each control place for a marking $\hat{\ell}(\tilde{M}) \in \mathcal{F}$ corresponding to configuration $\tilde{X}$ forbids the configuration $\tilde{X}$ and all larger configurations $\tilde{X}' \geqslant \tilde{X}$. From Theorem 9, the control place only prevents from reaching markings $\hat{\ell}(\tilde{M}) \in \mathcal{F}$. As a result, the control strategy corresponding to control places for all $\hat{\ell}(\tilde{M}) \in \mathcal{F}$ is maximally permissive. □

We now show that such a controller can be constructed from order 1 unfolding $\tilde{N}_1(M_0)$ including all its cut-off transitions and their output places. We first construct the control places that prevent reaching markings in $\mathcal{F}$ in $\tilde{N}_1(M_0)$.

Algorithm 13. **Control places for $\mathcal{F}$**

(1) Determine a reachable marking $\tilde{M}$ such that $\hat{\ell}(\tilde{M}) \in \mathcal{F}$ and such that no marking $\tilde{M}'$ with $\hat{\ell}(\tilde{M}') \in \mathcal{F}$ and $\mathrm{conf}(\tilde{M}') \subset \mathrm{conf}(\tilde{M})$ exists.
(2) If no such marking exists, then stop.
(3) Add to $\tilde{N}_1(M_0)$ the control place for $\tilde{M}$.
(4) Goto 1.

The net obtained by adding these control places to the order 1 unfolding is called $\tilde{N}_{1,c}(M_0)$. This net is not necessarily an occurrence net because the control places may contain more than one token.

Example 14. Given the net in Fig. 1, assume we want to forbid the set of markings $M \in R(N, M_0)$ such that $M(p_3) + M(p_4) + M(p_7) + M(p_8) = 2$. Clearly $\mathcal{F} = \{\{p_3,p_7\},\{p_4,p_7\},\{p_3,p_8\},\{p_4,p_8\}\}$, and it is not difficult to show that this forbidden set has property REACH for this net.

In the following table for each of the forbidden markings M we have shown the corresponding unfolding marking(s) $\tilde{M}$, the corresponding set of control transitions $\tilde{X}_c$ and finally the control place $\tilde{p}_c$. A symbol * (resp., **) in the last column denotes a configuration already forbidden by place $\tilde{p}_{c1}$ (resp., $\tilde{p}_{c2}$) hence no new place has to be added to the net for preventing it.

M	$\tilde{M}$	$\tilde{X}_c$	$\tilde{p}_c$
$\{p_3,p_7\}$	$\{\tilde{p}_3,\tilde{p}'_7\}$	$\{\tilde{t}_1,\tilde{t}'_4\}$	$\tilde{p}_{c1}$
$\{p_4,p_7\}$	$\{\tilde{p}_4,\tilde{p}_7\}$	$\{\tilde{t}_4\}$	$\tilde{p}_{c2}$
	$\{\tilde{p}'_4,\tilde{p}'_7\}$	$\{\tilde{t}_2,\tilde{t}'_4\}$	*
$\{p_3,p_8\}$	$\{\tilde{p}_3,\tilde{p}'_8\}$	$\{\tilde{t}_1,\tilde{t}'_5\}$	*
$\{p_4,p_8\}$	$\{\tilde{p}_4,\tilde{p}_8\}$	$\{\tilde{t}_5\}$	**
	$\{\tilde{p}'_4,\tilde{p}'_8\}$	$\{\tilde{t}_2,\tilde{t}'_5\}$	*

Note that place $\tilde{p}_{c2}$ contains no token because its corresponding set of control transitions is a singleton: this means that transition $\tilde{t}_4$ on tier 1 should never fire. ■

Definition 15. Let $\tilde{M}$ be a marking of an unfolding net such that $\text{conf}(\tilde{M}) = [\tilde{t}] \cup E$ and $[\tilde{t}] \cap E = \Phi$, i.e. $\text{conf}(\tilde{M})$ is an extension of $[\tilde{t}]$ denoted as $[\tilde{t}] \oplus E$ (Ezparza *et al.*, 2002). If $\tilde{t}$ is a cut-off transition with $\tilde{t}'$ as its mirror transition in $\tilde{N}_1(M_0)$, we define the mirror marking of $\tilde{M}$ as the marking of the configuration $[\tilde{t}'] \oplus E'$ where E' is the equivalent extension of E for $[\tilde{t}']$.

The concept of mirror marking can also be extended to a control places.

Definition 16. Let $\tilde{M}$ be a marking of $\tilde{N}_{1,c}(M_0)$ such that $\text{conf}(\tilde{M}) = [\tilde{t}] \cup E$ and $[\tilde{t}] \cap E = \Phi$. If $\tilde{t}$ is a cut-off transition with $\tilde{t}'$ as its mirror transition, the mirror marking of a control place p_c is $\tilde{M}(\tilde{p}_c) - \tilde{C}(\tilde{p}_c, \cdot)([\tilde{t}] - [\tilde{t}'])$.

According to the previous definition, after a cut-off transition fires, the control places should get back all those tokens that have been taken by the firing of $[\tilde{t}] - [\tilde{t}']$.

Algorithm 17. The control policy for $\mathcal{F}$ uses the net $\tilde{N}_{1,c}(M_0)$ and can be defined as follows.

(1) The plant and the net $\tilde{N}_{1,c}(M_0))$ are initialised with the respective initial marking.
(2) Compute a control pattern as follows: if $\tilde{T}_e$ is the set of transitions enabled in $\tilde{N}_{1,c}(M_0)$, the set of transitions that are enabled by the controller on the plant is $T_e = \ell(\tilde{T}_e)$.
(3) If a transition t fires in the plant, the unique transition $\tilde{t} \in \ell^{-1}(t)$ enabled in $\tilde{N}_{1,c}(M_0)$ is fired. After the firing of $\tilde{t}$, the marking of the unfolding is set to the related mirror marking if $\tilde{t}$ is a cut-off transition.
(4) Goto 2.

Theorem 18. The control policy of Algorithm 17 is maximally permissive.
Proof. Similar to the proof of Theorem 12, a marking $\tilde{M}$ of $\tilde{N}_1(M_0)$ is forbidden by control places of Algorithm 13 if and only if $\hat{\ell}(\tilde{M}) \in \mathcal{F}$. Since a marking $\tilde{M}$ obtained by a cut-off transition $\tilde{t}$ is replaced by its mirror marking $\tilde{M}'$, we need to prove that $\tilde{M}'$ is also permitted by control places. This is true since $\tilde{M}$ is accepted by control places which implies $\hat{\ell}(\tilde{M}) \notin \mathcal{F}$ and $\tilde{M} =_P \tilde{M}'$. □

6. CONCLUSIONS

In this paper we have used the technique of unfolding to design maximally permissive supervisors for safe Petri nets assuming that the specification is given by a set of forbidden markings with property REACH.

There are some lines for future research that are still open. In many cases it may be possible to find equivalent control structure to be added to the original net rather than to the unfolding. The approach may also be extended to nets with uncontrollable transitions.

7. REFERENCES

(Aghasaryan *et al.*, 1988) A. Aghasaryan, E. Fabre, A. Benveniste, R. Boubour, C. Jard, "Fault detection and diagnosis in distributed systems : an approach by partially stochastic Petri nets," *Discrete Events Dynamical Systems*, Vol. 8, p. 203-231, June 1998.

(Benveniste *et al.*, 2003a) A. Benveniste, E. Fabre, S. Haar, C. Jard, "Diagnosis of asynchronous discrete event systems, a net unfolding approach," *IEEE Trans. on Automatic Control*, Vol. 48, No. 5, pp. 714-727, May 2003.

(Benveniste *et al.*, 2003b) A. Benveniste, E. Fabre, S. Haar, "Markov nets: probabilistic models for distributed and concurrent systems," *IEEE Trans. on Automatic Control*, Vol. 48, No. 11, pp. 1936-1950, November 2003.

(Esparza *et al.*, 2002) J. Esparza, S. Römer, W. Vogler, "An improvement of McMillan's unfolding algorithm," *Formal Methods in System Design*, Vol. 20, pp. 285–310, 2002.

(Godefroid, 1996) P. Godefroid, *Partial-order methods for the verification of concurrent systems - an approach to the state-explosion problem*, Lecture Notes in Computer Science, Vol. 1032, Springer-Verlag, 1996.

(He and Lemmon, 2000a) K.X He, M.D. Lemmon, "Liveness verification of discrete-event systems modeled by n-safe ordinary Petri Nets," *Proc. 21st Int. Conf. on Application and Theory of Petri Nets* (Aarhus Denmark), Lecture Notes in Computer Science, Vol. 1825, pp. 227-243, Springer, 2000.

(He and Lemmon, 2000b) K.X He, M.D. Lemmon, "Liveness enforcing supervision of discrete-event systems modeled by *n*-safe Petri nets," *Proc. IFAC Int. Conf. on Control System Design* (Bratislava Slovakia), June 2000.

(He and Lemmon, 2002) K.X. He, M.D. Lemmon, "Liveness-enforcing supervision of bounded ordinary Petri nets using partial order methods," *IEEE Trans. on Automatic Control*, Vol. 47, No. 7, pp. 1042-1055, July 2002.

(Hellgren *et al.*, 1999) A. Hellgren, M. Fabian, B. Lennartson, "Deadlock detection and controller synthesis for production systems using partial order techniques," *Proc. of the 1999 IEEE Int. Conf. on Control Applications* (Kohala Coast, Hawaii, USA), Aug 1999.

(McMillan, 1995) K.L. McMillan, "A technique of state space search based on unfolding," *Formal Methods in System Design*, Vol. 6, No. 1, pp. 45-65, 1995.

(Murata, 1989) T. Murata, "Petri nets: properties, analysis and applications," *Proceedings of the IEEE*, Vol. 77, No. 4, pp. 541–580, April 1989.

(Neumair, 2002) C. Neumair, "Finite unfoldings of unbounded Petri nets," *Petri Net Newsletter* No. 63, pp. 5-10, October 2002.

(Ramadge and Wonham, 1989) P. Ramadge, W.M. Wonham, "Control of discrete event system," *Proceedings of the IEEE*, Vol. 77, No. 1, pp. 81-98, January 1989.

(Valmari, 1991) A. Valmari, "Stubborn sets for reduced state space generation," *Advances in Petri Nets '90*, Lecture Notes in Computer Science, Vol. 483, pp. 491-515, Springer, 1991.

(Valmari, 1994) A. Valmari, "State of the art report: stubborn sets," *Petri Net Newsletter* No. 46, pp. 6-14, April 1994.

(Xie and Giua, 2004) X. Xie, A. Giua, "Counterexamples to «Liveness-enforcing supervision of bounded ordinary Petri nets using partial order methods»," *IEEE Trans. on Automatic Control*, Vol. 49, 2004. To appear.

ELSEVIER
IFAC
PUBLICATIONS
www.elsevier.com/locate/ifac

MAXIMALLY PERMISSIVE MUTUALLY & GLOBALLY NONBLOCKING SUPERVISORS FOR DISCRETE EVENT SYSTEMS

Ratnesh Kumar [*] **Shigemasa Takai** [**] **Martin Fabian** [***]
Toshimitsu Ushio [****]

[*] *Department of Electrical & Computer Engineering, Iowa State University, Ames, IA 50014, USA*
E-mail: rkumar@iastate.edu
[**] *Department of Opto-Mechatronics, Wakayama University, 930 Sakaedani, Wakayama 640-8510, Japan*
E-mail: takai@sys.wakayama-u.ac.jp
[***] *Department of Signals and Systems, Chalmers University of Technology, SE-412 96 Göteborg, Sweden*
E-mail: fabian@s2.chalmers.se
[****] *Department of Systems Innovation, Osaka University, 1-3 Machikaneyama, Toyonaka, Osaka 560-8531, Japan*
E-mail: ushio@sys.es.osaka-u.ac.jp

Abstract: A supervisor is said to be mutually nonblocking with respect to a pair of specifications if upon completing a task in any of the specifications, it can continue on to complete the task in the other specification, i.e., the two specifications do not block each other. The notion of mutually nonblocking supervisor was introduced by Fabian and Kumar (2000). In this paper we present an algorithm of polynomial complexity for computing a maximally permissive mutually and globally nonblocking supervisor. The algorithm is based on a notion of *attractability*, and as a special case offers a new way of computing the maximally permissive nonblocking supervisors.

Keywords: Discrete event systems; Controllability; Nonblocking; Mutually nonblocking; Supervisory control

1. INTRODUCTION

A supervisory control theory for control of the qualitative behavior of discrete event systems was proposed by Ramadge and Wonham (1987). The control specification is a sublanguage of the language of the system under control, called a plant. A necessary and sufficient condition for the existence of a nonblocking supervisor is that the specification language be relative-closed and controllable (Ramadge & Wonham, 1987).

Many times the control specification is expressed as a set of specifications. When multiple specifications are present, one possible intent can be that all specifications must always be satisfied, i.e., the overall specification is the conjunct of the individual specifications. Another possible intent for providing multiple specifications can be that one of the specifications must always be satisfied, i.e., the overall specification is the disjunct of the individual specifications. In this case, the supervisor, besides being globally nonblocking (any generated trace of the controlled system can be extended to a marked trace generated by the controlled system), must also be mutually nonblocking (any generated trace of the controlled system satisfying a specification can be extended to a generated trace of the controlled system satisfying another specification). The notion of mutually nonblocking supervisors was introduced by Fabian and Kumar (2000), and is shown to have applications in feature interaction in telephony (Wong, Thistle, Hoang, & Malhame, 1995; Chen, Lafortune, & Lin, 2000), in protocol conversion (Kumar, Nelvagal, & Marcus, 1997), and in manufacturing (Fabian & Kumar, 2000).

When the supervisor is also required to be mutually nonblocking (besides being globally nonblocking) with respect to a set of specifications, then the additional condition of mutual nonblockingness with respect to the set of specifications is required for the controlled behavior (Fabian & Kumar, 2000). For the rest of the discussion we will only consider two specifications for the sake of simplicity. From the work of Fabian and Kumar (2000), the task of computing a maximally permissive globally and mutually nonblocking supervisor is equivalent to computing the supremal mutually nonblocking, relative-closed, and controllable sublanguage of the disjunct of the two specification languages. The fact that a unique such supremal sublanguage exists was established by Fabian and Kumar (2000). Further, a closed form formula for computing the supremal mutually nonblocking, relative-closed, and controllable sublanguage of the union of the two specification languages was obtained by Fabian and Kumar (2000), under the assumption that the two specifications are relative-closed, controllable, and "weakly nonconflicting".

The contribution of this paper is to present an algorithm for computing the supremal mutually nonblocking, relative-closed, and controllable sublanguage of the union of the two specifications without imposing any restriction on the specification languages. The reason for pursuing this work is that the assumption under which the closed form formula of Fabian and Kumar (2000) was obtained is quite restrictive: We show that under those assumptions the supremal merely equals the intersection of the two specifications, i.e., the maximally permissive behavior is restricted to that achievable under the conjunctive specifications setting.

Our algorithm is based on a notion of *controllable attraction* that is motivated by the notion of weak attraction introduced by Brave and Heymann (1990). A state is controllably attractable to a set of states, called an attractor, if under a suitable control all paths originating from the state eventually visit the attractor. The main difference with the notion of weak attraction of Brave and Heymann (1990) is that a finite-step attraction to the attractor is not required, i.e., ours is a weaker notion. We start with an automaton obtained by composing the plant and the two specifications, and then identify the set of all its maximal components that are strongly connected, control-invariant, and mutually nonblocking. Next we compute the region of controllable attraction of such components and remove all states (and their transitions) outside this region. The restricted automaton is shown to recognize the needed supremal sublanguage. Further, our algorithm can be specialized to the standard setting where mutual nonblockingness is not an issue and only a maximally permissive nonblocking supervisor needs to be computed. Thus the algorithm provides an alternative way of computing the supremal relative-closed and controllable sublanguage.

2. NOTATION AND PRELIMINARY

Given a set of events Σ, we use Σ^* to denote the set of all sequences of events of finite length including the zero length sequence, denoted ϵ. An element of Σ^* is called a trace, and a subset of it is called a language. For traces $s, t \in \Sigma^*$, $s \leq t$ is used to denote that s is a prefix of t. For a language $H \subseteq \Sigma^*$, the notation $pr(H)$, called the prefix closure of H, denotes the set of traces that are prefixes of traces in H.

A discrete event system is modeled as a state machine or an automaton, which is a five-tuple $G := (X, \Sigma, \alpha, x_0, X_m)$, where X is the set of states, Σ is the set of events, $\alpha : X \times \Sigma \to X$ is the (partial) state transition function, $x_0 \in X$ is the initial state, and $X_m \subseteq X$ is the set of marked states. The state transition function is recursively extended from events to traces. The generated language of G, denoted $L(G)$, is defined as $\{s \in \Sigma^* \mid \alpha(x_0, s) \text{ is defined}\}$. Similarly, the marked language of G, denoted $L_m(G)$, is defined as $\{s \in L(G) \mid \alpha(x_0, s) \in X_m\}$. Given a set of states $\hat{X} \subseteq X$, the state machine G *restricted to* $\hat{X}$, denoted $G|_{\hat{X}}$, is obtained by removing from G the states in $X - \hat{X}$ and all incoming/outgoing transitions. A state machine $\hat{G} := (\hat{X}, \Sigma, \hat{\alpha}, \hat{x}_0, \hat{X}_m)$ is said to *refine* the state machine G if $L(\hat{G}) = L(G)$ and $|\hat{X}| \geq |X|$.

For control purposes, the event set of a discrete event system to be controlled, called a plant, is partitioned into the set of controllable and the set of uncontrollable events. We use $\Sigma_u \subseteq \Sigma$ to denote the set of uncontrollable events. A supervisor is a map $f : L(G) \to 2^{\Sigma - \Sigma_u}$ such that it determines the set of controllable events $f(s) \subseteq \Sigma - \Sigma_u$ to be disabled following the execution of each trace $s \in L(G)$. The controlled plant is denoted G^f, and its generated and marked languages are obtained using:

$\epsilon \in L(G^f)$, $[s \in L(G^f), s\sigma \in L(G), \sigma \notin f(s)] \Leftrightarrow$

$[s\sigma \in L(G^f)]$; $L_m(G^f) := L(G^f) \cap L_m(G)$. It holds that $L_m(G^f) \subseteq L(G^f) = pr(L(G^f)) \neq \emptyset$. f is said to be (globally) nonblocking if $L(G^f) \subseteq pr(L_m(G^f))$.

Given a nonempty specification language $K \subseteq L_m(G)$, there exists a (globally) nonblocking supervisor if and only if K is controllable, i.e., $pr(K)\Sigma_u \cap L(G) \subseteq pr(K)$, and relative-closed, i.e., $pr(K) \cap L_m(G) \subseteq K$ (Ramadge & Wonham, 1987). The properties of controllability and relative-closure are preserved under union and hence there exists a supremal controllable and relative-closed sublanguage, denoted $supRC(K)$ of a given language K (Wonham & Ramadge, 1987).

When multiple, and without loss of generality, two specification languages $K_1, K_2 \subseteq L_m(G)$ are given, it is sometimes desired that the controlled system meet either of the specification without getting blocked (i.e., be globally nonblocking), and also always allow the possibility of completing either of the pending tasks (i.e., be mutually nonblocking). A supervisor $f : L(G) \to 2^{\Sigma-\Sigma_u}$ is said to be (K_1, K_2)-*mutually nonblocking* if

$$K_i \cap L_m(G^f) \subseteq pr(K_j \cap L_m(G^f))$$

for $i, j \in \{1, 2\}$. The notion of mutually nonblocking supervision was introduced by Fabian and Kumar (2000).

Definition 1. A language $H \subseteq K_1 \cup K_2$ is said to be (K_1, K_2)-*mutually nonblocking* if $H \cap K_i \subseteq pr(H \cap K_j)$ for $i, j \in \{1, 2\}$.

The following necessary and sufficient condition for the existence of a globally and mutually nonblocking supervisor was obtained by Fabian and Kumar (2000):

> Given a pair of specifications $K_1, K_2 \subseteq L_m(G)$, there exists a globally and mutually nonblocking supervisor f such that $L_m(G^f) \subseteq K_1 \cup K_2$ if and only if there exists a nonempty, controllable, relative-closed, and (K_1, K_2)-mutually nonblocking sublanguage of $K_1 \cup K_2$.

3. MAXIMALLY PERMISSIVE SUPERVISOR COMPUTATION

It was shown by Fabian and Kumar (2000) that the set of controllable, relative-closed, and mutually nonblocking languages is closed under union, so that the supremal such sublanguage of $K_1 \cup K_2$, denoted $supMRC(K_1 \cup K_2)$ exists.

3.1 *Formula for $supMRC(K_1 \cup K_2)$ in Specialized Setting*

A formula for computing $supMRC(K_1 \cup K_2)$ when K_1 and K_2 are controllable and mutually weakly nonconflicting was given by Fabian and Kumar (2000). The relative-closure property was ignored in obtaining this formula since "marking" supervisors were allowed. In what follows next we show that the formula remains valid even when non-marking supervisors are used so that the relative-closure property can no longer be ignored.

Recall that a pair of languages K_1, K_2 are mutually nonconflicting if $pr(K_1 \cap K_2) = pr(K_1) \cap pr(K_2)$ (Wonham & Ramadge, 1988). K_1, K_2 are called *mutually weakly nonconflicting* if $K_i, pr(K_j)$ are mutually nonconflicting (Fabian & Kumar, 2000). The following result presents the above mentioned formula.

> Given a pair of controllable and mutually weakly nonconflicting specifications $K_1, K_2 \subseteq L_m(G)$, the supremal mutually nonblocking and controllable sublanguage of $K_1 \cup K_2$ is given by Fabian and Kumar (2000): $(K_1 \cap pr(K_2)) \cup (pr(K_1) \cap K_2)$.

Theorem 1. Given a pair of controllable, relative-closed, and mutually weakly nonconflicting languages $K_1, K_2 \subseteq L_m(G)$, $supMRC(K_1 \cup K_2) = (K_1 \cap pr(K_2)) \cup (pr(K_1) \cap K_2) = K_1 \cap K_2$.

3.2 *Identifying Traces in $(K_1 \cup K_2) - supMRC(K_1 \cup K_2)$*

In this subsection we identify those "bad" traces that need to be removed from $K_1 \cup K_2$ in order to obtain $supMRC(K_1 \cup K_2)$ as the remainder. A trace is said to be bad if it violates the property of either mutual nonblockingness, or relative-closure, or controllability. The following iterative definition identifies the set of bad traces:

Definition 2. The set of bad traces of $K_1 \cup K_2$, denoted B, is defined iteratively as follows:

(1) $B_0 := L(G) - pr(K_1 \cup K_2)$; $k := 0$
(2) $B_{k+1} := B_k \cup \{s \in L(G) - B_k \mid \exists \sigma \in \Sigma_u : s\sigma \in B_k\}$
$\cup \{s \in L(G) - B_k \mid s \in K_i - pr((L(G) - B_k) \cap K_j)\}$
$\cup \{(L(G) - B_k) - pr(L_m(G) - B_k)\}$.
(3) If $B_{k+1} = B_k$, then stop and $B := B_k$; else $k := k + 1$ and go to step 2.

Thus all traces that do not belong to $pr(K_1 \cup K_2)$ are deemed bad in the initial iteration. In the

$(k+1)$th iteration, the set of bad traces consists of those traces that (i) are deemed bad in the kth iteration, or (ii) violate controllability, or (iii) violate mutual nonblockingness, or (iv) are not bad in the kth iteration but possess no extension that is marked and is not bad.

Next we show that the bad traces can be identified as bad states of a suitably refined model of the plant. Let $G := (X, \Sigma, \alpha, x_0, X_m)$ be an automaton model of the uncontrolled discrete event plant, and $S_i := (Y_i, \Sigma, \beta_i, y_{0,i}, Y_{m,i})$ be a trim acceptor of the specification K_i, which implies $L_m(S_i) = K_i$ and $L(S_i) = pr(K_i)$. Define the automata $\{\overline{S}_i := (Y_i \cup \{y_{d,i}\}, \Sigma, \overline{\beta}_i, y_{0,i}, Y_{m,i}), i = 1, 2\}$, where

$$\forall y \in Y_i \cup \{y_{d,i}\}, \sigma \in \Sigma :$$
$$\overline{\beta}_i(y, \sigma) := \begin{cases} \beta_i(y, \sigma) & \text{if } \beta_i(y, \sigma) \text{ defined,} \\ y_{d,i} & \text{otherwise.} \end{cases}$$

The automaton $\overline{S}_i$ is obtained by adding an unmarked "dump" state $y_{d,i}$ such that $y_{d,i} \notin Y_i$, and adding transitions from each state to the dump state on those events that are undefined at that state.

We refine the plant with respect to the two specifications by taking the synchronous composition, $G \| \overline{S}_1 \| \overline{S}_2 := (Z, \Sigma, \gamma, z_0, Z_m)$. Note that $Z = X \times (Y_1 \cup \{y_{d,1}\}) \times (Y_2 \cup \{y_{d,2}\})$, $z_0 = (x_0, y_{0,1}, y_{0,2})$, and we define the set $Z_m := (X \times Y_{m,1} \times (Y_2 \cup \{y_{d,2}\})) \cup (X \times (Y_1 \cup \{y_{d,1}\}) \times Y_{m,2})$ of marked states in a different way as conventional synchronous composition. Note that $L_m(\overline{S}_i) = L_m(S_i) = K_i$ and $L(\overline{S}_i) = \Sigma^*$. This implies that $L(G \| \overline{S}_1 \| \overline{S}_2) = L(G)$, i.e., $G \| \overline{S}_1 \| \overline{S}_2$ "refines" G. Also, from the definition of Z_m and the fact that $K_i \subseteq L_m(G)$ it follows that $L_m(G \| \overline{S}_1 \| \overline{S}_2) = K_1 \cup K_2$.

Using the definition of the bad traces we show that a trace s in $K_1 \cup K_2 = L_m(G \| \overline{S}_1 \| \overline{S}_2)$ is bad if and only if all traces leading to the same state as s in $G \| \overline{S}_1 \| \overline{S}_2$ are bad.

Theorem 2. Let $G := (X, \Sigma, \alpha, x_0, X_m)$ denote a plant, and $S_i := (Y_i, \Sigma, \beta_i, y_{0,i}, Y_{m,i})$ denote a trim acceptor of the specification $K_i \subseteq L_m(G)$ for $i = 1, 2$. Consider the definition of bad traces, B given in Definition 2. Suppose $s \in B$ and $t \in L(G)$ is such that the same state is reached in $G \| \overline{S}_1 \| \overline{S}_2$ by the executions of s and t. Then $t \in B$.

3.3 *Algorithm for Computing $supMRC(K_1 \cup K_2)$*

Theorem 2 establishes that the removal of bad traces in B from $K_1 \cup K_2$ can be achieved by the removal of "bad" states from $G \| \overline{S}_1 \| \overline{S}_2$. In this subsection we present an algorithm that accomplishes this. We first need to develop a few notions.

Definition 3. Given the state machine $G \| \overline{S}_1 \| \overline{S}_2 = (Z, \Sigma, \gamma, z_0, Z_m)$, a set of states $\hat{Z} \subseteq Z$, and a set of events $\hat{\Sigma} \subseteq \Sigma$, $\hat{Z}$ is said to be

(1) *strongly connected component* if for each pair of states $z_1, z_2 \in \hat{Z}$, there exists a trace $s \in \Sigma^*$ such that $\gamma(z_1, s) = z_2$ and $\gamma(z_1, t) \in \hat{Z}$ for all $t \leq s$.
(2) *$\hat{\Sigma}$-invariant component* if for any $z \in \hat{Z}$ and $\sigma \in \hat{\Sigma}$ whenever $\gamma(z, \sigma)$ is defined, $\gamma(z, \sigma) \in \hat{Z}$; *control-invariant component* if $\hat{Z}$ is a Σ_u-invariant component.
(3) *mutually nonblocking component* if $\hat{Z} \cap (X \times Y_{m,1} \times (Y_2 \cup \{y_{d,2}\}))$ and $\hat{Z} \cap (X \times (Y_1 \cup \{y_{d,1}\}) \times Y_{m,2})$ are both nonempty, i.e., $\hat{Z}$ contains marked states of each specification.
(4) *$\overline{Z}$-controllably attractable component* if $\gamma(\hat{Z}, \Sigma_u) \subseteq \overline{Z} \cup \hat{Z}$, $\gamma(\hat{Z}, \Sigma) \cap \overline{Z} \neq \emptyset$.

The notion of attractability was introduced by Brave and Heymann (1990), and Özveren, Willsky, and Antsaklis (1991), where a state is said to be *strongly attractable* (resp., *weakly attractable*) to a set of states if starting from the state it is guaranteed to reach the set of states in a finite number of transitions without (resp., with) control. The *region of strong* (resp., *weak*) *attraction* of a set of states consists of all states that are strongly (resp., weakly) attractable to the given set of states. Below we introduce a relaxed notion of attractability, which we call *controllably attractable*, by relaxing the requirement of convergence in a finite number of transitions.

Algorithm 1. Let $G := (X, \Sigma, \alpha, x_0, X_m)$ denote a plant, and $S_i := (Y_i, \Sigma, \beta_i, y_{0,i}, Y_{m,i})$ denote a trim acceptor of the specification $K_i \subseteq L_m(G)$ for $i = 1, 2$.

(1) Compute the refined plant $G \| \overline{S}_1 \| \overline{S}_2$ as explained above.
(2) Identify the maximal strongly connected, control-invariant, and mutually nonblocking components of $G \| \overline{S}_1 \| \overline{S}_2$, and let Z_0 denote the set of states of all such components that are contained in $Z_g := Z - (X \times \{y_{1,d}\} \times \{y_{2,d}\} \cup \{(x, y_1, y_2) \mid x \in X_m,\ y_i \notin Y_{m,i},\ i \in \{1, 2\}\})$.
(3) Compute the region of controllably attractable states of Z_0, denoted $\Omega(Z_0)$, using the following iterative computation:
    (a) $k := 0$.
    (b) $Z_{k+1} := Z_k \cup \{\hat{Z} \subseteq Z_g \mid \hat{Z}$ maximal strongly connected, and Z_k-controllably attractable component$\}$.
    (c) If $Z_{k+1} = Z_k$, then stop and $\Omega(Z_0) := Z_k$; else $k := k + 1$ and go to step 3(b).
(4) If $z_0 \notin \Omega(Z_0)$, then $supMRC(K_1 \cup K_2) = \emptyset$; else compute the states in $\Omega(Z_0)$ reachable from z_0, denoted $\overline{Z}$, and then $supMRC(K_1 \cup$

K_2) is the language accepted by $P := (G\|\overline{S}_1\|\overline{S}_2)\mid_{\overline{Z}}$, i.e., $G\|\overline{S}_1\|\overline{S}_2$ restricted to $\overline{Z}$.

The following theorem shows the correctness of Algorithm 1.

Theorem 3. Consider the notation of Algorithm 1.

(1) $L_m(P) = supMRC(K_1 \cup K_2)$.
(2) $supMRC(K_1 \cup K_2) \neq \emptyset$ if and only if $z_0 \in \Omega(Z_0)$.

Let m be the number of states in the plant model G, and for $i = 1, 2$, n_i be the number of states in the automaton for the specification K_i. The complexity of Algorithm 1 is $O(m^2 n_1^2 n_2^2)$.

Remark 1. The ideas of Algorithm 1 can be used to also compute the maximally permissive behavior of a globally nonblocking supervisor, i.e., the supremal relative-closed and controllable sublanguage. Let G denote the plant model and S denote a trim acceptor for the specification language $K \subseteq L_m(G)$. Assuming K to be relative-closed, we obtain an acceptor for $supRC(K)$, by applying Algorithm 1 with the following modifications:

(1) Replace $G\|\overline{S}_1\|\overline{S}_2$ by $G\|\overline{S}$.
(2) Define Z_0 to be the set of states contained in maximal strongly connected, and control-invariant components containing a state marked by the specification.

If the number of states in G and S are m and n respectively, then the complexity of this algorithm is $O(m^2 n^2)$.

The following example illustrates Algorithm 1.

Example 1. We consider an AGV working in a simple manufacturing system, shown in Fig. 1, that consists of one AGV station, one load station, and two machines Mn_1, Mn_2. The uncontrolled plant is modeled by an automaton shown in Fig. 2 (a), where a double circle is used to identify a marked state. The event labels in Fig. 2 (a) represent the following actions:

a: The AGV arrives at the load station from the AGV station,
b: The AGV arrives at the AGV station from the load station,
c: A type 1 part is loaded to the AGV,
d: A type 2 part is loaded to the AGV,
e: The AGV conveys a part to the machine Mn_1,
f: The AGV conveys a part to the machine Mn_2,
g: The AGV arrives at the load station from the machine Mn_1 or Mn_2.

Trim acceptors S_1 and S_2 of the specifications K_1 and $K_2 \subseteq L_m(G)$ are shown in Fig. 2 (b) and (c), respectively. The specification K_1 means that a type 1 part is always conveyed to Mn_1, while the specification K_2 means that a type 2 part is always conveyed to Mn_2.

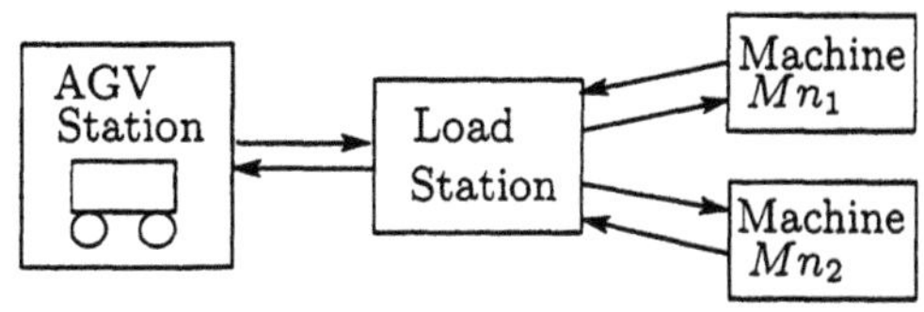

Fig. 1. A simple manufacturing system.

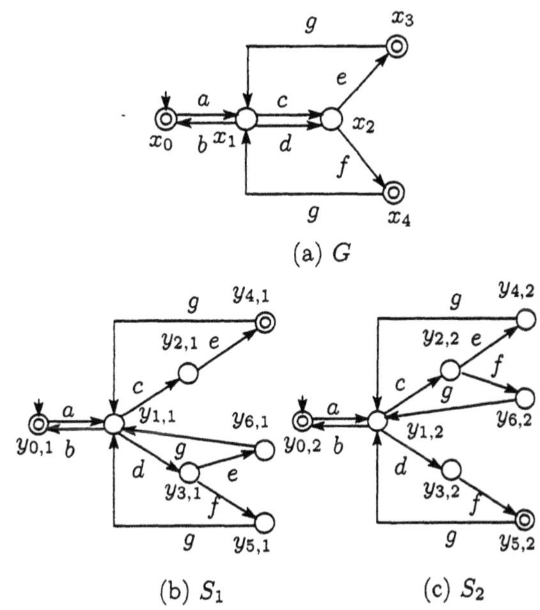

Fig. 2. The plant G and trim acceptors of K_1 and K_2.

The refined plant $G\|\overline{S}_1\|\overline{S}_2$ is shown in Fig. 3 (a). We assume that $\Sigma_u = \{c, d\}$. We compute $supMRC(K_1 \cup K_2)$ by Algorithm 1. A state in $G\|\overline{S}_1\|\overline{S}_2$ is not contained in Z_g if either the last two coordinates are dump, or the first coordinate is marked and the last two coordinates are unmarked. In Fig. 3 (a), such bad states are shaded dark. Fig. 3 (b) shows the reduced state machine $(G\|\overline{S}_1\|\overline{S}_2)\mid_{Z_g}$ obtained by removing the bad states. The unique strongly connected, control-invariant, and mutually nonblocking component contained in Z_g is $\{z_0, z_1, z_2, z_3, z_4, z_5\}$, where $z_0 = (x_0, y_{0,1}, y_{0,2})$, $z_1 = (x_1, y_{1,1}, y_{1,2})$, $z_2 = (x_2, y_{2,1}, y_{2,2})$, $z_3 = (x_2, y_{3,1}, y_{3,2})$, $z_4 = (x_3, y_{4,1}, y_{4,2})$, $z_5 = (x_4, y_{5,1}, y_{5,2})$. So we have $Z_0 = \{z_0, z_1, z_2, z_3, z_4, z_5\}$. It can be easily seen from Fig. 3 (b) that $Z_1 = Z_0$, and so we have $\Omega(Z_0) = Z_0$. Since $z_0 \in Z_0 = \Omega(Z_0)$, the automaton $(G\|\overline{S}_1\|\overline{S}_2)\mid_{Z_0} = P$ shown in Fig. 3 (c) accepts $supMRC(K_1 \cup K_2)$.

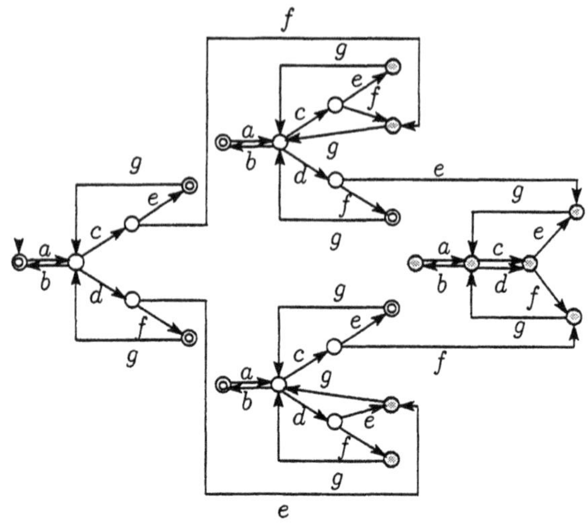

(a) $G \| \overline{S}_1 \| \overline{S}_2$

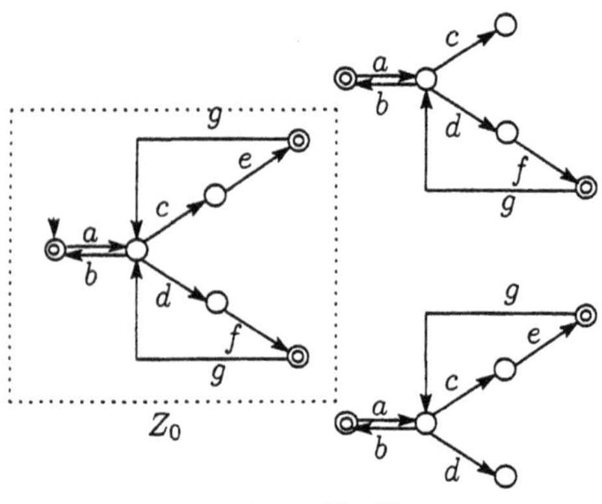

(b) $(G \| \overline{S}_1 \| \overline{S}_2) \, |_{Z_g}$

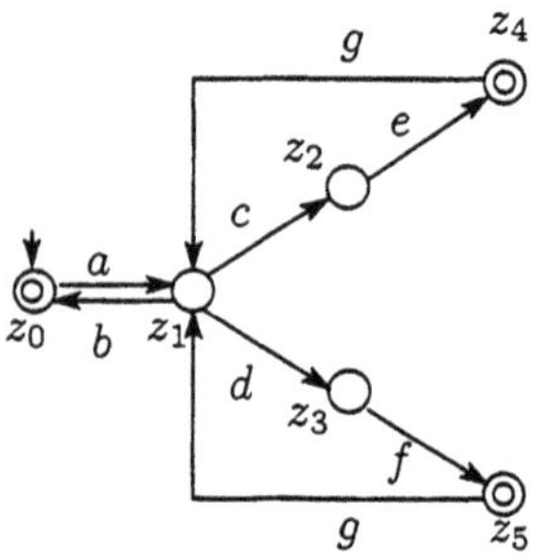

(c) P

Fig. 3. Refined plant $G \| \overline{S}_1 \| \overline{S}_2$, reduced plant $(G \| \overline{S}_1 \| \overline{S}_2) \, |_{Z_g}$, and automaton P.

4. CONCLUSION

We extended the prior work on modular control specified by the disjunct of a set of specifications. (Note there is another line of work on modular control specified by the conjunct of a set of specifications.) For the disjunctive modular control, mutual nonblockingness with respect to the specifications is one of the key issues (besides global nonblockingness with respect to the marked language of the plant). We presented an algorithm of polynomial complexity for computing the maximally permissive globally and mutually nonblocking supervisors. The algorithm can be specialized to offer a new way of computing the standard maximally permissive nonblocking supervisors.

ACKNOWLEDGMENTS

This work was supported in part by the National Science Foundation under the grants NSF-ECS-0218207, NSF-ECS-0244732 and NSF-EPNES-0323379, a DoD-EPSCoR grant from the Office of Naval Research under the grant N000140110621, a grant from the ICube initiative at the Iowa State University, and MEXT under Grant-in-Aid for Scientific Research (No. 15760321).

REFERENCES

Brave, Y., & Heymann, M. (1990). On stabilization of discrete event processes. *International Journal of Control, 51*(5), 1101–1117.

Chen, Y.-L., Lafortune, S., & Lin, F. (2000). Design of nonblocking modular supervisors using event priority functions. *IEEE Transactions on Automatic Control, 45*(3), 432–452.

Fabian, M., & Kumar, R. (2000). Mutually nonblocking supervisory control of discrete event systems. *Automatica, 36*(12), 1863–1869.

Kumar, R., Nelvagal, S., & Marcus, S. I. (1997). A discrete event systems approach for protocol conversion. *Discrete Event Dynamic Systems: Theory and Applications, 7*(3), 295–315.

Özveren, C. M., Willsky, A. S., & Antsaklis, P. J. (1991). Stability and stabilizability of discrete event dynamical systems. *Journal of ACM, 38*(3), 730–752.

Ramadge, P. J., & Wonham, W. M. (1987). Supervisory control of a class of discrete event processes. *SIAM Journal on Control and Optimization, 25*(1), 206–230.

Tarjan, R. (1972). Depth-first search and linear graph algorithms. *SIAM Journal on Computing, 1*(2), 146–160.

Wong, K. C., Thistle, J. G., Hoang, H.-H., & Malhame, R. P. (1995). Conflict resolution in modular control with applications to feature interaction. *Proceedings of the 34th IEEE Conference on Decision and Control*, New Orleans, LA, USA, (pp. 416–421).

Wonham, W. M., & Ramadge, P. J. (1987). On the supremal controllable sublanguage of a given language. *SIAM Journal on Control and Optimization, 25*(3), 637–659.

Wonham, W. M., & Ramadge, P. J. (1988). Modular supervisory control of discrete event systems. *Mathematics of Control, Signals, and Systems, 1*(1), 13–30.

ELSEVIER
IFAC
PUBLICATIONS
www.elsevier.com/locate/ifac

SUPREMAL NORMAL SUBLANGUAGES OF LARGE DISTRIBUTED DISCRETE-EVENT SYSTEMS

Jan Komenda *,[1] **Jan H. van Schuppen** **

* *Czech Academy of Sciences, Brno Branch, Zizkova 22, 616 62 Brno, Czech Republic*
** *CWI, P.O. Box 94079, 1090 GB Amsterdam, The Netherlands*

Abstract: Coalgebraic methods provide new results and insights for modular supervisory control of discrete-event systems (DES), where the overall system is composed of subsystems that are themselves partially observed DES. It is well known that the computation of supremal normal sublanguages is computationally very difficult. The attention of this paper is focused on complex distributed systems that are composed of a large number of small subsystems that are combined in a modular fashion. Conditions are derived under which supremal normal sublanguages commute with synchronous product, i.e. the computation of supremal normal sublanguages can be done locally. The coinduction proof principle is used to obtain our main result. *Copyright © 2004 IFAC*

Keywords: Modular discrete-event systems, Supremal normal sublanguages, Coalgebra, Coinduction

1. INTRODUCTION

The setting of this paper is that of the modular DES. A short historical overview of modular supervisory control of DES follows. The modular approach to the supervisory control of DES has been introduced in (Wonham and Ramadge, 1988) and (Ramadge and Wonham, 1989). The system is composed of local components (subsystems) that run concurrently (in parallel), i.e. the global system is the synchronous product of the local components. In the first papers on the topic the alphabets of the components were the same (Wonham and Ramadge, 1988), (Lin and Wonham, 1990). The general case of different local alphabets has been studied in (Willner and Heymann, 1991), where a very restrictive condition is imposed on events shared by several local alphabets: they must be controllable for all subsystems. This assumption has been generalized recently in (Wong and Lee, 2002) to the condition that the shared events must have the same control status for all subsystems that share a particular event.

Very little attention has been payed so far to the modular control with partial observations. A special case of modular supervisory control with partial observations is studied in (Rohloff and Lafortune, 2003).

This paper aims at developing methods for computing supremal normal sublanguages of large distributed discrete-event systems. In this paper we focus only on questions that do not consider the blocking issues. These blocking issues are very important for modular supervisory control, but

[1] Partial financial support of the EU Esprit LTR Project Control and Computation, ISO-2001-33520 and the grant No. 201/03/P077 of the Grant Agency of Czech Republic is gratefully acknowledged.

they have already been studied in modular control with full observations.

The coalgebraic approach has been recently used for decentralized DES. The following problems are of interest in modular supervisory control: can the control be exerted at the local level without violating the global control objectives or without affecting the optimality of the solution? If the answers to these questions are positive, then there is an exponential save on the computational complexity. In our coalgebraic framework these two problems can be paraphrased as follows: when does the supervised product commute with the synchronous product and when does the supremal normal and/or controllable sublanguage commute with the synchronous product?

Commutativity between supremal controllable sublanguages and the synchronous product has been studied in (Wong and Lee, 2002). Conditions under which supremal normal sublanguages commute with synchronous product, i.e. the computation of supremal normal sublanguages can be done locally, are presented in this paper.

2. COALGEBRA AND COINDUCTION

Coalgebras are categorical duals of algebras (the corresponding functor operates from a given set rather than to a given set). The theory of universal coalgebra (Rutten 2000) has been developed in analogy with the corresponding theory of universal algebra. It turns out that state-transition systems, i.e. in particular various types of automata are coalgebras. They can be studied using coalgebraic concepts, e.g. bisimulations and coinduction in contrast with their usual algebraic study based on concepts like congruence and induction. Final coalgebras give rise to coinductive definition and proof principles. These are heavily used throughout the paper.

2.1 Partial automata

In this section we recall from (Rutten 1999) partial automata as coalgebras with a special emphasis on the final coalgebra of partial automata, i.e. partial automaton of partial languages.

Let A be an arbitrary set (usually finite and referred to as the set of inputs or events). The empty string will be denoted by ε. Denote by $\Uparrow = \{\emptyset\}$ the one element set and by $2 = \{0, 1\}$ the set of Booleans. A partial automaton is a pair $S = (S, \langle o, t \rangle)$, where S is a set of states, and a pair of functions $\langle o, t \rangle \ : S \to 2 \times (\Uparrow + S)^A$, consists of an output function $o \ : \ S \to 2$ and a transition function $S \to (\Uparrow + S)^A$. The output function o indicates whether a state $s \in S$ is accepting (or terminating) : $o(s) = 1$, denoted also by $s \downarrow$, or not: $o(s) = 0$, denoted by $s \uparrow$. The transition function t associates to each state s in S a function $t(s) \ : \ A \to (\Uparrow + S)$. The set $\Uparrow + S$ is the disjoint union of S and $\Uparrow$. The meaning of the state transition function is that $t(s)(a) = \emptyset$ iff $t(s)(a)$ is undefined, which means that there is no a-transition from the state $s \in S$. Similarly, $t(s)(a) \in S$ means that the a-transition from s is possible and we define in this case $t(s)(a) = s_a$, which is denoted mostly by $s \xrightarrow{a} s_a$. This notation can be extended by induction to arbitrary strings in A^*. Assuming that $s \xrightarrow{w} s_w$ has been defined, define $s \xrightarrow{wa}$ iff $t(s_w)(a) \in S$, in which case $s_{wa} = t(s_w)(a)$ and $s \xrightarrow{wa} s_{wa}$.

A *homomorphism* between partial automata $S = (S, \langle o, t \rangle)$ and $S' = (S', \langle o', t' \rangle)$ is a function $f : S \to S'$ with, for all $s \in S$ and $a \in A$:

$$o'(f(s)) = o(s) \text{ and } s \xrightarrow{a} s_a \text{ iff } f(s) \xrightarrow{a} f(s)_a,$$

in which case $f(s)_a = f(s_a)$ (see diagram below).

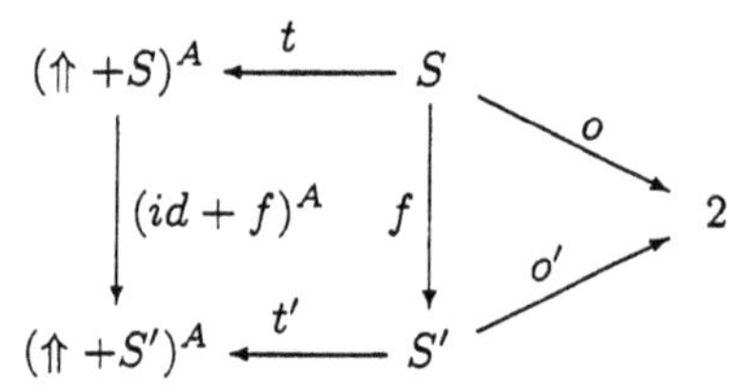

A partial automaton $S' = (S', \langle o', t' \rangle)$ is a *subautomaton* of $S = (S, \langle o, t \rangle)$ if $S' \subseteq S$ and the inclusion function $i : S' \to S$ is a homomorphism. It is important to notice that the coalgebraic concept of subautomaton corresponds to the notion of strict subautomaton in (Cho and Marcus, 1989). In the sequel we use always subautomata in the coalgebraic sense defined above, i.e. strict subautomata are meant.

Note that partial automaton as defined above is just a coalgebraic reformulation of what is understood to be a generator of a DES. Indeed, the transition function can be viewed in the coalgebraic form above, and the output function determines the subset of marked or final states (those whose output value is equal to 1).

A *simulation* between two partial automata $S = (S, \langle o, t \rangle)$ and $S' = (S', \langle o', t' \rangle)$ is a relation $R \subseteq S \times S'$ with, for all $s \in S$ and $s' \in S'$: $\langle s, s' \rangle \in R$

(*i*) $o(s) \leq o(s')$, i.e. $s \downarrow \Rightarrow s' \downarrow$, and
(*ii*) $\forall a \in A : s \xrightarrow{a} \Rightarrow (s' \xrightarrow{a}$ and $\langle s_a, s'_a \rangle \in R)$,

A *bisimulation* between two partial automata $S = (S, \langle o, t \rangle)$ and $S' = (S', \langle o', t' \rangle)$ is a relation $R \subseteq S \times S'$ with, for all $s \in S$ and $s' \in S'$: $\langle s, s' \rangle \in R$:

(*i*) $o(s) = o(s')$, i.e. $s \downarrow$ iff $s' \downarrow$
(*ii*) $\forall a \in A : s \xrightarrow{a} \Rightarrow (s' \xrightarrow{a}$ and $\langle s_a, s'_a \rangle \in R)$,
(*iii*) $\forall a \in A : s' \xrightarrow{a} \Rightarrow (s \xrightarrow{a}$ and $\langle s_a, s'_a \rangle \in R)$.

We write $s \sim s'$ whenever there exists a bisimulation R with $\langle s, s' \rangle \in R$. This relation is the union of all bisimulations, i.e. the greatest bisimulation also called bisimilarity. It is immediate from the definition of bisimulation that two states are bisimilar iff they can make the same transitions and they give rise to the same outputs:

Proposition 2.1. For any partial automaton $S = (S, \langle o, t \rangle)$ and any $s, s' \in S$: $s \sim s'$ iff
$\forall w \in A^* : s \xrightarrow{w} \iff s' \xrightarrow{w}$, in which case
$o(s_w) = o'(s'_w)$.

2.2 Final automaton of partial languages

In this subsection we define a partial automaton that is final among all partial automata and satisfies a proof principle called coinduction. The states of this final automaton represent minimal realizations of all possible behaviors (called partial languages) of all partial automata. Partial languages will be endowed with a (partial) automaton structure, which has the universal property of being final among all (partial) automata. The partial automaton of partial languages is defined using the Brzozowski notion of input derivative. Below we define the partial automaton of partial languages over an alphabet (input set) A, denoted by $\mathcal{L} = (\mathcal{L}, \langle o_{\mathcal{L}}, t_{\mathcal{L}} \rangle)$. More formally,
$\mathcal{L} = \{\Phi : A^* \rightarrow (\Uparrow +2) \mid dom(\Phi) =$
$\{w \in A^* \mid \Phi(w) \in 2\} \neq \emptyset$ is prefix-closed$\}$.
To each partial language Φ a pair $\langle V, W \rangle$ can be assigned: $W = dom(\Phi)$ and $V = \{w \in dom(\Phi) \mid \Phi(w) = 1 (\in 2)\}$. Conversely, to a pair $\langle V, W \rangle \in \mathcal{L}$, a function Φ can be assigned : $\Phi(w) = 1$ if $w \in V$, $\Phi(w) = 0$ if $w \in W$ and $w \notin V$, and $\Phi(w)$ is undefined if $w \notin W$. Therefore we can write :
$\mathcal{L} = \{(V, W) \mid V \subseteq W \subseteq A^*, W \neq \emptyset,$
and W is prefix-closed$\}$.

The transition function $t_{\mathcal{L}} : \mathcal{L} \rightarrow (1 + \mathcal{L})^A$ is defined using input derivatives. Recall that for any partial language $L = (L^1, L^2) \in \mathcal{L}$, $L_a = (L^1_a, L^2_a)$, where $L^i_a = \{w \in A^* \mid aw \in L^i\}$, $i = 1, 2$. If $a \notin L^2$ then L_a is undefined. Given any $L = (L^1, L^2) \in \mathcal{L}$, the partial automaton structure of $\mathcal{L}$ is given by:

$$o_{\mathcal{L}}(L) = \begin{cases} 1 & \text{if } \varepsilon \in L^1 \\ 0 & \text{if } \varepsilon \notin L^1 \end{cases}$$

and

$$t_{\mathcal{L}}(L)(a) = \begin{cases} L_a & \text{if } L_a \text{ is defined} \\ \emptyset & \text{otherwise} \end{cases}$$

Notice that if L_a is defined, then $L^1_a \subseteq L^2_a$, $L^2_a \neq \emptyset$, and L^2_a is prefix-closed. The following notational conventions will be used: $L \downarrow$ iff $\varepsilon \in L^1$, and $L \xrightarrow{w} L_w$ iff L_w is defined iff $w \in L^2$. The following two theorems are recalled from (Rutten 1999).

Theorem 2.2. Partial automaton $\mathcal{L} = (\mathcal{L}, \langle o_{\mathcal{L}}, t_{\mathcal{L}} \rangle)$ is final among all partial automata: for any partial automaton $S = (S, \langle o, t \rangle)$ there exists a unique homomorphism $l : S \rightarrow \mathcal{L}$. This homomorphism identifies bisimilar states: for $s, s' \in S$: $l(s) = l(s')$ iff $s \sim s'$.

Theorem 2.3. $\mathcal{L}$ satisfies the principle of coinduction: for all K and L in $\mathcal{L}$, if $K \sim L$ then $K = L$.

Coinduction is used as a proof and definition principle throughout this paper. The use of coinduction is limited to final coalgebras. Behavior equivalence of two elements of final coalgebra means that these are equal.

Proofs by coinduction consist in constructing appropriate relations: for instance a proof of equality of two elements of a final coalgebra consists in finding a bisimulation relation that relates them. Definition by coinduction of an operation on elements of a final coalgebra consists in defining the same coalgebraic structure on the operation (for instance we define binary operations on partial languages by defining derivatives and output functions further in this paper).

Recall from (Rutten 1999) the following coinductive definitions of the synchronous product. We assume that K is defined over the alphabet A_1 and L over A_2. Then the synchronous product $K \parallel L$ is a language over $A_1 \cup A_2$ with the following coinductive definition:

Definition 2.1.

$$(K \parallel L)_a = \begin{cases} K_a \parallel L_a & \text{if } a \in A_1 \cap A_2 \\ K_a \parallel L & \text{if } a \in A_1 \setminus A_2 \\ K \parallel L_a & \text{if } a \in A_2 \setminus A_1 \end{cases}$$

and $(K \parallel L) \downarrow$ iff $K \downarrow$ and $L \downarrow$.

3. MODULAR DES WITH PARTIAL OBSERVATIONS

A modular DES represented by a partial automaton S consists of local subsystems (modules) S_i, $i \in Z_n = \{1, \ldots, n\}$ such that the global DES S is the parallel composition (synchronous product) of S_i. We assume that each module S_i has only partial observation of its events, i.e. $A_i = A_{o,i} \cup A_{uo,i}$ is the decomposition of local events into locally observable and locally unobservable. The global system has observation set $A_o = \cup_{i=1}^n A_{o,i} \subseteq A = \cup_{i=1}^n A_i$. Additional notation is needed to set up our framework. Globally unobservable events are denoted by $A_{uo} = A \setminus A_o$ and locally unobservable events by $A_{uo,i} = A_i \setminus$

$A_{o,i}$. The projections of the global alphabet into local ones are denoted by $P_i : A^* \to A^*_{o,i}$. Partial observations in individual modules are expressed via local projections $P_i^{loc} : A_i^* \to A^*_{o,i}$, while global projection is denoted by $P : A^* \to A_o^*$. Local plant languages will be denoted by L_i, $i \in Z_n$ and local specification languages by K_i, $i \in Z_n$. We assume in our main theorem that $n = 2$, because the generalization to arbitrary n is quite straightforward. The following lemma concerning the relation between global and local observations is needed.

Lemma 1. If $A_{o,2} \cap A_1 = A_2 \cap A_{o,1}$ then for any $s, s' \in A^*$ we have:
if $P(s) = P(s')$ then for $i = 1, 2$: $P_i^{loc} P_i(s) = P_i^{loc} P_i(s')$.

An auxiliary concept that reflects the fact that due to partial observations it is not possible to distinguish between states is needed. The notation $\stackrel{\varepsilon}{\Rightarrow}$ is used for unobservable moves, i.e. $s \stackrel{\varepsilon}{\Rightarrow} s'$ iff $\exists \tau \in A^*_{uo} : s \stackrel{\tau}{\to} s_\tau = s'$.

Definition 3.1. (Observational indistinguishability relation on S.) A binary relation $Aux(S)$ on S, called *observational indistinguishability relation* is the smallest relation satisfying:

(i) $\langle s_0, s_0 \rangle \in Aux(S)$
(ii) If $\langle s, t \rangle \in Aux(S)$ then : ($s \stackrel{\varepsilon}{\Rightarrow} s'$ for some s' and $t \stackrel{\varepsilon}{\Rightarrow} t'$ for some t') $\Rightarrow \langle s', t' \rangle \in Aux(S)$
(iii) If $\langle s, t \rangle \in Aux(S)$ then $\forall a \in A_o$: ($s \stackrel{a}{\to} s_a$ and $t \stackrel{a}{\to} t_a$) $\Rightarrow \langle s_a, t_a \rangle \in Aux(S)$.

$Aux(S)$ can be characterized by the following lemma.

Lemma 2. For any $s, s' \in S$: $\langle s, s' \rangle \in Aux(S)$ iff there exist two strings $w, w' \in A^*$ such that $P(w) = P(w')$, $s = (s_0)_w$ and $s' = (s_0)_{w'}$.

4. SUPREMAL NORMAL SUBLANGUGES

It has been shown in (Wong and Lee, 2002) that a structural condition, called mutual controllability, is important for preserving optimality in the modular control with full observations (i.e. commutativity of the supremal controllable sublanguage with the synchronous product of partial languages). The natural question arises: With respect to which conditions this can be generalized to modular control with partial observations? Main problem of this paper is: when does the supremal normal sublanguage commutes with the synchronous product? Unfortunately, it is not possible to the best of our knowledge to define the supremal normal sublanguage by coinduction. The argument is similar to the corresponding argument for closed-loop languages under the antipermissive control policy (Komenda, 2004). Nevertheless, using suitable automata representations: state-partition automata (Yoo *et al.*, 2001) there is a monolithic algorithm for the computation of the supremal normal sublanguage. Algorithm 1 below looks almost like a coinductive definition and it will be used in the coinductive proof of our main theorem.

We use the representations of languages K and L by automata S_1 and S, where S_1 is a state-partition automaton (Yoo *et al.*, 2001). It is known how to construct such representations , see (Cho and Marcus, 1989) or (Yoo *et al.*, 2001). Let s_0 denote the common initial state of S_1 and S. The transition structure of S_1 and S is denoted by $\to_1$ and $\to$, respectively.

Normality of a specification language is an important property of partially observed DES, introduced in (Lin and Wonham, 1988).

Definition 4.1. (Normality.) Let $K, L \in \mathcal{L}$: $K \subseteq L$. K is said to be (L, P)-normal if $K^2 = L^2 \cap P^{-1}(P(K^2))$.

We recall below normal relations from (Komenda, 2002).

Definition 4.2. (Normal relation.) Given two (partial) automata $S_1 = (S_1, \langle o_1, t_1 \rangle)$ and $S = (S, \langle o, t \rangle)$ as above with common initial state $s_0 \in S$, a binary relation $N(S_1, S)$ on $S_1 \times S$ is called a *normal relation* if for any $\langle s, t \rangle \in N(S_1, S)$ the following items hold:

(i) $\forall a \in A : s \stackrel{a}{\to}_1 s_a \Rightarrow t \stackrel{a}{\to} t_a$ and $\langle s_a, t_a \rangle \in N(S_1, S)$
(ii) $\forall u \in A_{uo} : t \stackrel{u}{\to} t_u \Rightarrow s \stackrel{u}{\to}_1 s_u$.
(iii) $\forall a \in A_o : t \stackrel{a}{\to} t_a$ and $(\exists s' : \langle s, s' \rangle \in Aux(S_1) : s' \stackrel{a}{\to}_1 s'_a) \Rightarrow s \stackrel{a}{\to}_1 s_a$.

We have shown in (Komenda, 2002) that

Theorem 4.1. A (partial) language K is $(L, P)-$normal iff there exists a normal relation $N(S_1, S)$ on $S_1 \times S$ such that $\langle s_0, s_0 \rangle \in N(S_1, S)$.

Remark 4.2. We consider from now on an order relation on $\mathcal{L}$ induced by their second components only, i.e. we write $K \subseteq L \in \mathcal{L}$ iff $K^2 \subseteq L^2$. The same applies for supremum operation. Note that only the second condition (ii) of simulation relations must be checked to prove such defined inclusion of partial languages.

Now we are ready to formulate the algorithm for computation of supremal (L, P)−normal sublanguages.

Algorithm 1. Let automata S_1 and S representing K and L, respectively be such that S_1 is a subautomaton of S and S_1 is a state-partition automaton. Let us construct partial automaton $\tilde{S} = \langle \tilde{o}, \tilde{t} \rangle$ with $\tilde{t}$ denoted by $\rightarrow'$.
Define the auxiliary condition (*) as follows:
if $a \in A_{uo}$ then $\forall u \in A_{uo}^*$: $s_a \xrightarrow{u} \Rightarrow s_a \xrightarrow{u}_1$;
if $a \in A_o$ then $\forall s' \approx_{Aux(S_1)} s : s' \xrightarrow{a} \Rightarrow s' \xrightarrow{a}_1$, in which case also $\forall u \in A_{uo}^*$: $s'_a \xrightarrow{u} \Rightarrow s'_a \xrightarrow{u}_1$.

Below are the steps of the algorithm.
1. Put $\tilde{S} := \{s_0\}$.
2. For any $s \in \tilde{S}$ and $a \in A$ we put $s \xrightarrow{a}' s_a$ if condition (*) is satisfied and we put in the case $s \xrightarrow{a}'$ also $\tilde{S} := \tilde{S} \cup \{s_a\}$.
3. For any $s \in \tilde{S}$ we put $\tilde{o}(s) = o(s)$.

Let us denote by $\tilde{l} : \tilde{S} \rightarrow \mathcal{L}$ the unique (behavior) homomorphism given by finality of $\mathcal{L}$. It can be shown that

Theorem 4.3. $\tilde{l}(s_0)$ is the supremal (L, P)−normal sublanguage of K.

In our main theorem a condition similar to mutual controllability (Wong and Lee, 2002) is needed. We call it by analogy mutual normality.

Definition 4.3. Given partial languages $L_i = (L_i^1, L_i^2)$ and $L_j = (L_j^1, L_j^2)$, L_i and L_j are said to be mutually normal if

$$(P_i^{loc})^{-1} P_i^{loc}(L_i^2) \cap P_i(P_j)^{-1}(L_j^2) \subseteq L_i^2$$

and

$$(P_j^{loc})^{-1} P_j^{loc}(L_j^2) \cap P_j(P_i)^{-1}(L_i^2) \subseteq L_j^2.$$

The following lemma will be needed for the proof of our main theorem.

Lemma 3. Assume that $A_{o,2} \cap A_1 = A_2 \cap A_{o,1}$. Let $v \in A^*$ with $P_1(v) = v_1 \in A_1^*$ and $P_2(v) = v_2 \in A_2^*$. Let $v'_1 \in A_1^*$ be such that $P_1^{loc}(v_1) = P_1^{loc}(v'_1)$. Then there exists $v' \in A^*$ such that $P_1(v') = v'_1$, $P_2^{loc} P_2(v') = P_2^{loc}(v_2)$, and $P(v) = P(v')$.

Let us introduce the notation $\sup N(K, L, P)$ for the supremal (L, P)− normal sublanguage of K. Recall that local plant languages are denoted by L_i, $i \in Z_2$ and local specification languages by K_i, $i \in Z_2$. This means that we assume that the global specification language K is decomposable into local specification languages K_i, $i \in Z_2$, where $K = K_1 \parallel K_2$. Our main theorem follows. The coinductive proof principle and Algorithm 1 are used in the proof.

Theorem 4.4. If $A_{o,2} \cap A_1 = A_2 \cap A_{o,1}$, and L_1 and L_2 are mutually normal, then
$\sup N(K_1, L_1, P_1^{loc}) \parallel \sup N(K_2, L_2, P_2^{loc}) = \sup N(K_1 \parallel K_2, L_1 \parallel L_2, P)$.

Under very general conditions we have always one inclusion:

Corollary 1. If $A_{o,2} \cap A_1 = A_2 \cap A_{o,1}$, then we have $\sup N(K_1, L_1, P_1^{loc}) \parallel \sup N(K_2, L_2, P_2^{loc}) \subseteq \sup N(K_1 \parallel K_2, L_1 \parallel L_2, P)$.

Note that Theorem 4.4 is useful for the computation of (global) supremal normal sublanguages of large distributed plants. If the conditions of the theorem are satisfied, then it is sufficient to compute local supremal normal sublanguages and synchronize them.

The interest of this theorem should be clear: under the conditions that are stated it is possible to do the optimal (less restrictive) control synthesis with partial observations locally, which represents an exponential save on the computational complexity and makes in fact the optimal control synthesis of some large distributed plants feasible. Note that an extension of our results from $n = 2$ to an arbitrary number $n \in N$ of local modules is quite straightforward and thus omitted in this paper. The condition of mutual normality between any pair of local plants is required. In (Wong and Lee, 2002) there is a procedure how to arrange the mutual controllability for distributed plants, where this condition is not satisfied. It is conjectured that a similar procedure exists for mutual normality, possibly for special cases.

The condition of mutual normality is a structural condition, because it depends only on the (local) plant languages and not on the specification languages. The advantage is that for a given modular DES it must be checked only once and then it holds for arbitrary specification languages. In order to get more intuition about mutual normality we consider some extreme special cases of distributed DES. First of all, if all event alphabets are disjoint, the so called shuffe case, we notice that $P_i(P_j)^{-1}(L_j^2) = A_i^*$ for any $L_j^2 \subseteq A_j^*$. This means that the condition of mutual normality can not be satisfied. Thus one can not hope to find a universal procedure how to make a set of local plant languages mutually normal. The intuitive reason is that there is only a weak interconnection between local subsystems in this case. This is not surprising, because the observations of local agents are in this case completely independent and therefore there is a huge gap between local and global observations. On the other hand, it is obvious from the definition of mutual normality that in the case of full local observations, mutual normality is trivialy satified. Another extreme case occurs when all subsystems have the same event alphabets. Then all P_i's are identity mappings, i.e. the mutual normality becomes usual normality between two languages in a slightly more general sense (the assumption is lifted that one of the languages is a sublanguage of the other). This might justify why we call our condition mutual normality, it is a symmetric notion of normality. Note that mutual normality is not a necessary condition for commutativity between supremal normal sublanguages and the synchronous product as can be seen from simple examples.

5. CONCLUSION

Supremal normal sublanguages are important for the design of supervisory controllers with partial observations. For large distributed plants there is a problem of high computational complexity (Cho and Marcus 1989). Therefore the conditions under which the supremal normal sublanguage of the global plant can be computed using supremal normal sublanguages of the local plants are of a special interest. Coalgebraic framework for supervisory control of distributed systems with partially observed modules is used. The sufficient conditions obtained in this paper are discussed in some special cases.

There are many open problems left for a future investigation. For instance, all conditions we have obtained are only sufficient conditions. The question is whether they can be weakened, at least in special cases that occur in some relevant applications to be found. Another direction of future research is to study the conditions for commutativity between the synchronous product and the closed-loop languages using the antipermissive control policy. In this paper the blocking issues have not been considered. It is to be expected that for modular DES with partial observations the conditions for nonblocking are also the same as for modular DES with full observations (Wong and Lee, 2002), however it must still be proven.

Effective procedures for verification of our conditions must be found. For this purpose our coalgebraic approach helps through the relational characterization of all conditions that have been defined algebraicly (mutual normality, mutual controllability and decomposability). Since the new properties are similar to their classical counterparts, we believe that such effective procedures can be found. Even more, mutual controllability and mutual normality are likely to be much easier to verify than global controllability and normality, because the corresponding plant and specification languages involved are local, thus much smaller.

REFERENCES

Cassandras, S.G. and Lafortune, S. (1999). *Introduction to Discrete Event Systems*, Kluwer Academic Publishers, Dordrecht 1999.

Cho, H. and Marcus, S. I. (1989). Supremal and Maximal Sublanguages Arising in Supervisor Synthesis Problems with Partial Observations. *Math. Systems Theory*, **22**, 171-211, 1989.

Komenda, J. (2002). Computation of Supremal Sublanguages of Supervisory Control Using Coalgebra. Proceedings *WODES'02*, Workshop on Discrete-Event Systems, Zaragoza, 26-33.

Komenda, J. (2004). Coalgebra and coinduction in discrete-event control, submitted to *Discrete Event Dynamical Systems: Theory and Applications.*

Lin, F. and Wonham, W.M. (1988). On Observability. of Discrete-Event Systems. *Information Sciences*, **44** , 173-198.

Lin, F. and Wonham, W.M. (1990). Decentralized Control and Coordination of Discrete-Event Systems with Partial Observations, *IEEE Trans. Automatic Control*,**35**,1330-1337.

Rohloff, K. and Lafortune, S. (2003). The control and Verification of Similar Agents Operating in a Broadcast Network Environment. *In Proceedings CDC 2003*, Hawai, USA.

Rutten, J.J.M.M. (1999). Coalgebra, Concurrency, and Control. *Research Report CWI*, SEN-R9921, Amsterdam. Available also at `http://www.cwi.nl/~janr`.

Rutten, J.J.M.M. (2000). Universal Coalgebra: A Theory of Systems. *Theoretical Computer Science* **249(1)**, 3-80.

Ramadge, P.J. and Wonham, W.M. (1989). The Control of Discrete-Event Systems. *Proc. IEEE*, **77** :81-98.

Willner, Y. and Heymann, M. (1991). Supervisory control of concurrent discrete-event systems. *International Journal of Control*, **54(5)**, 1143-1166.

Wong, K.C. and Lee, S. (2002). Structural Decentralized Control of Concurrent Discrete-Event Systems. *European Journal of Control*, **0**, 1-15.

Wonham, W.M. and Ramadge, P.J. (1988) Modular supervisory control of discrete-event processes, *Mathematics of Control, Signal and Systems*, **1**, 13-30.

Yoo, T.S., Lafortune, S., and Lin, F. (2001). A Uniform Approach for Computing Supremal Sublanguages Arising in Supervisory Control theory. *Control group report CGR 02-04 Dept. of EECS, Univ. of Michigan*, Ann Arbor.

ELSEVIER
IFAC
PUBLICATIONS
www.elsevier.com/locate/ifac

EFFICIENT SUPERVISORY SYNTHESIS OF LARGE SYSTEMS

Arash Vahidi, Bengt Lennartson, Martin Fabian

The Control and Automation Lab. Department of Signals and Systems, Chalmers University of Technology, Sweden.
{ vahidi, bl, fabian } @s2.chalmers.se

Abstract: In this paper efficient computation of controllers in the context of Supervisory Control Theory (SCT) is discussed. A simple algorithm for efficient reachability search is presented to significantly improve controller synthesis performance, as the reachability search is the bottleneck of a controller synthesis algorithm. Simple measures to identify "hard" vs "easy" problems are also discussed and it is shown that a large state space is not always a guarantee for hardness! It is further demonstrated that well known handmade benchmark models that have been used over the years, can in fact be classified as easy, while hard problems often arise in real-world industrial systems. *Copyright © 2004 IFAC*

Keywords: Supervisory Control, Reachability Search, Symbolic Computation

1. INTRODUCTION

In the Supervisory Control Theory (SCT), *controller synthesis* is the task of computing a supervisor controller for a given problem (Wonham, 2002). This synthesis task has a clean and reliable theoretical meaning, while its implementation is problematic due to complexity issues. This is preventing SCT from having a major breakthrough in industry.

In this work, we will suggest a simple algorithm for more efficient reachability search that significantly improves the performance of controller synthesis. The algorithm uses a symbolic representation that compresses the size of the supervisor, reducing the need for supervisor reduction procedures (Vaz and Wonham, 1985).

Due to complexity issues, we cannot guarantee the efficiency of the algorithm for all problem instances. However, simple measures to identify hard and easy problems with good accuracy can be found. These simple measures can also be used as design support by identifying components that may affect the synthesis performance in a negative way.

2. PRELIMINARIES

A deterministic finite automaton is a fivetuple $A = \langle Q, \Sigma, \delta, q_i, Q_m \rangle$ where Q is the finite set of states and Σ is the set of events (the alphabet). Σ is divided into two disjoint subsets, the controllable events Σ_c and the uncontrollable events Σ_u. State transitions are described by the transfer function $\delta : Q \times \Sigma \rightarrow Q$, additionally, δ_u denotes the subset of δ associated with uncontrollable events. $q_i \in Q$ is the initial state and $Q_m \subseteq Q$ is the marked-states subset.

Composition between two or more automata is defined by the full synchronous operator $||$, originating from the early work of Hoare (Hoare, 1985). As a convention, we will use superscript indices to denote members of a composition, for example $A = A^1||...||A^n$.

In SCT, automata are often divided into logical groups such as *plants* and *specifications.* Let P and Sp denote the total composition of these two groups, respectively. In Supervisory Control Theory, the ultimate goal is to build a controlling device, a *supervisor*, which by limiting the behavior of plants guarantees that the specification plus some additional requirements (controllability and non-blocking) are met (Wonham, 2002).

2.1 Symbolic Computation

A *symbolic representation* of an object (for example a set or a relation) is a non-numeric representation that is more suited for computers than humans. *Symbolic computation* is the process of manipulating an object directly in its symbolic form. Throughout the rest of this paper, we will transparently use a *symbolic* representation of automata. We will use *Binary Decision Diagrams* to represent sets of states, transfer functions etc. for better performance. Therefore, when we discuss the size or the complexity of some object, we always refer to its symbolic form.

There are two important subjects on BDDs that the reader must know in order to understand the rest of this paper. First, a binary BDD operation $x \otimes y$ has under normal conditions the time complexity $O(|x| \times |y|)$, where $|x|$ and $|y|$ are the number of nodes in the BDDs x and y, respectively (Bryant, 1992). Second, symbolic computation with BDDs may suffer from *intermediate BDD size explosion* problem. In a series of BDD operations, even when the size of the BDD representing the final result is fairly small and manageable, the size of intermediate BDDs may very well exceed the space limit. This phenomenon often arises in reachability search where the size of the BDDs representing the intermediate set Q_k grows very fast (see Figure 3).

Although no further knowledge about BDDs is required for reading this paper, readers unfamiliar with BDDs and symbolic tools may be interested in consulting (Bryant, 1992; Hu, 1995) for some background information.

3. SAFE-STATE SYNTHESIS

It can be shown that a minimally restrictive supervisor that is both controllable and non-blocking, can be constructed as an automaton S such that $Q^S = Q^{P||Sp}$ while $L(S) \subseteq L(P||Sp)$ (some transitions are removed) (Wonham, 2002).

In this work, we will use the slightly different approach to controller synthesis where bad states are removed[1]. As long as the system is in one of the remaining states, controllability and non-blocking is guaranteed. The remaining states are therefore called the *safe states.* A safe-state supervisor can be created by building a supervisor candidate $S0 = P||Sp$, then removing states from Q^{S0} until it is both controllable and non-blocking. Let $Q^S \subseteq Q^{S0}$ denote the set of safe states. In practice, this set is the only part of the supervisor needed to be represented (as a BDD). $S0$ and S are never created. The transfer function δ^S and thus the language $L(S)$ can be constructed *online* by processing $\delta^{P||Sp}$ on the fly (see definition of $\Sigma_{enabled}$, Section 3.1). We call this approach an *online control scheme,*

Algorithm 1 presented here is a simple algorithm for computing the set of non-blocking and controllable safe states (Vahidi *et al.*, 2003). Given an explicit set of forbidden states Q_x, this algorithm computes the set of safe states.

Algorithm 1. SafeStateSynthesis
input : Q_x
let $k := 0$, $X_0 := Q_x$
repeat
  $k := k + 1$
  $Q' := RestrictedBackward(Q_m, X_{k-1})$
  $Q'' := UncontrollableBackward(Q - Q')$
  $X_k := X_{k-1} \cup Q''$
until $X_k = X_{k-1}$
return $Q - X_k$

Algorithm 2. RestrictedBackward
input : Q_m, Q_x
let $k := 0$, $Q_0 := Q_m - Q_x$
repeat
  $k := k + 1$
  $Q_k := Q_{k-1} \cup BackImage(Q_{k-1}, \delta) - Q_x$
until $Q_k = Q_{k-1}$
return Q_k

Algorithm 3. UncontrollableBackward
input : Q_x
let $k := 0$, $Q_0 := Q_x$
repeat
  $k := k + 1$
  $Q_k := Q_{k-1} \cup BackImage(Q_{k-1}, \delta_u)$
until $Q_k = Q_{k-1}$
return Q_k

Compared to existing algorithm such as (Hoffmann and Wong-Toi, 1992), this approach with the safe states can be made significantly more efficient,

[1] If the initial state is removed, then it is impossible to build a supervisor with the given requirements.

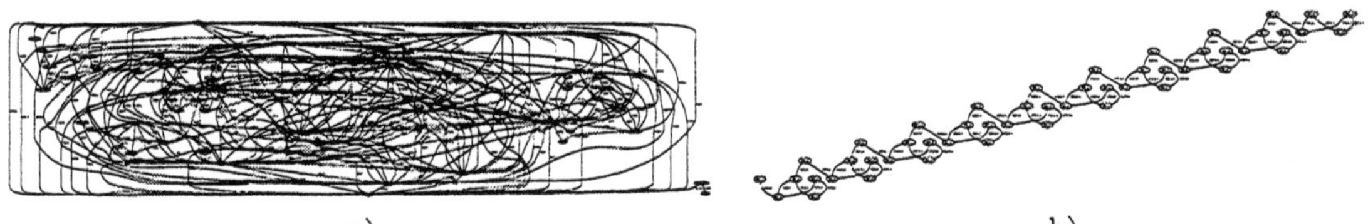

Fig. 1. PCG graphs for (a) the central locking example and (b) the transfer line (×10).

as we will see in the forthcoming pages. Notice that only backward reachability is needed (forward reachability can be completely ignored). The reachability searches turn however out to be the bottleneck of the algorithm. Therefore, in the remaining of this work, we only consider the performance of the reachability search.

3.1 Safe-State Supervision

Having computed the set of safe states Q^S, we can at any state $q_{current} \in Q^S$ compute the subset of enabled events:

$$\Sigma_{enabled} = \{ \sigma \mid \exists \langle \dot{q}^1, \cdots, \dot{q}^n \rangle \in Q^S. \forall_{1 \le i \le n}[\sigma \in \Sigma^i \wedge \delta^i(q^i_{current}, \sigma) = \dot{q}^i] \vee [\sigma \notin \Sigma^i \wedge q^i_{current} = \dot{q}^i] \}$$

This is equivalent to computing the control map $\Gamma : Q \rightarrow 2^\Sigma$. The control map can thus be computed on the fly, and without ever constructing the global transfer function δ. Since Q^S has often a very compact BDD representation ($10^2 - 10^4$ BDD nodes), our method also renders the need for supervisor reduction less critical.

4. COMMUNICATION COMPLEXITY IN AUTOMATA

In order to illustrate the complexity of a reachability search, we make use of *Process Communication Graphs*, adapted from (Aziz and Taziran, 1994). A PCG is a weighted undirected graph $PCG = \langle V, E, w \rangle$, where V is the set of vertices with a one to one mapping to a set of automata. $E \subseteq V \times V$, is the symmetric set of edges. Although not used in this paper, it should be mentioned that $w : E \rightarrow \mathbb{R}^+$ is a weight function that defines the intensity of *communication*. For Automata A and B, w may for example be defined as $w(A, B) = |\Sigma^A \cap \Sigma^B|$.

In Figure 1, process communication graphs for the central locking example[2] and the transfer line[3] ($TL(n)$ with $n = 10$ cells and 3 buffer places) are shown (for illustration, the are figures not meant to be read). If symbolic computation is used, the reachability search on the central locking turns out to be far harder (in terms of running time and memory usage) than the transfer line. A simple explanation is the high level of communication in the former model, as clearly seen in the figure.

We will also use the well known Dining Philosophers model, $DP(n)$ (Dijkstra, 1972). Another model that we will use is the *Pigeon Hole* model $PHP(n, m)$ based on the Pigeon Hole Principle[4], which despite the low number of reachable states has a very high communication complexity (refer to Tables 1 and 2).

The level-1 *dependency set* of an automaton A^x, denoted $D(A^x)$ is A^x itself plus the set of automata directly connected to A^x in its corresponding PCG[5]. The cardinality of such set equals $d(A^x) + 1$, that is, the degree of A^x plus one. We also define the *exclusive dependency set* $D^+(A^x) = D(A^x) - A^x$.

The level-2 *dependency set* of an automaton A^x, denoted $D^2(A^x)$ is defined as following:

$$D^2(A^x) = \{A^y | \exists A^z \in D(A^x).\ A^y \in D(A^z)\}$$

Assuming that some basic properties hold, for example all automata are minimized (and consequently all states in Q^x are reachable), the cardinality of the level-1 and level-2 dependency sets give often surprisingly good indications on how "hard" a reachability search will be. There are more accurate measures, but these two are easy to compute and precise enough for our purposes[6].

Statistics on the dependency sets for the discussed models is given in Table 1. To summarize, the central locking is conspired to be a "harder" model than the transfer line, despite the fact that it has far less reachable states (10^9 vs. 10^{21}). An illustrative way to explain this hardness is to refer

[2] This is a model of the alarm and locking system of a modern car. Unfortunately, no public references to this model exist.

[3] Used as a tutorial example in (Wonham, 2002).

[4] If n pigeons are inside m pigeon holes and $m < n$, then there must be at least one hole with more than one pigeon.

[5] To simplify the text, from this point no distinction is made between the automata and the corresponding PCG nodes.

[6] Here, we assume that the automata are "sound". Automata generated by random tend to have high dependency numbers but a very small number of reachable states, rendering them "easy" reachability search targets.

Table 1. Size of the level 1 and 2 dependency groups compared to the total number of automata.

	Central Locking	TL (10)	DP (25)	PHP $(n, n+1)$
Level1 min.	3%	4%	6%	53%
Level1 max.	51%	8%	6%	59%
Level1 avg.	19%	6%	6%	56%
Level2 min.	7%	8%	10%	100%
Level2 max.	67%	14%	10%	100%
Level2 avg.	35%	12%	10%	100%

to the average level 1 dependency value as a proof that one needs to do more work (i.e. consult a larger subset of the automata to verify that an event is fireable from a certain state) for each reachable state in the central locking model.

We also notice that bad design can easily be detected in a process communication graph. In Figure 2.a, the PCG for the Shoe Factory model [7] is shown. The model illustrates a clear and simple modular design of the plant. Figure 2.b on the other hand, is the same factory with the addition of a single automaton; a specification on how a single shoe is routed through the factory. Obviously, the new specification is designed poorly since it affects the communication complexity in such scale. In practice, supervisor synthesis of the shoe factory model with these type of specifications becomes intractable with as few as four shoes.

5. EFFICIENT SYNTHESIS BY EFFICIENT REACHABILITY SEARCH

A problem with composition of large sets of automata is that the total transfer function δ becomes extremely complex. If an algorithm uses a traditional state enumeration based approach, this is usually not a problem since the set of global states Q is considered to be a larger obstacle and the real bottleneck. However, if a symbolic approach is taken, the set of states is often compressed to such degree that representing the states is not a problem anymore (to some limit, of course). This does however not apply to δ, due to its complex structure.

Similar problems have been studied in the field of (symbolic) formal verification, where the corresponding transition relation T is often too large to be represented as a single monolithic relation. It has been suggested that by partitioning methods, one may split T into a set of less complex relations with a clear connection in between, e.g. $T = T^1 \otimes \cdots \otimes T^n$ (Burch *et al.*, 1991; Hu, 1995; Brezocnik *et al.*, 1996; Hojati *et al.*, 1996). In the following, we will show how this idea can be applied to the transfer function in a composite DES.

A simple *conjunctive partitioned* version of δ can be constructed by simple *event compensation*. This method will not be used in this work. Instead, we will construct a *disjunctive partitioned* version, denoted $\bar{\delta}^i$, such that

$$\delta(q, \sigma) = \bigcup_{A^i.\ \sigma \in \Sigma^i} \bar{\delta}^i(q, \sigma)$$

Assuming that $q = \langle q^1, \cdots, q^n \rangle$, let

$$\varsigma^{\hat{i},j}(q^j, \sigma) = \begin{cases} \delta^j(q^j, \sigma) & \text{if } \sigma \in \Sigma^i \cap \Sigma^j \\ q^j & \text{otherwise} \end{cases}$$

the disjunctive transfer function is

$$\bar{\delta}^i(q, \sigma) = \bigwedge_{A^j \in D(A^i)} \varsigma^{\hat{i},j}(q^j, \sigma) \wedge \underbrace{\bigwedge_{A^k \notin D(A^i)} q^k \leftrightarrow \dot{q}^k}_{\text{"keep"}}$$

Notice that the part marked by the word "keep" is not contributing to the complexity of the function and can in most cases be moved out of most equations. Notice further that it is trivial to remove all occurrences of σ from the function, transforming it into a much simpler transition relation $T^i \subseteq Q \times Q$. This representation is then used for more efficient symbolic computation.

Having a set of transfer functions/transition relations and the in between dependency map, we can now reformulate the reachability problem as the problem of calculating a *global fixpoint* of a set of functions. Of course, we could use a *monolithic* approach, that is, to compute the global fixpoint by applying all the transfer functions at the same time. But this requires building the global transfer function δ, which we wanted to avoid. Obviously, we need to find a way to compute the global fixpoint by calculating a series of dependent local fixpoints using the disjunctive transfer functions. This problem dates back to the classic work of the Polish logician Alfred Tarski in lattice theory (Tarski, 1955), and chaotic fixpoint computation (Cousot and Cousot, 1979). Based on these works and the static dependency map introduced by PCGs, we suggest a simple iteration strategy for reachability computation. Assuming that the average level-1 complexity is low, the disjunctive transfer function can be used for efficient reachability search by the *Workset* algorithm, which is inspired by the reachable marking vector exploration algorithm of Petri nets:

[7] The Shoe Factory model was inspired by Dawn Tilburys plenary talk at ACC 2002. Here we have used the model from (Huttunen and Härsjö, 2003).

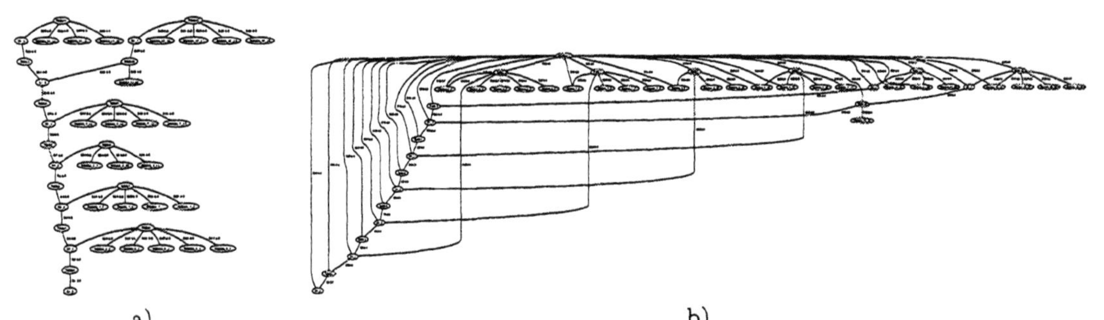

Fig. 2. The PCG for the shoe factory (a) plant only, (b) plant plus the specification for one shoe.

Algorithm 4. WorksetBackwardReachability
input : $Q_m, \Delta =$ set of $\tilde{\delta^i}$
let $W := \Delta$, $Q_0 := Q_m$, $K := 0$
repeat
  $\mathbb{H}$: Pick and remove a transition $\tilde{\delta^i} \in W$
  $k := k + 1$
  $Q_k = Q_{k-1} \cup BackwardReach(Q_{k-1}, \tilde{\delta^i})$
  if $Q_k \neq Q_{k-1}$ **then** $W := W \cup D^+(A^i)$
until $W = \emptyset$
return Q_k

Notice that we used *BackwardReach*, which is the "normal" reachability algorithm.

In the part marked by $\mathbb{H}$, a heuristic decision procedure is applied to choose the next automaton from the working set W. For simple problems with low level-1 dependency, we found a simple procedure based on *reinforcement learning* to be very reliable.

Table 2. Running times for reachability search by the *Workset* algorithm.

Model	States	Time [s]	
		Backward	Forward
ShoeF	2×10^{27}	2.7	3.3
ShoeF + 3 shoes	314783	5.4	5.1
DP(25)	3×10^{16}	0.8	0.9
DP(30)	6×10^{19}	1.1	1.1
TL(10)	1.2×10^{21}	0.4	0.3
TL(50)	2.3×10^{105}	59.0	39.9
TL(100)	5.3×10^{210}	617.3	342.4
PHP(6,7)	2.6×10^{5}	5.1	1.3
PHP(7,8)	4.8×10^{6}	89.4	27.2
PHP(8,9)	1×10^{8}	1129.3	445.8
Central Lock.	9.7×10^{8}	53.3	185.9

5.1 *Algorithm Efficiency*

In a monolithic search, a BDD operation related to the statement $[\cdots\ q \in Q_{k-1} \wedge \delta(q, \sigma) = \dot{q}\ \cdots]$ is carried out, which involves a binary operation between two BDDs representing δ (the global transfer function) and Q_{k-1} (intermediate set in the reachability search). We have already mentioned that δ has a very high complexity, this results in a BDD with a huge number of nodes (millions of nodes for medium sized set of automata). We have also mentioned the intermediate BDD size explosion problem. The plot labeled "Monolithic" in Figure 3 illustrates this problem for Q_{k-1} in the transfer line model. The above arguments leads to the fact that in a monolithic search, BDD operations with very large operands are carried out, making those operations slow and inefficient (recall the $O(|x| \times |y|)$ complexity).

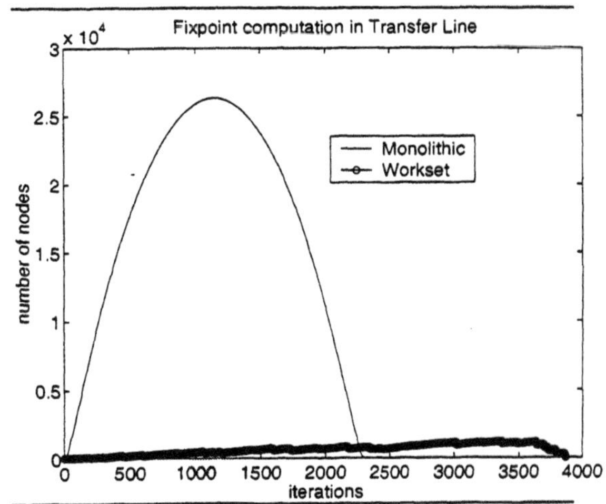

Fig. 3. Effect of the *Workset* algorithm on the intermediate BDDs, during reachability search in the transfer line model.

The *Workset* algorithm however, makes use of the disjunctive transfer functions $\tilde{\delta^i}$ which are in general orders of magnitudes smaller than δ. In addition, for reasons beyond the scope of this paper, the intermediate BDD peak have disappeared[8], resulting in much smaller Q_{k-1} (in terms of BDD nodes, see Figure 3) . Consequently, the *Workset* algorithm requires significantly less computational space and computation time is considerably shorter. This is clearly demonstrated in Table 2

[8] There are of course no guarantees that this always happens. The pattern of small intermediate BDDs has however been consistent in all cases we have studied so far.

[9]. With a traditional monolithic algorithm, only the reachable states of the PHP problem could be computed in reasonable time.

The size of the disjunctive transfer functions is directly dependent on the level-1 dependency. The number of iterations in the *Workset* algorithm is also dependent on this, since the exclusive dependency set D^+ is used in the algorithm. So, in general, the average level-1 dependency gives a good indication about the expected performance of the *Workset* algorithm.

6. SUMMARY AND CONCLUSIONS

We have suggested an alternative approach for supervisor synthesis where forward reachability search is not required, and the supervisor S is only represented as a subset of $Q^{P||Sp}$. Based on this approach, we have suggested an online control scheme. Neither the synthesis nor the control phase require building the global model of the system, $P||Sp$.

For this approach to be efficient, we have identified the reachability search to be the bottleneck and have proposed an algorithm for reachability search that performs significantly better than a monolithic algorithm. The algorithm is designed with symbolic computation in mind, and in order to predict its efficiency, we have suggested some measures that indicate the hardness of the reachability search.

Finally, we demonstrated the performance of the reachability algorithm on a set of academical and industrial examples. It is interesting to see that the hardness of many well-known benchmark examples do not scale as expected. Compare for example the forward search time for the transfer line with 5.3×10^{210} and the Central Locking model with less than 10^9 reachable states. Among the handmade/academical problems we studied so far, the Pigeon Hole is the only model that behaves more like an industrial problem in this regard. Therefore we encourage future use of PHP as a benchmark problem.

REFERENCES

Aziz, Adnan and Serdar Taziran (1994). BDD variable ordering for interacting finite state machines. In: *Proceedings of 31th Design Automation Conference*. pp. 283–288.

Brezocnik, Z., A. Casar and T. Kapus (1996). Efficient symbolic traversal algorithms using partitioned transition relations. In: *Proceedings of COST 247 International Workshop on Applied Formal Methods in System Design*. Slovenia. pp. 146–155.

Bryant, Randal E. (1992). Symbolic Boolean manipulation with ordered binary-decision diagrams. *ACM Computing Surveys* **24**(3), 293–318.

Burch, Jerry R., Edmund M. Clarke and David E. Long (1991). Symbolic model checking with partitioned transition relations. In: *Proceedings of 1991 Intl. Conf. on VLSI* (A. Halaas and P. B. Denyer, Eds.).

Cousot, P. and R. Cousot (1979). Constructive versions of Tarski's fixed point theorems. *Pacific Journal of Mathematics* **82**, 43–57.

Dijkstra, Edsger Wybe (1972). Hierarchical ordering of sequential processes. *Operating Systems Techniques* pp. 72–93.

Hoare, C.A.R. (1985). *Communicating Sequential Processes*. Prentice-Hall International Series in Computer Science.

Hoffmann, Gerard and Howard Wong-Toi (1992). Symbolic synthesis of supervisory controllers. In: *Proceedings of 1992 American Control Conference*. Chicago, IL, USA. pp. 2789–2793.

Hojati, R., S. Krishnan and R. Brayton (1996). Early quantification and partitioned transition relation. In: *Proceedings of IEEE International Conference on Computer Design*. pp. 12–19.

Hu, Alan John (1995). Techniques for Efficient Formal Verification Using Binary Decision Diagrams. PhD thesis. Carnegie Mellon University.

Huttunen, Arto and Michael Härsjö (2003). Control and simulation of a customer driven large-scale manufacturing system. Master thesis EX041/2003. Chalmers University of Technology, Sweden.

Tarski, Alfred (1955). A lattice-theoretical fixpoint theorem and its applications. *Pacific Journal of Mathematics* **5**, 285–309.

Vahidi, Arash, Bengt Lennartson and Martin Fabian (2003). Efficient supervisory synthesis with BDDs: Variable reduction, partitioning, and smoothing. *submitted to IEEE Transactions on Control Systems Technology*.

Vaz, Anthony Faria and William Murray Wonham (1985). On supervisor reduction in discrete-event systems. In: *Internation Journal of Control*. Vol. 44. pp. 475–491.

Wonham, William Murray (2002). *Notes on Control of Discrete-Event Systems*. University of Toronto.

[9] All computations were carried out on a standard desktop PC (1.8 GHz P4 with 512 MB memory of which only 128 MB was used) running Java JDK 1.4.2 on Microsoft Windows XP.

ELSEVIER
IFAC
PUBLICATIONS
www.elsevier.com/locate/ifac

APPROXIMATING THE MINIMAL SENSOR SELECTION FOR SUPERVISORY CONTROL

Samir Khuller [*] **Guy Kortsarz** [**]
Kurt R. Rohloff [***,1]

[*] *Department of Computer Science, University of Maryland College Park, MD 20742, USA*
samir@cs.umd.edu, www.cs.umd.edu/users/samir
[**] *Computer Science Department, Business and Science Building, Rutgers University, Camden, NJ, USA*
guyk@camden.rutgers.edu
[***] *Department of EECS, The University of Michigan 1301 Beal Avenue, Ann Arbor, MI 48109-2122, USA*
krohloff@eecs.umich.edu, www.eecs.umich.edu/umdes

Abstract: This paper discusses the problem of selecting a set of sensors of minimum cardinality that can be used for the synthesis of a supervisory controller. We show how this sensor selection problem is related to a type of directed graph *st*-cut problem that has not been previously discussed in the literature. Approximation algorithms to solve the sensor selection problem can be used to solve the graph cut problem and vice-versa. Polynomial time algorithms to find good approximate solutions to either problem most likely do not exist (under certain complexity assumptions), but a time efficient approximation algorithm is shown that solves a special case of these problems. *Copyright © 2004 IFAC*

Keywords: Discrete-Event Systems, Supervisory Control, Graph Theory, Approximation Algorithms, Observers, Computer-Aided Control System Design.

1. INTRODUCTION

When a controller operates on a system to match a specification, the controller may not need information from all sensors at its disposal. Therefore, for reasons of economy or simplicity, the control designer may desire the controller to use as few sensors as possible. There may be many sensor selections sufficient for the controller to match the specification, and the optimal selection may not be obvious. Unfortunately, the problem of finding this smallest cardinality sensor selection set is NP-complete (Yoo and Lafortune (2002)). Effective polynomial time approximation algorithms exist for many real-world NP-complete optimization problems (Ausiello et al. (1999)), and an approximation of this minimal solution may be acceptable. We explore the problem of approximating solutions to the sensor minimization problem.

In the next section the problem statement is formulated in the supervisory control framework of Lin and Wonham (1988) and necessary background information from computer science is given. We relate the sensor selection problem to a

[1] This paper reports on work commenced while the third author was under the direction of Prof. Jan H. van Schuppen at CWI in Amsterdam. Financial support in part for the investigation was made available by the European Commission through the project Control and Computation (IST-2001-33520) of the Information Society Technologies Program and by NSF grant CCR-0082784.

type of directed graph st-cut problem in Section 3 and we show some inapproximability results for our sensor selection and graph cutting problems in Section 4. We show a polynomial time approximation algorithm for a special case of the sensor selection and graph cutting problems in Section 5 and close with a brief discussion in Section 6.

2. THE SENSOR SELECTION PROBLEM AND APPROXIMATION ALGORITHMS

In our framework systems and specifications are modelled as the automata $G = (X^G, x_0^G, \Sigma, \delta^G)$ and $H = (X^H, x_0^H, \Sigma, \delta^H)$, respectively. The behavior generated by G is denoted by $\mathcal{L}(G)$ and the behavior generated by the controller S controlling G is denoted by $\mathcal{L}(S/G)$. The system S/G is said to match the specification H if $\mathcal{L}(S/G) = \mathcal{L}(H)$.

Controllable events ($\Sigma_c \subseteq \Sigma$) and observable events ($\Sigma_o \subseteq \Sigma$) are those events that can be respectively disabled or observed by the controller. Due to the controllability and observability theorem, Lin and Wonham (1988), there exists a controller S such that $\mathcal{L}(S/G) = \mathcal{L}(H)$ if and only if $\mathcal{L}(H)$ is controllable with respect to $\mathcal{L}(G)$ and $\Sigma \setminus \Sigma_c$ and $\mathcal{L}(H)$ is observable with respect to $\mathcal{L}(G)$, Σ_o and Σ_c. We assume that $\mathcal{L}(H)$ is always controllable. See Cassandras and Lafortune (1999) for a deeper introduction to supervisory control.

When designing a controller, the set of observable events can be chosen by selecting the appropriate respective sensors for those events. We call $\Sigma_o \subseteq \Sigma$ a *sufficient sensor selection* with respect to G, H and Σ_c if $\mathcal{L}(H)$ is observable with respect to $\mathcal{L}(G)$, Σ_o and Σ_c. We call the problem of finding a minimal cardinality set of observable events the *sensor selection problem.*

Problem 1. Sensor Selection: Given G, H and $\Sigma_c \subseteq \Sigma$, find a sufficient sensor selection Σ_o^{min} such that for any other sufficient sensor selection Σ_o, $|\Sigma_o^{min}| \leq |\Sigma_o|$.

For a survey of literature relevant to Problem 1, please consult Rohloff (2004).

Despite the NP-completeness of Problem 1 (Yoo and Lafortune (2002)), we may still be required to find a sufficient sensor selection Σ_o whose cardinality is approximately $|\Sigma_o^{min}|$. Fortunately, some NP-complete minimization problems have fairly accurate polynomial time approximation algorithms. This means sufficient and approximate solutions can be found for many difficult problems. However, not all NP-complete minimization problems are believed to have this property. We further discuss these approximation properties for Problem 1.

To better quantify what we mean by an approximation to Problem 1, suppose P is the set of instances of Problem 1. Let $p \in P$ be a specific problem instance corresponding to G, H and Σ_c. Suppose $\Sigma_o^{min}(p)$ is the solution of this problem instance and A is an algorithm that when given input p, returns $\Sigma_o^A(p)$ such that $\mathcal{L}(H)$ is observable with respect to $\mathcal{L}(G)$, $\Sigma_o^A(p)$ and Σ_c. We measure the utility of the approximation $\Sigma_o^A(p)$ with the ratio $|\Sigma_o^A(p)|/|\Sigma_o^{min}(p)|$. Problem 1 has an r-*approximation* if $\forall p \in P, |\Sigma_o^A(p)|/|\Sigma_o^{min}(p)| \leq r$. This r-approximation notation also holds for other approximation problems. For a deeper discussion of these topics, see Ausiello et al. (1999).

Other necessary background from computer science is that $DTIME(f(n))$ defines the set of all problems that can be solved by deterministic algorithms with time complexity in $O(f(n))$. Using standard computer science notation, n is the size of the encoding of the problem instance. Of interest is the class $DTIME(n^{\text{polylog } n})$. It is believed that $NP \not\subseteq DTIME(n^{\text{polylog } n})$, but this has not been proved (Arora and Lund (1997)).

3. THE GRAPH CUTTING PROBLEM

We now show how the sensor selection problem is related to a special type of directed graph st-cut problem. Examples of the constructions in this section are given in Rohloff (2004).

Suppose we are given an edge-colored directed graph $D = (V, A, C)$ where V is a set of vertices, $A \subseteq V \times V$ are directed edges, $C = \{c_1, \ldots, c_p\}$ is the set of colors and for $s, t \in V$, there is a path of directed edges from s to t. Each edge is assigned a color in C and let A_i be the edges having color c_i. Given $I \subseteq C$, let $A_I = \cup_{c_i \in I} A_i$. I is a colored st-cut if $(V, A \setminus A_I, C)$ has no path from s to t. This prompts us to define the colored cut problem.

Problem 2. Minimal Colored Cut: Given an edge colored directed graph $D = (V, A, C)$ and two vertices, $s, t \in V$, find a colored st-cut $I^{min} \subseteq C$ such that for any other colored st-cut $I \subseteq C$, $|I^{min}| \leq |I|$.

We now convert an instance of a colored cut problem to an instance of a sensor selection problem. Suppose we are given the edge colored directed graph $D = (V, A, C)$ and two vertices s, t. We first construct a system G.

If $C = \{c_1, \ldots, c_p\}$, let there be a set of events $\{\sigma_1, \ldots, \sigma_p\}$ such that for every color c_i, there is a corresponding event σ_i. Let γ be another event and define $\Sigma = \{\sigma_1, \ldots, \sigma_p, \gamma\}$. Define $X^G = V \cup \{s', s'', t'\}$ where s', s'', t' are states not in V. Let $x_0^G = s$. To define the state transition function, let v_1, v_2 be any vertices except s. To define δ^G, if $(v_1, v_2) \in A_i$, then $\delta^G(v_1, \sigma_i) = v_2$. If $(s, v_2) \in A_i$, then $\delta^G(s, \sigma_i) = v_2$ and $\delta^G(s'', \sigma_i) = v_2$. If $(v_1, s) \in A_i$, then $\delta^G(v_1, \sigma_i) = s''$. For simplicity we assume that $(s, s) \notin A$. Also, define $\delta^G(s, \gamma) = s'$ and $\delta^G(t, \gamma) = t'$. Let H be a copy of G except that $\delta^H(t, \gamma)$ is undefined. Let $\Sigma_c = \{\gamma\}$.

In the system above, γ must be enabled at s and be disabled at t. There is a control conflict if there is a path in G from s to t where no event is observed. Therefore, as system behavior progresses, if any event is observed, then γ can be disabled. Hence, a set of colors $I = \{c_a, \ldots, c_z\}$ is a colored cut for D if and only if selecting sensors for $\{\sigma_a, \ldots, \sigma_z\}$ makes the system observable. Therefore any approximation algorithm for the sensor selection problem can also be used with the same absolute effectiveness for the colored cut problem.

We now show the converse by converting an instance of Problem 1 to an instance of Problem 2. We start with a modified $\mathcal{M}_{\Sigma_o}$-machine method for testing observability (Tsitsiklis (1989)). Suppose we are given $H = (X^H, x_0^H, \Sigma, \delta^H)$, $G = (X^G, x_0^G, \Sigma, \delta^G)$, Σ_o and Σ_c such that we wish to test if $\mathcal{L}(H)$ is observable with respect to $\mathcal{L}(G)$, Σ_o and Σ_c. Let Σ' be a copy of the event set Σ where for every event $\sigma \in \Sigma$, there is a corresponding event $\sigma' \in \Sigma'$. We can now give the following definition: $\mathcal{M}_{\Sigma_o} = (X^{\mathcal{M}_{\Sigma_o}}, x_0^{\mathcal{M}_{\Sigma_o}}, \Sigma^{\mathcal{M}_{\Sigma_o}}, \delta^{\mathcal{M}_{\Sigma_o}})$ where $X^{\mathcal{M}_{\Sigma_o}} := X^H \times X^H \times X^G \cup \{d\}, x_0^{\mathcal{M}_{\Sigma_o}} := (x_0^H, x_0^H, x_0^G), \Sigma^{\mathcal{M}_{\Sigma_o}} := \Sigma \cup \Sigma'$. Let us define the set of conditions at state (x_1, x_2, x_3) that we call the $(*)$ conditions.

$$\left.\begin{array}{l} \delta^H(x_1, \sigma) \text{ is defined if } \sigma \in \Sigma_c \\ \delta^H(x_2, \sigma) \text{ is not defined} \\ \delta^G(x_3, \sigma) \text{ is defined} \end{array}\right\} \quad (*)$$

The nondeterministic transition function $\delta^{\mathcal{M}_{\Sigma_o}}$ is defined as follows.
For $\sigma \notin \Sigma_o$ and its Σ' equivalent, σ',

$$\delta^{\mathcal{M}_{\Sigma_o}}((x_1, x_2, x_3), \sigma') = \left\{ \begin{array}{l} (\delta^H(x_1, \sigma), x_2, x_3) \\ (x_1, \delta^H(x_2, \sigma), \delta^G(x_3, \sigma)) \end{array} \right\}$$

For $\sigma \in \Sigma$,

$$\delta^{\mathcal{M}_{\Sigma_o}}((x_1, x_2, x_3), \sigma) = \left\{ \begin{array}{l} (\delta^H(x_1, \sigma), \delta^H(x_2, \sigma), \delta^G(x_3, \sigma)) \\ d \quad \text{if } (*) \end{array} \right\}$$

For $\sigma \in \Sigma$, $\delta^{\mathcal{M}_{\Sigma_o}}(d, \sigma)$ is undefined, and $\delta^{\mathcal{M}_{\Sigma_o}}$ can be extended in the usual manner to be defined over strings of events.

The state d is reachable in $\mathcal{M}_{\Sigma_o}$ if and only if $\mathcal{L}(H)$ is observable with respect to $\mathcal{L}(G)$, Σ_o and Σ_c. The $\mathcal{M}_{\Sigma_o}$-machine here is modified from the original in Tsitsiklis (1989) in that Σ' transitions replace some Σ transitions.

Effectively $\mathcal{M}_{\Sigma_o}$ is a nondeterministic simulation of an observer's estimate of a system's behavior with respect to a specification. In a state (x_1, x_2, x_3) of $\mathcal{M}_{\Sigma_o}$, the first state is an observer's estimate of the specification state and the second and third states are the true states of the specification and system, respectively. In $\mathcal{M}_{\Sigma_o}$ a Σ transition occurs if an event occurrence in H and G is correctly predicted by the observer. A Σ' transition occurs if the prediction is not correct. Therefore, if an event is observed, it is predicted correctly and σ' transitions in $\mathcal{M}_{\Sigma_o}$ would be removed if σ were made observable. This implies that $\mathcal{M}_{\Sigma_o \cup \{\sigma\}}$ can be constructed from $\mathcal{M}_{\Sigma_o}$ by cutting all σ' transitions, and conversely, cutting all occurrences of σ' transitions in $\mathcal{M}_{\Sigma_o}$ corresponds to adding σ to Σ_o.

This realization shows that the sensor selection problem is really a type of colored cut problem. Suppose we consider $\mathcal{M}_\emptyset$ to be a colored directed graph as introduced above where the transition labels are really colors. A colored $x_0^{\mathcal{M}_\emptyset}$*d*-cut for $\mathcal{M}_\emptyset$ where we only select Σ' transitions for cutting corresponds to a sufficient sensor selection for observability to hold. This prompts the following proposition.

Proposition 3. $\mathcal{L}(H)$ is observable with respect to $\mathcal{L}(G)$, Σ_o and Σ_c if and only if $\Sigma'_o \subseteq \Sigma'$ is a colored $x_0^{\mathcal{M}_\emptyset}$*d*-cut $\mathcal{M}_\emptyset$.

However, the $\mathcal{M}_\emptyset$ cut problem is not in the same form as in Problem 2 because we are not allowed to cut Σ transitions from $\mathcal{M}_\emptyset$ in Proposition 3. To counter this difference, we use the following construction which performs a form of state condensation and hides the Σ transitions in $\mathcal{M}_\emptyset$.

Construct $\mathcal{M}_{\Sigma_o}$ from H, G, Σ_c and Σ_o. Define $X_x^{\mathcal{M}_{\Sigma_o}} = \delta^{\mathcal{M}_{\Sigma_o}}(x^{\mathcal{M}_{\Sigma_o}}, t)$ which represents all states that could be reached from $x^{\mathcal{M}_{\Sigma_o}}$ in $\mathcal{M}_{\Sigma_o}$ if only Σ transitions were allowed. The states in $X_x^{\mathcal{M}_{\Sigma_o}}$ would be reachable from $x^{\mathcal{M}_{\Sigma_o}}$ in $\mathcal{M}_{\Sigma_o}$ no matter what events were added to the observability set because only Σ'_o transitions can be cut. With this in mind, let us build the nondeterministic machine, $\tilde{\mathcal{M}}_{\Sigma_o}$ from $\mathcal{M}_{\Sigma_o}$. We assume that $d \notin X_{x_0}^{\mathcal{M}_{\Sigma_o}}$. Let $\tilde{\mathcal{M}}_{\Sigma_o} = (X^{\tilde{\mathcal{M}}_{\Sigma_o}}, x_0^{\tilde{\mathcal{M}}_{\Sigma_o}}, \Sigma^{\tilde{\mathcal{M}}_{\Sigma_o}}, \delta^{\tilde{\mathcal{M}}_{\Sigma_o}})$, where $X^{\tilde{\mathcal{M}}_{\Sigma_o}} := X^H \times X^H \times X^G \cup \{d\}, x_0^{\tilde{\mathcal{M}}_{\Sigma_o}} := (x_0^H, x_0^H, x_0^G)$ and $\Sigma^{\tilde{\mathcal{M}}_{\Sigma_o}} := \Sigma$.

The transition relation $\delta^{\tilde{\mathcal{M}}_{\Sigma_o}}$ is defined as follows. Suppose there exists $x^{\mathcal{M}_{\Sigma_o}}, y^{\mathcal{M}_{\Sigma_o}}, z^{\mathcal{M}_{\Sigma_o}} \in X^{\mathcal{M}_{\Sigma_o}}$ and $\sigma \in \Sigma$ such that $z^{\mathcal{M}_{\Sigma_o}} \in X_x^{\mathcal{M}_{\Sigma_o}}$, $\delta^{\mathcal{M}_{\Sigma_o}}(z^{\mathcal{M}_{\Sigma_o}}, \sigma') = y^{\mathcal{M}_{\Sigma_o}}$.

$$\delta^{\tilde{\mathcal{M}}_{\Sigma_o}}(x^{\mathcal{M}_{\Sigma_o}}, \sigma) = \begin{cases} y^{\mathcal{M}_{\Sigma_o}} & \text{if } d \notin X_y^{\mathcal{M}_{\Sigma_o}} \\ d & \text{if } d \in X_y^{\mathcal{M}_{\Sigma_o}} \end{cases}$$

Note that $\tilde{\mathcal{M}}_{\Sigma_o}$ is effectively a colored directed graph, with states as vertices, transitions as directed edges and transition labels for colors. This prompts one of the main results of this paper.

Theorem 4. Given a $\tilde{\mathcal{M}}_{\emptyset}$ machine ultimately constructed from H, G, Σ_c and $\emptyset$ as the set of observable events, $\mathcal{L}(H)$ is observable with respect to $\mathcal{L}(G)$, Σ_o and Σ_c if and only if Σ_o is a colored $x_0^{\tilde{\mathcal{M}}_{\Sigma_o}}$ d-cut in the colored directed graph $\tilde{\mathcal{M}}_{\emptyset}$.

With these conversions, any methods developed to approximate solutions to the colored cut problem could also be used to produce approximate solutions to the sensor selection problem and vice-versa.

4. INAPPROXIMABILITY RESULTS

To our knowledge the graph cutting problem has not been explored in the standard literature (from graph theory or computer science). Unfortunately, although many other types of graph cutting problems are computationally simple, it is shown here that solutions to Problem 2 are most likely difficult to approximate. Because of the above results, solutions to the sensor selection problem are similarly difficult to approximate.

Consider a bipartite graph $K = (V_1, V_2, E)$ where the sets V_1 and V_2 are partitioned into a disjoint union of q sets, $V_1 = \cup_{i=1}^{q} Y_i$ and $V_2 = \cup_{j=1}^{q} Z_i$. The sets $\{Y_1, Z_1, \ldots, Y_q, Z_q\}$ all have size N. E is the set of edges in K. The bipartite graph and the partitions of V_1 and V_2 induce a super-graph $\mathcal{H}$ where the vertices of $\mathcal{H}$ are the sets $\{Y_1, Z_1, \ldots, Y_q, Z_q\}$. Y_i and Z_j are connected by a (super)edge in $\mathcal{H}$ if and only if there exists $(a, b) \in Y_i \times Z_j$ that are adjacent in K. For our purposes, it is convenient to assume that $\mathcal{H}$ is d-regular.

Given $X \subseteq V_1 \cup V_2$, we say that the super-edge (Y_i, Z_j) in $\mathcal{H}$ is *covered* by X if there exists two nodes $a \in X \cap Y_i$ and $b \in X \cap Z_j$ such that $(a, b) \in E$. We can now formally introduce the MIN-REP problem.

Problem 5. MIN-REP: Given a bipartite graph K as introduced above, find a subset X of minimum size covering all superedges of $\mathcal{H}$.

The subset $X \cap Y_i$ is referred to as the *representatives* of Y_i in X (similarly for Z_j). We say that X is a set of *unique representatives* if $|X \cap Y_i| = 1$ and $|X \cap Z_j| = 1$ for all i, j.

In addition we will need is the following result from Kortsarz (1998) that shows the approximation difficulty of Problem 5. Consider the following theorem based on the well-known satisfiability problem (SAT) from computer science.

Theorem 6. Raz (1998). Let T be an instance of SAT. For any $0 < \epsilon < 1$, there exists a (quasi-polynomial) reduction of T to K, an instance of MIN-REP with n vertices so that if T is satisfiable, then there exists a set of unique representatives which cover all super-edges. If T is not satisfiable, then the size of any MIN-REP solution has at least $2q2^{\log^{(1-\epsilon)} n}$ vertices.

In the reduction in Theorem 6, if T is not satisfiable, then the average number of representatives needed per supernode is $\Omega(2^{\log^{(1-\epsilon)} n})$. Also, n is quasi-polynomial in the size of the SAT formula. This is used in Kortsarz (1998) to show the approximation difficulty of the MIN-REP problem.

Theorem 7. Kortsarz (1998). The MIN-REP problem admits no $2^{\log^{(1-\epsilon)} n}$ approximation for any $\epsilon > 0$ unless $NP \subseteq DTIME(n^{\text{polylog } n})$.

We can now show a conversion of the MIN-REP problem, $K = (V_1, V_2, E)$, to the colored cut problem, $D = (V, A, C)$ that establishes a similar connection between SAT and the colored cut problem and ultimately the sensor selection problem. For simplicity, we assume that D may be a multigraph, but by using a standard trick of subdividing edges, we can remove this assumption. We construct D in three parts.

First we construct the vertices V of D. Initially assign $V := V_1 \cup V_2$. For the edges $\{e_{ij}^1, \ldots, e_{ij}^b\}$ where $e_{ij}^k = (y^k, z^k)$ and $y^k \in Y_i$ and $z^k \in Z_j$, add a vertices $\{x_{ij}^1, \ldots, x_{ij}^b\}$ to V. For each superedge $e_{ij} = (Y_i, Z_j)$ in $\mathcal{H}$, add a vertex x_{ij}^{end} to V. Finally, add nodes s and t to V.

We now construct the edges A. Initially assign $A := E$, with all edges directed from the V_1 vertices to the V_2 vertices. Also add directed edges from s to all V_1 vertices. For some superedge $e_{ij} = (Y_i, Z_j)$, arbitrarily order the edges between Y_i and Z_j, $\{e_{ij}^1, \ldots, e_{ij}^b\}$. For each vertex in $z \in Z_j$, add an edge from z to x_{ij}^1. Add two parallel edges from x_{ij}^1 to x_{ij}^2, from x_{ij}^2 to x_{ij}^3 and so on until x_{ij}^b is reached. Then, add an edge from x_{ij}^b to x_{ij}^{end}. We call this construction of edges from the vertices in Z_j to x_{ij}^{end} the *chain* of superedge e_{ij}. Let X_{end} be the set of all x_{ij}^{end} vertices. Add additional edges from all of the X_{end} vertices to t. Observe that there is one chain associated with Z_j for every Y_i adjacent to it.

We finally color the edges of D. For each $y \in Y_i$, $z \in Z_j$ let there be colors c_y and c_z. For the paired edges $e_{ij}^k = (y^k, z^k)$ used to construct the chains, color one of the edges (x_{ij}^k, x_{ij}^{k+1}) in the chain c_{y^k} and the other c_{z^k}. The edges touching s and t are colored new distinct colors along with the edges originally in E.

Make the edges from s into V_1, from X_{end} into t and the edges of E *hard to cut* as follows. Suppose there are m vertices in the graph before edges are made hard to cut. An edge (x, x') can be made hard to cut by adding vertices $\{r_{xx'}^1, \ldots, r_{xx'}^{m^3}\}$. An edge is added from x to each $r_{xx'}^i$, and an edge is added from each $r_{xx'}^i$ to x'. Each of the new $2m^3$ edges is given a new distinct color. Now, cutting edge (x, x') corresponds to adding a very large number of new colors that also need to be cut, effectively making the cutting of (x, x') impossible in the solution.

Now that we have shown this construction, we discuss why this shows the colored cut problem is difficult to approximate. Consider a super-edge $e_{ij} = (Y_i, Z_j)$ and its chain. Let I be any feasible colored st-cut of D such that I does not contain colors for hard to cut edges.

Claim 8. There exists at least one parallel pair of edges in the chain corresponding to e_{ij} so that I contains both colors corresponding to this parallel pair of edges.

PROOF. Suppose that for each pair of parallel edges, at most one color is in I. This implies that even with I colored edges, there is a path from all Z_j vertices to t. All edges on paths to the Z_j from the vertices from s are hard to cut as are all paths from the X_{end} vertices to t. Therefore, if there is not a pair of parallel edges both cut in every chain, then I is not an st-cut. □

Let I be a colored st-cut for the colored cut problem constructed above.

Corollary 9. Let X be the vertices corresponding to the colored cut I. X is a feasible solution to the MIN-REP problem and $|X| = |I|$.

PROOF. Consider a superedge (Y_i, Z_j). There is a parallel pair of edges such that its colors are in I by Claim 8. The color of each edge corresponds to some vertex $y \in Y_i$ and $z \in Z_j$ such that $(y, z) \in E$. Hence, I defines in a natural way a collection $X \subseteq V$ that is a solution for MIN-REP. Therefore, the X collection covers all the superedges. Furthermore, the number of colors equals the number of vertices from X. □

Claim 10. Given any MIN-REP solution X for K, we can find a subset I of $|X|$ colors that solves the colored cut problem constructed from K.

PROOF. Given that $x \in X$, this defines a color c_x corresponding to x in the construction. Let I be the collection of all these colors. We show that I is a feasible solution for D.

Let $e_{ij} = (Y_i, Z_j)$. Consider the chain of e_{ij}. As e_{ij} is covered by X, there are $y \in X \cap Y_i$ and $z \in Z_j$ so that $(y, z) \in E$. By definition, there is a parallel pair of edges on this chain from x_{ij}^k corresponding to (y, z) and its colors are in I. The cutting of edges colored c_y and c_z cuts both of these parallel edges from D. It follows that there cannot be a path through the chain of e_{ij} to t. Since all superedges are covered by X, it follows that s cannot reach t after the edges colored by I are removed. Furthermore, $|I| = |X|$. □

The following is thus immediate.

Theorem 11. Let T be an instance of SAT. For any $0 < \epsilon < 1$, there exists a (quasi-polynomial) reduction of T to D, an instance of colored cut with n vertices so that if T is satisfiable, then there exists a set of $2q$ colors forming a feasible solution to the colored cut problem. If T is not satisfiable, then any colored cut solution has at least $2q2^{\log^{(1-\epsilon)} n}$ colors.

Corollary 12. The colored cut problem admits no $2^{\log^{(1-\epsilon)} n}$ approximation for any $\epsilon > 0$ unless $NP \subseteq DTIME(n^{\text{polylog } n})$.

Corollary 13. The sensor selection problem admits no $2^{\log^{(1-\epsilon)} n}$ approximation for any $\epsilon > 0$ unless $NP \subseteq DTIME(n^{\text{polylog } n})$.

It follows that if we can approximate solutions to either the colored cut or sensor selection problems with better than a $2^{\log^{(1-\epsilon)} n}$-approximation, then we have found a method for solving NP-complete problems in quasipolynomial time. This lower bound is generally considered to be a very poor lower bound in the computer science community. Indeed, as ϵ approaches 0, then $2^{\log^{(1-\epsilon)} n}$ approaches n.

5. ALGORITHM FOR A SPECIAL CASE

Now that we have shown that colored cuts and sensor selections are in a sense difficult to approximate, we show an approximation method for a special case with a reasonable approximation bound. Suppose we have an instance of the colored cut problem $D = (V, A, C)$ with vertices $s, t \in V$

such that there are no parallel edges in the same direction. K is a prespecified value which we will optimize in the following proof.

Algorithm 1.
Input: $D = (V, A, C), s, t \in V$ and K, an integer.
Output: A colored st-cut I for D.
Initialize: $I := \emptyset$, a set of colors.
As long as there is a st path of length at most K, add the colors associated with this path to I and remove all edges in D with colors in I.
After all paths of length at most K have been removed, let k be the distance from s to t.
Perform a depth-first-search to convert D to a layered directed graph D' of depth k.
Let V_i denote the set of vertices at distance i from s in D' and let A_i denote the the set of edges from V_i to V_{i+1}.
Select the set of edges A_j that uses the least number of colors and add those colors to I.
Return: I.

Theorem 14. Algorithm 1 gives a $|V|^{2/3}$ approximation for colored cut problems where D contains no parallel edges in the same direction.

PROOF. Let I^{min} represent the minimal colored cut. Suppose in the first phase of the algorithm, paths $\{P_1, \ldots, P_l\}$ are removed. At most lK colors are chosen during this step because there are at most K colors per path. The colors used by the paths are disjoint, but the optimal solutions needs to remove at most one color for each path. Hence, $l \leq |I^{min}|$. If $\{I_1, \ldots, I_l\}$ are the colors removed by each path, then $\Sigma_{i=1}^{l}|I_i| \leq K|I^{min}|$.

Let I_{A_j} represent the colors cut in the last part of the algorithm. Also, let $|A_p| = \min_i |A_i|$. It is known that $\Sigma_{i=1}^{k}|V_i| < N$ where k is the distance from s to t, $k \geq K$. Also, $A_i \leq |V_i||V_{i+1}|$. Therefore, $|I_{A_j}| \leq |I_{A_p}| \leq |A_p| \leq \left(\frac{N}{K}\right)^2$.

The optimal solution has at least one color from this cut that is different from the colors that have been added to I. $I = \Sigma_{i=1}^{l}|I_i| + |I_{A_j}| \leq K|I^{min}| + \left(\frac{N}{K}\right)^2$. By choosing $K = N^{2/3}$, we get a $N^{2/3}$ approximation. □

When a sensor selection problem is converted to a colored cut problem using the conversion outlined above, methods developed to solved the colored cut problem can be used to solve the sensor selection problem. In Algorithm 1, the restriction that D contains no parallel edges in the same direction is generally fairly restrictive for solving many interesting sensor selection problems. However, this method is still interesting and relevant from a graph theoretic and computer science point of view. We hope that by presenting this algorithm we might spur further research into approximation algorithms for other special cases of the colored cut and sensor selection problems. Our conversion of the sensor selection problem to the colored cut problem is helpful for analysis of the sensor selection problem as graph cutting problems are much more intuitive and this would helpfully aid further research on these problems.

6. DISCUSSION

We have shown how the theory of approximation algorithms can be applied in a formal manner for the analysis of the sensor selection problem in supervisory control. Although not discussed here, our contribution of using graph cutting methods to perform sensor selections allows us to force some events to be observable or unobservable depending on the construction of $\mathcal{M}_{\Sigma_o}$. Beyond the sensor selection problem, our methods could also be directly applied for the analysis of the corresponding actuator selection problem. There are also extensions to decentralized control and hybrid systems.

REFERENCES

S. Arora and C. Lund. Hardness of approximations. In D.S. Hochbaum, editor, *Algorithms for NP-Hard Problems*, pages 46–93. PWS Publishing, 1997.

G. Ausiello, P. Crescenzi, G. Gambosi, V. Kahn, A. Marchetti-Spaccamela, and M. Protasi. *Complexity and Approximation Combinatorial Optimization Problems and Their Approximability Properties.* Springer-Verlag, Berlin Heidelberg, 1999.

C.G. Cassandras and S. Lafortune. *Introduction to Discrete Event Systems.* Kluwer Academic Publishers, Boston, MA, 1999.

G. Kortsarz. On the hardness of approximating spanners. In *The First International Workshop on Approximation*, pages 135–146, 1998.

F. Lin and W. M. Wonham. On observability of discrete-event systems. *Information Sciences*, 44:173–198, 1988.

R. Raz. A parallel repetition theorem. *SIAM Journal of Computing*, 27(3):763–803, 1998.

K. Rohloff. *Computations on Distributed Discrete-Event Systems.* PhD thesis, The University of Michigan, Ann Arbor, 2004. Available at `http://www.eecs.umich.edu/umdes`.

J. Tsitsiklis. On the control of discrete-event dynamical systems. *Mathematics of Control, Signals and Systems*, 2:95–107, 1989.

T.-S. Yoo and S. Lafortune. NP-completeness of sensor selection problems arising in partially-observed discrete-event systems. *IEEE Trans. Auto. Contr.*, 47(9):1495–1499, 2002.

ELSEVIER
IFAC
PUBLICATIONS
www.elsevier.com/locate/ifac

ON SOLVABILITY OF AN AGENT-BASED CONTROL PROBLEM UNDER DYNAMIC ENVIRONMENT

Kunihiko Hiraishi *

* *School of Information Science,*
Japan Advanced Institute of Science and Technology
E-mail: hira@jaist.ac.jp

Abstract: In previous papers, we proposed a Petri-net-based formalism for agent-based control of discrete event systems. On this formalism, a control synthesis problem is shown to be undecidable. We discover reasonable assumptions on the model so that the problem becomes solvable. *Copyright © 2004 IFAC*

Keywords: discrete event systems, supervisory control, decentralized control

1. INTRODUCTION

Decentralized control of discrete events systems becomes an important research subject in the DES control theory. In previous papers[3,4], we proposed a new formalism together with a graphical notation for decentralized control of discrete event systems. The systems to be modeled are assumed to have the following properties: (i) The system is controlled by several control agents, each of which can observe and control only a subset of events. (ii) Control agents can communicate with each other. (iii) No global clock exists and all events in the system occur asynchronously. (iv) The set of controllable/observable events may vary dynamically in the environment. Communication links may also change. On these assumptions, a control synthesis problem is defined, and is shown to be undecidable in general. In this paper, we discover reasonable assumptions on the controllers and the communication model so that the problem becomes solvable.

2. PRELIMINARIES

A (deterministic) transition system is a 4 tuple $\mathcal{T} = (Q, \Sigma, \delta, q_0)$, where Q is the set of states, Σ is the set of events, $\delta : Q \times \Sigma \to Q$ is the transition function, and $q_0 \in Q$ is the initial state. $\mathcal{T}$ is called *finite* if Q is finite. The state transition function δ is defined as a partial function and we will write $\delta(q, \sigma)!$ to indicate that $\delta(q, \sigma)$ is defined. As usual, the domain of δ is extended to $Q \times \Sigma^*$. The set of trajectories generated by $\mathcal{T}$ is $L(\mathcal{T}) := \{s \in \Sigma^* \mid \delta(q_0, s)!\}$.

The accessible part of a transition system $\mathcal{T}$ is the transition system obtained by deleting all states which are not reachable from the initial state, and is denoted by $Acc(\mathcal{T})$. A transition system $\mathcal{T}$ is called *accessible* if $\mathcal{T} = Acc(\mathcal{T})$. Given two transition systems $\mathcal{T}_i = (Q_i, \Sigma_i, \delta_i, q_{0i})$, $i = 1, 2$, *parallel composition* of $\mathcal{T}_1$ and $\mathcal{T}_2$, denoted by $\mathcal{T}_1 \| \mathcal{T}_2$, as the accessible part of a transition system $(Q, \Sigma_1 \cup \Sigma_2, \delta, q_0)$, where $Q = Q_1 \times Q_2$, $\Sigma = \Sigma_1 \cup \Sigma_2$, $q_0 = (q_{01}, q_{02})$, and for each $q = (q_1, q_2) \in Q$ and $\sigma \in \Sigma$: $\delta(q, \sigma) := (\delta_1(q_1, \sigma), \delta_2(q_2, \sigma))$ if $\delta_i(q_i, \sigma)!(i = 1, 2)$ and $\sigma \in \Sigma_1 \cap \Sigma_2$; $(\delta_1(q_1, \sigma), q_2)$ if $\delta_1(q_1, \sigma)!$ and $\sigma \in \Sigma_1 - \Sigma_2$; $(q_1, \delta_2(q_2, \sigma))$ if $\delta_2(q_2, \sigma)!$ and $\sigma \in \Sigma_2 - \Sigma_1$; undefined otherwise.

We introduce Σ-*bisimilarity* on the set of transition systems. Intuitively, two transition systems are Σ-bisimilar if they show the same behavior with respect to occurrences of events in Σ. Let

$\mathcal{T}_i = (Q_i, \Sigma_i, \delta_i, q_{0i})$, $i = 1, 2$ be two transition systems and let $\Sigma \subseteq \Sigma_1 \cap \Sigma_2$. For $q, q' \in Q_i$ and $\sigma \in \Sigma$, we write $q \Rightarrow_{\mathcal{T}_i,\Sigma} q'$ if $\exists u \in (\Sigma_i - \Sigma)^* : q' = \delta_i(q, u)$, and write $q \overset{\sigma}{\Rightarrow}_{\mathcal{T}_i,\Sigma} q'$ if $\exists u, v \in (\Sigma_i - \Sigma)^* : q' = \delta_i(q, u\sigma v)$. Let R be a binary relation over $Q_1 \times Q_2$. We say that $\mathcal{T}_1$ *Σ-simulates* $\mathcal{T}_2$ for R if: (i) $q_{01} R q_{02}$, (ii) for each $q, q' \in Q_2$ and $\sigma \in \Sigma_2 - \Sigma$: if $\delta_2(q, \sigma) = q'$ and pRq, then there exists $p' \in Q_1$ such that $p \Rightarrow_{\mathcal{T}_1,\Sigma} p'$ and $p'Rq'$, (iii) for each $q, q' \in Q_2$ and $\sigma \in \Sigma$: if $\delta_2(q, \sigma) = q'$ and pRq, then there exists $p' \in Q_1$ such that $p \overset{\sigma}{\Rightarrow}_{\mathcal{T}_1,\Sigma} p'$ and $p'Rq'$. We say that $\mathcal{T}_1$ and $\mathcal{T}_2$ are *Σ-bisimilar* for R if $\mathcal{T}_1$ Σ-simulates $\mathcal{T}_2$ for R and $\mathcal{T}_2$ Σ-simulates $\mathcal{T}_1$ for R^{-1}.

3. AN AGENT-BASED CONTROL PROBLEM

3.1 An Instance of the Problem

Fig. 1 describes a simple AGV system. There are three zones $Z_i, i = 0, 1, 2$ each of which is controlled by a distributed controller C_i. There are communication links between C_0 and other controllers. Once the AGV enters the system, it visits M_1 and M_2 in any order, and after that it goes out. When the AGV moves to another zone, then the controller needs to approve it. The uncontrolled behavior and the target behavior of the AGV is given by transition systems $\mathcal{G}$ and $\mathcal{S}$, respectively. Then the objective is to design controllers so that the AGV *behaves like* $\mathcal{S}$.

Fig. 2 is a graphical notation describing the architecture of the system. It describes (i) the dynamic environment in which the object to be controlled behaves, and (ii) the set of actions available in each situation. The notation is based on PN^2 ("Petri Nets in a Petri Net")[3]. A PN^2 consists of two layers: the upper layer is called *the environment net*, and the labeled Petri nets in the lower layer are called *agent nets*. Controllers $C_i, i = 0, 1, 2$ and the AGV G are agent nets located in the environment net shown in Fig. 2. In this paper, we assume that each agent net is a transition system because transition systems can be considered as a class of labeled Petri nets. Each transition of an agent net can be synchronized with a transition of the environment net if they have the same label. Broken lines in the figure are *test arcs*, which is equivalent to two arcs with different directions. On the environment net, each agent net is treated as a token. This allows us to describe mobility and duplication of agents.

3.2 Problem Formulation

A discrete event system is modeled by a transition system $\mathcal{G} = (X, \Sigma, \delta_G, x_0)$, which is usually called

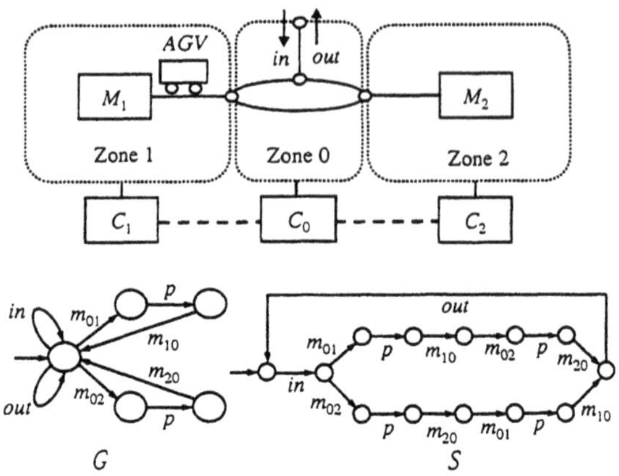

Fig. 1. An agent-based control problem.

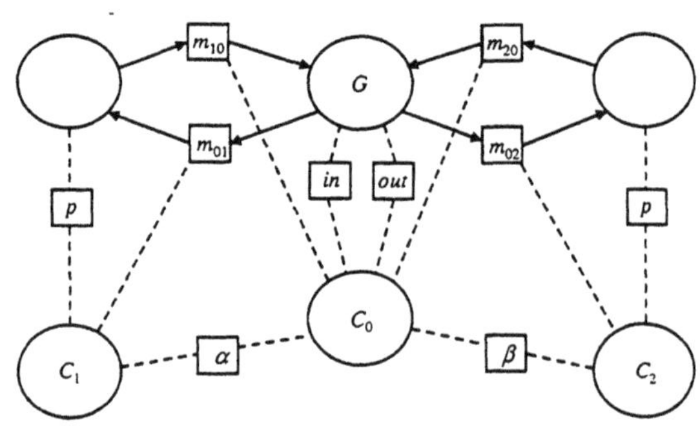

Fig. 2. A graphical notation of the system.

a plant. Each event in Σ is said to be *a plant event.* There are n controllers $Z = \{C_1, C_2, ..., C_n\}$, each of which can disable/observe a subset of events. As usual, let $\Sigma_{c,i} \subseteq \Sigma$ be the set of events which can be disabled by C_i, and let $\Sigma_{o,i} \subseteq \Sigma$ be the set of events which can be observed by C_i.

Each controller can manipulate another set of events, called *synchronization events*, each of which is related to *an autonomous action* of each controller or *a communication* between controllers. Each synchronization event is defined for a nonempty subset M of Z. Let $\widehat{\Sigma}_M$ denote the set of synchronization events defined for M. We assume that $\widehat{\Sigma}_M \cap \widehat{\Sigma}_{M'} = \emptyset$ if $M \neq M'$. Suppose that plant event a and synchronization event $\widehat{\sigma}$ are enabled. Since they are *independent*, i.e., the occurrence of one does not disable the other, both $a\widehat{\sigma}$ and $\widehat{\sigma}a$ may happen. This means that communication may be delayed relatively to occurrences of plant events.

Let $\widehat{\Sigma}$ denote the set of all synchronization events and let $\widehat{\Sigma}_i := \bigcup_{i \in M \subseteq Z} \widehat{\Sigma}_M$ for each controller C_i. Then each controller C_i is defined as a transition system $C_i = (Q_i, \Sigma_i, \delta_i, q_{i0})$, where $\Sigma_i = \Sigma \cup \widehat{\Sigma}_i$. Let $\Sigma_{all} := \Sigma \cup \widehat{\Sigma}$ be the set of all events.

In addition, we introduce the notion of *configuration*, which will be necessary for representing agent mobility. The sets of controllable/observable events and the set of possible synchronization events are assumed to vary according to evolution of the system. These sets depend on *the configuration* of the system. We assume that *there are finitely many configurations* and let Π be the set of them. For each configuration $\pi \in \Pi$, we define $\Sigma(\pi)$ (the set of plant events allowed to occur), $\Sigma_{c,i}(\pi) \subseteq \Sigma_{c,i}$ (the set of controllable events), $\Sigma_{o,i}(\pi) \subseteq \Sigma_{o,i}$ (the set of observable events), and $\widehat{\Sigma}_M(\pi) \subseteq \widehat{\Sigma}_M$, $\widehat{\Sigma}_i(\pi) \subseteq \widehat{\Sigma}_i$ (the set of possible synchronization events). The configuration changes according to a transition function $\delta_\Pi : \Pi \times \Sigma_{all} \rightarrow \Pi$, i.e., for each $\pi \in \Pi$ and $\sigma \in \Sigma_{all}$, $\delta_\Pi(\pi, \sigma)$ denotes the next configuration of π after an occurrence of σ. Let $\pi_0 \in \Pi$ be the initial configuration.

Example 3.1. The system described in 3.1 has three configurations $\pi_i, i = 0, 1, 2$ each of them is indicating that the AGV is in zone i. The behavior of the AGV is restricted by the environment, e.g, events *in* and *out* are allowed to occur only when the AGV is in zone 0. Controller $C_i, i = 1, 2$ can control and observe event p only when the AGV is in zone i.

We define the projection operator $P_i : \Pi \times \Sigma_{all}^* \rightarrow \Sigma_i^*$, which maps each global trajectory of the system to the observation by controller C_i, as follows: (i) $P_i(\pi, \varepsilon) := \varepsilon$ and (ii) for each $s \in \Sigma_{all}^*$ and $\sigma \in \Sigma_{all}$, $P_i(\pi, s\sigma) := P_i(\pi, s)\sigma$ if $\sigma \in \Sigma_{o,i}(\delta_\Pi(\pi, s)) \cup \widehat{\Sigma}_i(\delta_\Pi(\pi, s))$, $P_i(\pi, s\sigma) := P_i(\pi, s)$ otherwise. We simply write $P_i(s)$ to denote $P_i(\pi_0, s)$. We also define $P_i^{-1}(s) := \{s' \in \Sigma_{all}^* \mid P_i(s') = s\}$.

Let s be a global trajectory that has occurred in the system. Depending on the observation $P_i(s)$, each controller C_i specifies a set of events $\gamma_i \subseteq \Sigma_{c,i} \cup \widehat{\Sigma}_i$ to be disabled. Since the possible set of controllable/synchronization events varies depending on the current configuration π, only events in $\bigcup_i(\gamma_i \cap (\Sigma_{c,i}(\pi) \cup \widehat{\Sigma}_i(\pi)))$ are actually disabled[1].

Let $\mathcal{G}/Z$ denote the plant $\mathcal{G}$ under the control of decentralized controllers Z. Then the problem ABCDE (*Agent-Based Control under Dynamic Environment*) is defined as a problem of finding transition systems for controllers $C_i \in Z$ such that $\mathcal{S}$ and $\mathcal{G}/Z$ are Σ-bisimilar, where $\mathcal{S}$ is a transition system representing a desirable behavior such that $L(\mathcal{S}) \subseteq L(\mathcal{G})$. We remark that language equivalence is inadequate since the behavior of the entire system projected to Σ is generally nondeterministic. For a language L over Σ and $\Sigma' \subseteq \Sigma$, let $L|_{\Sigma'}$ denote the set of all strings obtained from strings in L by deleting all symbols not in Σ'. Even if $L(\mathcal{G}/Z)|_\Sigma = L(\mathcal{S})$, there may exist a trajectory in $L(\mathcal{G}/Z)$ that does not have a continuation as defined in $\mathcal{S}$. We show an example. Suppose that $L(\mathcal{S}) = Pref(\{ab\})$ and $L(\mathcal{G}/Z) = Pref(\{ab, \widehat{\sigma}a\})$, where $\Sigma = \{a, b\}$ and $Pref(\cdot)$ denotes the prefix closure of a language. Then $L(\mathcal{G}/Z)|_\Sigma = L(\mathcal{S})$ holds but b does not occur after $\widehat{\sigma}a$. Such a *blocking* in the controlled system is avoidable by introducing the notion of marked languages[9]. However, their effect is different form that of Σ-bisimilarity. Intuitively, Σ-bisimilarity can be checked locally, but conditions on marked languages cannot be.

3.3 Undecidability of the Problem

On decentralized control problems, there are several results on decidability. For decentralized control without communication, there are decidable cases (e.g., [6,7]) and undecidable cases[5,9]. Also for decentralized control with communication, there are decidable cases (e.g.[1,2]), and a negative result is shown in a work by Tripakis[10]. He proposed an architecture in which each controller sends every observed events to the other controller via a FIFO queue. The delay in communication is counted by the number of plant events that have occurred before the delivery of the message. Then the problem of finding decentralized controllers that satisfies a set of *responsiveness properties*[2] is undecidable if communication delay is unbounded. Using agents that simulate FIFO queues, the proposed formalism can simulate this architecture. Details are omitted here. Hence, we have the following result.

Theorem 3.1. Even if both $\mathcal{G}$ and $\mathcal{S}$ are finite, Problem ABCDE is undecidable.

4. SOLVABILITY OF THE PROBLEM

4.1 Finite-State Controllers

We will show a necessary and sufficient condition for the existence of controllers each of which is described by a finite transition system. Finiteness of controllers is necessary for them to be realized. We call such a controller *a finite-state controller*. Let $\mathcal{G} = (X, \Sigma, \delta_\mathcal{G}, x_0)$ be a finite transition system representing the plant and let $\mathcal{S} = (Z, \Sigma, \delta_\mathcal{S}, z_0)$ be a finite transition system describing the specification. To make the argument simpler, we assume that there are only two controllers.

[1] We assume a plant event is disabled if at least one of control agents disables it.

[2] Properties in the form $a \rightsquigarrow b$ (b occurs after every a). Such properties can be represented by a Σ-bisimilarity of transition systems.

Let $\mathcal{X} := \{\langle x, z, \pi\rangle \mid x \in X, z \in Z \cup \{ND\}, \pi \in \Pi\}$, where ND stands for *not defined*. Given some countable set I, we construct a finite transition system $\mathcal{N} = (Y, \Sigma_{all}, \delta_{\mathcal{N}}, y_0)$ representing a global behavior of the controlled system, called *an occurrence graph*, as follows:

- $Y \subseteq I \times \mathcal{X} \times 2^{\Sigma_{all}} \times 2^{\Sigma_{all}}$. Each state $(k, \langle x, z, \pi\rangle, \gamma_1, \gamma_2) \in Y$ is a global state of the controlled system, where $\gamma_i \subseteq \Sigma_{c,i} \cup \widehat{\Sigma}_i(\pi)$ is the set of events to be disabled by C_i. $y_0 = (0, \langle x_0, z_0, \pi_0\rangle, \gamma_{01}, \gamma_{02}) \in Y$.
- Let $y = (k, \langle x, z, \pi\rangle, \gamma_1, \gamma_2) \in Y$ and $\sigma \in \Sigma_{all}$. Then $\delta_{\mathcal{N}}(y, \sigma)$ is defined in the following two cases:

  (i) Suppose that $\sigma \in \Sigma(\pi)$. If $\sigma \notin \bigcup_i(\gamma_i \cap \Sigma_{c,i}(\pi))$ and $\delta_{\mathcal{G}}(x, \sigma)!$, then $\delta_{\mathcal{N}}(y, \sigma) = y'$ for $y' = (k', \langle x', z', \pi'\rangle, \gamma_1', \gamma_2') \in Y$ such that $x' = \delta_{\mathcal{G}}(x, \sigma)$; $z' = \delta_{\mathcal{S}}(z, \sigma)$ if $\delta_{\mathcal{S}}(z, \sigma)!$, $z' = ND$ otherwise; $\pi' = \delta_{\Pi}(\pi, \sigma)$; $\gamma_i' \subseteq \Sigma_{c,i} \cup \widehat{\Sigma}_i(\pi')$ $(i = 1, 2)$.

  (ii) Suppose that $\sigma \in \widehat{\Sigma}_M(\pi), \emptyset \neq M \subseteq Z$. If $\forall C_i \in M : \sigma \notin \gamma_i$, then $\delta_{\mathcal{N}}(y, \sigma) = y'$ for $y' = (k', \langle x, z, \pi'\rangle, \gamma_1', \gamma_2') \in Y$ such that $\pi' = \delta_{\Pi}(\pi, \sigma)$; $\gamma_i' \subseteq \Sigma_{c,i} \cup \widehat{\Sigma}_i(\pi')$ $(i = 1, 2)$.

We require that $\mathcal{N}$ is accessible. Different states may have the same components other than the elements in I. This means that the elements of I represent internal states of controllers.

An occurrence graph defines sets of disabled events as functions of trajectories that have occurred so far. Let $\gamma_i(s)$, $i = 1, 2$ denote the set of events disabled by C_i after a trajectory $s \in L(\mathcal{N})$. For each $y = (k, \langle x, z, \pi\rangle, \gamma_1, \gamma_2) \in Y$, let $y|_X := x$, $y|_Z := z$, and $y|_\Pi := \pi$.

- $\mathcal{N}$ is called *consistent* if for any $s, s' \in L(\mathcal{N})$, $P_i(s) = P_i(s')$ implies $\gamma_i(s) = \gamma_i(s')$. This condition is necessary for constructing each controller C_i from $\mathcal{N}$.
- State $y \in Y$ is called *legal* if for any $\sigma \in \Sigma$: (i) $\delta_{\mathcal{G}}(y|_X, \sigma)! \wedge \neg\delta_{\mathcal{S}}(y|_Z, \sigma)! \Rightarrow \neg\delta_{\mathcal{N}}(y, \sigma)$ (there are no states with ND) and (ii) $\delta_{\mathcal{G}}(y|_X, \sigma)! \wedge \delta_{\mathcal{S}}(y|_Z, \sigma)! \Rightarrow \exists u \in \widehat{\Sigma}^* : \delta_{\mathcal{N}}(y, u\sigma)!$. $\mathcal{N}$ is called *legal* if every state is legal.

Lemma 4.1. $\mathcal{N}$ is legal if and only if $\mathcal{N}$ and $\mathcal{S}$ are Σ-bisimilar.

Proof. Since $L(\mathcal{S}) \subseteq L(\mathcal{G})$, $\delta_{\mathcal{S}}(y|_Z, \sigma)!$ implies $\delta_{\mathcal{G}}(y|_X, \sigma)!$.

'Only if': Define a binary relation R over $Y \times Z$ by $\{(y, z) \mid y|_Z = z\}$. Since $y_0|_Z = z_0$, $y_0 R z_0$ holds. If $\delta_{\mathcal{S}}(z, \sigma) = z'$ and $y|_Z = z$ for $y \in Y$ and $\sigma \in \Sigma$, then by (ii) there exists $u \in \widehat{\Sigma}^*$ such that $\delta_{\mathcal{N}}(y, u\sigma) = y'$ and $y'|_Z = z'$. Therefore, $\mathcal{N}$ Σ-simulates $\mathcal{S}$ for R. Conversely, if $\delta_{\mathcal{N}}(y, \sigma) = y', \sigma \in \widehat{\Sigma}$, then $y|_Z = y'|_Z$ holds. If $\delta_{\mathcal{N}}(y, \sigma) = y', \sigma \in \Sigma$, then by (i) $\delta_{\mathcal{S}}(y|_Z, \sigma) = y'|_Z$. Hence, $\mathcal{S}$ Σ-simulates $\mathcal{N}$ for R^{-1}.

'If': Suppose that $\mathcal{N}$ and $\mathcal{S}$ are Σ-bisimilar for R. We first show that for any $y \in Y$, yRz if and only if $y|_Z = z$. This is proved by the following induction. (i) $y_0 R z_0$ and $y_0|_Z = z_0$. (ii) Assume that yRz and $y|_Z = z$. Since $\mathcal{S}$ Σ-simulates $\mathcal{N}$ for R^{-1}, if $\delta_{\mathcal{N}}(y, \sigma) = y', \sigma \in \widehat{\Sigma}$, then $y'Rz$ and $y'|_Z = y|_Z = z$; if $\delta_{\mathcal{N}}(y, \sigma) = y', \sigma \in \Sigma$, then $\delta_{\mathcal{S}}(z, \sigma) = z'$, $y'Rz'$ and $y'|_Z = z'$. Suppose that $\delta_{\mathcal{G}}(y|_X, \sigma)! \wedge \neg\delta_{\mathcal{S}}(y|_Z, \sigma)!$. Since $\mathcal{S}$ Σ-simulates $\mathcal{N}$ for R^{-1} and $yRy|_Z$, $\delta_{\mathcal{N}}(y, \sigma)$ is never defined and we obtain (i). Suppose that $\delta_{\mathcal{G}}(y|_X, \sigma)! \wedge \delta_{\mathcal{S}}(y|_Z, \sigma)!$. Since $\mathcal{N}$ Σ-simulates $\mathcal{S}$ for R and $yRy|_Z$, $y \stackrel{\sigma}{\Rightarrow}_{\mathcal{N},\Sigma} y'$ holds and we obtain (ii). □

Theorem 4.2. Suppose that both $\mathcal{G}$ and $\mathcal{S}$ are finite. Then there exist finite-state controllers that satisfy the requirement of Problem ABCDE if and only if there exists a consistent and legal occurrence graph $\mathcal{N}$ for some finite set I.

We describe an outline of the proof. 'Only if' is obvious because we can construct a finite, consistent and legal occurrence graph from the control actions by the controllers. Suppose that there exists a consistent and legal occurrence graph $\mathcal{N}$. Then we define controller C_i as the accessible part of a finite transition system $(H_i, \Sigma_{all}, \delta_i, h_{0i})$, where

- $H_i := 2^Y$,
- $\delta_i(h, \sigma) := \bigcup_{y \in h}\{\delta_{\mathcal{N}}(y, s) \mid \exists s \in \Sigma_{all}^* : P_i(y|_\Pi, s) = \sigma\} \neq \emptyset$,
- $h_{0i} := \{\delta(y_0, s) \mid \exists s \in \Sigma_{all}^* : P_i(\pi_0, s) = \varepsilon\}$.

From the consistency of $\mathcal{N}$, all elements in each state h of C_i have the same set of events to be disabled by controller C_i, which is defined as the disabling set at state h. Then the joint action of C_1 and C_2 restricts the behavior of $\mathcal{G}$ to $\mathcal{S}$.

As a countable set I, we use a finite index set $I_m := \{1, 2, \cdots, m\}$. Since I_m, $\mathcal{X}$ and Σ_{all} are finite, the number of possible states is also finite. Increasing m from 0, we can enumerate all possible occurrence graphs, and can check legality and consistency of them. That is, the set of all occurrence graphs is a recursively enumerable set. This gives a procedure for finding a finite controller. Because of the undecidability of the problem, the procedure may not terminate in general. However, if there exist finite controllers satisfying the requirement, then the procedure always finds them.

4.2 Communication Behavior

The undecidability arises from the fact that the behavior on communication is *unspecified* in the problem. Even if the languages of the plant and the specification are regular, the controlled behavior including communication may not. We introduce the notion of *communication behavior*. A communication behavior is a function that maps each trajectory observed so far to a set of synchronization events to be enabled.

For example, we consider the following communication behavior: each controller sends every observed event to other controllers. We call this *total communication*. As in the Tripakis's model[10], total communication is realized by FIFO queues. When a controller observes a plant event σ, the corresponding synchronization event $\widehat{\sigma}$ is appended to the tail of each queue. The only synchronization events to be enabled are those in the head of queues. Once a synchronization event in the head of a queue occurs, it is removed from the queue. As described in 3.2, synchronization events may be delayed relatively to plant events and therefore each controller cannot force these events to occur. In this sense, a communication behavior includes not only the communication policy of each controller, but also nondeterminism arising from the communication medium.

We can describe each communication behavior by a transition system. Suppose that there are two controllers $C_i, i = 1, 2$ with communication behaviors $\mathcal{W}_i = (W_i, \Sigma_{o,i} \cup \widehat{\Sigma}_i, \delta_{\mathcal{W}_i}, w_{0i}), i = 1, 2$. Then we construct a transition system $\mathcal{U} = (V, \Sigma_{all}, \delta_{\mathcal{U}}, v_0)$ representing the uncontrolled behavior including communication as follows:

- $V \subseteq W_1 \times W_2 \times X \times \Pi$. $v_0 = (w_{01}, w_{02}, x_0, \pi_0) \in V$.
- Let $v = (w_1, w_2, x, \pi) \in V$ and $\sigma \in \Sigma_{all}$. Then $\delta_{\mathcal{U}}(v, \sigma)$ is defined in the following two cases:
  (i) Suppose that $\sigma \in \Sigma(\pi)$. If $\delta_{\mathcal{G}}(x, \sigma)!$ and $\delta_{\mathcal{W}_i}(w_i, \sigma)!$ for every $\mathcal{W}_i$ such that $\sigma \in \Sigma_{o,i}$, then $\delta_{\mathcal{U}}(v, \sigma) = v'$ for $v' = (w'_1, w'_2, x', \pi') \in V$ such that $w'_i = \delta_{\mathcal{W}_i}(w_i, \sigma)$ if $\sigma \in \Sigma_{o,i}(\pi)$, $w'_i = w_i$ otherwise; $x' = \delta_{\mathcal{G}}(x, \sigma)$; $\pi' = \delta_{\Pi}(\pi, \sigma)$.
  (ii) Suppose that $\sigma \in \widehat{\Sigma}_M(\pi), \emptyset \neq M \subseteq Z$. If $\forall C_i \in M : \delta_{\mathcal{W}_i}(w_i, \sigma)!$, then $\delta_{\mathcal{U}}(v, \sigma) = v'$ for $v' = (w'_1, w'_2, x, \pi') \in V$ such that $w'_i = \delta_{\mathcal{W}_i}(w_i, \sigma)$ $(C_i \in M)$, $w'_i = w_i$ $(C_i \notin M)$; $\pi' = \delta_{\Pi}(\pi, \sigma)$.

Now we consider a decentralized control problem *without communication*, where the plant is $\mathcal{U}$, and each controller C_i observes events in $\Sigma_{o,i}(\pi) \cup \widehat{\Sigma}_i(\pi)$ and gives a control action $\gamma_i \subseteq \Sigma_{c,i}(\pi)$. The goal of the problem is to find a transition system for each controller $C_i \in Z$ such that $\mathcal{S}$ and $\mathcal{U}/Z$ are Σ-bisimilar. The following problem was shown to be decidable[7]: given a finite transition system $\mathcal{U}$ over Σ_{all}, and a finite transition system $\mathcal{E}$ over Σ_{all} such that $\mathcal{S}$ and $\mathcal{E}$ are Σ-bisimilar, find a transition system for each controller $C_i \in Z$ such that $L(\mathcal{U}/Z) = L(\mathcal{E})$. For the existence of such controllers, it must hold that $L(\mathcal{E}) \subseteq L(\mathcal{U}\|\mathcal{S})$. The difficulty in the problem is that there may be infinitely many such $\mathcal{E}$'s.

A communication behavior is called *rational* if it is given by a finite transition system that does not allow occurrence of infinite strings consisting only of synchronization events. Moreover, we say that observation by controller C_i *diverges* if $P_i^{-1}(P_i(s)) \cap L(\mathcal{G})$ is infinite for some $s \in L(\mathcal{G})$.

Theorem 4.3. Suppose that $\mathcal{G}$ and $\mathcal{S}$ are finite. If for each controller the communication behavior is rational and observation does not diverge, then Problem ABCDE is decidable.

Because of lack of space, we describe here only sketch of the proof. Let $\mathcal{E} = (Q, \Sigma_{all}, \delta_{\mathcal{E}}, q_0)$ be a finite transition system over Σ_{all} such that $L(\mathcal{E}) \subseteq L(\mathcal{U}\|\mathcal{S})$, and $\mathcal{E}$ and $\mathcal{S}$ are Σ-bisimilar. Given $\mathcal{U}$ as the plant and $\mathcal{E}$ as the specification, we construct an occurrence graph $\mathcal{N} = (Y, \Sigma_{all}, \delta_{\mathcal{N}}, y_0)$ with an index set Σ_{all}^* such that for each state $(s, \langle v, q, \pi\rangle, \Gamma_1, \Gamma_2) \in Y$, s is the sequence of events that has occurred so far. Let $y|_{trace} := s$. Note that all events are considered as plant events and no new synchronization events are introduced.

For each state $s \in L(\mathcal{E})$, let

$$\phi(s) := (\delta_{\mathcal{U}\|\mathcal{S}}((v_0, z_0), s), \nu(s)),$$

where $\delta_{\mathcal{U}\|\mathcal{S}}$ is the transition function of $\mathcal{U}\|\mathcal{S}$ and

$$\nu(s) := \{s'\sigma \in \Sigma_{all}^* \mid s' \in \widehat{\Sigma}^*, \sigma \in \Sigma, \delta_{\mathcal{E}}(q_0, ss'\sigma)!\}.$$

Since the communication behavior is rational, $\nu(s)$ is finite for any s. In the occurrence graph, whether state $\delta_{\mathcal{N}}(y_0, s)$ is legal or not can be determined by checking all states $\delta_{\mathcal{N}}(y_0, ss')$, $s' \in Pref(\nu(s))$. In this sense, Σ-bisimilarity is a local property. Let $\Phi := \{\phi(s) \mid s \in L(\mathcal{E})\}$. Since $\nu(s)$ is finite for any s, Φ is also.

For each trajectory $s \in L(\mathcal{N})$ and each controller C_i, let $E_i(s) := \{\phi(s') \mid \exists s' \in L(\mathcal{N}) : P_i(s) = P_i(s')\}$ be the estimate for Φ. Suppose that control actions applied to states $\delta_{\mathcal{N}}(y_0, s_1 s')$, $s' \in Pref(\nu(s_1))$ keep the graph legal. If $\phi(s_1) = \phi(s_2)$, then the same control actions applied to states $\delta_{\mathcal{N}}(y_0, s_2 s')$, $s' \in Pref(\nu(s_2))$ do also. As a result, if a solution of the problem exists, then we can always construct each controller as a feedback controller for 2^{Φ}, i.e., if $E_i(s_1) = E_i(s_2)$ then controller C_i gives the same control action to states $\delta_{\mathcal{N}}(y_0, s_1)$ and $\delta_{\mathcal{N}}(y_0, s_2)$.

$\mathcal{N}$ has the form of a rooted tree and we construct it in a breadth-first manner. We stop expanding the graph if one of the following conditions holds:

(1) $\mathcal{N}$ is found to be illegal, i.e., $\mathcal{N}$ cannot behaves like $\mathcal{E}$;
(2) $\mathcal{N}$ is found to be inconsistent w.r.t. the feedback policy for 2^{Φ};
(3) for each leaf y, there is an ancestor y' of it such that y' is found to be legal and $\forall C_i \in Z : E_i(y|_{trace}) = E_i(y'|_{trace})$.

If the above tree generation stops only by the condition (3), the resulting subtree is called *a closed subtree.* Once it is found, we can construct controllers that achieve the specification.

Since observation does not diverge, $E_i(\cdot)$ can be computed by exploring $\mathcal{N}$ to some finite depth. In addition, since the communication behavior is rational, whether a state is legal or not can be checked by exploring $\mathcal{N}$ to some finite depth.

Now we consider an algorithm that generates occurrence graphs nondeterministically for every possible $\mathcal{E}$'s. Though the set of all possible $\mathcal{E}$'s are infinite, we can enumerate all prefixes of them. The algorithm halts in the following two cases:

- a closed subtree is found;
- every nondeterministic branch of the algorithm falls into the case (1) or (2).

If the problem has a solution, then we will find a closed subtree. Otherwise, we will eventually know that the current subtree is illegal or inconsistent.

4.3 Instances of Rational Communication

We show two cases in which communication behavior is rational.

(***state-estimation-based controllers***) Given an occurrence graph $\mathcal{N}$, *the state estimate* by controller C_i after a trajectory $s \in L(\mathcal{N})$ is $\Psi_i(s) := \{\langle x(s'), z(s'), \pi(s')\rangle \in \mathcal{X} \mid s' \in L(\mathcal{N}), P_i(s) = P_i(s')\}$. Decentralized controllers are said to be ***state-estimation-based controllers*** if for any $s \in L(\mathcal{N})$: $\Psi_i(s) = \Psi_i(s')$ implies $\gamma_i(s) \cap \Sigma = \gamma_i(s') \cap \Sigma$ $(i = 1, \cdots, n)$, and each controller tries to send the current state estimate to all other controllers after every observation of plant events. The communication behavior of such a controller is described as a finite transition system with a state space $\{0,1\} \times 2^{\mathcal{X}}$, where $(0, \Psi)$ means that the current state estimate Ψ has not been sent yet and $(1, \Psi)$ means it has already been sent.

(***k-bounded-delay communication***[10]) Suppose that the communication behavior is total communication. Then the communication is said to be ***k-bounded-delay communication***, where k is a non-negative integer, if the number of plant events occurring between the time when a message is push to the queue and the time when it is pop from the queue is no greater than k. Then the maximal length of the queue is bounded by $k+1$, and therefore the communication behavior is represented by a finite transition system.

In addition, infinite strings over $\widehat{\Sigma}$ are not allowed to occur in both cases.

5. CONCLUDING REMARKS

We have studied solvability of an agent-based control problem under an assumption on communication. The next step is to develop an efficient synthesis algorithm. We should also consider minimization of communication, which was studied for communication without delay[8]

REFERENCES

[1] Barrett, G., and S. Lafortune (1998). On the synthesis of communicating controllers with decentralized information structures for discrete-event systems. Proc. IEEE CDC'98.

[2] Barett, G., and S. Lafortune (1998). A novel framework for decentralized supervisory control with communication. Proc. IEEE SMC'98.

[3] Hiraishi, K., (2001). A Petri-net-based model for the mathematical analysis of multi-agent systems. IEICE Trans., **Vol.E84-A**, No.11, pp.2829-2837.

[4] Hiraishi, K. (2003). PN^2: An elementary model for design and analysis of multi-agent systems. Proc. Coordination2002, Lecture Notes in Computer Science **2315**, Springer-Verlag, pp.220-235.

[5] Lamouchi, H., and J. Thistle (2000). Effective control synthesis for DES under partial observation. Proc. IEEE CDC, pp.22-28.

[6] Lin, F., and W. M. Wonham, Decentralized supervisory control of discrete-event systems, Information Sciences, **Vol.44**, pp.199-224 (1988).

[7] Rudie, K., and W. M. Wonham (1992). Think globally, act locally: decentralized supervisory control. IEEE Transactions on Automatic Control, **Vol.37**, No.11, pp.1692-1708.

[8] Rudie, K., S. Lafortune, and F. Lin (2003). Minimum communication in a distributed discrete event system. IEEE Transactions on Automatic Control, **Vol.48**, No.6, pp.957-975.

[9] Tripakis, S. (2001). Undecidable problems of decentralized observation and control. Proc 20th IEEE CDC, pp.4104-4109.

[10] Tripakis, S. (2002). Decentralized control of discrete event systems with bounded or unbounded delay communication. Proc. WODES'02.

ELSEVIER
IFAC
PUBLICATIONS
www.elsevier.com/locate/ifac

AN ERGODIC THEOREM FOR STOCHASTIC MAX-PLUS LINEAR SYSTEMS

Bernd Heidergott [*], **Jacob van der Woude** [**] and **Geert Jan Olsder** [**]

[*] *Department of Econometrics and Tinbergen Institute, Vrije Universiteit, De Boelelaan 1105, 1081 HV Amsterdam, The Netherlands*
[**] *Faculty of Electrical Engineering, Mathematics and Computer Science, Delft University of Technology, Mekelweg 4, 2628 CD Delft, The Netherlands*

Abstract: In this paper the asymptotic growth rate of stochastic max-plus linear systems is studied. For non-random irreducible max-plus linear systems the existence of this growth rate is well known. For stochastic systems conditions for the existence of the growth rate will be developed. The conditions are mild and natural, and weaker than the conditions used in literature until now. The interpretation of the conditions is that for the asymptotic growth rate to exist the stochastic max-plus linear system should be dominated from below by a non-random irreducible max-plus linear system. The latter condition need not to be true always, but should at least hold with some positive probability. The investigations in this paper are motivated by means of some examples.

Keywords: max-plus algebra, ergodic theory, Lyapunov exponent, irreducibility, subadditivity.

1. INTRODUCTION

Max-plus algebra has been successfully applied to many areas of discrete event systems. Typical examples of max-plus linear systems are the G/G/1-queue, networks of (finite) queues in tandem, Kanban systems, transportation networks, railway models, flexible manufacturing systems, fork/join queues or any parallel and/or series composition made by these elements, see (Heidergott, 2000) for more details.

Max-plus algebra is usually introduced as follows. Define $\varepsilon = -\infty$ and $\mathbb{R}_\varepsilon = \mathbb{R} \cup \{\varepsilon\}$, and set

$$x \oplus y = \max(x, y)\,, \quad x \otimes y = x + y\,,$$

for scalars $x, y \in \mathbb{R}_\varepsilon$, and set

$$[A \oplus B]_{ij} = a_{ij} \oplus b_{ij} = \max(a_{ij}, b_{ij})\,,$$

$$[A \otimes C]_{ij} = \bigoplus_{k=1}^{n} a_{ik} \otimes c_{kj} = \max_{k=1,\ldots,n}(a_{ik} + c_{kj})\,,$$

for matrices $A, B \in \mathbb{R}_\varepsilon^{m\times n}$ and $C \in \mathbb{R}_\varepsilon^{n\times p}$, see (Baccelli *et al.*, 1992; Cuninghame-Green, 1979).

This paper is devoted to the study of sequences $\{x(k)\}$ satisfying the recurrence relation

$$x(k+1) = A(k) \otimes x(k)\,, \quad k \geq 0\,, \qquad (1)$$

where $x(0) = x_0 \in \mathbb{R}^n_{\max}$ is the initial value and $\{A(k)\}$ is a sequence of $n \times n$ matrices over $\mathbb{R}_{\max}$. It is moreover assumed that $\{A(k)\}$ is a sequence of random elements in $\mathbb{R}^{n\times n}_{\max}$, defined on a common probability space, and that the matrices $A(k)$ for $k \geq 0$ are independent identically distributed (i.i.d.). The *asymptotic growth rate* of $\{x(k)\}$ is defined via the limits

$$\lim_{k\to\infty} \frac{x_j(k)}{k}, \tag{2}$$

for $j \in \underline{n}$, where, in general, the notation $\underline{m}$ stands for the set $\{1, \ldots, m\}$, for any integer $m \geq 1$. If the above limits exist and all have the same value, say λ, then this value is said to be the asymptotic growth rate of $\{x(k)\}$, or also the *Lyapunov exponent* of the sequence $\{A(k)\}$. When it comes to queueing networks, the Lyapunov exponent is the inverse throughput, and for systems operating according to a timetable, such as transportation networks, it is the (minimal) asymptotic period of the timetable.

Kingman's subadditive ergodic theorem applied to max-plus linear systems yields the existence with probability one of the limits of $\frac{1}{k}(\max_j x_j(k))$ and $\frac{1}{k}(\min_j x_j(k))$ for k to infinity. In the light of this result, the existence of the Lyapunov exponent follows when the aforementioned limits are shown to be equal. Finding the most general set of conditions to ensure the existence of this equality is a long standing problem. The first result in this direction can be found in (Baccelli *et al.*, 1992). There it is shown that a sufficient condition for the existence of the Lyapunov exponent is that (i) finite entries of $A(k)$ are non-negative, (ii) all diagonal elements of $A(k)$ are non-negative and (iii) the probability that $A_{ij}(k)$ is finite, hence non-negative, is either zero or one, independent of k. If $A(k)$ satisfies condition (iii), $A(k)$ is said to have a *fixed support* (see Definition 1 later on).

Conditions (ii) and (iii) impose the main restriction in the limit theorem of (Baccelli *et al.*, 1992). Vincent gives in (Vincent, 1997) a general condition for the existence of the Lyapunov exponent for matrices with fixed support. For matrices with denumerable state-space, Hong (Hong, 2001) provides conditions without assuming fixed support. His condition is that matrices B_i, with $i \in \underline{m}$, are in the support of $A(k)$ such that $B_1 \oplus B_2 \cdots \oplus B_m$ has only finite elements. Also for matrices with denumerable state-space, exploiting regenerative properties of max-plus linear systems, (Mairesse, 1997) derives yet another set of conditions.

The present paper establishes conditions for the existence of the Lyapunov exponent of sequences of square i.i.d. max-plus matrices with general state-space. More specifically, it establishes the existence of the Lyapunov exponent without requiring conditions (i) to (iii) to hold and without assuming that $A(k)$ has a denumerable state-space.

Example 1. Consider a closed system of n single–server queues in tandem, with infinite buffers. In the system, customers have to pass through the queues consecutively so as to receive service at each server. Initially, there is one customer present at each server. After service completion at the nth server, the customers return to the first queue for a new cycle of service. Let $\sigma_j(k)$ denote the kth service time at queue j and let $x_j(k)$ be the time of the kth service completion at node j, then the time evolution of the system can be described by a n-dimensional vector $x(k)$ following the homogeneous equation

$$x(k+1) = A(k) \otimes x(k),$$

where the matrix $A(k)$ looks like

$$\begin{pmatrix} \sigma_1(k+1) & \varepsilon & \varepsilon & \cdots & \sigma_1(k+1) \\ \sigma_2(k+1) & \sigma_2(k+1) & \varepsilon & \cdots & \varepsilon \\ \varepsilon & \sigma_3(k+1) & \sigma_3(k+1) & & \\ & \cdot & \cdot & \cdot & \\ & & \vdots & \vdots & \vdots \\ & & \sigma_{n-1}(k+1) & \sigma_{n-1}(k+1) & \varepsilon \\ \varepsilon & \cdots & \varepsilon & \sigma_n(k+1) & \sigma_n(k+1) \end{pmatrix}$$

for $k \geq 0$, see (Baccelli *et al.*, 1992) for details.

Example 2. (See (Baccelli and Hong, 2000)) Consider a cyclic tandem queueing network consisting of a single–server and a double–server, each with a deterministic service time. Service times at the single–server station equal σ, whereas service times at the double–server station equal σ'. Three customers circulate in the network. Initially, one customer is in service at station 1, the single–server, one customer is in service at station 2, the double–server, and the third customer is just about to enter station 2. The time evolution of this network is described by the sequence of vectors $x(k) = (x_1(k), \ldots, x_4(k))^\top$ ($^\top$ denotes transpose), where $x_1(k)$ is the kth beginning of service at the single–server station, $x_2(k)$ is the kth service completion at the single–server station, $x_3(k)$ is the kth beginning of service at the double–server station and $x_4(k)$ is the kth service completion at the double–server station. The system then follows

$$x(k+1) = D_2 \otimes x(k),$$

where

$$D_2 = \begin{pmatrix} \sigma & \varepsilon & \sigma' & \varepsilon \\ \sigma & \varepsilon & \varepsilon & \varepsilon \\ \varepsilon & 0 & \varepsilon & 0 \\ \varepsilon & \varepsilon & \sigma' & \varepsilon \end{pmatrix},$$

with $x(0)$ some appropriate initial vector.

Consider the cyclic tandem network again, but one of the servers of the double–server station has been broken down. The system is thus a tandem network with two single–server stations. Initially one customer is in service at station 1, one customer is in service at station 2, and the third customer is waiting at station 2 for service. This system follows

$$x(k+1) = D_1 \otimes x(k),$$

where

$$D_1 = \begin{pmatrix} \sigma & \varepsilon & \sigma' & \varepsilon \\ \sigma & \varepsilon & \varepsilon & \varepsilon \\ \epsilon & 0 & \sigma' & \varepsilon \\ \varepsilon & \varepsilon & \sigma' & \varepsilon \end{pmatrix},$$

with $x(0)$ some appropriate initial vector.

Assume that whenever a customer enters station 2, the second server of the double–server station breaks down with probability θ. Let $A_\theta(k)$ have distribution

$$P(\, A_\theta(k) = D_1\,) = \theta$$

and

$$P(\, A_\theta(k) = D_2\,) = 1 - \theta\,,$$

then

$$x_\theta(k+1) = A_\theta(k) \otimes x_\theta(k)$$

describes the time evolution of the system with breakdowns. The above recurrence relation models the sample path dynamics of the system with breakdowns. More details can be found in (Baccelli and Hong, 2000).

2. BASIC DEFINITIONS AND RESULTS

For a sequence of square matrices $\{A(k)\}$ set

$$\bigotimes_{k=l}^{m-1} A(k) \stackrel{\text{def}}{=} A(m-1) \otimes \cdots \otimes A(l+1) \otimes A(l)\,,$$

where $m > l$ and $\mathcal{E}$ otherwise. The above product will also be denoted as $A[m,l]$.

Let X be a random element in $\mathbb{R}_{\max}$ defined on a probability space $(\Omega, \mathcal{F}, P)$ modelling the underlying randomness. When defining the expected value of X, denoted by $\mathbb{E}[X]$, one has to take care of the fact that X may take value $\varepsilon = -\infty$ with positive probability. This is reflected in the following extension to $\mathbb{R}_{\max}$ of the usual definition of integrability of a random variable on $\mathbb{R}$, where $e = 0$. A stochastic variable $X \in \mathbb{R}_{\max}$ is called *integrable* if $X \oplus e = \max(X,0)$ and $X \oplus' e = \min(X,0)$ are integrable, and if $\mathbb{E}[X \oplus e]$ is finite. The expected value of X is then given by $\mathbb{E}[X] = \mathbb{E}[X \oplus e] + \mathbb{E}[X \oplus' e]$. This definition implies that $\mathbb{E}[X] = -\infty$ if $P(X = \varepsilon) > 0$. A random element A in $\mathbb{R}_{\max}^{n\times m}$ is called integrable, if its elements a_{ij} are integrable for all $i \in \underline{n}, j \in \underline{m}$. The expected value of A is given by the matrix $\mathbb{E}[A]$ with elements $\left[\mathbb{E}[A]\right]_{ij} = \mathbb{E}[a_{ij}\,]$.

To any $n \times n$ matrix A over $\mathbb{R}_{\max}$ a directed graph can be associated, called *communication graph*, denoted by $\mathcal{G}(A) = (\mathcal{N}(A), \mathcal{D}(A))$, as follows. The set of nodes of the graph is given by $\underline{n}$, i.e., $\mathcal{N}(A) = \{1, \ldots, n\}$, and a pair $(j,i) \in \underline{n} \times \underline{n}$ is an arc of the graph if $a_{ij} \neq \varepsilon$; in symbols $(j,i) \in \mathcal{D}(A) \Leftrightarrow a_{ij} \neq \varepsilon$. If (j,i) is an arc of the graph, then j is called the begin node and i the end node of the arc. A path in the graph is a sequence of consecutive arcs such that the end node of one arc is the begin node of the next arc. A cycle is a path of which the begin node of the first arc and the end node of the last arc are the same. The length of a circuit or path is the number of arcs it consists of. The cyclicity of a graph is the greatest common divisor of the lengths of all circuits of the graph. Graph $\mathcal{G}(A)$ is called strongly connected if there is a path from any node $j \in \underline{n}$ to any node $i \in \underline{n}$. In general, the communication graph of random matrix is random too, which motivates the following definition.

Definition 1. *The sequence $A(k)$ is said to have fixed support if the set of arcs of the communication graph of $A(k)$ is non-random and does not depend on k, in formula: for any $i,j \in \underline{n}$ it holds for all $k \geq 0$ that either $P(\,[A(k)]_{ij} = \varepsilon\,) = 0$ or $P(\,[A(k)]_{ij} = \varepsilon\,) = 1$, where $[A(k)]_{ij}$ denotes the entry of $A(k)$ in row i and column j.*

Note that a sequence $A(k)$ has fixed support if its 'finite/infinite structure' is independent of k. With the definition of fixed support at hand, a random matrix A is called *irreducible* if (i) it has fixed support, and (ii) the communication graph of any sample of A is strongly connected with probability one. Hence, for random matrices, irreducibility presupposes fixed support. Note that $A(k)$ in Example 1 has fixed support and is irreducible, provided that the service times have finite expected values. However, $A(k)$ as defined in Example 2 fails to have a fixed support and is thus reducible, i.e., not irreducible.

A matrix $A \in \mathbb{R}_{\max}^{n\times m}$ is called *regular* if A contains at least one element different from ε in each row. Regularity is a mere technical condition, for if A fails to be regular it contains redundant rows and any system modeled by $x(k+1) = A \otimes x(k)$ can as well be modeled by considering a reduced version of A which is then regular.

For $A \in \mathbb{R}_{\max}^{n\times m}$, the minimal finite entry of A, denoted by $||A||_{\min}$, is given by

$$||A||_{\min} = \min\{a_{ij} \mid (i,j) \in \mathcal{D}(A)\}\,,$$

provided that $\mathcal{D}(A) \neq \emptyset$ (recall that $\mathcal{D}(A)$ denotes the set of arcs in the communication graph of A). In the same vein, the maximal finite entry of $A \in \mathbb{R}_{\max}^{n\times m}$ is denoted by $||A||_{\max}$, which implies

$$||A||_{\max} = \max\{a_{ij} \mid (i,j) \in \mathcal{D}(A)\}\,,$$

provided $\mathcal{D}(A) \neq \emptyset$. Note that the case that $\mathcal{D}(A) = \emptyset$ is not interesting, because then the graph has no arcs and the underlying matrix has no finite entries. A direct consequence of the above definitions is that for any regular $A \in \mathbb{R}_{\max}^{n\times m}$

$$||A||_{\min} \leq ||A||_{\max}\,.$$

Notice that $||A||_{\min}$ and $||A||_{\max}$ can have negative values. It is easily checked that for regular $A \in \mathbb{R}_{\max}^{n\times m}$ and regular $B \in \mathbb{R}_{\max}^{m\times l}$ there holds for instance

$$||A||_{\min} + ||B||_{\min} \leq ||A \otimes B||_{\min} \tag{3}$$

and

$$\|A \otimes B\|_{\min} \leq \|A\|_{\min} + \|B\|_{\max} . \tag{4}$$

In the following $D^{\otimes k}$ will denote the kth power in max-plus sense of a given matrix D. Clearly, for this to make sense the matrix D should be square. Further, the inequality $A(k) \geq D$, given a sequence $\{A(k)\}$ and a matrix D of the same dimensions, should be interpreted componentwise.

Kingman's subadditive ergodic theorem, see (Kingman, 1973), implies the following limit theorem. See, for example, (Baccelli *et al.*, 1992) for a proof.

Result 1. (See (Baccelli *et al.*, 1992))*Let* $\{A(k)\}$ *be an i.i.d. sequence of integrable matrices such that* $A(k)$ *is regular with probability one. Then, for any finite and integrable initial condition* x_0 *it holds with probability one that*

$$\lambda^{\text{bot}} \overset{\text{def}}{=} \lim_{k\to\infty} \frac{||x(k)||_{\min}}{k}$$
$$\leq \lambda^{\text{top}} \overset{\text{def}}{=} \lim_{k\to\infty} \frac{||x(k)||_{\max}}{k}$$

and

$$\lambda^{\text{bot}} = \lim_{k\to\infty} \frac{1}{k}\mathbb{E}\left[||x(k)||_{\min}\right]$$
$$\leq \lambda^{\text{top}} = \lim_{k\to\infty} \frac{1}{k}\mathbb{E}\left[||x(k)||_{\max}\right].$$

The constant λ^{top} is called the *top* or *maximal Lyapunov exponent* of $\{A(k)\}$ and λ^{bot} is called the *bottom* or *minimal Lyapunov exponent* of $\{A(k)\}$.

3. MATRICES WITH FIXED SUPPORT

In this section i.i.d. sequences $\{A(k)\}$ of integrable and irreducible matrices are considered with the additional property that with probability one all finite elements are uniformly bounded from below by some constant. As it will be shown in the following theorem, the setting of this section implies that $\lambda^{\text{top}} = \lambda^{\text{bot}}$, which in particular implies convergence of $x_i(k)/k$, for all, $i \in \underline{n}$. The condition that all elements of $A(k)$, different from ε, are, with probability one, uniformly bounded from below by some constant, is not very restrictive when working, for example, with train networks. Then finite elements of $A(k)$ represent (sums of) travelling times on tracks, which by nature even are non-negative. The main technical result is provided in the following theorem.

Theorem 1. *Let* $D \in \mathbb{R}^{n\times n}_{\max}$ *be a non-random irreducible matrix whose graph has cyclicity* σ. *If* $A(k) \geq D$ *with probability one for any* k, *then integers* L *and* N *exist such that for any* $k \geq N$

$$\|x(k \cdot \sigma)\|_{\min} \geq \|x(k \cdot \sigma - L)\|_{\max} + L \cdot ||D||_{\min} ,$$

where the $\cdot$ *means multiplication in the usual sense.*

Proof Denote the communication graph of D by $\mathcal{G} = (\mathcal{N}, \mathcal{D})$ and let the graph be first of cyclicity one, i.e., $\sigma = 1$. Denote the number of (elementary) circuits in $\mathcal{G}$ by q and let β_i denote the length of circuit C_i, for $i \in \underline{q}$. Then the greatest common divisor of $\{\beta_1, \ldots, \beta_q\}$ is equal to one. A natural number N exists such that for all $\kappa \geq N$ there are integers $n_1, \ldots, n_q \geq 0$ such that $\kappa = n_1\beta_1 + \ldots + n_q\beta_q$; see (Brauer, 1942) for a proof.

Let l_{ij} denote the minimal length of a path from j to i containing *all* nodes of $\mathcal{G}$. Such paths exist because D is irreducible (and $\mathcal{G}$ is thus strongly connected). Let the maximal length of all these paths be denoted by l, i.e., $l = \max_{i,j\in\underline{n}} l_{ij}$.

Next, take L so large that $L - l \geq N$. Then for any $i, j \in \underline{n}$ there is a path from j to i of length L. Indeed, take any $i, j \in \underline{n}$ and choose a path, as mentioned above, from j to i containing *all* nodes of $\mathcal{G}$ and having minimal length l_{ij}. Clearly, the path has at least one node in common with each of the q circuits in $\mathcal{G}$. As $L - l_{ij} \geq N$, there are integers $n_1, \ldots, n_q \geq 0$ such that $L - l_{ij} = n_1\beta_1 + \ldots + n_q\beta_q$. Hence, adding n_1 copies of circuit C_1, and so on, up to n_q copies of circuit C_q, to the chosen path from i to j of length l_{ij}, a new path from j to i is created of length $l_{ij} + n_1\beta_1 + \ldots + n_q\beta_q = L$.

In graph theoretical terms, the element $[A(k, k - L)]_{ij}$ denotes the maximal weight of a path of length L from node j at 'time' $k - L$ to node i at 'time' k. Since $A[k, k - L] \geq D^{\otimes L}$ by assumption, it follows that for all $k \geq N$ and all $i \in \underline{n}$

$$x_i(k) = \bigoplus_{j=1}^{n} [A(k, k - L)]_{ij} \otimes x_j(k - L)$$
$$\geq \bigoplus_{j=1}^{n} \left[D^{\otimes L}\right]_{ij} \otimes x_j(k - L)$$
$$\geq \bigoplus_{j=1}^{n} ||D^{\otimes L}||_{\min} \otimes x_j(k - L) ,$$

implying that

$$||x(k)||_{\min} \geq ||x(k - L)||_{\max} + L \cdot ||D||_{\min} ,$$

for $k \geq N$. Note that (3) is used here.

One can show that if the graph of D has a general cyclicity σ, then $D^{\otimes\sigma}$ has cyclicity one (the proof is omitted for lack of space). Applying the arguments put forward above to $D^{\otimes\sigma}$, extends the obtained result to the case of a graph with a cyclicity greater than one. □

Theorem 1 provides the main technical means for establishing sufficient conditions for the equality of maximal, minimal and individual growth rates. The precise statement is provided in the following theorem.

Theorem 2. *Let $D \in \mathbb{R}_{\max}^{n \times n}$ be a non-random irreducible matrix and let $\{A(k)\}$ be a random sequence such that $A(k) \geq D$ with probability one for any k. For $x(k)$ defined in (1) it holds with probability one that the corresponding asymptotic growth rate of λ satisfies*

$$\lambda = \lim_{k \to \infty} \frac{1}{k} x_i(k),$$

for any $i \in \underline{n}$ and any finite initial state x_0.

Proof Assume that D has cyclicity σ, then according to Theorem 1 there are finite positive numbers L and N such that for $k \geq N$

$$||x(k \cdot \sigma; x_0)||_{\min} \geq ||x(k \cdot \sigma - L; x_0)||_{\max} + L \cdot ||D||_{\min},$$

where $x(l; x_0)$ is written to indicate the 'state' at 'time' l, when starting from $x(0) = x_0$. Dividing both sides of the above inequality by k and letting k tend to ∞ yields, after some manipulations, that

$$\lim_{k \to \infty} \frac{1}{k} ||x(k \cdot \sigma; x_0)||_{\min} \geq \lim_{k \to \infty} \frac{1}{k} ||x(k \cdot \sigma; x_0)||_{\max},$$

for any finite initial vector x_0. Note that the existence of both limits follows from Kingman's subadditive ergodic theorem, stated in Result 1.

Let $\mathbf{u} = (0, \ldots, 0)^\top$. Next it will be shown that the above limits are independent of the initial condition. Let $y \in \mathbb{R}_{\max}^n$ and notice that

$$\begin{aligned} ||x(k; \mathbf{u})||_{\min} + ||y||_{\min} &\leq ||x(k; y)||_{\min} \\ &\leq ||x(k; \mathbf{u})||_{\min} + ||y||_{\max}, \end{aligned}$$

for $k \geq 0$ (for a proof use (3) and (4)). Dividing the above row of inequalities by k and letting k tend to infinity yields

$$\lim_{k \to \infty} \frac{1}{k} ||x(k; \mathbf{u})||_{\min} = \lim_{k \to \infty} \frac{1}{k} ||x(k; y)||_{\min}$$

with probability one for any finite initial condition y. In a similar way one shows that

$$\lim_{k \to \infty} \frac{1}{k} ||x(k; \mathbf{u})||_{\max} = \lim_{k \to \infty} \frac{1}{k} ||x(k; y)||_{\max}$$

with probability one for any finite initial y.

Noticing that $x(k \cdot \sigma + l; x_0) = x(k \cdot \sigma; x(l; x_0))$ for $l \geq 0$, yields on the one hand, with k replacing $k \cdot \sigma + l$, that

$$\lim_{k \to \infty} \frac{1}{k} ||x(k; x_0)||_{\min} \geq \lim_{k \to \infty} \frac{1}{k} ||x(k; x_0)||_{\max}, \quad (5)$$

for any finite initial vector x_0. On the other hand, since $||x(k; x_0)||_{\max} \geq x_j(k; x_0) \geq ||x(k; x_0)||_{\min}$, for $j \in \underline{n}$, it follows that

$$\begin{aligned} \lim_{k \to \infty} \frac{1}{k} ||x(k; x_0)||_{\min} &\leq \lim_{k \to \infty} \frac{1}{k} x_i(k; x_0) \\ &\leq \lim_{k \to \infty} \frac{1}{k} ||x(k; x_0)||_{\max}, \end{aligned}$$

for $i \in \underline{n}$. Combining the above row of inequalities with (5) proves the claim. □

4. BEYOND FIXED SUPPORT

In this section possible relaxations of the conditions put forward in Theorem 2 are discussed. The main technical condition is the following.

(H) *There exists a non-random irreducible matrix D such that*

$$P(A(k) \geq D) \geq p,$$

for some $p \in (0, 1]$.

Condition **(H)** suffices to guarantee that the top and bottom Lyapunov exponent coincide. The precise statement is given in the following lemma.

Lemma 1. *Let $\{A(k)\}$ be an i.i.d. sequence of integrable matrices. If condition* **(H)** *holds, then the top and bottom Lyapunov exponent of $\{A(k)\}$ coincide.*

Proof By Theorem 1, there exists an integer L such that there is a path of length L from any node j to any node i in the graph of D with weight at least $L \cdot ||D||_{\min}$. Let σ denote the cyclicity. Consider the event that for some k it holds that

$$\forall l \in \underline{L}: \quad A(k \cdot \sigma - l) \geq D. \quad (6)$$

Hence,

$$\bigotimes_{l=1}^{L} A(k \cdot \sigma - l) \geq D^{\otimes L},$$

and in accordance with Theorem 1 it follows

$$||x(k \cdot \sigma)||_{\min} \geq ||x(k \cdot \sigma - L)||_{\max} + L \cdot ||D||_{\min}. \quad (7)$$

Notice that by assumption **(H)** the event characterized in (6) occurs at least with probability $p^L > 0$. Let $\{\tau_m\}$ be the sequence of 'times' k at which the event characterized in (6) occurs. The i.i.d. assumption implies that $\tau_m < \infty$ and $\lim_{m \to \infty} \tau_m = \infty$. By inequality (7),

$$||x(\tau_m \cdot \sigma)||_{\min} \geq ||x(\tau_m \cdot \sigma - L)||_{\max} + L \cdot ||D^{\otimes \sigma}||_{\min}$$

and dividing both sides of the above inequality by τ_m and letting τ_m tend to ∞ yields with probability one

$$\lim_{\tau_m \to \infty} \frac{1}{\tau_m} ||x(\tau_m \cdot \sigma)||_{\min} \geq \lim_{\tau_m \to \infty} \frac{1}{\tau_m} ||x(\tau_m \cdot \sigma)||_{\max}.$$

The existence of the top and the bottom Lyapunov exponent are guaranteed by Kingman's theorem and the above inequality for a subsequence of $x(k)$ is sufficient to establish equality of the top and bottom Lyapunov exponent. □

Lemma 1 allows to extend Theorem 2 to matrices that fail to have a fixed support. More precisely, the 'fixed support' condition can be replaced by the assumption that $A(k)$ with positive probability is bounded from below by a matrix with fixed support. Notice that D_2 in Example 2 is irreducible, and $\{A(k)\}$ thus satisfies condition **(H)** (take $D = D_2$). The extended version of Theorem 2 thus applies to this example.

Finally, the condition is considered that finite entries of $\{A(k)\}$ have to be bounded from below. If a random variable X has finite expected value, then $|X|$ has to be bounded by a constant on a set of positive measure. Thus, provided that element (i,j) of $A(k)$ is finite, $[A(k)]_{ij}$ is bounded from below by some constant with positive probability. Hence, integrability of $A(k)$ already implies that a matrix D exists with finite entries such that

$$P(\,A(k) \geq D) \geq p\,,$$

for some $p \in (0,1]$.

Let $\mathcal{D}(A(k))$ denote the set of arcs of the communication graph of $A(k)$ that occur with positive probability (note that $\mathcal{D}(A(k))$ is a finite set).

A sufficient condition for **(H)** is thus the following

(H)' The matrix $A(k)$ is integrable and $\mathcal{D}(A(k))$ contains at least one set of arcs such that the corresponding graph is strongly connected.

The resulting limit theorem is stated below.

Theorem 3. Let $\{A(k)\}$ be an i.i.d. sequence of irreducible and integrable matrices in $\mathbb{R}^{n\times n}_{\max}$ satisfying condition **(H)'**. For any finite integrable initial value x_0 it holds with probability one that

$$\lim_{k\to\infty} \frac{x_j(k)}{k} = \lim_{k\to\infty} \frac{1}{k}\mathbb{E}\big[x_j(k)\big] = \lambda\,,$$

for $j \in \underline{n}$.

Condition **(H)** (and consequently condition **(H)'**) can be relaxed as follows. The exist a finite number M and non-random matrices $D_i \in \mathbb{R}^{n\times n}_{\max}$, for $i \in \underline{M}$, such that $D_M \otimes \cdots \otimes D_2 \otimes D_1$ is irreducible and $P(A(k) \geq D_i) > 0$, for $i \in \underline{M}$.

In the previous the focus was on the existence of the overall asymptotic growth rate λ. For this the limits in (2) should all exist and have the same value. Sufficient for this was that the support of the sequence $\{A(k)\}$ with a certain positive probability is irreducible. However, if the latter fails to be true, i.e., when the sequence $\{A(k)\}$ with a probability one has a reducible support, then the limits in (2) may exist, but their values need not to be the same anymore. Then a vector of asymptotic growth rates exists. For non-random max-plus linear systems this vector is known as the cycle time vector. For stochastic max-plus linear systems the name stochastic cycle time vector may be used. To compute such a stochastic cycle time vector the approach of this paper has to be adapted and has to be combined with a decomposition of the graph of the support of the sequence $\{A(k)\}$ into irreducible subgraphs for each of which the ideas of this paper can be applied. A full description of this approach will be the subject of future work.

5. CONCLUSIONS

In this paper the asymptotic growth rate of a stochastic max-plus linear system is studied. For such systems it is proved that the asymptotic growth rate exists under some mild and natural conditions. The conditions are weaker than the conditions used in literature until now. The interpretation of the conditions is that for the asymptotic growth rate to exist the stochastic max-plus linear system should be bounded from below by a non-random irreducible max-plus linear system. The latter need not to be true always, but should at least hold with a probability larger than zero. The extension towards stochastic max-plus linear systems that with positive probability are dominated from below by a non-random reducible max-plus linear system will be reported in detail in a future work.

REFERENCES

Baccelli, F. and D. Hong (2000). Analytic expansions of (max,+) Lyapunov exponents. *Anls. Appl. Probab.* **10**, 779–827.

Baccelli, F., G. Cohen, G.J. Olsder and J.-P. Quadrat (1992). *Synchronization and Linearity*. John Wiley and Sons. New-York.

Brauer, A. (1942). On a problem of partitions. *Amer. J. Math.* **64**, 299–312.

Cuninghame-Green, R.A. (1979). *Minimax algebra*. vol. 166 of Lecture Notes in Economics and Mathematical Systems. Springer-Verlag. Berlin.

Heidergott, B. (2000). A characterization for (max,+)-linear queueing systems. *QUESTA* pp. 237–262.

Hong, D. (2001). Lyapunov exponents: When the top joins the bottom. Technical report no. 4198. INRIA Rocquencourt.

Kingman, J.F.C. (1973). Subadditve ergodic theory. *Annals of Probability* **1**, 883–909.

Mairesse, J. (1997). Products of irreducible random matrices in the (max,+) algebra. *Advances of Applied Probability* **29**, 444–477.

Vincent, J. (1997). Some ergodic results on stochastic iterative discrete event systems. *J. of Discrete Event Dynamic Systems* **7**, 209–232.

ELSEVIER
IFAC
PUBLICATIONS
www.elsevier.com/locate/ifac

MODELLING AND CONTROL OF URBAN BUS NETWORKS IN DIOIDS ALGEBRA

L. Houssin * **S. Lahaye** * **J.L. Boimond** *

* *Laboratoire d'Ingénierie des Systèmes Automatisés, 62, avenue Notre Dame du Lac, 49000 Angers, France*

Abstract: We aim at applying dioids algebraic tools to the study of urban bus networks. In particular, we show how they can be used for timetables synthesis.

Keywords: Discrete Event System, dioid, bus networks, timetables synthesis

1. INTRODUCTION

Discrete Event Dynamic Systems (DEDS) subject to synchronization phenomena can be modeled by linear equations in a particular algebraic structure called *dioid* or *idempotent semi-ring*. For about twenty years, this property has motivated the elaboration of a "new" linear system theory. Indeed, concepts such as state representation, transfer matrix, optimal control, correctors synthesis and identification theory have been transposed from conventional linear system theory to the considered algebraic structure (Baccelli et al., 1992). Applications of this theory have essentially concerned manufacturing systems (Menguy et al., 2000; Lahaye et al., 2003a), communication networks (Le Boudec and Thiran, 2001) and transportation networks (Braker, 1993; de Vries et al., 1998). In the latter, the focus has been on modelling, performance evaluation and stability analysis of railway networks.

For our concern, we are interested in applying dioids algebraic tools to the study of urban bus networks. The piloting of these networks can be seen as a two steps process. Firstly, a so-called *operating schedule* is defined for "mean conditions". In particular, timetables are defined at each stop to specify times at which buses should theoretically run. These timetables are used to inform passengers and, at particular stops, to re-synchronize buses. The second step consists in *regulation operations*: in reaction to current conditions (breakdown of a bus, modifications of traffic flows, *etc.*), a supervisor [1] may decide of adjustments from the operating schedule (transfer passengers, stop or reroute buses, *etc.*).

In (Lahaye et al., 2003b), we have proposed a first approach aiming at modelling some of these regulation operations. In practice, such a model may be used as a tool to aid decisions since it allows evaluating relevance of adjustments.

In this paper, the focus is rather on the first stage described above. In fact, we propose a model for urban bus networks functioning according to their operating schedule. We then show how dioid techniques [2] can be used for the synthesis of timetables. As described in (Ceder et al., 2000) this problem is crucial, and aims notably at maximizing connections between buses from different lines.

The outline of the paper is as follows. In §2, we recall elements of dioid theory as well as principles of DEDS description and control over dioids. In §3, we describe how bus networks operate, and a model is introduced. In §4, the method for timetables synthesis is proposed. An application to an elementary bus network is proposed in §5.

---

[1] Visualizing evolutions inside the network and communicating with bus drivers.

[2] The proposed approach is mainly based on residuation theory.

2. PRELIMINARIES

2.1 Elements of dioid theory

Definition 1. (Dioid). A dioid is a set $\mathcal{D}$ with two inner operations denoted $\oplus$ and $\otimes$. The sum is associative, commutative, idempotent ($\forall a \in \mathcal{D}$, $a \oplus a = a$) and admits a neutral element denoted ε. The product is associative, distributes over the sum and admits a neutral element denoted e. The element ε is absorbing for the product.

In a dioid $\mathcal{D}$, the equivalence : $a \succeq b \Leftrightarrow a = a \oplus b$ defines a partial order relation.

Definition 2. (Complete dioid). A dioid is said to be complete if it is closed for infinite sums and if product distributes over infinite sums too. The sum of all its elements is generally denoted $\top$.

A complete dioid has a structure of complete lattice (Baccelli et al., 1992, §4) , *i.e.* two elements in a complete dioid always have a *least upper bound* namely $a \oplus b$ and a *greatest lower bound* denoted $a \wedge b = \bigoplus_{\{x | x \preceq a,\ x \preceq b\}} x$.

Example 1. The set $\overline{\mathbb{Z}} = \mathbb{Z} \cup \{+\infty, -\infty\}$ endowed with the max operator as sum and the classical sum as product is a complete dioid, denoted $\overline{\mathbb{Z}}_{max}$ and usually called $(max, +)$ algebra, with $\varepsilon = -\infty$ and $e = 0$.

Theorem 1. The implicit equation $x = ax \oplus b$ defined over a complete dioid admits $x = a^*b$ as least solution with $a^* = \bigoplus_{i=0}^{+\infty} a^i$ and $a^0 = e$. The star operator $*$ is usually called *Kleene star.*

2.2 Residuation theory

A mapping f defined from a complete dioid $\mathcal{D}$ into a complete dioid $\mathcal{C}$ is said to be *isotone* if $a, b \in \mathcal{D}$, $a \preceq b \Rightarrow f(a) \preceq f(b)$. Residuation theory (Blyth and Janowitz, 1972) defines "pseudo-inverses" for some isotone mappings defined over ordered sets such as dioids (Cohen, 1998). More precisely, if the greatest element of set $\{x \in \mathcal{D} | f(x) \preceq b\}$ exists for all $b \in \mathcal{C}$, then it is denoted $f^\sharp(b)$ and $f^\sharp$ is called *residual* of f.

The isotone mapping $L_a : x \mapsto a \otimes x$ defined in a complete dioid is residuated. The greatest solution to $a \otimes x \preceq b$ exists and is equal to $L_a^\sharp(b)$, also denoted $\frac{b}{a}$ or $a \backslash b$. We furthermore have the following *formulæ* (see (Baccelli et al., 1992, §4.4)).

$$\frac{x}{ab} = \frac{a \backslash x}{b} \ (1) \quad \frac{x}{a^*} \preceq x \ (2) \quad \frac{a^* \backslash x}{a^*} = \frac{x}{a^*} \ (3)$$

Theorem 2. (Baccelli et al., 1992, Th. 4.56) Let $f : \mathcal{D} \to \mathcal{C}$ and $g : \mathcal{C} \to \mathcal{B}$. If f and g are residuated then $g \circ f$ is residuated and $(g \circ f)^\sharp = f^\sharp \circ g^\sharp$.

Definition 3. (Mapping restriction). Let $f : \mathcal{D} \to \mathcal{C}$ a mapping and $\mathcal{A} \subseteq \mathcal{D}$. We denote $f_{|\mathcal{A}} : \mathcal{A} \to \mathcal{C}$ the mapping defined by equality $f_{|\mathcal{A}} = f \circ Id_{|\mathcal{A}}$ where $Id_{|\mathcal{A}} : \mathcal{A} \to \mathcal{D}$ is the canonical injection.

A *constrained residuation problem* (Cohen, 1998) consists in finding the greatest solution to $f(x) \preceq b$ not in the whole $\mathcal{D}$ but only in a subset $\mathcal{A}$ of $\mathcal{D}$. More precisely, being given a residuated mapping f we aim at solving:

$$f_{|\mathcal{A}}(x) = f \circ Id_{|\mathcal{A}}(x) \preceq b. \qquad (4)$$

Proposition 1. If the canonical injection $Id_{|\mathcal{A}}$ is residuated, then the constrained residuation problem (4) admits an optimal solution denoted $f_{|\mathcal{A}}^\sharp(b)$.

Proof :

Thanks to theorem 2, if $Id_{|\mathcal{A}}$ is residuated, then so is $f \circ Id_{|\mathcal{A}}$, and the greatest solution to (4) is given by

$$f_{|\mathcal{A}}^\sharp(b) = (f \circ Id_{|\mathcal{A}})^\sharp(b) = Id_{|\mathcal{A}}^\sharp \circ f^\sharp(b) \qquad (5)$$

□

2.3 Representation of DEDS using dioids

Dioids algebra enables to model DEDS involving (only) synchronization phenomena. For instance, by dating each event, *i.e.* by associating with each event a discrete function x (resp. u and y for an input and an output) called *dater*[3], it is possible to get a linear state representation:

$$\begin{cases} x(k) = Ax(k-1) \oplus Bu(k) \\ y(k) = Cx(k). \end{cases} \qquad (6)$$

An analogous transform to $\mathcal{Z}$-transform (used to represent discrete-time trajectories in classical theory) can be introduced for daters. The γ-transform of a dater $x(k)$ is defined as the following formal power series: $x = \bigoplus_{k \in \mathbb{Z}} x(k)\gamma^k$. Indeed, variable γ can be interpreted as the backward shift operator in event domain. The set of formal power series in one variable γ and coefficients in $\overline{\mathbb{Z}}_{max}$ is a dioid denoted $\overline{\mathbb{Z}}_{max}[\![\gamma]\!]$.

State representation (6) becomes in $\overline{\mathbb{Z}}_{max}[\![\gamma]\!]$:

[3] $x(k)$ is the time of the $k+1$-th occurrence of event x.

$$\begin{cases} x = \gamma A x \oplus Bu \\ y = Cx. \end{cases} \quad (7)$$

Considering the earliest functioning of the system, we select the least solution of the first equation of (7) which is given according to theorem 1 by $x = (\gamma A)^* Bu$. This leads to $y = Hu = C(\gamma A)^* Bu$, in which $H = C(\gamma A)^* B$ is called the *transfer matrix*.

2.4 Just In Time control

The *Just In Time (JIT) control* of a DEDS consists in computing the latest input dates which cause output dates to be earlier than or equal to a given target. Let us define a dater specifying the desired outputs and z its γ-transform. From the representation in dioid algebra, this control problem is formulated as an inequality, and residuation theory leads to optimal solution u_{opt}:

$$u_{opt} = \bigoplus_{\{u|L_H(u)=Hu\preceq z\}} u = \frac{z}{H} = \frac{z}{C(\gamma A)^* B}. \quad (8)$$

3. MODELLING OF URBAN BUS NETWORKS

Many transportation systems may be studied as DEDS. With this point of view, their evolution is conditioned by events such as arrivals or departures of vehicles. In next sections, the focus is on urban bus networks. We first describe how they operate. Then, we propose a model in dioid $\overline{\mathbb{Z}}_{max}$.

3.1 Functioning of urban bus networks

As presented in (Hayat and Maouche, 1997), piloting of public transportation networks can be decomposed in two stages : the "operating schedule" and the "regulation".

Definition of an operating schedule The operating schedule is established with the aim of optimizing the offer of service according to objectives and constraints (bus fleet, line layouts, staff hours of work, *etc*). It is calculated for mean functioning conditions. In practical terms, this optimization results in:

(1) The definition of the bus routes and the bus stops for each line.
(2) The *choice of a level-of-service* for each line: distribution of resources (buses, drivers), definition of *the minimum and maximum headways* [4] (*i.e.* the expected minimum and maximum time separations between two buses).
(3) The *synthesis of timetables* defining times at which buses should theoretically run at each stop. Timetables are used to inform passengers and, at some stops, to re-synchronize buses.

Regulation This stage corresponds to adjustments or adaptations from the operating schedule in reaction to current functioning conditions. Common conditions leading to such adjustment operations are *disturbances*: breakdown of buses, modifications of traffic flows (for instance due to accidents), *etc.* A supervisor may then decide to transfer passengers, stop or reroute buses...

In the following, we are only interested in modelling the operating schedule. In §4, we furthermore show how dioid techniques can be used for the synthesis of timetables.

3.2 Modelling of a bus network

In this section, we propose a model for urban bus networks functioning according to their operating schedule. We assume that each line i includes n_i bus stops. The network is composed by M lines and $N = n_1 + \ldots + n_M$ stops. In the following, let $x_i(k)$ denote the departure time of the $(k+1)$-st bus at stop i. Without loss of generality, we assume that at the beginning of operation a bus departs from each stop [5]. Suppose that the bus coming from stop j reaches stop i. Then we have the following inequation: $x_i(k) \geq a_{ij} + x_j(k-1), k > 1$, in which a_{ij} denotes the travelling time from stop j to i. Let $x(k) = (x_1(k), x_2(k), \ldots, x_N(k))^t$, for the whole network this inequation can be written in max-algebraic matrix notation

$$x(k) \geq A \otimes x(k-1), \quad (9)$$

in which $A_{ij} = a_{ij}$ if stop j precedes stop i, otherwise $A_{ij} = \varepsilon$.

In practice, a public transportation network operates under a timetable which schedules the departure times of every bus. Buses respect timetables

[4] A minimum headway is defined to counteract the natural tendency of transit vehicles to bunch up. Thus, if a bus falls slightly behind schedule for any reasons, it will have more than the average number of passengers to pick up at the next station, which causes further delays. Thus, it keeps failing further behind schedule. Conversely, the bus behind it encounters fewer passengers than usual, allowing it to catch up with the preceding bus. Such bunching tends to appear if no minimum headway is defined. Symmetrically, a maximum headway specifies the desired level-of-service since it defines the minimum frequency of buses on the line.

[5] If no or several bus(es) initially depart from stops, then this results only in indexes modification. These cases can be dealt exactly as cases of places initially containing no or several token(s) for the modelling of timed event graph (Baccelli et al., 1992, §2.5.2).

only at specific stops[6] such as terminus or departure of lines or main stations. At the other stops, timetables are only used to inform passengers. We denote $u_l(k)$ the scheduled departure time for the $(k+1)$-st bus at stop l and S_u the set of specific stops, this leads to

$$x(k) \geq Bu(k), \tag{10}$$

in which B is a $N \times N$ matrix where $B_{ii} = e$ if stop $i \in S_u$, $B_{ij} = \varepsilon$ otherwise.

Considering inequations (9) and (10) and assuming that buses depart as soon as possible, a scheduled transportation network is modelled in dioid $\overline{\mathbb{Z}}_{max}$ by a state equation as in (6).

4. TIMETABLES SYNTHESIS OF URBAN BUS NETWORKS

4.1 Problem description

In a bus network, it may be profitable to minimize waiting times at a given stop for passengers arriving from another given stop. Such stops, generally close from a geographic point of view[7], belong to an itinerary which should be promoted[8] and are called *connection stops.*

With the aim of doing so, we consider a problem formulated in (Ceder et al., 2000)[9]. This problem consists in finding timetables which

- (i) maximize synchronizations at connection stops,
- (ii) ensure an expected level-of-service at particular stops of the network,
- (iii) satisfy planning constraints (*i.e.* minimum and maximum headways, see §3.1).

In order to compute timetables at each stop so as to synchronize bus departures at connection stops (objective mentioned at item (i)), we have to find input $\{u(k)\}_{k\in\mathbb{Z}}$ such that state $\{x(k)\}_{k\in\mathbb{Z}}$ satisfies:

$$x(k) = A'x(k-1) \oplus B'u(k) \tag{11}$$

where evolution matrix A' takes into account connections between lines and $B' = B$. More precisely, $A'_{ij} = A_{ij}$ if stop j precedes stop i on the same line; $A'_{ij} = w_{ij}$ denotes the walking time from stop j to stop i if i is in connection with j; $A'_{ij} = \varepsilon$ otherwise.

The desired level-of-service (mentioned at item (ii)) is specified for some particular stops (we denote S_o the set of these stops). More precisely, for such a stop $i \in S_o$, a dater $\{z_i(k)\}_{k\in\mathbb{Z}}$ can be defined: $z_i(k)$ specifies the latest date at which the $k+1$-st bus should depart from stop i. From equation (11), departure times at stops belonging to S_o are given by:

$$y(k) = C'x(k), \tag{12}$$

where C' is a $q \times N$ matrix ($q = Card(S_o)$) in which $C'_{ij} = e$ if i, j are indexes in y and x of a same stop in S_o, $C'_{ij} = \varepsilon$ otherwise.

We then aim at finding $\{u(k)\}_{k\in\mathbb{Z}}$ such that

$$y(k) = C'x(k) \preceq z(k). \tag{13}$$

According to item (iii), timetables given by dater $\{u(k)\}_{k\in\mathbb{Z}}$ must furthermore satisfy the planning constraints, *i.e.*

$$u(k+1) \geq \Delta^{min}u(k) \tag{14}$$
$$\Delta^{max}u(k) \geq u(k+1), \tag{15}$$

where Δ^{min} and Δ^{max} are diagonal matrices traducing minimum and maximum headways for each line. For all stops s of line i, the element Δ^{min}_{ss} (resp. Δ^{max}_{ss}) is equal to the minimum (resp. maximum) headway Δ^{min}_i (resp. Δ^{max}_i) of line i.

4.2 Formalization as a constrained residuation problem

From γ-transforms of equations (11) and (12), we obtain the expected *input/output behavior* (*cf.* § 2.3): $y = Hu = L_H(u)$, with $H = C'(\gamma A')^*B'$.

Equation (13) can then be written

$$y = L_H(u) \preceq z. \tag{16}$$

More precisely, we are interested in finding the greatest solution to this inequation. This solution allows buses to stay at stops belonging to S_u as late as possible while satisfying desired departures dates at stops in S_o (given by z). With this formulation, the problem given by equation (16) then appears to be a JIT control problem (*cf.* § 2.4).

Moreover, the searched solution u must satisfy inequations (14) and (15). Denoting $\mathbb{A}[\![\gamma]\!]$ the subset of $\overline{\mathbb{Z}}_{max}[\![\gamma]\!]$ composed of series satisfying γ-transforms of (14) and (15), we can rewrite the

[6] Where buses can park without blocking or disturbing traffic.
[7] Walking time between these stops is small (they are sometimes located at a same node of the network).
[8] Due to the important flow of users and/or for commercial reasons.
[9] Tools used in (Ceder et al., 2000) are essentially based on linear programming methods

considered problem as a constrained residuation problem (*cf.* § 2.2):

$$\begin{cases} L_H(u) \preceq z \\ u \in \mathbb{A}[\![\gamma]\!] \text{ (admissible domain).} \end{cases} \tag{17}$$

Canonical injection of the subset $\mathbb{A}[\![\gamma]\!]$ in $\overline{\mathbb{Z}}_{max}[\![\gamma]\!]$ is denoted $Id_{|\mathbb{A}}$ and $L_{H_{|\mathbb{A}}}$ is the mapping L_H restricted to the domain $\mathbb{A}[\![\gamma]\!]$. Regarding definition 3, the constrained problem is equivalent to solve:

$$L_{H_{|\mathbb{A}}}(u) = L_H \circ Id_{|\mathbb{A}}(u) \preceq z \,. \tag{18}$$

4.3 Solving the problem

From equation (18), we can apply proposition 1 if $Id_{|\mathbb{A}}$ is residuated:

$$L^{\sharp}_{H_{|\mathbb{A}}}(z) = Id^{\sharp}_{|\mathbb{A}} \circ L^{\sharp}_H(z). \tag{19}$$

The considered problem will then be solved in two stages:

(*i*) compute the solution to the "relaxed" problem $u_{opt} = L^{\sharp}_H(z)$ (*i.e* without planning constraints),
(*ii*) find the best approximation of (*i*) in admissible domain $u_{opt_{\mathbb{A}}} = Id^{\sharp}_{|\mathbb{A}}(L^{\sharp}_H(z))$.

To apply this reasoning, it remains to show that canonical injection $Id_{|\mathbb{A}}$ is residuated. With that goal, we first characterize the subset $\mathbb{A}[\![\gamma]\!]$.

Subset $\mathbb{A}[\![\gamma]\!]$ is composed of formal power series which satify planning constraints (14) and (15). Since Δ^{max} is diagonal, it is invertible and we obtain the two following inequations:

$$\begin{cases} \Delta^{min}\, u(k-1) & \preceq u(k) \\ (\Delta^{max})^{-1}\, u(k+1) & \preceq u(k), \end{cases}$$

The γ-transforms of these inequations lead to

$$\begin{cases} \Delta^{min}\, \gamma u & \preceq u \\ (\Delta^{max})^{-1}\, \gamma^{-1} u & \preceq u, \end{cases}$$

and $(\Delta^{min}\, \gamma \oplus (\Delta^{max})^{-1}\, \gamma^{-1})\, u \preceq u$. Since product is residuated, we equivalently have

$$u = \frac{u}{\Delta^{min}\, \gamma \oplus (\Delta^{max})^{-1}\, \gamma^{-1}} \wedge u.$$

The greatest solution is given by (see (Baccelli et al., 1992, Th. 4.7.3))

$$u = \frac{u}{(\gamma\Delta^{min} \oplus \gamma^{-1}(\Delta^{max})^{-1})^{*}}. \tag{20}$$

With the aim of abbreviating notations, we below denote $p = \gamma\Delta^{min} \oplus \gamma^{-1}(\Delta^{max})^{-1}$. Equality (20) allows the following definition of subset $\mathbb{A}[\![\gamma]\!]$:

$$A[\![\gamma]\!] = \{x \in \overline{\mathbb{Z}}_{max}[\![\gamma]\!] \mid x = \frac{x}{p^*}\}.$$

In order to show the residuability of canonical injection $Id_{|\mathbb{A}}$, we study the quotient of $\overline{\mathbb{Z}}_{max}[\![\gamma]\!]$ by a particular equivalence relation (this idea was inspired by (Cohen, 1993, Th. 30)).

Proposition 2. Let Π be the mapping from $\overline{\mathbb{Z}}_{max}[\![\gamma]\!]$ into itself defined by $\Pi : x \mapsto \frac{x}{p^*}$. We define the following equivalence relation

$$\{x \;\mathcal{R}\; y\} \iff \{\frac{x}{p^*} = \frac{y}{p^*}\}.$$

(1) Each equivalence class of $\overline{\mathbb{Z}}_{max}[\![\gamma]\!]_{/\mathcal{R}}$ contains one and only one element belonging to $\Pi(\overline{\mathbb{Z}}_{max}[\![\gamma]\!])$ and this element is explicitly given by $\frac{x}{p^*}$ for any x in the class.
(2) Element $\frac{x}{p^*}$ is the least element in $[x]_{\mathcal{R}}$, and it is the greatest element among those of $\Pi(\overline{\mathbb{Z}}_{max}[\![\gamma]\!])$ which are less than x.

Proof :

(1) For a given element $x \in \overline{\mathbb{Z}}_{max}[\![\gamma]\!]$, the element $\frac{x}{p^*}$ clearly belongs to $\Pi(\overline{\mathbb{Z}}_{max}[\![\gamma]\!])$ and since $\frac{p^* \backslash x}{p^*} = \frac{x}{p^*}$ (using (3)), it also belongs to $[x]_{/\mathcal{R}}$. Moreover, suppose that there is another element $x_1 \in [x]_{\mathcal{R}}$ which also belongs to $\Pi(\overline{\mathbb{Z}}_{max}[\![\gamma]\!])$: $\Pi(x'_1) = x_1 \Leftrightarrow x_1 = \frac{x'_1}{p^*} = \frac{p^* \backslash x'_1}{p^*} = \frac{x_1}{p^*}$, however $x_1 \mathcal{R} x$, so $\frac{x_1}{p^*} = \frac{x}{p^*}$ hence $x_1 = \frac{x}{p^*}$ which proves uniqueness of the element belonging to $[x]_{\mathcal{R}}$ and $\Pi(\overline{\mathbb{Z}}_{max}[\![\gamma]\!])$.
(2) Since $p^* \succeq e$, for all $y \in [x]_{\mathcal{R}}$, we have $\frac{x}{p^*} = \frac{y}{p^*} \preceq y$ (see (2)). Consequently $\frac{x}{p^*}$ is the least element of $[x]_{\mathcal{R}}$. For all $z \in \Pi(\overline{\mathbb{Z}}_{max}[\![\gamma]\!])$ such that $z \preceq x$, since Π is isotone, we obtain $\frac{z}{p^*} \preceq \frac{x}{p^*}$, hence $\frac{x}{p^*}$ is the greatest element of $\Pi(\overline{\mathbb{Z}}_{max}[\![\gamma]\!])$ which is less than x.

□

Note that for any element $x \in \Pi(\overline{\mathbb{Z}}_{max}[\![\gamma]\!])$ there exists y such that $x = \frac{y}{p^*}$, and we have $\frac{x}{p^*} = \frac{p^* \backslash y}{p^*} = \frac{y}{p^*} = x$ (using (3)), which shows that $\Pi(\overline{\mathbb{Z}}_{max}[\![\gamma]\!]) \subseteq \mathbb{A}[\![\gamma]\!]$. The converse is obvious and subsets $\Pi(\overline{\mathbb{Z}}_{max}[\![\gamma]\!])$ and $\mathbb{A}[\![\gamma]\!]$ are then identical.

Last assertion of proposition 2 consequently proves that canonical injection $Id_{|\mathbb{A}}$ is residuated and with $x' \in \overline{\mathbb{Z}}_{max}[\![\gamma]\!]$, its residual is

$$Id^{\sharp}_{|\mathbb{A}}(x') = \bigoplus_{\{x \in \mathbb{A}[\![\gamma]\!] \mid x \preceq x'\}} x = \frac{x'}{p^*}.$$

5. EXAMPLE

We now present an example of bus network composed of two lines including two connection stops (see the graph below). We assume that x_5 and x_7 are respectively in connection with x_1 and x_3 (null walking times for these connections). We then obtain the following matrix A'.

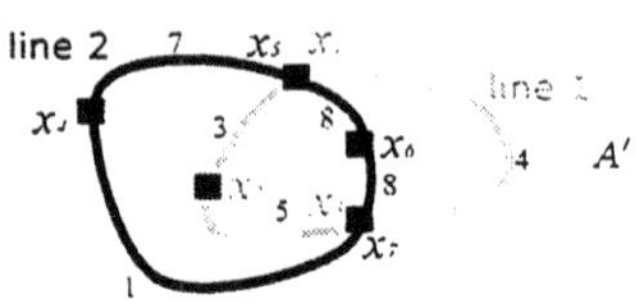

$$A' = \begin{pmatrix} \varepsilon & \varepsilon & 4 & \varepsilon & \varepsilon & \varepsilon & \varepsilon \\ 3 & \varepsilon & \varepsilon & \varepsilon & \varepsilon & \varepsilon & \varepsilon \\ \varepsilon & 5 & \varepsilon & \varepsilon & \varepsilon & \varepsilon & \varepsilon \\ \varepsilon & \varepsilon & \varepsilon & \varepsilon & \varepsilon & \varepsilon & 1 \\ e & \varepsilon & \varepsilon & 7 & \varepsilon & \varepsilon & \varepsilon \\ \varepsilon & \varepsilon & \varepsilon & \varepsilon & 8 & \varepsilon & \varepsilon \\ \varepsilon & \varepsilon & e & \varepsilon & \varepsilon & 8 & \varepsilon \end{pmatrix}$$

The desired output is specified for only one stop x_7 ($S_o = \{x_7\}$), this leads to $C = (\varepsilon\ \varepsilon\ \varepsilon\ \varepsilon\ \varepsilon\ \varepsilon\ e)$.

Moreover, the following tabular represents the desired level-of-service of x_7.

k	1	2	3	4	5
$z(k)$	10	20	24	28	40

Planning constraints are identical for both lines: $\triangle^{min} = 7$ and $\triangle^{max} = 10$.

We solve the problem in two stages (*cf.* item (i) and (ii) of §4.3). The first step consists in computing the solution of the relaxed problem u_{opt}. We show the computed dater u_{opt_6} of the generated timetables. Note that planning constraints are not satisfied.

k	1	2	3	4	5
$y(k)$	10	20	24	28	40
$u_{opt_6}(k)$	12	16	20	32	$\top$

Finally we find the best approximation of u_{opt} in $\mathbb{A}[\![\gamma]\!]$.

k	1	2	3	4	5
$y(k)$	7	14	21	28	38
$u_{opt_{\mathbb{A}6}}(k)$	6	13	20	30	40

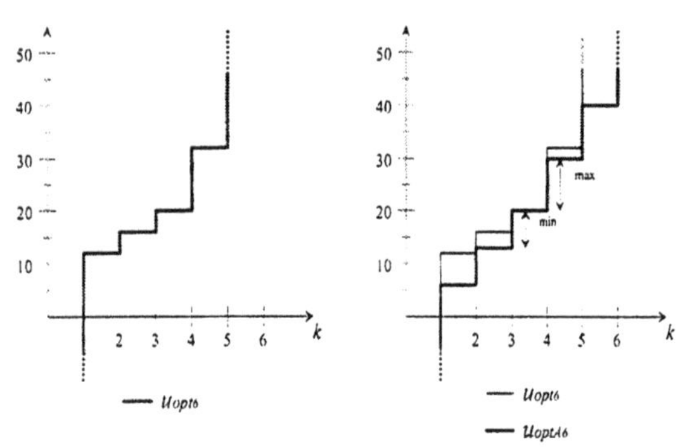

Fig. 1. Computation of dater $u_{opt_{\mathbb{A}6}}(k)$ in two stages

Remark 1. For the computation, we have used software tools dedicated to calculus in dioids (SW2004, 2004).

6. CONCLUSION

We show that dioids algebra can be suitable for bus networks modelling. First, we have described the functioning of such systems. A $(max, +)$ modelling is established to simulate such particular systems which are conditioned by timetables. We suggest a solution using residuation theory to the problem of timetables generation. We are currently trying to extend this work by considering "more general" synchronizations at connection stops (*e.g.* not occurring at each departure).

REFERENCES

F. Baccelli, G. Cohen, G. J. Olsder, and J. P. Quadrat. *Synchronization and Linearity.* Wiley, 1992.

T. S. Blyth and M. F. Janowitz. *Residuation Theory.* Pergamon press, 1972.

H. Braker. *Algorithms and applications in timed discrete event systems.* PhD thesis, Delft University of Technology, Dec 1993.

A. Ceder, B. Golany, and O. Tal. Creating bus timetables with maximal synchronization. *Transportation Research Part A: Policy and Practice*, 35:913–928, 2000.

G. Cohen. Two-dimensional domain representation of timed event graphs. In *Summer School on DES*, Spa, Belgium, June 1993.

G. Cohen. Residuation and Applications. In *Algèbres Max-Plus et applications en informatique et automatique : Ecole d'informatique théorique*, Noirmoutier, 1998. INRIA.

R. de Vries, B. de Schutter, and B. de Moor. On max-algebraic models for transportation networks. In *Proceedings of WODES'98*, Cagliary, Italy, 1998.

S. Hayat and S. Maouche. Régulation du trafic des autobus : amélioration de la qualité des correspondances. Technical Report LI-TU0192, INRETS, 1997.

S. Lahaye, L. Hardouin, and J. L. Boimond. Models combination in (max,+) algebra for the implementation of a simulation and analysis software. *Kybernetika*, 39(2):143–154, 2003a.

S. Lahaye, L. Houssin, and J. L. Boimond. Modelling of urban bus networks in dioids algebra. In *Proceedings of POSTA 2003*, pages 23–30, Roma, Italy, aug. 2003b. Springer Verlag.

J. Y. Le Boudec and P. Thiran. *Network Calculus.* Springer Verlag LNCS 2050, 2001.

E. Menguy, J.-L. Boimond, L. Hardouin, and J.-L. Ferrier. Just in time control of timed event graphs: Update of reference input, presence of uncontrollable input. *IEEE TAC*, 45(11):2155–2159, 2000.

SW2004.
`www.istia.univ-angers.fr/~hardouin/outils.html`, `www-rocq.inria.fr/maxplusorg/soft.html`., 2004.

ELSEVIER
IFAC
PUBLICATIONS
www.elsevier.com/locate/ifac

ON THE USE OF HYBRID SYSTEM IDENTIFICATION METHODS FOR MAX–PLUS–LINEAR SYSTEM IDENTIFICATION

Eberhard Münz, * **Gernot Schullerus** **
Volker Krebs *

* *Institut für Regelungs- und Steuerungssysteme,
Universität Karlsruhe (TH), Kaiserstraße 12,
D-76131 Karlsruhe, Germany,
Tel.: +49 (0)721/608-3179, Fax: +49 (0)721/608-2707,
e-mail:{muenz,krebs}@irs.uni-karlsruhe.de*

** *SEW-Eurodrive, Ernst-Blickle-Str. 42,
D-76646 Bruchsal, Germany,
Tel.: +49 (0)7251/75-5405, Fax: +49 (0)7251/75-505405,
e-mail: Gernot.Schullerus@sew-eurodrive.de*

Abstract: In recent years, different approaches for identification of hybrid systems and max–plus linear systems that are a special class of discrete event systems, have been developed. This paper introduces a new identification method for piecewise affine hybrid systems and shows that this approach is also easily applicable for identification of max–plus–linear systems. *Copyright© 2004 IFAC*

Keywords: System identification, max–plus–linear systems, hybrid systems.

1. INTRODUCTION

System identification is a powerful tool that can assist theoretical modelling by determining unknown parameters or adapt time–variant parameters of a process model obtained from physical principles. As well it can be used to experimentally obtain system models without having insight in the physical principles that govern the system behavior. As the resulting models are applied for system analysis, controller design, or fault diagnosis the need for new identification methods arises with the need of new model classes in control engineering.

In recent years, among others, two such model classes have gained significant attention: The max–plus–linear systems, a class of discrete event systems, and hybrid systems as a very powerful modelling formalism for complex systems with an interaction of time–driven and event–driven behavior. Consequently, identification methods have been developed for max–plus–linear systems (see (Menguy *et al.*, 2000; Schullerus and Krebs, 2002) and the references cited therein) and for hybrid systems (Roll, 2003; Ferrari-Trecate *et al.*, 2003; Münz *et al.*, 2003).

The aim of this paper is twofold:
(1) To introduce a new method for identification of hybrid system models with less computational burden than the existing approaches.
(2) To show that this method can be easily applied for identification of max–plus–linear systems as these systems can be represented by piecewise affine hybrid systems.

The paper is organized as follows: The next section briefly reviews the basic concepts of max–plus–linear systems considered in this paper, followed by a brief presentation of an existing identification procedure and its shortcomings that motivates the need for new algorithms. Then, section 3 discusses the class of piecewise affine systems and introduces a new identification algorithm for this system class. The application of this approach to max–plus–linear systems is presented in section 4 and illustrated in section 5 by the identification of the parameters of a max–plus–linear model of a manufacturing cell.

2. MAX–PLUS–LINEAR SYSTEM IDENTIFICATION

2.1 Max–Plus–linear systems

In the sequel discrete event systems are considered where the evolution of the events is governed by synchronization effects and no concurrency. The behavior of such systems can be modelled by the Petri net class of timed event graphs (see (Baccelli *et al.*, 1992) for details). To describe the timed event behavior of these systems, we denote the time instant when an internal event e_i occurs for the k–th time by the dater $x_i(k)$. Similarly, the input event times are denoted by $u_j(k)$. A relation between the event times $\underline{x}(k+1), \underline{x}(k) \in \mathbf{R}^n_{\max}$ and the input event times $\underline{u}(k+1) \in \mathbf{R}^p_{\max}$, where $\mathbf{R}_{\max} = \mathbf{R} \cup \{-\infty\}$, is given by the following equation (Baccelli *et al.*, 1992):

$$\underline{x}(k+1) = \underline{A}_0 \otimes \underline{x}(k+1) \oplus \underline{A}_1 \otimes \underline{x}(k) \oplus \underline{B}_0 \otimes \underline{u}(k+1) \,, \qquad (1)$$

where $\underline{A}_0, \underline{A}_1 \in \mathbf{R}^{n\times n}_{\max}$, $\underline{B}_0 \in \mathbf{R}^{n\times p}_{\max}$. The operators $\oplus$ and $\otimes$ are the addition and multiplication operators of the max–plus algebra and are defined by

$$x \oplus y = \max(x,y) \,, \qquad x \otimes y = x + y \qquad \forall\, x,y \in \mathbf{R}_{\max} = \mathbf{R} \cup \{-\infty\} \,.$$

The neutral elements of the max–plus addition and the max–plus multiplication are $-\infty = \varepsilon$ and 0, respectively. Note that ε is absorbing with respect to $\otimes$. Matrix addition and multiplication are defined similar to the conventional algebra:

$$\underline{P},\, \underline{Q} \in \mathbf{R}^{n\times p}_{\max} \,,\ (\underline{P} \oplus \underline{Q})_{ij} = (\underline{P})_{ij} \oplus (\underline{Q})_{ij} \,,$$

$$\underline{P} \in \mathbf{R}^{n\times p}_{\max},\ \underline{Q} \in \mathbf{R}^{p\times q}_{\max} \,,\ (\underline{P} \otimes \underline{Q})_{ij} = \bigoplus_{k=1}^{p} (\underline{P})_{ik} \otimes (\underline{Q})_{kj} \,.$$

Equation (1) can be used for parameter identification as will be shown in the subsequent section. However, it is not suited for system analysis or control design. If $(\underline{A}_0^n)_{ij} = \varepsilon$, $\forall\, i,j = 1,\ldots,n$, as will be assumed in the following considerations, (1) can be transformed into a recursive evolution equation

$$\underline{x}(k+1) = \underline{A} \otimes \underline{x}(k) \oplus \underline{B} \otimes \underline{u}(k+1) \,, \qquad (2)$$

where

$$\underline{A} = \underline{A}_0^* \otimes \underline{A}_1 \,,\ \underline{B} = \underline{A}_0^* \otimes \underline{B}_0 \ \text{ and } \underline{A}_0^* = \bigoplus_{k=0}^{n-1} \underline{A}_0^k \,.$$

The structural equivalence between (2) and the discrete time state space equation makes it possible to adapt well known concepts from system theory to this particular system class, provided the model and its parameter can be determined. The following section describes an identification procedure wich allows the determination of the model parameters using event time measurements. The model structure is assumed to be known.

2.2 Parameter estimation by minimization of an equation error

Consider the following parameter estimation problem:
Given the system model

$$\underline{x}(k+1) = \underline{A}_0 \otimes \underline{x}(k+1) \oplus \underline{A}_1 \otimes \underline{x}(k) \oplus \underline{B}_0 \otimes \underline{u}(k+1) \,,$$

the input event times $\underline{u}(k)$, $k = 1,\ldots,N$, and the measurements $\underline{x}(k)$, $k = 0,\ldots,N$, determine estimates $(\hat{\underline{A}}_0)_{ij}$, $(\hat{\underline{A}}_1)_{ij}$ and $(\hat{\underline{B}}_0)_{ij}$ for those elements $(\underline{A}_0)_{ij}$, $(\underline{A}_1)_{ij}$ and $(\underline{B}_0)_{ij}$ of the system matrices that are different from ε such that the equation error

$$\underline{\delta}(k+1) = \underline{x}(k+1) - \Big(\hat{\underline{A}}_0 \otimes \underline{x}(k+1) \oplus \hat{\underline{A}}_1 \otimes \underline{x}(k) \oplus \hat{\underline{B}}_0 \otimes \underline{u}(k+1)\Big) \qquad (3)$$

is minimized. It is assumed, that it is known which entries of the system matrices are equal to ε and which are not.
One solution to this problem in the case of time–invariant system parameters and no noise in the measured data is given by (Schullerus and Krebs, 2002) (cf. (Gallot *et al.*, 1997; Menguy *et al.*, 2000; De Schutter *et al.*, 2002) for other approaches and different model types)

$$(\hat{\underline{A}}_0)_{ij} = \min_{k=1}^{N} \left(x_i(k) - x_j(k)\right) \,,$$

$$(\hat{\underline{A}}_1)_{ij} = \min_{k=1}^{N} \left(x_i(k) - x_j(k-1)\right) \,,$$

$$(\hat{\underline{B}}_0)_{ij} = \min_{k=1}^{N} \left(x_i(k) - u_j(k)\right) \,.$$

It is tacitly assumed that the initial dater vector $\underline{x}(0)$ is known. However, if $\underline{x}(0)$ was unknown the estimation could be started with $k = 2$ and $N \geq 2$. Following the principles given in (Baccelli *et al.*, 1992; Cuninghame-Green, 1979; Menguy *et al.*, 2000) it can be shown that this solution, determined in the absence of noise for time–invariant parameters leads to an equation error $\underline{\delta}(k) = 0$, $\forall\, k = 1, \ldots, N$.

This is no longer true if the parameters are time–variant and the measured event times are corrupted by noise. To estimate the parameters of max–plus–linear systems in presence of noise we now propose to apply a new method used for the identification of piecewise affine hybrid systems presented in the following section.

3. IDENTIFICATION OF HYBRID SYSTEMS

To identify a large class of hybrid systems, piecewise affine discrete–time systems can be used. This has already been shown in (Münz *et al.*, 2003; Hodrus and Münz, 2003). If piecewise affine systems are used, the hybrid identification has to estimate the single submodels and their range of validity. For a particular class of these systems, (Roll, 2003) transformed the identification problem to a Mixed–Integer Linear Program (MILP) or to a Mixed–Integer Quadratic Program (MIQP) depending on the performance index used for identification. The problem investigated in this paper belongs to this class, but as it is stated in (Roll, 2003), the MILP resp. MIQP becomes very large and untractable very quickly for real problems.

Thus, an approach is presented, which is able to separate single models by minimizing a performance index, which is unconstrained. The major advantage of this procedure is, that very simple gradient based algorithms can be used to minimize this performance index. As a result of the simplicity, the algorithm is very fast, and practical applications are solvable.

To present this method, a short introduction into piecewise affine systems is given, and then the new identification approach is explained.

3.1 Piecewise affine hybrid systems

A piecewise affine system (sometimes referred to as piecewise linear system in literature) is obtained, if the domain of validity of a particular affine system is restricted to a particular domain of the state space; i.e. the whole system is defined on a subset Z of the cross product of the state space and the input space $Z \subseteq \mathbf{R}^n \times \mathbf{R}^m$. One single affine model

$$\dot{\underline{\xi}} = \underline{A}_i\underline{\xi} + \underline{B}_i\underline{\mu} + \underline{c}_i \tag{4}$$

is valid on a subset Z_i of the set Z, and the subsets Z_i form a complete and disjoint partition of the set Z, i.e. $\bigcup Z_i = Z$, $\bigcap Z_i = \emptyset$.

With these piecewise affine systems, it is possible to reproduce all hybrid phenomena mentioned in (Branicky, 1995), except jumps of the state. To reproduce all of these phenomena, either jump–equations or dirac impulses must be introduced in these time–continuous piecewise affine systems (Hodrus and Münz, 2003).

If piecewise affine systems in discrete time are considered, i.e. the differential equations (4) are replaced by difference equations, all hybrid phenomena can be reproduced. For simplicity of notation, the notations $\underline{A}_i$, $\underline{B}_i$ and $\underline{c}_i$ are used to denote such discrete time systems in the sequel.

In the case that the domains Z_i of the single submodels are known, it is rather easy to identify such piecewise affine systems in discrete time. To do so, the matrices $\underline{A}_i, \underline{B}_i$ and the vector $\underline{c}_i$ are combined to a parameter matrix $\underline{\Theta}_i$

$$\underline{\Theta}_i = \left[\underline{A}_i\ \underline{B}_i\ \underline{c}_i\right]^T.$$

The measured states $\underline{\xi}_k$ and inputs $\underline{\mu}_k$ which belong to the submodel i of the piecewise affine system, have to be arranged such that the equation

$$\underbrace{\begin{bmatrix} \underline{\xi}_k^T & \underline{\mu}_k^T & 1 \\ \vdots & \vdots & \vdots \\ \underline{\xi}_{k+N-1}^T & \underline{\mu}_{k+N-1}^T & 1 \end{bmatrix}}_{\underline{M}} \underline{\Theta}_i - \underbrace{\begin{bmatrix} \underline{\xi}_{k+1}^T \\ \vdots \\ \underline{\xi}_{k+N}^T \end{bmatrix}}_{\underline{Y}} = \underline{0}$$

is obtained. If noise is present, the equality to zeros does not hold, but it is possible to solve this equation in the Least–Squares sense by calculating the pseudo–inverse of $\underline{M}$. The parameter matrices $\underline{\Theta}_i$ are obtained from $\underline{\Theta}_i = \left(\underline{M}^T\underline{M}\right)^{-1}\underline{M}^T\underline{Y}$.

The identification approach for the piecewise affine systems where the domains Z_i are not known is somewhat different, but uses the same notations

3.2 Identification approach

The identification approach, which is able to identify the parameters of the submodels and the affiliation of the different data to the submodels simultaneously, proceeds in an iterative manner: some initial parameters for the submodels are obtained by minimizing a particular performance index. These initial estimates are enhanced step by step by correcting weight factors for each datum and minimizing the performance index again.

To explain this identification approach in detail, the noiseless case is considered. If each measured datum is error free, the distance between one submodel and the datum is zero

$$\forall j = 1, \ldots, N \; \exists i \in \{1, \ldots, M\} : \underline{m}_j^T \underline{\Theta}_i - \underline{y}_j^T = \underline{0} \ .$$

But not only the distance of the data to one of the submodels, but also one squared distance is zero. Furthermore, the product of all distances of one datum to all submodels is zero, since the product is zero, if at least one distance is zero. These considerations lead to the performance index

$$J = \sum_{j=1}^{N} \prod_{i=1}^{M} \left(\underline{m}_j^T \underline{\Theta}_i - \underline{y}_j^T \right)^T \left(\underline{m}_j^T \underline{\Theta}_i - \underline{y}_j^T \right) \ .$$

Now, the parameter matrices $\underline{\Theta}_i$ can be obtained analytically by choosing the elements of the $\underline{\Theta}_i$ such that $J = 0$. This is accomplished by minimizing the performance index J with respect to the parameter matrices $\underline{\Theta}_i$.
In the noise free case, it is shown in (Vidal *et al.*, 2003), that the parameters $\underline{\Theta}_i$ can be calculated analytically by using some results from Algebraic Geometry. If noise is present in the data, this is not possible any more. What can be done however in the presence of noise is to minimize the proposed performance index J. This leads to first estimates for the parameters in $\underline{\Theta}_i$; the affiliation of one datum to a submodel is indicated by the minimum distance of the considered datum to the different submodels.
Since the individual factors influence each other in the presence of noise a further enhancement of the performance index is necessary.
As mentioned before, the minimum distance of one datum to a submodel indicates, that one datum belongs to the corresponding submodel. The product of the other factors, consisting of squared distances to the other submodels, can be interpreted as weight for the smallest distance. This interpretation leads to the insight, that the parameter matrices $\underline{\Theta}_i$ are the result of a weighted Least-Squares estimate. Since the matrices $\underline{\Theta}_i$ should be the result of a not weighted Least-Squares algorithm, e. g. to avoid biases in the estimates, weights g_j are introduced such that the whole weight for the smallest distance from one submodel is set to 1. Without loss of generality, the first squared distance in addend k of the performance index J may be the smallest. Thus the product

$$w_k = \prod_{i=2}^{M} \left(\underline{m}_j^T \underline{\Theta}_i - \underline{y}_j^T \right)^T \left(\underline{m}_j^T \underline{\Theta}_i - \underline{y}_j^T \right)$$

can be interpreted as weight for the smallest distance. By multiplying addend k with $g_k = 1/w_k$ the overall weight for the smallest squared distance is set to 1 and a performance index with corrected weights is obtained

$$J_w = \sum_{j=1}^{N} g_j \cdot \prod_{i=1}^{M} \left(\underline{m}_j^T \underline{\Theta}_i - \underline{y}_j^T \right)^T \left(\underline{m}_j^T \underline{\Theta}_i - \underline{y}_j^T \right) \ .$$

The procedure consisting of the two steps, minimizing J_w, and correcting the weights g_k, is repeated until no data changes its affiliation to a submodel in two consecutive steps.
Afterwards, the data is grouped by the least squared distance to the different submodels and another Least-Squares estimate can be calculated in the different groups.
The most general procedure to obtain the domain of validity Z_i for each submodel, is to set up an environment around each datum in the set Z. An element ζ of Z belongs to the environment of a datum $\underline{m}_j$, iff the distance between the element ζ and all other data $\underline{m}_k$ is larger than the distance between datum $\underline{m}_j$ and ζ. The union set of all environments of data which belong to model i yields the set Z_i.
For the application of this method in the context of max-plus-linear systems, some additional a priori knowledge leads to a simplification of this general approach. This is discussed in the following section.

4. APPLICATION TO MAX-PLUS-LINEAR SYSTEMS

In (Heemels *et al.*, 2001) the equivalence of max-min-plus-scaling systems to piecewise affine systems has been demonstrated. As max-plus-linear systems are a subclass of max-min-plus-scaling systems, any system of max-plus-linear equations (1) can be represented by a set of piecewise affine systems as shown in the sequel. This discussion follows the considerations from (Baccelli *et al.*, 1992) in the context with the graphical representation of max-plus-polynomials.

Consider the max-plus-linear system

$$x(k+1) = a \otimes x(k) \oplus b \otimes u(k+1) \ . \qquad (5)$$

Recalling that the operator $\otimes$ represents conventional addition and the operator $\oplus$ corresponds to conventional maximization this equation can be rewritten as

$$x(k+1) \geq a + x(k), \quad x(k+1) \geq b + u(k+1),$$
$$(x(k+1) - a - x(k))\,(x(k+1) - b - u(k+1)) = 0 \ .$$

The corresponding piecewise affine system is represented in figure 1 with $a = 3$ and $b = 1$. The two affine parts represent the planes

$$x(k+1) = a + x(k) \qquad x(k+1) = b + u(k+1) \ .$$

So the identification of the parameters a and b of this max-plus-linear system is equivalent to the

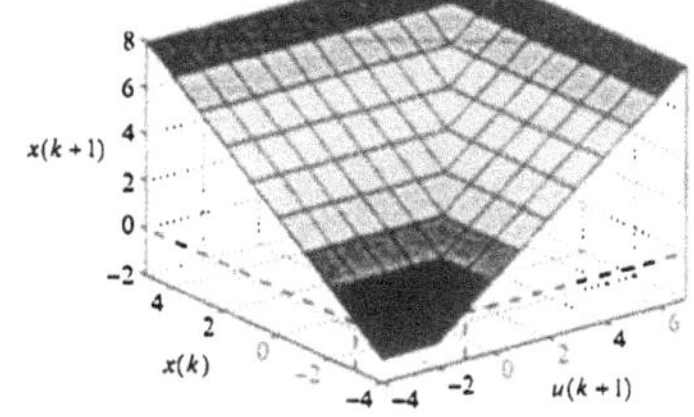

Fig. 1. Piecewise affine representation of the max–plus–linear system (5).

determination of the two plane equations. Once the planes are known, these parameters are easily obtained from the intersection with the plane $x(k+1) = 0$. The extension to the max–plus–linear system given by (1) is straightforward due to the fact that any row in this equation

$$x_i(k+1) = \bigoplus_{j=1}^{n} (\underline{A}_0)_{ij} \otimes x_j(k+1) \oplus \bigoplus_{j=1}^{n} (\underline{A}_1)_{ij} \otimes x_j(k) \oplus \bigoplus_{j=1}^{p} (\underline{B}_0)_{ij} \otimes u_j(k+1)$$

can be represented in a similar manner by as many hyperplanes as there exist entries different from ε in the i–th rows of $\underline{A}_0$, $\underline{A}_1$ and $\underline{B}_0$.

In the context of max–plus–linear systems, some parameters of the hyperplanes are known a priori: let M be the number of entries different from ε which must be estimated in one row of the system equation of a max–plus–linear model. As each of these parameters corresponds to one hyperplane, we need to determine M submodels in order to obtain estimates for these holding times with piecewise affine identification methods. Each submodel corresponds to an entry in the matrices $\underline{A}_0$, $\underline{A}_1$, and $\underline{B}_0$. Since the holding time in the max–plus–linear system corresponds to the affine part of the corresponding affine submodel, the linear coefficients in the affine submodels can be fixed to 1, iff the corresponding entry in $\underline{A}_0$, $\underline{A}_1$, or $\underline{B}_0$ is to be identified with this submodel, and to 0 otherwise; i.e. for each unknown holding time one submodel with one unknown parameter is obtained.

To estimate all holding times in a max–plus–linear system, a piecewise affine model has to be set up for each row, and the unknown holding times can be estimated, as it is described in section 3. The domains of validity Z_i of the different submodels are not of interest in this context, as the holding times represented by the affine parts of the affine submodels describe the behavior of a max–plus–linear system exactly.

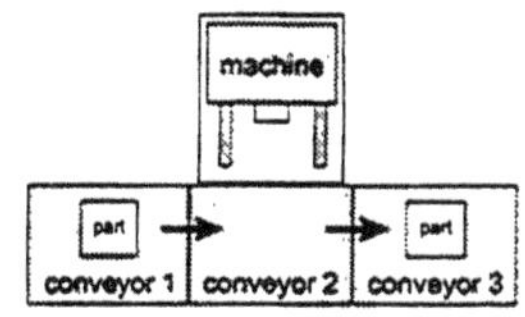

Fig. 2. Manufacturing cell.

5. EXAMPLE

Consider now a manufacturing cell shown in figure 2, where parts are delivered to the machine by a conveyor, machined and released to an additional conveyor. The capacity of each conveyor is limited to one part. The machine can process one part at the same time. The behavior of the plant can be described based on certain event times as follows: Let $x_1(k)$ be the time instant when the k–th part is loaded onto the conveyor 1. After τ_{21} time units, this part is ready to enter the machine. The dater $x_2(k)$ denotes the time when the k–th part enters the machine. After the machining operation which takes τ_{32} time units, the part is released to conveyor 2 at time $x_3(k)$ reaching a final position on conveyor 3 after τ_{43} time units. From this final position, the part is picked up at time $x_4(k)$. Conveyor 1 and 3 can receive a new part after τ_{12} and τ_{34} time units, respectively, whereas the machine must be prepared for a new operation for τ_{23} time units. The input event times $u_1(k)$ and $u_2(k)$ correspond to the time instants when a new part is available to be delivered for conveyor 1 or removed from conveyor 3, respectively.

Using the above reasoning and the initial state of the system as shown in figure 2, the following max–plus–linear equations hold for $\underline{x}(k)$:

$$\begin{aligned} x_1(k+1) &= \tau_{12} \otimes x_2(k+1) \oplus \tau_{11} \otimes u_1(k+1) \\ x_2(k+1) &= \tau_{21} \otimes x_1(k) \oplus \tau_{23} \otimes x_3(k) \\ x_3(k+1) &= \tau_{32} \otimes x_2(k+1) \oplus \tau_{34} \otimes x_4(k+1) \\ x_4(k+1) &= \tau_{43} \otimes x_3(k) \oplus \tau_{42} \otimes u_2(k+1) \end{aligned}$$

These equations are of the structure given by (1)

$$\underline{x}(k+1) = \underline{A}_0 \otimes \underline{x}(k+1) \oplus \underline{A}_1 \otimes \underline{x}(k) \oplus \underline{B}_0 \otimes \underline{u}(k+1)$$

$$\underline{A}_0 = \begin{bmatrix} \varepsilon & \tau_{12} & \varepsilon & \varepsilon \\ \varepsilon & \varepsilon & \varepsilon & \varepsilon \\ \varepsilon & \tau_{32} & \varepsilon & \tau_{34} \\ \varepsilon & \varepsilon & \varepsilon & \varepsilon \end{bmatrix}, \quad \underline{A}_1 = \begin{bmatrix} \varepsilon & \varepsilon & \varepsilon & \varepsilon \\ \tau_{21} & \varepsilon & \tau_{23} & \varepsilon \\ \varepsilon & \varepsilon & \varepsilon & \varepsilon \\ \varepsilon & \varepsilon & \tau_{43} & \varepsilon \end{bmatrix}$$

$$\underline{B}_0 = \begin{bmatrix} \tau_{11} & \varepsilon & \varepsilon & \varepsilon \\ \varepsilon & \varepsilon & \varepsilon & \tau_{42} \end{bmatrix}^T .$$

From the structure of the system, the entries that are equal to ε in the matrices $\underline{A}_0$, $\underline{A}_1$ and $\underline{B}_0$ are known. The other entries of these matrices, the unknown parameters of the system are now

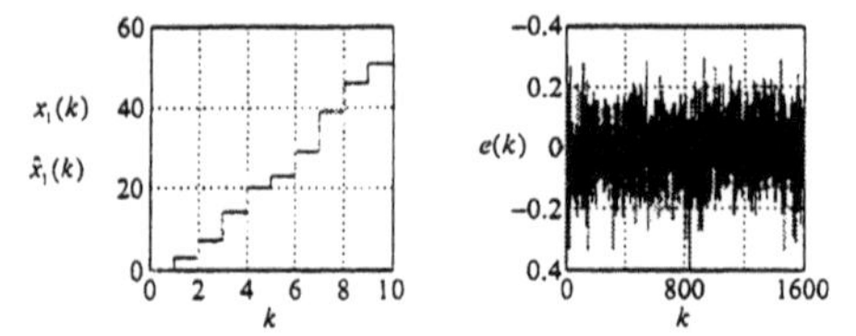

Fig. 3. Left: Simulation results obtained for the original system corrupted by the noise (–) and for the estimated system (– –) for $x_1(k)$. Both curves coincide in this scale. Right: Deviation $e(k)$ between the original data and the data obtained with the estimated parameters for $x_1(k)$.

identified using the method introduced in this paper.
To illustrate the results obtained with the new identification method, the system was simulated using the parameters

$$\tau_{12} = 2 \quad \tau_{21} = 1 \quad \tau_{23} = 1 \quad \tau_{11} = 0$$
$$\tau_{32} = 3 \quad \tau_{34} = 1 \quad \tau_{43} = 2 \quad \tau_{42} = 0$$

and choosing a sequence of 1600 input event times $\underline{u}(k)$. In order to demonstrate the performance of the identification algorithm in the presence of measurement noise the simulation data for $\underline{x}(k)$ was corrupted by gaussian noise with $\sigma^2 = 0.1$. Based on this data and the input event times, the following estimates were obtained for the entries of the system matrices:

$$\hat{\tau}_{12} = 1.8677 \quad \hat{\tau}_{21} = 1.0030, \quad \hat{\tau}_{32} = 3.0002$$
$$\hat{\tau}_{34} = 0.9897, \quad \hat{\tau}_{23} = 1.0035 \quad \hat{\tau}_{11} = 0.0045,$$
$$\tau_{43} = 1.9950 \quad \hat{\tau}_{42} = 0.0043.$$

Figure 3 illustrates simulation results, to compare the original data $\underline{x}$ used for parameter estimation with the data $\hat{\underline{x}}$ determined using the estimated system. The results are illustrated for $x_1(k)$ only. As the identification is very accurate, no differences are visible on the left side of figure 3. The right side illustrates the differences $e(k) = x_1(k) - \hat{x}_1(k)$ between the original and identified system. Similar results are obtained for the other components of $\underline{x}$.

6. CONCLUSIONS

The objective of this paper was the presentation of a new identification method for hybrid systems represented by piecewise affine systems and the application of this method for max–plus–linear system identification. The new identification method determines the different dynamic models for the different regions in state space and assigns the given data to these regions. The application to max–plus–linear systems is straightforward as these systems can be represented by piecewise affine models. Some results were illustrated in an example. The new method is applicable to a discrete event system class, and to hybrid systems, thus providing a new powerful identification tool.

REFERENCES

Baccelli, F., G. Cohen, G. J. Olsder and J.-P. Quadrat (1992). *Synchronization and Linearity.* John Wiley & Sons. Chichester.

Branicky, M.S. (1995). Studies in Hybrid Systems: Modeling, Analysis, and Control. PhD thesis. MIT.

Cuninghame-Green, R. (1979). *Minimax Algebra.* Vol. 166 of *Lecture Notes in Economics and Mathematical Systems.* Springer–Verlag. Berlin.

De Schutter, B., T. J. J. van den Boom and V. Verdult (2002). State space identification of max–plus–linear discrete event systems from input–output data. In: *Proceedings of the 41st IEEE Conference on Decision and Control.* Las Vegas. pp. 4024–4029.

Ferrari-Trecate, G., M. Muselli, D. Liberati and M. Morari (2003). A clustering technique for the identification of piecewise affine and hybrid systems. *Automatica* **39**, 205–217.

Gallot, F., J. L. Boimond and L. Hardouin (1997). Identification of Simple Elements in Max–Algebra: Application to SISO Discrete Event Systems Modelisation. In: *Proceedings of the 4th European Control Conference.* Brussels.

Heemels, W. P. M. H., B. De Schutter and A. Bemporad (2001). Equivalence of hybrid dynamical models. *Automatica* **37**(7), 1085–1091.

Hodrus, E. and E. Münz (2003). Hybride Phänomene in zeitdiskreter Darstellung. *at-Automatisierungstechnik.*

Menguy, E., J. L. Boimond, L. Hardouin and J. L. Ferrier (2000). A first step towards adaptive control for linear systems in max algebra. *Discrete Event Dynamic Systems* **10**(4), 347–368.

Münz, E., T. Hodrus and V. Krebs (2003). Top-down identification of hybrid characteristic maps. In: *Proceedings of ADHS.* Saint Malo.

Roll, J. (2003). Local and Piecewise Affine Approaches to System Identification. PhD thesis. Linköping University.

Schullerus, G. and V. Krebs (2002). A method for estimating the holding times in timed event graphs. In: *Proceedings of the 6th International Workshop on Discrete Event Systems (WODES'02).* IEEE Computer Society. Zaragoza. pp. 209–216.

Vidal, R., S. Soatto and S. Sastry (2003). An algebraic geometric approach to the identification of linear hybrid systems. In: *Proceedings of IEEE Conference on Decision and Control.* Maui, Hawaii USA.

ELSEVIER
IFAC
PUBLICATIONS
www.elsevier.com/locate/ifac

MODELLING AND CONTROL OF DISCRETE EVENT SYSTEMS USING SWITCHING MAX-PLUS-LINEAR SYSTEMS

T.J.J. van den Boom[1] and B. De Schutter[1],

Delft Center for Systems and Control, Delft University of Technology,
Mekelweg 2, 2628 CD Delft, The Netherlands,

Abstract: In this paper we consider the modelling and control of discrete event systems using switching max-plus-linear systems. In switching max-plus-linear systems we can switch between different modes of operation. In each mode the discrete event system is described by a max-plus-linear state equation with different system matrices for each mode. The switching allows us to change the structure of the system, to break synchronization and to change the order of events. We will give some examples of this type of system.
We define the model predictive control design problem for this type of discrete event system, and we show that solving this problem in general leads to a mixed integer optimization problem. *Copyright © 2004 IFAC*

1 INTRODUCTION

The class of discrete event systems essentially consists of man-made systems that contain a finite number of resources (such as machines, communications channels, or processors) that are shared by several users (such as product types, information packets, or jobs) all of which contribute to the achievement of some common goal (the assembly of products, the end-to-end transmission of a set of information packets, or a parallel computation) [1]. In general, models that describe the behavior of a discrete event system are nonlinear in conventional algebra. However, there is a class of discrete event systems — the max-plus-linear discrete event systems — that can be described by a model that is "linear" in the max-plus algebra [1, 4], which has maximization and addition as its basic operations. The max-plus-linear discrete event systems can be characterized as the class of discrete event systems in which only synchronization and no concurrency or choice occurs.

In this paper we will consider discrete event systems that can switch between different modes of operation. In each mode the system is described by a max-plus-linear state equation with different system matrices for each mode. The switching changes the structure of the system, and so allows us to break synchronization and to change the order of events. We define a control design problem for such a system to optimize the system's behavior. In general this will lead to a mixed integer optimization problem.

[1] {t.j.j.vandenboom,b.deschutter}@dcsc.tudelft.nl, http://www.dcsc.tudelft.nl

2 MAX-PLUS ALGEBRA AND SWITCHING MAX-PLUS-LINEAR SYSTEMS

2.1 Max-plus algebra

In this section we give the basic definition of the max-plus algebra and we present some results on a class of max-plus functions.
Define $\varepsilon = -\infty$ and $\mathbb{R}_\varepsilon = \mathbb{R} \cup \{\varepsilon\}$. The max-plus-algebraic addition ($\oplus$) and multiplication ($\otimes$) are defined as follows [1, 4]:

$$x \oplus y = \max(x, y) \qquad x \otimes y = x + y$$

for numbers $x, y \in \mathbb{R}_\varepsilon$, and

$$[A \oplus B]_{ij} = a_{ij} \oplus b_{ij} = \max(a_{ij}, b_{ij})$$

$$[A \otimes C]_{ij} = \bigoplus_{k=1}^{n} a_{ik} \otimes c_{kj} = \max_{k=1,\ldots,n} (a_{ik} + c_{kj})$$

for matrices $A, B \in \mathbb{R}_\varepsilon^{m\times n}$ and $C \in \mathbb{R}_\varepsilon^{n\times p}$.

2.2 Max-plus-linear systems

In [1, 4] it has been shown that discrete event systems in which there is synchronization but no concurrency can be described by a model of the form

$$x(k) = A(k) \otimes x(k-1) \oplus B(k) \otimes u(k) \ . \quad (1)$$

Systems that can be described by this model will be called max-plus-linear systems. The index k is called the event counter. For discrete event systems the state $x(k)$ typically contains the time instants at which the internal events occur for the kth time, the input $u(k)$ contains the time instants at which the input events occur for the kth time, and the output $y(k)$ contains the time instants at which the output events occur for the kth time.

2.3 Switching max-plus-linear systems

In this paper we will consider discrete event systems that can switch between different modes of operation. In each different mode $\ell = 1, \ldots, n_m$, the system is described by a max-plus-linear state equation

$$x(k) = A^{(\ell)}(k) \otimes x(k-1) \oplus B^{(\ell)}(k) \otimes u(k) \quad (2)$$

in which the matrices $A^{(\ell)}$, $B^{(\ell)}$ are the system matrices for the ℓ-th mode. The switching allows us to change the structure of the system, to break synchronization and to change the order of events. Note that each mode ℓ corresponds to a set of required synchronizations and an event order schedule, which leads to a model (2) with system matrices $(A^{(\ell)}, B^{(\ell)})$ for the ℓ-th model.

The moments of switching are determined by a switching mechanism. We define the switching variable $z(k)$, which may depend on the previous state $x(k-1)$, previous mode $\ell(k-1)$, the input variable $u(k)$ and an (additional) control variable $v(k)$:

$$z(k) = \Phi(x(k-1), \ell(k-1), u(k), v(k)) \in \mathbb{R}^{n_z} . \quad (3)$$

We partition $\mathbb{R}^{n_z}$ in n_m subsets $\mathcal{Z}^{(i)}$, $i = 1, \ldots, n_m$. The mode $\ell(k)$ is now obtained by determining in which set $z(k)$ is for event k. So if $z(k) \in \mathcal{Z}^{(i)}$, then $\ell(k) = i$. In some systems the switching mechanism will completely depend on the state $x(k-1)$ and input $u(k)$, in other examples $z(k)$ will be equal to $v(k)$ and so we can control the switching by choosing an appropriate $v(k)$.

3 EXAMPLES OF SWITCHING MAX-PLUS-LINEAR SYSTEMS

3.1 Production system

Consider the production system of Figure 1. This system consists of five machines M_1, M_2, M_3, M_4 and M_5. The raw material is fed to machine M_1 and M_2, where preprocessing is done. Both products now have to be finished in either unit M_3 and M_4, which basically perform

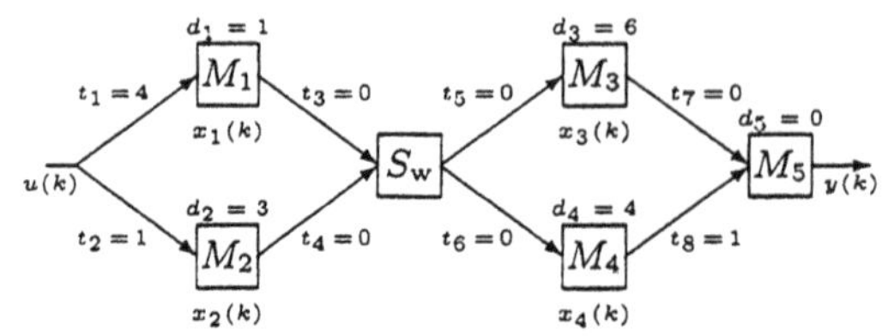

Figure 1: A production system.

the same task, but the processing time of M_3 is longer than M_4. Therefore, the products coming from machine M_1 and M_2 are directed to a switching device S_w, that feeds the first product in the kth cycle to the slower machine M_3 and the second product to the faster machine M_4. Finally, the products are assembled (instantaneously[1]) in machine M_5 and become available. We assume that each machine starts working as soon as possible on each batch, i.e., as soon as the raw material or the required intermediate products are available, and as soon as the machine is idle (i.e., the previous batch has been finished and has left the machine). We define $u(k)$ as the time instant at which the system is fed for the kth time, $y(k) = x_5(k)$ as time instant at which the kth product leaves the system and $x_i(k)$ as the time instant at which machine i starts for the kth time. The variable t_j for $j = 1, \ldots, 4$ is the transportation time, and d_i for $i = 1, \ldots, 5$ is the processing time on machine i. The system equations for x_1 and x_2 are given by

$$\begin{aligned} x_1(k) &= \max(x_1(k-1) + d_1, u(k) + t_1) \\ x_2(k) &= \max(x_2(k-1) + d_2, u(k) + t_2) \end{aligned}$$

If $x_1(k) + d_1 \leq x_2(k) + d_2$, the product of M_1 will be directed to M_3 and the product of M_2 will be directed to M_4, and we obtain:

$$\begin{aligned} x_3(k) &= \max(x_1(k) + d_1, x_3(k-1) + d_3) \\ &= \max(x_1(k-1) + 2d_1, x_3(k-1) + d_3, \\ &\qquad u(k) + d_1 + t_1) \\ x_4(k) &= \max(x_2(k) + d_2, x_4(k-1) + d_4) \\ &= \max(x_2(k-1) + 2d_2, x_4(k-1) + d_4, \\ &\qquad u(k) + d_2 + t_2) \\ x_5(k) &= \max(x_3(k) + d_3, x_4(k) + d_4 + t_8) \\ &= \max(x_1(k-1) + 2d_1 + t_1 + d_3, \\ &\qquad x_2(k-1) + 2d_2 + d_4 + t_8, \\ &\qquad x_3(k-1) + 2d_3, x_4(k-1) + 2d_4 + t_8, \\ &\qquad u(k) + d_1 + t_1 + d_3, \\ &\qquad u(k) + d_2 + t_2 + d_4 + t_8) \end{aligned}$$

[1] I.e. with a neglegible processing time.

For this first mode ($x_1(k)+d_1 \leq x_2(k)+d_2$) we obtain the system matrices

$$A^{(1)} = \begin{bmatrix} 1 & \varepsilon & \varepsilon & \varepsilon & \varepsilon \\ \varepsilon & 3 & \varepsilon & \varepsilon & \varepsilon \\ 1 & \varepsilon & 6 & \varepsilon & \varepsilon \\ \varepsilon & 6 & \varepsilon & 4 & \varepsilon \\ 8 & 11 & 12 & 9 & \varepsilon \end{bmatrix} \qquad B^{(1)} = \begin{bmatrix} 4 \\ 1 \\ 5 \\ 4 \\ 11 \end{bmatrix}$$

Similarly, for this second mode ($x_1(k) + d_1 > x_2(k) + d_2$) we obtain the system matrices

$$A^{(2)} = \begin{bmatrix} 1 & \varepsilon & \varepsilon & \varepsilon & \varepsilon \\ \varepsilon & 3 & \varepsilon & \varepsilon & \varepsilon \\ \varepsilon & 6 & 6 & \varepsilon & \varepsilon \\ 1 & \varepsilon & \varepsilon & 4 & \varepsilon \\ 7 & 12 & 12 & 9 & \varepsilon \end{bmatrix} \qquad B^{(2)} = \begin{bmatrix} 4 \\ 1 \\ 4 \\ 5 \\ 10 \end{bmatrix}$$

To decide the switching mechanism, we define the switching variable

$$\begin{bmatrix} z_1(k) \\ z_2(k) \end{bmatrix} = \begin{bmatrix} x_1(k) + d_1 \\ x_2(k) + d_2 \end{bmatrix}$$
$$= \begin{bmatrix} \max(x_1(k-1) + 2d_1, u(k) + d_1 + t_1) \\ \max(x_2(k-1) + 2d_2, u(k) + d_2 + t_2) \end{bmatrix}$$
$$= \begin{bmatrix} \max(x_1(k-1) + 2, u(k) + 5) \\ \max(x_2(k-1) + 6, u(k) + 4) \end{bmatrix}$$

and the sets

$$\mathcal{Z}^{(1)} = \{z \in \mathbb{R}^2 | z_1 \leq z_2\},$$
$$\mathcal{Z}^{(2)} = \{z \in \mathbb{R}^2 | z_1 > z_2\}.$$

Now the state space equation of our system is given by (2).

3.2 Railway network

Consider the railroad network of Figure 2 (see also [11]). There are 4 stations in this railroad network (A, B, C and D) that are connected by 5 single tracks (1/7, 2/4, 3, 5, 9) and one double track (tracks 6 & 8). There are three trains available. The first train follows the route $D \rightarrow A \rightarrow B \rightarrow D$, the second train follows the route $A \rightarrow B \rightarrow C \rightarrow A$, and the third train follows the route $D \rightarrow A \rightarrow C \rightarrow D$. We assume that there exists a periodic timetable that schedules the earliest departure times of the trains. The period of the timetable is $T = 60$ minutes. So if a departure of a train from station B is scheduled at 5.30 a.m., then there is also scheduled a departure of a train from station B at 6.30 a.m., 7.30 a.m., and so on.

Each track of the railway network has a number and a train allocated to it. For the sake of simplicity we will say "(virtual) train j" to denote the (physical) train on a specific track.

The number of tracks in the network is equal to 7, the number of physical trains in the network is equal to 3, and the number of virtual trains in the network is equal to 9. Let $x_j(k)$, $j = 1, \ldots, 9$ be the time instant at which train j departs from its station for the kth time. Let $d_j(k)$ be the departure time for this train according to the time schedule, and let $a_j(k)$ be the transportation time for this train j.

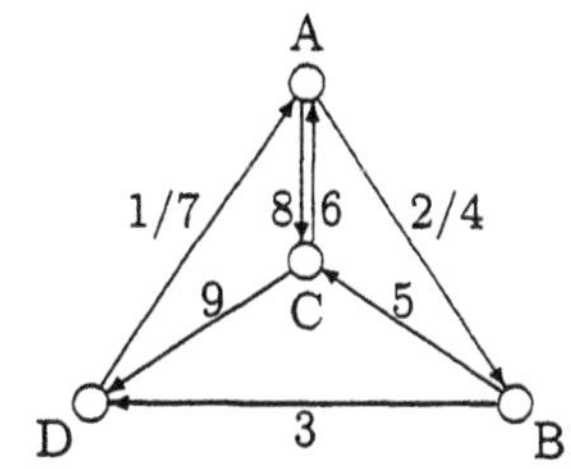

Figure 2: The railroad network of the example of Subsection 3.2.

Table 1 summarizes the information in connection with the nominal travelling times and the departure times. All the times are measured in minutes. The indicated departure times are the earliest departure times in the initial station of the track expressed in minutes after the hour. The first period starts at time $t = 0$. At the beginning of the first period the first train is in station A and the second train is in station B.

train	from -to	travel time	dep -arr	constraints
1	D-A	12	00-12	same train as 3^- connects to 9^- follows 7^-
2	A-B	12	15-27	same train as 1 connects to 6^- follows 4^-
3	B-D	20	30-50	same train as 2
4	A-B	12	19-31	same train as 6^- follows 2 connects to 7
5	B-C	10	34-44	same train as 4
6	C-A	25	47-12	same train as 5
7	D-A	12	04-16	same train as 9^- follows 1
8	A-C	25	19-44	same train as 7
9	C-D	10	47-57	same train as 8 connects to 5

Note: 3^- denotes train 3 in the previous cycle

Table 1: The nominal travelling times and the departure times for the railroad network.

The *continuity constraints* are that the trains on tracks 1, 2 and 3 are physically the same train, and the same holds for the trains on tracks 4, 5 and 6 and for the trains on tracks 7, 8 and 9.

Connection constraints are introduced to allow the passengers to change trains. In this network, train 1 has to wait for train 9 in the previous cycle with minimum connection time $c^{\min} = 3$. In the same way, train 2 waits for train 6 in the previous cycle, train 4 wait for train 7, and train 9 waits for train 5. The minimum stopping time of train j at station j to allow passenger to get off or on the train is fixed at $s^{\min} = 1$.

Follow constraints are introduced to guarantee sufficient separation time between two trains on the same track (moving in the same direction). In this network, train 4 is scheduled behind train 2 (train 4 follows train 2) with a minimum separation time $f^{\min} = 4$. In the same way, train 2 follows train 4 in the previous cycle, train 7 follows train 1, and train 1 follows train 7 in the previous cycle.

Each train departs as soon as all the connections are guaranteed (except for a connection when it is broken), the passengers have gotten the opportunity to change over and the earliest departure time indicated in the timetable has passed. We assume that in the first period all the trains depart according to schedule. The j-th state $x_j(k)$ is the time instant at which the train on track j departs from the initial station of the track for the kth time.

Now we write down the equations that describe the evolution of the $x_j(k)$'s. First we consider the train on track 1 and we determine $x_1(k)$, the time instant at which this train departs from station A for the kth time. The train has to wait at least until the train has arrived in station A for the $(k-1)$th time[2] and the passengers have got the time to get out of the train so we have $x_1(k) \geq x_3(k-1) + a_3(k-1) + 1$. Furthermore, the train on track 1 has to wait for the passengers of the train on track 9 in the $(k-1)$th cycle, which arrives in station B at time instant $x_9(k-1) + a_9(k-1)$. The passengers have $c^{\min} = 3$ minutes to change trains. Further the train on track 1 has to follow the train on track 7 in the previous cycle with a minimum separation time $f^{\min} = 4$. According to the timetable[3] the train on track 1 can only depart after time instant $00 + k\,60$. Hence, we have

$$\begin{aligned} x_1(k) &= \max(\, x_3(k-1) + a_3(k-1) + s^{\min}\,, \\ &\qquad x_7(k-1) + f^{\min}\,, \\ &\qquad x_9(k-1) + a_9(k-1) + c^{\min}\,, d_1(k)\,) \\ &= \max(\, x_3(k-1) + 21\,, x_7(k-1) + 4\,, \\ &\qquad x_9(k-1) + 13\,, k\,60\,) \end{aligned}$$

for $k = 1, 2, \ldots$ with $x_3(0) = x_9(0) = -\infty$.

Using a similar reasoning, we find that the other departure times are given by

$$\begin{aligned} x_2(k) &= \max(x_1(k) + 13, x_4(k-1) + 4, \\ &\qquad x_6(k-1) + 28, 15 + k\,60) \\ x_3(k) &= \max(x_2(k) + 13, 30 + k\,60) \\ x_4(k) &= \max(x_2(k) + 4, x_6(k-1) + 26, \\ &\qquad x_7(k) + 15, 19 + k\,60) \\ x_5(k) &= \max(x_4(k) + 13, 34 + k\,60) \\ x_6(k) &= \max(x_5(k) + 11, 47 + k\,60) \\ x_7(k) &= \max(x_9(k-1) + 11, x_1(k) + 4, 4 + k\,60) \\ x_8(k) &= \max(x_7(k) + 13, 19 + k\,60) \\ x_9(k) &= \max(x_8(k) + 26, x_5(k) + 13, 47 + k\,60) \end{aligned}$$

for $k = 1, 2, \ldots$ with $x_j(0) = -\infty$ for $j = 1, 2, \ldots, 9$.

Using successive substitution we can eliminate all right-hand terms with index k. By defining the appropriate matrix $A^{(1)}$ and by using the $(\otimes, \oplus)$-notation, we can rewrite the state equations as:

$$x(k) = A^{(1)}(k) \otimes x(k-1) \oplus d(k) \quad . \qquad (4)$$

In the nominal operation we have assumed that some trains should give pre-defined connections to other trains, and the order of trains on the same track is fixed. However, if one of the preceding trains has a too large delay, then it is sometimes better — from a global performance viewpoint — to let a connecting train depart anyway or to change the departure order on a specific track. This is done in order to prevent an accumulation of delays in the network. In this paper we consider the switching between different operation modes, where each mode corresponds to a different set of pre-defined or

[2]Under nominal operations the kth train on track 1 (e.g., the one that departs from station D at 10.00 a.m.) proceeds to the $(k-1)$th train on track 3 (which has departed from station B at 9.30 a.m.) and *not* to the kth train on track 3 (which will depart from station B at 10.30 a.m.).

[3]Note that in a undisturbed, well-defined time schedule the term $d_i(k)$ will be the largest. However, if due to unforeseen circumstances (an incident, a late departure, etc.) one of the trains has a delay, the corresponding term can become larger than the others, train i will depart later than the scheduled departure time $d_i(k)$ and will therefore also be delayed.

broken connections and a specific order of train departures. We allow the system to switch between different modes, allowing us to break train connections and to change the order of trains. Note that any broken connection or change of train order leads to a new model, similar to the nominal equation (4), but now with adapted system matrix $A^{(\ell)}$ for the ℓ-th model. We have the following system equation for the perturbed operation for $\ell = 2, \ldots, n_m$:

$$x(k) = A^{(\ell)}(k) \otimes x(k-1) \oplus d(k) \quad . \qquad (5)$$

In this railway network the switching variable $z(k)$ is equal to the control vector $v(k)$, and each entry of $v(k)$ corresponds to a specific control action, so a specific (scheduled) synchronization or specific (scheduled) event order. We assume $v(k)$ to be binary, where $v_i(k) = 0$ corresponds to the nominal case, and $v_i(k) = 1$ to the perturbed case (the synchronization is broken or the order of two events is switched). Each combination $v_1(k), \ldots, v_m(k)$ corresponds to a fixed routing schedule with a specific train order and specific connections.

4 THE MODEL PREDICTIVE CONTROL PROBLEM

Consider switching max-plus-linear model (2)–(3). We have two possible input signals, $v(k)$ and $u(k)$. Let $\mathcal{V}(k)$ and $\mathcal{U}(k)$ be the sets of possible future control actions $v(k)$ and $u(k)$, respectively. Sometimes values are predefined (e.g. in the railway system $u(k) = d(k)$) or not applicable (e.g. in the production system $v(k)$ is not used). Often $v(k)$ is assumed to be binary, and each entry corresponds to a specific control action (e.g. a specific scheduled synchronization or specific scheduled event order).

Just as in conventional Model Predictive Control (MPC) [10] we define a cost criterion J and we aim at computing the optimal input sequences $u(k), \ldots, u(k + N_p - 1)$, $v(k), \ldots, v(k + N_p - 1)$ that minimize a cost criterion $J(k)$, possibly subject to linear constraints on the inputs and the states, where N_p is the prediction horizon. The cost criterion reflects the input and output cost functions (J_{in} and J_{out}, respectively) in the event period $[k, k + N_p - 1]$:

$$J(k) = J_{out}(k) + \lambda J_{in}(k) \quad , \qquad (6)$$

where λ is a weighting parameter. The output cost function is usually chosen as

$$J_{out}(k) = \sum_{j=0}^{N_p-1} \| \hat{e}(k+j) \| \quad ,$$

where $\| \cdot \|$ is an appropriate norm and $\hat{e}$ is the due date error (e.g. for the production system the due date error of a product is given by $\hat{e}_i(k) = \max(y_i(k) - r_i(k), 0)$, where $r(k)$ is the desired due date of the product, and for the railway system the due date error is equal to the delay of a train, so $\hat{e}_i(k) = \max(x_i(k) - d_i(k), 0)$). The input cost function consists of two parts, $J_{in} = J_{in,u} + J_{in,v}$. The first part $J_{in,u}$ depends on $u(k+j)$, $j = k, \ldots, k + N_p - 1$ and is usually chosen as

$$J_{in,u}(k) = - \sum_{j=0}^{N_p-1} \| u(k+j) \| \quad ,$$

(see also [6]). The second part $J_{in,v}$ is a function of $v(k+j)$, $j = k, \ldots, k + N_p - 1$. For different applications, $J_{in,v}$ will have different appearances (e.g. for the production system $J_{in,v}(k) = 0$ and for the railway system we choose $J_{in,v}(k) = \sum_{j=0}^{N_p-1} \sum_{i=1}^{n_v} v_i(k+j)$).

Since the input signal $u(k)$ correspond to consecutive event occurrence times, we have the additional condition for $j = 0, \ldots, N_p - 1$:

$$\Delta u(k+j) = u(k+j) - u(k+j-1) \geq 0 \, .$$

Furthermore, in order to reduce the number of decision variables and the corresponding computational complexity we introduce a control horizon N_c ($\leq N_p$) and we impose the additional condition that the input rate should be constant from the point $k + N_c - 1$ on, so

$$\Delta u(k+j) = \Delta u(k + N_c - 1) \, ,$$

for $j = N_c, \ldots, N_p - 1$. The same often holds for the control variable $v(k)$, which will be assumed to be constant beyond control horizon N_c. This results in the constraint $v(k+j) = v(k + N_c - 1)$ for $j = N_c, \ldots, N_p - 1$, or equivalently

$$\Delta v(k+j) = v(k+j) - v(k+j-1) = 0 \, ,$$

for $j = N_c, \ldots, N_p - 1$. Now the MPC control problem for event step k can be defined as:

$$\min_{\{u(k) \in \mathcal{U}, v(k) \in \mathcal{V}(k)\}} J(k) \qquad (7)$$

subject to

$$x(k+j) = A^{(\ell(k))}(k) \otimes x(k+j-1) \oplus B^{(\ell(k))}(k) \otimes u(k+j) \qquad (8)$$

$$\Phi(x(k-1), \ell(k-1), u(k), v(k)) \in \mathcal{Z}^{(\ell(k))} \qquad (9)$$

$$\Delta u(k+j) \geq 0 \qquad (10)$$

$$\Delta v(k+m) = 0 \qquad (11)$$

$$\Delta u(k+m) - \Delta u(k + N_c - 1) = 0 \qquad (12)$$

$$A_c(k)\tilde{u}(k) + B_c(k)\tilde{y}(k) \leq c_c(k) \qquad (13)$$

for $j = 0, \dots, N_p - 1, \quad m = N_c, \dots, N_p - 1$

where (13) may represent additional linear constraints on the inputs and the outputs.

MPC uses a receding horizon principle. This means that after computation of the optimal control sequences $u(k), \dots, u(k + N_c - 1)$ and $v(k), \dots, v(k + N_c - 1)$, only the first control samples $u(k)$ and $v(k)$ will be implemented, subsequently the horizon is shifted one sample, and the optimization is restarted with new information of the measurements.

In principle we have all elements to solve the receding horizon control problem (7)–(13). In general we will have a mixed integer optimal control problem with both real parameters and binary parameters. Sometimes (e.g. the production system) the problem can be recast as a Extended Linear Complementary Problem (ELCP) and can be solved efficiently [7]. If the optimization is over a binary valued $v(k)$ (e.g. the railway problem) we obtain an integer optimization problem, which can be solved using genetic algorithms [5], tabu search [9], or a branch-and-bound method [3]. In some particular cases the problem can be recast as a Mixed Integer Linear Programming (MILP) or a Mixed Integer Quadratic Programming (MIQP), for which reliable algorithms are available [2, 8].

The application of the derived controller design method to a railway network is given in [11].

5 DISCUSSION

We have presented a new way to model a class of discrete event systems — the max-plus-linear discrete event systems — that can operate in different modes, in which the dynamics can be described by a model that is "linear" in the max-plus algebra. We have discussed two examples, a production system and a railway network. An MPC controller design technique has been derived for this type of systems. In general the resulting optimization problem requires a mixed integer optimization algorithm.

In future research we will study on the characterization of all discrete event systems that can be recast as a switching max-plus-linear system. Furthermore, we will try to find out what conditions are needed on J, Φ and $\mathcal{Z}^{(\ell(k))}$ to obtain particular optimization problems (ELCP, MILP, MIQP).

Acknowledgments

Research partially funded by the Dutch Technology Foundation STW project "Model predictive control for hybrid systems" (DMR.5675), by the European IST project "Modelling, Simulation and Control of Nonsmooth Dynamical Systems (SICONOS)" (IST-2001-37172), and by the NWO/STW VIDI project "Multi-agent control of large-scale hybrid systems" (DWV.6188).

References

[1] F. Baccelli, G. Cohen, G.J. Olsder, and J.P. Quadrat. *Synchronization and Linearity.* John Wiley & Sons, New York, 1992.

[2] A. Bemporad and M. Morari. Control of integrated logic, dynamics, and constraints. *Automatica*, 35(3):407–427, 1999.

[3] C. Cordier, H. Marchand, R. Laundy, and L.A. Wolsey. *bc-opt*: A branch-and-cut code for mixed integer programs. *Mathematical Programming, Series A*, 86(2):335–353, 1999.

[4] R.A. Cuninghame-Green. *Minimax Algebra*, volume 166 of *Lecture Notes in Economics and Mathematical Systems.* Springer-Verlag, Berlin, Germany, 1979.

[5] L. Davis, editor. *Handbook of Genetic Algorithms.* Van Nostrand Reinhold, New York, 1991.

[6] B. De Schutter and T.J.J. van den Boom. Model predictive control for max-plus-linear discrete event systems. *Automatica*, 37(7):1049–1056, July 2001.

[7] B. De Schutter and T.J.J. van den Boom. Model predictive control for max-min-plus-scaling systems — efficient implementation. In *WODES 2002, Zaragossa, Spain*, pages 343–348, 2002.

[8] R. Fletcher and S. Leyffer. Numerical experience with lower bounds for MIQP branch-and-bound. *SIAM Journal on Optimization*, 8(2):604–616, 1998.

[9] F. Glover and M. Laguna. *Tabu Search.* Kluwer Academic Publishers, Boston, 1997.

[10] J.M. Maciejowski. *Predictive control with constraints.* Prentice Hall, Pearson Education Limited, Harlow, UK, 2002.

[11] T.J.J. van den Boom and B. De Schutter. Modelling and control of railway networks. In *American Control Conference 2004, Boston, MA, USA*, 2004.

ELSEVIER
IFAC
PUBLICATIONS
www.elsevier.com/locate/ifac

CONTROL AND ROBUSTNESS ANALYSIS FOR (MAX, +)-LINEAR SYSTEMS

M. Lhommeau * L. Hardouin * C.-A. Maia ** R. Santos-Mendes ***

* *Laboratoire d'Ingénierie des Systèmes Automatisés 62, avenue Notre-Dame du lac 49000 ANGERS, FRANCE*
** *Depto. de Engenharia Elétrica - UFMG Av. Antônio Carlos 6627 - Pampulha 31270-010 Belo Horizonte - MG - BRAZIL*
*** *Faculdade de Engenharia Elétrica e de Computação, Universidade Estadual de Campinas P.O. Box 6101, 13083-970 Campinas, SP, BRAZIL*

Abstract: This paper deals with controller synthesis for (max,+) linear system. It aims at comparing the performances and the robustness of two control strategies introduced in (Cottenceau *et al.*, 2001) and (Maia *et al.*, 2003) respectively. In both strategies, the influences of the possible mismatches between the system and its model are analyzed. This work shows that the control strategy using simultaneously a precompensator and a feedback controller (introduced in (Maia *et al.*, 2003)) gives better performances. *Copyright © 2004 IFAC*

Keywords: Discrete event dynamic systems, Timed Petri nets, (max, +) algebra, Timed event Graphs, dioid, idempotent semiring, Control Systems.

1. INTRODUCTION

Timed Event Graphs (TEG) constitute a subclass of timed Petri nets in which each place has exactly one upstream and one downstream transition. It is well known that the timed/event behavior of a TEG, under earliest functioning rule [1], can be expressed by linear relations over some dioids, namely idempotent semiring (Baccelli *et al.*, 1992). Strong analogies then appear between the classical linear system theory and the (max, +)-linear system theory. In particular, the concept of *control* is well defined in the context of TEG study. It refers to the firing-control of the TEG input transitions in order to reach the desired performance. In the literature, an *optimal control* for TEG exists and is proposed in (Cohen *et al.*, 1989). For a given reference input, this open-loop structure control yields the latest input firing date in order to obtain the output before the desired date. One possible approach for the control of TEG is the model reference technique in which a given model describes the desired performance and the design goal is achieved through the calculation of a precompensator or a feedback controller (Cottenceau *et al.*, 2001; Luders and Santos-Mendes, 2002). The control strategies based on feedback control, although favoring stability, are limited in the sense that the reference model must satisfy certain restrictive conditions. Lately, a new technique for the design of controllers in which a precompensator and a feedback controllers are calculated simultaneously was introduced by (Maia *et al.*, 2003). This paper aims at comparing the performances and robustness

[1] *i.e.* a transition is fired as soon as it is enabled

of the above mentioned control methods. More precisely, we will compare performances regarding the just-in-time criterion and we will compare robustness, regarding possible mismatches between the system and its model. The paper is organized as follows. Section 2 introduces some algebraic tools concerning the dioid and residuation theories. Section 3 is devoted to recall some elements of DES representation over particular dioids and this section presents three control strategies. Section 4 is dedicated to the analysis of the performances and the robustness of these control strategies.

2. ALGEBRAIC PRELIMINARIES

A *dioid* $\mathcal{D}$ is an idempotent semiring, that is an algebraic structure with two internal operations denoted by $\oplus$ and $\otimes$. The neutral elements of $\oplus$ and $\otimes$ are represented by ε and e respectively. In a dioid, a partial order relation is defined by $a \succeq b$ iff $a = a \oplus b$ and $x \wedge y$ denotes the greatest lower bound between x and y. A dioid $\mathcal{D}$ is said to be complete if it is closed for infinite $\oplus$-sums and if $\otimes$ distributes over infinite $\oplus$-sums. Most of the time the symbol $\otimes$ will be omitted as in conventional algebra.

Theorem 1. ((Baccelli *et al.*, 1992), th. 4.75). The implicit equation $x = ax \oplus b$ defined over a complete dioid $\mathcal{D}$, admits $x = a^*b$ as least solution, where $a^* = \bigoplus_{i \in \mathbb{N}} a^i$ (Kleene star operator). It will be sometimes represented by the following mapping : $\mathcal{K} : \mathcal{D} \rightarrow \mathcal{D}, x \mapsto \bigoplus_{i \in \mathbb{N}} x^i$.

TEG control problems (Cohen *et al.*, 1989), stated in a just-in-time context, usually involves the inversion of isotone mappings[2], that is, one must find x such that $f(x) = y$ (where f is isotone). Residuation Theory (Blyth and Janowitz, 1972) deals with such problems stated in partially ordered sets.

Definition 2. (Residual and residuated mapping). A mapping $f : \mathcal{D} \rightarrow \mathcal{E}$ between two ordered sets is *residuated* if it is isotone, and if, for all $y \in \mathcal{E}$, the subset $\{x \in \mathcal{D} \mid f(x) \preceq y\}$ admits a maximal element, denoted $f^\sharp(y)$. The isotone mapping $f^\sharp : \mathcal{E} \rightarrow \mathcal{D}$ is called the *residual* of f. The residual $f^\sharp$ is the only isotone mapping satisfying the following properties :

$$f \circ f^\sharp \preceq \mathsf{Id} \text{ and } f^\sharp \circ f \succeq \mathsf{Id}, \qquad (1)$$

where Id is the identity mapping respectively on $\mathcal{D}$ and $\mathcal{E}$.

[2] f is an isotone mapping if it preserves order, that is, $a \preceq b \Longrightarrow f(a) \preceq f(b)$.

Lemma 3. ((Cohen, 1998)).

- If $f : \mathcal{D} \rightarrow \mathcal{E}$ and $g : \mathcal{E} \rightarrow \mathcal{F}$ are residuated mappings, then $f \circ g$ is also residuated and $(f \circ g)^\sharp = g^\sharp \circ f^\sharp$.
- If f is a residuated mapping from $\mathcal{D} \rightarrow \mathcal{E}$, then $f \circ f^\sharp \circ f = f$.

The mappings $L_a : x \mapsto a \otimes x$ and $R_a : x \mapsto x \otimes a$ defined over a complete dioid $\mathcal{D}$ are both residuated ((Baccelli *et al.*, 1992), p. 181). Their residuals are isotone mappings denoted respectively by $L_a^\sharp(x) = a \backslash x$ and $R_a^\sharp(x) = x / a$. Some useful dioid formulæ involving these residuals are given below.

$$a(a \backslash x) \preceq x \text{ and } (x / a)a \preceq x \qquad (2)$$

$$a(a \backslash (ax)) = ax \qquad (3)$$

$$a \backslash a = (a \backslash a)^* \qquad (4)$$

$$(a^*)^2 = a^* \qquad (5)$$

$$x \backslash (a^* x) = (a^* x) \backslash (a^* x) \qquad (6)$$

Definition 4. (Restricted mapping). Let $f : \mathcal{D} \rightarrow \mathcal{E}$ be a mapping and $B \subset \mathcal{E}$ with $f(\mathcal{D}) \subset B$. We will denote ${}_{B|}f : \mathcal{D} \rightarrow B$ the mapping defined by $f = i_B \circ_{B|} f$, where $i_B : B \rightarrow \mathcal{E}, x \mapsto x$ is the canonical injection.

Definition 5. (Closure mapping). An isotone mapping $f : \mathcal{D} \rightarrow \mathcal{D}$ defined on an ordered set $\mathcal{D}$ is a closure mapping if $f \succeq \mathsf{Id}$ and $f \circ f = f$.

Remark 6. According to (5), the Kleene star operator is a closure mapping since $a^* \succeq a$ and $(a^*)^* = a^*$.

Theorem 7. ((Cottenceau *et al.*, 2001)). Let $f : \mathcal{D} \rightarrow \mathcal{D}$ be a closure mapping. Then, ${}_{\mathsf{Im}f|}f$ is a residuated mapping whose residual is the canonical injection $i_{\mathsf{Im}f} : \mathsf{Im}f \rightarrow \mathcal{D}, x \mapsto x$.

Example 8. Mapping ${}_{\mathsf{Im}\mathcal{K}|}\mathcal{K} : \mathcal{D} \rightarrow \mathsf{Im}\mathcal{K}$ is a residuated mapping whose residual is $({}_{\mathsf{Im}\mathcal{K}|}\mathcal{K})^\sharp = i_{\mathsf{Im}\mathcal{K}}$. This means that $x = a^*$ is the greatest solution to inequality $x^* \preceq a^*$. Actually, this greatest solution achieves equality.

Theorem 9. ((Gaubert, 1992)). Let $f : \mathcal{D} \rightarrow \mathcal{D}$ be a residuated closure mapping, we have $f = f^\sharp \circ f$ and $f = f \circ f^\sharp$.

3. CONTROL METHOD

Firstly, let us consider the following $(\max, +)$-linear system

$$x(k) = Ax(k-1) \oplus Bu(k), \quad y(k) = Cx(k), \qquad (7)$$

where $x(k) \in \overline{\mathbb{Z}}_{\max}^{n \times 1}, u(k) \in \overline{\mathbb{Z}}_{\max}^{p \times 1}$ and $y(k) \in \overline{\mathbb{Z}}_{\max}^{m \times 1}$ are respectively the state, input and output vectors of the system. The matrices A, B and C are of proper sizes and have entries ranging over $\overline{\mathbb{Z}}_{\max}$. We know from (Baccelli *et al.*, 1992) that (7) represents the behavior of a class of discrete event systems called Timed Event Graphs (TEG). In the case of a TEG, x (resp. u and y) is a vector associated to the internal (resp. input and output) transitions, and $x_i(k)$ represents the kth firing dates of the internal transitions which are labelled x_i. Following the conventional approach, it is possible to define the transformation $x(\gamma) = \bigoplus_{k\in\mathbb{Z}} x(k)\gamma^k$ where γ is a backward shift operator in event domain (that is $y(\gamma) = \gamma x(\gamma) \Leftrightarrow \{y(k)\} = \{x(k-1), \forall k\}$, see (Baccelli *et al.*, 1992), p. 228). This transformation is analogous to the $\mathcal{Z}$-transform used in discrete-time classical control theory and the formal series $x(\gamma)$ is a synthetic representation of the trajectory $x(k)$. The set of the formal series in γ is a dioid denoted by $\overline{\mathbb{Z}}_{\max}[\![\gamma]\!]$. By using γ-transform, we obtain the following representation of (7) :

$$X(\gamma) = A\gamma X(\gamma) \oplus BU(\gamma), \quad Y(\gamma) = CX(\gamma),$$

where $U(\gamma), X(\gamma)$ and $Y(\gamma)$ are the γ-transform of u, x and y respectively. The implicit equation for the vector X, namely $X = A\gamma X \oplus BU$ which is solved (thanks to theorem 1) by $X = (A\gamma)^*BU$. Finally, we obtain the input-output representation (transfer matrix)

$$Y = HU \quad \text{with} \quad H(\gamma) = C(A\gamma)^*B. \tag{8}$$

Herein, three control strategies for the systems are presented, and their performances are compared in section 4. They are based on the Just-in-Time criterion and on the model reference approach (Cottenceau *et al.*, 2001). They can be described as follows : let $H \in (\overline{\mathbb{Z}}_{\max}[\![\gamma]\!])^{m\times p}$ be the transfer matrix of the plant, given by (8), and $G_{ref} \in (\overline{\mathbb{Z}}_{\max}[\![\gamma]\!])^{m\times p}$ be the reference model, *i.e.*, the desired transfer matrix for the controlled system, what are the controllers leading to the greatest controlled system lower than the reference model.

The precompensation problem is depicted Fig. 1.(a). It is an open-loop strategy. The relation between the input $V \in \overline{\mathbb{Z}}_{\max}[\![\gamma]\!]^p$, and the output Y is denoted G_c and the relation between V and U is denoted G_{uv}. They are given by

$$Y = G_cV = HPV \text{ and } U = G_{uv}V = PV.$$

The aim is to compute the greatest precompensator P such that $G_c \preceq G_{ref}$. The residual of mapping L_H gives the optimal solution, denoted by P_{op} [3],

$$P_{op} = H \backslash G_{ref}. \tag{9}$$

This means that, for a given external input [4] $V \in (\overline{\mathbb{Z}}_{\max}[\![\gamma]\!])^p$, the control input, given by $U = P_{op}V$, will be maximal. In fact, for any P such that $HP \preceq G_{ref}$, $P \preceq P_{op}$, therefore the isotony property assures that $U = PV \preceq P_{op}V$.

[3] In a manufacturing context, control $U = P_{op}V$ will delay as far as possible the input of raw material while ensuring that output of manufacturing part Y be lower that $G_{ref}V$

[4] In a manufacturing context, V represents the available catering of raw material and U represents the allowance of the raw material into the system.

a) The output feedback control strategy is depicted Fig. 1.(b). It is obviously more robust than the open-loop strategy. For a given feedback controller F_1, the closed-loop transfer relation, denoted G_{1_c}, between Y and V, and the transfer relation between U and V, denoted $G_{1_{uv}}$, are given by

$$Y = G_{1_c}V = H(F_1H)^*V \text{ and } U = G_{1_{uv}}V = (F_1H)^*V.$$

The problem is then to compute the greatest controller F_1 such that $G_{1_c} \preceq G_{ref}$.

This problem can be solved via residuation theory if some restrictions are imposed on the reference model. The two following results are due to (Cottenceau *et al.*, 2001).

Proposition 10. Let $H \in (\overline{\mathbb{Z}}_{\max}[\![\gamma]\!])^{m\times p}$ be the transfer function of a TEG. Let $M_H : \overline{\mathbb{Z}}_{\max}[\![\gamma]\!]^{p\times m} \to \overline{\mathbb{Z}}_{\max}[\![\gamma]\!]^{m\times p}, X \mapsto H(XH)^*$ be a mapping. This mapping represents the influence of an output feedback x on the closed-loop transfer dynamics. Consider $G \in \overline{\mathbb{Z}}_{\max}[\![\gamma]\!]^{m\times p}$, $D \in \overline{\mathbb{Z}}_{\max}[\![\gamma]\!]^{m\times m}$ and $N \in \overline{\mathbb{Z}}_{\max}[\![\gamma]\!]^{p\times p}$. Let us consider the following sets :

$$\mathcal{G}_1 = \{G \mid \exists D \text{ such that } G = D^*H\},$$
$$\mathcal{G}_2 = \{G \mid \exists N \text{ such that } G = HN^*\}.$$

The mapping ${}_{\mathcal{G}_1|}M_H$ and ${}_{\mathcal{G}_2|}M_H$ are both residuated. Their residuals are such that $({}_{\mathcal{G}_1|}M_H)^\sharp(X) = ({}_{\mathcal{G}_2|}M_H)^\sharp(X) = H\backslash X / H$.

Proposition 11. If $G_{ref} \in \mathcal{G}_1 \cup \mathcal{G}_2$, there exists a greatest realizable output feedback $F_{1_{op}}$ such that $M_H(F_{1_{op}}) \preceq G_{ref}$. This greatest controller is given by

$$F_{1_{op}} = H\backslash G_{ref} / H. \tag{10}$$

b) The model-reference control scheme proposed in the following is a generalization of the two strategies described above, that is, it uses both a precompensator and feedback controller (Maia *et al.*, 2003). Fig. 1.(c) illustrates the approach.

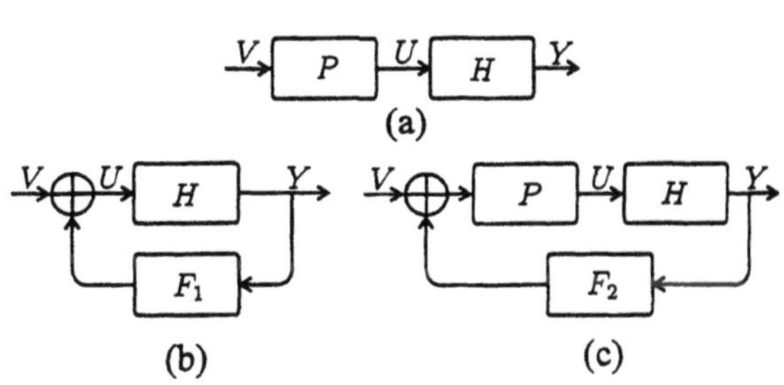

Fig. 1. Control structures.

By using theorem 1, one can obtain the closed-loop equations which relate U, V and Y:

$$Y = G_{2_c}V = HP(F_2HP)^*V, \quad (11)$$

$$U = G_{2_{uv}}V = P(F_2HP)^*V. \quad (12)$$

The problem can be stated as follows. *Given a reference model G_{ref}, what are the controller matrices P and F_2 which assure the greatest transfer function between U and V, i.e. $G_{2_{uv}}$, such that $G_{2_c} \preceq G_{ref}$?* Again, considering the Just-in-Time context, one seeks the controllers which satisfy the reference specification $G_{2_c} \preceq G_{ref}$ while delaying as much as possible the input trajectory (*e.g.* the entrance of products to be processed). Formally, the problem can be stated as follows:

$$\bigoplus_{P,\,F_2} G_{2_{uv}}(P, F_2) \quad (13)$$
$$\text{s.t. } G_{2_c} = HP(F_2HP)^* \preceq G_{ref}.$$

It is clear that $P = [\varepsilon]_{p\times p}$ is always a subsolution to the problem independently of the choice of F_2, meaning that the subsolution set is not empty. Furthermore, it is easy to notice that the strategies using exclusively a precompensator (by setting $F_2 = [\varepsilon]_{p\times m}$) or exclusively a feedback controller (by setting $P = I_{p\times p}$, where $I_{p\times p}$ is the identity matrix in dioid) are particular cases of this problem.

Proposition 12. ((Maia *et al.*, 2003)). For the proposed control scheme shown in Fig. 1.(c), the three following inequalities are equivalent:

$$HP(F_2HP)^* \preceq G_{ref}$$
$$P(F_2HP)^* \preceq H \backslash G_{ref}$$
$$HP(F_2HP)^* \preceq H(H \backslash G_{ref}).$$

Lemma 13. ((Maia *et al.*, 2003)). A solution to problem 13 must satisfy $P \preceq G_{2_{uv}} \preceq H \backslash G_{ref}$.

Proposition 14. ((Maia *et al.*, 2003)). A solution to the optimization problem proposed in (13) is given by :

$$P_{op} = H \backslash G_{ref}. \quad (14)$$

$$F_{2_{op}} = (HP_{op}) \backslash (HP_{op}) / (HP_{op}). \quad (15)$$

PROOF.

From lemma 13, $G_{2_{uv}}$ is maximum (it is equal to the upper bound) if $P = H \backslash G_{ref}$ and $F_2 = \varepsilon$. Then, the greatest F_2 for this value of P is given by the greatest subsolution of inequality $P_{op}(F_2HP_{op})^* \preceq H \backslash G_{ref}$, which in turn (by proposition 12) is equivalent to $HP_{op}(F_2HP_{op})^* \preceq H(H \backslash G_{ref}) = HP_{op}$. Moreover, from the residuation definition this inequality is equivalent to $(F_2HP_{op})^* \preceq (HP_{op}) \backslash (HP_{op})$. Equation (4) yields $((HP_{op}) \backslash (HP_{op}))^* = (HP_{op}) \backslash (HP_{op})$ then, thanks to corollary 8, $F_2HP_{op} \preceq (HP_{op}) \backslash (HP_{op})$. Finally, by solving this last inequality one obtains $F_{2_{op}} = (HP_{op}) \backslash (HP_{op}) / (HP_{op})$.

4. PERFORMANCES AND ROBUSTNESS ANALYSIS OF CONTROL METHODS

We will compare below the performances and the robustness of the control strategy given by proposition 11 and the one given by proposition 14. First, we must observe that unlike the first strategy, the second one does not restrict the reference model choice. Nevertheless, in order to compare performances of these strategies, we assume below that the controller $F_{1_{op}}$ exists and then that the reference model is such that $G_{ref} \in \mathcal{G}_1 \cup \mathcal{G}_2$ (see proposition 11).

4.1 Performance Comparison

Proposition 15. The control strategy given in proposition 14 leads to the same performances than the one obtained with the open-loop strategy, *i.e.*, the greatest closed-loop transfer functions $G_{2_{uv}}$ and G_{2_c} are equal to their upper bounds, that is, P_{op} and HP_{op} respectively. Formally, this means that

$$G_{2_c} = HP_{op}(F_{2_{op}}HP_{op})^* = HP_{op}$$
$$G_{2_{uv}} = P_{op}(F_{2_{op}}HP_{op})^* = P_{op}.$$

PROOF. This proposition follows directly from proposition 14, lemma 13 and from the observation that $G_{2_c} = HG_{2_{uv}}$ (see (11) and (12)).

Proposition 16. Let $G_{ref} \in \mathcal{G}_1 \cup \mathcal{G}_2$ be a reference model. The transfer relation between U and V are such that

$$G_{1_{uv}} = (F_{1_{op}}H)^* \preceq G_{2_{uv}} = P_{op}(F_{2_{op}}HP_{op})^* = P_{op}.$$

PROOF. We suppose that $G_{ref} \in \mathcal{G}_1$, that is $G_{ref} = D^*H$. Then, we have

$$G_{1_{uv}} = (F_{1_{op}}H)^* = ((H \backslash (D^*H) / H)\, H)^* \preceq (H \backslash (D^*H))^*$$

which follows from (2). But we also have that

$$G_{1_{uv}} \preceq (H \backslash (D^*H))^* = H \backslash (D^*H) = P_{op} = G_{2_{uv}}$$

by making use of (6) and (4).

Proposition 17. Let $G_{ref} \in \mathcal{G}_1 \cup \mathcal{G}_2$ be a reference model. The controlled system transfer $G_{1_c} = H(F_{1_{op}}H)^* \preceq G_{2_c} = HP_{op}(F_{2_{op}}HP_{op})^* = HP_{op}$.

PROOF. From proposition 16, we have $G_{1_{uv}} = (F_{1_{op}}H)^* \preceq G_{2_{uv}} = P_{op}(F_{2_{op}}HP_{op})^* = P_{op}$ and by isotony of product we obtain $G_{1_c} = HG_{1_{uv}} = H(F_{1_{op}}H)^* \preceq G_{2_c} = HG_{2_{uv}} = HP_{op}(F_{2_{op}}HP_{op})^* = HP_{op}$.

4.1.1. Summary These results mean that the pair of controllers $(P_{op}, F_{2_{op}})$:

- allows to obtain the same performances than the one obtained with the open-loop control;

- generates a control law greater than the one obtained by $F1_{op}$;
- leads to a controlled system transfer closer to the reference model G_{ref} than the one obtained by the controller $F1_{op}$.

4.2 Robustness analysis

a) In this section the aim is to analyse the robustness of the closed-loop control methods introduced previously.

First, we study the robustness of the controller given by proposition 11 (Fig. 1.(b)). We are looking for an upper bound denoted by $H1_{sup}$ to the set of systems which preserves the optimal closed-loop control objective, that is,

$$H1_{sup} = \sup \{X | X(F1_{op}X)^* = H(F1_{op}H)^*\}.$$

Then we characterize the set of systems which preserves the input output behavior. It means that the system can evolve in this set without alter the input-output performances of the closed-loop system.

Lemma 18. Let $Q_A : \mathcal{D} \to \mathcal{D}, X \mapsto X(AX)^*$ be a mapping defined over a complete dioid. Then ${}_{\mathsf{Im}Q_A|}Q_A$ is a residuated mapping and the residual is $({}_{\mathsf{Im}Q_A|}Q_A)^\sharp = i_{\mathsf{Im}Q_A}$, where $i_{\mathsf{Im}Q_A}$ is the canonical injection.

PROOF.

The mapping Q_A is a closure mapping, indeed $Q_A \circ Q_A(X) = X(AX)^*(AX(AX)^*)^* = X(AX)^*((AX)^+)^* = X(AX)^*(AX)^* = X(AX)^*$. Then proposition 7 gives the result.

Proposition 19. The system $H1_{sup} = H(F1_{op}H)^*$ is the greatest system which does not alter the closed-loop transfer relation, *i.e.*, $H1_{sup}(F1_{op}H1_{sup})^* = H(F1_{op}H)^*$.

PROOF.

According to lemma 18, we seek the greatest X such that $Q_{F1_{op}}(X) \preceq H(F1_{op}H)^*$. Lemma 18 yields $({}_{\mathsf{Im}Q_{F1_{op}}|}Q_{F1_{op}})^\sharp = i_{\mathsf{Im}Q_{F1_{op}}}$, and since $H(F1_{op}H)^* \in \mathsf{Im}Q_{F1_{op}}$, we have directly $H1_{sup} = H(F1_{op}H)^*$. Furthermore according to theorem 9, we have $Q_{F1_{op}} = Q_{F1_{op}} \circ Q^\sharp_{F1_{op}}$, which leads to equality $H1_{sup}(F1_{op}H1_{sup})^* = H(F1_{op}H)^* = H1_{sup}$.

Corollary 20. Whatever be the system behavior X such that $H \preceq X \preceq H1_{sup}$ the closed-loop transfer relation is equal to $H(F1_{op}H)^*$, *i.e.*, the input-output performances are not altered.

PROOF. Let X be a transfer relation such that $H \preceq X \preceq H1_{sup}$. Since the product and star operators are isotone, we have $H(F1_{op}H)^* \preceq X(F1_{op}X)^* \preceq H1_{sup}(F1_{op}H1_{sup})^*$, and proposition 19 leads to equality $H(F1_{op}H)^* = X(F1_{op}X)^* = H1_{sup}(F1_{op}H1_{sup})^*$.

b) We are now interested in the robustness analysis of the control method which equations are given in proposition 14. We are looking for an upper bound, denoted $H2_{sup}$, to the set of systems which preserves the optimal closed-loop control objective, that is

$$H2_{sup} = \sup \{X \mid XP_{op}(F2_{op}XP_{op})^* = HP_{op}(F2_{op}HP_{op})^*\}. \quad (16)$$

Proposition 21. The system $H2_{sup} = HP_{op}(F2_{op}HP_{op})^* \phi P_{op}$ is the greatest which satisfies (16), *i.e.*, $H2_{sup}P_{op}(F2_{op}H2_{sup}P_{op})^* = HP_{op}(F2_{op}HP_{op})^*$.

PROOF.

According to definition of the mappings $Q_{F2_{op}}$ and $R_{P_{op}}$, we first seek the greatest X such that $XP_{op}(F2_{op}XP_{op})^* \preceq HP_{op}(F2_{op}HP_{op})^*$, that is, $(Q_{F2_{op}} \circ R_{P_{op}})(X) \preceq (Q_{F2_{op}} \circ R_{P_{op}})(H)$. Since $(Q_{F2_{op}} \circ R_{P_{op}})(H) \in \mathsf{Im}Q_{F2_{op}}$ and as ${}_{\mathsf{Im}Q_{F2_{op}}|}Q_{F2_{op}}$ is a residuated mapping (see lemma 18), $X \preceq ({}_{\mathsf{Im}Q_{F2_{op}}|}Q_{F2_{op}} \circ R_{P_{op}})^\sharp \circ (Q_{F2_{op}} \circ R_{P_{op}})(H)$, and thanks to lemma 3, it follows $X \preceq R^\sharp_{P_{op}} \circ ({}_{\mathsf{Im}Q_{F2_{op}}|}Q_{F2_{op}})^\sharp \circ (Q_{F2_{op}} \circ R_{P_{op}})(H)$.

By recalling that $({}_{\mathsf{Im}Q_{F2_{op}}|}Q_{F2_{op}})^\sharp = i_{\mathsf{Im}Q_{F2_{op}}}$, we have $X \preceq R^\sharp_{P_{op}} \circ (Q_{F2_{op}} \circ R_{P_{op}})(H) = HP_{op}(F2_{op}HP_{op})^* \phi P_{op}$.

Now, we will show that this upper bound is solution of (16). From proposition 15 it follows $HP_{op}(F2_{op}HP_{op})^* \phi P_{op} = HP_{op} \phi P_{op}$, *i.e.*, $R^\sharp_{P_{op}} \circ Q_{F2_{op}} \circ R_{P_{op}}(H) = R^\sharp_{P_{op}} \circ R_{P_{op}}(H)$. Then, from lemma 3 it follows $Q_{F2_{op}} \circ R_{P_{op}} \circ R^\sharp_{P_{op}} \circ R_{P_{op}}(H) = Q_{F2_{op}} \circ R_{P_{op}}(H)$, which yields to $H2_{sup}P_{op}(F2_{op}H2_{sup}P_{op})^* = HP_{op}(F2_{op}HP_{op})^*$.

Corollary 22. Whatever be the system behavior X such that $H \preceq X \preceq H2_{sup}$ the closed-loop transfer relation is equal to $HP_{op}(F2_{op}HP_{op})^*$, *i.e.*, the input-output performances are not altered.

PROOF.

Let X be a transfer relation such that $H \preceq X \preceq H2_{sup}$. Since the product and star operators are isotone, we have $HP_{op}(F2_{op}HP_{op})^* \preceq XP_{op}(F2_{op}XP_{op})^* \preceq$

$H_{2_{sup}} P_{op}(F_{2_{op}} H_{2_{sup}} P_{op})^*$, and proposition 21 leads to equality $HP_{op}(F_{2_{op}} HP_{op})^* = XP_{op}(F_{2_{op}} XP_{op})^* = H_{2_{sup}} P_{op}(F_{2_{op}} H_{2_{sup}} P_{op})^*$.

4.3 Robustness evaluation

In the previous section, the upper bound of the system set which achieve the control objective is given for the both closed-loop control strategy. In order to compare these bounds we assume below that the optimal controller $F_{1_{opt}}$ exists, *i.e.* $G_{ref} \in \mathcal{G}_1 \cup \mathcal{G}_2$. Nevertheless, we recall that this restriction is not useful to ensure the existence of $F_{2_{op}}$.

Lemma 23. Consider a reference model $G_{ref} \in \mathcal{G}_1 \cup \mathcal{G}_2$. Let $F_{1_{op}}$ be the greatest controller such that $H(F_{1_{op}} H)^* \preceq G_{ref}$ and P_{op} the greatest precompensator such that $HP_{op} \preceq G_{ref}$. Then

$$H(F_{1_{op}} H)^* \preceq HP_{op} \preceq G_{ref}.$$

PROOF.

Since L_H is a residuated mapping (see definition 2), we have the following equivalences $H(F_{1_{op}} H)^* \preceq G_{ref} \iff (F_{1_{op}} H)^* \preceq H \circ\!\!\!\backslash G_{ref}$. Furthermore, by isotony of $\otimes$, we obtain $H(F_{1_{op}} H)^* \preceq H(H \circ\!\!\!\backslash G_{ref}) = HP_{op} \preceq G_{ref}$ in which the latter inequality follows from (14) and (2).

Lemma 24. If $G_{ref} \in \mathcal{G}_1 \cup \mathcal{G}_2$ the upper bound $H_{2_{sup}}$ is equal to HP_{op}, that is

$$H_{2_{sup}} = HP_{op}(F_2 HP_{op})^* \circ\!\!\!/ P_{op} = HP_{op} \circ\!\!\!/ P_{op} = HP_{op}.$$

PROOF. First assume that $G_{ref} \in \mathcal{G}_1$, *i.e.*, it exists D such that $G_{ref} = D^* H$. Then, thanks to (6),(3) and (4), it follows that $P_{op} = H \circ\!\!\!\backslash (D^* H) = (D^* H) \circ\!\!\!\backslash (D^* H) = ((D^* H) \circ\!\!\!\backslash (D^* H))^* = P_{op}^*$. Then, we have $HP_{op} \circ\!\!\!/ P_{op} = HP_{op}^* \circ\!\!\!/ P_{op}^* = R^{\sharp}_{P_{op}^*} \circ R_{P_{op}^*}$. By recalling that $R_{P_{op}^*}$ is a closure mapping (*i.e.*, $R_{P_{op}^*} \circ R_{P_{op}^*}(x) = xP_{op}^* P_{op}^* = R_{P_{op}^*}(x)$ and $R_{P_{op}^*} \succeq \mathsf{Id}$) and by using proposition 9 we have $HP_{op} \circ\!\!\!/ P_{op} = R_{P_{op}^*}(H) = HP_{op}^* = HP_{op}$. The proof, if $G_{ref} \in \mathcal{G}_2$ can be given in a similar way.

Proposition 25. If $G_{ref} \in \mathcal{G}_1 \cup \mathcal{G}_2$ then

$$H_{1_{sup}} \preceq H_{2_{sup}}.$$

This means that the pair of controllers $(P_{op}, F_{2_{op}})$ is more robust with regard to the system variations.

PROOF. Thanks to lemma 24, we have

$$H_{2_{sup}} = HP_{op} = HP_{op}^* = H\left(H \circ\!\!\!\backslash G_{ref}\right)^*.$$

From (2), we have $H \circ\!\!\!\backslash G_{ref} \succeq ((H \circ\!\!\!\backslash G_{ref}) \circ\!\!\!/ H)H$, then by isotony of the laws $\otimes$ and $\oplus$ it follows that

$$H_{2_{sup}} = H\left(H \circ\!\!\!\backslash G_{ref}\right)^* \succeq H\left((H \circ\!\!\!\backslash G_{ref} \circ\!\!\!/ H)H\right)^* = H_{1_{sup}}.$$

5. CONCLUSION

This paper compares the robustness and the performances of two control strategies for (max, +)-linear systems. More precisely, we show that the control proposed by (Maia *et al.*, 2003) gives a greatest control and ensures a greatest insensitivity to the mismatch between the system and the model used for the controller synthesis. The next step for the control proposed by (Maia *et al.*, 2003) aims at designing robust feedback controller when the system includes some parametric uncertainties which can be described by intervals (Lhommeau *et al.*, 2003).

REFERENCES

Baccelli, F., G. Cohen, G.-J. Olsder and J.-P. Quadrat (1992). *Synchronization and Linearity : An Algebra for Discrete Event Systems.* Wiley and Sons. http://maxplus.org/.

Blyth, T.-S. and M.-F. Janowitz (1972). *Residuation Theory.* Pergamon press.

Cohen, G. (1998). Residuation and applications. In: *Algèbres Max-Plus et applications en informatique et automatique, Ecole de printemps d'informatique théorique.* Noirmoutier, France.

Cohen, G., P. Moller, J.-P. Quadrat and M. Viot (1989). Algebraic Tools for the Performance Evaluation of Discrete Event Systems. *IEEE Proceedings: Special issue on Discrete Event Systems* **77**(1), 39–58.

Cottenceau, B., L. Hardouin, J.-L. Boimond and J.-L. Ferrier (2001). Model Reference Control for Timed Event Graphs in Dioid. *Automatica* **37**, 1451–1458.

Gaubert, S. (1992). Théorie des Systèmes Linéaires dans les Dioïdes. Thèse. École des Mines de Paris.

Lhommeau, Mehdi, Laurent Hardouin, Bertrand Cottenceau and Luc Jaulin (2003). Interval analysis in dioid : Application to robust controller design for timed event graphs. To appear in Automatica.

Luders, R. and R. Santos-Mendes (2002). Multivariable control of discrete event systems in dioids. In: *IFAC World Congress on Automatic Control.* Barcelona.

Maia, C.-A., L. Hardouin, R. Santos-Mendes and B. Cottenceau (2003). Optimal closed-loop control of timed event graphs in dioids. *IEEE Trans. on Automatic Control* **49**(12), 2284 – 2287.

ELSEVIER
IFAC
PUBLICATIONS
www.elsevier.com/locate/ifac

OPTIMAL CONTROL OF SWITCHING SURFACES IN HYBRID DYNAMIC SYSTEMS

Mauro Boccadoro * Magnus Egerstedt **,[1]
Yorai Wardi **,[1]

* boccadoro@diei.unipg.it
Dipartimento di Ingegneria Elettronica e dell'Informazione
Università di Perugia
06125, Perugia - Italy
** {magnus,ywardi}@ece.gatech.edu
Electrical and Computer Engineering
Georgia Institute of Technology
Atlanta, GA 30332

Abstract: This paper studies the problem of optimal switching surface design for hybrid systems. In particular, a formula is derived for computing the gradient of a given integral performance cost with respect to the switching surface parameters. The novelty of the work lies both in the problem formulation itself, and in the introduction of a discrete event variational system for propagating the perturbations across switching surfaces. *Copyright © 2004 IFAC*

Keywords: Optimal Control, Hybrid Systems, Switching Surfaces, Gradient Descent, Numerical Algorithms

1. INTRODUCTION

To view the switching surface parameters as free variables in an optimal control formulation for hybrid systems is an approach that previously has not been pursued. In this paper, we address this problem by computing the gradient of a given cost functional with respect to the switching surface parameters, which enables us to employ gradient-based numerical techniques for obtaining locally optimal solutions. The problem of optimal switching surface design draws its motivation from the many hybrid models where transitions between the continuous modes occur based on internal (state-based) events, as illustrated in Figure 1. Examples of such systems include autonomous navigation and obstacle avoidance for mobile robots (Arkin, 1998; Egerstedt, 2000), and production planning for manufacturing systems (Cassandras *et al.*, 2001; Boccadoro and Valigi, 2003). In this context, identifying optimal switching surfaces in the state space amounts to solving an optimal control problem in a closed loop fashion.

To control such hybrid systems is a problem that has received a significant amount of attention. In particular, the problem of finding optimal, continuous control laws, given a partition of the state space (Johansson and Rantzer, 1998) or a fixed set of switching surfaces (Shaikh and Caines, 2002; Sussmann, 1999; Xu and Antsaklis, 2002; Riedinger *et al.*, 1999) has been investigated. In (Giua *et al.*, 2001; Bemporad *et al.*, 2002) the authors addressed a timing optimization problem and derived homogeneous regions in the state space that determine the switchings. However, the

[1] The support from the Manufacturing Research Center at the Georgia Institute of Technology is gratefully acknowledged.

problem of optimal design of switching surfaces has not yet been addressed. In this paper, we assume that the switching surfaces are of the form $g : \Re^n \times \Re^m \to \Re$, where a transition occurs if $g(x, a) = 0$. Here, g is assumed to be continuously differentiable in both arguments, $x \in \Re^n$ is the state of the system, and $a \in \Re^m$ is a controllable parameter. *The main problem that will be considered in this paper is how to choose such parameters in order to minimize a suitable cost functional.*

This problem of optimal control of the switching surfaces will be approached from a numerical optimal control vantage point in the sense that the derivative of the cost with respect to the switching parameters will be computed. This derivative allows for gradient descent methods to be applied, and as a consequence, local optima can be obtained. The enabling result for this program to be successful is the computation of the derivative of the cost with respect to the switching parameters, and the program presents an extension to the work on optimal timing control (Egerstedt *et al.*, n.d.). The gradient will be shown to have a special structure, making it suitable as a basis for a numerical optimization algorithm. In fact, it will be based on the combination of a single, backwards costate computation (as established in (Egerstedt *et al.*, n.d.)) and a collection of forward integrations of the continuous dynamics as well as a discrete event system that describes, in an iterative fashion, how the perturbations propagate across the different switching surfaces.

This paper is organized as follows: In Section 2 we introduce the problem at hand, followed by a study of the single-switch case, in Section 3. In Section 4 a discrete event dynamics is introduced for the characterization of multiple switches trajectories. Section 5 presents the gradient computation, discusses its complexity, and includes an illustrative example.

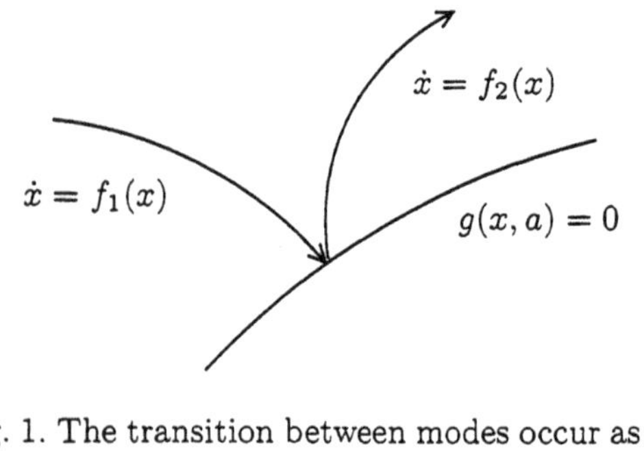

Fig. 1. The transition between modes occur as the state reaches a switching surface in the state space.

2. PROBLEM FORMULATION

In this paper we consider a switched system described by (see (Branicky *et al.*, 1998)):

$$\dot{x}(t) = f_i(x(t)) \tag{1}$$

$$i^+(t) = s(x(t), i(t)) \tag{2}$$

In particular, Eq. (1) describes the continuous dynamics, Eq. (2) describes the discrete event dynamics of the system: the switch from the *mode* indexed by i to mode i^+ occurs as the trajectory of the system, in the continuous state space, intersects a guard/threshold set described by a function of the kind $g_i(x, a_i) = 0$, i.e.,

$$s(x, i) = \begin{cases} i & g_i(x, a_i) \neq 0 \\ i^+ & g_i(x, a_i) = 0 \end{cases}$$

Note that the threshold functions g_i do not partition the state space in different regions each characterized by a certain mode, rather they dictate the occurrence of the ith switch and are "deactivated" thereafter. The functions g_i are assumed to be continuously differentiable in the x variable and the gradient w.r.t. the state variable $\partial g / \partial x$ will be denoted ∇g, a row vector. Also, the switching surfaces will be assumed to be always reachable. In formal terms:

Assumption 1.

$$\forall i \text{ and } \forall x \mid g_i(x, a_i) = 0,\ L_{f_i} g_i(x) \neq 0$$

where $L_{f_i} g_i(x) := \nabla g_i(x) f_i(x)$. The shape of the switching curves are determined by the form of the function g_i, given a priori, and by the value of the parameters a_i. These values constitute the control variables for the switched system. Consider a system evolution starting at $x(t_0) = x_0$, ruled by mode 1, which switches to mode 2 as the condition $g_1(x(t), a_1) = 0$ is satisfied, for some t. Similarly all subsequent switches from mode i to mode $i+1$ occur as the ith guard condition is met, i.e. when $g_i(x(t), a_i) = 0$. In the following, the ith switching instant will be denoted by t_i and the associated switching state as $x_i := x(t_i)$.

Here, the problem of tuning the values of parameters a_i, $i = 1, \ldots, N$ in order to minimize an integral cost function $J = \int_{t_0}^{T} L(x)$ is addressed. The final time T can be either a fixed final time or the time instant corresponding to the intersection with a terminal manifold $g_N(x, a_N) = 0$, i.e., in this case, system evolution terminates after $N-1$ switches at time $T = t_N$. The main contribution of this paper is to provide an expression for the derivatives dJ/da_i, which can be employed by a gradient descent algorithm for the (local) optimization of the switching surfaces. For simplicity, the case $m = 1$, will be considered, i.e., each switching surface is characterized by a single parameter, but it should be noted that similar arguments apply also in the general case $m > 1$.

3. PRELIMINARY RESULTS

Consider the trajectory of an autonomous system $\dot{x} = f(x)$, starting at time t_0 in x_0, hitting a surface $g(x, a) = 0$, $a \in \Re$, at time t_h, and incurring in a cost $J = \int_{t_0}^{t_h} L(x)$. Let $x_h = x(t_h)$, and let Φ denote the state transition matrix of the linearized system $\dot{z} = \frac{\partial f}{\partial x}(x(t))z(t)$.

Lemma 1.

$$\frac{dt_h}{da} = -\frac{1}{L_f g(x_h)} \frac{\partial g}{\partial a}(x_h) \tag{3}$$

Proof. Let a and $\tilde{a} = a + \Delta a$ denote the nominal and perturbed value for the switching surface parameter, which correspond to switches at time t_h in x_h or at time $\tilde{t}_h$ in $\tilde{x}_h$, respectively. Under Assumption 1 the perturbations $\Delta t_h := \tilde{t}_h - t_h$ and $\Delta x_h := \tilde{x}_h - x_h$ can be made arbitrarily small by proper small values of Δa, thus the following first order approximations can be made:

$$g(\tilde{x}_h, \tilde{a}) = g(x, a) + \frac{\partial g}{\partial x}\Delta x_h + \frac{\partial g}{\partial a}\Delta a + o(\Delta x_h, \Delta a)$$

$$\Delta x_h = f(x_h)\Delta t_h + o(\Delta t_h).$$

Eq. (3) follows from the switching conditions $g(x, a) = 0$, $g(\tilde{x}_h, \tilde{a}) = 0$, and considering that by Assumption 1 $L_f g(x_h) \neq 0$. □

Lemma 2.

$$\frac{dt_h}{dx_0} = -\frac{\nabla g(x_h)\Phi(t_h, t_0)}{L_f g(x_h)}. \tag{4}$$

Proof. Consider a perturbed initial condition $\tilde{x}_0 = x_0 + \Delta x_0$, resulting in a perturbed trajectory $\tilde{x}(\cdot)$, and assume that the intersection with the surface $g(x, a) = 0$ takes place at time $\tilde{t}_h = t_h + \Delta t_h$. A first order approximation of the condition to be satisfied by the perturbed trajectory at the intersection with the surface $g(x, a) = 0$ is given by:

$$\begin{aligned} 0 = g(\tilde{x}(\tilde{t}_h), a) = g(x(t_h), a) + \\ \nabla g(x(t_h)) \left(\Delta x(t_h) + f(x(t_h))\Delta t_h\right) + \\ o(\Delta x(t_h), \Delta t_h) \end{aligned} \tag{5}$$

where Δx denotes the variation in the nominal trajectory at a fixed time, hence:

$$\Delta x(t_h) = \Phi(t_h, t_0)\Delta x_0 + o(\Delta x_0) \tag{6}$$

Since $g(x(t_h), a) = 0$, substituting (6) in (5) yields the result. □

Lemma 3.

$$\frac{dJ}{dx_0} = p(t_0) \tag{7}$$

where the *costate* variable p satisfies the (backward) o.d.e.

$$\dot{p}^T = -p^T\frac{\partial f}{\partial x} - \frac{\partial L}{\partial x}, \tag{8}$$

$$p^T(t_h) = -\frac{L(x_h)\nabla g(x_h)}{L_f g(x_h)} \tag{9}$$

Also, if t_h is a *fixed final time* then $p^T(t_h) = 0$

Proof. This result is well known in optimal control theory (e.g. (Bryson and Ho, 1969)).

4. MULTIPLE SWITCHES

This section deals with the perturbations induced on the switching times and the corresponding states by small variations in the value of the threshold parameters. The manner in which these perturbations propagate across the switching surfaces will highlight the recursive nature of the solution, and how this is captured by introducing costate variables. The recursive propagation will take the form of a state-space-like discrete event system for the infinitesimal perturbations sampled in correspondence of the switches in the continuous dynamics. The variables $\tau_i(k)$ and $\chi_i(k)$ are introduced to represent the perturbations resulting at the kth switching instant due to variations in the ith threshold parameter a_i. In fact, by lifting the problem to a discrete event setting, we still capture all the information required to determine time and state perturbations at the next, $k+1$ switching surface from these new variables. Hence, they represent the *state* of the resulting discrete event system.

Since the switching surfaces are not time dependent, the hitting time at the kth switching surface, t_k depends on the parameters a_k of such surface and on the hitting time and state at the previous switching surface t_{k-1} and x_{k-1}.

$$t_k = t_k(t_{k-1}, x_{k-1}, a_k) \tag{10}$$

The kth switching state x_k, depends also on t_k, and is given by:

$$x_k = x_{k-1} + \int_{t_{k-1}}^{t_k} f_k(x(t))dt \tag{11}$$

Consider the infinitesimal approximation of the perturbations in the hitting times and states following a slight variation Δa_i on parameter a_i, assuming, for the moment, that this is independent by all the other parameters $a_j, j \neq i$. In this case, the variables $\tau_i \in \Re$ and $\chi_i \in \Re^n$ denote the following quantities:

$$\tau_i(k) = \frac{dt_k}{da_i}\Delta a_i + o(\Delta a_i) \tag{12}$$

$$\chi_i(k) = \frac{dx_k}{da_i}\Delta a_i + o(\Delta a_i); \tag{13}$$

According to this notation, differentiation of (10,11) give:

$$\tau_i(k) = \frac{\partial t_k}{\partial t_{k-1}}\tau_i(k-1) + \frac{\partial t_k}{\partial x_{k-1}}\chi_i(k-1) \quad (14)$$

$$\chi_i(k) = \frac{\partial x_k}{\partial t_k}\tau_i(k) + \frac{\partial x_k}{\partial t_{k-1}}\tau_i(k-1) + \frac{\partial x_k}{\partial x_{k-1}}\chi_i(k-1). \quad (15)$$

Since the equation $\dot{x} = f_k(x)$ is time-invariant in the interval $[t_{k-1}, t_k)$, it follows that $\partial t_k/\partial t_{k-1} = 1$. The term $\partial t_k/\partial x_{k-1}$ is the variation in the kth hitting time w.r.t. variations on x_{k-1} considering a fixed threshold surface, i.e., by Proposition 2:

$$\frac{\partial t_k}{\partial x_{k-1}} = -\frac{\nabla g_k(x_k)\Phi_k(t_k, t_{k-1})}{L_{f_k} g_k(x_k)} \quad (16)$$

Moreover, $dx_k/dt_k = f_k(x_k)$, since it is the equation $\dot{x} = f_k(x)$ in the interval $[t_{k-1}, t_k)$ that determines x_k (see Eq. 1); and $\partial x_k/\partial x_{k-1} = \Phi_k(t_k, t_{k-1})$. The term $\partial x_k/\partial t_{k-1}$ refers to a variation considering a fixed initial state x_{k-1} and fixed final time t_k; being the system time invariant it results: $\partial x_k/\partial t_{k-1} = -f_k(x_k)$. Finally, it results: $\partial x_k/\partial x_{k-1} = \Phi_k(t_k, t_{k-1})$. Having evaluated all the coefficients of Eqs. (14, 15), by substituting the former for the term $\tau_i(k)$ in the latter, it is possible to give a state space like expression for the iterative evolution of the variables τ_i and χ_i. In order to present this result with the least possible notational burden, the obvious arguments of functions f, g, and Φ and their (repeated) indexes will be dropped, adopting the following conventions: $f_k := f_k(x_k)$; $\nabla g_k := \nabla g_k(x_k)$; $L_k := L_{f_k} g_k(x_k)$ $\Phi_{k,k-1} = \Phi_k(t_k, t_{k-1})$.

$$\tau_i(k) = \tau_i(k-1) - \frac{\nabla g_k \Phi_{k,k-1}}{L_k}\chi_i(k-1) \quad (17)$$

$$\chi_i(k) = \left(I - \frac{f_k \nabla g_k}{L_k}\right)\Phi_{k,k-1}\chi_i(k-1) \quad (18)$$

4.1 Not independent threshold parameters

Assuming the a_i not independent (e.g. two distinct switches are determined by the same surface) adds forcing terms proportional to $\alpha_i(k) := \frac{da_k}{da_i}\Delta a_i + o(\Delta a_i)$ to the "autonomous" dynamics (17,18). In fact, (14) modifies as follows:

$$\tau_i(k) = \frac{\partial t_k}{\partial t_{k-1}}\tau_i(k-1) + \frac{\partial t_k}{\partial x_{k-1}}\chi_i(k-1) + \frac{\partial t_k}{\partial a_k}\alpha_i(k)$$

which yields:

$$\begin{pmatrix} \tau_i(k) \\ \chi_i(k) \end{pmatrix} = A(k)\begin{pmatrix} \tau_i(k-1) \\ \chi_i(k-1) \end{pmatrix} - \begin{bmatrix} \frac{\frac{\partial g_k}{\partial a}}{L_k} \\ \frac{\frac{\partial g_k}{\partial a}}{L_k} f_k \end{bmatrix} \alpha_i(k) \quad (19)$$

where $A(k)$ is the matrix whose members derive from Eqs. (17,18). The above dynamics describe how the perturbation in the hitting times due to a slight modification in the parameter a_i propagates from the ith hit. As initial conditions set $\tau_i(0) = 0$ and $\chi_i(0) = 0$; observe that the system describing the propagation of the perturbations remains at such a "state" zero until the first non null forcing term, i.e. $\alpha_i(i) = 1$, causes it to drift ($\alpha_i(k) = 0$ for $k < i$).

5. COMPUTATION OF THE GRADIENT OF THE COST FUNCTION

Consider a nominal trajectory $x(\cdot)$ characterized by N switches, occurring at time instants t_k and states x_k, and a perturbed trajectory $\tilde{x}(\cdot)$, due to a small variation in parameter a_i, with switches at time instants $\tilde{t}_k$ at states $\tilde{x}_k$. Let $\Theta_k := \max\{t_k, \tilde{t}_k\}$, $\theta_k := \min\{t_k, \tilde{t}_k\}$, and consider the piece of nominal and perturbed trajectories delimited by the time instant Θ_k and by their respective intersections with the next, $(k+1)^{st}$ switching surface. In this interval, both nominal and perturbed trajectories evolve according to the same mode $k+1$, and the corresponding variation in the cost function J, i.e.,

$$\int_{\Theta_k}^{\tilde{t}_{k+1}} L(\tilde{x}(s))ds - \int_{\Theta_k}^{t_{k+1}} L(x(s))ds, \quad (20)$$

can be referred to Lemma 3, where the perturbation on the initial conditions, according to (12,13) are given by:

$$\Delta x(\Theta_k) = \tilde{x}(\Theta_k) - x(\Theta_k) = \chi_i(k) - f_{k+1}(x_k)\tau_i(k), \quad (21)$$

independently by the relative ordering of t_k and $\tilde{t}_k$ (see Figure 2). Therefore, by (7,21) a first order approximation of Eq. (20) is:

$$p_{k+1}^T(t_k)(\chi_i(k) - f_{k+1}(x_k)\tau_i(k)) \quad (22)$$

where the costate variable evolves backwards according to:

$$\dot{p}_k^T = -p_k^T\frac{\partial f_k}{\partial x} - \frac{\partial L}{\partial x}, \quad (23)$$

with boundary conditions (see Eq. 9) :

$$p_k^T(t_k) = -\frac{L(x_k)\nabla g(x_k)}{L_{f_k} g_k(x_k)}, \quad (24)$$

A unique costate variable p can be defined by means of the switched dynamics (23), with reset conditions (24), for $k = 1, \dots N$, so that: $p^T(t_k^+) = p_{k+1}^T(t_k)$. Notice that the choice of time t_k instead of time Θ_k in (22) is justified since

$$p(\Theta_k) = p(t_k) + \dot{p}(t_k)\tau_i(k) + o(\Delta a_i),$$

so that Eq. (22) still constitutes a proper first order approximation to (20).

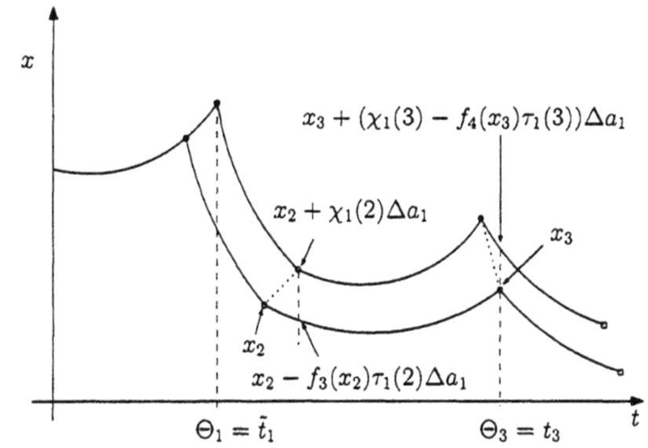

Fig. 2. An example of nominal (below) vs. perturbed (above) trajectories for one component of the state variables, denoted by x. A variation in parameter a_1 generates perturbations in the hitting times and states for all switches

For completing the derivation of dJ/da_i it is convenient to introduce the following function:

$$I_i(k) = \begin{cases} -\int_{t_k}^{\tilde{t}_k} L(x(s))ds, & t_k \le \tilde{t}_k \\ \int_{\tilde{t}_k}^{t_k} L(\tilde{x}(s))ds, & \tilde{t}_k < t_k \end{cases} \tag{25}$$

It is easy to see that independently by the relative ordering of t_k and $\tilde{t}_k$, it results:

$$I_i(k) = -L(x_k)\tau_i(k) \tag{26}$$

The variation in the cost function J due to a variation in a_i, is (see Figure 2):

$$\Delta J_i = \int_{\theta_i}^{\tilde{t}_N} L(\tilde{x}) - \int_{\theta_i}^{t_N} L(x) =$$
$$\int_{\theta_i}^{\Theta_i} L(\tilde{x}(s)) - L(x(s))ds + \sum_{k=i+1}^{N-1} I_i(k) +$$
$$\sum_{k=i}^{N-1} \left(\int_{\Theta_k}^{\tilde{t}_{k+1}} L(\tilde{x}(s))ds - \int_{\Theta_k}^{t_{k+1}} L(x(s))ds \right). \tag{27}$$

The only term of (27) which has not been characterized yet is: $\int_{\theta_i}^{\Theta_i}(L(\tilde{x}(s)) - L(x(s))ds$ which results to be a $o(\Delta a_i)$. According to this consideration, evaluating (27) at the light of (22,26), letting $\Delta a_i \to 0$, and recalling the definitions (12,13) yields:

$$\frac{dJ}{da_i} = \sum_{k=i}^{N-1} p^T(t_k^+)\left(\frac{dx_k}{da_i} - f_{k+1}(x_k)\frac{dt_k}{da_i}\right)$$
$$- \sum_{k=i+1}^{N-1} L(x_k)\frac{dt_k}{da_i}. \tag{28}$$

5.1 Computational Issues

Eq. (28) describes the generic term of the gradient dJ/da, $a := (a_1, \ldots, a_N)^T$. The computation of each term dJ/da_i needs:

i) the set of switching time instants t_k and switching states x_k, for $k = i, \ldots, N$; ii) the perturbation terms $\tau_i(k)$ $\chi_i(k)$, approximating $\frac{dt_k}{da_i}$ and $\frac{dx_k}{da_i}$ in (28), for $k = i, \ldots, N-1$; iii) the value of the costate variable p at times t_k^+, for $k = i, \ldots, N-1$. The computational cost for each of the above can be evaluated as follows:

I) A *single* forward integration of the state equations provides t_k, x_k, for $k = 1, \ldots, N$. Since those are nominal values they are independent by which parameter a_i is considered varying.

II) The characterization of the perturbation terms is provided via (19). Associated to the variations on each parameter a_i are the relative set of time and state perturbations $\tau_i(k)$ and $\chi_i(k)$, $k = i, \ldots, N$, whose computation needs *independent* forward integrations, one for each varying threshold parameter. Although the computational burden may seem large, it is worth noting that in some cases the number of switching surfaces is far less than the associated switchings (e.g. stability).

III) A *single* backwards integration for the costate dynamics (23) provides the needed values for the costate: in fact, such dynamics as well as the corner conditions (24) are independent of the varying parameter a_i.

5.2 Example

We present an example of switching surfaces optimization, using the gradient formulae derived (28) in a gradient descent algorithm. We consider the system, (after (Branicky, 1998)), $\dot{x} = A_i x$, characterized by two modes:

$$A_1 = \begin{pmatrix} -1 & -100 \\ 10 & -1 \end{pmatrix}; A_2 = \begin{pmatrix} 1 & 10 \\ -100 & 1 \end{pmatrix};$$

The system switches from mode 1 to mode 2 upon hitting the line $x_1 + a_1^2 x_2 = 0$ in the fourth quadrant; and from mode 2 to mode 1 upon hitting $a_2^2 x_1 + x_2 = 0$ in the second quadrant. The switching surfaces optimization is carried out by a gradient descent algorithm for the switching parameters. Following the procedure described in the preceding Section, the gradient dJ/da, $a = \{a_1, a_2\}$ is computed for the current, nominal instance of the switching parameters, and their value is iteratively optimized, by gradient descent, until some terminating condition for the algorithm is satisfied. As a cost function let

$$J = \frac{1}{2}\int_0^T ||x^2(t)||dt.$$

System evolution starts at $x_0 = (-1, 0)^T$ and terminates at a fixed final time $T = 0.3$. The optimization is achieved by a gradient descent algorithm with Armijo stepsize (Polak, 1997), and terminating condition $||dJ/da|| < 10^{-6}$. Figure

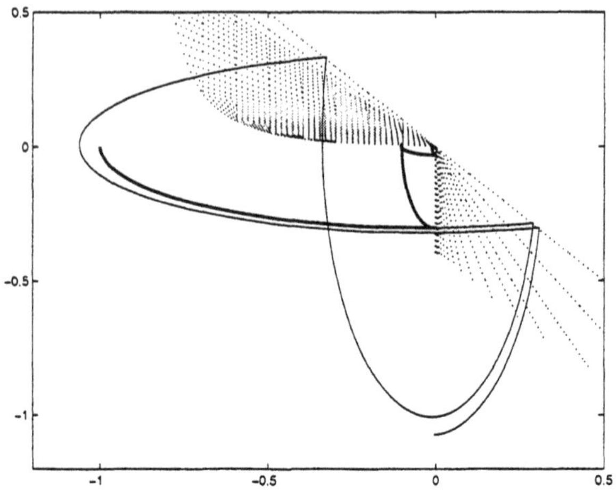

Fig. 3. Evolution of switching surfaces (dashed lines). System trajectories for nominal and terminal values of the switching parameters represented by thin and thick lines, respectively

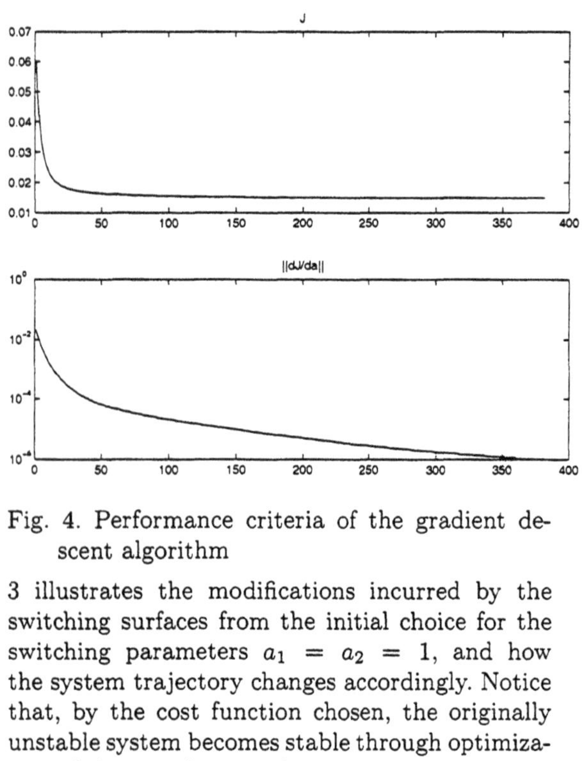

Fig. 4. Performance criteria of the gradient descent algorithm

3 illustrates the modifications incurred by the switching surfaces from the initial choice for the switching parameters $a_1 = a_2 = 1$, and how the system trajectory changes accordingly. Notice that, by the cost function chosen, the originally unstable system becomes stable through optimization of the switching surfaces.

REFERENCES

Arkin, R.C. (1998). *Behavior Based Robotics.* The MIT Press. Cambridge, MA.

Bemporad, A., A. Giua and C. Seatzu (2002). A master-slave algorithm for the optimal control of continuous-time switched affine systems. In: 41th *IEEE Conf. on Decision and Control (CDC 2002).* Las Vegas, USA. pp. 1976 –1981.

Boccadoro, M. and P. Valigi (2003). A modelling approach for the dynamic scheduling problem of manufacturing systems with non negligible setup times and finite buffers. In: *42nd IEEE Conference on Decision and Control (CDC '03).* Maui, Hawaii, USA.

Branicky, M. S. (1998). Multiple lyapunov functions and other analysis tools for switched and hybrid systems. *IEEE Trans. on Automatic Control* **43**(4), 475–482.

Branicky, M., V. Borkar and S. Mitter (1998). A unified framework for hybrid control: model and optimal control theory. *IEEE Trans. on Automatic Control* **43**(1), 31–45.

Bryson, A. E. and Y. C. Ho (1969). *Applied Optimal Control.* Ginn and Co.

Cassandras, C. G., D. L. Pepyne and Y. Wardi (2001). Optimal control of a class of hybrid systems. *IEEE Trans. on Automatic Control* **46**(3), 398–415.

Egerstedt, M. (2000). Behavior based robotics using hybrid automata. In: *Lecture Notes in Computer Science: Hybrid Systems III: Computation and Control.* Pittsburgh, PA. pp. 103–116.

Egerstedt, M., Y. Wardi and H. Axelsson (n.d.). Transition time optimization for switched systems. *IEEE Trans. on Automatic Control, submitted.*

Giua, A., C. Seatzu and C. Van Der Mee (2001). Optimal control of switched autonomous linear systems. In: 40th *IEEE Conf. on Decision and Control (CDC 2001).* Orlando, FL, USA. pp. 2472–2477.

Johansson, M. and A. Rantzer (1998). Computation of piecewise quadratic lyapunov functions for hybrid systems. *IEEE Trans. on Automatic Control* **43**(4), 555–559.

Polak, E. (1997). *Optimization: Algorithms and Consistent Approximations.* Springer.

Riedinger, P., F. Kratz, C. Iung and C. Zanne (1999). Linear quadratic optimization for hybrid systems. In: *Proceedings of the 38th IEEE Conference on Decision and Control.* Phoenix, AZ. pp. 3059–3064.

Shaikh, M. S. and P.E. Caines (2002). On trajectory optimization for hybrid systems: theory and algorithms for fixed schedules. In: 41th *IEEE Conf. on Decision and Control (CDC 2002).* Las Vegas, USA. pp. 1997–1998.

Sussmann, H.J. (1999). A maximum principle for hybrid optimal control problems. In: *Proceedings of the 38th IEEE Conference on Decision and Control.* Phoenix, AZ. pp. 425–430.

Xu, X. and P.J. Antsaklis (2002). An approach to optimal control of switched systems with internally forced switchings. In: *Proceedings of the American Control Conference.* Anchorage, USA. pp. 148 –153.

ELSEVIER
IFAC
PUBLICATIONS
www.elsevier.com/locate/ifac

DISCRETE EVENT CONTROL OF A PINBALL MACHINE

Gerwald Lichtenberg * Jörg Neidig **

* *University of Technology Hamburg-Harburg*
Institute of Control Engineering
Eissendorfer Str. 40, 21073 Hamburg
Lichtenberg@tu-harburg.de

** *Ruhr-Universität Bochum*
Institute of Automation and Computer Control
Universitätsstr. 150, 44780 Bochum
Neidig@atp.ruhr-uni-bochum.de

Abstract: The paper is concerned with the control of a pinball machine. The plant is a hybrid system because of the continuous movement of the ball and the only discrete sensors and actuators. As no additional optical sensors are used, a complicated discrete event control problem arises. With the help of an identified Semi Markov Model, an optimal controller is designed and tested. *Copyright © 2004 IFAC*

Keywords: Hybrid Systems, Semi Markov Models, Real Time Systems, Controllers

1. INTRODUCTION

The usual pinball game is well known. The human player looks at the playing field, watches the ball moving and presses some buttons that act on the flipper arms when a ball comes near one arm. A score display shows the actual score of the game depending on sensor events that have occurred during the ball movement. There are mechanical, magnetic or optical sensors on the playing field that can detect if the ball touches them or stays nearby.

This manual setup means control with full information: the actual position and the velocity of the ball in principle can be seen with the eyes, although this is no precise measurement. The basic block diagram of the control loop is shown in Figure 1. The difficulty of this game is mainly due to the fact that there are very few options to influence the ball with the flipper arms.

One idea to automate the human controlled system would be to install additional sensors that can track the position of the ball, such as cameras or other optical sensors. That approach would use the same class of information and would lead to severe problems mainly in the field of image processing.

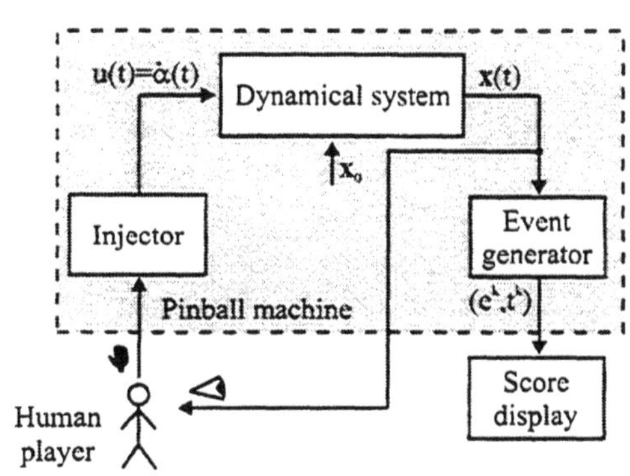

Fig. 1. Manual pinball control with full information

This is not the aim of this paper. Instead, the already installed sensors are used as the only ones giving information about the actual state. But there are no sensors in the working area of the flipper arms that might help to simplify the control task. The actual position only can be reconstructed by processing the sequences of former measured events.

Every algorithm used for reconstruction has to include a dynamical model of the pinball machine. As this is a real world experiment, it is clear that in no case the state can be determined exactly.

The laboratory experimental plant is a pinball machine *Indianapolis 500* from *Bally* that can be found in several playing halls around the world. This machine was introduced in 1996. It has 3 flipper arms and 35 sensors mounted to a playing field like shown in Figure 2.

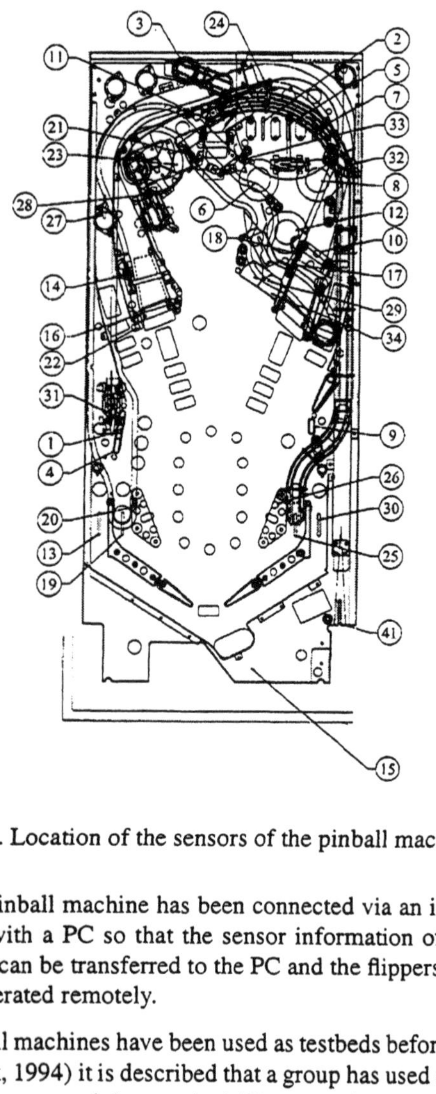

Fig. 2. Location of the sensors of the pinball machine

The pinball machine has been connected via an interface with a PC so that the sensor information of the game can be transferred to the PC and the flippers can be operated remotely.

Pinball machines have been used as testbeds before. In (Clark, 1994) it is described that a group has used such a game as a real time testbed. However, these systems have not been investigated as hybrid systems control problems like presented here.

In the field of hybrid systems control (Alur *et al.*, 1996), the problem of plants having discrete inputs and outputs while having a continuous states is addressed. In principle, a variety of modelling and design approaches for hybrid and discrete event systems can be applied, (Cassandras and Lafortune, 1999). Their capabilities can be tested using the control problem given here as a benchmark example.

In (Raisch, 1998) abstractions with an adjustable degree of accuracy are used for qualitative modelling of discrete time systems. This might be extended for the case of discrete event systems as it is formulated also in (Förstner and Lunze, 2000). The latter approach does not use information about probabilistic properties that can be introduced with the help of nondeterministic automata (Lichtenberg and Steele, 1996; Schröder, 2002). Especially for the case of discrete event systems, the concept of semi markov processes is seemingly interesting (Lunze *et al.*, 1999; Lunze, 2000).

Although Control of a Pinball machine has no direct applicability for engineering purposes (besides the possibility to let the machines run autonomously to attract potential players), there are industrial systems that show the same special behaviour that the plant can only be influenced by the inputs if certain conditions for the actual state hold. Combustion engines e.g. have upper and lower dead centers depending on the actual state of the crankshaft, valves and ignition.

The paper is organized as follows: In Section 2 the pinball machine is described formally as a hybrid system. In Section 3 the hybrid control problem is introduced and an approach to its solution is presented in Section 6 with the help of a semi markov model introduced in Section 4 which can be found as described in Section 5. In Section 7 a short overview of the real time implementation is given. Finally the results are summarized and the paper closes with an outlook.

2. THE HYBRID PLANT

This section is concerned with the plant to control shown as 3 subsystems in the upper part of Figure 3 that are composed in the block *Pinball machine*. The diagram of the pinball machine shows three subsystems *dynamical system, injector* and *event generator*. These will be described in detail in this section. Looking at the pinball machine from an outside point of view, it is a discrete event system. The input sequence as well as an output sequence only contain values of discrete sets.

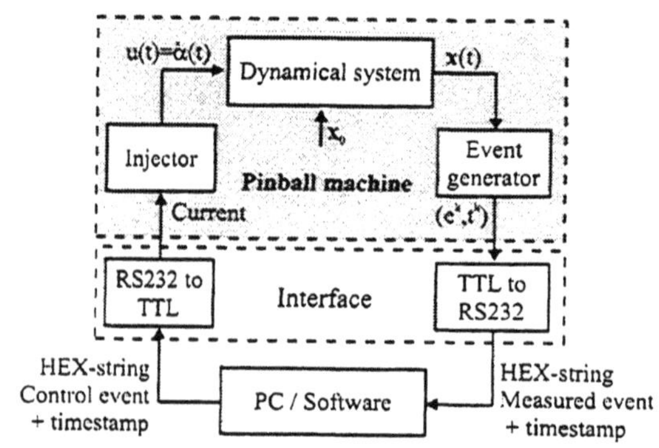

Fig. 3. Automatic pinball control

Continuous Dynamical System

A usual way to describe dynamical systems having continuous inputs and outputs is to set up state space models for the system dynamics. The state of the pinball machine can formally be described by a vector

$$\mathbf{x} = (s_1, s_2, \dot{s}_1, \dot{s}_2, \alpha_1, \alpha_2, \alpha_3)^T \quad , \qquad (1)$$

whereby s_1 and s_2 indicate the current position of the ball in the playfield and $\dot{s}_1$ and $\dot{s}_2$ the components of the current velocity vector. The angles α_i with $i \in \{1, 2, 3\}$ show the current position of the three flipper arms relative to their lower end position. This is a simplified state vector since we assume that the ball stays in the plain for all times and does not jump. The spin of the ball is also ignored.

The continuous dynamical system itself is a hybrid system, because the ball moves continuously on the playfield, but after every collision with a wall an autonomous jump takes place, (Branicky, 1995). This means that the absolute value and the direction of the velocity of the ball change abruptly.

The velocity also changes when the ball is hit by one of the flipper arms. This can be modelled neither as an autonomous nor a controlled jump, since the input of the system can be changed arbitrarily, but it only influences the system if other conditions hold. Regarding the pinball machine, it has to be modelled that the ball can only be influenced by the inputs if it is in a certain area given by the flipper geometry.

The dynamics of the system can be described by a nonlinear state space model

$$\dot{\mathbf{x}} = \boldsymbol{f}_n(\mathbf{x}, \mathbf{u}) + \boldsymbol{f}_a(\mathbf{x}) + \boldsymbol{f}_c(\mathbf{u}) + \boldsymbol{f}_s(\mathbf{x}, \mathbf{u}) \quad (2)$$

On the right side of eqn. (2) there is a vector field split into 4 components $\boldsymbol{f}_n$, $\boldsymbol{f}_a$, $\boldsymbol{f}_c$, $\boldsymbol{f}_s$, which are also vector fields. The first part $\boldsymbol{f}_n(\mathbf{x}, \mathbf{u})$ describes the continuous movement of the ball on the plane. The other components all are discontinuous. The second vector field

$$\boldsymbol{f}_a(\mathbf{x}) = \sum_{i=1}^{n_a} \boldsymbol{c}_{a_i}(\mathbf{x}) \cdot \delta(g_{a_i}(\mathbf{x})) \qquad (3)$$

describes autonomous jumps since the condition of the jump depends on the state $\mathbf{x}(t)$ only. The coefficient vectors $\boldsymbol{c}_{a_i}$ determine the jumping distance while the functions g_{a_i} give the jumping conditions.

The Dirac function δ implies jumping when the condition $g_{a_i}(\mathbf{x}) = 0$ holds. E.g. the system

$$\dot{x}(t) = x(t) + \delta(x(t) - x_s), \quad x(0) = x_0 \qquad (4)$$

causes a unit jump at time t_s for which the condition

$$x(t_s) = x_s \qquad (5)$$

is satisfied. Note that the time t_s is implicitly described by (4). All collisions of the ball with a wall are described by this component.

The third component

$$\boldsymbol{f}_c(\mathbf{x}) = \sum_{i=1}^{n_c} \boldsymbol{c}_{c_i}(\mathbf{u}) \cdot \delta(g_{c_i}(\mathbf{u})) \qquad (6)$$

describes controlled jumps accordingly. For the pinball machine $\boldsymbol{f}_c = \mathbf{0}$ holds. The last component

$$\boldsymbol{f}_s(\mathbf{x}, \mathbf{u}) = \sum_{i=1}^{n_s} \boldsymbol{c}_{s_i}(\mathbf{x}, \mathbf{u}) \cdot \delta(g_{s_i}(\mathbf{x})) \qquad (7)$$

introduces a special behaviour that can be found in the Pinball system. It models jumps that the system can undergo with state dependent influence of the inputs. Jumps of this class occur in the pinball machine whenever the ball is hitting a flipper arm while the arm is moving.

Event Generator

The event generator takes the continuous signal $\mathbf{x}(t)$ and generates a sequence of events. For a theoretical description $Q(i)$ is defined as the region in state space that is covered by the i-th sensor, (Nixdorf, 2003). The boundary of the region is symbolised by $\delta Q(i)$, as shown in Figure 4.

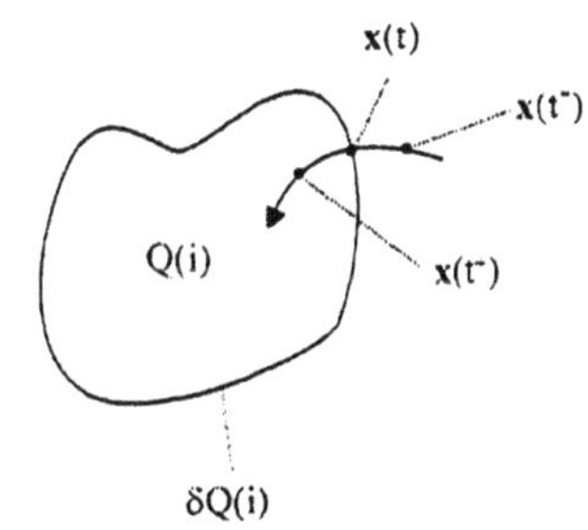

Fig. 4. Event Generation

The k-th event e_k takes place at time t_k, i.e.

$$\mathbf{x}(t_k) \in \delta Q(i) \qquad (8)$$

holds. With the times t^- and t^+ defined for $dt > 0$ as

$$t^- := \lim_{dt \to 0} t - dt \ , \quad t^+ := \lim_{dt \to 0} t + dt \ , \qquad (9)$$

two conditions for an event can be noted:

$$\mathbf{x}(t^+) \in Q(i) \ , \quad \mathbf{x}(t^-) \notin Q(i) \ . \qquad (10)$$

An example of a sequence of events of the pinball can be seen in Figure 8 in Section 7. Formally, an event sequence can be written as

$$W = ((w_1, t_1), (w_2, t_2), \ldots, (w_n, t_n)) \ , \qquad (11)$$

where the event times $t_1, t_2, \ldots, t_n \in \mathbb{R}^+$ and the event values $w_1, \ldots, w_n \in \mathcal{W}$ are combined and ordered.

For the Pinball machine *Indianapolis 500*, all 35 sensors on the playing field that can produce measurement events are shown in figure 2, i.e. the set $\mathcal{W}$ has 35 elements which can be uniquely labelled by symbols as e.g. *Center Lane*. One event that is very important for controller synthesis is the *Ball Lost* event, indicating that the ball has fallen through the lower hole of the playing field.

Injector

The injector transforms a discrete signal into a continuous one. In the case of the pinball machine the digital signal of the pressed button is converted to mechanical forces applied to the flipper arm. The input signal is of discrete event character and can be written as

$$V = ((v_1, t_1), (v_2, t_2), \ldots, (v_n, t_n)), \quad (12)$$

where

$$v_i \in \{UR_u, UR_d, LR_u, LR_d, LL_u, LL_d\} = \mathcal{V} \quad (13)$$

holds, i.e. only the 3 Flipper arms (upper right *UR*, lower right *LR* and lower left *LL*) can be turned up or down, symbolised by the index u or d, respectively. The injector converts the sequence (12) to a continuous signal applied as input to the continuous dynamical system.

3. HYBRID CONTROL PROBLEM

One principle aim of a controller is to keep the ball in the game i.e. in a subspace of the state space. The main problem hereby is the observation of the position (s_1, s_2) of the ball. Only when the ball hits a sensor, its position is exactly known. With the help of a dynamical model of the plant an approximation of the position at other times may be predicted from sequences (11), (12) of measurement and control events.

On the other hand the ways to influence the plant are quite poor. The controller only can make the flipper arms moving. Formally, the controller defines a certain input sequence (12) according to each measurement sequence (11) and, in general, asynchronous to it.

With respect to the goal that the ball should not be out of the game, for each event can be decided whether it belongs to the set $\mathcal{W}_0$ of forbidden events. For the pinball machine, a *Ball Lost* event is an element of that set.

The control problem for the Pinball machine should be solved with the help of a dynamical model. The problem is of the class that can be formulated as follows.

Safety Hybrid Control Problem

Given: Measurement event sequence

$$W = ((w_1, t_1), (w_2, t_2), \ldots, (w_n, t_n))$$

Set $\mathcal{W}_0$ of forbidden events

Find: Input event sequence

$$V = ((v_1, t_{n+1}), (v_2, t_{n+2}), \ldots, (v_m, t_{n+m}))$$

such that $w_{n+i} \notin \mathcal{W}_0 \; \forall \, i > 0$

4. SEMI MARKOV MODEL

It is clear that a state space model (2) could not be derived for the complete pinball machine, because an exact description of all walls and other configurations is too complicated. Stochastic approaches like Semi-Markov models seem to be more promising (Lunze, 2000). It's central element is a conditional probability density function

$$q_{hij}(\tau) = \text{Prob}\left(\begin{matrix} e_{k+1} = j \\ t_{k+1} \leq \tau \end{matrix} \middle| \begin{matrix} e_k = i \\ e_{k-1} = h \end{matrix} \right), \quad (14)$$

here given for a series of two past events. It holds the information with which probability the next event $e_{k+1} = j$ occurs within the time interval $t \leq \tau$ if the events $e_k = i$ and $e_{k-1} = h$ have occurred before. It is not distinguished between control and measurement events for the modelling of the pinball machine, so for any event $e_k \in \mathcal{V} \cup \mathcal{W}$ holds. This is due to the fact that these input and output events can occur in any order.

Using exactly two past events is not the only possible choice. But it is reasonable, because one past event does not even lead to a unique result if a quantitative model of the hybrid system would be known. On the other hand, using more past events would lead to a very complex Semi-Markov model. In principle, the more past events are considered in the condition part, the more exact the discrete event dynamics can be represented, (Raisch, 1998).

5. IDENTIFICATION

Approximations of (14) can be found with the help of experiments. Input and output event sequences can be stored in a data set and from this, the relative frequency can be computed and used as approximations of (14). The approximation becomes better with growing number of experiments.

Since the controller can only change the path of the ball by moving the flippers, an important information that needs to be calculated is the time when the ball enters the radius of the flipper arms. For an easy approach to controller design the transition times for this event *Ball in Flipper Radius* have to be introduced in the model in the same way as given in eqn. (14).

The basic problem now is, that this special event is not measurable with the standard setup. In order to determine the conditional probability for the event *Ball in Flipper Radius*, additional inductive sensors were placed above the flipper arms during the phase of identification and had been removed afterwards during the phase of control.

The Figures 5 and 6 show examples for results of the identification. The approximated conditional probability distribution (14) is shown for two different combination of past events. The total number of experiments in the 2 weeks identification phase was 3618, with a total number of $\approx$ 88200 events.

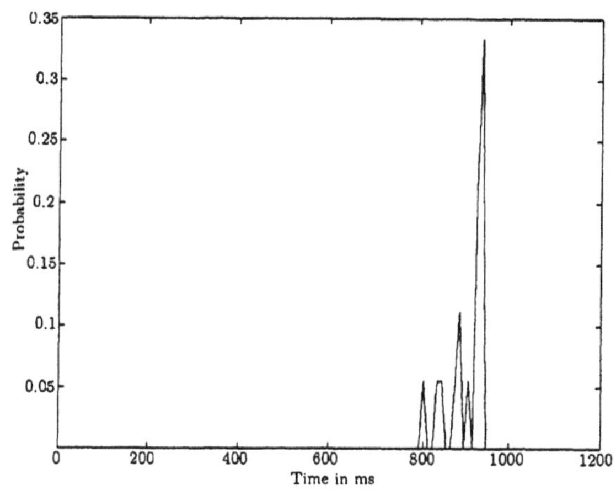

Fig. 5. Measured distribution (deterministic)

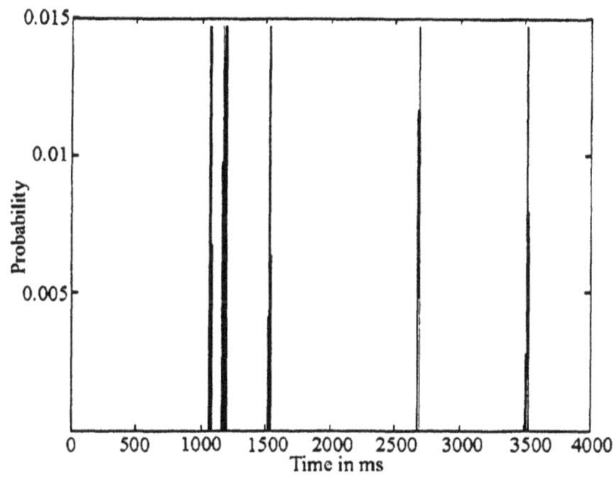

Fig. 6. Measured distribution (stochastic)

It is clearly visible, that the past events of Figure 5 lead to a next event that takes place between 8 and 9.5 seconds after. Thus it is relatively deterministic while as situation is quite stochastic in Figure 6.

The reason for this lies in the geometry of the playing field. The first example is from a loop where the movement of the ball is very restricted. The second example is from a region where so called "jets" change the velocity of the ball nearly arbitrarily. The identified distributions are not complete in a sense that they show zero probabilities at times where no event was measured but in principle can happen.

6. CONTROLLER DESIGN

With this extended model it can be determined, when a probability of the ball being in the radius of the flipper arms is greater then zero, assumed the actual and the past measurement events are known. Moreover, optimal times t_o for moving the flipper arms up can be predicted for each combination of present and past event with the help of the model (14) and using a cost function

$$J(t_0) = \int_{t_0}^{t_0+\Delta t} q_{hij}(\tau)d\tau \ . \qquad (15)$$

that reflects the last two events h and i as well as the event j = *Ball in Flipper radius* and the total time Δt of the upwards movement of the flipper arm. Classical continuous optimization algorithms are used off-line to compute the optimal times for all possible combinations (h, i) of two past events. The designed controller can not assure that the event *Ball Lost* never occurs, but lowers its probability.

7. REAL TIME EXPERIMENTS

For the experiments, the laboratory at the Technical University of Hamburg-Harburg was used. Whenever a sensor is activated a signal is sent to the internal processor of the pinball machine as an input. The inputs and outputs of this processor are checked for changes every millisecond by an interface. This can be seen as quasi-continuous measurements with respect to times between two events that are usually larger than 100 ms.

If there are changes, the interface generates an ASCII-string that contains information about these events. This string is send via a standard serial port to a PC and the desired input sequence for the movement of the flipper arms can be computed with programs like MATLAB. After the computation, it is sent from the PC to the pinball machine via the same interface. This leads to actions similar to manual pressing buttons of the Pinball Machine.

Thus, the closed loop system of Figure 3 can be operated autonomously, i.e. the PC plays a game on its own. Different control strategies can be tested and evaluated for example by determining the mean play times. The best results of the actual implemented controller which was designed with the help of the identified semi markov model are around one minute without a ball loss.

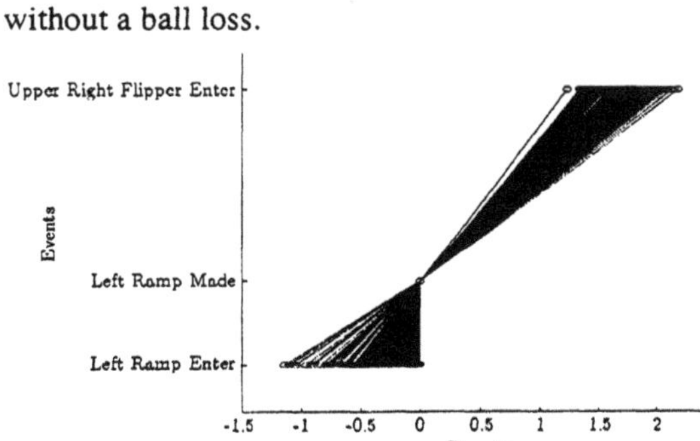

Fig. 7. Deterministic event sequence

In Figure 7, several measured event sequences are plotted in one picture and it is visible, that all sequences do have the same event codes but the time interval is variant. In this deterministic setup, the controller also works very well and hits the ball nearly every time.

On the other hand, there are other regions of state space, where the next event is not determined by the present and the past like shown in the Figures 8 and 9. Both pictures show event sequences that do have the same pair of events at time 0 and before (*Right Jet* and *Left Jet*) but the future as well as the past behavior is completely different. In this case, the controller is not able to hit the ball precisely and after these kind of situations, the ball often is lost.

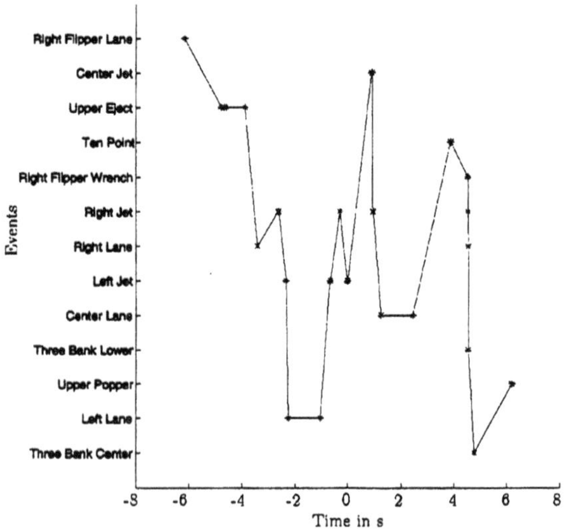

Fig. 8. Nondeterministic event sequence 1

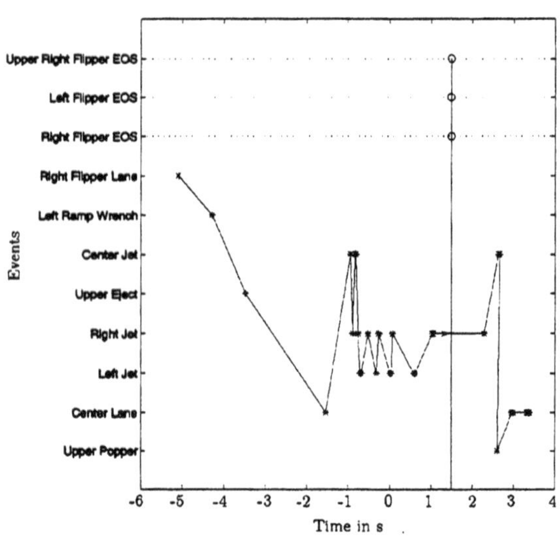

Fig. 9. Nondeterministic event sequence 2

8. CONCLUSION AND OUTLOOK

The paper shows that a classical Pinball Machine is a hybrid system that has an inherent control problem of great difficulty. Especially if the controller only accesses the event sequences measured by the on board sensors of the Pinball Machine, the reconstruction of the actual position of the ball is a non trivial problem.

The designed controller shows a performance which is good in deterministic areas of the state space whereas the more stochastic the dynamical behaviour gets, the chances to hit the ball are decreasing.

Other approaches to hybrid systems controller design will be evaluated in the future. Especially learning methods are promising, because the tactics of good players can be recorded while they play with the experimental Pinball machine.

The laboratory experiment is available at the Technical University of Hamburg-Harburg and easy to configure. Thus, it may be used for the formulation of real time benchmarks in Hybrid Systems Control. Other researchers are welcome to apply their modelling and design methods to this system.

REFERENCES

Alur, R., T. Henzinger and E. Sontag (1996). *Hybrid Systems III: Verification and Control*. Vol. LNCS 1066 of *Lecture Notes in Computer Science*. Springer-Verlag, Berlin Heidelberg.

Branicky, M. (1995). Studies in Hybrid Systems: Modeling, Analysis and Control. PhD thesis. MIT.

Cassandras, C. and S. Lafortune (1999). *Introduction to Discrete Event Systems*. Kluwer Academic Publishers.

Clark, D. (1994). An inexpensive realtime testbed - the pinball player project. *Proceedings of the IEEE Workshop on Real-Time Applications: Washington DC* pp. 86–88.

Förstner, D. and J. Lunze (2000). A discrete-event abstraction of continuous-variable systems with asynchronous inputs. In: *Proc. 3rd IMACS Symposium on Mathematical Modelling*. pp. 449–452.

Lichtenberg, G. and A. Steele (1996). An approach to fault diagnosis using parallel qualitative observers. In: *Workshop on Discrete Event Systems*. pp. 290–295.

Lunze, J. (2000). Diagnosis of quantised systems based on a timed discrete-event model. *IEEE Transactions on Systems, Man, and Cybernetics Part A* **SMC-30**(3), 322–335.

Lunze, J., B. Nixdorf and J. Schröder (1999). Deterministic discrete-event representations of linear continuous-variable systems. *Automatica* **35**, 395–406.

Nixdorf, B. (2003). *Discrete-Event Modelling and Control of Hybrid Systems*. Vol. 8 of *Fortschrittsberichte*. VDI-Verlag.

Raisch, J. (1998). Hierarchy of discrete abstractions for a hybrid plant. In: *Proceedings of the 3rd Conference on Automation of Mixed Processes*. pp. 55–62.

Schröder, J. (2002). *Modelling, State Observation and Diagnosis of Quantised Systems*. Lecture Notes in Control and Information Sciences. 1 ed.. Springer-Verlag Berlin, Heidelberg, New York.

ELSEVIER
IFAC
PUBLICATIONS
www.elsevier.com/locate/ifac

ON WEIGHTED TIME-OPTIMAL CONTROL FOR LINEAR HYBRID AUTOMATA USING QUANTIFIER ELIMINATION

Yan Pang * **Michael P. Spathopoulos** *

* *Department of Mechanical Engineering, University of Strathclyde, Glasgow, UK*
{ypang,mps}@mecheng.strath.ac.uk

Abstract: This paper considers the optimal control problem for linear hybrid automata. In particular, it is shown that the problem can be transformed into a constrained optimization problem whose constraints are a set of inequalities with quantifiers. Quantifier Elimination (QE) techniques are employed in order to derive quantifier free inequalities that are linear. The optimal cost is obtained using linear programming. The optimal switching times and optimal continuous control inputs are computed and used in order to derive the optimal hybrid controller. Our results are applied to an air traffic management example. *Copyright © 2004 IFAC*

Keywords: hybrid automata, quantifier elimination, time optimal control

1. INTRODUCTION

A mathematical model for hybrid systems is the *hybrid automaton*, which represents discrete components by finite-state machines and continuous components by real-numbered variables. An important subclass of hybrid systems is the so called rectangular hybrid automata (Preußig *et al.*, 1998), where in each mode the continuous dynamics are given as constant differential inclusions. When linear continuous dynamics with both control and d isturbance inputs are considered the so called linear hybrid automata (Xia *et al.*, 2002) are derived.

The optimal control problem, particularly the time-optimal or weighted time-optimal problem is considered in this paper. We derive a hybrid control strategy that drives the system from an initial state into the target set with minimal cost. Assuming that there exists a trajectory starting from the initial state to the target set, the constraints that the trajectory has to satisfy are converted into a set of inequalities with quantifiers. A set of quantifier-free inequalities with the switching times as variables is obtained using quantifier elimination tool QEPCAD (Collins and Hong, 1991). As the quantifier free inequalities are linear, linear programming is used in order to obtain the optimal cost and the associated optimal switching times. Based on the optimal switching times, the optimal continuous control inputs are derived and an optimal controller is constructed.

This paper is organized as follows. In section 2, we review the basic definitions and concepts related to linear hybrid automata. In section 3 we employ quantifier elimination for the optimal control of rectangular hybrid automata and design the corresponding optimal controller. In section 4, the optimal control problem for linear hybrid automata is solved. The air traffic management application is illustrated in section 5. Section 6 contains conclusions.

2. MODEL AND PROBLEM STATEMENT

2.1 *Hybrid Automata*

Hybrid automata are generalized finite-state machines where discrete transitions transfer the system between a finite number of modes Q. We use A to denote a hybrid automaton. In each discrete mode q, the system evolves continuously according to a set of differential equations :

$$\dot{x}_q = f_q(x(t), u(t), d(t)) \tag{1}$$

where $x \in X \subseteq \mathbb{R}^n$, $u(t) \in U$ is the continuous control input and $d(t) \in D$ is disturbance input. It is assumed that for every $q \in Q$, f_q is *globally lipschitz* to ensure the existence and uniqueness of solutions of the differential equation. Furthermore for each mode q there may be a set of state invariant constraints $Inv : Q \to 2^X$ that have to be fulfilled. Σ is the set of discrete events and the discrete transition set is $E \subseteq Q \times \Sigma \times Q$. We use guard $G : (Q \times \Sigma \times Q) \to 2^X$ to express the switching conditions that when $x \in G_{qq'}(\sigma)$ the system can be switched from q to q' by a discrete transition.

Depending on the characteristics of dynamics $\dot{x}_q = f_q(x, u, d)$ the following subclasses of hybrid automata can be defined:

- *Linear Hybrid Automata* (Xia *et al.*, 2002): $\dot{x}_q(t) = u(t) + d(t)$ where $u(t) \in U_q$ and $d(t) \in D_q$, $U_q = \{u | C_{U_q} u \leq h_{U_q}, C_{U_q} \in \mathbb{R}^{n_u \times n}, h_{U_q} \in \mathbb{R}^n\}$ and $D_q = \{d | C_{D_q} d \leq h_{D_q}, C_{D_q} \in \mathbb{R}^{n_d \times n}, h_{D_q} \in \mathbb{R}^n\}$. The continuous dynamics have both control and disturbance input.
- *Rectangular Hybrid Automata*(Preußig *et al.*, 1998): $\dot{x}_q(t) = u$ where $u \in [\underline{U}_q, \overline{U}_q], \underline{U}_q, \overline{U}_q \in \mathbb{R}^n$. The continuous dynamics are rectangular differential inclusions.
- *Integrator Hybrid Automata*(Xu and Antsaklis, 2003): $\dot{x}_q(t) = k_q, k_q \in \mathbb{R}^n$. The continuous variables evolve at a fixed speed.
- *Time Automata*(Alur *et al.*, 2001): $\dot{x}_q(t) = 1$. The continuous variables x act as clocks.

For these types of hybrid automata, we assume that all the invariant sets, guard sets and target set are convex polyhedral sets. It is assumed that the continuous states are the same before and after a discrete transition *i.e.* $q(t^-) \neq q(t^+)$, with $x(t^-) = x(t^+)$. Also, we assume that all the discrete events are controllable.

We use ξ to denote a hybrid trajectory. The set of hybrid trajectories starting from (q, x) is denoted by $L(A, (q, x))$ while $L(A)$ denotes the set of all hybrid trajectories generated by the automaton A. In the following $\xi(t)$ is used to denote the hybrid state of the trajectory at a given time instant t if no discrete transition happens at t. Otherwise, if a discrete transition happens at t, then $\xi(t^-)$ and $\xi(t^+)$ denotes the hybrid state before and after the transition respectively.

Definition 1. A hybrid controller is a map:

$$C : Q \times X \times T^+ \to 2^{\Sigma \times U} \tag{2}$$

where T^+ denotes time. Here, we assume that the controller has a clock which can record time. The controller issues both discrete inputs $C_d(q(t), x(t), t) \in 2^\Sigma$ and continuous inputs $C_c(q(t), x(t), t) \in 2^U$.

2.2 *Problem statement*

The *optimal control* problem we considered in this paper is as follows: Given a initial state (q_0, x_0) of a hybrid automaton A and a cost function J of a hybrid trajectory, find an optimal hybrid controller such that the controlled trajectory starting from (q_0, x_0) reaches the target set F with a minimal cost. The solvability of the above problem depends on the definition of the cost function and the characteristics of the given automaton A.

Given a trajectory $\xi \in L(A, (q_0, x_0)))$ and the target set $F \subset X \times Q$, the time that a trajectory ξ starting from (q_0, x_0) needs to reach F is:

$$T_f(\xi, F, (q_0, x_0)) = \begin{cases} t_f & if \quad \xi(0) = (q_0, x_0), \xi(t_f) \in F \wedge \\ & (\forall t \in [0, t_f), \xi(t) \in INV \wedge \xi(t) \notin F) \\ +\infty & otherwise \end{cases}$$

Given a trajectory $\xi \in L(A, (q_0, x_0))$, since we are only interested in the running cost incurred when the system evolves from (q_0, x_0) to F, we compute the cost along ξ during the time interval $[0, T_f(\xi, F, (q_0, x_0))]$. Let $e_j(\xi)$ denote the j-th discrete transition and $n(\xi)$ denote the total number of discrete transitions of ξ during the time interval $[0, T_f(\xi, F, (q_0, x_0))]$. Then, the cost along ξ is defined as:

$$J(\xi) = \begin{cases} \sum_{j=1}^{n(\xi)} J_d(e_j(\xi)) + \int_0^{t_f} J_c(q(t))dt & if \\ \qquad T_f(\xi, F, (q_0, x_0)) \text{ } is \text{ } finite \\ +\infty & otherwise \end{cases} \tag{3}$$

The running cost introduced above involves the following two cost functions:

- $J_d : E \to \mathbb{Q}$ discrete transition cost function.
- $J_c : Q \to \mathbb{Q}$ continuous transition cost function.

Basically we assign costs to both discrete and continuous transitions. Note that there is no penalty on the control input and on the continuous state.

From definition (3), it is clear that if there is no trajectory from (q_0, x_0) to F without violating the invariant constraints, then the optimal cost $\min_{\xi \in L(A,(q_0,x_0))} J(\xi) = +\infty$. In order to guarantee that the optimal cost is finite, a backward reachability analysis (Xia *et al.*, 2002) is performed starting from the target set F. If the target is reachable from (q_0, x_0), then the optimal cost is finite. Due to the fact that all discrete transitions are controllable, the procedure to find an optimal trajectory ξ^* which minimizes (3) is divided into the following steps :

(1) Find all discrete paths that connect q_0 with q_F, following the Dijkstra's algorithm (Martins *et al.*, 1999) on the discrete graph.
(2) Find the optimal trajectory which minimizes $\int_0^{t_f} J_c(q(t))dt$ for each discrete path.
(3) Among all optimal trajectories find the optimal hybrid trajectory ξ^* which minimizes (3).
(4) Design the optimal hybrid controller C^*.

Let $\Pi = \{\pi\}$ denote the set of all discrete paths from q_0 to q_F:

$$\Pi = \{\pi | \exists \sigma, \exists N \in \mathbb{N}, j = 0, ..., N-1, q_N = q_F : e_j = (q_j, \sigma, q_{j+1}) \in E \wedge \pi = \{q_0, ..., q_N\}\}$$

Let $l(\pi)$ be the number of discrete transitions in a path $\pi \in \Pi$. It is clear that the sum of discrete transition costs is constant along any given path π, $J_d^\pi := \sum_{j=1}^{l(\pi)} J_d(e_j^\pi)$. Then, the optimal solution of (3) can be represented as follows:

$$\min_{\pi \in \Pi}(J_d^\pi + \min_{t_j \geq 0} \sum_{j=0}^{l(\pi)} J_c(q_j^\pi)t_j) \tag{4}$$

where t_j denotes the time spent in q_j. Let $J_c^\pi := \min_{t_j \geq 0} \sum_{j=0}^{l(\pi)} J_c(q_j^\pi)t_j$ be the optimal continuous cost along a path π. Under the assumption that F is reachable from (q_0, x_0), the optimal cost $J^*(\xi)$ of $L(A, (q_0, x_0))$ can be found.

Algorithm 1. .

```
J_min := +∞
repeat i = 1, 2, ...
    π_i = ith − shortest − path(A)
    if (J_min < J_d^{π_i}) or (π_i = ∅) break
        J^{π_i} = J_d^{π_i} + J_c^{π_i}
        if(J^{π_i} < J_min)
            J_min = J^{π_i}
        end
    end
return J*(ξ) = J_min
```

The algorithm uses the results from (Martins *et al.*, 1999) to rank the *i-th* shortest discrete paths from the initial state to the target set in ascending order of the discrete cost, *i.e.* $J_d^{\pi_i} \leq J_d^{\pi_{i+1}}$ for $i = 1, 2, ...$ When $\pi_i = \emptyset$ then there are only $i-1$ paths in the graph of A.

Proposition 2. Under the assumption that the target set is reachable from (q_0, x_0), algorithm 1 terminates within finite iterations.

Proof: If $\pi_i = \emptyset$ for $i \in \mathbb{N}$, this means that the graph of A has a finite number of discrete paths, then it is obvious that algorithm 1 will terminate with finite iterations. On the other hand, as it is feasible to reach the target from (q_0, x_0), then $J^*(\xi)$ is finite. Since J_{min} is monotonously decreasing and $J_d^{\pi_i}$ is monotonously increasing during each iteration, it can be induced that there exists an integer i such that $J^{\pi_i} = J_d^{\pi_i} + J_c^{\pi_i}$ is optimal. □

From algorithm 1, the attainability of the optimal solution for the optimal control problem depends on one's ability to derive J_c^π. It is a *continuous cost* minimization problem along the fixed path π. This problem is considered in the rest of the paper.

Remark 3. Notice that if $J_c(q_j^\pi) = 1$ for all $j = 0, .., l(\pi)$ along the path π, the optimal control problem is transformed into a **time-optimal** control problem as follows: $t_{min} = J_c^\pi = \min_{t_j \geq 0} \sum_{j=0}^{l(\pi)} t_j$, and t_{min} is the minimal time that the hybrid automaton needs to reach the target set F starting from (q_0, x_0) along the path π. Otherwise the problem can be considered as a weighted time-optimal control problem. The overall cost is a balance between discrete transitions cost and time cost.

Let $G_{q_{i-1}, q_i}^\pi = \{x | C_i^\pi x \leq h_i^\pi\}$ denote the transition guard from q_{i-1}^π to q_i^π, with $i = 1, ..., l(\pi)$, where $C_i^\pi \in \mathbb{R}^{n_c \times n}$, $h_i^\pi \in \mathbb{R}^{n_c}$. Also, let $X_F = \{x | C_F x \leq h_F\}$ be the continuous part of the target set, with $C_F \in \mathbb{R}^{n_f \times n}$, $h_F \in \mathbb{R}^{n_f}$. For simplicity, we omit the symbol "π". Consider a discrete path π with $l(\pi)$ transitions. Our objective is to compute the minimal cost J_c^π. The time-optimal problem for integrator hybrid automata has been studied in (Xu and Antsaklis, 2003). Next, the optimal control for rectangular hybrid automata and linear hybrid automata is considered.

3. OPTIMAL SOLUTION FOR RECTANGULAR HYBRID AUTOMATA

This section deals with the optimal control problem for rectangular hybrid automata. Before proceeding with the main theorem, we discuss the relationship between the open and closed loop control. In each discrete state q, there are two types of controllers that generate the control input, namely the *open loop* controllers and the *closed loop* (feedback) controllers. Let $u_q \in [\underline{U}_q, \overline{U}_q]$ denote the control input generated by an open loop controller and let $u_q(t) \in [\underline{U}_q, \overline{U}_q]$ denote the control input (time-variant) generated by a closed loop controller. The relationship between these two controllers follows the *First Mean Value Theorem for Integrals*, see (Pang and Spathopoulos, 2004).

It follows that along a path π, for any trajectory ξ and a feedback control input $u(t)$, an alternative trajectory ξ' can always be found with open loop control input u, such that $J(\xi) = J(\xi')$ and $\xi(\sum_{i=0}^{j} t_j) = \xi'(\sum_{i=0}^{j} t_j)$ for $j = 0, ..., l(\pi)$ where t_j is the time the system spends in each location of the path π. Therefore, both type of controllers yield the same optimal cost. The advantage in the above consideration is that each open loop control u_{qi} can be treated as an existential ($\exists$) quantifier. Thus, *quantifier elimination* can be employed in order to get a set of linear inequalities for the variable t.

The objective of the real quantifier elimination is to *eliminate* "unwanted" variables from an algebraic description. The "unwanted" variables may represent unknown real quantities. Quantifiers give expressive power but do not enlarge the class of sets defined by quantifier-free formulas. This implies that given a formula including quantifiers

$$\varphi(x_1, ...x_n) \equiv \mathbf{Q_1}u_1, ...\mathbf{Q_m}u_m\psi(x_1, ...x_n, u_1, ...u_m)$$

where $\mathbf{Q_i} \in \{\exists, \forall\}$, there is always a logically equivalent quantifier free formula $\phi(x_1, ..., x_n)$ in the domain of the real numbers. A procedure computing such ϕ from φ is called real quantifier elimination.

Given a rectangular hybrid automaton A, let the initial state be $x_0 = (x_{01}, ..., x_{0n})$, the dynamics of a discrete state $q_j \in \pi$ be $\dot{x}_j = u_j = [u_{j1}, u_{j2}, ..., u_{jn}]^T$ for all $j = 0, ..., l(\pi)$, with control input $u_j \in [\underline{U}_j, \overline{U}_j]$, $\underline{U}_j, \overline{U}_j \in \mathbb{R}^n$.

Theorem 4. The optimal continuous cost along a path π for a rectangular automaton A with parameters defined above, can be computed by

$$\begin{aligned} J_c^{\pi} &= \min \ \lambda^T t \\ s.t. & \ \ \phi(t) \end{aligned} \qquad (5)$$

The quantifier free formula $\phi(t)$ that is linear in t is computed from $\varphi(t)$ by quantifier elimination in real domain where:

$$\varphi(t) \equiv \exists u_{11}, ..., \exists u_{1n}, \exists u_{21}, ..., \exists u_{2n}, ..., \exists u_{l(\pi)1}, ..., \exists u_{l(\pi)n}(\psi_1(u,t) \wedge \psi_2(u,t) \wedge \psi_3(t) \wedge \psi_4(u))$$

and:

$$\psi_1(u,t) = \bigwedge_{i=1}^{l(\pi)} C_i([x_{01}, ..., x_{0n}]^T + \sum_{j=0}^{i} [u_{j1}, ..., u_{jn}]^T t_j) \leq h_i(guard \ \ condition)$$

$$\psi_2(u,t) = C_F([x_{01}, ..., x_{0n}]^T + \sum_{j=0}^{l(\pi)} [u_{j1}, ..., u_{jn}]^T t_j) \leq h_F(target \ \ condition)$$

$$\psi_3(t) = \bigwedge_{j=1}^{l(\pi)} t_j \geq 0(nonnegative \ \ time \ \ condition)$$

$$\psi_4(u) = \bigwedge_{j=1}^{l(\pi)} \bigwedge_{r=1}^{n} \underline{U}_{jr} \leq u_{jr} \leq \overline{U}_{jr}(input \ \ restrictions)$$

Proof: See (Pang and Spathopoulos, 2004). □

It should be noted that since the two variables t and u are nonlinear in the formulas $\psi_1 \sim \psi_4$, it is impossible to solve the LP problem. Nevertheless, by considering the continuous input u as existential($\exists$) quantifier, we get an equivalent formula ϕ by eliminating the u quantifier. According to the quantifier elimination procedure, the quantifier free formula ϕ is formulated as *linear inequalities* whose closure is a convex set.

As mentioned early on, in order to guarantee the optimal performance, an optimal controller has to be imposed on the system. As the LP (5) converges to an optimal solution, we get the optimal control switching times t^*. Then the continuous optimal control law U^* can be obtained using the following two-step procedure:

- Step1: substitute the optimal time t^* derived from LP into the formulas ψ_1, ψ_2, ψ_4.
- Step2: simplify these formulas to eliminate redundant inequalities.

With the optimal times $t_0^*, ..., t_{l(\pi)}^*$ obtained from the LP problem 4, the optimal controller is:

$$C^*(q(t), x(t), t) = \begin{cases} (\epsilon, u_0^*) & if \ 0 \leq t < t_0^* \\ & \wedge q(t) = q_0 \wedge u_0^* \in U_0^* \\ (\sigma_{j,j+1}, \emptyset) & if \ t = \sum_{i=0}^{j} t_i^* \wedge q(t^-) = q_j \\ & \wedge (q_j, \sigma_{j,j+1}, q_{j+1}) \in E \\ & for \ j = 0, .., l(\pi) - 1 \\ (\epsilon, u_j^*) & if \ \sum_{i=0}^{j-1} t_i^* \leq t < \sum_{i=0}^{j} t_i^* \\ & \wedge q(t) = q_j \wedge u_j^* \in U_j^* \\ & for \ j = 1, .., l(\pi) \end{cases} \qquad (6)$$

It should be noted that the set U_j^*, obtained from above two-step procedure, is the optimal continuous input set for the discrete state q_j^{π}. The clock is set to 0 at the initial state (q_0, x_0). During the time $t \in [\sum_{i=0}^{j-1} t_i^*, \sum_{i=0}^{j} t_i^*)$ there is no discrete input and the continuous input is $u_j^* \in U_j^*$. At the optimal time $t = \sum_{i=0}^{j} t_i^*$ the discrete optimal controller generates a suitable discrete event to switch the system to the next discrete state q_{j+1} along the path π. At the same time the continuous optimal controller chooses a continuous input from U_{j+1}^* as soon as the system has been switched to the discrete state q_{j+1}.

4. OPTIMAL SOLUTION FOR LINEAR HYBRID AUTOMATA

If we consider linear hybrid automata with both control and disturbance inputs the problem becomes more difficult. The optimal control problem can not be solved using quantifier elimination when a feedback control is required. However, we can still get an *open loop* controller in the sense that the control input u is fixed at each discrete state q.

Due to the presence of disturbances, the open loop control is optimal in the *worst case* sense. The dynamics in a discrete state $q_j \in \pi$ become $\dot{x}_j(t) = u_j + d_j(t)$ for all $j = 0, ..., l(\pi)$, with control input $u_j \in U_j$, $U_j = \{u | C_{U_j} u \leq h_{U_j}, C_{U_j} \in \mathbb{R}^{n_u \times n}, h_{U_j} \in \mathbb{R}^n\}$ and disturbance $d_j(t) \in D_j$, $D_j = \{d | C_{D_j} d \leq h_{D_j}, C_{D_j} \in \mathbb{R}^{n_d \times n}, h_{D_j} \in \mathbb{R}^n\}$. Let d_{jm} $(m = 1, 2, ..., p_j)$ denote the vertices of the set D_j

Theorem 5. The worst case optimal continuous cost along a path π from an initial state x_0 for linear hybrid automata with both control and disturbance input can be computed as

$$\begin{aligned} J_c^\pi &= \min \ \lambda^T t \\ s.t. &\ \ \phi(t) \end{aligned} \tag{7}$$

The quantifier free formula $\phi(t)$ is computed from $\varphi(t)$ by quantifier elimination in real domain, where

$$\varphi(t) \equiv \exists u_{11}, ..., \exists u_{1n}, \exists u_{21}, ..., \exists u_{2n}, ..., \exists u_{l(\pi)1}, ..., \exists u_{l(\pi)n} (\psi_1(u,t) \wedge \psi_2(u,t) \wedge \psi_3(t) \wedge \psi_4(u))$$

and:

$$\psi_1(u,t) = \bigwedge_{i=1}^{l(\pi)} \overbrace{\bigwedge_{m=1}^{p_j} \cdots \bigwedge_{m=1}^{p_j}}^{j=0,..,i} (C_i(x_0 + \sum_{j=0}^{i} (u_j + d_{jm}) t_j) \leq h_i$$

$$\psi_2(u,t) = \underbrace{\bigwedge_{m=1}^{p_j} \cdots \bigwedge_{m=1}^{p_j}}_{j=0,..,l(\pi)} C_F(x_0 + \sum_{j=0}^{l(\pi)} (u_j + d_{jm}) t_j) \leq h_F$$

$$\psi_3(t) = \bigwedge_{j=1}^{l(\pi)} t_j \geq 0$$

$$\psi_4(u) = \bigwedge_{j=1}^{l(\pi)} C_{U_j} u_j \leq h_{U_j}$$

Proof: See (Pang and Spathopoulos, 2004). □

The optimal continuous control law U^* for linear hybrid automata can also be designed following the two-step procedure presented above. The optimal controller C^* is the same with the one derived for rectangular hybrid automata as in (6).

5. APPLICATION OF AIR TRAFFIC MANAGEMENT

In this section, the results of optimal control are applied on an Air Traffic Management System (ATMS) (Tomlin, 1996), (Pang and Spathopoulos, 2004). This two aircraft joining and collision

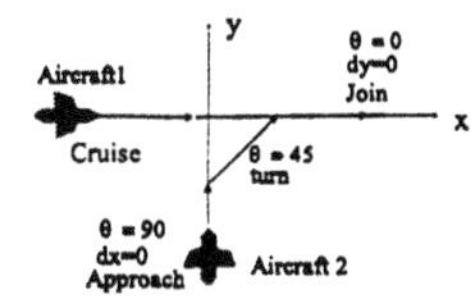

Fig. 1. One aircraft joining the path of another, in an approach to the airport runway

avoidance scheme can be seen as an optimal control problem. The scenario is a common one and illustrated in figure 5. Aircraft 2 is joining the path of Aircraft 1, in approach to the airport, and it is up to the controller to ensure that a minimal separation distance between the two aircraft is maintained and also minimize the cost during this procedure. Aircraft 1 has one mode of operation, called *Cruise*: it follows a straight path along the x-axis and has known bounds on its speed. Aircraft 2 is in three possible position modes: *Approach*, *Turn* and *Join*. Within the *Turn* and *Join* modes, Aircraft 2 can be in one of three possible speed modes, the controller effects both the transitions and the speed of the modes. The control scheme is as follows: if the x coordinate of Aircraft 2 is either greater than that of Aircraft 1 or more than 30 units behind that of Aircraft 1, then the Aircraft 2 should be in a *fast* mode. If the x coordinate of Aircraft 2 is less than 30 units behind that of Aircraft 1, then the Aircraft 2 should be in a slow mode. There are two fast modes and one *slow* mode: *Fast*1 is the mode in which Aircraft 2 is more than 30 units behind Aircraft 1. *Fast*2 is the mode in which Aircraft 2 is ahead of Aircraft 1.

The speed of Aircraft 1 is $\dot{x} = u_1$, where $u_1 \in [3, 4]$ in any position mode. For simplicity, we restrict the transitions between the *Fast* and *slow* modes if they are in the same position mode. Then the aim of this air traffic management is that Aircraft 2 join the path of Aircraft 1 safely with $x1 \geq 50 \wedge x2 \geq 50$. The safety requirement is that the distance between the two aircraft has to be kept over 10. That is roughly illustrated by the *forbidden* region $Bad = \{|x1 - x2| \leq 10 \wedge y2 \leq 10\}$.

It should be noted that the problem requires the composing the motions of both Aircraft 1 and Aircraft 2. The composed automaton has three discrete position modes of Aircraft 2. Its continuous states are in 3 dimensions: $(x1, x2, y2)$. We assume that the two aircraft communicate very well, and the Aircraft 1 can also change its speed when Aircraft 2 changes its position modes. Hence, by composing the two systems we get the target set F=($Join$,$F1 \cup F2$) for the composed hybrid system where the discrete state of Aircraft 2 is *Join* and the continuous sets are either $F1 = \{x1 - x2 \geq 10 \wedge x2 \geq 50\}$ or $F2 = \{x2 - x1 \geq 10 \wedge x1 \geq 50\}$

Path	π_1	$\pi 2$
q_0^π	$Appf$ $x2 = 0$ $-100 \leq y2 \leq -30$ $\dot{x}2 = 0$ $\exists \dot{y}2 \in [6, 7]$ $\exists \dot{x}1 \in [3, 4]$	$Appf$ $x2 = 0$ $-100 \leq y2 \leq -30$ $\dot{x}2 = 0$ $\exists \dot{y}2 \in [6, 7]$ $\exists \dot{x}1 \in [3, 4]$
q_1^π	$Turns$ $0 \leq x2 \leq 30$ $-30 \leq y2 \leq 0$ $x2 \leq x1 \leq 30 + x2$ $\exists \dot{x}2 \in [4, 5]$ $\dot{y}2 = \dot{x}2$ $\exists \dot{x}1 \in [3, 4]$	$Turnf2$ $0 \leq x2 \leq 30$ $-30 \leq y2 \leq 0$ $x2 \geq x1$ $\exists \dot{x}2 \in [5, 6]$ $\dot{y}2 = \dot{x}2$ $\exists \dot{x}1 \in [3, 4]$
q_2^π	$Joins$ $x2 \geq 30$ $y2 = 0$ $x2 \leq x1 \leq 30 + x2$ $\exists \dot{x}2 \in [4, 5]$ $\dot{y}2 = 0$ $\exists \dot{x}1 \in [3.5, 4]$	$Joinf2$ $x2 \geq 30$ $y2 = 0$ $x2 \geq x1$ $\exists \dot{x}2 \in [5, 6]$ $\dot{y}2 = 0$ $\exists \dot{x}1 \in [3, 4]$
$G_{q_0^\pi q_1^\pi}$	$x2 \leq x1 \leq 30 + x2$ $y2 = -30$	$x2 \geq x1$ $y2 = -30$
$G_{q_1^\pi q_2^\pi}$	$x2 = 30$ $y2 = 0$ $10 \leq x1 - x2 \leq 30$	$x2 = 30$ $y2 = 0$ $x2 - x1 \geq 10$
Target	$(Joins, F1)$	$(Joinf2, F2)$
J_d^π	$20 + 20 = 40$	$20 + 20 = 40$
Optimal time t^*	$t_0^* = 11.5457$; $t_1^* = 6$; $t_2^* = 4.9543$	$t_0^* = 10$; $t_1^* = 5$; $t_2^* = 7.5$
J_c^π	51	50
J^π	91	90

Table 1. Final results of two paths.

Mode	Controller
Appf	$\dot{x}1 = u_{10}^* = 3, \dot{y}2 = u_{20}^* = 7$
Turnf2	$\dot{x}1 = u_{11}^* = 4, \dot{x}2 = \dot{y}2 = u_{21}^* = 6$
Joinf2	$\dot{x}1 = u_{12}^* = 4, \dot{x}2 = u_{22}^* \in [5, 6]$

Table 2. Optimal controller for path π_2

The optimal control for this air traffic management is cast as follows: the two aircraft reach the target set F without entering the forbidden region *Bad* at minimal *cost*. Here, we associate each transition with the discrete cost $J_d = 20$. The continuous costs in *Approach* and *Join* position modes are $J_c(Approach) = J_c(Join) = 2$, and the continuous cost in the *Turn* position mode is $J_c(Turn) = 3$. Given an initial state in *Approach* mode: $x1(0) = -30, x2(0) = 0, y2(0) = -100$, there are only two discrete paths acceptable (the others generate blocking or lead the system to the forbidden region), namely $\pi_1 = \{Appf\ Turns\ Joins\}$ $\pi 2 = \{Appf\ Turnf2\ Joinf2\}$. The details of these paths are shown in table 1. It follows that the optimal cost is $J^{\pi_2} = 90$ with the optimal schedule t^* as shown in table 1. The optimal continuous inputs of Aircraft 1 and Aircraft 2 along the path π_2 are illustrated in table 2 and the optimal trajectories ξ^* in 3D are shown in figure 2.

6. CONCLUSIONS

We show how quantifier elimination can be applied in order to derive the appropriate optimal control strategy in the presence of continuous disturbances. A two aircraft joining and collision avoidance system is designed optimally using our method. The proposed linear hybrid automata

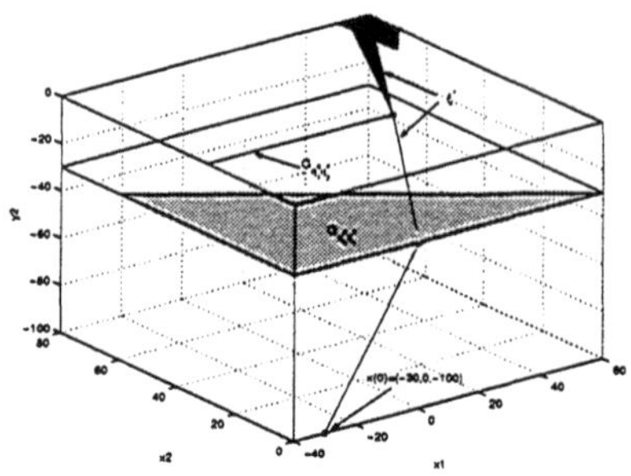

Fig. 2. Optimal trajectories in 3D.

have relative simple dynamics. Optimal control for the decidable classes of linear hybrid systems stated in (Lafferrire *et al.*, 2001) should be further investigated.

7. REFERENCES

Alur, R., S. La Torre and G.J. Pappas (2001). Optimal paths in weighted timed automata. In: *Hybrid Systems: Computation and Control.* LNCS 2034. Springer-Verlag.

Collins, G. and H. Hong (1991). Partial cylindrical algebriac decomposition for quantifier elimanation. *Journal of Symbolic Computation* **12**, 299–328.

Lafferrire, G., G. Pappas and S. Yovine (2001). Symbolic reachability computation for families of linear vector fields. *Journal of Symbolic Computation* **32**, 231–253.

Martins, E., M. Pascoal and J. Dos Santos (1999). Labelling algorithms for ranking shortest paths. Technical report. Department of Mathematics, Coimbra University. Portugal.

Pang, Y. and M. . Spathopoulos (2004). Reachability and optimal control for linear hybrid automata: A quantifier elimination approach. Technical report. University of Strathclyde (Submitted).

Preußig, J., S. Kowalewski, T. H. Hezinger and H. Wong-Toi (1998). An algorithm for the approximative analysis of simple rectangular automata. In: *Formal Techniques for Real-Time and Falt-Tolerant Systems.* LNCS. Spring-Verlag.

Tomlin, C. (1996). Verification of air traffic management protocol using hytech. Technical report. EE290A, Henzinger- Department of Electrical Engineering and Computer Science, University of California at Berkelay.

Xia, H., Y. Pang, A. Trontis and M. Spathopoulos (2002). Eventuality synthesis for controlled linear automata. In: *Proceedings of the American Control Conference.* Anchorage, Alaska.

Xu, X. and P.J. Antsaklis (2003). A linear programming approach to time optimal control of integrator switched systems with state constraints. In: *Proceedings of the 42nd Conference on Decision and Control.* Hawaii, USA.

ELSEVIER
IFAC
PUBLICATIONS
www.elsevier.com/locate/ifac

INTERMITTENT FAILURE DIAGNOSIS BASED ON DISCRETE EVENT MODELS

Correcher, A. García, E. Morant, F. Quiles, E.

Dept. of Syst. Eng. and Control, Universidad Politécnica de Valencia, Camino de Vera s/n, 46071 Valencia, (Spain)
E-mail: ancorsal@upvnet.upv.es, {egarciam, fmorant, equiles}@isa.upv.es

Abstract: This paper develops a methodology for diagnosing intermittent failures in industrial processes. Failure diagnosis methodologies based on discrete event systems are extended in order to diagnose this type of failures. A new structure for the supervision layer is designed, so information related to on-line diagnosis can be fused with failure historic patterns. Then, preventive maintaining can be assisted by the diagnosis system. A survey on the evolution of intermittent failure diagnosis over the past years is also included. *Copyright © 2004 IFAC*

Keywords: Intermittent failures, failure diagnosis, industrial processes.

1. INTRODUCTION

Production systems are becoming increasingly complex as safety and maintenance requirements are raised. These factors increase the need for new and better process supervision systems. This supervision requires specification of the variables monitoring, production scheduling and failure diagnosis.

Failures of production systems can be classified as permanent failures (PFs) or intermittent failures (IFs). Permanent failures are failures understood in the classical sense, i.e., malfunctions in some device integrated in the productive process that needs to be substituted or repaired. On the other hand, IFs can be recovered by process dynamics. An example of this type of failures is a pipe blocked that can be unblocked by the own internal pressure of the pipe.

This paper examines the problem of IFs from the point of view of discrete event systems (DES). Some limitations in IF diagnosis of the PhDs presented in (García, 2000) (Sampath, 1995) will be proved and solved.

The hypothesis stated in this paper is that the occurrence of intermittent failures in a device is a prelude of a permanent failure, as a consequence of a progressive degradation of its capabilities. In these cases, if IFs can be detected then the supervisor could select the most appropriate moment to stop and/or repair the system or the device. The best moment is that with the lowest economic cost.

2. IF APPROACHES

IFs occur in a number of fields and have been studied since the 1970's. The first field interested in this kind of failures was digital electronics. The initial studies of Savir (Savir, 1978) have been continued all over the world by many research groups and firms that develop circuit testing systems. In (Savir, 1978) two kind of IFs were distinguished: on one hand, stationary IFs (e.g.: connection looses) and transient IFs (e.g.: failures produced by electromagnetic interference). However, the last ones are completely random and were not considered. Only failures that always produce the same effects when they appear were studied. The diagnosis system developed compared the digital circuit output with a reference model free of failures.

In (Sorensen, et al., 1998) software used for the aircraft industry was presented. This paper outlines the importance of taking into account IFs in electronic devices because aging of devices will lead to IFs as a prelude of a PF. IFs due to connection problems have serious consequences over system reliability. This occurs because not repaired damaged units carry on injecting IFs in the system. These failures can be detected in other systems, generating false alarms and needless reparations. These situations are due to unsuitable testing of IFs, and often they are the main reason of considering that these electronic systems are "worn out".

Evolution of connection failures is shown in Fig. 1 (Sorensen, et al., 1998). This evolution occurs because connection failures are rarely repaired so its behaviour makes worse over time. When IFs begin (stage 1), they are reflected as small noise. Nevertheless, as the amplitude and duration of the fluctuations increases (stage 2), IFs start to occur. The effects of IFs are really severe in stage 3.

Other fields in which the importance of IFs have been studied are aircraft industry (Anderson and Aylward, 1994), (Salvatore, et al., 1994) and industrial processes (Madden and Nolan, 1999), (Baisheng, 2001), (Kui, et al. 2000).

However, electric contacts is, without any doubt, the industrial application where IF diagnosis has reached most advances. In this sense, the most studied phenomenon is fretting corrosion (Antler, 1999). Antler stated that fretting corrosion is caused by thermal or mechanical vibrations in the contacts. This induces relative movements between the two parts of the contact. As a consequence, these movements make the contact resistance to increase. These rises in resistance are intermittent and provoke intermittent connection failures. This phenomenon was widely studied in (Maul, et al., 2000, 2001, 2002).

The research group of the authors is working in behavioural patterns of IFs (Fig. 1) with the objective of detecting the IF stage. Then, strategies of preventive maintenance or corrective control actions can be carried out. These operations can be optimized to decrease the economic losses derived from unnecessary system shut downs. In (Correcher et al. 2003) IFs were modelled in order to generate its behavioural pattern.

3. ARCHITECTURE

In this paper, a specific architecture for diagnosing IFs has been designed. This architecture is based on two previous methodologies of DES failure diagnosis (García, 2000) (Sampath, 1995). See Fig. 2.

An IF diagnoser works as an on-line monitor of the process that also detects failures using its labels (next section will develop its design). These labels are used by a data acquisition system (Correcher et al. 2003) in order to extract the characteristics of the IF.

Once IF is characterized, it can be compared with behavioural pattern, in order to detect the stage of the IF (Fig. 1.) the scheduler performs this action. As a

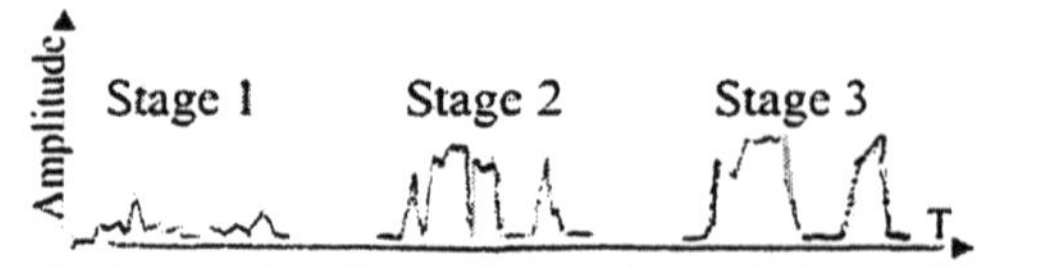

Fig. 1. Connection IF evolution through device life.

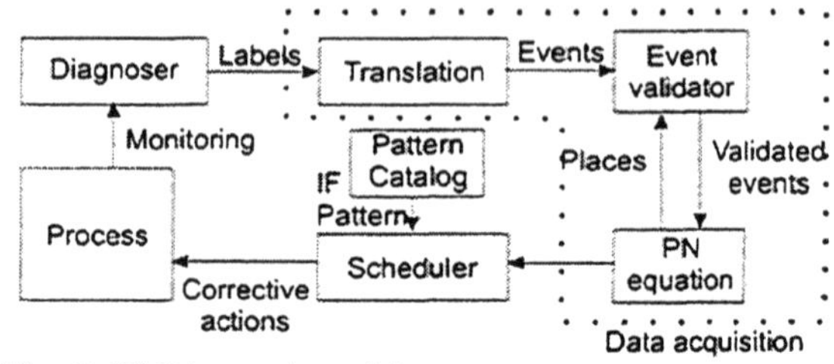

Fig. 2. IF Diagnosis architecture

result of the comparison, the scheduler takes corrective actions, for example, if the IF is placed over stages 1 or 2 the device will be able to perform much more operations, so no substitution is needed.

4. THE DIAGNOSER

Let us suppose a system G_G where IF can take place. The system can be modelled according to the diagnosis methodology based on DES (Sampath, 1995), (García, 2000) as:

$$G_G = (X_G, \Sigma_G, \delta_G, x_{0G}) \quad (1)$$

where X_G is the state set, Σ_G is the event set, δ_G is the transition function, and x_{0G} it is the initial state of G_G. This model, G_G is a finite automaton (Cassandras and Lafortune, 1999) that generates the event language L. L consists of all the valid event traces. Let us note as Σ^* the set of all the finite traces belonging to L. It will be assumed that the reader is familiar with the elements and the notation of the automaton theory (Cassandras y Lafortune, 1999).

The system model can be obtained by application of the synchronous composition operator to the individual models (2) of the devices (Sampath, 1995) (García, 2000).

$$G_j = (X_j, \Sigma_j, \delta_j, x_{0j}) \quad (2)$$

where X_j, Σ_j, δ_j, x_0 have the same meaning as in (1). The original methodology is extended by including transitions from faulty states to normal operation states, as it is shown in Fig. 3. This extension will allow the diagnoser to detect IFs. So that, the event set Σ_j is extended by addition of recovery events (RE) Σ_{rf}. These events, as the failure events Σ_f, will be considered unobservable. Behaviour of failure and recovery events is shown in Fig. 4, where some subsets can be defined:

- $X_{jF} \subseteq X_j$ subset of failure states.
- $X_{jFi} \subseteq X_{jF}$ subset where F_i is true.
- $X_{jN} \subseteq X_j$ subset of normal behaviour states.
- $X_{jRFi} \subseteq X_{jN.}$ states reachable from one state belonging to X_{jFi}. Formally: $X_{jFi}:=\{x \in X_{jN} / \exists x' \in X_{jFi}, \alpha \in \Sigma_{rf} / \delta_j(x', \alpha)=x\}$

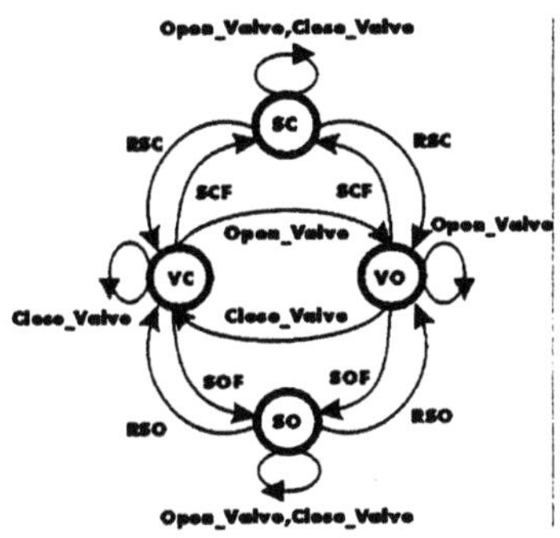

Fig. 3. Fault recovery on a valve discrete model. SC, SCF stand for stuck closed valve state and stuck closed valve failure; SO, SOF stand for stuck opened valve state and stuck closed valve failure; RSC, recovery from SC; RSO, recovery from SO; VC, valve closed and VO, valve opened

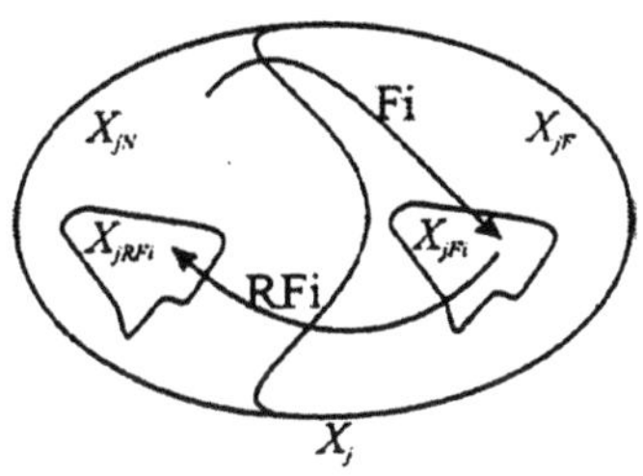

Fig. 4. System evolution through failure and recovery subsets.

Let us introduce some notation. The natural projection function $P: \Sigma^* \rightarrow \Sigma_o^*$ is defined in the usual way: $P(\varnothing)=\varnothing$; $P(\sigma)=\sigma$ if $\sigma\in\Sigma_o$; $P(\sigma)=\varnothing$ if $\sigma\in\Sigma_{uo}$; $P(s\sigma)=P(s)P(\sigma)$ $s,\sigma\in\Sigma^*$. The postlanguage of $s\in L$, L/s, is defined in the usual way, i.e., $t\in L/s$ if $st\in L$. Let us define $\Psi(\Sigma_{fi})$ as the set of traces that end in $\sigma_{fi}\in\Sigma_{fi}$, $\Psi(\Sigma_{fi}):=\{s\in L \Rightarrow s=s'\sigma_{fi},\ s, s'\in\Sigma^*,\ \sigma_{fi}, \in\Sigma_{fi}\}$. The length of the trace s is represented by $|s|$. Moreover, $\Sigma_x \in s$ $(s\in\Sigma^*)$, will represent that an event of Σ_x is included in trace s.

The diagnoser for system G is the finite state machine (Sampath, 1995):

$$G_d=(Q_d,\Sigma_0,\delta_d,q_0) \tag{3}$$

where Q_d, Σ_0, δ_d, and q_0, have the usual interpretation (1). Each diagnoser state is formed by a conjunction of original states from G and labels classifying these states. The label set is defined as:

$$\Delta=\{N\}\cup 2^{|\Delta f|} \tag{4}$$ [1]

N is to be interpreted with the meaning of "normal", and $\Delta_f=\{F_1, ..., F_m\}$ is the failure label set where F_i, $i\in\{1,...m\}$ as meaning that a failure of type F_i has occurred.

A state q_d of G_d is of the form $q_d=\{(x_1,l_1),...,(x_n,l_n)\}$ where $x_i\in X_0$ (X_0 is the accessible states set of X) and $l_i\in\Delta$, i.e., l_i is of the form $l_i=\{N\}$, $l_i=\{F_{i1}, F_{i2,...}, F_{ik}\}$, where $\{i_1,..., i_k\}\subseteq\{1,...,m\}$. A state q_d of G_d is said to be Fi-uncertain if $\exists(x,l_1),(y,l_2)\in q_d$ / $Fi\in l_1$ and $Fi\notin l_2$.

Initial state is always considered in normal working, so $q_0=(x_0, N)$. Starting from this state, the remaining states are built by using (later defined) the label propagation function (LP), remembering the partition of the event set:

$$\Sigma=\Sigma_o\cup\Sigma_{uo} \tag{5}$$

where Σ_o is the observable event set and Σ_{uo} is the unobservable event set. The present approach includes a new feature, it splits Σ_{uo} into three sets $\Sigma_{uo}=\Sigma_f\cup\Sigma_{rf}\cup\Sigma_{nuo}$, i.e., failure events, failure REs and unobservable events not belonging to the other two subsets. The inclusion of REs implies the development of new diagnosis tools.

We include in Σ_o at least one indicator for each failure in Σ_{fi}:

$$(\forall\sigma_{fi}\in\Sigma_f)\ (\exists I(\sigma_{fi})\in\Sigma_o)\ (\exists\, ni\in N)\ (\forall s\in\Psi(\Sigma_{fi}))$$
$$(\forall t\in L/s) \text{ if } [(|t|>ni) \text{ and } (\Sigma_{rfi}\notin t)] \Rightarrow I(\sigma_{fi})\in t.$$

Therefore, indicators are observable events that occur when a failure takes place but they does not during normal working. It can be outlined that only systems that include indicators are considered in this paper. This restriction is reasonable for ensuring failure observability. Moreover, industrial systems often include a high quantity of sensors, so it is reasonable to think that at least one of these sensors will be an indicator.

Proposition: Languages with failure recovery are not diagnosable in the sense of Sampath (1995).

Proof: A prefix closed and live language L is said to be diagnosable with respect to a projection P and a failure partition Π_{fi} of Σ_{fi}, in the sense of Sampath (1995) if
$(\forall i\in\Pi_{fi})\ (\exists\, ni\in N)\ (\forall s\in\Psi(\Sigma_{fi}))\ (\forall t\in L/s)$ / if $(|t|>ni)\Rightarrow w\in P^{-1}(P(st))\Rightarrow\Sigma_{fi}\in w$.

Let us choose $c=s_1s_2s_3\in L$, s_1, s_2, $s_3\in\Sigma^*$, and $c'=s_1\sigma_{fi}\, s_2\sigma_{rfi}s_3\in L$. Therefore, $P(c)=P(c')$. We can choose $c'=st$ with $s=s_1\sigma_{fi}$ $(s\in\Psi(\Sigma_{fi}))$ and $t=s_2\sigma_{rfi}s_3$, then $c\in P^{-1}(P(c'))$ but $\sigma_{fi}\notin c$.

Q.E.D.

So that, languages that include failure recovery are not diagnosable in the sense of Sampath (1995). Therefore, classical diagnosers do not guarantee diagnosability for systems with IFs. As derived from the proposition, a wider framework is needed to ensure the IF diagnosability.

An important element in the diagnoser construction is the label propagation function, LP. (Sampath, 1995) This function assigns to each state its corresponding label regarding the past events. LP must be extended in order to include IF recoveries.

[1] Note that A ("ambiguous") label (Sampath, 1995) is not considered. This assumption restricts the results of this paper to systems without ambiguities in the model.

Let us introduce this notation. $L(G,x)$ is defined as the set of traces belonging to Σ^* that start from $x \in G$, and $L_o(G,x) := \{s \in L(G,x) / s = u\sigma, u \in \Sigma_{uo}^*, \sigma \in \Sigma_o\}$.

Definition: Longest trace extraction function.
$\Gamma: \Sigma x \Sigma^* \rightarrow \Sigma^*$: Given $e \in \Sigma$, $s \in \Sigma^*$, therefore $\Gamma_e(s) = t$:

$(t = t'e)$, $t' \subseteq s$ if $(e \in s)$ and $(\neg \exists\ u \subseteq s$ such as $u = u'e$, $/ |u| > |t|)$

$(t = \varnothing)$ if $e \notin s$

Γ extracts the longest trace in s that ends in e. If the event does not belong to s, then $\Gamma_e(s) = \varnothing$. Note that if $|\Gamma_a(s)| = |\Gamma_b(s)|$ then ($a \notin s$ and $b \notin s$) or ($a = b$). Let us note as $\Gamma_\Sigma(s)$ the set of the longest traces in s that end in some event $e \in \Sigma$.

Definition: Extended label propagation function.
$ELP: X_0 x \Delta x \Sigma^* \rightarrow \Delta$: Given $x \in X_0$, $l \in \Delta$ and $s \in L_0(G,x)$
ELP propagates the label l over s, starting from x and following the dynamics of G according to $L(G,x)$. It is defined as:

$ELP(x,l,s)$:
$\{N\}$ if $\{l = \{N\}$ and $([\forall i\ \Sigma_{Fi} \notin s]$ or [if $\exists\ \Sigma_{Fi} \in s \Rightarrow |\Gamma_{\Sigma fi}(s)| > |\Gamma_{\Sigma rfi}(s)|]\}$ or $\{$if $Fi \in l \Rightarrow |\Gamma_{\Sigma rfi}(s)| > |\Gamma_{\Sigma fi}(s)|)\}$
$\{Fi\}$ if $|\Gamma_{\Sigma rfi}(s)| \leq |\Gamma_{\Sigma fi}(s)|$

This definition of ELP masks failures whose recovery takes place before reaching the next state.

5. DIAGNOSABILITY

The previous definitions for diagnosability (Sampath, 1995) do not take into account the physic reality of IFs neither the origins of recovery events in real signals. So, when diagnosability is defined, it avoids the existence of unobservable cycles of failures and REs. In this paper a more relaxed definition for diagnosability is proposed. We propose a definition for diagnosability that allows the existence of unobservable cycles in the model but only if the failure does not cause malfunction during the cycle. Therefore, a system will be considered diagnosable if any failure can be detected if it causes malfunction.

5.1. FR-diagnosability.

A prefix closed and live language L, is said to be FR-diagnosable (failure-recovery diagnosable) with respect to a failure partition Π_{fi} and a projection P if
$(\forall i \in \Pi_{fi})\ (\exists ni \in \mathbb{N})\ (\forall s \in \Psi(\Sigma_{fi}))\ (\forall t \in L/s / \Sigma_{rfi} \notin t)$ / if $(|t| > ni) \Rightarrow D$
D : if $w \in P^{-1}(P(st)) \Rightarrow \Sigma_{fi} \in w$ and $\forall u \subseteq w / u \in \Psi(\Sigma_{fi})$
$\exists u' \in \Psi(I(\Sigma_{fi})) \cap L/u / |\Gamma_{\Sigma fi}(uu')| > |\Gamma_{\Sigma rfi}(uu')|$.

These diagnosability conditions imply that all the traces with the same projection as one containing a failure (without later recovery) will contain the same failure. Moreover, when considering a sufficient long trace, the trace will contain an indicator. It can be outlined that this definition of diagnosability excludes failures with recovery that do not cause indicators between states. This definition corresponds with the physical reality of devices because if the recovery is fast enough the device will work normally.

5.2. Example: Relay.

Let us show *FR*-diagnosability with an example. The first stage consists of designing a discrete event model ($G_R = (X_R, \Sigma_R, \delta_R, x_{0R})$ as in (2)) of the device (Fig. 5). Two failures are considered: stuck closed and stuck open. This model can be combined with model form other devices in order to design the global model of the system (1).

When the relay gets stuck opened (CA) the contact resistance is high and current is interrupted. The same situation occurs when the control command *Roff* is given. So, this failure occurring during *Roff* activity, does not damage the relay operating so it will not be considered as a failure. Let us suppose that any of these traces occur:

C1 = {Roff Ron}
C2 = {Roff Fca Rca Ron}
C3 = {Roff Ron Roff Fca Ron}
C4 = {Roff Fca Rca Ron Roff Fca Ron}

The diagnoser will also observe the sensor readings from the relay and other events from other devices.

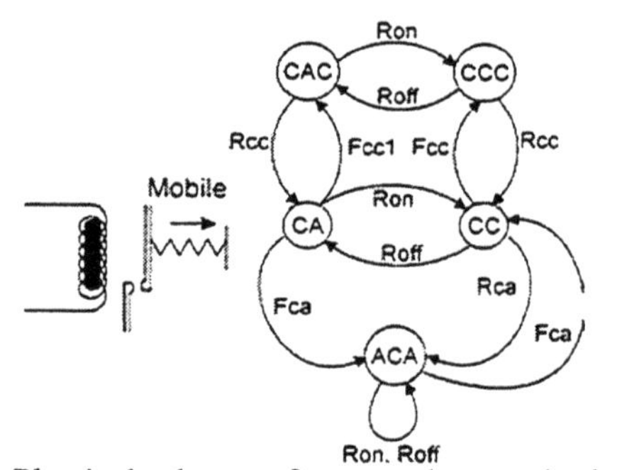

Fig. 5. Physical scheme of a normal opened relay and its event model. The meaning of each event and state is shown in table 1.

Table 1: Relay states and events

State	Meaning
CA	Circuit opened
CC	Circuit closed
ACA	Stuck opened
CAC	Stuck closed (short circuit) and CA
CCC	Stuck closed (short circuit) and CC
Event	
Ron	Close relay control command
Roff	Open relay control command
Fcc	Short circuit failure
Fca	Opened circuit failure
Rca	Short circuit failure recovery
Rcc	Opened circuit failure recovery
Fcc1	Short circuit failure (Spring broken)

Let us suppose that the relay includes a boolean current sensor ((*I(0)*, there is no current; *I(+)*, there is current). This sensor is the indicator for failures Fca and Fcc because its value will change when any of these failures occur (if no recovery). Let us build a diagnoser G_d for this system and let us choose s_1, $s_2 \in \Sigma_G^*$ such as: $Roff \in L/s_2$ $Ron \in L/s_1$. So that, the global system produces the traces *C1_s, C2_s, C3_s* and *C4_s*:

Roff I(0) s_1 *Ron I(+)* s_2
Roff I(0) s_1 *Fca Rca Ron I(+)* s_2
Roff I(0) s_1 *Ron I(+)* s_2 *Roff I(0)* s_2 *Fca Ron I(0)* s_1
Roff I(0) s_1 *Fca Rca Ron I(+)* s_2 *Roff I(0)* s_1 *Fca Ron I(0)* s_2

But the diagnoser can only detect their projections, *PC1_s, PC2_s, PC3_s* and *PC4_s*:

Roff I(0) $P(s_1)$*Ron I(+)*$P(s_2)$
Roff I(0) $P(s_1)$*Ron I(+)*$P(s_2)$
Roff I(0) $P(s_1)$*Ron I(+)*$P(s_2)$*Roff I(0)*$P(s_1)$*Ron I(0)* $P(s_2)$
Roff I(0) $P(s_1)$*Ron I(+)*$P(s_2)$*Roff I(0)*$P(s_1)$*Ron I(0)* $P(s_2)$

Therefore, the diagnoser is unable to distinguish between *PC1_s* and *PC2_s* with this sensor. Nevertheless, there is a failure in *PC2_s*. On the other hand, the operation of the relay is the same in both cases because current flows normally through the relay. FR-diagnosability definition allows the system to hide these situations.

Let us focus in traces *PC3_s* and *PC4_s*. The diagnoser observes the same projection and, therefore, it decides the same diagnosis. So that, the diagnoser will evolve to a state with a failure label afterwards *RonI(0)*. Nevertheless, PC3_*s* only contains a failure event and *PC4_s* contains two. The system carries on being FR-diagnosable because the recovery in the first failure in *PC4_s* occurs before the observable event *I(+)*. So, it maintains normal working as in the previous case.

This example shows that FR-diagnosability "filters" any failure produces in the system the same effect as the current control command. This limitation of the diagnosis system does not affect the normal working of the process. In fact, the function of the diagnosis system is to detect anomalous situations in the process and, in this case, the failure does not produce any effect.

5.3. *FR-diagnosability conditions.*

Let us introduce the concepts of cycle and *Fi*-uncertain cycle in G_d (Sampath, 1995). A state set $x_1, \ldots, x_n \in X$ form a cycle in G if $\exists s \in L(G, x_1)$ such as $s=\sigma_1\ldots\sigma_n$ and $\delta(x_l, \sigma_l)= x_{(l+1) \bmod n}$, $l=1,\ldots,n$. Moreover, a set of uncertain states $q_1, \ldots, q_n \in G_d$ is said to form a *Fi*-uncertain cycle if:

1) $q_1, \ldots, q_n$ form a cycle in G_d, with $\delta_d(q_l, \sigma_l)=x_{(l+1) \bmod n}$, where $\sigma_l \in \Sigma_o$, $l=1,\ldots,n$.
2) $\exists$ (x^k_l, l^k_l), $(y^r_l, l^{\prime r}_l) \in q_l$, $l=1\ldots n$, $k=1,\ldots\ldots,m$, $r=1,\ldots m'$ such as $\forall l,k,r$ $Fi \in l^k_l$ but $Fi \notin l^{\prime r}_l$ and $\{x^k_l\}$, $\{y^r_l\}$ form cycles in G.

<u>Theorem:</u> A prefix closed and live language *L* is FR-diagnosable iff there are not Fi-uncertain cycles in G_d for any failure.

Necessity: It will be proved by contradiction that if *L* is diagnosable, then the diagnoser does not have Fi-uncertain cycles in G_d for any failure.

Let us suppose that there exist a set of states $q_1, q_2, \ldots q_n \in Q_d$ that form a Fi-uncertain cycle. $\delta(q_l, \sigma_l)=q_{(l+1) \bmod n}$. So (x^k_l, l^k_l), $(y^r_l, l^{\prime r}_l) \in q_l$, $l=1\ldots n$, $k=1,\ldots\ldots,m$ and $r=1,\ldots m'$ form a cycle with $Fi \in l^k_l$ but $Fi \notin l^{\prime r}_l$. So

$$\begin{aligned} &\delta_d(x^k_l, s^k_l\sigma_l) = x^k_{(l+1)}, l=1,\ldots,n-1, k=1,\ldots,m, \\ &\delta_d(x^k_n, s^k_n\sigma_n) = x^{k+1}_l, \ k=1,\ldots,m-1, \\ &\delta_d(x^m_n, s^m_n\sigma_n) = x^1_l, \end{aligned} \tag{6}$$

and

$$\begin{aligned} &\delta_d(y^r_l, s'^r_l\sigma_l) = y^r_{(l+1)}, l=1,\ldots,n-1, k=1,\ldots,m', \\ &\delta_d(y^r_n, s'^r_n\sigma_n) = y^{r+1}_l, r=1,\ldots,m'-1, \\ &\delta_d(y^{m'}_n, s^{m'}_n\sigma_n) = y^1_l, \end{aligned} \tag{7}$$

As x^1_1 and $y^1_1 \in q_1$, then (by diagnoser construction) $\exists$ s_0 and s'_0 such as $\delta(x_0,s_0)=x^1_1$ and $\delta(y_0,s_0')=y^1_1$, moreover, $(P(s_0)=P(s'_0))$. Moreover, $Fi \in l^k_l$ but $Fi \notin l^{\prime r}_l$, then $\Sigma fi \in s_0$ and $|\Gamma_{\Sigma rfi}(s_0)|<|\Gamma_{\Sigma fi}(s_0)|$ or $Fi \in l_0$ and $|\Gamma_{\Sigma rfi}(s_0)| \leq |\Gamma_{\Sigma fi}(s_0)|$ and, on the other way, $\sigma_{Rfi} \in s'_0$ and $|\Gamma_{\Sigma rfi}(s)|>|\Gamma_{\Sigma fi}(s)|$ or $Fi \notin l_0$ and $|\Gamma_{\Sigma rfi}(s'_0)| \geq |\Gamma_{\Sigma fi}(s'_0)|$. Let us consider two traces, *w* and *w'*

$$w = s_0(s^1_1\sigma_1..s^1_n\sigma_n s^2_1\sigma_1..s^m_n\sigma_n)^{km'} \tag{8}$$
$$w' = s'_0(s'^1_1\sigma_1..s'^1_n\sigma_n s'^2_1\sigma_1..s'^{m'}_n\sigma_n)^{km} \tag{9}$$

for an arbitrary long *k*.

Then $P(w)=P(w')=P(s_0)P(\sigma_1\ldots\sigma_n)^{kmm'}$ and $\sigma_{fi} \notin w'$ (or $|\Gamma_{\Sigma rfi}(s_0)|>|\Gamma_{\Sigma fi}(s_0)|$) meanwhile $\sigma_{fi} \in w$ (or $|\Gamma_{\Sigma fi}(s_0)|<|\Gamma_{\Sigma fi}(s_0)|$). Let us choose $s \subseteq s_0$ such us $s \in \Psi(Fi)$ and let us choose $t \in L/s$ such us $w=st$. If choosing *k* sufficiently long it can be achieved that $|t|>n$ for any *n* natural. So $w' \in P^{-1}P(st)$ and $\sigma_{fi} \notin w$ or $|\Gamma_{\Sigma rfi}(s)|>|\Gamma_{\Sigma fi}(s)|$. Then, the chosen trace, *s*, contradicts the definition of *FR*-diagnosability for *Fi*, so *L* is not diagnosable.

Q.E.D.

Sufficiency: It will be proved by contradiction that if the diagnoser does not include *Fi*-uncertain cycles, then *L* is *FR*-diagnosable.

Let us suppose that *L* is not FR-diagnosable, then, $\exists$ $w, w' / w=s_0t, w'=s_0't'$ $w' \in P^{-1}(P(w))$ with $s_0 \in \Psi(\Sigma_{fi})$, $\Sigma_{rfi} \notin t$ and $(\Sigma_{fi} \notin w'$ or $\forall$ $u \subseteq w'$ / $u \in \Psi(\Sigma_{fi})$ $\neg\exists u' \in \Psi(I(\Sigma_{fi})) \cap L/u$ / $|\Gamma_{\Sigma rfi}(uu')|<|\Gamma_{\Sigma fi}(uu')|$. So that, $|\Gamma_{\Sigma rfi}(uu')| \geq |\Gamma_{\Sigma fi}(uu')|$. Let us chose *w* and *w'* from equations (8) and (9) so that $P(w)=P(w')=P(s_0)P(\sigma_1\ldots\sigma_n)^{kmm'}$.

Let us build the diagnoser derived from *w* and *w'*, and let us suppose, without lack of generality, that initial states are certain $q_0=(x_0, l_0)$ and $q'_0=(x_0', l'_0)$.

As $P(s_0)=P(s'_0)$, then, $q_1=((\delta(x_0,s_0),ELP(x_0, l_0, s_0)), (\delta(x'_0,s'_0), ELP(x'_0, l'_0, s'_0))$.

Therefore, $\delta(x_0,s_0)=x^1_1$ and $ELP(x_0,l_0,s_0)=Fi$ (since $s_0 \in \Psi(\Sigma_{fi})$) and where $\delta(x'_0,s'_0)= y^1_1$ and $ELP(x'_0,l'_0,s'_0)=l''^1_1$, with $Fi \notin l''^1_1$. So that, $q_1=\{(x^1_1,Fi), (y^1_1,l''^1_1)\}$, i.e., an uncertain state. Next states, $q_2...q_n$ are reached following equations (6) and (7), so they form a cycle in the same way as in the previous proof. On the other hand, *ELP* assigns the labels:

$$ELP(x^k_p, s^k_p\sigma_p, l^k_p) = Fi,\ p = 1,...,n-1, k = 1,...,m$$

$$ELP(x^k_n, s^k_n\sigma_n, l^k_n) = Fi,\ k = 1,...,m-1$$

$ELP(x^m_n, s^m_n\sigma_n, l^m_n) = Fi$, because, by hypothesis there is not recovery in w, moreover

$$ELP(y^r_p, s'^r_p\,\sigma_p, l'^r_p) = l'^r_{p+1},\ p = 1,...,n-1, k = 1,...,m'$$

$$ELP(y^r_n, s'^r_n\,\sigma_n, l'^r_n) = l'^{r+1}_p,\ k = 1,...,m'-1$$

$$ELP(y^{m'}_n, s'^{m'}_n\,\sigma_n, l'^{m'}_n) = l'^1_1$$

Taking into account that $Fi \notin l''^r_p$ since $|\Gamma_{\Sigma rfi}(s''^r_p')| \geq |\Gamma_{\Sigma fi}(s''^r_p')|$, the formed cycle is uncertain. This conclusion contradicts the starting hypothesis.

Q.E.D.

CONCLUSIONS

This paper extends the framework developed in (Sampath, 1995) and (Garcia, 2000) for diagnosing DES in order to include IF. The framework is based on the physical nature of IF failures so only real detectable failures are considered.

It has been proved that the previous methodologies do not support IF, therefore, a new definition for diagnosability (*FR*-diagnosability) has been defined and the conditions for the diagnosers to reach FR-diagnosability have been proved.

New diagnosers based on discrete event systems are designed. Moreover, they are embedded in a wider framework that includes not only failure diagnosis but a better facilities maintenance scheduling.

REFERENCES

Anderson, J. S. Aylward, (1994). "Lab Testing of Neural Networks for Improved Aircraft Onboard-Diagnostics on Flight-Ready Hardware" *LAMBDA NOTES*. Society of Reliability Engineers. Volume 26 - Issue 1.

Antler, M. (1999) Contact fretting of electronic connectors *IEICE transactions on electronics* **Vol. E82-C No. 1**

Baisheng Sun. (2001) Effects of electric contact failure on signal transmission in unmatched circuits. *Electrical Contacts, . Proceedings of the Forty-Seventh IEEE Holm Conference on ,* Page(s): 186 –191

Cassandras C.G. and S. Lafortune. (1999) "Introduction to Discrete Event Systems". Kluwer academic publishers.

Correcher, A. García, E., Morant, F., Quiles, E. and Blasco. R. (2003) Intermittent Failure Diagnosis in Industrial Processes. *Proceedings of the IEEE International Symposium on Industrial Electronics ISIE'2003* (ISBN 0-7803-7912-8) pp 1-6

David, R. and H. Alla. (1992) "Petri Nets and Grafcet. Tools for modelling discrete event systems". Prentice Hall International.

García, E. (2000) Descomposición modular de diagnosticadores de fallos basados en modelos de eventos discretos. PhD Thesis. Universidad Politécnica de Valencia.

Kui Li; Fang Yao; Jianguo Lu; Zhigang Li. (2000) Test and analysis of reliability for electromagnetic relay *Electrical Contacts, 2000. Proceedings of the Forty-Sixth IEEE Holm Conference on.* Page(s): 79 –82

Madden M. and Nolan P.J. (1999). "Monitoring and Diagnosis of Multiple Incipient Faults Using Fault Tree Induction", *IEE Proceedings – Control Theory and Applications* Journal, Institution of Electrical Engineers, London. **Vol. 146**, No. 2,

Maul, C. McBride, J. W., Swingler, J. (2000) On the Nature of Intermittences in Electrical Contacts, *20th International Conference on Electrical Contacts* Stockholm

Maul, C. McBride, J. W., Swingler, J. (2001) Intermittency phenomena in electrical connectors. *Components and Packaging Technologies, IEEE Transactions on*, **Volume: 24** Issue: 3 , Page(s): 370 –377

Maul, C. McBride, J. W. (2002) A model to describe intermittency phenomena in electrical connectors, *Electrical Contacts, 2002. Proceedings of the Forty-Eighth IEEE Holm Conference on ,* Page(s): 165 –174

Salvatore, J. Bavuso, Elizabeth Rothmann, Joanne Bechta Dugan, Kishor S. Trivedi, Nitin Mittal, Mark A. Boyd, Robert M. Geist and Mark D. Smotherman, (1994) HiRel: Hybrid Automated Reliability Predictor (HARP) Integrated Reliability Tool System (Version 7.0) HARP Introduction and User's Guide, *NASA TP-3452,* Vol. 1 , pp. 147

Sampath, M. (1995) A discrete event systems approach to failure diagnosis. PhD thesis. University of Michigan.

Savir. J. (1978) "Detection of intermittent faults in sequential circuits" Stanford University, Computer Systems Laboratory. T.R. 120. available in "ftp://reports.stanford.edu"

Sorensen, B. A. - Gary Kelly - Artur Sajecki - Paul W. Sorensen, (1994) "An analyzer for detecting aging faults in electronic devices" *AUTOTESTCON '94. IEEE Systems Readiness Technology Conference. 'Cost Effective Support Into the Next Century', Conference Proceedings. ,* 1994 Page(s): 417 –421. update (1998) available in www.usynaptics.com.

ELSEVIER
IFAC
PUBLICATIONS
www.elsevier.com/locate/ifac

HIERARCHICAL DISTRIBUTED DIAGNOSIS UNDER GLOBAL CONSISTENCY

R. Su and W.M. Wonham

Dept. Electrical & Computer Engineering
University of Toronto
Toronto, ON M5S 3G4, Canada
surong,wonham@control.utoronto.ca

Abstract: In this paper we first introduce the concept of the supremal global support to capture interaction among local components. Then we propose a hierarchical diagnosis approach, whose main objective is to achieve supremal global support at the bottom level.

Keywords: discrete-event systems, distributed diagnosis, hierarchical

1. INTRODUCTION

It is well known that in fault diagnosis for discrete-event systems (DES) a major challenge is the high space complexity of storing diagnosers and generating diagnostic results, which is typical in centralized approaches, e.g. (Sampath et al. (1998)). To reduce space complexity, researchers have turned their attention to distributed architecture, consisting of a set of simple modular components linked in a communication network, e.g. (Su et al. (2002)) (Fabre et al. (2002)) (Benveniste et al. (2003)). The resulting diagnoser consists of a set of relatively simple local diagnosers. Therefore, memory usage for storing the distributed diagnoser is low, but with a price of temporal delay introduced by inter-component communication. The inter-component interaction in a distributed approach is captured by an appropriate consistency property that is required to be achieved during communication. Considering that the communication procedure proposed in (Su et al. (2002)) (Fabre et al. (2002)) (Benveniste et al. (2003)) may not always terminate, in a companion paper (Su and Wonham (2004)) the authors provide a new algorithm to achieve the so-called *global consistency*. Nevertheless, it turns out that the algorithm in (Su and Wonham (2004)) may not be scalable and have high time complexity. To partially overcome these problems, in this paper we introduce hierarchical structure.

Hierarchical approaches have been studied in the AI community for many years, e.g. (Autio and Reiter (1998)) (Stroobandt and Campenhout (1996)) (Mozetic (1991)). Nevertheless, in those approaches each level possesses a centralized model instead of a distributed model. This is because the authors deal mainly with digital circuits whose behavior can be concisely expressed by simple logic expressions. So the size of a centralized model is not of great import; time complexity is the authors' major concern. As we have seen, a centralized model for automaton-based diagnosis usually leads to high space complexity. Another feature of AI hierarchical approaches is that their construction procedures for high-level abstract models usually are rather complicated, inasmuch as the models lack language structure, making the projection abstraction operation complicated. Thus checking consistency between two different levels is not a trivial matter. By contrast, we will see later that projection is not problematic for language-based fault diagnosis. Some work on distributed systems addresses hierarchical distributed diagnosis, e.g. (Su et al. (2001)) (Duarte and Nanya (1998)). Nevertheless, these approaches treat each local component simply as a node without any further evolution or input-out structure, and their objective is to determine which node fails to make a correct response. So this work is more about distributed fault detection than distributed fault diagnosis, which includes fault isolation, and its major concern is time complexity rather than space complexity.

This paper is organized as follows. In Section 2 we introduce the concept of hierarchical distributed model. Then in Section 3 we propose a hierarchical distributed diagnosis procedure and draw conclusions in Section 4.

2. HIERARCHICAL DISTRIBUTED MODEL

2.1 Supremal Global Support

Let I be an index set. For each $i \in I$ let Σ_i be an event set. As usual (Wonham (2003)), we use "$||$" to represent synchronous product. All natural projections in this paper obey the following notational rules: given an index set J with a predefined collection of event sets $\{\Sigma_j | j \in J\}$, we define

$$\Sigma_J := \bigcup_{j \in J} \Sigma_j$$

We use $P_{J,i}$ to mean the following natural projection,

$$P_{J,i} : \Sigma_J^* \to (\Sigma_J \cap \Sigma_i)^*$$

Let $\mathcal{L} = \{L_i \subseteq \Sigma_i^* | i \in I\}$ be a set of languages.

Definition 2.1. $\mathcal{L} = \{L_i | i \in I\}$ is *globally consistent* if $(\forall i \in I)\, L_i = P_{I,i}(||_{j \in I} L_j)$. □

Definition 2.2. Given $\mathcal{L} = \{L_i \subseteq \Sigma_i^* | i \in I\}$, a set $\{E_i \subseteq L_i | i \in I\}$ is a *global support* of $\mathcal{L}$ if it is globally consistent. Let $\Delta(\mathcal{L})$ be the set of all global supports of $\mathcal{L}$.

Clearly $\{E_i = \varnothing | i \in I\} \in \Delta(\mathcal{L})$. We define a binary relation "$\leq$" among elements in $\prod_{i \in I} \mathrm{Pwr}(\Sigma_i^*)$:

$$\{E_i | i \in I\} \leq \{\tilde{E}_i | i \in I\} \iff (\forall i \in I)\, E_i \subseteq \tilde{E}_i$$

$\mathcal{E} \in \Delta(\mathcal{L})$ is called the *supremal global support* of $\Delta(\mathcal{L})$, written $\mathrm{Sup}\Delta(\mathcal{L})$, if $(\forall \mathcal{E}' \in \Delta(\mathcal{L}))\, \mathcal{E}' \leq \mathcal{E}$.

Proposition 2.1. $\mathrm{Sup}\Delta(\mathcal{L}) = \{P_{I,i}(||_{j \in I} L_j) | i \in I\}$.

Definition 2.3. Let $\{\Sigma_i | i \in I\}$ be a collection of event sets. A *distributed reference model* $\mathcal{L}$ is a set of closed languages $\{L_i \subseteq \Sigma_i^* | i \in I\}$, where L_i is a *local component of* $\mathcal{L}$. There is a subset $\Sigma_{io} \subseteq \Sigma_i$ called the *observable event set of* i, which is not necessarily pairwise disjoint with other observable event sets, and another set $\Sigma_{if} \subseteq \Sigma_i$ called *fault event set of* i, which is pairwise disjoint with any other event set. □

Given a reference model $\mathcal{L} = \{L_i \subseteq \Sigma_i^* | i \in I\}$, for each $i \in I$ we have a local observable event set $\Sigma_{io} \subseteq \Sigma_i$. Let $P_{i,o} : \Sigma_i^* \to \Sigma_{io}^*$ be the natural projection for component i. The *symptom* of a distributed system is an I-tuple $\mathcal{O} = \{u_i \in \Sigma_{io}^* | i \in I\}$, where each u_i is a *local symptom* of component i. A distributed diagnosis problem under global consistency is proposed as follows.

Distributed Diagnosis under Global Consistency

1. **Local Computation:** $(\forall i \in I)\, M_i := P_{i,o}^{-1}(u_i) \cap L_i$
2. **Global Consistency:** $\mathcal{E} := \mathrm{Sup}\Delta(\{M_i | i \in I\})$ □

The motivation to formulate a distributed diagnosis problem as above can be simply explained as follows. Local computation is nothing but an estimation process, namely given observation u_i $(i \in I)$ we find all strings that can exhibit u_i. Once local estimates are obtained, we use supremal global support to capture "agreement" among these local estimates. In the companion paper (Su and Wonham (2004)) the authors provide an algorithm called *Computational Procedure for Global Consistency* (CPGC) to achieve the supremal global support. But it turns out that CPGC is rather time consuming when the index set I is large. So the main objective of this paper is to explore hierarchical structure to speed up CPGC.

2.2 Hierarchical Model Construction

In this paper we confine attention to a two-level hierarchy. It will be clear that the approach can be easily extended to a general hierarchy. Suppose we have a distributed reference model $\mathcal{L} = \{L_i \subseteq \Sigma_i^* | i \in I\}$. Let $\hat{I}$ be a new index set which is associated with a given partition $\{J_{\hat{k}} \subseteq I | \hat{k} \in \hat{I}\}$ on I. For each $i \in I$ let $\Sigma_{io} \subseteq \Sigma_i$ be the observable event set of component i. To preserve the consistency of observation, we assume that **if an event is observable in a local component, then it is observable in any local component that contains it as a local event**, namely

$$(\forall i, j \in I)(\forall \sigma \in \Sigma_i \cap \Sigma_j)\, \sigma \in \Sigma_{io} \iff \sigma \in \Sigma_{jo}$$

For notational consistency, elements and subsets of $\hat{I}$ have hat symbol, e.g. $\hat{i}, \hat{j}$, but no elements and subsets of I have it. For each $\hat{k} \in \hat{I}$, let $\Sigma_{\hat{k}o} := \cup_{i \in J_{\hat{k}}} \Sigma_{io}$. We call $\Sigma_{\hat{k}o}$ the *observable event set* of module $J_{\hat{k}}$. Let

$$\Sigma_{\hat{k}} := \Sigma_{\hat{k}o} \cup (\Sigma_{J_{\hat{k}}} \cap \Sigma_{I - J_{\hat{k}}}) \qquad (1)$$

be the *critical event set* of module $J_{\hat{k}}$. Thus, $\Sigma_{\hat{k}}$ contains the observable events of each component $i \in J_{\hat{k}}$ and all events shared by $J_{\hat{k}}$ and other modules $I - J_{\hat{k}}$. Recall that, with the same notational rules for natural projection described in Section 2.1, $P_{J_{\hat{k}},\hat{k}}$ represents the natural projection $P_{J_{\hat{k}},\hat{k}} : \Sigma_{J_{\hat{k}}}^* \to \Sigma_{\hat{k}}^*$. We define

$$L_{\hat{k}} := P_{J_{\hat{k}},\hat{k}}(||_{i \in J_{\hat{k}}} L_i) \qquad (2)$$

as an *abstract local module* of $\mathcal{L}$. The set $\hat{\mathcal{L}} := \{L_{\hat{k}} | \hat{k} \in \hat{I}\}$ is called the *abstract model* of $\mathcal{L}$.

Figure 1 depicts a simple paper acquisition model

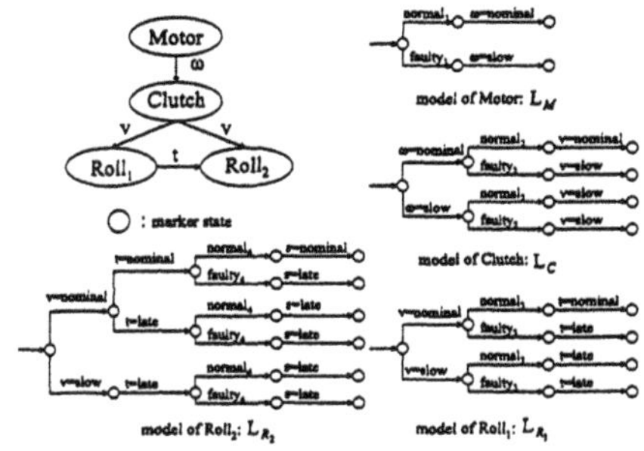

Fig. 1. Simple Paper Acquisition Model

(SPAM) which consists of one motor, one clutch and two rolls. The index set is $I = \{M, C, R_1, R_2\}$ and the corresponding distributed reference model is $\mathcal{L} = \{L_M, L_C, L_{R_1}, L_{R_2}\}$. To construct a high-level abstract model, we partition I into two subsets $J_1 = \{M, C\}$ and $J_2 = \{R_1, R_2\}$. Since there are no observable events in either Motor or Clutch, we have $\Sigma_{1o} = \varnothing$, namely the observable event set of module J_1 is empty. In module J_2, Roll$_1$ has no observable events but Roll$_2$ has observable events $\{s = nominal, s = late\}$. By expression (1) the critical event set of J_1 is

$$\Sigma_1 := \Sigma_{J_1} \cap \Sigma_{J_2} = (\Sigma_M \cup \Sigma_C) \cap (\Sigma_{R_1} \cup \Sigma_{R_2})$$

Similarly, we have for the critical event set of J_2

$$\Sigma_2 = \Sigma_{2o} \cup (\Sigma_{J_1} \cap \Sigma_{J_2})$$

By the notational rules for projections, we write

$$P_{J_1,1} : \Sigma_{J_1}^* \to \Sigma_1^* \text{ and } P_{J_2,2} : \Sigma_{J_2}^* \to \Sigma_2^*$$

Then by expression (2) we have abstract local modules

$$L_1 = P_{J_1,1}(L_M || L_C) \text{ and } L_2 = P_{J_2,2}(L_{R_1} || L_{R_2})$$

Figure 2 depicts L_1 and L_2, where the symbol v be-

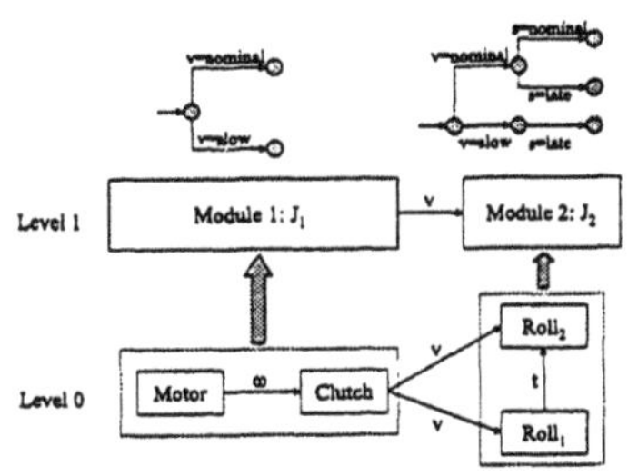

Fig. 2. Abstract Model for SPAM

tween two local modules in Figure 2 means that two modules share events assigning values to variable v, namely $v = nominal$ and $v = slow$. ◇

Suppose the symptom set is $\mathcal{O} = \{u_i \in \Sigma_{io}^* | i \in I\}$. Then we can perform local computation in each local component $i \in I$ to obtain the preliminary estimate,

$$M_i := P_{i,o}^{-1}(u_i) \cap L_i \tag{3}$$

Let $\mathcal{M} := \{M_i | i \in I\}$. For each $\hat{k} \in \hat{I}$ let

$$M_{\hat{k}} := P_{J_{\hat{k}},\hat{k}}(||_{i \in J_{\hat{k}}} M_i) \tag{4}$$

Thus, $\hat{\mathcal{M}} := \{M_{\hat{k}} | \hat{k} \in \hat{I}\}$ is an abstract model of $\mathcal{M}$. Let $P_{\hat{k},\hat{k}o} : \Sigma_{\hat{k}}^* \to \Sigma_{\hat{k}o}^*$ be the natural projection. We have the following result.

Proposition 2.2. $(\forall \hat{k} \in \hat{I})\, M_{\hat{k}} = L_{\hat{k}} \cap P_{\hat{k},\hat{k}o}^{-1}(||_{i \in J_{\hat{k}}} \{u_i\})$.

Prop. 2.2 tells us how to compute $\hat{\mathcal{M}}$ directly from the abstract model $\hat{\mathcal{L}}$ and the symptom set $\mathcal{O}$ without actually using expression (4). On the other hand, we can derive that

$$P_{I,\hat{I}}(||_{i \in I} M_i) = ||_{\hat{k} \in \hat{I}} M_{\hat{k}} \tag{5}$$

By Prop. 2.1 we have

$$\text{Sup}\Delta(\hat{\mathcal{M}}) = \{E_{\hat{i}} := P_{\hat{I},\hat{i}}(||_{\hat{k} \in \hat{I}} M_{\hat{k}}) | \hat{i} \in \hat{I}\}$$

Let $\text{Sup}\Delta(\mathcal{M}) = \{E_j | j \in I\}$. Then by (5) $\text{Sup}\Delta(\mathcal{M})$ and $\text{Sup}\Delta(\hat{\mathcal{M}})$ have the following relationship.

$$(\forall \hat{i} \in \hat{I})\, E_{\hat{i}} = P_{I,\hat{i}}(||_{j \in I} E_j) \tag{6}$$

The relation between two levels in a hierarchy can be summarized as a commutative diagram. Let

$$\mathbb{W} := \prod_{i \in I} \text{Pwr}(\Sigma_i^*) \quad \mathbb{O} := \prod_{i \in I} \Sigma_{io}^*$$
$$\hat{\mathbb{W}} := \prod_{\hat{k} \in \hat{I}} \text{Pwr}(\Sigma_{\hat{k}}^*) \quad \hat{\mathbb{O}} := \prod_{\hat{k} \in \hat{I}} \text{Pwr}(\Sigma_{\hat{k}o}^*)$$

Then Proposition 2.2, together with (2), (3), (4), (5) and (6), makes the following diagram commute, where

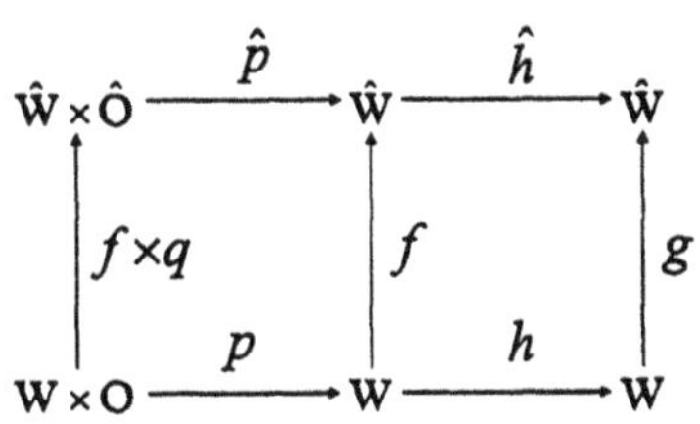

Fig. 3. Two-Level Hierarchical Computation

maps p, $\hat{p}$ are local estimations at low and high levels respectively (see (3) and Proposition 2.2); f, g, h, $\hat{h}$ and q are defined below.

(1) $f(\{L_i | i \in I\}) := \{P_{J_{\hat{k}},\hat{k}}(||_{j \in J_{\hat{k}}} L_j) | \hat{k} \in \hat{I}\}$.
(2) $g(\{E_i | i \in I\}) := \{E_{\hat{k}} = P_{I,\hat{k}}(||_{j \in I} E_j) | \hat{k} \in \hat{I}\}$.
(3) $h(\mathcal{M}) := \text{Sup}\Delta(\mathcal{M})$
(4) $\hat{h}(\hat{\mathcal{M}}) := \text{Sup}\Delta(\hat{\mathcal{M}})$
(5) $q(\{u_i | i \in I\}) := \{P_{J_{\hat{k}},\hat{k}o}(||_{j \in J_{\hat{k}}} \{u_i\}) | \hat{k} \in \hat{I}\}$

Given $\mathcal{L} \in \mathbb{W}$ and $\mathcal{O} \in \mathbb{O}$, our objective is to compute

$$\mathcal{E} = h(p(\mathcal{L}, \mathcal{O}))$$

But instead of using $h \circ p$, where "$\circ$" represents map composition, we first compute

$$\hat{\mathcal{E}} := \hat{h} \circ \hat{p} \circ (f \times q)(\mathcal{L}, \mathcal{O}) = \hat{h}(\hat{p}(f(\mathcal{L}), q(\mathcal{O})))$$

The reason is that computing supremal global support at the high level is usually easier than doing so at the bottom level, thanks to abstraction. The following result shows how to obtain $\mathcal{E}$ from $\hat{\mathcal{E}}$.

Proposition 2.3. Let f be defined as in Figure 3. Let $\mathcal{M} = \{M_i | i \in I\} \in \mathbb{W}, \hat{\mathcal{M}} = \{M_{\hat{k}} | \hat{k} \in \hat{I}\} \in \hat{\mathbb{W}}$ such that $f(\mathcal{M}) = \hat{\mathcal{M}}$. Suppose $\mathrm{Sup}\Delta(\mathcal{M}) = \{E_i | i \in I\}$ and $\mathrm{Sup}\Delta(\hat{\mathcal{M}}) = \{E_{\hat{k}} | \hat{k} \in \hat{I}\}$. Then we have that $(\forall \hat{k} \in \hat{I})(\forall i \in J_{\hat{k}})\, E_i = P_{J_{\hat{k}}, i}((||_{j \in J_{\hat{k}}} M_j) || E_{\hat{k}})$.

Once we have the supremal global support at the high level, say $\hat{\mathcal{E}}$, Prop. 2.3 tells us how to use $\hat{\mathcal{E}}$ to obtain the supremal global support $\mathcal{E}$ at the bottom level. Notice that concurrent computation can be done in different modules at the bottom level, which can substantially improve time complexity of achieving the supremal global support. This is the principal reason for introducing the hierarchical approach.

As an illustration, consider the following example depicted in Figure 4, where $I = \{1, 2, 3, 4\}$. Since all local components are prefix-closed languages, each state in Figure 4 is a marker state. Suppose the local

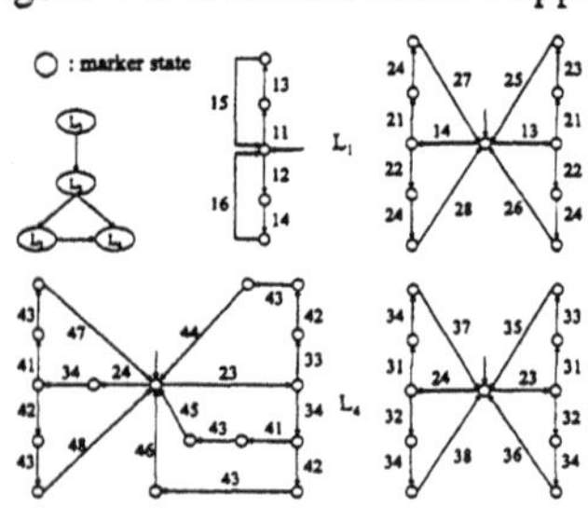

Fig. 4. Simple Example

observable event sets are $\Sigma_{1o} = \varnothing$, $\Sigma_{2o} = \varnothing$, $\Sigma_{3o} = \{33, 34\}$ and $\Sigma_{4o} = \{33, 34, 41, 42\}$. Let $\hat{I} := \{a, b\}$, where $J_a = \{1, 2\}$ and $J_b = \{3, 4\}$. According to expression (1) we have

$$\Sigma_a := \{23, 24\} \text{ and } \Sigma_b := \{23, 24, 33, 34, 41, 42\}$$

Based on the two critical event sets we build

$$L_a := P_{J_a, a}(L_1 || L_2) \text{ and } L_b := P_{J_b, b}(L_3 || L_4)$$

shown in Figure 5. Suppose the local symptom set at the bottom level is: $\mathcal{O} = \{\epsilon, \epsilon, 34, (34)(41)\}$, namely event 34 has been executed in L_3 and events 34, 41 in L_4. Then we can construct the local symptom set $\hat{\mathcal{O}} = \{U_a = ||_{j \in J_a}\{u_j\}, U_b = ||_{j \in J_b}\{u_j\}\}$ at the second level. So in L_a we have $U_a = \{\epsilon\} = \{\epsilon\} || \{\epsilon\}$ as the local observation, and in L_b we have $U_b = \{34\} || \{(34)(41)\} = \{(34)(41)\}$. By Proposition 2.2,

$$M_a = L_a || P^{-1}_{a, ao}(\{\epsilon\}), M_b = L_b || P^{-1}_{b, bo}(\{(34)(41)\})$$

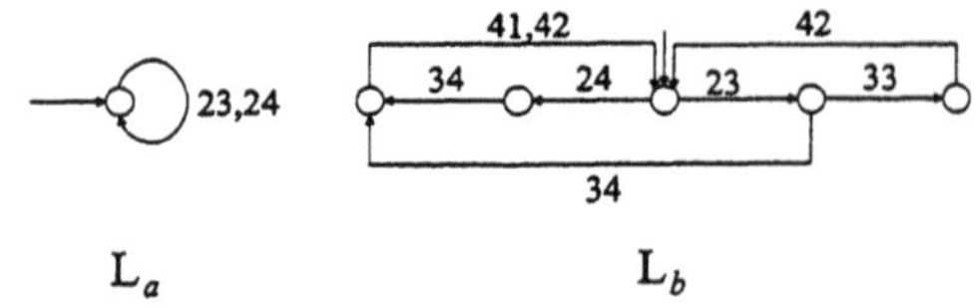

Fig. 5. Second Level Abstract Model $\hat{\mathcal{L}} = \{L_a, L_b\}$

Figure 6 depicts results of $\hat{\mathcal{M}} := \{M_a, M_b\}$. Then we

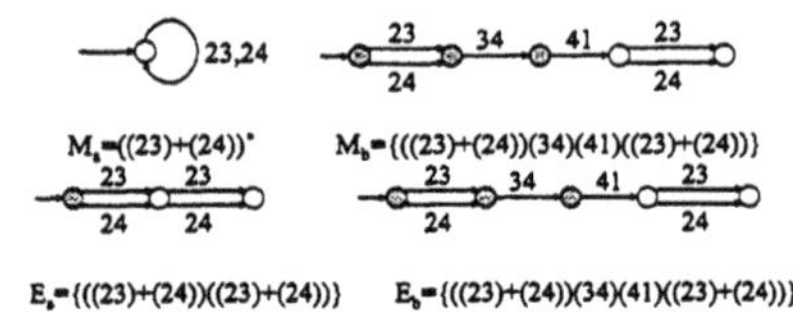

Fig. 6. M_a, M_b, E_a and E_b

compute $\hat{\mathcal{E}} := \{E_a, E_b\} = \mathrm{Sup}\Delta(\hat{\mathcal{M}})$, which are also shown in Figure 6. So far, we have finished computation at the second level. Now we descend to the first level and compute the supremal global support based on the local observation set $\mathcal{O} = \{\epsilon, \epsilon, 34, (34)(41)\}$ and messages E_a, E_b passed down from the second level. Since no observation is made from L_1 and L_2, we have $M_1 = L_1$ and $M_2 = L_2$. By Prop. 2.3, we can compute their local estimates based on E_a as follows:

$$E_1 = P_{J_a, 1}(M_1 || M_2 || E_a), E_2 = P_{J_a, 2}(M_1 || M_2 || E_a)$$

Figure 7 depicts the results of E_1 and E_2. Similarly,

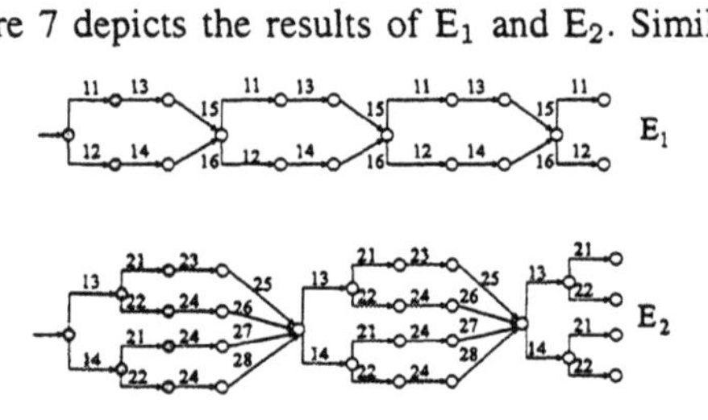

Fig. 7. E_1 and E_2

we can compute E_3 and E_4 by first computing

$$M_3 := L_3 \cap P^{-1}_{3,o}(\{34\}), M_4 := L_4 \cap P^{-1}_{4,o}(\{(34)(41)\})$$

then

$$E_3 = P_{J_b, 3}(M_3 || M_4 || E_b), E_4 = P_{J_b, 4}(M_3 || M_4 || E_b)$$

The left picture of Figure 8 depicts the results of M_3 and M_4. and the right picture of Figure 8 depicts the

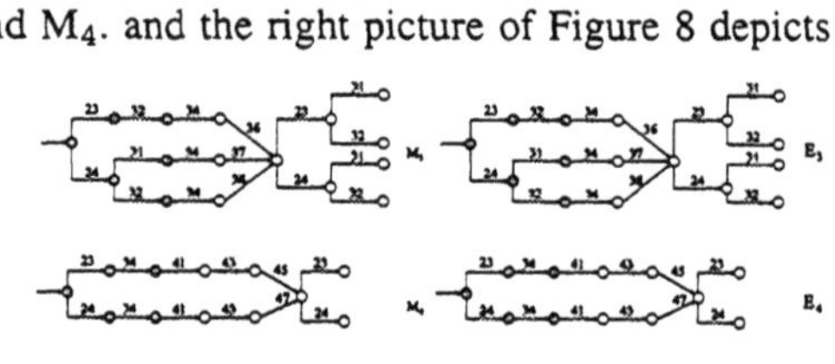

Fig. 8. M_3, M_4 (Left) and E_3, E_4 (Right)

results of E_3 and E_4. So far, we have the supremal global support $\mathcal{E} = \{E_1, E_2, E_3, E_4\}$ at the first level. On using the "brute-force" method (which, by the way, yields 1153 states and 3118 transitions for the size of $M_1 || M_2 || M_3 || M_4$), we confirm that all results generated from hierarchical distributed computation are correct. In this example, at the first level, computing E_1 and E_2 is independent of computing E_3 and E_4, making it possible to do concurrent computation with the two disjoint sets ($\{L_1, L_2\}$ and $\{L_3, L_4\}$).

So far we have described how to construct a two-level hierarchy and perform hierarchical computation in it. If we assign the symptom set at the bottom level as $\mathcal{O} := \{U_i := \{u_i\} \subseteq \Sigma_{io}^* | i \in I\}$, then we can easily check that all the results above hold in a general hierarchy; and between adjacent levels there is a commutative diagram according to Figure 3 and Propositions 2.2 and 2.3. As a summary, the main motivation for such hierarchical distributed computation is that computing the supremal global support at the high level is usually easier because of the relatively simple model at the high level due to abstraction. When computation reaches the bottom level, where the model is the most complicated, the problem of computing the supremal global support has been converted into many small problems where each problem only involves a single local module with a relatively small number of local components. This is nothing but another application of divide-and-conquer. Next, we introduce multi-resolution diagnosis.

3. MULTI-RESOLUTION DIAGNOSIS

In the previous section we have discussed how to use the divide-and-conquer strategy in a hierarchical structure to speed up computation of the supremal global support (see Proposition 2.3). But in reality we have another way to reduce computation, namely computation should only be carried out in a module that may potentially possess fault. This idea leads to the concept of multi-resolution diagnosis.

First, we need to figure out when we can decide that a module is absolutely fault free. To this end, at the bottom level $\mathcal{L}$, we define a normal distributed model.

$$(\forall i \in I)\, W_i := (\Sigma_i - \Sigma_{if})^* \cap L_i$$

Thus no string $s \in W_i$ contains a fault event. Clearly, W_i is also prefix-closed. For each $\hat{k} \in \hat{I}$ let $L_{J_{\hat{k}}} := ||_{i \in J_{\hat{k}}} L_i$ and $W_{J_{\hat{k}}} := ||_{i \in J_{\hat{k}}} W_i$. Then we define

$$L_{\hat{k},norm} := P_{J_{\hat{k}},\hat{k}}(W_{J_{\hat{k}}}) - P_{J_{\hat{k}},\hat{k}}(L_{J_{\hat{k}}} - W_{J_{\hat{k}}}) \quad (7)$$

For each string $s \in L_{\hat{k},norm}$ we have

$$P^{-1}_{J_{\hat{k}},\hat{k}}(s) \cap L_{J_{\hat{k}}} \subseteq W_{J_{\hat{k}}}$$

namely s contains no fault event. Thus, when s is executed in the abstract module $L_{\hat{k}}$ we know that every local component in the module $J_{\hat{k}}$ is absolutely normal. We call $L_{\hat{k},norm}$ the *truly normal abstract module* associated with $L_{\hat{k}}$. Let

$$\hat{\mathcal{L}}_{norm} := \{L_{\hat{k},norm} | \hat{k} \in \hat{I}\}$$

be the *truly normal abstract model of* $\mathcal{L}$. So when we build the two-level hierarchy, we build one abstract model $\hat{\mathcal{L}}$ and one truly normal abstract model $\hat{\mathcal{L}}_{norm}$, which will serve as the benchmark of normal behaviors. Now we describe how to use such a benchmark in hierarchical distributed diagnosis.

Multi-Resolution Diagnosis Procedure: (MRDP)

Given the model $\mathcal{L}$, suppose a symptom set is $\mathcal{O} \in \mathbb{O}$.

(1) At high level:

$$\hat{\mathcal{E}} := \hat{h} \circ \hat{p} \circ (f \times q)(\mathcal{L}, \mathcal{O}) = \{E_{\hat{k}} | \hat{k} \in \hat{I}\}$$

(2) At bottom level: $\mathcal{M} := p(\mathcal{L}, \mathcal{O})$

(3) For each $\hat{k} \in \hat{I}$, if $E_{\hat{k}} \not\subseteq L_{\hat{k},norm}$ then compute

$$(\forall i \in J_{\hat{k}})\, E_i := P_{J_{\hat{k}},i}((||_{j \in J_{\hat{k}}} M_j) || E_{\hat{k}})$$

The main feature of MRDP is that computation in a specific module, say $J_{\hat{k}}$, will be carried to the bottom level on each member $i \in J_{\hat{k}}$ if some behavior not belonging to normal has been detected in $L_{\hat{k}}$ at the higher level, namely $E_{\hat{k}} \not\subseteq L_{\hat{k},norm}$. Since at step 3 the supremal global support in $\hat{\mathcal{L}}$ has been achieved, by Propositions 2.2 and 2.3 we know that if computation at some module $J_{\hat{k}}$ is halted at the high level because no abnormality is detected, then computation in other local modules will not be affected. On the other hand, we may raise a question about MRDP: is it possible that some local component in a module has become faulty, namely some string containing a fault event has been executed, but computation is halted at the corresponding high-level abstract module? The following result answers the question.

Let $R_i : \Sigma_i^* \to \mathrm{Pwr}(\Sigma_{if})$ be a map of component i such that $(\forall s \in \Sigma_i^*)\, R_i(s) := \{\sigma \in \Sigma_{if} | \sigma \in s\}$, where $\sigma \in s$ means that σ appears at least once in s.

Proposition 3.1. Given a set $\{s_i \in \Sigma_i^* | i \in I\}$ with $||_{i \in I}\{s_i\} \neq \varnothing$, let $\mathcal{O} = \{P_{i,o}(s_i) | i \in I\}$. Then

$$(\forall \hat{k})(\forall i \in J_{\hat{k}})\, R_i(s_i) \cap \Sigma_{if} \neq \varnothing \Rightarrow E_{\hat{k}} \not\subseteq L_{\hat{k},norm}$$

where $E_{\hat{k}}$ is obtained by MRDP on $\mathcal{O}$. □

Proposition 3.1 says that, as long as the symptom set is generated by some feasible behavior $\{s_i | i \in I\}$, i.e. $||_{i \in I}\{s_i\} \neq \varnothing$, MRDP will not halt computation at the high level in a faulty module.

As an illustration, consider again the simple paper acquisition model (Figure 1). Besides the original abstract model (Figure 2), we construct another truly normal abstract model of the system. Figure 9 depicts normal behavior of each local component. We partition the system into two modules $J_1 :=$ {Motor, Clutch} and $J_2 := \{\mathrm{Roll}_1, \mathrm{Roll}_2\}$. Based on $\{W_i | i \in I\}$ we can compute that

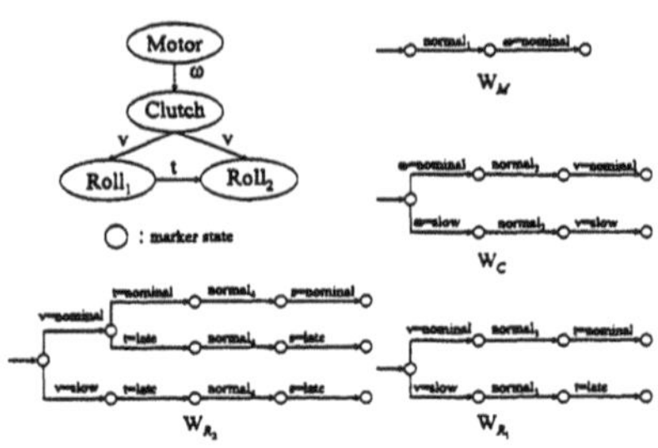

Fig. 9. $\{W_i | i \in I\}$ in SPAM

$$L_{1,norm} = (v = \text{nominal})$$
$$L_{2,norm} = (v = \text{nominal})(s = \text{nominal})$$

Case 1: Suppose at the bottom level, the observation is s = late in Roll_2 and empty in other local components. Then by the abstract model (Figure 2) we have that

$$E_1 = \{v = \text{nominal}, v = \text{slow}\}$$
$$E_2 = \{(v = \text{nominal})(s = \text{late}), (v = \text{slow})(s = \text{late})\}$$

Since $E_1 \nsubseteq L_{1,norm}$ and $E_2 \nsubseteq L_{2,norm}$, computation is carried to the bottom level in each local module.

Case 2: If the observation is s = nominal in Roll_2 and empty in other local components. Then we get that

$$E_1 = \{v = \text{nominal}\} = L_{1,norm}$$
$$E_2 = \{(v = \text{nominal})(s = \text{nominal})\} \subseteq L_{2,norm}$$

So by the proposed multi-resolution diagnosis procedure, no computation will be carried to the bottom level in any local module, and all components are declared *normal*.

4. CONCLUSIONS

In this paper we proposed a novel hierarchical distributed diagnosis approach. In this approach, at each single level of a pre-constructed hierarchy we solve a distributed diagnosis problem. Distributed diagnosis processes at different levels are linked together by information exchange between adjacent levels. Prop. 2.3 shows that after first achieving the supremal global support at the high level, we can apply concurrent computation to achieve the supremal global support at the low level. Furthermore, it turns out that the proposed hierarchical approach enables us to ignore any module that is confirmed to be *truly normal* at each intermediate level. Thus, computation to achieve the supremal global support at each level is only carried out in local modules that may contain faults. This selective computational procedure can further reduce time complexity of online diagnosis. Although the proposed hierarchical approach has to pay a price of extra memory to store high-level abstract models, our numerical computational results, which are not included in this paper due to limited space, indicate that the overall space complexity (which includes static memory to store models and dynamic memory to store intermediate computational results) is no worse (and in many cases much better) than a non-hierarchical approach. Finally, the simple procedure for constructing the high-level abstract models by using language-based natural projection and synchronous product is another advantage of our proposed hierarchical approach compared with other hierarchical approaches in the literature.

REFERENCES

K. Autio and R. Reiter. Structural abstraction in model-based diagnosis. In *Proc. 13th European Conference on Artificial Intelligence (ECAI'98)*, pages 269–273, Brighton, UK, August 1998.

A. Benveniste, S. Haar, E. Fabre, and C. Jard. Distributed monitoring of concurrent and asynchronous systems. In *Proc. ATPN-Workshop on Discrete Event Systems Control*, pages 97–142, Eindhoven, The Netherlands, June 2003.

E. Duarte and T. Nanya. A hierarchical adaptive distributed system-level diagnosis algorithm. *IEEE Trans. on Computers*, 47(1):34–45, 1998.

E. Fabre, A. Benveniste, and C. Jard. Distributed diagnosis for large discrete event dynamic systems. In *Proc. 15th IFAC World Congress*, Barcelona, Spain, July 2002.

Igor Mozetic. Hierarchical model-based diagnosis. *International Journal of Man-Machine Studies*, 35 (3):329–362, 1991.

M. Sampath, S. Lafortune, and D. Teneketzis. Active diagnosis of discrete-event systems. *IEEE Transactions on Automatic Control*, 40:908–929, July 1998.

D. Stroobandt and J. Van Campenhout. Hierarchical test generation with built-in fault diagnosis. In *Proc. 5th Asian Test Symp.*, pages 22–28, November 1996.

M.-S. Su, K. Thulasiraman, and A. Das. A multi-level adaptive distributed diagnosis algorithm for fault detection in a network of processors. In *Proc. 39th Annual Allerton Conf. on Communication, Control, and Computers*, Monticello, IL, October 2001.

R. Su and W.M. Wonham. A model of component consistency in distributed diagnosis. In *Proc. 2004 IFAC Workshop on DiscreteEvent Systems (WODES'04)*, Reims, France, September 2004.

R. Su, W.M. Wonham, J. Kurien, and X. Koutsoukos. Distributed diagnosis for qualitative systems. In *Proc. 6th International Workshop on Discrete Event Systems (WODES'02)*, pages 169–174, Zaragoza, Spain, October 2002.

W. M. Wonham. *Notes on Control of Discrete-Event Systems: ECE 1636F/1637S 2003-2004*. Systems Control Group, Dept. ECE, University of Toronto. URL: www.control.utoronto.ca/DES.html, 2003.

ELSEVIER
IFAC
PUBLICATIONS
www.elsevier.com/locate/ifac

BISIMULATION BASED DESIGN OF USER-INTERFACE FOR DISCRETE EVENT SYSTEMS

Masakazu Adachi * **Toshimitsu Ushio** *
Yoshitaka Ukawa *

* *Graduate School of Engineering Science,
Osaka University,
1-3 Machikaneyama, Toyonaka, Osaka, 560-8531, Japan*

Abstract: In human-machine systems, a user predicts machine's behavior using partial or abstracted information on the system's behavior through a user-interface. If the user-interface can not display sufficient information for user's prediction, an automation surprise may occur. In this paper, we consider interaction between a user and a machine through a user-interface, where both the user and the machine are modeled by automata and the user-interface is represented by a binary relation. First, automation surprises are classified into three cases: a blocking state, a mode confusion, and a refusal state. Next, we derive necessary and sufficient conditions for the nonexistence of the automation surprises using simulation and bisimulation relations between the machine model and the user model. Finally, we show that a user-interface and a user model without automation surprises can be designed by utilizing the bisimulation algorithm.

Keywords: Discrete event systems, user-interface, human-machine system, automation surprise, bisimulation,

1. INTRODUCTION

When a user controls and supervises mode-rich systems, a user-interface displays partial or abstracted information on a current state of a system or a machine, and the user operates the machine according to the information. In such human-machine systems, the user usually tracks or anticipates machine's behavior based on both information provided by a user-interface and a mental model of the underlying machine. But due to the incomplete information, machine's actual behavior may be different from user's awareness of the system configuration so that the user will be surprised by the difference. Such a faulty interaction is called an automation surprise (Sarter *et al.*, 1997). As the automated systems become more complex and large-scaled, a large number of mode indications is required in the user-interface, and machine's modes may change indirectly without user's intention according to their internal variables or environment. Consequently, if the user-interface and associated user's awareness are not adequate, the possibility of occurrence of automation surprises becomes higher. For example, Palmer reported two automation surprises in aircrafts (Palmer, 1995). In real-life situation, automation surprise may cause human errors so that a serious accident occurs. Hence, we need the methodology to realize a concise and reliable human-machine interaction.

Human-machine systems consist of the machine, the user, and the user-interface, where the machine is designed *a priori*. So, in order to obtain an adequate human-machine system without automation surprises, we need a model-based co-design of both a user-interface and a user model. Recently, systematic approaches to analysis of human-machine systems have been studied. Degani proposes a model-based analysis of human-machine systems (Degani, 1996). Parasuraman *et al.* propose a four-stage model (Parasuraman *et al.*, 2000). Moreover, techniques developed in computer science have been applied to verification of automation surprises. Rushby demonstrates applicability of model checking to exploration of automation surprises (Rushby, 2002). Degani and Heymann model both machine's behavior and user's action as discrete event systems, and propose an algorithm for generating an adequate user-interface which enables the users to perform specified tasks correctly (Heymann and Degani, 2002; Degani and Heymann, 2002). Their approach is based on a reachability problem in a composite model of the machine model and the user model. Oishi *et al.* introduce a concept of immediate observability in discrete event systems and apply it to reduction of a number of states in a user-interface (Oishi *et al.*, 2003).

In this paper, we address an alternative procedure for constructing human-machine systems without automation surprises by introducing a bisimulation relation of the machine model and the user model. First, we classify automation surprises into three cases: a blocking state, a mode confusion, and a refusal state. Then we characterize them by using simulation or bisimulation relations over the machine model and the user model. Second, we derive some properties which show relationship between the automation surprises and the bisimulation relation. Finally, we show that both a user-interface and a user model without automation surprises can be generated by utilizing the bisimulation algorithm.

2. MODEL OF HUMAN-MACHINE SYSTEMS

A mathematical model of a human-machine system is represented formally by composition of a machine model, a user-interface (information display), and a user model (a user manual) for operation as shown in Fig. 1. Usually, the machine model is given *a priori*, and both the user-interface and the user model are suitably designed in order to avoid automation surprises so that a friendly human-machine system is realized. First, we present mathematical representations of these components.

2.1 Machine model

We assume that a machine considered in this paper is a discrete event system modeled by a

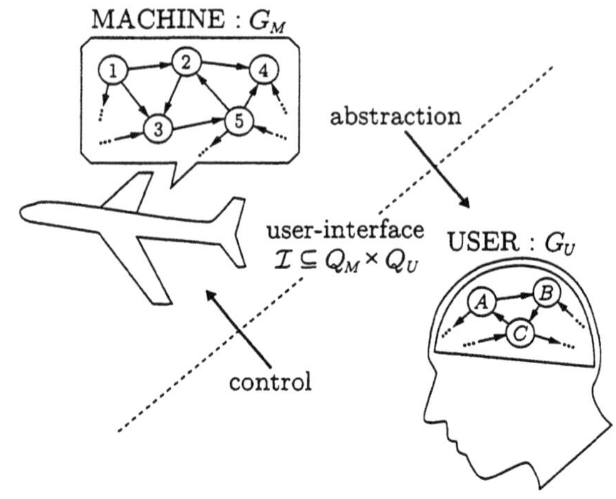

Fig. 1. Human-machine system.

deterministic finite automaton G_M:

$$G_M = (Q_M,\ \Sigma_M,\ \delta_M,\ q_{M,0}), \tag{1}$$

where Q_M is the set of states, Σ_M is the finite set of events, a partial function $\delta_M : Q_M \times \Sigma_M \to Q_M$ is the transition function, and $q_{M,0} \in Q_M$ is the initial state. Σ_M is partitioned into two subsets Σ_M^o and Σ_M^u as follows:

$$\Sigma_M = \Sigma_M^o \,\dot{\cup}\, \Sigma_M^u. \tag{2}$$

Σ_M^o is the set of observable events which correspond to commands given by the user. In other words, occurrence of every event in Σ_M^o is triggered by user's decision which is based on the user model. In this paper, events in Σ_M^o will be called observable events or commands alternatively. Σ_M^u is the set of unobservable events which are triggered by qualitative change of internal states of the machine. An example of machine model is shown in Fig. 2. In this figure, events α and $\beta \in \Sigma_M^o$ are commands given by the user and transitions by their occurrences are denoted by solid arrows while events μ and $\eta \in \Sigma_M^u$ represent qualitative change of the machine's states and transitions by them are denoted by dashed arrows. The user can observe occurrences of all observable events since they are triggered by the user. But, occurrences of unobservable events can be neither observed directly nor triggered.

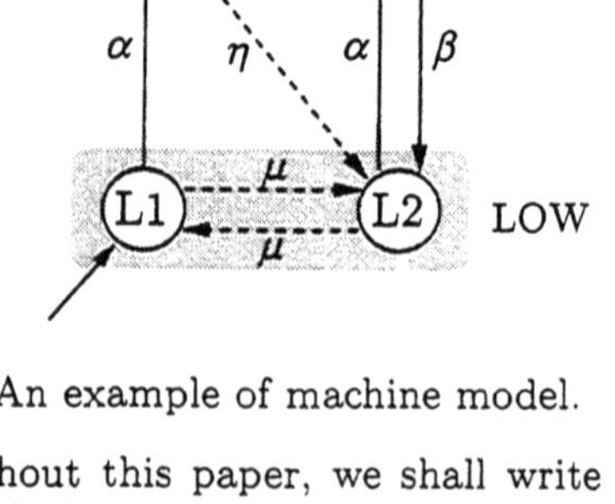

Fig. 2. An example of machine model.

Throughout this paper, we shall write $\delta_M(q,\sigma)!$ (*resp.* $\neg\delta_M(q,\sigma)!$) for any state $q \in Q_M$ and event

$\sigma \in \Sigma_M$ if $\delta_M(q,\sigma)$ is defined (*resp.* undefined). We also denote $q \xrightarrow{\sigma} q'$ if $\delta_M(q,\sigma) = q'$. These notations will be also used for other automata.

2.2 User-interface

The user-interface displays abstracted information on current states of the underlying machine and the user gives adequate commands based on both the information and the user model. Obviously, commands given by the user are useful information for user's perception of machine's current states. So, for simplicity, we assume that the event set Σ_M^o of observable events (commands) is also included in the information provided by the user-interface.

For example, in the machine model of Fig. 2, states L1 and L2 are abstracted as LOW, and H1 and H2 as HIGH. In this user-interface, the user knows whether the user-interface displays HIGH or LOW, but can not identify machine's internal state exactly. Thus, a state of the user-interface corresponds to a subset of machine's state set. We represent the correspondence as a binary relation as follows:

$$\mathcal{I} \subseteq Q_M \times Q_U, \tag{3}$$

where, Q_U is the set of states of the user-interface. Note that Q_U will be used as a set of states in the user model since the user gives commands based on states of the user-interface. In Fig. 2, the correspondence is given by $\mathcal{I}$ ={(L1, LOW), (L2, LOW), (H1, HIGH), (H2, HIGH)}. Note that, when a state transition by occurrence of an unobservable event changes a state of the user-interface, the user can recognize that some unobservable event occurs though the user can not identify which event occurs in general. In Fig. 2, the user can not know the occurrence of event μ at all while the user can know that some event occurs when event η occurs since the occurrence causes change of states of the user-interface.

2.3 User model

From the practical point of view, it is a difficult problem how to construct a user model (Norman, 1988). But, in this paper, we assume that the user can get an exact user model from a user manual and gives commands which are correct in the user model. All information the user can get is both the abstracted information of machine's states and commands through the user-interface. So, user's decision on given commands is based on states of the user-interface, which belong to the set Q_U introduced in the previous section, and user's prediction on the next state of the user-interface after giving a command in Σ_M^o is based on the user model. Moreover, since occurrences of unobservable events may cause changes of states of the user-interface, the user model must include such transitions, which can be described by nondeterministic ε-moves. Thus, the user model is described by the following nondeterministic finite automaton with ε-moves and the state set Q_U:

$$G_U = (Q_U, \Sigma_M^o \cup \{\varepsilon\}, \delta_U, q_{U,0}), \tag{4}$$

where ε is the empty string, a partial set-valued function $\delta_U : Q_U \times (\Sigma_M^o \cup \{\varepsilon\}) \rightarrow 2^{Q_U}$ is the transition function, and $q_{U,0} \in Q_U$ is the initial state. Since the unobservable events are not observed by the user, we introduce a projection $\Pi : \Sigma_M \rightarrow \Sigma_M^o \cup \{\varepsilon\}$:

$$\Pi(\sigma) = \begin{cases} \sigma & \text{if } \sigma \in \Sigma_M^o, \\ \varepsilon & \text{if } \sigma \in \Sigma_M^u. \end{cases} \tag{5}$$

Note that occurrences of unobservable events are classified into the following two cases:

(1) The occurrence of an unobservable event does not cause the change of the state in the user-interface, that is, for $(q_M, q_U) \in \mathcal{I}$, $q'_M \in Q_M$, and $\sigma_u \in \Sigma_M^u$ with $q_M \xrightarrow{\sigma_u} q'_M$, we have $(q'_M, q_U) \in \mathcal{I}$. In this case, the user can not know the occurrence of the unobservable event at all.
(2) The occurrence of an unobservable event causes the change of the state in the user-interface, that is, for $(q_M, q_U) \in \mathcal{I}$, $q'_M \in Q_M$, and $\sigma_u \in \Sigma_M^u$ with $q_M \xrightarrow{\sigma_u} q'_M$, we have $(q'_M, q_U) \notin \mathcal{I}$. In this case, the user can know that some event surely occurs.

Usually, the user manual does not mention about case (1) while it may mention case (2). We assume that, for any $q_U \in Q_U$, transition $q_U \xrightarrow{\varepsilon} q_U$ is defined in G_U. But, transition $q_U \xrightarrow{\varepsilon} q'_U$ with $q_U \neq q'_U$ is defined if the user manual notices it, otherwise undefined. In the following, for simplicity, we omit all self-loop transitions $q_U \xrightarrow{\varepsilon} q_U$ from figures of the user model.

In the user model, for simplicity, we assume that the user can predict the next state after giving commands uniquely (Degani and Heymann, 2002; Heymann and Degani, 2002). In other words, $|\delta_U(q_U, \sigma_o)| \leq 1$ for any $q_U \in Q_U$ and $\sigma_o \in \Sigma_M^o$, where $|\cdot|$ denotes the cardinality. However, transitions triggered by the empty string ε may be nondeterministic from Eq. (5).

Figure 3 shows an example of the user model. As mentioned before, machine's states are abstracted as LOW and HIGH in the user model of Fig. 3, where all self-loop transitions related to the empty string are omitted in G_U. Recall that α and $\beta \in \Sigma_M^o$ are commands (observable events) and μ and $\eta \in \Sigma_M^u$ are unobservable. In this example, the user manual is assume to notices that the ε-

move from HIGH to LOW, $\delta_U(\text{HIGH}, \varepsilon) = \text{LOW}$, can occur, otherwise it would be deleted.

In the following, a three-tuple $(G_M, G_U, \mathcal{I})$ of the machine model, the user model, and the binary relation will be called a human-machine system.

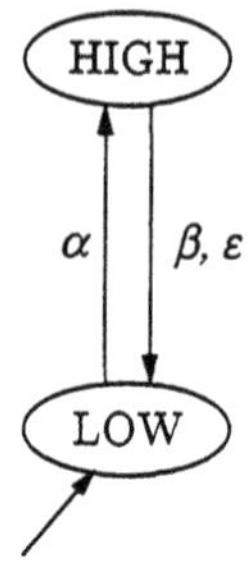

Fig. 3. An example of user model.

3. SIMULATION RELATION IN HUMAN-MACHINE SYSTEMS

In this section, we classify automation surprises into three cases, which will be defined formally by the binary relation $\mathcal{I}$. By applying a concept of simulation relation, we characterize each automation surprise. We make the following assumptions.

Assumption 1. In the human-machine system $(G_M, G_U, \mathcal{I})$, for all $(q_M, q_U) \in \mathcal{I}$, $q_M \xrightarrow{\sigma_o} q'_M$ $(\sigma_o \in \Sigma^o_M)$, there exits $q'_U \in Q_U$ such that $q_U \xrightarrow{\sigma_o} q'_U$ and $(q'_M, q'_U) \in \mathcal{I}$.

Assumption 2. In the human-machine system $(G_M, G_U, \mathcal{I})$, for every $q_M \in Q_M$, there exists $q_U \in Q_U$ such that $(q_M, q_U) \in \mathcal{I}$.

Intuitively, Assumption 1 means that, if a command $\sigma_o \in \Sigma^o_M$ is defined at a state in the machine model, the user can give the command at the corresponding state in the user-interface. Moreover, Assumption 2 means that the user-interface can work no matter which state the machine is in.

Automation surprises arise when some state transition in the machine is not transmitted to the user through the user-interface correctly or the machine can not follow the user's command. In the following, the inconsistencies of the state transitions caused by the commands and by the unobservable events in the machine are characterized by the pair $(q_M, q_U) \in \mathcal{I}$ and $(q_U, q_M) \in \mathcal{I}^{-1}$, respectively.

Definition 1. The human-machine system $(G_M, G_U, \mathcal{I})$ has a *blocking state* if the following condition holds: If $(q_M, q_U) \in \mathcal{I}$ and $q_M \xrightarrow{\sigma_u} q'_M$ for some $\sigma_u \in \Sigma^u_M$, then there does not exist $q'_U \in \delta_U(q_U, \varepsilon)$ such that $(q'_M, q'_U) \in \mathcal{I}$.

Definition 2. (Heymann and Degani, 2002; Degani and Heymann, 2002) The human-machine system $(G_M, G_U, \mathcal{I})$ has a *mode confusion* if the following condition holds: If $(q_U, q_M) \in \mathcal{I}^{-1}$ and $q_U \xrightarrow{\sigma_o} q'_U$ for some $\sigma_o \in \Sigma^o_M$, then there exits q'_M such that $q_M \xrightarrow{\sigma_o} q'_M$ and $(q'_U, q'_M) \notin \mathcal{I}^{-1}$

Definition 3. (Ukawa *et al.*, 2003) The human-machine system $(G_M, G_U, \mathcal{I})$ has a *refusal state* if $(q_U, q_M) \in \mathcal{I}^{-1}$ and $\delta_U(q_U, \sigma_o)! \wedge \neg\delta_M(q_M, \sigma_o)!$ for some $\sigma_o \in \Sigma^o_M$

Intuitive interpretations of the above definitions are as follows:

- A blocking state represents a situation that, when an unobservable event occurs in the machine, the corresponding ε-move is not defined in the user model adequately.
- We will say that a mode confusion occurs when the updated state of the user-interface after giving a command is different from a state predicted by the user model.
- A refusal state represents a situation that, when the user gives some command to the machine, it is prohibited from occurring in the machine. Practically, the machine beeps when the prohibited command is given.

For example, in Fig. 3, we consider a pair (HIGH, H1) $\in \mathcal{I}^{-1}$ of states in G_U and G_M. Suppose that the user gives the command β in order to change the state from HIGH to LOW in the user-interface. In the user model, the transition HIGH $\xrightarrow{\beta}$ LOW is defined, but the corresponding transition $\delta_M(\text{H1}, \beta)$ is not defined in the machine model. Thus, the command β is sometimes refused in the state HIGH . Obviously, the user-interface and the user model in Fig. 3 are inadequate.

Here, we define a simulation relation (Milner, 1989; Barrett and Lafortune, 1998) of the machine model by the user model.

Definition 4. For the human-machine system $(G_M, G_U, \mathcal{I})$, a simulation relation of G_M by G_U is a binary relation $\mathcal{I}$ such that:

1) For every $q_M \in Q_M$, there exists $q_M \in q_U$ such that $(q_M, q_U) \in \mathcal{I}$.
2) If $(q_M, q_U) \in \mathcal{I}$ and $q_M \xrightarrow{\sigma} q'_M$ $(\sigma \in \Sigma_M)$, then there exists $q'_U \in Q_U$ such that $q_U \xrightarrow{\Pi(\sigma)} q'_U$ and $(q'_M, q'_U) \in \mathcal{I}$.
3) If $(q_{M,0}, q_U) \in \mathcal{I}$, then $q_U = q_{U,0}$, and if $(q_M, q_{U,0}) \in \mathcal{I}$, then $q_M = q_{M,0}$.

If $\mathcal{I}$ is a simulation relation of G_M by G_U, we will say that G_U simulates G_M by $\mathcal{I}$, denoted by $G_M \preceq_{\mathcal{I}} G_U$.

Definition 5. For the human-machine system (G_M, G_U, $\mathcal{I}$), a bisimulation relation of G_M and G_U is a binary relation $\mathcal{I}$ such that $G_M \preceq_{\mathcal{I}} G_U$ and $G_U \preceq_{\mathcal{I}^{-1}} G_M$.

If $\mathcal{I}$ is a bisimulation relation of G_M and G_U, we will say that G_M and G_U are bisimilar by $\mathcal{I}$, denoted by $G_M \equiv_{\mathcal{I}} G_U$.

Remark 1. Without loss of generality, we can assume that G_M and G_U have the same initial state (e.g. "the machine does not work") in order to guarantee that the third condition of Definition 4 always holds.

Lemma 1. The human-machine system (G_M, G_U, $\mathcal{I}$) has no blocking states, if and only if $G_M \preceq_{\mathcal{I}} G_U$.

For a given machine model G_M and a user model G_U, let G_M^o and G_U^o be subautomata of G_M and G_U, respectively, by eliminating all transitions labeled by unobservable events or the empty string:

$$G_M^o = (Q_M,\ \Sigma_M^o,\ \delta_M^o,\ q_{M,0}), \quad (6)$$
$$G_U^o = (Q_U,\ \Sigma_M^o,\ \delta_U^o,\ q_{U,0}), \quad (7)$$

where transition functions $\delta_i^o : Q_i \times \Sigma_M^o \to Q_i$ ($i = U, M$) are defined by

$$\delta_i^o(q_i,\ \sigma) = \begin{cases} \delta_i(q_i,\ \sigma) & \text{if } \sigma \in \Sigma_M^o, \\ \text{undefined otherwise.} \end{cases}$$

Since all transitions labeled by observable events are deterministic, both G_M^o and G_U^o are deterministic. Both a mode confusion and a refusal state depend on commands, and we will show that the existence for such automation surprises can be checked by using both G_M^o and G_M^o.

Lemma 2. The human-machine system (G_M, G_U, $\mathcal{I}$) has neither mode confusions nor refusal states, if and only if $G_U^o \preceq_{\mathcal{I}^{-1}} G_M^o$.

Since G_M^o and G_U^o are subautomata of G_M and G_U, it is clear that if $G_M \preceq_{\mathcal{I}} G_U$, then $G_M^o \preceq_{\mathcal{I}} G_U^o$. From Lemmas 1 and 2, we have the following theorem.

Theorem 1. The human-machine system (G_M, G_U, $\mathcal{I}$) has no automation surprises, if and only if $G_M \preceq_{\mathcal{I}} G_U$ and $G_U^o \equiv G_M^o$.

4. DESIGN OF USER-INTERFACE

In the previous section, we show necessary and sufficient conditions for nonexistence of automation surprises. Based on the conditions, we propose a design method of both an adequate user-interface and a user model. In order to construct an adequate user-interface and a user model for a given machine model, it is sufficient to compute both a binary relation $\mathcal{I}$ and an automaton G_U satisfying Theorem 1. So, we propose a design algorithm for them by utilizing a quotient system (Henzinger, 1995).

Set Machine model G_M, equivalence relation $\sim^0$
set $Q_M/\sim^0 = \{Q_M \backslash \{q_{M,0}\}, q_{M,0}\}$
set $i = 0$, $G_U^0 = G_M/\sim^0$
while $\exists \mathcal{Q}, \mathcal{Q}' \in Q_M/\sim^i$ and $\sigma_o \in \Sigma_M^o$ such that $\emptyset \neq (\mathcal{Q} \cap Pre_{\sigma_o}(\mathcal{Q}')) \neq \mathcal{Q}$
  set $\mathcal{Q}_1 = \mathcal{Q} \cap Pre_{\sigma_o}(\mathcal{Q}')$, $\mathcal{Q}_2 = \mathcal{Q} \backslash Pre_{\sigma_o}(\mathcal{Q}')$
  refine $Q_M/\sim^{i+1} = (Q_M/\sim^i \backslash \{\mathcal{Q}\}) \cup \{\mathcal{Q}_1, \mathcal{Q}_2\}$
  set $G_U^{i+1} = G_M/\sim^{i+1}$
  set $i = i + 1$
end while
return G_U^i

Fig. 4. Bisimulation based design algorithm.

Definition 6. For a machine model $G_M = (Q_M,\ \Sigma_M,\ \delta_M,\ q_{M,0})$ and an equivalence relation $\sim \subseteq Q_M \times Q_M$, a quotient system of G_M is defined as $G_M/\sim = (Q_M/\sim,\ \Sigma_M^o \cup \{\varepsilon\},\ \delta_{M\sim},\ \tilde{q}_{M,0}/\sim)$, where

- $Q_M/\sim$ is a set of equivalence classes of Q_M by $\sim$,
- for every $\mathcal{Q}$ and $\mathcal{Q}' \in Q_M/\sim$, $\mathcal{Q}' \in \delta_{M\sim}(\mathcal{Q}, \Pi(\sigma))$ if there exist $q_M \in \mathcal{Q}$ and $q'_M \in \mathcal{Q}'$ such that $\delta_M(q_M,\ \sigma) = q'_M$, and
- $\tilde{q}_{M,0}/\sim = \{q \in Q_M \mid q \sim q_{M,0}\}$.

Note that $\tilde{q}_{M,0}/\sim = \{q_{M,0}\}$ by Remark 1. Let $G_U = G_M/\sim$, then we introduce a binary relation $\mathcal{I}_\sim$ over $Q_M \times Q_U$ induced by $\sim$ as follows:

$$\mathcal{I}_\sim = \{(q_M, q_U) \in Q_M \times Q_U \mid q_M \in q_U\}.$$

Then, we know that a user model G_U always simulates the machine model G_M by $\mathcal{I}_\sim$.

From Lemma 1, the binary relation $\mathcal{I}_\sim$ and the automaton G_U provide a user-interface and a user model with no blocking states. Moreover, $G_M^o \preceq_{\mathcal{I}_\sim} G_U^o$ is also true. Thus, from Theorem 1, an adequate user-interface and a user model without automation surprises can be obtained by a bisimilar relation $\mathcal{I}_\sim^*$, which is finer than $\mathcal{I}_\sim$, such that $G_M^o \equiv_{\mathcal{I}_\sim^*} G_U^o$. For $\sigma_o \in \Sigma_M^o$, let

$$Pre_{\sigma_o}(\mathcal{Q}') = \{q_M \in Q_M \mid q_M \xrightarrow{\sigma_o} q'_M, q'_M \in \mathcal{Q}'\}.$$

Then, by applying the bisimulation algorithm (Henzinger, 1995), we can construct an adequate user-interface and a user model without automation surprises (Fig. 4).

In general, a user model may be nondeterministic when we construct it as a quotient system of the machine model. It is undesirable that it is nondeterministic with respect to an observable event since the user can not predict the next state

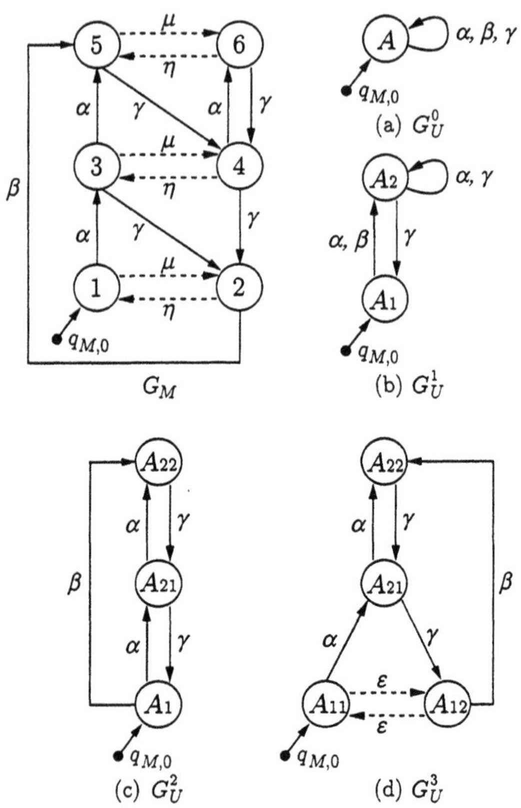

Fig. 5. Computing the user-interface and the user model.

uniquely after executing a command. The user model computed by the algorithm is deterministic with respect to any observable event.

We consider a machine model shown in Fig. 5, where $\Sigma_M^o = \{\alpha, \beta, \gamma\}$ and $\Sigma_M^u = \{\eta, \mu\}$. First, we set an initial binary relation as $\mathcal{I}_\sim^0 = \{(1, A), (2, A), (3, A), (4, A), (5, A), (6, A)\}$, where $A = \{1, 2, 3, 4, 5, 6\}$. We obtain the initial user model $G_U^0 = G_M/\sim^0$ as shown in Fig. 5(a). By Lemma 1, (G_M, G_U^0) has no blocking state on $\mathcal{I}_\sim^0$. However, for transition $A \xrightarrow{\gamma} A$, we have $A \neq (A \cap Pre_\gamma(A)) = \{3, 4, 5, 6\}$, which means G_M^o and G_U^o are not bisimilar. Next, the state A is partitioned into two states $A_1 = \{1, 2\}$ and $A_2 = \{3, 4, 5, 6\}$, and G_U^1 is shown in Fig. 5(b). After iterations, we finally have $G_M \preceq_{\mathcal{I}_\sim^3} G_U^3$ and $G_M^o \equiv_{\mathcal{I}_\sim^3} G_U^{3o}$ (Fig. 5(d)). Thus, an adequate user-interface and a user model are obtained.

5. CONCLUSIONS

In this paper, we studied interaction between a user and discrete event systems through a user-interface and classified automation surprises into three cases. First, we derive necessary and sufficient conditions for the nonexistence of the automation surprises using simulation and bisimulation relations between the machine model and the user model. Next, we propose a design algorithm for a user-interface and a user model without automation surprises. It is future work to extend the proposed algorithm to timed systems and hybrid systems.

REFERENCES

G. Barrett and S. Lafortune (1998). Bisimulation, the supervisory control problem and strong model matching for finite state machines. *Discrete Event Dynamic Systems* **8**, 377–429.

A. Degani (1996). Modeling Human-Machine Systems: On Modes, Error, and Patterns of Interaction. Ph.d. dissertation. Georgia Institute of Technologyy.

A. Degani. and M. Heymann (2002). Formal verification of human-automation interaction. *Human Factors* **44**(1), 28–43.

T. A. Henzinger (1995). Hybrid automata with finite bisimulations. In: *Proceedings of the 22nd International Colloquium on Automata, Languages, and Programming (ICALP)*. pp. 324–335. LNCS 944. Springer-Verlag.

M. Heymann and A. Degani (2002). On abstraction and simplification in the design of human-automation interaction. *NASA Technical memorandum 211397, NASA Ames Research Center, Moffett Field, CA.*

R. Milner (1989). *Communication and Concurrency.* Prentice Hall. New Jersey.

D. A. Norman (1988). *The Psychology of Everyday Things.* Basic Books. New York.

M. Oishi, I. Hwang and C. Tomlin (2003). Immediate observability of discrete event systems with application to user-interface design. In: *Proceedings of the 42nd IEEE Conference on Decision and Control.* pp. 2665–2672.

E. Palmer (1995). Oops, it didn't arm. - a case study of two automation surprises. In: *Proceedings of the 8th International Symposium on Aviation Psychology.* pp. pp. 227–232.

R. Parasuraman, T. B. Sheridan and C. D. Wickens (2000). A model for types and levels of human interaction with automation. *IEEE Transactions on Systems, Man, and Cybernetics-Part A: Systems and Humans* **30**(3), 286–297.

J. Rushby (2002). Using model checking to help discover mode confusions and other automation surprises. *Reliability Engineering and System Safety* **75**, 167–177.

N. Sarter, D. Woods and C. Billings (1997). Automation surprises. In: *Handbook of Human Factors and Ergonomics* (G. Salvendy, Ed.). pp. 1926–1943. Wiley. New York.

Y. Ukawa, T. Ushio and M. Adachi (2003). Formal detection of mode confusion in human-machine interaction. In: *Proceedings of the 1st International Symposium on Systems & Human Science.* Osaka, Japan. pp. 309–314.

ELSEVIER
IFAC
PUBLICATIONS
www.elsevier.com/locate/ifac

MODULAR SUPERVISION OF HYBRID SYSTEMS: A DES APPROACH

André B. Leal *, **José E. R. Cury** **

* *Depto. de Eng. Elétrica, Universidade do Estado de Santa Catarina, Brasil, E-mail: leal@joinville.udesc.br*
** *Depto. de Automação e Sistemas, Universidade Federal de Santa Catarina, Brasil, E-mail: cury@das.ufsc.br*

Abstract: This paper presents the formulation and solution of a modular supervisory control problem for a class of hybrid systems in which threshold-crossing events in the continuous state space force discrete state transitions. The continuous dynamics are determined by a discrete condition, which depends on the current discrete state of the system. The problem is to construct modular supervisors, each to satisfy a particular specification given in terms of sequences of threshold events, in such a way that the joint action of these supervisors results in an optimal control action. The supervisors for the hybrid system are obtained using the theory of supervisor synthesis for discrete event systems. A condition under which the proposed approach can be applied is given and an illustrative example is presented. *Copyright © 2004 IFAC*

Keywords: Hybrid systems; supervisory control; discrete-event systems; decentralized control.

1. INTRODUCTION

This paper reviews the supervisory control approach for hybrid systems (*HSs*), discussed in (González *et al.*, 2001), and extends it through the investigation of a modular control problem. The modular approach has two important motivations: decreasing the computational complexity if compared to the monolithic counterpart; and increasing flexibility. The class of *HSs* considered, called *threshold-event-driven hybrid systems*, is such that changes in the discrete state can occur only when continuous states variables encounter specified thresholds; the continuous dynamics are determined by a discrete condition, which depends on the current discrete state of the system. The objective is to synthesize modular supervisors, each to a particular specification, such that the joint action of these supervisors is optimal in the sense of being minimally restrictive. A similar control problem was studied in (Moor *et al.*, 2001), but while in that work the problem is stated within the *behavioral system theory* of Willems (1991), here it is presented using the formalism of condition/event (*C/E*) systems (Sreenivas and Krogh, 1991). This formalism provides a framework for defining continuous-time systems as the interconnection of subsystems with discrete-valued input and output signals and its use results in the following benefits: the hybrid system model can be easily built using a custom graphical user interface in the *Simulink/Matlab* environment; a discrete state model (approximation) for the hybrid system behavior can be computed running *CheckMate/Matlab*; and the synthesis of supervisors can be accomplished using the supervisory control toolbox developed by González *et al.* (2001) and extended in the context of this work to consider the approach introduced here. Furthermore, the considered approach deals explicitly with problems whose specifications are based on sequences of events generated when the continuous state crosses specified thresholds. This paper also discusses the use of a finite state automaton, called *approximating automaton* (Cury *et al.*, 1998), to describe the approximated logical behavior of the *HS*: it introduces a condition under which a solution to the supervisory control problem obtained for the approximating automaton is also a solution to the original problem, taking both the monolithic and the modular approach.

This paper is organized as follows. In Sect. 2 the supervisory control problem for *HSs* is revisited and some new concepts are presented. In Sect. 3 the modular control problem for *HSs* is formulated within the formalism of *C/E* systems and solved using the theory of supervisory control for discrete event systems (*DESs*). Section 4 discusses the use of a finite state automaton to describe the approximated logical behavior of the *HS*. Section 5 presents an illustrative example and Sect. 6 summarizes the contribution of this paper.

2. SUPERVISORY CONTROL OF HS

Consider the supervisory control scheme shown in Fig. 1 (González *et al.*, 2001). The hybrid plant $\mathcal{H}$

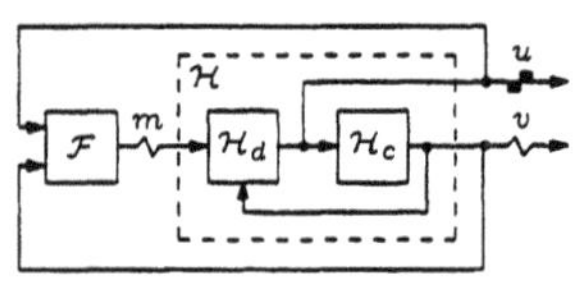

Fig. 1. Supervisory control scheme.

is a *C/E* system (Sreenivas and Krogh, 1991) composed of the interconnection of a continuous dynamic subsystem $\mathcal{H}_c$ and a discrete dynamic subsystem $\mathcal{H}_d$. The input signal for $\mathcal{H}_c$ is a condition signal $u(\cdot)$, a piecewise constant, right continuous signal, taking on discrete values in the finite set of conditions U. The space comprising all condition signals $u(\cdot)$ in $[0,\infty)$ is denoted by $\mathcal{U}$. As depicted in Fig. 2, $\mathcal{H}_c$ consists of two types of subsystems: a *switched continuous system (SCS)* and a *threshold event generator (TEG)*. The continuous state

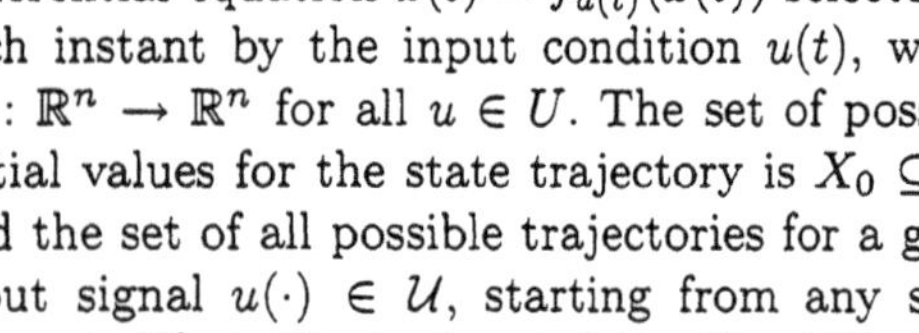

Fig. 2. Continuous dynamic subsystem $\mathcal{H}_c$.

trajectory, $x(\cdot)$, evolves in $\mathbb{R}^n$ according to the differential equation $\dot{x}(t) = f_{u(t)}(x(t))$ selected at each instant by the input condition $u(t)$, where $f_u : \mathbb{R}^n \to \mathbb{R}^n$ for all $u \in U$. The set of possible initial values for the state trajectory is $X_0 \subseteq \mathbb{R}^n$ and the set of all possible trajectories for a given input signal $u(\cdot) \in \mathcal{U}$, starting from any state in a set $X' \subset X_0$, is denoted by $\mathcal{X}_{u(\cdot)}(X')$. The function $g : X \to \mathbb{R}^m$ generates the continuous output signal $y(\cdot)$ from the state trajectory. Each component of y is compared to a threshold defined by a *threshold vector*, $T \in \mathbb{R}^m$, and a *zero detector* generates the event output signal, which is defined as $v(0) = \eta$ and for $t > 0$ and $i = 1, 2, ..., m$ as:

$$v_i(t) = \begin{cases} 1 & \text{if } \; y_i(t) - T_i = 0 \wedge (\exists \Delta > 0)(\forall \delta \in (0,\Delta)) : y_i(t-\delta) - T_i < 0 \\ 0 & \text{otherwise} \end{cases} \quad (1)$$

For the instants at which any $v_i(t) = 1$, it is said that a *threshold event* occurs. Note that for $t = 0$ it is considered the existence of an *initialization event* η. Thus, the event signal takes on nonzero values in the set $V^+ = \{\eta\} \cup V$ at isolated points in time, where $V = \{\{0,1\}^m - \{0\}^m\}$ is the *set of threshold events*. The space comprising all event signals $v(\cdot)$ in $[0,\infty)$ is denoted by $\mathcal{V}$.

The subsystem $\mathcal{H}_d$ is a purely discrete dynamic system which restricts the range of allowable input signals to $\mathcal{H}_c$, modeling physical constraints of the continuous subsystem. It is driven by two input event signals, $v(\cdot) \in \mathcal{V}$ and $m(\cdot) \in \mathcal{M}$, where $m(\cdot)$ is an external control input signal taking on values in $M = 2^U$. It is assumed that $\mathcal{H}_d$ can change the input condition signal for $\mathcal{H}_c$ only when a threshold event is observed. It is also assumed that $m(t) \neq \emptyset$ only when $v(t) \neq 0$, which makes discontinuities in $m(\cdot)$, $v(\cdot)$ and $u(\cdot)$ synchronous. The initialization event η is used to determine the range of allowable input signals to $\mathcal{H}_c$ at $t = 0$.

The continuous subsystem, $\mathcal{H}_c$, is defined as a subset of the *time synchronous cross product* of $\mathcal{V}$ and $\mathcal{U}$, denoted by $\mathcal{V} \otimes \mathcal{U}$, which represents the set of all pairs $(v(\cdot), u(\cdot)) \in \mathcal{V} \times \mathcal{U}$, such that discontinuities in $u(\cdot)$ occur only at instants when $v(\cdot)$ is nonzero. A pair $(v(\cdot), u(\cdot)) \in \mathcal{H}_c$ if and only if there exists a state trajectory $x(\cdot) \in \mathcal{X}_{u(\cdot)}(X_0)$ such that the resulting event signal is $v(\cdot)$. In a similar way, the discrete subsystem is defined as $\mathcal{H}_d \subseteq \mathcal{V} \otimes \mathcal{M} \otimes \mathcal{U}$. The hybrid plant $\mathcal{H} \subseteq \mathcal{V} \otimes \mathcal{M} \otimes \mathcal{U}$ is obtained by cascade and feedback connections of $\mathcal{H}_c$ and $\mathcal{H}_d$ (Sreenivas and Krogh, 1991).

The supervisor $\mathcal{F}$ observes the condition $u(\cdot)$ and event $v(\cdot)$ signals from $\mathcal{H}$ and applies a control input $m(\cdot)$ to $\mathcal{H}_d$ restricting the range of possible input conditions to $\mathcal{H}_c$. It is assumed that the supervisor outputs a nonempty value only when a threshold event is observed, which is in agreement with the synchronicity of signals stated before. A nonempty value for $m(t)$ should be interpreted as an updating (at time t) on the set of conditions that $\mathcal{H}_d$ can output to $\mathcal{H}_c$. It is also assumed that the feedback system of Fig. 1 does not lead to *chattering*. The *C/E* supervisor is defined as a causal system (Sreenivas and Krogh, 1991) $\mathcal{F} \subset \mathcal{H}$. Moreover, meaningful supervisors can disable any condition but can not disable all conditions that can be chosen by $\mathcal{H}_d$ for an active event, as follows.

Definition 1. A *C/E* supervisor $\mathcal{F}$ is said to be consistent for a plant $\mathcal{H}$ if $\forall (v(\cdot), m(\cdot), u(\cdot)) \in \mathcal{F}$, $(\forall t : v(t) \neq 0)\, u(t) \in m(t)$.

The closed-loop *C/E* system is $\mathcal{F}/\mathcal{H} \subseteq \mathcal{V} \otimes \mathcal{U}$, which is built by cascade and feedback connections of $\mathcal{F}$ and $\mathcal{H}$.

The logical behavior of *C/E* systems can be expressed in terms of languages. Thus, for a *C/E* system $\mathcal{D} \subseteq \mathcal{V} \otimes \mathcal{U}$, the language generated by $\mathcal{D}$, denoted by $\mathcal{L}(\mathcal{D})$, is defined over finite length strings obtained by the prefix-closure of all strings composed of sequences of values assumed by the pairs $(v(\cdot), u(\cdot)) \in \mathcal{V} \otimes \mathcal{U}$ at their points of discontinuity. As the signals $v(\cdot)$ and $u(\cdot)$ assume nonzero values in the discrete sets V^+ and U, respectively, then $\mathcal{L}(\mathcal{D}) \subseteq (V^+ \times U)^*$. In order to represent the open-loop behavior of $\mathcal{H}$, it is considered that $\forall (v(\cdot), m(\cdot), u(\cdot)) \in \mathcal{H}$, $m(t) = U$ for all $t \in [0, \infty) : v(t) \neq 0$. Thus, the open-loop language $\mathcal{L}(\mathcal{H})$ can be written as $\mathcal{L}(\mathcal{H}) \subseteq (V^+ \times U)^*$, as it does not depend on the signal $m(\cdot)$ from $\mathcal{F}$.

Next, the supervisor synthesis problem for hybrid systems (SSHS) can be stated. First, the projection of a language $K \subseteq (V^+ \times U)^*$ onto V^* is defined as $P_V(K) = \{t \in V^* : (\exists s \in K)\ P_V(s) = t\}$. Then, $P_V : (V^+ \times U)^* \to V^*$ erases the symbols η and the symbols $u \in U$ from strings in $(V^+ \times U)^*$.

Problem 1. (Supervisor Synthesis for HSs - SSHS) Let $\mathcal{H}$ be the *C/E* model for a hybrid plant and $E \subseteq V^*$ be a specification for the closed-loop behavior. Find a *C/E* supervisor $\mathcal{F}$, consistent for $\mathcal{H}$, such that $P_V[\mathcal{L}(\mathcal{F}/\mathcal{H})] \subseteq E$.

Next, the SSHS problem is translated to a purely discrete event control framework and a solution is proposed. It is then necessary to define two new sets. The *set of active events* in $K \subseteq (V^+ \times U)^*$ after $w \in \overline{K}$ is given by $V_K(w) = \{v \in V^+ : (\exists u \in U)\ w \circ vu \in \overline{K}\}$. The *set of active conditions* in K after $w \in \overline{K}$ and for a given $v \in V_K(w)$ is defined as $U_K(w, v) = \{u \in U : w \circ vu \in \overline{K}\}$. The *DES* model with control structure for $\mathcal{H}$ is given by $H = (L, \Gamma)$, where $L = \mathcal{L}(\mathcal{H})$ and the control structure is a map $\Gamma : L \to 2^{2^{V^+ \times U}}$ such that $\forall w \in L$, $\Gamma(w) = \{\gamma \in 2^{V^+ \times U} : (\forall v \in V_L(w))(\exists u \in U_L(w, v)) : vu \in \gamma\}$.

The *DES* supervisor F for the plant H is defined by the map $F : L \to 2^{V^+ \times U}$. The language $L(F/H)$ represents the behavior of the plant H under control of the supervisor F, and is recursively defined as: *(i)* $\epsilon \in L(F/H)$; and *(ii)* $w \circ vu \in L(F/H) \Leftrightarrow w \in L(F/H) \wedge w \circ vu \in L \wedge vu \in F(w)$.

Definition 2. A *DES* supervisor F is said to be consistent for a plant H if $\forall w \in L(F/H), F(w) \in \Gamma(w)$.

Thus, F is *consistent* for H if for any string $w \in L(F/H)$, and for every event v that can occur after w, there is at least one input condition u enabled by F, such that $w \circ vu \in L$.

González *et al.* (2001) present an algorithm to construct a *C/E* supervisor $\mathcal{F}_F$ from the *DES* supervisor F such that $L(F/H) = \mathcal{L}(\mathcal{F}_F/\mathcal{H})$. F can be seen as a discrete implementation of $\mathcal{F}_F$. This result indicates that the SSHS problem can be solved by solving an equivalent *DES* supervisory control problem, which is presented next.

Problem 2. (Supervisory Control Problem - SCP) Let H be the *DES* model for the hybrid plant $\mathcal{H}$ and $E \subseteq V^*$ a specification for the closed-loop behavior. Find a *DES* supervisor F, consistent for H, such that $P_V[L(F/H)] \subseteq E$.

Observe that while *nonblocking* is the required property for supervisors in classical modular control problems for *DES*, the feature to be met here is the capability of the supervisor to always react to physically possible threshold events, i.e., the consistency of the supervisor. The *vu-controllability* of languages, as defined in the following, plays an important role in the solution of Problem 2.

Definition 3. Given the languages $K \subseteq L \subseteq (V^+ \times U)^*$, K is said to be vu-controllable w.r.t. (with respect to) L if $(\forall w \in \overline{K})\ V_L(w) = V_K(w)$.

As the specifications are given in terms of languages in V^*, we introduce the following concept of *v-controllability*.

Definition 4. Given the languages $L \subseteq (V^+ \times U)^*$ and $E \subseteq P_V(L)$, E is said to be v-controllable w.r.t. L if there exists a language $K \subseteq L$ vu-controllable w.r.t. L and such that $P_V(K) = E$.

Consider the language $L = [\overline{v_1 u_2 \circ (v_3 u_1 + v_2 u_2)}] + [\overline{v_1 u_1 \circ (v_2 u_1 + v_2 u_2 \circ v_3 u_2)}]$. $E_1 = \overline{v_1 \circ v_2}$ is v-controllable w.r.t. L because there is a vu-controllable language $K_1 = \overline{v_1 u_1 \circ v_2 u_1}$ such that $P_V(K_1) = E_1$. The following propositions establish conditions for the existence of *DES* supervisors for H.

Proposition 1. Given a plant H and a language $K \subseteq L \subseteq (V^+ \times U)^*$, $K \neq \emptyset$, there exists a *DES* supervisor F such that $L(F/H) = K$ if and only if K is prefix-closed and vu-controllable w.r.t. L.

Proposition 2. Given a plant H and a language $E \subseteq V^*$, $E \neq \emptyset$, there exists a *DES* supervisor F such that $P_V[L(F/H)] = E$ if and only if E is prefix-closed and v-controllable w.r.t. L.

Consider an arbitrary language $K \subseteq L \subseteq (V^+ \times U)^*$ and let $\mathbb{C}_{vu}(K)$ be the set of all sublanguages of K that are vu-controllable w.r.t. L. According to González *et al.* (2001), the control structure Γ for L is closed under union for each $w \in L$. Therefore, a unique *supremal vu-controllable language* contained in K (denoted by $Sup\mathbb{C}_{vu}(K)$) always exists. Consider now a language $E \subseteq P_V(L) \subseteq V^*$ and let $\mathbb{C}_v(E)$ be the set of all sublanguages of E that are v-controllable w.r.t. L. The following proposition establishes results on the existence of a *supremal v-controllable language* and indicates how to compute it.

Proposition 3. Given the languages $L \subseteq (V^+ \times U)^*$ and $E \subseteq P_V(L)$, the set $\mathbb{C}_v(E)$ has a unique supremal v-controllable language, denoted by $Sup\mathbb{C}_v(E)$, which is given by $Sup\mathbb{C}_v(E) = P_V[Sup\mathbb{C}_{vu}(K)]$, where $K = P_V^{-1}(E) \cap L$.

Corollary 1. Given a plant H and a prefix-closed language $E \subseteq V^*$, $E \neq \emptyset$, there exists a supervisor F such that $P_V[L(F/H)] = Sup\mathbb{C}_V(E)$.

Returning to the SCP, the theory introduced before indicates that its solution requires the execution of the following steps: *(i)* Given $E \subseteq P_V(L)$, obtain $K = P_V^{-1}(E) \cap L$; *(ii)* Compute $Sup\mathbb{C}_{VU}(K)$ (Leal and Cury, 2002); *(iii)* The DES supervisor F is implemented by an automaton such that $L(F/H) = Sup\mathbb{C}_{VU}(K)$. The following results bring a formal solution to the SCP.

Theorem 1. The SCP is solvable if and only if $Sup\mathbb{C}_V(E) \neq \emptyset$.

Corollary 2. The SSHS problem is solvable if and only if the SCP is solvable. Furthermore, given the DES supervisor F as a solution to the SCP, the C/E supervisor $\mathcal{F}_F$ obtained by Algorithm 1 presented in (González *et al.*, 2001) is the corresponding solution to the SSHS problem.

3. MODULAR CONTROL OF HS

Consider the modular supervisory control scheme for HSs shown in Fig. 3. Given $m_1(\cdot), m_2(\cdot) \in \mathcal{M}$ two synchronous event signals taking on values in $M = 2^U$, we define the *synchronous intersection* $m_1(\cdot) \circledcap m_2(\cdot)$ of these signals as the event signal $m(\cdot)$ such that $m(t) = m_1(t) \cap m_2(t), t \in [0, \infty)$.

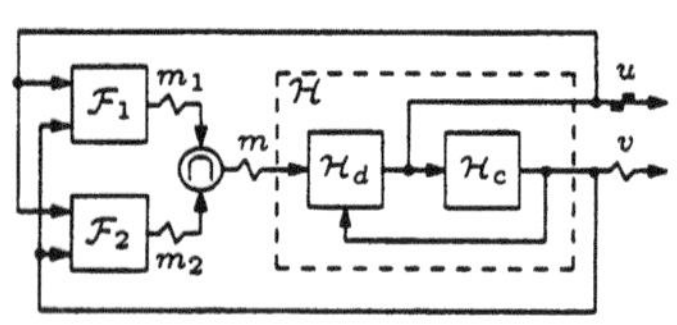

Fig. 3. Modular control architecture.

Definition 5. Let $\mathcal{F}_1, \mathcal{F}_2$ be consistent C/E supervisors for a hybrid plant $\mathcal{H}$. The conjunction of $\mathcal{F}_1$ and $\mathcal{F}_2$ is defined as a C/E system $\mathcal{F}_1 \wedge \mathcal{F}_2$ such that $(v(\cdot), m(\cdot), u(\cdot)) \in (\mathcal{F}_1 \wedge \mathcal{F}_2) \Leftrightarrow m(\cdot) = m_1(\cdot) \circledcap m_2(\cdot)$, where $(v(\cdot), m_i(\cdot), u(\cdot)) \in \mathcal{F}_i, i = 1, 2$.

Problem 3. (Modular Synthesis for HSs – MSHS) Let $\mathcal{H}$ be the C/E model for a hybrid plant and $E_1, E_2 \subseteq V^*$ be specifications for the closed loop behavior. Find C/E supervisors $\mathcal{F}_i$, consistent for $\mathcal{H}$, such that $P_V[\mathcal{L}(\mathcal{F}_i/\mathcal{H})] \subseteq E_i$ $(i = 1,2)$, in such a way that $\mathcal{F}_1 \wedge \mathcal{F}_2$ is a consistent supervisor for $\mathcal{H}$ and that $P_V[\mathcal{L}((\mathcal{F}_1 \wedge \mathcal{F}_2)/\mathcal{H})] \subseteq E_1 \cap E_2$.

The MSHS problem consists therefore of synthesizing C/E supervisors, each to satisfy a particular specification, and implementing global control by the conjunction of these supervisors. Each supervisor is designed as explained in the previous section. In the following, we employ the DES theory to propose a solution to this problem.

Definition 6. Let F_1 and F_2 be consistent DES supervisors for a plant $H = (L, \Gamma)$. The conjunction of F_1 and F_2 is defined as a map $(F_1 \wedge F_2) : L \to 2^{V^+ \times U}$ such that for $w \in L$, $vu \in (F_1 \wedge F_2)(w) \Leftrightarrow vu \in (F_1(w) \cap F_2(w))$.

Thus, after a string $w \in L$, a pair vu is enabled by $F_1 \wedge F_2$ if and only if vu is enabled by both F_1 and F_2 simultaneously.

The next definition is crucial for the establishment of conditions under which the conjunction of two supervisors leads to a consistent supervisor.

Definition 7. Two languages $K_1, K_2 \subseteq (V^+ \times U)^*$ are said to be interconsistent if:

$$(\forall w \in \overline{K_1} \cap \overline{K_2}) V_{K_1}(w) \cap V_{K_2}(w) = V_{K_1 \cap K_2}(w) \quad (2)$$

In words, K_1 and K_2 are *interconsistent* if after any string $w \in \overline{K_1} \cap \overline{K_2}$, the events that can occur in both languages can occur also in $K_1 \cap K_2$.

Proposition 4. Given a plant $H = (L, \Gamma)$, two specifications $K_1, K_2 \subseteq L$, and two DES supervisors F_1, F_2 such that $L(F_i/H) = K_i$ $(i = 1, 2)$, $F_1 \wedge F_2$ is a consistent supervisor for H if and only if K_1 and K_2 are interconsistent.

Let $L(F_i/H) = K_i$, $i = 1, 2$. It can be proved that if $F_1 \wedge F_2$ is consistent for H, then $L((F_1 \wedge F_2)/H) = K_1 \cap K_2$. Note that if K_1 and K_2 are non interconsistent, then, when a specific event v occurs after a string $w \in \overline{K_1} \cap \overline{K_2}$, there will not exist any condition enabled by both F_1 and F_2. Thus, the joint action of the supervisors will not provide an answer for the occurrence of v.

Theorem 2. Given a plant $H = (L, \Gamma)$ and two specifications $E_1, E_2 \subseteq P_V(L)$, there exist two DES supervisors, F_1 and F_2, such that: F_i is consistent for H and $P_V[L_m(F_i/H)] = Sup\mathbb{C}_V(E_i)$, $(i = 1, 2)$; $F_1 \wedge F_2$ is consistent for H; and $P_V[L((F_1 \wedge F_2)/H)] = Sup\mathbb{C}_V(E_1) \cap Sup\mathbb{C}_V(E_2) = Sup\mathbb{C}_V(E_1 \cap E_2)$, if and only if $Sup\mathbb{C}_{VU}[P_V^{-1}(E_1) \cap L]$ and $Sup\mathbb{C}_{VU}[P_V^{-1}(E_2) \cap L]$ are interconsistent.

Thus, if $Sup\mathbb{C}_{VU}[P_V^{-1}(E_i) \cap L]$, $i = 1, 2$, are interconsistent, then there exist consistent supervisors, F_i, whose conjunction results in a control action that is consistent and optimal in the sense of being equivalent to the solution $Sup\mathbb{C}_V(E_1 \cap E_2)$ obtained through the monolithic approach. Theorem 2 indicates that supervisors are built such that $L(F_i/H) = Sup\mathbb{C}_{VU}[P_V^{-1}(E_i) \cap L]$.

Proposition 5. Let F_1, F_2 be DES supervisors such that $L(F_1/H)$ and $L(F_2/H)$ are interconsistent. Let $\mathcal{F}_{F_1}$, $\mathcal{F}_{F_2}$ be C/E supervisors obtained from F_1, F_2 by Algorithm 1 in (González *et al.*, 2001). Then, $F_1 \wedge F_2$ and $\mathcal{F}_{F_1} \wedge \mathcal{F}_{F_2}$ are logically equivalent in the sense that $L((F_1 \wedge F_2)/H) = \mathcal{L}((\mathcal{F}_{F_1} \wedge \mathcal{F}_{F_2})/\mathcal{H})$.

Finally, a solution to the MSHS problem is stated.

Corollary 3. Let $H = (L, \Gamma)$ be the DES model for $\mathcal{H}$ and F_i $(i = 1, 2)$ be supervisors obtained according to Theorem 2. The C/E supervisors $\mathcal{F}_{F_i}$, equivalent to the DES supervisors F_i, lead to an optimal solution to the MSHS problem if and only if $L(F_1/H)$ and $L(F_2/H)$ are interconsistent.

4. USING FINITE STATE APPROXIMATIONS FOR HS

In general, it will not be possible to find a finite state realization of $\mathcal{H}$: one may not exist, or even if one does exist, it may not be feasible to compute it. However, in general the convergence of algorithms is only guaranteed over finite spaces. Thus, in the context of supervisors synthesis for hybrid systems, Cury *et al.* (1998) propose the use of a conservative approximation (finite state automaton) A for the logical behavior of $\mathcal{H}$, such that $L \subseteq L(A)$. The class of approximations introduced in the following will guarantee that a solution to the supervisory control problem obtained for the approximating automaton is also a solution to the original problem, taking both the monolithic and the modular approach.

Definition 8. Let H be the DES model for $\mathcal{H}$. A finite state automaton A is said to be a v-coherent approximation for H if $L \subseteq L(A)$ and

$$(\forall w \in L)(\forall v \in V_L(w)), U_{L(A)}(w,v) = U_L(w,v) \quad (3)$$

Generally, in the algorithms found in the literature, e.g. (Chutinan, 1999), the approximations are performed based on the possibilities of event occurrence and not on the choices of the conditions associated with them. Therefore, these approximations are v-coherent for the plant model. The following results can be presented now.

Proposition 6. Let A be a v-coherent approximation for H. If a DES supervisor F is consistent for A, then F is also consistent for H.

Proposition 7. Let A be a v-coherent approximation for H. If a prefix-closed language $K \subseteq L(A)$ is vu-controllable w.r.t. $L(A)$, then $K \cap L$ is vu-controllable w.r.t. L. Moreover, if a supervisor F is such that $L(F/A) = K$, then $L(F/H) = K \cap L$.

Let A be an approximation for H, $E \subseteq P_V[L(A)]$ be a specification for the closed-loop behavior, and F be a supervisor such that $P_V[L(F/A)] = Sup\mathbb{C}_V(E)$. From Prop. 3 we have $P_V[L(F/A)] = P_V[Sup\mathbb{C}_{VU}(K)]$, where $K = P_V^{-1}(E) \cap L(A)$ and then $L(F/A) = Sup\mathbb{C}_{VU}(K)$. From Prop. 7 we can state that if A is a v-coherent approximation for H, then $L(F/H) = Sup\mathbb{C}_{VU}(K) \cap L$ and therefore $P_V[L(F/H)] \subseteq Sup\mathbb{C}_V(E) \cap P_V(L) \subseteq E \cap P_V(L)$. We further know from Prop. 6 that if $P_V[L(F/A)] \neq \emptyset$ then $P_V[L(F/H)] \neq \emptyset$. Now, the following corollary can be stated.

Corollary 4. Let F be a DES supervisor designed to solve the SCP based on a v-coherent approximation A for H. Then F is a solution to the original SCP.

In the context of modular supervisory control of HSs, the following results can established.

Proposition 8. Let A be a v-coherent approximation for H and let F_1, F_2 be two DES supervisors designed for A. If $L(F_1/A)$ and $L(F_2/A)$ are interconsistent, then $L(F_1/H)$ and $L(F_2/H)$ are also interconsistent.

Corollary 5. Let F_i $(i = 1,2)$ be two DES supervisors designed to solve the MSHS problem in the DES framework and based on a v-coherent approximation A for H. Then the C/E supervisors $\mathcal{F}_{F_i}$ equivalent to the DES supervisors F_i are a solution to the original MSHS problem.

Suppose now that the modular supervisor synthesis is performed in such a way that $P_V[L(F_i/A)] = E_i$ $(i = 1,2)$, but that $L(F_1/A)$ and $L(F_2/A)$ are not interconsistent. The use of an improved approximating automaton A', obtained through the refinement of A (Cury *et al.*, 1998), for which $L \subseteq L(A') \subseteq L(A)$, may lead to supervisors F_i' such that $P_V[L(F_i'/A')] = E_i \cap P_V[L(A')]$ and whose languages $L(F_1'/A')$ and $L(F_2'/A')$ are interconsistent. Thus, if a given modular control problem has no solution for an approximation A, we may proceed the refinement of A in order to try to solve this problem.

5. EXAMPLE

This section presents an example in order to illustrate the proposed approach to the modular supervision of HSs. The system consists of 3 trains over cyclic tracks with shared sections, as depicted in Fig. 4. There are sensors over the tracks that register the crossing of the trains at locations A, B, C (train 1); D, E, F, G (train2); and H, I, J (train 3). Train 2 can travel at two speeds: *fast* and *slow* and trains 1 and 3 can travel only at the *fast* speed. An important characteristic of this system is that the *slow* mode for train 2 can be commanded only within sections $\overline{EF}$ and $\overline{GD}$. The purpose is to avoid collisions between trains.

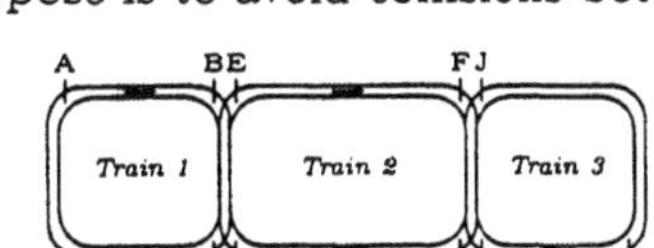

Fig. 4. Three trains over cyclic tracks.

The length of each track and the location of the sensors are as shown in Fig. 5. Points A, D and H were chosen as references for the measurement of the respective track lengths. At the beginning, train 1 is 68 km from A, train 2 is 13 km from D, and train 3 is 11 km from H. The value for the speed in the *slow* mode (train 2) is $\dot{x}_2 = 30$ km/h and for the *fast* mode is $\dot{x}_i = 60$ km/h, $i = 1,2,3$.

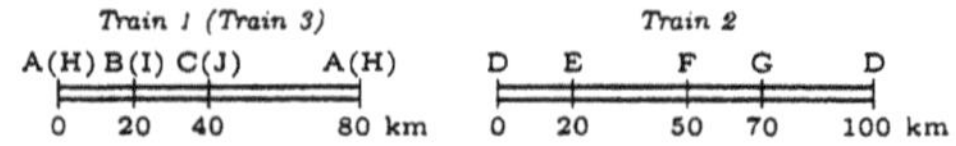

Fig. 5. Cyclic tracks.

The following steps are executed to synthesize the supervisors: *(i)* Build the hybrid system model; *(ii)* Generate a finite state *C/E* automaton that represents the approximated logical behavior of the hybrid system; *(iii)* Synthesize the modular supervisors and test whether their resulting closed-loop languages are interconsistent.

Step 1: To construct the model we use *CheckMate* (Chutinan, 1999). The crossing of a sensor position by one train is interpreted as the occurrence of a threshold event. In order to simplify, we consider that only one sensor position is crossed at a time and so we have a total of 10 possible threshold events. Furthermore, we use symbols to represent the threshold events. Symbol A, for example, represents the event $v(t) = [1\ 0\ 0\ 0\ 0\ 0\ 0\ 0\ 0\ 0]^{\mathrm{T}}$, which occurs at those instants when train 1 crosses position A. Consequently, we have that $V = \{A, B, C, \ldots, J\}$. Through simulation of the open-loop system behavior, in which the *fast* speed is always chosen, it can be seen that there are collisions between trains in the two shared sections. This paper proposes the use of two modular supervisors, one to avoid collisions between trains 1 and 2 (spec. E_1) and another to avoid collisions between trains 2 and 3 (spec. E_2). The specifications $E_1, E_2 \subseteq V^*$ are depicted in Fig. 6.

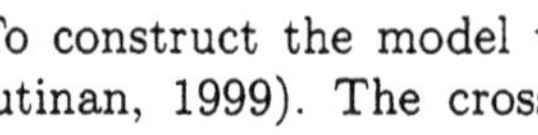

Fig. 6. Specifications $E_1, E_2 \subseteq V^*$.

Step 2: The calculation of a finite state approximation for the hybrid plant is accomplished by running *CheckMate*. The obtained automaton is then converted into a *C/E* automaton given in a format compatible with a computational tool called *Grail* (Raymond and Wood, 1994). For this example, a finite state machine with 182 states and 199 transitions is obtained, which in this case models the exact open-loop logical behavior.

Step 3: Following the supervisor synthesis procedure described in (Leal and Cury, 2002), and using a supervisory control toolbox developed for the *Grail* platform, a supervisor F_1 with 92 states and 97 transitions and a supervisor F_2 with 69 states and 73 transitions are obtained. Using a function developed in the context of this work, it can be verified that the languages implemented by these supervisors are *interconsistent*. Therefore, their joint action is optimal. The modular supervisors are not presented here, but an automaton for $F_1 \wedge F_2$ is depicted in Fig. 7 in order to illustrate their joint action. In this automaton, $u_1 = [f_1\ f_2\ f_3]^{\mathrm{T}}$ and $u_2 = [f_1\ s_2\ f_3]^{\mathrm{T}}$, where f_i and s_i refer to the *fast* and *slow* speed for train i, respectively. Note that a state aggregation was performed to show only those events associated with train 2, as this is the only one that can have its speed changed.

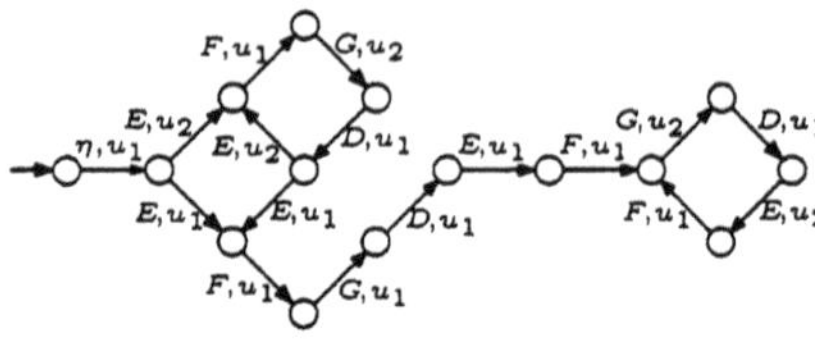

Fig. 7. Aggregated supervisor $F_1 \wedge F_2$.

6. DISCUSSION

This paper presented a modular approach to the supervisory control problem of hybrid systems and established a condition under which this approach can be used. It also introduced a condition under which one can use a finite state approximation for the logical behavior of the hybrid plant in order to find a computable solution to the MSHS problem.

ACKNOWLEDGEMENTS

The first author thanks PICDT/CAPES and the second thanks CNPq (300.953/93-3) and PRONEX (015/98) by their financial support.

REFERENCES

Chutinan, A. (1999). Hybrid System Verification Using Discrete Model Approximations. PhD Thesis. Dept. of Electrical and Computer Engineering, Carnegie Mellon University.

Cury, J. E. R., B. H. Krogh and T. Niinomi (1998). Synthesis of supervisory controllers for hybrid systems based on approximating automata. *IEEE Trans. Autom. Control* **43**(4), 564–568.

González, J. M. E., A. E. C. da Cunha, J. E. R. Cury and B. H. Krogh (2001). Supervision of event-driven hybrid systems: Modeling and synthesis. In: *Hybrid Syst.: Comp. and Contr.* Vol. 2034 of *LNCS*. Springer. pp. 247–260.

Leal, A. B. and J. E. R. Cury (2002). Controle modular de sistemas condição/evento. *XIV Congresso Brasileiro de Autom.* pp. 455–460.

Moor, T., J. M. Davoren and J. Raisch (2001). Modular supervisory control of a class of hybrid systems in a behavioural framework. In: *European Control Conf.* pp. 870–875.

Raymond, D. R. and D. Wood (1994). GRAIL: A C++ library for automata and expressions. *J. of Symbolic Computation* **17**(4), 341–350.

Sreenivas, R.S. and B.H. Krogh (1991). On condition/event systems with discrete state realizations. *DEDS: Theory & Appl.* **1**, 209–236.

Willems, J. C. (1991). Paradigms and puzzles in the theory of dynamical systems. *IEEE Trans. on Automatic Control* **36**, 259–294.

ELSEVIER
IFAC
PUBLICATIONS
www.elsevier.com/locate/ifac

MULTI-TASKING SUPERVISORY CONTROL OF DISCRETE-EVENT SYSTEMS

Max H. de Queiroz*, José E. R. Cury*, W. M. Wonham**

* *LCMI-DAS - Federal University of Santa Catarina,*
Florianópolis SC 88040-900 Brazil - Phone: 048 331-7602
{max, cury}@das.ufsc.br

** *Systems Control Group, Dept. of Electrical and Computer Engineering*
University of Toronto, ON, Canada M5S 3G4
wonham@control.utoronto.ca

Abstract: This paper presents an approach for functionally dealing with multiple tasks in the supervisory control of discrete-event systems (DES). The colored marking generator (CMG), a special type of Moore automaton, is introduced as a model that distinguishes classes of tasks in DES. Then, the main results of supervisory control theory are extended to this model, allowing the synthesis of more refined supervisors in control problems where the distinction of classes of task is necessary. Reversibility is also investigated as an alternative way of ensuring liveness of multiple tasks. Finally, an example illustrates the convenience of this approach. *Copyright © 2004 IFAC*

Keywords: Supervisory control, discrete-event systems, tasks, control system synthesis, automata.

1. INTRODUCTION

Supervisory control theory (SCT) as initiated by Ramadge and Wonham (1987) has been developed in recent decades as an expressive framework for the synthesis of control for discrete-event systems (DES). In SCT, the open-loop behavior of a DES, called plant, is modeled by a generator (Wonham, 2003), whose marked states represent completion of some task. The restrictions to be imposed on the plant can be expressed in terms of a language representing the admissible behavior. The Ramadge-Wonham framework provides computational algorithms for the synthesis of a minimally restrictive supervisor that constrains the behavior of the plant by disabling controllable events in such a way that it respects the admissible language and that it ensures nonblocking, i.e., there is always an event sequence available to reach a marked state.

While the admissible language can be viewed as a safety specification (assuring that nothing "bad" happens), nonblocking can be understood as a liveness specification that assures that the supervisor will not prevent the completion of a task (something good happens). Other classes of liveness specifications such as stability (Ozveren and Willsky, 1991) and fairness (Gohari, 2002) have also been investigated.

Situations where the liveness of multiple tasks is desired are common. In particular, interesting DES problems comprising multiple tasks arise in manufacturing and communication systems. In these situations, modeling the plant by a single generator can be problematic, for the completion of all classes of tasks would then be identified by the same marking. Especially for composite systems, the marking of each subsystem usually has an individual meaning. In that case, the global model, comprising the composition of all subsystems, identifies a string of events as a complete task if and only if it leads all subsystems to a marked state. Thus, according to the original approach of SCT, a nonblocking supervisor would assure that all subsystems would complete their tasks at the same global state. That constraint might be too conservative for some problems.

One possible approach to enforcing liveness of multiple tasks in a DES is the use of queues, as suggested by Wonham (2003). A queue, like any fairness rule, can be thought of as an extra restriction imposed on the behavior of a system that is potentially stronger than a liveness condition. As a result, this approach can lead to overly conservative solutions. The aim of the current work is to provide a convenient mechanism for the synthesis of minimally restrictive supervisors that assure the liveness of multiple tasks without enforcing any additional rule. These supervisors will disable an event if and only if it could start an uncontrollable sequence that leads the system out of the safe behavior or could render some task inaccessible. Note that such supervisors cannot force the completion of tasks, in conformance with the fact that they cannot directly force the occurrence of events.

Fabian and Kumar (1997) proposed a related approach, to synthesize mutually nonblocking supervisors for systems restricted by disjunctive specifications, which could be interpreted as different tasks. The results in the present paper have a different characteristic since here the tasks can be defined in the plant and the specification is taken as the conjunction of all rules. Thistle and Malhamé (1997) and Thistle et al. (1997) have studied similar issues arising with feature interactions in telephone networks, where the specification for a new feature defines a new marked language. They assumed that the specification is modeled by an automaton with multiple subsets of marked states (associated with different features) and extended the basic results of SCT to assure nonblocking of multiple marked languages. In this paper we propose a framework for multi-tasking supervisory control, where DES are modeled by Moore automata, whose outputs indicate completion of tasks. Since each task can be assigned to a marked language, some of our results coincide in many respects with those of Thistle and Malhamé (1997). However, the current work is developed in a more general context where the occurrence of tasks can be implicit in the plant or defined by specifications. Moreover, it promotes several additional contributions to the theory, which include the composition of multi-tasking DES, the generalization of the concept of $L_m(G)$-closure, and the study of reversibility.

Next, we introduce our approach for modeling DES with multiple tasks and discuss some properties and operations on this model. In the next section, we extend the main results of SCT to deal with multi-tasking generators. Further, we solve a problem of multi-tasking supervisory control. Finally, we investigate the use of reversibility as an alternative way of ensuring liveness of multiple tasks. Mathematical proofs are omitted for lack of space.

2. MULTI-TASKING DES

We say that a DES completes a task when it executes some sequence of events that accomplishes an objective of the control problem. For example, a machine might complete a task named "produce a part" whenever it reaches the initial state. Two tasks belong to the same class when they are related to objectives that are equivalent, i.e., that have the same meaning in the control problem. When a DES model incorporates multiple classes of tasks we call it a multi-tasking discrete-event system (MTDES).

2.1 Colored Behaviors

Let Σ^* be the set of all finite strings of elements in Σ, including the empty string ε. A language is a subset of Σ^*. The prefix closure of a language L is given by $\overline{L} := \{s \in \Sigma^* \mid \exists v \in \Sigma^* \wedge sv \in L\}$.

To distinguish multiple classes of tasks in a DES, we associate a color (label) to each class of task. Let Σ be the set of all events that can occur in the system and C be the set of all colors. For each color $c \in C$ we can assign a language $L_c \in Pwr(\Sigma^*)$ (power set of Σ^*) that represents the set of all sequences of events in Σ that can complete a task of the respective class. In this way, the colored behavior of a MTDES can be modeled by the set $\{(L_c, c), c \in C\}$. We define then a *colored behavior* $\Lambda_C \in Pwr(Pwr(\Sigma^*) \times C)$ as a set of pairs (language, color), with the restriction that any two distinct pairs have distinct colors. For a colored behavior Λ_C, the *language marked by* $c \in C$ is defined as $L_c(\Lambda_C) := L$ such that $(L, c) \in \Lambda_C$. The *language marked by* $B \subseteq C$ is defined by $L_B(\Lambda_C) := \cup_{b \in B} L_b(\Lambda_C)$. Let the colored behaviors $M_B \in Pwr(Pwr(\Sigma^*) \times B)$ and $N_C \in Pwr(Pwr(\Sigma^*) \times C)$. We say that $M_B \subseteq N_C$ if $B \subseteq C$ and $\forall b \in B, L_b(M_B) \subseteq L_b(N_C)$.

2.2 Colored Marking Generators

A direct way of representing a colored marking behavior is to model each language by a generator, so that a MTDES would be modeled as a set of pairs (generator, color). For convenience, a MTDES can be modeled by a special generator, whose states are marked by subsets of colors according to the classes of tasks to be achieved. Such a state machine, named *colored marking generator* (CMG), is formally defined as a 6-tuple $G := (Q,\Sigma,C,\delta,\chi,q_0)$, where Q is a set of states; Σ is a set of events; C is a set of colors; $\delta: Q \times \Sigma \rightarrow Q$ is a state transition function (extended for strings as usual); $\chi: Q \rightarrow Pwr(C)$ is a marking function; q_0 is the initial state. For a CMG G, we can define the eligible event function $\Gamma: Q \rightarrow Pwr(\Sigma)$, which associates each state $q \in Q$ to a subset of Σ with all events that can occur in q, i.e., $\Gamma(q) := \{\sigma \in \Sigma \mid \delta(q, \sigma)!\}$.

This model adds to the usual generator (Wonham, 2003) a set of colors C, representing all the classes of tasks that the system can execute, and a marking function χ, that assigns to each state of Q a subset (empty or not) of colors. Therefore, a CMG is basically a Moore automaton, whose outputs, represented by subsets of colors, define the classes of tasks that are completed after the corresponding sequences of events.

The *generated language* of a colored marking generator G, denoted by $L(G)$, is the set of all finite event strings that can be reached from the initial state q_0, that is, $L(G) := \{s \in \Sigma^* \mid \delta(q_0, s)!\}$. The marked language of a generator is the set of all strings that complete a task. Since a CMG usually comprises multiple classes of tasks, we can define a marked language for each class of task by the set of strings that reach states where the marking function returns a set containing the color associated with that class. Hence, $L_c(G)$, the *language marked by* $c \in C$, is formally defined by:

$$L_c(G) := \{s \in L(G) \mid c \in \chi(\delta(q_0, s))\}.$$

We can extend the concept of language marked by a color to a nonempty set of colors as the set of event strings that complete any of the respective tasks. Then, for the color set B, $\varnothing \subset B \subseteq C$, we define the *language marked by B* as:

$$L_B(G) := \{s \in L(G) \mid B \cap \chi(\delta(q_0, s)) \neq \varnothing\}.$$

The *colored behavior* of a CMG G, denoted by $\Lambda_C(G)$, is defined by $\Lambda_C(G) := \{(L_c(G), c), c \in C\}$.

Proposition 1: Let $\Lambda_C = \{(L_c, c), c \in C\}$ be a colored behavior. If Λ_C is regular (L_c is regular for all $c \in C$), there exists a finite-state colored marking generator G such that $\Lambda_C(G) = \Lambda_C$ and $L(G) = \cup_{c\in C}\overline{L_c}$.

We say that a state $q \in Q$ is *reachable* if it can be reached by a sequence of transitions starting from the initial state q_0, that is, if $\exists s \in \Sigma^*$ such that $\delta(q_0, s) = q$. G is reachable if q is reachable for every $q \in Q$ ($Q = Q_{ac}$). We denote by Q_{ac} the set of all reachable states of Q. The operation $Ac(G)$ eliminates all non-reachable states of G. A state $q \in Q$ is *weakly coreachable w.r.t. B* if there is a sequence of transitions that, starting from q, reaches a state that marks at least one color of B, i.e., if $\exists b \in B$, $\exists s \in \Sigma^*$ such that $b \in \chi(\delta(q, s))$. G is weakly coreachable w.r.t. B if q is weakly coreachable w.r.t. B for every $q \in Q$. A state $q \in Q$ is *strongly coreachable w.r.t. B* if, for every color b of B, there is a sequence of transitions that, starting from q, reaches a state that marks b., i.e., if $\forall b \in B$, $\exists s \in \Sigma^*$ such that $b \in \chi(\delta(q, s))$. Denote by $Q_{sco,B}$ the set of all states of Q that are strongly coreachable w.r.t. B. Clearly, $Q_{sco,B} = \cap_{b\in B} Q_{sco,\{b\}}$. G is strongly coreachable w.r.t. B if q is strongly coreachable w.r.t. B for every $q \in Q$, i.e., if $Q = Q_{sco,B}$. Note that it is possible that from some states $q \in Q_{sco,B}$ we can complete a task $b \in B$ only in states $q' \in Q - Q_{sco,B}$. Thus, we don't necessarily obtain a strongly coreachable generator by erasing from G all states that fail to be strongly coreachable. G is *strongly trim w.r.t. B* if it is reachable and strongly coreachable w.r.t. B, i.e., if $Q_{ac} = Q_{sco,B}$. In order to obtain a CMG strongly trim w.r.t. B from a given G, we have to eliminate all states of G that fail to be reachable and strongly coreachable w.r.t. B. Since the elimination of one state can alter the properties of another, in general this operation, represented by $STr(G, B)$, is performed in multiple iterations.

The synchronous composition of Moore automata requires the definition of an output function that represents the composition of the output functions for the concerned systems. We consider that, when a color is associated to different CMGs, the composed system marks the task represented by the color when all subsystems are in states marking this color. Hence, by the same way that a composed transition function synchronizes common events, the composed marking function synchronizes shared colors. Then, we define the *synchronous composition* of CMGs $G_1 = (Q_1, \Sigma_1, C_1, \delta_1, \chi_1, q_{01})$ and $G_2 = (Q_2, \Sigma_2, C_2, \delta_2, \chi_2, q_{02})$, with respective eligible event functions Γ_1 and Γ_2, as the CMG

$$G_1 \parallel G_2 := Ac(Q_1 \times Q_2, \Sigma_1 \cup \Sigma_2, C_1 \cup C_2, \delta, \chi, (q_{01}, q_{02})),$$

where:

$$\delta((q_1,q_2),\sigma) = \begin{cases} (\delta_1(q_1,\sigma),\delta_2(q_2,\sigma)), \text{ if } \sigma\in\Gamma_1(q_1)\cap\Gamma_2(q_2) \\ (\delta_1(q_1,\sigma),q_2), \text{ if } \sigma\in\Gamma_1(q_1)\setminus\Sigma_2 \\ (q_1,\delta_2(q_2,\sigma)), \text{ if } \sigma\in\Gamma_2(q_2)\setminus\Sigma_1 \\ \text{undefined otherwise} \end{cases}$$

and

$$\chi((q_1,q_2)) = [\chi_1(q_1)\cup(C_2 - C_1)]\cap[\chi_2(q_2)\cup(C_1 - C_2)].$$

The eligible event function Γ associated to $G_1 \parallel G_2$ is given by:

$$\Gamma((q_1,q_2)) = [\Gamma_1(q_1)\cup(\Sigma_2 - \Sigma_1)]\cap[\Gamma_2(q_2)\cup(\Sigma_1 - \Sigma_2)].$$

2.3 Blocking

The concept of blocking in a generator is related to the idea of executing a sequence of events that leads to a state from where it isn't possible to complete a task. When the generator comprises multiple colors, the idea of blocking allows two different interpretations: strong blocking, when it isn't possible to complete tasks of *any class*, and weak blocking, when it isn't possible to complete tasks of *at least one class*.

Given a nonempty subset of colors B, we say that a generator G is *weakly nonblocking w.r.t.* B if $L(G) = \overline{L_B(G)}$, that is, if any generated sequence of events is a prefix of at least one completed task or, equivalently, if $Ac(G)$ is weakly coreachable w.r.t. B. We say that a generator G is *strongly nonblocking w.r.t.* B if, $\forall b \in B$, $L(G) = \overline{L_b(G)}$, that is, if any generated sequence of events can be completed (not necessarily in the same way) to a task of all the classes represented by colors of B or, equivalently, if $Ac(G)$ is strongly coreachable w.r.t. B.

The idea of strong nonblocking can be extended to colored behaviors as follow. A colored behavior $\Lambda_C \in Pwr(Pwr(\Sigma^*) \times C)$ is strongly nonblocking w.r.t. $B \subseteq C$ whenever $\forall b \in B$, $\overline{L_b(\Lambda_C)} = \overline{L_C(\Lambda_C)}$, that is, when any incomplete string can complete tasks of all colors of B.

Proposition 2: There's a strongly trim CMG that models a colored behavior Λ_C if and only if Λ_C is strongly nonblocking.

Theorem 1: Let the set of strongly nonblocking behaviors contained in Λ_C be defined by $SNB(\Lambda_C, B) := \{M_C \subseteq \Lambda_C \mid \forall b \in B, \overline{L_b(M_C)} = \overline{L_C(M_C)}\}$. The set $SNB(\Lambda_C, B)$ has a supremal element $SupSNB(\Lambda_C, B)$ that represents the maximal strongly nonblocking behavior contained in Λ_C.

Proposition 3: Let G be a finite state CMG that models a regular colored behavior Λ_C. Then $SupSNB(\Lambda_C, B) = \Lambda_C(STr(G, B))$.

3. MULTI-TASKING SUPERVISORY CONTROL

Let the open loop behavior of a DES be modeled by a colored marking generator $G = (Q, \Sigma, C, \delta, \chi, q_0)$, with eligible event function Γ, whose alphabet is partitioned into controllable events $\sigma \in \Sigma_c$ (events that can be disabled) and uncontrollable events $\sigma \in \Sigma_u$. Let D be a set of important tasks for which we want liveness (strong nonblocking). It can be very convenient to allow the specification to define new classes of tasks ($D - C = E$), for it can bring new information (states) into the model. For example, the specification for a buffer between two machines can highlight the state where the buffer is empty, which could define a new task. Accordingly, let the specification be given by a colored behavior $\Lambda_D \in Pwr(Pwr(\Sigma^*) \times D)$ such that $\forall d \in D \cap C$, $\varnothing \subset L_d(\Lambda_D) \subseteq L_d(G)$, and $\forall d \in E$, $\varnothing \subset L_d(\Lambda_D) \subseteq L(G)$. Such specification contains sublanguages of the colored behavior of the plant and marks sublanguages of $L(G)$ with new colors. The objective of supervisory control is to generate an entity (designated supervisor) which, by disabling controllable events, prevents the controlled system from violating the languages of the specification. The supervisor should also guarantee that the controlled system is always capable of completing tasks of D.

3.1 Coloring Supervisor

In many problems, the specifications may include classes of tasks that are not explicit in the plant ($E \neq \varnothing$). In this case, the role of marking the new colors in the controlled system is assigned to the supervisor. A *coloring supervisor* is then defined as a mapping that associates to each sequence of events of the plant a set of enabled events and a set of new colors (of E) representing completed tasks, as illustrated in Fig. 1. Thus, a coloring supervisor S consists of a function S: $L(G) \to Pwr(\Sigma) \times Pwr(E)$.

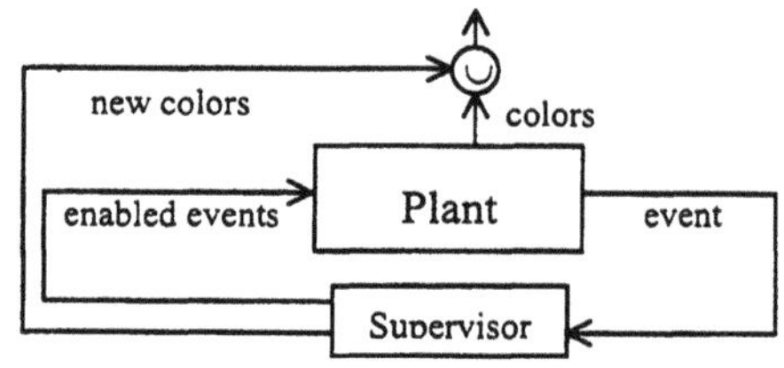

Fig. 1. Control architecture for coloring supervisor

Let $S(s) = (\gamma, \mu)$, and write $\Re(S(s)) = \gamma$ and $\Im(S(s)) = \mu$. The events that can occur after the occurrence of a string of events $s \in L(G)$ are given by $\Re(S(s)) \cap \Gamma(\delta(q_0, s))$. We say that a coloring supervisor S is *admissible* if it doesn't imply disabling uncontrollable events, that is, if $\forall s \in L(G)$, $\Sigma_u \cap \Gamma(\delta(q_0, s)) \subseteq \Re(S(s))$. The language generated by the controlled system S/G is defined by:

1. $\varepsilon \in L(S/G)$;
2. $s\sigma \in L(S/G) \Leftrightarrow (s \in L(S/G)) \wedge (s\sigma \in L(G)) \wedge (\sigma \in \Re(S(s)))$.

Besides marking with colors of C the complete tasks of the plant that remain after the supervision, the controlled system S/G marks with colors of E the tasks defined by the coloring supervisor S. Thus, the languages marked by S/G can be described by $\forall c \in C$, $L_c(S/G) := L(S/G) \cap L_c(G)$ and $\forall e \in E$, $L_e(S/G) := \{s \in L(S/G) \mid e \in \Im(S(s))\}$.

3.2 Existence of Supervisors

A supervisor S is *strongly nonblocking w.r.t* D if $\forall d \in D$, $\overline{L_d(S/G)} = L(S/G)$. A specification Λ_D is *D–closed w.r.t.* G if, for any color $d \in D \cap C$, every string, that is a prefix of $L_d(\Lambda_D)$ and marked by d in the plant, is also marked by d in the specification, that is, if $\forall d \in (D \cap C)$, $L_d(\Lambda_D) = \overline{L_d(\Lambda_D)} \cap L_d(G)$. A colored behavior Λ_D is *controllable w.r.t* G if the union of all languages $L_d(\Lambda_D)$ is controllable, i.e., if $\overline{L_D(\Lambda_D)}\Sigma_u \cap L(G) \subseteq \overline{L_D(\Lambda_D)}$.

Theorem 2: Necessary and sufficient conditions for the existence of an admissible coloring supervisor S

strongly nonblocking w.r.t. D such that $\Lambda_D(S/G) = A_D$ and $L(S/G) = \overline{L_D(A_D)}$ are:

1. controllability of A_D w.r.t. G;
2. strong nonblocking of A_D w.r.t. D;
3. D–closedness of A_D w.r.t. G.

The next proposition shows the existence of the least restrictive controllable behavior contained in A_D.

Proposition 4: Define the set of controllable behaviors contained in A_D by $C(A_D, G) := \{M_D \subseteq A_D$ such that $(\overline{L_D(M_D)}\Sigma_u \cap L(G) \subseteq \overline{L_D(M_D)})\}$. The set $C(A_D, G)$ has a supremal element $SupC(A_D, G)$.

The properties of strong nonblocking w.r.t. D and of D–closedness can be naturally assured by the construction of specification A_D. However, the controllability of A_D isn't always verified. In this case, we look for a strongly nonblocking supervisor S restricting A_D to $SupC(A_D, G)$. However, this is not always feasible, because $SupC(A_D, G)$ is not necessarily strongly nonblocking w.r.t. D, even if A_D is. The next theorem proves the existence of a supremal controllable and strongly nonblocking behavior contained in A_D. By Theorem 2, this behavior, if nonempty, can be generated by a minimally restrictive strongly nonblocking supervisor.

Theorem 3: Let the set of controllable and strongly nonblocking behaviors contained in A_D be defined by $CSNB(A_D, G, D) := C(A_D, G) \cap SNB(A_D, D)$ The set $CSNB(A_D, G, D)$ has a supremal element $SupCSNB(A_D, G, D)$.

We can show that $SupCSNB(A_D, G, D)$ can be computed for a regular A_D and a finite-state G by the iterative elimination of uncontrollable or not strongly coreachable states of a CMG modeling A_D. The following result shows that a supervisor for the maximal controllable and strongly nonblocking behavior can be more restrictive than a nonblocking supervisor for the maximal controllable language.

Proposition 5: In general, we have $L_D(SupCSNB(A_D, G, D)) \subseteq SupC(L_D(A_D), G)$.

The following proposition states that, if D–closedness of a given specification is verified, it also holds for the maximal controllable and strongly nonblocking behavior. Thus, when A_D is D–closed, $SupCSNB(A_D, G, D)$ is the least restrictive controllable, strongly nonblocking and D–closed behavior contained in A_D.

Proposition 6: If A_D is D–closed w.r.t. G, then $SupCSNB(A_D, G, D)$ is also D–closed.

4. EXAMPLE

A cat and a mouse share the maze presented in Fig. 2 (Ramadge and Wonham, 1989), where the arrows c_i signal the one-way passages exclusive for the cat and the arrows m_i signal the one-way passages exclusive for the mouse. The controller can close all passages, except c7. Fig. 2 indicates the room where food is available for the mouse and cat. The cat and the mouse are placed initially at the rooms indicated by the respective pictures. The controller should guarantee maximal freedom of movement excluding that both animals occupy the same room at the same time, but it should not prevent the cat and the mouse from eating (tasks **c** and **m**, respectively) and nor the system from returning to the initial state (task **i**).

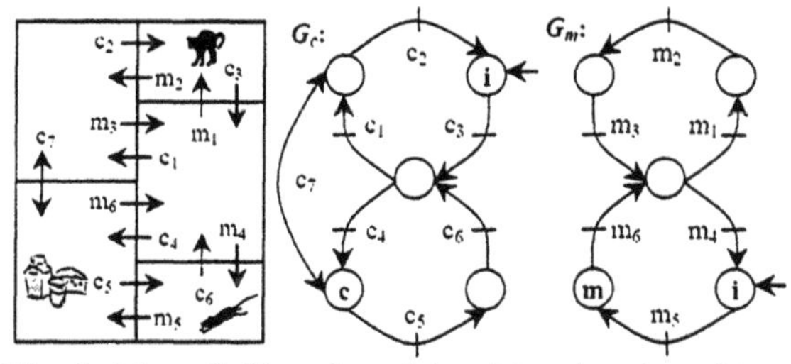

Fig. 2. Maze (left) and models with colored marking for cat (G_c) and mouse (G_m)

Let the alphabet $\Sigma = \{c_i, m_j: 1 \leq i \leq 7, 1 \leq j \leq 6\}$ and the set of colors $C = \{\mathbf{i}, \mathbf{c}, \mathbf{m}\}$. The movement of the cat and of the mouse is modeled by the system composed by the generators G_c and G_m on Σ and C, presented in Fig. 2, so that the global plant G is obtained by $G = G_c \parallel G_m$. In this example, the colors for which we want liveness are defined in the plant ($D = C$) and, thus, the supervisor doesn't need to introduce new colors ($E = \varnothing$). A generator for the maximal behavior A_D that respects the specifications is obtained from G, by erasing the states where both animals occupy the same room. The maximal behavior controllable and strongly nonblocking w.r.t. D contained in A_D is modeled by CMG of Fig. 3.

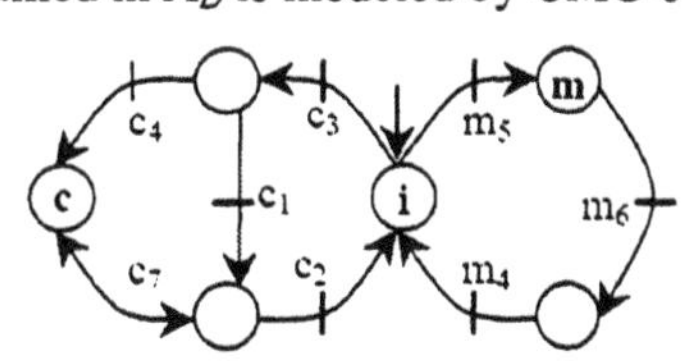

Fig. 3. CMG for $SupCSNB(A_D, G, \{\mathbf{i}, \mathbf{c}, \mathbf{m}\})$

This behavior coincides with the usual solution obtained by Ramadge and Wonham (1989), that considers as marked only the initial states. Therefore, the extra requirements for liveness of cat and mouse actually don't introduce additional restrictions to the solution in this particular problem. Nevertheless, the use of colored markings supports the verification of the liveness specifications.

5. REVERSIBILITY

A discrete-event system is called reversible when it is always possible to return from any reachable state to its initial state (Murata, 1989). This property is a natural requirement in many practical applications, particularly in manufacturing systems. Reversibility can be tested in an usual generator by marking only its initial state and verifying nonblocking. In the same way, in a colored marked generator we can define as a task the fact of reaching the initial state. By assuring strong nonblocking w.r.t. any set containing this task, we guarantee the reversibility of the DES. In this section we study the relation between reversibility and strong nonblocking with respect to multiple tasks.

Formally, a colored marking generator $G = (Q, \Sigma, C, \delta, \chi, q_0)$ is *reversible* if $\forall q \in Q_{ac}$, $\exists s \in \Sigma^*$ such that $\delta(q, s) = q_0$. For a nonempty CMG, we define the task $\mathbf{i} \in C$ as "reach the initial state", i.e., $\mathbf{i} \in \chi(q) \Leftrightarrow q = q_0$. So, G is reversible if G is nonblocking w.r.t. $\mathbf{i}$. A task $c \in C$ is *reachable* if for some reachable state $q \in Q_{ac}$, $c \in \chi(q)$. The set of reachable tasks $C_{ac} \subseteq C$ is given by $C_{ac} := \{c \in C \mid c$ is reachable$\}$. Clearly, $\mathbf{i} \in C_{ac}$.

Proposition 7: G is reversible if and only if G is strongly nonblocking w.r.t. C_{ac}.

The previous proposition states that we can test strong nonblocking with respect to a set of relevant tasks (including **i**) by verifying reversibility of the CMG and reachability of the respective tasks. For problems where reversibility is an important issue, this approach can be more efficient than testing for every reachable state of the CMG coreachability with respect to all tasks.

Proposition 8: Let $H_a = (X, \Sigma, D, \lambda, \kappa, x_0)$ be a CMG for a specification Λ_D. Let $\mathbf{i} \in D \subseteq C_{ac}$ be such that, for any $x \in X$, $\mathbf{i} \in \kappa(x) \Leftrightarrow x = x_0$. Then, $SupCSNB(\Lambda_D, G, D) = SupCSNB(\Lambda_D, G, \{\mathbf{i}\})$ whenever $SupCSNB(\Lambda_D, G, D) \neq \{(\varnothing, d) \mid d \in D\}$.

By the previous proposition, if a supervisory control problem admits a nonempty solution strongly nonblocking w.r.t. a set D, which includes **i**, it is equivalent to the optimal reversible solution. Therefore, in problems where reversibility of the specification is imposed, a natural way of computing *SupCSNB* is to compute the supremal controllable and reversible sublanguage and test reachability of all relevant tasks. This result could have been used directly in the solution of the cat and mouse problem, since reversibility is required for *SupCSNB* and all the other important tasks are reachable in *SupCSNB*. Thus, we could have achieved the same result by computing an optimal controllable supervisor that assures that the cat and the mouse can always be back in their respective initial rooms at the same time (task **i**). We can verify in the controlled behavior that the cat and the mouse can reach the food from the initial state. If the tasks **c** and **m** were not reachable from the initial state of the optimal reversible and controllable behavior, the problem would not admit any (nonempty) solution strongly nonblocking w.r.t. $\{\mathbf{i}, \mathbf{c}, \mathbf{m}\}$.

6. CONCLUSION

We have introduced a framework based on colored marking generators for dealing with multiple tasks in DES. In this model we can define strong nonblocking as a refinement of the usual property of nonblocking. Next, we extended the main results of SCT. The solution of a multi-tasking problem shows the convenience of the proposed methodology in the modelling as well as in the synthesis of (potentially) more refined supervisors. Finally, we presented some results on the property of reversibility that lead to an alternative approach for assuring liveness of multiple tasks. This approach could have been successfully applied to the cat and mouse problem, but certainly there are problems where reversibility would be too strong a constraint. Currently the authors are investigating the use of a modular approach (Wonham and Ramadge, 1988; Queiroz and Cury, 2000) to deal with state space explosion in multi-tasking supervisory control of composite DES.

AKNOWLEDGEMENTS

The authors thank the financial support provided by CNPq and CAPES.

REFERENCES

Fabian, M. and R. Kumar (1997). Mutually Nonblocking Supervisory Control of DES, *CDC'97*, San Diego, CA, USA.

Gohari-M., P. (2002) Fair Supervisory Control of DES, *Ph.D. Thesis*, Dept. of Elec. & Comp. Eng., Univ. of Toronto.

Murata, T. (1989). Petri Nets: Properties, Analysis and Application. *Proc. of the IEEE*, **77**, 541-580.

Ozveren, C.M. and A.S. Willsky (1991). Output stabilizability of discrete-event dynamic systems. *IEEE Trans. on Automatic Control*, **36**, 925-935.

Queiroz, M.H. de and J.E.R. Cury (2000) Modular Supervisory Control of Large Scale DES. *In: DES: Analysis and Control*, pp.103-110, Kluwer.

Ramadge, P.J. and W.M. Wonham (1989). The control of DES. *In: Proc. IEEE, Special Issue on DEDSs*, **77**, 81-98.

Thistle, J.G.R. and P. Malhamé (1997). Multiple Marked Languages and Noninterference in Modular Supervisory Control. *Proc. of the 35th Annual Allerton Conference*. 523-532.

Thistle, J.G.R.; P. Malhamé, H.-H. Hoang and S. Lafortune (1997), *Sup. control of distributed systems part I: modeling, spec., and synthesis.* Tech. Rep. EPM/RT-97/08, E. P. de Montreal.

Wonham, W.M. (2003). *Notes on Control of DES.* Dept. of Elec. & Comp. Eng., Univ. of Toronto. http://www.control.utoronto.ca/DES

Wonham, W.M. and P.J. Ramadge (1988). Modular supervisory control of discrete event systems. *Math. of Control, Signals & Systems*, **v. 1**, n. 1.

Published by Elsevier Ltd on behalf of IFAC

ELSEVIER
IFAC
PUBLICATIONS
www.elsevier.com/locate/ifac

MODULAR SUPERVISORY CONTROL OF A CLASS OF CONCURRENT DISCRETE EVENT SYSTEMS

B. Gaudin * H. Marchand *

* *Irisa, Campus universitaire de Beaulieu, Rennes.*

Abstract: In this paper, we are interested in the control of a particular class of Concurrent Discrete Event Systems defined by a collection of components that interact with each other. We investigate the computation of the supremal controllable language contained in the one of the specification. We do not adopt the decentralized approach. Instead, we have chosen to perform the control on some approximations of the plant derived from the behavior of each component. The behavior of these approximations is restricted so that they respect a new language property for discrete event systems called *partial controllability condition* that depends on the specification. It is shown that, under some assumptions, the intersection of these "controlled approximations" corresponds to the supremal controllable language contained in the specification with respect to the plant. This computation is performed without having to build the whole plant, hence avoiding the state space explosion induced by the concurrent nature of the plant.

Keywords: Discrete Event Systems, Supervision and control, Concurrent Systems, Partial controllability.

1. INTRODUCTION

In this paper, we are interested in the control of Concurrent Discrete Event Systems defined by a collection of components that interact with each other. *Supervisory control* [Ramadge and Wonham (1989)] consists in modifying a system (plant) such that the modified (or *controlled*) system satisfies a given property (or specification). Given a plant and a specification modeled by languages, in the Ramadge & Wonham theory, one important phase is the computation of the supremal controllable sub-language contained in a language that represents the expected behavior. However, although this computation is polynomial in the number of states of the plant and of the specification, it is well known that the plant size grows exponentially with the number of components that compose the plant. This renders the computation of the supervisors not practical because of the size of the generated state space which is often too important when dealing with large scale systems.

Several approaches have been recently investigated to deal with the complexity issue of the control of Concurrent Discrete Event Systems. Given a Concurrent Discrete Event System $G = G_1 \parallel \cdots \parallel G_n$ and a specification expressed by a language K, the problem is to compute the supremal controllable sublanguage of $K \cap \mathcal{L}(G)$ w.r.t. $\mathcal{L}(G)$ without having to build $\mathcal{L}(G)$. In [deQueiroz and Cury (2000); Åkesson et al. (2002)], the authors consider the control of a product plant (i.e. systems composed of asynchronous subsystems, not sharing common events)[1]. Given a specification, a local system is built from the components that are coordinated by the specification (i.e. all the components that share some events used to express it). It is then sufficient to compute the local supervisor ensuring the specification with respect to this local system in order to obtain the result on the whole system. Closely related to the decentralized theory, under the hypothesis that the specification is separable and that the shared events are controllable, the authors of [Willner and Heymann (1991)] provide a solution allowing to compute local modular supervisors S_i acting upon G_i and to operate the individually controlled plant S_i/G_i concurrently in such a way that the behavior of the controlled plant (i.e. $\mathcal{L}(\|_i\ S_i/G_i)$) corresponds to the supremal controllable sublanguage of K w.r.t. the plant $\mathcal{L}(G)$. The same methodology has been used in [Rohloff and Lafortune (2003)] for the control of concurrent plant for which the various components have an identical structure and under the constraints that the local supervisors S_i are only operating on a subset of the local events. See also [Abdelwahed and Wonham (2002); Jiang and Kumar (2000); Leduc et al. (2001)] for other works relating to the control of concurrent plant.

In this paper, compared to [Willner and Heymann (1991)], we adopt a dual approach. Instead of having one local supervisor per component that enforces local control actions with respect to the events of this component, we have chosen to perform the control on some approximations of the plant derived from the behavior of

[1] they actually first transform a concurrent plant in a modular plant.

each component. The behavior of these approximations is restricted so that they respect a new language property for discrete event systems called *partial controllability condition* that depends on K. Under some assumptions, it is shown that a supervisor can be derived from these "controlled approximations" such that the behavior of the controlled plant corresponds to the supremal controllable language contained in K w.r.t. the plant G. More details (and missing proofs) are available in [Gaudin and Marchand (2004)].

2. PRELIMINARIES

Model and Supervisory Control overview. The basic structures from which the plant is built are Finite State Machines (FSM), that are defined by a 4-tuple $G = \langle \Sigma, \mathcal{X}, x_o, \delta \rangle$, where Σ is the finite alphabet of G. $\mathcal{X}$ is the finite set of states, $x_o \in \mathcal{X}$ is the initial state, whereas δ is the partial transition function defined over $\Sigma \times \mathcal{X} \longrightarrow \mathcal{X}$. The notation $\delta(\sigma, x)!$ means that $\delta(\sigma, x)$ is defined, i.e., there is a transition labeled by an event σ out of state x in machine G. Likewise, for $x \in \mathcal{X}$ and $s \in \Sigma^*$, $\delta(s, x)$ denotes the state reached by taking the sequence of events defined by trace s from state x in machine G. The behavior of the system is described by the language $\mathcal{L}(G) \subseteq \Sigma^*$ generated by G. (i.e. $\mathcal{L}(G) = \{s \in \Sigma^* \mid \delta(s, x_o)!\}$.

Given $s, s' \in \Sigma^*$, we say that $s' \leq s$ whenever s' is a prefix of s (i.e. it exists $t \in \Sigma^*$ s.t. $s = s't$). We denote by $\overline{L}$ the prefix-closure of a language $L \subseteq \Sigma^*$ ($\overline{L} = \{s \in \Sigma^* \mid \exists s' \in L, s \leq s'\}$). Note that $\mathcal{L}(G)$, as defined above, is prefix-closed (i.e. $\mathcal{L}(G) = \overline{\mathcal{L}(G)}$). For $L \subseteq \Sigma^*$ and $\Sigma' \subseteq \Sigma$, we use $L(s, \Sigma')$ to denote the set of suffixes of L after s that belongs to Σ'^*, i.e. $L(s, \Sigma') = \{t \in \Sigma'^* \mid st \in L\}$.

Given a plant to be controlled, some of its events in Σ are said to be uncontrollable (Σ_{uc}), i.e., their occurrence cannot be prevented by the supervisor, while the others are controllable (Σ_c). First, we recall the definition of a *controllable language* [Ramadge and Wonham (1989)].

Definition 1. Let G be an FSM modeling the plant and $K \subseteq \mathcal{L}(G)$ the prefix-closed specification. Then K is controllable with respect to Σ_{uc} and G (or $\mathcal{L}(G)$) if $K\Sigma_{uc} \cap \mathcal{L}(G) \subseteq K$ •

We denote by $K^{\uparrow L,c}$ or $SupC(K, L, \Sigma_{uc})$ the supremal controllable sub-language of K w.r.t. Σ_{uc} and $\mathcal{L}(G) = L$ (see [Ramadge and Wonham (1989)]).

In some situations, it is also of interest to compute $K^{\downarrow L,c}$ the *infimal prefix-closed and controllable superlanguage of* K w.r.t. $\mathcal{L}(G)$ and Σ_{uc}, which basically corresponds to the smallest prefix-closed language that contains K and that is controllable w.r.t. Σ_{uc} and $\mathcal{L}(G) = L$. It can be shown (see e.g. [Cassandras and Lafortune (1999)]) that $K^{\downarrow L,c} = K\Sigma_{uc}^* \cap L$.

Concurrent DES. In this paper, a plant is composed of several components, sharing common events, i.e. a plant G is modeled as a collection of FSM $G_i = \langle \Sigma_i, \mathcal{X}_i, x_{oi}, \delta_i \rangle$. The global behavior of the plant is given by $G = G_1 \parallel \cdots \parallel G_n$, where the operation $\parallel$ is the classical *parallel composition* (i.e. $G_1 \parallel G_2$ represents the concurrent behavior of G_1 and G_2 with synchronization on the shared events). Now, given the set of FSM $(G_i)_{i \leq n}$ modeling G, we denote by Σ_s the set of shared events of G, i.e

$$\Sigma_s = \{\sigma \in \Sigma \mid \exists i \neq j,\ \sigma \in \Sigma_i \cap \Sigma_j\}. \quad (1)$$

Let $\Sigma' \subseteq \Sigma$, then $P_{\Sigma'} : \Sigma^* \rightarrow \Sigma'^*$ is the natural projection from Σ^* to Σ'^* that erases in a sequence of Σ^* all the events that do not belong to Σ'.This definition is easily extended to the projection of regular languages as follows: $P_{\Sigma'}(L) = \{s' \in \Sigma'^* \mid \exists s \in L,\ s' = P_{\Sigma'}(s)\}$. Given $L \subseteq \Sigma'^* \subseteq \Sigma^*$, the inverse projection is defined by $P_{\Sigma'}^{-1}(L) = \{s \in \Sigma^* \mid P_{\Sigma'}(s) \in L\}$. From an implementation point of view, if H denotes the FSM such that $\mathcal{L}(H) = L$, then the FSM modeling the inverse projection of L, noted H^{-1}, can be obtained from H by simply adding self-loops labeled by events in $\Sigma \setminus \Sigma'$ to each state of H.

Given a discrete event system $G = G_1 \parallel \cdots \parallel G_n$, with $\mathcal{L}(G_i) \subseteq \Sigma_i^*$, we simply denote by P_i the projection from Σ^* to Σ_i^* and by P_i^{-1} the inverse projection from Σ_i^* to Σ^*. Based on these operations, the language resulting from the parallel composition of FSM is characterized by:

$$\mathcal{L}(G) = P_1^{-1}[\mathcal{L}(G_1)] \cap \cdots \cap P_n^{-1}[\mathcal{L}(G_n)] \quad (2)$$

The following technical lemmas will be useful.

Lemma 1. [Gaudin and Marchand (2004)] Let $L \subseteq \Sigma'^*$ and $\Sigma' \subseteq \Sigma$, let $s \in L$ and $s' \in \Sigma^*$, then $ss' \in P_{\Sigma'}^{-1}(L) \Longrightarrow sP_{\Sigma'}(s') \in P_{\Sigma'}^{-1}(L)$. ◇

Lemma 2. ((3.1) of [Willner and Heymann (1991)]) Let $G = G_1 \parallel \cdots \parallel G_n$, $s \in \mathcal{L}(G)$, $i \in \{1, \ldots, n\}$ and $\sigma \in \Sigma_i \setminus \Sigma_s$. Then $s\sigma \in \mathcal{L}(G) \Longleftrightarrow s\sigma \in P_i^{-1}(\mathcal{L}(G_i))$. ◇

Control Problem formulation & Related Works Let $G = G_1 \parallel \cdots \parallel G_n$ be the plant to be controlled and $L_i = \mathcal{L}(G_i)$ be the language generated by the component G_i for $i \leq n$. The alphabet of G_i is partitioned into the controllable event set $\Sigma_{i,c}$ and the uncontrollable event set $\Sigma_{i,uc}$, i.e. $\Sigma_i = \Sigma_{i,uc} \mathbin{\dot\cup} \Sigma_{i,c}$. The alphabet of the global plant G is given by:

$$\Sigma = \bigcup_i \Sigma_i, \Sigma_c = \bigcup_i \Sigma_{i,c}, \text{ and } \Sigma_{uc} = \Sigma \setminus \Sigma_c.$$

Moreover, we assume that the following relation holds between the control status of shared events:

$$\forall i, j,\ \Sigma_{i,uc} \cap \Sigma_{j,c} = \emptyset \quad (3)$$

which simply means that the components that share an event agree on the control status of this event. Under this hypothesis, we have that

$$\Sigma_{uc} = \cup_i \Sigma_{i,uc}$$

Let $K \subseteq \Sigma^*$ be the expected behavior. The problem, we are interested in, is the Basic Supervisory Control Problem, i.e . the problem is to compute the supremal controllable sublanguage $(K \cap \mathcal{L}(G))^{\uparrow c}$ of $K \cap \mathcal{L}(G)$

w.r.t. $\mathcal{L}(G)$. However, knowing that the synthesis algorithms are polynomial in the number of states of G and that the size of the state space of G is exponential in the number of components of G, it is important to design algorithms that compute the controller by taking advantage of the structure of G without building it. Hence, the actual problem is to compute $(K \cap \mathcal{L}(G))^{\uparrow c}$ without computing neither $\mathcal{L}(G)$ nor $K \cap \mathcal{L}(G)$.

A Decentralized approach: The works of [Willner and Heymann (1991)] is closely related to the decentralized theory. The authors consider the control of Concurrent DES $G_1 \parallel \cdots \parallel G_n$. Given a language-based specification K, they provide some solutions allowing to compute local modular supervisors $\mathcal{S}_i$ on G_i (based on a notion of separable specification (See Definition 2)) and to operate the individually controlled system $\mathcal{S}_i/G_i$ concurrently in such a way that the controlled behavior corresponds to the supremal controllable sublanguage of $K \cap \mathcal{L}(G)$ w.r.t. $\mathcal{L}(G)$.

Definition 2. $\mathcal{L} \subseteq \Sigma^*$ is said to be separable w.r.t. $\{\Sigma_i\}_{i \leq n}$ with $\cup_{i \leq n}\Sigma_i = \Sigma$, whenever there exists a set of languages $\{\mathcal{L}_i\}_{i \leq n}$ (called generating set), s.t. $\mathcal{L}_i \subseteq \Sigma_i^*$ and $\mathcal{L} = \mathcal{L}_1 \parallel \cdots \parallel \mathcal{L}_n = \cap_i P_i^{-1}(\mathcal{L}_1)$.•

Based on this definition, [Willner and Heymann (1991)] shown that

Theorem 1. Let $G = G_1 \parallel \cdots \parallel G_n$, with $\mathcal{L}(G_i) \subseteq \Sigma_i^*$. and K the expected specification. If $\Sigma_s \subseteq \Sigma_c$ and K is separable w.r.t. $\{\Sigma_i\}_{i \leq n}$, then

$$\|_{i \leq n} SupC(P_i(K) \cap \mathcal{L}(G_i), \mathcal{L}(G_i), \Sigma_{i,uc}) = SupC(K \cap \mathcal{L}(G), \mathcal{L}(G), \Sigma_{uc}) \diamond$$

Hence, given a Concurrent DES G and a separable specification K, Theorem 1 shows that there exists a set of supervisors $\mathcal{S}_i$ acting upon G_i, such that $\|_{i \leq n} \mathcal{L}(\mathcal{S}_i/G_i) = (K \cap \mathcal{L}(G))^{\uparrow c}$.

If K is separable w.r.t. $\{\Sigma_i\}_{i \leq n}$ (which can be checked in $\mathcal{O}(m^{n+1})$, where m is the size of the FSM that generates K), then synthesizing the local supervisors requires the computation of the projection of K over Σ_i. In the worst case, the size of the FSM that generates $P_i(K)$ is in $\mathcal{O}(2^m)$. Hence, solving the supervisory control problem will require $\mathcal{O}(n.2^m.N)$ space where N is the size of each component.

Our approach: Our approach is different and is more related to the modular approach of [Wonham and Ramadge (1988)]. Indeed, the plant G can be described by the following parallel composition of FSM $G = \|_{i \leq n} G_i^{-1}$, where G_i^{-1} is the FSM such that $\mathcal{L}(G_i^{-1}) = P_i^{-1}(\mathcal{L}(G_i))$. In fact, each G_i^{-1} can be seen as an approximation of the plant G to be controlled.

Compared to [Willner and Heymann (1991)], we adopt a dual approach. Instead of controlling each component G_i (i.e. $\mathcal{L}(G_i)$) to enforce $P_i(K)$, we have chosen to control the approximations $\mathcal{L}(G_i^{-1})$ of the plant in order to enforce K. However, it is not sufficient to compute a supervisor $\mathcal{S}_i$ acting upon G_i^{-1} that restricts the behavior $\mathcal{L}(G_i^{-1})$ to the supremal controllable sublanguage of $K \cap \mathcal{L}(G_i^{-1})$ and to operate the controlled systems $\mathcal{S}_i/G_i^{-1}$ concurrently to obtain the supremal controllable sublanguage of $K \cap \mathcal{L}(G)$ (the result may be not supremal). So the idea of our method is to refine the notion of controllability.

The property that we ensure on each G_i^{-1} according to K is called *the partial controllability condition* and is defined in Section 3. As in the case of the controllability concept, it will be shown that there exists a supremal partially controllable sublanguage of $K \cap \mathcal{L}(G_i^{-1})$ w.r.t. K and G_i^{-1}, called $K_i^{\uparrow pc}$. It is then shown that $\cap_{i \leq n} K_i^{\uparrow pc} \subseteq (K \cap \mathcal{L}(G))^{\uparrow c}$ (Theorem 2) and that under some conditions on K the equality holds (Theorem 3 and 4). A comparison with the results of [Willner and Heymann (1991)] is then done.

3. PARTIAL CONTROLLABILITY PROPERTY

In this section, we introduce a new concept of controllability, named *Partial Controllability*, that will serve as the bases of the modular computation of supervisors for Concurrent discrete event systems.

3.1 Definition and useful properties

Definition 3. Let $M \subseteq L \subseteq \Sigma^*$ be prefix-closed languages. Let $\Sigma'_{uc} \subseteq \Sigma_{uc} \subseteq \Sigma$ be two sub-alphabets of Σ. Let $M' \subseteq M$ be a prefix-closed language. M' is partially controllable with respect to Σ'_{uc}, Σ_{uc}, M and L if

(i) M' is controllable w.r.t Σ'_{uc} and L.
(ii) M' is controllable w.r.t Σ_{uc} and M. •

In general, M is not partially controllable with respect to Σ'_{uc}, Σ_{uc}, M and L (e.g. if M is not controllable w.r.t. Σ'_{uc} and L). However, it can be shown that there exists a supremal sub-language of M that has this property.

Proposition 1. [Gaudin and Marchand (2004)]

Let $M \subseteq L \subseteq \Sigma^*$ be prefix-closed languages, $\Sigma'_{uc} \subseteq \Sigma_{uc}$. There exists a unique supremal language, denoted by $M^{\uparrow pc}$, which is partially controllable w.r.t Σ'_{uc}, Σ_{uc},M and L. Moreover

$$M^{\uparrow pc} = \overline{M^{\uparrow pc}} = \text{SupC}(\text{SupC}(M, \Sigma'_{uc}, L), \Sigma_{uc}, M) \diamond$$

Prop. 1 offers a practical way to compute the supremal partially controllable sub-language of M w.r.t. to Σ'_{uc}, Σ_{uc}, and L. It can be shown that its computation is in $\mathcal{O}(|\Sigma| N_M^2 N_L)$, where N_M is the size of the FSM encoding M and N_L the one of L.

4. CONTROL OF CONCURRENT DES

Given a Concurrent DES, $G = G_1 \parallel \cdots \parallel G_n$, and a control objective K, we want to compute a controllable sub-language of $K \cap \mathcal{L}(G)$ w.r.t. $\mathcal{L}(G)$ and Σ_{uc}, without having to build G itself.

4.1 Modular computation of a controllable sub-language of K w.r.t. $\mathcal{L}(G)$

Based on the concept of partial controllability applied on K and on the approximations of the plant $P_i^{-1}(\mathcal{L}(G_i))$

derived from each of its components, the next theorem provides a modular way to compute a sub-language of K that is controllable with respect to the plant.

Theorem 2. Let $G = G_1 \parallel \cdots \parallel G_n$, with G_i acting upon $\Sigma_i = \Sigma_{i,uc} \cup \Sigma_{i,c}$ and $L_i = \mathcal{L}(G_i)$. Let $K \subseteq \Sigma^*$ be a prefix-closed language modeling the expected behavior. For $i \leq n$, we note

- $K_i = K \cap P_i^{-1}(\mathcal{L}(G_i))$, and
- $K_i^{\uparrow pc}$ the supremal sublanguage of K_i partially controllable w.r.t. $\Sigma_{i,uc}$, Σ_{uc}, K_i and $P_i^{-1}(\mathcal{L}(G_i))$.

Then, $\bigcap_{i\leq n} K_i^{\uparrow pc}$ is controllable w.r.t. Σ_{uc} and $\mathcal{L}(G)$.

Proof : First, as K, $\mathcal{L}(G_i)$ are prefix-closed, languages $P_i^{-1}(\mathcal{L}(G_i))$ for $i \leq n$ are prefix-closed, and therefore K_i and $K_i^{\uparrow pc}$ are also prefix-closed. Now, according to Definition 1, we have to show that

$$(\bigcap_{i\leq n} K_i^{\uparrow pc})\Sigma_{uc} \cap \mathcal{L}(G) \subseteq \bigcap_{i\leq n} K_i^{\uparrow pc}$$

Let $s \in \bigcap_{i\leq n} K_i^{\uparrow pc}$ and $\sigma \in \Sigma_{uc}$ be such that $s.\sigma \in \mathcal{L}(G)$. We thus have to show that $s\sigma \in \bigcap_{i\leq n} K_i^{\uparrow pc}$. Without lost of generality we can assume that $\sigma \in \Sigma_{1,uc}$.

Since $s\sigma \in \mathcal{L}(G)$, $s\sigma \in P_1^{-1}(\mathcal{L}(G_1))$. Hence we have that $s\sigma \in K_1^{\uparrow pc}\Sigma_{1,uc} \cap P_1^{-1}(\mathcal{L}(G_1))$. Moreover, according to definition 3, $K_1^{\uparrow pc}$ is controllable w.r.t. $\Sigma_{1,uc}$ and $P_1^{-1}(\mathcal{L}(G_1))$, from which we can conclude that $s\sigma \in K_1^{\uparrow pc}$.

As $s\sigma \in K_1^{\uparrow pc}$, we have that $s\sigma \in K$. And as $s\sigma \in \mathcal{L}(G)$, then $\forall i \leq n$, $s\sigma \in P_i^{-1}(\mathcal{L}(G_i))$, which entails that $\forall i \leq n$, $s\sigma \in K_i = K \cap P_i^{-1}(\mathcal{L}(G_i))$. Hence, $\forall i \leq n$, $s\sigma \in K_i{}^{\uparrow pc}\Sigma_{uc} \cap K_i$. Now, according to definition 3, $\forall i \leq n$, $K_i^{\uparrow pc}$ is controllable w.r.t. Σ_{uc} and K_i. Hence $\forall i \leq n$, $s\sigma \in K_i^{\uparrow pc}$, which entails that $s\sigma \in \bigcap_{i\leq n} K_i^{\uparrow pc}$. ◇

4.2 Computation of $SupC((K \cap \mathcal{L}(G), \mathcal{L}(G), \Sigma_{uc})$

In the remainder of this paper we use $(K \cap \mathcal{L}(G))^{\uparrow c}$ to denote $SupC((K \cap \mathcal{L}(G), \mathcal{L}(G), \Sigma_{uc})$. According to Theorem 2, we have that $\bigcap_{i\leq n} K_i^{\uparrow pc} \subseteq (K \cap \mathcal{L}(G))^{\uparrow c}$. But, the equality does not hold in general. In this section, we present some conditions under which Theorem 2 gives access to the supremal solution.
Let us first introduce lemma 3. This lemma shows that whenever the shared events Σ_s are controllable, then $(K \cap \mathcal{L}(G))^{\uparrow c}$ verifies a part of the partial controllability condition.

Lemma 3. Let $G = G_1 \parallel \cdots \parallel G_n$ be the plant and $K \subseteq \Sigma^*$ a prefix-closed language modeling the expected behavior, then if $\Sigma_s \subseteq \Sigma_c$, then $\forall i \leq n$, $(K \cap \mathcal{L}(G))^{\uparrow c}$ is controllable w.r.t $\Sigma_{i,uc}$, $P_i^{-1}(\mathcal{L}(G_i))$.

Proof : Let $i \leq n$. Let us consider $s \in (K \cap \mathcal{L}(G))^{\uparrow c}$ and $\sigma \in \Sigma_{i,uc}$ such that $s\sigma \in P_i^{-1}(\mathcal{L}(G_i))$. We have to show that $s\sigma \in (K \cap \mathcal{L}(G))^{\uparrow c}$. Since $\sigma \in \Sigma_{i,uc}$ and $\Sigma_s \subseteq \Sigma_c$, we have $\sigma \in \Sigma_i \setminus \Sigma_s$. Moreover, $s \in P_i^{-1}(\mathcal{L}(G_i))$, $s \in \mathcal{L}(G)$ and $s\sigma \in P_i^{-1}(\mathcal{L}(G_i))$, which entails that $s\sigma \in \mathcal{L}(G)$ (Lemma 2). Now, as $\Sigma_{i,uc} \subseteq \Sigma_{uc}$, we have that $s\sigma \in (K \cap \mathcal{L}(G))^{\uparrow c}.\Sigma_{uc} \cap \mathcal{L}(G)$. Since $(K \cap \mathcal{L}(G))^{\uparrow c}$ is controllable w.r.t Σ_{uc} and $\mathcal{L}(G)$, this entails that $s\sigma \in (K \cap \mathcal{L}(G))^{\uparrow c}$. ◇

The next theorem states that whenever the shared event are controllable and the language of the control objective is included in the one of the plant then our methodology gives access to the supremal controllable sub-language of K w.r.t. L and Σ_{uc}.

Theorem 3. If $\Sigma_s \subseteq \Sigma_c$ and $K \subseteq \mathcal{L}(G)$, then with the notations of Theorem 2, $\bigcap_{i\leq n} K_i^{\uparrow pc} = K^{\uparrow c}$.

Proof : From Theorem 2, $\bigcap_{i\leq n} K_i^{\uparrow pc} \subseteq (K \cap \mathcal{L}(G))^{\uparrow c}$. As $K \cap \mathcal{L}(G) = K$ it is then sufficient to show that $K^{\uparrow c} \subseteq \bigcap_{i\leq n} K_i^{\uparrow pc}$ or, equivalently that $\forall i \leq n$, $K^{\uparrow c} \subseteq K_i^{\uparrow pc}$. To do so, let us pick up a $i \leq n$, and let us show that $K^{\uparrow c}$ is partially controllable w.r.t. $\Sigma_{i,uc}$, Σ_{uc}, K_i and $P_i^{-1}(L_i)$.

(i) First, according to lemma 3, $K^{\uparrow c}$ is controllable w.r.t $\Sigma_{i,uc}$ and $P_i^{-1}(\mathcal{L}(G_i))$.
(ii) Let us now show that $K^{\uparrow c}$ is controllable w.r.t. Σ_{uc} and K_i. Let us consider $s \in K^{\uparrow c}$, $\sigma \in \Sigma_{uc}$ such that $s\sigma \in K_i$, we have to prove that $s\sigma \in K^{\uparrow c}$. Since $K_i \subseteq \mathcal{L}(G)$, $s\sigma \in \mathcal{L}(G)$. We then have $s \in K^{\uparrow c}$, $\sigma \in \Sigma_{uc}$ and $s\sigma \in \mathcal{L}(G)$. Hence, because $K^{\uparrow c}$ is controllable w.r.t. Σ_{uc} and $\mathcal{L}(G)$, $s\sigma \in K^{\uparrow c}$.

This proves that $K^{\uparrow c}$ is partially controllable w.r.t. $\Sigma_{i,uc}$, Σ_{uc}, K_i, $P_i^{-1}(\mathcal{L}(G_i))$. Now, as $K_i^{\uparrow pc}$ is supremal and partially controllable w.r.t. $\Sigma_{i,uc}$, Σ_{uc}, K_i, $P_i^{-1}(\mathcal{L}(G_i))$, we can deduce that $\forall i \leq n$, $K^{\uparrow c} \subseteq K_i^{\uparrow pc}$. Finally, $K^{\uparrow c} \subseteq \bigcap_{i\leq n} K_i^{\uparrow pc}$. ◇

The interest of this method is that it avoids the building of the entire plant; hence reducing the complexity of the supervisory synthesis phase. Indeed, if $G = G_1 \parallel \cdots \parallel G_n$ is such that $|\mathcal{X}_{G_i}| = N$ and H, with $|\mathcal{X}_H| = m$, is the FSM modeling the language specification K, then the FSM modeling the *partial specification* K_i are in $\mathcal{O}(N.m)$. According to Section 3, the complexity to compute the supremal partially controllable sublanguage of each K_i is in $\mathcal{O}(N^2.m)$. Finally, the overall complexity is in $\mathcal{O}(n.N^2.m)$. This has to be opposed to the space complexity $\mathcal{O}(N^n.m)$ of computing $(K \cap \mathcal{L}(G))^{\uparrow c}$ on G, seen as a unique FSM.

Remark 1. To check that $K \subseteq \mathcal{L}(G)$ it is sufficient to check that $\forall 1 \leq i \leq n$, $K \subseteq P_i^{-1}(\mathcal{L}(G_i))$. Hence, it is not necessary to compute $\mathcal{L}(G)$.

In some situations, modeling the expected behavior by a language included in the one of the plant may lead to a language that is too large to be efficiently represented. Moreover, requiring the inclusion of languages makes that the specification of K may be itself relatively difficult to identify insofar as the language $\mathcal{L}(G)$ is not known. Theorem 4 gives another sufficient condition under which Theorem 2 gives access to the supremal solution. First, we need to introduce the notion of observable language.

Definition 4. Let K and M be two prefix-closed languages over Σ and $\Sigma', \Sigma'' \subseteq \Sigma$. K is said to be observable w.r.t. $P_{\Sigma'}$, Σ'' and M if $\forall s, s' \in K$, $\forall \sigma \in \Sigma''$, if $P_{\Sigma'}(s') = P_{\Sigma'}(s)$ and $s'\sigma \in K$, and $s\sigma \in M$, then $s\sigma \in K$. •

It is shown in e.g. [Cassandras and Lafortune (1999), Chap. 3.7] that the observability condition can be checked in $\mathcal{O}(m^2N)$, where m and N are the number of states of the FSM that generate K and M.

Definition 5. Let $K \subseteq \Sigma^*$ be a prefix-closed language and $G = G_1 \parallel \cdots \parallel G_n$ a concurrent system, then K is said to be G-observable if $\forall i \in \{1, \ldots, n\}$, $\forall s \in \Sigma^*$, $K_i(s, \Sigma_{uc})$ is observable w.r.t. $P_i, \Sigma_{i,uc}$ and $P_i^{-1}(\mathcal{L}(G_i))(s, \Sigma_{uc})$, where $K_i = K \cap \mathcal{L}(G_i)$, $\forall i \leq n$. •

The condition of G-observability states that after each trace s admissible in K_i, if there exists two admissible suffixes of s of uncontrollable events, say s_1 and s_2, that have the same projection over the local one (i.e. $P_i(s_1) = P_i(s_2)$), then if one can be extended by an uncontrollable event in a trace of K_i and the other one in a trace of the abstracted specification G_i^{-1}, then the two extended traces have to belong to the specification. Finally, note that the G-observability condition can be checked in $\mathcal{O}(nm^3N^4)$, where n is the number of components in G, N the state size of G_i and m the size of the FSM that generates K. Now, based on Definition 5, we have that

Theorem 4. Assume that $\Sigma_s \subseteq \Sigma_c$ and that K is G-observable, then $\bigcap_{i\leq n} K_i^{\uparrow pc} = (K \cap \mathcal{L}(G))^{\uparrow c}$ ◇

Proof : According to Theorem 2, $\bigcap_{i\leq n} K_i^{\uparrow pc} \subseteq (K \cap \mathcal{L}(G))^{\uparrow c}$ Thus, we have to show that

$$\forall i \leq n,\ (K \cap \mathcal{L}(G))^{\uparrow c} \subseteq K_i^{\uparrow pc}$$

To do so, let us consider $i \leq n$ and $((K \cap \mathcal{L}(G))^{\uparrow c})^{\downarrow K_i,c}$ (the infimal prefix-closed and controllable superlanguage of $(K \cap \mathcal{L}(G))^{\uparrow c}$ w.r.t. Σ_{uc} and K_i). We now prove that $((K \cap \mathcal{L}(G))^{\uparrow c})^{\downarrow K_i,c}$ is partially controllable w.r.t $\Sigma_{i,uc}$, Σ_{uc}, K_i and $P_i^{-1}(\mathcal{L}(G_i))$. Indeed, if $((K \cap \mathcal{L}(G))^{\uparrow c})^{\downarrow K_i,c}$ is partially controllable w.r.t $\Sigma_{i,uc}$, Σ_{uc}, K_i and $P_i^{-1}(\mathcal{L}(G_i))$, then we will have that $((K \cap \mathcal{L}(G))^{\uparrow c})^{\downarrow K_i,c} \subseteq K_i^{\uparrow pc}$. Moreover, as $(K \cap \mathcal{L}(G))^{\uparrow c} \subseteq ((K \cap \mathcal{L}(G))^{\uparrow c})^{\downarrow K_i,c}$, we will also have that $(K \cap \mathcal{L}(G))^{\uparrow c} \subseteq K_i^{\uparrow pc}$ and the proof will be done.

Let us now show that $((K \cap \mathcal{L}(G))^{\uparrow c})^{\downarrow K_i,c}$ is partially controllable w.r.t $\Sigma_{i,uc}$, Σ_{uc}, K_i and $P_i^{-1}(\mathcal{L}(G_i))$. According to Definition 3, we have to show that $((K \cap \mathcal{L}(G))^{\uparrow c})^{\downarrow K_i,c}$ is *(i)* controllable w.r.t. $\Sigma_{i,uc}$ and $P_i^{-1}(\mathcal{L}(G_i))$ and *(ii)* controllable w.r.t. Σ_{uc} and K_i. Item *(ii)* is obvious since by the definition of the infimal controllable language w.r.t. Σ_{uc} and K_i, $((K \cap \mathcal{L}(G))^{\uparrow c})^{\downarrow K_i,c}$ is controllable w.r.t. Σ_{uc} and K_i. Let us now prove the point *(i)*

Let us consider $s \in ((K \cap \mathcal{L}(G))^{\uparrow c})^{\downarrow K_i,c}$ and $\sigma \in \Sigma_{i,uc}$ such that $s\sigma \in P_i^{-1}(\mathcal{L}(G_i))$. We have

$$s\sigma \in ((K \cap \mathcal{L}(G))^{\uparrow c})^{\downarrow K_i,c}.\Sigma_{i,uc} \cap P_i^{-1}(\mathcal{L}(G_i))$$

Moreover, $((K \cap \mathcal{L}(G))^{\uparrow c})^{\downarrow K_i,c} = (K \cap \mathcal{L}(G))^{\uparrow c}.\Sigma_{uc}^* \cap K_i$. So s is of the form $s = s't$ with $s' \in (K \cap \mathcal{L}(G))^{\uparrow c}$ and $t \in \Sigma_{uc}^*$. Now as $s\sigma = s't\sigma \in P_i^{-1}(\mathcal{L}(G_i))$, we also have that $s'P_i(t\sigma) \in P_i^{-1}(\mathcal{L}(G_i))$ (Lemma 1).

We now have that $s' \in (K \cap \mathcal{L}(G))^{\uparrow c}$, $P_i(t\sigma) \in \Sigma_{i,uc}^*$ and $s'P_i(t\sigma) \in P_i^{-1}(\mathcal{L}(G_i))$. We then have that $s'P_i(t\sigma) \in (K \cap \mathcal{L}(G))^{\uparrow c}$ as $(K \cap \mathcal{L}(G))^{\uparrow c}$ is controllable w.r.t. $\Sigma_{i,uc}$ and $P_i^{-1}(\mathcal{L}(G_i))$ (Lemma 3). Finally, as $s'P_i(t\sigma) \in (K \cap \mathcal{L}(G))^{\uparrow c} \subseteq K_i$, we obtain $P_i(t\sigma) = P_i(t)\sigma \in K_i(s', \Sigma_{uc})$.

Moreover, as by hypothesis $s\sigma = s't\sigma \in P_i^{-1}(\mathcal{L}(G_i))$, we also have that $t\sigma \in P_i^{-1}(\mathcal{L}(G_i))(s', \Sigma_{uc})$. By definition of the projection, $P_i(t\sigma) = P_i(P_i(t\sigma))$. Overall, we have that $P_i(t) \in K_i(s', \Sigma_{uc})$, $t \in K_i(s', \Sigma_{uc})$, with $P_i(t) = P_i(P_i(t))$, $\sigma \in \Sigma_{uc}$, $P_i(t)\sigma \in K_i(s', \Sigma_{uc})$ and $t\sigma \in P_i^{-1}(\mathcal{L}(G_i))$. As $K_i(s', \Sigma_{uc})$ is observable w.r.t. P_i, $\Sigma_{i,uc}$ and $P_i^{-1}(\mathcal{L}(G_i))(s', \Sigma_{uc})$, we obtain that $t\sigma \in K_i(s', \Sigma_{uc})$. Hence $s\sigma$ $(= s't\sigma) \in K_i$ and it entails that

$$s\sigma \in ((K \cap \mathcal{L}(G))^{\uparrow c})^{\downarrow K_i,c}.\Sigma_{uc} \cap K_i$$

As $((K \cap \mathcal{L}(G))^{\uparrow c})^{\downarrow K_i,c}$ is (by definition) controllable w.r.t. Σ_{uc} and K_i, $s\sigma \in ((K \cap \mathcal{L}(G))^{\uparrow c})^{\downarrow K_i,c}$. Thus $((K \cap \mathcal{L}(G))^{\uparrow c})^{\downarrow K_i,c}$ is controllable w.r.t. $\Sigma_{i,uc}$ and $P_i^{-1}(\mathcal{L}(G_i))$.

Finally, $((K \cap \mathcal{L}(G))^{\uparrow c})^{\downarrow K_i,c}$ is partially controllable w.r.t $\Sigma_{i,uc}$, Σ_{uc}, K_i and $P_i^{-1}(\mathcal{L}(G_i))$, which concludes the proof. ◇

The complexity of checking the condition of G-observability and computing $(K \cap \mathcal{L}(G))^{\uparrow c}$ with our method is in $\mathcal{O}(n.m^2.N^2(m.N^2 + 1))$, which has to be opposed to $\mathcal{O}(N^n.m)$ (complexity with a monolithic approach).

The next proposition shows whenever K is separable then it respects the conditions of Theorem 4.

Proposition 2. [Gaudin and Marchand (2004)] Let $G = G_1 \parallel \cdots \parallel G_n$ be the plant to be controlled s.t. $\mathcal{L}(G_i) \subseteq \Sigma_i^*$. Let $K \subseteq \Sigma^*$ be the expected specification. If K is separable w.r.t. $\{\Sigma_i\}_{1\leq i\leq n}$ then K is G-observable, i.e. $\forall i \leq n$, $\forall s \in K_i = K \cap P_i^{-1}(\mathcal{L}(G_i))$, $K_i(s, \Sigma_{uc})$ is observable w.r.t. P_i, $\Sigma_{i,uc}$ et $P_i^{-1}(\mathcal{L}(G_i))(s, \Sigma_{uc})$. ◇

The previous proposition states that whenever the specification is separable then our methodology offers an alternative way to the one of [Willner and Heymann (1991)] to compute $(K \cap \mathcal{L}(G))^{\uparrow c}$. Indeed, the solution of [Willner and Heymann (1991)] gives access to a set of decentralized supervisors acting upon each local component of the plant whereas, in our case, the result is a centralized supervisor (See Section 4.3 above).

A contrario, the next example shows that a language that respects the G-observability condition, may be not separable.

Example 1. Let $G = G_1 \parallel G_2$ with $\mathcal{L}(G_1) = \overline{\{a_1.uc_1\}}$ and $\mathcal{L}(G_2) = \overline{\{a_2.uc_2\}}$. We have $\Sigma_{1,uc} = \{uc_1\}$ and $\Sigma_{2,uc} = \{uc_2\}$. Let $K = \overline{\{a_1.uc_1, a_2.uc_2\}}$ be a prefix-closed language over $\Sigma = \Sigma_1 \cup \Sigma_2$. One can check that K verifies the G-observability condition. However, K is not separable w.r.t. $\{\Sigma_1, \Sigma_2\}$, since $a_1 \in P_1(K)$ and $a_2 \in P_2(K)$, thus $a_1a_2 \in P_1(K) \parallel P_2(K)$ but $a_1a_2 \notin K$. ◇

4.3 *The Supervisor acting upon G.*

Let us now describe the way a supervisor can be extracted from the previously computed languages and how it can act upon G in order to achieve the control objective K. With the notations of Theorem 2, $\bigcap_{i \leq n} K_i^{\uparrow pc}$ is controllable with respect to Σ_{uc} and $\mathcal{L}(G)$. However, it is not of interest to perform the intersection between these languages and to derive a supervisor from the result (all the computational advantages of our method would be lost). Following the concept of modularity described in [Wonham and Ramadge (1988)], the supervisor $\mathcal{S}$ will be seen as an oracle taking its decision according to the history of the system and the so-called pc-supervisors $\mathcal{S}_i^{pc}$ derived from $K_i^{\uparrow pc}$. The supervisor architecture is summarized in Figure 1. From each $K_i^{\uparrow pc}$, we derive a "supervisor" $\mathcal{S}_i^{pc}$, which after a trace of G (which is also a trace of $P_i^{-1}(\mathcal{L}(G_i))$), delivers the set of events that extend s in a trace of $K_i^{\uparrow pc}$ (each of these pc-supervisor ensures on $P_i^{-1}(\mathcal{L}(G_i))$ the partial controllability property w.r.t. K and $P_i^{-1}(\mathcal{L}(G_i))$). Further following the modularity concept, the set of allowed events is given by $\mathcal{S}(s) = \mathcal{S}_1^{pc}(s) \cap \cdots \cap \mathcal{S}_n^{pc}(s)$. Finally, the sub-set of events that is allowed in G_i is given by $\mathcal{S}(s) \cap \Sigma_i$.

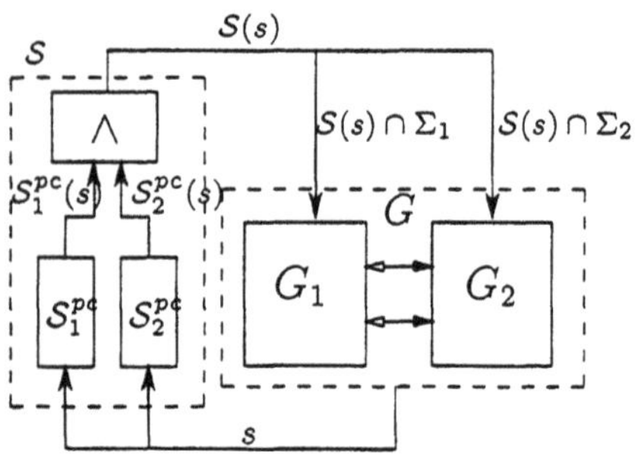

Fig. 1. Supervision Scheme

5. CONCLUSION

In this paper, we have investigated the Supervisory Control of Concurrent Discrete Event Systems. In particular, we proposed a modular method allowing to compute the supremal language of a specification K controllable w.r.t. to the plant G. From the plant G and each of its components G_i, we derive a set of approximations $\mathcal{L}(G_i^{-1})$ and we ensure by control that each of these approximations respects a new language property, called partial controllability condition that depends on K. It is then shown that whenever the original specification respects some conditions (either $K \subseteq \mathcal{L}(G)$, or K is G-observable) then a centralized supervisor can be extracted from the controlled approximations in such a way that the behavior of the controlled plant corresponds to the supremal controllable language contained in the one of the specification. Let us now emphasize some points that we did not mention so far but that are easy to deduce from the theorem (2,3,4). If the specification is given in a modular way e.g. $K \cap K'$, then in the case of prefix-closed languages, the modularity results in [Wonham and Ramadge (1988)] together with the results of Theorem 3, ensures that

$$\bigcap_{i \leq n} K_i^{\uparrow pc} \cap \bigcap_{i \leq n} K_i'^{\uparrow pc} = [(K \cap K') \cap \mathcal{L}(G)]^{\uparrow c}$$

as far as the conditions of Theorems 3 or 4 are satisfied by the two specifications.

If one want to change a component of G, e.g. replacing G_i by G_i', then as far as G_i' is expressed using the same alphabet as the one of G_i with the same partitioning between the controllable/uncontrollable event, then it is sufficient to recompute $K_i'^{\uparrow pc} = (K \cap P_i^{-1}(\mathcal{L}(G_i')))^{\uparrow pc}$ in order to obtain the new supervisor (note that only the conditions referring to G_i' has to be (re)-checked). Hence this methodology is suitable for reconfigurable plants.

So far we have been interested in the control of plant for prefix-closed specification. We are currently looking for results ensuring that the controlled plant is non-blocking while still avoiding the computation of the whole state space. Another point of interest would be to extend these techniques to the hierarchical model described in [Gaudin and Marchand (2003)].

REFERENCES

S. Abdelwahed and W. Wonham. Supervisory control of interacting discrete event systems. In *41th IEEE Conference on Decision and Control*, pages 1175–1180, Las Vegas, USA, December 2002.

C. Cassandras and S. Lafortune. *Introduction to Discrete Event Systems.* Kluwer Academic Publishers, 1999.

M.H. deQueiroz and J.E.R. Cury. Modular supervisory control of large scale discrete-event systems. In *Discrete Event Systems: Analysis and Control. Proc. WODES'00*, pages 103–110, 2000.

B. Gaudin and H. Marchand. Modular supervisory control of asynchronous and hierarchical finite state machines. In *European Control Conference, ECC 2003*, Cambridge, UK, September 2003.

B. Gaudin and H Marchand. Supervisory control of concurrent discrete event systems. Research Report 1593, IRISA, February 2004. available at http://www.irisa.fr/vertecs/Publis/Ps/1593.ps.

S. Jiang and R. Kumar. Decentralized control of discrete event systems with specializations to local control and concurrent systems. *IEEE Transactions on Systems, Man, and Cybernetics, Part B*, 30(5):653–660, October 2000.

K. Åkesson, Flordal, and M. Fabian. Exploiting modularity for synthesis and verification of supervisors. In *Proc. of the IFAC*, barcelona, Spain, July 2002.

R.J. Leduc, B.A. Brandin, W.M. Wonham, and M. Lawford. Hierarchical interface-based supervisory control: Serial case. In *Proc. of 40th Conf. Decision Contr.*, pages 4116–4121, December 2001.

P. J. Ramadge and W. M. Wonham. The control of discrete event systems. *Proceedings of the IEEE; Special issue on Dynamics of Discrete Event Systems*, 77(1):81–98, 1989.

K. Rohloff and S. Lafortune. The control and verification of similar agents operating in a broadcast network environment. In *42nd IEEE Conference on Decision and Control*, Hawaii, USA, 2003.

Y. Willner and M. Heymann. Supervisory control of concurrent discrete-event systems. *International Journal of Control*, 54(5): 1143–1169, 1991.

W. M. Wonham and P. J. Ramadge. Modular supervisory control of discrete event systems. *Mathematics of Control Signals and Systems*, 1:13–30, 1988.

ELSEVIER
IFAC
PUBLICATIONS
www.elsevier.com/locate/ifac

OPTIMAL CONTROL FOR A CLASS OF MANUFACTURING SYSTEMS WITH PRODUCTION RATE DEPENDENT FAILURE RATES [1]

Raffaele Chiovelli *, Francesco Martinelli * and Paolo Valigi **

* *Dipartimento di Informatica Sistemi e Produzione, Università di Roma Tor Vergata, I-00133 Rome, Italy*
** *Dip. di Ingegneria Elettronica e dell'Informazione, Università di Perugia, I-06125 Perugia, Italy*

Abstract: In this paper we consider a failure prone, single part type, single machine manufacturing system. The failure repair process of the machine is characterized by a non-homogeneous Markov chain, with a constant repair rate and a failure rate which is a non decreasing function $q_d(u)$ of the production rate. The stability issue is discussed with respect to the failure rate function considered. For a piecewise constant, two value function $q_d(u)$, the optimal policy is characterized and it is completely derived for a set of examples. This optimal policy is not a hedging point policy, which is known to be optimal only for particular failure rate functions $q_d(u)$.

Keywords: Production dependent failures, hedging point, manufacturing systems.

1. INTRODUCTION

The problem of production control for failure prone manufacturing systems has been considered since many years but it still presents several open questions. A complete analytical solution is available only for some particular cases, beginning from the one considered by (Kimemia and Gershwin, 1983; Akella and Kumar, 1986; Bielecki and Kumar, 1988), where the system comprises a single machine with a homogeneous Markov failure/repair process. The optimal policy for this problem is the so called *hedging point* policy, according to which the machine is operated at full rate until the inventory level hits a non-negative hedging level z^*, which is then maintained until the next failure event occurs. Extensions have been realized in several directions, to include multi-state machines, multi-part type systems, and backlog/inventory constraints, but usually assuming a homogeneous Markov failure/repair process. The problem becomes much more involved in the general case where the failure/repair process is non-homogeneous or not even Markov. The hedging point policy is not optimal in general, remaining optimal for the Markov case if the dependence of the failure rate on the production rate is affine (Hu *et al.*, 1994). As discussed in (Hu *et al.*, 1994) and in (Hu and Xiang, 1995), in the general case the optimal policy operates the machine increasing the inventory level as the age of the machine up-time increases.

Along this direction we consider a failure prone single machine system where the failure/repair process is described through a non-homogeneous Markov chain. It is non-homogeneous since the failure rate $q_d(u)$ depends on the production rate u. For a special case, where such a dependence is a

[1] Partially supported by MIUR (grants PRIN: prot. 2003090090, COFIN: Matrics and FIRB: Tiger)

two value piecewise constant function, we provide some structural properties of the optimal policy and completely characterize it on some examples.

Failure prone manufacturing systems can be studied in the context of hybrid systems, where the term *hybrid* is used for systems that combine time-driven and event-driven dynamics. The discrete part of the system considered in this paper is the state of the machine, while the continuous part is the buffer content, whose evolution depends on the current state of the machine. A continuous model is adopted for the part flow: the use of fluid networks to model discrete event queuing systems has been extensively used in literature since the '80 (Chen and Mandelbaum, 1991).

2. PROBLEM FORMULATION

According to a standard notation, let $x(t)$ denote the buffer content at time t, with $x(t) > 0$ representing an inventory surplus and $x(t) < 0$ a backlog of $-x(t)$. Let d be the constant demand rate to be met. The machine failure/repair process is described by means of a non-homogeneous Markov chain, comprising a failure state (denoted as state 0, or *down*) where the machine can not produce anything and a working state (denoted as state 1, or *up*) where the production rate $u(t)$ ranges in a finite interval $[0, \mu]$. The failure rate $q_d(u)$ depends on the production rate u of the machine while the repair rate q_u is a constant parameter. We assume that $q_d(u)$ is a non-decreasing function of u. The buffer level $x(t)$ at time t satisfies the following dynamical equation:

$$\dot{x} = u(t) - d \tag{1}$$

where $u(t) \equiv 0$ if at time t the machine is in the down state, and $u(t) \in [0, \mu]$ if at time t the machine is in the up state. The scheduling problem considered in this paper is the determination of the optimal control $u(t)$ minimizing the long-term average expected cost

$$J = \lim_{T\to\infty} \frac{1}{T} E\left[\int_0^T g[x(t), u(t)]dt\right], \tag{2}$$

where $g(x, u) = c_p x^+ + c_m x^-$, with $x^+ = \max\{0, x\}$, $x^- = \max\{0, -x\}$ and c_p, c_m non-negative constants. For constant failure rates q_d (Kimemia and Gershwin, 1983; Akella and Kumar, 1986; Bielecki and Kumar, 1988) and, more in general, for linear affine functions $q_d(u)$ (Hu *et al.*, 1994), the optimal policy $u^*(x)$ is a feedback hedging point policy, which works as much as possible until the backlog/inventory level hits a non-negative hedging level z^* to be determined. Formally:

$$u_{HP}(x) = \begin{cases} \mu & x < z^* \\ d & x = z^* \\ 0 & x > z^* \end{cases} \tag{3}$$

This policy is not optimal in general for the problem considered in this paper, as it will be illustrated below.

3. FEASIBILITY ANALYSIS

If $q_d(u) = q_d$ is a constant parameter, it is usual to take a feasibility assumption: namely $\frac{\mu q_u}{q_u + q_d} - d > 0$, which means that in the average the system is able to meet the demand, and the backlog remains bounded. We define in this case, where $q_d(u)$ is not constant with u, a feasibility function:

$$S(u) := \frac{u\, q_u}{q_u + q_d(u)} - d, \tag{4}$$

and remark that it is possible in the average to meet the demand if $S(u) > 0$ for at least a $u \in (0, \mu]$. In (Hu *et al.*, 1994) it was observed that, for several tool wear problems, it is of interest to consider failure rate functions $q_d(u) = a\, u^\beta + b$, where a, b and β are positive constants (see also (Drozda and C. Wick, 1983)). Under this assumption on the failure rate function it is straightforward to prove the following result.

Lemma 1. Assume $q_d(u) = a\, u^\beta + b$, with a, b and β positive constant. Then:

- if $0 \le \beta < 1$, it always exists $U_1 > 0$ such that $S(u) > 0$ for all $u > U_1$;
- if $\beta = 1$, it is possible to find a $U_1 > 0$ such that $S(u) > 0$ for all $u > U_1$ only if $q_u > a\, d$, otherwise $S(u) \le 0$ for all u;
- if $\beta > 1$, depending on system parameters, two different cases may occur:
  - (i) there exists U_1, U_2 such that $S(u) > 0$ for all $u \in (U_1, U_2)$;
  - (ii) $S(u) \le 0$ for all u.

Clearly, it is possible to meet the demand in the average only if it is possible to find such a U_1 (and a U_2 in the last case) and it happens that $\mu > U_1$.

4. THE TWO VALUE FAILURE RATE CASE

In the sequel of the paper we will consider a piecewise constant, two value, failure rate function $q_d(u)$, i.e.

$$q_d(u) := \begin{cases} q_{d1} & u \in (U, \mu] \\ q_{d2} & u \in (0, U] \end{cases} \tag{5}$$

where $q_{d1} \ge q_{d2}$ and $U \le \mu$. This function is considered to characterize a machine with a nominal failure rate q_{d2} which increases to a larger value q_{d1} if the machine is operated at a larger production rate. This is of practical interest but it is also intended as a preliminary step to solve the problem where $q_d(u)$ is a more general function.

For the problem to be of interest, the following Assumption will be considered in the following, namely that the demand may be met operating the machine with the large ($u > U$) and the small ($u \leq U$) production regime as well.

Assumption 1. Consider the problem with the two level failure rate function $q_d(u)$ given by (5). Then it is assumed that $S(U) > 0$ and $S(\mu) > 0$.

Remark 1. No assumption is taken on the order between $S(U)$ and $S(\mu)$. Since $q_{d1} \geq q_{d2}$ and $\mu \geq U$, both the cases $S(U) \geq S(\mu)$ and $S(U) < S(\mu)$ are possible.

The remaining of the paper is devoted to solve the following problem.

Problem 1. For the failure prone system considered above, with the two level failure rate function $q_d(u)$ given in (5), determine under Assumption 1 the policy u^* minimizing (2).

4.1 Hedging Point and Reduced Hedging Point policies.

Before investigating around the structure of the optimal policy, we compare in this section two in general non-optimal simple policies: a classical Hedging Point policy (HP), which operates the machine at the maximum possible rate μ until the inventory level reaches a safety stock z^*_μ (see eq. (3)) and a *Reduced Hedging Point* policy (RHP), which operates the machine at the smaller rate U until a safety stock z^*_U is attained. This could be of interest in applications where a frequent change of the operative regime of the machine may be undesirable and/or the inventory/backlog level may be inspected with difficulty and only approximately.

First of all observe that, under Assumption 1, $U > d$, hence, when the machine works with rate d, the failure rate is q_{d2}. Let $J_\mu(z)$ and $J_U(z)$ be the value of the performance index (2) if the safety stock is z and the machine is operated through the classical hedging point policy or the reduced one, respectively. Then define $z^*_U := \arg\min_z J_U(z)$ and $z^*_\mu := \arg\min_z J_\mu(z)$ as the optimal safety stock with respect to the two policies, with J^*_U and J^*_μ the corresponding optimal values. Then, using the same procedure followed in (Akella and Kumar, 1986; Bielecki and Kumar, 1988), it is possible to compute the steady state probability densities of the inventory/backlog level under both the two policies and the optimal values for the hedging points and the corresponding costs. We obtain the following expressions:

$$z^*_U := \begin{cases} 0 & \gamma_2 \geq \dfrac{c_m}{c_p + c_m} \\ \dfrac{1}{\alpha_2} \log\left[\dfrac{c_p + c_m}{c_p}(1 - \gamma_2)\right] & otherwise \end{cases}$$

$$J^*_U := \begin{cases} c_p z^*_U + \dfrac{c_p d}{q_{d2} + q_u} & \text{if } z^*_U > 0 \\ \dfrac{c_m q_{d2} U d}{(q_{d2} + q_u)\Delta_2} & \text{if } z^*_U = 0 \end{cases} \tag{6}$$

for the reduced hedging point policy (RHP) and

$$z^*_\mu := \begin{cases} 0 & \dfrac{\mu q_{d2}(c_p + c_m)}{(\Delta_1 + \mu q_{d2})c_p} \leq 1 \\ \dfrac{\log\left[\frac{\mu q_{d2}(c_p + c_m)}{(\Delta_1 + \mu q_{d2})c_p}\right]}{\alpha_1} & otherwise \end{cases}$$

$$J^*_\mu := \begin{cases} c_p z^*_\mu + \dfrac{c_p d(\mu - d)}{\Delta_1 + \mu q_{d2}} & \text{if } z^*_\mu > 0 \\ \dfrac{c_m \mu q_{d2}(\mu - d)d}{(\Delta_1 + \mu q_{d2})\Delta_1} & \text{if } z^*_\mu = 0 \end{cases} \tag{7}$$

for the standard hedging point policy (HP). In the expressions above we used part of the following notation, which will be also required in the sequel of the paper: $\Delta_1 := \mu q_u - d(q_u + q_{d1})$, $\Delta_2 := U q_u - d(q_u + q_{d2})$, $\alpha_1 := \Delta_1/(d(\mu - d))$, $\alpha_2 := \Delta_2/(d(U - d))$, $\gamma_1 := \Delta_1/[(\mu - d)(q_u + q_{d1})]$ and $\gamma_2 := \Delta_2/[(U - d)(q_u + q_{d2})]$. Observe that, under Assumption 1, $\Delta_i > 0$, $\alpha_i > 0$ and $\gamma_i \in (0, 1)$, $i = 1, 2$. Observe also that the expressions for HP are different from the classical expressions (see (Bielecki and Kumar, 1988) for example). This is because in this case the failure rate is different if the machine works at maximum rate μ or if it works at rate d. In general none of the two policies performs better than the other one: it depends on system parameters. In particular, based on (6) and (7), it is easy to see that as U increases from the minimum value for which $S(U) > 0$ to μ, there is a threshold value U_s where the RHP policy becomes more convenient with respect to the other one. The following Lemma states this result.

Lemma 2. For problem 1, there exists a threshold value U_s such that the HP policy performs better than the RHP policy for all $U < U_s$, while it is more expensive if $u > U_s$. The two policies give the same performance if $U = U_s$.

The threshold value U_s can be computed solving for U the equation $J^*_U = J^*_\mu$, where J^*_U and J^*_μ are given in (6) and (7) respectively. As an example, consider the system with the following parameters: $\mu = 6$, $d = 2.7$, $q_u = 4$, $q_{d1} = 3$,

$q_{d2} = 2$, $c_m = 4000$ and $c_p = 40$. For this system we have: $z^*_\mu = 7.4$ and $J^*_\mu = 318.6$. For all feasible U, i.e. $U \in (d\frac{q_{d2}+q_u}{q_u}, \mu]$, also the optimal RHP is not Just In Time (JIT), i.e. its optimal safety stock is always strictly positive. For this reason, to compute the threshold value U_s it is necessary to compare the first expression in (6) and (7). This gives $U_s = 4.9$. Figure 1 illustrates this result. This figure has been obtained by simulating the system under the HP policy (independent of U) and the RHP with U ranging from 4.4 to 5.65. Each value corresponds to the average of 5 simulations each one comprising 500000 state transitions. We want to remark that, for each value of U, the optimal hedging points z^*_U and z^*_μ are different, in general.

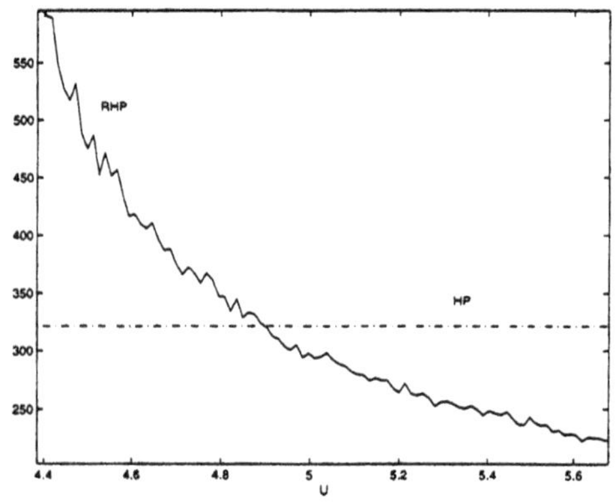

Fig. 1. A comparison between the HP policy (dashdotted) and the RHP policy (solid).

Remark 2. It is interesting to observe that the threshold value is different from the value $\bar{U}$ for which $\frac{\mu q_u}{q_u+q_{d1}} = \frac{U q_u}{q_u+q_{d2}}$, which in the example is $\bar{U} = 5.14$. This is because the HP policy works for finite amounts of time at rate d where the failure rate is q_{d2}, different from the failure rate q_{d1} at the maximum production μ.

4.2 The structure of the optimal policy

In this section, starting from the Hamilton Jacobi Bellman (HJB) equations reported below, we derive the structure of the optimal policy. The complete proof of optimality of such a policy will be only provided in the next section for a set of examples. The HJB equations in this case assume the following form, where $V_i(x)$ is the optimal cost-to-go function starting from a backlog/inventory level x with the machine in state i ($i = 0, 1$):

$$\min_{u\in[0,\,\mu]} \left\{ (u-d)\frac{dV_1}{dx} - q_d(u)[V_1(x) - V_0(x)] \right\} = J^* - g(x), \quad (8)$$

$$d\frac{dV_0}{dx} - q_u[V_1(x) - V_0(x)] = g(x) - J^*, \quad (9)$$

where J^* is the optimal value of the performance index (2). Based on (Rishel, 1975), if $V_i(x)$, $i = 0, 1$ are differentiable functions, a policy satisfying the HJB equations above is optimal. Lemma 3 gives a first key property of the optimal policy.

Lemma 3. At any time t, the optimal production control $u^*(t)$ takes values in the set $\{0, d, U, \mu\}$.

Proof. Let $\Delta V(x) := V_0(x) - V_1(x)$ and

$$W(x) := \frac{dV_1}{dx}(\mu - U) - \Delta V(x)(q_{d2} - q_{d1}).$$

Based on the equations (8)-(9), the optimal policy is given by:

$$u^*(x) = \begin{cases} 0 & \frac{dV_1}{dx} \geq 0 \text{ and } \Delta V(x) > 0 \\ \mu & \frac{dV_1}{dx} < 0 \text{ and } W(x) < 0 \\ U & \text{otherwise.} \end{cases} \quad (10)$$

The production rate d must be also considered since it may result on a switching surface among the conditions in (10). □

It is often assumed that V_1 is a convex function and it is also reasonable to expect that $\Delta V(x) > 0$ for all x, since the optimal cost-to-go should be larger if starting with the machine in the *down* state then in the *up* state. From these properties and from the observation of (10), it is possible to conceive the optimal policy as follows: the machine will work at maximum rate until the backlog/inventory level reaches a threshold value X and then switch to the lower production rate U until the backlog/inventory level hits an optimal hedging point Z. The optimal values X^* and Z^* for X and Z have to be evaluated. Formally:

$$u_{XZ}(x) = \begin{cases} \mu & x \leq X \\ U & x \in (X, Z) \\ d & x = Z \\ 0 & x > Z \end{cases} \quad (11)$$

Such a structure for the optimal policy agrees with the discussion reported in (Hu *et al.*, 1994). To prove the optimality of the policy (11), it is necessary to show that (11) satisfies the HJB equations (8) and (9). To proceed with the proof, it is necessary first of all to compute the expression of the optimal value J^*. To do this, we proceed as in previous literature (see for example (Liberopoulos and Caramanis, 1994)) and evaluate the steady state probability density functions pdf of the inventory/backlog level x when the policy (11) is applied. We get the following expressions for the pdf:

$$p_{XZ}(x) = \begin{cases} \dfrac{K_0 U e^{-\alpha_2(Z-x)}}{U-d} & x \in (X,Z) \\ \dfrac{K_0 \mu e^{-\alpha_2(Z-X)-\alpha_1(X-x)}}{\mu - d} & x < X \end{cases}$$

where $K_0 = K_0(X,Z)$ and $1/(K_0 d) = \frac{e^{\alpha_2(X-Z)}}{\Delta_1}\mu + \frac{1-e^{\alpha_2(X-Z)}}{\Delta_2}U + \frac{1}{q_{d2}}$. There is also a point mass probability $\gamma(X,Z) := K_0(X,Z)\frac{d}{q_{d2}}$ in $x = Z$. The index J in (2) with the machine operated through policy (11) is given by:

$$J(X,Z) = \int_{-\infty}^{Z} g(x)p_{XZ}(x)dx + \gamma(X,Z)g(Z), \quad (12)$$

and we will denote its optimal value as

$$J^* = J(X^*,Z^*) = \min_{X,Z:X\leq Z} J(X,Z).$$

While it is clear that $Z^* \geq 0$, as in the constant failure rate case, the optimal value of X can be both negative or positive. Policy RHP of Section 4.1 corresponds to $X^* = -\infty$, while policy HP to $X^* \equiv Z^*$.

5. EXAMPLES

In this section we provide a proof of optimality of policy (11) for a couple of interesting examples. This is a preliminary step to prove the optimality of policy (11), subject of ongoing research.

The proof technique is based on the HJB equations: the policy reported in (11), with an optimal selection of X and Z, is optimal if, integrating equations (8) and (9) with boundary condition on Z^* given by

$$c_p Z^* - J^* + q_{d2}\Delta V(Z^*) = 0, \quad (13)$$

with J^* the cost resulting from the application of such a policy, the obtained cost-to-go functions and the policy (11) satisfy equations (10).

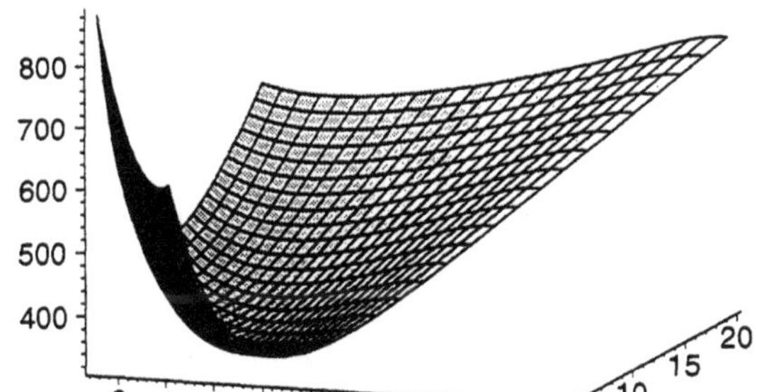

Fig. 2. Numerical evaluation of $J(X,Z)$ (through (12)) as a function of X and Z.

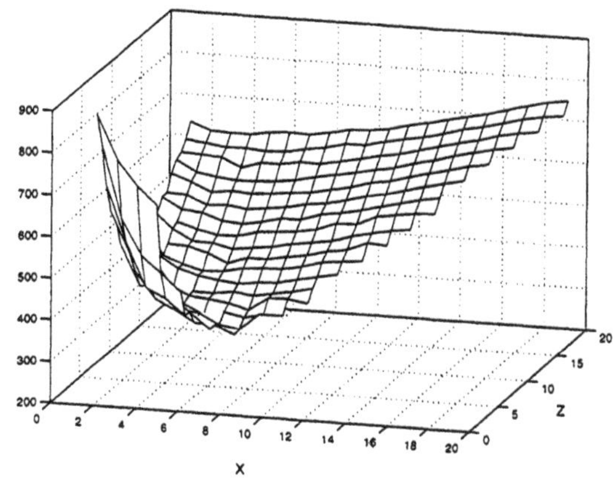

Fig. 3. Simulative evaluation of $J(X,Z)$ as a function of X and Z.

Consider the production system with the same parameter setting considered above, i.e. $\mu = 6$, $d = 2.7$, $q_u = 4$, $q_{d1} = 3$, $q_{d2} = 2$, $c_m = 4000$ and $c_p = 40$. We will consider two possible values for U, $U = 4.2$ and $U = 5$. The system with $U = 4.2$ is optimized with $X^* = 6.6$ and $Z^* = 7.5$ with an optimal cost of $J^* = 317.4$. In Fig. 2 it is reported the cost $J(X,Z)$ computed using (12) with Z and X ranging in the region $\{4 \leq Z \leq 20, 1 \leq X \leq Z\}$ (observe that $X \leq Z$). In Fig. 3 the same cost index $J(X,Z)$ has been evaluated as the average result of 5 simulations on a grid of values (X,Z) ranging in the same region, each simulation corresponding to 100000 state transitions. The optimal policy satisfies the HJB equations (8) and (9), in particular, solving (8) and (9) with $J^* = 317.4$, $X^* = 6.6$ and $Z^* = 7.5$, the optimal policy satisfies equation (10). In fact, as it is possible to see in Figs. 4, 5 and 6, the derivative of $V_1(x)$ is always negative for all $x \leq Z^*$ and, as it can be observed in Figs. 7, 8 and 9, the function $W(x)$ is positive for $x \in (X^*, Z^*)$ and is negative for $x < X^*$, with $W(X^*) = 0$.

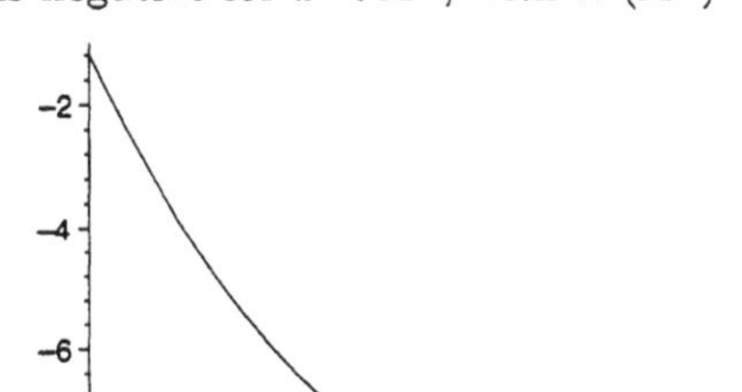

Fig. 4. The value function $V_1(x)$ for $x \in (X^*, Z^*)$.

The same discussion applies to the second case, where $U = 5$. In this case X^* is negative and given by $X^* = -1.03$ and $Z^* = 6.9$, with an optimal cost $J^* = 294.9$.

This allows to prove the optimality of policy (11) for the two examples considered. The generality of this result is the subject of ongoing research.

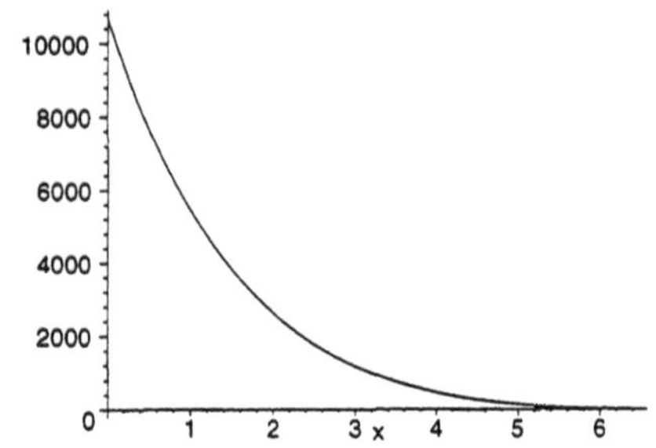

Fig. 5. The value function $V_1(x)$ for $x \in (0, X^*)$.

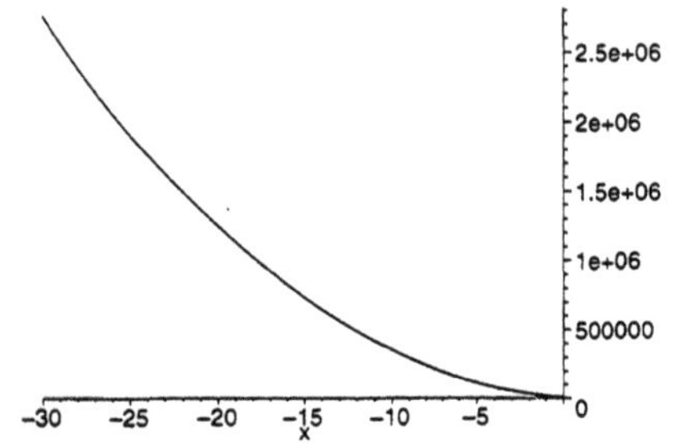

Fig. 6. The value function $V_1(x)$ for $x < 0$.

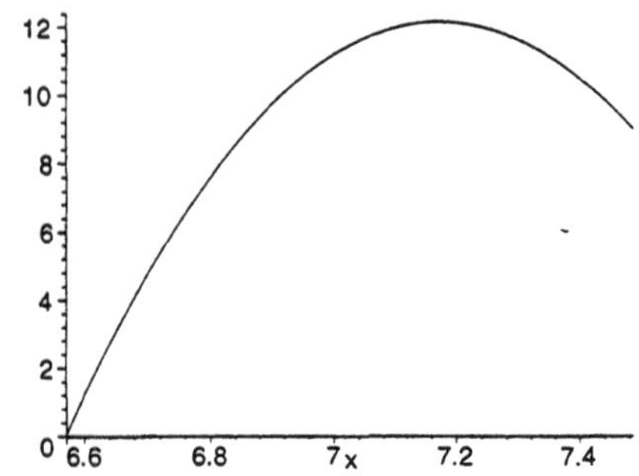

Fig. 7. Function $W(x)$ for $x \in (X^*, Z^*)$.

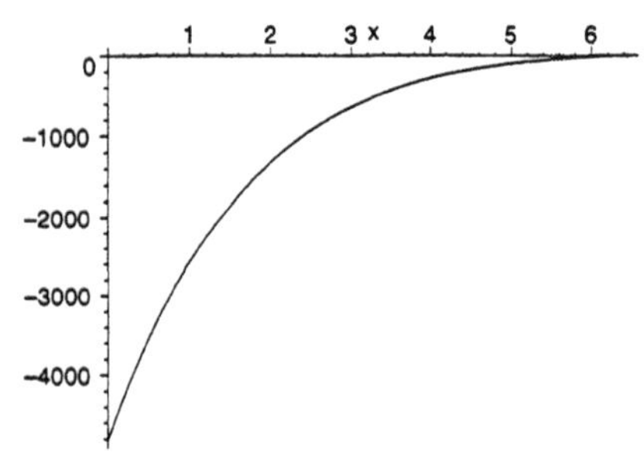

Fig. 8. Function $W(x)$ for $x \in (0, X^*)$.

6. CONCLUSIONS

A failure prone single machine manufacturing system has been considered in this paper. The failure/repair process is described through a Markov chain which is non-homogeneous due to the dependence of the failure rate $q_d(u)$ on the production rate u. The hedging point policy optimizes the case the failure rate function $q_d(u)$ is linear affine with u, while it is not optimal in general.

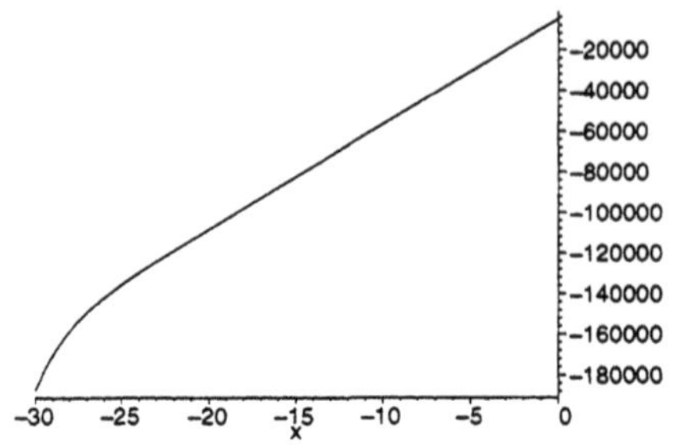

Fig. 9. Function $W(x)$ for $x < 0$.

For a two value failure rate function we provide in this paper some structural properties of the optimal policy and present a policy which is proved optimal on examples. The general proof is the subject of ongoing research. It is also intended as future work the generalization of these results to general failure rate functions $q_d(u)$, considering their discrete approximation.

REFERENCES

Akella, R. and P. R. Kumar (1986). Optimal control of production rate in a failure prone manufacturing system. *IEEE Transactions on Automatic Control* **AC-31**(2), 116–126.

Bielecki, T. and P. R. Kumar (1988). Optimality of zero-inventory policies for unreliable manufacturing systems. *Operations Research* **36**(4), 532–541.

Chen, H. and A. Mandelbaum (1991). Discrete flow networks: bottleneck analysis and fluid approximations. *Mathematics of Operations Research* **16**(2), 408–446.

Drozda, T.J. and Eds. C. Wick (1983). *Tools and Manufacturing Engineering Handbook, vol. I.* Dearborn, MI: Society Manufc. Eng.

Hu, J. Q., P. Vakili and G. X. Yu (1994). Optimality of hedging point policies in the production control of failure-prone manufacturing systems. *IEEE Transactions on Automatic Control* **AC-39**, 1875–1880.

Hu, J.Q. and Dong Xiang (1995). Monotonicity of optimal flow control for failure prone production systems. *Journal of Optimization Theory and Applications* **86**, 57–71.

Kimemia, J. and S. Gershwin (1983). An algorithm for the computer control of production in flexible manufacturing systems. *IIE Transactions* **15**(4), 353–362.

Liberopoulos, G. and M. Caramanis (1994). Production control of manufacturing systems with production-rate dependent failure rates. *IEEE Transactions on Automatic Control* **AC-39**(4), 889–895.

Rishel, R. (1975). Dynamic programming and minimum principles for systems with jump markov disturbances. *SIAM Journal of Control*.

ELSEVIER
IFAC
PUBLICATIONS
www.elsevier.com/locate/ifac

OPTIMAL ROUTING IN TWO PARALLEL QUEUES

Bruno Gaujal * **Emmanuel Hyon** **
Alain Jean-Marie ***

* *INRIA LIP, ENS Lyon, 46 allée d'Italie, F-69364 Lyon, France. E-mail Bruno.Gaujal@ens-lyon.fr*
** *LORIA UMR 7513, 615 route du jàrdin botanique, F-54606 Villers-les-Nancy, France. E-mail hyon@loria.fr*
*** *LIRMM, 161 Rue Ada, F-34392 Montpellier, France. E-mail ajm@lirmm.fr*

Abstract: This paper deals with the problem of the computation of the optimal routing sequence in a queuing system made of two parallel queues with exponential service times. The optimal policy (minimizing the expected waiting time) is a Sturmian sequence. An algorithm to compute the optimal routing sequence is proposed together with an extensive set of experiments which show several interesting features of the optimal policy with apparent discontinuities and a fractal behavior. Several good approximations using fast heuristics are given at the end.

Keywords: Open-Loop routing, parallel queues, Sturmian words.

1. INTRODUCTION

The problem of finding optimal routing policies in networks of queues is of uttermost importance as far as the new requirements of quality of service are considered. Indeed, increasing the bandwidth forever will not be enough to face all the problems (such as long delays). Load balancing policies between streams of customer in a network and admission control policies should be introduced to outperform the best effort policy currently applied. It turns out that this problem can be treated by the mean of the routing in parallel queues which models many situations arising in the communication networks. However, the problem of constructing the optimal routing policy is most often unsolved.

In the literature, the routing problem is often considered with closed-loop routing. This problem can then be expressed as a Markov decision process and the optimal policies are threshold policies (see Hajek (1984) for the full information case in two queues and Altman (2001) for a recent overview). However, the full information case is often unrealistic in practice. An alternative approach is to consider an open-loop control in which the only informations available to the decision maker are those available at time 0. For an infinite number of customers, an algorithm to compute the optimal policy when both services and queues are deterministic can be found in Gaujal and Hyon (2001) for parallel queues when the number of distinct services is two, in van der Laan (2003) for more than two distinct services and in Gaujal and Hyon (2002) for two parallel networks of tandem queues. When the number of customers is finite and both services and inter arrivals are exponential, an algorithm using dynamical programming methods is provided in Milito and Fernandez-Gaucherand (1995). But in the general case with an infinite number of customers only approximations of the optimal policy are given : Combe and Boxma (1994).

A long version of this paper with detailed proofs, comparisons with other methods as well as a rather extensive range of experimental results is available in Gaujal et al. (2004).

2. ROUTING SEQUENCES

We consider an arrival process $\mathcal{A}$ of intensity λ. The two queues have a FIFO discipline and the service times in Q_1 and Q_2 are random, independent and identically distributed sequences. The routing policy is determined by an infinite binary sequence m which will be called a *routing word* or a *routing sequence* in the following. Suppose that m is a word over the alphabet $\{0, 1\}$. If the ith letter of m: $m(i)$ is a one then the customer is sent in Q_1 otherwise the customer is sent in Q_2. The input processes in each queue are sampled processes of the global input process.

The *slope* $\mathfrak{s}(m)$ of m is the limit when n goes to infinity of $\sup_n \frac{m[n,1]}{n}$, where $m[n, 1]$ is the number of ones within the first n letters of m.

2.1 Stationary distribution

In order to find one optimal policy, we will compute the average performances of any periodic routing policy. This implies the knowledge of the average waiting time for each queue for the sampled process of both queue. This is why this section focuses on a single queue, say Q_1. To simplify the notations, the index of the queue (say 1) is omitted in this section. Moreover in this section, we assume that all the words are periodic, and, with a slight abuse of notation, m also denotes the smallest period of an infinite periodic word m. This smallest period has length ℓ and its number of ones is denoted a. The global input $\mathcal{A}$ is a Poisson process of intensity λ and the services are all independent and identically distributed with an exponential distribution of parameter μ.

The behavior of the number of customers in the system of this $G/M/1/\infty$ queue is given by a continuous-time Markov chain X_t which state space is equal to $\mathbb{N} \times \{1, \ldots, \ell\}$. The first entry represents the number of customers in the system at time t while the second entry represents the current letter of the routing word m (modulo ℓ).

Proposition 1. The continuous time Markov chain X_t is a Quasi Birth and Death process whose generator Q is given by

$$Q = \begin{bmatrix} C & A_0 & 0 & 0 & \ldots \\ A_2 & A_1 & A_0 & 0 & \ldots \\ 0 & A_2 & A_1 & A_0 & \ldots \\ 0 & 0 & A_2 & A_1 & \ldots \\ \vdots & \vdots & \vdots & \vdots & \ddots \end{bmatrix},$$

with $A_0 = \lambda J^+$, $A_1 = \lambda J^0 - (\lambda+\mu)I$, $A_2 = \mu I$ and $C = \lambda J^0 - \lambda I$, where I is the identity matrix of size ℓ and the matrices J^0 and J^+ are matrices which entries are respectively $(J^0)_{i,k} = (1 - m(i))\delta_{i,k+1}$ and $(J^+)_{i,k} = m(i)\delta_{i,k+1}$. The index k is taken modulo ℓ and the term $\delta_{i,k}$ is Kronecker's symbol (1 if $i = k$ and 0 otherwise).

Proof (sketch) Let us assume that we are in state (i, k) with $i > 0$ and $n \in \{1, \ldots, \ell\}$. Let us detail the states which could be reached from (i, k). The service rate being μ, the process jumps from (i, k) to $(i - 1, k)$ with rate μ as long as $i - 1$ remains non-negative.
The global arrival rate is λ and two cases can occur:
either $m(k) = 1$ (the current arrival is routed in the queue) and the process jumps in state $(i + 1, k + 1 \mod \ell)$,
or $m(k) = 0$ (the current arrival is not routed in the queue) and the process jumps in state $(i, k+1 \mod \ell)$. □

Let π be the invariant measure of the process X_t (when it exists). This probability satisfies

$$\pi Q = 0. \tag{1}$$

We now refine the notation by introducing block vectors π_n of dimension ℓ whose ith entry ($\pi_n(i)$) represents the stationary probability to have n customers in the system when we stand in the routing decision $m(i)$. Hence $\pi_k(1) + \cdots + \pi_k(\ell)$ is the stationary probability to have k customers in the system.

We will not try to compute π directly which can be quite hard, but we will rather determine its generating function. Some discussions on this choice can be found in Gaujal et al. (2004).

Definition 2. (Generating Function). Let $\overline{D}(0, 1)$ be the closed unit disk. The generating function of π is the function $\Pi(z)$ from $\overline{D}(0, 1)$ to $\mathbb{C}^\ell$ defined by

$$\Pi(z) = \sum_{n=0}^{\infty} z^n \pi_n.$$

Based on Theorem 1.3.2 of Neuts (1989), The Quasi Birth and Death process X_t is positive recurrent if and only if

$$a\lambda/\ell\mu < 1, \tag{2}$$

which we assume in the following.

The study can be restricted to one period of a periodic m (say the smallest period). Moreover it must be noticed that the performances of a shift of a finite word m repeated infinitely are equal to the performances of m, when the queue is stable. Indeed, one can consider that the sampling given by the shifted sequence is identical to the sampling given by m with an additional finite prefix which is negligible for asymptotic performances when the queue is stable.

2.2 The Kernel Method

This method can be decomposed in two major steps. First we determine a functional equation, depending on a parameter, which is satisfied by the generating function $\Pi(z)$. The second step uses the singularities of the kernel of the equation to compute some parameters of the equation which will then induce a complete computation of the generating function.

Lemma 3. Let us define $\rho = \lambda/\mu$. Let $K(z)$ be the $\ell \times \ell$ matrix $K(z)$ defined by

$$K(z) = \begin{bmatrix} 1-(1+\rho)z & \rho z^{1+m(1)} & & \\ & \ddots & \ddots & \\ & & 1-(1+\rho)z & \rho z^{1+m(\ell-1)} \\ \rho z^{1+m(\ell)} & & & 1-(1+\rho)z \end{bmatrix}.$$

Then the generating function satisfies the functional equation with kernel $K(z)$,

$$\Pi(z)K(z) = \pi_0(1-z). \qquad (3)$$

Proof (sketch) Using the global balance equation (1) we get the induction

$$\pi_0 C + \pi_1 A_2 = 0,$$
$$\pi_{n-1}A_0 + \pi_n A_1 + \pi_{n+1}A_2 = 0, \quad \forall n \geq 1.$$

This yields

$$\Pi(z)(z^2 A_0 + zA_1 + A_2) = \pi_0\mu(1-z).$$

Dividing by μ yields (3). □

Let us now study the zeros of $K(z)$. More precisely we will focus on the zeros inside the unit disk since $\Pi(z)$ is a power series with radius of convergence larger than one. Let us call $\Delta(z)$ the determinant of the matrix $K(z)$: $\Delta(z) = (-1)^{\ell+1}\rho^\ell z^{\ell+a} + (1-(1+\rho)z)^\ell$.

Lemma 4. If $\frac{a\rho}{\ell} < 1$ then $\Delta(z)$ has ℓ roots inside the unit disk.

Proof (sketch) This is a direct application of Rouché's theorem using the functions

$f(z) = \left(\frac{\rho+1}{\rho}\right)^\ell (z - \frac{1}{1+\rho})^\ell$, and $g(z) = f(z) - z^{\ell+a}$.
□

Theorem 5. If z_i is the ith root of $\Delta(z)$ in the unit disk and v_i is the right eigenvector of the eigenvalue 0 of $K(z_i)$, then π_0 is a solution of the system:

$$\begin{Bmatrix} (1-z_i)\pi_0 v_i = 0, \ 0 \leq i \leq \ell, z_i \neq 1 \\ \pi_0 \mathbf{1} = 1 - \frac{a\rho}{l}, \qquad \text{if } z_i = 1 \end{Bmatrix} \qquad (4)$$

where $\mathbf{1}$ is the column vector with all its components equal to 1.

Proof If $|z_i| < 1$, then it comes $(1-z_i)\pi_0 v_i = 0$. Note that the rank of the matrix $K(z_i)$ is $\ell - 1$ so that the vector v_i is unique up to one multiplicative constant. The case $z_i = 1$ has to be handled differently since $(1 - z_i) = 0$ and $K(1)\mathbf{1} = \mathbf{0}$. The equation for $z_i = 1$ comes from differentiating (3) at $z_i = 1$. □

If the set of equations (4) has a unique solution, this allows us to get the function $\Pi(z)$ as a vectorial analytical function in the closed unit disk, given by $\Pi(z) = (1-z)\pi_0 K(z)^{-1}$. However, this system may not have full rank, in which case our approach fails. Among the several thousand cases we considered in our experimental runs, they all had full rank. Let $\mathbb{E}(N_j(m))$ and $\mathbb{E}(W_j(m))$ be respectively the expected number of customers and the expected waiting time in the queue Q_j when the sample is made according m.

Then $\mathbb{E}(N_j(m)) = \frac{d\Pi(1)\mathbf{1}}{dz}|_{z=1}$ and by Little's law, the expected waiting time is $\mathbb{E}(W_j(m_j)) = \frac{l}{a\lambda}\mathbb{E}(N_j(m)) - \frac{1}{\mu}$.

This kernel method has been compared with the classical *matrix geometric* method using the most efficient (or considered so) iterative method, presented in Latouche and Ramaswami (1993). Here, this alternative approach is plagued by numerical stability problems (see Gaujal et al. (2004) for more details) which make it unusable.

3. OPTIMAL ROUTING

This section is dedicated to the search of one optimal routing policy when we consider an open-loop control in a system made of two parallel queues Q_1 and Q_2. We assume that the inputs in the whole system follow a Poisson process while the service times have exponential distributions with parameters μ_1 in Q_1 and μ_2 in Q_2. Our aim is to find one optimal routing policy, optimal in the sense that the average waiting time (or more generally any increasing convex function of the average waiting time) is minimized. In an open-loop routing since the routing does not depend on the state of the system, the routing sequence m is a binary word given before the system starts. This sequence will route the customers as described earlier in section 2.

The average waiting time when the routing sequence is m is denoted by $\mathbf{W}(m)$ given by

$$\mathbf{W}(m) = \lim \sup_{N \to \infty} \frac{1}{N} \sum_{k=1}^{k=N} W(m)_k,$$

where $W(m)_k$ denotes the waiting time of the kth customer.

3.1 Optimal allocations sequences

Definition 6. (Mechanical words). For a real number x, let $\lfloor x \rfloor$ (resp. $\lceil x \rceil$) be the largest (resp. smallest) integer smaller (resp. larger) than x.

The upper mechanical word with slope α is the infinite word $\overline{m}_\alpha$ where the nth letter ($n \geq 0$) is:

$$\overline{m}_\alpha(n) = \lceil (n+1) \times \alpha \rceil - \lceil n \times \alpha \rceil. \quad (5)$$

The lower mechanical word with slope α is the infinite word $\underline{m}_\alpha$ where the nth letter ($n \geq 0$) is:

$$\underline{m}_\alpha(n) = \lfloor (n+1) \times \alpha \rfloor - \lfloor n \times \alpha \rfloor. \quad (6)$$

The slope is a rational number if and only if the mechanical word is periodic.

Theorem 7. (Altman et al. (2000)). There exists an optimal policy which is a lower mechanical word.

This means that among all the binary policies the average waiting time is minimal for some lower mechanical word with a particular slope. However, this result does not give any clue on how to actually compute one optimal policy, that is, the slope α_{opt} of the optimal mechanical routing sequence. This will be done in the next section.

3.2 Optimal Ratio

For convenience and without loss of generality one can assume than Q_2 is the fastest queue ($\mu_2 > \mu_1$). The system is stable if the traffic intensity in the whole system ρ_s is strictly smaller than one *i.e.* $\rho_s = \frac{\lambda}{\mu_1+\mu_2} < 1$. The slope α of the routing word can vary in an interval that keeps the queue stable. This interval is called the interval of stability denoted by I_s. Owing to Equation (2), $I_s =]1 - \frac{\mu_2}{\lambda}, \frac{\mu_1}{\lambda}[\cap [0,1]$.

Using Theorem 7, only routing sequences which are mechanical are considered. When the routing sequence is $\underline{m}_\alpha$, randomizing an initial shift makes it a stationary and ergodic sequence and yields

$$\mathbf{W}(\overline{m}_\alpha) = \alpha \mathbb{E}(W_1(\overline{m}_\alpha)) + (1-\alpha)\mathbb{E}(W_2(\overline{m}_{1-\alpha})). \quad (7)$$

by using the ergodic theorem.

Our aim is now to compute the optimal slope α_{opt} such that

$$\alpha_{opt} = \arg\min_{\alpha \in I_s} \mathbf{W}(\overline{m}_\alpha).$$

Lemma 8. The function $\alpha \mapsto \mathbb{E}(W_j(\overline{m}_\alpha))$:
i) is increasing,
ii) goes to infinity when α approaches the frontiers μ_1/λ or $1 - \mu_2/\lambda$ of the stability interval, and
iii) is convex with respect to α.

Proof (sketch) The proof of *i)* uses coupling arguments.

ii) follows from Hajek (1985).

iii) can be found in Hordijk and van der Laan (2003). □

As a consequence, the function $\alpha \mapsto \mathbf{W}(\overline{m}_\alpha)$ admits an unique minimum in I_s.

3.2.1. Continuity issues One of the major open problem related to this work is the optimality of periodic policies. This mainly depends on the continuity of the function $(\lambda, \mu_1, \mu_2) \mapsto \arg\min W(\overline{m}_\alpha)$. If this function is continuous over some (possibly very small) domain, then, this implies that for some values of (λ, μ_1, μ_2), the slope α_{opt} is an irrational number which implies in turn that no periodic policy can be optimal.

3.3 Numerical Computations

The framework of the algorithm to compute α_{opt} is the following:

1- Compute the two rational numbers with the smallest denominators in I_s denoted by α and α', with $\alpha < \alpha'$ as well as two rational numbers β_l and β_u which are respectively a rational lower bound and a rational upper bound of α_{opt}.

2- Compute $\mathbf{W}(\overline{m}_\alpha)$ and $\mathbf{W}(\overline{m}_{\alpha'})$ using the kernel method. Since the cost function is convex, if $\mathbf{W}(\overline{m}_\alpha) \leq \mathbf{W}(\overline{m}_{\alpha'})$ then $\alpha_{opt} \leq \alpha'$ and one has to find a new point in $]\beta_l, \alpha'[$. Otherwise α becomes the lower bound and one has to find a new point in $]\alpha, \beta_u[$.

The remaining question is the choice of the new point in the search interval. Selecting the rational number with the smallest denominator in a zone "around" the middle of the current search interval does the trick because it provides a fast decrease of the search interval as well as a moderate increase of the size of the matrices involved in the computations (see Gaujal et al. (2004) for more details). The size of that zone is computed as hs, where s is the size of current interval and h is a parameter smaller than 1.

During our computations the parameter h has been chosen to be $h = 1/5$ for it provided the best convergence times on test beds.

The first series of computations is displayed in Figure 1. This figure presents the values of the optimal proportion α_{opt} (on the y-axis) as a function of the variation of the total load ρ_s (on the x-axis). In these computations, λ varies such that the total load of the system ρ_s ranges from 0 to 1 while the service times μ_1 and μ_2 are fixed. The value of μ_1 is taken to be $7/16$ and the value of $\mu_2 = 3\mu_1 = 21/16$. The arg min is computed with a precision of 10^{-4}. An interesting feature of

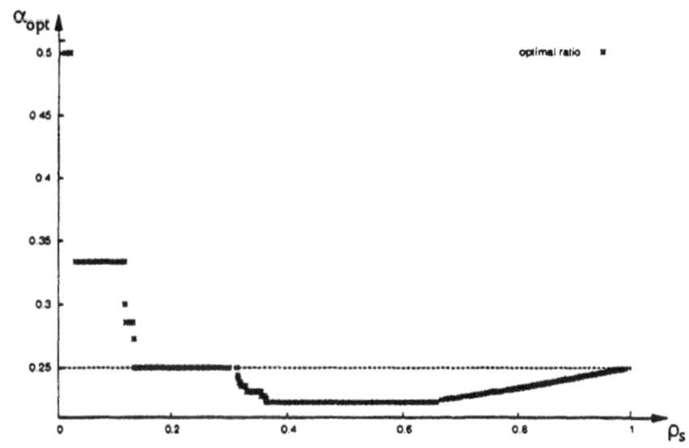

Fig. 1. Curve of $\rho \mapsto \alpha_{opt}$

α_{opt} is that it remains constant over long intervals. Moreover, the optimal values are most often rational numbers with small denominators such as $1/2$, $1/3$, $2/7$, $1/4$ and $2/9$ (this is not an artifact, since this curve is computed with a high precision). The dotted line on Figure 1, corresponds to the ratio $\alpha = 1/4$. This proportion corresponds to a simple load balancing policy which sends 3 times more customers in Q_2 than in Q_1. This seems natural since Q_2 is three times faster than Q_1. However, as seen in the figure this policy is not always optimal.

In the second set of experiments the arrival rate is set to $\lambda = 1$, and we let μ_1 vary (on the x-axis) and μ_2 vary (on the y-axis) while restricting our investigations to the stability domain I_s. The precision is chosen to be $1/500$. Figure 2 displays areas where the values of α_{opt} remain constant : each white cell represents the set of parameters that yield an identical optimal ratio. The largest cell represents the zone where the optimal policy is the *Round Robin* policy ($\alpha_{opt} = 1/2$). Surprisingly, this zone is larger than the well-known case where the two services are equal.

3.4 Heuristic routing algorithms

As shown above, computing the optimal routing policy is quite involved and requires a lot of computer power. In this section, we show some heuristics that provide good solutions very fast.

Here is the way to approximate α_{opt}.

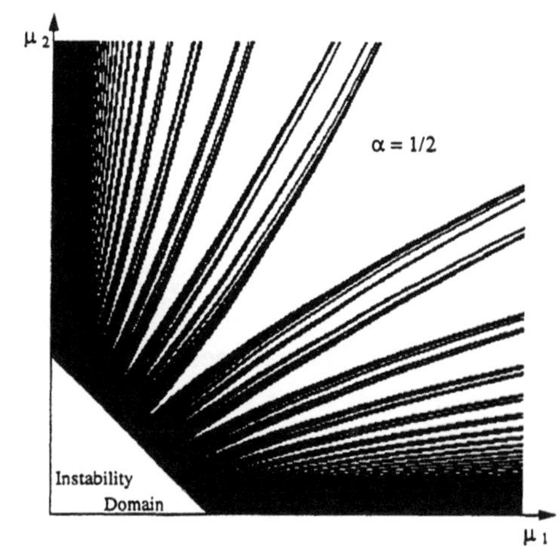

Fig. 2. Curve of α_{opt} when the intensity of the services vary

(1) Compute the optimal routing policy in a given class C of tractable distributions. This provides α_C, the probability that one packet is sent to Q_1.
(2) Use α_C as the routing density for a Sturmian routing.

Bernoulli Routing. The most immediate class of routing policies is Bernoulli (class B). Using classical properties of the expected waiting time in a $M/M/1$ queue (Gross and Harris (1985)) gives a direct way to compute the optimal α_B.

Gamma approximation. Another way to find an approximation of α_{opt} is by assuming that the input process in Q_1 has a Gamma distribution (class G) with parameters $(1/\alpha, \lambda)$ and a Gamma distribution with parameters $(1/(1-\alpha), \lambda)$ in Q_2 as done in Combe and Boxma (1994). Then we can use classical formulas for the expected waiting time for a $\Gamma/M/1$ queue (see Gross and Harris (1985)) in both queues and minimize the weighted sum, to obtain the best α, denoted α_G. Of course no routing policy can make the arrival process in both Q_1 and Q_2 Gamma-distributed, since $1/\alpha$ and $1/(1-\alpha)$ cannot be both integers (unless $\alpha = 2$). However, the optimal parameter α_G is a very good approximation of α_{opt} as seen next.

Erlang mixture approximation The third way to approximate α_{opt} is based on the following property of mechanical words Gaujal and Hyon (2001). The mechanical word $\overline{m}_\alpha$ can be factorized only using the two words u and v :

$$u = 1\overbrace{0\ldots0}^{s} \text{ and } v = 1\overbrace{0\ldots0}^{s-1},$$

with $s = \lfloor \alpha^{-1} \rfloor$. In addition, the frequency of u in $\overline{m}_\alpha$ is α_1 defined by $\alpha = 1/(s+\alpha_1)$. We consider the class of Erlang mixtures in queues Q_1 (class E) with an Erlang with parameters (s, λ) with probability $1-\alpha_1$ and an Erlang with parameters $(s+1, \lambda)$ with probability α_1. We optimize the parameters α_1 and s. The optimal parameters α_1 and s then provide the optimal α_E. Again here, the computations can be carried out using classical formulas based on Laplace transforms and no routing policy can actually realize the mixture input process in both queues.

3.4.1. Numerical Experiments Figure 3 displays the variations of the optimal ratios, obtained by the three methods. The parameters are those used in Figure 1.

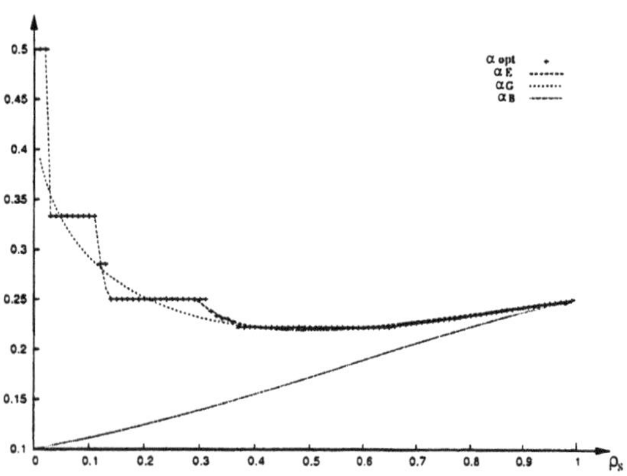

Fig. 3. Comparisons of the three heuristics

The ratios obtained by the Bernoulli routing method are always smaller than the exact values of α_{opt}. An intuitive explanation for that behavior is that sending two consecutive packets in the slowest queue has a high impact on the waiting time and a rather large probability in the Bernoulli case. Hence, this routing tends to send less packets in the slowest queue than it should. The ratios α_G and α_E are remarkably close to α_{opt}. For loads larger than 0.8, the differences with the exact optimal ratio $|\alpha_{opt} - \alpha_G|$ and $|\alpha_{opt} - \alpha_E|$ are smaller than 10^{-4} which is the precision of the computations of α_{opt} used in this Figure. Additionally, α_E remains constant over long interval like the optimal ratio α_{opt}.

Using α_B or α_G or α_E instead of α_{opt} for the routing policy always gives good performances, in terms of waiting times. The average relative difference for the expected waiting time is equal to 23% using α_B instead of α_{opt}, is equal to 1% with α_G and 0.1% with α_E!

4. CONCLUSION

In this paper, we present the computation of the optimal policy of open-loop routing in two exponential queues when the arrivals in the system follow a Poisson process. This requires a lot of computer power and the use of specific tricks to make it work efficiently. Some heuristics are used to approximate the policy which perform very well with a computational running time which can be very small.

However, some qualitative issues such as continuity are still not settled and the periodicity (or not) of the optimal policy remains open.

REFERENCES

E. Altman. *Markov Decision Processes, Models, Methods, Directions, and Open Problems*, chapter Applications of Markov Decision Processes in Communication Networks : a Survey, pages 488–536. Kluwer, 2001.

E. Altman, B. Gaujal, and A. Hordijk. Balanced sequences and optimal routing. *J. Assoc. Comput. Mach.*, 47:752–775, 2000.

M.B. Combe and O. Boxma. Optimization of static traffic allocation policies. *Theoritical Computer Science*, 125:17–43, 1994.

B. Gaujal and E. Hyon. Optimal routing policies in two deterministic queues. *Réseaux et systèmes répartis - Calculateurs Parallèles*, 13 (6):601–633, 2001.

B. Gaujal and E. Hyon. Optimal routing policies in deterministic queues in tandem. In *WODES*, pages 251–257. IEEE, 2002.

B. Gaujal, E. Hyon, and A. Jean-Marie. Optimal routing in two parallel queues with exponential service times. Technical Report 5109, INRIA, February 2004. http://www.inria.fr/rrrt/rr-5109.htm.

D. Gross and C.M. Harris. *Fundamentals of Queueing theory.* Wiley, 2nd edition edition, 1985.

B. Hajek. Optimal control of two interacting service stations. *IEEE Trans. Aut. Cont.*, 29: 491–499, 1984.

B. Hajek. Extremal splittings of point processes. *Mathematics of Operation Research*, 10(4):543–556, 1985.

A. Hordijk and D. van der Laan. Note on the convexity of the stationary waiting time as a function of the density. *Probability in the Engineering and Informational Sciences*, 17: 503–508, 2003.

G. Latouche and V. Ramaswami. A logarithmic reduction algoritm for quasi-birth-death processes. *Journal of Applied Probability*, 30:650–674, 1993.

R.A. Milito and E. Fernandez-Gaucherand. Open-loop routing of n arrivals to m parallel queues. *IEEE Trans. Aut. Cont.*, 40:2108–2114, 1995.

M.F. Neuts. *Structured stochastic matrices of M/G/1 type and their applications.* Marcel Dekker, 1989.

Dinard van der Laan. *The structure and performance of optimal routing sequences.* PhD thesis, Leiden University, 2003.

ELSEVIER
IFAC
PUBLICATIONS
www.elsevier.com/locate/ifac

JOB-SHOP SCHEDULING BY COMBINING REACHABILITY ANALYSIS WITH LINEAR PROGRAMMING

Sebastian Panek, **Olaf Stursberg**, and **Sebastian Engell**

Process Control Lab (BCI-AST)
University of Dortmund, 44221 Dortmund, Germany
{s.panek | o.stursberg | s.engell}@bci.uni-dortmund.de

Abstract: For optimization problems in planning and scheduling, techniques based on timed automata (TA) have the advantage of a simple and intuitive problem specification. In addition, search algorithms operating on the reachability tree of the TA have been proposed for determining the optimal (or a near-optimal) model evolution relatively efficiently. In order to improve the efficiency of the solution further, this paper describes the extension of a recently developed approach in which the search tree is pruned based on lower bounds for the cost that are computed by embedded linear programming. The paper shows how the use of a tailored linear program can considerably reduce the effort to compute the lower bounds such that the overall performance significantly increases.

Keywords: Automata, Optimization, Programming, Scheduling, Reachability.

1. INTRODUCTION

Scheduling, the task of assigning a given set of tasks or operations to a limited set of resources such that performance criteria are met, is of practical importance for a broad range of applications. These are as diverse as the task-processor assignment in real-time operating systems, the coordination of senders and receivers in telecommunication networks, or the assignment of production steps to machines or production units in processing systems. Commensurate with the importance in applications, research on scheduling has produced a variety of algorithmic techniques for the solution of scheduling problems. These include approaches based on mathematical programming (Groetschel *et al.*, 2001) as well as techniques that use constraint programming (Baptiste *et al.*, 2001) or genetic algorithms (Bruns, 1997). Among the first class of methods, mixed-integer linear programming (MILP) has been particularly successful (Kondili *et al.*, 1993; Kallrath, 2002), mainly because of an efficient exclusion (cutting) of possible solutions based on lower bounds for the performance costs. These bounds are computed by applying linear programming (LP) to problem instances in which integrality constraints are relaxed. It has to be noted, however, that the efficiency of the solution depends on a clever encoding of the relevant constraints in algebraic form (usually a time-consuming task).

In an alternative approach to the solution of scheduling problems, timed automata (TA) have been explored as the modeling basis. The techniques described in (Abdeddaim and Maler, 2001) and (Behrmann *et al.*, 2001) extend standard methods for reachability analysis of TA (see e.g., (Larsen *et al.*, 1997; Graf *et al.*, 2002)) by the consideration of costs, in order to determine an optimal evolution of the TA. The advantages of this approach are a very intuitive graphical specification of the problem, and the fact that the technique of reachability analysis can be extended relatively easy to a notion of costs. In (Panek *et al.*, 2004), it was proposed to combine the advantages of the TA-based approach with the MILP principle of computing lower bounds by linear programming, and to use these bounds to cut the reachability tree. This paper extends this idea in three directions: (a) for job-shop scheduling, a specialized formulation is introduced which is suitable to generate lower bounds of the cost-to-go efficiently; (b) an optimization algorithm with embedded LP solution and cutting of branches based on the lower bounds is presented; (c) new search heuristics are proposed which use the LP result to explore the search tree efficiently.

2. MODELING OF THE SCHEDULING TASK

The investigation in this paper is restricted to the case of job-shop scheduling with makespan minimization and without selection of alternative resources. (The extension of the approach to other types of scheduling problems seems possible, and is a matter of current investigation.) Job-shop scheduling is characterized by a set of jobs, each of which consists of an ordered set of operations, and each operation must be executed on a specific critical resource (or machine). Let $\mathcal{O} = \{o_1, \ldots, o_k\}$ denote the set of all operations, $\mathcal{R} = \{r_1, \ldots, r_m\}$ the set of resources, and the assignment of resources to operations $u : \mathcal{O} \rightarrow \mathcal{R}$. The durations of the operations are determined by a mapping $d : \mathcal{O} \rightarrow \mathbb{R}^+$. The set of jobs is denoted by $\mathcal{J} = \{j_1, \ldots, j_n\}$, $n \leq k$, and a mapping $\kappa : \mathcal{O} \rightarrow \mathcal{J}$ assigns operations to jobs.

The task of scheduling means to determine a *schedule* described by a mapping $s : \mathcal{O} \rightarrow \mathbb{R}^+$ which assigns a start time to each operation. A schedule is valid if it satisfies the following requirements: (a) while an operation o_i is executed on $u(o_i)$, no other operation o_j can be assigned to this resource: $\forall o_i \in \mathcal{O} : s(o_i) = t \Rightarrow \{o_j \in \mathcal{O} \mid u(o_i) = u(o_j) \wedge s(o_j) \in [t, t + d(o_i)]\} = \emptyset$; (b) operations that belong to the same job must be executed in the order defined by: $\kappa(o_i) = \kappa(o_j) \wedge i < j \Rightarrow s(o_i) + d(o_i) \leq s(o_j)$.

Let $\Sigma \subset \mathbb{R}^k$ be the set of all valid schedules. The length $L(s)$ of a schedule $s \in \Sigma$ is defined as $L(s) = \max\{s(o_i) + d(o_i) \mid o_i \in \mathcal{O}\}$. An optimal schedule s^* with respect to the overall time needed to finish all jobs, the *makespan*, is defined by $L(s^*) = \min\{L(s) | s \in \Sigma\}$.

Job-shop scheduling problems can easily be described by timed automata. For makespan optimization, it is sufficient to refer to the TA definition in (Alur and Dill, 1994); for more complex cost functions the use of priced timed automata according to (Behrmann *et al.*, 2001) is suitable. Consider the job $j_i \in \mathcal{J}$ with operations $(o_{i_1}, \ldots, o_{i_{n_i}}) \in \mathcal{O}$. One possibility to represent the scheduling problem is to define two locations called l_{i_j} and $\bar{l}_{i_j}$ for each operation o_{i_j}. In l_{i_j}, the job waits until the required resource $u(o_{i_j})$ becomes available, and in $\bar{l}_{i_j}$, the operation occupies the resource for the duration $d(o_{i_j})$. An additional final location f_i is defined to indicate that the job has finished. Clocks c_i are defined to monitor the durations for each individual job. Two types of transitions are introduced for an operation, one to start it $(l_{i_j} \overset{true, \alpha_{i_j}, c_i := 0}{\longrightarrow} \bar{l}_{i_j})$, and one to finish

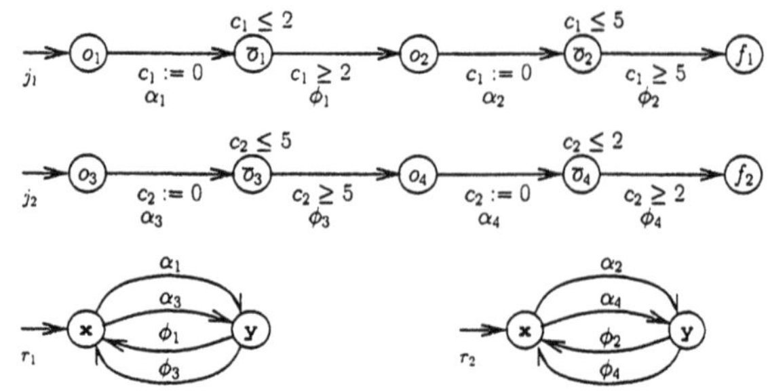

Fig. 1. TA model of a scheduling problem.

the operation $(\bar{l}_{i_j} \overset{c_i \geq d(o_{i_j}), \phi_{i_j}, \emptyset}{\longrightarrow} l_{i_{j+1}})$. The location invariants are defined by $I(l_{i_j}) = true$ and $I(\bar{l}_{i_j}) = (c_i \leq d(o_{i_j}))$. Furthermore, a label α_{i_j} is assigned to a transition which represents the allocation of a resource $u(o_{i_j})$, and a label ϕ_{i_j} marks a transition by which the resource is released. The automata of all jobs form a network and interact by the synchronization labels. To avoid that two operations occupy the same resource at the same time, the resource allocation has to be encoded mutually exclusively in the parallel composition of all automata. This can be achieved either by defining shared global binary variables (as in (Abdeddaim and Maler, 2001)), or by introducing an additional automaton for each resource. In the latter option, each of the resource automata has two locations and its transitions are synchronized with the job automata to model the resource allocation. To illustrate this concept, Fig. 1 shows an example with 2 jobs, each of which consists of 2 operations. In both cases, the first operation is executed on machine 1 and the second on machine 2. The notation for this example is as follows: set of operations $\mathcal{O} = \{o_1, o_2, o_3, o_4\}$, set of machines: $\mathcal{R} = \{r_1, r_2\}$, set of jobs: $\mathcal{J} = \{j_1, j_2\}$, job assignment: $\kappa(o_1) = j_1, \kappa(o_2) = j_1, \kappa(o_3) = j_2, \kappa(o_4) = j_2$, resource assignment $u(o_1) = r_1, u(o_2) = r_2, u(o_3) = r_1, u(o_4) = r_2$, durations: $d(o_1) = d(o_4) = 2, d(o_2) = d(o_3) = 5$. The synchronization of the job automata and resource automata is established by the labels $\alpha_1, \alpha_2, \alpha_3, \alpha_4$ to allocate resources, and $\phi_1, \phi_2, \phi_3, \phi_4$ to release the resources respectively.

3. THE OPTIMIZATION ALGORITHM

In order to minimize the makespan of job-shop scheduling problems, this paper proposes an algorithm that is motivated by the reachability algorithms for timed automata, as described e.g. in (Alur and Dill, 1994; Yovine, 1998). In addition to searching the reachability tree for feasible paths between the initial state (no job done) to the final state (all jobs completed), the objective of makespan minimization requires to determine the feasible path with minimal execution time. In (Abdeddaim and Maler, 2001; Behrmann *et al.*, 2001) procedures are described that combine the exploration of the reachability tree with the evaluation of a cost function to determine the cost-optimal path. The algorithm proposed here resembles these procedures, but adds one important feature: the generation of lower bounds for the cost-to-go by embedded linear programming and the pruning of the search tree based on these bounds. While introduced first in (Panek *et al.*, 2004), the idea is here modified with respect to the linear programming step and the use of priorities to steer the search towards the optimum.

The algorithm, which is shown in Fig. 2, represents a shortest-path algorithm for trees extended by functions to provide the lower bounds and priorities, and by the branch-and-bound principle well known from mixed-integer programming. The algorithm operates on a *waiting list* $\mathcal{W}$ and a *passed list* $\mathcal{P}$, the elements of which are quintuples (l, Z, b, π, h) where l denotes the location, h the history as a sequence of locations (and clock values) by which l was reached, Z is a cost zone with which l is reached given the history h, b an underestimate of the costs encountered for reaching the final state (all jobs done), and π a priority for selecting the quintuple from the waiting list. For job-shop scheduling with makespan minimization, the cost zone Z is simply the minimal time in which l is reached along h (Abdeddaim and Maler, 2001); for more general scheduling problems, the zones are polyhedral sets in the clock space. Furthermore, A is the composed TA model of the scheduling problem, M and M_r denote corresponding algebraic models (see below), L_f is the set of final locations, and Ω as the best solution found so far represents an upper bound for the optimal costs. The algorithm realizes cutting of

```
Ω := ∞; P := ∅
M := TRANSFORM(A); M_r := RELAX(M)
(b_0, x) := SOLVELP(M_r)
W := {(l_0, Z_0, b_0, π_0, −)}
WHILE W ≠ ∅
  (l, Z, b, π, h) := SELECTREM(W)
  IF l ∈ L_f THEN
    IF min(Z) < Ω THEN Ω := min(Z) END
  ELSEIF b ≤ Ω AND min(Z) < Ω THEN
    Succ := COMPSUCC(A, (l, Z, b, π, h))
    Succ' := REMOVEIDENT(P, Succ)
    FOR ALL (l', Z', −, π', h') ∈ Succ' DO
      M'_r := UPDATEM(M_r, (l', Z', −, π', h'))
      (b', x) := SOLVELP(M'_r)
      π' := COMPPRIORITY(x, M'_r, b')
      W := W ∪ {(l', Z', b', π', h')}
  END; END;
  P := P ∪ {(l, Z, b, π, h)}
END
```

Fig. 2. The search algorithm.

branches of the search tree over the lower bounds b and the upper bounds Ω. The following functions are embedded:

- The function TRANSFORM generates a completely algebraic model M that is equivalent to A with respect to the schedules it encodes. This transformation is described in detail in Sec. 4.
- RELAX replaces all binary variables of M (which encode the sequence of operations on the resources) by continuous variables that are defined on the interval $[0, 1]$. The result is the model M_r.
- SOLVELP solves the scheduling problem for the current model M_r' by linear programming (LP) and returns the objective value b' and the solution vector x which contains the values of the relaxed decision variables.
- Based on x, b', and the current model M_r', COMPPRIORITY assigns priorities π' to the paths starting from the current location. This issue is discussed in detail in Sec. 3.2.
- Using the priorities π', SELECTREM selects and removes the quintuple with the highest priority from the waiting list $\mathcal{W}$.
- For the selected quintuple (l, Z, b, π, h), COMPSUCC determines all successor states that are reachable from l. This step includes to update the value Z of each successor by adding the time spent in the predecessor to the accumulated time to reach the predecessor.
- REMOVEIDENT removes those quintuples from *Succ* for which the pair (l', Z') is already contained in *Succ* or in $\mathcal{P}$.
- UPDATEM updates the LP model such that all (originally relaxed) variables in M_r are fixed to those values that correspond to the history h into the current location l. If l is in the set L_f of final locations, all variables that encode the sequences are fixed to zero or one.

The input to this algorithm is the TA model, and the result is the cost-optimal path into the state in which all jobs are completed.

3.1 Computation of Lower Bounds

As is apparent from Fig. 2, the use of lower bounds b is essential to exclude those paths in an early stage which are not optimal. One possibility is to derive b from the problem structure. As discussed in (Abdeddaim and Maler, 2001), a possible lower bound for the makespan minimization in job-shop scheduling is to take the maximum of the sum of the durations of all operations that are not yet processed for a job or a resource. A drawback of this approach is that it does not consider conflicts of the operation-resource assignment, and thus may result in a very conservative bound which does not prune the search tree efficiently.

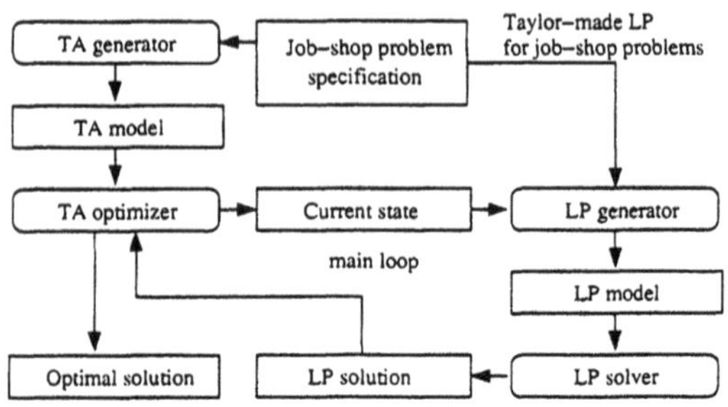

Fig. 3. Structure of the implementation.

The alternative advocated here is to generate bounds algorithmically by deriving algebraic models and solving them by LP. The further the algorithm in Fig. 2 explores the search tree in depth, the more efficient are the bounds for pruning the tree since the number of relaxed variables decreases.

In order to obtain models suitable for LP, (Panek *et al.*, 2004) describes a procedure in which an algebraic model is directly derived from the TA by: (a) introducing binary variables and linear algebraic (in)-equalities to encode the transitions; (b) using continuous and binary variables and appropriate algebraic equations to model the time-dependence of transitions; and (c) representing the synchronization between different automata by additional binary variables and inequalities. The solution of algebraic models requires to model explicitly the order by which the transitions can take place in time, i.e. all variables and algebraic equations have to be indexed over the set of time points. Due to this and due to the fact that the structure of the automata network (including synchronization) is mapped explicitly into the MILP model, the latter becomes relatively large, and the computation of b by LP can take more time as is gained by an efficient pruning of the reachability tree.

Therefore, a different approach is suggested here in which a tailor-made MILP model for the specific case of makespan minimization in job-shop scheduling is derived. While the functionality of TRANSFORM is restricted to this case, the advantage is a much more efficient LP solution which overall leads to a considerable decrease of the computation time. This mainly results from the fact that the new MILP formulation uses binary variables only to encode true degrees of freedom, but no additional ones to retain the automaton structure; see Sec. 4 for details. The MILP model is generated automatically from the scheduling problem specified by the user, i.e. it is not necessary for the latter to model the same problem in two different domains. The overall structure of a tool which realizes the algorithm in Fig. 2 with the embedded steps of MILP model generation and b-computation is shown in Fig. 3.

3.2 Search Heuristics

In order to determine the entry of the list $\mathcal{W}$ with which the algorithm should proceed (according to the function SELECTREM), several different strategies are conceivable. In the context of shortest-path algorithms, breadth-first, depth-first, and best-first (e.g. with respect to the minimal evolved absolute time) are the most prominent ones. In a series of experiments, it was found, however, that the following two heuristics are favorable for the considered scheduling problem:

- In each step, that entry in $\mathcal{W}$ is selected which has the lowest lower bound of cost-to-go b as returned by SOLVELP. This strategy is widely used in the mixed-integer domain, and it was observed for different examples that it works well also within the algorithm in Fig. 2. If more entries in $\mathcal{W}$ have the same lower bound, the one with the highest accumulated costs is chosen. It corresponds to the one for which the largest part of the schedule is already processed. This selection is mapped into the priorities π that are interpreted by the function SELECTREM.
- Alternatively, π is determined by COMPPRIORITY as a result of evaluating the solution vector x. The entries of the latter (real values in the interval $[0, 1]$) indicate a distance to the case that an operation is executed (1) or not (0). This information can be used in the following ways: (a) Over all transitions that leave the current location l, the priorities are assigned in decreasing manner if the distance of the x-entries to the value 1 increases. (b) For the complete path from l to a final location, the distances are accumulated, and the priorities decrease with an increasing accumulated distance.

4. EMBEDDED LINEAR PROGRAMMING

The key idea of generating concise MILP models (a relaxed version of which can be solved efficiently by LP) is to use binary variables only to encode the possible orders in which operations are processed on the same resource. This means that for each pair (o_i, o_j) of operations with $u(o_i) = u(o_j)$ one binary variable is defined and its value describes which operation is processed first. This approach, which has first been described in (Manne, 1960), is principally different from the state-task network or resource-task network formulations (Ierapetritou and Floudas, 1998; Kondili *et al.*, 1993). The latter use additional binary variables to model explicitly the execution of operations and the allocation of resources. (The benefits of the latter approaches are, however, that they are suitable to model more general scheduling problems.) Another difference with respect to the scheme proposed in (Panek *et al.*, 2004) is that the method proposed here does not use an a-priori discretization of the time axis (based on the start and end times of operations) but only introduces continuous variables that encode the time at which the operations are started. In detail, the scheduling problem described in Sec. 2 is represented in algebraic form as follows:

Parameters and constants:

- durations of the operations: $d(o) \in \mathbb{R}^+, o \in \mathcal{O}$;
- an overall time horizon $H \in \mathbb{R}^+$ which is a safe upper bound for the makespan.

Variables:

- Positive continuous variables store the start dates of the operations: $s(o) \in \mathbb{R}^{\geq 0}, o \in \mathcal{O}$.
- For each resource $r_i \in \mathcal{R}$, let $\mathcal{O}_i \subseteq \mathcal{O}$ denote the operations executed on r_i. For each pair o, o' of different elements in $\mathcal{O}_i$, a binary variable indicates the order in which the operations are executed: $p(o, o') = 1$ if o is executed before o', and $p(o, o') = 0$ otherwise.

Equations:

- Each operation must be started at or after time zero, and must be finished before H: $\forall o \in \mathcal{O}$: $s(o) \geq 0$, $s(o) + d(o) \leq H$.
- All operations belonging to one job must be executed in the specified order: $\forall o_i, o_q \in \mathcal{O}$ with $\kappa(o_i) = \kappa(o_q) \wedge i < q$: $s(o_i) + d(o_i) \leq s(o_q)$.
- If $\mathcal{O}_i$ denotes the set of all operations to be executed on r_i, then the simultaneous execution is excluded by: $p(o, o') + p(o', o) = 1 \; \forall o, o' \in \mathcal{O}_i$.
- The start and end times of the operations executed on r_i are ordered by the following conditions. $\forall o, o' \in \mathcal{O}_i$ with $o \neq o'$:
  $s(o) + d(o) - s(o') \leq H \cdot (1 - p(o, o'))$,
  $s(o) + d(o) - s(o') \geq -H \cdot p(o, o')$,
  $s(o') + d(o') - s(o) \leq H \cdot p(o, o')$,
  $s(o') + d(o') - s(o) \geq -H \cdot (1 - p(o, o'))$.
  These inequalities establish the link between the variables p and s.
- To formulate the minimization of the makespan, it is written as: $\min \Psi$, $\quad \forall o \in \mathcal{O} : \Psi \geq s(o) + d(o)$ subject to all (in-)equalities listed above.

The following extensions can be straightforwardly included in the MILP model, if this is required by the original formulation of the scheduling problem: (a) release dates (as earliest points of time to start o) and/or deadlines (the latest time to finish o) for the operations; (b) an additional machine allocation task if an operation is not uniquely assigned to one resource but can be operated on different ones; (c) durations that are not constant but depend on parallel operations; (d) the dates for finishing jobs are modeled as penalty terms in an extended cost function, rather than as hard constraints.

5. EXAMPLE AND RESULTS

To illustrate the procedure, the example introduced in Sec. 2 is considered. In order to obtain the corresponding MILP model, the following parameters, variables, and (in-)equalities are introduced:

- Estimated time horizon: $H = 100$,
- Positive real variables: $s(o_1)$, $s(o_2)$, $s(o_3)$, $s(o_4)$,
- Binary variables: $p(o_1, o_3)$, $p(o_3, o_1)$, $p(o_2, o_4)$, $p(o_4, o_2)$,
- Deadlines: $s(o_1) \leq 98$, $s(o_2) \leq 95$, $s(o_3) \leq 95$, $s(o_4) \leq 98$,
- Order specification: $s(o_1) + 2 \leq s(o_2)$, $s(o_3) + 5 \leq s(o_4)$,
- Mutual exclusion: $p(o_1, o_3) + p(o_3, o_1) = 1$, $p(o_2, o_4) + p(o_4, o_2) = 1$,
- Sequencing constraints:
  $s(o_1) + d(o_1) - s(o_3) \leq H \cdot (1 - p(o_1, o_3))$,
  $s(o_1) + d(o_1) - s(o_3) \geq -H \cdot p(o_1, o_3)$,
  $s(o_3) + d(o_3) - s(o_1) \leq H \cdot p(o_1, o_3)$,
  $s(o_3) + d(o_3) - s(o_1) \geq -H \cdot (1 - p(o_1, o_3))$,
  $s(o_2) + d(o_2) - s(o_4) \leq H \cdot (1 - p(o_2, o_4))$,
  $s(o_2) + d(o_2) - s(o_4) \geq -H \cdot p(o_2, o_4)$,
  $s(o_4) + d(o_4) - s(o_2) \leq H \cdot p(o_2, o_4)$,
  $s(o_4) + d(o_4) - s(o_2) \geq -H \cdot (1 - p(o_2, o_4))$.

The implementation of the approach (according to Fig. 3) generates this algebraic model from the TA shown in Fig. 1. The reachability tree produced by the algorithm in Fig. 2 is given in Fig. 4. The nodes are represented by quadruples (l_1, l_2, t, lb), where l_1 and l_2 are the names of the locations of both job automata, t is the absolute time (the accumulated cost Z), and lb is the lower bound obtained from solving the MILP model by LP. In order to find the optimal path from $o_1 o_3$ to $f_1 f_2$ quickly, the best-lower-bound heuristics has been applied. After finding the first path into $f_1 f_2$ with cost 9, the paths emerging from $o_1 \bar{o}_3$ and $f_1 \bar{o}_3$ (the right one) are not further explored but cut due to the lower bounds. Since the upper bound 9 is equal to the lower bound in the final state, the proven optimal solution has already been found.

In order to investigate the performance of the algorithm, different instances of the scheduling problem are optimized. In a first series (A), the number of jobs is varied from 2 to 5, each consisting of 2 operations, and 2 resources are available.

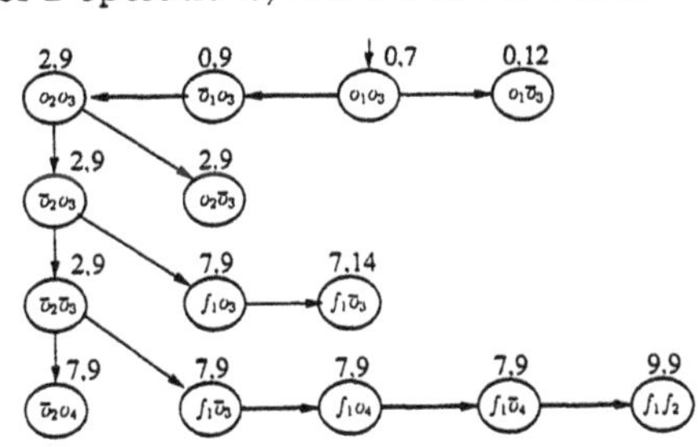

Fig. 4. Reachability tree of the example.

Table 1. Results for both series

		SP		BLB	
Jobs	series	states	sec.	states	sec.
2	A	39	0.03	12	0.01
3	A	433	0.5	52	0.14
4	A	7298	14.47	253	1.01
5	A	202914	579.41	11477	82.53
2	B	125	0.13	17	0.05
3	B	4159	7.56	50	0.2
4	B	247988	741.94	905	5.89
5	B	out of memory		20547	262.6

In a series B, the number of jobs varies in the same range but each job consists of 3 operations, and 3 resources are given. The durations are assigned as random numbers. For computing the optimal schedules, a standard shortest path algorithm without using lower bounds (SP) and the algorithm in Fig. 2 with the best-lower-bound heuristics (BLB) are compared. The table 1 contains the numbers of explored states and the computation times for these configurations. It should be remarked that *lazy* runs, i.e. executions in which operations start later than necessary (Abdeddaim and Maler, 2001), are sorted out by the algorithm. The results reveal that the algorithm presented in this paper reduces the computational effort considerably compared to the standard shortest path algorithm. A similar advantage has been observed in comparison to the approach in (Panek *et al.*, 2004). It has to be mentioned that the current implementation is a prototypical one, i.e. a further reduction of the computation times could be expected if, e.g., more advanced data structures were employed.

6. CONCLUSIONS

In comparison to existing methods for optimizing TA, it has been shown experimentally that the use of lower bounds obtained from LP can reduce the computational effort in the case of makespan minimization for job-shop problems considerably. In comparison to existing solution methods for solving MILP problems, the following two observations have to be balanced: On the one hand, the solution of the algebraic model described in Sec. 4 by existing mixed integer programming tools is faster than the algorithm in Fig. 2 for most of the experiments listed in Sec. 5. On the other hand, the specification of a given job-shop problem such that mixed-integer programming tools can read and efficiently solve it, is a cumbersome and time-consuming task. The specification with timed automata, however, is relatively easy even if many jobs and resources are considered. The authors believe that a fair comparison of different approaches should be based on the sum of the time for specifying the problem and solving it – in this respect the approach presented here is at least competitive.

Current work is focused on two issues: (a) to comprehensively evaluate the different search heuristics for a larger set of experiments; (b) to extend the tailor-made MILP models generated by the implementation to different classes of scheduling problems.

This work is financially supported within the EU-project AMETIST (IST-2001-35304).

REFERENCES

Abdeddaim, Y. and O. Maler (2001). Job-shop scheduling using timed automata. LNCS 2102. CAV'2001. Springer. pp. 478–492.

Alur, R. and D.L. Dill (1994). A theory of timed automata. *Theor. Comp. Science* **126**(2), 183–235.

Baptiste, P., C. Le Pape and W. Nuijten (2001). *Constraint-Based Scheduling: Applying Constraint Programming to Scheduling Problems.* Kluwer Acad. Publisher.

Behrmann, G., A. Fehnker, T.S. Hune, P. Petterson, K. Larsen and J. Romijn (2001). Efficient guiding towards cost-optimality in UPPAAL. In: *Proc. TACAS'01.* pp. 174–188.

Bruns, R. (1997). Scheduling. In: *Handbook of Evolutionary Comp.* (T. Baeck, D.B. Fogel and Z. Michalewicz, Eds.). Inst. of Physics Publ.. pp. F1.5:1–9.

Graf, S., M. Bozga and L. Mounier (2002). If-2.0: A validation environment for component-based real-time systems. In: *Proc. CAV'02.* Vol. 2404 of *LNCS.* Springer. pp. 343–348.

Groetschel, M., Krumke, S.O. and Rambau, J., Eds.) (2001). *Online Optimzation of Large Scale Systems.* Springer.

Ierapetritou, M. and C. Floudas (1998). Effective continuous-time formulation for short-term scheduling – multipurpose batch processes. *Ind. Eng. Chem. Res.* **37**, 4341–4359.

Kallrath, J. (2002). Combined strategic and operational planning - an MILP success story in chem. ind.. *OR Spectrum* **24**, 315–341.

Kondili, E., C.C. Pantelides and R.W.H. Sargent (1993). A general algorithm for short-term sched. of batch operations – MILP formulation. *Comp. Chem. Eng.* **17**, 211–227.

Larsen, K.G., P. Petterson and W. Yi (1997). Uppaal in a nutshell. *Int. J. Software Tools for Techn. Transfer* **1**(1), 134–152.

Manne, Alan S. (1960). On the job-shop scheduling problem. *Oper. Res.* **8**(2), 219–223.

Panek, S., O. Stursberg and S. Engell (2004). Optim. of timed automata models using mixed-integer programming. In: *Formal Mod. And Analysis of Timed Sys.*. Vol. 2791 of *LNCS.* Springer. pp. 73–87.

Yovine, S. (1998). Model checking timed automata. In: *Embedded Systems.* Vol. 1494 of *LNCS.* Springer. pp. 114–152.

ELSEVIER
IFAC
PUBLICATIONS
www.elsevier.com/locate/ifac

A DYNAMIC PROGRAMMING-BASED TECHNIQUE FOR MULTI-CLASS JOB SCHEDULING ON A SINGLE MACHINE

Davide Giglio and Riccardo Minciardi

DIST - Università di Genova
Via Opera Pia, 13
16145 - Genova, Italy

Abstract: Optimal scheduling of jobs on a single machine is considered in this paper. The jobs belong to different classes, and two jobs belonging to the same class are considered as completely equivalent. Due-dates are specified for each job of each class. The processing time of any job of a given class has to be fixed within an interval of possible values (the highest possible value is the nominal processing time). A deviation of the actual processing time from the nominal processing time induces a cost which has a linear dependence on such a deviation. The cost function to be minimized in the scheduling problem takes into account both the overall job tardiness and the sum of the execution time deviation costs. The decision variables are those concerning job sequencing (of course, taking into account the equivalence of jobs of the same class) and execution times. In the paper, the structure of closed-loop strategies is sought and determined.

Keywords: Scheduling algorithms, dynamic programming, optimization problems, manufacturing systems.

1. INTRODUCTION

Manufacturing systems are one of the application areas where scheduling techniques have their widespread diffusion (Blazewicz *et al.*, 1996; Pinedo, 1995). Actually, at a medium/short-term planning level, the assignment of jobs to the various resources, as well as the sequencing of their services over such resources, fall within the class of scheduling problems which are commonly dealt with in the scientific literature. However, some special features of scheduling problems in manufacturing environments have been so far rarely considered in the literature. Actually, such problems are generally characterized by at least two aspects that have to be carefully modelled. The first aspect concerns a high degree of repetitiveness of scheduling problem, in that generally jobs in a manufacturing process are not individually defined, but can be partitioned into different classes, being two jobs of the same class indistinguishable. The second aspect refers to the possibility of speeding up the execution of certain operations, when this appears to be necessary, owing to delays in the production timetable, or to other factors (e.g., the unforeseen arrival of new orders). Actually, such a reduction of the execution time (with respect to a pre-specified nominal one) may yield several kinds of additional cost. Such a cost may be due, for instance, to the necessity of using extra energy, or to the increase of the probability of a wrong operation execution, which may cause the necessity of repeating the operation execution or even eliminating the just processed component. Note that the introduction of the job processing times as decision variables makes the considered scheduling model affected by continuous decision variables (besides to integer ones), even in the case of *regular* optimization objectives (French, 1982). This fact deserves a special attention and has already noticed (Di Febbraro *et al.*, 2001; Di Febbraro *et al.*, 2002; Nowicki, 1994).

In this paper, the problem of scheduling jobs belonging to different classes on a single machine is considered. The objective is not just that of determining the optimal solution to such a problem, but also that of representing this solution in a closed-loop form. Namely, a solution is determined consisting in optimal decision strategies, capable of providing, at each stage, the values of the decision variables as functions of the present system state.

The rest of the paper is organized as follows. In the next section, the system model and the optimization problem are presented. In section 3, the structure of the optimal control strategy, for each decision stage, is determined through dynamic programming. Finally, some concluding remarks are presented in the last section.

2. THE MODEL AND THE OPTIMIZATION PROBLEM

Consider a single machine where N_1 jobs of type (class) P_1, N_2 jobs of type P_2, ..., and N_K jobs of type P_K have to be executed. All jobs belonging to the same class are completely equivalent. No precedence constraint has to be fulfilled, all jobs are available at time instant 0, and the execution of any job has to be non-preemptive. No set-up is required between the execution of jobs of different classes. A sequence of N_k due-dates for jobs of type P_k, namely $dd_{k,1}, dd_{k,2}, \dots, dd_{k,N_k}$, $k = 1, \dots, K$, is specified, as well as a set of coefficients, namely $\alpha_{k,1}, \alpha_{k,2}, \dots, \alpha_{k,N_k}$, $k = 1, \dots, K$, specifying the unitary tardiness penalties for each job.

The execution time of the j-th job of type P_k is a continuous variable $\tau_{k,j}$, $k = 1, \dots, K$, $j = 1, \dots, N_k$, whose value ranges from a lower bound τ_k^{low} up to a higher bound, namely τ_k^{nom}, which is also the nominal value of such an execution time. Note that the assumption that the highest possible execution time coincides with the nominal one may be justified by supposing that there is no economic advantage in slowing down the execution time of a job. Moreover, it is assumed that, once the execution time of a certain job has been chosen, it cannot be changed during the service.

The objective function to be minimized is

$$\sum_{k=1}^{K} \sum_{j=1}^{N_k} \Big[\alpha_{k,j} \cdot \max\{C_{k,j} - dd_{k,j}, 0\} + \beta_k \cdot (\tau_k^{\text{nom}} - \tau_{k,j}) \Big] \quad (1)$$

where β_k, $k = 1, \dots, K$, are weighting coefficients, and $C_{k,j}$, $k = 1, \dots, K$, $j = 1, \dots, N_k$, indicates the completion time of the j-th job of kind k. It is assumed that $\alpha_{k,j} > \beta_k$, for any $j = 1, \dots, N_k$, and for any $k = 1, \dots, K$. Actually, such an assumption simply states that the "marginal" tardiness cost is greater than the "marginal" cost related to the deviation of the actual execution time from the nominal value. Note that the cost function (1) takes into account, for each job of each class, the weighted tardiness and the weighted deviation from the nominal execution time. Then, the cost function to be optimized is of regular type (French, 1982), that is, there is no advantage in delaying any completion time if the other ones remain unchanged. Hence, there is an optimal solution of the scheduling problem where no idle time is inserted between the execution of two subsequent jobs.

The above model can be easily formalized through a state-space representation as it is apparent that the "system state" when a new decision has to be taken, i.e., at an instant t at which a job has been completed, is represented just by the $(K+1)$-tuple $\underline{x}(t) = [n_1(t), \dots, n_K(t), t]^T$, where $n_k(t)$, $k = 1, \dots, K$, is the number of jobs of type P_k already served at time instant t. The action $\underline{u}(t) = \underline{f}[\underline{x}(t)]$ that has to be taken at time instant t is represented by the choice of the class of the next job to be served, and by the execution time of this service. Let $\delta_k(t) \in \{0, 1\}$ denote a decision variable whose value is 1 if a job of kind P_k is selected , and 0 otherwise, and $\tau^{\text{ex}}(t)$ indicate the execution time. Obviously, $\sum_{k=1}^{K} \delta_k(t) = 1 \ \forall\, t$. Thus, $\underline{u}(t) = [\delta_k(t), k = 1, \dots, K, \tau^{\text{ex}}(t)]$.

As decisions are taken only at specific time instants corresponding to the completion of the service of a job (but for the last one), and to the initial time instant, the decision instants are actually discrete in time (although not equally spaced, and not generally multiple of a given value), and can be denoted as t_j, $j = 0, 1, \dots, \left(\sum_{k=1}^{K} N_k\right) - 1$. Of course $t_0 = 0$. At time instant t_j, $\left(\sum_{k=1}^{K} N_k\right) - j$ jobs are still to be served. In this way, the state equation of the system can be written as follows

$$\underline{x}(t_{j+1}) = \begin{bmatrix} n_1(t_{j+1}) \\ \vdots \\ n_K(t_{j+1}) \\ t_{j+1} \end{bmatrix} = \begin{bmatrix} n_1(t_j) + \delta_1(t_j) \\ \vdots \\ n_K(t_j) + \delta_K(t_j) \\ t_j + \tau^{\text{ex}}(t_j) \end{bmatrix} \quad (2)$$

$j = 0, 1, \dots, \left(\sum_{k=1}^{K} N_k\right) - 1$, where variables $\delta_k(t_j)$ and $\tau^{\text{ex}}(t_j)$ are respectively bounded by

$$\sum_{k=1}^{K} \delta_k(t_j) = 1 \quad (3)$$

$$\sum_{k=1}^{K} \left(\tau_k^{\text{low}} \cdot \delta_k(t_j) \right) \leq \tau^{\text{ex}}(t_j) \leq \sum_{k=1}^{K} \left(\tau_k^{\text{nom}} \cdot \delta_k(t_j) \right) \quad (4)$$

It is worth remarking that the state equation (2) has a linear form, the cost function (1) has a structure that can be easily converted into a linear one (via the introduction of auxiliary variables and auxiliary linear constraints), and all constraints affecting the problem variables are linear. Thus, the overall problem turns out to be a mixed-integer linear optimization problem. However, the solution of such a problem via mathematical programming techniques is not a satisfactory approach, owing to the difficulty of dealing with a high number of integer variables, and, above all, to the objective of determining optimal control strategies, instead of optimal control actions, as mentioned in the introduction.

3. THE OPTIMAL CONTROL STRATEGY

In the following, only two classes of equivalent jobs, that is, $K = 2$, will be considered, as the generalization to K classes is straightforward. In this case, the evolution of the system state, as regards the first two components of vector $\underline{x}(t_j)$, may be graphically represented through the diagram in Fig. 1, where, for each decisional stage, a variable number of possible states is allowed.

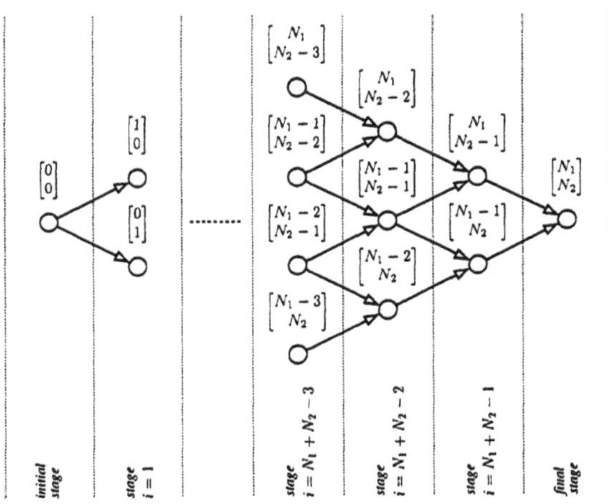

Fig. 1. A representation of the state diagram, when $K = 2$.

In order to provide the main result of this paper, it is convenient to introduce two preliminary definitions.

Definition 1. Given an integer number M, $M \geq 1$, and a positive real number ν, a function $f(x)$ is said to be a $\mathcal{PWL}(M, \nu)$-function (piece-wise linear-function, Fig. 2) if:

- $f(x)$ is a non-decreasing continuous piece-wise linear function of x, characterized by M break points; let γ_i, $i = 1, \ldots, M$, be the generic break point of $f(x)$, being $\gamma_{i+1} > \gamma_i$, $i = 1, \ldots, M-1$; let μ_0 be the slope of $f(x)$ in $(-\infty, \gamma_1)$, μ_i be the slope in (γ_i, γ_{i+1}) for any $i = 1, \ldots, M-1$, and μ_M be the slope in $(\gamma_M, +\infty)$; of course, $\mu_{i+1} \neq \mu_i$, $i = 1, \ldots, M-1$;
- $f(x) = 0$ for any $x \leq \gamma_1$, and thus $\mu_0 = 0$;
- $\mu_i \geq \nu$, $i = 1, \ldots, M$.

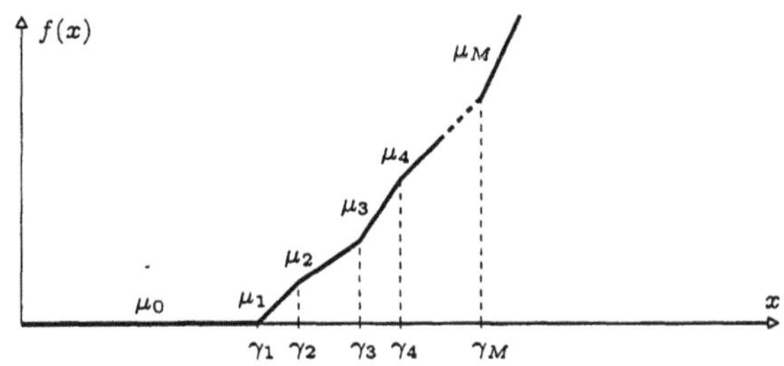

Fig. 2. $\mathcal{PWL}(M, \nu)$-function $f(x)$.

Definition 2. Given an integer number M, $M \geq 1$, and a positive real number ν, a function $f(x+t)$ is said to be a $\mathcal{PPWL}(M, \nu, t)$-function (parameterized piece-wise linear-function) if:

- $f(x+t)$ is a non-decreasing continuous piece-wise linear function of x, parameterized by t, and characterized by M break points; let $\gamma_i - t$, $i = 1, \ldots, M$, be the generic break point of $f(x+t)$; let μ_0 be the slope of $f(x+t)$ in $(-\infty, \gamma_1 - t)$, μ_i be the slope in $(\gamma_i - t, \gamma_{i+1} - t)$ for any $i = 1, \ldots, M-1$, and μ_M be the slope in $(\gamma_M - t, +\infty)$;
- $f(x)$ is a $\mathcal{PWL}(M, \nu)$-function.

In addition, three preliminary results are needed.

Lemma 3. Let $f(x+t)$ be a $\mathcal{PPWL}(M, \nu, t)$-function, and $g(x)$ be a linear decreasing function of x defined in interval $[x_1, x_2]$ and having a negative slope $-\nu$. Then, the function

$$x^\circ(t) = \min_{\substack{x \\ x_1 \leq x \leq x_2}} \left\{ f(x+t) + g(x) \right\} \quad (5)$$

is a non-increasing continuous piece-wise linear function of t defined as follows

$$x^\circ(t) = \begin{cases} x_2 & t < \gamma_1 - x_2 \\ -t + \gamma_1 & \gamma_1 - x_2 \leq t < \gamma_1 - x_1 \\ x_1 & t \geq \gamma_1 - x_1 \end{cases} \quad (6)$$

and the function

$$h(t) = f\left(x^\circ(t) + t\right) + g\left(x^\circ(t)\right) \quad (7)$$

is a $\mathcal{PWL}(P, \nu)$-function, where $P = M + 1$ if slope μ_1 of $f(x+t)$ is greater than ν, and $P = M$ otherwise, being γ_1 and μ_1 some of the parameters relevant to the definition of $f(x+t)$. Moreover, when $P = M + 1$, the break points o $h(t)$ are $\gamma_1 - x_2$, $\gamma_i - x_1$, $i = 1, \ldots, M$, whereas, when $P = M$, the break points are $\gamma_1 - x_2$, $\gamma_i - x_1$, $i = 2, \ldots, M$.

Lemma 4. Let $f(x+t)$ be a $\mathcal{PPWL}(M, \xi, t)$-function, and let $g(x+t)$ be a $\mathcal{PPWL}(N, \nu, t)$-function. Then the function

$$h(x+t) = f(x+t) + g(x+t) \qquad (8)$$

is a $\mathcal{PPWL}(P, \rho, t)$-function, where $\rho = \min\{\xi, \nu\}$ and $P \leq M + N$.

Lemma 5. Let $f(x)$ be a $\mathcal{PWL}(M, \xi)$-function, and let $g(x)$ be a $\mathcal{PWL}(N, \nu)$-function. Then the function

$$h(x) = \min\Big\{f(x), g(x)\Big\} \qquad (9)$$

is a $\mathcal{PWL}(P, \rho)$-function, where $\rho = \min\{\xi, \nu\}$ and $P \leq 2(M+N) - 1$.

PROOF. The proofs of Lemma 3, 4, and 5 can be directly obtained on the basis of geometric considerations, and are here omitted for the sake of brevity. □

Consider now state $[n_1, n_2, t_{n_1+n_2}]^T$, which is the state reached at time $t_{n_1+n_2}$, at which n_1 (n_2) jobs of class P_1 (P_2) have been executed, and let:

- $\tau^{\circ}_{1,n_1+1}(t_{n_1+n_2})$ ($\tau^{\circ}_{2,n_2+1}(t_{n_1+n_2})$) be the optimal execution time associated with the (n_1+1)-th ((n_2+1)-th) job of type P_1 (P_2), under the assumption $\delta_1(t_{n_1+n_2}) = 1$ ($\delta_2(t_{n_1+n_2}) = 1$);
- $\tilde{\mathcal{J}}^1_{n_1,n_2}(t_{n_1+n_2})$ ($\tilde{\mathcal{J}}^2_{n_1,n_2}(t_{n_1+n_2})$) be the optimal cost-to-go associated with state $[n_1, n_2, t_{n_1+n_2}]^T$, under the assumption $\delta_1(t_{n_1+n_2}) = 1$ ($\delta_2(t_{n_1+n_2}) = 1$);
- $\mathcal{J}^{\circ}_{n_1,n_2}(t_{n_1+n_2})$ denote the optimal *cost-to-go* associated with such a state.

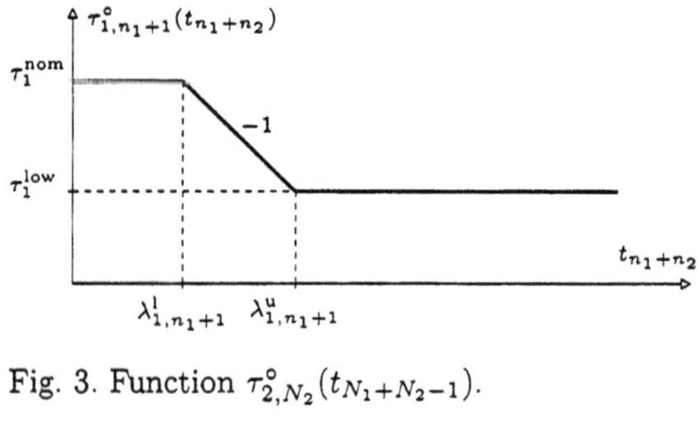

Fig. 3. Function $\tau^{\circ}_{2,N_2}(t_{N_1+N_2-1})$.

Then, the following theorem holds.

Theorem 6. For any n_1, n_2 such that $n_1 + n_2 \neq N_1 + N_2$ and $n_1 < N_1$ ($n_2 < N_2$):

- $\tau^{\circ}_{1,n_1+1}(t_{n_1+n_2})$ is a non-increasing continuous piece-wise linear function of $t_{n_1+n_2}$ (Fig. 3), defined as follows

$$\tau^{\circ}_{1,n_1+1}(t_{n_1+n_2}) = \begin{cases} \tau_1^{\text{nom}} & 0 \leq t_{n_1+n_2} < \lambda^{\text{l}}_{1,n_1+1} \\ -t_{n_1+n_2} + \lambda^{\text{l}}_{1,n_1+1} + \tau_1^{\text{nom}} & \lambda^{\text{l}}_{1,n_1+1} \leq t_{n_1+n_2} < \lambda^{\text{u}}_{1,n_1+1} \\ \tau_1^{\text{low}} & t_{n_1+n_2} \geq \lambda^{\text{u}}_{1,n_1+1} \end{cases} \qquad (10)$$

for some real $\lambda^{\text{l}}_{1,n_1+1}$ and $\lambda^{\text{u}}_{1,n_1+1}$ (an analogous result holds for $\tau^{\circ}_{2,n_2+1}(t_{n_1+n_2})$);
- $\tilde{\mathcal{J}}^1_{n_1,n_2}(t_{n_1+n_2})$ is a $\mathcal{PWL}(M, \nu)$-function, for some integer M and some real ν (an analogous result holds for $\tilde{\mathcal{J}}^2_{n_1,n_2}(t_{n_1+n_2})$);

Then, the optimal control strategy at state $[n_1, n_2, t_{n_1+n_2}]^T$ is given by

$$\delta_1(t_{n_1+n_2}) = \\ = \begin{cases} 1 & \text{if } \tilde{\mathcal{J}}^1_{n_1,n_2}(t_{n_1+n_2}) \leq \tilde{\mathcal{J}}^2_{n_1,n_2}(t_{n_1+n_2}) \\ 0 & \text{if } \tilde{\mathcal{J}}^1_{n_1,n_2}(t_{n_1+n_2}) > \tilde{\mathcal{J}}^2_{n_1,n_2}(t_{n_1+n_2}) \end{cases} \qquad (11)$$

$$\delta_2(t_{n_1+n_2}) = 1 - \delta_1(t_{n_1+n_2}) \qquad (12)$$

and by

$$\tau^{\text{ex}}(t_{n_1+n_2}) = \delta_1(t_{n_1+n_2}) \cdot \tau^{\circ}_{1,n_1+1}(t_{n_1+n_2}) + \\ + \delta_2(t_{n_1+n_2}) \cdot \tau^{\circ}_{2,n_2+1}(t_{n_1+n_2}) \qquad (13)$$

Besides, finally note that $\mathcal{J}^{\circ}_{n_1,n_2}(t_{n_1+n_2})$ is a $\mathcal{PWL}(M, \nu)$-function, for some integer M and some real ν.

PROOF. Consider state $[N_1, N_2, t_{N_1+N_2}]^T$ corresponding to the final stage. Clearly,

$$\mathcal{J}^{\circ}_{N_1,N_2}(t_{N_1+N_2}) = 0 \qquad (14)$$

since no job has to be executed anymore.

The final state can be reached both from state $[N_1, N_2 - 1, t_{N_1+N_2-1}]^T$ and from state $[N_1 - 1, N_2, t_{N_1+N_2-1}]^T$. The optimal cost-to-go at state $[N_1, N_2 - 1, t_{N_1+N_2-1}]^T$ is

$$\mathcal{J}^{\circ}_{N_1,N_2-1}(t_{N_1+N_2-1}) = \\ = \min_{\tau^{\text{ex}}(t_{N_1+N_2-1})} \Big\{\alpha_{2,N_2} \cdot \max\big\{t_{N_1+N_2-1} + \\ + \tau^{\text{ex}}(t_{N_1+N_2-1}) - dd_{2,N_2}, 0\big\} + \\ + \beta_2 \cdot \left(\tau_2^{\text{nom}} - \tau^{\text{ex}}(t_{N_1+N_2-1})\right) + \\ + \mathcal{J}^{\circ}_{N_1,N_2}(t_{N_1+N_2})\Big\} = \\ = \min_{\tau_{2,N_2}} \Big\{\alpha_{2,N_2} \cdot \max\big\{t_{N_1+N_2-1} + \\ + \tau_{2,N_2} - dd_{2,N_2}, 0\big\} + \beta_2 \cdot \left(\tau_2^{\text{nom}} - \tau_{2,N_2}\right) \\ + \mathcal{J}^{\circ}_{N_1,N_2}(t_{N_1+N_2})\Big\} \qquad (15)$$

Note that in writing (15), the only decision variable is the continuous variable $\tau^{\text{ex}}(t_{N_1+N_2-1}) = \tau_{2,N_2}$ as no decision about the class of the next job to execute has to be taken, since all jobs of class

P_1 have been previously completed. The rightmost term of equation (15) can be rewritten as follows

$$\min_{\tau_{2,N_2}} \left\{ s^{\alpha}_{2,N_2}(\tau_{2,N_2} + t_{N_1+N_2-1}) + s^{\beta}_{2,N_2}(\tau_{2,N_2}) \right\} \tag{16}$$

where (taking into account (14))

$$\begin{aligned} & s^{\alpha}_{2,N_2}(\tau_{2,N_2} + t_{N_1+N_2-1}) = \\ & \alpha_{2,N_2} \cdot \max\left\{t_{N_1+N_2-1} + \tau_{2,N_2} - dd_{2,N_2}, 0\right\} \\ & + \mathcal{J}^{\circ}_{N_1,N_2}(t_{N_1+N_2}) \end{aligned} \tag{17}$$

$$s^{\beta}_{2,N_2}(\tau_{2,N_2}) = \beta_2 \cdot (\tau_2^{\mathrm{nom}} - \tau_{2,N_2}) \tag{18}$$

are two functions of variable τ_{2,N_2}. Then note that, taking into account (14), $s^{\alpha}_{2,N_2}(\tau_{2,N_2} + t_{N_1+N_2-1})$ is a $\mathcal{PPWL}(1, \beta_2, t_{N_1+N_2-1})$-function (see Fig. 4). Instead, $s^{\beta}_{2,N_2}(\tau_{2,N_2})$, defined only for $\tau_2^{\mathrm{low}} \leq \tau_{2,N_2} \leq \tau_2^{\mathrm{nom}}$, is a decreasing linear function of τ_{2,N_2} (see Fig. 5).

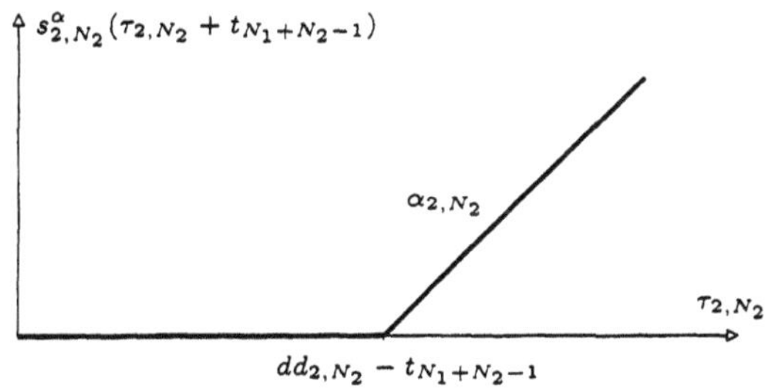

Fig. 4. Function $s^{\alpha}_{2,N_2}(\tau_{2,N_2}, t_{N_1+N_2-1})$.

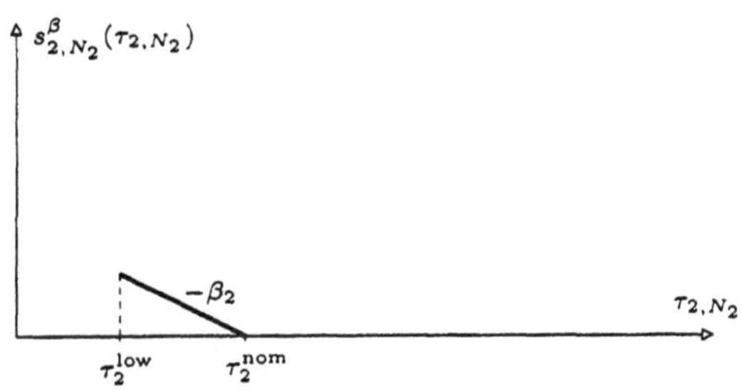

Fig. 5. Function $s^{\beta}_{2,N_2}(\tau_{2,N_2})$.

It is now possible to apply Lemma 3 and thus determine function $\tau^{\circ}_{2,N_2}(t_{N_1+N_2-1})$ (see Fig. 6), which expresses the optimal strategy at the considered stage, and the optimal cost-to-go $\mathcal{J}^{\circ}_{N_1,N_2-1}(t_{N_1+N_2-1})$ (see Fig. 7). Note that the optimal cost-to-go is a $\mathcal{PWL}(2, \beta_2)$-function, according to Lemma 3, and to the fact that it has been assumed $\alpha_{2,N_2} > \beta_2$.

Note that, obviously, the results plotted in Figs. 6 and 7 could also have been obtained directly from (16), (17), and (18). In fact, the interpretation of the results reported in Figs. 6 and 7 is obvious. Actually, when $t_{N_1+N_2-1} < dd_{2,N_2} - \tau_2^{\mathrm{nom}}$, the N_2-th job is in any case early, and then its optimal execution time is equal to the nominal value. Hence, the optimal cost is equal to 0. The optimal choice of the execution time

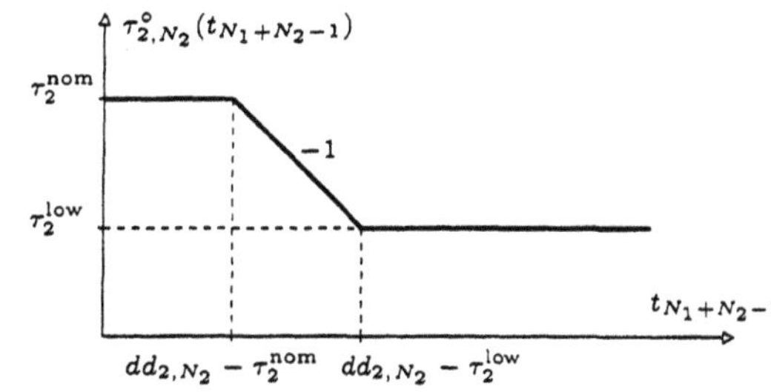

Fig. 6. Function $\tau^{\circ}_{2,N_2}(t_{N_1+N_2-1})$.

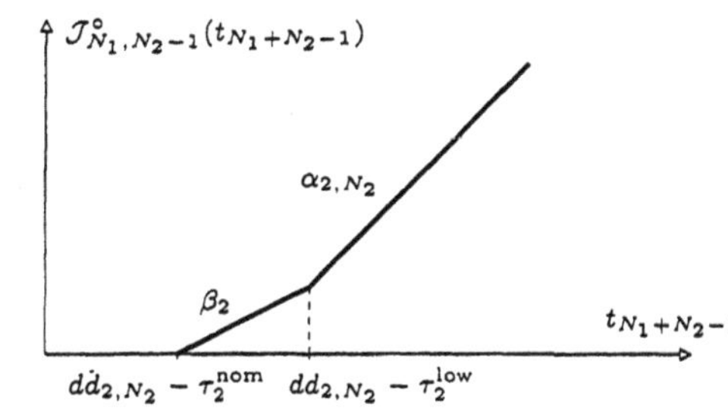

Fig. 7. Function $\mathcal{J}^{\circ}_{N_1,N_2-1}(t_{N_1+N_2-1})$.

when $dd_{2,N_2} - \tau_2^{\mathrm{nom}} \leq t_{N_1+N_2-1} \leq dd_{2,N_2} - \tau_2^{\mathrm{low}}$ makes the job just-in-time, as it has been assumed $\alpha_{2,N_2} > \beta_2$; in this case, the actual execution time is lower than its nominal value, and then a cost is paid, owing to the deviation of the optimal execution time from such a nominal value. Finally, when $t_{N_1+N_2-1} \geq dd_{2,N_2} - \tau_2^{\mathrm{low}}$, the job is in any case tardy, and then, as $\alpha_{2,N_2} > \beta_2$, in order to minimize the overall tardiness cost, the optimal execution time is equal to the smallest value, namely τ_2^{low}.

In a completely analogous way, it is possible to determine, with reference to state $[N_1 - 1, N_2, t_{N_1+N_2-1}]^T$, the optimal control strategy $\tau^{\circ}_{1,N_1}(t_{N_1+N_2-1})$ and the optimal cost-to-go $\mathcal{J}^{\circ}_{N_1-1,N_2}(t_{N_1+N_2-1})$. Of course, such an optimal cost-to-go is a $\mathcal{PWL}(2, \beta_1)$-function.

Consider now the optimal decision in state $[N_1 - 1, N_2 - 1, t_{N_1+N_2-2}]^T$. The optimal cost-to-go to be minimized is

$$\begin{aligned} & \mathcal{J}^{\circ}_{N_1-1,N_2-1}(t_{N_1+N_2-2}) = \\ & = \min_{\substack{\delta_k(t_{N_1+N_2-2}) \\ k=1,2 \\ \tau^{\mathrm{ex}}(t_{N_1+N_2-2})}} \Big\{ \delta_1(t_{N_1+N_2-2}) \cdot \\ & \cdot \Big(\alpha_{1,N_1} \cdot \max\left\{t_{N_1+N_2-2} + \tau^{\mathrm{ex}}(t_{N_1+N_2-2}) + \right. \\ & \left. - dd_{1,N_1}, 0\right\} + \beta_1 \cdot (\tau_1^{\mathrm{nom}} - \tau^{\mathrm{ex}}(t_{N_1+N_2-2})) + \\ & + \mathcal{J}^{\circ}_{N_1,N_2-1}(t_{N_1+N_2-1}) \Big) + \\ & + \delta_2(t_{N_1+N_2-2}) \cdot \\ & \cdot \Big(\alpha_{2,N_2} \cdot \max\left\{t_{N_1+N_2-2} + \tau^{\mathrm{ex}}(t_{N_1+N_2-2}) + \right. \\ & \left. - dd_{2,N_2}, 0\right\} + \beta_2 \cdot (\tau_2^{\mathrm{nom}} - \tau^{\mathrm{ex}}(t_{N_1+N_2-2})) + \\ & + \mathcal{J}^{\circ}_{N_1-1,N_2}(t_{N_1+N_2-1}) \Big) \Big\} \end{aligned} \tag{19}$$

The minimization in (19), has to be carried out with respect to the discrete decision variables $\delta_k(t_{N_1+N_2-2})$, $k = 1, 2$, and the continuous one $\tau^{\text{ex}}(t_{N_1+N_2-2})$.

Let us assume $\delta_1(t_{N_1+N_2-2}) = 1$ (which implies $\delta_2(t_{N_1+N_2-2}) = 0$). Then, the following minimization, with respect to the only decision variable $\tau^{\text{ex}}(t_{N_1+N_2-2}) = \tau_{1,N_1}$, has to be carried out

$$\begin{aligned}\min_{\tau_{1,N_1}} \Big\{ & \alpha_{1,N_1} \cdot \max\big\{t_{N_1+N_2-2}+ \\ & + \tau_{1,N_1} - dd_{1,N_1}, 0\big\} + \beta_1 \cdot (\tau_1^{\text{nom}} - \tau_{1,N_1}) \\ & + \mathcal{J}^{\circ}_{N_1,N_2-1}(t_{N_1+N_2-1})\Big\} = \\ = \tilde{\mathcal{J}}^{1}_{N_1-1,N_2-1} & (t_{N_1+N_2-1})\end{aligned} \tag{20}$$

In this case $t_{N_1+N_2-1} = t_{N_1+N_2-2} + \tau_{1,N_1}$, and then the optimal cost-to-go $\mathcal{J}^{\circ}_{N_1,N_2-1}(t_{N_1+N_2-1})$ represented in Fig. 7 is now expressed as

$$\begin{aligned}&\mathcal{J}^{\circ}_{N_1,N_2-1}(t_{N_1+N_2-1}) = \\ &\quad = \mathcal{J}^{\circ}_{N_1,N_2-1}(\tau_{1,N_1} + t_{N_1+N_2-2})\end{aligned} \tag{21}$$

being $\mathcal{J}^{\circ}_{N_1,N_2-1}(\tau_{1,N_1} + t_{N_1+N_2-2})$ a function of τ_{1,N_1}, parameterized by $t_{N_1+N_2-2}$.

Then, (20) can be rewritten as

$$\min_{\tau_{1,N_1}} \left\{ s^{\alpha}_{1,N_1}(\tau_{1,N_1} + t_{N_1+N_2-2}) + s^{\beta}_{1,N_1}(\tau_{1,N_1}) \right\} \tag{22}$$

where

$$\begin{aligned}& s^{\alpha}_{1,N_1}(\tau_{1,N_1} + t_{N_1+N_2-2}) = \\ & = \alpha_{1,N_1} \cdot \max\big\{t_{N_1+N_2-2} + \tau_{1,N_1} - dd_{1,N_1}, 0\big\} + \\ & \quad + \mathcal{J}^{\circ}_{N_1,N_2-1}(\tau_{1,N_1} + t_{N_1+N_2-2})\end{aligned} \tag{23}$$

$$s^{\beta}_{1,N_1}(\tau_{1,N_1}) = \beta_1 \cdot (\tau_1^{\text{nom}} - \tau_{1,N_1}) \tag{24}$$

Owing to Lemma 4, $s^{\alpha}_{1,N_1}(\tau_{1,N_1} + t_{N_1+N_2-2})$ is still a $\mathcal{PPWL}$-function. Then, as before, it is possible to apply Lemma 3 and obtain the function $\tau^{\circ}_{1,N_1}(t_{N_1+N_2-2})$ as well as the cost-to-go $\tilde{\mathcal{J}}^{1}_{N_1-1,N_2-1}(t_{N_1+N_2-1})$, which is still a $\mathcal{PWL}$-function.

In the same way, it is possible to determine the function $\tau^{\circ}_{2,N_2}(t_{N_1+N_2-2})$ and the cost-to-go $\tilde{\mathcal{J}}^{2}_{N_1-1,N_2-1}(t_{N_1+N_2-1})$ that correspond to the case $\delta_2(t_{N_1+N_2-2}) = 1$.

The optimal cost-to-go $\mathcal{J}^{\circ}_{N_1-1,N_2-1}(t_{N_1+N_2-2})$ corresponds, for any value of $t_{N_1+N_2-2}$, to

$$\begin{aligned}&\mathcal{J}^{\circ}_{N_1-1,N_2-1}(t_{N_1+N_2-2}) = \\ &\quad = \min\big\{\tilde{\mathcal{J}}^{1}_{N_1-1,N_2-1}(t_{N_1+N_2-2}), \\ &\qquad\quad \tilde{\mathcal{J}}^{2}_{N_1-1,N_2-1}(t_{N_1+N_2-2})\big\}\end{aligned} \tag{25}$$

Then, owing to Lemma 5, $\mathcal{J}^{\circ}_{N_1-1,N_2-1}(t_{N_1+N_2-2})$ is a $\mathcal{PWL}$-function.

The above argument can be applied for any decisional stage, i.e., at any of the states depicted in Fig. 1, and then it is possible to iteratively obtain the generic structure of the decision strategy indicated by theorem 6. □

4. CONCLUSIONS

The problem considered and solved in this paper can be considered as a first example of a class scheduling problems – quite common in several manufacturing environments – in which not all jobs are completely distinguishable and the processing times of operations are themselves matter of decision. It has been shown that it is possible to determine the structure of the optimal decision strategy for each stage, as a function of the system state, through a straightforward application of dynamic programming.

The extension of the proposed approach to different scheduling models is matter of current activity. First of all, the extension to the case where a set-up is required between the execution of jobs of different classes, has to be considered. Then, the case of quadratic (instead of linear) cost function has to be considered. Finally, the ultimate objective is that of extending the proposed approach (and the determination of closed-loop strategies) to the case in which the job completion requires a sequence of services on different machines.

REFERENCES

Blazewicz, J., K.H. Ecker, E. Pesch, G. Schmidt and J. Weglarz (1996). *Scheduling Computer and Manufacturing Processes.* Springer-Verlag.

Di Febbraro, A., R. Minciardi and S. Sacone (2001). Single machine scheduling with fixed lot-sizes and variable processing times. In: *Proc. of the 40th IEEE Int. Conf. on Decision and Control.* pp. 2349–2354.

Di Febbraro, A., R. Minciardi and S. Sacone (2002). Optimal control laws for lot-sizing and timing of jobs on a single production facility. *IEEE Transactions on Automatic Control* **47**(10), 1613–1623.

French, S. (1982). *Sequencing and Scheduling. An Introduction to the Mathematics of the Job-Shop.* Ellis Horwood Limited.

Nowicki, E. (1994). An approximation algorithm for a single-machine scheduling problem with release times, delivery times and controllable processing times. *European Journal of Operation Research* **72**, 74–81.

Pinedo, M. (1995). *Scheduling. Theory, Algorithms and Systems.* Prentice-Hall, Inc.

ELSEVIER
IFAC
PUBLICATIONS
www.elsevier.com/locate/ifac

AN URBAN TRAFFIC NETWORK MODEL VIA COLOURED TIMED PETRI NETS

Mariagrazia Dotoli, Maria Pia Fanti

Dipartimento di Elettrotecnica ed Elettronica, Politecnico di Bari
Via Re David 200, 70125 Bari, Italy

Abstract: This paper deals with modelling of traffic networks for real time control purposes. A modular framework based on coloured timed Petri nets is proposed to model the dynamics of signalized traffic network systems: places represent link cells and crossing sections, tokens are vehicles and token colours represent the routing of the corresponding vehicle. In addition, an ordinary timed Petri net models the signal timing plan controlling the area. The proposed modelling framework is applied to a real intersection located in Bari, Italy. A discrete event simulation of the controlled intersection validates the model and tests the signal timing plan obtained by an optimization strategy presented in the related literature. *Copyright © 2004 IFAC*

Keywords: modelling, road traffic, Petri-nets, discrete event systems, traffic control.

1. INTRODUCTION

Traffic signal on surface street networks plays a central role in traffic management. Most of the currently implemented traffic control systems are grouped into two principal classes (Patel and Ranganathan, 2001, Wey, 2000): 1) fixed-time systems and 2) vehicle actuated systems. In the first group, the control system determines fixed timing plans by using an off line optimization performed by computer programs. The second class of traffic control strategies employs actuated signal timing plans and performs an on-line optimization and synchronization of traffic signals. In the real-time optimization control strategy, detectors located on the intersection branches monitor traffic conditions and feed information on the actual system state to the real time controller. Moreover, the controller selects the duration of the green phases in the signal timing plan in order to optimize a performance index (Patel and Ranganathan, 2001, Wey, 2000, Dotoli, *et al.*, 2003, Barisone *et al.*, 2002).

Two main different approaches are developed in the related literature in order to model and analyze the behaviour of vehicles in Traffic Networks (TNs) and may be classified according to the level of modelling detail. First, microscopic models describe both the system entities and their interactions with high level of detail. However, these high-fidelity microscopic models and the resulting software are costly to develop and execute. Second, macroscopic models represent the traffic streams in an aggregate manner, e.g. by scalar values of flow rate, density and speed. These lower fidelity models are less costly to develop and maintain, but their representation may be inaccurate and not valid.

Petri net based models are suited to describe signal traffic control and concurrent activities in TNs (Giua, 1991, Di Cesare, *et al.*, 1994). In particular, Di Febbraro, *et al.*, (2002) present a traffic model in a timed Petri net framework, where tokens are vehicles and places are parts of lanes and intersections. However, in ordinary Petri nets tokens can not distinguish among different vehicles and their associated routes. Colours are introduced in (Di Cesare, *et al.*, 1994), where a different colour is assigned to each vehicle entering the system. However, the model methodology developed in (Di Cesare, *et al.*, 1994) does not use the coloured Petri net paradigm to simulate the system, that is substantially modelled with ordinary Petri nets.

This paper has two objectives. First, it presents a Coloured Timed Petri Net (CTPN) model to describe in a microscopic formulation the flow of vehicles in an urban TN system. The token colour represents the path which a vehicle has to follow and places are cells accommodating one vehicle only. Moreover, the paper models the TN traffic light and determines the green phases duration using an optimization method based on a macroscopic model (Dotoli, *et al.*, 2003, Barisone *et al.*, 2002). Second, the proposed CTPN model generates the code to simulate real intersections and to test the actuated control strategy proposed in (Dotoli, *et al.*, 2003, Barisone *et al.*, 2002). Hence, a case study is presented and the traffic flow in a real intersection located in the city of Bari, Italy, is simulated in different traffic scenarios.

The paper is organized as follows. Section 2 recalls the basics of CTPNs. Moreover, Section 3 defines the CTPN modelling the TN and the traffic light. Section 4 describes the case study and reports the results of a simulation performed under different traffic scenarios. Finally, Section 5 summarizes the conclusions.

2. BASICS OF COLOURED PETRI NETS

A Coloured Petri Net (CPN) is a bipartite directed graph represented by a quintuple $CPN=(P, T, Co, \boldsymbol{Pre}, \boldsymbol{Post})$, where P is a set of places, T is a set of transitions, Co is a colour function that associates with each element in $P\cup T$ a non empty ordered set of colours in the set of possible colours Cl (Jensen, 1992). Co maps each place $p_i \in P$ to the set of possible token colours $Co(p_i)=\{a_{i,1},a_{i,2},\ldots a_{i,ui}\}\subseteq Cl$, where $u_i=|Co(p_i)|$ is the number of possible colours of tokens in p_i. Note that we use symbol $|A|$ to denote the cardinality of the generic set A. Analogously, Co maps each transition $t_j \in T$ to the set of possible occurrence colours $Co(t_j)=\{b_{j,1},b_{j,2},\ldots,b_{j,uj}\}\subseteq Cl$ with $u_j=|Co(t_j)|$.

H is a weight function for an inhibitor arc that connects a transition to a place. More precisely, an inhibitor arc labelled $a\in N$ between $p_i \in P$ and $t_j \in T$ (i.e., $H(p_i,t_j)=a$) implies that as soon as there are a tokens in p_i, the arc inhibits the firing of t_j. Note that N is the set of non-negative integers.

Now, we define a non-negative multiset α over the set D as a mapping α: $D\rightarrow N$ (Jensen, 1992). Matrices ***Pre*** and ***Post*** are the pre-incidence and the post-incidence matrices, respectively. In particular, each element $\boldsymbol{Pre}(p_i,t_j)$ is a mapping from the set of occurrence colours of $t_j \in T$ to the set of non-negative multisets $\mathcal{N}(Co(p_i))$ over the set of colours of $p_i \in P$. More precisely, we denote $\boldsymbol{Pre}(p_i,t_j)$: $Co(t_j)\rightarrow \mathcal{N}(Co(p_i))$ as a matrix of $u_i \times u_j$ non-negative integers, whose generic element $Pre(p_i,t_j)(h,k)$ is equal to the weight of the arc from place p_i with respect to colour $a_{i,h}$ to transition t_j with respect to colour $b_{j,k}$.

Analogously, $\boldsymbol{Post}(p_i,t_j)$: $Co(t_j)\rightarrow \mathcal{N}(Co(p_i))$ for each $t_j \in T$ to each $p_i \in P$ corresponds to the set of directed arcs from T to P. Hence, we denote by $\boldsymbol{Post}(p_i,t_j)$ a matrix of $u_i \times u_j$ non-negative integers and the scalar $Post(p_i,t_j)(h,k)$ is the weight of the arc from transition t_j with respect to colour $b_{j,k}$ to place p_i with respect to colour $a_{i,h}$. Finally, the incidence matrix $\mathbf{C}$ of dimension $|P|\times|T|$ is defined as $\mathbf{C}(p_i,t_j)=\boldsymbol{Post}(p_i,t_j)-\boldsymbol{Pre}(p_i,t_j)$.

For each place $p_i \in P$, we define the marking $\boldsymbol{m}_i$ of p_i as a *non-negative multiset* over $Co(p_i)$. The mapping $\boldsymbol{m}_i$: $Co(p_i)\rightarrow N$ associates with each possible token colour in p_i a non-negative integer representing the number of tokens of that colour that is contained in p_i. In the following we denote $\boldsymbol{m}_i$ as a $(u_i \times 1)$ vector of non-negative integers, whose h-th component $m_i(h)$ is equal to the number of tokens of colour $a_{i,h}$ that are contained in p_i. The marking $\boldsymbol{M}$ of a CPN is an m-dimensional column vector of multisets, i.e., $\boldsymbol{M}=[\boldsymbol{m}_1\ldots \boldsymbol{m}_{|P|}]^T$. Finally, a coloured Petri net system <CPN,$\boldsymbol{M}_0$> is a CPN with initial marking $\boldsymbol{M}_0$.

Given a CPN and a node $x\in P\cup T$, the set $\bullet x=\{y\in P\cup T$: $(y,x)\in F\}$ is the preset of x and $x\bullet=\{y\in P\cup T$: $(x,y)\in F\}$ is the post-set of x. Now, a transition $t_j \in T$ is *enabled* with respect to colour $b_{j,k}$ at a marking $\boldsymbol{M}$ if and only if for each $p_i \in \bullet t_j$, we have:

$m_i(h)\geq Pre(p_i,t_j)(h,k)$ for $h=1,\ldots,u_i$.

If an enabled transition fires with respect to colour $b_{j,k}$, then we get a new marking $\boldsymbol{M}'$, where:

$m'_i(h)=m_i(h)+Post(p_i,t_j)(h,k)-Pre(p_i,t_j)(h,k)$ for $h=1,\ldots,u_i$ and for each $p_i \in \bullet t_j \cup t_j \bullet$,
$m'_i(h)=m_i(h)$ otherwise.

Now, to investigate the performance of the system it is convenient to extend the CPN with the time concept. To do this, we introduce a global clock, i.e., the clock values $\tau\in\Re^+$ represent the model continuous time. Moreover, the temporization of Petri nets can be achieved by attaching time either to places, to transitions (Desrochers and Al-Jaar, 1995) or to the expression functions of arcs (Jensen, 1992). Here we choose the second option and we consider timed transitions and immediate transitions. More precisely, the firing time of each transition t_j, namely FT_j, is a positive number specifying the deterministic duration of the firing of t_j. In this method, each token has a time stamp attached to it, in addition to token colours. The time stamp is described by the function s: $Co\rightarrow\Re+$ where $s(c)$ indicates the earliest delay after which the token of colour $c\in Co$ becomes available and can be removed by an enabled transition. Hence, as soon as the c-colour token arrives to the place p_i enabling transition t_j, $s(c)$ is set to FT_j.

Accordingly, after FT_j time instants, the enabled transition t_j becomes ready to fire. If FT_j is equal to zero, the transition is said to be an immediate transition.

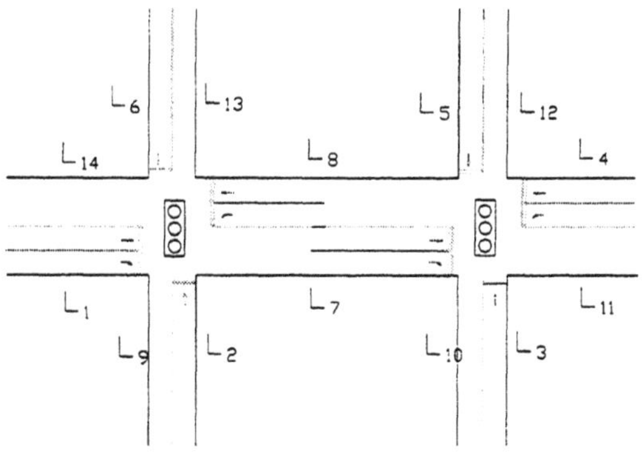

Fig. 1. Example of urban area comprising junctions pertaining to a common semaphoric cycle.

3. MODELLING A SIGNALIZED INTERSECTION AREA

3.1 The traffic network description.

In the proposed TN model, we consider the following fundamental components: signalized intersections, links and vehicles. More precisely, each link represents the space available between two adjacent intersections and may include one or several lanes. Hence, a generic signalized urban area comprises a number of junctions controlled by traffic lights pertaining to a common semaphoric cycle, including a set $L=\{L_i|\ i=1,...I\}$ of I links (see figure 1). A generic link L_i of length l_i with $i=1,...I$ has a finite capacity $C_i>0$ denoting the number of vehicles that a link can simultaneously accommodate. Hence, each link is divided in C_i cells with unit capacity. We assume that lane changing is allowed only in road.

Moreover, it is necessary to take into account the physical space that a vehicle crossing the intersection occupies. Such a physical space, named intersection cell, can be occupied by only one vehicle and may coincide with the whole intersection area in a very simple intersection or with a part of the physical space in a multi-lane intersection.

3.2 The coloured timed Petri net modelling the urban area.

In our model the CPN=(P, T, Co, H, ***Pre***, ***Post***) describes the TN in a modular representation. The CPN can be divided in two subnets: the first one describes the urban area comprising roads and junctions and the second describes the traffic lights pertaining to the common signal timing plan. Referring to the first subnet, a place $p_i \in P$ denotes a resource, i.e., a cell in a lane or in a crossing section, that a vehicle can acquire. More precisely, we have two types of places: places representing link cells and places corresponding to intersection cells.

In addition, transitions from $T_V \subset T$ model the flow of vehicles into the system between consecutive resources. We consider five types of transitions:

1) input transitions $t_{0,i} \in T_I$, modelling the input of a vehicle in a link from the outside of the considered intersection;

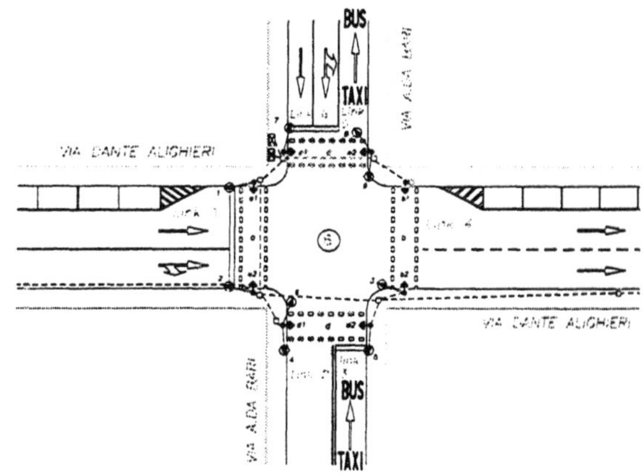

Fig. 2. Layout of a real traffic intersection located in the urban area of Bari, Italy.

2) output transitions $t_{i,0} \in T_O$, modelling the output of a vehicle from a link to the outside of the considered intersection;
3) flow transitions $t_j \in T_F$, modelling the flow of vehicles between two consecutive link cells;
4) intersection transitions $t_j \in T_C$, modelling vehicles entering or leaving a crossing section. The intersection transitions that model a vehicle entering the crossing sections are regulated as a multi-phase traffic light;
5) lane changing intersection transitions $t_j \in T_L$, modelling vehicles that change lane in road.

Transitions belonging to the set $T_V=T_I \cup T_O \cup T_F \cup T_C \cup T_L$ are timed transitions. The value of FT_j assigned to each $t_j \in T_F \cup T_L$ or $t_j \in T_C$ is equal to the average time interval in which a vehicle passes from a cell to the subsequent one or occupies an intersection cell, respectively, and depends on the vehicles speed. Moreover, the value of $t_{0,j} \in T_I$ ($t_{j,0} \in T_O$) is equal to the mean interval times in which vehicles can enter (leave) the system network. Obviously, these parameters can change during different traffic scenarios to model different congestion levels.

A coloured token in a place represents a vehicle. The colour of each token is the routing assigned to each vehicle indicating the different paths that a vehicle can follow starting from a particular position. For instance, with reference to figure 1, in link L_1 vehicles can have three different colours: a_1 if the vehicle has routing (L_1, L_7, L_{11}), a_2 if the vehicle route is (L_1, L_9) and a_3 if the vehicle route is (L_1, L_7, L_{10}). Finally, inhibitor arcs are introduced to model the finite capacities of each resource. More precisely, for each transition $t \in T_F \cup T_C \cup T_L$ and for each $p_i \in t^{\bullet}$, there exists an inhibitor arc between p_i and t, i.e., $H(p_i,t)=1$.
Comparing our model with the model proposed in Di Cesare, *et al.* (1994), we point out that the above assumptions allow us to obtain a concise and modular model, so that the development of the resulting simulation software is inexpensive.

To show the modelling technique, Example 1 describes a real traffic intersection located in the urban area of Bari, Italy, depicted in figure 2. We consider in the example a single junction. However, the modularity of the proposed model allows us to represent several intersections and to connect two consecutive links simply considering the output transitions of the first link coincident with the input transitions of the next link.

Example 1.
The real traffic intersection shown in figure 2 is composed by 6 links L_i (i=1,...,6) with length $l_1=l_4=l_5=l_6$=60m and $l_2=l_3$=40m, respectively. The capacities of the links are $C_1=C_4=C_5=C_6$=12 PCUs (Passenger Car Units) and $C_2=C_3$=8 PCUs, respectively, derived by assuming that one PCU is l_0=5m long. Figure 3 shows the CPN modelling the intersection. Places p_i, p'_i, with i=1,...,9 are a subset of places modelling cell links and places p_i, p'_i with i=10,11,12 model the intersection crossing area composed of six cells. For the sake of simplicity, the figure shows just two places for each lane in an input link and one place for each lane in an output link. Moreover, in this example we suppose that lane changing is not allowed.

With reference to figure 3, transitions $t_{0,1}$, $t_{0,2}$, $t_{0,3}$, $t_{0,4}$ and $t_{0,5}$ are input transitions and transitions $t_{1,0}$, $t_{2,0}$, $t_{3,0}$ and $t_{4,0}$ are output transitions. Transitions t_1, t'_1, t_3, t_6, t'_6 model vehicles entering a crossing section and are controlled by the traffic light Petri net. Vehicles travelling in the intersection are tokens of 6 colours. More precisely, colours a_1 and a_2 are associated with vehicles following the routings (L_1, L_4) and (L_1, L_2), respectively. Moreover, colours a_3, a_4 refer to vehicles following the routings (L_6, L_2) and (L_6, L_4), respectively. Finally, colours a_5, a_6 represent vehicles following the routings (L_3, L_5) and (L_3, L_4), respectively. Referring to the definitions of matrices $\boldsymbol{Post}(p_i,t_j)$ and $\boldsymbol{Pre}(p_i,t_j)$ given in section 2, we show as an example the following matrix entries, obviously the other ones are determined similarly:
$Pre(p'_{10},t'_7)(1,1)=1$ and $Pre(p'_{10},t'_7)(h,k)=0$ for $h,k=2,\dots,6$,
$Post(p'_{11},t'_7)(1,1)=1$ and $Post(p'_{11},t'_2)(h,k)=0$ for $h,k=2,\dots,6$,
$Post(p_7,t_2)(2,2)=1$, $Post(p_7,t_2)(3,3)=1$,
$Pre(p'_{10},t_2)(2,2)=1$, $Pre(p'_{10},t_2)(3,3)=1$,

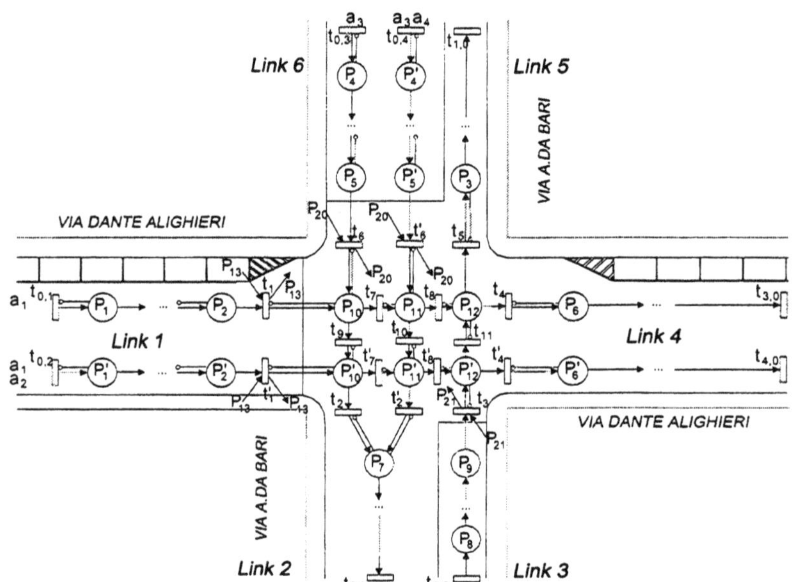

Fig. 3. The CTPN modelling the intersection in figure 2.

Table 1. The firing times of the transitions belonging to the CTPN modelling the intersection in figure 2.

FT_i [s]	Scenario 1		Scenario 2		Scenario 3		Scenario 4	
	Fixed	Dynamic	Fixed	Dynamic	Fixed	Dynamic	Fixed	Dynamic
Input transitions								
$FT_{0,1}$	6.7	6.7	5	5	4	/	5	5
$FT_{0,2}$	6.7	6.7	5	5	4	/	5	5
$FT_{0,3}$	6.7	6.7	5	5	4	/	2.9	2.9
$FT_{0,4}$	6.7	6.7	5	5	4	/	2.9	2.9
$FT_{0,5}$	26.7	26.7	20	20	16	/	20	20
Traffic light transitions								
FT_9	28	32	28	32	28	/	28	49
FT_{10}	4	4	4	4	4	/	4	4
FT_{11}	2	2	2	2	2	/	2	2
FT_{12}	40	36	40	36	40	/	40	19
FT_{13}	4	4	4	4	4	/	4	4
FT_{14}	30	30	30	26	30	/	30	10
FT_{15}	4	4	4	4	4	/	4	4
FT_{16}	0	0	0	0	0	/	0	0
FT_{17}	2	2	2	2	2	/	2	2

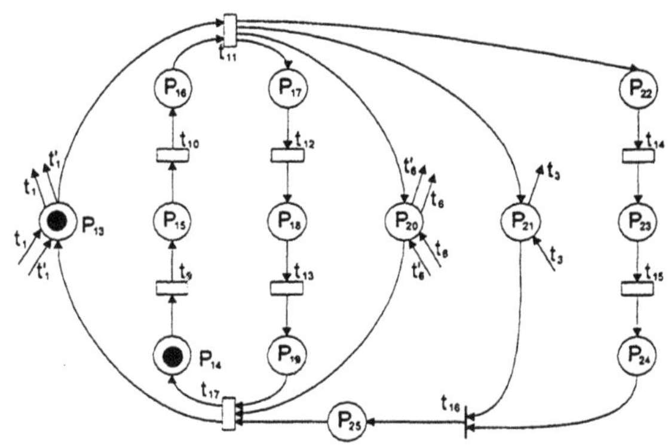

Fig. 4. The ordinary Petri net modelling the traffic light of the intersection in figure 2.

$Pre(p'_{10},t_2)(h,k)=0$ for $h,k=1,4,5,6$,
$Post(p_7,t_2)(h,k)=0$ for $h,k=1,4,5,6$.

The firing times of the input transitions $t_{0j} \in T_I$ for the CTPN modelling the intersection in figure 2 are reported in Table 1 for four different scenarios, corresponding to different congestion levels (see the sequel for details). In addition, all transitions $t_j \in T_O \cup T_F \cup T_C \cup T_L$ have a firing time $FT_j=l_0/v=0.45$ s, where $v=40$ km h^{-1} is assumed to be the vehicles average speed in the intersection.

3.3 The coloured timed Petri net modelling the traffic lights.

To model the traffic lights controller we use a formalism presented in the related literature (Di Febbraro, *et al.*, 2002). Since the Petri net modelling the traffic lights is rather simple, neutral colour tokens are considered. Moreover, places represent phases and transitions model the succession of red, yellow and green phases. To show the method to model a generic signal timing plan, we describe the Petri net representing the traffic light of the signalized intersection of Example 1, depicted in figure 4. To control the intersection in figure 2, a three-phase traffic light is used, modelled by three places representing the red, yellow and green phases, respectively. Hence, to model the phases of the cycle, nine places are necessary. In particular, a neutral colour token in place p_{13} enables transitions t_1 and t_1' to rule the vehicles of link 1. Moreover, places p_{14}, p_{15} and p_{16} describe the state of green, yellow and red, respectively. Moreover, the firing times FT_9, FT_{10}, FT_{11} referring to the corresponding transitions are the time intervals of green, yellow and all red (clearance time) of the traffic light of link 1. In a similar way, we describe the traffic light ruling link 6 and modelled by places p_{17}, p_{18}, p_{19} and p_{20} and the traffic light ruling link 3 and modelled by places p_{21}, p_{22}, p_{23} and p_{24}. Finally, the immediate transition t_{16} and place p_{25} are introduced to obtain for the traffic light in link 3 a different phase with respect to link 6, even if they start at the same time.

The start of the semaphoric cycle is with a token in p_{14} and in p_{13} (green for link 1). When the red phase begins for link 1 transition t_{11} fires and the green phases start for links 6 and 3 (tokens in p_{17} and p_{22}). The green phase for link 3 is shorter than the green phase of link 6: a token in p_{25} means that link 3 is in the red phase and link 6 is in the green or yellow phase. When transition t_{17} fires, the cycle begins again.

Finally, the firing times of the transitions for the CTPN modelling the traffic light in figure 4 are reported in Table 1 for the four traffic scenarios described in the sequel.

4. SIMULATION RESULTS

This section presents a simulation study performed for Example 1. The CTPN model of the signalized intersection reported in figure 3 was implemented and simulated in the Matlab environment. Indeed, the simplicity and modularity of the model suggest us to use an efficient software such as Matlab that allows us to simulate systems with a large number of places and transitions. Obviously, the CTPN simulation could be carried out in a generic commercial or freeware tool.

Starting from the initial marking $m_{13}=m_{14}=1$ and $m_i(h)=0$ for each $p_i \in P$ and for each colour $a_{i,h}$, the discrete event simulation program executes the CTPN in a fixed run time. Three simulation experiments are performed, each referring to a different traffic scenario. Each test is conducted for $K=3$ semaphoric cycles, of duration $C=80$s each, for a total run time $T=KC=240$s. In the first three experiments we use the fixed timing plans actually applied to the real intersection and independent from the traffic congestion. Table 1 reports the phases of the traffic light. In particular, the first case simulates under-saturated traffic: one vehicle every 6.7 seconds enters on average links L_1 and L_6 each, and one vehicle every 26.7 seconds enters L_3, respectively (see Table 1). The simulation model generates as an output the number of vehicles leaving each link L_i in the run time and the average number of vehicles queued in link L_i directed to link L_j. Figure 5 shows (light grey) the average number of vehicles queued in each input link L_i (i=1,3,6) and directed to an output link L_j (j=2,4,5). The second experiment simulates a medium congestion traffic condition: one vehicle every 5 seconds enters on average links L_1 and L_6 each, and one vehicle every 20 seconds enters L_3, respectively (see Table 1). Figure 5 (dark grey) reports the average number of vehicles queued in the input links, showing that these are characterized by a medium congestion, with queue sizes higher than those reported for scenario 1. In particular, L_1 and L_6 contain respectively 6.4 and 7.3 vehicles on average, then they are halfway to saturation, being $C_1=C_6=12$ PCUs. Finally, the third experiment simulates over-saturated traffic: one vehicle every 4 seconds enters on average links L_1 and L_6 each, and one vehicle every 16 seconds enters link L_3, respectively (see Table 1). Figure 5 (black) shows the average queue sizes. Now input link L_1 is oversaturated, with 12.5 vehicles on average and $C_1=12$ PCUs, while L_6 is almost saturated, with 10 vehicles on average and $C_6=12$ PCUs. Experimental evidence has proven that the simulation results are consistent with real traffic data in the considered scenarios.

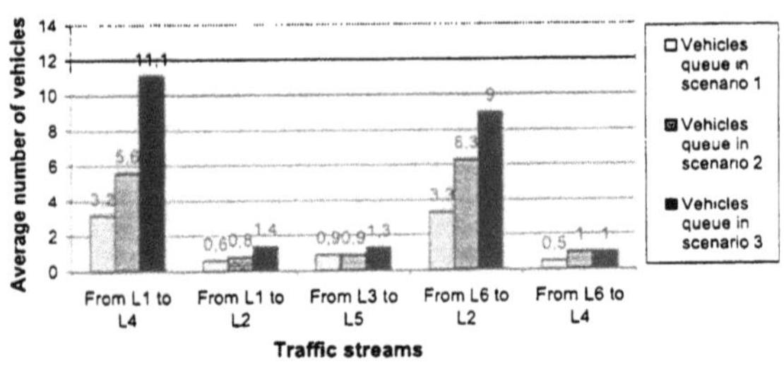

Fig. 5. Simulation results for the intersection under study with traffic scenarios 1, 2 and 3.

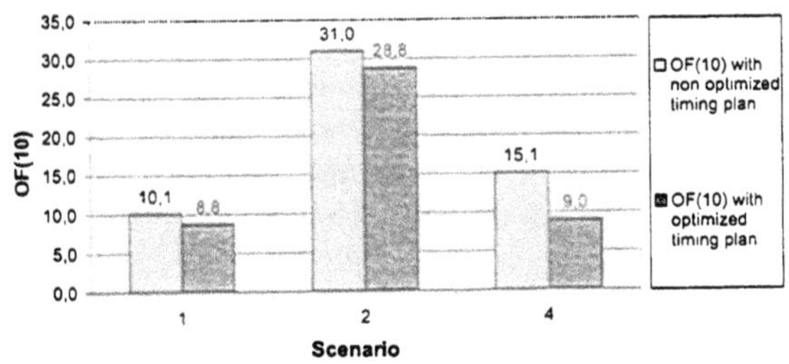

Fig. 6. Simulation results with traffic scenarios 1, 2 and 4 with original and optimized timing plan.

Now, in order to further validate the proposed model, with reference to Example 1, we apply a different signal timing plan obtained by the optimization method proposed in (Barisone, *et al.*, 2002, Dotoli, *et al.*, 2003). More precisely, the green phases of the connected traffic lights are determined dynamically, i.e., on the basis of traffic data, in order to minimize traffic congestion. In fact, the strategy is suited to be applied in an actuated control scheme. The minimized objective function is the following:

$$OF(K) = \sum_{i=\{1,3,6\}} \frac{1}{K}\left[\sum_{k=1}^{K} n_i(k)\right],$$

where $OF(K)$ is measured in K=10 cycles and denotes the average number of vehicles per cycle in the input links. In particular, the optimized signal timing plan differs from the original plan in the green phases allowing flow from L_1 to L_2 and L_4, from L_3 to L_4, from L_6 to L_2 and L_4 respectively. Moreover, Table 1 reports the values of FT_9, FT_{12} and FT_{14} of figure 4 obtained by the optimization method for scenarios 1 and 2 (dynamic).

Figure 6 shows that the simulation computes a lower number of vehicles in the queues if the intersection is controlled with the dynamically-optimized timing plan for K=10 cycles in scenarios 1 and 2. Comparison between the results obtained for the fixed and optimized timing plans under scenario 3 is neglected, since in such a case the TN is oversaturated. However, to stress the benefits of the optimization strategy, we consider a fourth traffic scenario, in which the L_1-L_4 direction is close to saturation and the L_3-L_6 direction is under-saturated (see Table 1 for the corresponding values of the input transitions firing times). Applying the optimization method yields FT_9=49, FT_{14}=10, FT_{12}=19 seconds, prioritizing the congested direction L_1-L_4. The simulation results with the original and optimized timing plans are depicted in figure 6. Clearly, the performance index OF(10) improves considerably controlling the network with the optimized plan. Hence, results show the benefits in applying an actuated traffic control strategy, that is able to modify the traffic lights phases with traffic congestion.

5. CONCLUSIONS

This paper provides a coloured timed Petri net model to describe a traffic network system focusing on real time control via a connected traffic light. The obtained microscopic model is able to give an accurate and valid representation to provide a simulation software not costly to develop. Moreover, the paper presents a real case study where simulation tests the traffic light phases derived by an optimization method proposed in the related literature. Simulation results give a confirmation of the model capability to correctly predict traffic performance measures.

REFERENCES

Barisone, A., D. Giglio, R. Minciardi and R. Poggi (2002) A macroscopic traffic model for real-time optimization of signalized urban areas. In: *Proc. 41st IEEE Int. Conf on Decision and Control*, Las Vegas, 900-903.

Desrochers, A.A. and R.Y Al-Jaar (1995). *Applications of Petri Nets in Manufacturing Systems*. IEEE Press, New York.

Di Cesare F., P. T. Kulp, Gile M. and G. List, (1994) The application of Petri nets to the modelling, analysis and control of intelligent urban traffic networks. *Application and Theory of Petri Nets, Lecture Notes in Computer Science*, **815**, Springer Verlag.

Di Febbraro, A., D. Giglio and N. Sacco (2002). On applying Petri nets to determine optimal offsets for coordinated traffic light timings. In: *Proc. 5th IEEE Int. Conf. on Intelligent Transportation Systems*, 773-778.

Dotoli, M., M. P. Fanti and C. Meloni (2003). Real time optimization of traffic signal control: application to coordinated intersections. In: *Proc. IEEE Int. Conf. on Systems, Man and Cybernetics*, Washington, 3288-3295.

Giua, A. (1991). *A traffic light controller based on Petri nets*. Rensselaer Polytechnic Institute, Troy, New York.

Jensen, K. (1992). *Colored Petri nets: basic concepts, analysis methods and practical use*, **1**. Springer, New York.

Patel, M. and N. Ranganathan (2001). IDUTC: An Intelligent Decision-Making System for Urban Traffic-Control Applications. *IEEE Trans. on Vehicular Technology*, **50**, 816-829.

Wey, W.-M. (2000). Model formulation and solution algorithm of traffic signal control in an urban network. *Computers, environment and urban systems*, **24**, 355-377.

ELSEVIER
IFAC
PUBLICATIONS
www.elsevier.com/locate/ifac

PNETLAB: A TOOL FOR THE SIMULATION, ANALYSIS AND CONTROL OF DISCRETE EVENT SYSTEMS BASED ON PETRI NETS

Francesco Basile * Ciro Carbone * Pasquale Chiacchio **

* *DIIIE, Università degli Studi di Salerno, Via Ponte Don Melillo, 1 - 84084 - Fisciano(SA), ITALY. E-mail: {fbasile, ccarbone}@unisa.it*

** *Dip. di Informatica e Sistemistica, Università degli Studi di Napoli Federico II, Via Claudio 21, 80125 - Napoli, ITALY.*

Abstract: Although supervisory control based on Petri Nets (PNs) appeared in literature in the early 90's, nowadays there is no tool which allows the analysis and simulation of the plant (modelled as a PN) together with the controller/supervisor that can be a PN itself (compiled supervisor case) or a program (interpreted supervisor case). A lot of commercial or freeware tools have been developed with the aim of analyzing and simulating a PN model, in timed or untimed case, and Colored Timed Petri Nets (CTPNs). In this paper we present PNetLab, a freeware tool developed by the Automatic Control Group of the University of Salerno. It allows the drawing of a PN model by a Java graphical user interface. The definition of a controller as a PN (CTPN) or as a standard C/C++ program is also possible. The simulation of the closed loop system and the token game in the timed and untimed case are performed. In addition, for the open loop PN model, the tool performs P-invariant, T-invariant, minimal siphon and trap computation. PNetLab can compute the coverability tree both for the open loop and closed loop system, and it offers a very suitable management of the transition conflicts. The tool functionality are shown via two detailed examples. *Copyright © 2004 IFAC*

Keywords: Petri Net, Supervisory Control, Simulation.

1. INTRODUCTION

Supervisory control theory for discrete event systems (DESs) was initiated by P.J.Ramadge and W.M. Wonham (1989). In their seminal work they represent both the plant — i.e., the system to be controlled — and the desired closed-loop behaviour, by regular languages. The specific problem addressed was to synthesize a controller, called *supervisor*, to achieve the largest subset of the desired language, disabling or enabling controllable events. The unwanted sequences may be related, for example, to safety requirements. Although regular languages have been an useful framework to start such DES control theory, they are limited in representing systems consisting of numerous interacting subsystems. For this reason, a control theory for DES modeled by Petri Net (PN) has been developed (Murata, 1989; DiCesare *et al.*, 1993), extending general PN models with the concept of *controllable* transitions.

In the supervisory control PN theory it is assumed that the set of transitions T of a net is partitioned into two disjoint subsets: T_{uc}, the set of uncontrollable transitions, and T_c, the set of controllable transitions.

In PN framework the state evolution of (rather than the traces of events generated by) a DES is considered. This approach, that we call *state-based*, is particularly attractive when Petri nets (PNs) are used to represent the plant and was used by Holloway *et al.* (1997), Li and Wonham (1994). Let us consider a PN system $\langle N, \mu_0 \rangle$ with m places, whose set of reachable markings is $R(N, \mu_0) \subseteq \mathbb{N}^m$. Assume we are given a set of legal markings $\mathcal{L} \subseteq \mathbb{N}^m$, and consider the basic control problem of designing a supervisor that restricts the reachability set of plant in closed loop to $\mathcal{L} \cap R(N, \mu_0)$. When controllable transitions are present, also legal markings from which it is possible to exit from $\mathcal{L}$ by firing only uncontrollable transitions has to be forbidden.

The CTPN setting extends the framework of PN by adding color, time and modular attributes to the net (Jensen, 1995). The color attribute is developed to deal with systems that have similar or redundant logical structures. The time attribute allows various time-based performance measures to be added in the system model. They have been widely used in the automation filed as modelling, simulation and specification language. Recently in the literature some results of supervisory control theory start to be extended in the CTPN context (Giua *et al.*, 2003).

Although supervisory control based on PNs appeared in literature in the early 90's, nowadays there is no tool which allows the analysis and simulation of the plant (modelled as a PN) together with the controller/supervisor that can be a PN itself (compiled supervisor case) or a program (interpreted supervisor case). A lot of commercial or freeware tools have been developed with the aim of analyzing and simulating a PN model, in timed or untimed case, and CTPNs.

In this paper we present PNetLab [1], a freeware tool developed by the Automatic Control Group of the University of Salerno. It allows the drawing of a PN model by a Java graphical user interface. The definition of a controller as a PN (CTPN) or as a standard C/C++ program is also possible. The simulation of the closed loop system and the token game in the timed and untimed case are performed. In addition, for the open loop PN model, the tool performs P-invariant, T-invariant, minimal siphon and trap computation. PNetLab can compute the coverability tree both for the open loop and closed loop system, and it offers a very suitable management of the transition conflicts. The tool functionality are shown via two examples.

2. PLACE TRANSITION NET BACKGROUND

A place/transition (P/T) net is a structure $\mathcal{N} = \langle P, T, \mathbf{Pre}, \mathbf{Post}\rangle$ where: P is a set of m *places* represented by circles; T is a set of n *transitions* represented by bars; $P \cap T = \emptyset$, $P \cup T \neq \emptyset$; **Pre** (**Post**) is the $\mid P \mid \times \mid T \mid$ sized, natural valued, pre-(post-)incidence matrix. For instance, $\mathbf{Pre}(p,t) = w$ ($\mathbf{Post}(p,t) = w$) means that there is an arc from p (t) to t (p) with weight w. The incidence matrix $\boldsymbol{C}$ of the net is defined as $\boldsymbol{C} = \mathbf{Post} - \mathbf{Pre}$. A *marking* is a $m \times 1$ vector $\boldsymbol{m} : P \to \mathbb{N}$ that assigns to each place of a P/T net a non-negative integer number of tokens. A P/T system or net system $\langle \mathcal{N}, \boldsymbol{m}_0\rangle$ is a P/T net $\mathcal{N}$ with an initial marking $\boldsymbol{m}_0$. A transition $t \in T$ is enabled at a marking $\boldsymbol{m}$ iff $\boldsymbol{m} \geq \mathbf{Pre}(\cdot, t)$. If t is enabled, then it may fire yielding a new marking $\boldsymbol{m}' = \boldsymbol{m} + \mathbf{Post}(\cdot,t) - \mathbf{Pre}(\cdot,t) = \boldsymbol{m} + \mathbf{C}(\cdot,t)$. The notation $\boldsymbol{m}[t > \boldsymbol{m}'$ means that an enabled transition t may fire at $\boldsymbol{m}$ yielding $\boldsymbol{m}'$. A *firing sequence* from $\boldsymbol{m}_0$ is a (possibly empty) sequence of transitions $\sigma = t_1 ... t_k$ such that $\boldsymbol{m}_0[t_1 > \boldsymbol{m}_1[t_2 > \boldsymbol{m}_2 ... [t_k > \boldsymbol{m}_k$. A marking $\boldsymbol{m}$ is reachable in $\langle \mathcal{N}, \boldsymbol{m}_0\rangle$ iff there exists a firing sequence σ such that $\boldsymbol{m}_0[\sigma > \boldsymbol{m}$. Given a net system $\langle \mathcal{N}, \boldsymbol{m}_0\rangle$ the set of reachable markings is denoted $R(\mathcal{N}, \boldsymbol{m}_0)$. The function $\boldsymbol{\sigma} : T \to \mathbb{N}$, where $\boldsymbol{\sigma}(t)$ represents the number of occurrences of t in σ, is called firing count vector of the fireable sequence σ. If $\boldsymbol{m}_0[\sigma > \boldsymbol{m}$, then we can write in vector form $\boldsymbol{m} = \boldsymbol{m}_0 + \boldsymbol{C}(\cdot, t) \cdot \boldsymbol{\sigma}$. This is known as the *state equation* of the system.

[1] Send an e-mail to the address fbasile@unisa.it to obtain the tool.

3. LOGICAL CONTROLLER CASE STUDY

Assume we are given a set of legal markings $\mathcal{L} \subseteq \mathbb{N}^m$, and consider the basic control problem of designing a supervisor that restricts the reachability set of plant in closed loop to $\mathcal{L} \cap R(\mathcal{N}, \boldsymbol{m}_0)$. Of particular interest are those PN state-based control problems where the set of legal markings $\mathcal{L}$ is expressed by a set of n_c linear inequality constraints called Generalized Mutual Exclusion Constraint (GMEC). A single GMEC is a couple $(\boldsymbol{l}, k)$ where $\boldsymbol{l} : P \to \mathbb{Z}$ is a $m \times 1$ weight vector and $k \in \mathbb{Z}$. Given the net system $\langle \mathcal{N}, \boldsymbol{m}_0\rangle$, a GMEC defines a set of markings that will be called *legal markings*: $\mathcal{M}(\boldsymbol{l}, \boldsymbol{k}) = \{\boldsymbol{m} \in \mathbb{N}^m \mid \boldsymbol{l}^T \boldsymbol{m} \leq k\}$. The markings that are not legal are called *forbidden markings*.

It is assumed that the set of transitions T of a net is partitioned into two disjoint subsets: T_{uc}, the set of uncontrollable transitions, and T_c, the set of controllable transitions. A controllable transition may be disabled by the supervisor — a controlling agent which ensures that the behavior of the system is within a legal behavior. In the paper we introduce the Control Variable function $CV : T_c \to \{0,1\}$; if $CV(t) = 0$, the controllable transition t is disabled. By T_{uc}^* we denote the set of all possible sequences of uncontrollable transitions. A controlling agent, called *supervisor*, must ensure that the forbidden markings will not be reached.

When the plant net is model is a Petri net in (Li and Wonham, 1994) it has been shown that if the critical subnet is acyclic, a maximally permissive control law is that a transition $t \in T_c$ has to be enabled under the net marking $\boldsymbol{m}$ if $\boldsymbol{m}[t > \boldsymbol{m}'$ and

$$\boldsymbol{l}^T \boldsymbol{m}' + \boldsymbol{l}^T \boldsymbol{C}_{uc} \boldsymbol{\sigma}_{uc}^* \leq k \tag{1}$$

where $\boldsymbol{\sigma}_{uc}^*$ is the solution of the following ILP

$$max_{\sigma_{uc}}\, \boldsymbol{l}^T \boldsymbol{C}_{uc} \boldsymbol{\sigma}_{uc}$$
$$s.t. \begin{cases} \boldsymbol{\sigma}_{uc} \geq \mathbf{0} \\ \boldsymbol{C}_{uc} \boldsymbol{\sigma}_{uc} \geq -\boldsymbol{m}' \end{cases} \tag{2}$$

where $\boldsymbol{C}_{uc}$ and $\boldsymbol{\sigma}_{uc}$ are respectively the the incidence matrix and the firing count vector of the uncontrollable subnet (i.e. obtained from plant net by removing uncontrollable transitions).

Consider the net in Fig. 1, a let $m_1 \leq b$ be a GMEC to be enforced on the net. From the control specification we have that $\boldsymbol{l} = \begin{bmatrix} 1\,0\,0\,0\,0\,0\,0\,0\,0\,0 \end{bmatrix}$, $b = [2]$. By applying (2) we have that under a certain net marking

m, $\forall t \in T_c$, $CV(t) = 1$ if $m[t > m'$ and the logical predicate

$$m'_1 + max(\sigma_2) \leq 2. \quad (3)$$

is verified.

A closed form solution is also possible in this case: under a certain net marking m, $\forall t \in T_c$, $CV(t) = 1$ if $m[t > m'$ and the logical predicate

$$m'_1 + min(m'_2 + m'_5, m'_3 + m'_6, m'_4 + m'_7) \leq 2. \quad (4)$$

is verified.

See algorithm 1 for an example of C++ implementation of the controller described in this example. Notice that the user from the controller can access to the current net marking and change the control variable function value.

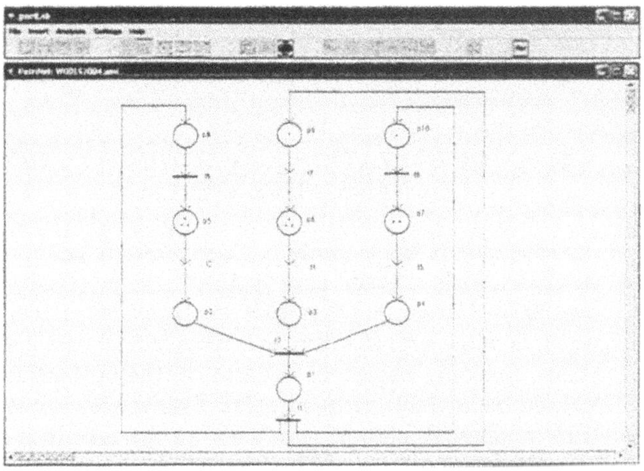

Fig. 1. PNetLab for the testing of the logical case study example (highlighted transitions are those one enabled by the controller).

4. COLORED PETRI NET BACKGROUND

CTPNs extends the framework of PNs by adding color, time and modular attributes to the net. The color attribute is developed to deal with systems that have similar or redundant logical structures. The time attribute allows various time-based performance measures to be added in the system model. We assume that the reader already knows CTPN concepts and terminology as introduced in (Jensen, 1995). Let us recall here the main characteristics of CPNs we deal with in the tool.

A multi-set m is a set which may contain multiple occurrences of elements of a non-empty set S. Formally, a multi-set is a function $m : S \rightarrow \mathbb{N}$, where $\mathbb{N}$ is the set of non-negative integers, which we represent as a formal sum

$$\sum_{s \in S} m(s)s.$$

We denote by: S_{MS} the set of all multi-sets over S; $Type(v)$ the type of a variable v; $Type(expr)$ the type of an expression $expr$; $Var(expr)$ the set of variables in an expression $expr$. In addition we define the binding of a set of variables V as the function b associating to each variable $v \in V$ an element $b(v) \in Type(v)$

Algorithm 1. Logical case study C++ Optimization Algorithm.

```
int nrow, ncol, nvgmec=1, i, j;
nrow=nPlaces(); // PNetLab function giving the net place number
ncol=nTrans(); // PNetLab function giving the net transition number
// Define the incidence matrix C and initialize it
int **C=NULL;
C=dmat_int(nrow, ncol, 0);
// Define the constraint matrix Lgmec and initialize it
int **Lgmec=NULL;
Lgmec=dmat_int(nvgmec, nrow, 0);
// Build the incidence matrix C by using
// the PNetLab function matInc(i,j)
for (i=0; i<nrow; i++)
     for (j=0; j<ncol; j++)
          C[i][j]=matInc(i, j);
// Build the constraint matrix Lgmec
Lgmec[0][0]=1;
// Define vector kgmec
int kgmec[]={2}; // Notice that in this example the GMEC is m(1)<=1
// Definition of the vector utran which specifies the controllable
// and uncontrollable transitions
int *utran=NULL;
utran=dvet_int(ncol, 0);
utran[0]=1; utran[5]=1;
utran[6]=1; utran[7]=1;

void Controllore::controllore(){
     // Load the current PN marking by using the PNetLab function m(i)
     int *marking=NULL;
     marking=dvet_int(nrow, 0);
     for(i=1; i <= nrow; i++)
         marking[i-1]=m(i);
     // Call function mpli which solves the (1) (and then
     // the ILP (2)) or, alternatively, the closed form solution (4)
     mpli(C, nrow, ncol, marking, Lgmec, kgmec, nvgmec, utran, CV);
}
```

and $expr < b >$ as the value obtained by evaluating an expression $expr$ in a binding b, i.e. substituting for each variable $v \in Var(expr)$ the value $b(v) \in Type(v)$ determined by the binding.

Now we recall the following definition (Jensen, 1995).

Definition 1. A CPN is a 9-tuple $CPN = (S, P, T, A, N, C, G, E, I)$ where:
S is a set of non-empty types, called color sets;
P is a finite set of places drawn by circles;
T is a finite set of transitions including immediate transitions drawn by black bars and timed transitions drawn by empty boxes;
A is a finite set of arcs such that $P \cap T = P \cap A = T \cap A = \emptyset$;
N is a node function defined from A to $P \times T \cup T \times P$;
C is a color function defined from P into S;
G is a guard function defined from T into expressions such that $\forall t \in T : [Type(G(t)) = B \wedge Type(Var(G(t))) \subseteq S]$, where B is the boolean type;
E is an arc expression function defined from A into expressions such that
$\forall a \in A : [Type(E(a)) = C(p)_{MS} \wedge Type(Var(E(a))) \subseteq S]$, where p is a place of $N(a)$;
I is an initialization function defined from P into closed expressions such that
$\forall p \in P : [Type(I(p)) = C(p)_{MS}]$. ◇

In order to define the binding of a transition t we introduce the following notation:

$$A(t) = \{a \in A | N(a) \in P \times \{t\} \cup \{t\} \times P\};$$

$$Var(t) = \{v | v \in Var(G(t)) \vee \exists a \in A(t) : v \in Var(E(a))\}.$$

A *binding* of a transition t is a function b defined on $Var(t)$ such that

$$v \in Var(t) : b(v) \in Type(v) \text{ and } G(t) < b >= 1,$$

i.e. the guard function is true. $B(t)$ denotes the set of all bindings for t.

A token is a pair (p, c) where $p \in P$ and $c \in C(p)$, while a binding element is a pair (t, b) where $t \in T$ and $b \in B(t)$. The set of all token elements is denoted by TE while the set of all binding elements is denoted by BE. A *marking* is a multi-set over TE while a *step* is a non-empty and finite multi-set over BE. The sets of all markings and steps are denoted by M and Y respectively. We assume $S = S_1 \times S_2 \times ... \times S_n$ and that a function between each set S_i and the integer number set $\mathbb{Z}$ is defined; this leads us to represent a token in a place p by a n-tuple $< t >= (c_1, c_2, \ldots, c_n)$, where c_j is an integer. We represent by (1) the black token, i.e. the token having no color.

A step Y is enabled in a marking M iff the following property is satisfied

$$\forall p \in P: \sum_{(t,b)\in Y} E(p,t) < b > \leq M(p).$$

When a step is enabled in a marking M_1 it can occur and, if the transition t fires, the net marking changes according to

$$M_2(p) = M_1(p) - \sum_{(t,b)\in Y} E(p,t) < b > + \sum_{(t,b)\in Y} E(t,p) < b > .$$

In this case we say that M_2 is reachable by M_1. This is written as $M_1[Y > M_2$.

The introduction of a timing specification is essential in order to use Petri net models for performance evaluation of distributed systems. We consider nets with deterministic timing and one phase firing rule, i.e. a timed enabling (called the service time of the transition) followed by an atomic firing. Service times of transitions are supposed to be mutually independent and time independent. A time function $f : B(t) \rightarrow \mathbb{R}$ is introduced; it is the time required by the timed transition t associated with a binding b to complete the firing.

In order to avoid the coupling between resolution of conflicts[2] and duration of activities, transitions in conflict are supposed to be immediate except for timed transitions modelling a watchdog timer or immediate transitions modelling a failure detection (Ajmone-Marsan *et al.*, 1995); only in these special cases it is allowed that a timed transition, modelling a watchdog timer, can interrupt another timed transition or that an immediate transition, modelling an error detection, can be in conflict with a timed transition.

In general, in a PN model there is no need of removing tokens from a place in the same order as they were added. We assume each place as being a queuing place with a FIFO policy.

The following expression functions, used in the arc expressions and guard functions, are defined:

[2] We say that two or more transitions are in conflict if they have common input places.

i) $pr(p, c_1, c_2, \ldots, c_k)$: for each token $< t >$ in the place p the function builds a new token formed only by the specified components $c_1, c_2, \ldots, c_k$ of the token $< t >$;
ii) $conc(< t_1 >, < t_2 >, ..., < t_k >)$: the function builds the new token $(< t_1 >, < t_2 >, ..., < t_k >)$ formed by the concatenation of the specified components;
iii) *id*: this function selects all the components of a token.

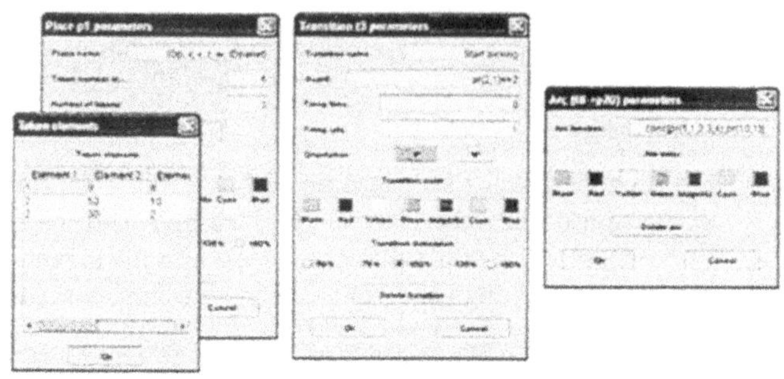

Fig. 3. Windows for the setting of place, transition and arc parameters.

5. AN OPTIMIZATION ALGORITHM FOR THE CRANE/AISLE SYSTEM

In this section we describe an optimization policy for a Crane/Aisle system already presented in (Basile *et al.*, 2003). We assume that each Stockage Unit arrives at the I/O point with the storage location already assigned. Thus, the only degree of freedom, in the optimization procedure, is the clever sequencing of the retrieval orders. The approach followed can be roughly described as follows: consider a list of N_C storage and retrieval orders; optimize the sequencing of the N_C retrievals; when the N_C Dual Commands have been performed consider a new list of N_C orders and repeat the same procedure.

To describe the algorithm we propose, let us denote by R the set composed of the retrieval location addresses, by S the ordered list of storage location addresses and by N_C the cardinality of R and S. Recall that, given two points $P_1 = (x_1, y_1)$ and $P_2 = (x_2, y_2)$, the Chebichev distance $d_C(P_1, P_2)$ between P_1 and P_2 is defined as

$$d_C(P_1, P_2) := max\{|x_1 - x_2|, |y_1 - y_2|\}. \quad (5)$$

We need to use Chebichev metric since the crane moves simultaneously in both directions.

Algorithm 2. Crane/Aisle Optimization.

begin
 repeat
  • Take the *first* element $s \in S$;
  • Select $r \in R$ with the minimum TB distance from s measured by the Chebichev metric;
  • Perform the DC cycle with storage in s and retrieval from r;
  • Update: $R = R - \{r\}$, $S = S - \{s\}$;
 until $(R = \emptyset)$;
end.

In algorithm 3 it is presented an implementation of the algorithm 2. In Fig. 2 it is shown how the user directly from PNetLab can edit the controller and can draw the net.

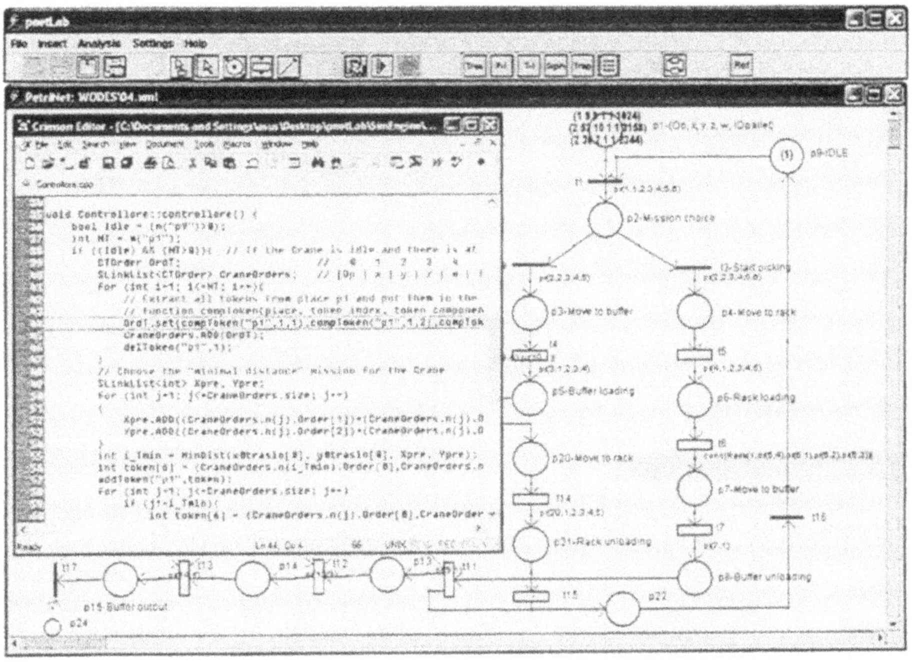

Fig. 2. PNetLab GUI showing the C++ optimization algorithm for the crane/aisle CTPN model.

Algorithm 3. Crane/Aisle C++ Optimization Algorithm.

```
void Controller::controller(){
  bool Idle = (m("p9")>0);
  int NT = m("p1");
  if ((Idle) && (NT>0)){ // If the Crane is idle and there is at least one mission...
     CTOrder OrdT;
     SLinkList<CTOrder> CraneOrders; // [Op | x | y | z | w | IDMission]
     for (int i=1; i<=NT; i++){
        // Extract all tokens from place p1 and put them in the CraneOrders list
        // by using the PNetLab function compToken(place, token index, token component)
        OrdT.set(compToken("p1",1,1),compToken("p1",1,2),compToken("p1",1,3),compToken("p1",1,4),...
           ...compToken("p1",1,5),compToken("p1",1,6));
        CraneOrders.ADD(OrdT);
        delToken("p1",1); // Remove the first token from place p1
     }
     // Choose the "minimal distance" mission for the Crane
     SLinkList<int> Xpre, Ypre;
     for (int j=1; j<=CraneOrders.size; j++)
     {
        Xpre.ADD((CraneOrders.n(j).Order[1])*(CraneOrders.n(j).Order[0]!=1));
        Ypre.ADD((CraneOrders.n(j).Order[2])*(CraneOrders.n(j).Order[0]!=1));
     }
     int i_Tmin = MinDist(xOtraslo[0], yOtraslo[0], Xpre, Ypre);
     int token[6] = {CraneOrders.n(i_Tmin).Order[0],CraneOrders.n(i_Tmin).Order[1],CraneOrders.n(i_Tmin).Order[2],...
        ...CraneOrders.n(i_Tmin).Order[3],CraneOrders.n(i_Tmin).Order[4],CraneOrders.n(i_Tmin).Order[5]};
     addToken("p1",token); // Append the token "token" at the end of the place p1
     for (int j=1; j<=CraneOrders.size; j++)
        if (j!=i_Tmin){
           int token[6] = {CraneOrders.n(j).Order[0],CraneOrders.n(j).Order[1],CraneOrders.n(j).Order[2],...
              ...CraneOrders.n(j).Order[3],CraneOrders.n(j).Order[4],CraneOrders.n(j).Order[5]};
           addToken("p1",token);
        }
  }
}
```

6. THE SIMULATOR ARCHITECTURE

In this section we present the architecture of a simulator prototype used to test optimization algorithms (see Fig. 4). PNetLab offers the possibility of compiling the Petri net model together with a given optimization algorithm, in order to simulate the closed-loop system and analyze its performance.

Starting from a PN (CTPN) drawn by using the graphical interface, the module `Build_Model` produces a formal representation in `.xml` format of the PN (CTPN) model in terms of pre-post-incidence matrixes (`PNmodel`), guard and arc functions (`Guard`, `ArcFunctions`), time functions associated to timed transitions (`TimeFunctions`) and a file (`NetParameters`) of model parameters. Then, two modules (`Arc_Expr` and `Trans_Expr`) are used to translate `ArcFunctions`, `Guard` and `TimeFunctions` in C++ source code files. Such files are compiled and linked together the optimization algorithm (files `Controller.cpp` and `Controller.h`) in order to pro-

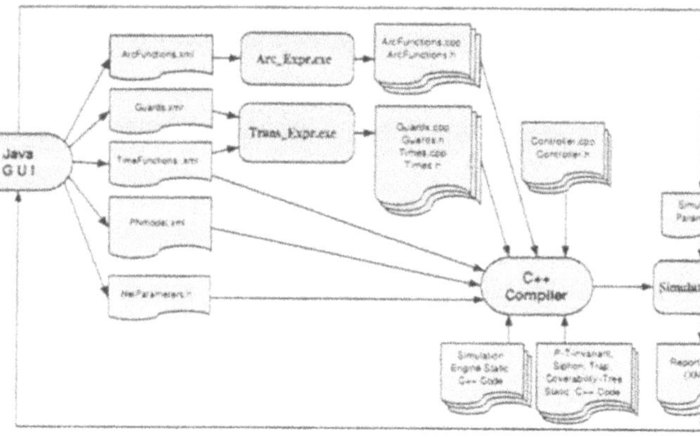

Fig. 4. PNetLab software architecture.

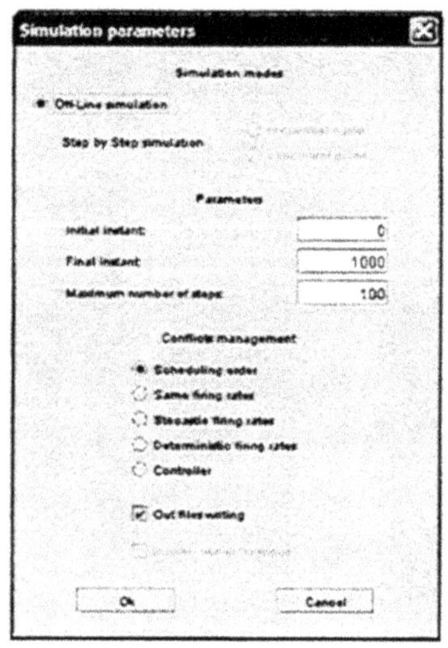

Fig. 5. PNetLab window for the setting of the simulation parameters.

duce a single executable file **Simulator**. By running **Simulator** it is possible to simulate the closed-loop system (Petri net model + Control algorithm) and analyze its performance. At the end of the simulation, the results are available in appropriate report files and can be shown in the graphical interface.

The simulation of the closed loop system and the token game in the timed and untimed case are possible. In addition, for the open loop PN model, the tool performs P-invariant, T-invariant, minimal siphon and trap computation. PNetLab can compute the coverability tree both for the open loop and closed loop system, and it offers a very suitable management of the transition conflicts.

In particular, the user can access to the list of transitions that are in effective conflict (two transitions are in effective conflict when the firing of one transition disables the other one) under the current net marking and decide which one of them has to be enabled. The conflict resolution, as shown in Fig. 5, can be managed by a control policy specified in the controller or by a predefined ones.

7. CONCLUSIONS

PNetLab, a freeware tool developed by the Automatic Control Group of the University of Salerno has been presented. The novel contributions of this tool are: a) the closed-loop and open loop analysis of Petri net based plant model; b) the possibility to specify a controller as a standard C/C++ program; c) a very suitable management of the transition conflicts. The main tool features has been shown via two case studies.

REFERENCES

Ajmone-Marsan, M., G. Balbo, G. Conte, S. Donatelli and G. Franceschinis (1995). *Modelling with Generalized Stochastic Petri Nets.* Wiley.

Basile, F., C. Carbone and P. Chiacchio (2003). An approach to control automated warehouse systems. *IEEE International Conference on Computation Engineering in Systems Applications (CESA'03), Lille, France.*

DiCesare, F., G. Harhalakis, J.M. Proth, M. Silva and F.B. Vernadat (1993). *Practice of Petri Nets in Manufacturing.* Chapman and Hall.

Giua, A., M.P. Fanti and C. Seatzu (2003). Generalized mutual exclusion constraints and monitors for colored petri nets. *2003 IEEE International Conference on Systems, Man and Cybernetics, Washington, DC* pp. 1860 – 1865.

Holloway, L.E., B.H. Krogh and A. Giua (1997). A survey of petri nets methods for controlled discrete event systems. *Discrete Event Dynamic Systems: Theory and Applications* **7**(7), 151–190.

Jensen, K. (1995). *Colored Petri nets. Basic concepts, analysis methods and pratical use. Volume 1.* Monographs on Theoretical Computer Science. Springer Verlag. New York.

Li, Yong and W.M. Wonham (1994). Control of vector discrete-event systems II - controller synthesis. *IEEE Trans. on Automatic Control* **39**(3), 512–531.

Murata, Tadao (1989). Petri nets: Properties, analysis and applications. *Proceedings of IEEE* **77**(4), 541–580.

P.J.Ramadge and W.M. Wonham (1989). The control of discrete event systems. *Proceedings of IEEE* **77**(1), 637–659.

ELSEVIER
IFAC
PUBLICATIONS
www.elsevier.com/locate/ifac

EXPLOITING PARTIAL SYMMETRIES IN WELL-FORMED NETS FOR THE REACHABILITY AND THE LINEAR TIME MODEL CHECKING PROBLEMS

Soheib Baarir **,2 **Serge Haddad** *,1
Jean-Michel Ilié **,2

* *LAMSADE, UMR CNRS 7024, Université Paris Dauphine, Place du Maréchal de Lattre de Tassigny, 75775 PARIS, FRANCE*
** *LIP6, Université PARIS VI, 8 rue du Capitaine Scott, 75015 PARIS, FRANCE*

Abstract : Taking advantage of the symmetries of a system is an efficient way to cope with the combinatory explosion involved by the verification process. Whereas numerous algorithms and tools efficiently deal with the verification of a symmetrical formula on a symmetrical model, the management of partial symmetries is still an open research topic. In this work, we present the design and the evaluation of two methods applicable on coloured Petri nets. These two methods are extensions of the symbolic reachability graph construction for the well-formed Petri nets. The first algorithm, called the extended symbolic reachability graph construction, tackles the reachability problem. The second one, called the symbolic synchronized product, checks a partially symmetric linear time formula on a net. The evaluations show that these two methods outperform the previous approaches dealing with partial symmetries. Furthermore they are complementary ones since the former while being less general gives better results than the latter when applied to the reachability problem. *Copyright © 2004 IFAC*

Keywords: Coloured Petri Net, Well-Formed Petri Net, Partial Symmetry, Symbolic Reachability Graph

1. INTRODUCTION

The coloured Petri nets formalism is an expressive model extending the representation of concurrency by Petri nets with a data management via the coloured domains and functions. However this expressiveness leads in practice to huge state graphs considerably restricting their use for the reachability and the model checking problems.

Therefore, a recurrent research topic is the building of a reduced graph equivalent to the original one w.r.t. some set of properties. Among the proposed approaches, the symmetry based method builds a symbolic reachability graph (SRG) where a node corresponds to a set of states leading to an equivalent behaviour up to some "admissible" colour permutation. In order to be applicable, such a method must detect the admissible permutations by a syntactical examination of the net. This requirement has motivated the introduction of the well-formed nets model which is

[1] haddad@lamsade.dauphine.fr
[2] {Souheib.Baarir,Jean-Michel.Ilie}@lip6.fr

equivalent to the coloured Petri net model but with a restricted syntax allowing the automatic computation of the SRG (Chiola *et al.* 1993). On this reduced graph, one solves the reachability problem and more generally the linear time temporal logic formula truth whenever the atomic propositions of the formula are symmetrical (e.g. $\mathbf{AND}_{c \in C(p)} m(p)(c) = 1$). (Clarke *et al.* 1996) establish the correctness of this checking in a general framework.

However, the previous approach suffers from two limitations. On the one hand, it is well-known that without process identities, many distributed problems do not have solutions. Indeed in distributed algorithms, identities comparisons break deadlock situations. Modelling such algorithms produces nets whose behaviour is symmetrical with the exception of a small set of transitions. Symmetrical methods are not able to efficiently handle partial symmetries since they require a symmetry upon the whole model.

On the other hand, many usual formulas have asymmetric propositions. For instance, the fairness formula "*If some process is waiting for a resource, then it will get it*" implies to distinguish a process (i.e. a colour in the net). By detecting the symmetries of the automaton related to the formula, (Emerson and Prasad Sistla 1996) efficiently handle this kind of formula. Unfortunately this technique fails when analyzing the automaton associated to the asymmetric formula: "*If some process is waiting for a resource then it will get it, provided none of the processes with higher identity will require the resource in the future*". Again viewed as a model, the automaton is partially symmetric. A more refined analysis of the automaton slightly extends the range of application but the main limitations remain (Ajami *et al.* 1998, Emerson and Trefler 1999).

Here, we present the efficient design of two methods for partially symmetric models and/or formulas in well-formed nets. The first method which solves the reachability problem and refines the one presented in (Haddad *et al.* 1995), may be summarized as follows: (1) partition the transitions in two subsets, the symmetrical and the asymmetrical ones; (2)build an Extended Symbolic Reachability Graph (ESRG), where each node corresponds to a set of symbolic states of the SRG which are developed as soon as an asymmetrical transition is fireable; (3) fire symbolically the symmetrical transitions from the representant of the node and asymmetrical ones from the developed symbolic markings ; (4) as soon as the symbolic markings associated to a node are no more relevant (i.e. the asymmetrical transitions have been fired and **all** the symbolic markings are reachable), they are deleted.

The second method solves the linear time model checking problem and applies on a symmetric well-formed net and an asymmetric Büchi automaton. It is an *instanciation* of a generic method presented in (Haddad *et al.* 2000). To any state of this automaton is associated a local partition of each colour class such that two colours inside the same partition are equivalent w.r.t. the atomic propositions labelling the state. Then a Symbolic Synchronized Product (SSP) is built where a state of this product is composed by a symbolic marking and a state of the automaton. Contrary to the SRG building, the decomposition of coloured classes into static subclasses is local to each state. Furthermore when building a state, we analyze its structure in order to minimize the decomposition into static subclasses whereas preserving the equivalence with the ordinary synchronized product.

The balance of the remaining paper is the following one. In the second section, we informally present the well-formed Petri nets and the standard model checking methods. Then we depict the principles of our methods. In the forth section, we evaluate them on two examples. At last, we conclude and we give perspectives to this work.

2. PRELIMINARIES

Below, we describe the main features of WFN. The reader can refer to (Chiola *et al.* 1993) for a formal definition :

- In a WFN, a colour domain is a cartesian product of colour classes which may be viewed as primitive domains. This product is possibly empty (e.g. a place which contains neutral tokens) and may include repetitions (e.g. a transition which synchronises two colours inside a class). A class is divided into static subclasses. The colours of a class have the same nature (processes, resources, etc.) whereas the colours inside a static subclass have the same potential behaviour (batch processes, interactive processes, etc.).
- In a WFN, a colour function is built by standard operations (linear combination, composition, etc.) on basic functions. There are three basic functions: a projection which selects an item of a tuple and is denoted by a typed variable (e.g. X, Y); a diffusion, a constant function which returns the bag composed by all the colours of a class or a subclass and is denoted S_C where C is the corresponding (sub)class; and a successor function which applies on an *orderered* class and returns the colour following a given colour.
- In a WFN, transitions are guarded by expressions (in fact even the coloured functions may

be guarded but we do not describe here this feature). An expression is a boolean combination of atomic predicates. An atomic predicate either identifies two variables $[X = Y]$ or restricts the domain of a variable to a static subclass $[X \in D]$.

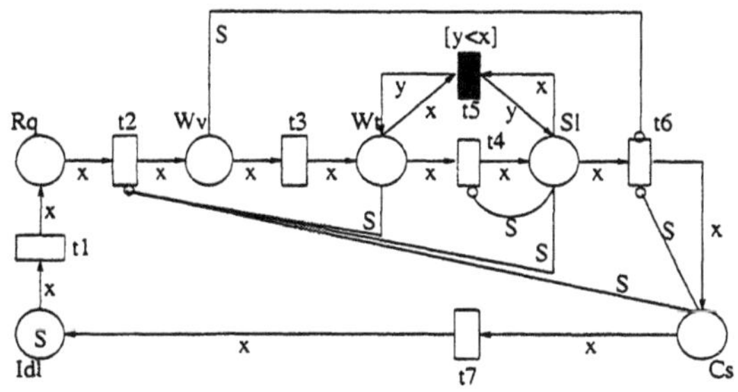

Figure 1. Modelling a distributed algorithm with priorities for a critical section

The WFN of Figure 1 models a distributed algorithm for critical section accesses. There is a single class C which represents the processes. The variables X and Y are typed by C. Since no confusion is possible, the constant S_C has been abbreviated by S. Let us note, for instance, that the colour domain of Rq is C and the colour domain of t_5 is $C \times C$. Roughly speaking, the distributed algorithm builds successive waves of requests (by firings of t_2 until some firing of t_3). Then one decides between the requests of a wave (inside the places Wv, Wt, Sl) according to the order of the processes identities. This tie break is done by the asymmetric transition t_5 which exchanges requests between Wt and Sl. This transition has priority over the other ones, thus t_6 is fireable when no more exchange is possible.

The class C is split into $D_1, \ldots, D_n$ static subclasses (one per colour). This decomposition is required to express the predicate $[Y < X]$ which is an abbreviation of $\text{OR}_{i<j}([Y \in D_i] \text{ AND } [X \in D_j])$.

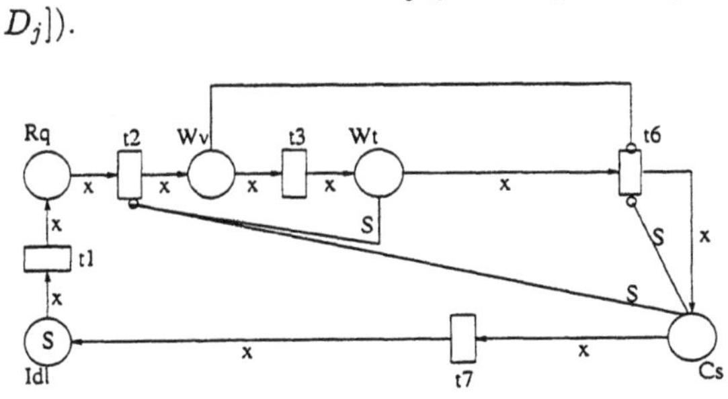

Figure 2. Modelling a distributed algorithm for a critical section

The WFN of Figure 2 represents a similar algorithm except that in a wave, the accesses to the critical section are not controlled. Since the potential behaviours of all processes are equivalent, the class is not splitted (i.e. it is composed by a single static subclass).

The symbolic reachability graph (SRG) building lies on a compact representation for a set of equivalent ordinary markings called a symbolic marking. In a symbolic marking, each static subclass is divided into dynamical subclasses. A dynamical subclass is only specified by its size, thus each consistent choice of colours for the dynamical subclasses leads to an ordinary marking. The implicit colours of a dynamical subclass are supposed to be *in the same state*. Consequently the marking of places is defined w.r.t. to dynamical subclasses instead of colours.

In order to build the SRG, a symbolic firing rule is applied. This rule is similar to the ordinary one except that the dynamical subclasses instanciating the firing must be reduced to singletons by an appropriate splitting. At last, different representations of a symbolic marking are possible. So a canonical one is defined and each time a new symbolic marking is built, it is canonized.

As an illustration, building the SRG of Figure 2 leads to exponential savings of space w.r.t. the ordinary reachability graph (RG). However, the SRG of Figure 1 has exactly the same size of the ordinary graph. Indeed the static decomposition of C forbids any saving by the standard SRG.

We briefly recall the principles of the standard model checking of a linear time temporal logic formula. At first, one builds the Büchi automaton of the negation of the formula. This automaton is then synchronized with the state graph i.e. a state of the synchronized product is composed by a state of the automaton and a state of the reachability graph which fulfills the atomic properties labelling the automaton state. There is an edge in the synchronized product iff the corresponding edges are present in the state graph and in the automaton. If a successful path is found on the synchronized product, the formula is invalidated. Usually, all the constructions are done on-the-fly.

As discussed in the introduction, using the SRG for a partially symmetric formula often requires to define "artificial" static subclasses. For instance, expressing the wave construction of the net of Figure 2 leads to the following LTL formula:

$$G(Rq_1 \text{ AND } Wt_2 \Rightarrow (\text{NOT } Cs_1 \text{ U } Cs_2)) \quad (1)$$

Here the two colours 1 and 2 must be singleton static subclasses in order to apply a synchronized product between the SRG and the automaton.

3. DESCRIPTION OF THE TWO METHODS

The common idea of both methods is to treat dynamically (on-the-fly) asymmetries.

Regarding the reachability problem, the asymmetry is due to the existence of some transitions whose treatment depends on colours. The ESRG method exploits the fact that these dependencies are only local so can be abstracted for other transitions.

Concerning model-checking, asymmetries can arise due the need to observe particular colours during the system runs. However, this particular observation is not relevant for all parts of the verification process. The SSP method takes profit from such observations to construct the more adapted group of symmetries.

The ESRG method: The first step of the method consists to partition the transitions in two subsets: the *asymmetric* transitions which involve a static subclass occurring either in some coloured function around the transition or in the guard and the other ones called the *symmetric* transitions.

Each node of the ESRG contains a symbolic marking which does not take into account the static subclasses (i.e. to each class corresponds a single subclass). Such symbolic marking is called *symmetrical representation (SR)*. When an asymmetrical transition is enabled in a node, it is completed with all the possible symbolic markings which refine the SR by taking into account the static subclasses. Such markings are called the *eventualities* of the SR. The firing of an asymmetrical transition is always processed from the eventualities of a node. The firing of a symmetrical transition is processed from the eventualities of a node if some of them (but not all) are present. A key point of the method is the removal of the eventualities of a node. It occurs when all the possible eventualities are present (i.e. the whole states represented by the SR are reachable) and all the asymmetric transitions fireable from the eventualities have been fired. We call this kind of node a *saturated* node. In order to accelerate this removal, we give priority to the firing of asymmetric transitions from nodes with all the eventualities present and then to the firing of symmetric transitions from saturated nodes since the reached node is immediately saturated.

From an implementation point of view, since the eventualities are also symbolic markings, no specific development is required for the firing rule. However a symbolic firing is more time consuming since both the symbolic marking and optionally the eventuality are computed and canonized.

The expected reduction of the space complexity of the ESRG building w.r.t. the SRG building is mainly related to the proportion of asymmetric transitions in the WFN. When all transitions are symmetric, the ESRG is the SRG. We expect that the reduction is maximal when the proportion of asymmetric transitions is small. Furthermore when this ratio increases, the reduction may remain important whenever the saturation effect is predominant.

The SSP method: The *Symbolic Synchronized Product* method (SSP) aims at synchronizing a partially symmetric Büchi automaton with the runs of a symmetrical system. The reader can refer to (Haddad *et al.* 2000) for the generalization to asymmetrical system specifications. The first step of the method consists to analyze the atomic propositions labelling a node of the Büchi automaton. *Local static subclasses* are associated to each node so that two colours inside a static subclass are equivalent w.r.t. the propositions (i.e. permuting the two colours let invariant the set of atomic propositions). Given a state b, we denote its local static decomposition by $ls(b)$.

A state of the synchronized product is defined by: a local decomposition into static subclasses LS, a symbolic marking defined upon these subclasses SM and a state of the automaton b. The atomic propositions of b are satisfied by all the ordinary markings associated to the symbolic marking.

The key point is the building of the successors of a node. We choose a successor for b, say b'. Then we apply a symbolic firing rule on SM and we obtain a new symbolic marking, say SM'. The next step consists to obtain the subset of ordinary markings of SM' which fulfill the atomic propositions of b'. We do it in a symbolic way by refining the static decomposition LS into $LS \cap ls(b')$. This refinement leads to replace SM' by a set of symbolic markings where we straightforwardly decide for each one if all of its ordinary markings satisfy the atomic propositions of b' (or none). We then check on the subset of remaining symbolic markings whether we can group some of them. The involved techniques are intricate and will not be described here.

Compared to the static method of (Ilié and Ajami 1997) for the model checking using the SRG, the expected reduction of space complexity depends on the degree of symmetry of the states of the Büchi automaton induced by the atomic propositions.

4. EVALUATION

We have implemented our symbolic methods by reusing the core implementation of the GreatSPN software. GreatSPN is a well-known software which computes the SRG of well formed nets, including the management of the symbolic marking representation (Chiola and Gaeta 1995).

Since both SSP and ESRG methods are based on the notion of *dynamic symmetries* we first implemented the module DySy, a *dynamic manager of symmetries*. Then we reused the GreatSPN core without significant changes in its internal structures, and we developed a modular implementation of the SSP and the ESRG methods. The computation of the SSP requires an external module which processes the synchronized product with the Büchi automata of an LTL formula, in order to perform the model checking. The LTL model checker called SPOT (see http://spot.lip6.fr), has been used (fig. 3).

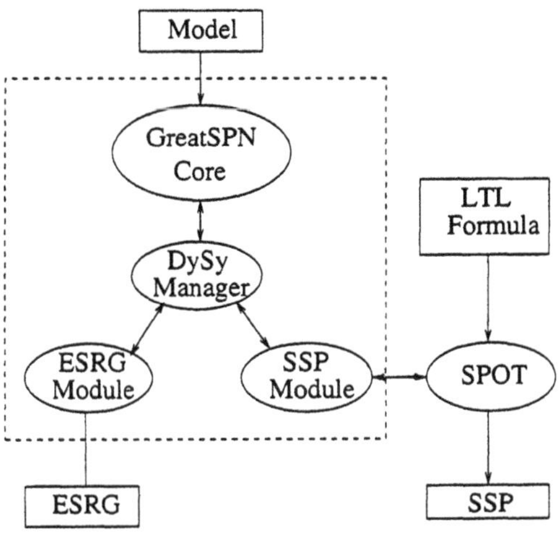

Figure 3. Architecture of the system

In the next subsections, the time spent as well as the memory consumed are measured for our two symbolic approaches, in comparison with previous ones. The memory consumption is measured in number of nodes.

It is worth noting that we used a 2 Ghz Intel Pentium IV machine, with 775 Mbytes memory size and working on Linux 9.1 Operating System.

4.1 ESRG Evaluation

Table 1 shows how ESRG and SRG behave on the asymmetrical model of Figure 1. On this example, the SRG is the RG because of the partition of the colors of the color class C in elementary static subclasses. In Table 1, columns denoted *CPU Time* shows the *CPU construction time* of each structure. The Column *SM* represent the number of symbolic markings in the SRG. The number of saturated and unsaturated ESMs are shown in the columns *(#ESM, #SAT)* and *(#ESM, #NOT SAT)*, respectively. The total number of stored eventualities are given away in the column *E*. The last column (denoted *Ratio*) highlights the obtained space gain.

Time evaluation :
Regarding the time spent for building each structure, we find similar values. This can be easily explained by the fact that the number of computed eventualities is equal to the number of *SM* in the SRG. Actually, the enabling of the asymmetrical transitions in the ESRG approach is tested on the eventualities.

Processes	SRG(RG)		ESRG				Ratio
	Cpu Time	# Symb. Mark. : SM	Cpu Time	# ESM # Sat	# ESM # Not Sat	# Event. : E	SM / (# ESM + E)
3	0	139	0	33	6	15	2
5	2	2709	1	96	20	325	6
7	41	50159	25	221	42	4665	10
9	1147	911017	939	394	72	56745	15
11	-----	16378179	54074	661	110	638285	25

Table 1. Evaluation of the ESRG method

Space evaluation :
There is a *super-linear* (Figure 4) gain in memory occupation because numerous ESM are saturated, and then are saved without their eventualities. We also observe that the number of eventualities, E, progresses much more slowly than the SM one. This is due to the fact that the progression of the number of eventualities depends on the number of unsaturated ESM, which are in a very small proportion compared to the total number of ESM ($\#SAT$ + $\#NOT\ SAT$). On a model with a less important proportion of asymmetric transitions, we would obtain an even greater reduction. We also notice that the last SM value (for 11 processes) in the Table 1 is estimated from the ESRG construction, because GreatSPN reaches its limit when computing the SRG of the asymmetric model with 10 processes.

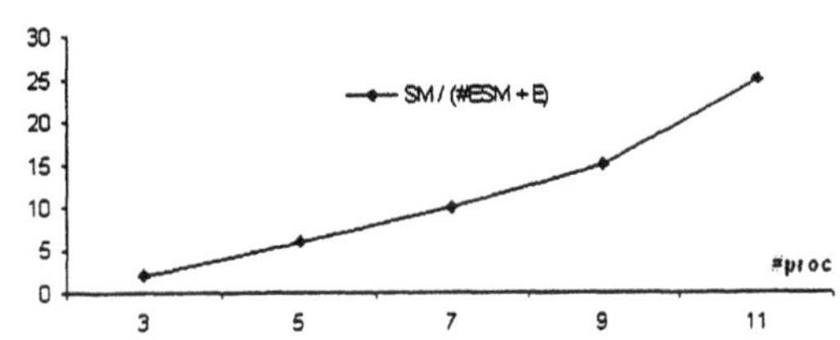

Figure 4. Ratio of SM w.r.t (#ESM + E)

4.2 Symbolic Synchronized Product Evaluation

In order to test our LTL verification method, we choose to check the formula 1 on the symmetric WFN of Figure 2 wherein the whole synchronized product structure is constructed, because of the truth of the formula 1 of section 2. Table 2 compares the SSP approach against both the standard synchronized product (SP-rg) and the symbolic one, obtained from the SP-srg presented in (Ilié and Ajami 1997). The columns designed by *CPU Time*, show the CPU construction time for each structure. The columns denoted *#States*, represent the number of states for each structure. The last two columns depicted *s-rg/s-ssp* and *s-srg/s-ssp* represent the obtained space gain.

Processes	SP-rg		SP-srg		SSP		Ratio	
	Cpu Time	#States : s-rg	Cpu Time	#States : s-srg	Cpu Time	#States : s-ssp	s-rg / s-ssp	s-srg / s-ssp
3	1	129	1	129	1	42	3	3
5	2	2655	2	739	4	155	17	4
7	53	51711	4	2197	12	392	131	5
9	920	974847	8	4871	31	801	1217	6
11	------	-------	14	9129	60	1430	------	7

Table 2. Evaluation of the SSP method.

Time evaluation :
Regarding the CPU time, the SSP building goes slightly slower than the SP-srg. Actually, the SSP requires more operations than the SP-srg, even if it contains less nodes. In contrast, the building of the SSP goes faster than the SP-rg because it is two order magnitude smaller (column *s-rg/s-ssp*).

Space evaluation :
A linear order reduction (Figure 5) is witnessed when comparing the space used by the SSP and the SP-srg (column *s-srg/s-ssp*). Generally, the reduction obtained by larger sets of ordinary markings than those allowed in the SRG is more important that the increase due to the fact that these sets may have a non empty intersection.

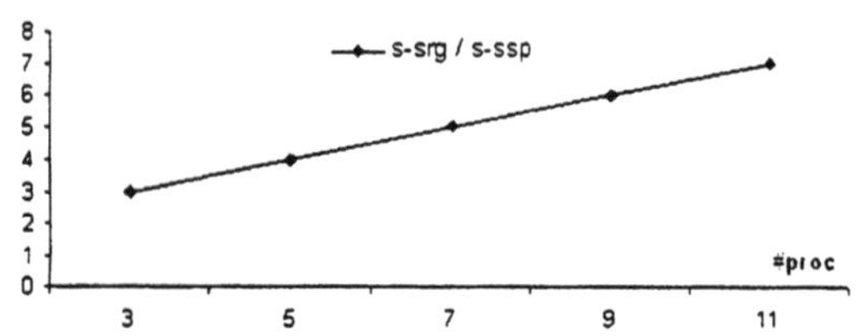

Figure 5. Ratio of s-srg w.r.t s-ssp

5. CONCLUSION AND PERSPECTIVES

In this work, we have presented the design and the evaluation of two verification methods exploiting partial symmetries and applicable on well-formed Petri nets. These two methods are extensions of the symbolic reachability graph construction for the well-formed Petri nets. The extended symbolic reachability graph construction tackles the reachability problem. The symbolic synchronised product checks a partially symmetric linear time formula on a net. On standard case studies, we have shown that these two methods outperform the previous approaches dealing with partial symmetries.

We plan to extend this work in two directions. On the one hand, we notice that the asymmetry is often introduced by the presence of guards on transitions like $X < Y$. Expressing such guards in the well-formed nets formalism implies to split the colour class of these variables into singleton static subclasses. So we want to handle in a dynamic way these guards (i.e. during the symbolic construction). On the other hand, as a method for the reachability problem, the symbolic synchronized product suffers from the following drawback: given two nodes of the product, it does not detect that the set of states associated to one of them is included in the set of states associated to other one. We are currently developping this additional test in the tool. However, taking advantage of this test for the model-checking problem requires a theoretical development on which we are working.

REFERENCES

Ajami, K., S. Haddad and J.-M. Ilié (1998). Exploiting Symmetry in Linear Temporal Model Checking: one Step Beyond. In: *Proc. of Tools and Algorithms for the Construction and Analysis of Systems TACAS'98, part of Theory and practice of Software ETAPS'98.* Vol. 1384 of *LNCS.* Springer Verlag. Lisbon - Portugal. pp. 52–67.

Chiola, G. and R. Gaeta (1995). Efficient Simulation of Parallel Architectures Exploiting Symmetric Well-formed Petri Net Models. In: *Sixth International Workshop on Petri nets and Performance Models.* IEEE Computer Society Press. Durham, NC, USA.

Chiola, G., C. Dutheillet, G. Franceschinis and S. Haddad (1993). Stochastic well-formed coloured nets for symmetric modelling applications. *IEEE Transactions on Computers* **42**(11), 1343–1360.

Clarke, E.M., R. Enders, T. Filkorn and S. Jha (1996). Exploiting Symmetry in Temporal Logic Model Chacking. *Formal Methods and System Design* **9**, 77–104.

Emerson, E. A. and R. J. Trefler (1999). From Asymmetry to Full Symmetry: New Techniques For Symmetry Reduction in Model Checking. In: *Proc of CHARMEÕ99.* Lecture Notes in Computer Science. Springer Verlag. Bad Herrenalb - Germany. pp. 142–156.

Emerson, E.A. and A. Prasad Sistla (1996). Symmetry and Model Checking. *Formal Methods and System Design* **9**, 307–309.

Haddad, S., J.M. Ilié and K. Ajami (2000). A model checking method for partially symmetric systems. In: *Proceedings of FORTE/PSTV'00.* Kluwer Academic Publishers. Pisa, Italy. pp. 121–136.

Haddad, S., J.M. Ilié, M. Taghelit and B. Zouari (1995). Symbolic Reachability Graph and Partial Symmetries. In: *Proc. of the 16th Intern. Conference on Application and Theory of Petri Nets.* Vol. 935 of *LNCS.* Springer Verlag. Turin, Italy. pp. 238–257.

Ilié, J.-M. and K. Ajami (1997). Model Checking through Symbolic Reachability Graph. In: *Proc. of Theory and Practice of Software Development TAPSOFT'97 - part of 7th CAAP.* Vol. 1214 of *LNCS.* Springer Verlag. Lille - France. pp. 213–224.

ELSEVIER
IFAC
PUBLICATIONS
www.elsevier.com/locate/ifac

AN APPLICATION EXAMPLE OF SYMBOLIC CALCULUS FOR SWN STRUCTURAL RELATIONS

L. Capra * M. De Pierro ** G. Franceschinis **

* *D.I.CO. Univerisità di Milano, Italia*
** *D.I., Università del Piemonte Orientale, Italia*

Abstract: Structural analysis techniques allow system properties to be efficiently verified, and may significantly improve the effectiveness of state-space based analysis. Symbolic approaches have been proposed to extend in an effective way structural analysis results from ordinary Petri nets to high level Petri nets but they apply to restricted sub-classes of high level Petri nets (e.g. Unary Regular nets). In the paper some salient points of a symbolic calculus of Stochastic Well-formed Nets structural relations are presented using an example. *Copyright ©2004 IFAC*

Keywords: Algebraic approaches, Calculus, Discret-event systems, Stochastic Petri-nets, Structural properties

1. INTRODUCTION

The introduction of High Level Petri Nets and in particular of the colored Petri Net (CPN) formalism stimulated the PN research community to develop new analysis techniques able to exploit the information conveyed by the more parametric representation of models.

The straightforward approach to apply standard PN analysis techniques to the unfolded net, on one hand allows to take advantage from the well known consolidated techniques developed for ordinary Petri Nets, but on the other hand has two drawbacks: the size of the unfolded net is often quite larger than the colored one, and the analysis techniques cannot take advantage of any kind of symmetry naturally encoded in the high level PN representation. Moreover, the unfolding breaks the parametric nature of the high level representation.

This paper concentrates on a structural analysis technique for a class of (timed) CPNs, the so-called Stochastic Well-formed Nets (SWN) whose definition can be found in (Chiola *et al.*, 1993*a*). Several scientific contributions, dealing with the manipulation of the colored PN higher level constructs in structural analysis techniques have been presented in the literature; among the others (Dutheillet and Haddad, 1993; Brgan and Poitrenaud, 1995; Barkaoui *et al.*, 1993) deal with problems close to those considered in this paper.

(Brgan and Poitrenaud, 1995) and (Barkaoui *et al.*, 1993) provide respectively stubborn set analysis for SWN and structural deadlock computation for CPN, both exploiting the symmetries encoded in the colored net, thus improving the efficiency of the methods with respect to the net unfolding alternative; the former however introduces constraints on both the definition of the structural relations and the SWN syntax. In (Dutheillet and Haddad, 1993) the basis for a symbolic approach to express structural relations in colored nets is proposed. The main contribution provided in this work is the possibility to represent structural relationships between transitions, such as conflict or causal connection, by means of algebraic manipulation of the color annotations that are part of the net description. They show how the theory may be practically applied to the class of Unary PN to compute extended conflict sets (ECS, (Chiola *et al.*, 1993*b*)), which represent sets of immediate transitions (transitions that fire in zero time, in contrast to timed transitions, which fire only after some enabling time) that are not independent (i.e. such that the firing of one transition in the set may affect the enabling status or the firing probability of the other transitions in the same set). This step is essential in the correct net-level definition of all the parameters needed to completely determine the stochastic process representing the behavior in time of stochastic Petri nets with immediate transitions (see (Teruel *et al.*, 2003)). In case high level stochastic Petri nets are considered (such as SWN), the dependency relation should be stated in a symbolic form, to characterize in a compact and parametric way the *color instances* of a given (immediate) transition that can influence the enabling/firing probability of other transition color instances.

This paper extends the structural analysis technique presented in (Dutheillet and Haddad, 1993) to SWN. The extension is based on a rather complex formal treatment of the color annotations appearing on the high level model: in this paper the formal aspects of the calculus, which are completely worked out in (De Pierro, 2004), are not detailed, but rather the problems and possible solutions are presented on a running example. The impact, with respect to the modeling task, of the symbolic representation of dependency relations, with the final goal of extending the guided procedure presented in (Teruel *et al.*, 2003) to the SWN formalism, and implementing it within the GreatSPN tool (`www.di.unito.it/~greatspn`), is also discussed.

2. SWN FORMALISM AND EXAMPLE

In this section the relevant features of the SWN formalism, together with the running example that will be used to illustrate the symbolic calculus, are introduced. For the sake of space the detailed SWN definition is not given, the reader may refer to (Chiola *et al.*, 1993*a*) for the details.

The SWN formalism. A SWN model is a tuple $(P, T, C, \mathcal{C}, W^+, W^-, H, \Phi, \Pi, \Omega, \mathbf{M}_0)$ where P is the set of places, T is the set of transitions, $C = \{C_1, \ldots, C_n\}$ is the set of basic color classes, each basic color class C_i is a finite non empty set of colors. $\mathcal{C}$ is a function associating a *color domain* to each place and transition; a color domain is defined as the Cartesian product of basic color classes. W^-, W^+ and H are respectively the input, output and inhibitor arc functions. An arc function $f(t,p)$ is defined as $f : \mathcal{C}(t) \rightarrow Bag(\mathcal{C}(p))$ where $Bag(A)$ denotes the set of all multisets that can be built on set A; arc function syntax takes the form of a sum of weighted and guarded tuples of *elementary functions* defined on the basic color classes. The allowed elementary functions are the *projection*, selecting one element of a transition instance color tuple, and the constant *diffusion*, always returning the set of all elements in a given basic color class. Φ, Π and Ω are T-indexed functions. Φ assigns a guard to each transition: guards are used to restrict the set of admissible color transition instances to those satisfying a given predicate. The predicate is a Boolean expression whose basic terms are *standard predicates*, representing simple conditions on the transition instance tuple elements. Π associates a priority level (in $\mathbb{N}$) to each transition instance: conflicts between transition instances with different priority level are deterministically resolved in favor of that with higher priority. Finally Ω assigns a *firing rate* to each timed transition instance and a *weight* to each immediate transition instance, used to probabilistically determine the conflict resolution policy between instances with equal priority. $\mathbf{M}_0$ is the initial marking and is such that $\mathbf{M}_0(p) \in Bag(\mathcal{C}(p))$; it is used to specify the multiset of colored tokens initially present in each place.

The example. The example, depicted in figure 1, is a portion of the SWN model presented in (Capra *et al.*, 1999), representing a fault-tolerant storing device called stable memory (SM). The SM is the basic component of an embedded FT architecture adopted on several plants of the Italian national electric power distribution network. The SM was designed to cope with transient memory faults caused by electromagnetic perturbations, and may be represented in abstract way as a reliable storing device, providing a read and a write operation. The SM implementation is based on temporal redundancy: only stable values are returned by the read operation. A value is stable only if it has been successfully written in the SM for three times in a row. The read/write operations have a parameter, the *group*, used to identify distinct storage areas. Each group is internally partitioned into a read-only address space (place RO) and a read/write address space (place WR). A three-position shift register implements the WR space of a group. A write operation (transition $Write$) causes the register shifting and the insertion of the new value in the first position; each write operation is followed by an equality voting (transition $Vote$) on the shift register contents. Unanimous voting results in copying the contents of the stable value, stored in the WR area, into the RO area. Place FS models the causal dependency between a write operation on a SM group and a voting on the same group. A read operation on the SM (transition *Read*) corresponds to reading the contents of the RO area of a group.

The running example of figure 1 has two basic color classes $GROUP$ denoting the mem-

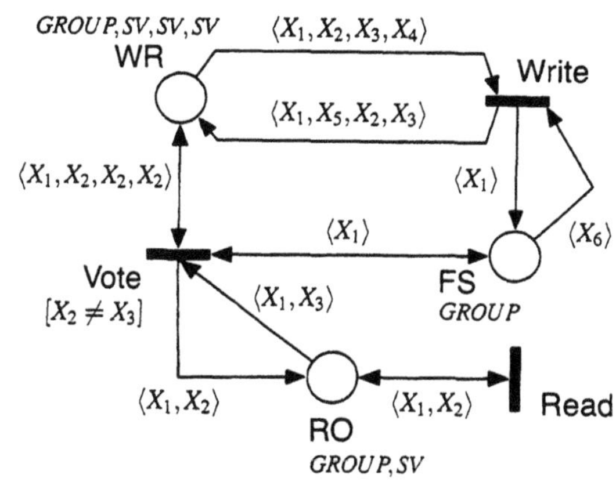

Fig. 1. Stable Memory SWN model.

ory groups, and SV denoting the possible values that can be written into the SM. Next the definition and interpretation of the place and transition color domains are discussed. A token $\langle g_1, v_1, v_1, v_2\rangle$ in place WR represents the fact that the shift register for group g_1 contains the values v_1, v_1 and v_2. A token $\langle g_1, v_1\rangle$ in place RO represents the last stabilized value v_1 copied in RO for group g_1. Place FS represents the last group on which a write operation has been performed. Transition *Read* has color domain $\mathcal{C}(Read) = GROUP \times SV$, meaning that a given value in SV of a given group is being read from RO. Transition $Vote$ has color domain $\mathcal{C}(Vote) = GROUP \times SV \times SV$, referring to the voting operation on a given group. The first occurrence of SV in $\mathcal{C}(Vote)$ (which can be selected using projection function X_2) represents the new stabilized value, while the second occurrence (X_3) represents the old value (stored in RO before the unanimous voting completion). Transition $Vote$ has an associated guard $[X_2 \neq X_3]$ restricting the set of allowed instances of $Vote$ to those with different second and third component. The meaning of this guard is that only a *new* stabilized value is copied from the write read into the read only address space of the required group (so that multiple copies in RO of the same value are prevented by the predicate). Finally, transition $Write$ has a rather complex color domain, namely $\mathcal{C}(Write) = GROUP \times SV \times SV \times SV \times SV \times GROUP$, where the first occurrence of $GROUP$ (X_1) corresponds to the target group of the write operation, the first three occurrences of SV (X_2, X_3, X_4) represent the shift register values before the write completion, the fourth occurrence of SV (X_5) represent the new value pushed into the shift register (first position) by the write operation; finally the second occurrence of $GROUP$ (X_6) represent the last written group (stored in FS) before the completion of the current write operation. All the arc functions appearing in the model are simple tuples of projection functions, that is each arc removes/adds into the connected place a single token whose elements are a subset of the transition instance elements.

A SWN characteristic is that due to the particular syntax adopted for color domains, arc functions and guards, behavioral symmetries can be automatically discovered and exploited to build an aggregate state space and corresponding stochastic process (continuous time Markov chain). In order for the underlying stochastic process to be completely defined however, the immediate transitions priority and weights must be properly specified by the modeler (see (Teruel *et al.*, 2003) for a discussion of this topic concerning the non colored formalism of GSPN). This requires that the modeler has got a precise knowledge of the possibly dependent immediate transition instances. In the example the transition instances $(Vote, \langle g1, v2, v1\rangle)$ is in structural conflict relation with $(Read, \langle g1, v1\rangle)$. This could be easily checked on the net unfolding where the uncolored transitions corresponding to the two transition instances share an input place, corresponding to the instance $(RO, \langle g1, v1\rangle)$ of colored place RO. Intuitively this structural conflict can arise in a marking where one token of color $\langle g1, v1\rangle$ is in place RO; in this marking both instances are enabled, but firing the latter disables the former.

Another relevant structural relation is the SCC highlighting the transition instances whose firing may enable other transition instances. It is easy to see that the instance $(Write, \langle g1, v2, v1, v1, v1, g1\rangle)$ is in SCC relation with $(Vote, \langle g1, v1, v3\rangle)$: in the unfolding an output place of the former is input place of the latter. It is obvious that if the basic color classes include a large number of colors, the number of instances for which the above relations hold may grow very large: representing them in a *symbolic form* allows to represent in a compact way all the relevant dependency patterns. The next sections discuss how such symbolic relations can be computed on the model structure.

3. SYMBOLIC STRUCTURAL RELATIONS

Symbolic structural relations have been defined in (Dutheillet and Haddad, 1993). A symbolic structural relation between nodes of a SWN is a functional description of a binary relation among the instances of the unfolded nodes. The acronyms used to denote symbolic relations are those utilized, for non-symbolic relations, in ordinary nets however a functional notation is added to distinguish between them. So, for example, if SC denotes the *structural conflict* in the unfolded net, then to denote the same relation but on the SWN between the nodes t, t', it will be written $SC(t, t')$. The meaning of such an expression is the following: $SC(t, t')$ is defined on $\mathcal{C}(t')$ and with values in

$\mathcal{P}(\mathcal{C}(t))$, the powerset on $\mathcal{C}(t)$, and assigns to each color instance c' of t' the set C composed by the color instances of t such that for each $(t,c) \in C$ it holds $(t,c)\ SC\ (t',c')$. A symbolic relation thus represents in a compact way a binary relation on the nodes of the unfolded net underlying a SWN. In (Dutheillet and Haddad, 1993) it has been shown that, when the high level PN satisfies to a given structured syntax, the symbolic relations such as $SC(t,t')$ may be *symbolically* derived. The formula of SC symbolic relation due to place p is:

$$SC_p(t,t') = \overline{W^-(t,p) - W^+(t,p)}^{\,t} \circ \overline{W^-(t',p)}$$

The first term $W^-(t,p) - W^+(t,p)$ is the functional description of the multiset of tokens actually removed from place p by the firing of an instance of t. The *support* of such expression $\overline{W^-(t,p) - W^+(t,p)}$ maps its argument into the set containing only one representative for each color in the original expression. The transpose $\overline{W^-(t,p) - W^+(t,p)}^{\,t}$ gives the reverse information i.e. the instances of t removing a given colored token from place p. The formal definitions are:
Definition 1. (Multiset Support). Let $a \in Bag(C)$ then the support of a denoted $\overline{a}$ is a function from $Bag(C)$ to $\mathcal{P}(C)$ such that $\overline{a} = \{c \in C : a(c) > 0\}$;
Definition 2. (Multiset Transpose). Let f be a function from $Bag(C)$ to $Bag(D)$ then the transpose of f denoted f^t is a function from $Bag(D)$ to $Bag(C)$ so defined:
$f^t(x)(y) = f(y)(x)\ \ \forall x \in D, \forall y \in C.$
Finally, composing $\overline{W^-(t,p) - W^+(t,p)}^{\,t}$ with the set $\overline{W^-(t',p)}$, the color instances of t removing from p some token needed by a given instance of t' are obtained.

Using similar arguments the functional description of the $SCC(t,t')$ symbolic relation when an output place p of t is an input to t' is:

$$\mathrm{SCC}_p(t,t') = \overline{W^+(t,p) - W^-(t,p)}^{\,t} \circ \overline{W^-(t',p)}$$

The above formulas summarize the issues of the calculus: it should manage symbols such as the arc functions with respect to the operations shown, i.e. the minus '-', the composition '$\circ$', the transposition 't' and finally the support '—'.

4. THE SYMBOLIC CALCULUS OF STRUCTURAL RELATIONS IN SWN.

In this section the symbolic computation of structural relations in SWN models is illustrated on the SM case example.

The calculus is based on the following normal form coming from an extension of SWN arc functions:

$$f_{D\to\mathcal{P}(E)} = [filter]\overline{\langle f_1, f_2, \ldots, f_n\rangle}[guard] \qquad (1)$$

where, D and E are SWN color domains and the f_i are expressed as intersection of elementary SWN functions which in turn are the basic symbols managed by the calculus. Filters are objects similar to SWN guards. Filters as well as guards are functions, so that (1) has a well-defined semantics in terms of function composition. Form (1) has been shown to be closed with respect to the SC and SCC calculus.

The first application here presented is the calculus of $SC(Vote, Read) : GROUP \times SV \rightarrow \mathcal{P}(GROUP \times SV \times SV)$, whose formula is

$$\mathcal{R}_1 = \overline{((\langle X_1, X_3\rangle - \langle X_1, X_2\rangle)[X_2 \neq X_3])^t} \circ \overline{\langle X_1, X_2\rangle}.$$

Eliminating the minus symbol. Many formulas of structural relations present the difference between tuples. If F and G are tuples, generally it holds $\overline{F - G} \neq \overline{F} - \overline{G}$, while it holds $\overline{F + G} = \overline{F} + \overline{G}$. Note that the left-side operators are on multisets, while the right-side operators are on sets. The proposed approach transforms a multi-set tuple difference into a formal sum. It is based on explicitly storing the difference result when it is negative and iteratively subtracting it from the addends of F. The $\mathcal{R}_1$ example is a quite simple case since the guarded terms represent disjoint multisets, later a more complex case is shown; it is straightforward to show that:

$$\overline{(\langle X_1, X_3\rangle - \langle X_1, X_2\rangle)[X_2 \neq X_3]} = \overline{\langle X_1, X_3\rangle}[X_2 \neq X_3].$$

Transposing a set function tuple. The transpose operation satisfies the distributive property with respect to the sum and the intersection, so without loss of generality transpose of complex expressions may be solved by considering transpose of elementary functions only. The algorithm relies on a table, here no reported, which stores the transpose of any SWN elementary function. The algorithm may be summarized as follows: (a) the filter and the guard of (1) are exchanged; (b) the transpose tuple is directly derived, using the table of elementary transposes, from both the original tuple and its color domain and codomain; repetitions of projections symbols translate into intersections of elementary transposes in the result tuple.

The transpose of $\overline{\langle X_1, X_3\rangle}[X_2 \neq X_3]^t : \mathcal{C}(RO) \rightarrow \mathcal{P}(\mathcal{C}(Vote))$, is $[X_2 \neq X_3]\overline{\langle X_1, S, X_2\rangle}$. Following is a comment on how the transpose tuple is built focusing on its 3rd component. Such component is tied to the transpose of function X_3 in $\overline{\langle X_1, X_3\rangle}$. Since X_3 occurs in the 2nd position of the original tuple, variable X_2 is used in the transpose tuple at the 3rd position. Note that the diffusion function S is used at 2nd position of the transpose tuple, since there is no occurrence of X_2 in the original tuple $\overline{\langle X_1, X_3\rangle}$. The result may be rewritten by applying some simple transformation rules in $\overline{\langle X_1, S - X_2, X_2\rangle}$.

Composing set function tuples. A critical point of the symbolic calculus is the transformation of the generic composition of tuples into form (1) introduced at the beginning of the section. Only the principles and the main issues to be dealt with are discussed here for the sake of space.

The composition algorithm solves $F \circ G$ by means of a table providing a collection of composition rules among basic functions, basic functions are somewhat more general than elementary functions. Generally, an elementary function X_i appearing in F at a given position, have to be composed with the basic function corresponding to the ith position of tuple G. The left distributive property of the intersection of elementary functions with respect to the composition may be exploited when necessary. In doing so, however, some syntactical transformation may be necessary on F and G, in order to cope with the problems of the operation: (a) possible repetitions of X_i in tuple F: the problem is addressed by stating appropriate composition rules, that allow only when necessary the effect of such repetitions to be expressed at the level of output filter of the resulting tuple; (b) possible filters in the expression $F \circ G$: the problem is addressed by giving equivalence transformation rules for function F, that allow the filters to be eliminated (this is achieved exploiting the properties of the intersection); (c) in general, letting f, g, h be elementary functions, $f \circ (g \cap h) \neq (f \circ g) \cap (f \circ h)$: the problem is addressed giving equivalence transformation rules for function G, that allow inner intersections to be expressed by means of guard.

Returning to $SC(Vote, Read)$ the simplification of $\mathcal{R}_1 = \overline{\langle X_1, S - X_2, X_2 \rangle} \circ \overline{\langle X_1, X_2 \rangle}$, applying the rules above, gives:

$\overline{\langle X \circ X_1, (S-X) \circ X_2, X \circ X_2 \rangle} = \overline{\langle X_1, S-X_2, X_2 \rangle}$ where X and X' denotes the unary identities on SV and $GROUP$ respectively. A symbolic reading of the $SC(Vote, Read)$ expression is that for any color instance (g_i, v_k) of $Read$, the set of conflicting color instances of $Vote$ are those having the form (g_i, v_j, v_k), where $v_j \neq v_k$. While each instance (g_i, v_j, v_h) or $(g_j v_h, v_l)$ with $g_j \neq g_i$ of $Vote$ such that $v_h \neq v_k$ are independent.

$SC(Write, Vote) : \mathcal{C}(Vote) \rightarrow \mathcal{P}(\mathcal{C}(Write))$ is the second example of symbolic formula discussed:

$$\mathcal{R}_2 = (\overline{\langle X_1, X_2, X_3, X_4 \rangle - \langle X_1, X_5, X_2, X_3 \rangle})^t \circ \overline{\langle X_1, X_2, X_2, X_2 \rangle}[X_2 \neq X_3] + (\overline{\langle X_6 \rangle - \langle X_1 \rangle})^t \circ \overline{\langle X_1 \rangle}[X_2 \neq X_3]$$

The first addend (i) of $\mathcal{R}_2$ expresses the conflict on place WR whereas the second one (ii) that on FS. In $\mathcal{D} = \overline{\langle X_1, X_2, X_3, X_4 \rangle - \langle X_1, X_5, X_2, X_3 \rangle}$ in (i), the subtrahend annuls the minuend given the predicate pr:$[X_2 = X_5 \wedge X_3 = X_2 \wedge X_4 = X_3]$. By rewriting $\mathcal{D}$ into the equivalent form $\mathcal{D}[pr]+\mathcal{D}[\neg pr]$ it holds $\mathcal{D} = \overline{\langle X_1, X_2, X_3, X_4 \rangle}[X_2 \neq X_5 \vee X_3 \neq X_2 \vee X_4 \neq X_3]$. Similarly $\overline{\langle X_6 \rangle - \langle X_1 \rangle}$ in (ii) translates into $\overline{\langle X_6 \rangle}[X_6 \neq X_1]$.

Applying the transposing rules on both (i) and (ii) sub-terms they respectively becomes:

$$\overline{\langle X_1, X_2, X_3, X_4 \rangle}\,[X_2 \neq X_5 \vee X_3 \neq X_2 \vee X_4 \neq X_3]^t = [X_2 \neq X_5 \vee X_3 \neq X_2 \vee X_4 \neq X_3]\overline{\langle X_1, X_2, X_3, X_4, S, S \rangle}$$

$$\overline{\langle X_6 \rangle}[X_6 \neq X_1]^t = [X_6 \neq X_1]\overline{\langle S, S, S, S, S, X_1 \rangle}$$

that may be rewritten as: $\overline{\langle S - X_1, S, S, S, S, X_1 \rangle}$.

Finally, by the composition rules, (i) becomes:

$$[X_2 \neq X_5 \vee X_3 \neq X_2 \vee X_4 \neq X_3]\overline{\langle X_1, X_2, X_3, X_4, S, S \rangle} \circ \overline{\langle X_1, X_2, X_2, X_2 \rangle}\,[X_2 \neq X_3] = [X_2 \neq X_5 \vee X_3 \neq X_2 \vee X_4 \neq X_3]\,\overline{\langle X_1, X_2, X_2, X_2, S, S \rangle}[X_2 \neq X_3] = \overline{\langle X_1, X_2, X_2, X_2, S - X_2, S \rangle}[X_2 \neq X_3]$$

because the 2nd,3rd and 4th elements are surely equal, so that only the 5th can be different from the 2nd and this is what $S - X_2$ expresses. Respectively, (ii) becomes:

$$\overline{\langle S - X_1, S, S, S, S, X_1 \rangle} \circ \overline{\langle X_1 \rangle}[X_2 \neq X_3] = \overline{\langle S - X_1, S, S, S, S, X_1 \rangle}[X_2 \neq X_3].$$

It is interesting to give an interpretation to the symbolic relations (i) and (ii) just derived: the former expresses the fact that the instances of $Write$ that may disable $Vote$ are those replacing the value in the first position of the WR shift register with a different value, when WR already contained a stabilized value for that group (the X_2, X_2, X_2 in 2nd,3rd and 4th position in the tuple expresses this), given that $Vote$ had not had a chance yet to copy the stabilized value in RO (this is expressed by the $[X_2 \neq X_3]$ predicate) The (ii) formulae instead expresses the fact that a write on a different group than the last one involved in a previous write operation, may be in conflict with the voting on the previously written group (whose identity is kept in place FS), thus preventing again a stabilized value to be copied to the Read Only (RO) storage area.

Following an approach very similar to that used for SC, also structural causal connection relations could be established. It is not difficult to see that some instances of $Write$ are in causal connection with some instances of $Vote$. The functional description of $SCC(Write, Vote)$ due to both places WR and FS is:

$$SCC_{WR \cup FS}(Write, Vote) = (\overline{\langle X_1, X_5, X_2, X_3 \rangle - \langle X_1, X_2, X_3, X_4 \rangle})^t \circ \circ \overline{\langle X_1, X_2, X_2, X_2 \rangle}[X_2 \neq X_3] + + (\overline{\langle X_1 \rangle - \langle X_6 \rangle})^t \circ \overline{\langle X_1 \rangle}[X_2 \neq X_3]$$

The first difference may be treated as done above giving as result:

$$(\overline{\langle X_1, X_5, X_2, X_3 \rangle - \langle X_1, X_2, X_3, X_4 \rangle})^t = (\overline{\langle X_1, X_5, X_2, X_3 \rangle}[X_2 \neq X_5 \vee X_3 \neq X_2 \vee X_4 \neq X_3])^t = [X_2 \neq X_5 \vee X_3 \neq X_2 \vee X_4 \neq X_3]\overline{\langle X_1, X_3, X_4, S, X_2, S \rangle}$$

Right composing the resulting expression with the $W^{-}(Vote, WR)$ the result is:

$$SCC_{WR}(Write, Vote) = \overline{[X_2 \neq X_5 \vee X_3 \neq X_2 \vee X_4 \neq X_3]\, \langle X_1, X_3, X_4, S, X_2, S\rangle} \circ \overline{\langle X_1, X_2, X_2, X_2\rangle [X_2 \neq X_3]} = \overline{[X_2 \neq X_5 \vee X_3 \neq X_2 \vee X_4 \neq X_3]\, \langle X_1, X_2, X_2, S, X_2, S\rangle}[X_2 \neq X_3] = \overline{\langle X_1, X_2, X_2, S - X_2, X_2, S\rangle}[X_2 \neq X_3]$$

$$SCC_{FS}(Write, Vote) = \overline{(\langle X_1\rangle - \langle X_6\rangle)^t} \circ \overline{\langle X_1\rangle}[X_2 \neq X_3] = \overline{((\langle X_1 \cap (S - X_6)\rangle))^t} \circ \overline{\langle X_1\rangle}[X_2 \neq X_3] = \overline{(\langle X_1, S, S, S, S, S\rangle \cap \langle X_1, S, S, S, S, S - X_6,\rangle)} \circ \overline{\langle X_1\rangle}[X_2 \neq X_3] = \overline{\langle X_1, S, S, S, S, S - X_6\rangle} \circ \overline{\langle X_1\rangle}[X_2 \neq X_3] = \overline{\langle X_1, S, S, S, S, S - X_6\rangle}[X_2 \neq X_3]$$

If it was established that *Read* has not priority over *Vote* then a form of conflict may arise between some instances of *Write* and some instances of *Read* (see propagation procedure in (Teruel *et al.*, 2003)). This dependency propagation requires to apply composition to express the relation between instances of *Write* in indirect conflict with *Read* through *Vote*. Its functional description is:

$$\mathcal{R}_3 = SCC(Write, Vote) \circ SC(Vote, Read) = \overline{\langle X_1, X_2, X_2, S - X_2, X_2, S\rangle}[X_2 \neq X_3] \circ \overline{\langle X_1, S - X_2, X_2\rangle}$$

since in the rightmost tuple the 2nd and 3rd component are always different the inner filter may be simplified and the formula becomes:

$$\mathcal{R}_3 = \overline{\langle X_1, X_2, X_2, S - X_2, X_2, S\rangle} \circ \overline{\langle X_1, S - X_2, X_2\rangle}.$$

The result of the above composition depends on the size of class SV. SV is the codomain of X_2 in the rightmost tuple. When the composition rules are being applied to the fourth component of the leftmost tuple the $(S - X_2) \circ (S - X_2)$ calculation results in (a) X_2 if $|SV| = 2$ or (b) S if $|SV| > 2$. The overall composition is thus:

(a) $\mathcal{R}_3 = [X_2 = X_3 \wedge X_2 = X_5 \wedge X_2 \neq X_4]\, \overline{\langle X_1, S - X_2, S - X_2, X_2, S - X_2, S\rangle} = \overline{\langle X_1, S - X_2, S - X_2, X_2, S - X_2, S\rangle}$

(b) $\mathcal{R}_3 = [X_2 = X_3 \wedge X_2 = X_5 \wedge X_2 \neq X_4]\, \overline{\langle X_1, S - X_2, S - X_2, S, S - X_2, S\rangle}$

The filter in the above results is due to X_2 in the leftmost tuple in conjunction with $S - X_2$.

5. CONCLUSIONS AND FUTURE WORK

In this paper the basic calculus needed to compute in a symbolic way structural relations in SWN models is introduced. The calculus has been applied to derive SC and SCC relations in the SWN model of a stable memory device. The proposed method exploits the particular color domain and arc function structure of SWN. The adopted syntax (a simple extension of SWN syntax), admitting the use of filters and of basic functions defined as intersection of SWN elementary functions, allows to compute the desired relations in a fully symbolic and closed way. A challenging point, currently addressed, is the simplification of a structural relation formula when several composition steps are required. Possible directions for future work are the implementation of the transpose and composition operations as well as the simplification procedure, and the application of these results in the context of net level specification of priorities and probabilities of immediate transitions of SWN models. The possibility of integrating the results presented within existing methods and algorithms based on the notion of stubborn sets is also intended to be explored. Finally, the study of the impact that the symbolic computation of structural conflicts, causal connection and mutual exclusion relations may have on the efficient update of the list of enabled transitions after a firing in the context of SWN simulation or reachability graph generation, is being considered.

ACKNOWLEDGEMENTS

This work has been partially funded by the MIUR FIRB project PERF (RBNE019N8N).

REFERENCES

Barkaoui, K., C. Dutheillet and S. Haddad (1993). An efficient algorithm for finding structural deadlocks in colored Petri nets. In: *Application and Theory of Petri Nets 1993, LNCS*. Vol. 691. Springer-Verlag. pp. 69–88.

Brgan, R. and D. Poitrenaud (1995). An efficient algorithm for the computation of stubborn sets of well formed Petri nets. In: *Application and Theory of Petri Nets, LNCS 935*. Springer-Verlag. pp. 121–140.

Capra, L., R. Gaeta and O. Botti (1999). SWN nets as a framework for the specification and the analysis of ft techniques adopted in electric plant automation. In: *Application and Theory of Petri Nets, LNCS*. Springer-Verlag.

Chiola, G., C. Dutheillet, G. Franceschinis and S. Haddad (1993*a*). Stochastic well-formed coloured nets for symmetric modelling applications. *IEEE TC* **42**(11), 1343–1360.

Chiola, G., M. A. Marsan, G. Balbo and G. Conte (1993*b*). Generalized stochastic Petri nets: A definition at the net level and its implications. *IEEE TSE* **19(2)**, 89–107.

De Pierro, M. (2004). *Ph.D. thesis: Structural analysis of conflicts and causality in GSPN and SWN*. Università di Torino - Italia.

Dutheillet, C. and S. Haddad (1993). Conflict sets in colored Petri nets. In: *In proc. of Petri Nets and Performance Models*. pp. 76–85.

Teruel, E., G. Franceschinis and M. De Pierro (2003). Well-defined generalized stochastic Petri nets: A net-level method to specify priorities. *IEEE TSE* **29(11)**, 962–973.

ELSEVIER
IFAC
PUBLICATIONS
www.elsevier.com/locate/ifac

SYNTHESIS OF CONTROLLERS USING COLOURED PETRI NETS AND THEORY OF REGIONS

Belhassen Zouari and Karima Ghedira

LIP2 Lab, Faculté des Sciences de Tunis
Le Belvédère – 1060 – Tunis, Tunisia. belhassen.zouari@fst.rnu.tn

Abstract: This paper addresses the forbidden state control problem of discrete event systems using coloured Petri nets. The problem is studied in the general framework of Ramadge and Wonham theory considering uncontrollable transitions and under non-blockingness requirement. The synthesis method computes a maximally permissive coloured Petri net controller. First, the legal behaviour the plant model should have is determined. Then, the theory of regions is applied in order to generate a maximally permissive controller. This controller is expressed in coloured Petri net terms and, is connected to the plant model. The proposed approach exploits the benefits of the parameterised modelling power of coloured Petri nets at the specification level of the plant model as well as at the synthesis level.

Keywords: Discrete event systems, Supervisory control, Petri nets, closed-loop controllers, Control system synthesis.

1. INTRODUCTION

The control theory pioneered by Ramadge and Wonham (1989) was introduced for modelling supervised discrete-event systems (DES) and synthesis of controllers. This theory was primarily studied in the context of automaton-based models. More recently, Petri nets have begun to receive attention as models for investigating the control problem as they represent a good trade-off between modelling power and analysis capabilities.

In a first approach, standard Petri nets are used for the plant model (i.e. the system to be controlled) and the control is ensured by a distinct model (Krogh, 1987; Giua and DiCesare, 1991; Sreenivas, 1997; Holloway *et al.*, 1996). In a second approach, the controller behaviour is modelled by a Petri net which is connected to the plant model (Ghaffari *et al.*, 2002; Zouari *et al.*, 2003). Ghaffari, *et al.* (2002) proposed an optimal design methodology of a Petri net controller for the forbidden state problem by considering uncontrollable transitions and under liveness requirement. The authors combine a Ramadge-Wonham-like approach, which is used to determine the automaton model of the controlled system, and the theory of regions (Badouel, *et al.*, 1995) to design a pure Petri net controller. This theory was also used in (Uzam, 2002) where an optimal deadlock prevention policy for flexible manufacturing systems is proposed.

In order to enhance the modelling power of Petri nets, the recent trends of the Petri net community are oriented to high-level models. Among these models, Coloured Petri nets (CP-nets) (Jensen and Rozenberg, 1991) seem to be powerful enough to describe complex systems in a manageable way. Nevertheless, high-level Petri nets have been rarely used just as well in the description of the plant model as in the synthesis of supervisors. Makungu, *et al.*, (1999) have developed a method for avoiding forbidden markings for a class of CP-nets where the process to be controlled is separated from the control logic.

This paper addresses the forbidden state control problem with uncontrollable transitions using bounded CP-nets and the theory of regions. The aim is to generate a controller, represented by a CP-net, which is connected to the CP-net of the plant model. A similar approach was introduced by Ghaffari, *et al.* (2002) where only ordinary Petri nets are used. In our approach, we start from a specification of the plant model in CP-nets and a specification of the control constraints expressed as a set of forbidden markings. In a first stage, the proposed approach determines the admissible behaviours of the system. Then, the theory of regions is used in order to derive a maximally permissive CP-net controller. Thanks to the high expression power of CP-nets, the generated controller is represented by only one place with an appropriate colour class and, the incidence colour functions necessary to the connection with the plant model. Hence the obtained closed-loop system is a CP-net. The CP-net model on which our study is based is the Well-Formed coloured Petri nets (WF-

nets) model of (Chiola, *et al.*, 1991). This model is equivalent in expressiveness to CP-nets, but is syntactically restricted by enforcing a particular structuring on colour classes and functions.
The remainder of the paper is organised as follows : Section 2 introduces the notation and definitions used in CP-nets. Section 3 presents the supervisory control theory and introduces an algorithm for the computation of the controlled system behaviour. Section 4 defines an adaptation of the theory of regions to CP-nets in order to design a CP-net controller.

2. HIGH LEVEL PETRI NETS

A natural extension of the ordinary Petri net formalism to allow the representation of larger systems is the introduction of 'colour' structures to identify tokens. Coloured (or in general High-level) Petri nets are particularly well-adapted to the modelling of parametric systems which behaviours depend on the basic structure of the model rather than on the cardinalities of the colour sets. In this section, we briefly present the different CP-net notions according to the syntax defined in the WF-net model.

2.1 Well-Formed Coloured Petri nets (WF-nets)

WF-nets have the same modelling power as CP-nets, although syntactically different. Before presenting a formal definition of the model, let us present the related basis notions :
A *multiset* is a set in which given elements may appear several times. Given a set A, $Bag(A)$ denotes the set of finite multisets on A. A multiset "a" can be represented as a sum: $a = \sum_{x\in A} a(x).x$ in which $a(x)$ gives the number of occurrences of the element x in the multiset "a".
Object classes are finite non-empty sets of objects or basic colours. A class may be viewed as a set made up of elements of the same type. We can distinguish particular type of classes, called *ordered classes*, for which an order relation is defined on its elements.
Colour domains may be defined as a Cartesian product of object classes and is associated with either a transition or a place. When associated with a transition, it defines the set of all its firing instances (coloured firings). When associated with a place, it defines the set of all its possible markings.
Colour functions are a pondered sum of tuples of *basic colour functions*. Colour functions are associated with the labels of WF-net arcs. These functions allow to specify the number of coloured tokens to be consumed and to be produced when firing a given transition. There are three basic colour functions : the *identity* function, used for the choice of any object in a class, specified by a variable X ; the *successor* function, used to specify the circular successor of an object in an ordered class, noted $\oplus X$; and the *diffusion* function, used to specify all the objects of a class C_i, noted All_{Ci}.
A *guard* is a Boolean function defined on a colour domain and which role is to restrict it to a sub-domain. When a guard is associated with the colour domain of a transition, it limits its possible firings. But a guard can also be associated with a colour function labelling an arc in order to indicate whether this arc is valid with respect to the guard value.
Let [g] be a guard and f be a colour function. The guarded function [g].f is defined by:
$\forall c\in C(t) \quad [g].f(c) = (\text{ if } g(c) \text{ then } f(c) \text{ else } 0)$
We can note that a guard can lead to the cancellation of an arc between a place and a transition.

A WF-net is a many-tuple
$N = <P, T, C_l, C, W^-, W^+, \Phi, M_0>$ where :
P is a finite set of places,
T is a set of transitions verifying $P\cap T=\phi$, $P\cup T\neq\phi$,
$C_l = \{C_1,...C_k\}$ is a set of object classes such that $\forall\, i\neq j \in \{1,...k\}$, $C_i\cap C_j=\phi$,
C is the colour domain function, defined from $P\cup T$ into the set of colour domains. An element c of $C(s)$ is a tuple $<c_1,\ldots, c_k>$ and is called a *colour* of s,
W^-, W^+ are the input and output functions (also called incidence functions) defined on $P\times T$, such that $W^-(p,t)$ and $W^+(p,t)$ are guarded colour functions representing linear applications mapping Bag(C(t)) onto Bag(C(p)), for all $(p,t) \in P\times T$,
Φ is a function which associates a guard with any transition. By default $\Phi(t)$ is *true* for any transition t.
M_0 the initial marking is a function defined on P, such that $M_0(p)\in Bag(C(p))$, for all $p\in P$.
$W=W^+-W^-$ indicates the incidence matrix, and W(p,.) is a line vector of such a matrix.

2.2 Definition of the firing rule

The dynamic behaviour of WF-nets is determined by the following firing :
A guarded transition t is enabled for a colour c and a marking M, denoted by $M[t,c\rangle$, iff
$\forall\, p\in P,\ M(p)\geq W^-(p,t)(c)$ and,
the guard associated with t is evaluated to true
The marking M' obtained after the firing of (t,c) is computed as :
$\forall\, p\in P,\ M'(p)=M(p)+W^+(p,t)(c)-W^-(p,t)(c)$
The notation $M[t,c\rangle M'$ is used to indicate this reachability relation. Using the firing rule, it is possible to construct a reachability graph R(N), whose nodes are the markings reachable from the initial marking, and whose arcs represent the reachability relation. Such an arc is labelled by the transition name and the associated colour involved in the reachability relation between two given nodes.

2.3 The dining philosopher problem

In order to illustrate the different steps of the proposed method, we will consider, throughout this paper, the well-known synchronisation problem of the dining philosophers. This system is modelled by the WF-net of Fig. 1. A set of n philosophers spend their lives thinking and eating. They share a common circular table laid with forks, one for each philosopher. Once a philosopher ph_i gets hungry, he tries to pick up both his fork (i) and that of his left-

hand neighbour (⊕*i*) if available. After eating, he puts down both forks and resumes thinking again.
Let $C_1=\{ph_1, ph_2, \ldots, ph_n\}$ be an ordered object class which represents just as well the identities of the philosophers as those of the forks. <X> and <⊕X> are respectively an identity function and a successor function defined on C_1. The initial marking is : $M_0(thinking)=M_0(free\_forks)=All_{C1}$.

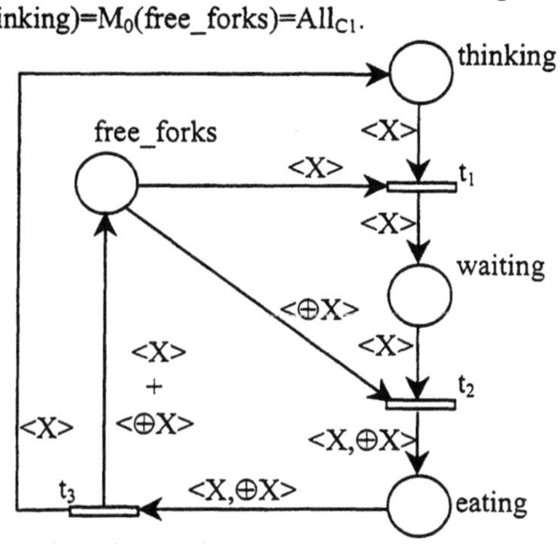

Fig. 1. Dining philosophers WF-net

3. DETERMINING THE CONTROLLED BEHAVIOUR

One of the main objectives of the supervisory control theory is to prevent a DES from reaching undesirable (forbidden) states. For this purpose, it is important to determine, in a first stage, which behaviours are allowed and which ones must be prohibited. This leads us to define the set of *admissible markings*. Then, we can show how a controller may act to achieve its purpose. In the following subsection, we present the problem hypothesis as well as the basic notions related to the supervisory control theory used in our approach.

3.1 Supervisory control problem

Given a DES, the specification of the supervisory control problem amounts to the following points :

- the CP-net of the *plant model* (i.e. the DES without control),
- a set of *forbidden states* (i.e. markings),
- a set of *final states*. A final state represents a proper termination of some operation or task and corresponds to a stable state of the system.

A forbidden state corresponds to an undesirable state because it either compromises the system safety or leads to a *blocking* situation (from which we can not reach a final state).
According to the supervisory control theory, the *uncontrollability* concept means that the transitions specified as uncontrollable are beyond any control procedure. Hence, we assume the transitions set T of the plan model is partitioned into two disjoint subsets: the set T_c of *controllable* transitions (that can be disabled if necessary), and the set T_u of *uncontrollable* transitions (that can not be disabled by the controller).

Furthermore, the *non-blocking* requirement is an aimed property of the controlled system which meaning is that from any reachable marking, it is possible to reach a final marking. The latter is not necessarily the initial marking.

In a first stage, we have to determine the *admissible behaviours* of the system which is represented by an appropriate graph, called the admissibility graph and denoted R_c. An admissible behaviour represents a behaviour of the controlled system under both safety specification and non-blocking requirement. The determination of the admissible behaviours is based on the reachability graph of the plant model. This graph will be treated to determine some markings called *dangerous markings* from which control has to be applied in order to avoid reaching forbidden states. A *dangerous marking* is defined as a node of the reachability graph, from which it exists at least one coloured transition that, when fired, inevitably leads to a forbidden marking. It is worth to note that a dangerous marking (that is not initially specified as forbidden), can be qualified as forbidden if all firing sequences from this marking, inevitably lead to a forbidden state. Moreover, the markings becoming unreachable from the initial marking and the non coreachable markings (i.e. markings from which we can not reach a final marking) will not appear in the graph R_c of admissible behaviours. Hence, we define the graph of admissible behaviours as the reachability graph with only *legal markings*. The set M_L of legal or admissible markings is the maximal set of reachable markings such that :

- a forbidden marking can not be a legal marking,
- from any legal marking, it is always possible to reach a final marking,
- from a legal marking, any coloured transition leading to a non legal marking is a controllable transition.

In a second stage, we have to identify the set Ω of *state-transitions* to be disabled by the control. These state-transitions are arcs of the reachability graph that bind legal markings to non legal markings. The role of the controller (reduced to one place) is to restrict the behaviour of the system within the limits of the legal behaviour. This must be achieved by only enabling and disabling the controllable transitions. Hence, a maximally permissive controller is a controller that ensures the admissible behaviours and forbids all state-transitions of Ω.

Let M be a marking, M_L the set of legal markings. The set Ω of state-transitions that we must forbid is formally defined as follows :

$$\Omega=\{ (M,t(c)) \,/\, M[t,c\rangle M' \wedge M\in M_L \wedge M'\notin M_L \}.$$

3.2 Computation of the controlled behaviour and forbidden state-transitions

In this section, we propose an algorithm which computes R_c and Ω. The idea is to use a colouring technique on the reachability graph (here, the meaning of 'colouring' is independent of CP-nets meaning) in order to identify the paths that lead to forbidden states. The algorithm begins by colouring the specified forbidden markings (*FS*) and their input arcs. Then, we use back propagation and forward propagation of colouring nodes and arcs according to the following rules:

- a node is dangerous if it exists at least one output coloured arc,
- a dangerous node becomes forbidden when all output arcs are coloured or if there exists at least an output coloured arc labelled by an uncontrollable transition,
- all input arcs and output arcs of a forbidden node must be coloured.

After that, if there are markings that can not be reachable from the initial marking, they become forbidden. Also, if the obtained graph contains a non coreachable node, then it becomes forbidden.
Let DS be the set of dangerous states, and TE be the set of state-transitions involving admissible markings to non admissible markings as well as non admissible markings to non admissible ones. The set Ω is included in TE.

Let *(t,c)* be the label of the arc *(x,f)*, where *x* is its origin node and *f* is its destination node. *(x,(t,c))* denotes a state-transition.

<u>*Algorithm1*</u> *Compute* R_c *and* Ω
<u>*Input*</u> *G: reachability graph of the plant model with G.nodes the set of nodes and G.arcs the set of arcs.*
MS: set of final states
FS: set of initially specified forbidden states
T_u: *set of uncontrollable transitions*
<u>*Initially*</u> $DS = \phi$; $TE = \phi$; $\Omega = \phi$; $R_c = G$;
<u>*Repeat*</u>
  <u>*Repeat*</u>
    Take a non-coloured node f from FS;
    Colour the node f;
    <u>*for*</u> *every input arc (x,f) where f is non-coloured* <u>*do*</u> *colour the arc (x,f);*
      $DS = DS \cup \{x\}$; $TE = TE \cup \{(x,(t,c))\}$;
      <u>*if*</u> $t \in T_u$ <u>*then*</u> $FS = FS \cup \{x\}$
      <u>*if*</u> *(x,j) is coloured for any j and* $x \notin MS$ <u>*then*</u> $FS = FS \cup \{x\}$
    <u>*if*</u> $M_0 \in FS$ <u>*then*</u> *exit //there is no solution*
    <u>*for*</u> *every output arc (f,x) non-coloured of f* <u>*do*</u>
      colour the arc(f,x);
      <u>*if*</u> *(x,j) is coloured for any j* <u>*then*</u> $FS = FS \cup \{x\}$
    // Eliminate the node f from R_c
    $R_c.nodes = R_c.nodes \setminus \{f\}$; $R_c.arcs = R_c.arcs \setminus \{(f,z)$ *and* (z,f) *for any* $z \in R_c.nodes\}$;
  <u>*Until*</u> *all nodes of FS are coloured*
    <u>*if*</u> $M_0 \in FS$ <u>*then*</u> *exit //there is no solution*
    <u>*for*</u> *every node m of* R_c*.nodes* <u>*do*</u>
      <u>*if*</u> *(m is not reachable from* M_0 *or m is non coreachable)* <u>*then*</u> $FS = FS \cup \{m\}$
<u>*Until*</u> *all nodes of FS are coloured*
  $DS = DS \setminus FS$;
  <u>*for*</u> *any element y of DS* <u>*do*</u>
    <u>*for*</u> *any element (x,(t,c)) of TE* <u>*do*</u>
      <u>*if*</u> *y==x* <u>*then*</u> $\Omega = \Omega \cup \{(x,(t,c))\}$
<u>*output*</u> R_c ,Ω;

One can easily verify that the algorithm terminates correctly. As we deal with finite reachability graphs, *FS* is finite, and the main loops of the algorithm will necessarily stop.

Let's consider the philosophers example of section 2.3 with n=3 (3 philosophers). Let's assume that transition t_2 is uncontrollable ($t_2 \in T_u$). The reachability graph G is given in Fig. 2.
Let M=(thinking, free_forks, waiting, eating) be the structure of the marking vector. Let's assume that the markings M_{13}=(0,0,All_{C1},0) and M_8=(ph_1+ph_2, ph_2, 0, <ph_3,ph_1>) are forbidden states. M_{13} corresponds to a deadlock and M_8 is just an undesirable state. So, we have MS={M_0} and FS={M_{13},M_8}.

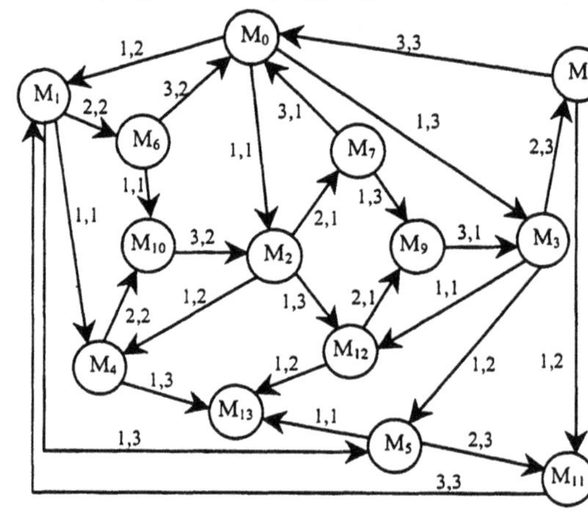

Fig. 2. Reachability graph G.
The state transition (t_x,ph_i) is denoted by x,i

If we apply algorithm 1, we obtain the admissibility graph presented in Fig. 3 and the following set of forbidden state-transitions Ω= {(M_0,t_1(ph_3)); (M_2,t_1(ph_3)); (M_4,t_1(ph_3)); (M_5,t_1(ph_1))}.

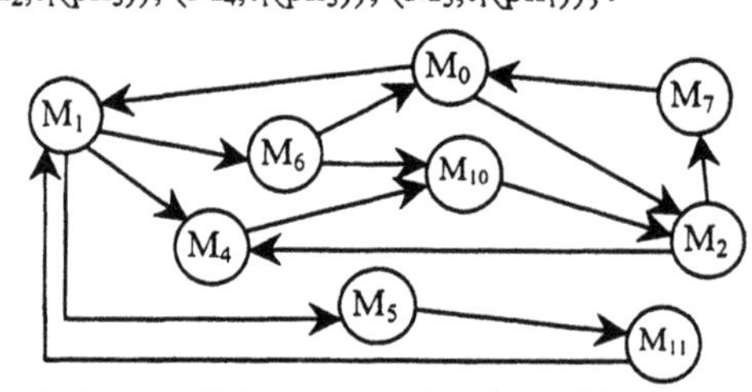

Fig. 3. Admissibility graph R_c where FS={M_{13},M_8}

4. THE CONTROLLER SYNTHESIS

In this section, we adapt the theory of regions to CP-nets in order to generate the aimed controller.
The theory of regions allows to formulate a linear programming system, on the basis of the reachability graph, in which the controller features (additional place, object class, colour functions, initial marking) are introduced as *unknown factors*. Solving this system provides the necessary information for the synthesis of the controller.

4.1 Formulating the admissibility features

Thanks to the parameterised modelling power of CP-nets, it is sufficient to add only one control place. This added place is connected with the original plant model for which it will act as a controller. The related information is determined by solving linear programming systems. These systems are formulated from the admissibility graph R_c and the set Ω of state transitions that must be disabled by the controller (which is also called the *set of event separation instances*). We obtain, at most, as many systems as the cardinality of the set of event separation instances. Each system is made up of three classes of equations/inequations called *the reachability conditions, the cycle equations* and *the event separation condition.*

The *reachability conditions* indicate that every admissible marking must remain reachable under control. Let p_c be the place representing the control place, and C_{pc} a new object class. The colour domain of p_c is C_{pc}. Each object of C_{pc} is related to one or more elements of Ω. In order to determine $M_0(p_c)$ the initial marking of p_c, and $W(p_c,.)$ the associated incidence vector, the place p_c must satisfy the reachability conditions (1). In other words, every marking M in the legal behaviour R_c must still be reachable with respect to p_c.

$$M(p_c)(v) = M_0(p_c)(v) + W(p_c,.)(v)\bar{\Gamma}_M \geq 0, \quad \forall M \in R_c \quad (1)$$

where $M(p_c)(v)$ is the occurrence of a given object v in the marking of p_c, $W(p_c,.)(v)$ is the incidence vector relatively to the object v, Γ_M is any non oriented path in R_c from M_0 to M, and $\bar{\Gamma}_M$ is its associated vector. $\bar{\Gamma}_M$ is indexed by transitions of T and is called the vector counting of Γ_M. Each line of $\bar{\Gamma}_M$ represents the sum of occurrence number of (t_i,c), for each colour c element of $C(t_i)$ in Γ_M. Formally:

$$\bar{\Gamma}_M = \begin{pmatrix} \Gamma_M^1 \\ \vdots \\ \Gamma_M^m \end{pmatrix} \text{ where } \Gamma_M^i \in Bag(C(t_i)) \text{ and } \Gamma_M^i = \sum_{c \in C(t_i)} \alpha_i(c).c\,;$$

$\forall i \in \{1,..,m\}$; where $\alpha_i(c)$ is the occurrence of (t_i,c) in Γ_M and which is determined from R_c.

Let's consider the admissibility graph of Fig. 3. The reachability condition related to the admissible marking M_{10} is : $M_0(p_c)(v) + W(p_c,t_1)(ph_2) + W(p_c,t_2)(ph_2) + W(p_c,t_1)(ph_1) \geq 0$.

Similarly, the *cycle equations* (2) indicate that the cycles of the admissibility graph R_c must remain reachable under control. In other words, the place p_c should satisfy (for every object v of C_{pc}):

$$W(p_c,.)(v).\bar{\gamma} = 0, \quad \forall \gamma \in S_c \quad (2)$$

where S_c is the set of cycles of admissibility graph R_c. $\bar{\gamma}$ is a vector counting of a cycle γ defined similarly as $\bar{\Gamma}_M$. We specify that according to well known results of graph theory, the cycle equations can be reduced to independent cycle equations of basis cycles (Schrijer, 1999).

In the example of Fig. 3. The oriented cycle (M_2, M_4, M_{10}) is represented by the equation :
$W(p_c,t_3)(ph_2) + W(p_c,t_1)(ph_2) + W(p_c,t_2)(ph_2) = 0$

Finally, an *event separation condition* (3) indicates that the control must prevent from firing a state transition of Ω. In other words, the place p_c and an occurrence object v of C_{pc}, must solve at least one event separation instance $(M,t(c))$ from Ω:

$$M_0(p_c)(v) + W(p_c,.)(v)\bar{\Gamma}_M + W(p_c,t)(c)(v) < 0 \quad (3)$$

We can deduce that the number of objects within C_{pc} is at most equal to, in practice much smaller than, the number of state-transitions to inhibit.

Let's take the philosophers example of Fig. 1 with n=3. Disabling the event separation instance $(M_2,t_1(ph_3))$ is ensured by determining the different occurrences of the object v of C_{pc} in the relation : $M_0(p_c)(v) + W(p_c,t_1)(ph_1)(v) + W(p_c,t_1)(ph_3)(v) < 0$

The formulation of the control problem in terms of (1), (2) and (3) is used for the synthesis of the controller as described in the next section.

4.2 The synthesis algorithm

The controller synthesis consists in solving numerous systems of type (1), (2) and (3) according to the algorithm described below. The latter builds, in an incremental manner, the controller components, namely the cardinality of C_{pc}, $W(p_c,.)$ the incidence vector associated with p_c, and $M_0(p_c)$ the initial marking of p_c. In each iteration, the proposed algorithm solves one system where a new object of C_{pc} is introduced as an unknown factor, and a new element of Ω is considered in (3).

Let R_c be the admissibility graph and Ω the set of event separation instances. Let $C_{pc} = \{v_1, \ldots, v_j\}$ (j is definitively determined at the end of the algorithm).

Step 0 : j is set to 1.
Step 1 : Generate the reachability conditions (1).
Step 2 : Compute the basis cycles of R_c.
Step 3 : Compute the independent cycle equations (2) using Gaussian elimination method.
Step 4 : Generate the event separation condition (3) for one element $(M,t(c))$ of Ω.
Step 5 : Solve the system made up of (1), (2) and (3) after replacing v by v_j. If there is no solution, the algorithm terminates as the legal behaviour can not be enforced. If a solution exists, $(M,t(c))$ is eliminated from Ω, and we verify whether it is also a solution of any other (3) (regarding the other elements of Ω). If it is the case, the related elements are also eliminated from Ω.
Step 6 : Here, we proceed to the *folding* of the partial solution $(W(p_c,t_i)(c)(v_j); \forall t_i \in T, \forall c \in C(t_i))$, obtained in step 5, in order to determine $W(p_c,.)(v_j)$. Indeed, the colour domain $C(t_i)$ is partitioned into k_i sets E_i^s $(s=1..k_i)$ such that: $c',c'' \in E_i^s$, iff $W(p_c,t_i)(c')(v_j) = W(p_c,t_i)(c'')(v_j) = \alpha_i^s$. Hence, for every transition t_i, the colour functions associated with the same set E_i^s are grouped. Thus, the folding is achieved as follows :

$$W(p_c,t_i)(v_j)=\sum_{s=1}^{k_i}\left[\bigvee_{c'\in E_i^s}\left[\bigwedge_{X\in \text{var}(t_i)}(X=X(c'))\right]\right].\alpha_i^s$$

Where $\text{var}(t_i)$ is the set of variables appearing on arcs that have t_i as source or destination.

Step 8 : If Ω is non empty, then j is set to j+1, and go to step 4 ; else the algorithm is terminated and a control solution is obtained which comes down to :

$$M_0(p_c)=\sum_{v\in C_{pc}} M_0(p_c)(v).v \quad ; \quad W(p_c,.)=\sum_{v\in C_{pc}} W(p_c,.)(v).v$$

The obtained controller, represented by C_{pc}, $M_0(p_c)$ and $W(p_c,.)$, is optimal as stated in the following theorem.

Theorem 1: The supervisory control problem can be optimally solved by adding one place of control to the plant model, if and only if there exists a solution $M_0(p_c)$, $W(p_c,.)$ satisfying equations (1), (2) and (3) for each event separation instance $(M,t(c))$ in Ω..

Proof: ($\Rightarrow$) Let N_c be the controlled net and $R(N_c)$ its reachability graph. We assume that $R_c=R(N_c)$. So for every place of N_c, (1) and (2) are satisfied. Also, for every $(M,t(c))$ element of Ω, $t(c)$ is forbidden by the control place p_c and some objects of C_{pc}. So (3) is satisfied for every element of Ω.
($\Leftarrow$) Since the place p_c and all the objects of C_{pc} satisfy (1) and (2), so all transitions appearing in the arcs of R_c are firable. They are also firable in $R(N_c)$. So we have $R_c \subseteq R(N_c)$. On the other hand, every (3) associated with an element of Ω is solved with the place p_c and an object v of C_{pc}. Thus, all the elements of Ω are prevented from R_c. So $R(N_c)=R_c$.

Let us apply this method to the philosophers example by considering the set Ω of Fig. 3. The CP-net of Fig. 4 represents the solution. The colour domain of the control place p_c is the object class $C_{pc}=\{v_1,v_2\}$. We can easily verify that the reachability graph of the controlled system is exactly the admissibility graph of Fig. 3, so the controller is optimal. At a first glance, it seems that the proposed controller tries to control uncontrollable transition t_2 via p_c. But it is not the case because the controller does not prevent the firing of t_2 when it is enabled by the other input places.

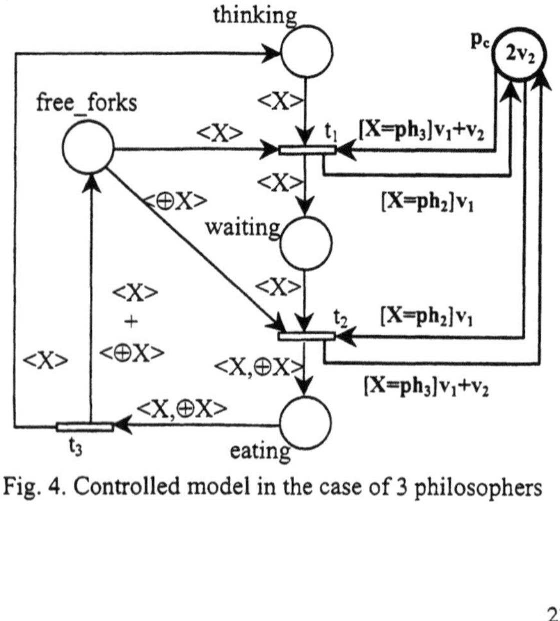

Fig. 4. Controlled model in the case of 3 philosophers

5. CONCLUSION

In this paper, we have presented an optimal design methodology of controllers for the forbidden state problem with non-blockingness requirement, and under the hypothesis of uncontrollable transitions. Our approach combines a Coloured Petri net model and the theory of regions. One of the benefits of using Coloured Petri nets instead of ordinary ones is the more compact and readable representation of the plant model as well as the controlled model. Indeed, the generated CP-net controller is reduced to only one place with an appropriate colour class, and the corresponding incidence colour functions. Improving the efficiency of the proposed method could be obtained by exploiting the notion of *symmetry* induced by the colour *structuring* of CP-nets, as it is done in (Chiola, *et al.*, 1991) for reducing the size of the reachability graph. Indeed, we have find that colour symmetries have a direct impact on *equivalent* solutions computed through the theory of regions. This would reduce the complexity of the controller synthesis procedure.

REFERENCES

Badouel, E, L. Bernardinello and P. Darondeau (1995). Polynomial algorithms for the synthesis of bounded nets. In: *Proc. of CAAP*, Aarhus.

Chiola, G, C. Dutheillet, G. Franceschinis and S. Haddad (1991). On well-formed coloured nets and their Symbolic reachability graph. In: *HLPN. Theory and application* (K. Jensen and G. Rozenberg (ed.), Springer Verlag.

Ghaffari, A, N. Rezg and X.L. Xie (2002a). Design of a live and maximally permissive Petri net controller using theory of regions. *IEEE Trans. on robotics and Automation.*

Giua, A. and F. DiCesare (1991). Blocking and controllability of Petri nets in supervisory control. In: *IEEE Trans on Automatic Control.*

Holloway, L.E., X. Guan and L. Zhang (1996). A generalization of state avoidance policies for controlled Petri nets. In: *IEEE Trans. On Automatic Control,* vol 41, n° 6, pp. 804-816.

Jensen, K. and G. Rozenberg, (1991). High-Level Petri Nets. Theory and application. *S. Verlag.*

Krogh, B.H.(1987). Controlled Petri nets and maximally permissive feedback logic. In: *Proc. 25th Allerton Conf. On Communication, Control, and Computing, Monticello,* pp. 317-326.

Makungu, M., M. Barbeau and R. St-Denis (1999). Synthesis of controllers of process Modeled as coloured Petri nets. In: *Journal Discrete Event Dynamic Systems Theory Applications Kluwer Academic Publishers,* vol. 9, n° 2, pp. 147-169.

Ramadge, P.J., W.M. Wonham (1989). The control of discrete event systems. In: *Proceedings IEEE,* vol.77, n°1, pp. 81-98.

Schrijer, A. (1999). Theory of Linear and Integer Programming. Wiley.

Sreenivas, R.S. (1997). On the existence of supervisory policies that enforces liveness in discrete-event dynamic systems modeled by controlled Petri nets. In: *IEEE Trans. on Automatic Control,* vol 42, n° 7, pp. 928-945.

Uzam, M., (2002). An Optimal Deadlock Prevention Policy for Flexible Manufacturing Systems Using Petri Net Models with Resources ad the Theory of Regions. *Int. Journal of Advanced Manufacturing Technology,* vol. 19, pp 192-208.

ELSEVIER
IFAC
PUBLICATIONS
www.elsevier.com/locate/ifac

INFINITESIMAL PERTURBATION ANALYSIS FOR MAKE-TO-STOCK MANUFACTURING SYSTEMS BASED ON STOCHASTIC FLUID MODELS

Christos Panayiotou * Christos G. Cassandras **,[1]

* *Dept. of Electrical and Computer Engineering, University of Cyprus.* christosp@ucy.ac.cy
** *Dept. of Manufacturing Engineering, Boston University, Brookline, MA 02446.* cgc@bu.edu

Abstract: In this paper we study Make-To-Stock manufacturing systems and generalize the results obtained in Panayiotou et al. (2002). Specifically, we use the same modeling framework to derive sample derivatives of the objective function of interest with respect to buffer capacities (hedging points). In the earlier work, the input processes (machine processing and demand) were assumed piecewise constant. In this work, we generalize the results to piecewise differentiable processes. The derived estimates are unbiased and non-parametric, in the sense that they require no knowledge of the distributions of the underlying random processes. However, unlike the earlier results, these estimators require knowledge of some rates at certain points in time.

1. INTRODUCTION

This paper revisits the Make-To-Stock (MTS) problem (see Paschalidis et al. (2004), Panayiotou et al. (2002) and references therein). The objective is to determine the optimal capacities (or hedging points) of the finished goods and work-in-process buffers such that a cost function (to be defined) is minimized. Solving such problems usually relies on gradient information thus, estimating the gradient of the cost function with respect to these hedging points becomes an essential task.

To derive sensitivity information, we adopt a Stochastic Fluid Model (SFM), a framework commonly used to deal with the analysis and control of manufacturing systems, as in Suri and Fu (1991). We emphasize that our goal is to make use of such a model for the purpose of *control and optimization* rather than *performance analysis*; one can often identify the solution of an optimization problem based on such a model that captures only those features of the underlying "real" system that are needed to lead to the right solution. Such observations have been made in several contexts (e.g., Cassandras et al. (2002) and references therein). We derive the sensitivity information through Perturbation Analysis (PA) methods (Cassandras and Lafortune (1999)) by appropriately adapting them to a SFM viewed as a hybrid system.

In earlier work, we derived such sensitivity estimates when the input processes (machine processing and demand) were piecewise constant. Under this assumption, the analysis was done using finite differences. It was assumed that either of the hedging points was increased by $\Delta\theta$, and through geometric arguments, we derived the change in the buffer contents and the change in the per-

[1] This work was supported in part by the National Science Foundation under Grants EEC-0088073 and DMI-0330171, by AFOSR under contract F49620-01-0056, and by ARO under grant DAAD19-01-0610.

formance measures of interest. In this work, we generalize those results to processes that are piecewise differentiable. The derived estimators still do not require any knowledge of the underlying stochastic processes, however, they require some rate information at certain points in time.

2. MODEL

In this paper we use the two machine model of Panayiotou et al. (2002) shown in Fig. 1. The upstream machine $M2$ can process fluid at a maximum rate $\mu_2(t)$ and is assumed to have an infinite supply. The output of $M2$ is placed in the work-in-process buffer W which has capacity θ_W. Machine $M1$ drains W at a maximum rate $\mu_1(t)$ and fills the finished part inventory buffer F. Finally, F is drained by the demand at a rate $\rho(t)$. The random functions $\mu_1(t)$, $\mu_2(t)$ and $\rho(t)$ are assumed piecewise differentiable w.p. 1. This is in contrast to Panayiotou et al. (2002) where they were assumed piecewise constant.

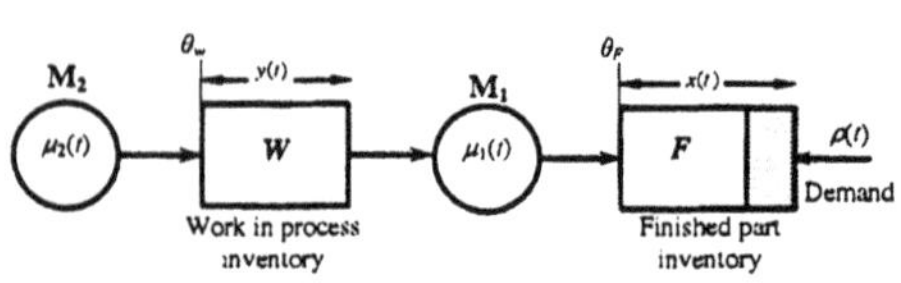

Fig. 1. System Model

Let $(x(t), y(t))$ denote the state of the system, where $x(t) \leq \theta_F, 0 \leq y(t) \leq \theta_W$ correspond to the amount of fluid present in F and W respectively. It is important to note that $x(t)$ can take negative values and these correspond to the amount of backlogged demand. Both, $x(t)$ and $y(t)$ depend on $\boldsymbol{\theta}$, however, to ease the notation, we omit $\boldsymbol{\theta}$. The dynamics of the system are given by

$$\frac{dx(t)}{dt} = \begin{cases} \gamma_1(t;\boldsymbol{\theta}) - \rho(t) & \text{if } x(t) < \theta_F, \\ 0 & \text{if } x(t) = \theta_F \end{cases} \quad (1)$$

$$\frac{dy(t)}{dt} = \begin{cases} \gamma_2(t;\boldsymbol{\theta}) - \gamma_1(t;\boldsymbol{\theta}) & \text{if } 0 < y(t) < \theta_W, \\ 0 & \text{if } y(t) = 0 \text{ or } y(t) = \theta_W \end{cases} \quad (2)$$

where $\gamma_1(t;\boldsymbol{\theta})$ is the inflow rate to buffer F and $\gamma_2(t;\boldsymbol{\theta})$ is the inflow rate to buffer W. The rates $\gamma_1(t;\boldsymbol{\theta})$ and $\gamma_2(t;\boldsymbol{\theta})$ capture the coupling between the dynamics of the two buffers and are given by

$$\gamma_1(t;\boldsymbol{\theta}) = \begin{cases} \rho(t) & \text{if } x(t) = \theta_F \\ \mu_1(t) & \text{if } x(t) < \theta_F \text{ and } y(t) > 0, \\ \mu_2(t) & \text{if } x(t) < \theta_F \text{ and } y(t) = 0 \end{cases} \quad (3)$$

$$\gamma_2(t;\boldsymbol{\theta}) = \begin{cases} \mu_2(t) & \text{if } y(t) < \theta_W \\ \mu_1(t) & \text{if } y(t) = \theta_W \text{ and } x(t) < \theta_F \\ \rho(t) & \text{if } y(t) = \theta_W \text{ and } x(t) = \theta_F \end{cases} \quad (4)$$

Fig. 2 shows a typical sample path for this system where we identify the following events. In general,

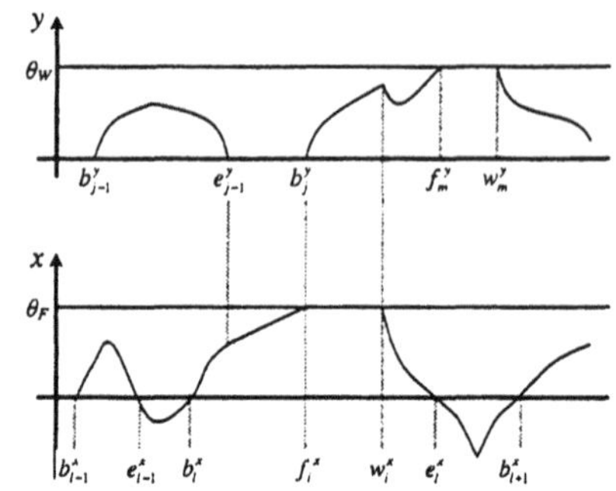

Fig. 2. Typical sample path

the time when these events occur depends on $\boldsymbol{\theta}$, however, to ease the notation, the explicit dependence is omitted.

β_i^q: The trajectory of $q \in \{x, y\}$ *ceases to be empty* for the ith time. This event occurs at time instants b_i^q.
ϵ_i^q: The trajectory of $q \in \{x, y\}$ *becomes empty* for the ith time. It occurs at time instant e_i^q.
ϕ_i^q: The trajectory of $q \in \{x, y\}$ *becomes full* for the ith time. It occurs at time instant f_i^q.
ω_i^q: The trajectory of $q \in \{x, y\}$ *ceases being full* for the ith time. It occurs at time instances w_i^q.

In the context of manufacturing systems of the type we study in this paper, one is typically interested in minimizing an expected cost function

$$J_T(\boldsymbol{\theta}) = \mathbb{E}\left[L_T(\boldsymbol{\theta})\right]$$

where, $\boldsymbol{\theta} = [\theta_F, \theta_W]$ and $L_T(\boldsymbol{\theta})$ is a sample function which in general has the form

$$L_T(\boldsymbol{\theta}) = \frac{1}{T}\int_0^T \left(c_F[x(t)]^+ + c_W y(t) + c_D[-x(t)]^+\right) dt$$

The first two terms correspond to the average inventory cost in F and W respectively and the third one corresponds to the average cost of backlogged demand. Furthermore, $[x]^+ = \max\{0, x\}$ and c_F, c_W, c_D are non-negative constants. We are interested in obtaining estimators of $dJ_T/d\theta$ based on observed data which are then used in stochastic approximation algorithms as in Panayiotou et al. (2002) and Paschalidis et al. (2004).

3. INFINITESIMAL PERTURBATION ANALYSIS (IPA)

In this section we derive sample derivatives for each of the components that make up the sample function $L_T(\boldsymbol{\theta})$, $\boldsymbol{\theta} = [\theta_F, \theta_W]$. So we define

$$X_T(\boldsymbol{\theta}) = \int_0^T x(t)dt, \; Y_T(\boldsymbol{\theta}) = \int_0^T y(t)dt \quad (5)$$

$$X_T^+(\boldsymbol{\theta}) = \int_0^T [x(t)]^+ \, dt, \; X_T^-(\boldsymbol{\theta}) = \int_0^T [-x(t)]^+ dt \quad (6)$$

In our earlier approach (Panayiotou et al. (2002)), we derived such IPA estimates under the assumption of piecewise constant input processes ($\mu_1(t)$, $\mu_2(t)$, and $\rho(t)$). In that work, we assumed that θ_q is increased by $\Delta\theta_q > 0$, $q \in \{F, W\}$ and determined the change in X_T and Y_T, ΔX_T and ΔY_T respectively. In this paper, we generalize the results for piecewise differentiable input processes. The derivation of the IPA derivative estimates is by differentiating the sample functions $X_T(\cdot)$, and $Y_T(\cdot)$ directly. First, we make the following assumptions and derive some preliminary results.

Assumptions:

A1: W.p. 1, the functions $\mu_1(t)$, $\mu_2(t)$ and $\rho(t)$ are piecewise analytic in the interval $[0, T]$.

A2: W.p. 1, except of (ϵ^y, ω^x), (ω^x, ω^y) and (ϕ^x, β^y) no two events can occur at the same time.

A3: W.p. 1, there does not exist an interval $(s, s+\tau)$, $\tau > 0$, such that $M_{n,m}(t) = 0$, where

$$M_{n,m}(t) = \mu_n(t) - \mu_m(t) \tag{7}$$

$n \neq m$, $n, m \in \{0, 1, 2\}$ and, by definition, $\mu_0(t) = \rho(t)$.

In A2, the exceptions are needed since in each pair, it is possible that the occurrence of the first event might also trigger the second event. All three assumptions are mild and are satisfied for general systems. A3 is rather technical. If it is violated, then one can still resort to one sided derivatives and get similar results, but we keep it to simplify our analysis. A1-A3 guarantee the existence of the sample derivatives $\frac{dL_T}{d\theta_F}$ and $\frac{dL_T}{d\theta_W}$.

3.1 Preliminary Results

In this section we are concerned with the evaluation of event time derivatives (with respect to either θ_F of θ_W) of some critical events. We derive a number of iterative relations for evaluating the required event time derivatives, but first, we introduce the following sample path decomposition.

3.1.1. Sample Path Decomposition In this section we decompose the sample paths of $(x(t), y(t))$ into *boundary periods* (BP) and *non-boundary periods* (NBP). BPs, are intervals where the buffer content is at some boundary, i.e., $x(t) = \theta_F$ and $y(t) = 0$ or $y(t) = \theta_W$. A NBP followed by a BP form a *cycle*. We also define the following variables. Let $v_{i,0}$ denote the beginning point of the ith cycle in the sample path of $x(t)$, therefore, it also denotes the beginning of the ith NBP of $x(t)$. This point coincides with the occurrence of the ω_{i-1}^x event (buffer F ceases to be full). $v_{i,1}$ denotes the end of the ith NBP (and the beginning of the ith BP) while $v_{i,2}$ denotes the end point of the ith BP. $v_{i,1}$ and $v_{i,2}$ coincide with the ϕ_i^x and ω_i^x events respectively. Similarly, for the $y(t)$ sample path, let $s_{i,0}$ denote the beginning of the ith NBP, $s_{i,1}$ the end of the ith NBP (same as the start of the ith BP) and $s_{i,2}$ the end of the ith BP. Points $s_{i,0}$ and $s_{i,2}$ occur due to either β^y or ω^y events (W ceases to be empty or full respectively). Point $s_{i,1}$ occurs due to either ϕ^y or ϵ^y events (W becomes full or empty respectively). Furthermore, we use the notation $C_i^x = [v_{i,0}, v_{i,2})$, $i = 1, \cdots, N_T^x$, $C_j^y = [s_{j,0}, s_{j,2})$, $j = 1, \cdots, N_T^y$, to denote the cycles in the sample paths of $x(t)$ and $y(t)$ respectively; N_T^x, N_T^y denote the random number of cycles in the corresponding sample path during the observation interval $[0, T]$.

3.1.2. Event Time Derivatives

Lemma 1. Let w_i^x denote the time when the ω_i^x event (F ceases to be full) has occurred. Then,

$$\frac{\partial w_i^x}{\partial \theta_F} = \begin{cases} \dfrac{\partial e_j^y}{\partial \theta_F} & \text{if } w_i^x = e_j^y \\ 0 & \text{otherwise} \end{cases} \quad \text{and}$$

$$\frac{\partial w_i^x}{\partial \theta_W} = \begin{cases} \dfrac{\partial e_j^y}{\partial \theta_W} & \text{if } w_i^x = e_j^y \\ 0 & \text{otherwise} \end{cases},$$

for all $i = 1, 2, \cdots$ and some $j = 1, 2, \cdots$ [2].

Lemma 2. Let w_i^y denote the time when the ω_i^y event (W ceases to be full) has occurred. Then [3],

$$\frac{\partial w_i^y}{\partial \theta_F} = \frac{\partial w_i^y}{\partial \theta_W} = 0, \quad \text{for all } i = 1, 2, \cdots.$$

Lemma 3. Let b_i^y denote the time when the ith β^y event (W ceases to be empty) has occurred. Then,

$$\frac{\partial b_i^y}{\partial \theta_F} = \begin{cases} \dfrac{\partial f_j^x}{\partial \theta_F} & \text{if } b_i^y = f_j^x \\ 0 & \text{otherwise} \end{cases} \quad \text{and}$$

$$\frac{\partial b_i^y}{\partial \theta_W} = \begin{cases} \dfrac{\partial f_j^x}{\partial \theta_W} & \text{if } b_i^y = f_j^x \\ 0 & \text{otherwise} \end{cases},$$

for all $i = 1, 2, \cdots$ and some $j = 1, 2, \cdots$ [4]. Also, $f_j^y = v_{j,1}$, $j = 1, 2, \cdots$.

Lemmas 1 and 3 show the coupling between the two sample paths. Next we derive four lemmas that show how perturbations in event times propagate from the beginning of the NBP to the end

---

[2] Using the decomposition of Sec. 3.1.1, $w_i^x = v_{i,2} = v_{i+1,0}$ for all $i = 1, 2, \cdots$ and $e_j^y = s_{n,1}$, $n \geq j = 1, 2, \cdots$ (n also includes all NBPs that end with a ϕ^y event).

[3] $w_i^y = s_{n,2} = s_{n+1,0}$ for all $n \geq i$ (n includes all BPs that end with a β^y event).

[4] $b_i = s_{n,2} = s_{n+1,0}$ where again $n \geq i$ since it may include NBPs that start with an event ω^y.

of the same NBP. The first two assume that no events occur during the NBP while the remaining two assume an arbitrary number of events. First, for notational economy we define the following indicator functions.

$$\mathbf{1}_F[t] = \mathbb{1}[x(t) = \theta_F], \quad \mathbf{1}_W[t] = \mathbb{1}[y(t) = \theta_W]. \quad (8)$$

Lemma 4. Let the interval $[s_{i,0}, s_{i,1}) \in [v_{j,0}, v_{j,1})$ or $[s_{i,0}, s_{i,1}) \in [v_{j,1}, v_{j,2})$, for some $i, j = 1, 2, \cdots$, then

$$\frac{\partial s_{i,1}}{\partial \theta_F} = \frac{M_{2,g}(s_{i,0})}{M_{2,g}(s_{i,1})} \frac{\partial s_{i,0}}{\partial \theta_F}$$

$$\frac{\partial s_{i,1}}{\partial \theta_W} = \frac{\mathbf{1}_W[s_{i,1}] - \mathbf{1}_W[s_{i,0}]}{M_{2,g}(s_{i,1})} + \frac{M_{2,g}(s_{i,0})}{M_{2,g}(s_{i,1})} \frac{\partial s_{i,0}}{\partial \theta_W}$$

where $M_{n,m}(t)$ is defined in (7) and $g = \mathbb{1}[x(t) < \theta_F]$, $t \in [s_{i,0}, s_{i,1})$.

Lemma 5. Let the interval $[v_{i,0}, v_{i,1}) \in [s_{j,0}, s_{j,1})$ or $[v_{i,0}, v_{i,1}) \in [s_{j,1}, s_{j,2})$, for some $i, j = 1, 2, \cdots$, then

$$\frac{\partial v_{i,1}}{\partial \theta_F} = -\frac{1 - \mathbf{1}_F[v_{i,0}]}{M_{g,0}(v_{i,1})} + \frac{M_{g,0}(v_{i,0})}{M_{g,0}(v_{i,1})} \frac{\partial v_{i,0}}{\partial \theta_F}$$

$$\frac{\partial v_{i,1}}{\partial \theta_W} = \frac{M_{g,0}(v_{i,0})}{M_{g,0}(v_{i,1})} \frac{\partial v_{i,0}}{\partial \theta_W}$$

where $M_{n,m}(t)$ is defined in (7) and $g = 1 + \mathbb{1}[y(t) = 0]$, $t \in [v_{i,0}, v_{i,1})$.

Lemmas 4 and 5 provide an iterative algorithm for determining the event time derivatives (with respect to either θ_F of θ_W) of the end point of a NBP given the event time derivatives of the starting point of a NBP, *when no other event occurs during the period.* Finally, we need to determine how a perturbation propagates from the beginning of a NBP to the end of a NBP when *several* other events occur during the interval. However, we first need to define the following notation. We concentrate in an interval $[v_{i,0}, v_{i,1})$ and identify a sequence of points $\xi_{i,j}$, $j = 0, \cdots, K_i^y + 1$ where $\xi_{i,0} = v_{i,0}$, $\xi_{i,K_i^y+1} = v_{i,1}$ and at any other point $\xi_{i,j}$, $y(\xi_{i,j})$ transitions either from empty to non-empty or vice versa. In other words, $\xi_{i,j}$ is a sequence with all β^y or ϵ^y events in the interval $[v_{i,0}, v_{i,1})$. Next, we also define the index function that takes values in $\{1, 2\}$

$$g_i(k) = \begin{cases} 1 \text{ if } g_i(k-1) = 2 \\ 2 \text{ if } g_i(k-1) = 1 \end{cases} \quad (9)$$

$$\text{with} \quad g_i(0) = \begin{cases} 1 \text{ if } y(\xi_{i,0}) > 0 \\ 2 \text{ if } y(\xi_{i,0}) = 0 \end{cases}$$

Lemma 6. The derivatives of $v_{i,1}$, $i = 1, 2, \cdots$ with respect to θ_F and θ_W are given by

$$v'_{i,1} = -\left(M_{g_i(K_i^y),0}(v_{i,1})\right)^{-1} \times \left(-\rho(v_{i,0})v'_{i,0} + \sum_{j=1}^{K_i^y} M_{g_i(j-1),g_i(j)}(\xi_{i,j})\xi'_{i,j}\right)$$

where the prime notation indicates the event time derivatives with respect to either θ_F or θ_W.

As mentioned above, all $\xi_{i,j}$ $j = 1, \cdots, K_i^y$ event times correspond to events in the $y(t)$ sample path. These either end a NBP (e.g., ϵ^y events) or start a new NBP (e.g., β^y events). For the first, we use either Lemma 4 or 7 below, to determine the event time derivative given the event time derivative of the start of the NBP. For the second, we use Lemmas 3 and 2 to determine the event time derivative of the NBP starting point.

Next, we analyze the interval $[s_{i,0}, s_{i,1})$ $i = 1, 2, \cdots$, and we identify a sequence of points $\zeta_{i,j}$, $j = 0, \cdots, K_i^x + 1$ where $\zeta_{i,0} = s_{i,0}$, $\zeta_{K_i^x+1} = s_{i,1}$ and at any other point $x(\zeta_{i,j})$ transitions either from full to non-full or vice versa. In other words, $\zeta_{i,j}$ is a sequence with all ϕ^x or ω^x events in the interval $[s_{i,0}, s_{i,1})$. Next, we also define the index function that takes values in $\{1, 2\}$

$$h_i(k) = \begin{cases} 0 \text{ if } h_i(k-1) = 1 \\ 1 \text{ if } h_i(k-1) = 0 \end{cases} \quad (10)$$

$$\text{with} \quad h_i(0) = \begin{cases} 0 \text{ if } x(\zeta_{i,0}) = \theta_F \\ 1 \text{ if } x(\zeta_{i,0}) < \theta_F \end{cases}$$

Lemma 7. The derivatives of $s_{i,1}$, $i = 1, 2, \cdots$ with respect to θ_F are given by

$$s'_{i,1} = -\left(M_{2,h_i(K_i^x)}(s_{i,1})\right)^{-1} \times \left(\mu_2(s_{i,0})s'_{i,0} + \sum_{j=1}^{K_i^x} M_{h_i(j-1),h_i(j)}(\zeta_{i,j})\zeta'_{i,j}\right).$$

The derivatives of $s_{i,1}$, $i = 1, 2, \cdots$ with respect to θ_W are given by

$$s'_{i,1} = -\left(M_{2,h_i(K_i^x)}(s_{i,1})\right)^{-1} \times \left(\mathbf{1}_W[s_{i,1}] - \mathbf{1}_W[s_{i,0}] + \mu_2(s_{i,0})s'_{i,0} + \sum_{j=1}^{K_i^y} M_{h_i(j-1),h_i(j)}(\zeta_{i,j})\zeta'_{i,j}\right)$$

Note that all $\zeta_{i,j}$ $j = 1, \cdots, K_i^y$ times correspond to events in the $x(t)$ sample path. These fall into two categories. (i) End points of BPs (or beginning points of NBPs) i.e., events ω^x. Therefore, the corresponding event time derivatives are determined by Lemmas 1. (ii) End points of NBPs of $x(t)$. In this case, use Lemma 5 or 6.

3.2 Sample Function Derivatives

Now we are ready to derive the derivatives of the sample functions of interest with respect to either θ_F or θ_W. First, using cycles, observe that $X_T(\boldsymbol{\theta})$ and $Y_T(\boldsymbol{\theta})$ can be written as

$$X_T(\boldsymbol{\theta}) = \sum_{i=1}^{N_T^x}\left(\int_{v_{i,0}}^{v_{i,1}} x(t)dt + \theta_F\,(v_{i,2} - v_{i,1})\right) \quad (11)$$

$$Y_T(\boldsymbol{\theta}) = \sum_{i=1}^{N_T^y}\left(\int_{s_{i,0}}^{s_{i,1}} y(t)dt + \theta_W(s_{i,2} - s_{i,1})\,\mathbf{1}_W[s_{i,1}]\right) \quad (12)$$

3.2.1. Evaluation of $\frac{\partial X_T(\boldsymbol{\theta})}{\partial \theta_F}$

Differentiating (11) with respect to θ_F we get

$$\frac{\partial X_T(\boldsymbol{\theta})}{\partial \theta_F} = \sum_{i=1}^{N_T^x}\left((v_{i,2} - v_{i,1}) + \int_{v_{i,0}}^{v_{i,1}} x'(t)dt\right) \quad (13)$$

where we use the prime notation to denote the derivative of a quantity with respect to θ_F. To obtain this result we used the facts that $v_{i,2} = v_{i+1,0}$, $i = 0, \cdots, N_T^x - 1$, $v'_{N_T^x,2} = T' = 0$ and $v'_{1,0} = 0' = 0$. During the ith NBP [5]

$$x(t) = \theta_F \mathbf{1}_F[v_{i,0}] + \int_{v_{i,0}}^{t} (\gamma_1(\tau;\boldsymbol{\theta}) - \rho(\tau))\, d\tau. \quad (14)$$

Again, we identify a sequence of points $\xi_{i,j}$, $j = 0, \cdots, K_i^y + 1$ where $\xi_{i,0} = v_{i,0}$, $\xi_{i,K_i^y+1} = v_{i,1}$ and at any other point $\xi_{i,j}$, $y(\xi_{i,j})$ transitions either from empty to non-empty or vice versa. Using this segmentation, we can rewrite $x(t)$

$$x(t) = \theta_F \mathbf{1}_F[v_{i,0}] + \sum_{j=0}^{K_i^y(t)} \int_{\xi_{i,j}}^{\xi_{i,j+1}} M_{g_i(j),0}(\tau)d\tau. \quad (15)$$

where $K_i^y(t)$ is the number of ϵ^y and β^y events in $[v_{i,0}, t)$ and $\xi_{K_i^y(t)+1} = t$ ($g_i(j)$ is given by (9) and $M_{i,j}(\cdot)$ by (7)). Next, we differentiate $x(t)$ with respect to θ_F:

$$x'(t) = \mathbf{1}_F[v_{i,0}] + \rho(\xi_{i,0})\xi'_{i,0} + \sum_{j=0}^{K_i^y(t)} \left[\mu_{g_i(j)}(\xi_{i,j+1})\xi'_{i,j+1} - \mu_{g_i(j)}(\xi_{i,j})\xi'_{i,j}\right] \quad (16)$$

where as earlier, $\xi'_{i,j}$ indicates the derivative of the time instant $\xi_{i,j}$ with respect to θ_F. Also, $\xi'_{i,K_i^y(t)+1} = t' = 0$. This is a piecewise constant function that changes value at points $t = \xi_{i,j}$, $j = 0, \cdots, K_i^y$ as more terms are added in the sum. The event time derivatives with respect to θ_F are determined by Lemmas 2, 3, 4 and 7.

[5] $\mathbf{1}_F[\cdot]$ is needed since $x(0) = 0$.

Theorem 1. The derivative of $X_T(\boldsymbol{\theta})$ with respect to θ_F is given by

$$\frac{\partial X_T(\boldsymbol{\theta})}{\partial \theta_F} = \sum_{i=1}^{N_T^x}(v_{i,2} - v_{i,1}) + \sum_{i=1}^{N_T^x}\sum_{j=1}^{K_i^y+1} x'(\xi_{i,j})\,(\xi_{i,j} - v_{i,0})$$

where $x'(\xi_{i,j})$, $j = 1, \cdots, K_i^y + 1$ is the derivative of $x(t)$ with respect to θ_F and is evaluated by (16) for all $i = 1, \cdots, N_T^x$.

3.2.2. Evaluation of $\frac{\partial X_T(\boldsymbol{\theta})}{\partial \theta_W}$

Differentiating (11) with respect to θ_W we get

$$\frac{\partial X_T(\boldsymbol{\theta})}{\partial \theta_W} = \sum_{i=1}^{N_T^x} \int_{v_{i,0}}^{v_{i,1}} x'(t)dt \quad (17)$$

where now the prime notation denotes the derivative of a quantity with respect to θ_W. Next, we evaluate $x'(t) = \frac{\partial x(t)}{\partial \theta_W}$. During the ith NBP, $x(t)$ is given by (15) where the same segmentation as in the previous case still holds. Next, we differentiate $x(t)$ with respect to θ_W to obtain

$$x'(t) = \rho(\xi_{i,0})\xi'_{i,0} + \sum_{j=0}^{K_i^y(t)} \left[\mu_{g_i(j)}(\xi_{i,j+1})\xi'_{i,j+1} - \mu_{g_i(j)}(\xi_{i,j})\xi'_{i,j}\right] \quad (18)$$

where now, $\xi'_{i,j}$ is the derivative of $\xi_{i,j}$ with respect to θ_W. This is again a piecewise constant function. The event time derivatives are again determined by Lemmas 2, 3, 4 and 7.

Theorem 2. The derivative of $X_T(\boldsymbol{\theta})$ with respect to θ_W is given by

$$\frac{\partial X_T(\boldsymbol{\theta})}{\partial \theta_W} = \sum_{i=1}^{N_T^x}\sum_{j=1}^{K_i^y+1} x'(\xi_{i,j})\,(\xi_{i,j} - v_{i,0})$$

where $x'(\xi_{i,j})$, $j = 1, \cdots, K_i^y + 1$ is the derivative of $x(t)$ with respect to θ_W and is evaluated by (18) for all $i = 1, \cdots, N_T^x$.

3.2.3. Evaluation of $\frac{\partial Y_T(\boldsymbol{\theta})}{\partial \theta_F}$

Differentiating (12) with respect to θ_F we get

$$\frac{\partial Y_T(\boldsymbol{\theta})}{\partial \theta_F} = \sum_{i=1}^{N_T^y} \int_{s_{i,0}}^{s_{i,1}} y'(t)dt \quad (19)$$

where we use the prime notation to denote the derivative of a quantity with respect to θ_F. Also, we used the fact that $s_{i,2} = s_{i+1,0}$. Next, we evaluate $y'(t) = \frac{\partial y(t)}{\partial \theta_F}$. During the ith NBP,

$$y(t) = \theta_W \mathbf{1}_W[s_0] + \int_{v_{i,0}}^{t} (\mu_2(\tau) - \gamma_1(\tau;\boldsymbol{\theta}))\, d\tau. \quad (20)$$

As in previous cases, in the interval $[s_{i,0}, s_{i,1})$ $i = 1, 2, \cdots$ we identify a sequence of points $\zeta_{i,j}$, $j = 0, \cdots, K_i^x + 1$ where $\zeta_{i,0} = s_{i,0}$, $\zeta_{K_i^x+1} = s_{i,1}$ and at any other point $x(\zeta_{i,j})$ transitions either from full to non-full or vice versa. Using this segmentation we rewrite $y(t)$

$$y(t) = \theta_W \mathbf{1}_W[s_0] + \sum_{j=0}^{K_i^x(t)} \int_{\zeta_{i,j}}^{\zeta_{i,j+1}} M_{2,h_i(j)}(\tau) d\tau. \quad (21)$$

where $K_i^x(t)$ is the number of ϕ^x and ω^x events in $[s_{i,0}, t)$. By convention we set $\rho(t) = \mu_0(t)$ and $h_i(j)$ is given by (10). Next, we differentiate $y(t)$ with respect to θ_F to obtain

$$y'(t) = -\mu_2(\zeta_{i,0})\zeta'_{i,0} + \sum_{j=0}^{K_i^x(t)} [\mu_{h_i(j)}(\zeta_{i,j+1})\zeta'_{i,j+1} - \mu_{h_i(j)}(\zeta_{i,j})\zeta'_{i,j}] \quad (22)$$

where as earlier, $\zeta'_{i,j}$ is the derivative of $\zeta_{i,j}$ with respect to θ_F. This is also a piecewise constant function. The event time derivatives with respect to θ_F are determined by the Lemmas 1, 5 and 6.

Theorem 3. The derivative of $Y_T(\boldsymbol{\theta})$ with respect to θ_F is given by

$$\frac{\partial Y_T(\boldsymbol{\theta})}{\partial \theta_F} = \sum_{i=1}^{N_T^y} \sum_{j=1}^{K_i^x+1} y'(\zeta_{i,j})\,(\zeta_{i,j} - s_{i,0})$$

where $y'(\zeta_{i,j})$, $j = 1, \cdots, K_i^x + 1$ is the derivative of $y(t)$ with respect to θ_F and is evaluated by (22) for all $i = 1, \cdots, N_T^y$.

3.2.4. Evaluation of $\frac{\partial Y_T(\boldsymbol{\theta})}{\partial \theta_W}$ Differentiating (12) with respect to θ_W we get

$$\frac{\partial Y_T(\boldsymbol{\theta})}{\partial \theta_W} = \sum_{i=1}^{N_T^y} \left((s_{i,2} - s_{i,1})\, \mathbf{1}_W[s_{i,1}] + \int_{s_{i,0}}^{s_{i,1}} y'(t) dt \right) \quad (23)$$

where now, the prime notation denotes the derivative of a quantity with respect to θ_W. Also, in the last step we used the fact that $s_{i,2} = s_{i+1,0}$. Next, we evaluate $y'(t) = \frac{\partial y(t)}{\partial \theta_W}$. During the ith NBP, $y(t)$ is given by (21) and we differentiate $y(t)$ with respect to θ_W to obtain

$$y'(t) = \mathbf{1}_W[s_{i,0}] - \mu_2(\zeta_{i,0})\zeta'_{i,0} + \sum_{j=0}^{K_i^x(t)} [\mu_{h_i(j)}(\zeta_{i,j+1})\zeta'_{i,j+1} - \mu_{h_i(j)}(\zeta_{i,j})\zeta'_{i,j}] \quad (24)$$

where as earlier, $\zeta'_{i,j}$ is the derivative $\zeta_{i,j}$ with respect to θ_W. The event time derivatives are determined by Lemmas 1, 5 and 6.

Theorem 4. The derivative of $Y_T(\boldsymbol{\theta})$ with respect to θ_W is given by

$$\frac{\partial Y_T(\boldsymbol{\theta})}{\partial \theta_W} = \sum_{i=1}^{N_T^y} \sum_{j=1}^{K_i^x+1} y'(\zeta_{i,j})\,(\zeta_{i,j} - s_{i,0})$$

where $y'(\zeta_{i,j})$, $j = 1, \cdots, K_i^x + 1$ is the derivative of $y(t)$ with respect to θ_W and is evaluated by (24) for all $i = 1, \cdots, N_T^y$.

Theorems 1-4 constitute an algorithm for evaluating the sample derivatives of interest with respect to θ_F and θ_W. These are similar with the ones derived in Panayiotou et al. (2002) but differ in that they include a number of "correction" terms that account for the fact that the input processes are no longer piecewise constants.

3.3 Unbiasedness

Theorem 5. This estimators obtained in Theorems 1-4 are unbiased estimators of the objective functions, i.e.,

$$\frac{\partial \mathbb{E}[X_T]}{\partial \theta_F} = \mathbb{E}\left[\frac{\partial X_T}{\partial \theta_F}\right], \quad \frac{\partial \mathbb{E}[X_T]}{\partial \theta_W} = \mathbb{E}\left[\frac{\partial X_T}{\partial \theta_W}\right],$$
$$\frac{\partial \mathbb{E}[Y_T]}{\partial \theta_F} = \mathbb{E}\left[\frac{\partial Y_T}{\partial \theta_F}\right], \quad \frac{\partial \mathbb{E}[Y_T]}{\partial \theta_W} = \mathbb{E}\left[\frac{\partial Y_T}{\partial \theta_W}\right].$$

REFERENCES

C. G. Cassandras and S. Lafortune. *Introduction to Discrete Event Systems.* Kluwer Academic Publishers, 1999.

C. G. Cassandras, Y. Wardi, B. Melamed, G. Sun, and C. G. Panayiotou. Perturbation analysis for on-line control and optimization of stochastic fluid models. *IEEE Transactions on Automatic Control*, AC-47(8):1234–1248, 2002.

C.G. Panayiotou, C.G. Cassandras, and P. Zhang. On-line inventory cost minimization for make-to-stock manufacturing systems. In *Proceedings of the IEEE American Control Conference*, pages 4469–4474, 2002.

I. Ch. Paschalidis, Y. Liu, C.G. Cassandras, and C. Panayiotou. Inventory control for supply chains with service level constraints: A synergy between large deviations and perturbation analysis. In *Annals of Operations Research, Special Volume on Stochastic Models of Production-Inventory Systems*, volume 126, pages 231–258, 2004.

R. Suri and B. R. Fu. On using continuous flow lines for performance estimation of discrete production lines. In *Proc. 1991 Winter Simulation Conference*, pages 968–977, 1991.

ELSEVIER
IFAC
PUBLICATIONS
www.elsevier.com/locate/ifac

PERFORMANCE EVALUATION OF NON MARKOVIAN STOCHASTIC DISCRETE EVENT SYSTEMS - A NEW APPROACH

Serge Haddad, Lynda Mokdad [*,1] **Patrice Moreaux** [**,1]

** LAMSADE, UMR CNRS 7024, Université Paris Dauphine, Place du Maréchal de Lattre de Tassigny, 75775 PARIS, FRANCE*
*** LAMSADE & CReSTIC, Université de Reims Champagne-Ardenne, REIMS, FRANCE*

Abstract In this work, we address the problem of transient and steady-state analysis of a stochastic discrete event system which includes non Markovian distributions with a finite support. Rather than computing an approximate distribution of the model (as done in the previous methods), we develop an exact analysis of an approximate model. The design of this method leads to a uniform handling for the computation of the transient and steady-state behaviour of the model. We have evaluated our method on a standard benchmark (the queuing model $M/D/S/K$). Our results demonstrate that: in most of the cases the solution of the approximate model converges quickly to the solution of the exact model, in the difficult cases (e.g. an heavy load on the queue) our method is more robust than the previous ones.

Keywords: Discrete event systems, Stochastic processes, Markovian processes, Periodic processes, Uniformization

1. INTRODUCTION

The transient and steady-state analysis of Markovian discrete event systems is now well established with numerous tools at the disposal of the modellers. The main open issue is the reduction of the space complexity induced by this analysis. However in a realistic system, the distribution of the occurrence (or the duration) of some events can not be described by a exponential law (e.g. the triggering of a time-out). Theoretically any "reasonable" distribution is approximated by a phase-type distribution enabling again a Markovian analysis (Cox 1955*b*). Unfortunately the continuous time Markov chain (CTMC) associated to this approximation is so huge that it forbids its analysis (indeed even its construction). Such a phenomenon often occurs when the non exponential distribution has a finite support (non null dirac, uniform, etc.); then a good phase-type approximation requires too much stages for its specification.

Hence the research has focused on alternative methods. In the case of a single realization of a non Markovian distribution at any time, successful methods have been proposed (German *et al.* 1995) both for the transient and steady state analysis. Let us cite, for instance, the method of supplementary variables (Cox 1955*a*, German and Lindemann 1994) or the method of the subordinated Markov chains (Ajmone Marsan and Chiola 1987).

The general case (i.e. simultaneous multiple realizations of such distributions) is more intricate. The method of supplementary variables is still theoretically applicable but the required space and the computation time limit its use to very small examples. An

[1] {haddad,mokdad,moreaux}@lamsade.dauphine.fr

alternative approach is described for non null Dirac distributions in (Lindemann and Schedler 1996). The stochastic process is observed at periodic moments of time ($\{i\Delta \mid i \in \mathbb{N}\}$) and this new process is expressed by a system of integro-differential equations and solved numerically. The steady-state distributions of these processes are identical and, with another computation, one obtains the transient distribution of the original process from some transient distribution of the transformed process. This method has been implemented in the DSPNexpress tool (Lindemann *et al.* 1999).

Here we propose to define an approximate model on which we perform an exact analysis. Let us first describe the kind of model we will analyse: its behaviour is described by a CTMC, a discrete time Markov chain (DTMC) and a time interval (say Δ). During an interval $(i\Delta, (i+1)\Delta)$ the behaviour is driven by the CTMC and at an instant $i\Delta$ a untimed probabilistic change of state is processed according to the DTMC.

Now we describe how we obtain such a model from a general model with non Markovian finite support distributions. At first, we transform such a distribution into a discrete distribution concentrated on the instants $i\Delta$ included in the support. This step can be done numerically in a straightforward way from the distribution function. So we will not develop it in the remaining paper. Let us note that Δ seems to be the approximation parameter but a better parameter is the ratio between Δ and the size of the support (moreover this indicator is the key factor for the complexity of our analysis).

In the approximate process, the Markovian events occur during the intervals $(i\Delta, (i+1)\Delta)$ and their logical occurrence is specified within the CTMC. Let us note that this is not an approximation since the set $\{i\Delta \mid i \in \mathbb{N}\}$ has a null measure. Non Markovian events always occur in $\{i\Delta \mid i \in \mathbb{N}\}$. Let us describe how they are scheduled. When a non Markovian event is enabled in an interval $(i\Delta, (i+1)\Delta)$ due to the occurrence of a Markovian event then its approximate distribution is interpreted as the number of points $i\Delta$ that must be met before its occurrence. Here we can choose whether we count the next point (an under-evaluation) or not (i.e. an overestimation). The impact of this choice will be discussed later. Thus the residual number of points to be met is included in the state. At any moment $i\Delta$, the current residual numbers are decreased. If some residues are null then the corresponding events occur with possibly some probabilistic choice in case of conflicts. The occurrence of these events may enable new non Markovian events. Such events are handled similarly except that the next point is always counted since now it corresponds to a complete interval.

This process is not exactly ergodic since the steady-state distribution depends on the relative position w.r.t. the points $i\Delta$. The transient analysis is done by successively computing the state distribution at the instants $\Delta, 2\Delta, \ldots$ applying a transient analysis of the CTMC during an interval Δ (via the uniformization technique (Gross and Miller 1984)) followed by a "step" of the DTMC. In order to smooth the effect of the discretization, we average the distribution on the last interval (with a variant of the uniformization). For the steady-state analysis, one computes the steady-state distribution at the instants $i\Delta$ and then starting from this distribution, one averages the steady-state distribution upon an interval.

The balance of the paper is the following one. In section 2 we present our approach. Section 3 reports evaluations of the method applied to a first model. At last, we conclude and give indications on future developments of our work.

2. A NEW APPROXIMATE METHOD

2.1 *Definition of the approximate process*

As discussed in the introduction, we choose a time interval Δ. We approximate the original stochastic process X with the process Y. Roughly speaking, the approximation is related to the non Markovian events which can only occur at times $t_h = h\Delta$. The impact of this approximation will be evaluated in the next section.

We then study the evolution of the stochastic process Y in each interval (t_h, t_{h+1}) and during the state changes at time t_h. We denote by t_h^- (t_h^+) the "time" before (after) the state change in t_h.

The process Y is defined by two components:

- the subordinated process in (t_h, t_{h+1}) associated to states at t_h^+ records only exponential events. It is then a CTMC defined by its generator $\mathbf{Q}$.
- the state changes at t_h are defined by a stochastic matrix $\mathbf{P}[i,j] = \Pr(X(h\Delta^+) = j \mid X(h\Delta^-) = i)$.

Thus Y is fully defined by its initial probability vector $\pi(0)$ and the matrices $(\mathbf{P}, \mathbf{Q})$. These three components depends obviously on Δ since for instance the state space includes the residual number of instants.

2.2 *Transient analysis*

Let $\pi(h\Delta+\tau)$ the probability vector of the process Y at time $h\Delta+\tau$. We have $\pi(h\Delta+\tau) = \pi(h\Delta^+)e^{\mathbf{Q}\tau}$ for $0 \le \tau < \Delta$, and $\pi((h+1)\Delta^-) = \pi(h\Delta^+)e^{\mathbf{Q}\Delta}$, so that $\pi((h+1)\Delta^+) = \pi(h\Delta^+)e^{\mathbf{Q}\Delta}\mathbf{P}$ and finally

$$\pi((h+1)\Delta^+) = \pi(0)(e^{\mathbf{Q}\Delta}\mathbf{P})^{h+1} \qquad (1)$$

Since we want to smooth the discretization effect, we define the approximate value $\pi^{(a)}(h\Delta)$ of $\pi_X(h\Delta)$ as the averaged value of the probabilities of the states of

Y in (t_h, t_{h+1}): $\pi^{(a)}(h\Delta) = \frac{1}{\Delta}\int_{h\Delta}^{(h+1)\Delta}\pi(\tau)d\tau$.
Using (1) we have then:

$$\pi^{(a)}(h\Delta) = \frac{1}{\Delta}\pi(0)\,(e^{\mathbf{Q}\Delta}\mathbf{P})^h \int_0^{\Delta} e^{\mathbf{Q}\tau}d\tau \qquad (2)$$

Finally, we are in general interested by performance measures defined on the states *of the system*, and not on the states of the stochastic process Y. Hence, all components of $\pi^{(a)}(t)$ corresponding to a given state of the original system (i.e. when forgetting the residual numbers) are summed up to compute performance measures (see the example in section 3).

2.3 Steady-state analysis

We also set the approximate value $\pi^{(a)}$ of π as the averaged value of the steady-state probabilities of the states of Y over (t_h, t_{h+1}) since this steady-state distribution depends on the relative position in the interval. If $\pi^{(\Delta)}$ is the steady-state distribution of the DTMC with initial distribution $\pi(0)$ and transition probabilities $\mathbf{P}^{(\Delta)} = e^{\mathbf{Q}\Delta}\mathbf{P}$ then

$$\pi^{(a)} = \frac{1}{\Delta}\pi^{(\Delta)} \int_0^{\Delta} e^{\mathbf{Q}\tau}d\tau \qquad (3)$$

By definition, $\pi^{(\Delta)}$ is the probability vector solution of $\pi = \pi.\mathbf{P}^{(\Delta)}$.

As in the transient case, all components of $\pi^{(a)}$ corresponding to a given state of the system are summed up to compute performance indices.

2.4 Numerical considerations

Formulae (2) and (3) for transient and steady-state probabilities involve vector-matrix products with possibly very large matrices, either $e^{\mathbf{Q}\Delta}$ or $I^{(\Delta)} = \int_0^{\Delta} e^{\mathbf{Q}\tau}d\tau$. Moreover, it is well known that, although $\mathbf{Q}$ is generally very sparse (see section 3 for an example), $e^{\mathbf{Q}\tau}$ is not sparse (the "fill in" phenomenon). The usual approach to compute these matrices is then based on series expansion and numerical summation until a required precision level. We have implemented a similar algorithm to compute the vector-matrix products directly using the series expansions.

When we need $e^{\mathbf{Q}\tau}$ we follows the uniformization approach (Stewart 1994). If $\mathbf{P}^{(u)} = \mathbf{I} + \frac{1}{u}\mathbf{Q}$ is the uniformized matrix of $\mathbf{Q}$ with rate $u > \max_i\{|q_{ii}|\}$, we have

$$e^{\mathbf{Q}\tau} = \sum_{k\geq 0} e^{-u\tau}\frac{(u\tau)^k}{k!}(\mathbf{P}^{(u)})^k \qquad (4)$$

The computation of $I^{(\Delta)} = \int_0^{\Delta} e^{\mathbf{Q}\tau}d\tau$ is based on (4). By definition,

$$I^{(\Delta)} = \sum_{k\geq 0}\left[\int_0^{\Delta} e^{-ut}\frac{(ut)^k}{k!}dt\right](\mathbf{P}^{(u)})^k$$

Using an integration by part,

$I_k = \int_0^{\Delta} e^{-ut}\frac{(ut)^k}{k!}dt$ may be expressed as

$$I_k = -\frac{e^{-u\Delta}(u\Delta)^k}{uk!} + I_{k-1} \quad \text{for } k \geq 1$$

Hence,

$$I_0 = -\frac{e^{-u\Delta}}{u} + \frac{1}{u}$$

$$I_k = \frac{1}{u}\left[1 - e^{-u\Delta}\sum_{h=0}^{h=k}\frac{(u\Delta)^h}{h!}\right]$$

and finally

$$I^{(\Delta)} = \frac{1}{u}\sum_{k\geq 0}\left[1 - e^{-u\Delta}\sum_{h=0}^{h=k}\frac{(u\Delta)^h}{h!}\right](\mathbf{P}^{(u)})^k \qquad (5)$$

As for $e^{(\mathbf{Q}\Delta)}$, we only need $I^{(\Delta)}$ through products $\frac{1}{\Delta}.W.I^{(\Delta)}$. Hence we compute these products iteratively to avoid the fill in. An analogous approach was used in (German 2001) for steady-state solution of Deterministic Stochastic Petri Nets (DSPN) but restricted to one deterministic event at any given time.

3. APPLICATION: THE $M/D/S/K$ QUEUE

In order to evaluate it, we apply our method to the M/D/S/K queue (Poisson arrival with rate λ, deterministic service time with duration D, S servers and finite capacity K). The $M/D/S/K$ queue is an interesting case study for several reasons. First, this model presents *multiple non exponential concurrent* activities (for $1 < S \leq K$). Second, no analytical results are available for the transient or even steady-state *distribution* of the queue length or other performance indices of this queueing model (for $S > 1$) unless $S = K$. It is thus significant to get quickly an approximation of these performance indices. Finally, due to the simplicity of the underlying process, we are able to easily compare our results with existing analytical or numerical tools for $S \leq 2$.

3.1 The approximate process

Let us recall that our approximation lies in the fact that non exponential events, i.e. the ends of (deterministic) service are recorded only at times $h\Delta$. For ease of notation, we set $\Delta = D/n$ so that n will be the indicator of the precision of the approximation in the rest of the paper.

S	K	$n=10$	$n=100$	$n=200$	$n=300$
1	10	101	911	1811	2711
2	10	597	46362	182712	409062
3	5	894	531984	4126459	13783434
3	10	2324	1416239	1416239	36736189

Table 1. Cardinalities of the state space of the approximate process

When we take into account these events, we may choose between two approximate strategies: either underestimate the (new) requested service time, choosing $n\Delta$ ("low" (L) choice) or in contrast overestimate this service time, choosing $(n+1)\Delta$ ("high" (H) choice). This leads to two series of parameters and approximate distributions that we denote by L or H superscript or $*$ for any of them; $n^{(*)}$ stands for n or $n+1$.

A state of the model is then $\langle c,x\rangle$ where

- c is the number of clients in the queue with $0 \leq c \leq K$;
- $x = (x_1, x_2, \ldots, x_{\min(c,S)})$ is the tuple of service states of each served client with $n^{(*)} \geq x_1 \geq x_2 \geq \ldots \geq x_{\min(c,S)}$.

The state $\langle 0,-\rangle$ corresponds to an empty queue and is the initial state of the system. States are ordered w.r.t. a kind of lexicographic order: the first state is $\langle 0,-\rangle$ and $\langle c,x\rangle < \langle c',x'\rangle$ iff $c < c'$ or $c = c'$ and $x = (x_1,\ldots,x_u) \leq x' = (x'_1,\ldots,x'_v)$ i.e. $u < v$ or $u = v$ and there is w such that $x_i = x'_i$ for all $i < w$ and $x_w < x'_w$ (lexical order between vectors).

There is no simple closed form for the cardinality of these state-spaces $S_n^{(L)}$ and $S_n^{(H)}$. They are obviously bounded by $(K+1)(n+1)^S$ but this is a very high bound. Table 1 gives the values of $S_n^{(H)}$ for some parameters.

In the $M/D/S/K$ case, the $\mathbf{Q}$ matrix defined in section 2 is very simple (in fact upper-triangular). From each state $\langle c,x\rangle$ (with $c < K$) the only exponential event is the Poisson arrival of a new client (rate λ), thus

$$\langle c',x'\rangle = \begin{cases} \langle c+1,x\rangle & (S \leq c < K) \\ \langle c+1,(n^{(*)}x_1,\ldots,x_{\min(c,S)})\rangle & (c < S) \end{cases}$$

If $c = K$ there is no transition from $\langle c,x\rangle$.

The $\mathbf{P}$ matrix records state changes due to non exponential events at times $h\Delta$. It has one only non null term per row (with $c > 0$) corresponding to decrease of remaining service times and possible ends of service followed by starts of service of waiting clients ($\mathbf{P}$ is lower-triangular). Let $e = |\{x_i \mid x_i = 1\}|$ be the number of servers which will end their service at next $h\Delta$ time in $s = \langle c,x\rangle$. Then $b = \max(\min(e, c-S), 0)$ is the number of servers that will start a new service at $h\Delta$ from s (such a server may have been busy or idle before $h\Delta$). Then, from $\langle c,x\rangle$, there is a transition with probability 1 to $\langle c',x'\rangle$ equal to

$$\langle c-e,(n^{(*)},\ldots,n^{(*)},x_1-1,\ldots,x_{\min(c,S)-e}-1)\rangle$$

with b occurrences of $n^{(*)}$ on the left side of x'. There is no transition from $\langle 0,-\rangle$ ($\mathbf{P}[1,1] = 0.0$).

3.2 Solving the approximate process

We compute approximate distributions (see section 2) for increasing values of n until they are as close as the required precision. Transient distributions are computed for discrete time horizons $h\Delta$. In all the paper, we use the L1 norm $\| \pi_2 - \pi_2 \| = \sum_i |\pi_1[i] - \pi_2[i]|$ to compare two probability distributions π_1 and π_2.

Let us denote $\pi_n^{(L)}(h)$ (resp. $\pi_n^{(H)}(h)$) the distribution at time $h\Delta$ computed for a given elementary time step $\Delta = D/n$ by taking approximate service delay $n\Delta$ (resp. $(n+1)\Delta$). We observed that each of the sequences $(\pi_n^{(L)}(h))_n$ and $(\pi_n^{(H)}(h))_n$ converges to $\pi^{(L)}(h)$ and $\pi^{(H)}(h)$ but that $\| \pi_n^{(L)}(h) - \pi_n^{(H)}(h) \|$ does not necessarily converge to 0. Moreover, several comparisons showed that depending on the parameters, one of the two sequences converges faster than the other and also that one of the two limits is closer to the exact distribution than the other, but we did not found any clear rule to choose between the two approximate distributions. These behaviours lead us to define the approximate distribution for n as a weighed sum of $\pi_n^{(L)}(h)$ and $\pi_n^{(H)}(h)$ based on their respective convergence rate:

$$\pi_n^{(a)}(h) = \frac{1}{d_n^{(L)} + d_n^{(H)}} \left(d_n^{(L)} \pi_n^{(L)}(h) + d_n^{(H)} \pi_n^{(H)}(h) \right)$$

with $d_n^{(*)} = \frac{1}{\|\pi_n^{(*)}(h) - \pi_{n'}^{(*)}(h)\|}$. and n' is the previous value used in the iterative approximation process. The steady-state approximate distribution is defined similarly.

For a given n, in both transient and steady-state cases, we first compute the matrices $\mathbf{P}_n^{(L)}, \mathbf{P}_n^{(H)}$, $\mathbf{Q}_n^{(L)}, \mathbf{Q}_n^{(H)}$. The initial probability vector is defined as $\pi_n(0)[1] = 1$ and $\pi_n(0)[j] = 0$ for all $j > 1$.

In the transient case (see (2)) we first compute the vector

$$\mathbf{V} = \frac{1}{\Delta} \pi_n(0) \left(e^{\mathbf{Q}_n^{(*)}\Delta} (\mathbf{P}_n^{(*)})^h \right)$$

then the product

$$\pi_n^{(*)}(h) = \mathbf{V}. \int_0^{\Delta} e^{\mathbf{Q}_n^{(*)}\tau} d\tau$$

Note that we never compute neither $e^{\mathbf{Q}_n^{(*)}\Delta}$ nor the integral matrices to avoid fill in.

In the steady-state case (see (3)), we first compute the solution $\pi_n^{(*,\Delta)}$ of $\pi = \pi.\mathbf{P}_n^{(*,\Delta)}$ with $\mathbf{P}_n^{(*,\Delta)} =$

ρ	time horizon 10D	50D	100D
0.1	40 - 10^{-4}	40 - 10^{-4}	40 - 10^{-4}
1.0	120 - 10^{-4}	140 - 10^{-3}	150 - 10^{-3}
10.0	110 - N/A	145 - N/A	150 - N/A

Table 2. Minimal n for $d \leq 10^{-3}$ and precision of the approximation for $S = 2, K = 5$

$e^{\mathbf{Q}_n^{(*)}\Delta}\mathbf{P}_n^{(*)}$ using an iterative method. Then we compute the product

$$\pi_n^{(*)} = \frac{1}{\Delta}\pi_n^{(*,\Delta)} \int_0^{\Delta} e^{\mathbf{Q}_n^{(*)}\tau} d\tau$$

From the two vectors $\pi_n^{(L)}(h), \pi_n^{(H)}(h)$ (or their steady-state versions) we derive $\pi_n^{(a)}(h)$ and we iterate (increasing n by some step value) until $\| \pi_n^{(a)}(h) - \pi_{(n')}^{(a)}(h) \|$ reaches a given precision level ε.

3.3 *Numerical results*

We did several comparisons to validate our method and study its behaviour w.r.t. to the convergence and the precision of the approximation. All our computations were done on a Pentium-PC 2.6Ghz, 512MB, with Scilab 2.7.2 (INRIA 2002) under SuSE Linux. For all these measures, we have fixed $\varepsilon = 10^{-3}$, $K = 5, 10$. As we will show, the behaviour of the implementation depends heavily on the ratio $\rho = \lambda D/S$ which can be seen as the load indicator of the queue. Hence we have fixed $\lambda = 1.0$ in all experiments and we vary ρ in $\{0.1, 0.5, 1.0, 5.0, 10.0\}$, that is to say from a very low to a very high loaded system. For $S = 1$ we used several tools, mainly based on SPN: DSPNexpres (Lindemann 2003), GreatSPN (P.E. Group 2002), SPNica and TimeNET (German 2000). We also developed an exact solution based on the supplementary variable approach (Cox 1955*a*, German and Lindemann 1994) and we compared our approximation with this method too. For $S = 2$, we used DSPNepress-NG, the last version of DSPNexpress, the only analytical tool available to the best of our knowledge for two concurrent deterministic events (see (Lindemann 1998, Lindemann *et al.* 2000) for a detailed presentation). We do not report detailed results do to space limitation. We simply indicate the main trends we have observed.

3.3.1. Transient analysis In the transient case, we did computations for time horizons $t = 10D, 50D$ and $100D$ which correspond to $h = 10n, 50n$ and $100n$ for a given value of n (which varies during the computation of the approximation). To evaluate the convergence of the approximation, we computed the minimal n such that $d = \| \pi_n^{(a)}(h) - \pi_{(n-1)}^{(a)}(h) \| \leq 10^{-3}$. Results for $S = 2$ are given in Table 2. In all cases, the minimal n was between 100 and 200 and is increasing slowly with t. For $K = 10$, we have observed however that for $\rho \gg 1$, the variation on $\pi_n^{(a)}(h)$ first decreases ($n \leq 300$) but then increased with n. This indicates clearly that for large n, numerical computations become unstable. In all cases, we observed a difference lower than 10^{-3} between our approximation and the result given by DSPNexpress-NG.

3.3.2. Steady-state analysis We did the same experiments for the steady-state case. As suspected by the transient analysis, the required n to reach a fixed precision depends largely on ρ. In fact, since we force the end of service time at $n\Delta$ only, our method provides an almost exact result when either the servers are almost always idle ($\rho \ll 1$) or busy ($\rho \gg 1$). We recognize this behaviour both in the higher values of n and the lower precision. This property of our method is also illustrated by the results (not reported here due to lack of space) for the special case $S = K$ for which the exact result is known to be equal to the $M/M/S/S$ queue. In this infinite server case, every end of service is immediately followed by a beginning of service when the system is heavily loaded. In these cases we have obtained small values of n to get the approximation. In contrast, for $\rho \approx 1$, the convergence is slow ($n \approx 200 - 300$) and the approximation is not so good (precision=10^{-3}). For $K = 5$ (resp. $K = 10$) and $\rho = 4.5$ (resp. $\rho > 1.75$), we have noticed that DSPNexpress-NG gives erroneous probabilities (i.e. negative values) so that we could not compare our result with an "exact" numerical value.

4. CONCLUSIONS AND PERSPECTIVES

We have presented a new approximate method for stochastic processes with multiple non Markovian concurrent activities. Contrary to the other methods, we have approximated the model and applied an exact method rather than the opposite way to do. The key factor for the quality of this approximation is that the occurrence of Markovian events are not approximated as it would be in a naive discretization process. Furthermore, the design of its analysis is based on robust numerical methods (i.e. uniformization) and the steady-state and transient cases are handled similarly.

The examples have shown that our method is robust. For instance when analysing an heavy queue the precision is still correct (i.e. 10^{-3} whereas DSPNexpress-NG (a very efficient tool in general) gives negative probabilities!

We are currently extending our method in two directions. The first one is to adapt the approach to models exhibiting very different durations (stiff models). In such cases, the determination of Δ uniquely based on the smallest deterministic delay would lead to intractable matrices and vectors.

The second extension is related to the introduction of tensorial expressions for the matrices in order to manage the space complexity associated to systems with many non Markovian concurrent activities.

Acknowledgments We would like to thank the authors of the different solvers we have used for providing us the last version of their software.

REFERENCES

Ajmone Marsan, M. and G. Chiola (1987). On Petri nets with deterministic and exponentially distributed firing times. In: *Advances in Petri Nets 1987* (G. Rozenberg, Ed.). pp. 132–145. Number 266 In: *LNCS*. Springer–Verlag.

Cox, D. R. (1955*a*). The analysis of non-Markov stochastic processes by the inclusion of supplementary variables. *Proc. Cambridge Philosophical Society (Math. and Phys. Sciences)* **51**, 433–441.

Cox, D. R. (1955*b*). A use of complex probabilities in the theory of stochastic processes. *Proc. Cambridge Philosophical Society* pp. 313–319.

German, R. (2000). *Performance analysis of communication systems*. John Wiley & Sons,Ltd. Chichester, England.

German, R. (2001). Iterative analysis of Markov regenerative models. *Performance Evaluation* **44**, 51–72.

German, R. and C. Lindemann (1994). Analysis of stochastic Petri nets by the method of supplementary variables. *Performance Evaluation* **20**(1–3), 317–335. special issue: Peformance'93.

German, R., D. Logothesis and K.S. Trivedi (1995). Transient analysis of Markov regenerative stochastic Petri nets: A comparison of approaches. In: *Proc. of the 6th International Workshop on Petri Nets and Performance Models*. IEEE Computer Society Press. Durham, NC, USA. pp. 103–112.

Gross, D. and D.R. Miller (1984). The randomization technique as a modeling tool an solution procedure for transient markov processes. *Operations Research* **32**(2), 343–361.

INRIA (2002). Scilab home page: `http://www.scilab.org`.

Lindemann, C. (1998). *Performance Modelling with Deterministic and Stochastic Petri Nets*. John Wiley & Sons.

Lindemann, C. (2003). DSPNexpress home page: `http://www.dspnexpress.de.`.

Lindemann, C., a. Klemm, M. Lohmann and O. Walhorst (2000). Quantitative system evaluation with DSPNexpress 2000. In: *Proc. of the* 2nd *Int. Workshop on Software and Performance (WOSP)*. Ottawa, canada. pp. 12–17.

Lindemann, C., A. Reuys and A. Thümmler (1999). DSPNexpress 2.000 performance and dependability modeling environment. In: *Proc. of the* 29th *Int. Symp. on Fault Tolerant Computing*. Madison, Wisconsin.

Lindemann, C. and G.S. Schedler (1996). Numerical analysis of deterministic and stochastic Petri nets with concurrent deterministic transitions. *Performance Evaluation* **27–28**, 576–582. special issue: Proc. of PERFORMANCE'96.

P.E. Group (2002). GreatSPN home page: `http://www.di.unito.it/~greatspn`.

Stewart, W. J. (1994). *Introduction to the numerical solution of Markov chains*. Princeton University Press, USA.

ELSEVIER
IFAC
PUBLICATIONS
www.elsevier.com/locate/ifac

ANT COLONY OPTIMIZATION FOR MULTI-STATE SERIES-PARALLEL SYSTEM EXPANSION-SCHEDULING

Mustapha Nourelfath(1), **Abdelhakim Khatab**(2)

(1) *Université du Québec en Abitibi-Témiscamingue (UQAT)*
Network Organization Technology Research Center (CENTOR)
Canada

(2) *INSA de Lyon*
Département Informatique
France

Abstract: The demand increases with time in many industrial cases, and the design problem concerns expansions to adapt the manufacturing production system capacity to the demand. This paper uses an ant colony meta-heuristic optimization method to solve the expansion planning problem for multi-state series-parallel systems. A universal generating function technique is applied to evaluate the availability of multi-state series-parallel systems. *Copyright © 2004 IFAC*

keywords: Optimization, Performance evaluation, Expansion scheduling, Ant colony, Redundancy, Manufacturing production systems.

1. INTRODUCTION

1.1 Problem description

The redundancy optimization problem is a well known combinatorial optimization problem where the design goal is achieved by discrete choices made from elements available on the market. The problem is to find the minimal cost configuration of a series-parallel system under availability constraints. The system is considered to have a range of performance levels from perfect functioning to complete failure, and it is called a *multi-state* system (MSS). The MSS consists of n components C_i ($i = 1, 2, \ldots, n$) in series. For each component C_i there are various versions, which are proposed by the suppliers on the market. Elements are characterized by their cost, performance and availability according to their version. These elements can represent machines in a manufacturing system. Each component C_i contains a number of elements connected in parallel.

In many industrial cases the demand increases with time, and the design problem concerns expansions (or reinforcements) to adjust the manufacturing system capacity to the demand. In the redundancy optimization problem it is sufficient to ask *what* should be done and *where*. In designing the expansion plan an answer is needed also to the question of *when*. This paper deals with the expansion planning problem for multi-state series-parallel systems. The system study horizon is divided into several periods. At each period the demand distribution is predicted in the form of a cumulative demand curve. Elements are included at any period to adapt the total system capacity to the demand. The elements of the system are characterized by their cost, productivity and availability. These elements are chosen from a list of products available on the market. The objective is to minimize the cost of each investment over its study period while satisfying availability constraints at each separate period.

1.2 Prior literature

The classical reliability theory is based on the binary assumption that the system is either working perfectly or completely failed. In this paper, because the system and its components are considered to have a range of performance levels, MSS reliability theory will be used. A good and extensive recent review of the literature can be found for example in (Ushakov, Levitin and Lisnianski, 2002) or (Levitin and Lisnianski, 2001).

The problem of total investment-cost minimization, subject to reliability or availability constraints, is well known as the redundancy optimization problem (ROP). The ROP for series parallel systems is NP-hard (Chern, 1992) and has been studied in many different forms as summarized in (Tillman, Hwang and Kuo, 1977), and more recently in (Kuo and Prasad, 2000). In most existing works on ROP, it is usually assumed that a system may experience only two possible sates. The ROP for the multi-state reliability was introduced in (Ushakov, 1987). In (Lisnianski, Levitin, Ben-Haim and Elmakis, 1996) and (Levitin, Lisnianski, Ben-Haim and Elmakis, 1997), genetic algorithms were used to find the optimal or nearly optimal power system structure. In (Levitin, 2000) the multistage expansion problem for multi-state series-parallel systems is solved by using a genetic algorithm as an optimization tool. In (Nourelfath and Nahas, 2003), the authors have used quantized Hopfield networks to solve the reliability optimization problem. In (Nourelfath and Dutuit, 2004), the authors combine universal moment generating function technique with stochastic Petri nets to solve the redundancy optimization problem for multi-state systems under repair policies.

This paper uses an *ant colony* optimization approach to solve the expansion planning problem for multi-state series-parallel systems. The idea of employing a colony of cooperating agents to solve combinatorial optimization problems was recently proposed in (Dorigo, Maniezzo and Colorni, 1996). The ant colony approach has been successfully applied to the classical traveling salesman problem (Dorigo and Gambardella, 1997a; Dorigo and Gambardella, 1997b), to the quadratic assignment problem (Maniezzo and Colorni, 1999), and to scheduling (Bauer, Bullnheimer, Hartl and Strauss, 2000; Den Besto, Stützle and Dorigo, 2000). Ant colony shows very good results in each applied area. It has been recently adapted for the reliability design of binary state systems (Liang and Smith, 2001). The ant colony has also been adapted with success to other combinatorial optimization problems such as the vehicle routing problem (Bullnheimer, Hartl and Strauss, 1997), telecommunication networks management (Di Caro and Dorigo, 1998), graph coloring (Costa and Hertz, 1997), constraint satisfaction (Schoofs and Naudts, 2000) and Hamiltonian graphs (Wagner and Bruckstein, 1999). The ant colony method has not yet been used for the expansion planning problem for MSS.

1.3 Approach and outline

The problem formulated in this paper lead to a set of complicated combinatorial optimization problems. The total number of different solutions to be examined is very large, even for rather small problems. An exhaustive examination of the enormous number of possible solutions is not feasible given reasonable time limitations. In (Levitin, Lisnianski, Ben-Haim and Elmakis, 1997) and (Levitin, 2000), the authors applied a genetic algorithm to solve similar problems. Because of the search space size of the expansion planning problem for MSS and lack of dominant heuristic, it is a good candidate for other meta-heuristic approaches including the focus of this paper, the ant colony optimization (or simply ACO). The ACO approach is adapted to find optimal or nearly optimal solutions to be obtained in a short time. The newer developed meta-heuristic method solves the problem without the limitation on the diversity of versions of elements in parallel.

During the optimization process, artificial ants will have to evaluate the availability of a given selected structure of the series-parallel system. To do this, a fast procedure of availability estimation is developed. This procedure is based on a modern mathematical technique: the *z*-transform or universal moment generating function which was introduced in (Ushakov, 1986). It was proven to be very effective for high dimension combinatorial problems: see e.g. (Ushakov, Levitin and Lisnianski, 2002), (Levitin and Lisnianski, 2001) and references therein. The universal moment generating function is an extension of the ordinary moment generating function (Ross, 1993). The method developed in this paper allows the availability function of reparable series-parallel MSS to be obtained using a straightforward numerical procedure.

The remainder of this paper is organized as follows. In section 2, the expansion planning problem is formulated. In section 3, the ant colony optimization method is adapted to solve the expansion planning problem. In section 4, illustrative examples and numerical results are presented in which the optimal choice of elements in a system is found. Conclusions are drawn in section 5.

2. EXPANSION PALANNING PROBLEM

2.1. Series-parallel systems

Let consider a series-parallel system containing n components C_i (i = $1, 2, \ldots, n$) in series. Every component C_i contains a number of different elements connected in parallel. For each component i, there are a number of element versions available in the market. For any given system component, different versions and number of elements may be chosen. For each component i, elements are characterized according to their version v by their cost (C_{iv}), availability (A_{iv}) and performance (Σ_{iv}). The structure of system component i can be defined by the numbers of parallel elements (of each version) k_{iv} for $1 \le v \le V_i$, where V_i is a number of versions available for element of type i. The entire system structure is defined by the vectors $\boldsymbol{k}_i = \{k_{iv_i}\}$ $(1 \le i \le n,\ 1 \le v \le V_i)$. For a given set of vectors $\boldsymbol{k}_1$, $\boldsymbol{k}_2$, ..., $\boldsymbol{k}_n$ the total cost of the system can be calculated as:

$$C = \sum_{i=1}^{n} \sum_{v=1}^{V_i} k_{iv} C_{iv} \tag{1}$$

2.2. Availability of reparable multi-state systems

The series-parallel system is composed of a number of failure prone elements, such that the failure of some elements leads only to a degradation of the system performance. This system is considered to have a range of performance levels from perfect functioning to complete failure. In fact, the system failure can lead to decreased capability to accomplish

a given task, but not to complete failure. An important MSS measure is related to the ability of the system to satisfy a given demand.
When applied to electric power systems, reliability is considered as a measure of the ability of the system to meet the load demand (D), i.e., to provide an adequate supply of electrical energy (Σ). This definition of the reliability index is widely used for power systems: see e.g., (Ross, 1993; Murchland, 1975; Levitin, Lisnianski, Ben-Haim and Elmakis, 1998). The Loss of Load Probability index (LOLP) is usually used to estimate the reliability index (Billinton and Allan, 1990). This index is the overall probability that the load demand will not be met. Thus, we can write $R = Probab(\Sigma \geq D)$ or R = 1-LOLP with LOLP = $Probab(\Sigma < D)$. This reliability index depends on consumer demand D.
For reparable MSS, a multi-state steady-state availability E is used as $Probab(\Sigma \geq D)$ after enough time has passed for this probability to become constant. In the steady-state the distribution of states probabilities is given by equation (2), while the multi-state stationary availability is formulated by equation (3):

$$P_j = \lim_{t\to\infty}[Probab(\Sigma(t) = \Sigma_j)] \tag{2}$$

$$E = \sum_{\Sigma_j \geq D} P_j \tag{3}$$

If the operation period T is divided into M intervals (with durations $T_1, T_2, \ldots, T_M$) and each interval has a required demand level (D_1, D_2, ..., D_M, respectively), then the generalized MSS availability index A is:

$$A = \frac{1}{\sum_{j=1}^{M} T_j} \sum_{j=1}^{M} Probab(\Sigma \geq D_j)\ T_j \tag{4}$$

We denote by $\boldsymbol{D}$ and $\boldsymbol{T}$ the vectors $\{D_j\}$ and $\{T_j\}$ $(1 \leq j \leq M)$, respectively. As the availability A is a function of $\boldsymbol{k}_1, \boldsymbol{k}_2, \ldots, \boldsymbol{k}_n$, $\boldsymbol{D}$ and $\boldsymbol{T}$, it will be written $A(\boldsymbol{k}_1, \boldsymbol{k}_2, \ldots, \boldsymbol{k}_n, \boldsymbol{D}, \boldsymbol{T})$. In the case of a power system, the vectors $\boldsymbol{D}$ and $\boldsymbol{T}$ define the cumulative load curve (consumer demand). In general, this curve is known for every power system.

2.3. Optimal design of the initial structure

At the initial period, the problem can be formulated as a multi-state system redundancy optimization one: find the minimal cost system configuration $\boldsymbol{k}_1, \boldsymbol{k}_2, \ldots, \boldsymbol{k}_n$ that meets or exceeds the required availability A_0. That is,

$$\text{Minimize } C = \sum_{i=1}^{n} \sum_{v=1}^{V_i} k_{iv} C_{iv} \tag{5}$$

$$\text{Subject to } A(\boldsymbol{k}_1, \boldsymbol{k}_2, \ldots, \boldsymbol{k}_n, \boldsymbol{D}, \boldsymbol{T}) \geq A_0 \tag{6}$$

2.4. Determination of the expansion plan

To provide the desired level of productivity and reliability, the manufacturing system should be expanded at different periods of the horizon study. Each period θ $(1 \leq \theta \leq \Theta)$ begins $\Gamma(\theta)$ years after the base period (period 0) and is characterized by its demand distribution which is represented by cumulative demand curve.
The entire system structure at period θ can be defined by the vectors:

$\boldsymbol{k}_i(\theta) = \{k_{iv_i}(\theta)\}$ $(1 \leq i \leq n,\ 1 \leq v \leq V_i)$.

For a given set of vectors $\boldsymbol{k}_1(\theta), \boldsymbol{k}_2(\theta), \ldots, \boldsymbol{k}_n(\theta)$ the total cost of the system at period θ can be calculated in present values as:

$$C(\theta) = \frac{1}{(1+\mathrm{IP})^{\Gamma(\theta)}} \sum_{i=1}^{n} \sum_{v=1}^{V_i} k_{iv}(\theta) C_{iv}(\theta) \tag{7}$$

where IP represent the annual interest rate.

The universal z-transform is a modern mathematical technique introduced in (Ushakov, 1986). This method, convenient for numerical implementation, is proved to be very effective for high dimension combinatorial problems assuming unlimited repair or maintenance resources. For the sake of space, the presentation of the UMGF method is omitted. The unfamiliar reader is referred for example to (Ushakov, 1986).

3. THE ANT COLONY OPTIMIZATION APPROACH

3.1. The ACO principle

Ants lay down in some quantity an aromatic substance, known as *pheromone*, in their way to food. The pheromone quantity depends on the length of the path and the quality of the discovered food source. An ant chooses a specific path in correlation with the intensity of the pheromone. The pheromone trail evaporates over time if no more pheromone is laid down. Other ants can observe the pheromone trail and are attracted to follow it. Thus, the path will be marked again and will therefore attract more ants. The pheromone trail on paths leading to rich food sources close to the nest will be more frequented and will therefore grow faster. In that way, the best solution has more intensive pheromone and higher probability to be chosen. The described behavior of real ant colonies can be used to solve combinatorial optimization problems by simulation: artificial ants searching the solution space simulate real ants searching their environment. The objective values correspond to the quality of the food sources. The ACO approach associates pheromone trails to features of the solutions of a combinatorial problem, which can be seen as a kind of adaptive memory of the previous solutions. In addition, the artificial ants are equipped with a local heuristic function to guide their search through the set of feasible solutions. Solutions are iteratively constructed in a randomized heuristic fashion biased by the pheromone trails left by the previous ants. The pheromone trails are updated after the construction of a solution, enforcing that the best features will have a more intensive pheromone.

3.2. Applying the ACO meta-heuristic to the EPP

3.2.1. The general algorithm

To apply the ACO meta-heuristic to a combinatorial optimization problem, it is convenient to represent the problem by a graph $G = (\varsigma, \Lambda)$, where ς are the nodes and Λ is the set of edges. To represent our ROP as such a graph, the set of nodes ς is given by components and elements, and edges connect each component to its available elements. Some nodes are added to represent positions where additional component was not used. As in (Liang and Smith, 2001), these nodes are called blanks nodes and have attributes of zero. The obtained graph is partially connected. Ants cooperate by using indirect form of communication mediated by pheromone they deposit on the edges of the graph G while building solutions.
Informally, our algorithm works as follows: m ants are initially positioned on a node representing a component. Each ant represents one possible structure of the entire system. This structure is represented by K_i elements in parallel for n different components. The K_i elements can be chosen in any combination from the V_i available type of elements. Each ant builds a feasible solution (called a tour) to the ROP problem by repeatedly applying a stochastic greedy rule, i.e., the *state transition rule*. While constructing its solution, an ant also modifies the amount of pheromone on the visited edges by applying the *local updating rule*. Once all ants have terminated their tour, the amount of pheromone on edges is modified again (by applying the *global updating rule*). Ants are guided, in building their tours, by both heuristic information (they prefer to choose "less expansive" edges), and by pheromone information. Naturally, an edge with a high amount of pheromone is a very desirable choice. The pheromone updating rules are designed so that they tend to give more pheromone to edges which should be visited by ants. The overview of the algorithm is reported in figure 1.

```
Initialize
 Loop /* at this level each loop is
 called an iteration */
   A colony of ants are positioned on a
   node representing a component
   Loop /* at this level each loop is
   called a step */
     Each ant applies a state transition
     rule to incrementally build a
     solution and a local pheromone
     updating rule
   Until all ants built a complete
   solution
   Evaluate solutions of all ants in
   the colony, rank them and
   record the best feasible one
   A global pheromone updating rule is
   applied
Until a stopping criterion is reached
```

Fig. 1. Overview of the algorithm

In the following we discuss the state transition rule, the global updating rule, and the local updating rule.

3.2.2. State transition rule

In the above algorithm, at each step of the construction process, ants use problem-specific heuristic information (denoted by η_{ij}) and pheromone trails (denoted by τ_{ij}) to select K_i elements in each component. An ant positioned on node i (representing a component C_i) chooses the element j by applying the rule given by:

$$j = \begin{cases} \arg\max_{m \in AC_i} \left\{ [\tau_{im}]^{\alpha} \cdot [\eta_{im}]^{\beta} \right\} & \text{if } q \leq q_0 \\ J & \text{otherwise} \end{cases} \quad (8)$$

and J is a random variable selected according to the probability distribution given by:

$$p_{ij} = \begin{cases} \dfrac{[\tau_{ij}]^{\alpha} \cdot [\eta_{ij}]^{\beta}}{\sum_{m \in AC_i} [\tau_{im}]^{\alpha} \cdot [\eta_{im}]^{\beta}} & \text{if } j \in AC_i \\ 0 & \text{otherwise} \end{cases} \quad (9)$$

where:
α and β are parameters that control the relative weight of the pheromone (τ_{ij}) and the local heuristic (η_{ij}), respectively. AC_i is the set of available element choices for component i. q is a random number uniformly distributed in [0 .. 1]; and q_0 is a parameter ($0 \leq q_0 \leq 1$).
The parameter q_0 determines the relative importance of exploitation versus exploration: every time an ant in node i has to choose an element j, it samples a random number $0 \leq q \leq 1$. If $q \leq q_o$ then the best edge, according to equation (8), is chosen (exploitation), otherwise an edge is chosen according to equation (9) (biased exploration). The state transition rule resulting from equations (8) and (9) is a *pseudo-random-proportional rule*.
The heuristic information used is $\eta_{ij} = 1/(1+c_{ij})$ where c_{ij} represents the associated cost of element j for component i. In equation (9) we multiply the pheromone on edges by the corresponding heuristic value. In this way we favor the choice of edges which are weighted with smaller costs and which have a greater amount of pheromone. That is, elements with smaller cost have greater probability to be selected.

3.2.3. Global updating rule

Once all ants have built a complete system, pheromone trails are updated. Only the globally best ant (i.e., the ant which constructed the best design solution from the beginning of the trial) is allowed to deposit pheromone. A quantity of pheromone $\Delta\tau_{ij}$ is deposited on each edge that the best ant has used. The quantity $\Delta\tau_{ij}$ is given by $1/TC_{best}$, where TC_{best} is the total cost of the design feasible solution constructed by the best ant. Therefore, the global updating rule is:

$$\tau_{ij} \leftarrow (1-\rho).\tau_{ij} + \rho.\Delta\tau_{ij} \quad (10)$$

where $0 < \rho < 1$ is the pheromone decay parameter representing the evaporation of trail.

Global updating is intended to allocate a greater amount of pheromone to less expansive design solution. Equation (10) dictates that only those edges belonging to the globally best solution will receive reinforcement.

3.2.4. Local updating rule

While building a solution of the EPP problem of MSS, ants choose elements by visiting edges on the graph *G*, and change their pheromone level by applying the following local updating rule:

$$\tau_{ij} \leftarrow (1-\rho).\tau_{ij} + \rho.\tau_{o} \quad (11)$$

where ρ is a coefficient such that $(1-\rho)$ represents the evaporation of trail; and τ_o is the initial value of trail intensities.

The application of the local updating rule, while edges are visited by ants, has the effect of lowering the pheromone on visited edges. The pheromone reduction is small but sufficient to lower the attractiveness (or desirability) of precedent edge. This favors the exploration of edges not yet visited, since the visited edges will be chosen with a lower probability by the other ants in the remaining steps of an iteration of the algorithm. Thus, by discouraging the next ant from choosing the same element during the same cycle, ants never converge to a common solution and premature convergence is avoided.

The heuristic described above has been applied succefully to a power station coal transportation system which supplies a boiler. The example of a system to be optimized was taken from (Lisnianski, Levitin, Ben-Haim and Elmakis, 1996) where a genetic algorithm is used to solve the ROP for MSS. The drawback of the algorithm presented in (Lisnianski, Levitin, Ben-Haim and Elmakis, 1996) is that it allows each system component to contain only elements of the *same* version. The proposed ACO algorithm overcomes this limitation and offers a practical way to solve efficiently the problem extended to the case where different elements can be placed in parallel. In the next section, numerical results are given. Note that because the expansion planning problem has been formulated as a sequence of redundancy optimization problems, the proposed ant colony approach will be applied to one redundancy optimization problem for illustration.

4. EXAMPLE

A large industrial example has been studied by the proposed method. Note that details are omitted for the lack of space. The studied example is a power station coal transportation system supplies a boiler (Lisnianski, Levitin, Ben-Haim and Elmakis, 1996). Feeder 1 loads the coal from the bin to conveyor 1. Convoeyor 1 transports the coal to the stacker-reclaimer which lifts the coal up to the burner level. Feeder 2 loads conveyor 2 which supplies the burner feeding system of the boiler. Each element of the system is considered as a unit with total failures. The characteristics of the products available on the market for each type of equipment are given in Levitin, Lisnianski, Ben-Haim and Elmakis, 1997). The nominal capacities are given as a percentage of the nominal boiler capacity. The data of the piecewise cumulative boiler demand curve are also given in Levitin, Lisnianski, Ben-Haim and Elmakis, 1997). The operation periods T_i are long enough that the steady state is attained.

The objective is to define the minimal cost system configuration which provides the requested level of availability $A \geq A_0$. In the experiments the numeric parameters of the ACO algorithm are set to the following values: $\alpha = 1$, $\beta = 2$, $q_0 = 0.9$ and $\rho = 0.95$. These values were obtained by a preliminary optimization phase. The ACO is not very sensitive to changes in these values and tested well for quite a range of them. Ten ants are used in each iteration. The stopping criterion is when the number of iterations reaches 450 or the best feasible solution ant has not changed for 100 consecutive iterations. The best solutions are obtained very quickly. The best solution is obtained after 450 cycles. The search space visited by the ants is then composed of 4500 solutions (10 ants $\times$ 450 cycles) instead of the huge space size of an exhaustive search (about 10^{27}). Such a comparison between the number of solutions constructed by the proposed ACO method and an exhaustive search clearly highlights the goodness of the proposed ACO meta-heuristic with respect to the computing time. From a comparison with optimal solutions for a system with identical parallel elements obtained by the genetic algorithm suggested in (Lisnianski, Levitin, Ben-Haim and Elmakis, 1996), we found that the algorithm suggested in this paper allows for much more effective solutions to be obtained. In fact, with different parallel elements the same avalability level is achieved with lower cost. For example, the cost of the system configuration with different parallel elements obtained for A_0 = 0.975 is about 20 % less than the cost of the configuration obtained with identical parallel elements. However, in a latter version of genetic algorithm (GA), the identical parallel elements limitation is relaxed (Levitin, Lisnianski and Elmakis, 1997) (Levitin and Lisnianski, 2001). In problems with non-identical element versions, the new version of GA can obtain the optimal solution, which is shown in table 8. This solution is better than the one obtained by ACO. Detailed comparison of ACO, GA and other methods remains a perspective of this work.

5. CONCLUSION

The expansion planning problem, a very interesting problem usually encountered in industry, has been solved in this paper. It is formulated as a sequence of redundancy optimization problems for multi-state systems, and solved efficiently by developing an ant colony optimization method. It remains that when considering the expansion planning problem as a sequence of separate redundancy optimization problems, only partial optimal solutions are obtained. This might not lead to the optimization of the whole plan because, in general, partial optimal solutions cannot guarantee an overall optimal solution. Another drawback we would like to move is that the solution methodology in this paper can be applied only to problems in which the planning horizon and the capacity/reliability requirements are defined deterministically. Finally, the comparison of ACO, GA and others meta-heuristics (such as Tabu search, GRASP, Iterated Local Search, Variable Neighborhood Search and Simulated Annealing) remains also a perspective to investigate in further work.

ACKNOWLEDGMENTS

The authors would like to thank the Natural Sciences and Engineering Research Council of Canada (NSERC), and the "Fonds de Recherche sur la Nature et la Technologie" of Quebec (NATEQ).

REFRENCES

(Ait-Kadi and Nourelfath, 2003) D. Ait-Kadi, D., M. Nourelfath (2003). Availability optimization of fault-tolerant production systems. *Journal of Decision Systems*, Vol. 12, Issue 1.

(Bauer, Bullnheimer, Hartl and Strauss, 2000) Minimizing Total Tardiness on a Single machine using Ant Colony Optimization. *Central European Journal of Operations research*, Vol. 8, no. 2, 125-141.

(Bullnheimer, Hartl and Strauss, 1997) Applying the Ant System to the vehicle Routing problem. *2nd Metaheuristics International Conference (MIC-97),* Sophia-Antipolis, France, pp. 21-24.

(Billinton and Allan, 1990) Reliability evaluation of power systems. *Pitman.*

(Chern, 1992) On the Computational Complexity of Reliability redundancy Allocation in a Series System. *Operations Research Letters*, Vol. 11, 309-315.

(Costa and Hertz, 1997) Ants Can Colour Graphs. *Journal of the Operational Research Society*, Vol. 48, 295-305.

(Dallery and Gershwin, 1992) Manufacturing Flow Line Systems: A Review of Models and Analytical Results. *Queueing Systems theory and Applications*, Special Issue on Queueing Models of Manufacturing Systems, Vol. 12, No. 1-2, pp. 3-94.

(Den Besto, Stützle and Dorigo, 2000) Ant Colony Optimization for the Total Weighted tardiness Problem. *Proceeding of the 6th International Conference on Parallel Problem Solving from Nature (PPSNVI)*, LNCS 1917, Berlin, 611-620.

(Di Caro and Dorigo, 1998) Mobile Agents for Adaptive Routing. *Proceedings for the 31st Hawaii International Conference On System Sciences*, Big Island of Hawaii, pp. 74-83.

(Dorigo, Maniezzo and Colorni, 1996) The Ant System: Optimization by a colony of cooperating agents. *IEEE Transactions on Systems, Man and Cybernetics- Part B*, Vol. 26, No.1, pp. 1-13.

(Dorigo and Gambardella, 1997a) Ant Colony System: A Cooperative Learning Approach to the Traveling Salesman Problem", *IEEE Transactions on Evolutionary computation*, Vol. 1, No.1, pp. 53-66.

(Dorigo and Gambardella, 1997b) Dorigo, M. and L. M. Gambardella. Ant Colonies for the Travelling Salesman Problem. *Bio Systems*, Vol. 43, 73-81.

(Kuo and Prasad, 2000) An Annotated Overview of System-reliability Optimization. IEEE Trans. on Reliability, Vol. 49, no. 2, 176-187.

(Levitin and Lisnianski, 2001) A new approach to solving problems of multi-state system reliability optimization. *Quality and Reliability Engineering International*, vol. 17, No. 2, pp. 93-104.

(Levitin, 2000) Multi-state series-parallel system expansion-scheduling subject to availability constraints. *IEEE Transactions on Reliability* vol. 49, No. 1, pp.71-79.

(Levitin, Lisnianski, Ben-Haim and Elmakis, 1997) Structure optimization of power system with different redundant elements. *Electric Power Systems Research*, vol. 43, No. 1, pp.19-27.

(Levitin, Lisnianski, Ben-Haim and Elmakis, 1998) Redundancy optimization for series-parallel multi-state systems. *IEEE Transactions on Reliability*, vol. 47, No. 2, pp.165-172.

(Liang and Smith, 2001) An Ant Colony Approach to Redundancy Allocation. Submitted to *IEEE Transactions on Reliability.*

(Lisnianski, Levitin, Ben-Haim and Elmakis, 1996) Power system structure optimization subject to reliability constraints. *Electric Power Systems Research*, vol. 39, No. 2, pp.145-152.

(Maniezzo and Colorni, 1999) The Ant System Applied to the Quadratic Assignment Problem. *IEEE Transactions on Knowledge and Data Engineering*, Vol. 11, no. 5, 769-778.

(Murchland, 1975) Fundamental concepts and relations for reliability analysis of multi-state systems. *Reliability and Fault Tree Analysis*, ed. R. Barlow, J. Fussell, N. Singpurwalla. SIAM, Philadelphia.

(Nourefath, Ait-Kadi and Soro, 2003) M. Nourefath, D. Ait-Kadi and W.-I. Soro. Availability modelling and optimization of reconfigurable manufacturing systems. *Journal of Quality in Maintenance Engnn* 2003;9(3):284-302.

(Nourefath and Nahas, 2003) M. Nourefath and N. Nahas. Quantized Hopfield networks for reliability optimization. *Reliability Engineering and System Safety*, Volume 81, Issue 2, pages 191-196, 2003.

(Nourefath and Dutuit, 2004) M. Nourefath and Y. Dutuit. A combined approach to solve the redundancy optimization problem for multi-state systems under repair policies. *Reliability Engineering and System Safety*, in press, 2004.

(Ross, 1993) Introduction to probability models. *Academic press.*

(Schoofs and Naudts, 2000) Schoofs, L. and B. Naudts, "Ant Colonies are Good at Solving Contraint Satisfaction Problem," *Proceeding of the 2000 Congress on Evolutionary Computation*, San Diego, CA, July 2000, 1190-1195.

(Tillman, Hwang and Kuo, 1977) Tillman, F. A., C. L. Hwang, and W. Kuo, "Optimization Techniques for System Reliability with Redundancy – A review," *IEEE Trans. on Rel.*, Vol. R-26, no. 3, 1977, 148-155.

(Ushakov, Levitin and Lisnianski 2002) Multi-state system reliability: from theory to practice. *Proc. of 3 Int. Conf. on mathematical methods in reliability, MMR 2002*, Trondheim, Norway, pp. 635-638.

(Ushakov, 1987) Optimal standby problems and a universal generating function. *Sov. J. Computing System Science*, Vol. 25, N 4, pp 79-82.

(Ushakov, 1986) Universal generating function. *Sov. J. Computing System Science*, Vol. 24, N 5, pp 118-129.

(Wagner and Bruckstein, 1999) Wagner, I. A. and A. M. Bruckstein. Hamiltonian(t)-An Ant inspired heuristic for Recognizing Hamiltonian Graphs. *Proceeding of the 1999 Congress on Evolutionary Compuation*, Washington, D.C., 1465-1469.

ELSEVIER
IFAC
PUBLICATIONS
www.elsevier.com/locate/ifac

THE THINNING PROBLEM

Spyros A. Reveliotis * Jin Young Choi *

* *School of Industrial & Systems Engineering*
Georgia Institute of Technology
{spyros,choijy68}@isye.gatech.edu

Abstract: The problem addressed in this work is to confine a set of sequential processes, that interact through a finite set of resources, to a subset of their feasible behaviors, while optimizing some performance criterion. A canonical characterization of the problem is provided for the case that the optimized performance criterion is the maximization of the system throughput.

Keywords: Resource Allocation Systems, Deadlock Avoidance, Performance Modelling and Analysis

1. INTRODUCTION

A general statement of the problem addressed in this work is as follows: Given a set of sequential processes, each of which can be executed through a number of *process plans* – i.e., sequences of processing stages with each stage requesting the exclusive allocation of some resource set – and a *"budget"* on the *"active"* process plans for each process, select the process plans to be activated with respect to each process, so that some performance index is optimized. The entire set of process plans available for each process is represented by a digraph with its nodes defined by the set of processing stages, and its edges indicating all the possible transitions among these stages. Hence, the problem can be posed as the selection of a number of edges from each digraph, so that the restriction of the system operation on the process plans expressed by the selected edges optimizes the considered performance index. Since, this edge selection process is equivalent to the removal of the remaining edges from the corresponding process graphs, the entire problem can be perceived as a *"thinning"* operation of the original process graphs.

The subsequent development of this work focuses on the particular criterion of *maximizing the system throughput*, and provides a canonical characterization of the corresponding thinning problem in the form of a *Mixed Integer Programming (MIP)* formulation. The presented formulation constitutes a synthesis of recently obtained results on liveness-enforcing supervision of resource allocation systems (RAS) (Ezpeleta *et al.*, 2002; Park and Reveliotis, 2001) with the performance modelling framework of Markov Decision Processes (MDP) (Puterman, 1994). From a practical standpoint, its applicability is limited by the fact that the numbers of variables and constraints involved are a super-polynomial function of the size of the underlying system configuration. However, it provides a rigorous basis for the eventual development of other approximating methods with reduced computational complexity; the development of such computationally efficient methods for the thinning problem is part of our current investigations.

The rest of the paper is organized as follows: Section 2 introduces the class of Disjunctive / Conjunctive(D/C-)RAS, which models the resource allocation dynamics considered in this work, and it summarizes the currently available results on the development of liveness-enforcing supervisors for these systems. Section 3 proceeds to the modelling of the timed dynamics of the logically controlled D/C-RAS, and it provides a linear programming (LP) formulation for the performance control problem, that is based on concepts and techniques borrowed from the field of Continuous-Time Markov Decision Processes (CT-MDP) (Puterman, 1994). Section 4 builds upon this background in order to provide a Mixed Integer Programming (MIP) formulation of the thinning problem, itself. Finally, Section 5 concludes the paper and discusses directions for future work. A more extensive write-up of this work, that includes also a number of elucidating examples, can be found, under the same title, at *http://www.isye.gatech.edu/~spyros*.

2. DISJUNCTIVE / CONJUNCTIVE RESOURCE ALLOCATION SYSTEMS AND LIVENESS-ENFORCING SUPERVISION

D/C-RAS The class of *Disjunctive / Conjunctive Resource Allocation Systems (D/C-RAS)*, considered in this work, is formally defined as a 5-tuple $\Phi = < \mathcal{R}, C, \mathcal{P}, \mathcal{A}, \mathcal{T} >$ where: (i) $\mathcal{R} = \{R_1, \ldots, R_m\}$ is the set of the system *resource types*. (ii) $C : \mathcal{R} \rightarrow Z^+$ – the set of strictly positive integers – is the system *capacity* function, characterizing the number of identical units from each resource type available in the system. Resources are considered to be *reusable*, i.e., each allocation cycle does not affect their functional status or subsequent availability, and therefore, $C(R_i) \equiv C_i$ constitutes a system *invariant* for each i. (iii) $\mathcal{P} = \{\Pi_1, \ldots, \Pi_n\}$ denotes the set of the system *process types* supported by the considered system configuration. Each process type Π_j is a composite element itself, in particular, $\Pi_j = < \mathcal{S}_j, \mathcal{G}_j >$, where: (a) $\mathcal{S}_j = \{\Xi_{j1}, \ldots, \Xi_{j,l(j)}\}$ denotes the set of *processing stages* involved in the definition of process type Π_j, and (b) $\mathcal{G}_j$ is an *acyclic digraph* with its node set, V_j, being bijectively related to the set $\mathcal{S}_j$. Let $V_j^{\nearrow}$ (resp., $V_j^{\searrow}$) denote the set of *source* (resp., *sink*) nodes of $\mathcal{G}_j$. Then, any *path* from some node $v_s \in V_j^{\nearrow}$ to some node $v_f \in V_j^{\searrow}$ defines a *process plan* for process type Π_j. (iv) $\mathcal{A} : \bigcup_{j=1}^n \mathcal{S}_j \rightarrow \prod_{i=1}^m \{0, \ldots, C_i\}$ is the *resource allocation function* associating every processing stage Ξ_{jk} with a *resource allocation request* $\mathcal{A}(j,k) \equiv A_{jk}$. More specifically, each A_{jk} is an m-dimensional vector, with its i-th component indicating the number of resource units of resource type R_i necessary to support the execution of stage Ξ_{jk}. Obviously, in a well-defined RAS, $A_{jk}(i) \leq C_i, \ \forall j, k, i$. Furthermore, the resource set A_{jk}, required for the execution of a particular processing stage Ξ_{jk}, is allocated exclusively and non-preemptively to each process instance, and it is released by it only upon the allocation of the resources required for the execution of the subsequent stage. (v) $\mathcal{T} : \bigcup_{j=1}^n \mathcal{S}_j \rightarrow \mathcal{D}$ is the *timing function*, corresponding to each processing stage Ξ_{jk} a distribution D_{jk} that characterizes the statistics of the the processing time t_{jk}, experienced during the execution of stage Ξ_{jk}. Finally, $|\Phi| \equiv |\mathcal{R}| + |\bigcup_{j=1}^n \mathcal{S}_j| + \sum_{i=1}^m C_i$ will be referred to as the *size of* Φ.

For reasons that will become obvious in the subsequent development, each of the aforementioned graphs $\mathcal{G}_j$ is augmented with an additional node v_{j0}, and a new set of edges $\{(v_{j0}, v_{jk}) : v_{jk} \in V_j^{\nearrow}\} \cup \{(v_{jk}, v_{j0}) : v_{jk} \in V_j^{\searrow}\}$. The resulting graph will be denoted by $\bar{\mathcal{G}}_j$. Notice that $\bar{\mathcal{G}}_j$ is strongly connected, with each cycle in it containing node v_{j0}. Node v_{j0} essentially models the RAS environment, i.e., that state occupied by the process before entering or upon leaving the system.

A logical characterization of the RAS behavior A formal characterization of the logical behavior generated by the D/C-RAS can be provided by means of the modelling framework of *Finite State Automaton (FSA)* (Hopcroft and Ullman, 1979). The D/C-RAS *state* $s(t)$ at time t is a vector of dimensionality equal to the total number of distinct processing stages, such that each of its components $s(t;q)$ corresponds to a processing stage Ξ_{jk} and indicates the number of process instances executing stage Ξ_{jk} at time t. To simplify the notation, the following discussion omits the dependence of state s on time t. The information contained in the RAS state is sufficient for the determination of the distribution of the resource units to the various processing stages, as well as of the *slack* (or *idle*) resource capacity in the system; in particular, the *slack* capacity, $\delta_i(s)$, of resource R_i at sate s, can be computed as $\delta_i(s) \equiv C_i - \sum_{q=1}^D s(q(j,k)) \cdot A_{jk}(i)$. The set S of *feasible* resource allocation states for the considered RAS is defined by $S \equiv \{s \in (Z_0^+)^D : \delta_i(s) \geq 0, \ \forall i = 1, \ldots, m\}$. The finiteness of the resource capacities implies that $card(S) \equiv |S| < \infty$. However, in general, $|S|$ will be a *super-polynomial* function of the RAS size.

The set of *events*, E, that can change the system state, comprises: (i) the events e_{jk}^l, $j = 1, \ldots, n$, $k \in \{1, \ldots, l(j) : \Xi_{jk} \in V_j^{\nearrow}\}$, corresponding to the *loading* of a new instance of process type Π_j into the system, that is to follow a process plan starting with stage Ξ_{jk}, (ii) the events e_{jkh}^a, $j = 1, \ldots, n$, $k, h = 1, \ldots, l(j)$, $k \neq h$, corresponding to the *advancement* of a process instance executing stage Ξ_{jk} to a successor stage Ξ_{jh}, and (iii) the events e_j^u, corresponding to the *unloading* of a finished process instance of type Π_j. Without loss of generality, it is assumed that, during a single state transition, only one of these events can take place. The resulting transition, however, is *feasible* only if the additionally requested set of resources can be obtained from the system slack capacity.

A natural definition for the *initial* state is $s_0 \equiv \mathbf{0}$ i.e., the state in which the system is idle and empty of any process instances. Since the main logical concern is the establishment of non-blocking behavior, the set of *marked* states is defined as

$S_m \equiv \{s_0\}$. Hence, the *marked language* $\mathcal{L}_m$ of this automaton corresponds to *"complete runs"*.

The above FSA-based model of the RAS behavior can be expressed graphically by the *State Transition Diagram (STD)*, i.e., a digraph $\mathcal{G}$ with *nodes* corresponding to the FSA states, and *edges* corresponding to the feasible state transitions. Of particular interest is the STD subgraph induced by the nodes s that are reachable from node s_0; this subgraph is denoted by S_r and is characterized as the *reachable* subspace of the considered RAS.

Liveness and Liveness-Enforcing Supervision for D/C-RAS A major concern in the logical control of RAS is the establishment of *live* – or *deadlock-free* or *non-blocking* – behavior. *Deadlocks* are defined as RAS states where there is a set of process instances, such that each of its processes, in order to advance, requests the allocation of resources currently held by some other process(es) in the considered set. Their development results from (i) the fact that processes will hold upon their allocated resources in a non-preemptive manner and (ii) the arbitrary structure of the process routes that can give rise to cyclical patterns of resource requests among the various executing processes.

In the FSA-based modelling of the RAS operation, deadlocks are represented by the formation of *strongly connected components* in the system reachable space, S_r, which, however, are not *co-accessible*, i.e., the empty state, s_0, is not reachable from them through any sequence of feasible transitions. Hence, a *correct Liveness-Enforcing Supervisor (LES)*, Δ, tries to restrict the system operation to a *strongly connected component of S_r which contains the empty state s_0*. The RAS subspace that is reachable under – or *admissible* by – some LES Δ will be denoted by $S_r(\Delta)$. Given a D/C-RAS configuration, an applied LES is characterized as *optimal*, if the corresponding admissible subspace is the *maximal* strongly connected component of S_r which contains the empty state s_0. The set of states admitted by the optimal LES, Δ^*, is characterized as (the set of) *reachable safe* states, and it is denoted by S_{rs}. The complement of S_{rs} with respect to S_r is denoted by S_{ru}, and it constitutes the system *reachable unsafe* region.

In the C/D-RAS operational context, the optimal LES, Δ^*, is well-defined, and it is effectively computable through an *one-step lookahead* scheme that admits a tentative resource allocation if and only if the resulting state is safe. However, the corresponding state safety problem is NP-complete (Araki *et al.*, 1977). In the light of this result, the research community has sought the development of sub-optimal LES that are implementable in polynomial complexity with respect to the underlying RAS size, and yet, efficient, i.e., they manage to admit a large part of S_{rs}. This idea has been formalized by the concept of *Polynomial Kernel (PK-)* LES (Reveliotis *et al.*, 2001). From an implementational standpoint, the design of PK-LES requires the identification of a property $\mathcal{H}(s)$, $s \in S$, s.t. (i) the complexity of testing $\mathcal{H}()$ on the RAS states is *polynomial with respect to the RAS size*; (ii) $\mathcal{H}(s_0)$ = TRUE; and (iii) the subspace $\{s \in S_r : H(s) = \texttt{TRUE}\}$ is strongly connected. Two conditions $\mathcal{H}()$ leading to correct PK-LES for D/C-RAS are provided in (Park and Reveliotis, 2001) and (Ezpeleta *et al.*, 2002).

3. D/C-RAS PERFORMANCE MODELLING AND CONTROL

The assumption of exponentially distributed processing times In order to study the performance of the D/C-RAS, the FSA model developed in Section 2 must be augmented with the timing aspects of the system operation. In the D/C-RAS model, these aspects are captured by the distributions D_{jk} characterizing the processing times experienced during the execution of stage Ξ_{jk}. For reasons that will become obvious in the subsequent analysis, it is assumed that D_{jk} is an *exponential* distribution, with parameter λ_{jk}. Although, technically, this assumption can be considered as a restriction of the model and its applicability, the derived results can still be applied in an approximate sense to D/C-RAS involving more general distributions D_{jk}, by substituting the relevant non-exponential stages by some appropriately structured acyclic process sub-nets modelling a *phase-type* distribution that approximates D_{jk} to some satisfactory level; the details of this approximation can be found in (Papadopoulos *et al.*, 1993).

The refined RAS state space Under the assumption of exponentially distributed processing times, the temporal aspects of the RAS behavior can be captured through the following refinement of the basic FSA model introduced in Section 2: (i) Each state component $s(q(j,k))$ is expanded to a 3-dimensional vector $[s^p(q(j,k)), s^f(q(j,k)), s^b(q(j,k))]$. Components $s^p(q(j,k))$ and $s^f(q(j,k))$ indicate the numbers of process instances of stage Ξ_{jk} that are, respectively, in *processing* or *finished*. Component $s^b(q(j,k))$ is a *binary*-valued component, with the value of 1 indicating that the advancement of the finished processes in stage Ξ_{jk} is currently *blocked*. Furthermore, an additional component $s^b(q(j,0))$ is added for each process type Π_j. This component is also *binary*-valued, and its value of 1 indicates that the loading of any further instances of process type Π_j is currently *blocked*. (ii) At any point in time, there should be at least one process instance under processing in the system, i.e., the system operation should be *globally non-idling*. The feasibility of this requirement is guaranteed by (a) the correctness of the applied LES, and (b) the further assumption of an *infinite* number of instances from each process type waiting for loading in the considered D/C-RAS; this last assumption is consistent with the posed objective of throughput maximization. (iii) The completion of an active process instance at a certain state immediately resets all the state components of $s^b()$ type to 0, i.e., it unblocks all the process loading and advancing events. (iv) A state containing unblocked loading and advancing events is followed by a sequence of *immediate* transitions each of which either (a) advances an unblocked finished process instance to its next stage, or (b) loads a new process instance of some unblocked process type, or (c) sets an $s^b()$ state component to its blocked state. Process instances j_j having completed a *"sink"* processing stage Ξ_{jk} in their corresponding graph $\mathcal{G}_j$ cannot be blocked, but they must be unloaded from the system. The advancement of any other process instance or the loading of a new process instance must be admissible by the applied LES.

The underlying Semi-Markov process Based on the above, states containing unblocked finished process instances or unblocked loading events present zero sojourn times, and therefore, they are characterized as *vanishing*. The remaining states containing only process instances that are in processing or blocked, have a finite sojourn time that is determined by the *"exponential race"* for completion of all the process instances that are in processing; i.e., the processing times of these states are exponentially distributed with a rate equal to $\sum_{(j,k)}[\lambda_{jk} \cdot s^p(q(j,k))]$. These states will be characterized as *tangible*. Let $S_{rv}(\Delta)$ and $S_{rt}(\Delta)$ denote respectively the sets of reachable vanishing and tangible states, under the supervision of LES Δ. Then, the RAS performance control problem can be expressed as *the determination of a probability distribution for each vanishing state, that will regulate the execution of the corresponding LES-admissible immediate events*. The resulting problem is a *semi-Markov decision process (semi-MDP)* (Puterman, 1994).

The induced CT-MDP The above problem formulation can be further simplified by focusing on the stochastic process which is induced by the vanishing states that result *directly* from tangible states, upon the finishing of some process instance. Let us denote this particular subset of vanishing states as $S^0_{rv}(\Delta)$. The possible state transitions out of $s \in S^0_{rv}(\Delta)$, to another state $s' \in S^0_{rv}(\Delta)$, are determined by (i) the set of immediate transition sequences σ emanating from s, and (ii) the transitions enabled in the tangible states s'' that result from the execution of these σ-sequences. More specifically, let $\Sigma(s)$ denote the set of all the immediate LES-admissible transition sequences associated with state $s \in S^0_{rv}(\Delta)$. Then, $\Sigma(s)$ defines the set of *options* associated with state s, since the enabling of the process loading and advancing events in each of these transition sequences is completely in the jurisdiction of the RAS controller. The set of states $s' \in S^0_{rv}(\Delta)$ that result from the selection of some option $\sigma \in \Sigma(s)$ at state s, is determined by the set of potential finishing events that are enabled in the tangible state s'' such that $s'' \stackrel{\sigma}{\leftarrow} s$; i.e., for any given option σ, the resulting states s' correspond to the finishing of some process instance of stage Ξ_{jk} with $(s'')^p(q(j,k)) > 0$, and the corresponding *transition probabilities* are determined by

$$p(s'|s,\sigma) = \frac{\lambda_{jk} \cdot (s'')^p(q(j,k))}{\sum_{(j',k')}[\lambda_{j'k'} \cdot (s'')^p(q(j',k'))]}$$

The *sojourn time* associated with the transition resulting from the selection of option σ at state s is exponentially distributed with *mean value*

$$\tau(s,\sigma) = \frac{1}{\sum_{(j,k)}[\lambda_{jk} \cdot (s'')^p(q(j,k))]}$$

where s'' is defined as above. Thus, by focusing on the $S^0_{rv}(\Delta)$ subspace, the original semi-MDP problem can be transformed to an equivalent *Continuous-Time MDP (CT-MDP)*.

Characterizing the objective function of the CT-MDP To completely define this CT-MDP problem, it remains to specify the *(immediate) gain* $g(s,\sigma)$ associated with each state-option pair (s,σ), and the *objective* under consideration. The characterization of these two issues is determined by our focus on the long-run system throughput as the performance criterion of interest. Hence, $g(s,\sigma)$ is defined as

$$g(s,\sigma) = \sum_{(j,k)} \Big(\frac{\lambda_{jk} \cdot (s'')^p(q(j,k))}{\sum_{(j',k')}[\lambda_{j'k'} \cdot (s'')^p(q(j',k'))]} \cdot I_{\{\Xi_{jk} \text{ is a "sink" stage of } \mathcal{G}_j\}} \Big)$$

i.e., it is equal to the probability that some process instance will be completed and unloaded during the transition resulting from the state-option pair (s,σ). The long-run system throughput obtained by initializing the RAS in some reachable LES-admissible state s^0 and applying some *policy* Γ for governing the selection of the options σ to be enabled at each state s, is characterized by the *average expected gain* $\bar{g}^\Gamma(s^0)$:

$$\bar{g}^\Gamma(s^0) = \liminf_{t\to\infty} \frac{1}{t} E^\Gamma_{s^0}\Big[\sum_{n=0}^{\infty} \int_0^t g(s_n,\sigma_n) \cdot \delta(t - t_n)\Big]$$

In the above equation, t_n denotes the time of the n-th transition in the induced CT Markov chain, assuming that the RAS operation is started at state s^0 in time $t_0 = 0$, and it is controlled by policy Γ. The expectation $E^\Gamma_{s^0}[\ ]$ is taken over all the corresponding sample paths of the aforementioned CT-Markov chain, and the function $\delta(t)$ is Kronecker's delta function. Our objective is to identify a policy Γ^* such that

$$\bar{g}^{\Gamma^*}(s) = \sup_\Gamma\{\bar{g}^\Gamma(s)\}, \quad \forall s \in S^0_{rv}(\Delta)$$

An LP formulation for the performance control problem The observation of the LES constraints during the construction of the state space of the aforementioned CT-MDP guarantees that there exists a deterministic policy that can take the RAS from any given state $s \in S^0_{rv}(\Delta)$ to any other state $s' \in S^0_{rv}(\Delta)$. Hence, the considered CT-MDP is a *communicating* model and the optimal average expected gain $\bar{g}^{\Gamma^*}(s)$ is constant for all initial states $s \in S^0_{rv}(\Delta)$ ((Puterman, 1994), Section 8.3). In the following, this constant optimal gain is denoted by $\bar{g}^*$. Its value and an optimal policy Γ^* can be obtained by solving the following *linear programming (LP)* formulation in variables $x(s,\sigma)$, $s \in S^0_{rv}(\Delta)$, $\sigma \in \Sigma(s)$ (Puterman, 1994):

$$\max \sum_{s \in S^0_{rv}(\Delta)} \sum_{\sigma \in \Sigma(s)} g(s,\sigma)x(s,\sigma) \quad (1)$$

s.t.

$$\forall s \in S^0_{rv}(\Delta), \quad \sum_{\sigma \in \Sigma(s)} x(s,\sigma) - \sum_{s' \in S^0_{rv}(\Delta)} \sum_{\sigma \in \Sigma(s')} p(s|s',\sigma)x(s',\sigma) = 0 \quad (2)$$

$$\sum_{s \in S^0_{rv}(\Delta)} \sum_{\sigma \in \Sigma(s)} \tau(s,\sigma)x(s,\sigma) = 1 \quad (3)$$

$$x(s,\sigma) \geq 0, \quad \forall s \in S^0_{rv}(\Delta),\ \forall \sigma \in \Sigma(s) \quad (4)$$

Variables $x(s,\sigma)$ can be interpreted as the *rates* with which the RAS finds itself in state s selecting option σ in that state. Therefore, the value of the optimal gain, $\bar{g}^*$, is equal to the optimal objective value for this formulation, and an optimal policy Γ^* is provided by a *stationary randomizing* policy, where the *option selection probabilities* $\xi_{\Gamma^*(s)}(\sigma)$, for selecting option σ at state s, are determined by

$$\xi_{\Gamma^*(s)}(\sigma) = \frac{x(s,\sigma)}{\sum_{\sigma' \in \Sigma(s)} x(s,\sigma')} \quad (5)$$

Incorporating additional throughput-ratio constraints In case of RAS with more than one process types, the definition of the process gains and the problem objective function provided above, imply that the ultimate objective is the *unconditional* maximization of the cumulative throughput across all process types. In most practical situations, the throughput with respect to the various process types will be required to observe some *ratio constraints*, i.e.,

$$TH_j = \rho_j TH_1, \quad \forall j \geq 2 \quad (6)$$

The resulting problem is a *constrained* CT-MDP problem ((Puterman, 1994), Section 8.9). In order to differentiate the throughput attained with respect to each process type, the constrained version of the problem will employ a set of gains $g_j(s,\sigma)$, one set for each process type Π_j, with

$$g_j(s,\sigma) = \sum_k \left(\frac{\lambda_{jk} \cdot (s'')^p(q(j,k))}{\sum_{(j',k')}[\lambda_{j'k'} \cdot (s'')^p(q(j',k'))]} \cdot I_{\{\Xi_{jk} \text{ is a "sink" stage of } \mathcal{G}_j\}} \right)$$

The long-term throughput $\bar{g}^*$ and an optimal policy Γ^* for the constrained version of the RAS performance control problem can be obtained by an LP formulation similar to that of Equations 1–4. The new objective function for this formulation is

$$\max \sum_{s \in S^0_{rv}(\Delta)} \sum_{\sigma \in \Sigma(s)} g_1(s,\sigma)x(s,\sigma) \quad (7)$$

and the constraint set is augmented with the following constraint that expresses the throughput ratio requirement of Equation 6:

$$\forall j \geq 2, \quad \sum_{s \in S^0_{rv}(\Delta)} \sum_{\sigma \in \Sigma(s)} g_j(s,\sigma)x(s,\sigma) - \rho_j \sum_{s \in S^0_{rv}(\Delta)} \sum_{\sigma \in \Sigma(s)} g_1(s,\sigma)x(s,\sigma) = 0 \quad (8)$$

4. THE THINNING PROBLEM

Problem statement In the context of the RAS modelling framework presented in Sections 2 and 3, the thinning problem is posed as the selection of a subset of edges, $\mathcal{E}^S_j$, from each process graph $\mathcal{G}_j$, with $|\mathcal{E}^S_j| = \mu_j$, such that the throughput of the resulting RAS is maximized. A detailed formal characterization of this general statement can be obtained as follows:

Consider the graph $\mathcal{G}_j$ associated with some process type Π_j of some D/C-RAS Φ, and a selection of edges from this graph, $\mathcal{E}^S_j \subseteq \mathcal{E}_j$. The subgraph $\mathcal{G}_j(\mathcal{E}^S_j)$ that encodes all the process plans for process type Π_j involving only transitions corresponding to edges in $\mathcal{E}^S_j$, can be computed by (i) considering the subgraph $\bar{\mathcal{G}}_j(\mathcal{E}^S_j)$ of $\bar{\mathcal{G}}_j$, defined by all the nodes v_{jk}, $k = 0, 1, \ldots, l(j)$, and the edges in $\mathcal{E}^S_j$; (ii) trimming $\bar{\mathcal{G}}_j(\mathcal{E}^S_j)$ with respect to node v_{j0} to obtain $\bar{\mathcal{G}}^{Tr}_j(\mathcal{E}^S_j)$; and (iii) deleting node v_{j0} and its incident edges from $\bar{\mathcal{G}}^{Tr}_j(\mathcal{E}^S_j)$. Then, given a D/C-RAS Φ and sets $\mathcal{E}^S_j \subseteq \mathcal{E}_j$ for each process type Π_j, let $\Phi(\mathcal{E}^S_1, \ldots, \mathcal{E}^S_n)$ denote the D/C-RAS obtained by substituting each process graph $\mathcal{G}_j$ by $\mathcal{G}_j(\mathcal{E}^S_j)$, in the definition of Φ. Of particular interest for the pursued formulation are all the D/C-RAS $\Phi(\mathcal{E}^S_1, \ldots, \mathcal{E}^S_n)$ for which (i) the edge sets $\mathcal{E}^S_j$ have $|\mathcal{E}^S_j| = \mu_j$, where $\mu_j \leq |\mathcal{E}_j|$ is externally defined, and (ii) $\mathcal{G}_j(\mathcal{E}^S_j) \neq NULL$, $\forall j$. In the following, the distinct edge selections defining these D/C-RAS will be denoted by $\{\mathcal{S}_1, \ldots, \mathcal{S}_N\}$, and the corresponding D/C-RAS by $\Phi(\mathcal{S}_u)$, $u = 1, \ldots, N$; obviously $N \leq \prod_{j=1}^n \binom{|\mathcal{E}_j|}{\mu_j}$.

In the context of this formalism, the thinning problem can be posed as follows: Given a D/C-RAS Φ, a correct LES Δ, target throughput ratios ρ_j for $j = 2, \ldots, |\mathcal{P}|$,[1] and parameters $\mu_j \leq |\mathcal{E}_j|$, $j = 1, \ldots, |\mathcal{P}|$, find an edge selection $\mathcal{S}^*$ such that

$$\mathcal{S}^* = \arg\max_{\mathcal{S}}\{\bar{g}_1^*(\Phi(\mathcal{S}); \Delta)\} \quad (9)$$

s.t.

$$|\mathcal{E}_j^S| = \mu_j \ \wedge \ \mathcal{G}_j(\mathcal{E}_j^S) \neq NULL, \ \forall \mathcal{E}_j^S \in \mathcal{S} \quad (10)$$

A MIP formulation The maximal throughput $\bar{g}_1^*(\Phi(\mathcal{S}); \Delta)$, for D/C-RAS $\Phi(\mathcal{S})$, can be computed through the formulation of Equations 7, 2–4 and 8, by forcing to zero every variable $x(s,\sigma)$ that corresponds to a transition leading to a state s' which does not belong to $S_r(\Phi(\mathcal{S}); \Delta)$, the reachable Δ-admissible subspace for the D/C-RAS $\Phi(\mathcal{S})$. In the light of Equation 3, this requirement of forcing to zero every variable $x(s,\sigma)$ that corresponds to a transition leading to a state $s' \notin S_r(\Phi(\mathcal{S}); \Delta)$, can be expressed by the addition of the following constraint in the LP formulation of Equations 7, 2–4 and 8:

$$\forall s \in S_{rv}^0(\Delta), \ \forall \sigma \in \Sigma(s), \ \ x(s,\sigma) \leq \quad (11)$$
$$\{\min_{(s,\sigma):\ s \in S_{rv}^0(\Delta),\ \sigma \in \Sigma(s)} \tau(s,\sigma)\}^{-1} I_{\{s' \in S_r(\Phi(\mathcal{S});\Delta)\}}$$

Equation 11 subsequently suggests the following formulation of the thinning problem: Consider an enumeration $\{\mathcal{S}_1, \ldots, \mathcal{S}_N\}$ of all the edge selections satisfying Constraint 10, and associate a *binary* variable I_u with each selection $\mathcal{S}_u$, such that $I_u = 1$ indicates that the finally selected edge sets $\mathcal{E}_j^S$ are those in selection $\mathcal{S}_u$. Obviously,

$$\sum_u I_u = 1 \quad (12)$$

In addition,

$$\forall s \in S_{rv}^0(\Delta), \ \forall \sigma \in \Sigma(s), \ \ x(s,\sigma) \leq \quad (13)$$
$$\{\min_{(s,\sigma):s \in S_{rv}^0(\Delta),\sigma \in \Sigma(s)} \tau(s,\sigma)\}^{-1} \sum_{u: s' \in S_r(\Phi(\mathcal{S}_u);\Delta)} I_u$$

i.e., the transition corresponding to the selection of option σ in state s can be activated in the final solution, only if this solution engages an edge selection $\mathcal{S}_u$ such that the resulting state s' is in $S_r(\Phi(\mathcal{S}_u); \Delta)$. The complete formulation of the considered thinning problem is defined by Equations 7, 2, 3 4, 13, 12, and the additional requirement that I_u is a binary variable, for all u. This formulation is a *Mixed Integer Program (MIP)* and, in principle, it can be solved through available commercial solvers. However, notice that the size of both variable sets, $x(s,\sigma)$ and I_u, is a super-polynomial function of $|\Phi|$ and, in general, these sets will grow very fast. Therefore, for practical purposes, it is important to develop alternative approximating schemes to it; this is part of our current investigations.

[1] In the case of D/C-RAS with a single process type, this data set is vacuous, and the unconstrained version of the throughput maximization problem must be employed.

5. CONCLUSIONS

This paper introduced the "thinning" problem and provided a detailed formal characterization of it. The derived formulation takes the form of a mixed integer program, and, in principle, it can be solved through well-developed computational algorithms. A severe limitation of the approach, though, is that the number of variables and constraints grows super-polynomially with respect to the underlying RAS size. Therefore, future work will seek to exploit the derived formulation towards the development of efficient approximating solution algorithms. An additional contribution of the presented results is the employment of the CT-MDP modelling framework for the systematic characterization of the interaction of logical and performance control problems arising in the operational context of D/C-RAS.

REFERENCES

Araki, T., Y. Sugiyama and T. Kasami (1977). Complexity of the deadlock avoidance problem. In: *2nd IBM Symp. on Mathematical Foundations of Computer Science*. IBM. pp. 229–257.

Ezpeleta, J., F. Tricas, F. Garcia-Valles and J. M. Colom (2002). A banker's solution for deadlock avoidance in fms with flexible routing and multi-resource states. *IEEE Trans. on R&A* **18**, 621–625.

Hopcroft, J. E. and J. D. Ullman (1979). *Introduction to Automata Theory, Languages and Computation*. Addison-Wesley. Reading, MA.

Papadopoulos, H. T., C. Heavy and J. Browne (1993). *Queueing Theory in Manufacturing Systems Analysis and Design*. Chapman & Hall. New York, NY.

Park, J. and S. A. Reveliotis (2001). Deadlock avoidance in sequential resource allocation systems with multiple resource acquisitions and flexible routings. *IEEE Trans. on Automatic Control* **46**, 1572–1583.

Puterman, M. L. (1994). *Markov Decision Processes: Discrete Stochastic Dynamic Programming*. John Wiley & Sons.

Reveliotis, S. A., M. A. Lawley and P. M. Ferreira (2001). Structural control of large-scale flexibly automated manufacturing systems. In: *The Design of Manufacturing Systems* (C. T. Leondes, Ed.). pp. 4-1 – 4-34. CRC Press.

ELSEVIER
IFAC
PUBLICATIONS
www.elsevier.com/locate/ifac

A COMPOSITION ORIENTED PERSPECTIVE ON CONTROLLABILITY OF LARGE SCALE DES

Florian Wenck, Jan H. Richter

Department of Process Automation, Technische Universität Hamburg-Harburg, Schwarzenbergstraße 95, 21073 Hamburg, Germany
wenck@tu-harburg.de

Abstract: In this paper the concept of controllability is reviewed within the context of modularly designed and modelled large-scale discrete-event systems. Modular methods are considered the natural way of modelling real manufacturing systems, for example. The problem of conserving controllability under compositional operators is introduced. Sufficient conditions for the conservation of controllability under composition are presented. *Copyright © 2004 IFAC*

Keywords: Discrete-event systems, controllability, large-scale systems, system analysis, automation

1. INTRODUCTION

Discrete-event systems (DES) are characterized by the occurrence of discrete state transitions induced by events. Large-scale DES are often treated using modular approaches, mainly for two reasons. First, reduction in computational complexity is desired. Second, for large-scale systems, modular plant modelling is the only feasible approach. Regarding their underlying physical structure, real complex systems in manufacturing should be treated as a composition of subsystems, such as shops or workcells.

The computational complexity of supervisor design increases exponentially with the number of modules present in a global plant. The complexity issue has been examined in detail by Gohari and Wonham (2000). The original modular approach proposed by Ramadge and Wonham (1988) introduces the concept of modular design to allow for modular specifications, defined on the global plant. Hence the global plant has to be expressed explicitly for the computation of supremal controllable sublanguages. The approach is refined in (de Queiroz and Cury, 2000*a*) and (de Queiroz and Cury, 2000*b*), who introduce the concept of local modularity to avoid the explicit computation of the global plant.

The notion of controllability for specification languages is key to the solution of supervisory control problems. Composition operators are the basis of any module-oriented system representation. The focus of this work is on the conservation of controllability under the application of compositional operators.

The remainder of this paper is organized as follows. Section 2 lays the foundations for the study. A collection of compositional operators is introduced and explained in Section 3. In Section 4, the meaning of conserving controllability under composition is explained and results are provided for some of the operators. Section 5 illustrates these ideas by an example. Finally, a conclusion is made in Section 6.

2. PRELIMINARIES

In this work, DES are treated using the Supervisory Control Theory (SCT) framework introduced by Ramadge and Wonham (1989). Plants are modeled as deterministic finite automata (dfa).

Let Σ denote a finite alphabet of events and the Kleene-closure Σ^* the set of finite event chains including the empty chain ε. The prefix closure of a language $L \subseteq \Sigma^*$ is denoted by $\overline{L}$.

A plant model dfa is represented by a generator $G = (X, \Sigma, \delta, x_0)$ with state set X, event alphabet Σ, partial transition function $\delta : X \times \Sigma \to X$ and initial state x_0. Marker states are not relevant in the context of this paper and thus omitted. The automaton generates the regular language $L(G)$, expressing the uncontrolled plant behavior. The event alphabet is partitioned into the subsets of controllable and uncontrollable events: $\Sigma = \Sigma_c \cup \Sigma_u$. A specification language $K \subseteq L(G)$ represents the legal plant behavior.

The standard criterion for controllability of a pair (K, G) is used: K is controllable with respect to G if and only if

$$\overline{K}\Sigma_u \cap L(G) \subseteq \overline{K}. \tag{1}$$

Concerning the composition of automata representing DES, the classification of operators mentioned by Heymann (1990) is roughly followed. Some of the operators defined there are equivalent to one of Wonham's (2002) synchronous product, shuffle and meet, which could be used as a basis as well. To get all the operators presented in (Heymann, 1990) as special cases of Wonham's (2002) system of operators, the event sets of the respective generators have to be changed from case to case. This renders a comparison between operators more difficult. A review of compositional operators based on Heymann (1990) follows.

3. COMPOSITION OF DES

In the SCT context, composition means synchronization of deterministic finite automata. The basis for the definition of all following composition operators are two Generators $G_1 = (X_1, \Sigma_1, \delta_1, x_{01})$ and $G_2 = (X_2, \Sigma_2, \delta_2, x_{02})$ with disjoint state sets $X_1 \cap X_2 = \emptyset$ but generally overlapping event sets. The result of any composition is a generator $G = G_1 ||_{(\cdot)} G_2 = (X, \Sigma, \delta, x_0)$ with the state set $X = X_1 \times X_2$, event set $\Sigma = \Sigma_1 \cup \Sigma_2$ and initial state $x_0 = (x_{01}, x_{02})$, where $(\cdot)$ is one of {PSC, BSC, PSPC, SYPC, LPC, SPC}, defined below. Each operator is defined by a distinct transition function presented subsequently, with $\sigma \in \Sigma$ a single event, and $x \in X$ a state. The standard notation $\delta(x, \sigma)!$ is used to denote that $\delta(x, \sigma)$ is defined. The notation $\neg\delta(x, \sigma)!$ embodies the opposite.

Definition 3.1 (Prioritized Synchronous Comp. (PSC)) For each of G_1 and G_2, a priority set $\Psi_1, \Psi_2 \subseteq (\Sigma_1 \cup \Sigma_2)$ is defined.

$$\delta(x, \sigma) =$$

$$\begin{cases} \delta_1(x_1,\sigma) \times \delta_2(x_2,\sigma) & \text{if } \delta_1(x_1,\sigma)! \\ & \wedge\, \delta_2(x_2,\sigma)! \\ \delta_1(x_1,\sigma) \times \{x_2\} & \text{if } \delta_1(x_1,\sigma)! \\ & \wedge \neg\delta_2(x_2,\sigma)! \wedge \sigma \notin \Psi_2 \\ \{x_1\} \times \delta_2(x_2,\sigma) & \text{if } \neg\delta_1(x_1,\sigma)! \\ & \wedge\, \delta_2(x_2,\sigma)! \wedge \sigma \notin \Psi_1 \\ \emptyset & \text{otherwise.} \end{cases}$$

In PSC, each generator insists on participating in the execution of every event in its priority set. Such an event can only occur in that generator or in both simultaneously. The composed generator language $L(G)$ is a subset of the Kleene-closure of the source alphabets' union: $L(G) \subset (\Sigma_1 \cup \Sigma_2)^*$. PSC, introduced by Heymann (1990), is the most general compositional operator considered here.

Definition 3.2 (Biased Synchronous Comp. (BSC)) BSC is obtained as a specialization of PSC by setting $\Psi_1 = \emptyset$, $\Psi_2 = \Sigma_1 \cup \Sigma_2$. The generator G_2 is called master, G_1 is called follower.

$$\delta(x, \sigma) =$$

$$\begin{cases} \delta_1(x_1,\sigma) \times \delta_2(x_2,\sigma) & \text{if } \delta_1(x_1,\sigma)! \\ & \wedge\, \delta_2(x_2,\sigma)! \\ \{x_1\} \times \delta_2(x_2,\sigma) & \text{if } \neg\delta_1(x_1,\sigma)! \\ & \wedge\, \delta_2(x_2,\sigma)! \\ \emptyset & \text{otherwise.} \end{cases}$$

In BSC, introduced by Lafortune and Chen (1990), the master (a button, for example) may execute state transitions whenever it can, completely independent from the follower. The follower (a valve, for example) moves synchronously, if possible. The language generated by G is exactly the master generator language: $L(G) = L(G_2)$, see (Lafortune and Chen, 1990).

Definition 3.3 (Partially Strict Product Comp. (PSPC)) Another specialization of PSC is obtained from defining a single common priority set $\Psi_1 = \Psi_2 = \Psi \subseteq (\Sigma_1 \cup \Sigma_2)$.

$$\delta(x, \sigma) =$$

$$\begin{cases} \delta_1(x_1,\sigma) \times \delta_2(x_2,\sigma) & \text{if } \delta_1(x_1,\sigma)! \\ & \wedge\, \delta_2(x_2,\sigma)! \\ \delta_1(x_1,\sigma) \times \{x_2\} & \text{if } \delta_1(x_1,\sigma)! \\ & \wedge \neg\delta_2(x_2,\sigma)! \wedge \sigma \notin \Psi \\ \{x_1\} \times \delta_2(x_2,\sigma) & \text{if } \neg\delta_1(x_1,\sigma)! \\ & \wedge\, \delta_2(x_2,\sigma)! \wedge \sigma \notin \Psi \\ \emptyset & \text{otherwise.} \end{cases}$$

PSPC, mentioned in (Heymann, 1990), enforces the generators G_1, G_2 into lockstep regarding events in Ψ only. The language generated by G is a subset of the Kleene-closure of the source alphabets' union: $L(G) \subset (\Sigma_1 \cup \Sigma_2)^*$.

Definition 3.4 (Synchronous Product Comp. (SYPC)) PSPC becomes Wonham's synchronous product when $\Psi = \Sigma_1 \cap \Sigma_2$, assuming $\Sigma_1 \cap \Sigma_2 \neq \emptyset$.

$$\delta(x,\sigma) =$$

$$\begin{cases} \delta_1(x_1,\sigma)\times\delta_2(x_2,\sigma) & \text{if } \delta_1(x_1,\sigma)! \\ & \wedge\, \delta_2(x_2,\sigma)! \\ \delta_1(x_1,\sigma)\times\{x_2\} & \text{if } \delta_1(x_1,\sigma)! \\ \wedge\neg\delta_2(x_2,\sigma)! \wedge \sigma\notin(\Sigma_1\cap\Sigma_2) & \\ \{x_1\}\times\delta_2(x_2,\sigma) & \text{if } \neg\delta_1(x_1,\sigma)! \\ \wedge\,\delta_2(x_2,\sigma)! \wedge \sigma\notin(\Sigma_1\cap\Sigma_2) & \\ \emptyset & \text{otherwise.} \end{cases}$$

For SYPC, $L(G) = P_1^{-1}L(G_1) \cap P_2^{-1}L(G_2)$ (see (Wonham, 2002)), where P_i denotes the natural projection onto Σ_i, $i = 1, 2$.

Definition 3.5 (Loose Product Comp. (LPC))
The loose product composition is a specialization of PSPC by setting $\Psi = \emptyset$.

$$\delta(x,\sigma) =$$

$$\begin{cases} \delta_1(x_1,\sigma)\times\delta_2(x_2,\sigma) & \text{if } \delta_1(x_1,\sigma)! \\ & \wedge\,\delta_2(x_2,\sigma)! \\ \delta_1(x_1,\sigma)\times\{x_2\} & \text{if } \delta_1(x_1,\sigma)! \\ & \wedge\neg\delta_2(x_2,\sigma)! \\ \{x_1\}\times\delta_2(x_2,\sigma) & \text{if } \neg\delta_1(x_1,\sigma)! \\ & \wedge\,\delta_2(x_2,\sigma)! \\ \emptyset & \text{otherwise.} \end{cases}$$

LPC allows the generators G_1, G_2 to run concurrently. The language generated by G is a subset of the Kleene-closure of the source alphabets' union: $L(G) \subseteq (\Sigma_1 \cup \Sigma_2)^*$.

Definition 3.6 (Strict Product Comp. (SPC))
The strict product composition is a specialization of PSPC by setting $\Psi = \Sigma_1 \cup \Sigma_2$, that is, by synchronizing on all events.

$$\delta(x,\sigma) =$$

$$\begin{cases} \delta_1(x_1,\sigma)\times\delta_2(x_2,\sigma) & \text{if } \delta_1(x_1,\sigma)! \\ & \wedge\,\delta_2(x_2,\sigma)! \\ \emptyset & \text{otherwise.} \end{cases}$$

SPC forces the generators G_1, G_2 into complete lock-step operation. The language generated by G is simply the intersection of the source languages: $L(G) = L(G_1) \cap L(G_2)$. It is equivalent to Wonham's meet operator if $\Sigma_1 = \Sigma_2$. Note that $\Sigma_1 \cap \Sigma_2 = \emptyset$ results in an empty generator, thus $\Sigma_1 \cap \Sigma_2 \neq \emptyset$ is assumed.

The operators were introduced by repeated specialization from PSC. They are all special cases of PSC. This can be expressed by introducing an embracing relation on the set of composition operators. The relation then induces a partial order on the set of the above defined operators, which is depicted as a Hasse diagram in Figure 1 (Kowalewski, 1996, p. 174).

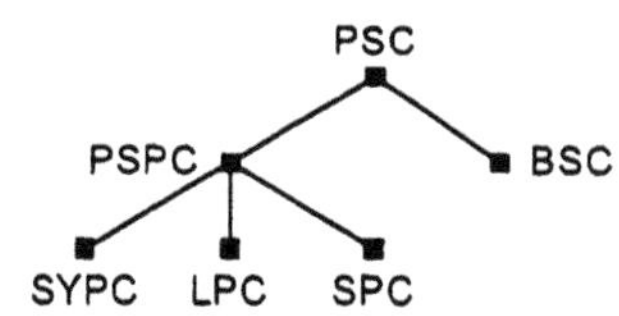

Figure 1. Hasse diagram illustrating the partial order of composition operators.

4. CONSERVATION OF CONTROLLABILITY UNDER COMPOSITION

In Section 4.1, the problem of the conservation of controllability under the application of compositional operators is further specified. In Sections 4.3 and 4.4, sufficient conditions are given for the controllability of the global plant specification, which is defined in Sections 4.1 and 4.2.

4.1 Problem Definition

It is known from the general SCT (Wonham, 2002) that the set of all controllable sublanguages $C(L(G))$ of a given language $L(G)$ is closed under arbitrary unions: $(K_1 \in C(L(G))) \wedge (K_2 \in C(L(G))) \Rightarrow K_1 \cup K_2 \in C(L(G))$. The same holds for intersections, provided that K_1 and K_2 are non-conflicting: $(\overline{K_1 \cap K_2} = \overline{K}_1 \cap \overline{K}_2) \wedge (K_1 \in C(L(G))) \wedge (K_2 \in C(L(G))) \Rightarrow K_1 \cap K_2 \in C(L(G))$. In any case, the original specification languages and their unions and intersections are controllable with respect to a common, distinct language $L(G)$.

Now consider the situation where G is a composed system, consisting of two components ($n = 2$). K_1 and K_2 are defined on modules G_1 and G_2, respectively. One possible way of analyzing the controllability of unions and intersections of modular specifications in that context would be to check the controllability of the modular specifications against the composed plant. In case of successful testing the above results about unions and intersections apply immediately. This approach however involves the explicit computation of both the composed system and the combined specifications.

The objective of this paper is to avoid, when possible, the explicit computation of the composed system and the combined specification in order to judge the controllability of the union or intersection of the modular specifications. For this purpose the following common assumptions are introduced.

Let G_1, G_2 be generator modules as defined in Section 3, and G be the composed generator. Let $K_1 \subseteq L(G_1)$, $K_2 \subseteq L(G_2)$ be controllable specifications local to G_1 and G_2:

$$\overline{K}_1\Sigma_{u1} \cap L(G_1) \subseteq \overline{K}_1 \qquad (2)$$
$$\overline{K}_2\Sigma_{u2} \cap L(G_2) \subseteq \overline{K}_2$$

Let possible combinations of modular specifications be intersection $K = K_1 \cap K_2$ and union $K = (K_1 \cup K_2) \cap L(G)$. Intersecting two different plants' local specifications means that when considered together, they may only perform common tasks. Taking the union respectively means that they may each perform any specified operation individually and independently from each other. The restriction to $L(G)$ is necessary for the specification to be consistent with the context of the composed system.

Problem 1. Given $n = 2$ local specifications K_1, K_2, locally controllable with respect to their $n = 2$ generator modules G_1, G_2 as noted above, decide if their intersection and union are globally controllable with respect to the composed plant $G = G_1||_{(\cdot)}G_2$, with the composition operator $||_{(\cdot)}$ as in Section 3.

Sufficient conditions to ensure the controllability of the combined specification are provided below for the SPC and BSC operators. A brief discussion of the four remaining operators follows.

All subsequent proofs are based on basic set operation properties like distributivity of concatenation, union and intersection and the use of identities like $\overline{K_1 \cap K_2} \subseteq \overline{K}_1 \cap \overline{K}_2$.

4.2 Uncontrollable Event Alphabet

To derive the following statements, exact analytic expressions for the uncontrollable event alphabets $\Sigma_{\mathrm{u}}^{(\cdot)}$, $(\cdot) \in \{\mathrm{BSC}, \mathrm{SPC}\}$ of the resulting generator language $L(G)$ are necessary for both used compositional operators. The estimate $\Sigma_{\mathrm{u}} \subseteq (\Sigma_{\mathrm{u1}} \cup \Sigma_{\mathrm{u2}})$ is not accurate enough. The result is operator dependent.

For BSC, the result

$$\Sigma_{\mathrm{u}}^{\mathrm{BSC}} = \Sigma_{\mathrm{u2}} \tag{3}$$

with G_2 as master is self-explanatory. For SPC, consider that only common events can be part of the resulting language $L(G)$. Common events can be all controllable or all uncontrollable in both respective component systems, or controllable in one component and uncontrollable in the other. In the latter case, controllability is assumed to win over uncontrollability. Thus $\Sigma_{\mathrm{u}} = (\Sigma_{\mathrm{u1}} - \Sigma_{\mathrm{c2}}) \cap (\Sigma_{\mathrm{u2}} - \Sigma_{\mathrm{c1}})$, with the first bracketed term marked by horizontal stripes, the second marked by vertical stripes in Figure 2. By inspection it is clear that the result is

$$\Sigma_{\mathrm{u}}^{\mathrm{SPC}} = \Sigma_{\mathrm{u1}} \cap \Sigma_{\mathrm{u2}}. \tag{4}$$

4.3 Strict Product Composition (SPC)

As mentioned in Definition 3.6, $G = G_1||_{\mathrm{SPC}}G_2$, $L(G) = L(G_1) \cap L(G_2)$, and by Section 4.2, $\Sigma_{\mathrm{u}}^{\mathrm{SPC}} = \Sigma_{\mathrm{u1}} \cap \Sigma_{\mathrm{u2}}$.

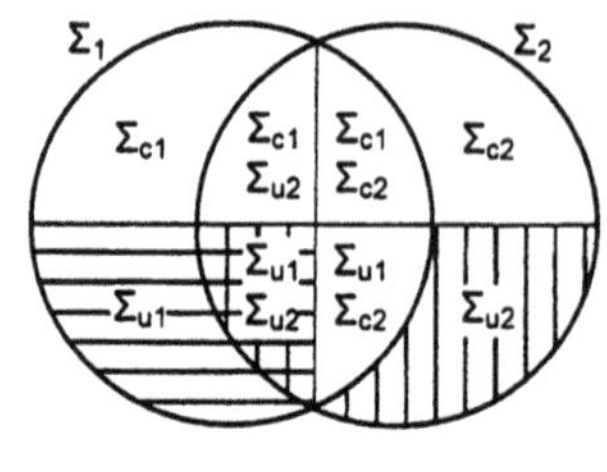

Figure 2. Venn diagram illustrating the deduction of the uncontrollable event alphabet $\Sigma_{\mathrm{u}}^{\mathrm{SPC}}$.

Proposition 4.1 ($K = K_1 \cap K_2$ for SPC)
K is controllable w. r. t. G ($\overline{K}\Sigma_{\mathrm{u}}^{\mathrm{SPC}} \cap L(G) \subseteq \overline{K}$) if

(1) $\overline{K_1 \cap K_2} = \overline{K}_1 \cap \overline{K}_2$, i.e. K_1, K_2 are non-conflicting

PROOF.

$$\begin{aligned}
& \overline{K}\Sigma_{\mathrm{u}}^{\mathrm{SPC}} \cap L(G) && (5)\\
\subseteq\ & (\overline{K}_1 \cap \overline{K}_2)(\Sigma_{\mathrm{u1}} \cap \Sigma_{\mathrm{u2}}) \cap (L(G_1) \cap L(G_2)) \\
=\ & \underbrace{(\overline{K}_1(\Sigma_{\mathrm{u1}} \cap \Sigma_{\mathrm{u2}}) \cap L(G_1))}_{\subseteq \overline{K}_1} \\
& \cap \underbrace{(\overline{K}_2(\Sigma_{\mathrm{u1}} \cap \Sigma_{\mathrm{u2}}) \cap L(G_2))}_{\subseteq \overline{K}_2} \\
\subseteq\ & \overline{K}_1 \cap \overline{K}_2 \\
=\ & \overline{K_1 \cap K_2} \text{ if } K_1, K_2 \text{ are non-conflicting} \\
=\ & \overline{K}
\end{aligned}$$

The condition in Proposition 4.1 is used for the last but one equation. All other steps are based on well-known identities. □

Proposition 4.2 ($K = (K_1 \cup K_2) \cap L(G)$ for SPC)
K is controllable w. r. t. G ($\overline{K}\Sigma_{\mathrm{u}}^{\mathrm{SPC}} \cap L(G) \subseteq \overline{K}$) if

(1) $\overline{K_1 \cap L(G)} = \overline{K}_1 \cap \overline{L}(G)$, i.e. K_1, $L(G)$ are non-conflicting
(2) $\overline{K_2 \cap L(G)} = \overline{K}_2 \cap \overline{L}(G)$, i.e. K_2, $L(G)$ are non-conflicting

PROOF.

$$\begin{aligned}
& \overline{K}\Sigma_{\mathrm{u}}^{\mathrm{SPC}} \cap L(G) && (6)\\
\subseteq\ & [(\overline{K}_1 \cup \overline{K}2) \cap \overline{L(G)}](\Sigma_{\mathrm{u1}} \cap \Sigma_{\mathrm{u2}}) \\
& \cap L(G_1) \cap L(G_2) \\
=\ & \underbrace{[\overline{K}_1 \cap \overline{L(G)}](\Sigma_{\mathrm{u1}} \cap \Sigma_{\mathrm{u2}}) \cap L(G_1) \cap L(G_2)}_{a)} \\
& \cup \underbrace{[\overline{K}_2 \cap \overline{L(G)}](\Sigma_{\mathrm{u1}} \cap \Sigma_{\mathrm{u2}}) \cap L(G_1) \cap L(G_2)}_{b)}
\end{aligned}$$

Term a) simplifies to $\overline{K}_1(\Sigma_{\mathrm{u1}} \cap \Sigma_{\mathrm{u2}}) \cap \overline{L(G)}(\Sigma_{\mathrm{u1}} \cap \Sigma_{\mathrm{u2}}) \cap L(G_1) \cap L(G_2) = \overline{K}_1(\Sigma_{\mathrm{u1}} \cap \Sigma_{\mathrm{u2}}) \cap \underline{L(G_1)} \cap \underline{\overline{L(G)}(\Sigma_{\mathrm{u1}} \cap \Sigma_{\mathrm{u2}}) \cap L(G_1) \cap L(G_2)} \subseteq \overline{K}_1 \cap \overline{L(G)} = \overline{K_1 \cap L(G)}$. The last equation is based on the first condition in Proposition 4.2. By a similar argument and using the second condition in Proposition 4.2, term b) simplifies to $\overline{K}_2(\Sigma_{u1} \cap \Sigma_{u2}) \cap \overline{L(G)}(\Sigma_{u1} \cap \Sigma_{u2}) \cap L(G_1) \cap L(G_2) = \overline{K}_2(\Sigma_{u1} \cap \Sigma_{u2}) \cap L(G_1) \cap \overline{L(G)}(\Sigma_{u1} \cap \Sigma_{u2}) \cap L(G_1) \cap L(G_2) \subseteq \overline{K}_2 \cap \overline{L(G)} = \overline{K_2 \cap L(G)}$. Then, Equation 6 continues as $\overline{K}\Sigma_u \cap L(G) \subseteq \overline{(K_1 \cap L(G))} \cup \overline{(K_2 \cap L(G))} = \overline{(K_1 \cup K_2) \cap L(G)} = \overline{K}$, as required. □

The non-conflicting test in Proposition 4.1 involves the language K explicitly, however not the global plant. The non-conflicting test in Proposition 4.2 involves the language $L(G)$ explicitly, however not the combined specification. Both propositions have a benefit: The computation of K for intersection is easy, and the global generator need not be computed. For union, the computation of the global generator is necessary but the combined specification need not be computed in addition.

4.4 Biased Synchronous Composition (BSC)

Here, $G = G_1 ||_{BSC} G_2$, $L(G) = L(G_2)$ by Definition 3.2, $\Sigma_u^{BSC} = \Sigma_{u2}$ by Section 4.2.

Proposition 4.3 ($K = K_1 \cap K_2$ for BSC)
K is controllable w. r. t. G ($\overline{K}\Sigma_u^{BSC} \cap L(G) \subseteq \overline{K}$) if

(1) $\overline{K}_1\Sigma_{u2} \cap L(G_2) \subseteq \overline{K}_1$, i. e. K_1 is controllable w. r. t. G_2.
(2) $\overline{K_1 \cap K_2} = \overline{K}_1 \cap \overline{K}_2$, i. e. K_1, K_2 are non-conflicting

PROOF.

$$\begin{aligned} & \overline{K}\Sigma_u^{BSC} \cap L(G) \\ \subseteq \quad & (\overline{K}_1 \cap \overline{K}_2)\Sigma_{u2} \cap L(G_2) \\ = & \underbrace{(\overline{K}_1\Sigma_{u2} \cap L(G_2))}_{a)} \cap \underbrace{(\overline{K}_2\Sigma_{u2} \cap L(G_2))}_{b)} \end{aligned} \tag{7}$$

Term b) simplifies to $\overline{K}_2\Sigma_{u2} \cap L(G_2) \subseteq \overline{K}_2$ immediately. Term a) is included in $\overline{K}_1$ if K_1 is controllable w. r. t. G_2, the first condition in Proposition 4.3. Thus Equation 7 continues as $\overline{K}\Sigma_u \cap L(G) \subseteq \overline{K}_1 \cap \overline{K}_2 = \overline{K_1 \cap K_2} = \overline{K}$ if K_1, K_2 are non-conflicting, the second condition in Proposition 4.3, which completes the proof. □

Proposition 4.4 ($K = (K_1 \cup K_2) \cap L(G)$ for BSC)
K is controllable w. r. t. G ($\overline{K}\Sigma_u^{BSC} \cap L(G) \subseteq \overline{K}$) if

(1) $\overline{K}_1\Sigma_{u2} \cap L(G_2) \subseteq \overline{K}_1$, i. e. K_1 is controllable w. r. t. G_2.
(2) $\overline{K_1 \cap L(G_2)} = \overline{K}_1 \cap \overline{L}(G_2)$, i. e. K_1, $L(G_2)$ are non-conflicting

Note for the following proof that here $K = (K_1 \cap L(G)) \cup (K_2 \cap L(G)) = (K_1 \cap L(G_2)) \cup K_2$.

PROOF.

$$\begin{aligned} & \overline{K}\Sigma_u^{BSC} \cap L(G) \\ & \subseteq [(\overline{K}_1 \cap \overline{L}(G_2)) \cup \overline{K}_2]\Sigma_{u2} \cap L(G_2) \\ = \quad & \underbrace{[(\overline{K}_1 \cap \overline{L}(G_2))\Sigma_{u2} \cap L(G_2)]}_{a)} \\ & \cup \underbrace{[\overline{K}_2\Sigma_{u2} \cap L(G_2)]}_{b)} \end{aligned} \tag{8}$$

Term b) simplifies to $\overline{K}_2\Sigma_{u2} \cap L(G_2) \subseteq \overline{K}_2$ immediately. Term a) continues as $\overline{K}_1\Sigma_{u2} \cap \overline{L}(G_2)\Sigma_{u2} \cap L(G_2) = (\overline{K}_1\Sigma_{u2} \cap L(G_2)) \cap (\overline{L}(G_2)\Sigma_{u2} \cap L(G_2)) \subseteq \overline{K}_1 \cap \overline{L}(G_2)$ if K_1 is controllable w. r. t. G_2, the first condition in Proposition 4.4. Now Equation 8 continues as $\overline{K}\Sigma_u \cap L(G) \subseteq (\overline{K}_1 \cap \overline{L}(G_2)) \cup \overline{K}_2 = \overline{(K_1 \cap L(G_2)) \cup K_2} = \overline{K}$ if $(K_1, L(G_2))$ is non-conflicting, the second condition in Proposition 4.4, which completes the proof. □

The benefit for intersection of specifications is the same as for SPC, saving the explicit computation of $L(G)$. For union, the situation is also similar to that of SPC, but the computation of $L(G)$ is especially simple.

4.5 Remaining operators: PSC, PSPC, SYPC, LPC

Explicit and useful analytic expressions for the resulting languages $L(G)$ from the application of the compositional operators PSC, PSPC, and LPC are not easily found. Such expressions were key to the above deductions for SPC and BSC. For this reason no such results are presented here. This is a problem for further consideration.

Remark 1. The case $K = K_1 ||_{SYPC} K_2$, $G = G_1 ||_{SYPC} G_2$ was treated in (de Queiroz and Cury, 2000*a*) for the special case $\Sigma_1 \cap \Sigma_2 = \emptyset$. It was shown that K is controllable w. r. t. G iff K_i is controllable w. r. t. G_i, $i = 1, 2$.

5. EXAMPLE

The ideas developed above are now illustrated by an example. Consider two plant generators $G_1 = (X_1, \Sigma_1, \delta_1, x_{01})$, $G_2 = (X_2, \Sigma_2, \delta_2, x_{02})$ with $\Sigma_1 = \{\alpha, \beta, \gamma, \delta, r\}$, $\Sigma_2 = \{\alpha, \beta, \gamma, r\}$, $\Sigma_{u1} = \Sigma_{u2} = \{\beta\}$, state set and transition structures as in Figure 3. Given are locally controllable specifications $K_1 = \{\alpha\beta, \alpha\delta r\}$ and $K_2 = \{\alpha\beta, \alpha\gamma\}$. This results in $K_1 \cap K_2 = \{\alpha\beta\}$, $K_1 \cup K_2 = \{\alpha\beta, \alpha\gamma, \alpha\delta r\}$.

The composed generator G_{SPC} is shown in Figure 4(a). Observe that G_{SPC} is blocking, but would not be so if state c of G_2 were marked, and that here $\Psi = \Sigma_1$. The intersected specification $K = K_1 \cap$

$K_2 = \{\alpha\beta\}$ is controllable if the condition of Proposition 4.1 holds: 4.1 – (1) K_1, K_2 are non-conflicting as $\overline{K}_1 \cap \overline{K}_2 = \underline{\{\varepsilon, \alpha, \alpha\beta, \alpha\delta, \alpha\delta r\} \cap \{\varepsilon, \alpha, \alpha\beta, \alpha\gamma\}} =$ $\{\varepsilon, \alpha, \alpha\beta\} = \overline{\{\alpha\beta\}} = \overline{K_1 \cap K_2}$ ✓. Since K_1, K_2 are each controllable w. r. t. their own plant, by Proposition 4.1 their intersection $K = K_1 \cap K_2 = \{\alpha\beta\}$ is controllable w. r. t. G_{SPC}. This is easily verified by inspection of Figure 4(a). Note that the composed plant was not used for the decision.

If $K = (K_1 \cup K_2) \cap L(G_{\mathrm{SPC}}) = \{\alpha\beta, \alpha\gamma\}$, Proposition 4.2 is relevant. By condition 4.2 – (1) $(K_1, L(G_{\mathrm{SPC}}))$ is non-conflicting because $\overline{K}_1 \cap \overline{L}(G_{\mathrm{SPC}}) = \{\varepsilon, \alpha, \alpha\beta, \alpha\delta, \alpha\delta r\} \cap \overline{L}(G_{\mathrm{SPC}}) = \{\varepsilon, \alpha, \alpha\beta\} = \overline{\{\alpha\beta, \alpha\delta r\} \cap L(G_{\mathrm{SPC}})} = \overline{K_1 \cap L(G_{\mathrm{SPC}})}$ ✓. 4.2 – (2) $(K_2, L(G_{\mathrm{SPC}}))$ is non-conflicting because $\overline{K}_2 \cap \overline{L}(G_{\mathrm{SPC}}) = \{\varepsilon, \alpha, \alpha\beta, \alpha\gamma\} \cap \overline{L}(G_{\mathrm{SPC}}) = \{\varepsilon, \alpha, \alpha\beta, \alpha\gamma\} = \overline{\{\alpha\beta, \alpha\gamma\} \cap L(G_{\mathrm{SPC}})} = \overline{K_2 \cap L(G_{\mathrm{SPC}})}$ ✓. K is controllable, which is easily verified by inspection. Observe that in this case, the composed plant language was required for the last two conditions, but the combined specification language was not used.

Finally the composed generator G_{BSC}, with G_1 master and G_2 follower, is shown in Figure 4(b). Checking the conditions of Proposition 4.3 yields: 4.3 – (1) Controllability of K_2 w. r. t. G_1 is easily verified by inspection ✓, 4.3 – (2) the same as in 4.1 – (1). Since K_1, K_2 are each controllable w. r. t. their own plant, by Proposition 4.3, their intersection $K = K_1 \cap K_2 = \{\alpha\beta\}$ is controllable w. r. t. G_{BSC}, which is easily verified by inspection of Figure 4(b) as well.

When $K = (K_1 \cup K_2) \cap L(G_{\mathrm{BSC}}) = \{\alpha\beta, \alpha\gamma, \alpha\delta r\}$, Proposition 4.4 is applied, with condition 4.4 – (1) the same as in Proposition 4.3. 4.4 – (2) $(K_2, L(G_{\mathrm{BSC}}))$ is non-conflicting because $\overline{K}_2 \cap \overline{L}(G_2) = \underline{\{\varepsilon, \alpha, \alpha\beta, \alpha\gamma\} \cap \overline{L}(G_2)} = \{\varepsilon, \alpha, \alpha\beta, \alpha\gamma\} = \overline{\{\alpha\beta, \alpha\gamma\}} = \overline{K_2 \cap L(G_1)}$ ✓. Thus, K is controllable w. r. t. G_{BSC}, as easily verified using Figure 4(b).

6. CONCLUSION

The notion of controllability was reviewed in the context of composed systems. Results on the conservation of controllability under the application of two compositional operators were presented and illustrated by an example. The taken approach fits naturally into a modular modelling and control perspective appropriate for dealing with large systems. Further considerations are necessary concerning the remaining operators PSC, PSPC, SYPC and LPC, with focus on PSC as the most general operator. Furthermore, the computational benefit and expansion to multi-component systems ($n > 2$) remains to be investigated in detail.

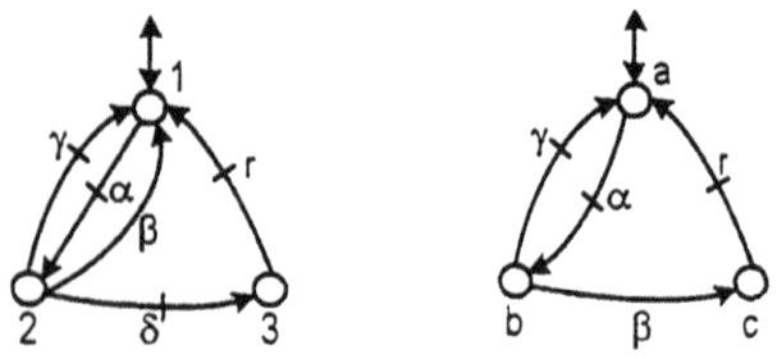

(a) Generator G_1. (b) Generator G_2.

Figure 3. Plant modules used to obtain the composed plant.

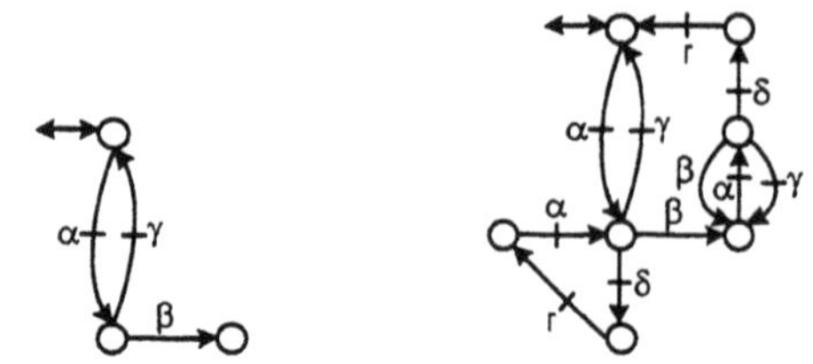

(a) Generator G_{SPC}. (b) Generator G_{BSC}.

Figure 4. Automata graphs of composed plants.

In addition, investigations regarding the conservation of other relevant properties under composition, such as normality, nonblocking or optimality, represent potentially useful expansion.

REFERENCES

de Queiroz, M.H. and J.E.R. Cury (2000*a*). Modular Control of Composed Systems. In: *Proceedings of the American Control Conference*. Chicago, USA.

de Queiroz, M.H. and J.E.R. Cury (2000*b*). Modular Supervisory Control of Large Scale Discrete Event Systems. In: *Discrete Event Systems: Analysis and Control* (R. Boel and G. Stremersch, Eds.). pp. 103–110. Kluwer Academic Publishers.

Gohari, P. and W.M. Wonham (2000). On the Complexity of Supervisory Control Design in the RW Framework. *IEEE Trans. on Systems, Man and Cybernetics – Part B* **30**(5), 643–652.

Heymann, M. (1990). Concurrency and Discrete Event Control. *IEEE Control Systems Magazine* **10**(4), 103–112.

Kowalewski, S. (1996). *Modulare diskrete Modellierung verfahrenstechnischer Anlagen zum systematischen Steuerungsentwurf*. Shaker Verlag, Aachen, Germany.

Lafortune, S. and E. Chen (1990). The Infimal Closed Controllable Superlanguage and Its Application in Supervisory Control. *IEEE Trans. on Automatic Control* **35**(4), 398–405.

Ramadge, P.J. and W.M. Wonham (1988). Modular Supervisory Control of Discrete-Event Systems. *Mathematics of Control, Signals and Systems* **1**(1), 13–30.

Ramadge, P.J. and W.M. Wonham (1989). The Control of Discrete Event Systems. *Proc. IEEE, Special Issue on Discrete Event Dynamic Systems* **77**(1), 81–98.

Wonham, W.M. (2002). *Notes on Control of Discrete-Event Systems*. Dept. of Elec. and Comp. Eng., University of Toronto, Toronto, Canada.

ELSEVIER
IFAC
PUBLICATIONS
www.elsevier.com/locate/ifac

ON THE SET OF CERTAIN CONFLICTS OF A GIVEN LANGUAGE

Robi Malik

Department of Computer Science, University of Waikato
Hamilton, New Zealand

Abstract: Two concurrent processes are said to be in conflict if they can get trapped in a situation where they both are waiting or running endlessly, forever unable to complete their common task. In the design of reactive systems, this is a common fault which can be very subtle and hard to detect. This paper studies conflicts in more detail and characterises the most general set of behaviours of a process which certainly leads to a conflict when accepted by another process running in parallel. It shows how this set of certain conflicts can be used to simplify the automatic detection of conflicts and thus the verification of reactive systems.

Keywords: Discrete event systems, large-scale systems, formal verification, deadlock, termination.

1. INTRODUCTION

Conflicts are a common fault in the design of concurrent programs, and have long been studied in the context of discrete event systems (Dietrich *et al.*, 2002; Ramadge and Wonham, 1989; Wong *et al.*, 2000). Two concurrent processes are in conflict if they can reach a state, from which no terminal state can be reached anymore. This includes both the possibility of *deadlock*, where processes are stuck and unable to continue at all, and *livelock*, where processes continue to run forever without achieving any further progress.

Since discrete event systems are used to model complex, safety-critical systems (Brandin and Charbonnier, 1994; Leduc and Wonham, 1995; Malik, 2003), tools to detect conflicts in very large systems are needed. Existing model checking techniques (Clarke *et al.*, 1999) can be used, but are limited by the state explosion problem. This paper studies the special nature of conflicts in deeper detail and proposes a method to facilitate the automatic detection of conflicts in modular systems.

A common way of fighting state explosion is to exploit modular system structures (Leduc *et al.*, 2000; Wong and Wonham, 1998). Often it suffices to analyse only subsystems of a large system to establish that the entire system satisfies a property of interest (Åkesson *et al.*, 2002; Brandin *et al.*, 2004). Yet, the nature of conflict is more sophisticated: even if each individual subsystem is free from conflict, the system composed from these subsystems may be in conflict (Wong *et al.*, 2000).

In this situation, it is possible to use abstractions of subsystems that preserve possible conflicts with other subsystems. Structural (Leduc *et al.*, 2000; Lee and Wong, 2002) and priority-based (Wong *et al.*, 2000) approaches have been suggested. These techniques give some insight how systems can be designed to be free from conflict, but they rely on design carried out by users, and therefore cannot be performed automatically. Other techniques have been discovered in the context of model checking (Clarke *et al.*, 1999; Dams *et al.*, 1997), but they are not tailored for the special needs of conflict analysis and therefore are limited in power or cannot be applied directly.

Recently, a heuristic algorithm for detecting a special case of conflicts has been developed (Malik, 2003). The experience gained from this algorithm can be used to find abstractions of discrete event systems that preserve all possible conflicts. This paper introduces the set of certain conflicts of a language, a set of behaviours that can always be treated specially for verification purposes, considerably improving performance.

Section 2 introduces the necessary notations from supervisory control theory (Ramadge and Wonham, 1989; Wonham, 1999) including conflicts. Section 3 defines the set of certain conflicts and shows how it can simplify the task of checking a complex system for conflicts. Section 4 presents a fixed point algorithm to compute the set of certain conflicts, and section 5 gives some concluding remarks.

2. NOTATION AND PRELIMINARIES

2.1 Languages

Event sequences and languages are a simple means to describe discrete system behaviours. Their basic building blocks are *events*, which are taken from a finite *alphabet* Σ. Then, Σ^* denotes the set of all finite *strings* of the form $\sigma_1\sigma_2\cdots\sigma_k$ of events from Σ, including the *empty string* ε. A *language* over Σ is any subset $\mathcal{L} \subseteq \Sigma^*$.

The *catenation* of two strings $s, t \in \Sigma^*$ is written as st. A string $t \in \Sigma^*$ is called a *prefix* of $s \in \Sigma^*$, written $t \sqsubseteq s$, if $s = tu$ for some $u \in \Sigma^*$. The *prefix-closure* $\overline{\mathcal{L}}$ of a language $\mathcal{L} \subseteq \Sigma^*$ is the set of all prefixes of strings in $\mathcal{L}$, i.e. $\overline{\mathcal{L}} = \{t \in \Sigma^* \mid t \sqsubseteq s \text{ for some } s \in \mathcal{L}\}$. A language $\mathcal{L}$ is called *prefix-closed* if $\mathcal{L} = \overline{\mathcal{L}}$.

Projection is used to hide events that are not of interest. For $\Omega \subseteq \Sigma$, the *natural projection* $P_\Omega \colon \Sigma^* \to \Omega^*$ removes all events not in Ω from a string. This is extended to languages by $P_\Omega(\mathcal{L}) = \{P_\Omega(s) \mid s \in \mathcal{L}\}$. The *inverse projection* adds events not in Ω to strings of a language in all possible ways, $P_\Omega^{-1}(\mathcal{L}) = \{s \in \Sigma^* \mid P_\Omega(s) \in \mathcal{L}\}$.

2.2 Discrete Event Systems

A *discrete event system (DES)* over Σ is a pair $\mathbf{G} = (\mathcal{L}, \mathcal{M})$, where $\mathcal{L} \subseteq \Sigma^*$ is a prefix-closed language, and $\mathcal{M} \subseteq \mathcal{L}$ (Cassandras and Lafortune, 1999). The language $\mathcal{L}$ contains all possible sequences of events that can be executed by the system, while $\mathcal{M}$ contains only event sequences corresponding to completed tasks. A DES $\mathbf{G}$ is completely characterised by these two languages, which are also referred to as $\mathcal{L}(\mathbf{G})$ and $\mathcal{M}(\mathbf{G})$, respectively.

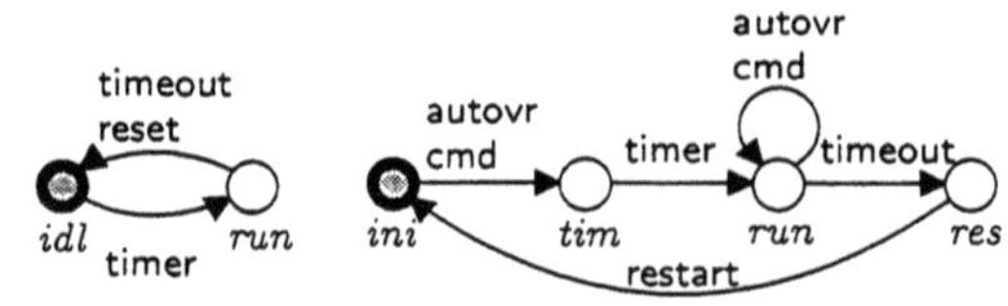

Fig. 1. Two DES represented as state transition diagrams.

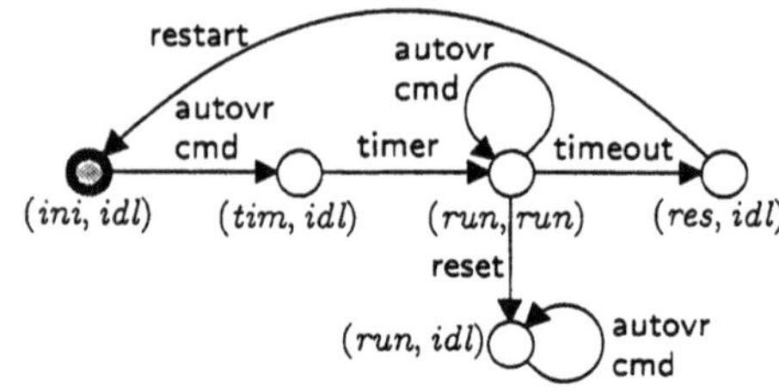

Fig. 2. The synchronous product of the DES in fig. 1.

Discrete event systems can be represented visually using state transition diagrams. Fig. 1 shows two examples of this representation. The left DES represents a timer which, after having been started by event timer, can either produce a timeout or be reset before being started again. Initial states have a thick border, and terminal states are coloured gray. In this example, the states *idl* and *ini* are both initial and terminal states.

2.3 Synchronous Product

When multiple DES run in parallel, lock-step synchronisation is used. In order to execute an event in a certain state, all participating DES must be able to accept that event (Hoare, 1985). It follows that the behaviour $\mathbf{G}_1 \parallel \mathbf{G}_2$ of two DES running in parallel is characterised by the intersection of their languages (Wonham, 1999).

This kind of synchronisation requires $\mathbf{G}_1$ and $\mathbf{G}_2$ to use the same event alphabet Σ. Otherwise, the event alphabet Σ_1 of $\mathbf{G}_1$, e.g., can be extended to the common alphabet $\Sigma = \Sigma_1 \cup \Sigma_2$ using inverse projection. The events not considered in $\mathbf{G}_1$ are assumed to be possible in every state and not to cause any state change.

Given such an extended DES, it is interesting which events are actually used to characterise the behaviour, and which may just have been introduced by the above procedure. An event set $\Omega \subseteq \Sigma$ is called *adequate* for $\mathcal{L} \subseteq \Sigma^*$ if $P_\Omega^{-1}(P_\Omega(\mathcal{L})) = \mathcal{L}$, i.e., if the language can be reconstructed from its projection. Ω is called adequate for a DES G, if Ω is adequate for $\mathcal{L}(G)$ and $\mathcal{M}(G)$.

In the state transition diagram representation, the synchronous product of DES is obtained from the Cartesian product of all involved state spaces. Fig. 2 shows the synchronous product of the two DES in fig. 1, after extending both DES to use the events autovr, cmd, reset, and restart which occur in only one of the original DES.

2.4 Conflicts

Given a DES $\mathbf{G}$, it is desirable that every behaviour possible in $\mathcal{L}(\mathbf{G})$ can be extended to a complete task in $\mathcal{M}(\mathbf{G})$. A DES satisfying this property is called *nonblocking*; otherwise it is said to be *blocking* (Wonham, 1999). The DES in fig. 2 is blocking, because no terminal state can be reached from state (*run*, *idl*).

Blocking becomes more interesting when multiple DES are running in parallel—in this case the term *conflicting* is used instead of blocking. Two DES $\mathbf{G}_1$ and $\mathbf{G}_2$ are called *nonconflicting* if $\mathbf{G}_1 \parallel \mathbf{G}_2$ is nonblocking, i.e.,

$$\mathcal{L}(\mathbf{G}_1) \cap \mathcal{L}(\mathbf{G}_2) = \overline{\mathcal{M}(\mathbf{G}_1) \cap \mathcal{M}(\mathbf{G}_2)}. \quad (1)$$

Otherwise $\mathbf{G}_1$ and $\mathbf{G}_2$ are called *conflicting.* The above (1) extends the traditional definition of nonconflicting languages (Wonham, 1999) to potentially blocking DES.

Conflicts are difficult to analyse in a modular way (Wong *et al.*, 2000). For example, although each of the two DES in fig. 1 is nonblocking, they are conflicting when run in parallel (fig. 2). In computation tree logic (CTL) (Clarke *et al.*, 1999), the property of being nonblocking or nonconflicting is expressed as

$$\mathbf{AG}\,\mathbf{EF}\; marked\_state, \quad (2)$$

where *marked_state* is a propositional formula identifying the states in which all involved DES have completed their tasks. This formula is neither in $\forall$CTL* nor in $\exists$CTL*, which explains why many known abstraction techniques (Clarke *et al.*, 1999) cannot be used for this kind of property.

3. FINDING CONFLICTS IN LARGE SYSTEMS

3.1 The Problem

The aim of this paper is to find efficient algorithms to determine whether a large system of concurrent DES is blocking or not. The straightforward approach to do this is to construct and examine a synchronous product such as

$$\mathbf{G}_1 \parallel \mathbf{G}_2 \parallel \cdots \parallel \mathbf{G}_n .$$

The check is done by exploring all reachable states and checking whether a terminal state can be reached from every reachable state. Using symbolic representations such as BDDs (Bryant, 1986) or IDDs (Zhang and Wonham, 2002), this approach has been used to analyse very large models. Yet, the technique always remains limited by the amount of memory available to store representations of the synchronous product.

An alternative approach avoids building the entire synchronous product by analysing smaller subsystems first. Assume, for example, that the blocking DES in fig. 2 is the first component $\mathbf{G}_1$ of a large system of concurrent DES. If it can be shown that all components of the system can execute the event sequence cmd timer reset, e.g., then it already follows that the entire system is blocking. In order to determine whether this is the case without constructing the entire synchronous product, an *incremental language inclusion check* (Brandin *et al.*, 2004) can be used.

Further to the example, assume that the events timer and reset are known never to be disabled by any of the other components of the system. In this case, the system is known to be blocking if $\mathbf{G}_1$ can ever enter state (*tim*, *idl*): if $\mathbf{G}_1$ is in this state, the entire system can execute the event sequence timer reset and put $\mathbf{G}_1$ into the blocking state (*run*, *idl*).

Therefore, the states (*tim*, *idl*), (*run*, *run*), and (*run*, *idl*) are considered as states of *certain conflicts* of $\mathbf{G}_1$: any system that is to be nonconflicting with $\mathbf{G}_1$ has to prevent $\mathbf{G}_1$ from entering these states. In the following, these ideas are presented in a more formal way using languages.

3.2 The Set of Certain Conflicts

If a DES $\mathbf{G}$ interacts with other systems, then its event set Σ can be split into a set Υ of events that are used only by $\mathbf{G}$ and are irrelevant to all other systems, and a set Ω of events that can be shared between $\mathbf{G}$ and other systems.

$$\Sigma = \Omega \dot{\cup} \Upsilon \quad (3)$$

As seen in the above example, if $\mathbf{G}$ is blocking, then there exists a set of strings accepted by $\mathbf{G}$ which, when accepted by another DES $\mathbf{H}$ running in parallel with $\mathbf{G}$, causes $\mathbf{G}$ and $\mathbf{H}$ to be conflicting. These strings constitute the set of certain conflicts.

Definition 1. Let $\mathbf{G}$ be a DES over Σ, and $\Omega \subseteq \Sigma$. A string $s \in \Omega^*$ is said to be *Ω-blocking* for $\mathbf{G}$, if, for every DES $\mathbf{H}$ over Σ such that Ω is adequate for $\mathbf{H}$ and $s \in \mathcal{L}(\mathbf{H})$, the DES $\mathbf{G}$ and $\mathbf{H}$ are conflicting. Otherwise s is said to be *Ω-nonblocking* for $\mathbf{G}$.

To be Ω-blocking is a property of strings—for s to be Ω-blocking, *every* DES $\mathbf{H}$ with the mentioned properties has to be conflicting with $\mathbf{G}$.

Definition 2. For a DES $\mathbf{G}$ over Σ, and $\Omega \subseteq \Sigma$,

$$\textsc{Conf}(\mathbf{G},\Omega) = \{\, s \in \Omega^* \mid s \text{ is } \Omega\text{-blocking for } \mathbf{G} \,\};$$
$$\textsc{NConf}(\mathbf{G},\Omega) = \{\, s \in \Omega^* \mid s \text{ is } \Omega\text{-nonblocking for } \mathbf{G} \,\}.$$

$\textsc{Conf}(\mathbf{G},\Omega)$ is the set of *certain conflicts* of $\mathbf{G}$ with respect to Ω. Its complement $\textsc{NConf}(\mathbf{G},\Omega)$ is the set of all strings s of shared events for which there exists another system which accepts s and is nonconflicting with $\mathbf{G}$.

Let $\mathbf{G}_1$ be the DES in fig. 2, and assume $\Upsilon = \{\mathsf{reset}, \mathsf{timer}\}$. Then every DES $\mathbf{H}$ that has the events in Υ selflooped in all states and accepts autovr or cmd must be conflicting with $\mathbf{G}_1$. Thus,

$$\textsc{Conf}(\mathbf{G}_1,\Omega) = \{\mathsf{autovr},\mathsf{cmd}\}\,\Omega^*. \qquad (4)$$

The following result shows how the set of certain conflicts can simplify conflict analysis: another system can only be nonconflicting with a DES $\mathbf{G}$ if the behaviour of that system remains confined by $\textsc{NConf}(\mathbf{G},\Omega)$. This gives a necessary condition for complex systems to be nonconflicting.

Proposition 3. Let $\mathbf{G}$ and $\mathbf{H}$ be two DES over Σ, and let $\Omega \subseteq \Sigma$ be adequate for $\mathbf{H}$. If $\mathbf{G}$ and $\mathbf{H}$ are nonconflicting then

$$P_\Omega(\mathcal{L}(\mathbf{H})) \subseteq \textsc{NConf}(\mathbf{G},\Omega). \qquad (5)$$

Proof. Assume $\mathbf{G}$ and $\mathbf{H}$ are nonconflicting, and let $s \in P_\Omega(\mathcal{L}(\mathbf{H}))$. Since Ω is adequate for $\mathbf{H}$, it follows that $s \in P_\Omega^{-1}(P_\Omega(\mathcal{L}(\mathbf{H}))) = \mathcal{L}(\mathbf{H})$. Hence, s is Ω-nonblocking for $\mathbf{G}$, and $s \in \textsc{NConf}(\mathbf{G},\Omega)$ by definition. □

Consider once more the example subsystem $\mathbf{G}_1$ in fig. 2, which is part of a larger system $\mathbf{G}_1 \parallel \cdots \parallel \mathbf{G}_n$. The events in $\Upsilon = \{\mathsf{timer}, \mathsf{reset}\}$ are not used in $\mathbf{G}_2,\ldots,\mathbf{G}_n$. Therefore, the above result says that, if the behaviour of $\mathbf{G}_2,\ldots,\mathbf{G}_n$ can be shown not to be contained in the language $\textsc{NConf}(\mathbf{G}_1,\Omega)$, i.e. if it is not the case that

$$\mathcal{L}(\mathbf{G}_2 \parallel \cdots \parallel \mathbf{G}_n) \subseteq P_\Omega^{-1}(\textsc{NConf}(\mathbf{G}_1,\Omega)), \quad (6)$$

then the entire system must be conflicting.

The example discussed above was taken from a faulty model of an actual automotive central locking system (KorSys project), which consists of 54 components and has a reachable state space of 7.5×10^8 states. The two components in fig. 1 that are responsible for the conflict were found and combined automatically, by appropriate heuristics, and an incremental language inclusion check quickly revealed that the event cmd can indeed be executed by the entire system, after executing only one other event. The error was that the event reset in the timer model of fig. 1 had been forgotten when modelling the rest of the system. This example demonstrates that proposition 3 provides a very powerful means to detect conflicts in large systems.

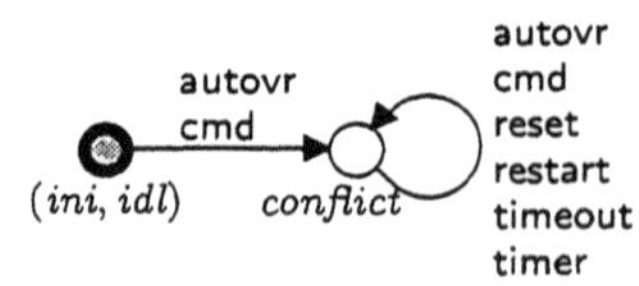

Fig. 3. A DES which is conflict equivalent to fig. 2.

3.3 Simplifying Subsystems

If condition (6) is true, it remains unknown whether the composed system $\mathbf{G}_1 \parallel \cdots \parallel \mathbf{G}_n$ is conflicting or not. In this case, the information about the certain conflicts of $\mathbf{G}_1$ can still be used to simplify conflict analysis. The component $\mathbf{G}_1$ can be replaced by a simpler model $\mathbf{G}_1'$, and then the system $\mathbf{G}_1' \parallel \mathbf{G}_2 \parallel \cdots \parallel \mathbf{G}_n$ can be considered for analysis.

The prerequisite for such an approach to work is that the replacement $\mathbf{G}_1'$ preserves all possibilities of $\mathbf{G}_1$ being in conflict with the rest of the system. This suggests to use conflicts to define a testing semantics along the lines of (Brinksma *et al.*, 1995).

Definition 4. Two DES $\mathbf{G}_1$ and $\mathbf{G}_2$ over Σ are called *conflict equivalent* with respect to $\Omega \subseteq \Sigma$ if, for every DES $\mathbf{H}$ such that Ω is adequate for $\mathbf{H}$, $\mathbf{G}_1$ and $\mathbf{H}$ are nonconflicting if and only if $\mathbf{G}_2$ and $\mathbf{H}$ are nonconflicting.

The DES in fig. 3 is conflict equivalent to the DES in fig. 2: any DES $\mathbf{H}$, which does not use the events timer and reset, is conflicting with either of these two DES if and only if it ever accepts one of the events autovr or cmd. Therefore, fig. 2 can be replaced by the smaller fig. 3 before starting a conflict check of the entire system. Furthermore, other components of the system may be replaced in a similar way.

The simplification of fig. 2 to fig. 3 was based on the observation that, as far as conflicts are concerned, all the strings that bear a certain conflict for a DES can be treated in the same way. In a state transition diagram, this can be done by collapsing all states representing certain conflicts into a single state where all events are possible.

The following two results justify this idea. Proposition 5 states that strings from the set of certain conflicts can be added to the language $\mathcal{L}(\mathbf{G})$ of a DES without affecting its possible conflicts.

Proposition 6 states that certain conflicts do not represent proper completed tasks, and therefore can be removed safely from the language $\mathcal{M}(\mathbf{G})$. Fig. 3 can be obtained from fig. 2 by first applying proposition 6 and then proposition 5.

Proposition 5. Let $\mathbf{G}$ be a DES over Σ, and $\Omega \subseteq \Sigma$. Then $\mathbf{G}$ is conflict equivalent to $\mathbf{G}'$ with respect to Ω, where

$$\mathcal{L}(\mathbf{G}') = \mathcal{L}(\mathbf{G}) \cup P_\Omega^{-1}(\textsc{Conf}(\mathbf{G},\Omega));$$
$$\mathcal{M}(\mathbf{G}') = \mathcal{M}(\mathbf{G}).$$

Proof. Let $\mathbf{H}$ be a DES over Σ such that Ω is adequate for $\mathbf{H}$.

First assume that $\mathbf{G}'$ and $\mathbf{H}$ are conflicting. Then there exists $s \in \mathcal{L}(\mathbf{G}') \cap \mathcal{L}(\mathbf{H})$ such that $s \notin \overline{\mathcal{M}(\mathbf{G}') \cap \mathcal{M}(\mathbf{H})} = \overline{\mathcal{M}(\mathbf{G}) \cap \mathcal{M}(\mathbf{H})}$ since $\mathcal{M}(\mathbf{G}') = \mathcal{M}(\mathbf{G})$. If $s \in \mathcal{L}(\mathbf{G})$, it follows immediately that $\mathbf{G}$ and $\mathbf{H}$ are conflicting. Otherwise, if $s \notin \mathcal{L}(\mathbf{G})$, note that $s \in \mathcal{L}(\mathbf{G}')$ and therefore $s \in P_\Omega^{-1}(\textsc{Conf}(\mathbf{G},\Omega))$. Then $s' = P_\Omega(s) \in \textsc{Conf}(\mathbf{G},\Omega)$, and $s' \in \mathcal{L}(\mathbf{H})$ since $s \in \mathcal{L}(\mathbf{H})$ and Ω is adequate for $\mathbf{H}$. By definition of $\textsc{Conf}(\mathbf{G},\Omega)$, $\mathbf{G}$ and $\mathbf{H}$ are conflicting.

Now assume that $\mathbf{G}'$ and $\mathbf{H}$ are nonconflicting. To prove that $\mathbf{G}$ and $\mathbf{H}$ are nonconflicting, let $s \in \mathcal{L}(\mathbf{G}) \cap \mathcal{L}(\mathbf{H}) \subseteq \mathcal{L}(\mathbf{G}') \cap \mathcal{L}(\mathbf{H})$. Since $\mathbf{G}'$ and $\mathbf{H}$ are nonconflicting, there exists $t \in \Sigma^*$ such that $st \in \mathcal{M}(\mathbf{G}') \cap \mathcal{M}(\mathbf{H}) = \mathcal{M}(\mathbf{G}) \cap \mathcal{M}(\mathbf{H})$. Thus, $\mathbf{G}$ and $\mathbf{H}$ are nonconflicting. □

Proposition 6. Let $\mathbf{G}$ be a DES over Σ, and $\Omega \subseteq \Sigma$. Then $\mathbf{G}$ is conflict equivalent to $\mathbf{G}'$ with respect to Ω, where

$$\mathcal{L}(\mathbf{G}') = \mathcal{L}(\mathbf{G});$$
$$\mathcal{M}(\mathbf{G}') = \mathcal{M}(\mathbf{G}) \cap P_\Omega^{-1}(\textsc{NConf}(\mathbf{G},\Omega)).$$

Proof. Let $\mathbf{H}$ be a DES over Σ such that Ω is adequate for $\mathbf{H}$.

First assume that $\mathbf{G}$ and $\mathbf{H}$ are nonconflicting. To prove that $\mathbf{G}'$ and $\mathbf{H}$ are nonconflicting, let $s \in \mathcal{L}(\mathbf{G}') \cap \mathcal{L}(\mathbf{H}) = \mathcal{L}(\mathbf{G}) \cap \mathcal{L}(\mathbf{H})$. Since $\mathbf{G}$ and $\mathbf{H}$ are nonconflicting, there exists $t \in \Sigma^*$ such that $st \in \mathcal{M}(\mathbf{G}) \cap \mathcal{M}(\mathbf{H}) \subseteq \mathcal{L}(\mathbf{H})$. Since Ω is adequate for $\mathbf{H}$, $P_\Omega(st) \in \mathcal{L}(\mathbf{H})$ is Ω-nonblocking for $\mathbf{G}$. By definition of $\textsc{NConf}(\mathbf{G},\Omega)$ it follows that $P_\Omega(st) \in \textsc{NConf}(\mathbf{G},\Omega)$. Thus, $st \in \mathcal{M}(\mathbf{G}) \cap P_\Omega^{-1}(\textsc{NConf}(\mathbf{G},\Omega)) \cap \mathcal{M}(\mathbf{H}) = \mathcal{M}(\mathbf{G}') \cap \mathcal{M}(\mathbf{H})$, i.e., $\mathbf{G}'$ and $\mathbf{H}$ are nonconflicting.

Now assume that $\mathbf{G}'$ and $\mathbf{H}$ are nonconflicting. To prove that $\mathbf{G}$ and $\mathbf{H}$ are nonconflicting, let $s \in \mathcal{L}(\mathbf{G}) \cap \mathcal{L}(\mathbf{H}) = \mathcal{L}(\mathbf{G}') \cap \mathcal{L}(\mathbf{H})$. Since $\mathbf{G}'$ and $\mathbf{H}$ are nonconflicting, there exists $t \in \Sigma^*$ such that $st \in \mathcal{M}(\mathbf{G}') \cap \mathcal{M}(\mathbf{H}) \subseteq \mathcal{M}(\mathbf{G}) \cap \mathcal{M}(\mathbf{H})$. Thus, $\mathbf{G}$ and $\mathbf{H}$ are nonconflicting. □

4. A FIXED POINT ALGORITHM

This section provides an operational characterisation and an algorithm to compute the set of certain conflicts. The set of certain conflicts can be characterised as a fixed point of the following operator which manipulates languages over Ω.

Definition 7. Let $\mathbf{G}$ be a DES over Σ, and $\Omega \subseteq \Sigma$. Then define the mapping

$$\Xi_{\mathbf{G},\Omega} : 2^{\Omega^*} \to 2^{\Omega^*};$$
$$X \mapsto P_\Omega(\mathcal{L}(\mathbf{G}) \setminus \overline{\mathcal{M}(\mathbf{G}) \setminus P_\Omega^{-1}(X)})\, \Omega^*.$$

This operator is designed to compute the set of certain conflicts of the DES $\mathbf{G}$ iteratively. Iteration starts with the empty language $X_0 = \emptyset$ which is extended repeatedly, until the complete set $\textsc{Conf}(\mathbf{G},\Omega)$ is obtained. The first step yields

$$X_1 = \Xi_{\mathbf{G},\Omega}(\emptyset) = P_\Omega(\mathcal{L}(\mathbf{G}) \setminus \overline{\mathcal{M}(\mathbf{G})})\, \Omega^*,$$

the set of all strings that resemble, considering only the shared events from Ω, strings that cannot be extended to a completed task, plus any continuations of such strings. The set of certain conflicts can be shown to be the least fixed point of this iteration.

Proposition 8. Let $\mathbf{G}$ be a DES over Σ, and $\Omega \subseteq \Sigma$. Then $\textsc{Conf}(\mathbf{G},\Omega) = \operatorname{lfp} \Xi_{\mathbf{G},\Omega}$.

The proof uses the fixed point theory of (Tarski, 1955) and is technical. While the operator $\Xi_{\mathbf{G},\Omega}$ can be shown to be monotonic, it is not necessarily continuous, and therefore the iteration is not guaranteed to converge in all cases.

If the DES to be analysed can be represented by a finite-state transition diagram, then the iteration does converge and terminate in a finite number of steps. In this case, it is convenient to use sets of states X_i' instead of languages X_i in the algorithm. In the example of fig. 2, the iteration terminates with the second step, producing

$$X_0' = \emptyset;$$
$$X_1' = \{(run, idl), (run, run), (tim, idl)\};$$
$$X_2' = X_1'.$$

Not all state transition diagrams can be manipulated in such a simple way. A single state may be reachable via different sequences of events which may need to be treated separately.

Such states can be split on the fly or in a preprocessing step. It is possible to compute a representation $\mathbf{P}$ of the language $P_\Omega(\mathcal{L}(\mathbf{G}))$ first, using subset construction (Hopcroft *et al.*, 2001), and then use the synchronous product $\mathbf{G} \parallel \mathbf{P}$ as input for the algorithm. This shows that the procedure outlined above can indeed be used to compute the set of certain conflicts for a finite-state DES.

5. CONCLUSION

The set of *certain conflicts* for a discrete event system has been characterised as the largest set of behaviours which, when accepted by another system running in parallel, definitely cause a livelock or deadlock. This set can help to detect conflicts in complex models of reactive systems without exploring their complete state space, and to simplify the analysis of conflicts in general.

While the abstraction techniques proposed give very powerful reductions for systems that actually have conflicts, i.e. deadlocks or livelocks, the effect on nonconflicting systems remains limited. Although some simplification is possible, it remains necessary to compose all components of a system in order to prove that it actually is nonconflicting.

In the future, the author plans to study abstraction techniques which also deal with behaviours that are not certain conflicts, and obtain better reductions for nonconflicting systems. The goal is to develop efficient techniques for verifying complex systems to be conflicting or nonconflicting.

REFERENCES

Åkesson, K., H. Flordal and M. Fabian (2002). Exploiting modularity for synthesis and verification of supervisors. In: *Proc. 15th IFAC World Congress on Automatic Control.* Barcelona, Spain.

Brandin, Bertil A., Robi Malik and Petra Malik (2004). Incremental verification and synthesis of discrete-event systems guided by counter-examples. *IEEE Trans. Contr. Syst. Technol.* **12**(3), 387–401.

Brandin, Bertil and François Charbonnier (1994). The supervisory control of the automated manufacturing system of the AIP. In: *Proc. Rensselaer's 4th Int. Conf. Computer Integrated Manufacturing and Automation Technology.* Troy, NY, USA. pp. 319–324.

Brinksma, Ed, Arend Rensink and Walter Vogler (1995). Fair testing. In: *Proc. 6th Int. Conf. Concurrency Theory, CONCUR '95* (Insup Lee and Scott A. Smolka, Eds.). Vol. 962 of *LNCS.* Springer. Philadelphia, PA, USA. pp. 313–327.

Bryant, Randal E. (1986). Graph-based algorithms for Boolean function manipulation. *IEEE Trans. Comput.* **35**(8), 677–691.

Cassandras, C. G. and S. Lafortune (1999). *Introduction to Discrete Event Systems.* Kluwer.

Clarke, Jr., Edmund M., Orna Grumberg and Doron A. Peled (1999). *Model Checking.* MIT Press.

Dams, Dennis, Rob Gerth and Orna Grumberg (1997). Abstract interpretation of reactive systems. *ACM Trans. Programming Languages and Systems* **19**(2), 111–149.

Dietrich, P., R. Malik, W. M. Wonham and B. A. Brandin (2002). Implementation considerations in supervisory control. In: *Synthesis and Control of Discrete Event Systems* (B. Caillaud, P. Darondeau, L. Lavagno and X. Xie, Eds.). Kluwer. pp. 185–201.

Hoare, C. A. R. (1985). *Communicating Sequential Processes.* Prentice-Hall.

Hopcroft, John E., Rajeev Motwani and Jeffrey D. Ullman (2001). *Introduction to Automata Theory, Languages, and Computation.* Addison-Wesley.

Leduc, R. J. and W. M. Wonham (1995). PLC implementation of a DES supervisor for a manufacturing testbed. In: *Proc. 33rd Allerton Conf. Communication, Control and Computing.* Monticello, Illinois. pp. 519–528.

Leduc, R. J., B. A. Brandin and W. M. Wonham (2000). Hierarchical interface-based nonblocking verification. In: *Proc. Canadian Conf. Electrical and Computer Engineering.* pp. 1–6.

Lee, Sang-Heon and Kai C. Wong (2002). Structural decentralised control of concurrent discrete-event systems. *European J. Control* **8**, 477–491.

Malik, Petra (2003). From Supervisory Control to Nonblocking Controllers for Discrete Event Systems. PhD thesis. University of Kaiserslautern. Kaiserslautern, Germany.

Ramadge, Peter J. G. and W. Murray Wonham (1989). The control of discrete event systems. *Proc. IEEE* **77**(1), 81–98.

Tarski, Alfred (1955). A lattice-theoretical fixpoint theorem and its applications. *Pacific J. Math.* **5**(2), 285–309.

Wong, K. C. and W. M. Wonham (1998). Modular control and coordination of discrete-event systems. *Discrete Event Dynamic Systems: Theory and Applications* **8**(3), 247–297.

Wong, K. C., J. G. Thistle, R. P. Malhame and H.-H. Hoang (2000). Supervisory control of distributed systems: Conflict resolution. *Discrete Event Dynamic Systems: Theory and Applications* **10**, 131–186.

Wonham, W. M. (1999). Notes on control of discrete event systems. Systems Control Group, Dept. of Electrical Engineering, University of Toronto, Ontario, Canada; at `http://www.control.utoronto.ca/` under "Research".

Zhang, Z. H. and W. M. Wonham (2002). STCT: An efficient algorithm for supervisory control design. In: *Synthesis and Control of Discrete Event Systems* (B. Caillaud, P. Darondeau, L. Lavagno and X. Xie, Eds.). Kluwer. pp. 77–100.

ELSEVIER
IFAC
PUBLICATIONS
www.elsevier.com/locate/ifac

MAXIMALLY PERMISSIVE CONTROLLERS IN ALL CONTEXTS

Stéphane Riedweg, Sophie Pinchinat

IRISA-INRIA, F-35042, Rennes, France
{sriedweg,pinchina}@irisa.fr

Abstract: We propose a logical formalism for the supervisory control specifications where the maximal permissiveness is explicitly handled, independently of the plant. The approach relies on a modal logic with quantified propositions and additionally provides a powerful machinery for decision and synthesis procedures.

Keywords: Supervision, Discrete-events systems, Optimal control, Temporal logic

1. INTRODUCTION

The control theory for discrete event systems was initiated by (Ramadge and Wonham, 1989), followed by (Thistle and Wonham, 1994) and many others. More recently, temporal logics specifications were considered by (Kupferman *et al.*, 2000) and (Arnold *et al.*, 2003) in order to solve a larger class of problems.

Typically, control problems require to find a supervision of the plant such that its (operational) behavior meets some desired properties, often called *the control objectives.* In regard to the objectives, they most often concern standard properties such as *admissibility*, *non-blocking*, *safety* etc. Among those, the fact that the controllers should allow any sound event, hence fulfilling the *maximal permissiveness property*, is generally taken for granted. As studied in depth by (Ramadge and Wonham, 1989) and others, the theory of regular languages offers clear results on the subject: the existence of a controller implies the existence of a unique maximally permissive one. The proposed algorithms hence synthesize precisely this optimal solution. Unfortunately, this cozy situation does not extends to larger frameworks: for example, the ω-closure assumption is necessary for ω-regular definable objectives (as in (Thistle and Wonham, 1994)), the set of control patterns of (Golaszewski and Ramadge, 1987) must be union-closed, etc. Worse is the case of branching-time logic specifications as in (Kupferman *et al.*, 2000; Arnold *et al.*, 2003) where neither the unicity and nor the existence of a maximally permissive controller are guaranteed anymore. For instance, there is no optimal supervision of the plant $(a+b)^\omega$ which eventually disables event b. Surprisingly, to our knowledge, the maximal permissiveness property has never been approached on its own, like a possible parameter of the supervisory control problems, likely because regular language semantics withstand this subject.

In this paper, we propose a logical formalism which allows an explicit handling of maximally permissiveness, in addition to standard supervisory control specifications. The approach relies on a logical framework and provides a powerful machinery for decision procedures and effective synthesis.

The logic is an extension of the propositional mu-calculus (Kozen, 1983; Arnold and Niwinski, 2001), called the *Quantified mu-calculus* ($\mathrm{Q}L_\mu$), as originally proposed by (Riedweg and Pinchinat, 2003*a*). The extension feature relies on the use of quantifications over (atomic) propositions. The resulting logic remains decidable as well as its model-checking thanks to its equivalence with infinite tree-automata, as already the case for the mu-calculus (see (Emerson and Lei, 1986)). For this reason, effective procedures to synthesis supervisory controllers are made possible by computing a finite model of a (infinite) tree-automaton (see also (Riedweg, 2003)).

In this paper, we do not aim at explaining the synthesis procedure, but instead we shall take it for granted from (Riedweg and Pinchinat, 2003*a*). Rather, we demonstrate how the logical formalism elegantly specifies supervisory control problems with an explicit mention of the *maximal permissiveness property* inside. Moreover, we make a proof that no other framework known in the literature can reach this expressive power.

The paper is organized as follows : Section 2 introduces the logic $\text{Q}L_\mu$. Section 3 explains how control objectives can be specified. Next, Section 4 dedicates to the maximal permissiveness specification with a proof of correctness. In Section 5, expressiveness issues are discussed, followed by a section of concluding remarks.

2. QUANTIFIED MU-CALCULUS

Models are deterministic finite state machines with marked states, called *processes* in Definition 2, as plants in supervisory control problems normally are. Since branching-time temporal statements will be considered, the natural notion of behavior is the execution tree, namely the (possibly infinite) unfolding of the finite state machine. Given that controllers aim at pruning the execution tree, the proposed logic can state in particular the existence of some pruning of the tree that satisfies a (mu-calculus definable) desired property.

The prunings are simply represented by placing a fresh atomic proposition on the tree to delimit the remaining subtree after pruning: in this way, an edge of the tree where the source node is positively labeled by the proposition whereas the target node is not is meant to be pruned.

Technically, an atomic proposition p is added to the unfolding of a process by composing the process with a particular one, namely a *p-labeling process* $\mathcal{E}$ as in Definition 3; controllers are actually derived from those. To get the tree pruning induced from the proposition done, we prune the labeling process according to Definition 5 to get $\mathcal{E}_{(\to p)}$, before we compose it with the system.

So that, as stated by Proposition 7, wondering whether a given pruning/control $\mathcal{S} \times \mathcal{E}_{(\to p)}$ of the system verifies some desired property α reduces to wonder whether the corresponding p-labeling by $\mathcal{S} \times \mathcal{E}$ is appropriated satisfies some adjustment $\alpha{*}p$ of the original property.

We assume given a finite set of events $\Sigma = \{a, b, \dots\}$, a set of atomic propositions $AP = \{p, p', c, c', \dots\}$, and a set of variables $Var = \{X, Y, \dots\}$.

Definition 1. **(Syntax of $\text{Q}L_\mu$)** We fist recall the syntax of the pure mu-calculus, written L_μ. The set of formulas of L_μ is defined by the following grammar:

$$\top \mid p \mid X \mid \neg\beta \mid \langle a\rangle\beta \mid \beta \vee \beta' \mid \mu X.\beta(X)$$

where $a \in \Sigma$, $p \in AP$ and $X \in Var$.

Fix-points formulas $\mu X.\beta(X)$ can properly interpreted (in Definition 4) whenever each occurrence of X in $\beta(X)$ is under an even number of negation symbols $\neg$.

The quantified mu-calculus, written $\text{Q}L_\mu$, extends L_μ as follows. The set of formulas of $\text{Q}L_\mu$ is defined by:

$$\exists p.\alpha \mid \neg\alpha \mid \alpha \vee \alpha' \mid \beta$$

where $p \in AP$ and $\beta \in L_\mu$.

Freely extending the classical terminology of the mu-calculus to the quantified mu-calculus, we name *sentences* all the formulas where each occurrence of a variable X is binded by a fix-point symbol μ. Also, we write $\bot$, $[a]\alpha$, $\alpha \wedge \alpha'$, $\nu X.\alpha(X)$, and $\forall p.\alpha$ respectively for $\neg\top$, $\neg \langle a\rangle\neg\alpha$, $\neg(\neg\alpha \vee \neg\alpha')$, $\neg\mu X.\neg\alpha(\neg X)$ and $\neg\exists p.\neg\alpha$, as well as $\xrightarrow{a}$, $[\,]\alpha$ and $\alpha \Rightarrow \alpha'$ respectively for $\langle a\rangle\top$, $\bigwedge_{a\in\Sigma}[a]\alpha$ and $\neg\alpha \vee \alpha'$.

Lastly, for $\beta \in L_\mu$, $\text{INV}(\beta)$ is a notation $\nu X.[\,]X \wedge \beta$: according to Definition 4 further, it states "from now on, the property β always holds".

Since in general, fixed-point operators and quantifiers do not commute, we forbid the use of quantifications inside fixed-point terms, all the more since the formulas will have enough expressive power for control problems specifications, as shown in the next section.

The semantics of the quantified mu-calculus (Definition 4) relies on models called *processes*:

Definition 2. **(Processes)** Given a finite set $\Gamma \subseteq AP$, a *process on* Γ is a tuple $\mathcal{S} = \langle S, s^0, t, L\rangle$, where S is a set of states, $s^0 \in S$ is the initial state, $t : S \times A \to S$ is a partial function called the *transition function* and $L : S \to 2^\Gamma$ labels states by propositions.

A process $\mathcal{S}$ is *finite* if S is finite and it is *complete* if $t(s, a)$ is defined everywhere.

Processes are normally subject to a natural abstraction according to their execution tree isomorphism class. Two processes with isomorphic execution trees are called *bisimilar*.

As announced, processes are composed in a synchronous manner, as to diminish their behavior: The *(synchronous) product* of $\mathcal{S}_1 = \langle S_1, s_1^0, t_1, L_1 \rangle$ on Γ_1 and $\mathcal{S}_2 = \langle S_2, s_2^0, t_2, L_2 \rangle$ on Γ_2 (with disjoint Γ_1 and Γ_2) is $\mathcal{S}_1 \times \mathcal{S}_2 = \langle S_1 \times S_2, (s_1^0, s_2^0), t, L \rangle$ on $\Gamma_1 \cup \Gamma_2$ where:

(1) $t((s_1, s_2), a) = (s_1', s_2')$ whenever $s_1' = t_1(s_1, a)$ and $s_2' = t_2(s_2, a)$), and
(2) $L(s_1, s_2) = L_1(s_1) \cup L_2(s_2)$.

Definition 3. **(Labeling Processes)** Given $p \in AP$, a *p-labeling process* is a complete process on $\{p\}$. We let Lab_p be the set of p-labeling processes, and we use $\mathcal{E}$, $\mathcal{E}'$ for typical elements. A *labeling of $\mathcal{S}$ by p*, or a *p-labeling of $\mathcal{S}$*, is a product $\mathcal{S} \times \mathcal{E}$, where $\mathcal{E} \in Lab_p$.

The logic can now be interpreted. Given a process $\mathcal{S}$, a formula α denotes a subset of the states, those which "satisfy" it. The semantics is given by induction over α, hence the need to interpret variable formulas: the standard manner consists in setting a valuation $val : Var \to \mathcal{P}(S)$ which defines the subset of states denoted by the formulas $X \in Var$. The pure mu-calculus formulas semantics is standard.

Definition 4. **(Semantics of $\mathrm{Q}L_\mu$)** Given a process $\mathcal{S} = \langle S, s^0, t, L \rangle$ and $val : Var \to \mathcal{P}(S)$, the interpretation of a $\mathrm{Q}L_\mu$-formula α is

$$[\![\alpha]\!]_{\mathcal{S}}^{[val]} \subseteq S$$

defined inductively by:

$$[\![\top]\!]_{\mathcal{S}}^{[val]} = S, \qquad [\![p]\!]_{\mathcal{S}}^{[val]} = \{s \in S \mid p \in L(s)\},$$
$$[\![X]\!]_{\mathcal{S}}^{[val]} = val(X), \; [\![\neg\alpha]\!]_{\mathcal{S}}^{[val]} = S \setminus [\![\alpha]\!]_{\mathcal{S}}^{[val]},$$
$$[\![\alpha \vee \alpha']\!]_{\mathcal{S}}^{[val]} = [\![\alpha]\!]_{\mathcal{S}}^{[val]} \cup [\![\alpha']\!]_{\mathcal{S}}^{[val]},$$
$$[\![\langle a \rangle \alpha]\!]_{\mathcal{S}}^{[val]} = \{s \in S \mid \exists s', t(s,a) = s', s' \in [\![\alpha]\!]_{\mathcal{S}}^{[val]}\},$$
$$[\![\mu X.\alpha(X)]\!]_{\mathcal{S}}^{[val]} = \bigcap \{V \subseteq S \mid [\![\alpha]\!]_{\mathcal{S}}^{[val(V/X)]} \subseteq V\},$$

and $[\![\exists p.\alpha]\!]_{\mathcal{S}}^{[val]}$ is the set of states $s \in S$, for which there exists $\mathcal{E} = \langle E, \varepsilon^0, t', L' \rangle \in Lab_p$, such that

$$(s, \varepsilon^0) \in [\![\alpha]\!]_{\mathcal{S}\times\mathcal{E}}^{[val']}$$

where $val'(X) \subseteq S \times E$ is $val(X) \times E$, in order to be sound.

Since the interpretation of a sentence α is independent of the valuation val, we simply write $[\![\alpha]\!]_{\mathcal{S}}$, and we use $\mathcal{S} \models \alpha$ whenever the initial state of $\mathcal{S}$ belongs to $[\![\alpha]\!]_{\mathcal{S}}$. To sum up, the assertions "$\mathcal{S} \models \alpha_1 \wedge \alpha_2$", "$\mathcal{S} \models \neg\alpha$", ... respectively mean "$\mathcal{S} \models \alpha_2$ and $\mathcal{S} \models \alpha_2$", "not $\mathcal{S} \models \alpha$" (also written "$\mathcal{S} \not\models \alpha$"),... Finally, "$\mathcal{S} \models \exists p.\alpha$" expresses that "there exists a p-labeling process $\mathcal{E}$ such that $\mathcal{S} \times \mathcal{E} \models \alpha$".

Apart from formulas of the form $\exists p.\alpha$, the semantics is the classical mu-calculus definition.

Because the mu-calculus is bisimulation invariant (formulas have the same truth value when interpreted over bisimilar processes), so is the logic $\mathrm{Q}L_\mu$ since the synchronous product which the interpretation of $\exists p.\alpha$ relies on preserves the execution tree isomorphisms.

In the rest of this section we establish how to adjust $\mathrm{Q}L_\mu$ formulas according to a fixed proposition when the latter issued to prune the model. We first formalize what the pruning is.

Definition 5. **(Pruning)** Given any process $\mathcal{S} = \langle S, s^0, t, L \rangle$ on Γ and $p \in \Gamma$, the *p-pruning of $\mathcal{S}$* is $\mathcal{S}_{(\to p)} = \langle S, s^0, t', L' \rangle$ on $\Gamma \setminus \{p\}$, where:

(1) for all $s \in S$ and $a \in \Sigma$, $t'(s,a) = t(s,a)$ if $p \in L(t(s,a))$, otherwise $t'(s,a)$ is undefined, and
(2) $L'(s) = L(s) \setminus \{p\}$.

Definition 6. **(Adjustment)** For any sentence $\alpha \in \mathrm{Q}L_\mu$ and for any $p \in AP$, the *p-adjustment of α* is $\alpha * p \in \mathrm{Q}L_\mu$, inductively defined by (by convention, $*$ has highest priority) :

$$\top * p = \top, \quad p' * p = p', \quad X * p = X,$$
$$(\langle a \rangle \alpha) * p = \langle a \rangle (p \wedge \alpha * p), \; (\neg\alpha) * p = \neg\alpha * p,$$
$$(\alpha \vee \alpha') * p = \alpha * p \vee \alpha' * p, \; (\mu X.\alpha) * p = \mu X.\alpha * p,$$
$$(\exists p.\alpha) * p = \exists p.\alpha * p.$$

Definition 5 and 6 together lead to the following fundamental result:

Proposition 7. Given a sentence $\alpha \in \mathrm{Q}L_\mu$, $\Gamma \subseteq AP$, $p \notin \Gamma$, a process $\mathcal{S}$ on Γ and $\mathcal{E} \in Lab_p$, we have:

$$\mathcal{S} \times (\mathcal{E}_{(\to p)}) \models \alpha \text{ if and only if } \mathcal{S} \times \mathcal{E} \models \alpha * p.$$

PROOF. Since, the processes $\mathcal{S} \times (\mathcal{E}_{(\to p)})$ and $(\mathcal{S} \times \mathcal{E})_{(\to p)}$ are isomorphic, they must satisfy the same formulas in each isomorphic states: $[\![\alpha]\!]_{\mathcal{S}\times(\mathcal{E}_{(\to p)})} = [\![\alpha]\!]_{(\mathcal{S}\times\mathcal{E})_{(\to p)}}$. An easy proof by induction on α gives the following more general result to conclude: $[\![\alpha]\!]_{(\mathcal{S}\times\mathcal{E})_{(\to p)}} = [\![\alpha * p]\!]_{\mathcal{S}\times\mathcal{E}}$.

As shown by Proposition 7, objects of type $\mathcal{E}_{(\to p)}$ indicate what controllers should be like. The soundness of the coming logical specifications can be argued by using two kinds of binary relations between processes. The first kinds are equivalence relations: given a process $\mathcal{S}$, two p-labeling processes $\mathcal{E}_1$ and $\mathcal{E}_2$ will be equivalent if they cause the same p-labeling on $\mathcal{S}$. They will be used in the next section.

Definition 8. **(The $\equiv_{\mathcal{S}}$-equivalences)** Given a process $\mathcal{S}$ and two c-labeling processes $\mathcal{E}_1$ and $\mathcal{E}_2$, *$\mathcal{E}_1$ and $\mathcal{E}_2$ induce the same c-labeling on $\mathcal{S}$*, written $\mathcal{E}_1 \equiv_{\mathcal{S}} \mathcal{E}_2$, whenever $\mathcal{S} \times \mathcal{E}_1$ and $\mathcal{S} \times \mathcal{E}_2$ are bisimilar.

The next binary relation between processes we consider here is the preorder of *simulation*; its main use further on is to compare more or less permissive behaviors.

Definition 9. **(Simulation)** Given two processes $\mathcal{S}_1 = \langle S_1, s_1^0, t_1, L_1\rangle$ and $\mathcal{S}_2 = \langle S_2, s_2^0, t_2, L_2\rangle$, a *simulation of* $\mathcal{S}_1$ *by* $\mathcal{S}_2$ is a binary relation $R \subseteq S_1 \times S_2$ such that $(s_1^0, s_2^0) \in R$ and, for any $(s_1, s_2) \in R$ and for any $a \in \Sigma$, we have: if $t_1(s_1, a)$ is defined, then $t_2(s_2, a)$ is defined and $(t(s_1, a), t_2(s_2, a)) \in R$.
We write $\mathcal{S}_1 \preceq \mathcal{S}_2$ whenever there exists a simulation of $\mathcal{S}_1$ by $\mathcal{S}_2$. Basically, $\mathcal{S}_1 \preceq \mathcal{S}_2$ tells that $\mathcal{S}_1$ has more executions than $\mathcal{S}_2$.

One easily checks that $\preceq$ is a preorder. Note that if two processes $\mathcal{S}_1$ and $\mathcal{S}_2$ are bisimilar, then $\mathcal{S}_1 \preceq \mathcal{S}_2$ and $\mathcal{S}_2 \preceq \mathcal{S}_1$; the relation $\preceq$ ignoring the atomic propositions, the reciprocal does not hold in general. We let $\mathcal{S}_1 \prec \mathcal{S}_2$ mean $\mathcal{S}_1 \preceq \mathcal{S}_2$ and not $\mathcal{S}_2 \preceq \mathcal{S}_1$.

We now establish two useful technical lemmas.

Lemma 10. For any process $\mathcal{S}$, we have: $\mathcal{S} \preceq \mathcal{S}_{(\neg c)}$ if and only if $\mathcal{S} \models [\,]\texttt{INV}\,(p)$.

PROOF. Let S and S' be respectively the set of states of $\mathcal{S}$ and $\mathcal{S}_{(\neg p)}$. Hence $S' \subseteq S$. Actually one proves that $\mathcal{S} \preceq \mathcal{S}_{(\neg p)}$ by check the simulation $\{(s, s) \in S' \times S'\}$.

Lemma 11. For any p-labeling process $\mathcal{E}$: $\mathcal{S} \preceq \mathcal{S} \times \mathcal{E}_{(\neg p)}$ if and only if $\mathcal{S} \times \mathcal{E} \models [\,]\texttt{INV}\,(p)$.

PROOF. By Lemma 10 for $\mathcal{S} \times \mathcal{E}$, $\mathcal{S} \times \mathcal{E} \models [\,]\texttt{INV}\,(p)$ if and only if $\mathcal{S} \times \mathcal{E} \preceq (\mathcal{S} \times \mathcal{E})_{(\neg p)}$. Since $\mathcal{E}$ is complete, $\mathcal{S} \preceq \mathcal{S} \times \mathcal{E}$ and because $(\mathcal{S} \times \mathcal{E})_{(\neg p)}$ is bisimilar to $\mathcal{S} \times \mathcal{E}_{(\neg p)}$, we obtain $\mathcal{S} \preceq \mathcal{S} \times \mathcal{E} \preceq (\mathcal{S} \times \mathcal{E})_{(\neg p)} \preceq \mathcal{S} \times \mathcal{E}_{(\neg p)}$. By transitivity, $\mathcal{S} \preceq \mathcal{S} \times \mathcal{E}_{(\neg p)}$. The reciprocal is easy.

3. BASIC CONTROLLERS

The current section shows how basic controllers can be specified in the logic $\mathrm{Q}L_\mu$. The correctness of the specifications is also given.

¿From now on, we assume given $\Gamma \subseteq AP$ and a process $\mathcal{P}$ on Γ which represents a plant: usually, the set of events Σ divides between *uncontrollable* events $\Sigma_{uc} \subseteq \Sigma$ on the one hand and *controllable* events, by complementary, on the other hand.

We suppose that the reader is familiar with the notion of *controller of* $\mathcal{P}$ *for some control objective*, as proposed in the reference work of (Ramadge and Wonham, 1989). In our setting, since the control objectives are mu-calculus, and even quantified mu-calculus, definable properties, we agree that *a controller of* $\mathcal{P}$ *for* $\alpha \in \mathrm{Q}L_\mu$ is a process $\mathcal{C}$ with the following:

(1) at any time, it allows all uncontrollable events:
$\mathcal{C} \models \texttt{Admi} \stackrel{\text{def}}{=} \texttt{INV}\,(\bigwedge_{u\in\Sigma_{uc}} \stackrel{u}{\rightarrow})$, and
(2) it achieves the objectives inasmuch as:
$\mathcal{P} \times \mathcal{C} \models \alpha$.

In the following, we fix a fresh atomic proposition c and we let the following mu-calculus formula:

$$\texttt{Controller}(c) \stackrel{\text{def}}{=} c \wedge \texttt{INV}\,((\neg c) \Rightarrow \texttt{INV}\,(\neg c)) \wedge \texttt{INV}\,(c \Rightarrow \bigwedge_{u\in\Sigma_{uc}} [u]c)$$

$\texttt{Controller}(c)$ is interpreted on a c-labeling of $\mathcal{P}$, where c represents reachable states under control: a c-labeling of $\mathcal{P}$ satisfies $\texttt{Controller}(c)$ whenever, under control,

(1) the initial state is reachable, since c-labeled,
(2) successor states of unreachable states are neither reachable, and
(3) the successors of reachable states via uncontrollable events remain reachable.

Proposition 12. For any $\mathcal{E} \in Lab_c$ s.t. $\mathcal{P} \times \mathcal{E}$ satisfies $\texttt{Controller}(c)$, there exists $\mathcal{E}' \in Lab_c$ which defines a control, in the sense that $\mathcal{E}'_{(\neg c)} \models \texttt{Admi}$, and s.t. $\mathcal{E}' \equiv_{\mathcal{P}} \mathcal{E}$.

PROOF. Let $\mathcal{E} = \langle E, e^0, t, L\rangle \in Lab_c$ and let us assume that $\mathcal{P} \times \mathcal{E} \models \texttt{Controller}(c)$. Define $\mathcal{E}' \in Lab_c$ by $\mathcal{E}' = \langle E, e^0, t', L\rangle$ where

$$t'(e, a) = \begin{cases} e & \text{if } (a \in \Sigma_{uc}) \text{ and } (c \in L(e)) \\ & \quad\text{and } (c \notin L(t(e, a))) \\ t(e, a) & \text{otherwise} \end{cases}$$

By construction, $\mathcal{E}' \models c \wedge \texttt{INV}\,(c \Rightarrow \bigwedge_{u\in\Sigma_{uc}}[u]c)$; hence $\mathcal{E}'_{(\neg c)} \models \texttt{INV}\,(\bigwedge_{u\in\Sigma_{uc}} \stackrel{u}{\rightarrow})$. The fact that $t'(e, a)$ differs from $t(e, a)$ only when $a \in \Sigma_{uc}$, $c \in L(e)$ and $c \notin L(t(e, a))$, and since by assumption, $\mathcal{P} \times \mathcal{E} \models \texttt{INV}\,(\bigwedge_{u\in\Sigma_{uc}} c \Rightarrow [u]c)$, we conclude that $\mathcal{E}' \equiv_{\mathcal{P}} \mathcal{E}$.

Propositions 7 and 12 together lead to Theorem 13 below, provided we agree upon the following:

$$\texttt{Solution}(c, \alpha) \stackrel{\text{def}}{=} \alpha{*}c \wedge \texttt{Controller}(c)$$

Theorem 13. For any $\alpha \in \mathrm{Q}L_\mu$,

there exists a controller of $\mathcal{P}$ for α
if and only if

$$\mathcal{P} \models \exists c.\texttt{Solution}(c, \alpha) \qquad (1)$$

PROOF. The specification above is correct as we show now: assume there is an controller $\mathcal{C}$ for α. We will define $\mathcal{E} \in Lab_c$ such that $\mathcal{E}_{(\to c)}$ and $\mathcal{C}$ are isomorphic: we assign c to each state of $\mathcal{C}$ and we complete the resulting process with a sink state. Then $\mathcal{P} \times \mathcal{E}$ satisfies $c \wedge \texttt{INV}((\neg c) \Rightarrow \texttt{INV}(\neg c))$ and since $\mathcal{C}$ is a controller, $\mathcal{P} \times \mathcal{E}$ satisfies $\texttt{INV}(c \Rightarrow \bigwedge_{u \in \Sigma_{uc}} [u]c)$. Since $\mathcal{C}$ is a controller for α on $\mathcal{P}$ and since $\mathcal{E}_{(\to c)}$ and $\mathcal{C}$ are isomorphic, by Proposition 7, $\mathcal{P} \times \mathcal{E}$ satisfies $\alpha{*}c$.

Assume now that $\mathcal{P}$ satisfies the formula (1). Then there exists $\mathcal{E} \in Lab_c$ such that $\mathcal{P} \times \mathcal{E}$ satisfies $(\alpha * c)$ and $\texttt{Controller}(c)$. Since $\mathcal{P} \times \mathcal{E} \models \texttt{Controller}(c)$, there exists (Proposition 12) a labeling process $\mathcal{E}'$ such that $\mathcal{E}' \equiv_{\mathcal{P}} \mathcal{E}$ and $\mathcal{E}'_{(\to c)}$ is an controller. Moreover, $\mathcal{E}' \equiv_{\mathcal{P}} \mathcal{E}$ and $\mathcal{P} \times \mathcal{E} \models \alpha{*}c$ imply $\mathcal{P} \times \mathcal{E}' \models \alpha{*}c$. Then, take $\mathcal{E}'_{(\to c)}$ to conclude (by Proposition 7).

4. MAXIMALLY PERMISSIVE CONTROLLERS

We first give the definition for *Maximal Permissiveness* in a fully natural way. Next we specify it in $\mathrm{Q}L_\mu$, with the arguments for its correctness.

Definition 14. **(Maximally Permissive Controller)** Given a formula $\alpha \in \mathrm{Q}L_\mu$, we say that the controller $\mathcal{C}$ of $\mathcal{P}$ for α is *maximally permissive* if for any other controller $\mathcal{C}'$ of $\mathcal{P}$ for α, we do not have $\mathcal{S} \times \mathcal{C} \prec \mathcal{S} \times \mathcal{C}'$.

Essentially, since the labelings of the plant can be compared regarding the size of the remaining subtree after pruning, so can the underlying controllers, hence our capability to express optimality in the logic: given any two fresh atomic propositions c and c', we define the formulas $c \sqsubseteq c'$ and $c \sqsubset c'$ by:

$$c \sqsubseteq c' \stackrel{\text{def}}{=} ([\,]\texttt{INV}(c'))*c$$

$$c \sqsubset c' \stackrel{\text{def}}{=} c \sqsubseteq c' \wedge \neg(c' \sqsubseteq c)$$

Formula $c \sqsubseteq c'$ contains a c-adjustment to express that c' holds all along c-labeled executions.

Proposition 15. For all $\mathcal{E}, \mathcal{E}' \in Lab_c$, we have:

$$\mathcal{P} \times \mathcal{E}_{(\to c)} \preceq \mathcal{P} \times \mathcal{E}'_{(\to c')} \text{ iff } \mathcal{P} \times \mathcal{E} \times \mathcal{E}' \models c \sqsubseteq c'$$

PROOF. Assume that $\mathcal{P} \times \mathcal{E} \times \mathcal{E}'$ satisfies $c \sqsubseteq c'$. By Proposition 7, this is equivalent to say that $\mathcal{P} \times \mathcal{E}' \times \mathcal{E}_{(\to c)}$ satisfies $[\,]\texttt{INV}(c')$. Applying now Lemma 11 for $\mathcal{P} \times \mathcal{E}_{(\to c)}$ leads to $\mathcal{P} \times \mathcal{E}_{(\to c)} \preceq \mathcal{P} \times \mathcal{E}_{(\to c)} \times \mathcal{E}'_{(\to c')}$. Because clearly $\mathcal{P} \times \mathcal{E}_{(\to c)} \times \mathcal{E}'_{(\to c')} \preceq \mathcal{P} \times \mathcal{E}'_{(\to c')}$, we conclude by transitivity. Reciprocally, assume $\mathcal{P} \times \mathcal{E}_{(\to c)} \preceq \mathcal{P} \times \mathcal{E}'_{(\to c')}$; by composing both terms with $\mathcal{E}_{(\to c)}$, we obtain $\mathcal{P} \times \mathcal{E}_{(\to c)} \preceq \mathcal{P} \times \mathcal{E}'_{(\to c')} \times \mathcal{E}_{(\to c)}$. Now similar arguments to previous ones can be put forward to conclude.

Theorem 16. For any $\alpha \in \mathrm{Q}L_\mu$, there exists a maximally permissive controller of $\mathcal{P}$ for α if and only if

$$\begin{aligned}\mathcal{P} \models \exists c.\ &\texttt{Solution}(c,\alpha)\ \wedge \\ &\forall c'.[\,c \sqsubset c' \Rightarrow \neg\texttt{Solution}(c',\alpha)\,]\end{aligned} \quad (2)$$

PROOF. $\Leftarrow$) By Theorem 13 and because $\mathcal{P} \models \exists c.\texttt{Solution}(c,\alpha)$, there exists $\mathcal{E} \in Lab_c$ s.t. $\mathcal{E}_{(\to c)}$ is a controller on $\mathcal{P}$ for α and

$$\mathcal{P} \times \mathcal{E} \models \forall c.c \sqsubset c' \Rightarrow \neg\texttt{Solution}(c',\alpha) \quad (3)$$

Equation 3 with Proposition 15 means for all process $\mathcal{E}' \in Lab_{c'}$, if $\mathcal{P} \times \mathcal{E}_{(\to c)} \prec \mathcal{P} \times \mathcal{E}'_{(\to c')}$ then $\mathcal{P} \times \mathcal{E} \times \mathcal{E}' \not\models \texttt{Solution}(c',\alpha)$. Now, because $\mathcal{E}$ is complete and since c does not occur in $\texttt{Solution}(c',\alpha)$, we can assert: $\mathcal{P} \times \mathcal{E}' \not\models \texttt{Solution}(c',\alpha)$. By Theorem 13, $\mathcal{E}'_{(\to c')}$ necessarily cannot be a controller of $\mathcal{P}$ for α. By Definition 9, $\mathcal{E}_{(\to c)}$ is a maximally permissive controller of $\mathcal{P}$ for α.

$\Rightarrow$) Assume a maximally permissive controller for α on $\mathcal{P}$, say $\mathcal{C}$. Let us label all states of $\mathcal{C}$ by c, and complete the result to get some $\mathcal{E} \in Lab_c$ s.t. $\mathcal{E}_{(\to c)}$ equals $\mathcal{C}$. By Theorem 13 (and actually its proof arguments), $\mathcal{P} \times \mathcal{E} \models \texttt{Solution}(c,\alpha)$. Moreover, since $\mathcal{C}$ is maximally permissive, for all $\mathcal{E}' \in Lab_{c'}$, we have $\mathcal{P} \times \mathcal{E} \times \mathcal{E}' \models \texttt{Solution}(c',\alpha) \Rightarrow \neg(c \sqsubset c')$. Which concludes.

5. EXPRESSIVENESS

We focuse now on Formula (2), where α is a parameter, and we partially answer expresiveness. Recall that Formula (2) represents a new kind of specification which expresses the maximal permissiveness requirement for controllers.

Theorem 17. Let α be formula in L_μ. **There cannot exist** formulas β and γ in L_μ, or even in $\mathrm{Q}L_\mu$, such that for all plant $\mathcal{P}$:

$\mathcal{C}$ is a maximally permissive controller of $\mathcal{P}$ for α
if and only if
$\mathcal{C} \models \gamma$ and $\mathcal{P} \times \mathcal{C} \models \beta$

PROOF. Take $\alpha = \mu X.[\,]X \vee p$, written $\mathbf{AF}p$ in the following, as in the classic logic CTL. $\mathbf{AF}p$ tells that along any execution, the proposition p will eventually hold.

Assume some pair (β,γ) exists. Take $\mathcal{P}$ as in Fig.1, where the two events a and b of Σ are both controllable. Clearly, the single state process with an a-loop is a maximally permissive controller for $\mathbf{AF}p$. By assumption, $\mathcal{C} \models \gamma$ and $\mathcal{P} \times \mathcal{C} \models \beta$.

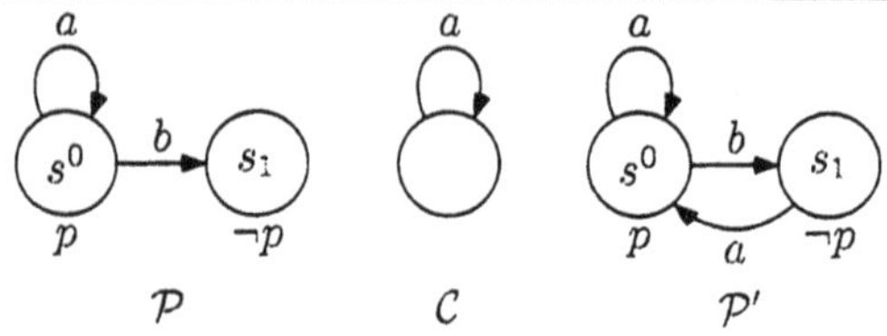

Fig. 1. $\mathcal{P}$, the controller $\mathcal{C}$ and $\mathcal{P}'$

Now define $\mathcal{P}'$ as $\mathcal{P}$ plus an a-transition back to the initial state, as drawn in Figure 1. Because $\mathcal{P} \times \mathcal{C}$ and $\mathcal{P}' \times \mathcal{C}$ are bisimilar, we also have $\mathcal{P}' \times \mathcal{C} \models \beta$. Since by assumption $\mathcal{C} \models \gamma$, $\mathcal{C}$ is a solution of (β, γ) and yet $\mathcal{C}$ is not a maximally permissive controller of $\mathcal{P}'$ for $\mathbf{AF}p$: indeed, $\mathcal{C}$ cuts the transition from s^0 to s_1 of $\mathcal{P}'$, which should be kept. This leads to a contradiction.

As a consequence of Theorem 17, while with the logic $\mathrm{Q}L_\mu$ the maximal permissiveness is expressed in a uniform manner, its statement in the framework of (Arnold *et al.*, 2003) must be designed according to the plant, hence an ad hoc solution.

6. CONCLUSION

The presented work proposes a logical framework for the specification of control problems, which additionally allows to perform synthesis. The key ideas are built on an extension of the propositional mu-calculus using quantified atomic propositions. The resulting is expressive enough to specify maximally permissive controllers for any mu-calculus definable control objective, independently of the plant. Moreover, we showed that other existing logical frameworks, like (Arnold *et al.*, 2003), cannot express maximal permissiveness on its own, but would rather need a system-dependent specification. For the sake of simplicity, we have only proposed here a fully observational setting, but the results still hold in the more general case of partial observation, as explained in (Riedweg and Pinchinat, 2003*b*), as well in a forthcoming paper.

As explained in the paper, the maximal permissiveness property is definitely a strong argument in favor of our approach and demonstrates a fundamental feature of the quantified logic framework. Indeed, contrary to other approaches, it is possible to express properties outside the remaining system after the synchronous product of the plant with the controller. As a consequence, many other kinds of optimality criteria can be considered, such as the unicity of a solution, but also, "contextual" properties: one can specify that deadlocks are forbidden unless they already exist in the plant, priority between events, fairness of the control, forcing events, etc.

REFERENCES

Arnold, A. and D. Niwinski (2001). *Rudiments of mu-calculus*. North-Holland.

Arnold, Vincent and Walukiewicz (2003). Games for synthesis of controllers with partial observation. *Theoretical Computer Science* **1**, 7–34.

Emerson, E. A. and C. Lei (1986). Efficient model checking in fragments of the propositional mu-calculus. In: *Proc. 1st IEEE Symp. Logic in Computer Science, Cambridge, Massachussets.*. pp. 267–278.

Golaszewski, C. H. and P. J. Ramadge (1987). Control of discrete event systems with forced events. In: *Proc. of 26th Conf. Decision and Control*. Los Angeles, CA, USA. pp. 247–251.

Kozen (1983). Results on the propositional μ-calculus. *Theoretical Computer Science* **27(3)**, 333–354.

Kupferman, O., P. Madhusudan, P. S. Thiagarajan and M. Y. Vardi (2000). Open systems in reactive environments: Control and synthesis. In: *Proc. 11th Int. Conf. on Concurrency Theory*. Vol. 1877 of *LNCS*. Springer-Verlag. pp. 92–107.

Ramadge and Wonham (1989). The control of discrete event systems. In: *Proc. of the IEEE*. Vol. 77(1). pp. 81–98.

Riedweg (2003). Logiques pour le Contrôle d'Automatismes Discrets. PhD thesis. Université de Rennes I.

Riedweg and Pinchinat (2003*a*). Quantified mu-calculus for control synthesis. In: *Mathematical Foundations of Computer Science*. Vol. 2747 of *LNCS*. pp. 642–651.

Riedweg, S. and S Pinchinat (2003*b*). Quantified loop mu-calculus for control synthesis under partial observation. Technical Report R4949. Irisa.

Thistle and Wonham (1994). Control of of infinite behavior of finite automata. *SIAM Control and Optimization* **32(4)**, 1075–1097.

ELSEVIER
IFAC
PUBLICATIONS
www.elsevier.com/locate/ifac

HIERARCHICAL CONTROL FOR STRUCTURAL DECENTRALIZED DES

Klaus Schmidt, Johann Reger, Thomas Moor

Lehrstuhl für Regelungstechnik, Universität Erlangen-Nürnberg
Cauerstraße 7, D-91058 Erlangen, Germany
klaus.schmidt@rt.eei.uni-erlangen.de

Abstract: In this contribution, we consider structural decentralized DES and supplement the existing control architecture with a two-level hierarchy. For the proposed overall system, we prove hierarchical consistency and that the closed-loop behavior is nonblocking. A comprehensive example demonstrates the computational benefit of our method.

Keywords: supervisory control, structural decentralized control, hierarchical control.

1. INTRODUCTION

In the past decade a great variety of ideas have been studied to reduce the complexity of synthesis algorithms for the supervisory control of discrete event systems (Rudie and Wonham, 1992; Jiang et al., 2001; Yoo and Lafortune, 2000; Hubbard and Caines, 2002; da Cunha et al., 2002; Zhong and Wonham, 1990; Lee and Wong, 2002; Wong and Wonham, 1996). A key ingredient of promising approaches is to assume or to impose a particular control architecture, such that computationally expensive product compositions of individual subsystems can be either avoided altogether or at least postponed to a more favorable stage in the design process. Our contribution builds on two recent results from this category, namely structural decentralized control and hierarchical control.

The decentralized control architecture proposed by Lee and Wong (2002) addresses plant models that are composed of a number of subsystems, which are coupled via shared events. Specifications are given for each subsystem individually, and the task is to synthesize individual supervisors. As the subsystems are coupled, synthesis will in general need to refer to the synchronous product of all subsystems. Conditions under which such product can be avoided are given in Lee and Wong (2002).

In hierarchical architectures (Zhong and Wonham, 1990; da Cunha et al., 2002; Hubbard and Caines, 2002), controller synthesis is based on a plant abstraction (high-level model), which is supposed to be less complex than the original plant model (low-level model). Technically, abstractions can be defined as language projections. While from worst case scenarios projections are known to be of exponential computational complexity, application relevant cases with polynomial complexity are identified in Wong (1997). An important question is how to derive the plant abstraction, such that a high-level controller can be implemented by available low-level control actions (hierarchical consistency). A characterization of this property is given in Zhong and Wonham (1990).

In this paper, we consider the decentralized setting of Lee and Wong (2002), where the overall system is modelled by the synchronous product of the individual subsystems. As an abstraction, we propose the natural language projection based on the shared events. We then show that this abstraction does comply with hierarchical consistency as defined in Zhong and Wonham (1990). Furthermore, the high-level model can be computed by the synchronous product of the projections of the individual subsystems (rather than the projection of the product of the individual subsystems). Whenever the projections of the subsystems behave computationally nicely, this change of order promises a substantial computational benefit. This is demonstrated by an example.

The outline of the paper is as follows. Basic notations and definitions of supervisory control theory are recalled in Section 2. Section 3 and Section 4 introduce structural decentralized and hierarchical control of discrete event systems, respectively. In Section 5, both methods are combined so as to form a decentralized and hierarchical control architecture. A comprehensive example in Section 6 illustrates our contribution.

2. PRELIMINARIES

We recall basic facts from supervisory control theory. (Wonham, 2001; Cassandras and Lafortune, 1999).

For a finite alphabet Σ, the set of all finite strings over Σ is denoted Σ^*. We write $s_1 s_2 \in \Sigma^*$ for the concatenation of two strings $s_1, s_2 \in \Sigma^*$. We write $s_1 \leq s$ when s_1 is a *prefix* of s, i.e. if there exists a string $s_2 \in \Sigma^*$ with $s = s_1 s_2$. The empty string is denoted $\varepsilon \in \Sigma^*$, i.e. $s\varepsilon = \varepsilon s = s$ for all $s \in \Sigma^*$. A *language* over Σ is a subset $H \subseteq \Sigma^*$. The *prefix closure* of H is defined by $\overline{H} := \{s_1 \in \Sigma^* | \exists s \in H \text{ s.t. } s_1 \leq s\}$. A language H is *prefix closed* if $H = \overline{H}$.

The *natural projection* $p_i : \Sigma^* \to \Sigma_i^*$, $i = 1,2$, for the (not necessarily disjoint) union $\Sigma = \Sigma_1 \cup \Sigma_2$ is defined iteratively: (1) let $p_i(\varepsilon) := \varepsilon$; (2) for $s \in \Sigma^*$, $\sigma \in \Sigma$, let $p_i(s\sigma) := p_i(s)\sigma$ if $\sigma \in \Sigma_i$, or $p_i(s\sigma) := p_i(s)$ otherwise. The set-valued inverse of p_i is denoted $p_i^{-1} : \Sigma_i^* \to 2^{\Sigma^*}$, $p_i^{-1}(t) := \{s \in \Sigma^* | p_i(s) = t\}$. The *synchronous product* $H_1 || H_2 \subseteq \Sigma^*$ of two languages $H_i \subseteq \Sigma_i^*$ is $H_1 || H_2 = p_1^{-1}(H_1) \cap p_2^{-1}(H_2) \subseteq \Sigma^*$.

A *finite automaton* is a tuple $G = (X, \Sigma, \delta, x_0, X_m)$, where X is the finite set of *states*; Σ is the finite alphabet of *events*; $\delta : X \times \Sigma \to X$ is the partial *transition function*; $x_0 \in X$ is the *initial state*; and $X_m \subseteq X$ is the set of *marked states*. We write $\delta(x, \sigma)!$ if δ is defined at (x, σ). In order to extend δ to a partial function on $X \times \Sigma^*$, recursively let $\delta(x, \varepsilon) := x$ and $\delta(x, s\sigma) := \delta(\delta(x, s), \sigma)$, whenever both $x' = \delta(x, s)$ and $\delta(x', \sigma)!$. $L(G) := \{s \in \Sigma^* : \delta(x_0, s)!\}$ and $L_m(G) := \{s \in L(G) : \delta(x_0, s) \in X_m\}$ are the *closed* and *marked language* generated by the finite automaton G, respectively. For a formal definition of the synchronous composition of two automata G_1 and G_2 we refer to e.g. Cassandras and Lafortune (1999) and note that $L(G_1 || G_2) = L(G_1) || L(G_2)$.

When $L(G)$ represents the plant behavior in a supervisory control context, we write $\Sigma = \Sigma_c \cup \Sigma_u$, $\Sigma_c \cap \Sigma_u = \emptyset$, to distinguish *controllable* (Σ_c) and *uncontrollable* (Σ_u) events. A *control pattern* is a set γ, $\Sigma_u \subseteq \gamma \subseteq \Sigma$, and the set of all control patterns is denoted $\Gamma \subseteq 2^\Sigma$. A *supervisor* is a map S: $L(G) \to \Gamma$, where $S(s)$ represents the set of enabled events after the occurrence of string s; i.e. a supervisor can disable controllable events only. The language $L(S/G)$ generated by G under supervision S is iteratively defined by (1) $\varepsilon \in L(S/G)$ and (2) $s\sigma \in L(S/G)$ iff $s \in L(S/G), \sigma \in S(s)$ and $s\sigma \in L(G)$. Thus, $L(S/G)$ represents the behavior of the *closed-loop system*. To take into account the marking of G, let $L_m(S/G) := L(S/G) \cap L_m(G)$. The closed-loop system is *nonblocking* if $\overline{L_m(S/G)} = L(S/G)$, i.e. if each string in $L(S/G)$ is the prefix of a marked string in $L_m(S/G)$.

A language H is said to be controllable w.r.t. $L(G)$ if there exists a supervisor S such that $\overline{H} = L(S/G)$. The set of all languages that are controllable w.r.t. $L(G)$ is denoted $C(L(G))$ and can be characterized by $C(L(G)) = \{H \subseteq L(G) | \ \exists S \text{ s.t. } \overline{H} = L(S/G)\}$. Furthermore, the set $C(L(G))$ is closed under arbitrary union. Hence, for every *specification* language E there uniquely exists a *supremal controllable sublanguage* of E w.r.t. $L(G)$, which is formally defined as $\kappa_{L(G)}(E) := \cup\{K \in C(L(G)) | \ K \subseteq E\}$. A supervisor S that leads to a closed-loop behavior $\kappa_{L(G)}(E)$ is said to be *maximal permissive*. A maximal permissive supervisor can be realized on the basis of a generator of $\kappa_{L(G)}(E)$. The latter can be computed from G and a generator of E. The computational complexity is of order $O(N^2M^2)$, where N and M are the number of states in G and the generator of E, respectively.

A language E is L_m-*closed* if $\overline{E} \cap L_m = E$ and the set of $L_m(G)$-closed languages is denoted $F_{L_m(G)}$. For specifications $E \in F_{L_m(G)}$, the plant $L(G)$ is nonblocking under maximal permissive supervision.

3. STRUCTURAL DECENTRALIZED DES

Structural decentralized DES as proposed in Lee and Wong (2002) are composed of subsystems, realized by finite state automata G_i, $i = 1, 2, \ldots, n$ with respective alphabets Σ_i. The synchronization of each two subsystems G_i and G_j is organized via shared events $\Sigma_i \cap \Sigma_j$.

Definition 3.1. A *decentralized control system* (DCS) consists of subsystems, modelled by finite state automata $G_i, i = 1, \ldots, n$ over the respective alphabets Σ_i. The overall system is defined as $G := ||_{i=1}^n G_i$ over the alphabet $\Sigma := \bigcup_{i=1}^n \Sigma_i$. The controllable and uncontrollable events are $\Sigma_{i,c} := \Sigma_i \cap \Sigma_c$ and $\Sigma_{i,u} := \Sigma_i \cap \Sigma_u$, respectively, where $\Sigma_c \cup \Sigma_u = \Sigma$ and $\Sigma_c \cap \Sigma_u = \emptyset$. For brevity and convenience, let $L := L(G)$, $L_m := L_m(G)$, $L_i := L(G_i)$, and $L_{i,m} := L_m(G_i)$.

For our applications we assume local specifications $E_i \in F_{L_{i,m}} \subseteq \Sigma_i^*, i = 1, \ldots, n$ for each subsystem G_i. Relative to the overall alphabet Σ, the E_i become $(p_i)^{-1}(E_i)$, where $p_i : \Sigma^* \to \Sigma_i^*$ denotes the natural projection. Taking into account the language L of G, the global specification is $E = \bigcap_{i=1}^n (p_i)^{-1}(E_i) \cap L$.

There are two approaches for generating a supervisor implementing the given set of specifications: the synthesis of a monolithic supervisor S for the overall specification E leading to the closed language $L(S/G) = \overline{\kappa_L(E)}$ of the supervised system, and the synthesis of local supervisors S_i for G_i and E_i, $i = 1,\ldots,n$ resulting in $||_{i=1}^n L(S_i/G_i) = \bigcap_{i=1}^n (p_i)^{-1}(\kappa_{L_i}(E_i)) \cap L$; for the second approach see Figure 1.

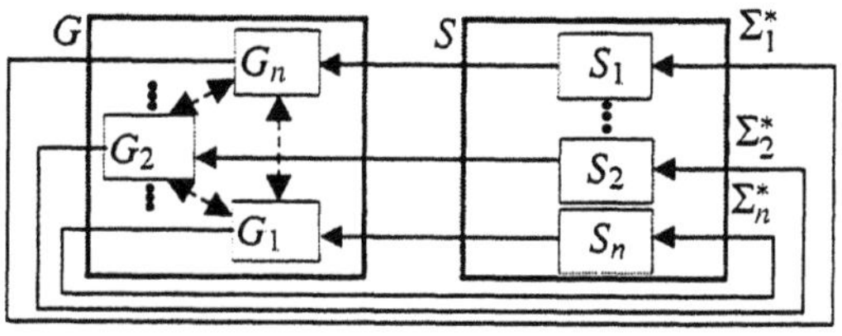

Fig. 1. Structural Decentralized Control Structure

In Lee and Wong (2002), conditions under which both approaches lead to the same overall result are developed. Thus, for $i = 1,\ldots,n$ the following holds:

$$\bigcap_{i=1}^{n}(p_i)^{-1}(\overline{\kappa_{L_i}(E_i)}) = \overline{\kappa_L(E)}, \tag{1}$$

$$p_i(\overline{\kappa_L(E)}) \text{ is nonblocking w.r.t. } L_{i,m}. \tag{2}$$

Theorem 3.1. (Structural Decentralized DES Lee and Wong (2002)). Using notation as in Definition 3.1, denote the natural projection $p_i^{ij} : (\Sigma_i \cup \Sigma_j)^* \to \Sigma_i^*$. Suppose that $E_i \in F_{L_{i,m}}$ and that for $i, j = 1,\ldots,n, i \neq j$

(i) $L_{i,m}$ marks $\Sigma_i \cap \Sigma_j$, i.e.

$$\Sigma_i^*(\Sigma_i \cap \Sigma_j) \cap \overline{L_{i,m}} \subseteq L_{i,m}(\Sigma_i \cap \Sigma_j) \tag{3}$$

(ii) L_i and L_j are mutually controllable, i.e.

$$\overline{L_i}(\Sigma_{j,u} \cap \Sigma_i) \cap p_i^{ij}((p_j^{ij})^{-1}(\overline{L_j})) \subseteq \overline{L_i} \tag{4}$$

Then (1) and (2) hold. [1]

In order to discuss the computational complexity, let N and M denote bounds for the number of states in G_i and generators of E_i, respectively. Then N^n and M^n are the respective bounds for the number of states in the overall system. The computational complexity of the monolithic synthesis procedure is $O(N^{2n}M^{2n})$, whereas it is $O(nN^2M^2)$ for the local synthesis approach. It is clear that this benefit does not come for free. The approach in Lee and Wong (2002) requires the computation of the language projection (with exponential worst case complexity) in condition (4). However, $G = ||_{i=1}^n G_i$ need not be computed.

4. HIERARCHICAL CONTROL

We consider the event-based hierarchical control scheme in Zhong and Wonham (1990) (Figure 2 (a)).

The detailed plant model G and the supervisor S^{lo} form a low-level closed-loop system, indicated by Con^{lo} (control action) and Inf^{lo} (feedback information). Similarly, the high-level closed loop consists of an abstract plant model G^{hi} and the supervisor S^{hi}. The two levels are interconnected via Com^{hilo} and Inf^{lohi}. The former allows S^{hi} to impose high-level control on S^{lo}, the latter drives the abstract plant G^{hi} in accordance to the detailed model. From the perspective of the high-level supervisor, the forward path sequence Com^{hilo}, Con^{lo} is usually designated "command and control", while the feedback path sequence Inf^{lohi}, Inf^{hi} is identified with "report and advise". Formally, the high-level abstraction is defined as follows.

Definition 4.1. (Hierarchical Abstraction). Let $G = (X, \Sigma, \delta, x_0, X_m)$ be a finite automaton and $\Sigma^{hi} \subseteq \Sigma$ a set of high-level events. A *reporter map* [2] is a map $\theta : \Sigma^* \to (\Sigma^{hi})^*$ such that (1) $\theta(\varepsilon) = \varepsilon$ and (2) either $\theta(s\sigma) = \theta(s)$ or $\theta(s\sigma) = \theta(s)\sigma^{hi}$, where $\sigma \in \Sigma$, $\sigma^{hi} \in \Sigma^{hi}$. The high-level language is defined by $L^{hi} := \theta(L(G))$. The high-level marking is chosen s.t. $L_m^{hi} \subseteq L^{hi}$, where L_m^{hi} is required to be regular. The canonical recognizer of L_m^{hi} is denoted G^{hi}, and hence, $L(G^{hi}) = L^{hi}$, $L_m(G^{hi}) = L_m^{hi}$. Finally, high-level controllable and uncontrollable events are denoted Σ_c^{hi} and Σ_u^{hi}, respectively, where $\Sigma^{hi} = \Sigma_c^{hi} \cup \Sigma_u^{hi}$, $\Sigma_c^{hi} \cap \Sigma_u^{hi} = \emptyset$.

For our control architecture, not only the high-level events and the reporter map, but also the choice of controllable and uncontrollable events Σ_c^{hi} and Σ_u^{hi} are essential. For the synchronous composition of subsystems, we will show in Section 5 that this choice can be based on shared events.

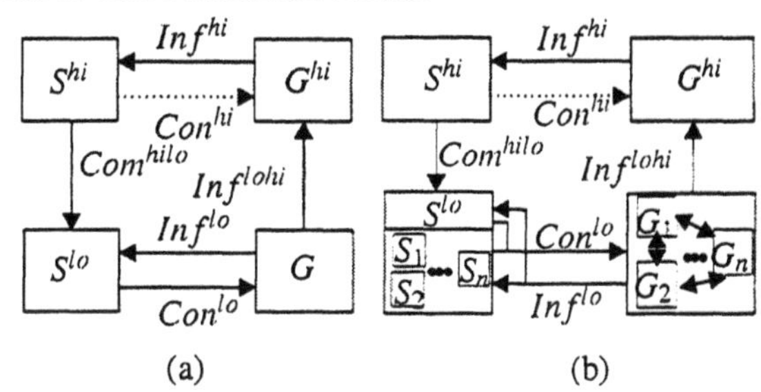

Fig. 2. Hierarchical Control Schemes

For our further discussion, we need to define the interconnection of low- and high-level supervisors with the plant.

Definition 4.2. (Hierarchical Control System). Referring to the notation in Definition 4.1, a *hierarchical control system (HCS)* consists of G, G^{hi}, S^{hi} and S^{lo}, where the *high-level supervisor* S^{hi} and the *low-level supervisor* S^{lo} fulfill the following conditions: $S^{hi}: L^{hi} \to \Gamma^{hi}$ with the high-level control patterns $\Gamma^{hi} := \{\gamma | \Sigma_u^{hi} \subseteq \gamma \subseteq \Sigma^{hi}\}$; and $S^{lo} : L(G) \to \Gamma$ with $\theta(L(S^{lo}/G)) \subseteq L(S^{hi}/G^{hi})$.

Given a high-level specification $E^{hi} \in F_{L_m^{hi}}$, we can synthesize S^{hi} such that $L(S^{hi}/G^{hi}) = \kappa_{L^{hi}}(E^{hi})$ with a nonblocking high-level closed-loop. At this stage, the remaining task is to implement high-level control actions for the low-level plant by means of S^{lo}.

[1] Note also that in this case $E \in F_{L_m}$.

[2] In the sequel, we focus our attention on the reporter map $\theta = p^{hi}$, where p^{hi} is the natural language projection into $(\Sigma^{hi})^*$.

Definition 4.3. (Hierarchical Control Problem). Given G, G^{hi}, S^{hi}, find a low-level supervisor S^{lo} as in Definition 4.2 such that the low-level controlled language of the HCS, $L(S^{lo}/G)$, is nonblocking.

On the one hand, there may not exist a low-level supervisor S^{lo} such that $\theta(L(S^{lo}/G)) = L(S^{hi}/G^{hi})$, and we end up with a strict subset relation. This is interpreted as an overoptimistic high-level supervisor expecting low-level behavior that is not possible. On the other hand, if the above equation turns out true for some S^{lo}, it provides a powerful tool to show that the the low-level is nonblocking. This is one motivation for the following notion of hierarchical consistency.

Definition 4.4. (Hierarchical Consistency (HC)). The hierarchical control system in Definition 4.2 is *hierarchically consistent* if the following applies:

$$\theta(L(S^{lo}/G)) = L(S^{hi}/G^{hi}). \qquad (5)$$

There has been a variety of event-based contributions which tackle the complexity problem associated with the hierarchical abstraction and which provide appropriate definitions of the high-level marking as well as high-level controllable events in order to prove hierarchical consistency. Zhong and Wonham (1990) introduces the notion of strict-output-control-consistency to guarantee a consistent low-level controller fulfilling a high-level specification. Blocking is not considered in this approach. da Cunha et al. (2002) utilizes a generalized model for controlled DES in the high-level to obtain hierarchical consistency without adding complexity by refining the hierarchy. Moor et al. (2003) uses Willems' I/O behaviours to establish a nonblocking hierarchical control architecture. Other state-based approaches (Hubbard and Caines, 2002) use state-aggregation for hierarchical abstraction.

5. HIERARCHICAL CONTROL FOR DECENTRALIZED SYSTEMS

We discuss how a combination of the decentralized approach in Section 3 with the hierarchical approach in Section 4 can improve the computational efficiency of supervisor synthesis; see Figure 2 (b) for our proposed overall scheme. As our detailed plant model, we consider a structural decentralized DES $G_1, \ldots G_n$, $G = ||_{i=1}^n G_i$, subject to local feedback control by supervisors $S_1, \ldots, S_n$. The abstraction G^{hi} of the low-level closed loop is based on the observation of shared events Σ^{hi} via Inf^{lohi}. A high-level supervisor S^{hi} is designed to control G^{hi}, where effect on the low level is taken via Con^{hilo} and a low-level supervisor S^{lo}.

Definition 5.1. A Hierarchical and Decentralized Control System (HDCS) consists of the following entities.

- A detailed plant model is a decentralized control system $G := ||_{i=1}^n G_i$ with subsystems G_i over respective alphabets Σ_i, $i = 1, \ldots, n$ and $\Sigma := \bigcup_{i=1}^n \Sigma_i$. Relevant languages are $L := L(G)$, $L_m := L_m(G)$, $L_i := L(G_i)$, and $L_{i,m} := L_m(G_i)$, controllable and uncontrollable events are $\Sigma_{i,c} = \Sigma_i \cap \Sigma_c$ and $\Sigma_{i,u} = \Sigma_i \cap \Sigma_u$ as in Definition 3.1.
- Local low-level controllers are denoted $S_i\colon L_i \to \Gamma_i$, where Γ_i are the respective control patterns. Low-level closed-loop languages are denoted $L_i^c := L(S_i/G_i)$, $L_{i,m}^c := L_i^c \cap L_{i,m}$, $L^c := ||_{i=1}^n L_i^c$, $L_m^c := ||_{i=1}^n L_{i,m}^c = L^c \cap L_m$. Also let G^c be a generator such that $L^c = L(G^c)$, $L_m^c = L_m(G^c)$.
- For the abstraction the reporter map: $\theta := p^{hi}$ is used, where $p^{hi} : \Sigma^* \to (\Sigma^{hi})^*$ denotes the natural projection and $\Sigma^{hi} := \bigcup_{i,j,i \neq j}(\Sigma_i \cap \Sigma_j)$. The high-level language is $L^{hi} := \theta(L^c)$ and[3]

$$L_m^{hi} := \{s^{hi} \in L^{hi} |\ (p^{hi})^{-1}(s^{hi}) \cap L_m^c \neq \emptyset\},$$

with canonical recognizer G^{hi} s.t. $L^{hi} = L(G^{hi})$, $L_m^{hi} = L_m(G^{hi})$. High-level controllable events are defined as $\Sigma_c^{hi} := \Sigma_c \cap \Sigma^{hi}$ and $\Sigma_u^{hi} := \Sigma_u \cap \Sigma^{hi}$.
- The high-level supervisor is denoted $S^{hi}\colon L^{hi} \to \Gamma^{hi}$ with the high-level closed-loop language $L(S^{hi}/G^{hi})$. A valid low-level supervisor $S^{lo}\colon L^c \to \Gamma$ must fulfill $\theta(L(S^{lo}/G^c)) \subseteq L(S^{hi}/G^{hi})$.

Lemma 5.1. (High Level Plant). Assume the control architecture of Definition 5.1 and let $L_i^{hi} = p^{hi}(L_i^c)$. Then the high level language is

$$L^{hi} = p^{hi}(||_{i=1}^n L_i^c) = ||_{i=1}^n L_i^{hi}.$$

Proof (Sketch): We use induction and the identity $p_0(H_1||H_2) = p_0(H_1)||p_0(H_2)$ for languages H_1 and H_2 with alphabets Σ_{H_1} and Σ_{H_2} and the natural projection $p_0 : (\Sigma_{H_1} \cup \Sigma_{H_2})^* \to (\Sigma_0)^*$ with $\Sigma_0 \subseteq (\Sigma_{H_1} \cup \Sigma_{H_2})$ and $(\Sigma_{H_1} \cap \Sigma_{H_2}) \subseteq \Sigma_0$ as in Wonham (2001). □

The previous lemma enables the computation of the hierarchical abstraction by only using the hierarchically abstracted subsystems and it is not necessary to compute the full low-level plant automaton. Consequently the computational effort reduces from computing $||_{i=1}^n L_i^c$ and subsequent projection $p^{hi}(||_{i=1}^n L_i^c)$ to first computing $p^{hi}(L_i^c)$ and then evaluating $||_{i=1}^n p^{hi}(L_i^c)$. While technically the computational complexity is of the same order, a significant computational benefit shows in applications where realizations of $p^{hi}(L_i^c)$ have less states than those of L_i^c.

For our further discussion, we assume that the low-level subsystems fulfill conditions (i) and (ii) in Theorem 3.1 and that the low-level supervisors have been synthesized to enforce a local specification $E_i \in F_{L_{i,m}}$; i.e. $L_i^c := L(S_i/G_i) = \kappa_{L_i}(E_i)$. Thus, as a result of Theorem 3.1, the low-level is nonblocking, i.e. $L^c = \overline{L_m^c}$.

The next theorem gives a possible choice of the low-level supervisor such that the proposed control architecture is hierarchically consistent and nonblocking.

Theorem 5.1. (Main Result): Let the hierarchical control architecture for structural decentralized DES be

[3] By construction L_m^{hi} is regular.

defined as in Definition 5.1 and define the low-level supervisor S^{lo} for each $s \in L^c$ as

$$S^{lo}(s) := S^{hi}(p^{hi}(s)) \cup (\Sigma - \Sigma^{hi}).$$

Then S^{lo} is valid according to Definition 4.2, the HDCS is hierarchically consistent and the controller is maximally permissive. In particular, S^{lo} solves the hierarchical control problem in Definition 4.3

Proof: We use induction to show that S^{lo} is valid, i.e. $p^{hi}(L(S^{lo}/G^c)) \subseteq L(S^{hi}/G^{hi})$. Obviously $\varepsilon \in L(S^{lo}/G^c)$ and $p^{hi}(\varepsilon) = \varepsilon \in L(S^{hi}/G^{hi})$. Pick any s and $\sigma \in \Sigma$ with $s\sigma \in L(S^{lo}/G^c)$ and $p^{hi}(s) \in L(S^{hi}/G^{hi})$. Then either $\sigma \in \Sigma^{hi}$ or $\sigma \in (\Sigma - \Sigma^{hi})$. In the first case, $\sigma \in S^{lo}(s)$ implies $\sigma \in S^{hi}(p^{hi}(s))$ and, hence, $p^{hi}(s\sigma) = p^{hi}(s)\sigma \in L(S^{hi}/G^{hi})$. In the second case $p^{hi}(s\sigma) = p^{hi}(s) \in L(S^{hi}/G^{hi})$.

We prove hierarchical consistency by induction. As validity has been established above, we are left to show $p^{hi}(L(S^{lo}/G^c)) \supseteq \kappa_{L^{hi}}(E^{hi})$. It is clear that $\varepsilon \in p^{hi}(L(S^{lo}/G^c))$ and $\varepsilon \in \kappa_{L^{hi}}(E^{hi})$. Let $s^{hi} \in \kappa_{L^{hi}}(E^{hi})$ and $s^{hi} \in p^{hi}(L(S^{lo}/G^c))$ and assume for $\sigma^{hi} \in \Sigma^{hi}$, $s^{hi}\sigma^{hi} \in \kappa_{L^{hi}}(E^{hi})$. Then $\sigma^{hi} \in S^{hi}(s^{hi})$ and by definition of the low-level supervisor S^{lo}, $\forall s \in (p^{hi})^{-1}(s^{hi}) \cap L^c$ we know that $\sigma^{hi} \in S^{lo}(s)$. Further on, as $s^{hi}\sigma^{hi} \in L^{hi}$ it follows that $\exists s' \in L^c$ s.t. $s' \in (p^{hi})^{-1}(s^{hi})$ and $s'\sigma^{hi} \in L^c$. But then $s'\sigma^{hi} \in L(S^{lo}/G^c)$ and thus $p^{hi}(s'\sigma^{hi}) = p^{hi}(s')\sigma^{hi} = s^{hi}\sigma^{hi} \in p^{hi}(L(S^{lo}/G^c))$.

For proving nonblocking behavior it must be shown that $L(S^{lo}/G^c) = \overline{L(S^{lo}/G^c) \cap L^c_m}$. As $L(S^{lo}/G^c) \supseteq \overline{L(S^{lo}/G^c) \cap L^c_m}$ is obvious, we only prove $L(S^{lo}/G^c) \subseteq \overline{L(S^{lo}/G^c) \cap L^c_m}$. Assume $s \in L(S^{lo}/G^c)$ but $s \notin \overline{L(S^{lo}/G^c) \cap L^c_m}$. As $L(S^{lo}/G^c) \subseteq \overline{L^c_m}$, $s \in \overline{L^c_m}$ holds, and thus $s \notin \overline{L(S^{lo}/G^c) \cap L^c_m}$ requires that $\forall u \in \Sigma^*$ s.t. $su \in L^c_m$, $\exists \sigma^{hi} \in \Sigma^{hi}$ and $u', u'' \in \Sigma^*$ s.t. $u = u'\sigma^{hi}u''$ and $\sigma^{hi} \notin S^{lo}(su')$ and $\forall \bar{u}' \leq u'$, $s\bar{u}' \notin L^c_m$. [4] Now let u be such that $su \in L^c_m$ and let σ^{hi}, u', u'' as above. Then it holds that $p_i(su') \in L_{i,m}$ for i such that $\sigma^{hi} \in \Sigma_i$ because of (3) and $\forall i$ s.t. $\sigma^{hi} \notin \Sigma_i$ (let $i_1, \ldots, i_m$ be the corresponding indices), $\exists u_i \in (\Sigma_i - \Sigma^{hi})^*$ such that $p_i(s)u_i \in L^c_{i,m}$ as $L^c_i = \overline{L^c_{i,m}}$. Thus $su'u_{i_1}u_{i_2}\ldots u_{i_m} \in L^c_m$ and thus $s \in \overline{L(S^{lo}/G^c) \cap L^c_m}$.

For proving maximal permissiveness we assume that $\exists \tilde{S}^{lo}$ as in Definition 4.2 s.t. $L(S^{lo}/G^c) \subset L(\tilde{S}^{lo}/G^c) \subseteq (p^{hi})^{-1}(L(S^{hi}/G^{hi}))$ with S^{lo} from Theorem 5.1. Then $\exists s \in L(\tilde{S}^{lo}/G^c)$ with $s^{hi} = p^{hi}(s) \in L(S^{hi}/G^{hi})$ but $s \notin L(S^{lo}/G^c)$. Because of HC $\exists s' \in L(S^{lo}/G^c)$ with $p^{hi}(s') = s^{hi}$ and $s = s'u$ with $u \in (\Sigma - \Sigma^{hi})^*$. But we know that $(\forall s'u' < s'u)$, $(\Sigma - \Sigma^{hi}) \in S^{lo}(su')$ and thus $s = s'u \in L(S^{lo}/G^c)$, which contradicts the assumption and S^{lo} is maximally permissive. □

By the above theorem, the complexity of synthesis for the high-level specification becomes $O((N^{hi})^{2n}(M^{hi})^2)$ compared to $O(N^{2n}(M^{hi})^2)$ for the local monolithic synthesis; where N, N^{hi}, M^{hi} denote number of states in G^c, G^{hi} and the canonical recognizer of E^{hi}, respectively. Again, in the case of $N^{hi} < N$ we expect a computational benefit.

6. EXAMPLE

We consider the two cooperating machine cells G_1 and G_2 given in Figure 3. Both machines have 6 states.

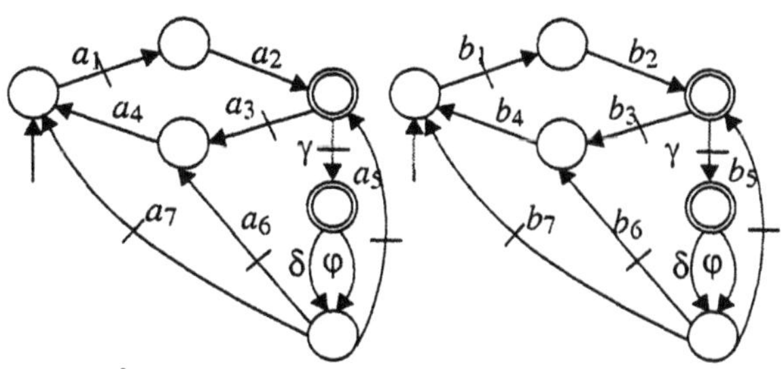

Fig. 3. local machines G_1 and G_2

The event γ represents the start of the cooperation of the two machines, δ indicates that the cooperation terminated successfully and φ represents failure of the cooperation. The machines evolve independently while $a_1, \ldots, a_7$ or $b_1, \ldots, b_7$ occur. Thus the shared events are $\Sigma_1 \cap \Sigma_2 = \{\gamma, \delta, \varphi\}$ and the uncontrollable shared events are $\{\delta, \varphi\}$.

6.1 Hierarchical Method

Following the lines of Lee and Wong (2002) (3) holds as all states before shared events are marked. The mutual controllability condition (4) for $i, j = 1, 2$ is also valid. For example, from Figure 3 and Figure 4 [5] it can readily be observed that $\overline{L_1}$ is controllable w.r.t. $(p_2^{12})^{-1}(p_1^{12}(\overline{L_2}))$ and the event set $\Sigma_{2u} \cap \Sigma_1$. [6]

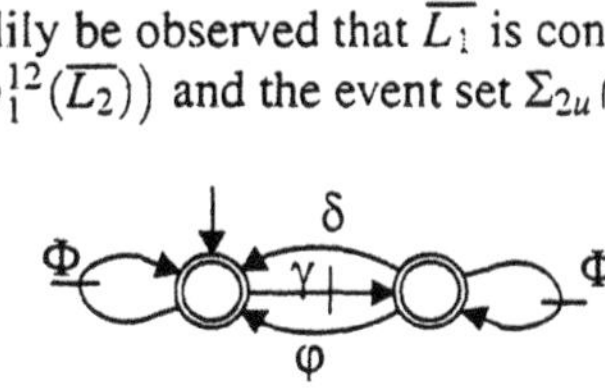

Fig. 4. $(p_2^{12})^{-1}(p_1^{12}(\overline{L_{2,m}}))$

Thus $G = G_1 || G_2$ constitutes a structural-decentralized control system and we identify G as the low-level plant of a HDCS. The low-level specifications E_1 and E_2 in Figure 5 are controllable w.r.t. their respective subsystems and are identical to the low-level controlled subsystems G_1^c and G_2^c.

Now, in addition to low-level supervision, the high-level specification E^{hi} in Figure 6 is implemented by employing our hierarchical control method. E^{hi} states that after the occurrence of two successive failures in cooperation no more cooperation should take place.

[4] This means all extensions of s to a marked string must be disabled and disabling can only occur if a high-level event is disabled by S^{lo}.

[5] $\Phi = \{a_1, a_2, a_3, a_4, a_5, a_6, a_7\}$. The corresponding transition is labeled with a tick as Φ contains controllable events.

[6] The reverse relation is obvious because of symmetry.

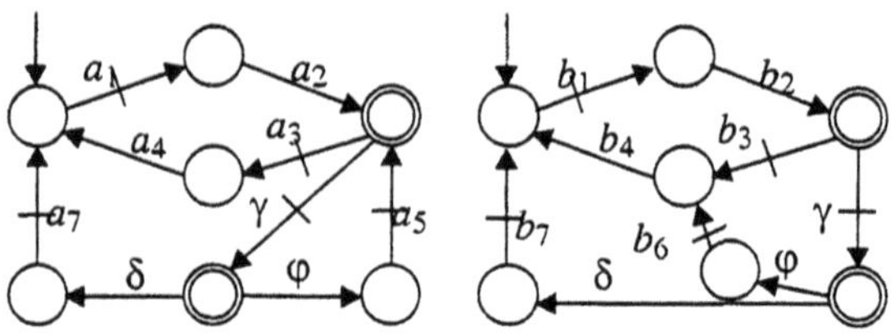

Fig. 5. low-level specifications E_1 and E_2

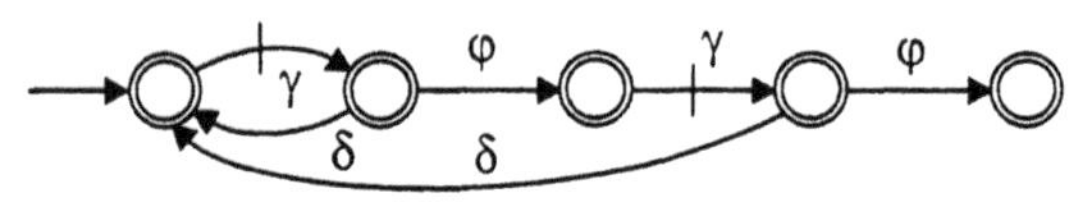

Fig. 6. high-level specification E^{hi}

For the abstraction, $\Sigma^{hi} = \{\gamma, \delta, \varphi\}$ and Figure 7 shows recognizers G_1^{hi} and G_2^{hi} of the respective supervised subsystems G_1^c and G_2^c. It turns out that, in this case, the computation of the projection is easy as the cooperating part and the independent part of the two machines are clearly separated. The high-level plants have only two states. By computing the synchronous composition of the abstracted subplants G_1^{hi} and G_2^{hi}, we obtain the two-state high-level plant $G^{hi} = G_1^{hi}||G_2^{hi}$. It is evident that this is easier than computing $p^{hi}(L_1^c||L_2^c)$, where $G_1^c||G_2^c$ has 26 states.

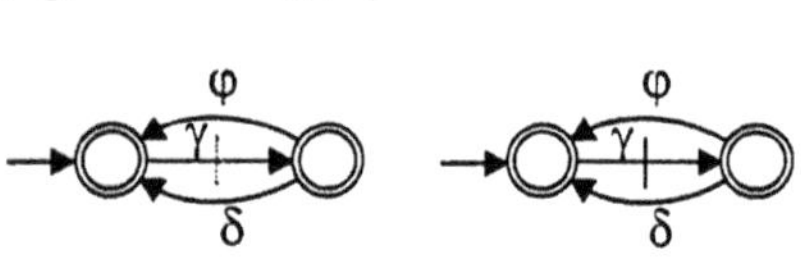

Fig. 7. high-level abstractions G_1^{hi} and G_2^{hi}

It can easily be verified that E^{hi} is controllable w.r.t. $L(G^{hi})$ and the high-level controlled language $L(S^{hi}/G^{hi})$ equals E^{hi} with a 5-state recognizer. Because of Theorem 5.1 the low-level supervisor $S^{lo}(s) = S^{hi}(p^{hi}(s)) \cup (\Sigma - \Sigma^{hi})$ realizes the desired high-level behavior and implements non-blocking and maximally permissive behavior in the low level.

6.2 Comparison with other methods

For the standard RW control synthesis the overall plant and specification automata have 26 states and 14 states, respectively and the supervised plant has 77 states. For applying the method of Zhong and Wonham (1990); da Cunha et al. (2002) it is also necessary to compute the overall plant automaton. The reporter map for the high-level abstraction is defined by labeling the low-level strings and for Zhong and Wonham (1990) there is additional computational effort for verifying hierarchical consistency, whereas in da Cunha et al. (2002), a supervisory control problem must be solved for each low-level subsystem, corresponding to a high-level string via the reporter map. For the technique of Lee and Wong (2002) the additional high-level specification E^{hi} can be decomposed into two local specifications $p_i(E^{hi})$, $i = 1,2$ which leads to low-level specifications with 8 states, each, and supervised low-level plants with 17 states both, versus 7 states for the case without high-level control. Thus, compared to our method the high-level specification must be decomposed, which leads to relatively large low-level supervisors.

7. CONCLUSIONS

Recent work in control of DES has been focused on exploiting structural properties for reducing the computational complexity of DES controller synthesis. Our approach embeds a decentralized approach Lee and Wong (2002) in a hierarchical control scheme and it was shown that our architecture guarantees nonblocking behavior of the controlled system. Furthermore we pointed out that hierarchical consistency need not be verified as it is directly implied by the proposed control architecture. The computational benefit of our method was illustrated by an example. Ongoing work aims for weaker conditions guaranteeing nonblocking behavior and the demonstration of our results by a laboratory case study of a manufacturing system.

REFERENCES

C.G Cassandras and S. Lafortune. Introduction to discrete event systems. *Kluwer*, 1999.

A.E.C. da Cunha, J.E.R. Cury, and B.H. Krogh. An assume guarantee reasoning for hierarchical coordination of discrete event systems. *WODES*, 2002.

P. Hubbard and P.E. Caines. Dynamical consistency in hierarchical supervisory control. *IEEE TAC*, 2002.

S. Jiang, V. Chandra, and R. Kumar. Decentralized control of discrete event systems with multiple local specializations. *Proc. ACC*, 2001.

S-H. Lee and K.C. Wong. Stuctural decentralised control of concurrent discrete-event systems. *EJC*, 2002.

T. Moor, J. Raisch, and J.M. Davoren. Admissibility criteria for a hierarchical design of hybrid control systems. In *Proc. ADHS'03*, 2003.

K. Rudie and W.M. Wonham. Think globally, act locally: decentralized supervisory control. *IEEE TAC*, 1992.

K. Wong. On the complexity of projections of discrete-event systems., 1997. URL `citeseer.nj.nec.com/wong98complexity.html`.

K.C. Wong and W.M. Wonham. Hierarchical control of discrete-event systems. *Discrete Event Dynamic Systems*, 1996.

W.M Wonham. Notes on control of discrete event systems. *Department of Electrical & Computer Engineering, University of Toronto*, 2001.

T. Yoo and S. Lafortune. A generalized framework for decentralized supervisory control of discrete event systems. *WODES*, 2000.

H. Zhong and W.M. Wonham. On the consistency of hierarchical supervision in discrete-event systems. *IEEE TAC*, 1990.

ELSEVIER
IFAC
PUBLICATIONS
www.elsevier.com/locate/ifac

VALIDATING FAULT-TOLERANT BEHAVIORS OF SYNCHRONOUS SYSTEM SPECIFICATIONS BY DISCRETE CONTROLLER SYNTHESIS [1]

Emil DUMITRESCU, Alain GIRAULT
Eric RUTTEN

INRIA, Montbonnot, FRANCE

Abstract: We demonstrate the utility of discrete controller synthesis to formally assess the fault-tolerance capabilities of a dependable system from the early design stages. We start with an executable specification in order to yield a new fault-tolerant executable specification. Then, we obtain manually the final distributed implementation and we formally verify its conformity to the initial fault-tolerant specification. *Copyright © 2004 IFAC*

1. INTRODUCTION

This work addresses the fault-tolerance issue in discrete-event synchronous systems design. Fault occurrences must be taken into account from the early design stages of dependable systems. First, the designer must think about the functionalities of the system that may fail. Then, for each possible failure, a tolerance policy must be specified in order to limit error propagation and to ensure the correct behavior of the system, despite the failure. At the functional level, a fault-tolerance policy is described as an additional behavior that should be enabled whenever a failure occurs. Its purpose is the total or partial compensation of the loss of functionality/performance caused by the failure.

Given the complexity of a design, programming a fault-tolerance policy is a difficult task. Indeed, the designer must think of all possible evolutions that may follow the failure of a component, and distinguish which ones must be avoided. Such an approach is not realistic, as it amounts to finding all possible execution sequences that satisfy a correctness requirement of a design. In this context, we propose a formal analysis framework, allowing designers to deal with fault tolerance at the functional level of the design executable specification. The main interest of working with executable specifications is that formal techniques are much more efficient at this level, as specifications are usually more concise.

[1] This work was supported by the *ARTIST* European project (IST-2001-34820).

We propose to use the *discrete controller synthesis* (DCS) technique (Ramadge and Wonham, 1989) in its symbolic approach (Marchand *et al.*, 2000) to generate automatically a behavior which implements fault-tolerance. We compose it with the design's executable specification, in order to yield a new *fault-tolerant specification.* If such a behavior does not exist, the design is not fault-tolerant: there is no means to prevent failure propagations or inconsistent executions. The executable specification should hence be redesigned.

We claim that most fault-tolerant behaviors achieve a *total or partial service continuity*, according to the initial formal specification of the design. Thus, we specify *the nominal service of a design*, and require that this service be maintained for any expected fault occurrence. DCS should then ensure the continuity of this service.

The resulting fault-tolerant specification is a non-deterministic combination of fault tolerant behaviors. To derive a deterministic fault-tolerant implementation, a definitive choice must be made among the existing behaviors. Moreover, the implementation must meet the initial architectural constraints. In practice, distributed control solutions are often required, because the target architecture is distributed. To handle distribution, we chose to address the implementation step *manually.* Indeed, the automatic generation of *communicating* distributed controllers is the only practical alternative to the manual distribution, which is not appropriate and we given our fault-tolerance goal, which prohibits the introduction of additional communication.

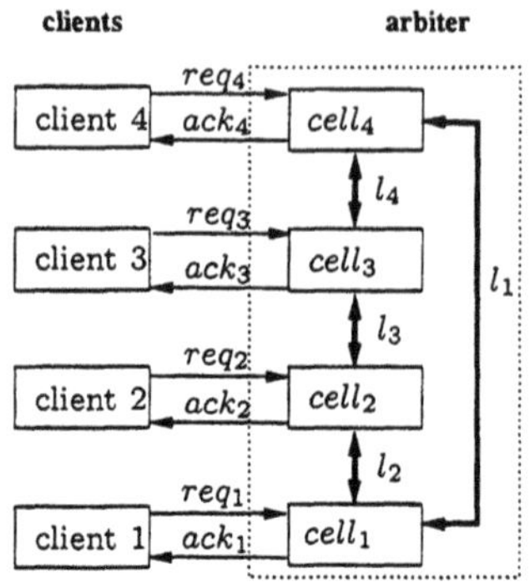

Fig. 1. Architecture of the distributed arbiter

Once an implementation has been manually obtained, its conformity to the initial fault-tolerant specification can be formally assessed. We propose to use the *refinement checking* technique (Grumberg and Long, 1991) in order to achieve this last step.

In the sequel, we present our general design approach and illustrate its different steps on a running example: the functional specification of a fault-tolerant distributed arbiter. Then, we discuss the results of our method as well as the manual implementation aspects, and we conclude on the advantages of our approach.

2. THE DISTRIBUTED ARBITER

The arbiter model we use as a running example is inspired from (McMillan, 1993). It deals with four clients (Figure 1) accessing a shared resource. Each client i ($i = 1..4$) requests access by asserting the input signal req_i. The arbiter replies by asserting the proper acknowledge signal ack_i only. The arbiter must achieve two basic functionalities, mutual exclusion and liveness, formally expressed with the temporal logic CTL (Clarke and Emerson, 1981):

- **Mutual Exclusion (ME)**: at any moment, at most one acknowledge signal may be asserted: $ME : AG(not(ack_1.ack_2{+}ack_1.ack_3{+}ack_1.ack_4 + ack_2.ack_3 + ack_2.ack_4 + ack_3.ack_4))$
- **Liveness (LIV)**: for each client i, if a request is asserted, then it is eventually acknowledged: $LIV_i : AG(req_i \rightarrow AF(ack_i))$

Incoming requests are not memorized. Thus, to satisfy LIV_i, we must assume that each client is persistent: it maintains req_i at least until receiving ack_i. In order to implement liveness, a token mechanism is used. Each cell is periodically given priority with respect to the others. Each token is valid for only one time unit (synchronous execution cycle) and then expires even though it has not been consumed. At most one token may circulate through the cells. The token is transmitted via the communication links l_i ($i = 1..4$) established between consecutive cells. The cell i may acknowledge its client if it is holding the token. However, if $cell_i$ is not holding the token, then it may still acknowledge its client according to a weak priority policy: the cells are statically ordered, from $cell_1$, which has the highest priority to $cell_4$. If the cell holding the token does not have an incoming request, then other cells must be able to acknowledge their requests, provided mutual exclusion is respected. Thus, if $cell_i$ receives a request while not holding the token, then it may acknowledge this request only if there are no more prioritary requests (cells $j = 1..i-1$) and if none among the less prioritary cells ($k = i+1..4$) is about to acknowledge its corresponding request while holding the token.

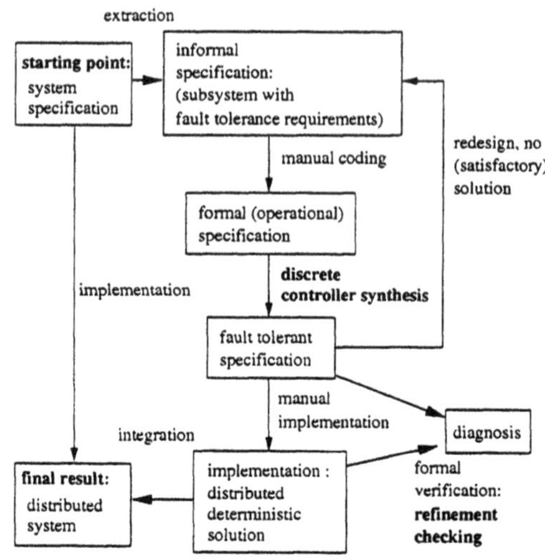

Fig. 2. Overview of the validation method

3. OVERVIEW OF OUR METHOD

Figure 2 presents our design and formal validation flow. A component specification with fault-tolerance requirements is first extracted from the global system specification. This specification contains a *failure hypothesis* that specifies the modules that are likely to fail, as well as a *failure model* that specifies the behavior of a faulty module. The specification is formalized through a *manual coding* step. At this stage, the designer only concentrates on the component's nominal functionalities to be implemented.

A *fault tolerance policy* is automatically generated by the DCS tool, to ensure service continuity despite failures. The result of this step is a new *fault tolerant specification*: a constrained model able to avoid all unwanted configurations triggered by failures. If no control solution exists, the component should be redesigned as it is not fault-tolerant.

The constrained model cannot be directly implemented, as it is usually non-deterministic and monolithic (not suitable for distribution). This is why we choose to *manually implement* the fault-tolerant components. However, the implementation can be formally assessed to ensure that all the implemented behaviors are part of the fault-tolerant specification.

3.1 Faults, errors, and failures

This is the fundamental causal principle in the fault pathology of dependable systems (J.C. Laprie, 1992; Jalote, 1994). Faults are physical hazardous events. A fault may yield an error. An error may yield a failure of one or more components. A failure of one component can be viewed as a fault at the system level. Each possible failure is permanent.

We limit our study to *error processing*: we assume there exists a reliable external unit, which reports an error if and only if a physical fault has occurred. Failure management may involve both hardware and software or temporal redundancy. In our context, we rely on the manual introduction of hardware redundancy, followed by the automatic generation of redundant software, which exploits both the existing and the redundant hardware.

The design of dependable systems calls for a dedicated specification and validation procedure. Three key points must be taken into account: the failure hypothesis, the failure model and the fault tolerance policy.

3.2 Defining a failure hypothesis

A failure hypothesis states which components of the system may fail. If more than one component is likely to fail, *failure configurations* are a common way to express subsets of components that may fail together. According to this hypothesis, the remaining components are supposed to be reliable: they never fail, or if such a failure occurs, the whole system fails. The failure hypothesis needs to be confirmed by a stochastic analysis step, in order to find the probability for each failure configuration. In this document, we assume that all failure configurations specified are equally probable.

3.3 Defining a failure model

When a component fails, specify what this failure implies. This amounts to defining a behavior that is triggered by this failure. For instance, when a component fails (processor, communication link, sensor, etc.), it may stop reacting to its environment. This means either sending a constant value, or sending random values to the environment (Byzantine behavior), or sending no value at all (fail-silent behavior).

3.4 Defining a fault tolerance policy

Ideally, a fault-tolerant system should maintain its functionalities and its performance (nominal service) even though some of its components are down. In practice, this assumption is too strict and expensive to implement. Thus, if a failure occurs, the nominal service may be replaced by a *degraded operating mode*. When the system runs inside a degraded mode, only a subset of its initial functional requirements are still met. We wish to achieve fault-tolerance by establishing such a degraded mode when a failure occurs. In our context, this is done by building appropriate constraints that are used to control the system's behavior from outside, in order to ensure a minimal service. This approach only applies to systems that are *controllable*: for such systems, a set of dedicated *controllable inputs* is provided for service maintenance purposes. The system behavior is constrained by driving the controllable inputs appropriately. All remaining input variables are called *uncontrollable*. The DCS technique is an excellent choice for automatically producing such controlling constraints.

In our context, adding controllability to a design amounts to adding supplementary hardware (input variables and additional states and transitions) for backup purposes. This amounts to specifying a hardware redundancy mechanism. On the other hand, the resulting synthesized controller, if it exists, represents a "program" that appropriately uses the redundant hardware in order to establish the degraded operating mode.

DCS needs a *control objective* expressing fault-tolerance: what the system should always do, despite failure occurrences.

4. SPECIFYING THE DEPENDABILITY

4.1 Failure hypothesis

We assume that only the communication links l_i between cells may fail. When link l_i fails, it stops transmitting tokens to the cell i. Each link is fail-silent and its failure is permanent. Moreover, we assume that the probability to have a failure affecting more than one link is negligible.

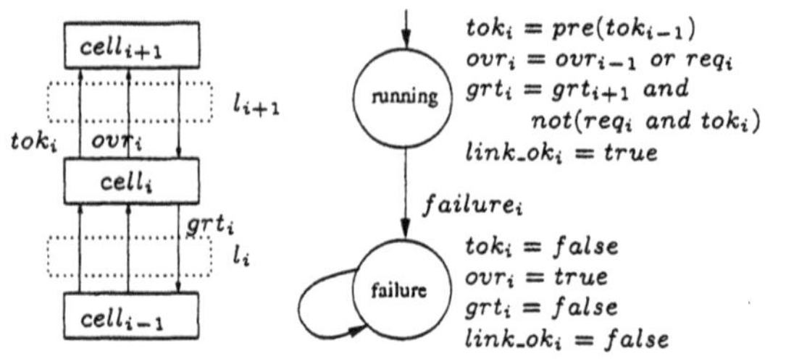

Fig. 3. Abstract model of a communication link l_i

4.2 Link failure model

An abstract model for each communication link is given in Figure 3. Each link may be either functional, in which case it transmits correctly the tokens and priority management information, or in a failure state, in which case it stops transmitting forever. The transition to the failure state is triggered by the environment and corresponds to a physical damage event.
The static priority between the cells is defined using an override/grant mechanism. Cell i overrides cell $i+1$ iff req_i is active or if cell $i-1$ overrides cell i. The first cell in the chain (cell 1) cannot be overridden. On the other hand, cell i grants cell $i-1$ iff it is granted itself by cell $i+1$ and if it is not holding both an incoming request and a token. The last cell in the chain (cell 4) is always granted.

Each cell i is aware of the state (running or failure) of the communication link l_i it is connected to. When link l_i is down, cell i stops receiving tokens. Thus, it stops complying with the specification LIV_i. Moreover, a link failure invalidates the static priority scheme, as no cell can be aware anymore of other requests of higher/lower priority.

4.3 Fault tolerance policy

When link l_i is down, the token keeps moving until it reaches l_i, and then it is lost. From that moment, no other request will ever be acknowledged. Thus, the mutual exclusion requirement is still trivially satisfied, but the response time is no longer guaranteed for any cell. Also, the static priority scheme is disabled.

Fault-tolerance must address the response time problem, while ensuring mutual exclusion. When a link fails, we wish to appropriately constrain the arbiter behavior, so that the response time becomes acceptable. However, in its current design state, our arbiter can hardly be controlled: the incoming request or error signals should not be constrained, as they should be able to be set at any moment.

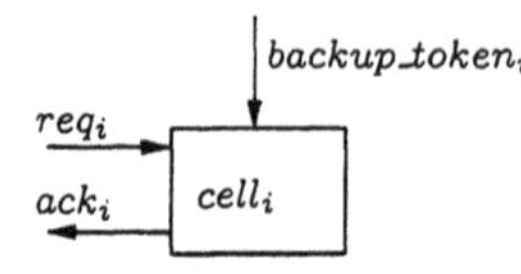

Fig. 4. Cell i interface with hardware redundancy

Besides, the only way to constrain the behavior of a reactive system is via its input signals. Thus, in order to achieve fault-tolerance, we must redesign our arbiter interface and behavior, to make it more "controllable".

Adding controllability. When a link is down, the arbiter stops complying with its specification as soon as the token is lost. Thus, a possible solution involves adding a *backup token* insertion point mechanism (Figure 4). The arbiter environment can reinsert a new token by adequately controlling this supplementary input. However, if l_i is operational, backup token reinsertion has no effect.

We redesign the arbiter by assuming the existence of an external global *supervisor*. By asserting the $backup\_token_i$ signal, this supervisor inserts a new token inside the arbiter. By deasserting the $backup\_token_i$ signal, it prevents a new token from being inserted. Backup tokens should only be inserted when at least one link is down. However, they should not be inserted too often, as at most one token should circulate inside the arbiter. Finally, token insertion should anticipate incoming requests, so that the response time remains reasonable.

Achieving liveness Up to now, DCS handles safety and non-blocking objectives. In order to address liveness, we have chosen to replace the concept of "response in finite time" by a *bounded time* response requirement, which is more easily solvable by currently available DCS algorithms. This requires finding a reasonable bound on the response time for each ack_i signal. Since there are four cells, and since at each cycle a new cell receives the token, we set the response bound to four cycles. Thus, the liveness requirement adapted to bounded time becomes:

$$LIV_i' : AG(req_i \rightarrow ack_i + AX(ack_i + AX(ack_i + AX(ack_i + AX(ack_i)))))$$

The synthesized controller satisfying ME and LIV_i' ($i = 1..4$), has the following features:

- it observes every variable in the system;
- it is only allowed to control the inputs $backup\_token_i$ ($i = 1..4$) in order to constrain the arbiter to achieve its functional requirement;

- if link l_i fails, the controller will re-establish the global system coherence by reinserting backup tokens through cell i, while preventing such token insertions whenever this would violate the ME requirement.

5. VALIDATION OF THE FT BEHAVIOR

We have modeled our arbiter and we have validated its fault-tolerance capability, by actually synthesizing a controller which ensures all the requirements stated in Section 4. For example, if link l_1 fails, the following two scenarios are possible:

① req_1 is asserted. If req_2, req_3, and req_4 are asserted and if links l_2, l_3, and l_4 are operational, then the controller enforces the fair acknowledgment of all the requests, by inserting a backup token through cell 1. The token insertion takes into account the response delay bound of each cell. Thus, a possible future violation of the delay response bound is *anticipated* and avoided, by enforcing the insertion of a token at the appropriate moment.
② If cell j ($j = 2..4$) holds a token, the controller disables the insertion of a backup token until the currently existing token is lost inside link l_1.

The control which has been synthesized is called *maximally permissive*: controllable inputs are only constrained w.r.t. the control objectives. Indeed, if link l_i is operational, the value of the controllable input *backup_token*$_i$ does not influence the behavior of the arbiter. If link l_i is down, the input *backup_token*$_i$ only needs to be constrained in two specific scenarios:

① As long as the ring contains a token, *backup_token*$_i$ is set to $false$.
② When the ring becomes empty (the token gets lost by link l_i), the controller enforces the introduction of a new backup token only when a possibility exists that a request is not acknowledged within the 4-cycles time bound.

The resulting controlled arbiter offers a choice among several available fault-tolerant behaviors. Because of its non-deterministic nature, we consider this result as a newly generated *executable specification* which embeds a set of fault-tolerant behaviors.

6. IMPLEMENTING AND VERIFYING THE FAULT-TOLERANT ARBITER

Implementating a *deterministic* fault-tolerant arbiter involves choosing *one* behavior among those embedded inside the fault-tolerant specification. The architecture of the resulting controller must comply with the initial failure hypothesis. Thus, we assume that the controller cannot fail. Unfortunately, the controller achieves *global control and observation*; hence, such an assumption is not realistic w.r.t. the distributed architecture of the arbiter. Besides, the implementation of the controlled arbiter must also meet the architectural constraint: *each cell i should be locally controlled*, and the only variables the local controller is allowed to monitor should be the internal variables of cell i. The automatic generation of local controllers achieving global control objectives is a more difficult task, known as *decentralized controller synthesis* (Lin and Wonham, 1988). Distributed control solutions often require supplementary communication between the local controllers. In our context, assuming *reliable* communication between local controllers would contradict our initial failure hypothesis, which states that communication between cells is not reliable. On the other hand, the distributed controller synthesis problem without communication between local controllers has been shown to be undecidable (Tripakis, 2002).

As a consequence, we build local controllers *manually*. By construction, our arbiter has several interesting properties. First of all, the controllable inputs are "don't cares" as long as no link failure has occurred. Then, when link l_i is down, only its corresponding *backup_token*$_i$ needs to be controlled. By running a sufficient set of simulation scenarios, we isolate a set of traces that are allowed by the controlled arbiter specification and that can be played on *backup_token*$_i$. All these traces are composed of the following successive parts: (1) an arbitrary length prefix, corresponding to the normal operation; all along this prefix, *backup_token*$_i$ can be set to any value; (2) once a failure occurs, a bounded-length prefix corresponds to the token propagation until it is lost; (3) finally, a periodic suffix corresponds to the periodic reinsertion of a new token once the previous one is lost.

Such a behavior is *local* to each cell and can be reproduced by a simple manually built deterministic finite state machine, such as the one presented in Figure 5. Hence, by composing the arbiter with four instances of this state machine, fault-tolerance is achieved, provided that the fault hypothesis holds.

We formally assess the correctness of the above implementation by using the *refinement checking* technique (Grumberg and Long, 1991). This technique is dedicated to assessing implementations

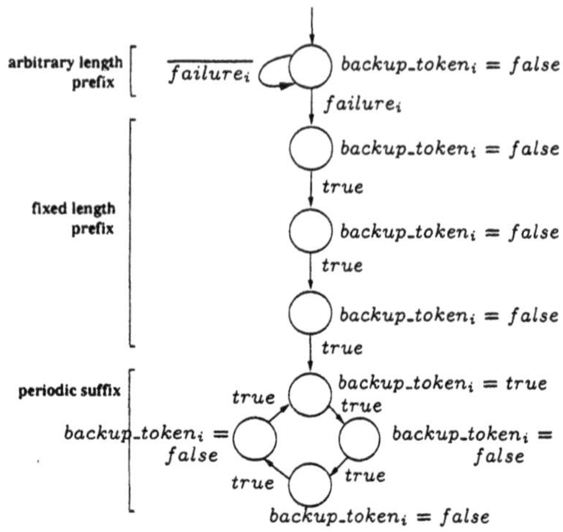

Fig. 5. Possible implementation of a local controller for cell i

against a non-deterministic operational specification. Its application in our context is efficient for two reasons: first, if an implementation *refines* a specification, then it inherits every ∀−CTL property satisfied by the specification (Grumberg and Long, 1991). Our fault-tolerance objectives are compliant to the ∀−CTL subset. Thus, we can prove that the implementation is fault-tolerant. Second, refinement checking can be applied compositionally, and its complexity does not exceed the reachable state space computation of the implementation (Henzinger *et al.*, 1998). In our experimental framework, we carried out this step manually, as refinement checking is not an implemented feature of the control synthesis tool we have used.

7. DISCUSSION AND CONCLUSION

Discrete Controller Synthesis (DCS) is a formal technique that builds a controller able to constrain the behavior of a specification so that it satisfies some given properties. To do this, the synthesized controller can only constrain the so called *controllable* inputs of the specification. We have used DCS as a design tool for assessing the fault tolerant capabilities of an executable specification. At the specification level, this assessment is only behavioral: "is my specification able to deal with faults, and if yes, how?". The synthesized controller answers this question. The benefit of our technique comes from the ability to specify the nominal service to be maintained despite fault occurrences, and to automatically generate the behavior that maintains this service. This issue cannot be addressed using model checking tools, as they do not distinguish between controllable and uncontrollable input events. To the best of our knowledge, the only available model checking tool that could also solve this question is MOCHA (Henzinger *et al.*, 1998). However, this tool only checks the existence of a solution, without constructing it.

The whole experimentation part has been performed using the SIGALI (Marchand *et al.*, 2000) DCS tool, as well as the MATOU (Maraninchi *et al.*, 2000) mode automata environment.

Further improvements of this work include the automatic distribution of the fault-tolerant control solution. Another important research direction is the development of compositional methods dedicated for control synthesis, to efficiently deal with combinational explosion.

REFERENCES

Clarke, E.M. and E.A. Emerson (1981). Design and synthesis of synchronization skeletons using branching time temporal logic. In: *Logic of Programs.* Vol. 131 of *LNCS.* Springer.

Grumberg, O. and D.E. Long (1991). Model checking and modular verification. In: *CONCUR'91.* Vol. 527 of *LNCS.* Springer.

Henzinger, T. A., S. Qadeer and S. K. Rajamani (1998). You assume, we guarantee: Methodology and case studies. In: *CAV'98.* pp. 440–451.

Jalote, P. (1994). *Fault-Tolerance in Distributed Systems.* Prentice Hall.

J.C. Laprie, et al. (1992). *Dependable Computing and Fault-Tolerant Systems.* Chap. Dependability: Basic Concepts and Terminology. Springer.

Lin, F. and W. M. Wonham (1988). Decentralized supervisory control of discrete-event systems. *Information Sciences.*

Maraninchi, F., Y. Rémond and Y. Raoul (2000). Matou: An implementation of mode-automata into DC. In: *Compiler Construction.* Springer.

Marchand, H., P. Bournai, M. Le Borgne and P. Le Guernic (2000). Synthesis of discrete-event controllers based on the signal environment. *Discrete Event Dynamic System: Theory and Applications* **10**(4), 325–346.

McMillan, K.L. (1993). *Symbolic Model Checking.* Kluwer Academic Publishers. Norwell Massachusetts.

Ramadge, P.J. and W. M. Wonham (1989). The control of discrete event systems. *IEEE Special Issue on Dynamics of Discrete Event Systems.*

Tripakis, S. (2002). Decentralized control of discrete event systems with bounded or unbounded delay communication. In: *WODES'02.* Zaragoza, Spain.

ELSEVIER
IFAC
PUBLICATIONS
www.elsevier.com/locate/ifac

ON THE SUITABILITY OF OBJECT-ORIENTED MODELS FOR INDUSTRIAL LOGIC CONTROLLERS

Marcello Bonfè * Cesare Fantuzzi **

* *Università di Ferrara, Dip. Ingegneria*
Via Saragat 1, Ferrara, Italy.
** *Università di Modena e Reggio Emilia, DISMI*
Via Allegri 15, Reggio Emilia, Italy.

Abstract: The paper presents a discussion on the practical issues of applying object-oriented modeling and formal verification techniques to the design of Manufacturing Systems logic controllers. In particular, the semantical aspects of specification languages like UML and Statecharts, which are widely used in many areas of Software Engineering, are analysed in terms of their adequacy for the industrial domain. Then, the paper presents the results of an application of these modeling languages to a practical case, showing that model checking techniques can also be adopted, given an adequate and domain-specific semantics of the object-oriented design model. *Copyright © 2004 IFAC*

Keywords: Manufacturing systems, Logic controllers, Discrete-event systems, Verification

1. INTRODUCTION

The application of systematic approaches and formal methods to the design of logic control systems for industrial applications has been the subject of many academical projects, as described in (Frey and Litz, 2000). However, despite the excellent results achieved in these projects, their impact on the industrial community have been quite limited, because of several reasons. First of all, modeling languages typically adopted for Discrete Event Systems (DES) need most of the times to be adapted to the application domain, in order to ease the implementation on Programmable Logic Controllers (PLCs) of control system models. Moreover, design models must be graphically represented in modular and hierarchical forms, in order to be of practical use in real cases. Another key issue, in the industrial practice, is the design and representation of exceptions handling mechanisms. In that sense, it is worth to remark that the visual language of Statecharts (Harel, 1987), which supports state hierarchy and concurrency, have a significative advantage over languages like Grafcet (David and Alla, 1992) or similar, which derives from the possibility to define explicitly *inter-level transitions* (i.e. transitions crossing the boundaries of states hierarchy).

Finally, the verification and implementation of logic control designs on target PLCs is still an activity, in the industry, subject to improvements. The correctness of PLC programs is typically not verified with formal techniques, but tested on the field with many "try-and-error" steps. Of course, the use of intuitive graphical specification languages would help the designer in the description of the desired control logic, whose correcteness can be formally verified at model level and then translated in an executable program.

The aim of the paper is to discuss the benefits and critical issues related to the use of popular specification languages, namely Unified Modeling Language (UML) (Object Management Group, 2001) and Statecharts, to design industrial logic control systems, with particular emphasis on practical results of a development project at Tetra Pak Carton Ambient SpA, manufacturer of packaging machines for liquid foods, that have supported the authors' research. The paper describes in Section 2 the subset of the UML notation that can be of use in the application domain, specifying also how it should be extended. Then Section 3 shows how Model Checking techniques can be applied to the Object-Oriented model, in order to verify its correctness before starting the real implementation phase. The paper ends with the illustration, in Section 4, of the case of study subject of the joined project between the authors and Tetra Pak Carton Ambient SpA, plus some concluding remarks.

2. THE OBJECT-ORIENTED SPECIFICATION LANGUAGE

The benefits of the O-O approach in analysis and design of software systems, deriving from its cardinal principles of *modularity*, *abstraction* and *information hiding*, are well-known to scientists of all engineering disciplines. Some basics of the O-O approach or, at least, *Object-Based* approach can be found also in the international standards related to industrial controllers programming, like the IEC 61131-3 (International Electrotechnical Commission, 2002) and the IEC 61499-1 (International Electrotechnical Commission, 2000). In fact, both documents emphasize the definition of modular software architectures based on Function Blocks (FBs), which are reusable components, defined as types and used as instances, with an interface of Input/Output (I/O) "signals" and encapsulated algorithms and data (see Figure 1).

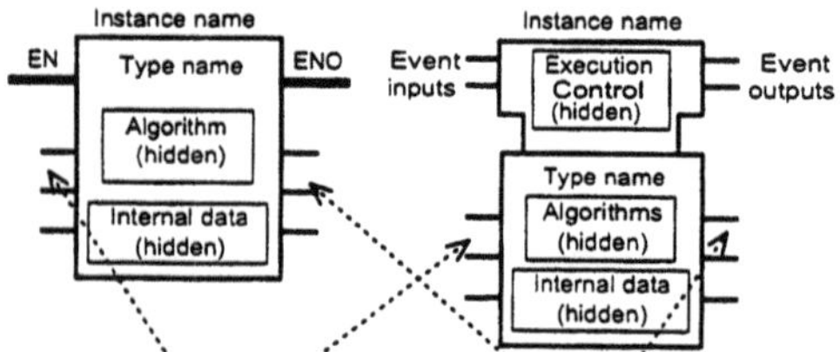

Fig. 1. Function Blocks as defined in IEC 61131-3 (left) and IEC 61499-1 (right)

Any design methodology for industrial controllers that aims for building modular programs should emphasize the use of these software structures. In fact, FBs can be viewed as means to implement controllers of *mechatronic objects*. However, the role of FBs is not necessarily to model the machine control system, therefore industrial control engineers would benefit from the support of a specification language like UML, for several reasons. First of all, it would help them in the description at a higher level of abstraction of system components, showing properties which are not always visible in FB networks. Moreover, the knowledge of O-O modeling techniques will help control engineers in the adoption of newer technologies, simply mapping existing design models to different implementation frameworks. Finally, UML is supported by several "off-the-shelf" CASE tools, that could be extended to automatically generate customized code. The UML language has also some defects: it is not a purely formal language and some of its semantical aspects are not adequate for the application domain. Therefore, we propose in the following a domain-specific extension of a subset of UML.

Structural Model:

Structural views of an O-O system are described with UML by means of *Class Diagrams*. Links between classes in a diagram can represent *association*, *simple* or *strong aggregation* (also called *composition*), and *generalization*, which involves inheritance mechanisms. The UML language itself is defined by means of class diagrams representing an extensible *meta-model*, which means that domain-specific concepts can be added to the language by means of *stereotyped* elements. This feature is particularly useful, because the concept of *mechatronic object* is quite different from that of *software objects*. The class of a mechatronic object, in fact, should have a signal-based interface, rather than operation-based, for the reasons discussed before. Moreover, since mechatronic objects have a tigthly aggregated physical counterpart, it would be also useful to model explicitly the hardware connections with sensors and actuators.

Therefore, in order to support the design activity described in Section 4, the following UML stereotypes have been introduced. Classifiers for mechatronic objects are defined as ≪`mechatronic`≫ classes, which have an interface of publicly visible properties, in their turn stereotyped as ≪`input`≫ or ≪`output`≫. A ≪mechatronic≫ class cannot have any public operations, while private operations may be used to model complex data-processing activities. The part of a ≪mechatronic≫ class related to the connection with physical components is specified with the help of classes stereotyped as ≪`hardware`≫. Classes of this kind are always related by means of a *composition* link with a ≪mechatronic≫ class and their ≪input≫ and ≪output≫ properties represent the hardware I/O ports as a private part of the ≪mechatronic≫ class. With this approach, the behavior of a ≪mechatronic≫ class, specified by an attached Statechart, represents the behavior of the controller, while the behavior of a ≪hardware≫ class can model the plant. Figure 2 shows the graphical representation of the proposed stereotypes in a UML Class Diagram (notice that inputs of a ≪hardware≫ class are outputs of the controller and viceversa).

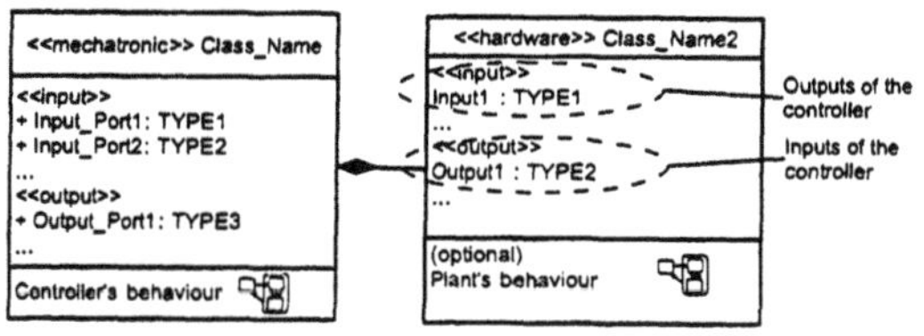

Fig. 2. UML stereotypes for mechatronic object models

A complete structural model for a complex machine would be defined by one or more Class Diagrams, in which *part/whole* relationships between modules are modeled by *composition* links between mechatronic classes. The physical interpretation of objects in this framework suggests the definition of well-formedness rules prescribing that there must always be a single *"top-level"* class (i.e. the machine), that simple (shareable) *aggregation* links cannot be used, and that composition links must be qualified with fixed multiplicity, since dynamic creation of objects is not admissible. With regards to *generalization* links, they clearly involve inheritance, which is not admitted in both IEC 61131-3 and IEC 61499-1. However, in a framework where the behavior of objects is specified by Statecharts, also the "structural" inheritance mechanisms of O-O programming languages (e.g. *interface conformity*) are unuseful. Actually, *behavioral conformity* of classes with a state-based specified behavior is a much more interesting and challenging issue, which involves formal techniques to ensure *trace containment* or *simulation*, in the sense described in (Harel and Kupferman, 2002). We believe that the model checking framework described in Section 3 is a starting point to introduce the concept of *behavioral inheritance* for mechatronic objects (an issue for future works).

Behavioral Model:

With the proposed approach, functional requirements of the manufacturing machine are mapped into a modular input/output model of the controller, specified with the association of a Statechart diagram to each ≪mechatronic≫ class of the model, a behavioral description admitted also by the UML language. The graphical notation of Statecharts has been recognized intuitive and powerful also by the engineers involved in the project described in Section 4, thanks to hierarchy and inter-level transitions. With regard to textual expressions, like transition labels or state actions, a redefinition in terms of an IEC 61131-3 compatible syntax has been considered more adequate for the application domain. This choice, would ease the development of automatic code generation procedures for industrial controllers. In particular, we propose to label transitions with a string in the form:

```
trigger[guard]/actions
```

where events in the `trigger` must be the *rising edge* of the result of a boolean expression (with the syntactical rules of IEC 61131-3), over attributes of the ≪mechatronic≫ class and ≪output≫ attributes of contained instances, or the same kind of attributes explicitly typed as `EVENT`. The *falling edge* of a boolean expression is also admissible, provided that the expression is written in negated form. The `guard` must also be a valid boolean expression, and `actions` will follow the same rules of the similar string included in a state action, which is specified by a textual expression like:

```
when / actions IF[guard]
```

Here, `when` is a qualifier that can be `entry` or `exit`, `guard` is an optional boolean expression that may prevent the action from being executed, if it evaluates to false, and `actions` is an ordered list of operations that can: *set* or *reset* a boolean variable (denoted with `SET(..)` or `RESET(..)`), *assign* the value of an expression to a variable of a non-boolean data-type, denoted with `Var_name := Expression`, or *emit* an attribute typed as `EVENT` (denoted with `EMIT(..)`).

3. VERIFICATION OF DESIGN MODELS

The languages proposed so far are not "formal" in a strict sense. In order to apply common verification techniques for finite state systems it is necessary to define an adequate semantics for the specification language and then translate design models into the input form of a formal verification tool. The tool adopted in our project has been *SMV* (Symbolic Model Checker) (McMillan, 1993), developed at Carnegie Mellon University, which verifies properties expressed in the temporal logic *CTL* for finite state models described as *Kripke structures*. In order to formalize a given mechatronic objects model, we will restrict the discussion to a framework of centralized PLC-like controllers. In this context, the "original" Statecharts semantics (see (Harel and Naamad, 1996)) is more adequate than the UML semantics (based on event queues), especially assuming the *synchronous time model.* The formalization of stereotyped UML models follows an approach similar to the one described in (Chan *et al.*, 1998), but in addition we assume that the global system is modular and sequential (modules perform their own step according to a given "calling" order).

The instantiation of the "top-level" class of an O-O model generates (deterministically) a *mechatronic system* $M_s = (M, t, \gamma)$, where M is a set of *mechatronic modules*, $t \in M$ is the top-level one and $\gamma : M \rightarrow 2^M$ is a function that retrieves for each module the ordered set of its components. Each $M_i \in M$ is a tuple (S_i, T_i, P_i, r_i), with a set S_i of *states*, a set T_i of *transitions*, a set $P_i = P_i^I \cup P_i^O \cup P_i^{HI} \cup P_i^{HO}$ of "port" variables, including visible interface (I/O superscripts) and private interface with hardware parts (denoted with HI/HO), and a root state $r_i \in S_i$. The topology of a Statechart is defined by typing each $s \in S_i$ as an AND-state, OR-state or basic, and by the functions $def(s)$ (retrieving the default state of each OR-state), $chldn(s)$ (retrieving the substates of s), and $chldn^*(s)$, the reflexive-transitive closure of $chldn(s)$. A *configuration* C is a maximally consistent subset of S_i and $compl(X)$ retrieves a configuration which is the *default completion* of a consistent set of states X. We also assume to have a language $\mathcal{L}_B$ to describe the set of boolean expressions over variables in P_i, a language $\mathcal{L}_A$ to describe the set of assignments over variables in $P_i^O \cup P_i^{HO}$ and a function $edge(exp)$ whose boolean result is true if the value $exp \in \mathcal{L}_B$ changes from false to true with regard to the previous evaluation. The details of each transition $tr \in T_i$ are specified by: $src(tr) \in S_i$, the source state, $trig(tr) \in \mathcal{L}_B$, the trigger expression that must be evaluated with the edge function, $grd(tr) \in \mathcal{L}_B$, the guard expression, $act(tr) \in 2^{\mathcal{L}_A}$, a set of assignment actions, and $targ(tr) \in S_i$, the target state. The scope of tr is the smallest OR-state containing both $src(tr)$ and $targ(tr)$, while $maxsrc(tr)$ is the unique child of the scope of tr such that $src(tr) \in chldn^*(maxsrc(tr))$. A transition is enabled if the predicate $en(tr) = in(src(tr)) \wedge edge(trig(tr)) \wedge grd(tr)$ is true, but is firable only if an additional predicate $conflict(tr)$ is false. The latter is true if any transition with a priority higher that that of tr is enabled. Our priority rules enforce an explicit order between transitions with the same source state and give higher priority to transitions exiting states at a higher level in the hierarchy. To conclude, we assume that states $s \in S_i$ have associated entry and exit actions, defined as tuples (w, a, g), where $w \in \{entry, exit\}$, $a \in 2^{\mathcal{L}_A}$ is a set of assignments and $g \in \mathcal{L}_B$ is a guard expression.

The global states of a mechatronic system M_s are defined by all the possible configurations of the modules and all the possible values taken by their variables. The initial state of M_s is the one in which each module is in the configuration $C_i = compl(r_i)$ and each variable takes its initial value. The transition relation is defined by the following interpretation algorithm, which assumes that TD is the totally ordered set of subscripts of $M_i \in M_s$, whose initial value TD^0 is univocally derived from the knowledge of t and γ:

Algorithm 1. **Mechatronic System Step**:

(1) Update simultanously the value of variables in sets P_i^{HI}, for each $M_i \in M_s$.
(2) Pick up the next element i of TD and repeat until $TD = \emptyset$:
   (a) compute the set $F_i = (tr \in T_i | en(tr) \wedge \neg conflict(tr))$ of firable transitions;
   (b) compute the next configuration

$$C_i' = compl((C_i - \bigcup_{tr \in F_i} chldn^*(maxsrc(tr))) \cup \bigcup_{tr \in F_i} targ(tr));$$

   (c) execute exit actions related to exited states, actions related with each $tr \in F_i$ and entry actions related to entered states
(3) Update simultanously the value observed by the external environment of variables in sets P_i^{HO}, for each $M_i \in M_s$.
(4) Set $TD = TD^0$ and go to Step 1).

Given this semantics, the translation of the design model into an SMV program can take advantage of the modularity features supported also by the SMV language. Therefore, mechatronic modules will be defined as:

```
MODULE Mech_Class(EN, Inputs, READ_HW_INPUTS)
VAR
Exec : {IDLE, ACTIVE, EXECUTING1,
       EXECUTING2, ..., FINISHED};
SUBModule1 : Mech_Class1(..);
...
```

where EN is a boolean input that forces the "activation" of the module and READ_HW_INPUTS is a variable that permits the synchronous update of all hardware inputs. The state variable Exec takes the enumerated values according to the sequence of execution followed by the controller: when ACTIVE, the given module performs its own "step", when EXECUTINGi one of its sub-modules is activated. The other state variables and the transition relation will be defined according to the Statechart specification, encoding the hierarchy as follows:

```
VAR
Root : {State1, State2, ..., StateN};
SUBState1 : {State11, ..., State1N};
```

and evaluating the configuration and the set of enabled and actually firable transitions with assignments like:

```
DEFINE
  IN-State1 := (Root = State1);
  IN-State11 := IN-State1 & (SUBState1 = State11);
  EN-Trans1 := IN-SourceState & Trigger & Guard;
  CONFLICT-Trans1 := EN-Trans2 | EN-Trans3 | ..;
  FIRABLE-Trans1  := EN-Trans1 & !CONFLICT-Trans1 &
                     (Exec = ACTIVE);
ASSIGN
  init(Exec) := IDLE;
  init(Root) := State1; -- Default ...
  next(Root) := case
                 FIRABLE-Trans1 : State2; ...
                 1 : Root;
                esac;
  next(Hw_In1) := case
                   READ_HW_INPUTS : {0,1};
                   1: Hw_Input1;
                  esac;
```

Notice that, in this example, the value of a hardware input is updated non-deterministically when the controller is in the "read inputs" phase, which mimic the fact that the plant can have any unpredictable behavior. However, the SMV program can be easily extended to include the specification for ≪hardware≫ classes, if they are explicitly modeled with Statecharts.

Once that the design model has been formalized, the SMV tool will prove that the system holds correctness properties, expressed with CTL formulas, or will provide, if possible, a counter-example (i.e. a computational path) proving that the system violates a given property. Useful CTL formulas would be, for example, those expressing state reachability:

`SPEC EF ModX.IN-StateY`, which means "it exists (E) a path in which `StateY` is reached sometime in the future (F)", or absence of deadlocks:

`SPEC AG (ModX.IN-StateY -> EF (!ModX.IN-StateY))`,

which means "for all paths (A), it is always true (G) that if `StateX` is activated, there exists a future time (EF) in which it is deactivated". Even though many other CTL patterns or templates can be provided, they would certainly be difficult to handle for control engineers, that may prefer to specify system's requirements with graphical tools. In a UML based methodology, Sequence Diagrams can be adopted for this purpose, provided that they are extended with means to distinguish between *possible* and *mandatory*, scenarios and that rules to translate them automatically into CTL formulas.

4. A CASE OF STUDY

The proposed specification languages have been successfully adopted in a development project at Tetra Pak Carton Ambient SpA to design the logic control system of a Tetra Brik Aseptic® packaging machine for liquid foods (see Figure 3).

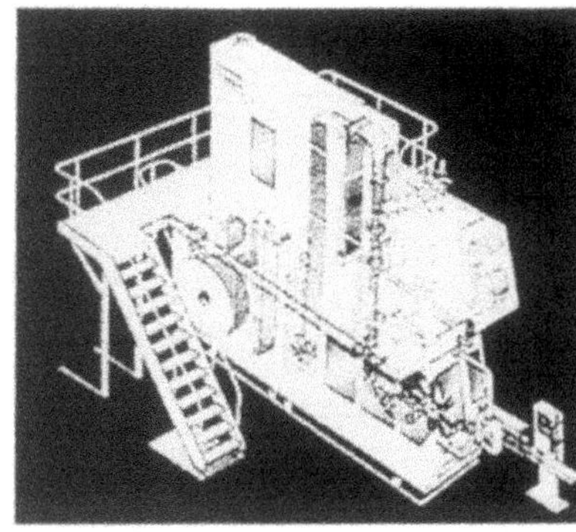

Fig. 3. A Tetra Brik Aseptic® packaging machine

The machine case of study is a complex system that packages liquid products with a so-called *vertical filling&forming* process: the packaging material is unrolled, sterilized in a bath of H_2O_2 (a very critical part of the machine), then folded and sealed longitudinally in order to form a tube, which is filled with the product. Finally, the package is cut, sealed transversally and folded to take a "brick" shape.

The methodology adopted to design the logic controller has followed a "top-down" approach: the machine has been decomposed into main modules, aggregated to the "top-level" class, then each module has been further decomposed into significative components, as described by Class Diagrams. The most simple of these modules, which is adequate for the illustrative purpose of the paper, is the one called Automatic Strip Splicing Unit (ASSU).

The ASSU is in charge of unreeling bobbins of plastic strip, necessary in the transversal sealing of the package, and manage the splicing of the strip, in order to ensure its continuity. The strip flowing from one of the bobbins runs over a set of pulleys, called *dancer*, which permits to decouple the unreeling speed with the speed of the other machine components pulling the strip. Whenever necessary, the strip in the "running" bobbin is spliced with that in the other bobbin simply pushing forward and heating the *sealing head*. Considering the functional requirements and the physical components, the structural specification for this module has been described by the Class Diagram shown in Figure 4.

The behavioral specification of the controller is divided into three statecharts: one for the "Splicing Unit" composite class (Figure 5, actions omitted), one for the "Sealing Head" and one for the "Motion System" (omitted). As said before, the design pattern adopted forces the two sub-parts of this module to interact only with the state machine of the composite class, therefore the latter is in charge of the synchronization of its components, especially during the splicing operation.

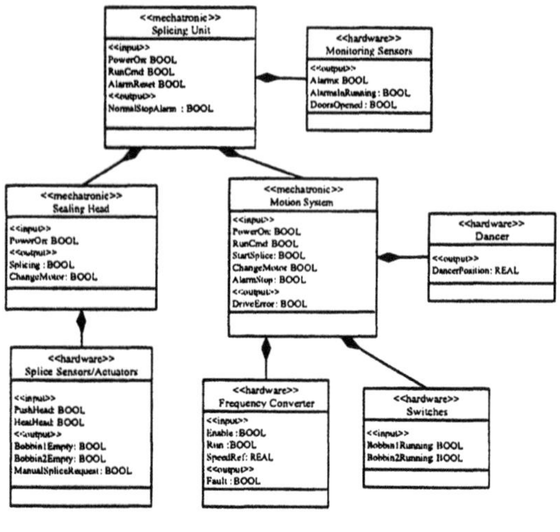

Fig. 4. Class Diagram of the Splicing Unit control system

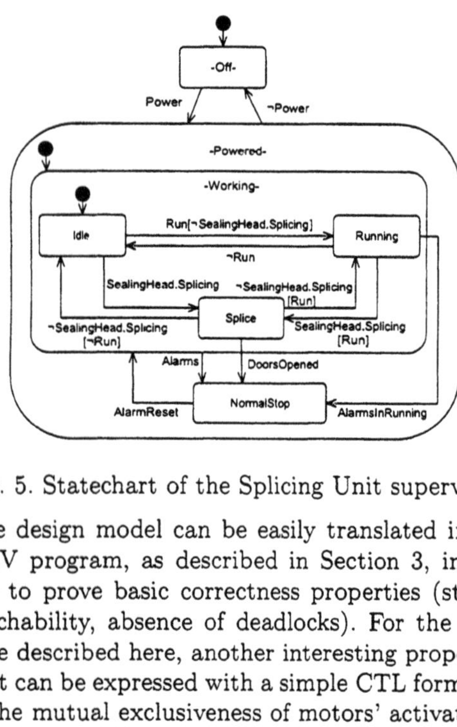

Fig. 5. Statechart of the Splicing Unit supervisor

The design model can be easily translated in an SMV program, as described in Section 3, in order to prove basic correctness properties (states reachability, absence of deadlocks). For the module described here, another interesting property, that can be expressed with a simple CTL formula, is the mutual exclusiveness of motors' activation, specified as follows:

```
SPEC AG !(MotSys.Bobbin1Running & MotSys.Bobbin2Running)
```

To conclude, the controller has been fully implemented on a commercial PLC, translating the design specification in a modular program, coded in Ladder Diagram language according to translation rules (not discussed here) that follow the same semantical interpretation described in Section 3.

5. CONCLUSION

The paper has described an application of O-O modeling techniques, supported by an extended subset of UML, to the design and verification of a complex manufacturing system logic controller. Technicians and engineers involved in the machine project have been able to use efficiently the modeling language without a background in formal methods, while model checking techniques have been applied thanks to a translation of O-O models into the input language of the SMV tool. Furthermore, design specifications have been easily implemented in a modular and efficient PLC program, thanks to the adaptation of the modeling language to the application domain.

ACKNOWLEDGEMENT: The authors fully acknowledge Tetra Pak Carton Ambient SpA for its support and cooperation during the research work described in the paper.

REFERENCES

Chan, W., R.J. Anderson, P. Beame, S. Burns, F. Modugno, D. Notkin and J.D. Reese (1998). Model checking large software specifications. *IEEE Trans. on Software Engineering* **24**(7), 498–520.

David, R. and H. Alla (1992). *Petri Nets and Grafcet: Tools for modelling discrete events systems.* Prentice–Hall.

Frey, G. and L. Litz (2000). Formal methods in PLC programming. In: *Proc. IEEE Conf. on Systems Man and Cybernetics (SMC) 2000.* pp. 2431–2436.

Harel, D. (1987). Statecharts: a visual formalism for complex systems. *Science of Computer Programming* **8**, 231–274.

Harel, D. and A. Naamad (1996). The STATEMATE semantics of Statecharts. *ACM Transactions on Software Engineering and Methodologies* **5:4**, 293–333.

Harel, D. and O. Kupferman (2002). On object systems and behavioral inheritance. *IEEE Trans. on Software Engineering* **28**(9), 889–903.

International Electrotechnical Commission (2000). IEC 61499-1. Function Blocks for Industrial Process Measurement and Control - Part 1: Architecture. Public Available Specification (PAS).

International Electrotechnical Commission (2002). IEC 61131-3. Programmable Controllers - Part 3: Programming Languages (2nd Edition). Final Draft International Standard (FDIS).

McMillan, K.L. (1993). *Symbolic Model Checking: an Approach to the State Explosion Problem.* Kluwer Academic Publishers.

Object Management Group (2001). UML, v.1.4, OMG specification. Document N. formal/2001-09-67. www.omg.org/uml.

ELSEVIER
IFAC
PUBLICATIONS
www.elsevier.com/locate/ifac

DISCRETE EVENT SYSTEMS APPROACH TO THE VERIFICATION OF THE INFORMATION FLOW PROPERTIES IN SECURE PROTOCOLS

Nejib Ben Hadj-Alouane * Stéphane Lafrance ** Feng Lin * John Mullins ** Moez Yeddes ***

* *Department of Applied Computer Sciences National School of Information Sciences. University of Manouba, Tunisia. Emails:*
{Nejib.BenHadjAlouane, Moez.yeddes}@ensi.rnu.tn
** *Department of Computer Engineering. École Polytechnique de Montréal. Montréal, Quebec, Canada. Emails:*
{Stephane.Lafrance, John.Mullins}@polymtl.ca
*** *Department of Electrical and Computer Engineering Wayne State University. Detroit, MI 48202, U.S.A. Email:* flin@ece.eng.wayne.edu

Abstract: This paper introduces a new algorithmic approach to the problem of checking the intransitive non-interference (INI) using discrete event systems (DES) tools and concepts. INI is an information flow property widely used in formal verification of computer systems and security protocols. First a new property called iP-observability (observability based on a purge function) is introduced to capture INI. An equivalence between iP-observability and P-observability (observability as used in DES) is then established. This paper also presents an algorithm to transform the automaton modelling the system/protocol into an automaton where P-observability can be checked, which is equivalent to verifying INI for the original system. Since P-obervability can be checked with a polynomial complexity, this algorithmic approach can effectively verify the important security property of INI.

Keywords: Interference, Intransitive Non-interference, Security Policies, Formal Verification, Observability.

1. INTRODUCTION

The theory of supervisory control of discrete event systems (DES) was introduced over fifteen years ago [15], [12], [4]. Since then, the properties of controllability and observability have been used as tools to characterize and solve many problems with diverse application domains: control and supervision applied to manufacturing [3], verification of communication protocols [16], and database systems [11], just to name a few. In this paper, we focus our attention on the application of the theory of DES to the area of computer system's and protocol's security.

Security is a crucial property of system behavior. The need for secure protocols has seen a rapid growth during the past decade due to, on the one hand, the development and wide spread use of computer networks and distributed systems, and on the other hand, the globalization of various forms of electronic communication, collaboration and trade. Within this global context, security policies and protocols are mainly seen as tools for providing secure information exchange. The main task of cryptographic protocols is to guarantee, among other issues, the confidentiality of transmitted data over a network, the authentication of the agents participating in a protocol run, and the integrity of the data transmitted over a network. A subclass of cryptographic protocols, namely *e-commerce protocols*, have the additional task of guaranteeing properties addressing anonymity of the participating parties, atomicity of money and goods, non-repudiation from parties and fair exchange issues.

Although the information-security community has not yet reached a consensus on the exact meaning of the terms "security" or "confidentiality", it is quite clear that both require a strict control over the information flowing between different agents manipulating objects and acting within systems with multiple security levels [1]. In this regard, several information-flow security properties have been proposed. *Non-interference* [6], [7], [13], [18], with its different generalizations and forms, is one such property. Intuitively, non-interference captures any causal dependency from a high-level action h to a lower-level action l. By causal dependency we mean that the dependent action cannot occur without the occurrence of the preceding action. Such a dependency creates an insecure channel called *covert channel*, having the capability to transmit high-level information concerning h to any low-level agent observing l.

In practice, however, many security problems go beyond the scope of simple non-interference. In particular, the problem of confidentiality in multi-level security systems, where the relation over the set of security levels capturing allowed information flows, is not transitive. *Intransitive non-interference* has been treated in the literature [8], [10], [17], [14]. The paradigmatic example for intransitive non-interference is a three-level (private, public and downgrade) security system modelling a cryptosystem. Such a system has the requirement of insuring non-interference from a secret data m and encryption key k to the public level, unless it has been previously *downgraded* through the cryptosystem. Clearly here, encryption creates a causal dependency between the secret pair (m, k) and the declassified data $\{m\}_k$ (m encrypted by k) since any variation of m or k is reflected in $\{m\}_k$.

Literature has showed that intransitive information flow properties are suitable for expressing security properties dealing with information transmitted by cryptographic protocols. The general schema of these (local) properties is based on a process-algebraic notion of observability [2]. The (global) operational characterization of this schema is called the *Unwinding Theorem*, which states that in any hostile environment (the detailed specifications depend on the type of security property), no condition observable from a point of view H will be observable from a point of view L, unless this observation has been previously downgraded from H to L (through a downgrading level).

Observability has played an important role within the DES theory. However, its various uses have been tied to the notion of a static projection capturing the fact that a fixed subset of events is observable and the rest are not. An important aim of this paper consists of extending the expressiveness of DES to capture information flow specifications based on a purge function, rather then a projection. The purge function erases events from a given sequence (or trace), based on an allowable information flow lattice defined over the security levels of a given system. Intransitive non-interference is then expressed as observability with respect to a purge function. In the case of a three level security system, a transformation is given to characterize the problem of verifying intransitive non-interference in terms of regular observability (with respect to a projection).

2. INTRANSITIVE NON-INTERFERENCE IN MULTI-LEVEL SYSTEMS

The concept of *non-interference* is introduced by Goguen and Meseguer [7] as a basis for specifying and analyzing security issues in computer systems and protocols. The basic idea behind non-interference can be simply stated as follows: the behavior of a given entity is said not to interfere with the behavior of a second entity whenever no action performed by the first can influence subsequent outputs seen by the second.

Rushby [17] gives a formalization of *non-interference*, in terms of input/output automata, and introduces the concept of *intransitive non-interference* (INI). Information flow property INI extends *non-interference* and enable the specification of a generalized class of security policies dealing with channel control mechanisms. Roughly speaking, channel control mechanisms require the following type of specification: given a system with three channels H, L and D, information is allowed to flow from H to L only after passing trough D, but never directly (intransitivity). Here, channel D is seen as a downgrading channel (for example an encryption mechanism). In terms of non-interference, the event stream generated by H is allowed to interfere with the event stream generated by L, only through D events.

In the remainder of this section, we formally define the notions required to capture intransitive non-interference as presented by Rushby [17]. Our treatment is based on event automata, instead of Rushby's input/output automata; it is well-known, however, that these two types of automata are equivalent.

We are given a set $\mathbb{D}$ of *security domains* and a set of events Σ partitioned over these domains. The operator dom : $\Sigma \to \mathbb{D}$ is used to capture this partition: to every domain $U \in \mathbb{D}$, the set $\Sigma_U = \{\sigma \in \Sigma \mid \mathrm{dom}(\sigma) = U\}$ specifies the events associated with U. The domains are interpreted to represent the security channels for which we will define non-interference requirements. We also consider an *interference relation* $\rightsquigarrow \subseteq \mathbb{D} \times \mathbb{D}$ defined over $\mathbb{D}$: given domains U, U', the intended meaning of the relation $\rightsquigarrow$ is such that the domain U is allowed to interfere with the domain U' whenever $U \rightsquigarrow U'$. We write $U \not\rightsquigarrow U'$ whenever $(U, U') \notin \rightsquigarrow$.

We assume that our system, i.e., the combined behavior associated with all the domains, is modelled by a language $K \subseteq \Sigma^*$. Moreover, K is generated by the finite automaton $G = (\Sigma, X, \delta, x_0)$, that is, $K = \mathcal{L}(G)$ (in particular, K is prefix-closed).

Intuitively, intransitive non-interference can be understood from the following example.

Example 1. Consider a system with two domains: one is a high security domain (classified), the other is a low security domain (unclassified); e.g., $\mathbb{D} = \{H, L\}$. The non-interference relation is such that:

$$\rightsquigarrow = \{(L, H)\}.$$

Let $h \in \Sigma_H$ and $l \in \Sigma_L$, and consider the two automata given in Figure 1 and Figure 2 as possible specifications for system behavior. In the case of automaton G_1 (Figure 1), the behavior of the system poses a problem: at the initial state, the L domain cannot execute the event l; but in its second state, after the execution of h by the H domain, the L domain can now execute l. Thus, the behavior generated by the H domain interferes with the behavior of the L domain (classified information is being leaked to the unclassified levels). This is contrary to the specification of the domain structure given above (i.e., $H \not\rightsquigarrow L$). Such a dependency does not exist in automaton G_2 (Figure 2).

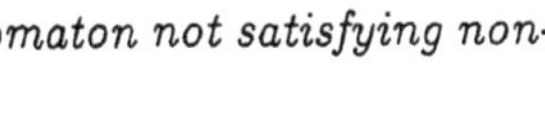

Fig. 1. *Automaton not satisfying non-interference.*

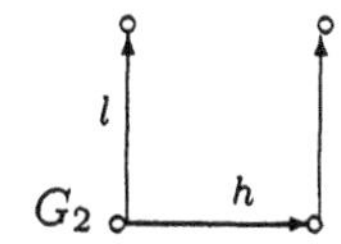

Fig. 2. *Automaton satisfying non-interference.*

Next, consider a three domain system, $\mathbb{D} = \{H, D, L\}$, where D is a downgrading domain. The non-interference relation is such that:

$$\rightsquigarrow = \{(H,D),(D,L),(D,H),(L,D),(L,H)\},$$

i.e., only $H \rightsquigarrow L$ is not allowed ($H \not\rightsquigarrow L$). We consider the automata given in Figure 3 and Figure 4, and assume $h \in \Sigma_H$, $l \in \Sigma_L$ and $d \in \Sigma_D$.

Automaton G_3 (Figure 3) poses a problem with intransitive non-interference, since the event l is not possible following the trace hd, but becomes possible following hdh.

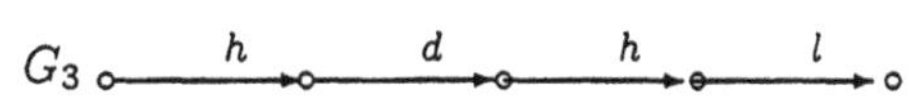

Fig. 3. *Automaton not satisfying intransitive non-interference.*

However, the system specified by automaton G_4 (Figure 4) satisfies intransitive non-interference. For instance, note that the fact that ll is possible following hd and is not possible following h does not constitute an intransitive non-interference problem. According to the domain structure, H is allowed to interfere with L through D events. In other words, we only have the explicit requirement to preserve confidentiality of H with respect to L in between D events, but not across them.

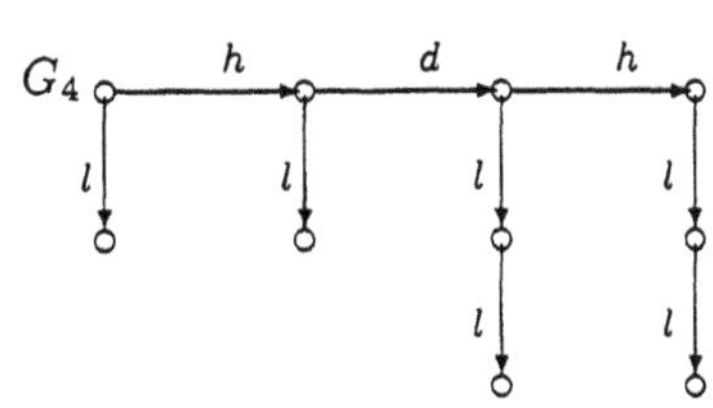

Fig. 4. *Automaton satisfying intransitive non-interference.*

To formally capture INI, a reduction function iP, called *intransitive purge*, is introduced by Rushby [17]. The purge of a trace, with respect to a given domain, removes from the trace all events from other domains that are not allowed to interfere with the given domain. The function iP is defined using the following auxiliary function: $sources : \Sigma^* \times \mathbb{D} \to \mathbb{D}$ given as follows: $sources(\epsilon, U) = \{U\}$, and $sources(\sigma s, U) = sources(s, U) \cup \{\mathrm{dom}(\sigma)\}$ if $(\exists V \in \mathrm{sources}(s, U))\mathrm{dom}(\sigma) \rightsquigarrow V$, and $sources(\sigma s, U) = sources(s, U)$.

Intuitively, function sources(.,.) captures the set of domains which are allowed to interfere throughout the execution of a trace. This set of domains is determined backwards (i.e., starting from the end of the trace). Moreover, the fact that a given domain V is in sources(s, U) either means that $V = U$ or there is a subsequence $\sigma_1, \sigma_2, ..., \sigma_n$ of the trace s, such that dom$(\sigma_1) \rightsquigarrow$ dom$(\sigma_2) \rightsquigarrow ... \rightsquigarrow$ dom(σ_n) with $V =$ dom(σ_1) and dom$(\sigma_n) \rightsquigarrow U$.

We then consider the following intransitive purge function

$$iP(\epsilon, U) = \epsilon, \quad \text{and}$$
$$iP(\sigma s, U) = \begin{cases} \sigma iP(s, U) & \text{if } \mathrm{dom}(\sigma) \in sources(\sigma s, U) \\ iP(s, U) & \text{otherwise} \end{cases}$$

Informally, purge function $iP(.,.)$ is a string reduction function such that $iP(s, U)$ consists of the subsequence obtained from s by removing every event belonging to any domain that should not interfere with the domain U (i.e. domains which are not in sources(s, U)).

It should be noted that iP is left congruent, but not right congruent (we do not prove these facts as they are not used in the developments in this paper). More importantly, given two traces s and s', with s a prefix of s', s by itself may be purged differently than s as part of s'. In addition, iP retains an important property of projections, namely, for any given trace s, $iP(iP(s)) = iP(s)$; this can be easily proved, as $sources(s) = sources(iP(s))$.

Based on the above definition of $sources(.,.)$ and $iP(.,.)$, intransitive non-interference (INI) is defined as follows, which is essentially the definition of Rushby [17].

Definition 1. (Intransitive Non-Interference). A language K satisfies *INI* if,

$$\forall_{U \in \mathbb{D}} \forall_{s \in K} \forall_{\sigma \in \Sigma_U} \quad s\sigma \in K \iff iP(s, U)\sigma \in K.$$

The following result leads to a characterization of INI as a variant of the observability property using the iP reduction.

Lemma 1. Language K satisfies INI if and only if, for every $U \in \mathbb{D}$, for every $s_1, s_2 \in K$ and for every $\sigma \in \Sigma_U$,

$$iP(s_1, U) = iP(s_2, U) \wedge s_1\sigma \in K) \implies s_2\sigma \in K.$$

Now consider the following property, similar in form to the observability property used in the theory of discrete event systems [12]. Nevertheless iP is not a projection but rather a reduction as defined above.

Definition 2. (iP-observability). Let $M \subseteq \Sigma^*$ and $\Sigma_U \subseteq \Sigma$. A prefix-closed sublanguage $K \subseteq M$ is *iP-observable* w.r.t. (M, Σ_U) if for every $U \in \mathbb{D}$, for every $s_1, s_2 \in K$ and for every $\sigma \in \Sigma_U$, $(iP(s_1, U) = iP(s_2, U) \wedge s_1\sigma \in K \wedge s_2\sigma \in M) \implies s_2\sigma \in K$.

Theorem 1. Language K satisfies INI if and only if K is iP-observable w.r.t (Σ^*, Σ_U), for all $U \in \mathbb{D}$.

3. P-OBSERVABILITY VS IP-OBSERVABILITY

Asssumption: For the remainder of this paper, we restrict our attention to systems and protocols with only three security domains H, L and D:

$$\rightsquigarrow = \{(H, D), (D, L), (D, H), (L, D), (L, H)\},$$

Within the context of a three domains system, only domain L can pose an interference problem. Therefore, the corollary given below follows from Theorem 1.

Corollary 1. Language K satisfies INI if and only if K is iP-observable w.r.t. (Σ^*, Σ_L).

Furthermore, we write $iP(s)$ instead of $iP(s, L)$ since we have to verify the iP-observability of K only w.r.t. (Σ^*, Σ_L).

The property of iP-observability is novel. However, it will be unfruitful and unnecessary to develop a new theory for iP-observability, because iP-observability is closely related to P-observability (i.e., observability based on the projection P) that was introduced by Lin and Wonham [12].

In this section, we will show that by properly defining a new language K_{iP} from the original language K, we can transform iP-observability of K into P-observability of K_{iP}. In this way, the verification of the INI property amounts to checking P-observability, for which we have efficient algorithms.

3.1 iP-Quotient Language

This section presents a language K_{iP} obtained by transforming the language K. Recall that $\Sigma = \Sigma_H \cup \Sigma_L \cup \Sigma_D$ and $K = \mathcal{L}(G)$.

First, we start by introducing the notion of a minimal subtrace for a given trace, used in the construction of the transformed language K_{iP}.

Intuitively, a minimal subtrace, with respect to G, of a given trace in $\mathcal{L}(G)$, is obtained by removing any events involved in loops in G. In this manner, we can obtain a subtrace (still in $\mathcal{L}(G)$) "representing the original trace" that does not go through any loop in G. Our goal is to use the set of minimal subtraces which is finite, to represent a possible infinite sublanguage of $\mathcal{L}(G)$.

For our purposes, we adopt the following constructive definition, which associates with every given trace a unique subtrace.

Definition 3. For every trace $s \in K$, **the minimal subtrace** $\underline{s}$ **(with respect to** G**)** is obtained from s by repeatedly, starting from the first event of the trace, removing any subsequence of contiguous events generated by a loop in G.

Explicitly, if $s = \sigma_1 \ldots \sigma_{i-1}\sigma_i \ldots \sigma_j\sigma_{j+1} \ldots \sigma_n$ and if $\sigma_1 \ldots \sigma_j$ is the smallest prefix of s that has a loop at the end, i.e. $\delta(x_0, \sigma_1 \ldots \sigma_{i-1}) = \delta(x_0, \sigma_1 \ldots \sigma_{i-1}\sigma_i \ldots \sigma_j)$, then the subsequence $\sigma_i \ldots \sigma_j$, involving the loop, is removed to obtain the new trace $s' = \sigma_1 \ldots \sigma_{i-1}\sigma_{j+1} \ldots \sigma_n$. Note that $s' \in K = \mathcal{L}(G)$. The above process is repeated until the minimal subtrace $\underline{s}$ is obtained.

The set of all the minimal subtraces of K, based on G, can be given as the language generated by an acyclic automata $G' = (\Sigma, X, \delta', x_0)$ that is the depth-first expansion of G. G' has the same state space, initial state, and event set as G. The following algorithm implements a depth-first expansion on G, to compute δ', the transition function of G'. This Algorithm can be found in [9].

The following lemma states an immediate property of minimal subtraces used in results that follow.

Lemma 2. $\forall_{s \in K} \forall_{t \in \Sigma^*} \quad (st \in K \Leftrightarrow \underline{s}t \in K)$.

The event set Σ_{iP} of the transformed automaton G_{iP} can be given as follows based on the above notion of minimal subtraces.

Definition 4. $\Sigma_{iP} = \Sigma_{L_{iP}} \cup \Sigma_{H_{iP}} \cup \Sigma_{D_{iP}}$ where,

- $\Sigma_{L_{iP}} = \{[\sigma] \mid \sigma \in \Sigma_L\}$;
- $\Sigma_{H_{iP}} = \{[\sigma] \mid \sigma \in \Sigma_H\}$;
- $\Sigma_{D_{iP}} = \{[s'\sigma] \mid s' = \underline{s}, s\sigma \in K, \sigma \in \Sigma_D\}$.

Note that [.] is used to denote the new events of G_{iP} based on the symbols of G's events.

In the above definition, the sets $\Sigma_{L_{iP}}$ and $\Sigma_{H_{iP}}$ are computed in an obvious way. The event set $\Sigma_{D_{iP}}$ are computed by finding all traces in G' that can be appended with a Σ_D event: $\Sigma_{D_{iP}} = \{[s'\sigma] \mid s' \in \mathcal{L}(G') \wedge \sigma \in \Sigma_D \wedge s'\sigma \in \mathcal{L}(G)\}$

Not that $\Sigma_{D_{iP}}$ can be computed using a modified version of G' and by backward reaching with a worst case complexity of $O(|\Sigma|^{|X|})$.

Definition 5. Consider the operator $\langle\cdot\rangle : K \rightarrow \Sigma_{iP}^*$ defined inductively as follows: $\langle\epsilon\rangle = \epsilon$, and

$$\langle s\sigma\rangle = \begin{cases} \langle s\rangle[\sigma] & \text{if } \sigma \in \Sigma_L \cup \Sigma_H \\ [\underline{s}\sigma] & \text{if } \sigma \in \Sigma_D \end{cases}$$

The operator $\langle\cdot\rangle$ transforms a trace $s = \sigma_1 ... \sigma_i\sigma_{i+1} ... \sigma_n$ from K, where σ_i is the first event in Σ_D from the right, into a new trace $\langle s\rangle = [\underline{\sigma_1 ... \sigma_i}][\sigma_{i+1}] ... [\sigma_n]$ in Σ_{iP}^*. Note that $[\underline{\sigma_1 ... \sigma_i}] \in \Sigma_{D_{iP}}$ and $[\sigma_{i+1}], ..., [\sigma_n] \in \Sigma_{H_{iP}} \cup \Sigma_{L_{iP}}$. If $s \notin K$, then $\langle s\rangle$ is undefined.

We now define the iP-quotient language of K as follows.

Definition 6. For $K \subseteq \Sigma^*$, consider the following language over Σ_{iP}^* defined as follows:

$$K_{iP} = \{\langle t\rangle \in \Sigma_{iP}^* \mid t \in K\}.$$

We sometimes also use $\langle K\rangle$ to denote this language, that is, $\langle K\rangle = K_{iP}$. The "inverse" of $\langle\cdot\rangle$, denoted by $\widetilde{\cdot}$, is defined below.

Definition 7. Consider the operator $\widetilde{\cdot} : \Sigma_{iP}^* \rightarrow \Sigma^*$ defined inductively as follows: $\widetilde{\epsilon} = \epsilon$ and

$$\widetilde{s\sigma} = \begin{cases} \widetilde{s}\sigma' & \text{if } \sigma = [\sigma'] \in \Sigma_{L_{iP}} \cup \Sigma_{H_{iP}} \\ s'\sigma' & \text{if } \sigma = [s'\sigma'] \in \Sigma_{D_{iP}} \end{cases}$$

The operator $\widetilde{\cdot}$ transforms a trace

$s = [\sigma_1'] ... [\sigma_m'][s'\sigma][\sigma_1] ... [\sigma_n]$ from Σ_{iP}^* into a trace $\widetilde{s} = s'\sigma\sigma_1 ... \sigma_n$ in Σ^*. Note that we define $\widetilde{\cdot}$ from Σ_{iP}^* to Σ^*, not just from K_{iP} to Σ^*. If $s \in K_{iP}$, then s must have the form $s = [s'\sigma][\sigma_1] ... [\sigma_n]$, where $[s'\sigma] \in \Sigma_{D_{iP}}, [\sigma_i] \in \Sigma_{L_{iP}} \cup \Sigma_{H_{iP}}$.

Now we can present the main result of this section: iP-observability of K is equivalent to P-observability of K_{iP}.

Theorem 2. K is iP-observable w.r.t. (Σ^*, Σ_L) if and only if K_{iP} is P-observable w.r.t. $(\Sigma_{iP}^*, \Sigma_{L_{iP}})$.

3.2 iP-Quotient Automaton

To verify the iP-observability of K_{iP}, an automaton generating K_{iP} needs to be constructed. This automaton is called iP-quotient automaton and is defined as follows.

Definition 8. Let $G = (\Sigma, X, \delta, x_0)$ and $K = \mathcal{L}(G)$. The *iP-quotient* automaton of G is defined as

$$G_{iP} = (\Sigma_{iP}, X, \delta_{iP}, x_0)$$

where the transition function $\delta_{iP} : X \times \Sigma_{iP} \rightarrow X$ is given by $\delta_{iP}(x, \sigma) =$

$$\begin{cases} \delta(x, \sigma') & \text{if } \sigma = [\sigma'] \in \Sigma_{L_{iP}} \cup \Sigma_{H_{iP}}, \\ \delta(x_0, s\sigma') & \text{if } \sigma = [s\sigma'] \in \Sigma_{D_{iP}}, \text{ and } x = x_0, \\ \text{undefined} & \text{otherwise.} \end{cases}$$

Theorem 3. If G does not have self-loops of Σ_D events at the initial state, then $K_{iP} = \mathcal{L}(G_{iP})$.

Example 2. Consider the state machine G of Figure 5, with $h \in \Sigma_H$, $d \in \Sigma_D$, and $l \in \Sigma_L$. In this example, we use Definition 8 and the corresponding implicit algorithm, to transform G into its iP-quotient, G_{iP}.

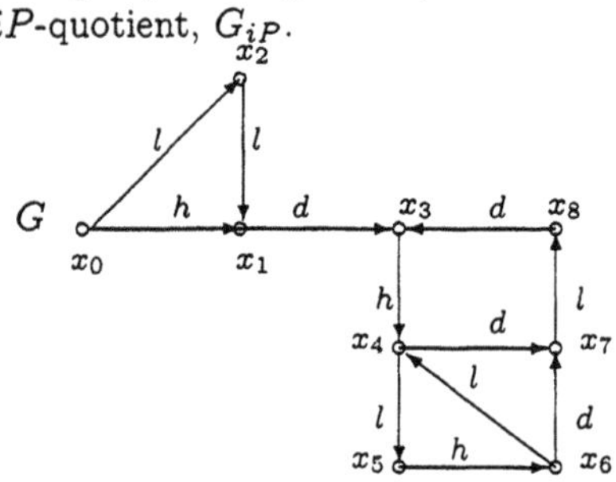

Fig. 5. *A machine G with loops.*

First consider the computation of the minimal traces of G needed in order to determine $\Sigma_{D_{iP}}$. The acyclic automaton corresponding to minimal traces is given in [9].

From automaton G' and the formula:

$$\Sigma_{D_{iP}} = \{[s'\sigma] \mid s' \in \mathcal{L}(G') \wedge \sigma \in \Sigma_D \wedge s'\sigma \in \mathcal{L}(G)\}$$

we obtain: $\Sigma_{D_{iP}} = \{[hd], [hdhd], [hdhlhd], [hdhdld], [hdhlhdld], [lld], [lldhd], [lldhlhd], [lldhdld], [lldhlhdld]\}$

The G_{iP} automaton is given in [9].

4. CONCLUSION

In this paper, we used the observability theory of discrete event systems to formulate and provide an algorithmic approach for the problem of checking the property of intransitive non-interference of security policies. This property plays a key role in the area of security, because it is used to model information flow and the lack of information flow within muli-level security systems and protocols.

REFERENCES

[1] D.E. Bell and L.J. LaPadula. Secure computer system: unified exposition and multics interpretation. Technical Report MTR-2997, Mitre Corp., Bedford, Mass., USA, June 1976.

[2] G. Boudol. Logic and models of concurrent systems. *Notes on algebraic calculi of processes, NATO ASI. Springer*, F-13:261–303, 1985.

[3] B. A. Brandin, W. M. Wonham, and B. Benhabib. Manufacturing cell supervisory control—a timed discrete event system approach. In *Proceedings of the IEEE Conference on Robotics and Automation*, Nice, France, May 1992.

[4] C.G. Cassandras and S. Lafortune. *Introduction to Discrete Event Systems.* Kluwer Academic Publishers, Boston, MA, 1999.

[5] J. Clark and M. Jacob. A survey of authentication protocols. Technical report, Available at www.cs.york.ac.uk/jac/papers/, 1997.

[6] R. Focardi and R. Gorrieri. A classification of security properties for process algebras. *Journal of Computer Security*, 3(1):5–33, 1994/1995.

[7] J.A. Goguen and J. Meseguer. Security policies and security models. In *Proceedings 1982 IEEE Symposium on Research in Security and Privacy*, pages 11–20, April 1982.

[8] J.A. Goguen and J. Meseguer. Unwinding and inference control. In *Proceedings of the 1984 IEEE Symp. on Research in Security and Privacy*, pages 75–285, Oakland, CA, 1984. IEEE Computer Society.

[9] N. Hadj-Alouane, S. Lafrance, F. Lin, J. Mullins, and M. Yeddes. Characterizing Intransitive Non-Interference in Security Policies with Observability. *Preprint*, available at www.ece.eng.wayne.edu/~ flin.

[10] J.T. Haigh and W.D. Young. Extending the noninterference version of mls for sat. *IEEE Trans. on Software Engineering*, 13(2):141–150, 1987.

[11] S. Lafortune. Modeling and analysis of transaction execution in database systems. *IEEE Transactions on Automatic Control*, vol. 33, N5, May 1988, pp. 439-447.

[12] F. Lin and W. M. Wonham. On observability of discrete-event systems. *Information Sciences*, 44:173–198, 1988.

[13] J. McLean. A general theory of composition for a class of possibilistic properties. *IEEE Trans. on Software Engineering*, 22(1):53–66, 1996.

[14] S. Pinsky. Absorbing covers and intransitive non-interference. *In Proceedings of the IEEE Symp. on Research in Security and Privacy. Oakland, CA*, pages 102–113, May 1995.

[15] P. J. Ramadge and W. M. Wonham. Supervisory control of a class of discrete-event processes. *SIAM Journal of Control and Optimization*, 25(1):206–230, 1987.

[16] K. Rudie and W. M. Wonham. Protocol verification using discrete-event systems. In *Proceedings of the 31st IEEE Conference on Decision and Control*, pages 3770–3777, Tucson, Arizona, December 1992.

[17] J. Rushby. Noninterference, transitivity and channel-control security policies. Technical Report CSL-92-02, SRI International, Menlo Park CA, USA, december 1992.

[18] P.Y.A Ryan and S.A Schneider. Process algebra and non-interference. *In Proceedings of CSFW-12. Mordano. Italy. IEEE*, June 1999.

ELSEVIER
IFAC
PUBLICATIONS
www.elsevier.com/locate/ifac

DESIGNING DEPENDABLE LOGIC CONTROLLERS USING ALGEBRAIC SPECIFICATIONS

Jean-Marc ROUSSEL[(1)], **Jean-Marc FAURE**[(1)(2)]

(1)LURPA, ENS de CACHAN, 61 avenue du président Wilson,
F-94235 CACHAN Cedex, France.
(2)SUPMECA, 3 rue Fernand Hainaut, F-93407 Saint-Ouen Cedex, France.

Abstract: A formal method to design a logic control system from specifications given in natural language is presented. To prevent designer's faults coming from a specifications' misinterpretation, the specifications are translated into algebraic statements. A part of the paper is devoted to the presentation of the formal framework that underlies this formal design method: algebra $\mathbb{I}$. The operations and relations in this algebra allow one to state formally specifications, to detect and to remove inconsistencies within the set of specifications. From the consistent set of specifications obtained, control laws can be generated. Application of this formal design method to an example permits to illustrate its main advantages. *Copyright © 2004 IFAC*

Keywords: Formal methods, Logic control systems, Algebraic approach, Dependabilility, Programmable Logic Controller (PLC).

1. INTRODUCTION

The fast-paced development of information and communication technologies has led to the introduction of an increasing number of automatic operations within everyday objects as well as within production systems for goods and services. Whole components of our daily lives and of the economy at large thereby rely upon the successful operations of control systems.

This evolution, related to the increasing demand on the part of society to better control technological risks, explains the significant development efforts devoted to methods that enable guaranteeing, as of the design phase, that the control system meets all requirements imposed by the application. A control system can in fact only be qualified as dependable if no flaw due to a misinterpretation of its specifications has been introduced during design. By adopting the taxonomy proposed in (Laprie, 1992), these methods have thus staked a position in the removal of faults caused by the control system designer.

Only the control of discrete event systems (DES), which represent a sizable share of all industrial systems (all the more so given the focus on safety functions), will be considered herein. Under such conditions, two main categories of methods for fault removal during design have been identified:

- A posteriori verification of the design result;
- (Semi-)automatic design.

The first approach consists of letting the control system designer develop control laws based on the requirements contained in the set of specifications and then of automatically analysing a formal representation of these control laws. Such an analysis relies on formal techniques that are either analysis techniques for state automata (i.e. model-checking techniques (Bérard, *et al.*, 1999)) or symbolic calculus techniques (i.e. theorem-proving techniques (Roussel and Denis, 2002)).
The second approach, qualified as synthesis by some authors (Ramadge and Wonham, 1989; Zaytoon and Carré-ménétrier, 2001), is intended directly to deduce the control laws from the specifications, without any involvement of a designer (or at least in limiting involvement to a strict minimum). This approach necessitates, in exchange, a formal modelling set-up as

regards both the specifications and the rules for manipulating the representative formal models.

The work presented herein lies within this latter category and is aimed at contributing to fault removal during the design of a control system for a logic DES, by means of proposing a method that enables deducing from specifications expressed using natural language a complete and consistent formal description of the control laws. This formal model may then be easily implemented in either hardware or software form.
Section 2 is devoted to presenting this method's main objective and introduces the formalism used to ground the methodology. This formalism will then be discussed in section 3; the various operations and relations admitted that provide a formal basis for the design method will be explored therein. This construction serves section 4 to describe the proposed design method for a dependable control system. Section 5 illustrates a simple example using this method.

2. OBJECTIVE

The starting point for the proposed design method is the set of specifications inherent in the control system, as expressed with natural language. These specifications describe the expected behaviour of the control system, in the form of vivacity constraints (what the control system must accomplish) and safety constraints (what the system must not accomplish), and may include constraints coming from actuator and sensor technology choices. All of these constraints are to be expressed in the form of logic assertions, i.e. propositions that must be true for the desired control system (c.f. Section 5.1). At this point, however, it is important to highlight the following points:

- The control specifications of a logic system can make reference to logic variable *states*, to state changes of these logic variables (*events*), or to *physical time* values. The formalism that supports the design method is to be endowed with the capability of expressing these three types of variables.
- A set of specifications does not necessarily have to be consistent. The design method must therefore be capable of detecting possible inconsistencies in the specifications and then of proposing solutions that enable resolving these problems.

Control laws are to be designed on the basis of this list of assertions. For a dynamic logic control system with n boolean input variables $U_1(t)$ to $U_n(t)$ and m boolean output variables $Y_1(t)$ to $Y_m(t)$, the target control laws must specify at each time t the output values as functions of input values, which leads to (Cassandras and Lafortune, 1999):

$$\begin{cases} Y_1(t) = f_1(U_1(t), \ldots, U_n(t)) \\ \ldots \\ Y_m(t) = f_m(U_1(t), \ldots, U_n(t)) \end{cases}$$

The search for solutions to this system of m equations generally requires a reformulation of the problem in the form of a state model, thereby necessitating the introduction of other variables, state variables X_j recording the system evolution. Regardless of the advantages inherent in this approach, it displays the disadvantage of merely providing specific solutions, at a given point in time, as exemplified in the form of «Y_i becomes true when U_k becomes false and if variables X_2, X_5 and X_8 are false while variables X_3 and X_{10} true», but never a general solution that holds true regardless of the date considered. Moreover, modelling a state automaton-based dynamic system corresponds to an imperative design approach, whereas control system specifications are in most cases given in declarative form. The transition from one of these representation modes to the other always requires a major effort.

For both reasons, the proposed design method relies upon a special formalism, named algebra Π, that has been developed by our research team. The basic elements of this formalism consist of the time functions $U_i(t)$, $Y_i(t)$. The operations that enable composing these functions lead to defining an algebraic structure in which the simultaneous manipulation of boolean variable states, state changes of these variables (events) and physical time values is possible.
In sum, the proposed design method for dependable logic control systems calls for developing, on the basis of specifications given in the form of assertions in natural language, a set of control laws in the form of time functions. This method requires formalizing specifications into relations within the algebra Π (figure 1) as well as checking the consistency of the set of assertions and generating the expected control laws from the consistent set of specifications obtained.

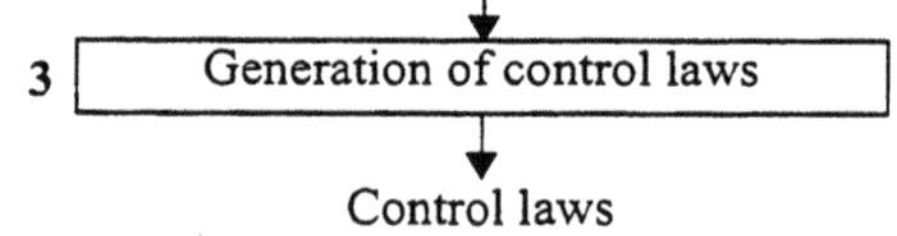

Fig. 1 Method overview

A large part of the method (consistency checking and control laws generation) might be automated with a software tool able to manipulate algebra Π statements. Conversely inconsistencies removal will always require the designer's competence.

3. FORMAL FRAMEWORK: A BOOLEAN ALGEBRA FOR BINARY SIGNALS

3.1 Binary signal modelling

Algebra $\mathbb{I}$ shall provide a formal framework to represent and to manipulate boolean variable states, events and physical delays. When defining this algebra, the main idea was to consider binary signals, i.e. functions describing the evolutions of boolean values in time. These evolutions are usually represent by timing diagrams. Though this representation is quite useful for control engineers, it is not at all based on a sound formalism. Hence the first step in the definition of the algebra $\mathbb{I}$ is aimed at giving a formal definition of binary signals: piecewise-continuous functions from $\mathbb{R}^{+*}$ ($\mathbb{R}^{+*}$ means $(0, +\infty)$) to $\mathbb{B} = \{0, 1\}$. The elements of $\mathbb{I}$ are consequently formally defined in the following way:

$$\mathbb{I} = \{f : \mathbb{R}^{+*} \to \mathbb{B} \mid \forall t \in \mathbb{R}^{+*}: (\exists \varepsilon_t > 0 : (\forall(\varepsilon_1, \varepsilon_2) \in (0, \varepsilon_t)^2, f(t-\varepsilon_1) = f(t-\varepsilon_2)))\}$$

Figure 2 shows an example of a function element of $\mathbb{I}$. Attention shall be paid to the right-continuity used at the dates t_1 and t_3 and to the double-discontinuity at the dates t_2 and t_4, mandatory to model events.

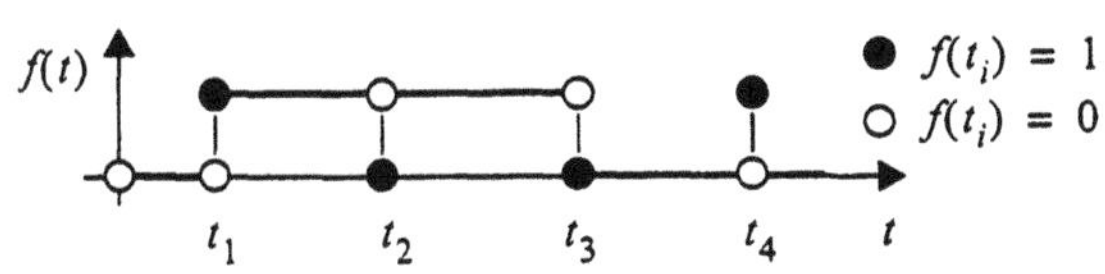

Fig. 2 Graphical representation of a binary signal

$\mathbb{I}$ contains two special elements $1*$ (the one element) and $0*$ (the zero element) defined as follows:

$$1* : \begin{array}{l} \mathbb{R}^{+*} \to \mathbb{B} \\ 1*(t) \to 1 \end{array} \qquad 0* : \begin{array}{l} \mathbb{R}^{+*} \to \mathbb{B} \\ 0*(t) \to 0 \end{array}$$

3.2 Structure of boolean algebra

To compose the elements of $\mathbb{I}$, three closed operations have been defined:

AND operation	OR operation	NOT operation
$\mathbb{I}^2 \to \mathbb{I}$	$\mathbb{I}^2 \to \mathbb{I}$	$\mathbb{I} \to \mathbb{I}$
$(f, g) \to (f \cdot g)$	$(f, g) \to (f + g)$	$f \to \bar{f}$

Where $\forall t \in \mathbb{R}^{+*}$,

$$(f \cdot g)(t) = f(t) \wedge g(t)$$
$$(f + g)(t) = f(t) \vee g(t)$$
$$\bar{f}(t) = \neg f(t)$$

$(\mathbb{I}, \cdot, +, \bar{}, 1*, 0*)$ is a boolean algebra (Grimaldi 2000) (Roussel and Faure 2002). The three basis operations (AND, OR, NOT) enable to combine binary signals only in a combinatory way, i.e. to obtain a signal whose value at each date is obtained from the values of the operands at the same time. Sequential and timed operations are to be defined to describe more complex behaviours, such as those included in the specifications of industrial control systems. Only the definitions of a sequential operation (RS) and of a timed operation (TON) are given hereinafter. Other operations are presented in (Roussel, *et al.*, 2004):

TON operation	RS operation
$\mathbb{I} \to \mathbb{I}$	$\mathbb{I}^2 \to \mathbb{I}$
$f \to d/f$	$(s, r) \to RS(s, r)$

Where $\forall t \in \mathbb{R}^{+*}$,

$$(d/f)(t) = \begin{cases} 0 & \forall t < d \\ (\forall t_1 \in (t-d, t], (f(t_1) = 1)) & \forall t \geq d \end{cases}$$

$$RS(s, r)(t) = (s(t) \wedge \neg r(t)) \vee [\exists t_1 < t \mid ((s(t_1) = 1) \wedge (\forall d \in [t_1, t], (r(d) = 0)))]$$

3.3 Equality and partial ordering relations

An obvious relation in the algebra $\mathbb{I}$ is the equality between two signals, denoted $=$, which states that the values of these signals are equal, whatever the considered date. Moreover as $(\mathbb{I}, \cdot, +, \bar{}, 1*, 0*)$ is a boolean algebra, the relation $\leq$ defined as follows is a partial ordering relation (Grimaldi, 2000).

If $(f, g) \in \mathbb{I}$, define $f \leq g$, if $f \cdot g = f$

As the relation $\leq$ is a partial ordering relation, this relation is reflexive, antisymmetric and transitive. This relation can be also stated in the following way:

$$\forall t \in \mathbb{R}^{+*}, f(t) \wedge \neg g(t) = 0 \text{ or } \neg f(t) \vee g(t) = 1$$

That means in natural language: «for all dates t_i such as $f(t_i)$ is true, the value of $g(t_i)$ is true too.»

For all f, g elements of $\mathbb{I}$, the 6 following relations have been proved equivalent:

$f \leq g$	$f \cdot g = f$	$\bar{f} + g = 1*$
$\bar{g} \leq \bar{f}$	$f + g = g$	$f \cdot \bar{g} = 0*$

The following results, enabling to split a partial ordering relation into two equivalent ones, have also been proved:

$$[(f+g) \leq h] \equiv \left[\begin{cases} f \leq h \\ g \leq h \end{cases}\right] \qquad [f \leq (g \cdot h)] \equiv \left[\begin{cases} f \leq g \\ f \leq h \end{cases}\right]$$

This partial ordering relation is the cornerstone of the design method of dependable control systems as described in the next section.

4. CONTRIBUTION OF ALGEBRA $\mathbb{I}$ TO THE DESIGN OF CONTROL SYSTEMS

4.1 Specifications formalization

The operations and relations of algebra $\mathbb{I}$ enable one to state formally specifications given in the form of assertions in natural language and including boolean variable states, events and physical delays. Some usual assertions that can easily be obtained from control systems specifications and the equivalent formal relations are shown in table 1.

4.2 Generation of a control law

Consider an elementary single output control system whose output O is assumed to be specified by the fol-

Table 1: Assertions and formal relations

Assertions given in natural language	Formal relations
Values of f and g signals are always equal.	$f = g$
Values of f and g signals are never simultaneously true.	$f \cdot g = 0^*$
At each time, at least one of the values of the f and g signals is true.	$f + g = 1^*$
When the value of signal f is true, the value of signal g is true.	$f \leq g$
It is sufficient that the value of signal f is true to get the value of signal g true.	$f \leq g$
Value of the signal f must be true to obtain the value of signal g true.	$f \leq g$ or $\overline{g} \leq \overline{f}$
Value of the signal f is never true more than 3 seconds.	$3s/f = 0^*$

lowing two relations:

$$\begin{cases} (1a) & A \leq O \\ (1b) & B \leq \overline{O} \end{cases}$$

where A, B and O are binary signals. A and B can be two inputs of the control system or complex statements using the operations of algebra Π. The searched control law shall relate O to an expression involving A and B. The solution of each ones of these relations is quite easy to obtain:

- $O = A + f_1$ with $f_1 \in \Pi$, for the first relation (1a). This relation sets indeed that $O(t)$ must be true when $A(t)$ is true, no matter what its value is when $A(t)$ is false.
- $O = \overline{B} \cdot f_2$ with $f_2 \in \Pi$, for the second one (1b). $O(t)$ must be false when $B(t)$ is true, no matter what its value is when $B(t)$ is false.

Relations set analysis

When analysing the set of the two relations, three cases can be pointed out:

- No solution if the statement $A \cdot B \neq 0^*$ holds. In that case, there is indeed at least one date t_i such as $(A \cdot B)(t_i) = 1$, that leads to inconsistency ($O(t_i)$ should be at the same time true and false).
- Unique solution: $O = A$, if the statement $A = \overline{B}$ holds.
- Infinite number of solutions (the output is not completely specified) if $A \cdot B = 0^*$ and $A \neq \overline{B}$. These solutions can be written in the form: $O = (A + \overline{B} \cdot f_1)$ or $O = (A + f_2) \cdot \overline{B}$ with $f_1 \in \Pi$ and $f_2 \in \Pi$.

The previous analysis showed how inconsistency and incompleteness can be detected thanks to relations of algebra Π. The inconsistency of a set of relations must be removed to generate a control law; this shall be performed by introducing priority levels between relations as explained in what follows. Incompleteness does not prevent to generate a control law; all you have to do is to choose a possible solution.

Control laws

To establish a control law, whatever the value of $A \cdot B$, the three following situations are to be considered:

- Situation # 1: No inconsistency ($A \cdot B = 0^*$). The consistent specification is:

$$\begin{cases} (1a) & A \leq O \\ (1b) & B \leq \overline{O} \end{cases} \qquad A \cdot B = 0^*$$

  The form of the control law is: $O_1 = (A + \overline{B} \cdot f_1)$ or $O_1 = (A + f_2) \cdot \overline{B}$ with $(f_1, f_2) \in \Pi^2$.
- Situation # 2: Initial inconsistency ($A \cdot B \neq 0^*$). The relation (1a) is dominant beside the relation (1b), denoted (1b) « (1a). The consistent specification is:

$$\begin{cases} (1a) & A \leq O \\ (1b) & B \leq \overline{O} \\ (1b) \ll (1a) & \end{cases} \text{ equivalent to } \begin{cases} (1a) & A \leq O \\ (1b) & \overline{A} \cdot B \leq \overline{O} \end{cases}$$

  The form of the control law is: $O_2 = (A + \overline{B} \cdot f)$ with $f \in \Pi$.
- Situation # 3: Initial inconsistency ($A \cdot B \neq 0^*$) The relation (1b) is dominant beside the relation (1a). The consistent specification is:

$$\begin{cases} (1a) & A \leq O \\ (1b) & B \leq \overline{O} \\ (1a) \ll (1b) & \end{cases} \text{ equivalent to } \begin{cases} (1a) & A \cdot \overline{B} \leq O \\ (1b) & \overline{A} \leq \overline{O} \end{cases}$$

  The form of the control law is: $O_3 = \overline{B} \cdot (A + f)$ with $f \in \Pi$.

Attention shall be paid that for the dates t_i such as $A(t_i) \vee B(t_i) = 0$, the value $O(t_i)$ of the solution O can be true or false according to the choice made for f, f_1, f_2. To obtain deterministic control laws, i.e. control laws involving only the A and B signals and no specific designer's choices, one of the following solutions is to be chosen:

- At these dates t_i, $O(t_i) = 0$, that leads to:

$$O_1 = A \qquad O_2 = A \qquad O_3 = A \cdot \overline{B}$$

- At these dates t_i, $O(t_i) = 1$, that leads to:

$$O_1 = \overline{B} \qquad O_2 = A + \overline{B} \qquad O_3 = \overline{B}$$

- At these dates t_i, $O(t_i)$ keeps the last value determined by the assertions set (memory behaviour), that leads to:

$$O_1 = RS(A, B) \qquad O_2 = SR(A, B) \quad O_3 = RS(A, B)$$

The last solution will be adopted herein for it seems closer to industrial concerns and practice.

5. APPLICATION EXAMPLE

This section is aimed at dealing with a simple example: the control of an automatic gate for a car park. The inputs and the outputs of the control system to design are given in figure 3.

5.1 *Control system specifications*

The expected behaviour of the control system with regard to the application requirements may be expressed by the set of assertions given hereafter. Among these 7

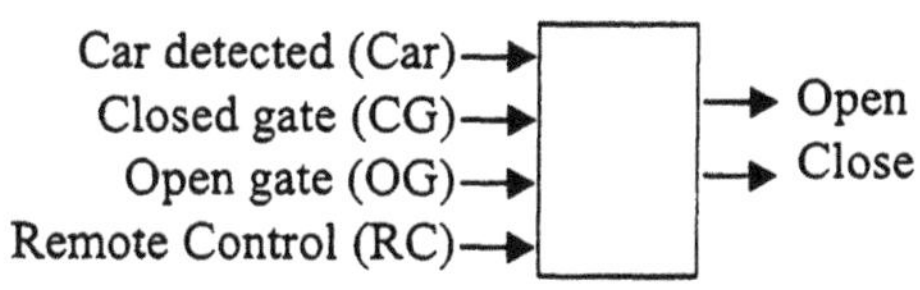

Fig. 3 Inputs and outputs of control system to design

assertions, the first three ones (A1 to A3) are related to vivacity requirements (what must be done to perform the expected task), the assertion A4 expresses a safety requirement. The assertion A5 and A6 express constraints coming from actuators features and the last one (A7) is an assumption on the correct operation of the sensors (the problem of sensors monitoring will not be dealt with in this study).

A1 When the remote control is activated, the gate opens.

A2 When the gate is open during 3 seconds without a request of the user or the detection of a car, the gate closes.

A3 While the gate is not totally closed, the detection of a car causes the reopening of the gate.

A4 The gate must never be simultaneously controlled to open and to close.

A5 An open gate can't be controlled to open.

A6 A closed gate can't be controlled to close.

A7 The gate is never simultaneously open and closed.

5.2 Specifications formalisation

The previous set of assertions can be translated into a set of formal relations (table 2) that include input signals (Car, CG, OG, RC) and output signals (Open, Close). The main objective of the design method is to state, from this set of relations, the control laws that link the output signals to the input signals. For systems more complex than this example, internal signals may also be defined.

Table 2: Formal specifications of the control system

Assertions	Assertions written on $\amalg$
A1	$RC \le Open$
A2	$3s/(OG \cdot \overline{(RC + Car)}) \le Close$
A3	$\overline{CG} \cdot Car \le Open \cdot \overline{Close}$
A4	$Open \cdot Close = 0^*$
A5	$OG \cdot Open = 0^*$
A6	$CG \cdot Close = 0^*$
A7	$OG \cdot CG = 0^*$

5.3 Consistency checking

When dealing with a set of formal relations involving several output signals, consistency checking comprises three steps:

- Assertions labelling
- Dependency analysis
- Consistency checking for each output signal.

Consistency of a set of relations depends indeed not only on the consistency of the relations defining a given output signal, as previously shown in Section 4.2, but also on the lack of cross-references in the relations set, e.g. the output O_i is defined from the internal signal IS_j that is itself defined from O_i. The first two steps are aimed at checking cross-references while the last step looks for inconsistencies in the specification of each signal.

Assertions labelling

This step aims to rank each of the relations of the set of specifications into one of the following categories:

- Relations stating formally assertions that involve only input signals. These relations are assumptions.
- Relations stating formally assertions that involve only one output or one internal signal and input signals. Each of these relations will be used to build the control law of the given output or the internal signal from input signals.
- Relations stating formally assertions that involve several outputs or internal signals. In that case, either the relation can be decomposed in elementary relations (relations comprising only one output or internal signal) as stated in Section 3.3, that leads to the previous case, or the designer has to decide which signal will be function of the other ones (e.g. assertion A4 in this example).

From this analysis it is possible to determine the assertions that must be employed to elaborate an output signal as well as which other signals the considered signal depends on. The result of this analysis for the control system of the automatic gate is presented in table 3.

Table 3: Dependency relations derived from assertions labelling

Output signal	Assertions to be used	Signals involved in the assertions
Open	A1, A3, A5	Car(A3), CG(A3), OG(A5), RC(A1)
Close	A2, A3, A4, A6	Car(A2, A3), CG(A3, A6), OG(A2), RC(A2), Open(A4)

Dependency analysis

From previous results a dependency graph can be built. If this graph does not include any cycle, no cross-reference lies within the assertions set. Conversely, any cycle found when analysing the dependency graph enables to point out specifications inconsistency. In that case, part of the assertions set must be modified in order to eliminate the inconsistencies. The table 3 shows that the set of assertions of the studied example does not include any inconsistency.

Consistency checking for each output signal

The principle of this step has been already described in Section 4.2. To remove inconsistencies, three priorities between assertions must be introduced. The table 4 shows the result of consistency checking.

Table 4: Result of consistency checking

Signal	Initial specification	Priority
Open	(A1) $RC \le Open$ (A3) $\overline{CG} \cdot Car \le Open$ (A5) $OG \le \overline{Open}$	(A1) « (A5) (A3) « (A5)
Close	(A2) $3s/(OG \cdot \overline{RC} \cdot \overline{car}) \le Close$ (A3) $\overline{CG} \cdot Car \le \overline{Close}$ (A4) $Open \le \overline{Close}$ (A6) $CG \le \overline{Close}$	(A2) « (A4)

5.4 Control laws

Applying the synthesis technique indicated in the Section 4.2 leads to the following control laws:

$$\begin{cases} Open = RS[(RC + \overline{CG} \cdot Car), OG] \\ Close = RS[3s/(OG \cdot \overline{(RC + Car)}), (CG + Car + Open)] \end{cases}$$

This set of formal statements can be easily translated into a Programmable Logic Controller (PLC) program written in a standardized language (IEC, 1993), like Ladder Diagram, a widespread programming language for PLCs (figure 4). It matters to highlight that the instruction lines of this program are ordered according to the dependency constraints previously obtained. The output Close, for instance, must be elaborated once the output Open is computed.

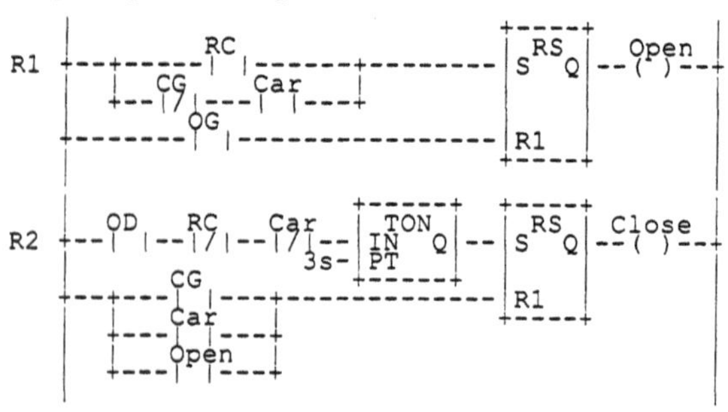

Fig. 4 PLC program developed from the control laws

Bridging the gap between control laws design and control software development is therefore an indirect benefit of the proposed design method.

6. CONCLUSION

Formal design methods may contribute efficiently to improve control systems dependability by preventing designer's faults. The first advantage of the presented formal method is the ability to detect and to remove inconsistencies within the specifications of a given control system; these inconsistencies may take place in the specifications of a single output or of a set of outputs. From the consistent set of specifications obtained, control laws are stated by formal manipulations in algebra $\mathbb{I}$, the formal framework that underlies this design method. At last, the control laws may be used to obtain a control software compliant with the specifications. The whole design method has been applied successfully to several industrial cases.

The main weakness of the method lies in the need to manipulate formal statements that are not user friendly for automation engineers. To overcome this problem, a software tool supporting the method has been developed. This tool embeds operations, relations and theorems of algebra $\mathbb{I}$ and performs automatically consistency checking, for a single signal as well as for a set of signals, and control laws generation if the relations set is consistent. When an inconsistency is detected, the tool warns the designer who must then introduce dominance, in case of inconsistency in the specification of a single output, or dependency constraints, in case of cross-reference. Steps 2 and 3 of the method (figure 1) have therefore been automated. Further prospect concerns the development of a library of formal relations stating formally assertions commonly found in industrial systems specifications. Using this library will contribute to automate step 1 of the method and consequently will ease industrial acceptance of this design method.

REFERENCES

BÉRARD, B., M. BIDOIT, A. FINKEL, F. LAROUSSINIE, A. PETIT, L. PETRUCCI, P. SCHNOEBELEN (1999). *Systems and Software Verification: Model-Checking techniques and tools,* Heidelberg, Springer-Verlag Editions.

CASSANDRAS, C.G., S. LAFORTUNE (1999). *Introduction to Discrete Event Systems,* Boston, Kluwer Academic Publishers.

GRIMALDI, R.P. (2000). *Discrete and Combinatorial Mathematics: An Applied Introduction,* New-York, Addison-Wesley Editions, 4th edition.

INTERNATIONAL ELECTROTECHNICAL COMMISSION (1993). IEC 61131-3, *Programmable controllers - programming languages.*

LAPRIE, J.C. (1992). *Dependability: basic concepts & terminology,* Springer-Verlag Editions

RAMADGE, P.J., W.M. WONHAM (1989). The control of discrete-event systems. *Proceedings of the IEEE,* **77**, 81-97

ROUSSEL, J.-M., B. DENIS (2002). Safety properties verification of ladder diagram programs, *Journal Européen des Systèmes Automatisés,* Hermès Editions, **36-7**, 905-917.

ROUSSEL, J.-M., J.-M. FAURE (2002). An algebraic approach for PLC programs verification. In: *Proceedings of 6th International Workshop on Discrete Event Systems (WODES'02),* Zaragoza, Spain, 2-4 October, 2002, pp. 303-308.

ROUSSEL, J.-M., J.-M. FAURE, J.-J. LESAGE , A. MEDINA (2004), Algebraic approach for dependable logic control systems design. In: *International Journal of Production Research,* **42-13**, to appears

ZAYTOON, J., V. CARRÉ-MÉNÉTRIER, (2001), Synthesis of control implementation for discrete manufacturing systems. In: *International Journal of Production Research,* **39-2**, 329-345

ELSEVIER
IFAC
PUBLICATIONS
www.elsevier.com/locate/ifac

ABSTRACTION OF CONTINUOUS SYSTEM TRAJECTORIES INTO TIMED AUTOMATA

Arnaud HÉLIAS[1,2], François GUERRIN[1] and Jean-Philippe STEYER[2]

[1]*CIRAD, Gdor team, Station de la Bretagne, BP 20, 97408 Saint-Denis, Reunion Island, France, guerrin@cirad.fr, arnaud.helias@inria.fr*
[2]*INRA, Laboratory of environmental biotechnology, av. des étangs, 11100 Narbonne, France, steyer@ensam.inra.fr*

Abstract: This paper deals with the representation of continuous system dynamics into a timed discrete-event formalism to the end of system analysis. The continuous model of the system is first approximated using intervals and then translated into the timed automata formalism by comparison with thresholds defined on the state variables' domains. The detection of thresholds crossing is characterised by two time instants corresponding respectively to the earliest and latest crossing dates. This approach is briefly illustrated with real data obtained from a 1 m^3 wastewater treatment pilot plant.

Keywords: Model approximation, Discrete system, Continuous system, Intervals, Thresholds, Automata, Waste treatment.

1. INTRODUCTION

A classical approach to analyse the interaction between continuous and discrete elements within a system is to represent it with a unique formalism. In the Qualitative reasoning and Hybrid dynamical system communities, this problem was often addressed. However, Struss (2002) underlined three main difficulties: (i) explosion in the number of discrete states , (ii) rounding errors of intermediate variables and (iii) thresholds detection. Moreover, time is not often clearly represented. This study is based on a timed discrete-event representation similar to those developed by Henzinger *et al.* (1998), Kowalewski *et al.* (1999), Lunze (1999), Supavatanakul *et al.* (2003); however it differs from several points:

- as the continuous system dynamics is assumed to be partially known, the resulting imprecision is represented by numerical intervals bounding both initial states and input variables values;
- thresholds, partitioning the state-variable domains according to expert knowledge are used as a basis of the state space discretisation;
- to avoid combinatorial explosion, timed automata are purposely built to answer specific questions, according to some initial state and the properties to be checked.

The paper is organised as follows. After a brief recall on timed automata, one of the most widespread timed discrete formalism, the continuous system under study is described in Section 2. The discretisation procedure is then detailed in Section 3 and illustrated by an application example in Section 4. Finally, this approach is briefly discussed in Section 5 and conclusions are drawn in Section 6.

2. SYSTEM FORMAL DESCRIPTION

Timed automata

Introduced by Alur and Dill (1994), a timed automaton is composed by a finite state machine and the expression of continuous time. This formalism allows timed constraints to be introduced using variables named *clocks*.

Clock. $\mathrm{x} \in \mathrm{X}$, with X a finite set, is a clock whose value grows linearly with time (*i.e.*, $\mathrm{x}(t+\varepsilon)=\mathrm{x}(t)+\varepsilon$ with ε a positive time lapse and x's derivative always equal to 1).

Clock constraint. A clock constraint is an atomic constraint conjunction. Taking x and y two clocks, $c \in \mathbb{Q}$ a constant and # a symbol relation from the set $\{<,\le,=,\ge,>\}$, the set $\Phi(X)$ of the atomic constraints φ is defined by the grammar $\varphi := x\#c \mid x-y\#c \mid \varphi_1 \wedge \varphi_2$. These constraints are possibly associated to the locations or edges of the automaton.

Timed automaton. A timed automaton G is defined by a 6-tuple (S,P,X,E,A,Inv) with:

- S a finite set of locations with $s_0 \in S$ the initial location;
- P a mapping which associates to each $s \in S$, a set of atomic propositions valid at this location;
- X a finite set of clocks;
- E a finite set of labels;
- $A \subseteq S \times E \times \Phi(X) \times 2^X \times S$ a finite set of edges; each edge a is a 5-tuple (s,e,φ,δ,s') with $s \in S$ the start location and $s' \in S$ the destination location, $\varphi \in \Phi(X)$ a clock constraint named *guard* that must be satisfied to trigger the discrete transition and $\delta \subseteq X$ the clock subset to be reinitialised during the transition;
- $Inv(s): S \to \Phi(X)$ a mapping that associates to each location a timed constraint φ', named *invariant*. The system can remain in the same location as long as the invariant is true.

Continuous system

We consider a non-linear piecewise continuous system described by an ordinary differential equation (ODE) system such as:

$$\begin{cases} \dot{\xi} = f(\xi,\zeta) \\ \xi(t_0) = \xi_0 \end{cases} \quad (1)$$

where $\xi \in \Xi \subseteq \mathbb{R}^n$, with Ξ a n-dimensional state space, $\zeta = g(p,t)$, $\zeta \in Z \subseteq \mathbb{R}^m$, with Z a m-dimensional input space, and $p \in P$ a parameter. It is assumed that we have only a partial knowledge of both the initial state $\xi(t_0)$ and the input variables values ζ that can be estimated by:

$$\forall i, \xi_i^-(t_0) \le \xi_i(t_0) \le \xi_i^+(t_0) \quad (2)$$

$$\forall j, \forall t, \zeta_j^-(t) \le \zeta_j(t) \le \zeta_j^+(t) \quad (3)$$

with $i \in \{1,\dots,n\}$, $j \in \{1,\dots,m\}$, $(\xi_i^-,\xi_i^+) \in \Xi_i^2$ and $(\zeta_j^-,\zeta_j^+) \in Z_j^2$ ($\Xi_i \subset \Xi$ and $Z_j \subset Z$).

Uncertainty representation. We define $\zeta^<$ and $\zeta^>$ according to the influence of the interval bounds on the derivative's values (taking $k = \{1,\dots,m\} \setminus \{j\}$):

$$\begin{aligned} f_i(\xi,\zeta_k,\zeta_j^-) \le f_i(\xi,\zeta) \le f_i(\xi,\zeta_k,\zeta_j^+) \Rightarrow \zeta_j^< = \zeta_j^-, \zeta_j^> = \zeta_j^+ \\ f_i(\xi,\zeta_k,\zeta_j^-) \ge f_i(\xi,\zeta) \ge f_i(\xi,\zeta_k,\zeta_j^+) \Rightarrow \zeta_j^< = \zeta_j^+, \zeta_j^> = \zeta_j^- \end{aligned} \quad (4)$$

From (4) we can rewrite (1) as the double ODE system:

$$\begin{cases} \dot{\xi}^- = f(\xi^-,\zeta^<) \\ \xi^-(t_0) = \xi_0^- \\ \\ \dot{\xi}^+ = f(\xi^+,\zeta^>) \\ \xi^+(t_0) = \xi_0^+ \end{cases} \quad (5)$$

With the property $\forall t, \forall i, \xi_i^-(t) \le \xi_i(t) \le \xi_i^+(t)$, Eq. (5) can approximate the system (1) by introducing uncertainty on the initial state (Eq. (2)) and the system inputs (Eq. (3)) if the *cooperativity* property is verified on the system's dynamics (Smith 1995). Cooperativity simply states that the off-diagonal elements of the Jacobian matrix of a dynamic system are positive or equal to zero (additional details can be found in Gouzé *et al.*, (2000) or Moor and Raisch (2002)).

Threshold

To represent the continuous system dynamics in a discrete formalism, the first step is to discretise the state space. For this, each ξ_i's domain is partitioned according to thresholds into a finite number of intervals that can be considered as qualitative states. Thresholds are defined from expert knowledge; *e.g.*, a stock level can be qualified as "insufficient", "low", "medium", "high" and "critical". L_i denotes the ordered set of thresholds in the ith dimension, that is, with $i \in \{1,\dots,n\}$ and $\Omega_i \in \mathbb{N}$:

$$(L_i,<) = \{ l_{i,0},\dots,l_{i,\Omega_i} \} \quad (7)$$

The continuous state space Ξ is divided according to the $\prod_i \Omega_i$ thresholds by the mapping:

$$D_X : \Xi \to \{1,\dots,\Omega_1\} \times \dots \times \{1,\dots,\Omega_i\} \quad (8)$$

with the assumption that $\xi_i \in [l_{i,0}, l_{i,\Omega_i}], \forall i$. A rectangular partition is thus obtained.

3. DISCRETE APPROXIMATION

The idea is (i) not to consider as discrete states the cells partitioning the state space themselves but, rather, the cells' faces and (ii) determine the state transitions from the simulation output of the continuous system within each cell as was done by Kowalewski *et al.*, 1999. These authors defined a grid on each cell's face and simulated all the trajectories from or bounded to each grid-point. In comparison, our approach depends upon each considered initial state and focuses on handling uncertainty by propagating the interval $[\xi^-,\xi^+]$ within the cells by simulating Eq. (5).

Trajectories within a cell

Thresholds crossing. Let τ be the time-point when the system (1) reaches v, a threshold value defined in the ith dimension, *i.e.*, $\xi_i(\tau) = v$, $v \in \mathbb{R}$. Because ξ_i is estimated by the interval $\left[\xi_i^-, \xi_i^+\right]$, the time window within which crossing the v threshold may occur can be described by τ^- and τ^+ such that $\xi_i^-(\tau^-) = \xi_i^+(\tau^+) = v$. From Eq. (5), the following implications can then be set:

$$\left(\dot{\xi}_i^-\left(\tau^-\right) > 0\right) \wedge \left(\dot{\xi}_i^+\left(\tau^+\right) > 0\right) \Rightarrow \tau^+ < \tau^- \quad (9)$$

$$\left(\dot{\xi}_i^-\left(\tau^-\right) < 0\right) \wedge \left(\dot{\xi}_i^+\left(\tau^+\right) < 0\right) \Rightarrow \tau^+ > \tau^- \quad (10)$$

Timed automaton representation. Let τ_{min}=min(τ^-,τ^+), τ_{max}=max(τ^-,τ^+) and F the mapping that associates a label to each threshold crossing:

$$F(v) \rightarrow \begin{cases} "v^{\Delta}" \text{ if } \tau_{min} = \tau^+ \\ "v^{\nabla}" \text{ if } \tau_{min} = \tau^- \end{cases} \quad (12)$$

with the symbols Δ and ∇ denoting threshold crossing with increasing and decreasing trends respectively.

τ_{min} and τ_{max} being approximated on $\mathbb{Q}$ (*cf.* the clock constraint definition), the system trajectory between two thresholds w and v is represented within the time window $[\tau_{min}, \tau_{max}]$. This window can be modelled in the timed automata formalism as follows (*cf.* Fig. 1):

- s_1 and s_2 are two locations with associated labels "w^{Δ}" and "v^{Δ}" respectively,
- x is a clock,
- $x \leq \tau_{max}$ is the invariant of s_1,
- $(s_1, \varnothing, x \geq \tau_{min}, \varnothing, s_2)$ is the edge.

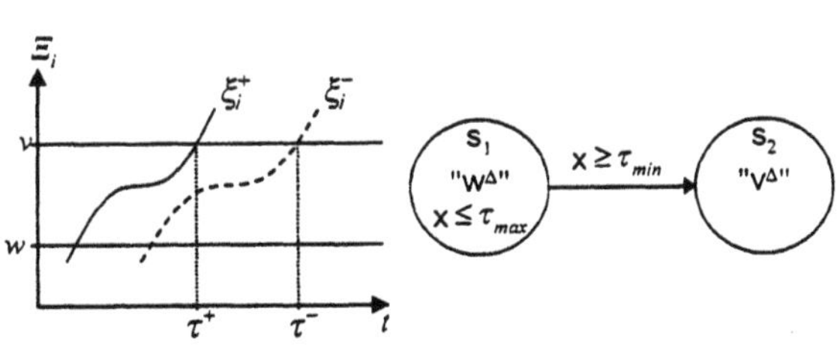

Fig. 1. Timed automaton for describing a system's trajectory from thresholds w to v with an increasing trend.

Generalisation

For sake of clarity, the index i is omitted in notations for all the variables in the following sections. Applying the above described approach to the lower bound of Eq. (5) in the ith dimension allows one to obtain T^-, the ordered set of time instants when ξ^- crosses the thresholds and Σ^-, the ordered set of threshold crossings:

$$\left(T^-, \leq\right) = \left\{\tau_j^- \,\middle|\, \xi^-(\tau_j^-) = l_{\omega}\right\} \quad (13)$$

$$\left(\Sigma^-, \leq\right) = \left\{\sigma_j^- \,\middle|\, \sigma_j^- = \xi^-(\tau_j^-)\right\} \quad (14)$$

with $j = 1,\dots,J$, $J = \text{card}\left(T^-\right) = \text{card}\left(\Sigma^-\right)$ and $\omega \in \{1,\dots,\Omega\}$. In the following, the mapping that associates τ_{min} or τ_{max} to each τ_j^- according to the sign of the derivative is denoted A_{T^-}:

$$A_{T^-} : \tau_j^- \mapsto \begin{cases} \tau_{min} \text{ if } \dot{\xi}^-\left(\tau_j^-\right) < 0 \\ \tau_{max} \text{ if } \dot{\xi}^-\left(\tau_j^-\right) > 0 \end{cases} \quad (15)$$

A similar approach can be used on the upper bound to obtain T^+, Σ^+ and associated labels by the mapping A_{T^+}.

Building the timed automaton associated to the system's trajectory in the ith dimension

From Eqs. (13-15), the following sets can be defined:

- B, the ordered set of time instants when occur threshold crossings by ξ^- and ξ^+ with label τ_{min} ($k_B = 1,\dots,K_B$, $K_B = \text{card}(B) = \text{card}(D)$):

$$(B,\leq) = \left\{b_{k_B} \in T^- \,\middle|\, A_{T^-}(b_{k_B}) = \tau_{min}\right\} \cup \left\{b_{k_B} \in T^+ \,\middle|\, A_{T^+}(b_{k_B}) = \tau_{min}\right\} \quad (16)$$

- C, the ordered set of time instants when occur threshold crossings by ξ^- and ξ^+ with label τ_{max} ($k_C = 1,\dots,K_C$, $K_C = \text{card}(C)$):

$$(C,\leq) = \left\{c_{k_C} \in T^- \,\middle|\, A_{T^-}(c_{k_C}) = \tau_{max}\right\} \cup \left\{c_{k_C} \in T^+ \,\middle|\, A_{T^+}(c_{k_C}) = \tau_{max}\right\} \quad (17)$$

- D, the ordered set of threshold crossings associated to the elements of B:

$$(D,\leq) = \left\{d_{k_B} \in \Sigma^- \,\middle|\, \xi^-(b_{k_B}) = d_{k_B}\right\} \cup \left\{d_{k_B} \in \Sigma^+ \,\middle|\, \xi^+(b_{k_B}) = d_{k_B}\right\} \quad (18)$$

- Σ_0^{∇}, the set of threshold crossings with decreasing trends at the initial time ($k_{\Sigma^{\nabla}} = 1,\dots,K_{\Sigma^{\nabla}}$, $K_{\Sigma^{\nabla}} = \text{card}\left(\Sigma_0^{\nabla}\right)$):

$$\Sigma_0^\nabla = \left\{\sigma^\nabla_{0,k_{\Sigma^\nabla}} \mid \left(\sigma^\nabla_{0,k_{\Sigma^\nabla}} \in L\right) \wedge \left(\xi^- < \sigma^\nabla_{0,k_{\Sigma^\nabla}} < \xi^+\right)\right\}$$
$$\cup \left\{\sigma^\nabla_{0,k_{\Sigma^\nabla}} \mid \left(\sigma^\nabla_{0,k_{\Sigma^\nabla}} \in L\right) \wedge \left(\xi^- = \sigma^\nabla_{0,k_{\Sigma^\nabla}}\right) \wedge \left(\dot{\xi}^- \le 0\right)\right\} \quad (19)$$
$$\cup \left\{\sigma^\nabla_{0,k_{\Sigma^\nabla}} \mid \left(\sigma^\nabla_{0,k_{\Sigma^\nabla}} \in L\right) \wedge \left(\xi^+ = \sigma^\nabla_{0,k_{\Sigma^\nabla}}\right) \wedge \left(\dot{\xi}^+ \ge 0\right)\right\}$$

- Σ_0^Δ the set of threshold crossings with an increasing trend at the initial time ($k_{\Sigma^\Delta} = 1,\dots,K_{\Sigma^\Delta}$, $K_{\Sigma^\Delta} = \mathrm{card}\left(\Sigma_0^\Delta\right)$):

$$\Sigma_0^\Delta = \left\{\sigma^\Delta_{0,k_{\Sigma^\Delta}} \mid \left(\sigma^\Delta_{0,k_{\Sigma^\Delta}} \in L\right) \wedge \left(\xi^- < \sigma^\Delta_{0,k_{\Sigma^\Delta}} < \xi^+\right)\right\}$$
$$\cup \left\{\sigma^\Delta_{0,k_4} \mid \left(\sigma^\Delta_{0,k_4} \in L\right) \wedge \left(\xi^- = \sigma^\Delta_{0,k_{\Sigma^\Delta}}\right) \wedge \left(\dot{\xi}^- \ge 0\right)\right\} \quad (20)$$
$$\cup \left\{\sigma^\Delta_{0,k_{\Sigma^\Delta}} \mid \left(\sigma^\Delta_{0,k_{\Sigma^\Delta}} \in L\right) \wedge \left(\xi^+ = \sigma^\Delta_{0,k_{\Sigma^\Delta}}\right) \wedge \left(\dot{\xi}^+ \le 0\right)\right\}$$

Clock set. The clock set is a singleton since the automaton is obtained from only one simulation of the continuous model.

Location set. Let us denote:

- S, a mapping that associates a location s to each element of D, Σ_0^∇ and Σ_0^Δ. From S, one gets $\mathsf{S} = \{\mathsf{s}_k\}_{k=1,\dots,K} \cup \mathsf{s}_0$ with $K = K_B + K_{\Sigma^\nabla} + K_{\Sigma^\Delta}$ and s_0, the location corresponding to the initial state of the continuous system.
- P, a mapping that associates a proposition to each element of S denoting which threshold is crossed and the trend (with $\omega \in \Omega$ and with $\mathsf{P}(\mathsf{s}_0) \to \mathsf{l}_0$):

$$\mathsf{P}(\mathsf{s}_k) :\to \begin{cases} \left(\mathsf{S}(d_{k_B}) \to \mathsf{s}_k\right) \wedge \left(d_{k_B} = l_\omega\right) \wedge \left(d_{k_B} = \Sigma^-\right) \Rightarrow \mathsf{l}^\nabla_\omega \\ \left(\mathsf{S}(d_{k_B}) \to \mathsf{s}_k\right) \wedge \left(d_{k_B} = l_\omega\right) \wedge \left(d_{k_B} = \Sigma^+\right) \Rightarrow \mathsf{l}^\Delta_\omega \\ \left(\mathsf{S}\left(\sigma^\nabla_{0,k_{\Sigma^\nabla}}\right) \to \mathsf{s}_k\right) \wedge \left(\sigma^\nabla_{0,k_{\Sigma^\nabla}} = l_\omega\right) \Rightarrow \mathsf{l}^\nabla_\omega \\ \left(\mathsf{S}\left(\sigma^\Delta_{0,k_{\Sigma^\Delta}}\right) \to \mathsf{s}_k\right) \wedge \left(\sigma^\Delta_{0,k_{\Sigma^\nabla}} = l_\omega\right) \Rightarrow \mathsf{l}^\Delta_\omega \end{cases} \quad (21)$$

Forward edges. Edges that can reach a new location at the earliest time point τ_{min} are named *Forward edges*. Let $\mathsf{s}_n, n \in K_B$, a location corresponding to a threshold crossing with increasing trend, and its associated proposition l^Δ_ω. The preceding location can reflect the following cases (cf. Fig. 2):

a) crossing a lower threshold with increasing trend,

b) crossing the same threshold with decreasing trend,

c) the initial state whenever no threshold has already been crossed and (i) the threshold is within the initial state interval, or (ii) the initial state interval lies below the threshold.

The same reasoning can be made the other way around about threshold crossing with a decreasing trend (*i.e.* about a location with a proposition l^∇_ω).

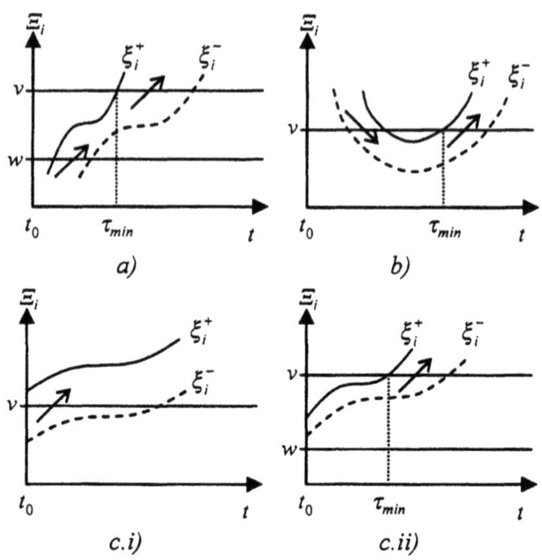

Fig. 2. Different cases of threshold crossing with increasing trend.

Based on this, a set of locations can be obtained which satisfy one of these conditions. The preceding location is the element to which is associated the nearest τ_{min} of the τ_{min} of s_n. Formally, a_1, the forward edge of s_n is defined by the 5-tuple $(\mathsf{s}_m, \varnothing, \varphi, \varnothing, \mathsf{s}_n)$ with $\varphi := \mathsf{x} \ge b_n$ [1]. Let the ordered set $\mathsf{S}' \subset \mathsf{S}$, the union of the different cases described by the Fig. 2:

$$(\mathsf{S}_1', \le) = \begin{cases} \left(\mathsf{P}(\mathsf{s}_n) = \mathsf{l}^\Delta_\omega\right) \Rightarrow \left\{\mathsf{s}_k \mid \left(\mathsf{P}(\mathsf{s}_k) = \mathsf{l}^\Delta_{\omega-1}\right)\right\} \\ \left(\mathsf{P}(\mathsf{s}_n) = \mathsf{l}^\nabla_\omega\right) \Rightarrow \left\{\mathsf{s}_k \mid \left(\mathsf{P}(\mathsf{s}_k) = \mathsf{l}^\nabla_{\omega+1}\right)\right\} \end{cases} \quad (23)$$

$$(\mathsf{S}_2', \le) = \begin{cases} \left(\mathsf{P}(\mathsf{s}_n) = \mathsf{l}^\Delta_\omega\right) \Rightarrow \left\{\mathsf{s}_k \mid \left(\mathsf{P}(\mathsf{s}_k) = \mathsf{l}^\nabla_\omega\right)\right\} \\ \left(\mathsf{P}(\mathsf{s}_n) = \mathsf{l}^\nabla_\omega\right) \Rightarrow \left\{\mathsf{s}_k \mid \left(\mathsf{P}(\mathsf{s}_k) = \mathsf{l}^\Delta_\omega\right)\right\} \end{cases} \quad (24)$$

$$\mathsf{S}_3' = \left(\mathsf{s}_n = \mathsf{S}\left(\sigma^\nabla_{0,k_3}\right)\right) \vee \left(\mathsf{s}_n = \mathsf{S}\left(\sigma^\Delta_{0,k_4}\right)\right) \Rightarrow \{\mathsf{s}_0\} \quad (25)$$

$$\mathsf{S}_4' = \begin{cases} \left(\mathsf{s}_n = \mathsf{S}\left(d_{k_1}\right)\right) \wedge \left(\mathsf{P}(\mathsf{s}_n) = \mathsf{l}^\Delta_\omega\right) \wedge \\ \left(l_{\omega-1} < \xi^-(t_0) \le \xi^+(t_0) \le l_\omega\right) \Rightarrow \{\mathsf{s}_0\} \\ \left(\mathsf{s}_n = \mathsf{S}\left(d_{k_1}\right)\right) \wedge \left(\mathsf{P}(\mathsf{s}_n) = \mathsf{l}^\nabla_\omega\right) \wedge \\ \left(l_\omega < \xi^-(t_0) \le \xi^+(t_0) \le l_{\omega+1}\right) \Rightarrow \{\mathsf{s}_0\} \end{cases} \quad (26)$$

$$(\mathsf{S}', \le) = \mathsf{S}_1' \cup \mathsf{S}_2' \cup \mathsf{S}_3' \cup \mathsf{S}_4' \quad (27)$$

with $k = 1,\dots,n-1$ (*i.e.*, only reachable locations before s_n are taken). The preceding location $\mathsf{s}_m \subseteq \mathsf{S}'$ is the latest that could be reached. The set of locations of S forward edges is named A_1.

A similar approach can be used about the latest time points (*i.e.*, threshold crossing at τ_{max}) to determine

[1] If the location is an initial state threshold crossing, then the guard is of the form $\mathsf{x} = t_0$.

the location's invariants. Moreover, when a location has no next location, a *Backward edge* is defined. Figure 3 shows this case: the system crosses a threshold successively twice, and the edge with guard $x = \tau^+_{v_2}$ is a backward edge. Doing this, the Inv mapping, that associates the invariants to the locations, and A_2, the set of backward edges, are defined. A complete description of this approach can be found in Hélias, (2003).

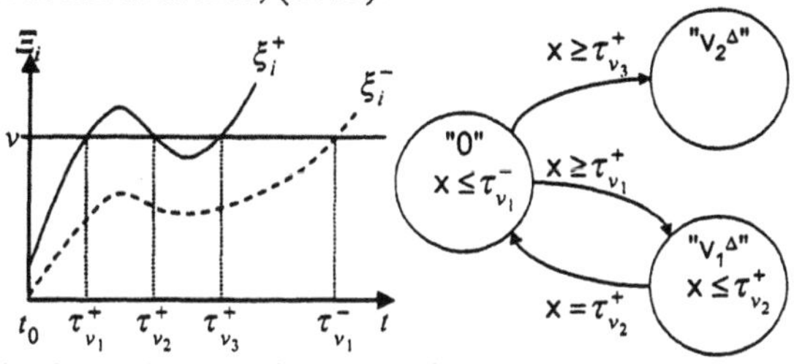

Fig. 3. Backward edge example.

From (16-27), the continuous system dynamics in the ith dimension is approximated by the timed automaton $G_i = (S_i, P_i, X_i, E_i, A_i, \mathsf{Inv}_i)$ where:

- $S_i = \{s_{k_i}\}_{k_i=1,\dots,K_i} \cup s_{0,i}$ is a set of locations,
- P_i is a mapping that determines, for each location, the threshold crossing and its trend,
- X_i is a singleton composed of the clock x_i,
- E_i, the set of labels, is an empty set,
- $A_i = A_{1,i} \cup A_{2,i}$ is the set of edges,
- Inv_i is a mapping that associates the invariants to the locations.

4. EXAMPLE: APPROXIMATION OF A WASTEWATER ANAEROBIC DIGESTION MODEL

To briefly illustrate this approach, the procedure previously described is used on an anaerobic digestion process. A more complete description of this example can be found in Hélias *et al.*, (2004). The aim is to predict possible dysfunctions of a wastewater treatment plant by a discrete approximation of a biological reaction model in front of expert partial knowledge of the system's dynamics. Anaerobic digestion is a biological process used for carbon removal from wastewater. The principle is to transform organic matter into biogas (*i.e.*, that is a methane (CH_4) and carbon dioxide (CO_2) mixture) in absence of oxygen. This complex process can be modelled in two steps: the acidogenic and methanogenic phases. A simplified version of the mass balance analytical model of our process (Bernard *et al.* , 2001) is used:

$$\begin{cases} \dot{S}_1 = D(S_{1in} - S_1) - k_1\mu_1 X_1 \\ \dot{S}_2 = D(S_{2in} - S_2) + k_2\mu_1 X_1 - k_3\mu_2 X_2 \\ \dot{Z} = D(Z_{in} - Z) \\ \dot{C} = D(C_{in} - C) + k_4\mu_1 X_1 + k_P\mu_2 X_2 \end{cases} \quad (28)$$

with $k_P = k_5 - k_6 \dfrac{P_{CO_2}}{P_T - P_{CO_2}}$. S_1, S_2, Z, and C are the organic matter concentration, the volatile fatty acids concentration, the total alkalinity and the total inorganic carbon, respectively. k_1 to k_6 are the yield coefficients for the reactions and μ_1 and μ_2 are the growth rates of the bacterial biomass X_1 according to S_1 (in the acidogenic phase) and bacterial biomass X_2 according to S_2 (in the methanogenic phase) respectively. These growth rates are assumed to follow a Monod-type kinetics, *i.e.* $\mu = \mu_{max} \dfrac{S}{S + K_S}$ with μ_{max} the maximum bacterial growth rate and K_S the half-saturation constant. Finally, P_T is the total pressure in the reactor, P_{CO_2} is the partial pressure of CO_2, D is the dilution rate according to input flow and S_{1in}, S_{2in}, Z_{in} and C_{in} the different influent concentrations. Because of a relation holding between P_T and P_{CO_2}, the cooperativity property holds in this model. Thus, intervals bounding influent and biomass concentrations can be used to handle imprecision, following the approach described in Section 2.. This model is run with parameters measured on an anaerobic digestion pilot plant (Steyer *et al.*, 2002).

The inputs of the process (*i.e.*, influent concentrations) and the initial state are thus represented by intervals. Qualitative states resulting from expert knowledge (*e.g.*, S_2 can be assigned labels "normal", "high" or "critical") also available are used in this approach. The plant operation is apprehended on a short time period (a few days) by combining information on different variables. For example, an important organic overload (*i.e.* dysfunction due to an inhibition of the reaction by S_2) can be associated to a "low" Z value and a "high" S_2 value according to thresholds specifically set in these variables' domains. To check the possible occurrence of this situation, we use the model-checking tool on the obtained timed automaton. This approach has been used to check about fifteen operating states during a four days period (Hélias *et al.*, 2004). The complete abstraction procedure (system simulation, threshold crossing detection, timed automata creation), with 11 thresholds defined on 4 variable domains, takes less than ten seconds on a personal computer with a CPU running at 1.7 GHz. This procedure has been implemented by combining the *Matlab*® environment and the *Kronos* model-checker (Yovine, 1997). The resulting automaton is composed of 24 locations and 66 edges.

5. DISCUSSION

The final size of the automaton is set according to the system's dimension, the number of thresholds and the crossing number. Consequently, this approach is inappropriate when:

- the system oscillates around a threshold value,
- a large number of thresholds are defined.

The idea is not to create an automaton to check all the possible properties of the continuous system but, rather, to create the automaton specifically suited to check a given property according to purposely defined thresholds.

Whereas in the classical definition of timed automata the clock constraints compare the clock values using rational numbers, integers must be used in the *Kronos* model-checker. Consequently, in our approach, the threshold crossing time values must be rounded. Thus, if the time unit used in the model is not carefully chosen according to the system dynamics, some additional imprecision may be introduced.

6. CONCLUSION

This paper has described a procedure aimed at representing a continuous system's dynamics as a timed automaton, where:

- some elements (inputs, initial state) are estimated by intervals, which comes down to reformulate the system as extremal ODE systems (upper and lower bounds ξ^- and ξ^+ approximating the real dynamics),
- dynamical envelopes stemming from the simulation of this extremal ODE systems are translated into the timed automata formalism (discrete model with continuous time) according to expert defined thresholds on the state variables' domain.

The principal characteristics of this approach are:

- taking threshold crossings themselves as qualitative discrete states (disregarding the interval between two consecutive thresholds);
- defining the discrete state transitions from the piecewise simulation of the extremal ODE systems within consecutive thresholds;
- starting the procedure from (i) an initial state, (ii) imprecise values of inputs and (iii) threshold definitions, all defined in accordance to the property to be checked in order to reduce the state explosion.

This approach was briefly illustrated on an anaerobic digestion model. The timed discrete-event representation makes model-checking possible to check properties of the continuous system's dynamics. An interesting extension would be to couple this representation to the discrete part of a wastewater treatment process in order to check properties on mixed continuous and discrete elements.

ACKNOWLEDGEMENTS

This work was partially supported by the Région Réunion and the European Social Fund.

REFERENCES

R. Alur and Dill D (1994). The theory of timed automata. *Theor. Comp. Sc.*. **126**: pp. 183-235.

O. Bernard, Hadj-Sadok Z., Dochain D., Genovesi A. and Steyer J.-P (2001). Dynamical model development and parameter identification for anaerobic wastewater treatment process. *Biotech. and Bioengineering*. **75**(4): pp. 424-438.

J.L. Gouzé, Rapaport A. and Hadj-Sadok M.Z (2000). Interval observers for uncertain biological systems. *Ecological Modelling*. **133**(1-2): pp. 45-56.

A. Hélias (2003). *Agrégation/abstraction de modèles pour l'analyse et l'organisation de réseaux de flux : application à la gestion des effluents d'élevage à la Réunion*. PhD thesis. ENSAM. Montpellier (France). 224 p.

A. Hélias, Guerrin F. and Steyer J.-P (2004). Abstracting continuous system behaviours into timed automata: application to diagnosis of an anaerobic digestion process. In Proc. *DX-2004 15th International Workshop on Principles of Diagnosis*. Carcassonne (France).

T.A. Henzinger, Ho P.-H. and Wong-Toi H (1998). Algorithmic Analysis of Nonlinear Hybrid Systems. *IEEE Trans. on Automatic Control*. **43**(4): pp. 540-554.

S. Kowalewski, Engell S., Prußig J. and Stursberg O (1999). Verification of logic controller for continuous plants using timed condition/event-system models. *Automatica*. **35**(3): pp. 505-518.

T., Moor, and Raisch, J (2002). Abstraction based supervisory controller synthesis for high order monotone continuous systems. *LNCIS* **279:** pp. 247-265.

J. Lunze (1999). A timed discrete-event abstraction of the continuous variable systems. *International Journal of Control*, **72**(13), pp. 1147-1164.

H.L. Smith (1995). *Monotone dynamical systems: An introduction to the theory of competitive and cooperative systems*. Mathematical Surveys and Monographs. **41**. 174 p.

J.-P. Steyer, Bouvier J.-C., Conte T., Gras P. and Sousbie P (2002). Evaluation of a four year experience with a fully instrumented anaerobic digestion process. *Water Science and Technology*. **45**(4-5): pp. 495-502.

P. Struss (2002). Automated Abstraction of Numerical Simulation Models - Theory and Practical Experience. in *Model Based Systems and Qualitative Reasoning for Intelligent Tutoring Systems*. San Sebastian (Spain). pp. 161-168.

P. Supavatanakul, Falkenberg C. and Lunze J (2003). Identification of timed discrete event models for diagnosis. in *DX-2003 14th International Workshop on Principles of Diagnosis*. Washington DC (USA).

S. Yovine (1997). Kronos: A verification tool for real-Time Systems. *Journal of Software Tools for Technology Transfer*. **1**(1-2): pp. 123-133.

ELSEVIER
IFAC
PUBLICATIONS
www.elsevier.com/locate/ifac

REACHABILITY ANALYSIS USING PROXIMITY BASED AUTOMATA *

Jim Kapinski, Klaus Schmidt, and Bruce H. Krogh

Carnegie Mellon University
Dept. of Electrical and Computer Engineering
Pittsburgh, PA 15213 USA
{jpk3|klaussch|krogh}@andrew.cmu.edu

Abstract: This paper introduces *proximity based automata* (PBA) to perform reachability analysis for discrete-time, continuous-state systems. The notion of proximity of states in the continuous valued state space is used to evaluate the response of the system efficiently. A PBA is constructed using ellipsoidal techniques and tools for solving optimization problems constrained by linear matrix inequalities. Several empirical studies demonstrate how trade-offs between conservativeness and computational complexity can be exploited by varying parameters in the procedure for constructing the PBA. *Copyright © 2004 IFAC*

Keywords: reachability, hybrid systems, verification, dynamic systems

1. INTRODUCTION

There has been much interest in recent years in developing methods for formally verifying hybrid systems (Alur et al., 1996; Dang and Maler, 1998; Silva et al., 2002). Typically, the goal is to verify that a hybrid system satisfies some safety property for an entire range of initial conditions and system parameters. The most computationally expensive aspect of verifying hybrid systems is the computation of the set of reachable states for a system due to variations in operating conditions characterized by system parameters, initial conditions and ranges of inputs (controls and disturbances). Many different types of reachability have been studied (Chutinan and Krogh, 2003; Dang and Maler, 1998; Henzinger et al., 1997; Jönsson, 2002; Mitchell et al., to appear), but these techniques only provide methods for estimating the set of reachable states in the continuous-valued state space and there is no systematic way of exploring multiple paths in the reachable space due to inputs.

This work extends the concept of systematic simulation introduced in Kapinski et al. (2003) by representing the set of reachable states of the infinite state system with a finite state machine, called a *proximity based automaton* (PBA). The definition of a PBA is based on the notion that trajectories from regions that are near each other in the state space will evolve in a similar manner. We use ellipsoidal methods to perform reachability analysis for dynamic systems with bounded inputs (Kurzhanski and Vályi, 1997). These methods provide an efficient way of approximating reachable sets using semidefinite programming techniques (Boyd et al., 1994). The notion of proximity is exploited by merging PBA states that fulfill specified nearness conditions. The PBA can also be used to perform refinement of the

* This research was supported in part by Ford, the US Defense Advance Projects Research Agency (DARPA) contract nos. F33615-00-C-1701 and F33615-02-C-4029, US Army Research Office (ARO) contract no. DAAD19-01-1-0485, the US National Science Foundation (NSF) contract no. CCR-0121547, and the Institute of Control Engineering and Automation at the University of Erlangen-Nuremberg.

reachable set approximation and the construction easily extends to hybrid systems. Further, we improve on systematic simulation by introducing a computationally efficient grid based method for testing proximity of states in the PBA.

Some researchers have investigated reachability of hybrid systems with discrete-valued inputs (Cury et al., 1998; Raisch and O'Young, 1998; Moor et al., 2004). We consider systems where the continuous set of available control inputs is covered by a finite collection of sets. The information contained in the PBA is used to perform refinement of the representation if necessary. We exploit the trade-off between conservativeness and computational complexity by varying parameters of the reachability analysis.

2. THE REACHABILITY PROBLEM

A discrete-time dynamic system (DDS) consists of a state space $X \subseteq \mathbb{R}^n$, a set of initial states $X_0 \subseteq X$, an input set $U \subseteq \mathbb{R}^m$, which covers control inputs as well as disturbance inputs, and an update equation $f : \mathbb{R}^n \times \mathbb{R}^m \to \mathbb{R}^n$. In this paper, we assume the dynamics of the DDS are linear; that is, there are matrices A and B such that $f(x, u) = Ax + Bu$.

Definition 2.1. A sequence $(x_0, u_0, x_1, u_1, \ldots)$ is a *run* of a DDS $\mathcal{H}$ if $x_0 \in X_0$ and for all $k \geq 0$, $x_{k+1} = f(x_k, u_k)$ and $u_k \in U$. $\mathcal{R}_\mathcal{H}$ denotes the set of all runs of $\mathcal{H}$, and $\mathcal{R}_\mathcal{H}(k)$ denotes the set of all runs of $\mathcal{H}$ of length less than or equal to k.

Definition 2.2. Given a DDS $\mathcal{H}$, a proximity based automaton (PBA) for $\mathcal{H}$ is a tuple, $P = (Q, \Sigma, E, Q_0, \lambda, \rho_X, \rho_U)$, where

- Q - a finite set of states;
- Σ - a finite set of symbols;
- $E \subseteq Q \times Q$ - a set of transitions;
- $Q_0 \subseteq Q$ - a set of initial states;
- $\lambda : E \to 2^\Sigma$ - a labelling function that assigns a set of input symbols to each transition;
- $\rho_X : Q \to C_{\mathbb{R}^n}$ - a function that assigns a region in $\mathbb{R}^n$ to each state in P. $C_{\mathbb{R}^n}$ denotes the set of compact, connected subsets of $\mathbb{R}^n$;
- $\rho_U : \Sigma \to C_{\mathbb{R}^m}$ - a function that assigns a region in $\mathbb{R}^m$ to each symbol in Σ,

and $\forall (q, q') = e \in E, \sigma \in \lambda(e), u \in \rho_U(\sigma)$ and $x \in \rho_X(q), \exists x' \in \rho_X(q')$ such that $f(x, u) = x'$.

Remark 2.1. Note that Definition 2.2 can easily be extended to piecewise-linear discrete time systems with a finite set of discrete modes $\mathcal{I}$ and a partition $\mathcal{X} = \{X_i\}_{i \in \mathcal{I}}$ of the continuous state space $\mathbb{R}^n$, where for each mode the dynamics are $f_i(x, u) = A_i x + B_i u$. Then the function $\rho_X : Q \to C_{\mathbb{R}^n} \times \mathcal{I}$ assigns a region in $\mathbb{R}^n$ and a mode in $\mathcal{I}$ to each state in Q.

Definition 2.3. The *language of the PBA P*, denoted $\mathcal{L}_P$, is the set of all input sequences $\pi = u_0 u_1 \cdots u_k$ such that:

$\exists \omega = q_0 q_1 \cdots q_{k+1} \ni \forall 0 \leq i \leq k, \sigma \in \lambda(q_i, q_{i+1})$ with $u_i \in \rho_u(\sigma)$.

The language of a PBA P will be a subset of U^*. Also, we define U^k as the set of all input sequences of length less than or equal to k.

Definition 2.4. Let P be a PBA as in Definition 2.2 and let $\mathcal{H}$ be a DDS. P is said to be a conservative estimate of the reachset $\mathcal{R}_\mathcal{H}(k)$ if $\mathcal{R}_\mathcal{H}(k) \subseteq \bigcup_{q \in Q} \rho_X(q)$.

Lemma 2.1. If the language of the PBA P contains U^k, then the PBA is a conservative estimate of the reachset for up until time k.

Proofs appear in Kapinski (2004).

Definition 2.5. An *invariant PBA* is one whose language is U^*.

Example 2.1. To illustrate the concept of a PBA, consider the servo system shown in Figure 1, which has first-order plant dynamics and a first-order reference signal filter, where u is an input reference signal, x_1 is the position, x_2 is the filtered version of u, and $e = x_2 - x_1$ is the error. The system matrices for the sampled version of the system, with $g = 10$, $\tau = 0.1$, and a sample period of 0.1, are

$$A = \begin{bmatrix} 0.368 & 0.368 \\ 0.000 & 0.368 \end{bmatrix} \quad B = \begin{bmatrix} 0.264 \\ 0.632 \end{bmatrix}.$$

The set of initial conditions is given by a circle, centered at zero, with $\|x\|^2 \leq \frac{1}{10}$.

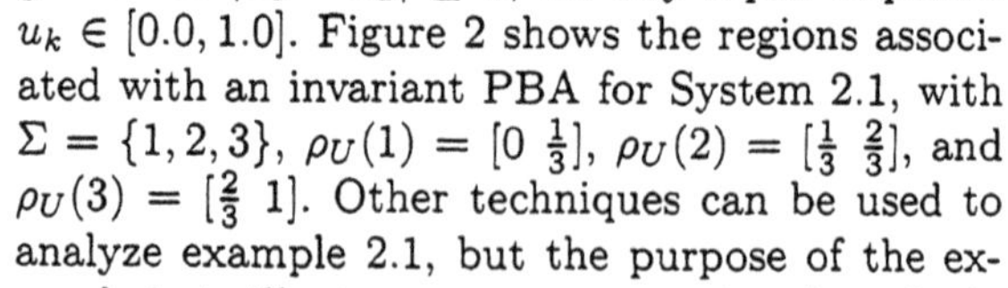

Fig. 1. A simple servo system.

Suppose we wish to verify that the servo system error does not exceed 1, that is, we want to guarantee $|x_1 - x_2| \leq 1$, for any input sequence $u_k \in [0.0, 1.0]$. Figure 2 shows the regions associated with an invariant PBA for System 2.1, with $\Sigma = \{1, 2, 3\}$, $\rho_U(1) = [0\ \frac{1}{3}]$, $\rho_U(2) = [\frac{1}{3}\ \frac{2}{3}]$, and $\rho_U(3) = [\frac{2}{3}\ 1]$. Other techniques can be used to analyze example 2.1, but the purpose of the example is to illustrate our computational method.

This example shows the general safety problems we want to address with the PBA in this paper. Given a region, $Fail \subset \mathbb{R}^n$, a DDS $\mathcal{H}$ is *k-step safe with respect to Fail* if $\mathcal{R}_\mathcal{H}(k) \cap Fail = \emptyset$ and *safe wrt. Fail* if $\mathcal{R}_\mathcal{H} \cap Fail = \emptyset$.

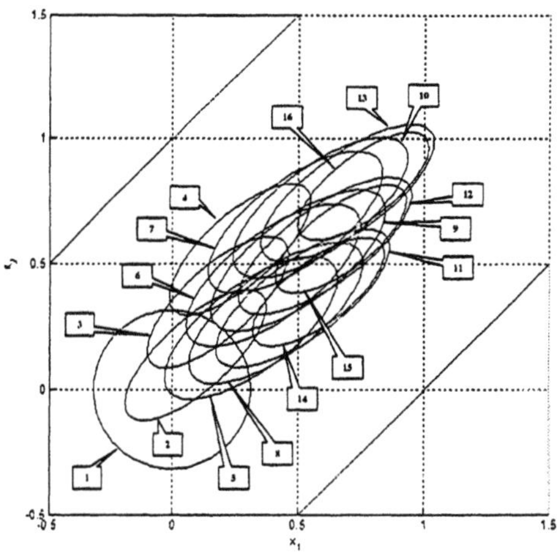

Fig. 2. The set of regions for the PBA of System 2.1. Each region is labelled with the corresponding state of the PBA.

3. REACHABILITY ANALYSIS USING PBA'S

The PBA reachability procedure (PRP), shown in Figure 3, constructs a PBA for a DDS. The procedure propagates regions forward using the reachability operator $Next$. If the reachability estimate resulting from $Next$ satisfies a nearness criterion for a region that is already associated with the PBA, the procedure merges the two regions; if the reachability estimate does not satisfy the nearness criterion for any region associated with P, a new state is created and associated with the new region. The result is that at each iteration the PBA represents all of the possible behaviors of the underlying system until time k. The procedure terminates if either there are no more states to explore or after K iterations, where K represents the termination time step.

The procedure requires a dynamical system description, a set of initial regions S, which includes the set of initial conditions X_0, a set of input symbols Σ, and a function $\rho_U : \Sigma \rightarrow C_{\mathbb{R}^m}$.

The PRP uses three functions: *Jointest*, *Join*, and $Next$. $Jointest(N, P)$ is a function that determines when two regions will be merged. It returns a state within PBA P that has a region associated with it that satisfies some nearness criterion to region N if such a state in P exists, otherwise it returns the empty set. In our experiments, we partition the space and merge regions when their center points are in the same element of the partition.

$Join(N, N')$ is a function that returns a region containing N and N'. Our implementation uses ellipsoidal representations; $Join(N, N')$ is an LMI that solves for the minimum volume ellipsoid that contains ellipsoids N and N' (Boyd et al., 1994).

```
/* Create PBA P^temp */
For each N ∈ S
  Insert q_N into Q^temp
  ρ_X^temp(q_N) := N
Q_0^temp := Q^temp
Q_W^temp := Q^temp
Loop K times
  Copy machine P^temp to P
  Q_W := ∅
  For each q ∈ Q_W^temp
    For each σ ∈ Σ
      /* Compute new region */
      N' := Next(ρ_X^temp(q), σ)
      q̂ := Jointest(N', P)
      If q̂ ≠ ∅
        /* New region should join with
           existing region */
        If N' ⊈ ρ_X(q̂)
          Ñ := Join(N', ρ_X(q̂))
          ρ_X(q̂) := Ñ
          Insert q̂ into Q_W
        Insert (q, q̂) into E
        λ(q, q̂) := λ(q, q̂) ∪ {σ}
      Else
        /* Create new state which is
           associated with new region */
        Insert q_N' into Q
        Insert q_N' into Q_W
        ρ_X(q_N') := N'
        Insert (q, q_N') into E
        λ(q, q_N') := σ
```

Fig. 3. PBA Reachability Procedure

$Next(N, \sigma)$ returns a region N', which contains the set of states that are reachable from region N in one time increment given some input in $\rho_U(\sigma)$, i.e. $f(N, \rho_U(\sigma)) \subseteq N'$. We compute the minimum volume ellipsoid containing $f(N, \rho_U(\sigma))$ using the ellipsoidal technique developed by Kurzhanski and Vályi (1997).

The PRP is initiated with a set of regions S. The set of initial conditions X_0 should be included in S, but additional regions may be added. It is often possible to add regions to S so that greater connectivity in the PBA array results from the PRP. We call this process *seeding the reachset*. This is beneficial because it can help reduce the amount of computation time spent on the PRP, and in some cases, it results in the PRP terminating with the final PBA being invariant. We have found that ellipsoids that are invariants given one input or given one mode are good seeding regions. (For a survey of methods for computing invariant regions, see Blanchini (1999).)

Lemma 3.1. For a DDS $\mathcal{H}$ and a given region $Fail \subset \mathbb{R}^n$, if $X_0 \subseteq Inv$ for some $Inv \subset \mathbb{R}^n$, $f(Inv, U) \subseteq Inv$, and $Inv \cap Fail = \emptyset$, then $\mathcal{H}$ is safe with respect to $Fail$.

If a region Inv, as described in lemma 3.1, can be identified, then the DDS is safe, and no further analysis needs to be performed.

The PRP generates a PBA whose language is guaranteed to contain U^K. Due to the merging of regions during the PRP procedure, which creates cycles in the finite state representation of the system, the PBA will possibly contain other strings, some of which are of infinite length.

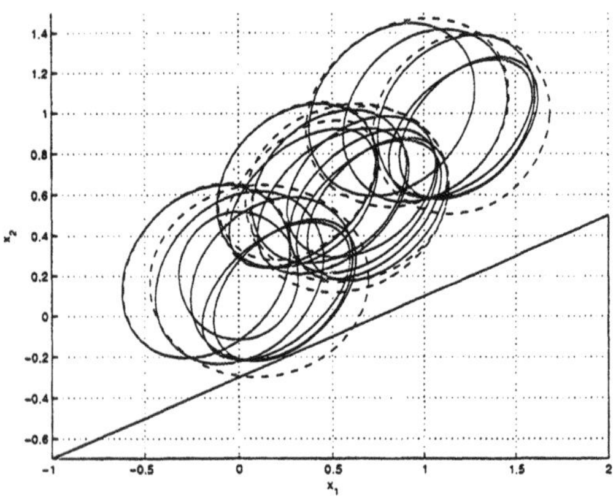

Fig. 4. Example of the reachset estimate for two different merging conditions. The 3D example is projected into the first two dimensions. The dashed lines correspond to a coarser merging condition grid and the solid lines correspond to a finer merging condition grid.

Proposition 3.1. If the PRP terminates with $Q_W^K = \emptyset$, then the PBA created is an invariant PBA.

In the case where $Jointest(N, P)$ is always empty, merging is never performed, and the complexity of the PRP is $O(n^3|\Sigma|^K)$, where the n^3 term is due to the linear algebra computations necessary for the propagation of ellipsoids (Kurzhanski and Vályi, 1997; Golub and Loan, 1996). If $Jointest(N, P)$ always returns a region in P for N to merge with, the complexity of the PRP is $O(n^{6.5}|\Sigma|K)$, where the $n^{6.5}$ term is due to the LMIs used to compute bounding ellipsoids (Vandenberghe et al., 1998). The main challenge in using the PRP method is to choose the *Jointest* function such that a sufficient amount of merging occurs so as not to incur the exponential complexity, but not so much merging occurs so as to cause the approximation to become overly conservative.

4. EXPERIMENTS

In using the PRP, choosing an appropriate *Jointest* function is essential for achieving a proper balance between computation time and conservativeness. For our PRP we use an implementation of *Jointest* based on a partition of the state space. Each ellipsoid in the PBA is associated with the partition element containing the center point of the ellipsoid. Two ellipsoids are merged if their center points lie in the same partition element. It is expected that computation time will decrease as the merging condition is made more liberal, but conservativeness will increase. In our case, the merging condition being more liberal corresponds to the size of the partition elements being larger.

We would like to trade off conservativeness and computation time depending on the problem. Figure 4 illustrates this for an example that has three state variables and three inputs. (Details of this example are available from the authors.) The diagonal line in the figure indicates the failure region. The union of the solid ellipsoids represent the reachset estimate when the PRP is used with a grid size of 0.2. The union of the dashed ellipsoids represents the reachset found with a grid size of 0.4. In both cases, the PRP was performed for $K = 4$. Figure 4 shows that the reachset estimate using the grid size of 0.4 is too conservative, and the reachset estimate using the grid size of 0.2 is sufficient for the 4-step verification problem.

Several experiments were performed to determine the effect of merging condition, system stability, and state space dimension on computation time and conservativeness using the PRP. Our procedure was implemented in MATLAB, and all experiments were performed on an 800 MHz, Pentium III machine with 120MB of RAM running Windows ME.

The first experiment investigates how the computation time changes for different grid sizes and spectral radii of the system A matrix. We used randomly generated A matrices with spectral radii of 0.2, 0.4, 0.6, 0.8, 0.9, and 0.95 and used grid sizes between 0.1 and 1 with a granularity of 0.1. Figure 5 shows results for this experiment.

It can be seen that the computation time decreases as grid size increases and as spectral radius decreases. Note that, as the figures show particular examples, the computational effort is not monotone with respect to the joining condition, but that the general trend is what we expect, that is, that the computation time decreases as the merging condition becomes more liberal.

In the next experiment, system dynamics with spectral radii of 0.4 and system dimensions varying from 2 to 5 were randomly created. Figure 5 shows that the computation time for performing the PRP decreases as the grid size increases. Also, the increase in computation time in going from the 2D system to the 5D system is less than one order-of-magnitude.

The last experiment compares the computation time per time step for the case where no merg-

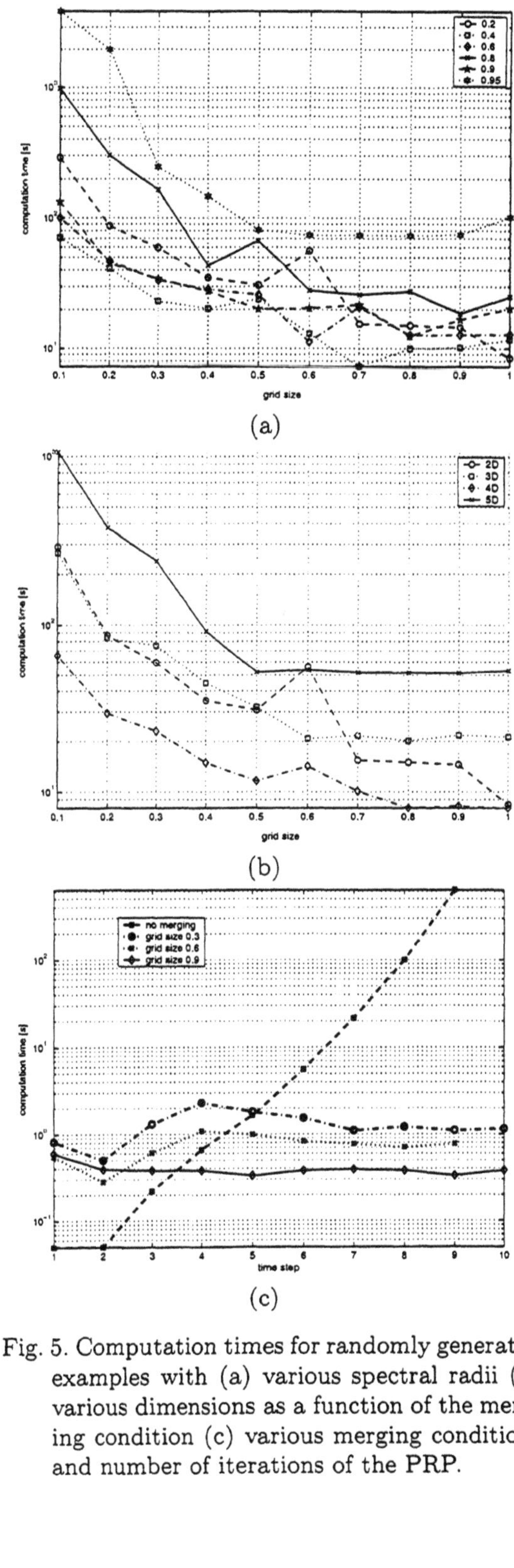

Fig. 5. Computation times for randomly generated examples with (a) various spectral radii (b) various dimensions as a function of the merging condition (c) various merging conditions and number of iterations of the PRP.

ing takes place with the case with merging for different grid sizes for a two-dimensional system with a spectral radius of 0.4. Figure 5 suggests an exponential growth of the computation time if we do not merge any ellipsoids and it also shows that the computation time per time step remains at a nearly constant level until the exploration terminates in 9 and 10 time steps for grid size 0.6 and 0.3, 0.9, respectively.

5. REFINEMENT

Section 4 showed that the PRP could achieve less conservative approximations of the reachable set if the merging condition was conservative enough. Refining the PBA accomplishes the same task without repeating the entire reachability process. The refinement procedure consists of taking regions that are the result of a merging operation and lead to the failing region and *unmerging* them, that is, taking regions that result from a merging operation and breaking them into the two regions that were merged. In this way, the PBA that results from the PRP can be refined so that the conservativeness introduced by the merging operation is reduced.

Consider again example 2.1. Suppose that we wish to show that $|x_1 - x_2| \leq .424$ for all input sequences. The reachset was first seeded with three regions, and the PRP was performed with a merging condition grid size of 1. The ellipsoids in Figure 6 (a) illustrate the result of the first two steps of the reachability procedure. Note that the regions associated with the PBA intersect the failing region. Since this intersection may be due to the conservativeness introduced from the merging operation, a refinement procedure was performed on the PBA.

Figure 6 (b) shows the final result of the refinement procedure. Note that the failing ellipsoid from the previous figure is removed and replaced with the one step reachable sets from the failing region, the initial condition set, and the seeding regions. This final collection of regions satisfies the safety condition.

6. CONCLUSIONS

This paper presents a method for constructing a finite-state structure, the proximity based automaton (PBA), to guide reachability analysis for discrete-time continuous systems. The PBA is a discrete state, input driven representation of the system dynamics. Empirical studies confirm the efficiency that can be achieved with the PBA and illustrate the tradeoffs to be made between conservativeness and computation time.

We are currently developing other applications of the PBA, including the use of the PBA representation to perform safe input sequence synthesis (Schmidt et al., 2004). The PBA can also be extended to hybrid systems. This makes it possible to compute reachability for nonlinear systems by approximating them with a piecewise affine switched mode system. The approximation error, based on the Lipschitz constant, can then be compensated for by adding a disturbance to the system dynamics, similar to the method proposed by Asarin et al. (2003). Then the PRP can be

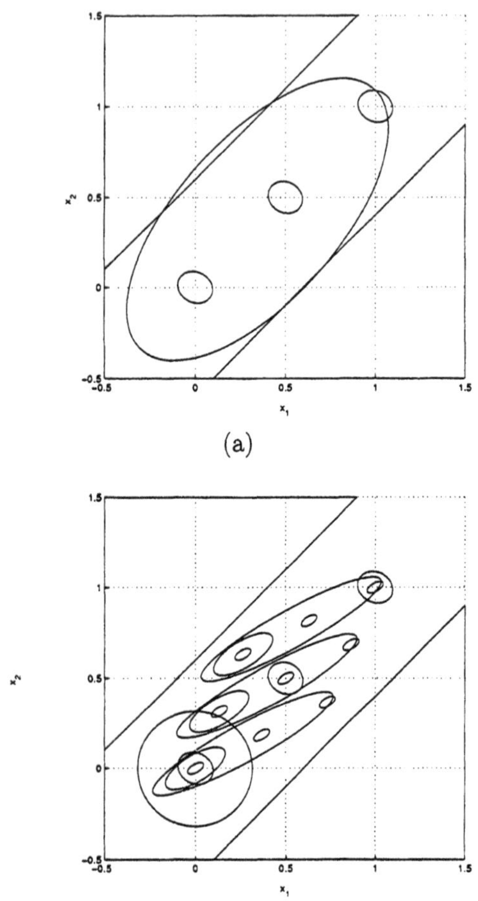

(a)

(b)

Fig. 6. Result of (a) the PRP with a liberal merging condition and (b) the final step in the refinement procedure for the servo system example. Note that the reachset no longer intersects *Fail*.

applied to the system to compute a conservative approximation of the reachable set.

REFERENCES

R. Alur, T.A. Henzinger, and P.-H. Ho. Automatic symbolic verification of embedded systems. *IEEE Trans. on Software Engineering*, 22(3):181–201, Mar 1996.

E. Asarin, T. Dang, and A. Girard. Reachability analysis of nonlinear systems using conservative approximations. In *Hybrid Systems: Computation and Control 2003*, pages 20–35. Springer-Verlag, 2003.

F. Blanchini. Set invariance in control. *Automatica*, 35:1747–1767, 1999.

S. Boyd, L.E. Ghaoui, E. Feron, and V. Balakrishnan. *Linear Matrix Inequalities in System and Control Theory*, volume 15 of *SIAM Studies in Applied Mathematics*. SIAM, 1994.

A. Chutinan and B. Krogh. Computational techniques for hybrid system verification. *IEEE Transactions on Automatic Control*, 48(1):64–75, Jan. 2003.

J.E.R. Cury, B.H. Krogh, and T. Niinomi. Synthesis of supervisory controllers for hybrid systems based on approximating automata. *IEEE Transaction on Automatic Control*, 43(4):564–569, April 1998. a preliminary version of this paper also appeared in *Proc. 34th IEEE CDC*, pp. 1461-1466.

T. Dang and O. Maler. Reachability via face lifting. In *Hybrid Systems: Computation and Control 1998*, pages 96–109. Springer-Verlag, 1998.

G. H. Golub and C. F. Van Loan. *Matrix Computations*. Johns Hopkins Press, Baltimore, 1996.

T. A. Henzinger, P. Ho, and H. Wong-Toi. Hytech:a model checker for hybrid systems. In *Software Tool for Technology Transfer 1*, pages 110–122. Springer-Verlag, 1997.

U. Jönsson. Robustness of trajectories with finite time extent. *Automatica*, 38:1485–1497, 2002.

J. Kapinski. *Control System Verification and Synthesis Using Proximity Based Automata*. PhD thesis, Carnegie Mellon University, 2004.

J. Kapinski, O. Maler, O. Stursberg, and B. H. Krogh. On systematic simulation of open continuous systems. In *Hybrid Systems: Computation and Control, 6th International Workshop, HSCC'03*, pages 283–297. Springer-Verlag, 2003.

A. B. Kurzhanski and I. Vályi. *Ellipsoidal Calculus for Estimation and Control*. Birkhä user, Boston, 1997.

I. Mitchell, A. Bayen, and C. Tomlin. Computing reachable sets for continuous dynamic games using level set methods. *IEEE Trans. on Automatic Control*, to appear.

T. Moor, J. Raisch, and S.D. O'Young. Discrete supervisory control of hybrid systems based on l-complete approximations. *Journal of Discrete Event Dynamic Systems*, pages 83–107, 2004.

J. Raisch and S. D. O'Young. Discrete approximation and supervisory control of continuous systems. *IEEE Transactions on Automatic Control*, 43(4):569–573, April 1998.

K. Schmidt, J. Kapinski, and B. H. Krogh. Control input synthesis for hybrid systems using informed search. In *Computer Aided Control System Design Conference (to appear)*, 2004.

B.I. Silva, O. Stursberg B.H. Krogh, and S. Engell. An assessment of the current status of algorithmic approaches to the verification of hybrid systems. In *Proceedings of the 40th IEEE Conference on Decision and Control*, volume 3, pages 2867–2874, 2002.

L. Vandenberghe, S. Boyd, and W. Shao-Po. Determinant maximization with linear matrix inequality constraints. *SIAM Journal on Matrix Analysis and Applications*, 19(2):499–533, April 1998.

ELSEVIER
IFAC
PUBLICATIONS
www.elsevier.com/locate/ifac

ACTL STRONG NEGATION AND ITS APPLICATION TO HYBRID SYSTEM VERIFICATION

Alongkrit Chutinan* Zhi Han Bruce H. Krogh****

* *Fabrinet Co., Ltd., 294 Moo 8, Vibhavadi Rangsit Rd., Kookot, Lumlookka, Patumthanee 12130, Thailand*
ac4c@ece.cmu.edu
** *Dept. of Electrical and Computer Engineering, Carnegie Mellon University, Pittsburgh, PA 15213, USA*
{zhih|krogh}@ece.cmu.edu

Abstract: Model checking procedures for verifying properties of hybrid dynamic systems are based on the construction of finite-state abstractions. If the property is not satisfied by the abstraction, the verification is inconclusive and the abstraction needs to be refined so that a less conservative model can be checked. If the hybrid system does not satisfy the property, this verify-refine procedure will usually not terminate. This paper introduces the concept of strong negation for ACTL as an auxiliary condition that can be verified to obtain a conclusive negative verification result from a finite-state abstraction in certain cases. A method for identifying a sufficient set of states to be refined when the verification is inconclusive is also presented. The concepts are illustrated with an example. *Copyright © 2004 IFAC*

Keywords: Verification, hybrid system, model checking, abstraction, refinement

1. INTRODUCTION

A standard approach to hybrid system verification is to construct a finite-state abstraction (usually a quotient of the hybrid system) and apply model checking. That is, a conservative discrete model is constructed so that methods for verifying properties of discrete state systems can be applied to hybrid dynamic systems. The main restriction in this approach is that only *universal* properties can be verified, such as properties specified by the temporal logic ACTL. If the specification is true for the abstraction, it is true for the hybrid system since all possible hybrid trajectories are accounted for in the conservative abstraction. If the model checker returns a negative result, one is assured the specification is false for the hybrid system only if the abstraction is a bisimulation of the hybrid system (Alur et al., 2000). As this is rarely the case, the only recourse is to refine the abstraction and try again with the hope that eventually an abstraction will be constructed for which the specification is true (Chutinan and Krogh, 2003).

Although the literature reports many successful applications of this approach to hybrid systems with non-trivial continuous dynamics (at least one success per tool), it is disconcerting that the refinement process will never terminate if the hybrid system really doesn't satisfy the specification. The concept of strong negation introduced in this paper aims to offer some relief in this situation. In certain cases, checking the strong negation condition will allow an abstraction-based verification tool to terminate with a conclusive negative result, even if the abstraction is not a bisimulation of the hybrid system.

If no conclusive result is found, the abstraction must be refined. Often counterexamples (CEs) are useful in the refinement procedure. The counterexample-guided refinement method for software systems has been studied in (Clarke et al., 2000; Glusman et al., 2003). For verification of hybrid systems, methods based on the same idea have been proposed in (Clarke et al., 2003) and (Alur et al., 2003). In this paper we show that the strong negation condition can help to reduce the search space in the counterexample-guided abstraction refinement procedure.

Methods based on abstractions which 'strongly preserve' certain class of properties have been proposed for μ-calculus (Dams et al., 1997), for fractions of CTL* (Dams et al., 1993) and for modal transition systems (Godefroid et al., 2001; Godefroid and Jagadeesan, 2002). Those techniques define two types of transitions in the abstract system: `must-` (or constrained) transitions and `may-` (or free) transitions. By distinguishing the two types of transitions in the model checking algorithm, conclusive negative results can be obtained for the concrete system. This technique is not applicable for verification of hybrid systems, however, since the discrete abstraction of a hybrid system is based on reachability analysis for the continuous dynamics (Alur et al., 2000; Chutinan and Krogh, 2001), where, in the terminology of (Godefroid et al., 2001), all the transitions are `may-` transitions. To construct the `must-` transitions for a hybrid dynamical system, the post condition of each abstract state would have to be computed, which involves the most expensive computation in the verification procedure.

The following section introduces notation, definitions and background material. Section 3 presents the concept of strong negation for ACTL formulas and how it can be used to demonstrate an ACTL expression is not satisfied based on verification of an abstraction of a hybrid system. Section 4 concerns the construction of subgraphs for the abstraction that identify the collection of sets that need to be refined when the verification of a quotient system is inconclusive. Section 5 illustrates the application of strong negation with an example. The concluding section summarizes the contributions of this paper and discusses directions for further work.

2. PRELIMINARIES

2.1 Quotient Transition Systems

Definition 1. A labeled transition system *T is a tuple $T = (Q, \rightarrow, Q_0, \mathcal{L}, L)$, where Q is the set of* states, *$\rightarrow \subseteq Q \times Q$ is the set of* transitions, *$Q_0 \subseteq Q$ is the set of* initial states, *$\mathcal{L}$ is a countable set of* labels, *and $L : Q \rightarrow \mathcal{L}$ is the* labelling function.

Abstraction-based verification of hybrid systems is based upon a transition-system semantics for hybrid automata (Henzinger, 1996). A *quotient transition system* is constructed from a partition of the transition system state space. The partition must be consistent with the labelling function and the initial states in the following sense.

Definition 2. Given a labelled transition system $T = (Q, \rightarrow, Q_0, \mathcal{L}, L)$, a partition $\mathcal{P}$ of Q is said to be consistent *if and only if for all $P \in \mathcal{P}$ and for all $q, q' \in P$, $L(q) = L(q')$ and $q \in Q_0 \iff q' \in Q_0$.*

Definition 3. (Quotient Transition System) (QTS) Given a labelled transition system $T = (Q, \rightarrow, Q_0, \mathcal{L}, L)$ and a consistent partition $\mathcal{P}$ of Q, the quotient transition system *of T is defined as $T/\mathcal{P} = (\mathcal{P}, \rightarrow_{\mathcal{P}}, Q_0/\mathcal{P}, \mathcal{L}, L_{\mathcal{P}})$, where*

(i) for all $P, P' \in \mathcal{P}$, $P \rightarrow_{\mathcal{P}} P'$ iff there exist $q \in P$ and $q' \in P'$ such that $q \rightarrow q'$;
(ii) $Q_0/\mathcal{P} = \{P \in \mathcal{P} \mid P \subseteq Q_0\}$; and
(iii) for all $P \in \mathcal{P}$, $L_{\mathcal{P}}(P) = L(q)$ for some $q \in P$.

In words, a partition state in the QTS can make a transition to another partition state if one of its element states can make a transition to another element state in the destination partition state.

2.2 Computation Tree Logic (CTL)

CTL is a well-established temporal logic formalism for transition systems (Clarke et al., 1999). The specification is formulated by attaching a finite set of *atomic propositions AP* to each state in the transition system. Thus, we consider a labelled transition system of the form $T = (Q, \rightarrow, Q_0, 2^{AP}, L)$ where $L : Q \rightarrow 2^{AP}$ is a labeling function that assigns to each state in Q a set of atomic propositions that are true for that state. We assume that the transition relation $\rightarrow$ is *total*; that is, for all states $q \in Q$ there exists a state $q' \in Q$ such that $q \rightarrow q'$.

A CTL formula consists of *temporal operators* and the *path quantifiers*. A temporal operator specifies the system evolution along a single path of the computation tree. There are five basic temporal operators, $\mathbf{X}$(*next time*), $\mathbf{F}$(*in the future*), $\mathbf{G}$(*globally*), $\mathbf{U}$(*until*), and $\mathbf{R}$(*release*). The universal and existential path quantifiers, $\mathbf{A}$ and $\mathbf{E}$, specify that the path assertion holds along *all* and *some* paths in the computation tree, respectively. Well-formed CTL formulas are defined recursively in terms of its subformulas. For a detailed treatment of CTL formulas, the reader is referred to (Clarke et al., 1999). In this paper, we consider a class of CTL formulae called ACTL, which allows only universal assertions (Clarke et al., 1999). The syntax of an ACTL formula φ is given by

$$f \rightarrow ap \mid \neg ap \mid f \vee f \mid f \wedge f \mid \mathbf{AX}\, f \mid \mathbf{AF}\, f \\ \mid \mathbf{AG}\, f \mid \mathbf{A}\, f\, \mathbf{U}\, f \mid \mathbf{A}\, f\, \mathbf{R}\, f.$$

We write $T, q \models f$ to mean that the CTL formula f holds at state q in T, and let $SAT(T, f) = \{q|T, q \models f\}$ denote the set of states for which the CTL formula f is true. This set can be computed using fixed-point iterations that are guaranteed to terminate for finite-state systems (Clarke et al., 1999). T will be dropped from these notations when the labelled transition system is clear from the context. The labelled transition system T is said to satisfy a CTL specification f, denoted $T \models f$, if and only if $Q_0 \subset SAT(T, f)$.

2.3 Verification Using Quotient Transition Systems

The following proposition is the basis for verification of ACTL specifications based on quotient transition systems.

Proposition 1. (Chutinan and Krogh, 2001) Given a labeled transition system T, an ACTL formula φ, and a consistent partition $\mathcal{P}$, if $T/\mathcal{P}, P \models \varphi$ then $T, q \models \varphi$ for all $q \in P$.

If model checking a QTS does not give a conclusive result, the partition can be refined to see if verification can be achieved with a less conservative QTS.

Definition 4. (Refinement of Partitions). Given two partitions $\mathcal{P}_1$ and $\mathcal{P}_2$ of Q, $\mathcal{P}_2$ is said to be a refinement of $\mathcal{P}_1$ if and only if for all $P_2 \in \mathcal{P}_2$, there exists $P_1 \in \mathcal{P}_1$ such that $P_2 \subseteq P_1$.

If $\mathcal{P}_2$ is a refinement of $\mathcal{P}_1$, for any $P_1 \in \mathcal{P}_1$ either there exist a $P_2 \in \mathcal{P}_2$ where $P_1 = P_2$ or there exist a collection of sets $\{P_2^1, \ldots, P_2^k\} \subset \mathcal{P}_2$ such that $P_1 = \cup_{i=1}^k P_2^i$. We say state P_1 is refined by $\mathcal{P}_2$ for the second case. The refinement induces a surjective mapping $\alpha : \mathcal{P}_2 \rightarrow \mathcal{P}_1$ such that

$$\alpha(P_2) = P_1 \Rightarrow P_1 \supseteq P_2$$

When the partitions $\mathcal{P}_1, \mathcal{P}_2$ are consistent partitions, the function α defines an *abstraction relation* between the quotient transition systems $T/\mathcal{P}_1$ and $T/\mathcal{P}_2$ (Clarke et al., 2003).

To identify a set of states that should be refined when a QTS does not satisfy an ACTL specification, we will use the following definition of a counterexample from (Clarke et al., 2002).

Definition 5. (Counterexample). Let T be a labelled transition system, and φ be an ACTL formula. A counterexample of T for the specification φ is a labelled transition system C such that

(i) $C \models \neg\varphi$
(ii) T simulates C (Clarke et al., 1999), and therefore $T \models \neg\varphi$.

We will consider counterexamples constructed from subgraphs of the given transition system, where the graph G of a transition system $T = (Q, \rightarrow, Q_0, \mathcal{L}, L)$ is $G_T = (Q, \rightarrow)$, the set of states and transition relation for T. The labelling function and initial states are not included in the definition of the graph of T.

3. ACTL STRONG NEGATION

This section introduces the concept of ACTL strong negation and shows how it can be used to demonstrate an ACTL expression is false based on QTS model checking. The strong negation of an ACTL formula is constructed as follows.

Definition 6. Given an ACTL formula f, its strong negation $!f$ *is defined by the following recursive rules.*

(i) $!ap = \neg ap$	*(v)* $!\mathbf{AX}\, f = \mathbf{AX}\, !f$
(ii) $!(\neg ap) = ap$	*(vi)* $!\mathbf{AF}\, f = \mathbf{AG}\, !f$
(iii) $!(f \vee g) = !f \wedge !g$	*(vii)* $!\mathbf{AG}\, f = \mathbf{AF}\, !f$
(iv) $!(f \wedge g) = !f \vee !g$	*(viii)* $!\mathbf{A}\, f\, \mathbf{U}\, g = \mathbf{A}\, !f\, \mathbf{R}\, !g$
	(ix) $!\mathbf{A}\, f\, \mathbf{R}\, g = \mathbf{A}\, !f\, \mathbf{U}\, !g$

Given an ACTL formula f, its strong negation $!f$ is also an ACTL formula. This fact can be proved easily by induction on the length of the ACTL formula. This is in contrast to the regular negation $\neg f$, which may be an existential formula. The strong negation $!f$ is called "strong" in the sense that it is the negation of f that applies to all paths in the computation tree. For any state q, the following proposition states that $q \models !f$ implies $q \models \neg f$. (Proofs of all propositions are given in (Chutinan et al., 2004).)

Proposition 2. Given a transition system $T = (Q, \rightarrow, Q_0, 2^{AP}, L)$ and an ACTL formula f, for any state $q \in Q$, if $q \models !f$ then $q \models \neg f$.

Proposition 3. Given a transition system T, its quotient transition system $T/\mathcal{P}$ constructed from a consistent partition $\mathcal{P}$, and an ACTL specification f, if there exists $P_0 \in Q_0/\mathcal{P}$ such that $T/\mathcal{P}, P_0 \models !f$, then $T \not\models f$.

In words, if the strong negation of an ACTL specification is true for *any* initial state of a QTS, the transition system does not satisfy the specification. Therefore, by augmenting the specification used for model checking a QTS to include the strong negation of the specification, the verification process for a hybrid system can terminate with a conclusive negative result in the event that the strong negation of the given ACTL specification is true. The verification procedure using strong negation is illustrated in Fig. 1. In the step "verify $\varphi \vee !\varphi$", the term $!\varphi$ is true iff $Q_0/\mathcal{P} \cap S(T/\mathcal{P}, !\varphi) \neq \emptyset$. This condition is easily verified by standard model checking since $!\varphi$ is an ACTL expression.

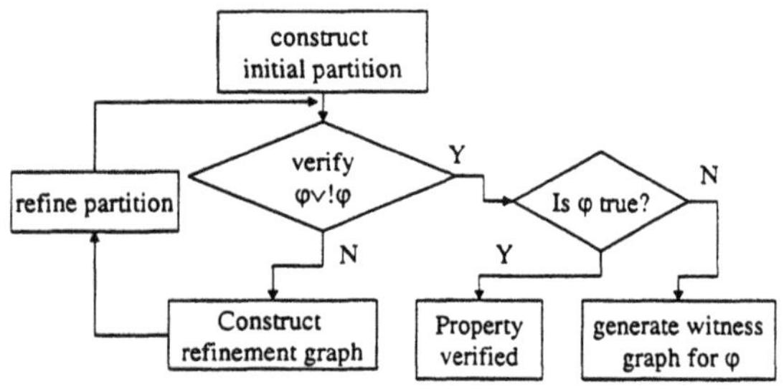

Fig. 1. Hybrid system verification procedure. ($!\varphi$ is true iff $Q_0 \cap S(T, !\varphi) \neq \emptyset$)

4. COUNTEREXAMPLES AND REFINEMENT

This section presents a method for identifying the set of QTS states to be refined when model checking of the QTS is inconclusive. We show that by incorporating the strong negation condition the set of to-be-refined states can be reduced. Our approach is motivated by counterexample-guided refinement procedures proposed for finite-state transition systems in (Clarke et al., 2000) and (Glusman et al., 2003), for hybrid systems in (Alur et al., 2003) and (Clarke et al., 2003). In these papers, the QTS states to be refined at each iteration are identified from a single given counterexample. Our goal is to identify efficiently *all* states that need to be refined to guarantee that a conclusive result can be obtained, if a conclusive result is feasible based on refinement of the current QTS. We achieve this by constructing a subgraph for the QTS such that refinement of any state not included in the subgraph will not aid in the verification; that is, by focusing on particular subgraph (called the *refinement graph*), we eliminate states that can be ignored in the refinement procedure.

We first consider the construction of a QTS subgraph, called the *witness graph*, which is a graph that represents all possible witnesses for a given ECTL formula φ_E. The proposed procedure to obtain the witness graph is given in (Chutinan et al., 2004). This procedure is a modification of the witness algorithm commonly used in model checking tools to generate counterexamples. (Clarke et al., 2002, 1999). The difference is that the set of trajectories for our witness graph contains *all* witnesses for the given ECTL formula. The witness graph procedure recursively unwind the subformulas f of φ_E and construct the graph from the states in $SAT(T, f)$.

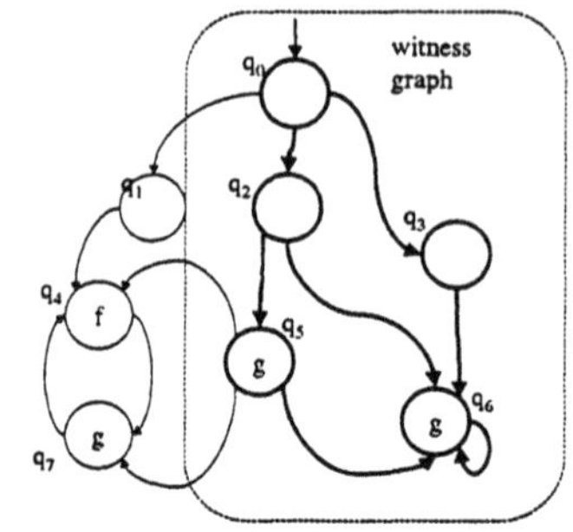

Fig. 2. Witness graph for $EFEG\neg f \vee EFEG\neg g$.

Example 1. (witness graph). Figure 2 illustrates the definition of the witness graph. The states labelled f and g denote that the atomic proposition f and g are true for the states. The ACTL specification $AGAFf \wedge AGAFg$ denotes "On every path in the future, f and g are true infinitely often." The specification is false for the transition system, the witness graph of the negation $\varphi_E = EFEG\neg f \vee EFEG\neg g$ are shown as the subgraph with bold lines.

The following proposition shows that the witness graph constructed above serves as a counterexample (CE) of the transition system, i.e., the specification is valid if and only if the witness graph of the negation is empty.

Proposition 4. The witness graph for $\varphi_E = \neg\varphi_A$ is empty iff $T \models \varphi_A$.

A refinement procedure constructs a new partition where one or more states is partitioned. An intuitive implication from Prop. 4 is that the states to be refined can be restricted to the witness graph. (This is proved in (Chutinan et al., 2004).) In fact, the set of states to be refined can be reduced even further by considering the strong negation condition. Consider a subformula f_A of φ_A, we know that its strong negation $!\varphi_A$ is invariant under refinement. Thereby refining these states does not change the value of the subformula. From the property of the strong negation condition, the corresponding sub-formula f_E of the negation $\varphi_E = \neg\varphi_A$ is also invariant under refinement. Thus any state for which f_E is true remains inside the witness graph after refinement. We can exclude such states from the refinement procedure by incorporating strong negation. The set of states to be refined is reduced from the witness graph to the *refinement graph* which is defined by the procedure $RGraph(T, Q_0 \cap S'(T, f), f)$, where f is an ECTL formula,

$$S'(T,f) = \begin{cases} \emptyset & \text{, if } f \in AP; \\ SAT(T,f) \backslash SAT(T, !(\neg f)) & \text{, if } f \notin AP. \end{cases} \quad (1)$$

and *RGraph* is constructed from the *witGraph* procedure in (Chutinan et al., 2004) using $S'(T,f)$ to replace $SAT(T,f)$.

The function S' differs from $SAT(T,f)$ in that it excludes the set of states where $!(\neg f_E)$ is true. It can be easily seen that

$$RGraph(T, Q_0 \cap S'(T,f), F) \subseteq witGraph(T, Q_0 \cap SAT(T,f), f)$$

since the $S'(T, f_E)$ function is a subset of $SAT(T, f_E)$ function. To show that the refinement graph contains all the states that need to be refined, we have the following proposition.

Proposition 5. (Reduced refinement states). Given a transition system T, an ACTL specification φ_A, consistent partitions $\mathcal{P}_1$, $\mathcal{P}_2$, where $\mathcal{P}_2$ refines $\mathcal{P}_1$ with refinement function a, and $T/\mathcal{P}_1 \not\models \varphi_A$, $T/\mathcal{P}_2 \models \varphi_A$. Let $\mathcal{P}_2'$ be the partition containing the sets P_2' defined for each $P_2 \in \mathcal{P}_2$ as

$$P_2' = \begin{cases} \alpha(P_2), & \text{if } \alpha(P_2) \notin RGraph(T/\mathcal{P}_1, Q_0/\mathcal{P}_1 \cap \\ & \quad S'(T/\mathcal{P}_1, \varphi_E), \varphi_E) \\ P_2, & \text{otherwise,} \end{cases}$$

where $\varphi_E = \neg\varphi_A$, then $T/\mathcal{P}_2' \models \varphi_A$.

By taking advantage of the refinement graph, the set of states to be refined is reduced by the strong negation condition compared with (Chutinan and Krogh, 2001; Clarke et al., 2003). The reduction of states to be refined is illustrated in the following example.

Example 2. Consider again the transition system in example 1. The refinement graph for the same specification is shown in Fig. 3. The witness graph in Fig. 2 contains five states to refine, whereas by incorporating the strong negation condition, the number of states of the witness graph is reduced to three.

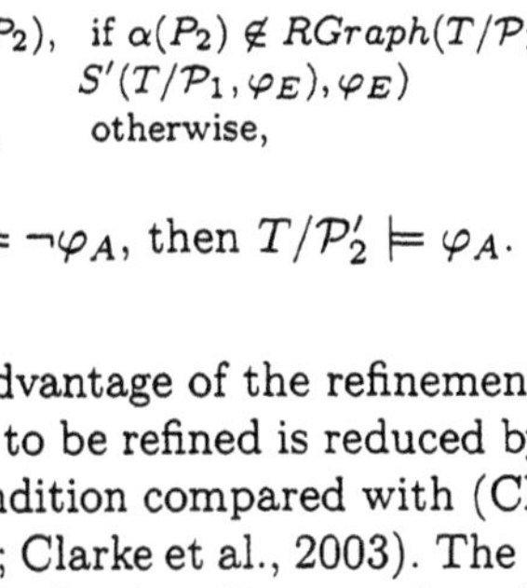

Fig. 3. Refinement graph for $(AGAFf \wedge AGAFg)$.

5. EXAMPLE

This section demonstrates an application of the ACTL strong negation through a simple application example. The example is implemented in an abstraction-based hybrid system verification tool called CHECKMATE (Chutinan and Krogh, 2001, 2003).

We consider a simplified automotive powertrain model with the transmission held at the first gear. The continuous dynamics of the system is $\dot{x} = Ax + b \in \mathbb{R}^5$. The state variables comprising the state vector x are T_s, the drive shaft torque,v, the vehicle speed, ω_t, the engine speed, *tps*, the throttle position, and *sin_grade*, the sine of the road grade, respectively. The derivatives of *tps* and *sin_grade* are set to zero since these are constant parameters. Our verification objective is to ensure that the system satisfies the specification (to be discussed below) over a given set of initial states. A detail discussion with verification result with CHECKMATE can be found in (Chutinan and Butts, 2002).

Starting from an initial condition, we would like to ensure that the throttle position provides the vehicle with enough power to overcome the steepness of the road grade. For a trajectory starting from any initial condition, if the vehicle reaches the speed of 1 m/s within 1.5 seconds, then the trajectory is a *success*. If the vehicle reaches the speed of -1 m/s the trajectory is a *failure*. If the time limit of 1.5 seconds has expired before the vehicle reaches the speed of 1 m/s, then the trajectory is a *time_limit*. Following the discussions above, we write the ACTL specification for our problem as

$$\mathbf{AF}\ success \wedge \mathbf{AG}\ \neg time\_limit.$$

The reachability analysis in CHECKMATE yields the quotient transition system shown in Figure 4. We note that the quotient transition system in Figure 4 is actually an *approximate quotient transition system (AQTS)* because CHECKMATE uses reachability approximations to compute the transitions between the partition states (Chutinan and Krogh, 2001). There are three special states in the AQTS corresponding to the terminal states, *success* and *failure* and *time limit*.

The QTS constructed with strong negation checking is shown in Fig. 4. Initial State 2 of the quotient transition system in satisfies the strong negation. Thus, we conclude by Proposition 3 that the original powertrain system does not satisfy the specification and there is no need to refine the quotient transition system. As a comparison CHECKMATE continue to refine the partition. The partition after the 10 iterations is shown in Fig.

5. The two boundaries from closed-form solution are superimposed which partitioned the states into *success*, *time_limit* and *failure* sets. The refinement of CheckMate is getting close to the boundaries, however, it will never terminate since the system does not satisfies the specification. The application of ACTL strong negation in this case has helped avoid unnecessary computations.

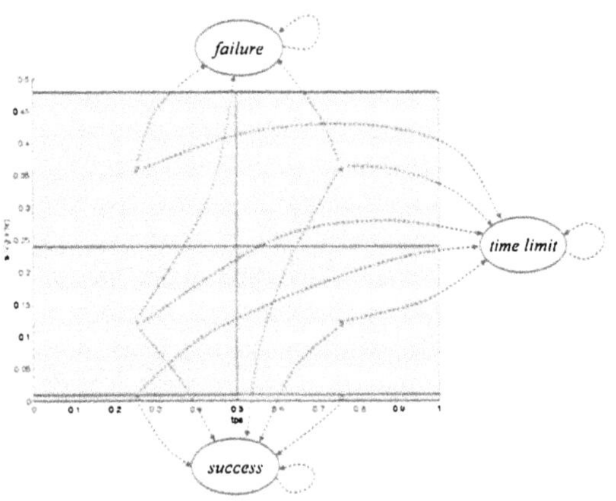

Fig. 4. QTS constructed using strong negation

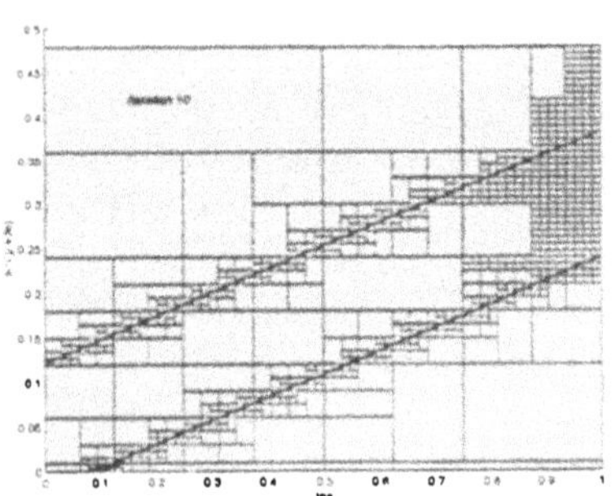

Fig. 5. CheckMate Partition after iteration 10

6. DISCUSSION

This paper introduces the concept of ACTL strong negation that can be applied to a conservative abstraction to draw the conclusion that the underlying system fails a universal specification. As demonstrated in Section 5, ACTL strong negation can help avoid unnecessary refinement computations for the abstraction with respect to the ACTL specification. ACTL strong negation can be used to complement any verification procedure that involves abstraction refinement methods such as the *counterexample-guided refinement* (Clarke et al., 2000, 2003). The procedures described in this paper have been implemented in a new version of CheckMate, an abstraction-based model checker for hybrid systems (Chutinan and Krogh, 2003).

REFERENCES

R. Alur, T. Dang, and F. Ivancic. Counter-example guided predicate abstraction of hybrid systems. In *9th International Conference on Tools and Algorithms for the Construction and Analysis of Systems (TACAS)*. Springer, 2003.

R. Alur, T. A. Henzinger, G. Lafferriere, and G. J. Pappas. Discrete abstractions of hybrid systems. In F. W. Vaandrager and J. H. Van Schuppen, editors, *Proceedings of the IEEE*, pages 971–84, 2000.

A. Chutinan and K. R. Butts. Dynamic analysis of hybrid system models for design validation. Technical report, University of California, 2002.

A. Chutinan, Z. Han, and B. H. Krogh. Actl strong negation and its application to hybrid verification. Technical report, Department of ECE, Carnegie Mellon University, 2004.

A. Chutinan and B. H. Krogh. Verification of infinite-state dynamic systems using approximate quotient transition systems. *IEEE Transactions on Automatic Control*, 46 (9):1401–1410, September 2001.

A. Chutinan and B. H. Krogh. Computational techniques for hybrid system verification. *IEEE Transactions on Automatic Control*, 48(1):64–75, January 2003.

E. Clarke, A. Fehnker, Z. Han, B. H. Krogh, J. Ouaknine, O. Stursberg, and M. Theobald. Abstraction and counterexample-guided refinement in model checking of hybrid systems. *International Journal of Foundations of Computer Science (IJFCS): Special Issue on Verification and Analysis of Infinite State Systems*, 2003.

E. Clarke, O. Grumberg, S. Jha, Y. Lu, and H. Veith. Counterexample-guided abstraction refinement. In *Computer-Aided Verification (CAV)*, volume 1855 of *LNCS*, pages 154–169. Springer, 2000.

E. Clarke, S. Jha, Y. Lu, and H. Veith. Tree-like counterexamples in model checking. In *17th Annual IEEE Symposium on Logic in Computer Science (LCS)*, pages 19– 29, 2002.

E. M. Clarke, O. Grumberg, and D. A. Peled. *Model Checking*. The MIT Press, Cambridge, MA, 1999.

D. R. Dams, R. Gerth, and O. Grumberg. Generation of reduced models for checking fragments of ctl. In Costas Courcoubetis, editor, *Computer Aided Verification*, Berlin, 1993. number 697 in Lecture Notes in Computer Science, Springer-Verlag.

D. R. Dams, R. Gerth, and O. Grumberg. Abstract interpretation of reactive systems. *ACM Transactions on Programming Languages and Systems*, 19(2):253–91, March 1997.

M. Glusman, G. Kamhi, S. Mador-Haim, R. Fraer, and M. Vardi. Multiple-counterexample guided iterative abstraction refinement: an industrial evaluation. In *9th International Conference on Tools and Algorithms for the Construction and Analysis of Systems (TACAS)*. Springer, 2003.

P. Godefroid, M. Huth, and R. Jagadeesan. Abstraction-based model checking using modal transition systems. In *Proceedings of CONCUR'2001 (12th International Conference on Concurrency Theory)*, Aalborg, 2001. Lecture Notes in Computer Science, Springer-Verlag.

P. Godefroid and R. Jagadeesan. Automatic abstraction using generalized model checking. In *Proceedings of CAV'2002 (14th Conference on Computer Aided Verification)*, Copenhagen, 2002. Lecture Notes in Computer Science, Springer-Verlag.

T. A. Henzinger. The theory of hybrid automata. In *Proceedings of the 11th Annual Symposium on Logic in Computer Science*, pages 278–292. IEEE Computer Society Press, 1996. Invited tutorial.

ELSEVIER
IFAC
PUBLICATIONS
www.elsevier.com/locate/ifac

DIAGNOSIS OF MODULAR DISCRETE EVENT SYSTEMS [1]

**Olivier Contant Stéphane Lafortune
Demosthenis Teneketzis**

*Department of EECS, The University of Michigan
1301 Beal Avenue, Ann Arbor, MI 48109-2122 USA
{olivier, stephane, teneket}@eecs.umich.edu
http://www.eecs.umich.edu/umdes/*

Abstract: The diagnosis of unobservable fault events in discrete event systems modeled in a modular manner by means of parallel composition is considered. The notion of modular diagnosability is introduced for these systems. Necessary and sufficient conditions for this new notion of diagnosability are given. A new algorithm for verifying modular diagnosability is presented. The main feature of this algorithm is the exploitation of the modular structure of the system in order to save on computational efforts. *Copyright © 2004 IFAC*

Keywords: Discrete-event systems, Distributed models, Fault diagnosis, Failure detection, Observers, Detection algorithms.

1. INTRODUCTION

Monitoring and diagnosis methodologies that are based upon discrete-event models of dynamic systems have proved successful in many application areas including document processing systems, heating, ventilation, and air-conditioning systems, intelligent transportation systems, chemical process control, and telecommunication networks. Most of the discrete-event fault diagnosis methodologies that have been studied in the literature require a "monolithic" model of the system under consideration for diagnosability analysis and/or construction of the diagnostic protocol; see (Lafortune *et al.*, 2001) and the references therein. Notable exceptions include the approaches developed in (Holloway and Chand, 1994; Ricker and Fabre, 2000; Fabre *et al.*, 2002; García *et al.*, 2002; Su *et al.*, 2002; Debouk *et al.*, 2002; Genc and Lafortune, 2003) where the modular structure of the underlying system is exploited by the respective monitoring or diagnostic protocols. This paper has a similar objective, although the approach adopted is different from those in the above references.

We consider systems that are modeled by the parallel composition of a set of automata. Each individual automaton models a component or subsystem of the overall system. The coupling of these system components is captured by the use of common events. We are concerned with systems where the construction of the complete system model (namely, the *monolithic* model obtained by performing the parallel composition of the set of individual automata) is computationally difficult or intractable, making the use of fault diagnosis methodologies such as the "Diagnoser Approach" initiated in (Sampath *et al.*, 1995) impractical. Therefore, it is necessary to develop *modular approaches* that are computationally tractable for

[1] This research is supported in part by NSF grants ECS-0080406, CCR-0082784, and CCR-0325571, and by a grant from the Xerox University Affairs Committee. Useful discussions with R. Debouk are gratefully acknowledged.

analyzing the diagnosability properties of the system and for synthesizing appropriate diagnostic protocols. By "modular approaches" we mean approaches that exploit the structure of the system as captured by the individual automata and their respective sets of common events.

The first contribution of this paper is to present a new notion of diagnosability that is explicitly geared towards systems with modular representations. Necessary and sufficient conditions for modular diagnosability are presented and discussed. The second contribution of this paper is the development in Section 4 of a novel algorithm for verifying modular diagnosability. Due to space limitations, all proofs and several discussions have been omitted; for a complete exposition of this work, the reader is referred to (Contant *et al.*, 2004).

2. SYSTEM MODEL

Let I be the total number of components or subsystems in the given modular system under consideration. Let $T = \{1, \ldots, I\}$ and $S \subseteq T$. Elements i of T are often termed "sites" hereafter. We use the notation

$$G_S = (X_S,\ \Sigma_S,\ \delta_S,\ x_{S_0},\ X_{S_m}) \qquad (1)$$

to denote the automaton with state space X_S, set of events Σ_S, (partial) transition function δ_S, initial state x_{S_0}, and set of marked states X_{S_m}. When $S = T$, G_S denotes the global (or complete) system model. When $S = \{i\}$, with $i \in T$, G_S denotes individual component model i. When $S \subset T$, $S \neq \{i\}$, $i \in T$, G_S denotes the partial system model comprised of the individual automata in the set S, which we will call the "system G_S" hereafter. In all of the above cases, system G_S accounts for the normal and failed behavior of the components in S. $G_S = \|_{z \in S}\, G_z$ is obtained by composing the individual automata G_z, $z \in S$, using the parallel composition operation.

The behavior of the system G_S is described by the prefix-closed language $\mathcal{L}(G_S)$ generated by G_S. $\mathcal{L}(G_S)$ is assumed to be live. This means that there is a transition defined at each state x in X_S. The liveness assumption on $\mathcal{L}(G_S)$ is made for the sake of simplicity. With slight modifications, all the main results of this paper hold true when the liveness assumption is relaxed. Some of the events in Σ_S are observable, i.e., their occurrence can be observed by sensors, while the rest are unobservable. We use the notation Σ_{os} and Σ_{uos} to represent the set of observable and unobservable events of G_S, respectively, where $\Sigma_{os} = \Sigma_S \setminus \Sigma_{uos}$. Let Σ_{fs} denote the set of fault events in the system G_S. Without loss of generality, we assume that $\Sigma_{fs} \subseteq \Sigma_{uos}$, since an observable fault event can be diagnosed trivially.

Due to space limitations, it has been necessary to assume that the reader is familiar with basic notions[2] in languages and automata theory and with the notation and main concepts of the Diagnoser Approach of (Sampath *et al.*, 1995), such as diagnosers, certain and uncertain states, and indeterminate cycles. We add the following definitions regarding notation. The notation $\Sigma = \Sigma' \dot{\cup} \Sigma''$ indicates that the event set Σ is the disjoint union of Σ' and Σ'', i.e., $\Sigma' = \Sigma \setminus \Sigma''$. Let $R \subset T$. The event set Σ_R is partitioned as $\Sigma_R = \Sigma_{CM_R} \dot{\cup} \Sigma_{PV_R}$, where $\Sigma_{CM_R} = \Sigma_R \cap [\cup_{z \in T \setminus R} \Sigma_z]$ represents the set of common events in G_R and where Σ_{PV_R} represents the set of private events in G_R. We define $\Sigma_{CM} = \cup_{i \in T} \Sigma_{CM_i}$ the set of all common events. We use the notation $\Sigma_{CM_{o_R}} = \Sigma_{CM_R} \cap \Sigma_{o_R}$ to represent the set of common and observable events. The notation $CoAc(G_S)$ represents the automaton obtained from G_S by retaining only those states x of G_S that are coaccessible, namely, that can reach a (marked) state in X_{S_m}. This notation will often be specialized to $CoAc(G_S, M_x)$ where M_x will be a particular label associated with the marked states of G_S. In this case, $CoAc(G_S, M_x)$ will be the coaccessible part of G_S with respect to the marked states of G_S that are labeled with M_x.

It will be necessary in many instances to explicitly identify the event set associated with an automaton. By default, the event set Σ_S associated with automaton G_S will be $\Sigma_S := \{\sigma \in s : s \in \mathcal{L}(G_S)\}$, namely, all the events that appear in all the traces in $\mathcal{L}(G_S)$. In special cases it is required to define a larger set of events associated with G_S than the default one. In this regard, the notation (G_S, Σ) will denote the automaton G_S together with the event set Σ such that $\Sigma \supseteq \Sigma_S$. For example, (G_i, Σ_S) implies that the original event set Σ_i of G_i is now augmented by the set $\Sigma_S \setminus \Sigma_i$. While (G_i, Σ_S) has the same language properties as (G_i, Σ_i), it has different behavior when performing parallel composition (or, more generally, any operation using sets of events) with other modules as it will prevent the occurrence of events in $\Sigma_S \setminus \Sigma_i$.

We define

$$Obs(G_S, \Sigma_S^{obs}) = (X_S^{obs},\ \Sigma_S^{obs},\ \delta_S^{obs},\ x_{S_0}^{obs}) \qquad (2)$$

to be the *observer* of G_S with respect to the set of observable events $\Sigma_S^{obs} \cap \Sigma_{os}$, where $\Sigma_S^{obs} \subset \cup_{z \in T} \Sigma_z$. (See (Cassandras and Lafortune, 1999) for the complete definition of an observer.) For example, $Obs(G_i, \Sigma_{CM_l})$, $l \neq i$, is the observer of G_i with respect to $\Sigma_{CM_l} \cap \Sigma_{o_i}$, the observable events of G_i that are common with those of G_l.

[2] See, e.g., Chapter 2 of (Cassandras and Lafortune, 1999).

Let Σ_X and Σ_Y be any sets of events. We define two projection operators relative to these sets, $P_{\{\Sigma_X,\Sigma_Y\}}$ for the usual natural projection and $R_{\{\Sigma_X,\Sigma_Y\}}$ for the so-called "reverse" projection. Specifically, the natural projection $P_{\{\Sigma_X,\Sigma_Y\}} : \Sigma_X^* \rightarrow \Sigma_Y^*$ is defined in the usual manner:

$$P_{\{\Sigma_X,\Sigma_Y\}}(\epsilon) := \epsilon \tag{3}$$

$$P_{\{\Sigma_X,\Sigma_Y\}}(\sigma) := \begin{cases} \sigma & \text{if } \sigma \in \Sigma_Y \\ \epsilon & \text{if } \sigma \notin \Sigma_Y \end{cases} \tag{4}$$

$$P_{\{\Sigma_X,\Sigma_Y\}}(s\sigma) := P_{\{\Sigma_X,\Sigma_Y\}}(s)P_{\{\Sigma_X,\Sigma_Y\}}(\sigma) \quad \text{for } s \in \Sigma_X^*, \sigma \in \Sigma_X. \tag{5}$$

In contrast, the *reverse projection* $R_{\{\Sigma_X,\Sigma_Y\}}$ is applied to traces of events from Σ_X^* and produces "inverse" traces from Σ_Y^* as is usually done in inverse projection operations. More precisely, $R_{\{\Sigma_X,\Sigma_Y\}} : \Sigma_X^* \rightarrow 2^{\Sigma_Y^*}$ is defined as follows:

$$R_{\{\Sigma_X,\Sigma_Y\}}(s) = \{t \in \Sigma_Y^* : P_{\{\Sigma_Y,\Sigma_X\}}(t) = s\}. \tag{6}$$

The natural and reverse projections can also be defined with respect to a particular language L. For $L \subseteq \Sigma_Y^*$,

$$P^L_{\{\Sigma_X,\Sigma_Y\}}(s) = \{t \in L : P_{\{\Sigma_X,\Sigma_Y\}}(s) = t\}, \tag{7}$$

and

$$R^L_{\{\Sigma_X,\Sigma_Y\}}(s) = \{t \in L : P_{\{\Sigma_Y,\Sigma_X\}}(t) = s\}. \tag{8}$$

We conclude this section by stating a key assumption that will be required for the results presented in the remainder of the paper: *all common events are observable*, i.e., $\forall i \in T$, $\Sigma_{CM_i} \subseteq \Sigma_{o_i}$. This implies that all faults are private events (since they are unobservable).

3. MODULAR DIAGNOSABILITY

We define the notion of modular diagnosability as follows.

Definition 1. : **Modular Diagnosability**
Let $T = \{1, ..., I\}$, $S \subseteq T$, $G_S = \|_{z \in S} G_z$, and $S^- \subseteq S$. The language $\mathcal{L}(G_S)$ is modularly diagnosable w.r.t. $(\Sigma_{o_z} : z \in S)$ and $(\Sigma_{f_z} : z \in S^-)$ if $\forall i \in S^-$, $\forall f \in \Sigma_{f_i}$, $\forall s \in \mathcal{L}(G_S)$ s.t. s ends with f, $\exists n \in \mathbb{N}$ s.t. $\forall t \in \mathcal{L}(G_S)/s$, $\| P_{\{\Sigma_S,\Sigma_{o_i}\}}(t) \| \geq n \Rightarrow D(st) = 1$. The diagnosability condition function D is given by

$$D(st) = \begin{cases} 1 & \text{if } \omega \in R^{\mathcal{L}(G_S)}_{\{\Sigma_{oS},\Sigma_S\}}[P_{\{\Sigma_S,\Sigma_{oS}\}}(st)] \Rightarrow \\ & f \in \omega, \\ 0 & \text{otherwise.} \end{cases} \tag{9}$$

To draw comparisons between our modular diagnosability algorithm, presented in Section 4, and any other potential approach, we give necessary and sufficient conditions for modular diagnosability based on monolithic constructions of the system model and diagnoser for a given set S of system components. G_{dS} and G'_S denote, respectively, the diagnoser of G_S and the nondeterministic automaton built from G_S by eliminating unobservable events; both are defined in (Sampath *et al.*, 1995).

Definition 2. : F^{M_i}**-indeterminate cycle**
A set of F-uncertain states $q_1, q_2, \ldots, q_n \in Q_{dS}$ is said to form an F^{M_i}-indeterminate cycle in G_{dS} if the following condition C1 is satisfied.

C1) States $q_1, q_2, \ldots, q_n \in Q_{dS}$ form a cycle in G_{dS} with $\delta_{dS}(q_u, \sigma_u) = q_{u+1}$, $u = 1, \ldots, n-1$, $\delta_{dS}(q_n, \sigma_n) = q_1$ where $\sigma_u \in \Sigma_{oS}$, $u = 1, \ldots, n$ and $\exists l \in \{1, \ldots, n\}$ s.t. $\sigma_l \in \Sigma_{o_i}$.
Considering the states $q_1, q_2, \ldots, q_n \in Q_{dS}$, $\exists$ (x_u^k, ℓ_u^k), $(y_u^r, \tilde{\ell}_u^r) \in q_u$, $u = 1, \ldots, n$, $k = 1, \ldots, m$, and $r = 1, \ldots, m'$ such that:
(i) $[(F \in \ell_u^k) \wedge (F \notin \tilde{\ell}_u^r)]$, for all u, k, and r, where F represents the label associated with the fault event $f \in \Sigma_{f_i}$, $i \in S$,
(ii) the sequences of states $\{x_u^k\}$, $u = 1, \ldots, n$, $k = 1, \ldots, m$, and $\{y_u^r\}$, $u = 1, \ldots, n$, $r = 1, \ldots, m'$, form cycles in G'_S with $(x_u^k, \sigma_u, x_{(u+1)}^k) \in \delta_{G'_S}$, $u = 1, \ldots, n-1$, $k = 1, \ldots, m$, $(x_n^k, \sigma_n, x_1^{k+1}) \in \delta_{G'_S}$, $k = 1, \ldots, m-1$, $(x_n^m, \sigma_n, x_1^1) \in \delta_{G'_S}$, and $(y_u^r, \sigma_u, y_{(u+1)}^r) \in \delta_{G'_S}$, $u = 1, \ldots, n-1$, $r = 1, \ldots, m'$, $(y_n^r, \sigma_n, y_1^{r+1}) \in \delta_{G'_S}$, $r = 1, \ldots, m'-1$, $(y_n^{m'}, \sigma_n, y_1^1) \in \delta_{G'_S}$.

Remark 3. The symbol "M_i", in the notation "F^{M_i}-indeterminate cycle", stands for "Module G_i". F^{M_i}-indeterminate cycles differ slightly from the F-indeterminate cycles introduced in (Sampath *et al.*, 1995) in two respects. First, we require that there exists at least one observable event from module G_i in the cycle of states $q_1, q_2, \ldots, q_n \in Q_{dS}$: cf. "$\exists l \in \{1, \ldots, n\}$ s.t. $\sigma_l \in \Sigma_{o_i}$" in Condition C1. Second, we require that the label F in hypothesis (i) of Condition C1 represents the label associated with the fault event $f \in \Sigma_{f_i}$, i.e., the fault event f originates from module G_i.

Theorem 4. : Consider the language $\mathcal{L}(G_S)$ generated by automaton $G_S = \|_{z \in S} G_z$. $\mathcal{L}(G_S)$ is modularly diagnosable w.r.t. $(\Sigma_{o_z} : z \in S)$ and $(\Sigma_{f_i} : i \in S)$, iff there are no F^{M_i}-indeterminate cycles in the diagnoser G_{dS}.

The main difference between the modular and the monolithic (in the sense of (Sampath *et al.*, 1995)) definitions of diagnosability concerns the type of traces that need to be considered. Modular diagnosability focuses only on traces where events from module G_i, which is the module where the fault originates, occur with some regularity. Consequently, the notion of modular diagnosability is weaker than the notion of monolithic diagnosability since more languages will satisfy this definition than the monolithic one.

Our primary motivation for defining modular diagnosability is to ensure that after a fault occurs in one of the system modules, detection and isolation of that fault is only required for continuations that involve events from the given module. It is reminiscent of the familiar "persistency of excitation" condition in system identification. In other words, continuations that entirely exclude the module where the fault originates from cannot lead to a violation of modular diagnosability. (Recall that the approach that we propose assumes that faults do not bring the system, or any of its modules, to a halt.)

4. TEST FOR MODULAR DIAGNOSABILITY

We propose a novel approach that tests modular diagnosability by incorporating incrementally, in a systematic manner, subsystems into the test. We prove that our approach provides the correct answer to the question "Is $\mathcal{L}(G_T)$ modularly diagnosable w.r.t. $(\Sigma_{o_z} : z \in T)$ and $(\Sigma_{f_z} : z \in T)$?" in a finite number of steps. We proceed as follows. In Section 4.1 we present the algorithm; in Section 4.2 we state its properties; in Sections 4.3 and 4.4 we present a discussion of the key steps of the algorithm and online diagnosis, respectively.

4.1 Modular Diagnosability Algorithm

We present the detailed statement of our Modular Diagnosability Algorithm (MDA). For the sake of clarity, MDA is broken into three algorithms. Algorithm 1 is the core of MDA; it calls Algorithm 2 to perform preliminary steps involving indeterminate cycles that could lead to a violation of modular diagnosability. Algorithm 1 also calls Algorithm 3 where the incremental analysis of each indeterminate cycle is performed.

Algorithm 1. ***MDA***

1.) Let $T = \{1, \ldots, I\}$. Construct the local diagnosers G_{d_i}, $i \in T$, and search for indeterminate cycles. If, $\forall i \in T$, $\mathcal{L}(G_i)$ is monolithically diagnosable w.r.t. Σ_{o_i} and Σ_{f_i}, i.e., none of the local diagnosers G_{d_i} have F-indeterminate cycles, then stop and declare $\mathcal{L}(G_T)$ modularly diagnosable w.r.t. $(\Sigma_{o_z} : z \in T)$ and $(\Sigma_{f_z} : z \in T)$. Else, go to Step 2.

2.) Let $ND = T \setminus D$, where $D = \{z : \mathcal{L}(G_z)$ is monolithically diagnosable w.r.t. Σ_{o_z} and $\Sigma_{f_z}\}$. Call **Preliminary Function**. For each local diagnoser G_{d_i}, $i \in ND$, and for each sequence of traces SEQ_x, $x \in \{1, ..., X_i\}$, perform the Steps 2-a to 2-d:

2-a.) Mark with the label M_x, $x \in \{1, ..., X_i\}$, $i \in ND$, the states $q \in Q^x$ in G_{d_i}. The label M_x stands for "State of G_{d_i} part of the indeterminate cycle associated with SEQ_x".

2-b.) Construct

$$G_{CMi} = Obs(G_{d_i}, \Sigma_{CM_i}). \tag{10}$$

A state of G_{CMi} is marked with label M_x if one or more of its state components are marked with M_x.

2-c.) Construct

$$G_{ICM_x} = CoAc(G_{CMi}, M_x). \tag{11}$$

The resulting event set of machine G_{ICM_x} is denoted by Σ_{ICM_x}. Enlarge the set of events Σ_{ICM_x} by adding $\Sigma_{CM_i} \setminus \Sigma_{ICM_x}$ to it. The machine G_{ICM_x} and its newly associated event set Σ_{CM_i} are hereafter represented by the notation $(G_{ICM_x}, \Sigma_{CM_i})$.

2-d.) Call **Reachability Function** with argument $\{i, SEQ_x, G_{ICM_x}, \Sigma_{ICM_x}, \Sigma_{CM_i}\}$. If the Reachability Function returns "Reachable" then stop and declare $\mathcal{L}(G_T)$ not modularly diagnosable w.r.t. $(\Sigma_{o_z} : z \in T)$ and $(\Sigma_{f_z} : z \in T)$.

3.) Stop and declare $\mathcal{L}(G_T)$ modularly diagnosable w.r.t. $(\Sigma_{o_z} : z \in T)$ and $(\Sigma_{f_z} : z \in T)$. $\diamond$

Algorithm 2. - ***Preliminary Function***

I.) For each $i \in ND$, do the following:

i.) Call EIC_z, $z \in \{1, \ldots, Z\}$, the elementary[3] indeterminate cycles in G_{d_i} and t^z_y, $y \in \{1, \ldots, Y_z\}$, their corresponding sequences of events, where Z represents the total number of elementary indeterminate cycles in diagnoser G_{d_i}, $i \in ND$, and Y_z represents the total number of sequences of events that satisfy the indeterminate cycle definition for the particular EIC_z.

ii.) Number and name $SEQ_1, SEQ_2, \ldots, SEQ_{X_i}$ all sequences of events t^z_y, $y = 1, \ldots, Y_z$ and $z = 1, \ldots, Z$. To each SEQ_x, $x \in \{1, \ldots, X_i\}$, associate its corresponding F_m-indeterminate cycle, $m \in \{1, \ldots, M\}$, its corresponding sequence of states $Q^x = q_1 \ldots q_{N_x}$, and its corresponding sequence of events t^z_y, $z \in \{1, \ldots, Z\}$, $y \in \{1, \ldots, Y_z\}$, that form the cycle.

[3] A cycle is called elementary if no state appears more than once in it.

II) Return to MDA with SEQ_x and Q^x, $\forall x \in \{1,\ldots,X_i\}$, $\forall i \in ND$. ◇

Algorithm 3. - **Reachability Function** $\{i, SEQ_x, G_{ICM_x}, \Sigma_{ICM_x}, \Sigma_{CM_i}\}$

A.) Let $c := 1$, $B_c^x = \{i\}$, $S_c = B_c^x$, and

$$\bar{s}^x = P_{\{\Sigma_{o_i}, \Sigma_{CMo_i}\}}(SEQ_x). \quad (12)$$

Let $c := c+1$,

$$B_c^x = \{l : [\Sigma_{CM_l} \cap \Sigma_{ICM_x} \neq \emptyset \vee (l \in B_{c-1}^x)], l \in T\}, \quad (13)$$

and

$$S_c = B_c^x \setminus B_{c-1}^x. \quad (14)$$

B.) Construct

$$G_{mod_c^x} = (G_{ICM_x}, \Sigma_{CM_i}) \parallel (\|_{l \in B_c^x, l \neq i} G_{CMl}). \quad (15)$$

If there does not exist in $G_{mod_c^x}$ a cycle of states labeled M_x then return to MDA; otherwise denote by $s_1^c, s_2^c, \ldots, s_P^c$ the sequences of events that describe such (elementary) cycles and go to step C.

C.) $\forall p \in \{1,\ldots,P\}$, let

$$\bar{s}_p^c = P_{\{\Sigma_{CMo_S}, \Sigma_{CMo_i}\}}(s_p^c). \quad (16)$$

If $\exists p \in \{1,\ldots,P\}$ such that $\bar{s}^x = \bar{s}_p^c$ or if $\exists s'^x, s''^x$, and $p \in \{1,\ldots,P\}$ such that $\bar{s}^x = s'^x s''^x$ and $s''^x s'^x = \bar{s}_p^c$, then go to Step D; otherwise return to MDA.

D.) Construct

$$\widetilde{G}_c = CoAc(G_{mod_c^x}, M_x). \quad (17)$$

Let $\widetilde{\Sigma}_c$ be the event set of $\widetilde{G}_c$.

E.) Let $c := c+1$ and

$$B_c^x = \{l : [(\Sigma_{CM_l} \cap \widetilde{\Sigma}_{c-1} \neq \emptyset) \vee (l \in B_{c-1}^x)], l \in T\}. \quad (18)$$

Define

$$S_c = B_c^x \setminus B_{c-1}^x. \quad (19)$$

If $S_c \neq \emptyset$ then go to step B; otherwise declare the indeterminate cycle associated with SEQ_x "Reachable" and return to MDA. ◇

4.2 Properties of MDA

Theorem 5. MDA returns an answer in a finite number of steps.

Theorem 6. MDA returns the correct answer, namely, whether $\mathcal{L}(G_T)$ is or is not modularly diagnosable w.r.t. $(\Sigma_{o_z} : z \in T)$ and $(\Sigma_{f_z} : z \in T)$.

4.3 Discussion

To give more insight into MDA, we discuss the key steps of its operation. The procedure starts by building local diagnosers for each module of the system and checking if they are monolithically diagnosable or not. Clearly, if each individual module is (monolithically/modularly) diagnosable, then the complete system is both monolithically and modularly diagnosable. Therefore, we need only focus on the modules that are not diagnosable in order to find out if a corresponding violation of modular diagnosability occurs or not when the given module is coupled with the rest of the system. To do so, we concentrate on the traces that form indeterminate cycles in local diagnosers. We need to test these traces one by one and determine if they survive in the diagnoser of the complete system, without constructing this monolithic diagnoser.

The testing procedure starts by selecting one indeterminate cycle in a given (non-diagnosable) local diagnoser and isolating all the so-called "troublesome traces", namely, the traces that lead to and form indeterminate cycles in local diagnosers; there could be more than one troublesome trace depending on the accessibility of the indeterminate cycle in the transition structure of the local diagnoser. For each troublesome trace, we select all other modules that contain an event common with the ones in the troublesome trace, build machines (observers for common events - cf. Step 2-b of Algorithm 1) for each module selected, perform the parallel composition of these machines, and finally check if the indeterminate cycle under consideration survives (cf. Step B of Algorithm 3). If it does not survive at this stage then it will not survive if we were to construct the monolithic diagnoser. However, if it does survive, then we need to consider the effect of other modules, namely those that have common events with the result of the above parallel composition. This is the heart of the incremental procedure performed in Algorithm 3. We iterate using essentially the same steps as described above - cf. the loop formed by Steps B through E of Algorithm 3.

The incremental procedure in Algorithm 3 ceases to add local modules and consequently stops when either (i) it has been determined that the indeterminate cycle under consideration is not reachable in the complete system - if this holds for all indeterminate cycles then the monolithic system is modularly diagnosable or (ii) no other module is added in the incremental process at Step E of Algorithm 3 - in which case the monolithic system is not modularly diagnosable. Note that the latter conclusion can be reached *without* having to consider all modules in the set T. This potential computational gain depends on the structure of the machine $G_{mod_c^x}$ and its co-accessibility properties with respect to the indeterminate cycle under consideration, as determined in Steps C and D of Algorithm 3.

The main feature exploited within MDA is the incremental addition of modules by considering only those that are necessary to reach a decision on the modular diagnosability of the monolithic system. Depending on the structure of the system, MDA may consider a smaller number of modules rather than all of T when performing parallel composition operations. Nevertheless, the worst case can possibly occur and yield $|B_c^x| = |T|$, which implies that every system module is considered in the parallel composition for obtaining $G_{mod_c^x}$. However, it should be emphasized that the parallel compositions performed in MDA involve only machines with common events (observers) built from the individual modules. Therefore the resulting automata are likely to have a smaller state spaces than corresponding ones built using entire system modules. It should also be emphasized that in MDA, we construct diagnosers only for the individual G_i modules. The process of building diagnosers may result in a large state space growth in the worst case, since subsets of states of the model under consideration must be accounted for. Therefore, it is clearly advantageous to build diagnosers for local subsystems as opposed to building the diagnoser of the monolithic system G_T. The same advantage holds for online diagnosis, cf. Section 4.4. Practical experience with models of modular discrete event systems will be key to gaining insight into the potential computational advantages of MDA over a monolithic approach. This is the object of current research.

Finally, the whole procedure followed in MDA not only exploits the modular structure of the given system, but also may provide insight into causes of non diagnosability and possible remedies for it through coupling of system modules with one another. Thus MDA could be a useful tool in modular system design.

4.4 Online Diagnosis

We can perform *online diagnosis* by simply running the local diagnosers G_{d_i}, $i \in T$, at each local site. When the system is modularly diagnosable, we know from the property of modular diagnosability that even if the local diagnoser G_{d_i} contains an indeterminate cycle, the local observations at site i will never stay forever in this cycle when the complete system G_T is functioning.

REFERENCES

Cassandras, C. G. and S. Lafortune (1999). *Introduction to Discrete Event Systems.* Kluwer Academic Publishers.

Contant, O., S. Lafortune and D. Teneketzis (2004). Diagnosis of distributed discrete event systems: Architecture and modular verification approach. Technical Report CGR 04-04. College of Engineering Control Group Reports, University of Michigan.

Debouk, R., R. Malik and B. Brandin (2002). A modular architecture for diagnosis of discrete event systems. In: *Proc. 41st IEEE Conf. on Decision and Control.* Las Vegas, NV, USA. pp. 417–422.

Fabre, E., A. Benveniste and C. Jard (2002). Distributed diagnosis for large discrete event dynamic systems. In: *Proc. 15th IFAC World Congress.* Barcelona, Spain.

García, E., F. Morant, R. Blasco-Gimnez, A. Correcher and E. Quiles (2002). Centralized modular diagnosis and the phenomenon of coupling. In: *Proc. of the 2002 International Workshop on Discrete Event Systems - WODES'02.* Zaragoza, Spain. pp. 161–168.

Genc, S. and S. Lafortune (2003). Distributed diagnosis of discrete-event systems using Petri nets. In: *Proc. 2003 International Conf. on Application and Theory of Petri Nets.* Eindhoven, The Netherlands. pp. 316–336.

Holloway, L. E. and S. Chand (1994). Time templates for discrete event fault monitoring in manufacturing systems. In: *Proc. 1994 American Control Conference.* Baltimore, MD, USA. pp. 701–706.

Lafortune, S., D. Teneketzis, M. Sampath, R. Sengupta and K. Sinnamohideen (2001). Failure diagnosis of dynamic systems: An approach based on discrete event systems. In: *Proc. 2001 American Control Conference.* Arlington, VA, USA. pp. 2058–2071.

Ricker, L. S. and E. Fabre (2000). On the construction of modular observers and diagnosers for discrete event systems. In: *Proc. 39th IEEE Conf. on Decision and Control.* Sydney, Australia. pp. 2240–2244.

Sampath, M., R. Sengupta, K. Sinnamohideen S. Lafortune and D. Teneketzis (1995). Diagnosability of discrete event systems. *IEEE Trans. Automatic Control* **40**(9), 1555–1575.

Su, R., W. M. Wonham, J. Kurien and X. Koutsoukos (2002). Distributed diagnosis for qualitative systems. In: *Proc. of the 2002 International Workshop on Discrete Event Systems - WODES'02.* Zaragoza, Spain. pp. 169–174.

ELSEVIER
IFAC
PUBLICATIONS
www.elsevier.com/locate/ifac

DISTRIBUTED CONTEXTUAL DIAGNOSIS FOR VERY LARGE SYSTEMS

René K. Boel and George Jiroveanu

SYSTeMS Research Group, Ghent University
Technologiepark 914, Zwijnaarde 9052, Belgium

Abstract: The occurrence of a fault usually influences immediately a (small) part of a large system, while the behavior in the rest of the system deteriorates only after some delay. Exploiting this, we propose in this paper a distributed diagnosis algorithm using a contextual analysis of Petri Nets. Using a backward reasoning technique we derive the minimal context (*what must have happened*) that explains the observed events. The method can tackle Petri Nets models with uncertain initial markings, as is the case when in a distributed architecture, two Petri Nets share some places where the tokens can pass unobservably from one net to another. We show that local diagnosers communicating at fixed points in time lead to scalable diagnosers that diagnose a fault before it can deteriorate the overall plant.

Keywords: diagnosis, Petri Nets, large systems, backward

1. DIAGNOSIS - INTRODUCTION

The goal of diagnosis of a plant is to decide, based on limited observation of the behavior of the plant whether a fault has occurred for sure. Model-based diagnosis of a system combines the observed symptoms and the knowledge of the plant structure in order to generate an *"alarm"* as soon as possible after the occurrence of the fault.

The diagnosis of Discrete Event Systems (DES), with faults represented explicitly by unobservable transitions occurring at unpredictable times (Baroni and Lamperti, 1999; Sampath et al., 1994) is usually performed in two steps. First, the set of all possible traces that match the system description and the observation are derived and then it is checked whether these traces contain faulty transitions.

To avoid the overwhelming computationally complexity of a centralized structure due to the size and complexity of the reachable set of traces, a feasible solution is to design a network of diagnoser-agents (d-agents in short), where each d-agent, based on its local knowledge (the local model and local observation) generates a *preliminary diagnosis* of a sub-system of the plant. The d-agents reconcile their local preliminary diagnostics with each other by exchanging messages, recovering the *conclusions* of a centralized diagnoser (Febre et al., 2003). We consider here a general framework for communication between the d-agents: d-agents communicate only at fixed points in time (e.g. after completion of a working cycle), not necessarily after each observation (Kurine et. al, 2002). Hence, d-agent i must be able to compute a preliminary diagnosis, using only a sequence of locally observed events.

In this paper we consider that each d-agent has a partial knowledge of the system, knowing the Petri Net (PN) model of a sub-system, where the observation (received *via* local sensors) is represented by observable transitions in the PN model. The PN models of sub-system i and j have a common border, consisting of *shared places.* Tokens may pass from one PN model to another *via* the shared places. In this paper we relax the assumption (Genc and Lafortune, 2003) that the *common border* is observable (tokens can enter or leave the shared places *only* by firing locally observable events). We only assume that any path from where a token enters a subsystem $\mathcal{N}_i$ (from a neighbour) to a place where a tokens leaves $\mathcal{N}_i$ has at least one observable event.

Since the border is unobservable, the main difficulty for computing a local preliminary diagnosis, is that a d-agent has no information on the external tokens that have crossed the border (the marking is uncertain (Giua and Seatzu, 2002)). To avoid the consideration of the maximum number of external tokens that may have crossed the border, we propose here an algorithm, based on backward analysis, that derives the *minimal context* of a local observation. By minimal context we mean what *must have happened before* for the observation to be allowable. The minimal context calculates a minimal number of external tokens that is needed for explaining the local observations. The agents agree by communicating whether the tokens could have crossed the border or not.

This paper is organized as follows. Section 2 introduces PN notation used throughout the paper. In Section 3 we introduce the notion of minimal context for (a sequence of) observed events. Section 4 describes centralized CD while in Section 5 the distributed CD is presented. The paper concludes with some final remarks.

2. PETRI NETS

A Petri Net is a structure $\mathcal{N} =< \mathcal{P}, \mathcal{T}, Pre, Post >$ where $\mathcal{P}$ denotes the set of $\sharp\mathcal{P}$ places, $\mathcal{T}$ denotes the set of $\sharp\mathcal{T}$ transitions, and $Pre(p,t) : \mathcal{P} \times \mathcal{T} \to \mathbb{N}$ and $Post(t,p) : \mathcal{T} \times \mathcal{P} \to \mathbb{N}$ are the *pre- and post-incidence function* that specify the arcs. We use the standard notations: $p^\bullet$, ${}^\bullet p$ for the set of input, respectively output transitions of a place; similarly ${}^\bullet t$ and $t^\bullet$ denote the set of input places to t, and the set of output places of t respectively. A *marking* M of a PN is represented by a $\sharp\mathcal{P}$-vector, $M : \mathcal{P} \to \mathbb{N}$, that assigns to each place of $\mathcal{N}$ a non-negative number of tokens.

The set $\mathcal{L}_\mathcal{N}(M_0)$ of all possible traces of PN, $< \mathcal{N}, M_0 >$, with initial marking M_0 is defined as follows. A transition t is *enabled* at the marking M if $M \geq Pre(\cdot,t)$. Firing, an enabled transition t consumes $Pre(p,t)$ tokens in the input places ${}^\bullet t$ and produces $Post(t,p)$ tokens in the output places $t^\bullet$. The next marking is $M' = M + Post(t,\cdot) - Pre(\cdot,t)$. A trace τ in $< \mathcal{N}, M_0 >$ is defined as: $\tau = M_0 \xrightarrow{t_1} M_1 \xrightarrow{t_2} \ldots \xrightarrow{t_k} M_k$, where for $i = 1 \ldots k$, $M_{i-1} \geq Pre(t_i)$. $M \xrightarrow{\tau} M_k$ denotes that the enabled sequence τ may fire at M yielding M_k.

The transition set $\mathcal{T}$ is partitioned into the disjoint subsets of observable $\mathcal{T}_o$, and unobservable $\mathcal{T}_{uo}$ transitions. Denote by $\mathcal{T}^*$ the Kleene closure of the set $\mathcal{T}$ and by ϵ the empty string. Let $s \in \mathcal{L}_\mathcal{N}(M_0) \subseteq \mathcal{T}^*$. The projection $\Pi : \mathcal{L}_\mathcal{N}(M_0) \to \mathcal{T}_o^*$ is defined as: i) $\Pi(\epsilon) = \epsilon$; ii) $\Pi(t) = t\ if\ t \in \mathcal{T}_o$; iii) $\Pi(t) = \epsilon\ if\ t \in \mathcal{T}_{uo}$; iv) $\Pi(st) = \Pi(s)\Pi(t)\ for\ s \in \mathcal{L}_\mathcal{N}(M_0)\ and\ t \in \mathcal{T}$.

For a given PN $\mathcal{N}$, $\mathcal{T}_F$ denotes the set of faulty events, where $\mathcal{T}_F \subset \mathcal{T}_{uo}$. $\mathcal{T}_F$ can be partitioned regarding the kinds of faults that may happen in the process as $\mathcal{T}_F = \mathcal{T}_{F_1} \cup \mathcal{T}_{F_2} \ldots \cup \mathcal{T}_{F_m}$.

Bellow Σ_σ is the set of transitions belonging to a string σ (the alphabet of σ) and $\wp(M_0)$ is the set of all partial initial markings of the initial marking M_0 (i.e. $M \in \wp(M_0)$ iff $\forall p : M(p) \leq M_0(p)$).

3. CONTEXTUAL DIAGNOSIS

3.1 The minimal context of an observed event

The minimal context for explaining an observation is derived in this paper using backward reasoning. Computationally, backward search starts from an observation (t^o), trying to find a minimal subset of the initial marking from where a minimal trace that terminates with t^o is possible. The advantage of backward reasoning is that it avoids computations in a very large search space.

Definition 1. Given $< \mathcal{N}, M_0 >$, let $t^o \in \mathcal{T}_o$ be the first observed event. The *set of minimal contexts* of t^o is: $MinC(M_0, t^o) = \{MinC^1(t^o), \ldots MinC^m(t^o)\}$ with $m =\mid MinC(M_0,t^o) \mid$, $(1 \leq k \leq m)$:

1) $MinC^k(t^o) =< M_0, MinM^k, MinE^k >$
2) $MinM^k \in \wp(M_0)$
3) $MinE^k \in \mathcal{L}_\mathcal{N}(MinM^k)$, $MinE^k = \sigma^k_{uo}t^o$ with $\sigma^k_{uo} \in \mathcal{T}^*_{uo}$
4) $\forall M' \subset MinM^k$ then $MinE^k \notin \mathcal{L}_\mathcal{N}(M')$
5) $\forall \sigma'_{uo}$ obtained from σ^k_{uo} deleting a transition $t \in \Sigma_{\sigma^k_{uo}}$ then $\sigma'_{uo}t^o \notin \mathcal{L}_\mathcal{N}(MinM^k)$

A minimal context $MinC^k(t^o) \in MinC(M_0,t^o)$ specifies the smallest marking $MinM^k$ needed for deriving the minimal explanation $MinE^k$ of the first occurrence of the first observed event t^o.

Due to size limitation of the paper we only present (*Algorithm 1*) the idea behind the computation of the set of minimal contexts (see (Jiroveanu and Boel,2004) for the complete algorithm).

Algorithm 1

1) $M_{int} = Pre(\cdot,t^o)$; $E_{int} = t^o$;
2) *if* $M_{int} \in \wp(M_0)$ then save $< M_{int}, E_{int} >$ as $MinC^k$
3) *else* explain how M_{int} was reached by considering for any $m(p^*) \neq 1$ in M_{int}, that $t^* \in^\bullet p^* \cap \mathcal{T}_{uo}$ has backfired

4) $M_{int} = (M_{int} \uplus Post(t^{*}, \cdot)) \backslash Pre(\cdot, t^{*})$, $E_{int} = t^{*} E_{int}$ and go to 2

Notice that $MinE^k$ contains only those unobservable transitions that *must have happened* before t^o. Hence, for $MinE^k$, we have that $\forall t_{uo} \in \Sigma_{\sigma^k_{uo}}$, $t_{uo} \prec t^o$ (t_{uo} must have happened before t^o) with $\prec$ the precedence relation. Then let $\mathcal{L}_C(t^o)$ denote the set of all minimal traces $MinE^k$ that explain the observed event t^o:
$\mathcal{L}_C(t^o) = \{MinE^k \mid MinE^k = \sigma^k_{uo} t^o, \ \ \sigma^k_{uo} \in \mathcal{T}^*_{uo}\}$

By firing $\mathcal{L}_C(t^o)$, starting from M_0 we obtain the set of estimated markings:

$$\mathcal{M}_C = \left\{ M^k_C \mid \exists MinE^k \in \mathcal{L}_C(t^o) \ s.t. \ M_0 \xrightarrow{MinE^k} M^k_C \right\}$$

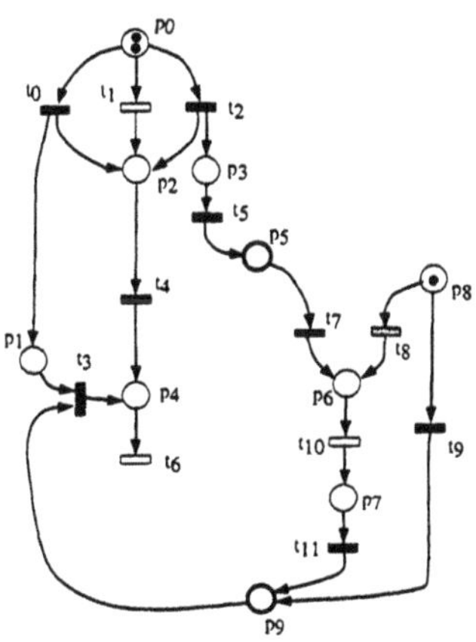

Fig. 1.

If we expand each estimated state $M^k_C \in \mathcal{M}_C$ by firing all the allowable unobservable strings, we obtain the extended language $\mathcal{L}_{C_{ex}}$ and the extended set of estimated states $\mathcal{M}_{C_{ex}}$ as:

$$\mathcal{L}_{C_{ex}}(t^o) = \{MinE^k s^k_{uo} \mid s^k_{uo} \in \mathcal{L}_{\mathcal{N}}(M^k_C) \cap \mathcal{T}^*_{uo}\}$$
$$\text{and } \mathcal{M}_{C_{ex}} = \left\{ M_{C_{ex}} \mid M_0 \xrightarrow{MinE^k s^k_{uo}} M_{C_{ex}} \right\}$$

For the observed transition t^o, by forward search, we derive $\mathcal{L}_S(t^o) = \{E^k \in \mathcal{L}(\mathcal{N}, M_0) \mid \Pi_{\mathcal{T}^o} E^k = t^o\}$ that is the set of strings E^k that explain the observed event t^o, considering in each explanation E^k of t^o unobservable transitions that *could have happened* before, after, or concurrently with t^o. By firing $\mathcal{L}_S(t^o)$ from M_0, we obtain the set $\mathcal{M}_S = \left\{ M^k_S \mid M_0 \xrightarrow{E^k} M^k_S \right\}$ that represents the set of estimated markings, derived by a standard forward search, observing t^o.

Proposition 1. $\mathcal{L}_C(t^o) \subseteq \mathcal{L}_{C_{ex}}(t^o) \subseteq \mathcal{L}_S(t^o)$ and $\mathcal{M}_C \subseteq \mathcal{M}_S = \mathcal{M}_{C_{ex}}$.

Example 1. Consider PN $\mathcal{N}$ in Fig. 1 with the observable transitions t_6, t_{10}. Let the first observed event be t_{10}. We have:

$MinC^{1_1} : MinM^{1_1} = \left\{ M^{1_1}(p_0) = 1 \right\}$, $MinE^{1_1} = t_2 t_5 t_7 \mathbf{t}_{10}$

$MinC^{2_1} : MinM^{2_1} = \left\{ M^{1_1}(p_8) = 1 \right\}$, $MinE^{2_1} = t_8 \mathbf{t}_{10}$

$M^{1_1}_C = \left\{ M^{1_1}(p_0) = 1, M^{1_1}(p_2) = 1, M^{1_1}(p_7) = 1, M^{1_1}(p_8) = 1 \right\}$

$M^{2_1}_C = \left\{ M^{1_1}(p_0) = 2, M^{1_1}(p_8) = 1 \right\}$

3.2 The minimal context of a sequence of observed transitions

Algorithm 1 must now be extended to deal recursively with a sequence of n observed events. ($\tau^o_n = t^o_1 \ldots t^o_n$). The first observed event is *explained* as before, deriving the set of minimal contexts $MinC(M_0, t^o_1)$. The second observed event t^o_2, is explained using as "initial conditions" a state $M^{k_1}_C$ obtained considering each minimal context $MinC^{k_1}(t^o_1) \in MinC(M_0, t^o_1)$ ($1 \leq k_1 \leq m_1$; $m_1 = \mid MinC(M_0, t^o_1) \mid$).

Definition 2. For $< \mathcal{N}, M_0 >$ and an observed sequence τ^o_n, the *set of minimal contexts* denoted by $MinC(M_0, \tau^o_n)$ is defined recursively as follows:

1) $MinC(M_0, \tau^o_n) = \{MinC^1(t^o_n) \ldots MinC^{m_n}(t^o_n)\}$ where $m_n = \mid MinC(M_0, \tau^o_n) \mid$
2) $MinC^{k_n}(t^o_n) =< M^{k_{n-1}}_C, MinM^{k_n}, MinE^{k_n} >$
3) $MinE^{k_n} \in \mathcal{L}_{\mathcal{N}}(M^{k_{n-1}}_C)$, $MinE^{k_n} = \sigma^{k_n}_{uo} t^o_n$

Then for $MinC^{k_n}(t^o_n) \in MinC(M_0, \tau^o_n)$ we have:

$$M_0 \xrightarrow{MinE^{k_1}} M^{k_1}_C \ldots M^{k_{n-1}}_C \xrightarrow{MinE^{k_n}} M^{k_n}_C \quad (1)$$

By computing $MinE^{k_1}, \ldots, MinE^{k_n}$ we avoid the enumeration of all possible traces, obtained by interleaving the concurrent unobservable transitions. Each $\tau^{k_n} = MinE^{k_1} \ldots MinE^{k_n}$ is an arbitrary minimal trace explaining τ^o_n. To avoid confusion, we use a vector notation $\mathbf{MinE}^{k_n}$ where each element $MinE^{k_q} = \sigma^{k_q}_{uo} t^o_q$ ($1 \leq q \leq n$) gives the minimal explanation of the observed event t^o_q (the events that *must have happened* before t^o_q).

$$\mathbf{MinE}^{k_n} = [MinE^{k_1}, \ldots, MinE^{k_n}]$$

Then by firing $\mathbf{MinE}^{k_n}$ from M_0 (see Algorithm 2 bellow) we obtain the minimal state $\mathbf{M}^{k_n}_C$ in a vector form: $M_0 \xrightarrow{\mathbf{MinE}^{k_n}} \mathbf{M}^{k_n}_C$.

Algorithm 2

1) $rank = 0$ for each token considered in M_0
2) fire $MinE^{k_1} \ldots MinE^{k_n}$ from M_0 with the rule:
- if $t \in \mathcal{T}_{uo}$ then make the rank of the newly produced tokens equal to the maximum rank of the tokens that were consumed
- if $t \in \Sigma_{\tau^o_n}$ then make the rank of the newly produced tokens equal the rank (the position) of the observed transition in τ^o_n (e.g. t^o_q produces tokens in $(t^o_q)^{\bullet}$ with rank q)

$\mathbf{M}^{k_n}_C$ is a vector of size $n+1$ (n the number of observed events), whose elements $M^{k_n}_C[q]$ contain tokens that have rank q: $\mathbf{M}^{k_n}_C = [M^{k_n}_C[0], \ldots, M^{k_n}_C[n]]$. In words, it means that the marking $\mathbf{M}^{k_n}_C$ is partitioned, s.t. an element $M^{k_n}_C[q]$ contains tokens that were produced at a time later or equal to the firing time of the observed event t_q^o.

Definition 3. Given the minimal explanation $\mathbf{MinE}^{k_n}$ of τ_n^o, *the concurrent behavior* $\Upsilon^{k_n}(\tau_n^o)$ is:

$$\Upsilon^{k_n}(\tau_n^o) = [\Upsilon^{k_n}[0], \ldots, \Upsilon^{k_n}[n]]$$

where $\Upsilon^{k_n}[q] = \mathcal{L}(\mathcal{N}, \uplus_{h=0}^{h=q} M_C^{k_n}[h]) \cap \mathcal{T}_{uo}^*$ $(0 \leq q \leq n)$ gives the set of unobservable words s_{uo}^q that are allowable from the marking obtained by the union of all the tokens having rank lower or equal than q. It turns out that $\forall s_{uo}^q \in \Upsilon^{k_n}[q]$, s_{uo}^q can happen concurrently with t_{q+1}^o or (obviously) any other observed transition $t_{q+2}^o, \ldots t_n^o$).

Example 2. For the observed sequence $\tau^o = t_{10} t_6$ we have:
$MinC^{1_2}$: $MinM^{1_2} = \{M^{1_2}(p_0) = 2\}$,
$\mathbf{MinE^{1_2}} = [MinE^{1_1}, MinE^{1_2}]$ with $MinE^{1_2} = t_0 t_4 t_6$
$MinC^{2_2}$: $MinM^{2_2} = \{M^{2_2}(p_0) = 2\}$,
$\mathbf{MinE^{2_2}} = [MinE^{1_1}, MinE^{2_2}]$ with $MinE^{2_2} = t_0 t_9 t_3 t_6$
$MinC^{3_2}$: $MinM^{3_2} = \{M^{3_2}(p_0) = 2\}$,
$\mathbf{MinE^{3_2}} = [MinE^{1_1}, MinE^{3_2}]$ with $MinE^{3_2} = t_1 t_4 t_6$
$MinC^{4_2}$: $MinM^{4_2} = \{M^{4_2}(p_0) = 1\}$,
$\mathbf{MinE^{4_2}} = [MinE^{1_1}, MinE^{4_2}]$ with $MinE^{4_2} = t_4 t_6$
$MinC^{5_2}$: $MinM^{5_2} = \{M^{5_2}(p_0) = 1, M^{5_2}(p_8) = 1\}$,
$\mathbf{MinE^{5_2}} = [MinE^{2_1}, MinE^{5_2}]$ with $MinE^{5_2} = t_0 t_4 t_6$
$MinC^{6_2}$: $MinM^{6_2} = \{M^{6_2}(p_0) = 1, M^{6_2}(p_8) = 1\}$,
$\mathbf{MinE^{6_2}} = [MinE^{2_1}, MinE^{6_2}]$ with $MinE^{6_2} = t_0 t_9 t_3 t_6$
$MinC^{7_2}$: $MinM^{3_2} = \{M^{3_2}(p_0) = 1, M^{7_2}(p_8) = 1\}$,
$\mathbf{MinE^{7_2}} = [MinE^{2_1}, MinE^{7_2}]$ with $MinE^{7_2} = t_1 t_4 t_6$
$MinC^{8_2}$: $MinM^{8_2} = \{M^{8_2}(p_0) = 1, M^{8_2}(p_8) = 1\}$,
$\mathbf{MinE^{8_2}} = [MinE^{2_1}, MinE^{8_2}]$ with $MinE^{8_2} = t_4 t_6$

For $MinC^{8_2}$ we have $\mathbf{M_C^{8_2}} = [M_C^{8_2}[0], M_C^{8_2}[1], M_C^{8_2}[2]]$ where: $M_C^{8_2}[0] = \{M(p_0) = 1\}$, $M_C^{8_2}[1] = \{M(p_7) = 1\}$, $M_C^{8_2}[2] = \{M(p_2) = 1\}$.
Then we have:
$\Upsilon^{8_2}[0] = \mathcal{L}_{\mathcal{N}}(M_C^{8_2}[0]) \cap \mathcal{T}_{uo}^*$,
$\Upsilon^{8_2}[0] = \{t_0, t_1, t_2, t_1 t_4, t_0 t_4, t_2 t_4, t_2 t_5, t_2 t_5 t_7\}$ etc.

4. CENTRALIZED CONTEXTUAL DIAGNOSIS (CD)

The contextual-diagnosis of $< \mathcal{N}, M_0 >$, given τ_n^o, means first the computation of $\mathcal{L}_C(\tau_n^o)$, and then CD is obtained as a set of possible contextual-diagnostics by projecting $\mathcal{L}_C(\tau_n^o)$ onto $\mathcal{T}_F$.

Definition 4. The CD of $< \mathcal{N}, M_0 >$, given τ_n^o is a set of possible diagnostics:

$$CD_n = \{CD_n^{1_n} \ldots C_n^{m_n}\}$$

where $CD_n^{k_n} =< CD_{n-1}^{k_{n-1}}, MinC^{k_n}, DR_n^{k_n} >$ $(k_n = 1 \ldots m_n)$. $CD_{n-1}^{k_{n-1}}$ is the previous contextual-diagnostic (after observing t_{n-1}^o), $MinC^{k_n}$ is the minimum context for explaining t_n^o and $DR_{CD_n}^{k_n}$ is the new CD result, obtained by projecting $\mathbf{MinE}^{k_n}$ on to set of faulty transitions $\mathcal{T}_F$: $DR_{CD_n}^{k_n} = \Pi_F()\mathbf{MinE}^{k_n})$ (Fig. 2-left).

Assumption 1. We make the natural assumption that there are no reachable states M^* from M_0 where only faulty transitions can be fired (the faults would then be predictable).

Proposition 2. For $< \mathcal{N}, M_0 >$ under Assumption 1 and given an arbitrary observed sequence τ_n^o, whenever a standard diagnoser (SD) detects that a fault F_i has happened, a contextual diagnoser (CD) does the same. We write that $DR_{SD} \sim_F DR_{CD}$ (Fig. 2) where the standard definitions of $\{N, F, UF\}$ are considered (Sampath et al. 1994).

Proof 1. Contextual-diagnosis does not consider the set of unobservable strings $\Upsilon^{k_n}(\tau_n^o)$ that may be executed concurrently with the observed transitions in τ_n^o considering that $\mathbf{MinE}^{k_n}$ explains τ_n^o. By Assumption 1 the proof comes immediately.

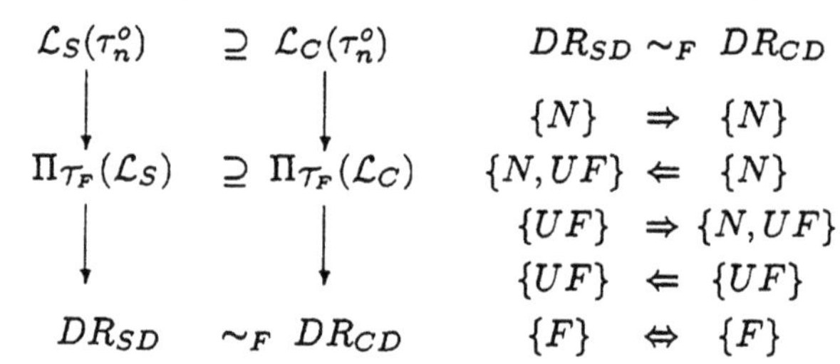

Fig. 2.

Remark 1. The unobservable strings $s_{uo} \in \Upsilon^{k_n}(t_n^o)$ allowable from an estimated state $\mathbf{M}_C^{k_n} \in \mathcal{M}_C(\tau_n^o)$ are *not inferable* form the recorded observation τ_n^o. *Not inferable* means that both assumptions "s_{uo} has happened" and "s_{uo} has not happened" can be true for minimal explanation of τ_n^o given by $\mathbf{MinE}^{k_n}$. It turns out that the consideration of these strings is much more a prognosis problem than a diagnosis one since the diagnostic of a system should provide information about events for which one infers for sure that they happened before, based on available observation.

The *CD evolution graph* is a tree, $\Omega(\mathcal{N}, \tau_n^o) = (CD, \mathcal{E}, \zeta)$, whose set of vertices CD is a multi set $CD = \{CD_0, CD_1, \ldots CD_n\}$, where $CD_q \in CD$ $(1 \leq q \leq n)$ is a set of diagnostics derived after is observed t_q^o in τ_n^o: $CD_q = \{CD_q^{1_q} \ldots CD_k^{m_q}\}$. The set of edges $\mathcal{E}$, is represented by the set of minimal evolutions $MinE_q^{k_q} = \sigma_{uo}^{k_q} t_q^o$ $(1_q \leq k_q \leq m_q)$. The transition function $\zeta : CD \times \mathcal{E} \rightarrow CD$ gives for each diagnoser state $CD_q^{k_q}$, and a feasible minimal explanation $MinE_{q+1}^{k_{q+1}}$, the *next diagnoser state* $CD_{q+1}^{k_{q+1}}$. A trace in *diagnosis evolution graph* is: $CD_0 \xrightarrow{MinE^{k_1}(t_1^o)} \ldots CD_{n-1}^{k_{n-1}} \xrightarrow{MinE^{k_n}(t_n^o)} CD_n^{k_n}$

Example 3. Consider that t_1 and t_8 models two kind of faults F_1 and F_2 respectively. We have:
$CD_{1_1} =< \emptyset, MinC^{1_1}, DR^{1_1} = F_2 >$
$CD_{2_1} =< \emptyset, MinC^{2_1}, DR^{2_1} = N >$
$CD_{1_2} =< CD_{1_1}, MinC^{1_2}, DR^{1_2} = N >$
$CD_{3_2} =< CD_{1_1}, MinC^{3_2}, DR^{3_2} = \{F_1, F_2\} >$
$CD_{5_2} =< CD_{2_1}, MinC^{5_2}, DR^{5_2} = N >$ and so on.

5. DISTRIBUTED CONTEXTUAL DIAGNOSIS (DCD)

We consider the following settings :
i) there are two d-agents Ag_i and Ag_j, s.t. for $\nu = i, j$, Ag_ν *knows* $\mathcal{N}_\nu$ and *records* $\tau^o_{n_\nu} \in \mathcal{T}^*_{o_\nu}$
ii) $\mathcal{N} = \mathcal{N}_i \uplus \mathcal{N}_j$, with $SH_{ij} = \mathcal{P}_i \cap \mathcal{P}_j$; $SH_{ij} = \{p^1_{ij}, \ldots, p^{n_p}_{ij}\}$, and $\mathcal{T}_i \cap \mathcal{T}_j = \emptyset$
iii) $SH_{ij} = IN_i \cup OUT_i = IN_j \cup OUT_j$
iv) $IN_\nu = \{p \in SH_{ij} \mid \exists t \in \mathcal{T}_\nu \ s.t. p \in^\bullet t\}$
v) $OUT_\nu = \{p \in SH_{ij} \mid \exists t \in \mathcal{T}_\nu \ s.t. p \in t^\bullet\}$
vi) communication allowed from time to time

We make the natural assumption that there are no delays in receiving the observation and the locally observed events have time tags attached (according to a global clock) s.t. $\forall t \in \mathcal{T}^i_{uo} \cup \mathcal{T}^j_{uo}$, if $t \prec t^o_{k_i}$ and $t^o_{k_i} \prec_o t^o_{k_j}$ then $t \prec t^o_{k_j}$ (the global order is preserved by $\prec_o$).

Assumption 2. There are no unobservable traces that allow a token to enter $\mathcal{N}_i$ *via* an input place $p' \in IN_i$ and then another token to leave $\mathcal{N}_i$ *via* an exit place $p'' \in OUT_i$ without the execution of one observable event.

Without having any information about the tokens that may have entered $\mathcal{N}_\nu$ ($\nu = i, j$) a local-agent Ag_ν must be able to compute a preliminary diagnostic of a locally observed sequence of events $\tau^o_{n_\nu}$. This can be done by the extension of *Algorithm 1*, the only difference is that instead of finding the minimal explanation in the minimal subset of M_0, we search for the minimal subset of M^i_0 (the marking of places $\mathcal{P}_i \backslash IN_i$) together with the minimal marking of the shared places $IN_i = OUT_j$, denoted $MinIN_i$ (the minimum number of external tokens required).

Let Ag_ν ($\nu = i, j$) have, after processing n_ν observed events, m_{n_ν} ($\mid CD^i_{n_\nu}(M^\nu_0, \tau^o_{n_\nu}) \mid = m_{n_\nu}$) possible diagnostics where

$$CD^{k_{n_\nu}}_{n_\nu} =< CD^{k_{n-1_\nu}}_{n-1_\nu}, MinC^{k_{n_\nu}}, DR^{k_{n_\nu}}_{CD_\nu} > \text{ with}$$
$$MinC^{k_{n_\nu}} =< MinM^{k_{n_\nu}}, MinIN^{k_{n_\nu}}_i, MinE^{k_{n_\nu}} >$$

$MinIN^{k_{q_\nu}}_\nu$ is the marking of the places IN_ν of $\mathcal{N}_\nu$ for explaining the q_ν^{th} event in the locally observed sequence $\tau^o_{n_\nu} = t^o_{1_\nu} \ldots t^o_{q_\nu} \ldots t^o_{n_\nu}$. The upper index k_{q_ν} is a counter in the $(q_\nu)^{th}$ layer of the preliminary CD_{n_ν} of Ag_ν ($CD^{k_{n_\nu}}_{n_\nu}$).

Hence the minimal traffic *via* IN_i places (minimal number of external tokens) needed for having the minimal explanation $\mathbf{MinE}^{k_{n_i}}$ is $\mathbf{MinIN}^{k_{n_i}}_i = [MinIN^{k_{1_i}}_i, \ldots, MinIN^{k_{n_i}}_i]$. In words, for explanation $\mathbf{MinE}^{k_{n_i}}$ to be minimal at least $MinIN^{k_{1_i}}_i$ tokens must have entered (to IN_i) *before* $t^o_{1_i}$, $MinIN^{k_{2_i}}_i$ tokens *before* $t^o_{2_i}$ an so on.

Now it remains to check the consistency of (*to coordinate*) any two pairs $(CD^{w_{n_i}}_{n_i}, CD^{z_{n_j}}_{n_j})$ in the Cartesian product $(m_{n_i} \times m_{n_j})$. For two arbitrary paths in Ω_i and Ω_j we have before communication:

$$M^i_0 \uplus \mathbf{MinIN}^{w_{n_i}}_i \xrightarrow{\mathbf{MinE}^{w_{n_i}}} \mathbf{M}^{w_{n_i}}_C \qquad (1.1)$$
$$M^j_0 \uplus \mathbf{MinIN}^{z_{n_j}}_j \xrightarrow{\mathbf{MinE}^{z_{n_j}}} \mathbf{M}^{z_{n_j}}_C \qquad (1.2)$$

A local preliminary diagnostics $CD^{z_{n_j}}_{n_j}$ of Ag_j can be a feasible diagnostic of $\mathcal{N}_j$ if there is a preliminary diagnostic $CD^{w_{n_i}}_{n_i}$ of Ag_i that can *provide*: $\mathbf{MinIN}^{z_{n_j}}_j = [MinIN^{z_{1_j}}_j, \ldots, MinIN^{z_{n_j}}_j]$. It means that $CD^{z_{n_j}}_{n_j}$ of Ag_j can be a diagnostic only if $MinIN^{z_{1_j}}_j$ would have been the minimal marking of IN_j at a time before $t^o_{1_j}$, $MinIN^{z_{2_j}}$ would have been the minimal marking of IN_j at a time before $t^o_{2_j}$ and so on. Hence the tokens of $MinIN^{z_{1_j}}_j$ must have left $\mathcal{N}_i$ *via* OUT_i before $t^o_{1_j}$, the tokens of $MinIN^{z_{2_j}}_j$ must have left $\mathcal{N}_i$ *via* OUT_i before $t^o_{2_j}$ etc.. Hence $\mathbf{MinIN}^{z_{n_j}}_j$ represents the *image* of the local observation $(\tau^o_{n_j})$ of Ag_j on the border with Ag_i.

Hence beside explaining $\tau^o_{n_i}$ by $\mathbf{MinE}^{w_{n_i}}$, Ag_i must explain also the sequence of virtual transitions $\tau^v_{n_j} = t^v_{1_j} \ldots t^v_{n_j}$ (where $t^v_{k_j}$ "produces" $MinIN^{k_j}_j$ for $\mathcal{N}_j$). Having the marking $\mathbf{M}^{w_{n_i}}_C$, Ag_i calculates the minimal explanations of $\tau^v_{n_j}$ $\mathbf{MinE}^{z_{n_j}}_i(x_{n_j})$ $(x_{n_j} = 1 \ldots max^{z_{n_j}}_{w_{n_i}})$ as unobservable strings in $\Upsilon^{w_{n_i}}(\tau^o_{n_i})$. Then $\mathbf{MinE}^{w_{n_i}}$ is extended by $\mathbf{MinE}^{z_{n_j}}_i(x_{n_j})$ providing a local explanation of both observations $\tau^o_{n_i}$ and $\tau^o_{n_j}$. It means that if $CD^{w_{n_i}}_{n_i}$ is feasible, then $CD^{w_{n_i}}_{n_i}$ can be coordinated with $CD^{z_{n_j}}_{n_j}$ ($CD^{w_{n_i}}_{n_i}$ is not known to be feasible since it requires $\mathbf{MinIN}^{w_{n_i}}$ to be explained by Ag_j).

The coordination function ψ_i (relative to Ag_i) is :

$$\psi_i : CD_{n_i}(M^i_0, \tau^o_{n_i}) \times CD_{n_j}(M^j_0, \tau^o_{n_j}) \rightarrow \{true,\ false\}$$

For $\tau^o = t^o_{1_i} \ldots t^o_{q_i} t^o_{1_j} t^o_{q+1_i} \ldots t^o_{2_j} \ldots t_{n_i}$ where $\tau^o_{n_i} = t^o_{1_i} \ldots t^o_{q_i} t^o_{q+1_i} \ldots t^o_{n_i}$ and $\tau^o_{2_j} = t^o_{1_j} t^o_{2_j}$, the coordination procedure does the following:

Coordination procedure (as implemented by Ag_i)
1) take first element $MinIN^{z_{1_j}}_j$ of $\mathbf{MinIN}^{z_{n_j}}_j$
2) find the last $t^o_{q_i}$ of Ag_i s.t. $t^o_{q_i} \prec_o t^o_{1_j} \prec_o t^o_{q+1_i}$.
3) make a virtual transition $t^v_{1_j}$ and virtual arcs s.t. the tokens given by $MinIN^{z_{n_j}}_{1_j}$ are consumed by firing $t^v_{1_j}$
4) for explaining $MinIN^{z_{1_j}}_{1_j}$ use the first $q + 1_i$ elements

of the vector $\mathbf{M}_C^{w_{n_i}}$ (denoted $\mathbf{M}_C^{w_{n_i}}(q_i)$)

5) compute $MinC(\mathbf{M}_C^{w_{n_i}}(q_i), t^v_{z_{1_j}})$

6) for each $MinC_i^{z_{1_j}}(x_{1_j}) \in MinC(\mathbf{M}_C^{w_{n_i}}(q_i), t^v_{z_{1_j}})$:

- $\mathbf{M}_{n_i}^{wq_i} \xrightarrow{MinE_i^{z_{1_j}}(x_{1_j})} \mathbf{M}_C^{w_{n_i}}(z_{1_j}, x_{1_j})$

7) go to 1 - explain $MinIN_{2_j}^{z_{2_j}}$ for $\mathbf{M}_C^{w_{n_i}}(z_{1_j}, x_{1_j})$.

If Ag_i has found in $\Upsilon^{w_{n_i}}(\tau^o_{n_i})$ $max_{z_{n_j}}^{w_{n_i}}$ explanations ($\mathbf{MinE}^{w_{n_i}}(x_{n_j})$) of $\mathbf{MinIN}_j^{z_{n_j}}$ ($1 \le x_{n_j} \le max_{z_{n_j}}^{w_{n_i}}$) while Ag_j did similarly, finding $max_{w_{n_i}}^{z_{n_j}}$ explanations $\mathbf{MinE}^{z_{n_j}}(y_{n_i})$ ($1 \le y_{n_i} \le max_{w_{n_i}}^{z_{n_j}}$) of $\mathbf{MinIN}^{w_{n_i}}$ in $\Upsilon^{z_{n_j}}(\tau^o_{n_j})$, it results that there where made $max_{z_{n_j}}^{w_{n_i}} \times max_{w_{n_i}}^{z_{n_j}}$ *bridges* between the preliminary diagnosis $CD_{n_i}^{w_{n_i}}$ and $CD_{n_j}^{z_{n_j}}$.

Protocol 1(Ag$_i$ considered)

1) for each $CD_{n_i}^{w_{n_i}} \in CD_{n_i}(M_0^i, \tau_{n_i}^{w_{n_i}})$ *send* $\mathbf{MinIN}_i^{w_{n_i}}$

2) *receive* $\mathbf{MinIN}_j^{z_{n_j}}$ for $CD_{n_j}^{z_{n_j}} \in CD_{n_j}(M_0^j, \tau_{n_j}^{z_{n_j}})$

3) *send* $\psi_i(CD_{n_i}, CD_{n_j})$ || *receive* $\psi_j(CD_{n_j}, CD_{n_i})$

After communication, and *coordination procedure* (1.1) and (1.2) are extended adding $\mathbf{MinE}^{z_{n_j}}(x_{n_j})$ and $\mathbf{MinE}^{w_{n_i}}(y_{n_i})$ respectively

$$M_{0_i} \uplus \mathbf{MinIN}_j^{w_{n_i}} \xrightarrow{\mathbf{MinE}^{w_{n_i}} \oplus \mathbf{MinE}^{z_{n_j}}(x_{n_j})} \mathbf{M}_C^{w_{n_i}}(z_{n_j}, x_{n_j})$$

$$M_{0_j} \uplus \mathbf{MinIN}_i^{z_{n_j}} \xrightarrow{\mathbf{MinE}^{z_{n_j}} \oplus \mathbf{MinE}^{z_{n_j}}(y_{n_i})} \mathbf{M}_C^{z_{n_j}}(w_{n_i}, y_{n_i})$$

$\mathbf{MinE}^{w_{n_i}} \oplus \mathbf{MinE}^{z_{n_j}}(x_{n_j})$ is a vector having $n_i + n_j$ elements, obtained by interleaving the elements of $\mathbf{MinE}^{w_{n_i}}$ and $\mathbf{MinE}^{z_{n_j}}(x_{n_j})$ according to the order of the observed transitions in $\mathcal{N}_i$ and $\mathcal{N}_j$ given by the global order ($\tau^o_{n_i+n_j}$).

Proposition 3. Two local diagnosis $CD_{n_i}^{w_{n_i}} \in CD_{n_i}(M_0^i, \tau^o_{n_i})$ and $CD_{n_j}^{z_{n_j}} \in CD_{n_j}(M_0^j, \tau^o_{n_j})$ are consistent, written $\psi(CD_{n_i}^{w_{n_i}}, CD_{n_j}^{z_{n_j}}) = true$ iff:

$$\psi_i(CD_{n_i}^{w_{n_i}}, CD_{n_j}^{z_{n_j}}) = true \land \psi_j(CD_{n_j}^{z_{n_j}}, CD_{n_i}^{w_{n_i}}) = true$$

Theorem 1. The centralized standard diagnosis result (DR_{SD}) of the overall plant $< \mathcal{N}, M_0 >$, is F-equivalent ($\sim_F$) with the result of the distributed contextual diagnosis (DR_{DCD}) immediately after communicating according to *Protocol1*.

$$DR_{SD} \sim_F DR_{DCD}$$

Proof 2. First we prove that $DR_{DCD} \equiv DR_{CD}$. The proof uses double induction using Assumption 2, considering cases: $\tau^o_{n_i} = \epsilon, \tau^o_{n_j} = t^o_{1_j}$; $\tau^o_{n_i} = \epsilon, \tau^o_{n_j} = t^o_{1_j} \dots t_{q_j}$; $\tau^o_{n_i} = t^o_{1_i}, \tau^o_{n_j} = t^o_{1_j} \dots t_{q_j}$ and $\tau_o = t^o_{1_i} t^o_{1_j} \dots t_{q_j}$; $\tau^o_{n_i} = t^o_{1_i} \dots t_{q_i}, \tau^o_{n_j} = t^o_{1_j} \dots t_{q_j}$ and $\tau_o = t^o_{1_i} \dots t^o_{q_i} t^o_{1_j} \dots t_{q_j}$ etc.. Then Prop. 2.

Example 4. Consider that the PN in Fig. 1 is split in two PNs, the left one $\mathcal{N}_1$ with places $p_0 - p_4$ and the right one $\mathcal{N}_2$ with places $p_6 - p_8$. The common border is made by the shared places p_5 and p_9. Consider that the left agent (Ag_1 that *knows* $\mathcal{N}_1$) receives from the plant the sequence $\tau^1_3 = t_6 t_6 t_6$ while Ag_2 receives from $\mathcal{N}_2$, $\tau^2_1 = t_{10}$.

Ag_1 derives four minimal explanations:

$\mathbf{MinE}_1^{1_3} = [s_1, s_1, s_5]$; $MinIN_1^{1_3} = \{m(p_9) = 2\}$; $DR_1^{1_3} = N$

$\mathbf{MinE}_1^{2_3} = [s_1, s_2, s_5]$; $MinIN_1^{2_3} = \{m(p_9) = 1\}$; $DR_1^{1_3} = N$

$\mathbf{MinE}_1^{3_3} = [s_1, s_5, s_3]$; $MinIN_1^{3_3} = \{m(p_9) = 1\}$; $DR_1^{1_3} = F_1$

$\mathbf{MinE}_1^{4_3} = [s_1, s_5, s_4]$; $MinIN_1^{4_3} = \{m(p_9) = 1\}$; $DR_1^{1_3} = N$

$s_1 = t_0 t_3 t_6$, $s_2 = t_0 t_3 t_6$, $s_3 = t_1 t_4 t_6$, $s_3 = t_2 t_4 t_6$, $s_5 = t_4 t_6$.

Ag_2 derives two minimal explanations:

$MinE_2^{1_1} = t_7 t_{10}$; $MinIN_2^{1_1} = \{m(p_5) = 1\}$; $DR_2^{1_1} = N$

$MinE_2^{1_1} = t_8 t_{10}$; $MinIN_2^{2_1} = \{m(p_5) = 0\}$; $DR_2^{2_1} = F_2$

Considering the global order of the received observation $\tau^o = t_6 t_6 t_6 t_{10}$, the only solution is obtained for the pair ($MinC_1^{4_3}, MinC_2^{1_1}$), by extending $\mathbf{MinE}_1^{4_3}$ with t_5 (since a token is required by Ag_2), and adding t_9 at any time before t_{10} to $MinE_2^{1_1} = t_7 t_{10}$.

6. CONCLUSIONS

In this paper we have introduced the contextual analysis method for diagnosis of large systems modeled as PNs. We have shown that contextual diagnosis (CD) is equivalent with standard diagnosis (SD) with respect to fault detection, and the contextual analysis of PN gives a more compact representation of the set of possible traces, allowing the reduction of the information exchanged in distributed architectures. The method is very suitable when the diagnostic is used for taking actions (e.g. fault isolation), that are initiated only when the diagnostic is certain that a fault has occurred.

Acknowledgment: This research was partially supported by IAP Program initiated by the Belgium State, Prime Minister's Office for Science, Technology and Culture. G. Jiroveanu is supported by BOF - Ghent University.

REFERENCES

P.Baroni and G.Lamperti (1999) Diagnosis of large active systems. *Artificial Intelligence*, 110

M. Sampath, R. Sengupa et.al (1994) Diagnosability of DES - *Report*, The Univ. of Michigan

S. Genc and S. Lafortune (2003) Distributed diagnosis for DES using Petri Nets - *ATPN'03*

E. Febre, A. Benvensite et al. (2003) Distributed diagnosis for Large Discrete Event Dynamic Systems - *ATPN'03*, Eindhoven

J. Kurine, X. Koutsoukos et al. (2002) Distributed diagnosis for qualitative systems *WODES'02*

G. Jiroveanu and R. K. Boel (2004) Contextual Analysis of Petri Nets for distributed applications - *MTNS'04*, Leuven

A. Giua and C. Seatzu (2002) Observability of Place/Transition Nets *IEEE-T on AC*, 47(9)

ELSEVIER
IFAC
PUBLICATIONS
www.elsevier.com/locate/ifac

COMPLEXITY REDUCTION IN STATE OBSERVATION OF STOCHASTIC AUTOMATA

Jan Lunze *

* *Ruhr-University Bochum*
Institute of Automation and Computer Control
44780 Bochum
email: Lunze@esr.rub.de

Abstract: If an automaton has a set of unobservable states, then the observation result cannot be improved as long as the automaton moves within this set. It is, therefore, reasonable to reduce the automaton and, hence, the complexity of the state observation algorithm by replacing the set of unobservable states by a single state. The paper shows that for stochastic automata it is impossible to perform this reduction in such a way that the observation result for the remaining states remains unchanged. Consequently, the state observation algorithm applied to the reduced automaton does not generate the same probability distribution as for the original automaton. The paper introduces the reduced requirement that the observation results should be complete in the sense that all states in which the complete automaton can reside with positive probability is associated with a positive probability by the reduced state observer. A sufficient condition for satisfying this requirement is given.

Keywords: State observation, stochastic automaton, complexity reduction.

1. INTRODUCTION

Discrete-event systems with input v, output w and state z are considered, which are described by a stochastic automaton. A state observer is used to reconstruct the state $z(k_e)$ at the current time k_e from the known sequences

$$V(0...k_e) = (v(0), v(1), ..., v(k_e))$$
$$W(0...k_e) = (w(0), w(1), ..., w(k_e))$$

of input and output symbols (Fig. 1). For a given time horizon k_e, $z(k_e)$ is the last element of the immeasurable state sequence $Z(0...k_e) = (z(0), z(1), ..., z(k_e))$ that the automaton follows. As Fig. 1 shows, the state observer processes the pair $(V(0...k_e), W(0...k_e))$, which is called the input/output (I/O) pair.

State observation problem	
Given:	Automaton
	Input sequence $V(0...k_e)$
	Output sequence $W(0...k_e)$
Find:	Current state $z(k_e)$

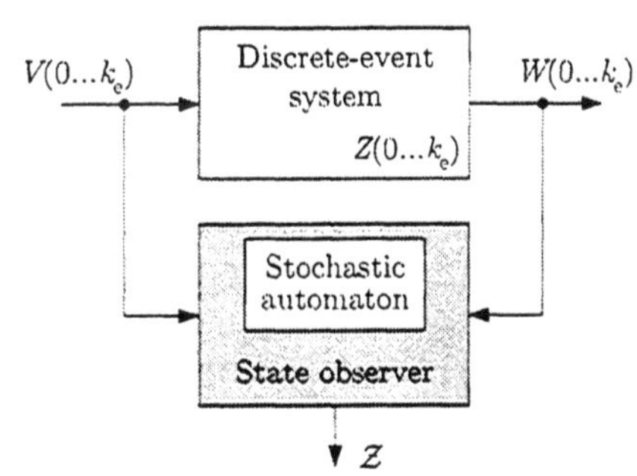

Fig. 1: State observation of stochastic automata

As the stochastic automaton is a nondeterministic system, the observation result is generally not a single state but a set $\mathcal{Z}(k_e)$ that includes the last elements of all the state sequences $Z(0...k_e)$ that the automaton can follow under the input sequence $V(0...k_e)$ while generating the output sequence $W(0...k_e)$. The state observation algorithm described in (Lunze, Schröder 2001) and summarised in Section 3 does not only yield the set $\mathcal{Z}$ but also the a-posteriori state probability distribution

$$\text{Prob}\,(z(k_e) = z \,|\, k_e) :=$$
$$\text{Prob}\,(z(k_e) = z \,|\, V(0...k_e), W(0...k_e))$$

describing the probability that the state at time k_e is given by z provided that the I/O pair $(V(0...k_e), W(0...k_e))$ occurred. Hence, the set $\mathcal{Z}$ satisfies the relation

$$\mathcal{Z}(k_e) = \{z \mid \text{Prob}\,(z(k_e)=z \mid k_e) > 0\}. \quad (1)$$

This paper concerns the question how the complexity of the state observation algorithm can be reduced by simplifying the stochastic automaton that represents the discrete-event system under investigation. The idea is as follows. The observation algorithm processes the current I/O pairs $(v(k_e), w(k_e))$ for increasing time horizon k_e step-by-step. This I/O pair improves the observation result only if the automaton does not move within a set $\bar{\mathcal{N}}_z$ of stochastically unobservable states (cf. Definition 2). Hence, during the movement within the set $\bar{\mathcal{N}}_z$ the observer yields the same result as a simulator (that does not know the system output). That is why the movement within $\bar{\mathcal{N}}_z$ can be ignored during the observation, which means that this state set can be replaced by a single state $\bar{z}$ (Fig. 2). In this way, all sets of unobservable states are removed from the automaton and, consequently, the complexity of the state observation algorithm is reduced.

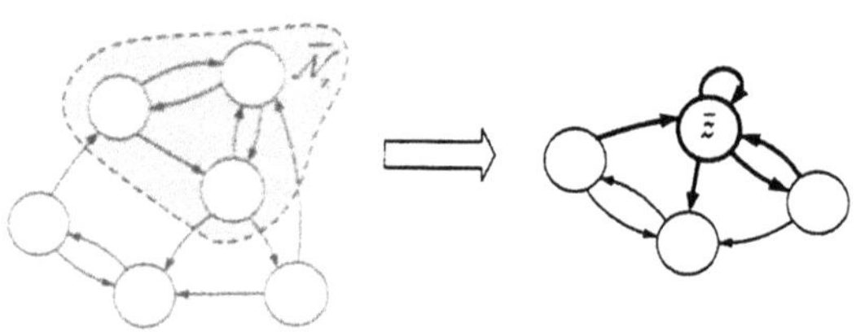

Fig. 2: Idea of model reduction

The main questions to be answered when using this approach are the following:

- Is such a reduction, in general, possible without changing the observation result?
- How can the sets of unobservable states be found?
- How can the state transition probabilities among the "old" automata states and the new state $\bar{z}$ (bold lines in Fig. 2) be determined?

These questions will be answered in this paper.

A similar idea has been described in (Zad et al. 2003) for finite state machines within the context of fault diagnosis. Based on the investigation of the diagnosability of nondeterministic automata the model used for diagnosis has been reduced by lumping states together if the transitions among these states do not provide additional information on the fault condition. It has been shown that this model reduction does not change the diagnostic result. In contrast to this result, the following investigation will show that for stochastic automata such a model reduction is more involved and cannot be applied without changing the observation result.

In Section 2 the stochastic automaton is introduced in the form used here. Section 3 summarises the observation algorithm elaborated in (Lunze, Schröder 2001). The observability of stochastic automata is investigated in Section 4. Then the coincidence of observation results obtained by means of two different stochastic automata is defined in Section 5. The counterexample presented in Section 6 shows that, in general, both observation results do not coincide. Therefore, Section 7 introduces reduced requirements on the model simplification, which should ensure that the observation result remains "complete" in the sense defined there. Finally, Section 8 gives a method for reducing the complexity of the stochastic automaton and, hence, the observation algorithm.

2. STOCHASTIC AUTOMATA

The stochastic automaton

$$\mathcal{S}(\mathcal{N}_z, \mathcal{N}_v, \mathcal{N}_w, L, \text{Prob}\,(z(0)))$$

is described by the state set $\mathcal{N}_z$, the set $\mathcal{N}_v$ of input symbols, the set $\mathcal{N}_w$ of output symbols, the behavioural relation L and the a-priori state probability distribution $\text{Prob}\,(z(0))$. The behavioural relation $L : \mathcal{N}_z \times \mathcal{N}_w \times \mathcal{N}_z \times \mathcal{N}_v \rightarrow [0, 1]$ describes the probability

$$L(z', w \mid z, v) =$$
$$\text{Prob}\,(z(k_e+1)=z', w(k_e)=w \mid z(k_e)=z, v(k_e)=v)$$

that the automaton with state z at time k_e jumps towards the state z' at time $k_e + 1$ and generates the output w provided that it has the input value v. Since it represents a conditional probability distribution, the symbol $L(z', w \mid z, v)$ is used instead of $L(z', w, z, v)$.

3. STATE OBSERVATION ALGORITHM

The state observation starts with the given a-priori state probability distribution

$$\text{Prob}\,(z(0) \mid -1) := \text{Prob}\,(z(0)) \quad (2)$$

and recursively determines, for the current I/O pair $(w(k_e) = \tilde{w},\ v(k_e) = \tilde{v})$, the probability distributions $\text{Prob}\,(z(k_e)=\bar{z} \mid k_e)$ and $\text{Prob}\,(z(k_e+1)=\bar{z} \mid k_e)$ due to the following relations

$$\text{Prob}\,(z(k_e)=\bar{z} \mid k_e) = \quad (3)$$

$$\frac{\sum_{z'} L(z',\tilde{w} \mid \tilde{z},\tilde{v}) \cdot \mathrm{Prob}\,(z(k_e)=\tilde{z} \mid k_e-1)}{\sum_{z,z'} L(z',\tilde{w} \mid z,\tilde{v}) \cdot \mathrm{Prob}\,(z(k_e)=z \mid k_e-1)}$$

$$\mathrm{Prob}\,(z(k_e+1)=\tilde{z}' \mid k_e)= \tag{4}$$

$$\frac{\sum_{z} L(\tilde{z}',\tilde{w} \mid z,\tilde{v}) \cdot \mathrm{Prob}\,(z(k_e)=z \mid k_e-1)}{\sum_{z,z'} L(z',\tilde{w} \mid z,\tilde{v}) \cdot \mathrm{Prob}\,(z(k_e)=z \mid k_e-1)}.$$

provided that the denominator does not vanish (for more details cf. (Lunze, Schröder 2001) and the textbook (Lunze 2003)). This algorithm is applied for increasing time horizon k_e:

Algorithm *State observation of stochastic automata*

Given:	Automaton $\mathcal{S}$
Initialise:	$\mathrm{Prob}\,(z(0) = z \mid -1) :=$ $\mathrm{Prob}\,(z(0) = z)$ for all $z \in \mathcal{N}_z$ $k_e := 0$
Loop:	1. Wait for the next I/O pair $(\tilde{v},\ \tilde{w})$ 2. Determine the probability distributions $\mathrm{Prob}\,(z(k_e)=\tilde{z} \mid k_e)$ for all $\tilde{z} \in \mathcal{N}_z$ by eqn. (3) $\mathrm{Prob}\,(z(k_e+1)=\tilde{z} \mid k_e)$ for all $\tilde{z} \in \mathcal{N}_z$ by eqn. (4) 3. $k_e := k_e + 1$ Continue with step 1
Result:	$\mathrm{Prob}\,(z(k_e)=\tilde{z} \mid k_e)$ for $k_e = 0, 1, \ldots$.

Note that the state observation algorithm can be adapted to a given discrete-event system by simply using the stochastic automaton that describes the given system in the best way. This opens the possibility to reduce the complexity of the state observer by using a stochastic automaton with the smallest possible state set that represents the given system with sufficient precision.

For the comparison of the results obtained by applying the observation algorithm to the original automaton or to a reduced automaton, the former is called the *complete observer* whereas the latter is called the *reduced observer*.

4. OBSERVABILITY OF STOCHASTIC AUTOMATA

The model reduction described later is based on an observability test. The definition of observability introduced in (Lunze, Schröder 2001) is motivated by a comparison of the observation result

$$\mathrm{Prob}\,(z(k_e) \mid V(0\ldots k_e), W(0\ldots k_e))$$

with the result $\mathrm{Prob}\,(z(k_e) \mid V(0\ldots k_e))$ that is obtained by generating the state sequence for given input sequence $V(0...k_e)$ without using the output sequence $W(0...k_e)$ (which is referred to as the simulation result). If the stochastic automaton is unobservable, both results coincide for arbitrary time horizon k_e:

$$\begin{aligned}&\mathrm{Prob}\,(Z(0\ldots k_e) \mid V(0\ldots k_e), W(0\ldots k_e))\\ &= \mathrm{Prob}\,(Z(0\ldots k_e) \mid V(0\ldots k_e)).\end{aligned} \tag{5}$$

This is the case if the behavioural relation L can be decomposed as follows:

Definition 1. (Lunze, Schröder 2001) A stochastic automaton is called **stochastically unobservable** if the behavioural relation L can be represented as product of some functions

$$F: \mathcal{N}_z \times \mathcal{N}_z \times \mathcal{N}_v \to [0,1]$$
$$G: \mathcal{N}_w \times \mathcal{N}_v \to [0,1]$$

such that

$$L(z',w \mid z,v) = F(z' \mid z,v) \cdot G(w \mid v) \tag{6}$$

holds for all $z', z \in \mathcal{N}_z$, $v \in \mathcal{N}_v$ and $w \in \mathcal{N}_w$.

Usually, a given stochastic automaton is not "completely" unobservable in the sense that eqn. (6) is satisfied for all z, z', v and w. However, unobservability occurs as long as the automaton moves within a subset $\bar{\mathcal{N}}_z$ of the state set $\mathcal{N}_z$. Hence, the definition above is applied to this subset $\bar{\mathcal{N}}_z$ as follows:

Definition 2. A stochastic automaton is called **stochastically unobservable within a set** $\bar{\mathcal{N}}_z \subset \mathcal{N}_z$ of at least two states ($|\bar{\mathcal{N}}_z| \geq 2$) if

$$\bar{L}(z',w \mid z,v) = \bar{F}(z' \mid z,v) \cdot \bar{G}(w \mid v) \tag{7}$$

holds for all $z, z' \in \bar{\mathcal{N}}_z$, $v \in \mathcal{N}_v$, and $w \in \mathcal{N}_w$ with

$$\bar{L}(z',w \mid z,v) = \frac{L(z',w \mid z,v)}{\sum_{z' \in \bar{\mathcal{N}}_z} \sum_w L(z',w \mid z,v)} \tag{8}$$

$$\bar{F}(z' \mid z,v) = \sum_w \bar{L}(z',w \mid z,v) \tag{9}$$

$$\bar{G}(w \mid v) = \sum_{z' \in \bar{\mathcal{N}}_z} \bar{L}(z',w \mid z,v). \tag{10}$$

A stochastic automaton is called stochastically observable if there does not exist any set of stochastically unobservable states. There may exist more than one set within which the automaton is stochastically unobservable.

5. COINCIDENCE OF THE OBSERVATION RESULTS

The idea of the complexity reduction of the state observer is to replace sets $\tilde{\mathcal{N}}_z$ of unobservable states by single states $\bar{z}$ and to apply the observation algorithm to the reduced automaton $\hat{\mathcal{S}}(\hat{\mathcal{N}}_z, \mathcal{N}_v, \mathcal{N}_w, \hat{L}, \mathrm{Prob}(\hat{z}(0)))$.

Without loosing generality, the following considerations are restricted to automata with only one unobservable state set $\tilde{\mathcal{N}}_z$. For them, $\hat{\mathcal{N}}_z = (\mathcal{N}_z \backslash \tilde{\mathcal{N}}_z) \cup \{\bar{z}\}$ holds. Note that both automata share the state set $\bar{\mathcal{N}}_z = \mathcal{N}_z \cap \hat{\mathcal{N}}_z$. For all $z \in \bar{\mathcal{N}}_z$ the a-priori state probability is the same for the original and the reduced automaton:

$$\mathrm{Prob}(\hat{z}(0) = z) = \mathrm{Prob}(z(0) = z) \quad (11)$$
$$\text{for all } z \in \bar{\mathcal{N}}_z.$$

Assume that the relation

$$L(z', w \mid z, v) = \hat{L}(z', w \mid z, v) \ \text{ for } z', z \in \bar{\mathcal{N}}_z \quad (12)$$

holds for all $v \in \mathcal{N}_v$ and $w \in \mathcal{N}_w$ whereas for all other pairs of z and z' the value of the behavioural relation $\hat{L}(z', w \mid z, v)$ has to be chosen yet. The rest of the paper deals with this choice.

Clearly, the aim of choosing $\hat{L}$ is to make sure that the observation results obtained by the complete or the reduced observer, respectively, coincide. To define clearly what this means, denote the observation result produced by the reduced observer $\hat{\mathcal{S}}$ by $\hat{\mathrm{Prob}}(z(k_e) = z \mid V(0...k_e), W(0...k_e))$.

Definition 3. The observation results obtained by means of the complete or the reduced observer, respectively, are said to **coincide** if the relation

$$\hat{\mathrm{Prob}}(z(k_e) = z \mid V(0...k_e), W(0...k_e)) \quad (13)$$
$$= \mathrm{Prob}(z(k_e) = z \mid V(0...k_e), W(0...k_e))$$
$$\text{for all } z \in \bar{\mathcal{N}}_z$$

holds for arbitrary time horizon k_e and arbitrary I/O pair $(V(0...k_e), W(0...k_e))$.

6. A COUNTEREXAMPLE

The following example shows that a model reduction without loosing the coincidence of the observation results is, in general, impossible. Hence, the seemingly straightforward idea of the complexity reduction yields a reduction of the precision of the observation result.

Consider the automaton whose graph is shown in Fig. 3 and whose initial state is known to be the state 1. The set $\tilde{\mathcal{N}}_z = \{2, 3\}$ is unobservable according to Definition 2 which can be checked by

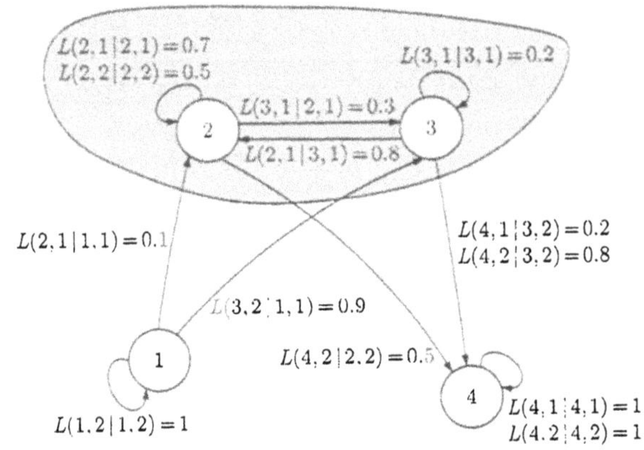

Fig. 3: Example of an incompletely observable stochastic automaton

simple calculations. The set $\tilde{\mathcal{N}}_z$ is replaced in the automaton graph by the state $\bar{z} = 5$ as shown in Fig. 4. Then, $\bar{\mathcal{N}}_z = \{1, 4\}$ holds.

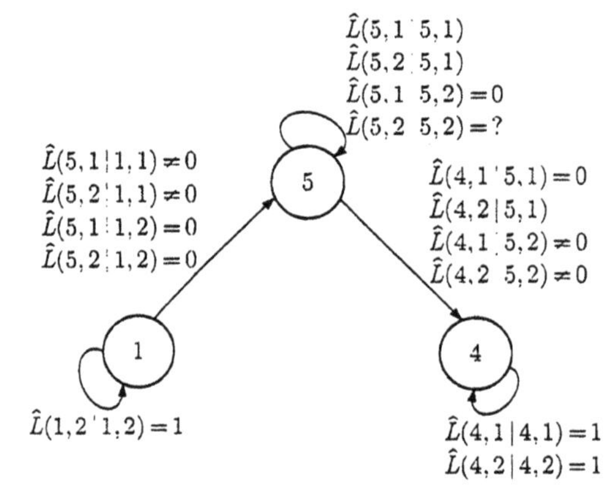

Fig. 4: Simplified model of the automaton shown in Fig. 3

The values of the behavioural relation $\hat{L}$, which belong to the state transitions $1 \to 5$, $5 \to 5$ and $5 \to 4$ should be fixed in such a way that for an arbitrarily given I/O pair (V, W) the observation results coincide and, hence, the relation

$$\hat{\mathrm{Prob}}(z(k_e) = 5 \mid V(0...k_e), W(0...k_e)) =$$
$$\mathrm{Prob}(z(k_e) = 2 \mid V(0...k_e), W(0...k_e)) +$$
$$\mathrm{Prob}(z(k_e) = 3 \mid V(0...k_e), W(0...k_e))$$

is satisfied. The following considerations, which omit the detailed calculations, show that this is impossible. They lead to the values of the behavioural relation $\hat{L}$ given in the figure.

First, use the I/O pair $(V(0...1) = (1,\ 2), W(0...1) = (2,\ 2))$. The complete observer yields

$$\mathrm{Prob}(z(1) = 3 \mid 1) = 1$$
$$\mathrm{Prob}(z(2) = 4 \mid 1) = 1,$$

where all other probabilities for $k_e = 1$ vanish. For the reduced observer the relation

$$\hat{\text{Prob}}\,(z(2)=5\,|\,1) = \frac{\hat{L}(5,\,2\,|\,5,\,2)\cdot \hat{\text{Prob}}\,(z(1)=5\,|\,0)}{(\hat{L}(4,\,2\,|\,5,\,2)\cdot \hat{\text{Prob}}\,(z(1)=5\,|\,0) + \hat{L}(5,\,2|5,\,2)\cdot \hat{\text{Prob}}\,(z(1)=5\,|\,0))}$$

should vanish. This requires to set

$$\hat{L}(5,\,2\,|\,5,\,2) = 0. \tag{14}$$

Second, consider $(V(0...2) = (1,\,2), W(0...2) = (1,\,2))$. The complete observer yields

$$\text{Prob}\,(z(2)=2\,|\,1) = 0.5$$
$$\text{Prob}\,(z(2)=4\,|\,1) = 0.5,$$

and the reduced observer

$$\hat{\text{Prob}}\,(z(2)=5\,|\,1) = \frac{\hat{L}(5,\,2\,|\,5,\,2)\cdot \hat{\text{Prob}}\,(z(1)=5\,|\,0)}{(\hat{L}(4,\,2\,|\,5,\,2)\cdot \hat{\text{Prob}}\,(z(1)=5\,|\,0) + \hat{L}(5,\,2\,|\,5,\,2)\cdot \hat{\text{Prob}}\,(z(1)=5\,|\,0))},$$

which have to be equal to 0.5. Hence,

$$\hat{L}(5,\,2\,|\,5,\,2) \neq 0 \tag{15}$$

has to be valid. Equations (14), (15) contradict. Consequently, the reduced automaton cannot be chosen in such a way that the observation results obtained by the complete and the reduced observers coincide.

Corollary 1. The observation result of the reduced observer does, in general, not coincide with the observation result of the complete observer.

7. COMPLETENESS OF THE OBSERVATION RESULT

The counterexample shows that a complexity reduction is impossible without loosing the precision of the observation result. This is true although the set $\tilde{\mathcal{N}}_z$ of states, which is replaced by a single state $\bar{z}$ during the model reduction step is unobservable and, hence, the movement of the automaton through this state set does not generate information that improves the observation result.

Hence, the question arises what aims should be followed by the model reduction. This paper proposes to reduce the model in such a way that the completeness of the observation result is retained. Completeness means that both observers yield the same set $\mathcal{Z}(k_e)$ of states in which the automaton is known to be at time k_e with positive probability. For the complete observer, this set is given by eqn. (1) and for the reduced observer by

$$\hat{\mathcal{Z}}(k_e) = \{z\,|\,\hat{\text{Prob}}\,(z(k_e)=z\,|\,k_e) > 0\}.$$

Definition 4. The result obtained by the reduced observer is said to be **complete** if the relation

$$\hat{\mathcal{Z}}(k_e) \supseteq \mathcal{Z}(k_e) \cap \tilde{\mathcal{N}}_z \tag{16}$$

holds for arbitrary time horizon k_e and arbitrary I/O pairs (V, W).

That is, the reduced observer should associate all those states of the common state set $\tilde{\mathcal{N}}_z$ with a positive probability, that are associated with positive probability by the complete observer.

8. MODEL REDUCTION FOR STATE OBSERVATION

The following theorem states requirements on the behavioural relation $\hat{L}$ of the reduced automaton such that the observation result is complete.

Theorem 1. The observation result obtained by the reduced observer is complete if the behavioural relation $\hat{L}$ of the reduced automaton satisfies the following requirements for all $z, z' \in \tilde{\mathcal{N}}_z$ $v \in \tilde{\mathcal{N}}_v$, $w \in \mathcal{N}_w$:

$$\begin{aligned} L(\bar{z}, w\,|\,z, v) > 0 &\Rightarrow \hat{L}(\bar{z}, w\,|\,z, v) > 0 \\ L(z', w\,|\,\bar{z}, v) > 0 &\Rightarrow \hat{L}(z', w\,|\,\bar{z}, v) > 0 \\ \hat{L}(\bar{z}, w\,|\,\bar{z}, v) > 0 & \end{aligned} \tag{17}$$

Proof. Due to (11), for the initialisation of the complete and the reduced observers the relation

$$\text{Prob}\,(z(0)=z\,|\,-1) = \hat{\text{Prob}}\,(z(0)=z\,|\,-1) \;\text{ for all }\; z \in \tilde{\mathcal{N}}_z \tag{18}$$

holds, where as before the state probability of the reduced observer is marked with a hat. Consider $k_e = 0$. If for the complete observer eqn. (3) yields, for given $v(0) = \tilde{v}$ and $w(0) = \tilde{w}$, a positive probability for some $\tilde{z} \in \tilde{\mathcal{N}}_z$

$$\text{Prob}\,(z(0)=\tilde{z}\,|\,0) = \frac{\sum_{z'} L(z', \tilde{w}\,|\,\tilde{z}, \tilde{v})\cdot \text{Prob}\,(z(0)=\tilde{z}\,|\,-1)}{\sum_{z,z'} L(z', \tilde{w}\,|\,z, \tilde{v})\cdot \text{Prob}\,(z(0)=z\,|\,-1)} > 0$$

then the same holds true for the reduced observer

$$\hat{\text{Prob}}\,(z(0)=\tilde{z}\,|\,0) = \frac{\sum_{z'} \hat{L}(z', \tilde{w}\,|\,\tilde{z}, \tilde{v})\cdot \hat{\text{Prob}}\,(z(0)=\tilde{z}\,|\,-1)}{\sum_{z,z'} \hat{L}(z', \tilde{w}\,|\,z, \tilde{v})\cdot \hat{\text{Prob}}\,(z(0)=z\,|\,-1)} > 0$$

because there is at least one pair z, z' for which the product $L(z', \tilde{w}\,|\,z, \tilde{v})\cdot \text{Prob}\,(z(0)=z\,|\,-1)$

is positive. Due to eqns. (17) and (18) then the product $\hat{L}(z', \tilde{w} \mid z, \tilde{v}) \cdot \hat{\text{Prob}}(z(0) = z \mid -1)$ is positive as well. Similarly, if eqn. (4) yields a positive probability

$$\text{Prob}(z(1) = \bar{z}' \mid 0) = \frac{\sum_z L(\bar{z}', \tilde{w} \mid z, \tilde{v}) \cdot \text{Prob}(z(0) = z \mid -1)}{\sum_{z,z'} L(z', \tilde{w} \mid z, \tilde{v}) \cdot \text{Prob}(z(0) = z \mid -1)} > 0$$

then

$$\hat{\text{Prob}}(z(1) = \bar{z}' \mid 0) = \frac{\sum_z \hat{L}(\bar{z}', \tilde{w} \mid z, \tilde{v}) \cdot \hat{\text{Prob}}(z(0) = z \mid -1)}{\sum_{z,z'} \hat{L}(z', \tilde{w} \mid z, \tilde{v}) \cdot \hat{\text{Prob}}(z(0) = z \mid -1)} > 0$$

holds true. Hence, the completeness condition (16) is satisfied for $k_e = 0$. This argument can be applied for increasing time horizon k_e, which proves the theorem. □

The question that remains to be answered is how to choose the behavioural relation $\hat{L}(z', w \mid z, v)$ for $z = \bar{z}$ and/or $z' = \bar{z}$. Different proposals can be made and their usefulness depends on the movement of the automaton through the set $\bar{\mathcal{N}}_z$ of unobservable states, which in turn depends upon the input sequence V.

If the state probability distribution tends to become approximately uniform after some time steps, the following proposal for selecting $\hat{L}$ is reasonable. Assume that for the current I/O pair $(\tilde{v}, \tilde{w})$ the automaton goes with positive probability to some state $z' \in \bar{\mathcal{N}}_z$. Then the Chapman-Kolmogorov equation leads for the time k_e considered

$$\text{Prob}(z(k_e) = z') = \sum_{z \in \bar{\mathcal{N}}_z} L(z', \tilde{w} \mid z, \tilde{v}) \cdot \frac{1}{\bar{N}},$$

because $\text{Prob}(z(k_e - 1) = z) = \frac{1}{\bar{N}}$ holds with $\bar{N}$ denoting the number of states in the unobservable set $\bar{\mathcal{N}}_z$. For the reduced automaton

$$\hat{\text{Prob}}(z(k_e) = z') = \hat{L}(z', \tilde{w} \mid \bar{z}, \tilde{v})$$

holds because the automaton is assumed in the current considerations to be in the state $\bar{z}$ with probability one. Hence, $\hat{L}$ should be chosen such that

$$\hat{L}(z', \tilde{w} \mid \bar{z}, \tilde{v}) = \sum_{z \in \bar{\mathcal{N}}_z} L(z', \tilde{w} \mid z, \tilde{v}) \cdot \frac{1}{\bar{N}} \quad (19)$$

holds. Finally,

$$\hat{L}(\bar{z}, \tilde{w} \mid \bar{z}, \tilde{v}) = 1 - \sum_{z' \in \bar{\mathcal{N}}_z} \hat{L}(z', \tilde{w} \mid \bar{z}, \tilde{v}) \quad (20)$$

has to hold for all $\tilde{v} \in \mathcal{N}_v$ and $\tilde{w} \in \mathcal{N}_w$. The reduced automaton is completely described by eqns. (11), (12), (19) and (20).

9. CONCLUSIONS

The paper has considered the complexity reduction of the state observation problem for stochastic automata by replacing sets of unobservable states by single states and, hence, reducing the automaton used in the observation algorithm. An important result is that such a complexity reduction introduces, in general, a bias of the observation result. Hence, it has been proposed to require the completeness of the observation result obtained by means of the reduced model. The condition given in Theorem 1 ensures this property. Furthermore, a proposal to choose the behavioural relation of the reduced automaton for the state changes between the remaining states of the original automaton and the new state representing the set of unobservable states has been made.

The results can be directly extended to fault diagnosis, because the diagnostic algorithm developed in (Lunze, Schröder 2001) is based on the state observation algorithm investigated here. For the diagnosis of quantised systems (cf. (Blanke et al. 2003), (Schröder 2002)) the model reduction has direct consequences on the choice of the state quantiser. If the stochastic automaton obtained from abstraction has for a given quantised system unobservable state sets, then the corresponding state partitions generated by the quantiser can be lumped together and, hence, the quantiser and the automaton be simplified.

Acknowledgement. Thanks are due to Christian Falkenberg for his valuable comments on the counterexample.

10. REFERENCES

[1] Blanke, M.; M. Kinnaert, J. Lunze, M. Staroswiecki: *Diagnosis and Fault-Tolerant Control.* Springer-Verlag, Heidelberg 2003.

[2] Lunze, J.: *Automatisierungstechnik.* Oldenbourg-Verlag, München 2003.

[3] Lunze, J. and J. Schröder: State observation and diagnosis of discrete-event systems described by stochastic automata. *Discrete Event Dynamic Systems: Theory and Applications,* **11**, 2001, 319-369.

[4] Schröder, J.: *Modelling, State Observation and Diagnosis of Quantised Systems.* Springer-Verlag, Berlin 2002.

[5] Zad, S. H., R. H. Kwong, W.M. Wonham: Fault diagnosis in discrete-event systems: framework and model reduction, *IEEE Trans. Autom. Control* **AC-48**, 1199-1212, 2003.

ELSEVIER
IFAC
PUBLICATIONS
www.elsevier.com/locate/ifac

NEW COLOURED REDUCTIONS FOR SOFTWARE VALIDATION

Sami Evangelista * Serge Haddad **
Jean-François Pradat-Peyre *

* *CEDRIC-CNAM Paris 292, rue St Martin, 75003 Paris*
** *LAMSADE-CNRS UMR 7024 Université Paris 9*
Place de Lattre de Tassigny 75775 Paris Cedex 16

Abstract: Structural model abstraction is a powerful technique for reducing the complexity of a state based enumeration analysis. We present in this paper accurate reductions for high-level Petri nets based on new ordinary Petri nets reductions. These reductions involve only structural and algebraical conditions. They preserve the liveness of the net and any LTL formula that does not observe the reduced transitions of the net. The mixed use of structural and algebraical conditions significantly enlarges their application area. Furthermore the specification of the transformation is parametric with respect to the cardinalities of coloured domains.

Keywords: Software Validation, Reductions, High-level Petri nets

1. INTRODUCTION

The use of formal methods in software design may be decomposed in two steps: a modelling stage which must lead to a model as close as possible to the analysed software and a verification stage involving properties expression and model checking via adequate algorithms.

Two kinds of verification techniques can be used. The state enumeration based methods lead to a complete verification but the analysis is restricted by the combinatory explosion factor. The structural methods are generally efficient but they do not ensure the complete correctness of the modelled system.

Thus an attractive trade-off would be to first perform structural abstractions in order to obtain a simplified model on which an enumeration based method can more easily be applied. The model may be abstracted in two ways: data abstraction and operation abstraction. Here we will focus on the latter one which merges consecutive instructions into a virtual atomic one. Such a transformation drastically reduces the combinatory explosion due to the elimination of the intermediate states. In the context of (high-level) Petri nets this abstraction is called a net reduction. A reduction is characterised by some application conditions, a transformation rule and the properties for which the initial and the reduced models are equivalent. In order to obtain reductions with a broad range of applications while preserving a large set of properties, we base our coloured reductions on new efficient ordinary Petri nets reductions (see Evangelista et al. (2004a)) and we use the following approach to extend them to coloured models. We characterise some properties of the coloured functions labelling an arc which ensure that the unfolding of this arc will be appropriate for the conditions involved in ordinary reductions. We exhibit coloured flows which lead to the satisfaction of the algebraic conditions of ordinary reductions. We show how the use of composition, inverse and transpose of mappings enables us to handle the transformation of the labelling of arcs in the reduced net. Given a subclass of the Well-formed nets, Chiola et al. (1990), we specify reductions at a syntactic level in order to efficiently check the conditions and apply the transformations. We will not describe this part which can be found in Evangelista et al. (2004b). Compared to previous works concerning high-level nets reductions, Colom et al. (1986), Genrich (1990), Haddad (1990), our new coloured reductions lie on accurate application conditions (since they are based on efficient ordinary Petri nets reductions) and then permit to reduce more realistic models. Moreover, this analysis does not need to fix a value for the parameters of the model (which is not the case for methods that reduce the reachability graph) and can be followed by any other analysis method.

The paper is organised as follows. In the next section, we recall the basics of coloured Petri nets with a focus on the coloured functions. In the third section, we first demonstrate that existing reductions do not cover typical patterns of concurrent programming. Then we show how the analysis of coloured functions and coloured invariants helps to accurately characterise behavioural conditions on the net. At last, we formally develop the post-agglomeration. In the fourth section, an example illustrates the power of these new reductions.

2. DEFINITIONS AND NOTATIONS

Coloured Petri nets handle tokens that are typed (or coloured) upon non empty finite sets called colour domains; a marking is then a multi-set over a colour domain and we denote $Bag(C)$ the set of multi-sets over C (the related definitions can be found in the appendix).

DEFINITION *2.1.* A coloured net is a 5-tuple $CN = \langle P, T, C, W^+, W^- \rangle$ with:

- P a non empty and finite set of places;
- T a non empty and finite set of transitions (disjoint of P);
- C is the colour mapping from $P \bigcup T$ to ω where ω is a set of finite and non empty sets. An item of $C(s)$ is called a colour of s and $C(s)$ denotes the colour domain of s.
- W^+ (resp. W^-) is the post (resp. pre) incidence mapping that associates to each place p and each transition t a colour mapping form $C(t)$ to $Bag(C(p))$. We note $W = W^+ - W^-$.

We note $\epsilon = \{\bullet\}$ the domain reduced to the single value $\bullet$ (the neutral token); so, ordinary Petri nets can be viewed as particular coloured Petri nets (the unique and common colour domain is ϵ).

DEFINITION *2.2.* A marking is a mapping that associates to each place p a value in $Bag(C(p))$. We note m_0 the initial marking of a net. A transition t is fireable for an instance $c_t \in C(t)$ from a marking m (denoted by $m[t, c_t\rangle$) if

$$\forall p \in P, m(p) \geq W^-(p,t)(c_t)$$

The firing of t, c_t from m leads to the marking m' ($m[t, c_t\rangle m'$) defined by $\forall p \in P, m'(p) = m(p) + W(p,t)(c_t)$. A marking m' is reachable from a marking m if there exists a sequence $t_1, c_1, \ldots, t_k, c_k$ such that $m[t_1, c_1\rangle m_1$, $m_1[t_2, c_2\rangle m_2$, ..., $m_{k-1}[t_k, c_k\rangle m'$. We denote by $Reach(CN, m_0)$ the set of all reachable markings from m_0. As usual, an infinite sequence is a firing sequence iff all its finite prefixes are firing sequences.

To each coloured net corresponds a unique Petri net which is called the *underlying* Petri net. This net is composed by the set of places, $p[c_p]$ where $p \in P$ and $c_p \in C(p)$ and the set of transitions $t[c_t]$, $t \in T$, $c_t \in C(t)$. The pre and the post conditions are defined by the instantiation of colour function. This unfolded net is defined in the appendix.

We now introduce the coloured flows and invariants. These invariants can be used to characterise specific behaviours like, for instance, mutual exclusion. In order to obtain a sound definition of flows, we extend by linearity a function from C to $Bag(D)$ to a function from $Bag(C)$ to $Bag(D)$.

DEFINITION *2.3.* A flow $\mathcal{F}$, on a domain $C(\mathcal{F})$, is a vector over P, noted as the formal sum $\mathcal{F} = \sum_{p \in P} \lambda_p.\mathcal{F}_p.p$, where $\forall p \in P$, $\lambda_p \in \mathbb{Z}$ and $\mathcal{F}_p$ a mapping from $Bag(C(p))$ to $Bag(C(\mathcal{F}))$ such that: $\forall t \in T, \sum_{p \in P} \lambda_p.\mathcal{F}_p \circ W(p,t) = 0$ [1].
$\mathcal{F}$ induces the invariant:
$\forall m \in Reach(CN, m_0)$,
$\sum_{p \in P} \lambda_p.\mathcal{F}_p(m(p)) = \sum_{p \in P} \lambda_p.\mathcal{F}_p(m_0(p))$
An invariant $\mathcal{F}$ is **positive** if $\forall p \in P$, $\lambda_p \geq 0$. It is **binary** if $\forall c \in C(\mathcal{F})$, $\sum_{p \in P} \lambda_p.\mathcal{F}_p(m_0(p))(c) = 1$. It is a **synchronisation** invariant if $\forall c \in C(\mathcal{F})$, $\sum_{p \in P} \lambda_p.\mathcal{F}_p(m_0(p))(c) = 0$.

When no confusion is possible (i.e. the initial marking is given), we will not distinguish the flow and its corresponding invariant. We want to analyse the structure of the underlying Petri net using the structure and the functions of the coloured Petri net. This requires to characterise and manipulate coloured functions. The following definition and notations are enough for our purposes.

DEFINITION *2.4.* Let f be a mapping from $Bag(C)$ to $Bag(C')$.

- ${}^t(f)$ is the mapping defined from $Bag(C')$ to $Bag(C)$ by ${}^t(f)(c')(c) = f(c)(c')$

[1] 0 denotes here the null mapping from $C(t)$ to $Bag(C(\mathcal{F}))$

- $\overline{f}$ is defined from $\mathcal{P}(C)$ to $\mathcal{P}(C')$ by $\overline{f}(D) = \{c' \in C' \mid \exists d \in D, f(d)(c') \neq 0\}$ where $\mathcal{P}(C)$ denotes the power set of C. Note that the linearity is preserved by this transformation (substituting $\cup$ to $+$) and that $\overline{f}$ may be viewed as a function from C to $\mathcal{P}(C')$.

DEFINITION *2.5.* Let f and g be two linear mappings from $\mathcal{P}(C)$ to $\mathcal{P}(C')$. We note $f \sqsubseteq g$ if $\forall c \in Bag(C), f(c) \subseteq g(c)$.

Below we list particular mappings. An **orthonormal** mapping is a colour domain permutation, a **unitary** mapping produces at most one token per colour, a **projection** is a canonic mapping from $Bag(C \times D)$ to $Bag(C)$, an **ortho-projection** is the composition of an orthonormal mapping with a projection. f is a **quasi-one to one** mapping if $\forall c \neq d\ \overline{f}(c) \cap \overline{f}(d) = \emptyset$. f is a **quasi-onto** mapping from $Bag(C)$ to $Bag(D)$ if $\overline{f}(C) = D$. The complete definitions are given in the appendix.

3. COLOURED AGGLOMERATIONS

We suppose in the sequel that the set of transitions of the net is partitioned as: $T = T_0 \uplus H \uplus F$. The underlying idea of this decomposition is that the couple (H, F) defines transitions sets that are causally dependent: an occurrence of $f \in F$ in a firing sequence may always be related to a previous occurrence of some $h \in H$ in this sequence.

We have extended two kinds of agglomerations: the pre and the post agglomeration. Informally stated, in the **pre-agglomeration** scheme, firing a transition $h \in H$ is only useful for firing any transition of $f \in F$. Thus in the reduced net, the firings of h are postponed until the corresponding firing of f. In the **post-agglomeration** scheme, the firing of any transition $f \in F$ is mainly conditioned by the firing of the transitions of H. Thus, in the reduced net, one fires f immediately after the firing of some $h \in H$.

3.1 An introducing example

In the following coloured net (see Fig.1), the transition h models the update of a variable modelled by the place $V1$: the value X is replaced by the value $G1(X)$ where $G1$ models a mapping from C to C. Initially, this variable has the value $x0$. Similarly, the transition f models the update of a second variable modelled by the place $V2$. Generally, this model does not have the same behaviour as the one of the model depicted in Fig.2 where the two updates are performed simultaneously.

However, there exist many cases for which these two models are equivalent. In particular, as soon

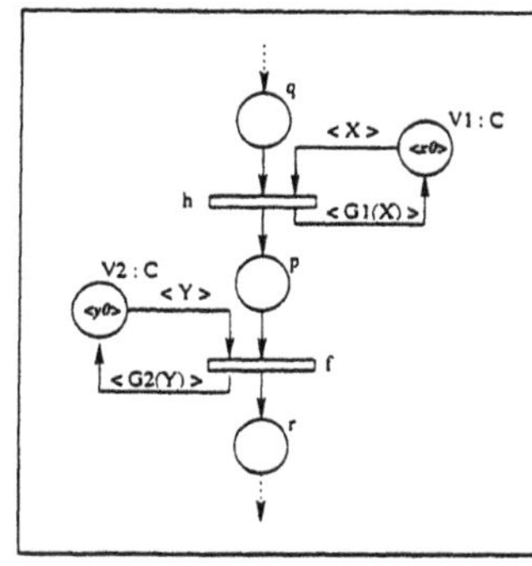

Fig. 1. Updating variables sequentially

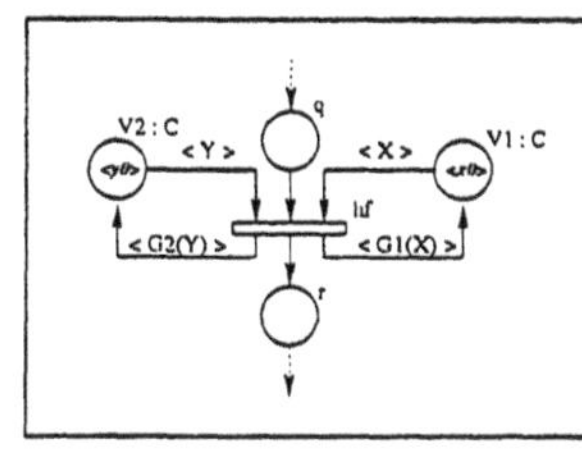

Fig. 2. Updating variables atomically

as we can prove that either the value of $V1$ does not change when q or p are marked or the value of $V2$ does not change when p is marked, the two models share a large set of properties.

Indeed let us suppose that, in the first model, the value of $V1$ does not change when q is marked. The variable $V1$ cannot be modified when p is marked. So, we can delay the update of the variable $V1$ until we are ready to perform the update of the variable $V2$ without modifying the properties of the model. This corresponds to the scheme of the pre-agglomeration: h can be delayed until f is fireable. In the second case, updating $V2$ after having waited in state p or updating $V2$ just after having updated $V1$ is equivalent since value of $V2$ cannot change when p is marked. This corresponds to the scheme of the post-agglomeration: f is fireable as soon as h is fired.

Nevertheless, whereas these behaviours correspond to the scheme of the pre or of the post agglomeration, none of the previously defined reductions cover such behaviours. The present work is based on reductions for ordinary Petri nets that we proposed in Evangelista et al. (2004a). Such reductions cover a large range of patterns by introducing algebraical conditions whereas the previously defined ones solely lie on structural conditions. However the extension of the conditions and the transformation of these reductions to high-level nets require careful analysis of the coloured functions labelling the arcs of the net. Due to the lack of space, we focus in this paper on the post-agglomeration; complete results can be found in Evangelista et al. (2004b).

3.2 Exploiting coloured functions and invariants

The structure of a coloured net does not necessarily reflect the structure of the underlying Petri net since we have to take into account colour mappings. Especially, we need to follow colour transformation using composition or transposition of colour mappings. Let us consider the following coloured Petri net and suppose that, given a

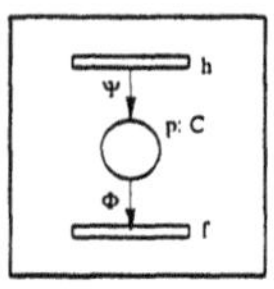

Fig. 3. Colour mapping manipulation illustration

colour $c_f \in C(f)$, we want to compute the colours $\{c_h \in C(h)\}$ such that the firing of h for a colour c_h may help the firing of f for the instance c_f (by producing useful tokens in place p). We have to start from c_f and to find the instances of p that are linked to $f[c_f]$. By definition, this set is $\overline{\Phi}(c_f)$. Then we have to find instances of h that are linked to a place $p[c_p]$, $c_p \in \overline{\Phi}(c_f)$. These instances are the set $\{c_h \in C(h), \overline{\Psi}(c_h) \cap \overline{\Phi}(c_f) \neq \emptyset\}$. By definition of the transposition of a function, this set is $\overline{{}^t\Psi}(\overline{\Phi}(c_f))$. Thus, the set of colours we look for is $(\overline{{}^t\Psi} \circ \overline{\Phi})(c_f)$. In an opposite way, the set of instances of f that are causally dependent of an instance c_h of h, are $(\overline{\Psi} \circ \overline{{}^t\Phi})(c_h)$.

Let us consider now the following coloured Petri net where p is an ordinary place (see Fig.4).

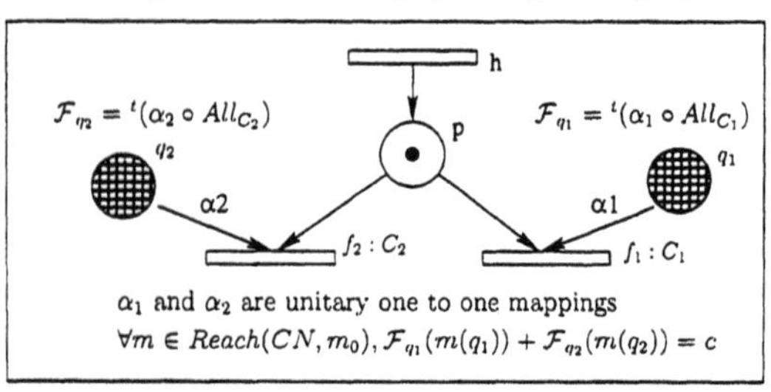

Fig. 4. An invariant controlling f_1 and f_2

Let us prove that there is always an instance of f_1 or of f_2 that is fireable when p is marked.

- The interpretation of the invariant $\mathcal{F}$ is the following one: there is at least one token either in the place q_1 or in q_2 whose colour is either in the set $\overline{\alpha_1}(\overline{All_{C_1}}(\bullet))$ or in $\overline{\alpha_2}(\overline{All_{C_2}}(\bullet))$.
- Since ${}^t\alpha_i\ _{(i=1,2)}$ is an unitary quasi-one to one mapping, each firing instance (f_i, c_i) requires, when $\alpha_i(c_i) \neq 0$, in addition to the token in p, exactly a token in the place q_i which colour is in the singleton $\overline{\alpha_i(c_i)}$.
- Combining these two facts, an instance (f_i, c_i) for some i is always fireable when p is marked.

Remark that this reasoning is still valid if we only require that $\forall i, \overline{\mathcal{F}_{q_i}} \sqsubseteq {}^t(\overline{\alpha_i} \circ \overline{All_{C_i}})$.

3.3 Post-agglomeration hypotheses

We present the four conditions of the post-agglomeration: the **potentially post-agglomerability**, the **HF-interchangeability**, the **F-independence** and the **F-continuation**.

The **potentially post-agglomerability** ensures that in any fireable sequence the number of occurrences of H is greater or equal than the number of occurrences of F.

DEFINITION *3.1.* (Hypothesis R1). A coloured net is potentially post-agglomerable (p-post-agglomerable) if $\exists\ H \subset T$, $F \subset T$, $p \in P$ such that

(1) ${}^\bullet p = H$, $p^\bullet = F$ and $m_0(p) = 0$
(2) $\forall f \in F$, $C(f) = C(p) \times C_f$ and $W^-(p, f)$ is an ortho-projection from $C(p) \times C_f$ to $C(p)$;
(3) $\forall h \in H$ $W^+(p, h)$ is a unitary quasi-onto mapping such that ${}^t(W^+(p, h))$ is a quasi-onto mapping

The first point ensures that place p models an intermediate state between the firing of a transition in H and the firing of a transition in F. The second one ensures that any firing of a transition f requires exactly one token in p. The last point guarantees that all instances of any firing of $h \in H$ produces a token in the place p and that any coloured token of $C(p)$ may be produced by a firing of some transition $h \in H$.

The **HF-interchangeability** hypothesis mainly restricts either the set H or F to be a singleton in order to avoid the case where $h \in H$ and $f \in F$ are live in the original net whereas the transition hf is not live in the reduced net.

DEFINITION *3.2.* (Hypothesis R2). A p-post-agglomerable coloured net is HF-interchangeable if one of these conditions is fulfilled:

(1) $H = \{h\}$ and $W^+(p, h)$ is orthonormal
(2) $F = \{f\}$, $C(f) = C(p)$ (thus $W^-(p, f)$ is orthonormal)

In the following, we will assume w.l.o.g. that depending on the item of the above definition either $W^+(p, h)$ is the identity function and $W^-(p, f)$ is a projection or $W^-(p, f)$ is the identity function. Indeed applying the reduction called orthonormalization leads to this situation (see Haddad (1990)).

The **F-independence** hypothesis ensures that when the place p is marked, no transition that can produce tokens useful for the firing of a transition in F can be fired.

DEFINITION *3.3.* (Hypothesis R3). A p-post-agglomerable coloured net is F-independent if $\forall f \in F$, $\forall q \in ({}^\bullet f \setminus \{p\})$, $\forall t \in {}^\bullet q \setminus F$, $\exists p_t \in {}^\bullet t$, such that

(1) there exists a binary coloured positive invariant $\mathcal{F} = \sum_{r \in P} \mathcal{F}_r.r$ on a domain D

(2) let us note
$$\phi = \overline{{}^t(W^+(q,t))} \circ \overline{W^-(q,f)} \circ \overline{{}^t(W^-(p,f))}$$
$$\psi = \overline{{}^t(W^-(p_t,t))} \circ \overline{{}^t(\mathcal{F}_{p_t})} \circ \overline{\mathcal{F}_p}$$
then $\phi \sqsubseteq \psi$

Furthermore, if there exists a binary positive invariant $\mathcal{F}'$ on the domain $C(p)$ such that ${}^t\mathcal{F}'_p$ is a quasi-onto mapping then the net is **strongly** F-independent.

These two points ensure that the transitions of ${}^\bullet({}^\bullet f[c_f])$ (dashed transitions of figure Fig. 5) are not fireable when the related instance $p[c_p]$ of place p is marked. This behaviour is obtained by the use of a binary positive invariant that ensures a mutual exclusion of the place $p[c_p]$ with place $p_t[c_{p_t}]$ which are pre-conditions of these transitions t. The mapping ψ allows us to highlight the instances of the transition t that are linked to an instance of p_t covered by the a positive invariant (in the unfolded net) which covers a given instance of p.

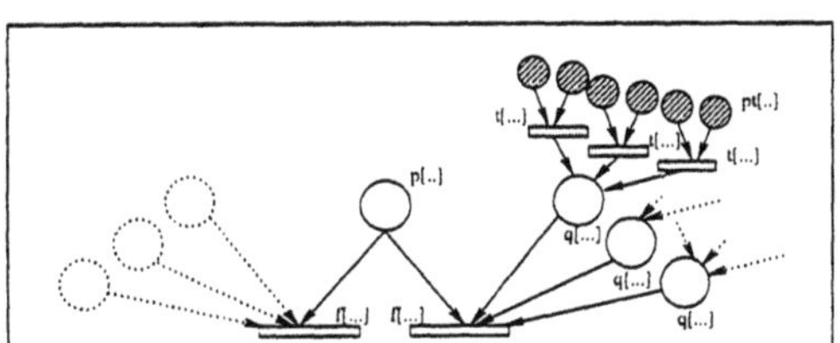

Fig. 5. A subset of $\phi(c_p), c_p \in C(p)$

The third hypothesis, the F-**continuation**, means that an excess of occurrences of $h \in H$ can always be reduced by subsequent firings of transitions of F (when the place p is marked, a transition of F is necessarily fireable).

DEFINITION *3.4.* (Hypothesis R4). A p-post-agglomerable net is F-continuable if either $\exists f \in F$ such that ${}^\bullet f = \{p\}$ or $\exists\ F_s \subset F$ such that:

(1) $\forall f \in F_s$, ${}^\bullet f = \{p, p_f\}$,

(2) $\forall f \in F_s$, ${}^t(W^-(p_f, f))$ is a quasi-one to one mapping,

(3) there exists a flow on $C(p)$ with
$$\mathcal{F} = \sum_{f \in F_s} \mathcal{F}_{p_f}.p_f - \lambda.\mathcal{F}_p.p$$
$\forall f \in F_s$, $W^-(p_f, f) \circ \langle X_{C(p)}, All_{C_f}\rangle = {}^t\mathcal{F}_{p_f}$
and such that

(a) either $\lambda = 0$ and $\mathcal{F}$ induces a binary positive invariant

(b) or $\lambda = 1$ and $\mathcal{F}$ induces a synchronisation invariant

3.4 Post-agglomeration transformation

We define now the transformation associated to the coloured post-agglomeration. The reduced net is the same as the original one except that we merge any transition of H with transitions of F (we form couples (hf)).

DEFINITION *3.5.* The reduced net (CN_r, m_{0r}) obtained from a coloured net (CN, M_0) by a coloured post-agglomeration is defined by:

- $P_r = P$ and $T_r = T \setminus (H \cup F) \cup (H \times F)$; we note hf a new transition $(h, f) \in H \times F$.
- $\forall p' \in P_r, m_{0r}(p') = m_0(p')$
- $\forall t \in T_r \setminus (H \times F)$, $\forall p' \in P_r$, $W_r^-(p', t) = W^-(p', t)$ and $W_r^+(p', t) = W^+(p', t)$
- $\forall hf \in (H \times F)$, $\forall q \in P_r$
  - $W_r^-(q, hf) = max(\Gamma_1, \Gamma_2)$
  - $W_r^+(q, hf) = \Upsilon + W_r^-(q, hf))$

  where
  when $H = \{h\}$, then, $C(hf) = C(f)$ and
  - $\Gamma_1 = W^-(q,h) \circ W^-(p,f)$,
  - $\Gamma_2 = W^-(q,f) - W(q,h) \circ W^-(p,f)$
  - $\Upsilon = W(q,f) + W(q,h) \circ W^-(p,f)$

  when $F = \{f\}$, then $C(hf) = C(h)$ and
  - $\Gamma_1 = W^-(q,h)$
  - $\Gamma_2 = W^-(q,f) \circ W^+(p,h) - W(q,h)$
  - $\Upsilon = W(q,f) \circ W^+(p,h) + W(q,h)$

This transformation preserves the Petri net liveness and the properties related to the maximal or infinite sequences (e.g. deadlock, fairness, mutual-exclusion,etc.). The corresponding theorem can be found in the appendix.

3.5 An example

The following coloured net (Fig. 6) models the allocation of resources ($C2$) to processes ($C1$). When receiving a request from a process X, the server chooses a free resource Y, sends this resource identity to the process and stores locally (in place **Taken**) this allocation (`ts1`). Upon reception of a resource release request (`t3`) the server services this request (`ts2`) when the request corresponds to an already stored allocation (a token `(X,Y)` is in the place **Taken**).

Let us describe the reduction process on this net in order to look for possible deadlocks. At first, we delete the implicit places `Server` and `Att2`. Then we apply a post-agglomeration of the transition **t2** with the transition `t3` and a post-agglomeration of the transition `ts2` with the transition **t4**. Note that these agglomerations would be still possible with the reductions of Haddad (1990). But on the reduced net of Fig.7 no reductions previously defined are applicable. However, the post-agglomeration of h = `ts1` with f = `t2` around

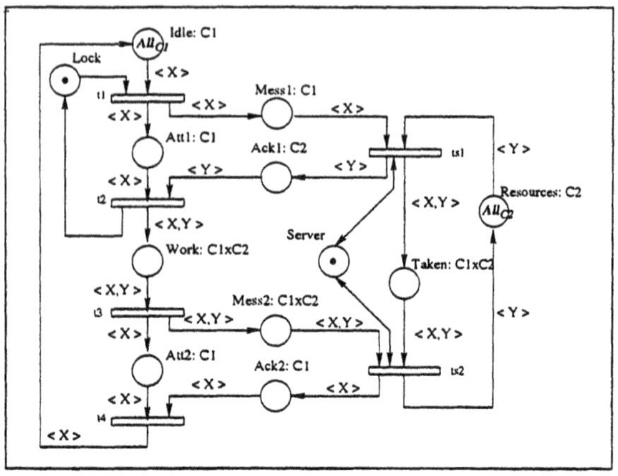

Fig. 6. Typed resource allocation

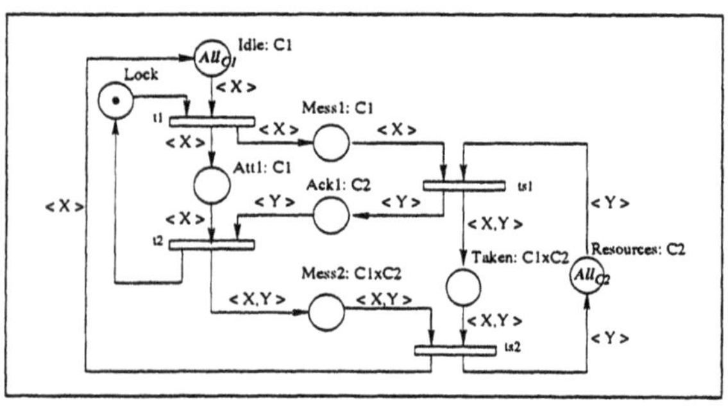

Fig. 7. No more standard reductions

the place p = **Ack1** is applicable. Indeed, $\mathcal{C}(f) = \mathcal{C}(p) \times C2$, $W^-(p, f)$ is an ortho-projection, $H = \{h\}$, $W^+(p, h)$ is an ortho-projection so the net is p-post agglomerable (but does not verify the HF-interchangeability).

The unique transition related to the F-independency hypothesis is the transition $t1$ (which may put token in place `Att`). The flow $\mathcal{F} = \langle All_{C1}\rangle.Lock + \langle All_{C1}\rangle.Mess1 + \langle All_{C1}\rangle.Ack1$ on $C1$ induces the binary positive invariant: $\forall m \in Reach(N, m_0)$, $m(Lock) + \sum_{c \in C1} m(Mess1)(c) + \sum_{c \in C2} m(Ack1)(c) = 1$. Using notation of the hypothesis conditions, $\phi = All_{C1}$ and $\psi = All_{C1}$. So, $\phi \sqsubseteq \psi$ and since ${}^t(All_{C1}) = All_{C1}$ is a quasi-onto mapping the net is strongly F-independent.

At last, the flow $\mathcal{F} = \langle All_{C2}\rangle.Mess1 + \langle All_{C2}\rangle.Ack1 - \langle All_{C2}\rangle.Att1$ induces the synchronisation invariant $\forall m \in Reach(N, m_0)$, $\sum_{c \in C1} m(Mess1)(c) + \sum_{c \in C2} m(Ack1)(c) - \sum_{c \in C1} m(Att11)(c) = 0$ which fulfils the conditions of the F-continuation hypothesis (`t2` is fireable as soon as `Att1` is marked). Applying this reduction leads to the net depicted Fig.8. Finally the implicit places `Mess2` and `Att1` are deleted and then a post-agglomeration of `ts1` with `ts2` followed by a post-agglomeration of `t1` with `ts1` produces the net depicted Fig.9 where all the places being implicit may be deleted. As the final net is live, the original one is also live (see Theorem 1 in appendix).

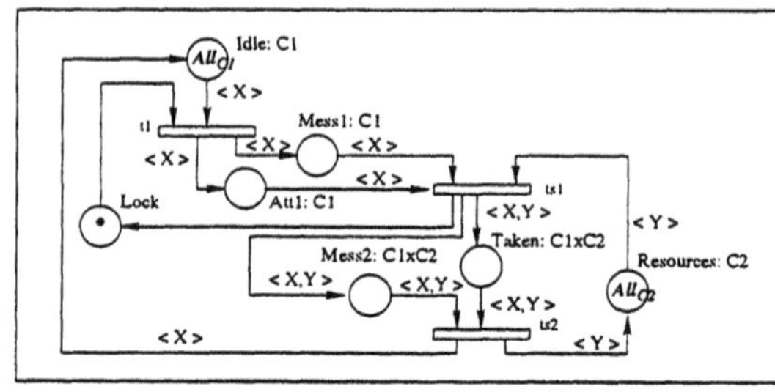

Fig. 8. Application of a post-agglomeration

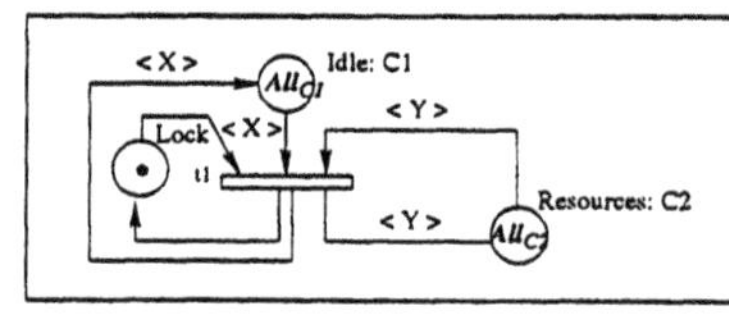

Fig. 9. A live net

4. CONCLUSION

We have presented in this paper new coloured reductions and we have precisely defined one of them: the post-agglomeration. These reductions are based on accurate conditions using linear invariants that cover more realistic concurrent software behaviours (compared to initial conditions which were only based on the structure of the model). We are integrating a syntactic version of these reductions in the Quasar tool, a framework for verifying concurrent programs (see http://quasar.cnam.fr).

REFERENCES

C. Chiola, C. Dutheillet, G. Franceschinis, and S. Haddad. On well-formed colored nets and their symbolic reachability graph. In *ICATPN*, Paris-France, June 1990.

J.M. Colom, J. Martinez, and M. Silva. Packages for validating discrete production systems modeled with Petri nets. In *IMACS-IFAC Symposium*, Lille, France, 1986.

S. Evangelista, S. Haddad, and J.F. Pradat-Peyre. New coloured reductions for software validation. Technical Report 655, CEDRIC, CNAM, Paris, 2004a.

S. Evangelista, S. Haddad, and J.F. Pradat-Peyre. Petri nets reductions for software validation. Technical report, CEDRIC, CNAM, Paris, 2004b.

H.J. Genrich. Equivalence transformations of prt-nets. In Jensen and Rozenberg, editors, *High-level Petri Nets, Theory and Application*, volume 424, pages 426–453. Springer-Verlag, 1990.

S. Haddad. A reduction theory for colored nets. In Jensen and Rozenberg, editors, *High-level Petri Nets, Theory and Application*, volume 424 of *LNCS*, pages 399–425. Springer-Verlag, 1990.

ELSEVIER
IFAC
PUBLICATIONS
www.elsevier.com/locate/ifac

MONITOR DESIGN FOR COLORED PETRI NETS WITH UNCONTROLLABLE AND UNOBSERVABLE TRANSITIONS

Alessandro Giua * Carla Seatzu *

* *Dip. Ing. Elettrica ed Elettronica, U. di Cagliari, Italy*
Email: {giua,seatzu}@diee.unica.it

Abstract: In this paper we deal with the problem of designing a supervisory controller that enforces certain specifications on the marking of a Colored Petri net (CPN). In particular, we consider colored Generalized Mutual Exclusion Constraints (GMEC) that represent in a compact way several linear constraints that limit the weighted sum of colored tokens in subsets of places. In a previous work we have shown that, when all transitions are controllable and observable with respect to all colors, these constraints can be enforced by a colored monitor place that can be added to the net to obtain the closed-loop system. The novel contribution of this paper is that of showing how these results can be extended to the case of transitions that are uncontrollable and/or unobservable with respect to certain colors. In particular, we show that the parametrization and the tabular procedure proposed by Moody and Antsaklis for uncolored Petri nets can still be used to compute – when it does exist – a less permissive GMEC that can be forced by an admissible colored monitor place. *Copyright © 2004 IFAC*

Keywords: Petri nets, colored Petri nets, generalized mutual exclusion constraints, monitor places.

1. INTRODUCTION

In (Fanti *et al.*, 2003) we showed how the control approach based on the construction of monitor places presented in (Giua *et al.*, 1992) for place/transition nets can be extended to the more general case of CPN (Jensen, 1992). A colored GMEC may represent in a compact way several constraints, and can be unfolded into a set of uncolored GMEC. In (Fanti *et al.*, 2003) it was shown that, when all transitions are controllable and observable, a GMEC can still be enforced by adding a monitor place p_c, and it was provided a systematic procedure to compute the incidence matrix defining such a monitor place, as well as its initial marking. When all transitions are controllable and observable with respect to (wrt) all colors, the monitor place minimally restricts the behavior of the closed-loop system, in the sense that it prevents only those transition firings that yield forbidden markings.

In the presence of transitions that are uncontrollable and/or unobservable wrt certain colors, it may well be the case that the monitor designed for a given GMEC is not admissible. In other words, it may occur that the monitor either disables a transition wrt an uncontrollable color, or observes the firing of a transition wrt an unobservable color. We show that in such a case it may be possible to construct a less permissive, but admissible monitor, extending to the case of colored nets the parametrization and the tabular procedure proposed by Moody and Antsaklis (1998) for uncolored Petri nets (PN).

It is important to observe that the computational complexity of the proposed approach is the same as that of computing the equivalent admissible GMEC and the corresponding monitor places for the unfolded net. There is still some advantages in using our approach with CPN rather than unfolding the net and use the standard approach in (Moody and Antsaklis, 1998).

— In many application cases it is much more convenient to deal with a CPN model rather than a PN model, so as to keep all the advantages deriving from the greater expressive power of CPN and the possibility of using other techniques developed to analyze and simulate CPN. In such a case an alternative to our procedure would be that of unfolding the net, converting the colored GMEC into a set of uncolored GMEC, then compute the admissible monitor places, and finally convert them into a single colored monitor place.

Our approach provides a systematic procedure to do this within the framework of CPN.

— Moreover, when computing an admissible GMEC using the proposed tabular procedure, we do not need to compute the unfolding of the whole CPN, but we only consider the incidence matrix of the unfolded subnet containing those transitions and those colors wrt whom transitions are either uncontrollable or unobservable.

— Finally, the proposed approach is a first step towards the formulation of a symbolic procedure for the computation of admissible monitor places for arbitrary high-level PN.

2. MULTISETS

Definition 2.1. Let D be a set. A *multiset* (resp., *non negative multiset*) $\boldsymbol{\alpha}$ over D is defined by a mapping $\alpha : D \rightarrow \mathbb{Z}$ ($\alpha : D \rightarrow \mathbb{N}$) and may be represented as $\boldsymbol{\alpha} = \sum_{d \in D} \alpha(d) \otimes d$ where the sum is limited to the elements such that $\alpha(d) \neq 0$.
Let $\mathcal{Z}(D)$ (resp., $\mathcal{N}(D)$) denote the set of all multisets (resp., non negative multisets) over D.
The multiset $\boldsymbol{\varepsilon}$ is the empty multiset such that for all $d \in D$, $\varepsilon(d) = 0$. ■
Now, given two sets D and D', let $\boldsymbol{F} : D \rightarrow \mathcal{Z}(D')$ be a function that associates to each element $d \in D$ a multiset on D':

$$\boldsymbol{F}(d) = \sum_{d' \in D'} F(d, d') \otimes d' \in \mathcal{Z}(D').$$

Definition 2.2. Given two sets D and D', a function $\boldsymbol{F} : D \rightarrow \mathcal{Z}(D')$, and a multiset $\boldsymbol{\alpha} \in \mathcal{Z}(D)$,

$$\boldsymbol{F}(\boldsymbol{\alpha}) \triangleq \boldsymbol{F} \circ \boldsymbol{\alpha} \triangleq \sum_{d \in D} \alpha(d) \boldsymbol{F}(d) = \sum_{d \in D} \sum_{d' \in D'} \alpha(d) F(d, d') \otimes d' \in \mathcal{Z}(D')^{\cdot}$$

■

We finally observe that it is possible to give a matrix representation of multisets and of functions over multisets.
Given two sets D and D', let us arbitrary order their elements as follows: $D = \{d_1, \ldots, d_k\}$ and $D' = \{d'_1, \ldots, d'_{k'}\}$.
A multiset $\boldsymbol{\alpha} \in \mathcal{Z}(D)$ can be represented by a vector:

$$\boldsymbol{\alpha} = \begin{bmatrix} \alpha(d_1) \\ \alpha(d_2) \\ \vdots \\ \alpha(d_k) \end{bmatrix} \in \mathbb{Z}^k.$$

Thus, given a function $\boldsymbol{F} : D \rightarrow \mathcal{Z}(D')$ for all $d \in D$ we can write

$$\boldsymbol{F}(d) = \begin{bmatrix} F(d, d'_1) \\ F(d, d'_2) \\ \vdots \\ F(d, d'_{k'}) \end{bmatrix} \in \mathbb{Z}^{k'}.$$

while its extension $\boldsymbol{F} : \mathcal{Z}(D) \rightarrow \mathcal{Z}(D')$ can be represented by the matrix

$$\boldsymbol{F} = [\, \boldsymbol{F}(d_1) \; \boldsymbol{F}(d_2) \; \ldots \; \boldsymbol{F}(d_k) \,] \in \mathbb{Z}^{k' \times k}$$

and finally the multiset $\boldsymbol{F}(\boldsymbol{\alpha}) = \boldsymbol{F} \circ \boldsymbol{\alpha}$ can be computed with the usual matrix-vector product denoted by $\cdot$, i.e.,

$$\boldsymbol{F}(\boldsymbol{\alpha}) = \boldsymbol{F} \circ \boldsymbol{\alpha} = \boldsymbol{F} \cdot \boldsymbol{\alpha} = \begin{bmatrix} \sum_{i=1}^{k} \alpha(d_i) F(d_i, d'_1) \\ \sum_{i=1}^{k} \alpha(d_i) F(d_i, d'_2) \\ \vdots \\ \sum_{i=1}^{k} \alpha(d_i) F(d_i, d'_{k'}) \end{bmatrix} \in \mathbb{Z}^{k'}.$$

Finally, given a multiset $\boldsymbol{\alpha} \in \mathcal{Z}(D')$, where $D' \in \{d'_1, \ldots, d'_{k'}\}$, with the notation $\boldsymbol{\alpha}'$ we denote a multiset represented by a row vector, i.e.,

$$\boldsymbol{\alpha}' = \left[\, \alpha(d'_1) \; \ldots \; \alpha(d'_{k'}) \,\right] \in \mathbb{Z}^{1 \times k'}.$$

The multiset $\boldsymbol{\alpha}' \circ \boldsymbol{F}$ represented by a row vector, can be computed with the usual matrix-vector product, i.e.,

$$\boldsymbol{\alpha}' \circ \boldsymbol{F} = \boldsymbol{\alpha}' \cdot \boldsymbol{F} =$$

$$\left[\sum_{i=1}^{k'} \alpha(d'_i) F(d'_i, d_1) \; \ldots \; \sum_{i=1}^{k'} \alpha(d'_i) F(d'_i, d_k) \right] \in \mathbb{Z}^{1 \times k'}.$$

3. COLORED PETRI NETS

A *Colored Petri Net* (CPN) is a bipartite directed graph represented by a quintuple $N = (P, T, Co, \boldsymbol{Pre}, \boldsymbol{Post})$ where P is the set of places, T is the set of transitions, $Co : P \cup T \rightarrow Cl$ is a color function that associates to each element in $P \cup T$ a non empty ordered set of colors in the set of possible colors Cl.
Therefore, for all $p_i \in P$, $Co(p_i) = \{a_{i,1}, \ldots, a_{i,u_i}\} \subseteq Cl$ is the ordered set of possible colors of tokens in p_i, and u_i is the number of possible colors of tokens in p_i. Analogously, for all $t_j \in T$, $Co(t_j) = \{b_{j,1}, \ldots, b_{j,v_j}\} \subseteq Cl$ is the ordered set of possible occurrence colors of t_j, and v_j is the number of possible occurrence colors in t_j.
We assume that $m = |P|$ and $n = |T|$.
Matrices $\boldsymbol{Pre}$ and $\boldsymbol{Post}$ are the pre-incidence and the post-incidence $m \times n$ dimensional matrices respectively. Each element $\boldsymbol{Pre}(p_i, t_j)$ (the same reasoning applies to $\boldsymbol{Post}$) is a mapping from the set of occurrence colors of t_j to a non negative multiset over the set of colors of p_i, namely, $\boldsymbol{Pre}(p_i, t_j) : Co(t_j) \rightarrow \mathcal{N}(Co(p_i))$, for $i = 1, \ldots, m$ and $j = 1, \ldots, n$. We denote $\boldsymbol{Pre}(p_i, t_j)$ as a matrix of $u_i \times v_j$ non negative integers, whose generic element $Pre(p_i, t_j)(h, k)$ is equal to the weight of the arc from place p_i wrt color $a_{i,h}$ to transition t_j wrt color $b_{j,k}$.
The incidence matrix $\boldsymbol{C}$ is an $m \times n$ matrix, whose generic element $\boldsymbol{C}(p_i, t_j) : Co(t_j) \rightarrow \mathcal{Z}(Co(p_i))$, for $i = 1, \ldots, m$ and $j = 1, \ldots n$. In particular $\boldsymbol{C}(p_i, t_j) = \boldsymbol{Post}(p_i, t_j) - \boldsymbol{Pre}(p_i, t_j)$.
For each place $p_i \in P$, we define the *marking* $\boldsymbol{m}_i$ of p_i as a *non negative multiset* over $Co(p_i)$. The mapping $m_i : Co(p_i) \rightarrow \mathbb{N}$ associates to each possible token color in p_i a non negative integer representing the number of tokens of that color that is contained in place p_i, and $\boldsymbol{m}_i = \sum_{d \in Co(p_i)} m_i(d) \otimes d$.
Here we denote $\boldsymbol{m}_i$ as a column vector of u_i non negative integers, whose h-th component $m_i(h)$ is equal to the number of tokens of color $a_{i,h}$ that are contained in p_i.
Finally, the marking $\boldsymbol{M}$ of a CPN is an m-dimensional column vector of multisets whose i-th entry is equal to $\boldsymbol{m}_i$.
A colored Petri net system $\langle N, \boldsymbol{M}_0 \rangle$ is a colored Petri net N with initial marking $\boldsymbol{M}_0$.
A transition $t_j \in T$ is *enabled* wrt color $b_{j,k}$ at a marking $\boldsymbol{M}$ if and only if for each place $p_i \in P$ and for all $h = 1, \ldots, u_i$, we have $m_i(h) \geq Pre(p_i, t_j)(h, k)$.
The firing of transition t_i wrt color b_{ij} follows the standard rules of CPN (Giua and Seatzu, 2004).

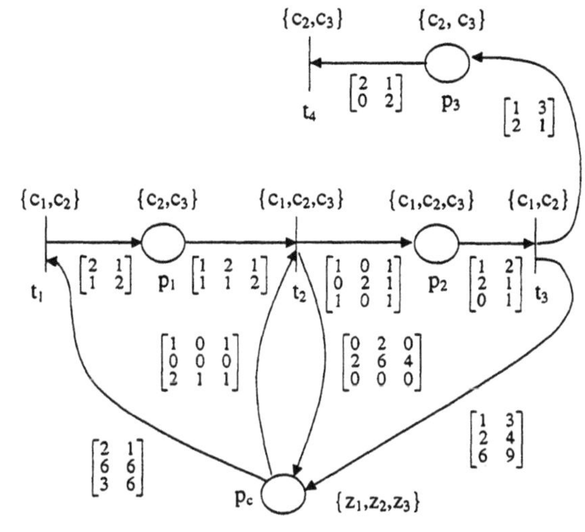

Fig. 1. The closed-loop colored Petri net of Example 1.

Example 3.1. Let us consider the CPN in Figure 1.a apart from place p_c and all connected arcs. The set of colors is $Cl = \{c_1, c_2, c_3\}$. Places p_1 and p_3 may only contain tokens of colors c_2 and c_3, while place p_2 may contain tokens of any color in Cl. Finally, transitions t_1 and t_3 may only fire wrt to colors c_1 and c_2, while transition t_2 may fire wrt any color in Cl, and transition t_4 may fire wrt colors c_2 and c_3. Given the structure of the net, the only non null matrices $\boldsymbol{Pre}$ and $\boldsymbol{Post}$ are those reported Figure 1.a using the matrix notation. ■

Finally, let us introduce the following definitions.

Definition 3.2. Given a CPN $N_p = (P, T, Co, \boldsymbol{Pre}_p, \boldsymbol{Post}_p)$, a transition $t_j \in T$ is *uncontrollable wrt color* $b_{j,k} \in Co(t_j)$ if its firing wrt $b_{j,k}$ cannot be inhibited by an external action. Thus, the set $Co(t_j)$ may be partitioned as $Co(t_j) = Co^{uc}(t_j) \cup Co^{c}(t_j)$ where $Co^{c}(t_j)$ $(Co^{uc}(t_j))$ is the set of colors wrt whom transition t_j is controllable (uncontrollable).

If $Co^{uc}(t_j) = \emptyset$ we say that t_j is *controllable*; if $Co^{c}(t_j) = \emptyset$ we say that t_j is *uncontrollable*. ■

Definition 3.3. Given a CPN $N_p = (P, T, Co, \boldsymbol{Pre}_p, \boldsymbol{Post}_p)$, a transition $t_j \in T$ is *unobservable wrt color* $b_{j,k} \in Co(t_j)$ if its firing wrt $b_{j,k}$ cannot be measured by an external action. Thus, the set $Co(t_j)$ may be partitioned as $Co(t_j) = Co^{uo}(t_j) \cup Co^{o}(t_j)$ where $Co^{o}(t_j)$ $(Co^{uo}(t_j))$ is the set of colors wrt whom transition t_j is observable (unobservable).

If $Co^{uo}(t_j) = \emptyset$ we say that t_j is *observable*; if $Co^{o}(t_j) = \emptyset$ we say that t_j is *unobservable*. ■

4. GMEC IN COLORED PETRI NETS

Now, we recall the notion of colored GMEC (Fanti *et al.*, 2003).

Definition 4.1. A GMEC is a couple $(\boldsymbol{W}, \boldsymbol{k})$ where $\boldsymbol{W} = [\boldsymbol{w}_1 \; \ldots \; \boldsymbol{w}_m]$, $\boldsymbol{k} \in \mathcal{Z}(D)$ and for all i, $\boldsymbol{w}_i : Co(p_i) \to \mathcal{Z}(D)$, while D is a set of colors different from $Co(p_i)$, $i = 1, \ldots, m$. Thus $\boldsymbol{W}$ can also be represented by a matrix with $|D|$ rows and $\sum_{i=1}^{m} |Co(p_i)|$ columns, and

$$\mathcal{M}(\boldsymbol{W}, \boldsymbol{k}) = \left\{ \boldsymbol{M} = \begin{bmatrix} \boldsymbol{m}_1 \\ \vdots \\ \boldsymbol{m}_m \end{bmatrix} \;\middle|\; \boldsymbol{m}_i \in \mathcal{N}(Co(p_i)), \; \boldsymbol{W} \circ \boldsymbol{M} \triangleq \sum_{i=1}^{m} \boldsymbol{w}_i \circ \boldsymbol{m}_i \leq \boldsymbol{k} \right\}.$$

is the set of legal markings. ■

Note that here we are extending the $\circ$ operator to the case of scalar product of vectors of multisets.

Example 4.2. Let us consider again the CPN in Figure 1.a apart from place p_c and all connected arcs. Assume $D = \{z_1, z_2, z_3\}$. Moreover, let $\boldsymbol{w}_1 = [\boldsymbol{w}_1(c_2) \; \boldsymbol{w}_1(c_3)]$, $\boldsymbol{w}_1(c_1) = 1 \otimes z_1 + 2 \otimes z_2$, $\boldsymbol{w}_1(c_3) = 2 \otimes z_2 + 3 \otimes z_3$, $\boldsymbol{w}_2 = [\boldsymbol{w}_2(c_1) \; \boldsymbol{w}_2(c_2) \; \boldsymbol{w}_2(c_3)]$, $\boldsymbol{w}_2(c_1) = 1 \otimes z_1 + 2 \otimes z_2 + 2 \otimes z_3$, $\boldsymbol{w}_2(c_2) = 2 \otimes z_3$, $\boldsymbol{w}_2(c_3) = 1 \otimes z_1 + 3 \otimes z_3$, $\boldsymbol{w}_3(c_2) = \boldsymbol{w}_3(c_3) = \boldsymbol{\varepsilon}$ and $\boldsymbol{k} = 3 \otimes z_1 + 5 \otimes z_2 + 6 \otimes z_3$. Therefore,

$$\boldsymbol{W} \circ \boldsymbol{M} \triangleq \sum_{i=1}^{m} \boldsymbol{w}_i \circ \boldsymbol{m}_i = \begin{bmatrix} 1 & 0 \\ 2 & 2 \\ 0 & 3 \end{bmatrix} \cdot \begin{bmatrix} m_1(c_2) \\ m_1(c_3) \end{bmatrix} + $$

$$+ \begin{bmatrix} 1 & 0 & 1 \\ 2 & 0 & 0 \\ 2 & 2 & 3 \end{bmatrix} \cdot \begin{bmatrix} m_2(c_1) \\ m_2(c_2) \\ m_2(c_3) \end{bmatrix} \leq \begin{bmatrix} 3 \\ 5 \\ 6 \end{bmatrix} \implies$$

$$\mathcal{M}(\boldsymbol{W}, \boldsymbol{k}) = \{ \; \boldsymbol{M} = \begin{bmatrix} \boldsymbol{m}_1 \\ \boldsymbol{m}_2 \\ \boldsymbol{m}_3 \end{bmatrix} \; | \; \boldsymbol{m}_i \in \mathcal{N}(Co(p_i)),$$
$$m_1(c_2) + m_2(c_1) + m_2(c_3) \leq 3,$$
$$2m_1(c_2) + 2m_1(c_3) + 2m_2(c_1) \leq 5$$
$$3m_1(c_3) + 2m_2(c_1) + 2m_2(c_2) + 3m_2(c_3) \leq 6\} .$$

■

5. MONITORS FOR COLORED PETRI NETS

Definition 5.1. Given a CPN system $\langle N_p, \boldsymbol{M}_{p,0} \rangle$, with $N_p = (P, T, Co, \boldsymbol{Pre}_p, \boldsymbol{Post}_p)$, and a GMEC $(\boldsymbol{W}, \boldsymbol{k})$ with $\boldsymbol{k} \in \mathcal{Z}(D)$, the *monitor* that enforces this constraint is a new place p_c with $Co(p_c) = D$, to be added to N_p. The resulting system is denoted $\langle N, \boldsymbol{M}_0 \rangle$, with $N = (P \cup \{p_c\}, T, Co, \boldsymbol{Pre}, \boldsymbol{Post})$. Then N will have incidence matrix

$$\boldsymbol{C} = \begin{bmatrix} \boldsymbol{C}_p \\ \boldsymbol{C}_c \end{bmatrix}, \quad \boldsymbol{C}_c = -\boldsymbol{W} \circ \boldsymbol{C}_p. \qquad (1)$$

We are assuming that there are no selfloops containing p_c in N, hence $\boldsymbol{Pre}$ and $\boldsymbol{Post}$ may be uniquely determined by $\boldsymbol{C}$. The initial marking of $\langle N, \boldsymbol{M}_0 \rangle$ is

$$\boldsymbol{M}_0 = \begin{bmatrix} \boldsymbol{M}_{p,0} \\ \boldsymbol{m}_{c,0} \end{bmatrix}, \quad \boldsymbol{m}_{c,0} = \boldsymbol{k} - \boldsymbol{W} \circ \boldsymbol{M}_{p,0}. \quad (2)$$

We assume that the initial marking $\boldsymbol{M}_{p,0}$ of the system satisfies the constraint $(\boldsymbol{W}, \boldsymbol{k})$. ■

In the case of controllable and observable transitions we proved the following result (Fanti *et al.*, 2003).

Theorem 5.2. Let $\langle N_p, \boldsymbol{M}_{p,0} \rangle$ be a CPN system, and $(\boldsymbol{W}, \boldsymbol{k})$ a colored GMEC. Let $\langle N, \boldsymbol{M}_0 \rangle$ be the system with the addition of the monitor place p_c.

(1) The monitor place p_c enforces the GMEC $(\boldsymbol{W}, \boldsymbol{k})$ when included in the closed-loop system $\langle N, \boldsymbol{M}_0 \rangle$.

(2) The monitor place p_c minimally restricts the behavior of the closed-loop system $\langle N, \boldsymbol{M}_0 \rangle$, in the sense that it prevents only transition firings that yield forbidden markings.

6. UNCONTROLLABLE AND UNOBSERVABLE GMEC

In this section we first introduce the definition of controllable and observable GMEC. Then we provide necessary and sufficient conditions for its controllability and observability.

Definition 6.1. Given a CPN system $\langle N_p, \boldsymbol{M}_{p,0} \rangle$, with $N_p = (P, T, Co, \boldsymbol{Pre}_p, \boldsymbol{Post}_p)$, and a GMEC $(\boldsymbol{W}, \boldsymbol{k})$ with $\boldsymbol{k} \in \mathcal{Z}(D)$, let p_c with $Co(p_c) = D$ be the corresponding monitor place with (pre-)incidence matrix $(\boldsymbol{Pre}_c)$ $\boldsymbol{C}_c$.

— The monitor place p_c is said *structurally controllable* (or simply, *controllable*) if for all transitions $t_j \in T$ and for all colors $b_{j,k} \in Co^{uc}(t_j)$, $\boldsymbol{Pre}_c(\cdot, b_{j,k}) = \boldsymbol{\varepsilon}$, i.e., the monitor place is controllable if the weights of the input arcs to transitions that are uncontrollable wrt certain colors, are null wrt those colors.

— The monitor place p_c is said *structurally observable* (or simply, *observable*) if for all transitions $t_j \in T$ and for all colors $b_{j,k} \in Co^{uo}(t_j)$, $\boldsymbol{C}_c(\cdot, b_{j,k}) = \boldsymbol{\varepsilon}$, i.e., the monitor place is observable if the weights of the input and the output arcs to transitions that are uncontrollable wrt certain colors, are null wrt those colors. ■

A monitor place that is controllable and observable is said *admissible*.

Analogously, we say that a GMEC is *controllable* and *observable* (*admissible*) if the corresponding monitor place is controllable and observable (*admissible*).

Proposition 6.2. Given a CPN system $\langle N_p, \boldsymbol{M}_{p,0} \rangle$ with incidence matrix $\boldsymbol{C}_p$, let $(\boldsymbol{W}, \boldsymbol{k})$ with $\boldsymbol{k} \in \mathcal{Z}(D)$ be a GMEC that we want to force, and p_c be the corresponding monitor place with incidence matrix $\boldsymbol{C}_c$. Let $\boldsymbol{C}_p^{uc}$ be the matrix that we obtain selecting from $\boldsymbol{C}_p$ the only columns relative to those transitions and those colours such that the considered transitions are uncontrollable wrt the considered colors.

(i) The GMEC and the monitor place are controllable *if and only if*

$$\boldsymbol{C}_c^{uc} = -\boldsymbol{W} \circ \boldsymbol{C}_p^{uc} \geq \underbrace{[\varepsilon \ \dots \ \varepsilon]}_{\sum_{j:\ t_j \in T} |Co^{uc}(t_j)|}.$$

(ii) The GMEC and the monitor place are observable *if and only if*

$$\boldsymbol{C}_c^{uo} = -\boldsymbol{W} \circ \boldsymbol{C}_p^{uo} = \underbrace{[\varepsilon \ \dots \ \varepsilon]}_{\sum_{j:\ t_j \in T} |Co^{uo}(t_j)|}.$$

Proof. See (Giua and Seatzu, 2004). □

Example 6.3. Let us consider again the CPN system in Figure 1.a apart from place p_c and all connected arcs. Assume that transition t_2 is uncontrollable wrt colors c_2 and c_3. In such a case the GMEC $(\boldsymbol{W}, \boldsymbol{k})$ defined in the previous Example 4.2 is not controllable being

$$\boldsymbol{C}_c^{uc} = -\boldsymbol{W} \circ \boldsymbol{C}_p^{uc} = \begin{array}{c} \begin{array}{cc} c_2 & c_3 \end{array} \\ \begin{bmatrix} 2 & -1 \\ 6 & 4 \\ -1 & -1 \end{bmatrix} \end{array} \begin{array}{c} \\ z_1 \\ z_2 \\ z_3 \end{array} \not\geq [\varepsilon \ \varepsilon]$$

Now, assume that transition t_2 is unobservable wrt color c_1, and transition t_3 is unobservable wrt color c_2. Being

$$\boldsymbol{C}_c^{uo} = -\boldsymbol{W} \circ \boldsymbol{C}_p^{uo} = \begin{array}{c} \begin{array}{cc} c_1 & c_2 \end{array} \\ \begin{bmatrix} -1 & 3 \\ 2 & 4 \\ -2 & 9 \end{bmatrix} \end{array} \begin{array}{c} \\ z_1 \\ z_2 \\ z_3 \end{array} \neq [\varepsilon \ \varepsilon]$$

the GMEC $(\boldsymbol{W}, \boldsymbol{k})$ is not observable ■

7. MONITORS FOR UNCONTROLLABLE AND UNOBSERVABLE GMEC

As well known from the uncolored PN theory, given an uncontrollable or unobservable GMEC, the maximal permissive supervisor may not be a monitor place (Giua *et al.*, 1992). An important contribution in this area is due to Moody and Antsaklis (1998) who proposed a very simple and efficient procedure to compute a family of more restrictive *controllable and observable* GMEC that force the original GMEC. Clearly, these GMEC are not in general maximally permissive. Here we show how the procedure by Moody and Antsaklis can be easily extended to the case of CPN.

Proposition 7.1. Given a GMEC $(\boldsymbol{W}, \boldsymbol{k})$ with

$$\boldsymbol{W} = \begin{bmatrix} \boldsymbol{w}_1 & \dots & \boldsymbol{w}_m \end{bmatrix}, \quad \boldsymbol{w}_i : Co(p_i) \to \mathcal{Z}(D), \quad i = 1, \dots, m,$$

$$\boldsymbol{k} \in \mathcal{Z}(D), \quad D = \{z_1, \ \dots, \ z_q\},$$

let us consider the modified GMEC $(\tilde{\boldsymbol{W}}, \tilde{\boldsymbol{k}})$ where

$$\tilde{\boldsymbol{W}} = \begin{bmatrix} \boldsymbol{s}_1' \circ \boldsymbol{W} \\ \vdots \\ \boldsymbol{s}_q' \circ \boldsymbol{W} \end{bmatrix} + \boldsymbol{R},$$

$$\boldsymbol{s}_i \in \mathcal{N}(D), \quad \boldsymbol{s}_i(z_j) \in \mathbb{N}^+ \text{ if } i = j, \quad \boldsymbol{s}_i(z_j) = 0 \text{ if } i \neq j,$$

$$\boldsymbol{R} = \begin{bmatrix} \boldsymbol{r}_1 & \dots & \boldsymbol{r}_m \end{bmatrix}, \quad \boldsymbol{r}_i : Co(p_i) \to \mathcal{N}(D), \quad i = 1, \ \dots, \ m,$$

$$\tilde{\boldsymbol{k}} = \begin{bmatrix} \boldsymbol{s}_1' \circ (\boldsymbol{k} + \boldsymbol{1}) \\ \vdots \\ \boldsymbol{s}_q' \circ (\boldsymbol{k} + \boldsymbol{1}) \end{bmatrix} - \boldsymbol{1}$$

where $\boldsymbol{1} \in \mathcal{N}(D)$ is the multiset of ones over D. It holds $\mathcal{M}(\tilde{\boldsymbol{W}}, \tilde{\boldsymbol{k}}) \subseteq \mathcal{M}(\boldsymbol{W}, \boldsymbol{k})$.

Proof. If $\boldsymbol{M} \in \mathcal{M}(\tilde{\boldsymbol{W}}, \tilde{\boldsymbol{k}})$, then $\tilde{\boldsymbol{W}} \circ \boldsymbol{M} \leq \tilde{\boldsymbol{k}}$, or equivalently, by definition,

$$\begin{bmatrix} \boldsymbol{s}_1' \circ \boldsymbol{W} \\ \vdots \\ \boldsymbol{s}_q' \circ \boldsymbol{W} \end{bmatrix} \circ \boldsymbol{M} + \boldsymbol{R} \circ \boldsymbol{M} \leq \begin{bmatrix} \boldsymbol{s}_1' \circ (\boldsymbol{k} + \boldsymbol{1}) \\ \vdots \\ \boldsymbol{s}_q' \circ (\boldsymbol{k} + \boldsymbol{1}) \end{bmatrix} - \boldsymbol{1}. \quad (3)$$

Therefore,

$$
\begin{bmatrix} s_1' \circ \boldsymbol{W} \\ \vdots \\ s_q' \circ \boldsymbol{W} \end{bmatrix} \circ \boldsymbol{M} \leq \begin{bmatrix} s_1' \circ (\boldsymbol{k}+1) \\ \vdots \\ s_q' \circ (\boldsymbol{k}+1) \end{bmatrix} - 1 - \boldsymbol{R} \circ \boldsymbol{M}
$$

$$
\leq \begin{bmatrix} s_1' \circ (\boldsymbol{k}+1) \\ \vdots \\ s_q' \circ (\boldsymbol{k}+1) \end{bmatrix} - 1 \tag{4}
$$

being $\boldsymbol{R} \circ \boldsymbol{M} \geq \varepsilon$. Now, let us rewrite the weighting matrix $\boldsymbol{W}$ as

$$
\boldsymbol{W} = \begin{bmatrix} \bar{\boldsymbol{w}}_1' \\ \vdots \\ \bar{\boldsymbol{w}}_q' \end{bmatrix}.
$$

Because each multiset s_i has only one non-zero (positive) element in correspondence to the i-th position, the inequality (4) can be rewritten as

$$
s_i(z_i)\ \bar{\boldsymbol{w}}_i' \circ \boldsymbol{M} \leq s_i(z_i)\ (k(z_i)+1) - 1, \quad i = 1, \ldots, q.
$$

Now, because $s_i(z_i) \in \mathbb{N}^+$, it holds

$$
\bar{\boldsymbol{w}}_i' \circ \boldsymbol{M} \leq k(z_i) + 1 - \frac{1}{s_i(z_i)} < k(z_i) + 1, \quad i = 1, \ldots, q.
$$

Being $\bar{\boldsymbol{w}}_i' \circ \boldsymbol{M} \in \mathbb{Z}$ for all $i = 1, \ldots, q$, it holds that $\bar{\boldsymbol{w}}_i' \circ \boldsymbol{M} \leq k(z_i)$, for $i = 1, \ldots, q$, or equivalently $\boldsymbol{W} \circ \boldsymbol{M} \leq \boldsymbol{k}$, i.e., $\boldsymbol{M} \in \mathcal{M}(\boldsymbol{W}, \boldsymbol{k})$, thus proving the statement. □

A modified GMEC can be easily determined using the Algorithm 7.2 that is an extension of the algorithm presented by Moody and Antsaklis in (1998).

Note that with no ambiguity in the notation we often refer to the matrix representation of multisets and of matrices of multisets. Moreover, we denote as $\boldsymbol{0}_{k \times q}$ the zero matrix of dimension $k \times q$ and with $\boldsymbol{0}_k$ the zero column vector of dimension k.

If the algorithm stops with $k = q+1$ and $flag = 0$, then consider the transformed GMEC $(\tilde{\boldsymbol{W}}, \tilde{\boldsymbol{k}})$ defined as follows

$$
\tilde{\boldsymbol{W}} = \begin{bmatrix} s_1' \circ \boldsymbol{W} \\ \vdots \\ s_q' \circ \boldsymbol{W} \end{bmatrix} + \boldsymbol{R}, \quad \tilde{\boldsymbol{k}} = \begin{bmatrix} s_1' \circ (\boldsymbol{k}+1) \\ \vdots \\ s_q' \circ (\boldsymbol{k}+1) \end{bmatrix} - 1
$$

that is controllable and observable being

$$
-\tilde{\boldsymbol{W}} \circ \boldsymbol{C}_p^{uc} \geq \underbrace{[\,\varepsilon \;\; \cdots \;\; \varepsilon\,]}_{n_{uc}}
$$

$$
-\tilde{\boldsymbol{W}} \circ \boldsymbol{C}_p^{uo} = \underbrace{[\,\varepsilon \;\; \cdots \;\; \varepsilon\,]}_{n_{uo}}
$$

Moreover, by Proposition 7.1, the controllable and observable monitor $\tilde{p}_c$ corresponding to the GMEC $(\tilde{\boldsymbol{W}}, \tilde{\boldsymbol{k}})$, if physically implementable, is sub-optimal (or even optimal) for the GMEC $(\boldsymbol{W}, \boldsymbol{k})$.

Example 7.3. Let us consider the CPN system in Figure 1.a. apart from place p_c and all connected arcs. Assume that transition t_2 is uncontrollable wrt colors c_2 and c_3 and unobservable wrt color c_1, while transition t_3 is unobservable wrt color c_2. Let us apply Algorithm 7.2 to compute an admissible GMEC $(\tilde{\boldsymbol{W}}, \tilde{\boldsymbol{k}})$. In such a case $n_{uc} = n_{uo} = 2$, and $M = 7$. Moreover, at step 1 of Algorithm 7.2 we define

$$
\boldsymbol{V} = \begin{bmatrix} -2 & 1 \\ -6 & -4 \\ 1 & 1 \end{bmatrix}, \quad \boldsymbol{U} = \begin{bmatrix} 1 & -3 \\ -2 & -4 \\ 2 & -9 \end{bmatrix},
$$

$$
\boldsymbol{R} = [\,\varepsilon\ \varepsilon\ \varepsilon\ \varepsilon\ \varepsilon\ \varepsilon\ \varepsilon\,], \quad s_i = \varepsilon,
$$

$i = 1, \ldots, 3$. Thus, for $k = 1$ we consider the table

$$
\boldsymbol{A} = \left|\begin{array}{cc|cc|ccccccc|c}
-2 & -1 & -1 & 0 & 1 & 0 & 0 & 0 & 0 & 0 & 0 & 0 \\
-1 & -2 & -1 & 0 & 0 & 1 & 0 & 0 & 0 & 0 & 0 & 0 \\
0 & 1 & 1 & -2 & 0 & 0 & 1 & 0 & 0 & 0 & 0 & 0 \\
2 & 1 & 0 & -1 & 0 & 0 & 0 & 1 & 0 & 0 & 0 & 0 \\
0 & 1 & 1 & -1 & 0 & 0 & 0 & 0 & 1 & 0 & 0 & 0 \\
0 & 0 & 0 & 3 & 0 & 0 & 0 & 0 & 0 & 1 & 0 & 0 \\
0 & 0 & 0 & 1 & 0 & 0 & 0 & 0 & 0 & 0 & 1 & 0 \\
\hline
-2 & 1 & 1 & -3 & 0 & 0 & 0 & 0 & 0 & 0 & 0 & 1
\end{array}\right|
$$

To make null the element $v(2) = 1$ we update the last row of $\boldsymbol{A}$ as: $\boldsymbol{A}(8,\cdot) = \boldsymbol{A}(8,\cdot) + \boldsymbol{A}(1,\cdot)$. To make null the element $u(2) = -3$ we update the last row of $\boldsymbol{A}$ as: $\boldsymbol{A}(8,\cdot) = \boldsymbol{A}(8,\cdot) + \boldsymbol{A}(6,\cdot)$.

Now, $\boldsymbol{v} \leq [0\ \ 0]'$ and $\boldsymbol{u} = [0\ \ 0]'$, thus we update $s_1(1) = 1$, $\boldsymbol{R}(1,\cdot) = [1\ 0\ 0\ 0\ 0\ 1\ 0]$, and repeat the procedure for $k = 2$.

An admissible GMEC is $(\tilde{\boldsymbol{W}}, \tilde{\boldsymbol{k}})$ where

$$
\tilde{\boldsymbol{w}}_1 = \begin{bmatrix} 2 & 0 \\ 2 & 2 \\ 1 & 4 \end{bmatrix} \quad \tilde{\boldsymbol{w}}_2 = \begin{bmatrix} 1 & 0 & 1 \\ 2 & 0 & 2 \\ 2 & 2 & 3 \end{bmatrix}
$$

$$
\tilde{\boldsymbol{w}}_3 = \begin{bmatrix} 1 & 0 \\ 0 & 6 \\ 3 & 0 \end{bmatrix} \quad \tilde{\boldsymbol{k}} = \begin{bmatrix} 3 \\ 5 \\ 6 \end{bmatrix}
$$

Thus, the resulting monitor place is defined by

$$
\tilde{C}_c = \begin{array}{c} \\ \left[\begin{array}{cc|ccc|cc|cc}
\overset{c_1}{-4} & \overset{c_2}{-2} & \overset{c_1}{0} & \overset{c_2}{4} & \overset{c_3}{0} & \overset{c_1}{0} & \overset{c_2}{0} & \overset{c_2}{2} & \overset{c_3}{0} \\
-6 & -6 & 0 & 6 & 2 & -10 & 0 & 6 & 12 \\
-6 & -9 & 0 & 2 & 2 & 3 & 0 & 6 & 0
\end{array}\right] \end{array}
\begin{array}{l} z_1 \\ z_2 \\ z_3 \end{array}
$$

Let $\boldsymbol{M}_{p,0} = [\varepsilon\ \ 1 \otimes c_1 + 1 \otimes c_3\ \ \varepsilon]'$ that satisfies the modified GMEC $(\tilde{\boldsymbol{W}}, \tilde{\boldsymbol{k}})$. The initial marking of $\tilde{p}_c$ should be taken equal to $\tilde{\boldsymbol{m}}_{c,0} = \tilde{\boldsymbol{k}} - \tilde{\boldsymbol{W}} \circ \boldsymbol{M}_{p,0} = \tilde{\boldsymbol{k}} - \sum_{i=1}^{3} \tilde{\boldsymbol{w}}_i \circ \boldsymbol{m}_{p,0,i} = 1 \otimes z_1 + 1 \otimes z_2 + 1 \otimes z_3$. ■

8. CONCLUSIONS

In this paper we first recalled some results in (Fanti *et al.*, 2003) where we shown that the classic PN control approach based on GMEC and monitor places can be extended to the case of CPN. Then, we focused our attention to the case in which not all transitions are controllable and observable thus a colored monitor designed for a given colored GMEC may not be admissible (it either disables a transition wrt an uncontrollable color, or observes the firing of a transition wrt an unobservable color). We shown how it may be possible to construct a less permissive, but admissible monitor, extending to the case of CPN the parametrization and the tabular procedure proposed by Moody and Antsaklis (1998) for uncolored PN.

REFERENCES

Fanti, M.P., A. Giua and C. Seatzu (2003). Generalized mutual exclusion constraints and monitors for colored petri nets. In: *Proc. 2003 IEEE*

Algorithm 7.2 (Design of a suboptimal monitor place for CPN).
Let us consider a CPN system $\langle N_p, \boldsymbol{M}_{p,0} \rangle$ with incidence matrix $\boldsymbol{C}_p$. Let $\boldsymbol{C}_p^{uc}$ ($\boldsymbol{C}_p^{uo}$) be the matrices obtained selecting from $\boldsymbol{C}_p$ the only columns relative to those transitions and those colors such that the considered transitions are uncontrollable (unobservable) wrt those colors.
Let $(\boldsymbol{W}, \boldsymbol{k})$ with $\boldsymbol{k} \in \mathcal{Z}(D)$, $D = \{z_1, \ldots, z_q\}$, be the GMEC that we want to force.
Finally, let $n_{uc} = \sum_{j:t_j \in T} |Co^{uc}(t_j)|$, $n_{uo} = \sum_{j:t_j \in T} |Co^{uo}(t_j)|$, and $M = \sum_{i=1}^{m} |Co(p_i)|$.

1. Let $\boldsymbol{V} := \boldsymbol{W} \circ \boldsymbol{C}_p^{uc}$, $\boldsymbol{U} := \boldsymbol{W} \circ \boldsymbol{C}_p^{uo}$,

$$\boldsymbol{R} := \underbrace{[\,\boldsymbol{\varepsilon} \;\ldots\; \boldsymbol{\varepsilon}\,]}_{M} \quad s_i := \boldsymbol{\varepsilon}, \quad i = 1, \ldots, q, \quad \boldsymbol{\varepsilon} \in \mathcal{Z}(D), \quad k := 1, \quad flag := 0.$$

2. While $k \leq q$ **and** $flag = 0$, **do**
if $\boldsymbol{V}(k, \cdot) \leq \boldsymbol{0}'_{n_{uc}}$ and $\boldsymbol{U}(k, \cdot) = \boldsymbol{0}'_{n_{uo}}$, **then let** $k := k + 1$, **else**
begin

2.1. let $\boldsymbol{A} := \begin{array}{|c|c|c|c|} \boldsymbol{C}_p^{uc} & \boldsymbol{C}_p^{uo} & \boldsymbol{I}_M & \boldsymbol{0}_M \\ \hline \boldsymbol{v} & \boldsymbol{u} & \bar{\boldsymbol{r}}_1 & r_2 \end{array}$ where $\boldsymbol{v} := \boldsymbol{V}(k, \cdot)$, $\boldsymbol{u} := \boldsymbol{U}(k, \cdot)$, $\bar{\boldsymbol{r}}_1 := \boldsymbol{0}'_M$, $r_2 := 1$.

2.2. Let $\mathcal{J}_{uc} := \{j \mid \boldsymbol{v}(j) > 0\}$ be the set of indices of columns of $\boldsymbol{A}$ corresponding to positive elements of $\boldsymbol{v}$.
Let $\mathcal{J}_{uo}^{+} := \{j + n_{uc} \mid \boldsymbol{u}(j) > 0\}$ be the set of indices of columns of $\boldsymbol{A}$ corresponding to positive elements of $\boldsymbol{u}$.
Let $\mathcal{J}_{uo}^{-} := \{j + n_{uc} \mid \boldsymbol{u}(j) < 0\}$ be the set of indices of columns of $\boldsymbol{A}$ corresponding to negative elements of $\boldsymbol{u}$.
Let $\mathcal{J} := \mathcal{J}_{uc} \cup \mathcal{J}_{uo}^{+} \cup \mathcal{J}_{uo}^{-}$. **If** $\mathcal{J} = \emptyset$, then **goto** 2.4.

2.3 If $\mathcal{J} \neq \emptyset$, then **choose** a value $\bar{\jmath} \in \mathcal{J}$.

If $\bar{\jmath} \in \mathcal{J}_{uc}$ then try to reduce $v(\bar{\jmath})$ to zero using the following procedure.

(a) **Let** $\mathcal{I} := \{i \mid C_p^{uc}(i, \bar{\jmath}) < 0\}$ be the set of row indices of the elements of $C_p^{uc}(\cdot, \bar{\jmath})$ that are negative.
If $\mathcal{I} = \emptyset$ it is not possible to get null the positive element $v(\bar{\jmath})$,
then **let** $flag := 1$ and **goto** 2.4.

(b) **Choose** an index $\bar{\imath} \in \mathcal{I}$ and compute $d := \text{l.c.m.}\{-C_p^{uc}(\bar{\imath}, \bar{\jmath}), v(\bar{\jmath})\}$
the least common multiple among the two elements $-C_p^{uc}(\bar{\imath}, \bar{\jmath})$ and $v(\bar{\jmath})$.

(c) **Update** the $(M+1)$–th row of $\boldsymbol{A}$ as follows:

$$\boldsymbol{A}(M+1, \cdot) := \frac{d}{v(\bar{\jmath})} \boldsymbol{A}(M+1, \cdot) + \frac{d}{C_p^{uc}(\bar{\imath}, \bar{\jmath})} \boldsymbol{A}(\bar{\imath}, \cdot).$$

By construction $v(\bar{\jmath})$ is null.

If $\bar{\jmath} \in \mathcal{J}_{uo}^{+}$ then try to reduce $u(\bar{\jmath} - n_{uc})$ to zero using the following procedure.

(a) **Let** $\mathcal{I}^{-} := \{i \mid C_p^{uo}(i, \bar{\jmath} - n_{uc}) < 0\}$ be the set of row indices of the elements of $C_p^{uo}(\cdot, \bar{\jmath} - n_{uc})$ that are negative.
If $\mathcal{I}^{-} = \emptyset$ it is not possible to get null the positive element of $\boldsymbol{u}(\bar{\jmath} - n_{uc})$,
then let $flag := 1$ and goto 2.4.

(b) **Choose** an index $\bar{\imath} \in \mathcal{I}^{-}$ and compute $d^{+} := \text{l.c.m.}\{-C_p^{uo}(\bar{\imath}, \bar{\jmath} - n_{uc}), u(\bar{\jmath} - n_{uc})\}$.

(c) **Update** the $(M+1)$–th row of $\boldsymbol{A}$ as follows:

$$\boldsymbol{A}(M+1, \cdot) := \frac{d^{+}}{u(\bar{\jmath} - n_{uc})} \boldsymbol{A}(M+1, \cdot) + \frac{d^{+}}{C_p^{uo}(\bar{\imath}, \bar{\jmath} - n_{uc})} \boldsymbol{A}(\bar{\imath}, \cdot).$$

By construction $u(\bar{\jmath} - n_{uc})$ is null.

If $\bar{\jmath} \in \mathcal{J}_{uo}^{-}$ then try to reduce $u(\bar{\jmath} - n_{uc})$ to zero using the following procedure.

(a) **Let** $\mathcal{I}^{+} := \{i \mid C_p^{uo}(i, \bar{\jmath} - n_{uc}) > 0\}$ be the set of row indices of the elements of $C_p^{uo}(\cdot, \bar{\jmath} - n_{uc})$ that are positive.
If $\mathcal{I}^{+} = \emptyset$ it is not possible to get null the negative element $u(\bar{\jmath} - n_{uc})$,
then let $flag := 1$ and **goto** 2.4.

(b) **Choose** an index $\bar{\imath} \in \mathcal{I}^{+}$ and compute $d^{-} := \text{l.c.m.}\{C_p^{uo}(\bar{\imath}, \bar{\jmath} - n_{uc}), -u(\bar{\jmath} - n_{uc})\}$.

(c) **Update** the $(M+1)$–th row of $\boldsymbol{A}$ as follows:

$$\boldsymbol{A}(M+1, \cdot) := \frac{d^{-}}{u(\bar{\jmath} - n_{uc})} \boldsymbol{A}(M+1, \cdot) + \frac{d^{-}}{C_p^{uo}(\bar{\imath}, \bar{\jmath} - n_{uc})} \boldsymbol{A}(\bar{\imath}, \cdot).$$

By construction $u(\bar{\jmath} - n_{uc})$ is null.

2.4. If $flag = 1$ then the algorithm stops without providing a solution,
else
the last row of the table is in the form $|\boldsymbol{v}|\boldsymbol{u}|\bar{\boldsymbol{r}}_1|r_2|$
where $\boldsymbol{v} \leq \boldsymbol{0}_{n_{uc}}$, $\boldsymbol{u} = \boldsymbol{0}_{n_{uo}}$, $\boldsymbol{r}_1 \in \mathbb{N}^m$ and $r_2 \in \mathbb{N}$.
Let $s_k(k) := r_2$, $\boldsymbol{R}(k, \cdot) := \bar{\boldsymbol{r}}_1$,
If $k < q$, then **let** $k := k + 1$ and **goto** 2.2.

endif

end ■

Int. Conf. on Systems, Man, and Cybernetics. pp. 1860–1865.

Giua, A. and C. Seatzu (2004). Monitor design for colored petri nets with uncontrollable and unobservable transitions. *Technical Report, University of Cagliari, Italy.* http://www.diee.unica.it/ seatzu/CPN.pdf.

Giua, A., F. DiCesare and M. Silva (1992). Generalized mutual exclusion constraints on nets with uncontrollable transitions. In: *Proc. 1992 IEEE Int. Conf. on Systems, Man, and Cybernetics*. pp. 974–979.

Jensen, K. (1992). *Coloured Petri Nets. Basic concepts, analysis methods and practical use. Volume 1: Basic Concepts.* Springer Verlag.

Moody, J.O. and P.J. Antsaklis (1998). *Supervisory control of discrete event systems using Petri nets.* Kluwer Academic Publishers.

ELSEVIER
IFAC
PUBLICATIONS
www.elsevier.com/locate/ifac

ON THE LINEARIZABILITY OF DISCRETE TIMED EVENT GRAPHS WITH MULTIPLIERS USING (MIN,+) ALGEBRA

Samir Hamaci, Jean-Louis Boimond,
Sébastien Lahaye, Mustapha Mostefaoui

LISA, 62 avenue Notre Dame du Lac,
49000 Angers - France

e-mail:
[Hamaci,Boimond,Lahaye,Mostefaoui]@istia.univ-angers.fr

Abstract: We deal with the reduction of timed event graphs with multipliers to timed event graphs without multipliers which allows (min, +) linear representations of graphs. The proposed linearization method uses an algebra defined on a set of operators and endowed with the pointwise minimum as addition and the composition as multiplication. The method considers a change of variables which does not increase the number of transitions of graph, but the graph is supposed consistent and conservative and its (unique) elementary T-semiflow has at least one component equal to one. *Copyright © 2004 IFAC*

Keywords: Timed weighted marked graphs, timed event graphs with multipliers, linearization, dioid, (min, +) algebra.

1. INTRODUCTION

Timed Event Graphs (TEG's) are a subclass of Petri nets well adapted to model synchronization and saturation phenomena. Moreover their behaviors can be modeled by linear equations in (min, +) algebra, or in an equivalent algebra, which allows analogy with conventional linear system theory (Baccelli *et al.*, 1992). We deal here with a broader class of graphs, in which weights (in $\mathbb{N}$) are associated with arcs. The use of weights is appropriate to model a large number of systems, for example, when the achievement of a specific task requires several units of a same resource, or when an assembly operation requires several units of a same piece.

The study of these graphs, called Timed Event Graphs with Multipliers (TEGM's), or timed weighted marked graphs, requires a particular algebra, analogous to the (min, +) algebra but defined on a set of operators endowed with the pointwise minimum as addition and the composition of operators as multiplication (Cohen *et al.*, 1998). Due to the presence of weights assigned to arcs, models are not linear which prevents from using performance analysis or control methods developed in the linear system setting (Baccelli *et al.*, 1992), (Lahaye, 2000), (Menguy, 1997). We propose here a method of linearization which allows to reduce particular TEGM's into TEG's.

We consider that TEGM's are consistent and conservative, so they are strongly connected and have an unique elementary T-semiflow. To our knowledge, relevant works on linearization of such graphs lead to a duplication of transitions. The system transformation proposed in (Munier, 1993) under *single* server semantics hypothesis, or in (Nakamura and Silva, 1999) under *infinite* server semantics hypothesis, leads to a TEG with $|\theta|$

transitions ($|\theta|$ is the 1-norm of the elementary T-semiflow of the corresponding TEGM). Another approach stated in $(\min, +)$ algebra is proposed in (Trouillet *et al.*, 2002) when each elementary circuit of graph contains at least one *normalized* transition (*i.e.*, a transition for which its corresponding elementary T-semiflow component is equal to one). This method also increases the number of transitions. We improve it in the sense that, the number of transitions is not increased and only the presence of one normalized transition is required in the graph. Note that such a method allows the use of a multi *server* semantics.

This paper is organized as follows. We briefly define the class of TEGM's in Section 2. An algebra of operators is introduced in Section 3 to model these graphs: an operator μ allowing multipliers on arcs is added to operators γ, δ usually used to model TEG's (without multipliers) (Cohen *et al.*, 1995), (Cohen *et al.*, 1998). The proposed linearization method of TEGM's is presented in Section 4. Before concluding, we give a short example in which the cycle time is computed from the linearized graph.

2. RECURRENT EQUATIONS OF TEGM'S

Let us briefly introduce the definitions and some basic properties of Petri nets, see (Murata, 1989) for a more exhaustive description.

A *timed Petri net with multipliers* is a valued bipartite graph given by a 5-tuple (P, T, M, m, τ).

(1) P is the finite set of places, T is the finite set of transitions.
(2) $M \in \mathbb{N}^{P\times T \cup T\times P}$. Given $p \in P$ and $q \in T$, the multiplier M_{pq} (resp. M_{qp}) specifies the weight of the arc from transition q to place p (resp. from place p to transition q).
(3) $m \in \mathbb{N}^P$: m_p assigns an initial number of tokens to place p.
(4) $\tau \in \mathbb{N}^P$: τ_p gives the minimal time a token must spend in place p before it can contribute to the enabling of its downstream transitions.

A vector $\theta \in \mathbb{N}^T$ such that $\theta \neq 0$ and $W\theta = 0$ is a *T-semiflow* (W is the incidence matrix). A vector $Y \in \mathbb{N}^P$ such that $Y \neq 0$ and $Y^t W = 0$ is a *P-semiflow*.

A T-semiflow θ is *elementary iff* it is *canonical* (*i.e.*, the greatest common divisor of its components is equal to one) and has a *minimal support* (*i.e.*, no other T-semiflow θ' is such that $\{q \in T | \theta'(q) > 0\} \subset \{q \in T | \theta(q) > 0\}$).

An *event graph* is a Petri net such that each place has exactly one upstream and one downstream transition.

The TEGM's we consider in the following are *consistent* (*i.e.*, there exists a T-semiflow θ covering all transitions: $\{q \in T | \theta(q) > 0\} = T$) and *conservative* (*i.e.*, there exists a P-semiflow Y covering all places: $\{p \in P | Y(p) > 0\} = P$).

Such graphs verify the next properties (Teruel *et al.*, 1992):

- A TEGM allows a live and bounded initial marking m *iff* it is consistent and conservative.
- A consistent TEGM is strongly connected *iff* it is conservative.
- A consistent TEGM has a unique elementary T-semiflow.
- The product of multipliers along any circuit of a conservative TEGM is equal to one.

We denote by $^\bullet q$ (resp. $q^\bullet$) the set of places upstream (resp. downstream) transition q. Similarly, $^\bullet p$ (resp. $p^\bullet$) denotes the set of transitions upstream (resp. downstream) place p.

The *earliest* functioning rule of TEGM's is defined as follows. A transition q fires as soon as all its upstream places $\{p \in {^\bullet q}\}$ contain enough tokens (M_{qp}) having spent at least τ_p units of time in place p. When the transition q fires, it consumes M_{qp} tokens in each upstream place $p \in {^\bullet q}$ and produces $M_{p'q}$ tokens in each downstream place $p' \in q^\bullet$.

With each transition is associated a *counter variable* n_q: $n_q(t)$ denotes the cumulated number of firings of transition q up to time t ($t \in \mathbb{Z}$).

We suppose in the following that all the tokens in a graph are "frozen" before time 0 and are available at time 0 which is a classical assumption in Petri nets theory. Hence, with each counter variable of a TEGM is added a counter variable corresponding to an impulse input e (*i.e.*, $e(t) = 0$ for $t \leq 0$ and $e(t) = +\infty$ for $t > 0$), see (Lahaye, 2000, §5.2) for more details.

Assertion 1. The counter variables of a TEGM (under the earliest firing rule) satisfy the following *transition to transition* equation:

$$n_q(t) = \min_{p \in {^\bullet q},\, q' \in {^\bullet p}} \lfloor M_{qp}^{-1}(m_p + M_{pq'} n_{q'}(t - \tau_p)) \rfloor . \tag{1}$$

Note the presence of inferior integer part to preserve integrity of Eq. (1). In general, a transition q may have several upstream transitions $\{q' \in {^{\bullet\bullet} q}\}$ which implies that the associated counter variable is given by the min of *transition to transition* equations obtained for each upstream transition.

Example 1. The counter variable associated with transition q described in Fig. 1 satisfies the following equation:

$$n_q(t) = \min(\lfloor M_{qp}^{-1}(m_p + M_{pq'}n_{q'}(t - \tau_p))\rfloor, e(t)).$$

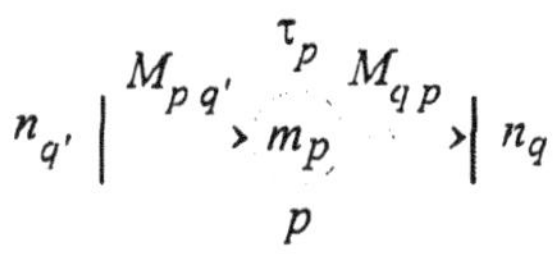

Fig. 1. A simple TEGM.

Example 2. Let us consider TEGM depicted in Fig. 2.

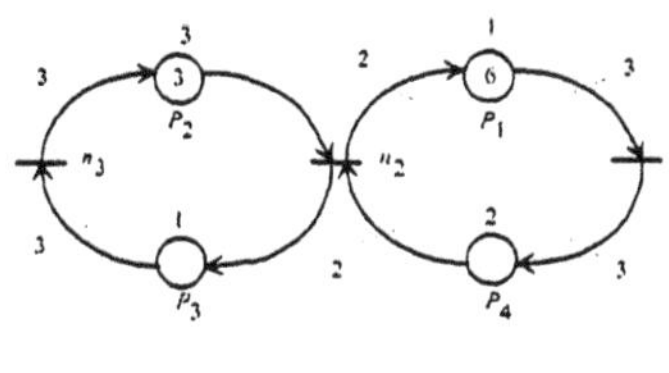

Fig. 2. A TEGM.

Counter variables verify the following equations:

$$\begin{cases} n_1(t) &= \min(\lfloor \dfrac{6 + 2n_2(t-1)}{3} \rfloor, e(t)), \\ n_2(t) &= \min(\lfloor \dfrac{3n_1(t-2)}{2} \rfloor, 3 + 3n_3(t-3), e(t)), \\ n_3(t) &= \min(\lfloor \dfrac{n_2(t-1)}{3} \rfloor, e(t)). \end{cases}$$

The marking graph given in Fig. 3 shows that a periodic behavior is reached after a transient behavior.

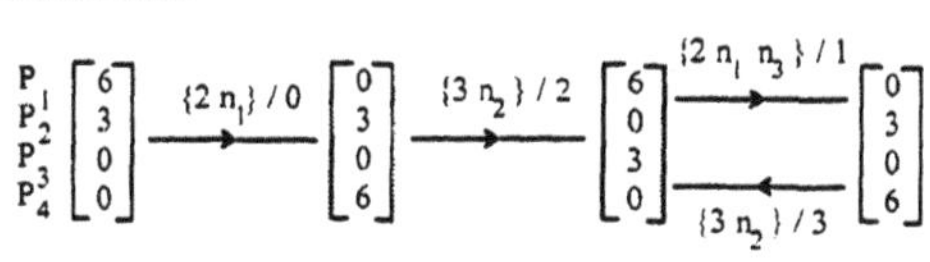

Fig. 3. Marking graph.

3. DIOID, OPERATORIAL REPRESENTATION OF TEGM'S

Let us briefly introduce the definitions and some basic properties of dioid algebra, see (Baccelli *et al.*, 1992) for a more exhaustive description.

A *dioid* $(\mathcal{D}, \oplus, \otimes)$ is a semiring in which $\oplus$ is idempotent ($\forall a,\ a \oplus a = a$). Neutral elements of $\oplus$ and $\otimes$ are denoted ε and e respectively.

A dioid is *commutative* when $\otimes$ is commutative. The symbol $\otimes$ is often omitted. Due to idempotency of $\oplus$, a dioid can be endowed with a natural order relation defined by $a \preceq b \Leftrightarrow b = a \oplus b$ (the least upper bound of {a,b} is equal to $a \oplus b$).

A dioid $\mathcal{D}$ is *complete* if every subset A of $\mathcal{D}$ admits a least upper bound denoted $\bigoplus_{x \in A} x$, and if $\otimes$ distributes at left and at right over infinite sums. The greatest element noted T of a complete dioid $\mathcal{D}$ is equal to $\bigoplus_{x \in \mathcal{D}} x$. The greatest lower bound of every subset X of a complete dioid always exists and is noted $\bigwedge_{x \in X} x$.

Example 3. The set $\mathbb{Z} \cup \{\pm\infty\}$, endowed with min as $\oplus$ and usual addition as $\otimes$, is a complete dioid noted $\overline{\mathbb{Z}}_{\min}$ and usually called (min, +) algebra with neutral elements $\varepsilon = +\infty$, $e = 0$ and $T = -\infty$.

Example 4. Starting from a 'scalar' dioid $\mathcal{D}$, let us consider $p \times p$ matrices with entries in $\mathcal{D}$. The sum and product of matrices are defined conventionally after the sum and product of scalars in $\mathcal{D}$. Let A and B two matrices in $\mathcal{D}^{p \times p}$, then

$$(A \oplus B)_{ij} = A_{ij} \oplus B_{ij},$$

$$(A \otimes B)_{ij} = \bigoplus_{k=1}^{p} A_{ik} \otimes B_{kj}.$$

This set of matrices, endowed with these two operations, is also a dioid denoted $\mathcal{D}^{p \times p}$.

Let us give definitions of *signal* and *set of signals* to represent counter variables. A *signal* is an increasing map from $\mathbb{Z}$ to $\mathbb{Z} \cup \{\pm\infty\}$. We denote $S = (\mathbb{Z} \cup \{\pm\infty\})^{\mathbb{Z}}$ the set of signals.

This set is endowed with a kind of module structure, called *min-plus semimodule*, the two associated operations are:

- pointwise minimum of time functions to add signals: $\forall t \in \mathbb{Z}$,
$$(u \oplus v)(t) = u(t) \oplus v(t) = \min(u(t), v(t));$$
- addition of a constant to play the role of external product of a signal by a scalar: $\forall t \in \mathbb{Z}, \forall \lambda \in \mathbb{Z} \cup \{\pm\infty\}$,
$$(\lambda.u)(t) = \lambda \otimes u(t) = \lambda + u(t).$$

An *operator* is a map from a set of signals to a set of signals. An operator H is *linear* if it preserves the min-plus semimodule structure, *i.e.*, for all signals u, v and constant λ,

$$H(u \oplus v) = H(u) \oplus H(v) \quad \text{(additive property)},$$

$$H(\lambda \otimes u) = \lambda \otimes H(u) \quad \text{(homogeneity property)}.$$

We will introduce in the following three operators defined from S to S, which are used for the modeling of TEGM's:

(1) $\gamma^{\nu} : n(t) \mapsto n(t) + \nu$ in which $\nu \in \mathbb{Z} \cup \{\pm\infty\}$,
(2) $\delta^{\tau} : n(t) \mapsto n(t - \tau)$ in which $\tau \in \mathbb{Z} \cup \{\pm\infty\}$,
(3) $\mu_r : n(t) \mapsto \lfloor r \times n(t) \rfloor$ in which $r \in \mathbb{Q}^+$ (r is equal to a ratio of elements in $\mathbb{N}$).

An algebra of operators is used in (Cohen *et al.*, 1998) to model continuous TEGM's. A similar algebra is proposed here to model discrete TEGM's.

We denote by $\mathcal{D}_{\min}$ the (noncommutative) dioid of finite sums of operators $\{\mu_r, \gamma^{\nu}\}$ endowed with pointwise min ($\oplus$) and composition ($\otimes$) operations, with neutral elements equal to $\varepsilon =$

$\mu_{+\infty}\gamma^{+\infty}$ and $e = \mu_1\gamma^0$ respectively. Thus, an element in $\mathcal{D}_{\min}$ is a map $p = \bigoplus_{i=1}^{k} \mu_{r_i}\gamma^{\nu_i}$ defined from S to S such that

$$\forall\, t \in \mathbb{Z},\, p\,(n(t)) = \min_{1\le i\le k} (\lfloor r_i(\nu_i + n(t))\rfloor).$$

Let a map $g : \mathbb{Z} \to \mathcal{D}_{\min},\ \tau \mapsto g(\tau)$ in which $g(\tau) = \bigoplus_{i=1}^{k_\tau} \mu_{r_i^\tau}\gamma^{\nu_i^\tau}$. We define the power series $G(\delta)$ in the indeterminate δ with coefficients in $\mathcal{D}_{\min}$ by:

$$G(\delta) = \bigoplus_{\tau\in\mathbb{Z}} g(\tau)\delta^\tau.$$

The set of these formal power series endowed with the two following operations:
$F(\delta) \oplus G(\delta)$:
$(f \oplus g)(\tau) = f(\tau) \oplus g(\tau) = \min(f(\tau), g(\tau))$,
$F(\delta) \otimes G(\delta)$:
$(f \otimes g)(\tau) = \bigoplus_{i\in\mathbb{Z}} f(i) \otimes g(\tau - i) = \inf_{i\in\mathbb{Z}}(f(i) + g(\tau - i))$,
is a dioid noted $\mathcal{D}_{\min}[\![\delta]\!]$, with neutral elements $\varepsilon = \mu_{+\infty}\gamma^{+\infty}\delta^{-\infty}$ and $e = \mu_1\gamma^0\delta^0$.

Elements of $\mathcal{D}_{\min}[\![\delta]\!]$ allow modeling the transfer between two transitions of a TEGM. A formal series of $\mathcal{D}_{\min}[\![\delta]\!]$ can also represent a signal n as $N(\delta) = \oplus_{\tau\in\mathbb{Z}} n(\tau)\ \delta^\tau$, simply due to the fact that it is also equal to $n \otimes e$ (by definition of neutral element e of $\mathcal{D}_{\min}$). For example, the graph depicted in Fig. 1 is represented by equation $N_q(\delta) = \mu_{M_{qp}^{-1}}\gamma^{m_p}\delta^{\tau_p}\mu_{M_{pq'}}N_{q'}(\delta) \oplus E(\delta)$ where $N_q(\delta)$ and $N_{q'}(\delta)$ denote elements of $\mathcal{D}_{\min}[\![\delta]\!]$ associated with transitions q and q' respectively.

Assertion 2. The counter variables of a TEGM satisfy the following equation:

$$N_q(\delta) = \bigoplus_{p\in{}^\bullet q,\, q'\in{}^\bullet p} \mu_{M_{qp}^{-1}}\gamma^{m_p}\delta^{\tau_p}\mu_{M_{pq'}}N_{q'}(\delta). \tag{2}$$

Example 5. The TEGM depicted in Fig. 2 admits the following representation:

$$\begin{pmatrix} N_1 \\ N_2 \\ N_3 \end{pmatrix} = \begin{pmatrix} \varepsilon & \mu_{1/3}\gamma^6\delta^1\mu_2 & \varepsilon \\ \mu_{1/2}\delta^2\mu_3 & \varepsilon & \gamma^3\delta^3\mu_3 \\ \varepsilon & \mu_{1/3}\delta^1 & \varepsilon \end{pmatrix} \begin{pmatrix} N_1 \\ N_2 \\ N_3 \end{pmatrix} \oplus \begin{pmatrix} E \\ E \\ E \end{pmatrix}.$$

Let us express some properties of operators γ, δ, μ.

Proposition 1. Let $a,\, b \in \mathbb{N}$, we have:

(1) $\gamma^a\delta^b = \delta^b\gamma^a$, $\mu_a\delta^b = \delta^b\mu_a$ (commutative properties).
(2) $\mu_{a^{-1}}\mu_b = \mu_{(a^{-1}b)}$.
(3) Let $N(\delta)$ such that, $\forall t \in \mathbb{Z}$, $n(t)$ is a multiple of a then $\mu_{a^{-1}}\gamma^b N(\delta) = \gamma^{\lfloor a^{-1}b\rfloor}\mu_{a^{-1}}N(\delta)$.
(4) $\gamma^b\mu_a = \mu_a\gamma^{a^{-1}b}$, or equivalently, $\mu_a\gamma^b = \gamma^{ab}\mu_a$.

Proof: Point 1 is obvious.
Point 2: $\mu_{a^{-1}}\mu_b N(\delta)$ corresponds to $\lfloor a^{-1}\lfloor b\,n(t)\rfloor\rfloor = \lfloor a^{-1}b\,n(t)\rfloor$ which leads to $\mu_{(a^{-1}b)}N(\delta)$.
Point 3: $\mu_{a^{-1}}\gamma^b N(\delta)$ corresponds to $\lfloor a^{-1}(b + n(t))\rfloor = \lfloor a^{-1}b\rfloor + a^{-1}n(t)$ since $n(t)$ ($\in \mathbb{Z} \cup \{\pm\infty\}$) is a multiple of a, which leads to $\gamma^{\lfloor a^{-1}b\rfloor}\mu_{a^{-1}}N(\delta)$.
Point 4: $\gamma^b\mu_a N(\delta)$ corresponds to $b + \lfloor a\,n(t)\rfloor = \lfloor a(a^{-1}b + n(t))\rfloor$ which leads to $\mu_a\gamma^{a^{-1}b}N(\delta)$.

4. LINEARIZATION OF TEGM'S

The presence of integer part in operator μ induces a nonlinearity in Eq. (2) used to represent a TEGM. So, as far as possible, we seek to represent a TEGM with linear equations in order to apply standard results of linear system theory developed in the dioid setting, which leads to transform a TEGM into a TEG (represented without operator μ).

4.1 Principle of linearization

A consistent TEGM has a unique elementary T-semiflow in which components are in $\mathbb{N}^*$. The principle of linearization consists in normalizing the T-semiflow, *i.e.*, to obtain a same value for all the T-semiflow components (Trouillet, 2003).

The proposed method is based on a diagonal change of counter variables. Let $\theta = (\theta_1\ \theta_2\ ...\ \theta_n)^t$ the T-semiflow of a TEGM ($n = card(T)$). We note LCM the *Least Common Multiple* of θ_q, for $q = 1, ..., n$.

Proposition 2. Let us consider the following change of counter variables: $x_i(t) = \lambda_i\ n_i(t)$, $\forall i = 1, ..., n$ such that $\lambda_i\ \theta_i = LCM$. We obtain the following counter variables by performing this change of variables in Eq.(2):

$$X_q(\delta) = \bigoplus_{p\in{}^\bullet q,\, q'\in{}^\bullet p} \mu_{\lambda_q}\mu_{\lambda_q^{-1}}\ \gamma^{\lambda_q M_{qp}^{-1} m_p}\ \delta^{\tau_p}\ X_{q'}(\delta),$$

or equivalently,

$$x_q(t) = \min_{p\in{}^\bullet q,\, q'\in{}^\bullet p}(\lfloor\lambda_q\lfloor\lambda_q^{-1}(\lambda_q M_{qp}^{-1}m_p + x_{q'}(t - \tau_p))\rfloor\rfloor).$$

Proof: We have $X_q(\delta) = \mu_{\lambda_q}N_q(\delta)$ (with $\lambda_q \in \mathbb{N}^*$), so we deduce from Eq. (2) that:

$$X_q(\delta) = \bigoplus_{p\in{}^\bullet q,\, q'\in{}^\bullet p} \mu_{\lambda_q}\mu_{M_{qp}^{-1}}\ \gamma^{m_p}\delta^{\tau_p}\ \mu_{M_{pq'}}\ N_{q'}(\delta),$$

i.e., thanks to proposition 1.4:

$$X_q(\delta) = \bigoplus_{p\in{}^\bullet q,\, q'\in{}^\bullet p} \mu_{\lambda_q}\mu_{M_{qp}^{-1}}\mu_{M_{pq'}}\ \gamma^{M_{pq'}^{-1}m_p}\ \delta^{\tau_p}\ N_{q'}(\delta),$$

i.e., thanks to proposition 1.2:

$$X_q(\delta) = \bigoplus_{p\in {}^\bullet q,\, q'\in {}^\bullet p} \mu_{\lambda_q}\mu_{(M_{qp}^{-1}M_{pq'})}\gamma^{M_{pq'}^{-1}m_p}\,\delta^{\tau_p}\,N_{q'}(\delta).$$

From the definition of θ (*i.e.*, $\theta \neq 0$ and $W\theta = 0$ (Munier, 1993)) we easily deduce that $M_{pq'}\theta_{q'} = M_{qp}\theta_q$. Moreover, knowing that $\theta_q = \lambda_q^{-1}LCM$ and $\theta_{q'} = \lambda_{q'}^{-1}LCM$, we deduce that:
$M_{qp}^{-1}M_{pq'} = \lambda_q^{-1}\lambda_{q'}$ and $M_{pq'}^{-1} = \lambda_{q'}^{-1}\lambda_q M_{qp}^{-1}$.
Hence, by using propositions 1.2 and 1.4, we have:

$$\begin{aligned} X_q(\delta) &= \bigoplus_{p\in {}^\bullet q,\, q'\in {}^\bullet p} \mu_{\lambda_q}\mu_{(\lambda_q^{-1}\lambda_{q'})}\gamma^{\lambda_{q'}^{-1}\lambda_q M_{qp}^{-1}m_p}\delta^{\tau_p}N_{q'}(\delta) \\ &= \bigoplus_{p\in {}^\bullet q,\, q'\in {}^\bullet p} \mu_{\lambda_q}\mu_{\lambda_q^{-1}}\gamma^{\lambda_q M_{qp}^{-1}m_p}\delta^{\tau_p}\mu_{\lambda_{q'}}N_{q'}(\delta), \end{aligned}$$

which leads to the result since $X_{q'}(\delta) = \mu_{\lambda_{q'}}N_{q'}(\delta)$.

Let us note that the commutation of the operator $\mu_{\lambda_q^{-1}}$ with $\gamma^{\lambda_q M_{qp}^{-1}m_p}$ to linearize this equation is not always possible (see proposition 1.3).

A transition q is said to be *normalized* if its corresponding elementary T-semiflow component, θ_q, is equal to one (Trouillet *et al.*, 2002).

Proposition 3. : Let q' a normalized transition, we have: $\forall a \in \mathbb{N}, \forall q = 1, ..., n,$

$$\mu_{\lambda_q}\mu_{\lambda_q^{-1}}\gamma^a X_{q'}(\delta) = \gamma^{\lambda_q \lfloor \lambda_q^{-1}a\rfloor}X_{q'}(\delta).$$

Proof: We have $x_{q'}(t) = LCM\, n_{q'}(t)$ by definition of q', so $x_{q'}(t)$ is a multiple of λ_q ($\lambda_q \in \mathbb{N}^*$), $\forall q = 1, ..., n$. Hence, thanks to proposition 1.3, we deduce that:
$\mu_{\lambda_q^{-1}}\gamma^a X_{q'}(\delta) = \gamma^{\lfloor \lambda_q^{-1}a\rfloor}\mu_{\lambda_q^{-1}}X_{q'}(\delta)$.
So, thanks to proposition 1.4, we deduce that:
$\mu_{\lambda_q}\mu_{\lambda_q^{-1}}\gamma^a X_{q'}(\delta) = \gamma^{\lambda_q\lfloor \lambda_q^{-1}a\rfloor}\mu_{\lambda_q}\mu_{\lambda_q^{-1}}X_{q'}(\delta)$.
Finally, we prove the result by noting that $\mu_{\lambda_q}\mu_{\lambda_q^{-1}}X_{q'}(\delta) = X_{q'}(\delta)$ since $\mu_{\lambda_q}\mu_{\lambda_q^{-1}}X_{q'}(\delta)$ corresponds to $\lfloor \lambda_q \times \lfloor \lambda_q^{-1}x_{q'}(t)\rfloor\rfloor = x_{q'}(t)$ because $x_{q'}(t)$ is a multiple of λ_q.

We give in the following the state model obtained after the change of counter variables described in proposition 2. Consider the vector X composed of the counter variables described in proposition 2. Let $X^t = (X_{\overline{N}}^t\, X_N^t)$ in which X_N is composed with counter variables associated with normalized transitions, and $X_{\overline{N}}$ is composed with the other counter variables. The counter variables corresponding to impulse input e added with each counter variable X_i - to represent that all tokens are "frozen" before time 0 and are available at time 0 - are considered separately from the other counter variables. More precisely, they are simply added with each counter variable X_i, since $\mu_{\lambda_q}\mu_{\lambda_q^{-1}}E(\delta) = E(\delta)$. We have finally:

$$\begin{pmatrix} X_{\overline{N}} \\ X_N \end{pmatrix} = \begin{pmatrix} F & G \\ H & L \end{pmatrix}\begin{pmatrix} X_{\overline{N}} \\ X_N \end{pmatrix} \oplus \begin{pmatrix} E \\ E \end{pmatrix}, \qquad (3)$$

knowing that such an equation admits the following earliest solution[1]:

$$\begin{cases} X_{\overline{N}} &= F^*\,G\,X_N \oplus F^*\,E, \\ X_N &= (HF^*\,G \oplus L)\,X_N \oplus HF^*\,E \oplus E, \end{cases} \qquad (4)$$

in which $F^* = e \oplus F \oplus F^2 \oplus \cdots$.

Proposition 4. A consistent and conservative TEGM is linearizable without increasing the number of its transitions, if it contains at least one normalized transition.

Proof: By performing the change of counter variables described in proposition 2, the graph depicted in Fig. 1 is represented by equation $X_q(\delta) = \mu_{\lambda_q}\mu_{\lambda_q^{-1}}\,\gamma^{\lambda_q M_{qp}^{-1}m_p}\,\delta^{\tau_p}\,X_{q'}(\delta) \oplus E(\delta)$. Let us suppose that transition q' is normalized, then we obtain by using proposition 3:
$X_q(\delta) = \gamma^{\lambda_q\lfloor M_{qp}^{-1}m_p\rfloor}\,\delta^{\tau_p}\,X_{q'}(\delta) \oplus E(\delta)$,
i.e., a linear transfer between transitions q' and q (corresponding to a TEG). More generally, such a linearization procedure can be applied to all transitions, since any transition can be linked with a normalized transition (see Eq. (4)), due to the strongly connected property of TEGM we consider.

4.2 Example, computation of the cycle time

The TEGM depicted in Fig. 2 admits the elementary T-semiflow $\theta = (2\;3\;1)^t$ in which θ_3 is associated with a normalized transition.

From proposition 2, we define the state vector $X = (X_1\,X_2\,X_3)^t$ in which $X_i(\delta) = \mu_{\lambda_i}N_i(\delta)$, $i = 1, 2, 3$ with $\lambda_1 = 3, \lambda_2 = 2, \lambda_3 = 6 (= LCM)$. We obtain the following expressions of matrices F, G, H, L with $X_{\overline{N}} = (X_1\,X_2)^t$, $X_N = X_3$ (see Eq. (3)):

$$F = \begin{pmatrix} \varepsilon & \mu_3\mu_{1/3}\gamma^6\delta^1 \\ \mu_2\mu_{1/2}\delta^2 & \varepsilon \end{pmatrix}, G = \begin{pmatrix} \varepsilon \\ \mu_2\mu_{1/2}\gamma^6\delta^3 \end{pmatrix},$$

$H = (\varepsilon\;\mu_6\mu_{1/6}\delta^1)$, $L = \varepsilon$. We deduce the following expressions:

$F^*E = \begin{pmatrix} (\gamma^6\delta^3)^* \\ \delta^2(\gamma^6\delta^3)^* \end{pmatrix}$ (we have $\forall a, b \in \mathbb{N}$, $\mu_{\lambda_q}\mu_{\lambda_q^{-1}}\gamma^a\delta^b E(\delta) = \gamma^{\lambda_q\lfloor\lambda_q^{-1}a\rfloor}\delta^b E(\delta)$),

$F^*G\,X_N = \begin{pmatrix} \gamma^{12}\delta^4(\gamma^6\delta^3)^* \\ \gamma^6\delta^3(\gamma^6\delta^3)^* \end{pmatrix} X_N$ (thanks to proposition 3),

---

[1] The implicit equation $x = ax \oplus b$ admits a^*b as least solution.

which allows obtaining the following earliest solution by using Eq. (4):

$$\left\{ \begin{array}{l} \begin{pmatrix} X_1 \\ X_2 \end{pmatrix} = \begin{pmatrix} \gamma^{12}\delta^4(\gamma^6\delta^3)^* \\ \gamma^6\delta^3(\gamma^6\delta^3)^* \end{pmatrix} X_3 \oplus \begin{pmatrix} (\gamma^6\delta^3)^* \\ \delta^2(\gamma^6\delta^3)^* \end{pmatrix} E, \\ \quad X_3 \quad = \gamma^6\delta^4(\gamma^6\delta^3)^* \, X_3 \oplus \delta^3(\gamma^6\delta^3)^* E. \end{array} \right.$$

Let us express these equations in usual counter setting (dioid $\overline{\mathbb{Z}}_{\min}$), we have, $\forall t \in \mathbb{Z}$:

$$\left\{ \begin{array}{l} x_1(t) = 6\, x_1(t-3) \oplus 12\, x_3(t-4) \oplus e(t), \\ x_2(t) = 6\, x_2(t-3) \oplus 6\, x_3(t-3) \oplus e(t-2), \\ x_3(t) = 6\, x_3(t-4) \oplus e(t-3). \end{array} \right.$$

These equations are quite (min,+) linear. It turns out that the TEG depicted in Fig. 4 is a possible representation of the previous equations.

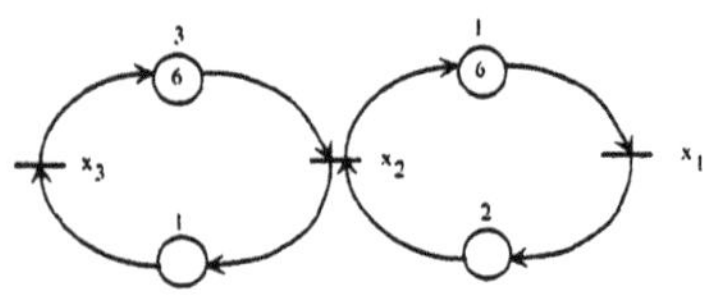

Fig. 4. TEG (linearized TEGM).

The expressions of the counter variables associated with the TEGM are finally deduced from whose of associated with the TEG by using the relations $x_i(t) = \lambda_i \, n_i(t)$, $i = 1, 2, 3$.

Let us compute, for example, the cycle time of the previous TEGM from the corresponding TEG. The computation of the cycle time of a strongly connected TEG is classical (Gondran and Minoux, 1985). Let us denote by C the set of elementary circuits of the graph, $T(c), M(c)$ the sum of holding times and the total number of tokens initially contained in the circuit c, then the cycle time, denoted TC, is equal to:

$$\max_{c \in C} \frac{T(c)}{M(c)}.$$

We deduce from this expression that the cycle time of the TEG depicted in Fig. 4 (which has two elementary circuits: (x_1, x_2) and (x_2, x_3)) is equal to:

$$\max\left(\frac{3}{6}, \frac{4}{6}\right) = \frac{2}{3},$$

which equivalently means that the firing frequency of transitions x_1, x_2, x_3 is equal to $3/2$.

So, from the relation $x_i(t) = \lambda_i \, n_i(t)$, $i = 1, 2, 3$, we easily deduce the firing frequencies, denoted f_1, f_2, f_3, of transitions n_1, n_2, n_3 of the TEGM; we have :

$$f_1 = \frac{3}{2} \times \frac{1}{3}, f_2 = \frac{3}{2} \times \frac{1}{2}, f_3 = \frac{3}{2} \times \frac{1}{6},$$

which implies a cycle time of the TEGM equal to 4 thanks to the following relation:

$$TC = \frac{\theta_i}{f_i} \text{ for } i = 1,\ 2,\ 3.$$

5. CONCLUSION

We propose a method to linearize a consistent and conservative TEGM when its elementary T-semiflow has at least one component equal to one. This method does not increase the number of transitions. Future works aim at studying the fact that some properties of TEGM's (such as structure) may be preserved after linearization, as in the considered example.

REFERENCES

Baccelli, F., G. Cohen, G.J. Olsder and J.-P. Quadrat (1992). *Synchronization and Linearity: An Algebra for Discrete Event Systems.* Wiley.

Cohen, G., S. Gaubert and J.-P. Quadrat (1995). Asymptotic Throughtput of Continuous Timed Petri Nets. In: *Proceedings of the IEEE-CDC.* New Orleans.

Cohen, G., S. Gaubert and J.-P. Quadrat (1998). Timed-Event Graphs with Multipliers and Homogeneous Min-Plus Systems. *IEEE TAC* **43(9)**, 1296–1302.

Gondran, M. and M. Minoux (1985). *Graphes et algorithmes.* Eyrolles.

Lahaye, S. (2000). Contribution à l'étude des systèmes linéaires non stationnaires dans l'algèbre des dioïdes. PhD thesis. Université d'Angers. France.

Menguy, E. (1997). Contribution à la commande des systèmes linéaires dans les dioïdes. PhD thesis. Université d'Angers. France.

Munier, A. (1993). Régime asymptotique optimal d'un graphe d'événements temporisé généralisé : Application à un problème d'assemblage. In: *RAIPO-APII.* Vol. 27(5). pp. 487–513.

Murata, T. (1989). Petri Nets: Properties, Analysis and Applications. In: *Proceedings of the IEEE.* Vol. 77(4). pp. 541–580.

Nakamura, M. and M. Silva (1999). Cycle Time Computation in Deterministically Timed Weighted Marked Graphs. In: *IEEE-ETFA.* pp. 1037–1046.

Teruel, E., P. Chrzastowski-Wachtel, J.M. Colom and M. Silva (1992). On Weighted T-Systems. In: *Advances in Petri nets, Lecture Notes in Computer Science.* Vol. 616. pp. 348–367.

Trouillet, B. (2003). Contribution à la commande des systèmes linéaires dans les dioïdes. PhD thesis. Ecole Centrale de Lille. France.

Trouillet, B., A. Benasser and J.-C. Gentina (2002). On the Modeling of the Dynamical Behavior of Weighted T-Systems. In: *APII-JESA.* Vol. 36(7). pp. 931–944.

ELSEVIER
IFAC
PUBLICATIONS
www.elsevier.com/locate/ifac

OFF-LINE SCHEDULING OF REAL TIME APPLICATIONS WITH VARIABLE DURATION TASKS.

Stéphane Pailler [1] **Annie Choquet-Geniet** [2]

LACS(LISI) - ENSMA
Téléport 2 - 1 Avenue Clément Ader BP 40109
86960 FUTUROSCOPE cedex, FRANCE

Abstract: We propose a model oriented scheduling methodology for highly coupled real time applications. We show that variations in the computation times of tasks may hazard the safeness of the controlled process. For the sake of reliability we take conditional instructions of tasks' code explicitly into account, in order to reduce the potential failures. We fisrt adapt the task's temporal model to this context, then we model applications using autonomous Petri nets which run under the earliest firing rule with terminal marking set. We define two concepts of schedulability: the local schedulability and the global one and we define the concept of scheduling graph. Finally, we show how to obtain a scheduling graph from the graph of global schedulability. *Copyright © 2004 IFAC*

Keywords: off-line scheduling, Petri Nets, conditional instructions, scheduling graph.

1. INTRODUCTION

A constantly increasing number of either autonomous or assisted processes (cars, nuclear power station, plane, space probe) are controlled by real time systems interacting with their own environment. Yet these systems are not safe from failures that may put human beings' lives in danger or jeopardize substantial economic value. That is why, it is necessary to develop methods to minimize those risks by validating the real time applications that control the processes. In most cases, the failures come from disregarded temporal constraints. Indeed, a real time application is defined as a multitask application in which each task is subjected to the inherent temporal constraints of the controlled processes. Thus, the main problem will be to choose a scheduling policy which would distribute each active job of task on the processor(s) so that the temporal constraints are respected. Two techniques can achieve that. The first one consists in executing a scheduling algorithm, which aim is to define at every moment the task to execute: it is the online scheduling. The second one consists in a previous analysis of the application and of the computation of valid schedules of the task executions within a temporal window of suitable size. Whatever the method, the schedulability analysis relies on a temporal model of the task (Liu and Layland, 1973; Stankovic *et al.*, 1998) allowing to represent each task by its temporal constraints: R_i, the date of the first activation, D_i the relative deadline of the execution of the task, T_i the period and C_i the computation time of the task calculated on the target hardware. Many works have made validation tools achievable by relying on heterogeneous scheduling contexts: hard and soft temporal constraints, the presence

[1] pailler@ensma.fr
[2] ageniet@ensma.fr

of real time primitives (resources and synchronizations which make the scheduling problem NP-Hard (Dertouzos and Mok, Dec. 1989), the reduction of response time, of jitter etc). Yet, a very few tools are actually used or usable in the industrial world. Indeed, the inaccuracy of execution time due to both the task's code itself and the processor that executes it, questions the relevance of the classic tools of schedulability analysis. It is the reason why we propose an off-line schedulability analysis methodology that takes the fluctuations of duration of tasks into account by relying on works that deal with the determination of the worst case of execution times. For that purpose, we have to extend the temporal model of Liu Layland in order to take the variations coming from the conditional instructions of the tasks (an important cause of variation) into account. We used the modeling of real time application by Petri nets (Choquet-Geniet *et al.*, 1996; Grolleau and Geniet, 2000) to analyze the feasibility of highly coupled applications (i.e. applications with a lots of resources and synchronizations) , then we show how to describe valid behaviors of the application, by means of scheduling graphs and finally, we explain how to get those graphs from the analysis of the Petri Net.

In a first part, we look into the difficulties of computation time estimation and the scheduling problems that it generates. Then in a second part, we introduce a new temporal model of task which takes the conditional instructions of the code into account and we reformulate the scheduling problem according to this new context. In a third part, we present the modeling of the application by autonomous colored Petri nets under the earliest firing rule and we show how to extract a scheduling graph describing a feasible behavior of the application.

2. SCHEDULING INSTABILITY PHENOMENA

In most real time applications, to find out a scheduling policy respecting constraints of the system proves to be a major issue. As an answer to that problem, we generally rely on a temporal model (Liu and Layland, 1973) that describes the temporal constraints of each task of the application in order to find and then validate a scheduling policy. But among the parameters that model a task, the execution time is probably the most complex to estimate. Those durations enable us to quantify the time allotted to the CPU to process each job of the tasks. It is thus an upper bound which is reckoned up from the task's program code and the features of the hardware. Each tool of scheduling analysis being based on a pattern of the tasks, that upper bound (Worst Case Execution Time) represents then an input data whose accuracy of estimation will determine the quality of the scheduling policy's validation. Therefore, researches on the WCET must take the two following preconditions into account (Puschner and Koza, 1989). First, the reckoned WCET must be safe, ie. it has to be an upper bound whatever the execution context may be. Second, the estimation of WCET has to be as accurate as possible or instability phenomena may occur as shown in figure 1: the reduction of the duration of one of the job of the task can cause a temporal fault in spite of the validation of the scheduling policy (Deadline Monotonic[3]). This kind of instability phenomena had been highlighted by (Graham, 1969).

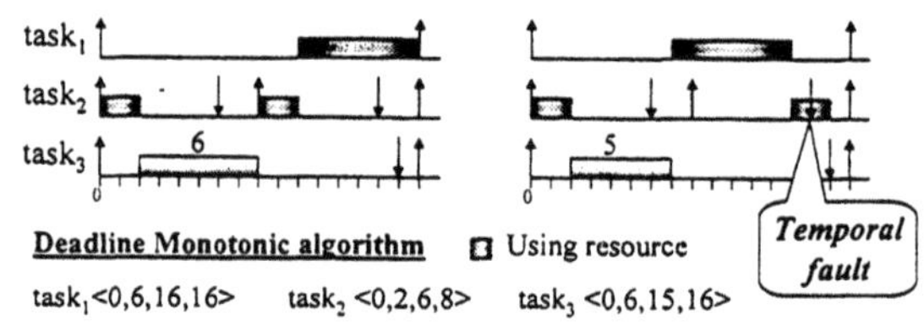

Fig. 1. An example of a scheduling failure due to a computation time reduction. If task 3 has an effective computation time equal to 5 instead of 6, task 1 can get the resource at time 7 before task 2 is released, causing task 2 to miss its deadline.

However the current methods of calculation of WCET do not all take the importance of real time primitives into account. In a context of highly coupled application, the traditional estimation of the WCET does not always correspond to the paths of execution where all the real time primitives are activated. However, as far as the analysis of schedulability is concerned, it is necessary to consider all the real time primitives that can occur. Consequently it is essential to consider the estimation of the WCET, to which should be added every real time primitive that can be activated in the various paths of the execution of the task. It seems then obvious that the resulting temporal model will be strongly over constrained what reduces the number of potential scheduling and complicates validation. To overcome that problem we suggest a method that allows to take the various paths of execution of a same task into account in order to isolate each real time primitive. Thus a task will be modeled for the analysis of schedulability by a set of execution paths. The existence of these paths is partly due to the different forms of conditional instructions present in the source code. That is why we propose to extend the temporal model of Liu Layland in

[3] DM (Deadline Monotonic) refers to the priority allocation algorithm: the task with the shortest deadline (the smallest value of D) is assigned the highest priority.

order to explicitly consider the IF THEN ELSE relevant instructions that is to say those which make it possible to isolate temporal primitives or sufficiently different execution times. Thereafter we will more specifically study the influence of the conditional tests on the schedulability analysis while considering that the methods of WCET analysis allow to obtain our model of conditional task.

3. CONDITIONAL TASKS AND SCHEDULING

3.1 The extended temporal model

We consider real time applications compounded of synchronous periodic tasks ($r_i = 0$). However let's point out that our method can take asynchronous periodic tasks into account, but it does at the cost of a combinatory explosion as far as the accessibility graph construction and the scheduling graph extraction are concerned. Optimization criteria need to be found to solve the problem.

Our objective is to refine the functional block description in order to take conditional instructions into account, which requires to modify the temporal model of tasks. We extend the Liu-Layland model in order to describe all the durations of the conditional branches: a task is represented by three deterministic parameters (the date of the first release, the relative deadline and the period) and by a *multiset* E of durations, each one corresponding to a possible behavior of the task. This new pattern of tasks is very close to the multiframe model of (Mok and Chen, 1996). However, unlike the later, we don't need the knowledge of the series of execution time for each task since each job of task may be processed with whatever execution time of the multiset. Let us note that if there is no conditionals instruction, the multiset E contains a single duration, which corresponds to the usual model.

In this temporal model, we have left the association between the real time primitives and the execution times aside. However, this information is not lost since it will be entirely integrated in the modeling of the application by Petri nets.

3.2 Scheduling graph

Considering on line scheduling strategies, the presence of conditional instructions matters only for the application validation, but most of these on line algorithms can be used without any adaptation even when conditional instructions are involved. On the contrary, off line scheduling strategies must be adapted. Indeed, if we consider tasks without conditional statements (durations are deterministic), the objective of an off line scheduling strategy is to build one or more valid schedules. This cannot be used if conditional statement are involved, because the various choices in conditional instructions will induce different behaviors of the application, which could not be described in a single schedule. In order to describe the behavior of a conditional application, we introduce scheduling graphs: a scheduling graph is a graph where each branch corresponds to a schedule of the application obtained by considering for each task only one of their paths. We call *split sub-application* each of these applications.

We then reformulate the scheduling problem, and define two concepts of schedulability:

- An application is said to be *locally schedulable* if each of its split sub-application is schedulable, i.e. there is for each one of them a valid schedule,
- An application is said to be *globally schedulable* if there is at least one valid scheduling graph, i.e. all deadlines are met whatever the conditional choices.

A globally schedulable application is obviously locally schedulable (but the converse does not hold if the application uses real time primitives). This comes from the fact that each branch of a scheduling graph is a valid schedule for one split sub-application.

4. MODELLING BY MEANS OF PETRI NETS

The schedulability analysis methodology which we proposed relies on (Choquet-Geniet *et al.*, 1996; Grolleau and Geniet, 2000; Grolleau, 1999). It consists in modeling the application by a constrained marking colored Petri nets, under the earliest firing rule with terminal marking set. The feasible schedules are then obtained through the construction of the state graph. The model includes two parts: the task system which is obtained through a classical modeling of the functional description of the application, and a clock system which models time (see figure 2). We have adopted a discrete modeling of time (Kopetz, 1992; Fohler, 1994): an external clock (RTC) counts the time in each place Time$_i$, which acts as a local clock used to release periodically the related tasks. Let us note that each transition corresponds to an action of duration one time unit, and that all transitions of the task system are in competition for obtaining the processor. It follows that, according to the earliest firing rule, at each time, one single valid transition is fired.

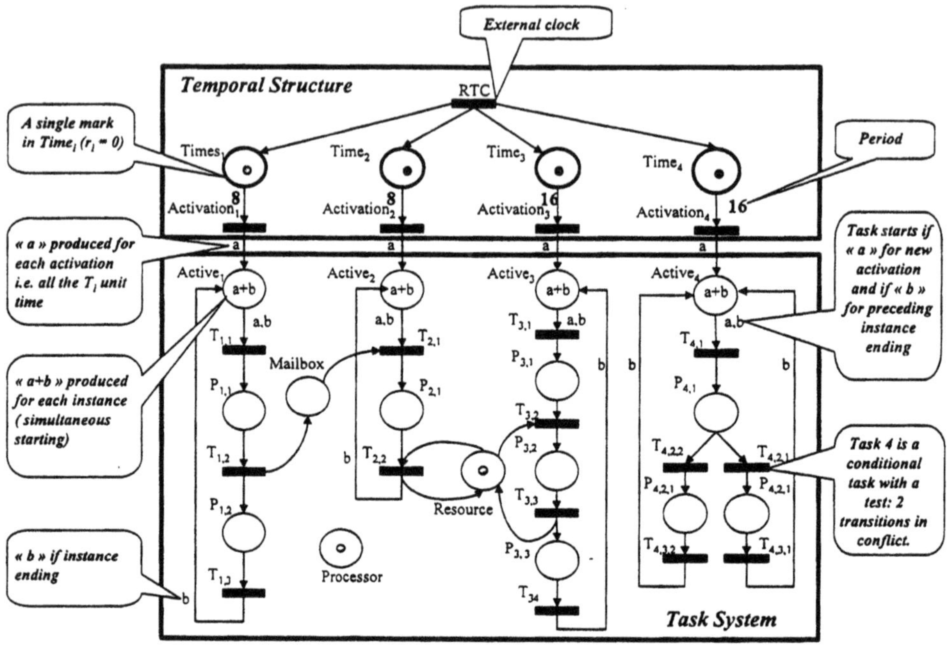

Fig. 2. Petri Nets model for an application of 4 tasks with a conditional task. The links toward processor are not shown.

It results that only conservative schedules[4] can be produced. But for the sake of scheduling power since there is no optimal conservative scheduling algorithm in a scheduling context with resource and synchronization, we need also to consider non work conserving schedules. For that purpose, we introduce in the task system a further task, called idle task, which models the inactivity of the processor. When transitions of this task fire, the processor remains idle what allows to produce non work conserving schedules too. The computation time of the idle task is P(1-U) [5]. Morever, the knowledge of the number of idle times during the metaperiod P help us for the detection of temporal fault during marking graph construction. When conditional tasks are involved, the utilization factor U is no more deterministic and the duration of the idle task becomes variable too. In order to match the previously used rule (a transition of the task system fires each unit time), we have adapted our model so that the duration of the idle task could be dynamically modified during the simulation of the Petri Nets. To allow these time units exchanges, the idle task is represented by a special place, holding as many tokens as idle time has been assumed within the metaperiod. For that purpose, we have chosen a method which has the advantage to minimize the number of backtracking operations during the construction of the accessibility graph. It consists in assuming that the maximal durations are always chosen within the multiset. Consequently, the duration of the idle task is assumed to be minimal. When a conditional task is computed, if the conditional test doesn't correspond to the maximum execution path, the duration of the idle task has to be increased of the difference between the worst and effective duration (cf. figure 3).

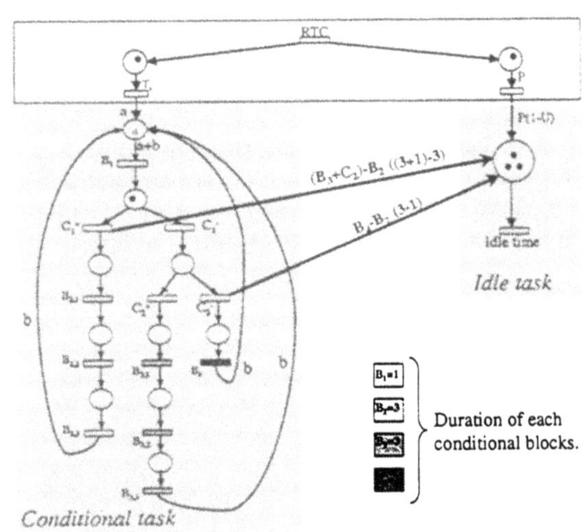

Fig. 3. Idle time exchange between the Idle task and a Conditional task with maximal duration policy. The multiset of the Conditional task is E={6,7,5}, so the number of token of the Idle task is calculated with max(E)=7. Each computation of the Conditional task with a different duration increases the number of idle times.

[4] in a conservative (or work-conserving) schedule, a task never intentionally waits

[5] where U is the processor utilization factor of the application and P is the LCM of the periods of the tasks

5. OFF-LINE ANALYSIS

5.1 Construction of global accessibility graph

Once the model constructed, we are interested in its exploitation in order to obtain valid scheduling graphs. From this Petri net, we only need to build a P depth final marking graph since the application is cyclic with a period of P (metaperiod). This means that, we have the same states at a t-depth and at t+k×P (k>0) in the marking graph. We then remove from the marking graph all states without successor, of depth less than P. They correspond to states of the application where fatal scheduling decisions have been made: temporal faults can no more be avoided. We call *accessibility graph* the reduced graph which contains only valid states. This graph contains all the possible valid schedules for the different splitted sub-applications. It can thus be used for studying the local schedulability. For the sake of global schedulability, the graph must be further reduced; indeed, we must insure that at each node, when a edge labelled by one alternative of the conditional statement exists, the edge labelled by the other alternative exists too. This means at a given depth in the graph, if there is a conditional test then it must exist a succesion of states until the depth P, corresponding to the metaperiod of the application, independently of the positive or negative result of the test. In this case only, the global schedulability is considered and scheduling graphs can be extracted. In order to purge the marking graph from lonely alternative paths, the basic idea is that each time, a choice is made of a path in a task, the choice of the other path corresponding to the other alternative in the conditional instruction, must be possible. Else, the first path must be removed. This operation can be made directly during the construction of the accessibility graph if we chose the construction in depth methods. The resulting graph is then called a *global accessibility graph.*

5.2 Extraction of a scheduling graph

The last step of the analysis consists then in the extraction of the scheduling graphs. The main difficulty is to deal with the size of the scheduling graph which depends on the number of the possible execution paths of the tasks.

Property: We consider n tasks defined by $< r_i, E_i, D_i, T_i >$ for $1 \leq i \leq n$, which do not use real time primitives, so:

- The number of possibles behaviors in a scheduling graph is
$$\prod_{i=1}^{n}(\#E_i)^{\frac{P}{T_i}},$$
- The global accessibility graph which contains all of the possible scheduling graphs is entirely hold in a hyper-cube which has
$$\sum_{i=1}^{n}(\Phi(E_i)) + 1 \text{ dimensions},$$

 # represents the number of elements of the multiset E_i,

$\Phi(E_i)$ *is the maximum number of interleave conditional tests from all the behaviors of the task* i.

Nevertheless, we can note that in the case of highly coupled applications, the size of the graph decreases significantly. For a better result, we can choose to act in two ways. First, when we build the global accessibility graph, we can apply some constraints like successor constraints or define a threshold for jitter or response time for a task or a set of tasks (Grolleau, 1999). These kind of criterion are called *a priori constraints* and allow us to reduce the size of the global accessibility graph which is the most important drawback of our method. The second method is used to choose from this reduced graph a optimized scheduling graph with some precise criterion : the better response time, the minimum jitter...(cf. figure 4). Anyway, both methods must be refined in the case of conditional task. When a task could have several behaviors, we can choose for example to only take into account the best response time for all of the different behaviours or at the opposite the worse. The result will be surely different. By default, we have choosen to consider the most pessimistic way for all the criterion.

Now that we can schedule periodic tasks with variable duration and by that way control the number of idle time in a scheduling graph, we have the possibility to optimize their position in the schedule. In the case some aperiodic tasks occur, we can look into the positionning of each idle time to simulate the on-line algorithms like Polling Server, Deferrable Server, Priority Exchange (Lehoczky *et al.*, 1987)... It can be made by a special criteria that uses a weighting function (like sinusoidal function with a period equal to the period of the simulated server). It forces each idle time to occur periodically and then may reduce their response time. We can then extract scheduling graphs that can be used for mixed scheduler in the case of both periodic and aperiodic tasks occurrences.

6. CONCLUSION

We have proposed a method for scheduling analysis of highly coupled real time applications which uses a Petri net modeling. We have shown that the variations of execution time have to be taken into account. Indeed, the use of the WCET does not give inevitably good results. The indeterminism of the current processors, the use of the optimized

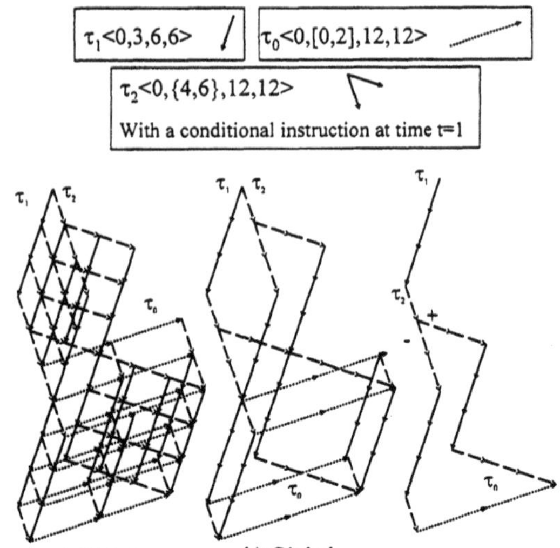

a) Global accessibility graph

b) Global accessibility graph with constraint of successor

c) A scheduling graph

Fig. 4. (a) The global d'accessibilit graph; (b) a reduce graph with constraint of successor; (c) a scheduling graph which is extract from b .

compilers, the programming with particular instructions which increases the number of different behaviors from each task, makes it impossible to get an exact measurement of the execution times (Puschner, 2002). In spite of these difficulties, the analysis of WCET and schedulability strongly interests industrialists (Nielsen *et al.*, 2002). It is the reason why we have to try to benefit from works on the WCET to integrate in our method the relevant conditional instructions that can be found in the code of tasks. After extending the temporal model of (Liu and Layland, 1973) to conditional tasks, we have defined two concepts of schedulability, coming directly from the implications of conditional tasks: local and global schedulability. We have shown that there is no equivalence between both. In addition, we have adopted the concept of idle task, because the duration of this task is closely related to the duration fluctuations of the application tasks. We have proposed a modeling of the idle task that allows on one hand a production of non work conserving schedules and on the other an optimization in the construction of the marked graph (the idle task enables to considerably reduce the number of non valid states). We have shown how to extract a scheduling graph with some optimization criteria for response time or jitter minimizing. The next step of this work is to extend our method to sporadic tasks, induced by conditional blocks within which new tasks are requested. Our method must be adapted in order to take into account this new kind of tasks model. Our goal is then, thanks to efficient idle time managing, to define a mixed scheduler for both periodic and sporadic tasks.

REFERENCES

Choquet-Geniet, A., D. Geniet and F. Cottet (1996). Exhaustive computation of the scheduled task execution sequences of a real-time application. In: *Proc. of the 4th International Symposium on FTRTFTS.* Uppsala, Sweden. pp. 246–262.

Dertouzos, M.L. and A.K. Mok (Dec. 1989). Multiprocessor on-line scheduling of hard real-time tasks. *IEEE TSE* **15(12)**, 1497–1506.

Fohler, G. (1994). Flexibility in statically scheduled hard real-time sys-ems. PhD thesis. University of Wien.

Graham, R. (1969). Bounds on multiprocessing timing anomalies. In: *SIAM Journal of Applied Mathematics 17, no. 2.* pp. 416–429.

Grolleau, E. (1999). Ordonnancement temps-réel hors-ligne optimal à l'aide de réseaux de Petri en environnement mono-processeur et multiprocesseur. PhD thesis. ENSMA - Université de Poitiers.

Grolleau, E. and A. Choquet Geniet (2000). Off line computation of real-time schedules by means of petri nets. In: *WODES2000.* Analysis and Control. Kluwer Academic Publishers. Ghent, Belgium. pp. 309–316.

Kopetz, H. (1992). Sparse time versus dense time in distributed real-time systems. In: *12th Int. Conf. On Distributed Computing Systems.* Japon. pp. 460–467.

Lehoczky, J.P., L. Sha and J.K. Strosnider (1987). Enhanced periodic responsiveness in hard real time environments. In: *Proceeding of the RTSS, IEEE.* San Jose. pp. 261–270.

Liu, C.L. and J.W. Layland (1973). Scheduling algorithms for mutltiprogramming in real-time environnement. *Journal of the ACM* **20(1)**, 46–61.

Mok, A. and D. Chen (1996). A multiframe model for real time tasks. In: *in Proceedings of the 17th RTSS.* pp. 22–29.

Nielsen, M.R., E. Conquet and J.L. Terraillon (2002). The european space agency's involvment and interest is wcet and scheduling analysis. In: *2nd Intl. Workshop on WCET Analysis 14th Euromicro Conference on Real- Time Systems.*

Puschner, P. (2002). Is worst-case execution-time analysis a non problem? - towards new software and hardware architectures. In: *2nd Intl. Workshop on WCET Analysis 14th Euromicro Conference on Real- Time Systems.*

Puschner, P. and C. Koza (1989). Calculating the maximum execution time of real-time programs. *Journal of Real-Time Systems* **1(2)**, 159–176.

Stankovic, J.A., M. Spuri, K. Ramamritham and G. C. Buttazzo (1998). *Deadline scheduling for real-time systems.* Vol. ISBN 0 7923 8269 2.

ELSEVIER
IFAC
PUBLICATIONS
www.elsevier.com/locate/ifac

A BRANCH & BOUND PROCEDURE TO OPTIMIZE LOADING SEQUENCES OF A SIMPLE SUPPLY CHAIN UNDER APERIODIC DEMAND

S.E. Merzouk, O. Grunder and M. El Bagdouri

Laboratory S.e.T. University of Technology Belfort Montbéliard, France
UTBM, site de Belfort 90000, Belfort Cedex.
Phone : +33(0)3 84 58 33 19 Fax : +33(0)3 84 58 3342
{salah-eddine.merzouk, olivier.grunder, mohammed.elbagdouri}@utbm.fr

Abstract The purpose of this paper is to propose an exact procedure to optimize the lots delivery scheduling in a simple supply chain under just in time sitting. This problem is mainly characterized by the fact that the products have to be delivered to the customer before given due dates and as late as possible. Some interesting mathematical properties are given for this problem in case of there is a single transporter to deliver the goods. These results greatly improve the search of the optimal solution by an appropriate branch and bound procedure. Comparative experimental results illustrate the efficiency of the proposed algorithm *Copyright © 2004 IFAC*

Keywords: optimal load flows, optimization problem, scheduling algorithm, optimal control, synchronization, transportation control, tree search.

1. INTRODUCTION

The Supply Chain can be defined as a set of actors (suppliers, producers, distributors...) which contribute to manufacture, sale and provision of finished products for the customers. The logistic is the management of business operations, such as the acquisition, storage, transportation and delivery of goods along the supply chain. In today's increasing global and competitive marketplace, it is imperative that members of supply chain work together in an effort to minimize overall times and costs of production and transportation (F.E. Vergara & al, 2002). Thus, the problem of optimization production and transport sequences was largely studied these last years. Indeed, the Economic Lot and delivery Scheduling Problem (*ELDSP*) was studied by (J. Hahm & C. A Yano, 1992) and an extension of the problem is introduced to hold account the capacity of the transporter. The purposed model was also used in (F.Elizabet Vergara *et al.*, 2002). They generalized this model to a linear supply chain (customers and suppliers in cascade) and they developed a genetic algorithm to find the best scheduling sequence of various products flows through the production line. (M. Khouja, *et al.* 1998) also used the genetic algorithms to solve the sequences of production scheduling problem of several products in the same production line. (S.K Goyal & A.Z Szendrovits, 1986) developed a model which assumes that a constant lot size is produced through a fixed sequence of manufacturing stages with a single set-up and without interruption at each stage. In (M.A Hoque & S.K Goyal, 2000), a delivery policy of batches with equal and non equal sizes was developed under a constant demand for also a single vendor and single buyer. (S.Y Chou & S.L Chang, 2001) developed heuristics to find the minimal cost in the context of only one order of production and several deliveries. A dynamic programming was introduced by (K. H Kim & J. B Kim, 2002) in order to select a model for trucks transporting components between two members of logistic chain.

However, all of this research maintained the assumption of a deterministic and constant logistic demand.

In (I. Elmahi, *et al.*, 2002), a max-plus algebra based model is proposed to control a simple supply chain under the assumption that the load sequence of the transporter is given. If the load sequence become an input variable of the model, all the possible load sequences have to be considered to find the optimal one.

The purpose of this paper is to propose an exact procedure to optimize the lots delivery scheduling in a simple supply chain under just in time sitting. The main contribution of this work is to demonstrate some interesting mathematical properties for this problem in case of there is a single transporter to deliver the goods. These results greatly improve the search of the optimal solution by an appropriate branch and bound procedure.

This paper is organized as follow: In section 2, we present the studied system and the constraints related to its dynamic. The mathematical model of this system and its properties are provided in section 3. We propose in the following section an efficient Branch and Bound Procedure to find the optimal solution of the studied problem. Comparative experimental results show the efficiency of the proposed algorithm in section 5.

Section 6 wraps up the paper with conclusions and perspectives

2. PROBLEM STATEMENT

The system studied in this work is a simple supply chain made up of three actors: one supplier, one customer and one transporter (Fig. 1). The customer needs a number of products which must be available in its stock at given due dates, that is to say that the products can arrive eventually earlier compared with their due dates but never later. The products are manufactured by the supplier who has to respect the requirements of the customer. Once manufactured, they can be transported from the supplier toward the customer by the transporter. Several parameters characterize the transporter such as the capacity of loading, the travel duration between the customer and the supplier and the loading time and unloading time of each product from the transporter.

Fig. 1. The studied system.

The studied system can be encountered in several application fields, in particular in the automobile industry where the time factor is the primary importance for the suppliers but on all for the assembly facilities.

The purpose of this study is to optimize the lots delivery scheduling in this chain under just in time sitting. This problem is mainly characterized by the fact that the products have to be delivered to the customer before given due dates and as late as possible. Under these assumptions, one has to find the optimal lot sequence that minimizes the global advance of the products, that is to say the sum of the differences between the due date and the real arrival date for all products.

The following section gives a mathematical model of this system and its properties.

3. MATHEMATICAL MODEL

The notations used in this paper are given as follows:

- np : total number of the requested products
- $ch_{\max}$: loading capacity of the vehicle.
- t_g : travel duration from supplier to customer.
- t_r : time duration to return back at the supplier.
- $T = t_g + t_r$: transport time.
- t_l : loading time per product.
- t_u : unloading time per product.
- Y_d : vector of due dates, where :

$$Y_d = (y_d(k), k = 1, \cdots np)$$

  $y_d(k)$: the due date of the kth product.

- Y_r : vector of the real arrival dates, where:

$$Y_r = (y_r(k), k = 1, \cdots np)$$

  $y_r(k)$: the real arrival date of the kth product. If the kth and $k+1^{th}$ products are in the same lot, we have: $y_r(k+1) = y_r(k) + t_u$

- s_p : partial sequence for the last p products.

$$s_p = \{s_p^1, s_p^2, \cdots, s_p^k\}$$

  s_p^i : number of loaded products in the ith lot of s_p.

- $A_g(s_p)$: the global advance of the partial sequence s_p
- $A_{av}(s_p)$: the average advance of the partial sequence s_p

3.1 Definitions

Definition 1: We define a loading sequence s_{np} as a series $(s_{np}^i)_{i=1,np}$ which satisfies:

a) $\forall i \in [1, np]$, $0 \le s_{np}^i \le ch_{\max}$.

b) $(\exists i \in [1, np] / s_{np}^i = 0) \Rightarrow (\forall j > i, s_{np}^j = 0)$

c) $\sum_{i=1}^{np} s_{np}^i = np$.

Definition 2: A partial sequence s_p is a sequence of p products such as $p < np$.

The products of any partial sequence always correspond to the last ones required by the customer. Thus, the first product of the partial sequence s_p corresponds to the product $np - p + 1$ needed by the customer.

Definition 3: The global advance of a given partial sequence is defined as the sum of the differences between the due date and the real arrival date for all products.

$$A_g(s_p) = \sum_{k=np-p+1}^{np} (y_d(k) - y_r(k))$$

Definition 4: The average advance for a given partial sequence s_p is defined as: $A_{av}(s_p) = \dfrac{Ag(s_p) + A_b}{p+1}$

with:

$$A_b = [y_d(np - p) - (y_r(np - p + 1) - T - t_u - s_p^1 t_l)]$$

The next section provides the expressions of the real arrival dates for each product and a proposition to compare partial sequences of this problem.

3.2 Model analysis

We first formulate the real arrival dates of the last lot of a partial sequence with the following lemma.

Lemma 1

Given a partial sequence s_p , the real arrival date of the first product of the last lot is:

$$y_r(np - s_p^k + 1) = \min_{i=1}^{s_p^k} [y_r(np - s_p^k + i) - (i-1) \times t_u]$$

Proof: The real arrival dates of the last lot have to satisfy:

$$\begin{cases} y_r(np - s_p^k + 1) \le y_d(np - s_p^k + 1) \\ y_r(np - s_p^k + 2) \le y_d(np - s_p^k + 2) \\ \vdots \\ y_r(np) \le y_d(np) \end{cases}$$

$$\Rightarrow \begin{cases} y_r(np-s_p^k+1) \le y_d(np-s_p^k+1) \\ y_r(np-s_p^k+1)+t_u \le y_d(np-s_p^k+2) \\ \vdots \\ y_r(np-s_p^k+1)+(s_p^k-1)\cdot t_u \le y_d(np) \end{cases}$$

Consequently, we have:

$$\begin{cases} y_r(np-s_p^k+1) \le y_d(np-s_p^k+1) \\ y_r(np-s_p^k+1) \le y_d(np-s_p^k+2)-t_u \\ \vdots \\ y_r(np-s_p^k+1) \le y_d(np)-(s_p^k-1)\cdot t_u \end{cases}$$

$$\Rightarrow y_r(np-s_p^k+1) \le \min_{i=1}^{s_p^k}\left[y_r(np-s_p^k+i)-(i-1)\times t_u\right]$$

The real arrival dates of other products in this lot are obtained by adding the unloading time t_u.

From this result, we give the expression of the real arrival dates of the previous lots in the following lemma:

Lemma 2

Given a partial sequence $s_p=\{s_p^1,s_p^2,\cdots,s_p^k\}$, the real arrival date for the 1st product in s_p^j ($j<k$) is:

$$y_r(np-\sigma+1)=\min\begin{Bmatrix} \min_{i=1}^{s_p^j}\left[y_d(np-\sigma+i)-(i-1)\times t_u\right], \\ y_r(np-\sigma+s_p^j+1)-T- \\ s_p^{j+1}\cdot t_l-s_p^j\cdot t_u \end{Bmatrix}$$

with $\sigma=\sum_{l=j}^{k} s_p^l$

Proof: The real arrival dates in the previous lots have to satisfy the same inequality as in the lemma 1. Moreover, the following constraint has to be satisfied:

$$y_r(np-\sigma+s_p^j) \le y_r(np-\sigma+s_p^j+1)-t_u-s_p^{j+1}\cdot t_l-T$$

Indeed, the two lots have to be separated with a necessary temporal interval which allows the transporter to go to the supplier, load the products and then return toward the customer.

thus:

$$y_r(np-\sigma+1) \le \begin{bmatrix} y_r(np-\sigma+s_p^j+1)-t_u-s_p^{j+1}\cdot t_l \\ -T-(s_p^j-1)\cdot t_u \end{bmatrix}$$

$$y_r(np-\sigma+1)=\min\begin{Bmatrix} \min_{i=1}^{s_p^j}\left[y_d(np-\sigma+i)-(i-1)\times t_u\right], \\ y_r(np-\sigma+s_p^j+1)-T- \\ s_p^{j+1}\cdot t_l-s_p^j\cdot t_u \end{Bmatrix}$$

This second lemma uses a backward recursion to determine the real arrival date of the first product in the previous lots. Thus, using this lemma, all real arrival dates for all products in the previous lots are calculated. These dates are the latest possible for a given sequence; they thus allow to minimize the total advance of this sequence.

The following lemmas 3 and 4 are provided to compare two partial sequences of the problem. They will be used later in this paper to improve the efficiency of the branch and bound search.

Lemma 3

Given two partial sequences s_p and s'_p such as $y_r(np-p+1)-s_p^1\cdot t_l \ge y'_r(np-p+1)-s'^1_p\cdot t_l$, and a lot of x products, we consider the following partial sequences:

$s_{x+p}=\{x,s_p^1,s_p^2,\cdots,s_p^k\}$ and $s'_{x+p}=\{x,s'^1_p,s'^2_p,\cdots,s'^k_p\}$.

Let $y_r(i)$ and $y'_r(i)$ be the real arrival dates of the ith product in s_{x+p} and s'_{x+p} respectively, then

$$\forall i\in[np-p-x+1,np-p]\quad,\quad y_r(i)\ge y'_r(i)$$

Proof: Let formulate the real arrival dates of the first product for the two partial sequences:

- s_{x+p}:

$$y_r(np'-x+1)=\min\begin{Bmatrix} \min_{i=1}^{x'}\left[y_d(np'-x+i)-(i-1)\times t_u\right], \\ y_r(np'+1)-T-s_p^1.t_l-xt_u \end{Bmatrix}$$

with $np'=np-p$

- s'_{x+p}:

$$y'_r(np'-x+1)=\min\begin{Bmatrix} \min_{i=1}^{x'}\left[y_d(np'-x+i)-(i-1)\times t_u\right] \\ y'_r(np'+1)-T-s'^1_p.t_l-xt_u \end{Bmatrix}$$

We note $m=\min_{i=1}^{x'}\left[y_d(np'-x+i)-(i-1)\times t_u\right]$

We have to consider 4 cases:

1) $y_r(np'-x+1)=m$
$y'_r(np'-x+1)=m$

$\Rightarrow \forall i\in[np-p-x+1,np-p],\quad y_r(i)=y'_r(i)$

2) $y_r(np'-x+1)=m$
$y'_r(np'-x+1)=y'_r(np'+1)-T-s'^1_p.t_l-xt_u$

So $y'_r(np'-x+1)<m=y_r(np'-x+1)$

$\Rightarrow \forall i\in[np-p-x+1,np-p],\quad y_r(i)\ge y'_r(i)$

3) $y_r(np'-x+1)=y_r(np'+1)-T-s_p^1\cdot t_l-x.t_u$
$y'_r(np'-x+1)=m$

thus:

$$y'_r(np'-x+1)=M\le y'_r(np'-x+1)-T-s'^1_p\,t_l-x\cdot t_u$$

Then:

$$y_r(np'-x+1)\le y'_r(np'-x+1)$$
$$\Rightarrow y_r(np'+1)-s_p^1\cdot t_l\le y'_r(np'+1)-s'^1_p\cdot t_l$$

This last inequality is contradictory with the assumptions of the lemma. This is impossible.

4) $y_r(np'-x+1)=y_r(np'+1)-T-s_p^1.t_l-xt_u$
$y'_r(np'-x+1)=y'_r(np'+1)-T-s'^1_p.t_l-xt_u$

We have $y_r(np'+1) - s_p^1 \cdot t_l \geq y'_r(np'+1) - s''^1_p \cdot t_l$

Then $y_r(np'-x+1) \geq y'_r(np'-x+1)$

$\Rightarrow \forall i \in [np-p-x+1, np-p], \quad y_r(i) \geq y'_r(i)$

This lemma means that if two different partial sequences are completed by the same lot and the real arrival date of the first product in the first partial sequence is greater or equal to the real arrival date of the first product in the second partial sequence, then, the first partial sequence is a better solution than the second one.

Definition 5: We define $\zeta(s_p)$ the set of solutions (complete sequences) constructed from the partial sequence s_p by adding in upstream s_p necessary lots to satisfy the customer request. We denote $s^*(s_p)$ the sequence belonging to $\zeta(s_p)$ which global advance is the smallest.

$$\forall s \in \zeta(s_p) \quad , \quad A_g(s) \geq A_g(s^*(s_p))$$

Proposition 1:
Let s_p and s'_p be two partial sequences for the same number of products p.
Assume that:

$$\begin{cases} A_g(s_p) \leq A_g(s'_p) \\ y_r(np-p+1) - s_p^1 \times t_l \geq y'_r(np-p+1) - s'^1_p \times t_l \end{cases}$$

then

$$A_g(s^*(s_p)) \leq A_g(s^*(s'_p))$$

Proof: We suppose that: $A_g(s^*(s_p)) > A_g(s^*(s'_p))$

This means that:

$$\forall q \in \zeta(s_p) \quad , \quad A_g(q) > A_g(s^*(s'_p))$$

The sequence $s^*(s'_p)$ is made up of two parts: the first one is the partial sequence s'_p. The second one is another partial sequence which composed of lots sc^i_{np-p} necessary to satisfy the total request of the customer (fig 2). We note this last one sc_{np-p} with $sc_{np-p} = \{sc^1_{np-p}, sc^2_{np-p}, \cdots, sc^l_{np-p}\}$.

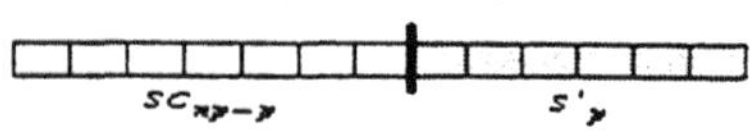

Fig. 2. loading sequence $s^*(s'_p)$.

We complete the partial sequence s_p with the partial sequence sc_{np-p} as shown in figure 3. We note this obtained sequence r_{np}.

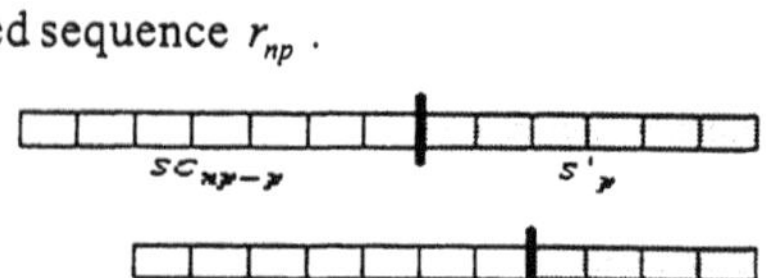

Fig. 3. Completion of partial sequences.

The lemma 3 applied to r_{np} and $s^*(s'_p)$ gives:

$$\forall i \in [1, np-p] \quad , \quad y_r(i) \geq y'_r(i)$$

Then: $y_d(i) - y_r(i) \leq y_d(i) - y'_r(i)$

$$\Rightarrow \sum_{i=1}^{np-p} (y_d(i) - y_r(i)) \leq \sum_{i=1}^{np-p} (y_d(i) - y'_r(i))$$

$$\Rightarrow \begin{bmatrix} \sum_{i=1}^{np-p} (y_d(i) - y_r(i)) + \\ A_g(s_p) \end{bmatrix} \leq \begin{bmatrix} \sum_{i=1}^{np-p} (y_d(i) - y'_r(i)) + \\ A_g(s'_p) \end{bmatrix}$$

$$\Rightarrow A_g(r_{np}) < A_g(s^*(s'_p))$$

The last inequality means that we find one sequence belonging to $\zeta(s_p)$ whose the global advance is smallest than $A_g(s^*(s'_p))$. Hence, the assumption of $A_g(s^*(s_p)) > A_g(s^*(s'_p))$ was false. Consequently, we have:

$$A_g(s^*(s_p)) \leq A_g(s^*(s'_p))$$

The mathematical model given above allows to compute the latest real arrival dates for a given sequence. If the load sequence become an input variable of the model, all the possible load sequences have to be considered to find the optimal one. As the size of problem increases exponentially with the number of products, it becomes impossible to explore all the search space. We propose in the following section, a Branch and Bound procedure based on the above results which allows, in addition, to find the optimal sequence in very reducing time.

4. BRANCH AND BOUND PROCEDURE

The BBP belongs to the exact resolution methods which thus guarantee the optimality of the found solution. The BBP has techniques to detect the failures as soon as possible and thus obtaining a considerable gain in computing time. It allows to reduce the search space by an implicit enumeration of complete branches of tree search (Hao J.K, *et al.*, 1999).

The tops of the tree search correspond to subsets which are divided into levels. There is only one top with level zero which represents the root of the arborescence. A minimal evaluation of the branch (called limit) is needed to construct the tops of the other levels. Otherwise say, all the other tops which obtained from the current node will have an evaluation greater or, at best, equal to this limit.

The BBP is based on two elementary techniques which act considerably on the quality and the speed of the algorithm: The first one is the evaluation of the solution during its construction. The second one is the separation which consists in stopping the construction of a solution as s oon as the evaluation of this last one become greater than the limit evaluation.

The loading sequences optimization problem can be modelled in the shape of tree search. Each node models the quantity of products loaded in a given lot. With each node, we associate an evaluation corresponding to the advance of the partial sequence of products delivered until this top. The algorithm has to be able to stop as soon as possible the construction of a branch which will end inevitably to a bad solution.

The proposition 1 is used by the developed BBP to explore the search arborescence and find the optimal loading sequence. In following section, we explain how the BBP proceeds to choose the best lot among the various possibilities in the same level.

4.1 Branching scheme

The search arborescence holds at most np levels corresponding to the maximal number of trips necessary to satisfy the customer request. In each level, all the nodes that correspond to the various possibilities of loading remaining products, are considered. Thus, the first level of tree contains n_1 nodes where $n_1 = \min(np, ch_{\max})$. These nodes correspond to the last lot in the sequence.

In the next levels, the number of nodes depends of the path from the root node to the parent node of the previous level. So on, in each level, the number of choices is calculated using:

$$n_j = \min(np'_j, ch_{\max}) \text{ with } np'_j = np - \sum_{l=j-1}^{k} s_{np}^{l}$$

For the last lot, we have to select among a set of nodes the best one. Thus, the corresponding average advance is calculated for each loading. This advance is computed using definition 4. Hence, the selected loading is the one with the smallest average advance. The following bounding scheme is used by the BBP to stop the exploration of a given branch. It is based essentially on the proposition 1 given above.

4.2 Bounding scheme

If the choice of the last lot is equal to s_{np}^{k}, the best partial sequence to deliver the last s_{np}^{k} products is to group them in the same lot. Indeed, to calculate the average advance for a given partial sequence, we need to calculate the global advance for this partial sequence and then, the real arrival date for product preceding the sequence. In proposition 1, we have proved that for a same number of products, the partial sequence with the smallest global advance and greatest real arrival date will give inevitably the optimal solution. Thus, it is not necessary to explore the other possibilities for the same number of products. Otherwise say, this partial sequence is the best manner to deliver that number of products.

Once the last loading is chosen, we obtain with the same manner the previous loading. The different average advances for the partial sequences $s_{np}^{k-1} - s_{np}^{k}$ are computed where s_{np}^{k-1} is the different possibilities of loading remaining products $s_{np}^{k-1} \in [1, \min(np - s_{np}^{k}, ch_{\max})]$. And so on, all the other previous lots are obtained.

During the exploration of tree search, the best partial sequence of each number of products is registered. The best global advance and the latest real arrival dates are compared with the global advance and real arrival date of the current branch. If they not satisfy the two conditions of the proposition 1, the exploration of this branch will stop immediately. Otherwise, the current branch becomes the new best partial sequence. So on, we obtain the best loading sequence for all products.

To explain this scheme, the following example is given:

Let to deliver 8 products ($np = 8$) from the supplier to the customer. The vehicle capacity is $ch_{\max} = 5$ and the vector of due dates is:

$$Y_d = \{100, 101, 102, 120, 122, 130, 131, 132\}$$

The temporal parameters are:

$$t_g = t_r = 5 \Rightarrow T = 10 \quad , \quad t_l = t_u = 1$$

We start by finding the last lot. So, the supplier can deliver 1, 2, 3, 4 or 5 products.

For 1 product, the real arrival date is $y_r(8) = 132$, then, the average advance is:

$$A_{av}(1) = \frac{Ag(1) + [y_d(7) - (y_r(8) - T - t_u - t_l)]}{2}$$

$$A_{av}(1) = \frac{0 + [131 - (132 - 10 - 1 - 1)]}{2} = 5.5$$

Thus, in the same manner, we obtain the following table 1:

<u>Table 1 global and average advance for the different sizes of the last lot.</u>

Products number	Real arrival dates	Global advance	Average advance
2	$y_r(7) = 131$ $y_r(8) = 132$	0	4
3	$y_r(6) = 130$ $y_r(7) = 131$ $y_r(8) = 132$	0	1.5
4	$y_r(5) = 122$ $y_r(6) = 123$ $y_r(7) = 124$ $y_r(8) = 125$	21	6.8
5	$y_r(4) = 120$ $y_r(5) = 121$ $y_r(6) = 122$ $y_r(7) = 123$ $y_r(8) = 124$	25	3.8

We conclude that the best manner to deliver the last three products is to take them on the same lot. Then the last trip is equal to 3.

In the previous trip, we repeat the same process to calculate the number of loaded products. The selected partial sequence is $2-3$ and the average advance corresponding is $A_{av}(2-3) = 1.66$.

So on, we find finally that the best sequence to deliver the 8 products is $3-2-3$. The global advance for this sequence is $A_g(3-2-3) = 11$. The following figure 4 resumes the different leafs explored by the BBP and the corresponding average advance for each one.

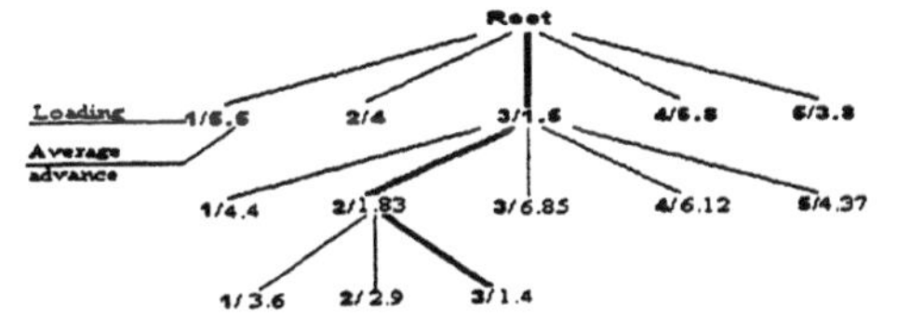

Fig. 4: Explored nodes for the example.

We see that we didn't explore all possibilities in each level. Indeed, the average advance for each leaf allows algorithm to choice the way to follow during the exploration of tree search. The proposition 1 makes it possible to stop the construction of a given branch as soon as its evaluations end inevitably to a bad solution.

5. EXPERIMENTAL RESULTS

To test the performance of the developed BBP, we have considered several problems with different small sizes and different parameters. For each problem size, 10 problem instances with different data ($ch_{max}, T, \ldots$) are generated. The minimum, the maximum and the average computation times of the branch and bound algorithm are reported in the table 2, as well as the computation time needed for an enumeration procedure. In this experimentation study, we have not found examples which deal with the same kind problems, it's why the examples that we presented here are generated randomly.

Table 2 Average run time for BBP and enumeration procedure.

Products number	Run time			
	BBP			Enum.
	Min time	Max time	Average time	
20	15 ms	16 ms	15 ms	30 ms
30	15 ms	16 ms	15 ms	14 mn 48sec
40	15 ms	32 ms	20 ms	>1 day
50	15 ms	47 ms	25 ms	>150 days
80	31 ms	157 ms	88 ms	> 8.10^5 yrs
100	47 ms	1.15 sec	250 ms	> 10^9 yrs
200	187 ms	46.4 sec	16.7 sec	> 10^{29} yrs
300	2.12 sec	1.2min	34.5 sec	> 10^{49} yrs

Note that these results are obtained with a computer which has a P4 processor with a frequency equal to 2.66GHz. In addition, it is interesting to note the difference between running time of the BBP as opposed to the enumeration procedure. As the table shows, as the problems become larger, the BBP quickly becomes much faster than the enumeration procedure. For example, for 40 products, the BBP took on average 20 ms versus more than one day for the enumeration procedure. However, we note that the running time for a given number of products differs of a problem to another. It is explaining by the fact that data ($ch_{max}, T, \ldots$) differ of a problem to another.

This BBP is also applied to problem with large size in order to find the limit of the procedure. We find that problems of 1000 products can be resolved easily by this BBP. The running time to find the optimal solution is very tolerable. That time which borders 48mn 11sec is roughly the same one as that carried out by heuristics like the "genetic algorithms" to find an approached solution.

6. CONCLUSION

In this paper, we have proposed a very efficient optimization procedure for the lots delivery scheduling problem in a simple supply chain under a just in time criterion. This problem is mainly characterized by the fact that the products have to be delivered to the customer before given due dates and as late as possible. Under these assumptions, one has to find the optimal lot sequence that minimizes the global advance of the products. Some interesting mathematical properties are given for this problem which act considerably on the speed of the algorithm. Indeed, these results applied to a branch and bound procedure, allow to find the optimal solution of this problem in a very reducing computing time. We have provided some comparative experimental results which illustrate the efficiency of the proposed algorithm Indeed, this BBP open the field to ensure the exactitude of the found solution of problems with large size.

However, this study has to be generalized to take in account supply chains with several suppliers and customers and large fleet of transporters. The transport, the loading, the unloading and the holding costs has also to be integrated in the evaluation function.

REFERENCES

Chou S.Y, Chang S.L & Yang W.D (2001). *Optimal multiple delivery schedule for demand in logistic model.* Int. Journal of production Economics. Vol 73, p 241 – 249.

Elizabet Vergara F., Khouja M.& Michalewicz Z. (2002). *An evolutionary algorithm for optimizing material flow in supply chains.* Computers & Industrial Engineering. Vol 43 p 407 – 421.

Elmahi I., Grunder O. & Elmoudni A. (2002). *A Petri net approach for the evaluation and command of supply chain using the Max-Plus algebra.* Proceedings of the 2002 IEEE International Conference on Systems, Man and Cybernetics. 6-9 October 2002, IEEE SMC 2002. Tunisie.

.K Goyal & A.Z Szendrovits, (1986). *A constant lot size model with equal and unequal sized batch shipments between production stages.* Engeneering Costs and Production Economics, 10, pp 203-210

Hahm, J and Yano C. A. (1992). *The economics lot delivery scheduling problem: The single item case.* Int. Journal of production Economics. 28. pp 235-252.

Hao, J. K Galinier, P., Habib M. (1999). *Méthaheuristiques pour l'optimisation combinatoire et l'affectation sous contraintes.* Revue d'Intelligence Artificielle.

Hoque, M.A and Goyal, A. (2000). *An optimal policy for a single-vendor single-buyer integrated production-inventory system with capacity constraint of the transport equipment.* Int. Journal of production Economics. 65. pp 305-315.

Kim, J.B and Kim K.H. (2002) *Determining load pattern for the delivery of assembly components under JIT systems.* Int. Journal of Production Economics. 77. pp 25-38.

Khouja M., Michalewicz Z., Wilmot M. (1998). *The use of genetic algorithms to solve the economic lot size scheduling problem.* European Journal of Operation research. Vol 110. p 509-524.

ELSEVIER
IFAC
PUBLICATIONS
www.elsevier.com/locate/ifac

USING TIMED AUTOMATA FOR THE VERIFICATION OF IEC 61499 APPLICATIONS

Marius Stanica
Hervé Guéguen

Supélec- IETR
Avenue de la Boulais BP81127
35511Cesson-Sévigné cedex, France

Abstract: This paper presents a methodology for the modelling of IEC 61499 function blocks behaviour using timed automata. The aim is to be able to associate basic timed automata with each functional block and to compose them in order to verify the execution of a network of functional blocks. The method and the verification that can be made are exemplified on a distributed control application. *Copyright © 2004 IFAC*

Keywords: functional blocks, IEC 61499, verification, timed automata.

1. INTRODUCTION

The IEC 61499 (IEC TC65/WG6, 2003) standard specifies a model of applications and re-usable software components for the design of distributed industrial process control applications. This model is based on function blocks that are widely used in this domain and is then based on a dataflow approach. However, in order to further the standardisation of the function blocks and their reuse on various platforms the execution control flow is made explicit. For each bloc the control of the execution of the algorithms, according to input control events and the state of the block, is specified by a state machine. The global application is specified by connecting the data inputs and outputs of the blocks in order to define the data flows but also their control inputs and outputs for the control flows. It can be noticed that the standard is mainly concerned with the control part and that the specification of the algorithms is out of its scope.

When connecting such components it is difficult to determine the global behaviour of the execution and it is useful to get a model on which formal verifications can be made in order to guarantee the properties of the execution. In this paper an approach that used timed automata (Alur and Dill, 1990) and the UPPAAL tool (Larsen, *et al.*, 1998) to provide such verifications is presented.

The IEC 61499 standard is briefly introduced in section 2 and some related work on verification of function blocks applications is presented in section 3. The IEC 61499 structure of an example of control application is given in section 4 to complete this presentation of the standard. The proposed formalisation of the behaviour with timed automata is then presented in section 5 and its application to the example of section 4 is described in section 6 where some examples of properties that can be verified are also presented.

2. IEC 61499 SPECIFICATION.

The specification of an application consistent with the IEC 61499 standard is based on two distinct steps. First, each component (function block) is specified and then the global application is build by connecting the blocks.

A block is specified, on the one hand by the algorithms part that specifies how the input data are used to generate the output data. This part is only considered from an external point of view in the standard that only includes the input and output data and algorithms declarations. On the other hand the block is also specified by its control part that defines how and when these algorithms are executed according to control input events.

This control is specified by a state machine called execution control chart (ECC). Each state of the ECC (called an EC state) can be associated with zero or more EC actions. Each EC action is associated with zero or one algorithm and zero or one output control event. The EC transitions are associated with zero or one input control event and zero or one Boolean condition on data. It can be cleared when its associated event occurs, its condition is true and its source state is active. When a transition is cleared its target EC state is activated, the algorithms associated with its EC actions are then executed, the output data value are changed and the output events are set to true. The standard specifies that this is a run to completion reaction; it means that if a transition can be cleared from the new active state, it is cleared when the algorithms execution is finished and the algorithms associated with the new active state are executed. The reaction stops when no more transition can be cleared, all the output events that where set during the reaction are then issued outside the block.

When an input control event occurs the input data are sampled and the reaction of the block is launched. All reactions are event driven, that is, if the previous reaction was stopped because the condition associated with a transition without event was false, and the input data have changed to values such that this condition becomes true, the block will not react until a new input control event occurs. If an input event occurs when the block is busy in a reaction the event is postponed and a new reaction is launched as soon as the current one finishes.

The application is then build with these blocks by connecting the data and the control events. Each output data may be connected to at most one input data and each control output event may be connected to at most one input control event. When the value of an output data is changed the value of the connected input data is immediately changed and when an output event is issued the connected input event immediately occurs.

The execution control of the application is then propagated through the blocks by the event connections. As the standard aims at dealing with distributed applications some specific interface blocks that propagate the data and control flows over networks are specified. Finally other specific blocks are specified to generate control events.

3. RELATED WORK.

A lot of work has been done in the field of verification of control and PLC applications (see e.g. (De Smet *et al.* 2001) or (Mader and Wupper, 2000)). Generally this work considers the problem of the verification of programs specified in one of the IEC 61131-3 languages and their consistency with the behaviour of the plant. The work presented here is concerned with the formalization of the execution control and this problem of validation of the application is not considered as it can be seen below. This part is, then, more specifically devoted to work on IEC 61499 standard.

Vyatkin and Hanisch (2000) have proposed a formalisation based on Net/Condition Event Systems, which are an extension of Petri nets. This extension introduces the concept of module: a Petri net specifies the behaviour of a module that can be composed with others modules by condition and event outputs and inputs. The condition signals correspond to places that can be tested and the event signals corresponds to transitions that can be used to synchronize other transitions. The verification of models expressed with this formalism can be performed with the tools SESA and VEDA. In order to model IEC 61499 semantic, modules modelling each input and output events, an invocation manager, an execution manager and the ECC are introduced. As the structure of the model is very closed to the one of the block and of its execution and because of the composition mechanism this lead to rather complicated models.

Wurmus (2002) also presents an approach based on a variant of Petri nets (CNET) that is less complete and based on a practical implementation. The verification is then oriented to the synchronisation of the blocks and the absence of deadlocks.

Faure *et al* (2002) propose a model based on the synchronous dataflow language Signal and its associated verification tool (Leguernic *et al*, 1991) to verify IEC 61499. This approach that uses the notion of clocks and constraints on clocks of Signal to synchronize the input and the ECC takes into account the sequential aspects of the execution but not its timed aspects that are not considered.

4. THE EXAMPLE APPLICATION.

The example considered to complete this presentation and later to exemplify the methodology of formal modelling and verification is based on the control of the flow and temperature of an air stream. The general structure of the model of the plant and the controller is given figure 1.

With this structure of the controller it is possible to use a distributed application: the control of the flow (controller C1) is implanted in one resource and the control of the temperature (controllers C2 and C3) in an other one. A communication link is used to implement the data flow between C1 and C3.

The IEC 61499 specification of the control of the flow and the temperature are shown figure 2 and

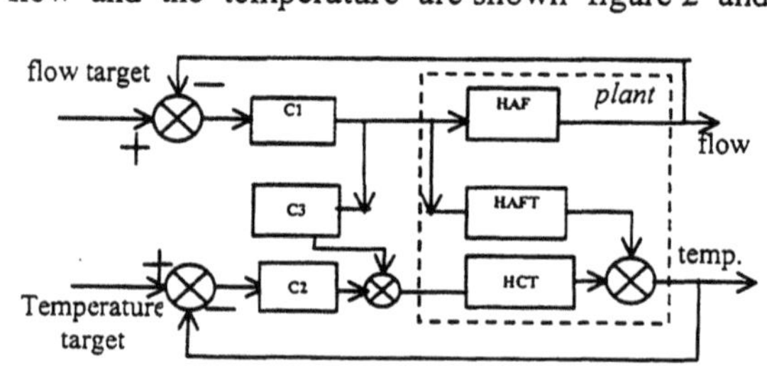

Fig. 1. The control application.

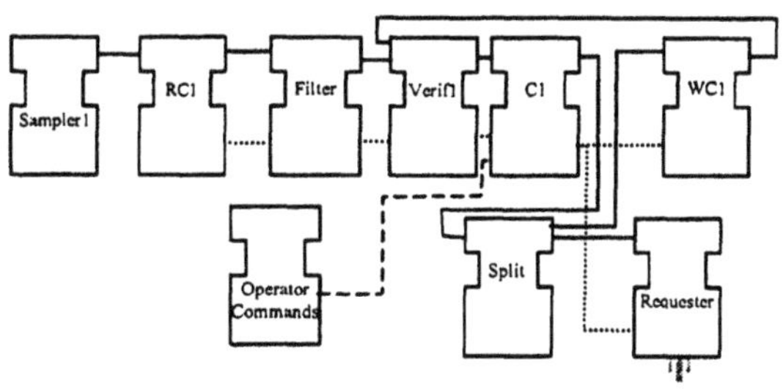

Fig. 2. IEC specification for airflow control.

figure 3, where the dotted lines represent the data flows and continuous lines the control event flows. The block *C1*, *C2* and *C3* are the implementation of the corresponding controllers of figure 1. The blocks *RC1(2)* and *WC1(2)* are interface blocks that perform the reading and the writing of plant variables by means of the AD/DA board. The blocks *Requester* and *Responder* are a couple of distribution blocks defined by the standard to perform the propagation of the data and events flows between two resources. The *Split* block is used to split the event issued by the bloc *C1* to input events of the blocks *WC1* and *Requester* in order to launch their simultaneous execution. The blocks *Sampler1*(2) are timer blocks that periodically generate an event that is the source of a reaction of the application. The functions of the *Filter* and *Adder* blocks are clearly given by their name. It can be noticed that in this specification the execution of the *adder* and consequently of the *WC2* blocks is launched either by the executions of the *C2* or *C3* blocks that are asynchronous. Finally the Verifl(2) blocks are used to insure that a new execution of C1(2) is not launched before the value issued by the previous one is written on the output of the system.

5. FORMAL MODELING OF IEC 61499 COMPONENTS.

This formal modelling is based on IEC TC65/WG6 (2000) and is consistent with this document that first specified the behaviour of blocks and applications. Its main aim is to define a modular translation schema in order that a formal model may be associated with each functional block and that the global model of an application may be mainly built by composing these basic models. An other important aim is to define a model that is fully consistent with the external behaviour of the block but not necessary to its internal structure. For example, as the specification of the algorithms are not in the scope of the standard, they are not taken

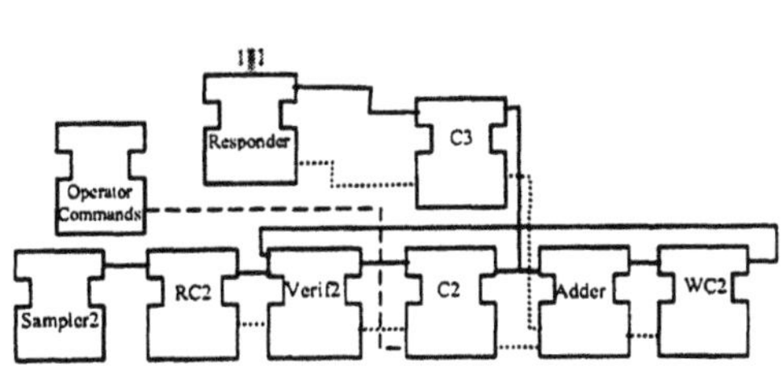

Fig. 3. IEC specification for the temperature control.

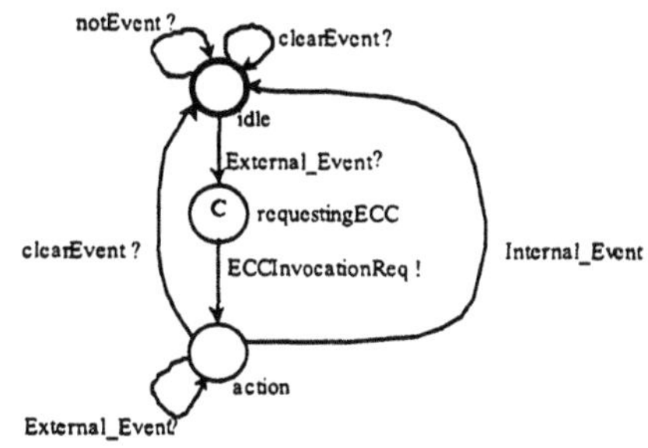

Fig. 4. General model for input event.

into account in the modelling and consequently neither are the data that are abstracted. This abstraction means firstly that algorithms will be modelled by their processing time and secondly that a different reaction will be considered for each value (true or false) of conditions on the data. Of course, this abstraction introduces some spurious reactions but most verification results are still worthwhile.

5.1. The function block.

The basic component of the specification is the function block (FB) and of course as far as control is concerned the main part is the ECC. However it is not sufficient to consider this ECC, especially because modularity is emphasized and each FB will be associated with a set of automata: one for the ECC, one for each input event and two additional ones for synchronization.

The input events. When an input occurs a reaction of the ECC is requested but if the ECC is busy the event has to be memorised and may be used later. The model that is associated with each input event of a block is then given figure 4. This automaton has 2 main locations and two main synchronisation channels. The channel *External_event* is used to synchronize with the outside of the block when the event occurs and the channel *Internal_event* is used to synchronize with the ECC reaction (see figure 6) when the event is taken into account. The channel *ECCInvocationReq* is used to request the reaction of the block (see figure 7). The location *requestingECC* is then a committed[1] location that is introduced in UPPAAL in order to synchronize the *External_event* and the *ECCInvocationReq*. The channel *notEvent* is used by the model of the ECC to determine the reaction that is to be launched when an invocation is requested and the *clearEvent* channel is used to reset the events that where not used when this invocation is processed. The location *action* is then active when an external event has occurred and no reaction has been launched since then.

All the channels are parameters of the model and have to be specified when creating an instance for each input event. Most of them are specified at the

[1] In UPPAAL a committed location is a location that can not remain active as time elapses.

FB level and the *External_Event* is specified at the application level.

The ECC. The automaton that models the ECC is the core of the modelling as it is synchronized with all the others. As the ECC is a state machine its modelling may seem quite straightforward, however, as the semantic is based on a run to completion view, the basic element is not really the EC state but the *reaction to an invocation.* Such a reaction consists of a sequence of deactivations and activations of EC state, algorithm executions, outputs event setting leading to a stable state that is a state that will remain active until a new input event occurs. The output events are all issued at the end of the reaction. This concept of reaction has to be introduced because some transitions are not associated to any event and is very similar to transient evolutions in Grafcet (Guéguen, and Bouteille, 2000; Marcé, and Le Parc, 1992).

The modelling of the ECC is based on three types of elements or patterns. First the waiting locations and the invocation synchronization, second the reaction selection part and third the reaction modelling

The source of a reaction is an EC state that can be stable. Generally these states are the source of at least one transition with an associated event. In the first step, a location is then associated to such EC state. For example if the ECC of FB *C1* (that is given figure 5) is considered, a location will be associated with EC state *Start,* where the FB can wait for an event but not with EC states *C1* or *Init*. As a reaction is launched by an invocation request these locations are the source of a unique transition synchronized on the *InvokeECC* channel of the FB with the automaton of figure 7.

The target of this transition is a set of committed locations and transitions that is more or less complex according to the number of relevant input events. For each event that is associated with a transition which source is the considered EC state a committed location and two downstream transitions that are synchronized with the automaton of the event (see figure 4) are introduced in order to test whether the event has occurred or not. These patterns are composed in a sequential way and all the relevant events are tested one after the other. This part selects the reaction, if any, that must be launched and at the end of this selection part committed locations and synchronized transitions are introduced to reset all the input events. For the ECC example of figure 5 this selection part consists of the locations *Decision1* and *Decision2* of the automaton of figure 6 and the resetting part consists of locations *Clear1*, *Clear2* and *Clear3* and the transitions synchronized on the channel *resetEvsC1* with the automaton of figure 7.

The modelling of the reaction itself is also based on three parts. The first one is the synchronization with the scheduler of the resource that controls which FB executes its algorithms and is composed of one location and one transition. For the example this part

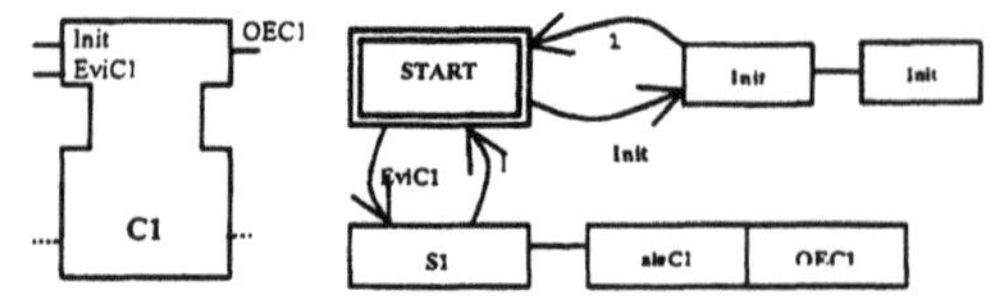

Fig. 5. Function block and ECC structure for *C1* controller.

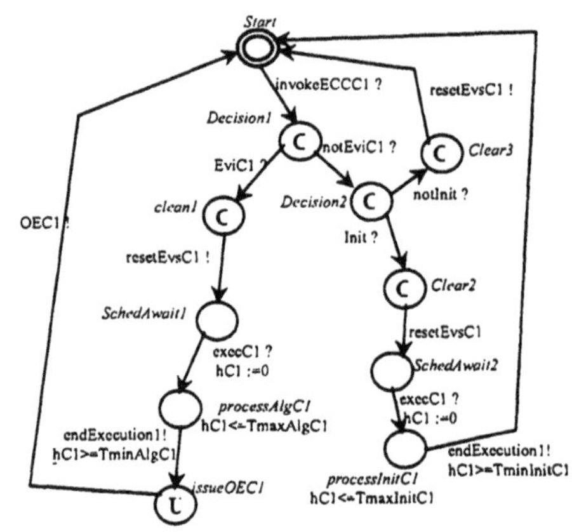

Fig. 6. ECC timed automaton model for the *C1* function block of the application.

consists of locations *SchedAwait1* for the reaction to event *EviC1* and *SchedAwait2* for the reaction to event *Init* and the transitions synchronized on channel *ExecC1*. The second part is also composed of one location and one transition and is used to model the execution of all algorithms of the reaction. This is a timed location that is associated with a lower and upper value of the execution time of the reaction that are specified by the invariant and guard concepts of UPPAAL (see for example location *processAlgC1*). The last part of the reaction is then the issuing of output events. According to UPPAAL constraints, in this part an urgent location and a transition synchronized on the corresponding channel is introduced for each output event (see for example location *issueOE1* in figure 6). The target of the last transition of this sequence is then the location modelling the stable EC state reached by the reaction. A new synchronization with the invocation request can then take place.

ECC invocation request mechanism. The ECC invocation model automaton is introduced for each FB to memorise the occurrence of events and to launch the reaction selection and the input event resetting even if the occurred events are not relevant. It has the general structure in figure 7 where the synchronisation channels are parameters that are instantiated for each FB.

Event input resetting model. This automaton is introduced because there is no broadcast mechanism in UPPAL. It is used to broadcast the resetting event

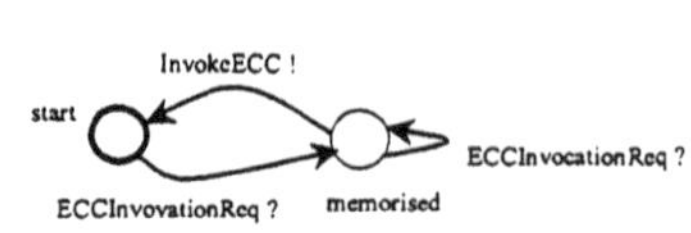

Fig. 7. ECC invocation request automaton.

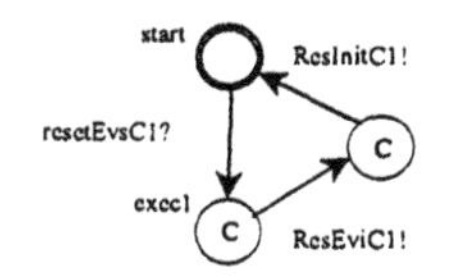

Fig. 8. Event input resetting model.

issued by the ECC model at the end of selection part to all input event models of the FB. It has then a transition synchronized on the *ClearAllEvent* channel of the block and one synchronized on the *ClearEvent* channel of each input event and is very simple as it can be seen on figure 8 for the example.

5.2. The application and resource modelling.

The previous set of automata is attached to each function block to model its execution control. Composing these sets of automata with synchronisation channels and, building a new automaton to model the scheduling, then model a specific application.

Control flow connection. In the function block diagram the control flow is specified by connecting an output event to an input event. In the formal model this connection is realised by instantiating to the same synchronisation channel, the channel of the output event used in the automaton modelling the ECC (e.g. figure 6.) and the *External_Event* channel of the input events model specified figure 4.

Scheduling function. An automaton that ensured that only one location modelling the execution of the algorithms of the functional blocks is active models the scheduling. In the current version of this work, this automaton is simple and only implements this function (e.g. figure 9). This corresponds to a non-pre-emptive and without priorities scheduling policy that could be improved in further work for example by using stopwatch automata (Abdeddaïm and Maler, 2002) for pre-emptive scheduling.

6. FORMAL VERIFICATION.

This modelling approach can be applied to the example of figures 2 and 3 in order to check some properties of the possible executions. Actually some

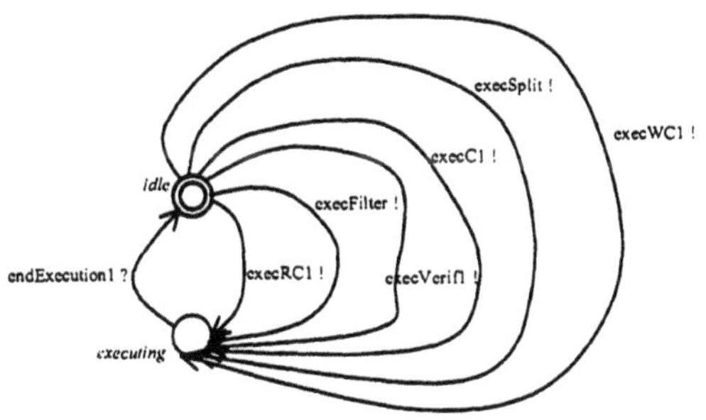

Fig. 9. Scheduler of algorithms for the air flow control loop resource.

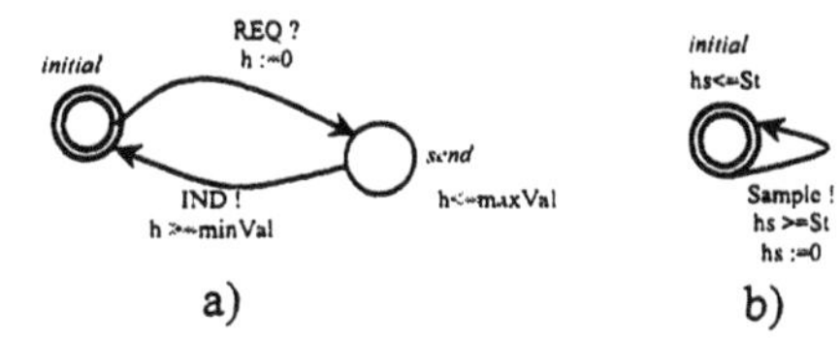

Fig. 10. Timed automaton model of a) the communication blocks, b) the sampler blocks.

blocks are not modelled as described above as they are specific. The first ones are the *Requester-Responder* blocks, and at a first level of analysis this communication link is abstracted by the delay introduced in the event flow and modelled by the automaton of figure 10.a. The second ones are the *sampler* blocks that have no input event and are modelled by the automaton of figure 10.b. For this application, the timed automata network is then composed of 55 automata that are the instantiations of only 12 UPPAAL templates; as for example 14 instances of the automaton of figure 4 are included in the model.

This automata network can be analysed and a first property that can be checked is that there is no deadlock in the model. An other important property is that the control algorithms are executed; it means that the location *processAlg1* of figure 6 and the relative locations for the other controllers become active. This can be verified by checking with UPPAAL that the predicate E<>(*C1.processAlg1*) is true. Actually if the *Verif* FB specified figure 11 is used this property is false as the output event *OE* is never issued because of the loop in the control flow. In order to correct this error that is introduced by the reuse of a block in a bad context, this block has to be substituted by the one specified figure 12.
An other important property that can be checked is that the sampling period is consistent with the processing time of the application and that the output relative to a sampling event is really applied before a new sampling event is issued. This property can be verified by checking that the predicate A[]not(RC1.processAlg and Verif1.S1) is true.

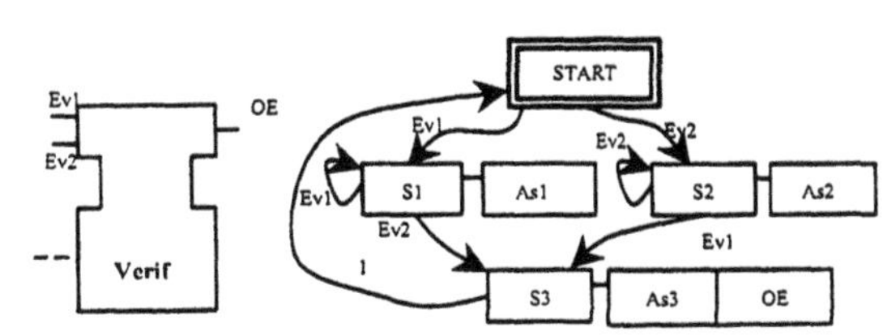

Fig. 11. *Verif* functional block specification

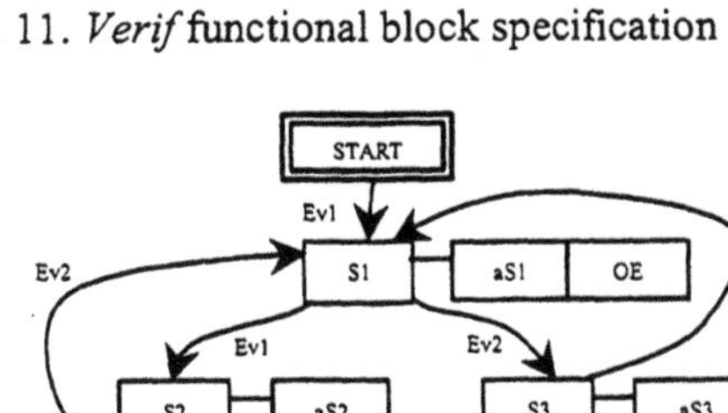

Fig. 12. Alternative ECC for the *Verif* functional block.

7. CONCLUSIONS AND FURTHER RESEARCH.

This paper presents a formalisation of the behaviour of the execution of IEC 61499 function blocks. As it may be quite difficult to understand the global behaviour of an application built by composing such blocks this formalisation and the associated verification tool are useful. The proposed formalisation is based on a modular modelling and uses timed automata in order to take into account the execution time of the algorithms. This behaviour modelling of IEC 61499 function block is basically based on the concept of reaction rather than state in order to get simpler models. Finally some examples of properties that can be verified and their relationship with the application are given.

It may seem that this formalisation is rather complex but the automata are very simple and most of them are specified as instances of generic templates. A tool has been developed that translates the standardised XML specification of an application to the timed automata network in order to make it easier. From the verification complexity point of view it may be noticed that most locations are untimed and that there is only one clock for each function block.

Of course some points have to be considered further. The abstraction of data and its effects on verification results is among the most important ones. For example this formalisation is useful to detect some deadlocks that are structural but it can not detect deadlocks linked to conditions on data. An other point is the relationships that can be made between the properties of the execution that can be checked with this approach and the properties of the application of control of plant that can be verified with other approaches. The scheduling of the application and the improvement of the modelling of more complex scheduling algorithms is a point that has been studied in other contexts and that has also to be included in this work.

Finally the confrontation of these principles to real industrial applications would be very reach. Of course the number of blocks would introduce some complexity but it is not sure that the ECC of each block would itself be very complex.

REFERENCES

Abdeddaïm, Y. and O., Maler (2002). Preemptive Job-Shop Scheduling using Stopwatch Automata. *Tools and Algorithms for Construction and Analysis of Systems,* pages 113-126.

Alur, R. and D., Dill (1990). A Theory of Timed Automata, *17th International Colloquium on Automata, Languages, and Programming. ICALP*. LNCS 443.

De Smet O., J.J. Lesage, J.M. Roussel, (2001). Formal verification of industrial control systems, *10th IFAC Symposium on Information Control Problems in Manufacturing, , INCOM'2001*, CDRom paper

Faure, J.M. , C. Schnakenbourg and J.J. Lesage (2002). Towards IEC 61499 function blocks diagram verification, *Conference on System, Man and Cybernetics, SMC 2002*, Hammamet (Tunisie), CRRom paper n°TA1C2.

Guéguen, H. and N., Bouteille (2000). Extension of Grafcet to structure behavioural specifications. *Control Engineering Practice*. Vol 9/7 pp 743-756.

IEC TC65 WG6 (2000). Function Blocks for Industrial Process Measurements and Control Systems. *Voting Draft.*

IEC TC65 WG6 (2003). Function Blocks for Industrial Process Measurements and Control Systems. *Committee Draft for Vote.*

Larsen, K. G., P. Pettersson and W. Yi (1997). UPPAAL in a Nutshell. *Springer International Journal of Software Tools for Technology Transfer*, volume 1, issue 1+2, pp 134-152

Leguernic P, M. Leborgne, T Gautier and C. Lemaire (1991) Programming real –time applications with Signal. *Proceedings of the IEEE*, Vol. 79, N 9 pp 1321-1336.

Mader A. and H. Wupper (2000). What is the method in applying formal methods to PLC applications? *ADPM2000, Hybrid Dynamical Systems*, pp 165-170.

Marcé, L. and P., Le Parc, (1992). Defining the Semantics of Languages for Programmable Controllers with Synchronous Processes. *WRTP '92.*

Vyatkin, V. and H.M., Hanisch (2000). Modelling of IEC 61499 Function Blocks. A clue to their verification. *11th Workshop on Supervising and Diagnostics of Machining Systems.*

Wurmus, H. (2002). Cnet – Komponentenbasierter Entwurf verteilter Steuerungsysteme mit Petri-Netzen. *PhD Dissertation, Hannover.*

ELSEVIER
IFAC
PUBLICATIONS
www.elsevier.com/locate/ifac

SEPARATED APPROACH FOR "ACTIVE" MONITORING OF DISCRETE EVENT SYSTEMS

Moamar Sayed Mouchaweh, Alexandre Philippot, Slim Triki, Bernard Riera

Centre de Recherche en STIC (CReSTIC)
Université de Reims-Champagne-Ardennes
Moulin de la Housse, BP 1039
51687 Reims, FRANCE
{moamar.sayed-mouchaweh ; alexandre.philippot ; slim.triki ; bernard.riera }@univ-reims.fr

Abstract: In this paper, we propose an approach to realize "active" monitoring for Discrete Event Systems. The "active" monitoring means that human operators have a good perception of process state and detect failures by their own observation of process running. This approach uses one model of the operative part to design two complementary monitoring tools. The first tool is used to follow the normal operating function. Indeed only sensors are not sufficient to guarantee an accurate perception of the process state. The idea is to supply high-level information to human operator or automated system in order to improve the situation awareness. However, it is important to supply reliable information. For that, the second complementary tool detects failures of sensors and consequently generates an alarm. This approach is a separated one since it models the behavior of the operative part independently of the command part. Additionally this approach is simple to be implemented and it better controls the combinatory explosion. An experimental prototype based on this approach is developed. Finally two examples of implementation are studied to illustrate the functionality of this approach. *Copyright © 2004 IFAC*

Keywords: Monitoring, Discrete Event Systems, Automata, Human Machine Systems.

1. INTRODUCTION

The monitoring gathers all the functions allowing to detect if a process works under abnormal operating state, to localize the responsible elements and to identify the reasons and consequences. Thus the monitoring comprises three essential functions (Riera, 2003, Combacau, 2002a, b): detection, localisation and diagnosis. The supervision includes, besides the monitoring, the control or decision function. The goal of the decision function is to determine the necessary procedures to return the process back to a normal operating state. Each of these functions is realised by means of a model. The choice of this model is based on the type of application (continuous process, manufacturing, batch) and on the level of process automation (partially or completely). Indeed the increasing complexity of process entails the automation to be partially. Consequently the results of the monitoring are intended for the Human Operator who takes the final decision. An abnormal process state can be detected by human operator either with an alarm (i.e. passive detection) or by his own observation of process running (i.e. active detection).

The actual researches aim developing powerful Human-Machine Interface libraries to help human operators in realizing an active monitoring (Riera, 2001, 2003) especially in the case of continuous systems. In other hand, it is quite different for the Discrete Event Systems (DES) where variables are Boolean and behavior is relatively simple. The failures, in the DES, can appear in the command part or in the operative part. If we considered that the command part is perfect then the failures in the operative part can be happened because of: sensors, actuators and process. These failures can be multiple, permanent or periodic.

In this paper, we propose an approach to realize "active" monitoring for DES. This approach uses one model of the operative part to design two complementary monitoring tools. The first tool is used to follow the normal operating function. Indeed only sensors are not sufficient to guarantee an accurate perception of the process state. The idea is to supply high-level information to human operator or automated system in order to improve the situation awareness. However, it is important to supply reliable information. For that, the second complementary tool detects failures of sensors and consequently generates an alarm. An experimental prototype based in this approach is developed.

2. DISCRETE EVENT SYSTEMS MONITORING APPROACHES

The detection of abnormal operating state is realized by comparing in real time the observed current state with the one coming from the model of operative part. This comparison can be done according to three criterions: logical, sequential and temporal ones (Combacau, 2002a). Thus an alarm is generated as soon as logical, sequential or temporal inconsistence is appeared. Consequently, to monitor a process, a reliable model of the operative part must be built in basing upon the knowledge of DES.

There are generally three approaches to monitor DES (Combacau, 2002a, b, Ramadge, 1989, Zaytoon, 2001). The first is the integrated approach where a general model of the process and the command part is constructed by using finite state automata. This approach is based on the interdiction of events that can entail dangerous functioning (Ramadage, 1989, Zaytoon, 2001, Valette, 1994). Thus the integrated approach needs the definition of all the possible scenarios of the DES. This fact entails the necessity to a perfect knowledge about the normal and abnormal functioning modes, which leads often to the combinatory explosion. In the second approach, the monitoring is completely separated from the command part, thus it is called the separated approach. In this approach the monitoring can be seen as an observatory of the exchange between the operative part and the command part. The results of this observation are sent either to an automated system or to a human operator. We can find several applications of this approach in (Zamaï, 1997). The third approach is the mixed one, which combines the two last approaches. Indeed, in this approach the monitoring is considered as mixed because the reasoning is done separately from the command while the command part is used to order the monitoring (Combacau, 2002b). Anyway this approach stays dependent on the command part. In all these approaches, the modelling of operative part is essential to realize a reliable monitoring.

An "active" monitoring based on separated approach is proposed in the next subsection. The separated approach was selected for the following reasons: firstly to obtain a monitoring independent of the command part, secondly to better control the combinatory explosion and lastly to obtain information more intense about the causes of failures.

3. SEPARATED APPROACH TO REALIZE AN "ACTIVE" MOINTORING

An "active" monitoring requires an efficiency perception of the structure of the operative part (Vicente, 1992). Thus a follow up tool, which follows the normal functioning of the operative part, is necessary. But the results provided by this tool can be erroneous if the information coming from sensors is not correct. Thus another complementary tool is necessary to detect failures of sensors and consequently generates an alarm. Indeed only sensors are not sufficient to guarantee an accurate perception of the process state. Thus the goal of the use of these two complementary tools is to supply high-level information to human operator or automated system in order to improve the situation awareness.

The Boolean variables Input/Output (I/O) of the Command Part (CP) of an automated system is divided into the following subsets: $S = \{S1, S2\}$ and $E=\{E1, E2, E3\}$. $S1$ is the outputs sent to the Monitoring Part (MP), these outputs characterize the logical state of the OP (sensors, actuators, switches, ...). $S2$ denotes the outputs sent to the OP, they can be divided into two types: $S21$ which are the orders applied on the actuators of the OP and $S22$ which are the used to activate other composites such as a lamp or an indicator. $E1$ denotes the commands coming from MP and $E2$ the information provided by the set of sensors and other elements of the OP. The information provided by the sensors used and activated by the CP form the subset $E21$ while the other elements not activated by the CP form the subset $E22$. $E3$ corresponds to the set of sensors not used by the CP but installed in the OP. The functioning of the proposed approach is done as following: the following tool observes the physical state of OP in using $S21$, $E2$ and $E3$. Each output of $S21$ produces a set of activation and deactivation of the sensors and consequently a change in the information provided by $E2$ and $E3$. The observation of the MP characterizes the physical state of the OP. This observation is used by the detection tool to determine if there is an inconsistence or not. In the case of inconsistence, an alarm is generated to indicate sensors failures. The fig. 1 shows the flowchart of the functioning of this approach.

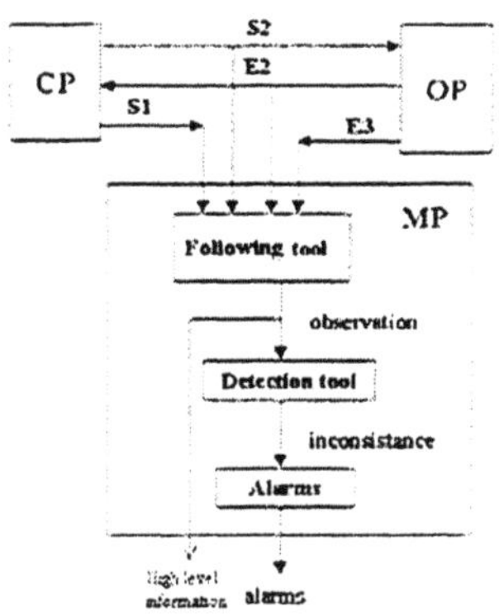

Fig. 1. Monitoring principle

To explain how this approach detects sensors failures, let us take the following example of a wagon that can move to the right or to the left (cf fig. 2), we suppose that the wagon has no inertia and there is no intersection between the axes of movement. The normal functioning is the following: after the activation of the switch of the departure of cycle, the wagon moves to the right till the position *c* then it returns back to the position *b*, then returns to the position *c* and finally it returns to the position *a*. The outputs *S21* are the commands Right (R) and Left (L). The subset *E21* comprises the information provided by the sensors *a*, *b* and *c* while *E22* comprises the switch of departure of cycle (dcy). If the normal functioning is the cycle *aca* then *E3* includes only the sensor *b*.

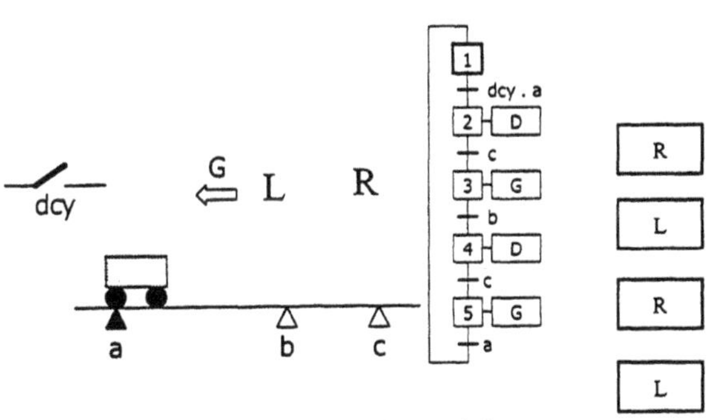

Fig. 2. Illustration example of a mobile wagon

The modeling of the OP is based on the reaction of the two subsets *E21* and *E3* to the commands *S21*. The following subsection illustrates how modeling the OP of the wagon example.

3.1 Follow up tool of the OP

The follow up tool is based on one oriented graph for each possible reaction (*L* and *R*). Each graph is represented by a pair (Q,A) with: *Q* the set of dots (also called nodes) and *A* is the set of arcs functions. Each reaction entails a change in one or more of sensors outputs according to a certain sequence. A combination, σ, of *m* values of sensors outputs is assigned to each node. We consider that only one change can happen for each reaction. Thus, according to a certain node and for *m* sensors we can have maximally *m* possible events. In other words, we can have *m* possible sensors combinations for the next node. As an example, if the actual node is the node 1, which has the combination (100) for the three sensors *a*, *b* and *c*, then the next node 2 can have one of the following combinations: (101), (011) or (000). Additionally we consider two nodes different if they indicate a different physical state. As an example the node which indicates that the wagon is located between *a* and *b* is different of the node which indicates that the position of the wagon is between *b* and *c* even they have the same combination of sensors values $\sigma_1 = \sigma_2 =$ (0 0 0).

If at the instant t_n and as a response to the command $(S21)_i$, the actual node is $q_{n,i}$, then the next node $q_{n+1,i}$ can be determined in estimating all the possible next nodes $\hat{T}_{n+1,i}$ and in calculating the adjacent matrix AD_i corresponding to the command $(S21)_i$:

$$T_{n+1,i} = diag\left(\hat{T}_{n+1,i}\cdot\left((S21)_{n,i}\cdot T^{tr}_{n,i}\cdot AD_i\right)\right) \qquad (1)$$

The value one in the line *l* and the colon *f* in the matrix AD_i indicates a link going from the node *l* towards the node *f*. Each line of this matrix can have maximally a number of elements with the value one equal to the number of sensors. One element maximally in $T_{n,i}$ is equal to one to indicate that one node is active at each reaction. $T_{n,i}$ is a vector of *N* lines, where *N* is the number of nodes. Consequently the matrix AD_i is of dimension $N\,x\,N$. $\hat{T}_{n+1,i}$ is a vector of *N* lines, it indicates all the possible future nodes, thus it can have several ones. $T_{n+1,i}$ is a vector of *N* lines and it indicates the future node corresponding to the index of line where the element of value 1 is located. $(S21)_i$ is equal to one if the command *i* is active and 0 if not.

Let us take the example of the fig. 2, and let us suppose that the wagon, is located in the position *a* at the instant $t = 0$. We have two oriented graphs, one for the command Right (*R*) and the other for the command Left (*L*) as it is illustrated in the fig. 3.

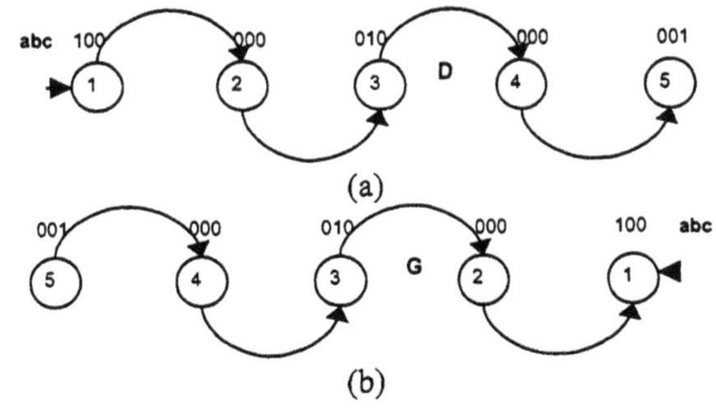

Fig. 3. Graphs corresponding (a) to the right reaction (R), (b) to the left reaction (L)

There are five different nodes for each reaction: $Q_R = \{q_1, q_2, q_3, q_4, q_5\}$ for reaction right and $Q_L = \{q_5, q_4, q_3, q_2, q_1\}$. The node q_1 indicates that the wagon is located in the position *a*, q_2 between *a* and *b*, q_3 in the position *b*, q_4 between *b* and *c* and finally q_5 in the position *c*. The initial node q_0 is the first one q_1. The set of combinations for the reactions R and L are respectively: $\{\sigma_1 = (100), \sigma_2 = (000), \sigma_3 = (010), \sigma_4 = (000)\ \sigma_5 = (001)\}$. The set of arcs A_i for each reaction $i = R$ or L is calculated as following: $\{(q_n, \sigma_{n+1}, q_{n+1})\}$ for n = 1.. 5, thus we have for the example of wagon: $A_R = \{(1, (000), 2), (2, (010, 3), (3, (000), 4), (4, (001), 5)\}$ and $A_L = \{(5, (000), 4), (4, (010), 3), (3, (000), 2), (2, (100), 1)\}$. The adjacent matrix AD_i is formed of lines. Each line shows if there is a link between two successive nodes or not. As an example: if the first line is [0 1 0 0 0], the value 1 in the line 1 and colon 2 indicates a link going from the node q_1 to the node q_2. Consequently the adjacent matrix for the two reactions R and L are respectively:

$$AD_R = \begin{pmatrix} 0 & 1 & 0 & 0 & 0 \\ 0 & 0 & 1 & 0 & 0 \\ 0 & 0 & 0 & 1 & 0 \\ 0 & 0 & 0 & 0 & 1 \\ 0 & 0 & 0 & 0 & 0 \end{pmatrix} \quad AD_L = \begin{pmatrix} 0 & 0 & 0 & 0 & 0 \\ 1 & 0 & 0 & 0 & 0 \\ 0 & 1 & 0 & 0 & 0 \\ 0 & 0 & 1 & 0 & 0 \\ 0 & 0 & 0 & 1 & 0 \end{pmatrix}$$

We can see simply that there is no link for the node 5 since it is the last one in the adjacent matrix corresponding to the reaction R as well as for the node 1 in the adjacent matrix of the reaction L.

Let us suppose that the wagon is located in the position a and the command R is applied, so the next node is calculated as following: the command is the action towards the right $(S12)_D = 1$, since the wagon is located in the position 1 then $T^{tr}_{n,D} = [1\ \ 0\ \ 0\ \ 0\ \ 0]$. When the wagon moves from the position 1, the sensors provide the following combination $\sigma = (0\ 0\ 0)$. In looking for all the nodes that have this combination, we find the nodes 2 and 4, consequently $\hat{T}^{tr}_{n,D} = [0\ \ 1\ \ 0\ \ 1\ \ 0]^{tr}$. $T_{n+1,D}$ is calculated in using the equation 1:

$$T_{n+1,D} = diag\left(\begin{bmatrix} 0 \\ 1 \\ 0 \\ 1 \\ 0 \end{bmatrix} . \left(1.[1\ 0\ 0\ 0\ 0] . \begin{bmatrix} 0 & 1 & 0 & 0 & 0 \\ 0 & 0 & 1 & 0 & 0 \\ 0 & 0 & 0 & 1 & 0 \\ 0 & 0 & 0 & 0 & 1 \\ 0 & 0 & 0 & 0 & 0 \end{bmatrix}\right)\right)$$

$$T_{n+1,D} = diag\left(\begin{bmatrix} 0 \\ 1 \\ 0 \\ 0 \\ 0 \end{bmatrix} . [0\ 1\ 0\ 0\ 0]\right)$$

$$T_{n+1,D} = diag\left(\begin{bmatrix} 0 & 0 & 0 & 0 & 0 \\ 0 & 1 & 0 & 0 & 0 \\ 0 & 0 & 0 & 0 & 0 \\ 0 & 0 & 0 & 0 & 0 \\ 0 & 0 & 0 & 0 & 0 \end{bmatrix}\right) = [0\ 1\ 0\ 0\ 0]$$

Thus the next node is q_2, which corresponds to the reality.

Indeed the observation of the probable position of the wagon can be simply and clearly visualized to a human operator and consequently it allows an efficiency perception of the OP. The fig. 4 shows an example of this visualization of the wagon position between a and b which corresponds to the node 4.

Fig. 4. Following the evolution of the wagon position

The sensors are not sufficient to supply a reliable perception of the OP. For this reason, a detection tool is used to look for inconsistencies in the image constructed by the follow up tool. Indeed using separately one tool provides only a partial perception of the OP. For this reason we use conjointly the two tools. In the following subsection we present the complementary detection tool, which uses the observation provided by the following tool.

3.2 Detection tool

When a new event is occurred in response to a command i, the following tool computes the next estimated node $\hat{q}_{n+1,i}$ and the next calculated node $q_{n+1,i}$ in using the equation 1. In comparing these two nodes, the detection tool can determine one of the following cases:

1) the next estimated node is equal to the calculated one: $\hat{q}_{n+1,i} = q_{n+1,i}$. This case corresponds to a consistence one and thus to normal functioning,
2) $\hat{q}_{n+1,i} = null$ which indicates the impossibility to find a next estimated node corresponding to the actual event. This case corresponds to a logical inconsistence resulting from a failure in the output of one or more of sensors. This case can also corresponds to a new node which is not considered by the following tool. We suppose that all the nodes corresponding to a normal functioning are considered by the following tool. Surly when $\hat{q}_{n+1,i} = null$, $q_{n+1,i} = null$,
3) $\hat{q}_{n+1,i} \neq q_{n+1,i}$, this case corresponds to a sequential inconsistence due to sensors failure. Indeed this case signifies that due to a sensors failure, the next node must be $q_{n+1,i}$ and not $\hat{q}_{n+1,i}$. Thus the sensors give incorrect perception of the physical situation of the OP to the CP.

Let us explain these two cases of inconsistence in using the wagon example.

Case 1: logical inconsistence

Let suppose that when the wagon arrives at the position b a failure in the sensor c is occurred in changing its output from 0 to 1. Here the follow up tool provides the observation $\hat{q}_{n+1,i} = null$ and $q_{n+1,i} = null$ which indicates a logical inconsistence since all the nodes corresponding to a normal functioning are considered by the follow up tool (cf. Fig. 5). An alarm will consequently be generated.

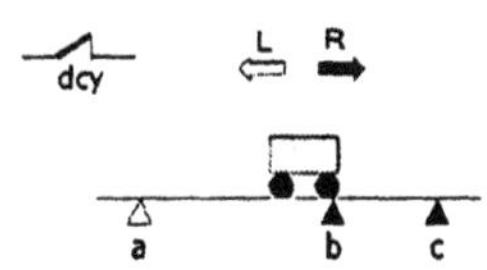

Inputs			Outputs		Next estimated state	Current state	Next calculated state	Message
a	b	c	R	L	$\hat{q}_{n+1,R}$	$q_{n,R}$	$q_{n+1,R}$	
0	1	1	1	0	3	3	3	Logical Inconsistence Alarm

Fig. 5. Detection of a logical inconsistence

Case 2: sequential inconsistence

Let suppose that the wagon, after being in the position a, is moving to the right and it is located at this instant between a and b (node 2). Let suppose that because of a failure the sensor c changes its output to the value 1. Thus σ_5 = (001). The next estimated node is $\hat{q}_{n+1,D} = 5$ while the next calculated node is $q_{n+1,D} = 3$. This case corresponds to a sequential inconsistence and consequently an alarm is generated as it is shown in the fig. 6.

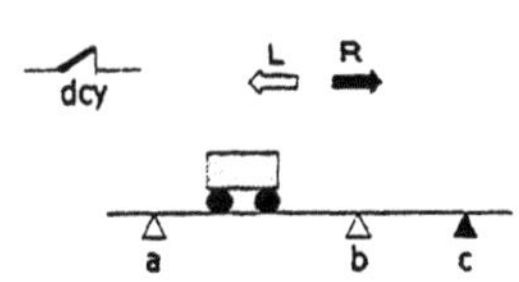

Inputs			Outputs		Next estimated state	Current state	Next calculated state	Message
a	b	c	R	L	$\hat{q}_{n+1,R}$	$q_{n,R}$	$q_{n+1,R}$	
0	0	1	1	0	5	2	3	sequential inconsistence

Fig. 6. Detection of a sequential inconsistence

4. EXPERIMENTAL PROTOTYPE AND APPLICATION EXAMPLE

An experimental prototype is developed to evaluate the proposed monitoring approach. This prototype is based upon: SCADA software, simulator allowing modelling the OP and CP, and Matlab to explain the functions of the modelling and detection tools. The communication between Matlab and the SCADA software is realized in using Excel software with a Dynamic Data Exchange (DDE) link.

This prototype was used [illegible] the wagon example of the fig. 2 and the results were presented in the fig. 5 and fig. 6.

Another example is used to evaluate the performances of the separated approach, presented in this paper, in using the experimental prototype. This example presents a pneumatic gripper, which aspirates a pinion from a drawer to put it down in another one (cf fig. 7). Gripping is carried out by aspiration, the displacements are done by double-acting cylinder controlled by a monostable valve for the vertical movement and by a bistable valve for the horizontal movement. The complete characteristics are given in http://www.lurpa.ens-cachan.fr/cosed/.

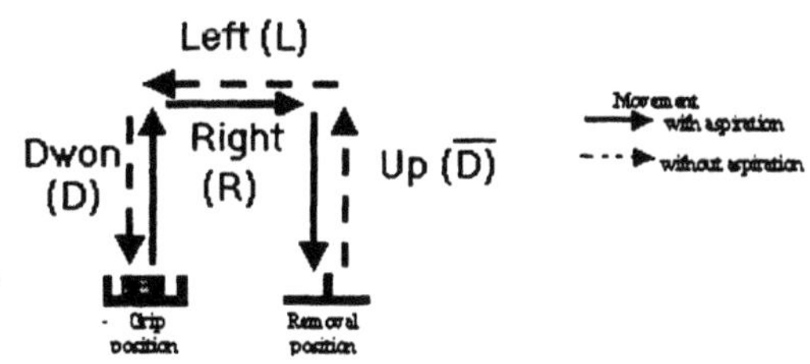

Fig. 7. Gripper example

5 sensors: a, b, c, d and e are used. a, b and c are used for the horizontal movement: a left position, b in the middle and c in the right. d and e are used in the vertical movement: d up position and e bottom position. The commands are: Left (L), Right (R), Down (D) and up ($\overline{D}$). The graphs for this example are shown in the fig. 8. We can find 15 nodes indicating all the possible states for the gripper. As an example the node 1 corresponds to the gripper in position up and in the left, which means the combination of the nodes 1 and 6 in the graphs.

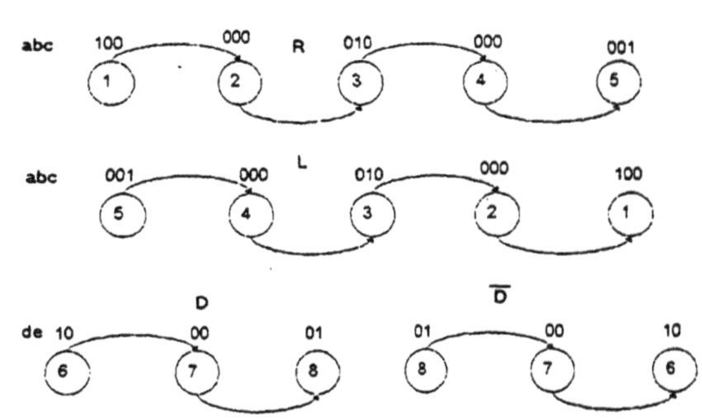

Fig. 8. Follow up tool graphs for gripper example

Let us suppose that the gripper is located in the bottom at the right (node 15), the sensors combination corresponding to this node is σ_{15} = ($abcde$ = 00101). When the commands L and $\overline{D}$ are applied, the gripper moves in diagonal displacement and the sensors combination observed is σ_{15} = (00000). We can see here that there is no information provided by the sensors. The application of the equation 1 allows determining the next state (node 9). In the fig. 9, this case is represented by the experimental prototype with the yellow lines and the green rectangle, which indicates a diagonal movement.

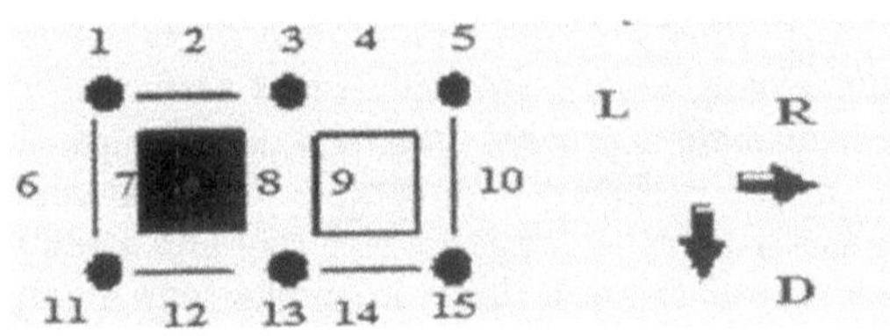

Fig. 9. Representation of diagonal movement by the experimental prototype

Let us suppose that when the gripper started to move from the node 15 diagonally, after the application of the commands L and $\bar{D}$, a failure in the sensor b has occurred and thus instead to have the sensors combination (00000), we have (01000). The estimated node corresponding to this event is $\hat{q}_{n+1,L,\bar{D}}$ = 8 while the calculated state resulting of the application of the equation 1 is $q_{n+1,L,\bar{D}}$ = 9. This indicates a sequential inconsistence and leads to generate an alarm. An example of a logical inconsistence can be seen in the fig. 10. We can see two activated sensors in the same time. This leads to obtain an estimated state null and thus to generate an alarm.

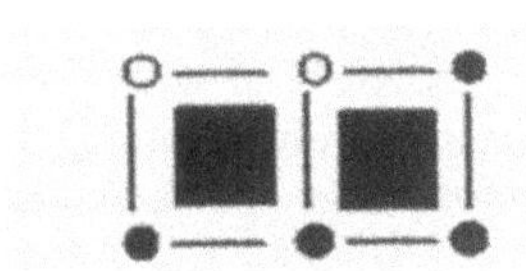

Fig. 10. Logical inconsistence case

5. CONCLUSION

A new separated approach to realize an active monitoring is presented in this paper. This approach uses two combinatory tools: follow up tool and detection tool. The follow up tool is used to follow the normal functioning of the Operative Part in using a discrete finite oriented graph for each command. The observation of the follow up tool is not reliable since sensors cannot guarantee a reliable perception of the OP. For this reason the detection tool is used to look for inconsistencies in the perception provided by the follow up tool. Indeed using separately one tool provides only a partial perception of the OP.

Generating alarms cannot provide information about the context of the failure, and consequently it leads to a passive monitoring. In other hand using only the follow up tool can give a perception not reliable about the OP and consequently lead to bad decisions taken either by human operators or by automated systems.

For this reason we use conjointly the two tools.
This approach presents different advantages: firstly it is based on a separated approach, which makes it independent of the CP, and thus it better controls the combinatory explosion. Secondly the information resulting of this approach can be simply and clearly visualized to a human operator and consequently it allows an efficiency perception of the OP. Finally his approach can be simply implemented. However this approach cannot detect all the sensors failures and thus it must be complementary with the others ones which treat the other types of sensors failures (logical and temporal). In addition this method is intuitive and thus must be constructed by hand.

There are different kinds and levels of models of OP. Some of them take into consideration the case where the inputs are modified not only by the actuators but also with other factors. An example of this case is the installation of a detector of existence of a piece in the OP. This case will be treated with the approach presented in this paper and it will be discussed in future paper. In addition the use of graph enables us to solve the problem of wagon inertia and the change simultaneously of two or more of input variables. We are treating this case and it will be presented in future paper.

REFERENCES

Combacau, M. (2002a). Approche discrète de la surveillance des systèmes de production, Rapport LAAS No01450, Maîtrise des risques et sûreté de fonctionnement des systèmes de production, Hermes Science, N°ISBN 2-7462-0402-9, Chapitre 11, pp.335-362.

Combacau M. (2002b). L.Kouiss, A.Toguyeni. Supervision d'atelier, Rapport LAAS No01569, Fondements du pilotage des systèmes de production, Traité IC2 Productique, Hermes Science, N°ISBN 2-7462-0513-0, pp.121-154.

Ramadge, P. J. G., Wonham, W. M. (1989). The control of discrete event systems. Proc. IEEE, 77,pp.81-98, 1989.

Riera, B., Debernard, S. (2003). Supervisory Systems for process control, Handbook of Cognitive Task Design, Lawrence Erlbaum Associates, pp.255-282.

Riera, B. (2001). Specifications, design and evaluation of an advanced human adapted supervisory system, Cognition, Technology and Work, Springer-Verlag London Limited, 3, pp.53-65.

Valette, R., Künzle, A. L. (1994). Réseaux de Petri pour la détection et le diagnostic. Journées Nationales : Sûreté, Surveillance, Supervision du G.R. Automatique, Paris, pp.17-18.

Vicente, K.J., Rasmussen, J. (1992). Ecological Interface Design : Thereotical foundations IEEE Transactions on Systems, Man, and Cybernetics, SMC 22, N°4, pp.589-606.

Zamaï, E. (1997). Architecture de surveillance-commande pour les systèmes à événements discrets complexes. Rapport LAAS No97361, Doctorat, Université Paul Sabatier, Toulouse, N°2760, 154 pages.

Zaytoon, J. (2001). Systèmes dynamiques hybrides, sous la direction de J. Zaytoon, traité IC2, Hermès-Paris, 378 pages.

ELSEVIER
IFAC
PUBLICATIONS
www.elsevier.com/locate/ifac

ON-LINE SYNTHESIS APPROACH BASED ON A STRUCTURED PLANT MODELLING

A. Philippot, A. Tajer, F. Gellot, V. Carré-Ménétrier

Centre de Recherche en STIC (CReSTIC), Moulin de la Housse B.P. 1039, 51687 REIMS Cedex 2, FRANCE. Tel: +33.3.26.91.32.26, Fax: +33.3.26.91.31.06
{alexandre.philippot, abdelouahed.tajer, francois.gellot, veronique.carre}@univ-reims.fr

Abstract: This paper is based on an approach of synthesis and optimal control. For a given Grafcet, a plant and constraints models, this method must respect the process evolution. We developed two improvements of this approach to be used on complex real systems. The new results obtain concern the plant modelling and the on-line synthesis. Formal algorithms and an example are presented in the paper. *Copyright © 2004 IFAC*

Keywords: Automata, Grafcet, Control synthesis, Plant modelling, Supervisory control theory, On-line synthesis.

1. INTRODUCTION

The objective of our work is to generate the required correct controller implementation for a given Grafcet (IEC 60848, 2002). This controller must to respect a number of safety and liveness constraints related to the modelled plant. The method is based on the use of a supervisor and a controller. The supervisor enforces the given safety constraints, whereas the controller directs the system to accomplish a specific set of tasks. An intuitive presentation of the different steps involved is given below, the formal algorithms can be found in (Carré-Ménétrier, *et al.*, 1999).

The first step consists of modelling the plant behaviour, the safety and liveness constraints using automata. The specification of the control behaviour is described by Grafcet. The second step is used to generate the optimal controller by an intersection. This step is realised between the controller, the plant and constraints models (Carré-Ménétrier, *et al.*, 1999). The maximum reactive non-blocking execution controller is then generated. The last step consists of implementing the correct controller by taking into account the constraints. The proposed implementation architecture executes the correct controller and correspondingly updates the state of Grafcet. This step allow to visualise the execution of the control model and to highlight the corrections induced by the synthesis framework.

We showed the applicability of this method on several examples (Carré-Ménétrier, *et al.*, 1999), (Philippot, *et al.*, 2003). However, to be practicable on complex real systems, this approach encounters two problems. The first is lied in the plant modelling which is realised without methodology. The second in computational complexity of the oof-line synthesis. The aspects related to the plant modelling are presented around an example (fig. 1) in Section 2 by an adaptation of Chandra modelisation (Chandra and Kumar, 2001). Section 3 is dedicated to a particular synchronisation algorithm which proposes an off-line synthesis approach. The contribution of the present paper is the establishment of the on-line synthesis formalization in Section 4.

2. METHOD OF PLANT MODELLING

Our previous works showed that some knowledge of plant behaviour is required to be able to identify and to treat deadlocks. It is necessary to make the image of the plant explicit, by connecting an appropriate abstract model for each of its elements.

Hence, a modular approach is required to extract a simple model for each of the elements of the plant. Such a model can be derived using automata that accept the control actions, and react by changing the logical values of Grafcet inputs. The behaviour of the plant is modelled by means of the spontaneous event generators of the supervisory control theory (Wonham and Ramadge, 1987). However, this requires the use of an adequate methodology for the modelling of the plant, which is proposed in this paper. This methodology can be considered as an adaptation of Chandra's approach (Chandra and Kumar, 2001) to be concistence with ours. The activation and deactivation of Grafcet actions correspond to controllable events Σ_c. This because of the possibility to prevent their occurrence by an appropriate conditioning of Grafcet actions. Uncontrollable events, Σ_u, are initiated by the plant. These events (Σ_c and Σ_u), which cannot be disabled by control action, are associated with the rising and falling edges of Grafcet inputs. This interpretation can be considered as an adaptation of Balemi's scheme (Balemi, *et al.*, 1993) to the case of Grafcet.

The adapted methodology is based on occurrence rules and precedence relations. These rules define the interactions between controllable or uncontrollable events. The precedence relations specify the links between the uncontrollable events. The user define the rules and relations which are translated into automata compatible with our synthesis approach.

2.1. Occurrence Rules and precedence relations

To determine the occurrence rules, it is necessary : (i) to fix the initial conditions of the system from the inputs vector of the controller model, (ii) to determine all the events related to the element of plant, (iii) to define with rules, the influence of the activation and the deactivation of the controllable events on the uncontrollable events. Each rule is a "cause/consequence". The cause relates the controllable event to the consequence (uncontrollable event). For each occurrence rule with the same cause, the chronology between the consequent uncontrollable events are established with precedence relations.

For the pneumatic gripper example (fig. 1), the complete model consists of 3 automata. They describe respectively the horizontal movement, the vertical movement and the movement of aspiration. These models take account of the cylinders technology. For the horizontal movement, there are two controllable events *GO_RIGHT* and *GO_LEFT*, and two uncontrollable events *left* and *right*. In initial situation, the cylinder is on the left and no order is sent to the plant. To define the occurrence rules, the consequences of the sent

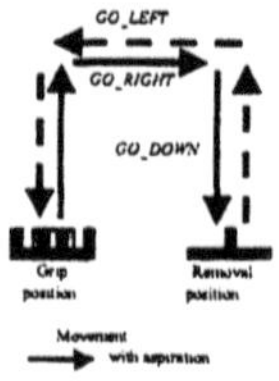

A pneumatic gripper system illustrated this paper. This system takes a pinion in a drawer and put down the pinion on an axis. The catch is carried out by a venturi. The complete characteristics are given in http://www.lurpa.ens-cachan.fr/cosed/.

Fig. 1. Characteristics of the pneumatic gripper

orders must be observed. The *GO_RIGHT* order activation makes possible the displacement of the cylinder to move towards the line and to leave the left position. The chronology of the uncontrollable events is the following one : deactivation of *left* then activation of *right*. The parameters are presented in table 1.

Initial Conditions	Occurrence rules	Precedence relations
$\downarrow$*GO_RIGHT*	$\uparrow$*GO_RIGHT* → $\downarrow$*left* $\uparrow$*GO_RIGHT* → $\uparrow$*right*	$\downarrow$*left precede* $\uparrow$*right*
$\downarrow$*GO_LEFT*	$\uparrow$*GO_LEFT* → $\downarrow$*right* $\uparrow$*GO_LEFT* → $\uparrow$*left*	$\downarrow$*right precede* $\uparrow$*left*
$\downarrow$*right*	$\downarrow$*GO_RIGHT* → $\uparrow$*right*	*nothing*
$\uparrow$*left*	$\downarrow$*GO_LEFT* → $\uparrow$*left*	*nothing*

Table 1. Parameters of the horizontal movement

2.2. Plant automaton construction

The plant automaton construction is based on rules and relations which is carried out according to 4 following steps :

1. To build the automaton with 2^n states (n is the number of controllable events) describing all the possible evolutions of the controllable events starting from the initial conditions.
2. To add to this automaton the uncontrollable events resulting from occurrence rules,
3. To build a precedence automaton starting from the precedence relations and initial situation between the uncontrollable events,
4. To make the synchronous cross product of the automaton resulting from the occurrence rules and the precedence automaton.

This method of modelling is carried out for each part of the system. The complete model of plant is obtained then by asynchronous composition of all the elementary models.

For the horizontal movement, the one step consists in establishing a 4 states automaton. The evolutions between states are related to activation or deactivation of the *GO_RIGHT* and *GO_LEFT* orders (fig. 2.a). The second step respects the occurrence rules defined by the user and permits to supplement this automaton by the uncontrollable events *left* and *right* (fig. 2.b). With the precedence relations, the third step gives the second automaton

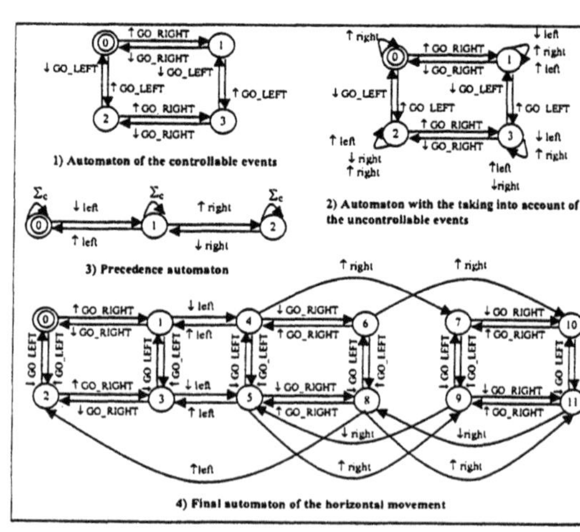

Fig. 2. Horizontal movement automaton

presented figure 2.c. Each state of this automaton authorizes all the events related to the plant element model. A synchronous cross product between the automaton resulting from step 2 and the precedence automaton resulting from step 3 allow to obtain the final automaton (fig. 2.d). The movement is then described in a complete way independently of the controller through 12 states and 35 transitions.

The automata of the vertical movement and aspiration movement are established in an identical way. The plant automaton is obtained by asynchronous product cross of these elementary models and corresponds to an automaton of 288 states and 2184 transitions.

2.3. Modelling of the constraints

We can use this methodology for the modelling of the constraints. We need only to extend the definitions of the occurrence rules and the precedence relations given previously. The occurrence rules define a link between the controllable events and the uncontrollable events and these events can be "cause" or "consequence". The precedence relations relate the same types of events. (Philippot *et al*, 2004).

We defined two types of constraints. The safety constraints between controllable events themselves use directly precedence relations. The liveness constraints establish a link between controllable events and uncontrollable events. Consequently, the modelling depends on the type of constraints to model.

The plant and constraints models are used to obtain the automaton *SUP*. This automaton corresponding to the supremal language of the supervised plant by applying the synthesis algorithm proposed in (Kumar, 1991). The transition structure of *SUP* represents the maximum permissible behaviour of the controlled plant with respect to the given constraints. At the same time, the automaton of stable situations, *GSS*, of Grafcet, deterministic and reactive, is extracting by applying the algorithm given in (Roussel, 1994).

3. PRINCIPE OF OFF-LINE SYNTHESIS

The synthesis framework is based on the use of a controller and a supervisor. A dedicated intersection algorithm extracts the sequences of events that can be generated both by the controller, *GSS*, and by the supervisor, *SUP*. This algorithm results in an event-based automaton *SYNC*. A reduction algorithm allow to remove the deadlocks of *SYNC* and to generate the control-execution model.

The intersection step generate the automaton *SYNC* representing the behaviour common to the automata *SUP* and *GSS*. The evolutions of the control model, *GSS*, are authorised by the supervisor, *SUP*. Each state in *GSS* is associated with the orders of the corresponding situation of Grafcet. The transitions of *GSS* are given in terms of logical expressions combining input variables and their edges. On the other hand, *SUP* is an elementary automaton whose transitions correspond to simple events. The intersection algorithm goes through the events that can be accepted by *GSS* and *SUP* to construct *SYNC*. Only the control evolutions that are acceptable by the supervisor are retained in automaton *SYNC*. The states of *SYNC* relating transitions are grouped into a region, which corresponds to the given Grafcet situation. The intra-region transitions correspond to controllable events and the inter-region transitions represent uncontrollable events. The interleaving of parallel orders, inside a region, is used here to provide an event-based semantics to *SYNC*. This, in view of simplifying the formal development of the intersection procedure.

The reduction step is used to generate the maximum reactive non-blocking execution controller by removing all the blocking situations from *SYNC*. The resulting automaton (*OPT*) represents the most permissive sub-set of the behaviour of Grafcet. *OPT* is both non-blocking, and satisfies the given safety and liveness specifications. Each state of this automaton corresponds to an aggregation of the states of a region of *SYNC* after the removal of blocking evolutions. The states of *OPT* correspond to the stable situations of the controller to be implemented. They contain information about the active Grafcet steps. The outputs are simultaneously activated and deactivated in the corresponding situation. The transitions of *OPT* correspond to uncontrollable events whose effect is to drive the controller from one stable situation to another. In the worst case, the size of the automaton *OPT* is proportional to $|GSS|\times|SUP|\times 2^{\Sigma u/2}$, where |GSS| and |SUP| represent the number of states of the automaton *GSS* and *SUP* respectively.

Here, the off-line synthesis approach results in an exhaustive control implementation model. This one requires a very large memory in the case of complex systems. An alternative on-line synthesis approach has therefore been proposed to reduce the size of such an exhaustive model. The on-line synthesis proceeds by successive calculations of a partial control implementation model for a temporal window.

4. ON-LINE SYNTHESIS

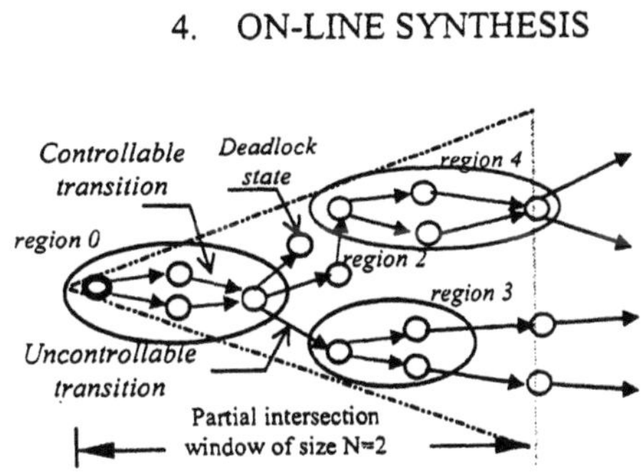

Fig. 3. Semantic model of a partial intersection

In 1992, (Chung, *et al.*, 1992) implements an on-line scheme which is basis of an N-step ahead projection of the behaviour of the process. For us, the on-line synthesis is based on the three steps of the off-line approach among which the first (modelling, synthesis and extraction) is identical to the off-line. Consequently, the procedure of control generation is based on a partial intersection step between *SUP* and *GSS* for a number of anticipated future evolutions. The step of treatment of blocking situations results from this partial intersection. A partial automaton represents the anticipated control implementation model of anticipated future evolutions.

4.1. Principle of the partial intersection.

The partial intersection procedure consists to construct the common behaviour of *GSS* and *SUP* starting from their current state from anticipated future evolutions. The method give in result an automaton $SYNC^N$ where N represents the step of calculation. N is the future possible reactions of the plant from courant situation of the system. For a partial intersection window where the size is limited by the value of N, the automaton $SYNC^N$, like *SYNC*, is constituted by regions, intra-regions transitions and inter-regions transitions. The $SYNC^N$ construction is in real time on a limited part of the controller and gives an unfinished automaton (fig. 3). Thus, while being interested only in the number of uncontrollable events Σ_u, the automaton $SYNC^N$ is constructed compared to futures reachable situations of Grafcet.

Consequently, for a region reached, all actions associate to the corresponding Grafcet situation are taken into account. The states of $SYNC^N$ with no output transition are blocking states.

4.2. Formalization of the partial intersection

The automaton $SYNC^N$ is defined by a 9-uplet (Σ, ST, TR, st_0, r_0, f_0, BLOC, EM, RD) with:

- $\Sigma = \Sigma_c \cup \Sigma_u$, $\Sigma_c = \uparrow Z \cup \downarrow Z$ et $\Sigma_u = \uparrow E \cup \downarrow E$. In the continuation, $\uparrow z$ corresponds to an element of $\uparrow Z$, $\downarrow z$ with an element of $\downarrow Z$, $\uparrow e$ with an element of $\uparrow E$ and $\downarrow e$ with an element of $\downarrow E$.

- ST is the whole of the states of $SYNC^N$. ST is partitioned in subsets called regions. A state $st \in ST$ corresponds to a single configuration of the 3-uplet (q, y, E). q is a state of *SUP*, y a state of *GSS* and E the whole of the logical values (0 or 1) of the inputs of Grafcet. All the states corresponding to a situation of Grafcet and a single configuration of its inputs variables are grouped in a region. These regions can be grouped in a window of partial intersection whose size depends on N.

- TR: $ST \times \Sigma \rightarrow ST$ is a partial function representing the transitions from *SYNC*. A transition is defined by a triplet (st, σ, st') $\in$ TR.

- $st_0 \in ST$ is the initial state which is given by (q_0, y_0, E_0). q_0 is the initial state of *SUP*, y_0 the initial state of *GSS* and E_0 corresponds to the whole of the initial values of the inputs of Grafcet.

- $r_0 \subset ST$ is the initial region corresponding to the initial situation y_0 of *GSS* and to the initial values E_0 of its inputs variables. This region includes the initial state st_0.

- BLOC $\subset$ ST contains the whole of the blocking states, i.e. those not having outgoing transitions.

- EM $\subset$ ST contains the whole of the states and the marked regions.

- RD $\subset$ ST contains the whole of the already developed regions.

The automaton $SYNC^N$ is generated by means of the following iterative algorithm (Philippot, 2002). When no window of intersection can be created anymore, the procedure is terminated.

Step 1: To develop the window of initial partial intersection f_0 starting from the initial region r_0.

Step 2: To develop a window of partial intersection unspecified f_c.

This algorithm receives as an input the automaton *SUP*, *GSS* and the initial structure of the automaton $SYNC^N$, given by (Σ, {st_0}, $\varnothing$, st_0, {st_0}, {st_0}, $\varnothing$, $\varnothing$, $\varnothing$). SUP = (Σ, Q, Δ, q_0) corresponds to the supervisor generated by the synthesis step. The automaton GSS = (E, S, Y, T, y_0) is the equivalent of the Grafcet. In the following, the current state of $SYNC^N$ is characterised by the current state of *SUP*, the current situation of *GSS*, as well as the current valuation of the input variables.

The iterative procedure corresponding to step (❶) is given in figure 4. This procedure creates the states and the transitions of the initial region r_0. In the second step (❷), as long as the number of uncontrollable events running *Ncou* is lower or equal to N, it is about:

A) To develop the inter-regions transitions and to create the first state of each regions downstream. If $Ncou = N\text{-}1$ then the states and the regions upstream of the transition are safeguarded in the set of marked states *EM*. The regions downstream are gathered in the set of developed regions *RD* to avoid the developments for the next step.

B) To safeguard the states blocking in *BLOC*.

C) To develop each region downstream according to a principle similar to ❶, these new regions belong to the initial window of partial intersection.

The algorithm associated at the development of an unspecified window breaks up into 2 steps (fig. 5). The first step (❶) consists in recovering the information already developed in the preceding partial intersection. It is about:

A) To recover the states and the regions marked in the intersection preceding and pertaining to *EM*.

B) To recover the already developed regions assemble in *RD*.

C) To initialise sets *EM* and *RD* for the future safeguards for the development of another unspecified window f_c.

The second step (❷) corresponds to ❷ algorithm of development of the initial window (fig. 4).

```
Ncou = 0
❶ /* To develop the initial region */
∀st = (q, y0, E0) ∈ r0 such as r0 ∈ f0, ∀(q, Σ, q') ∈ Δ such as (Σ = ↑Z ∧ Z ∈ Zy0) Do :
    i) ST = ST ∪ st' with st' = (q', y0, E0); ii) TR = TR ∪ {(st, Σ, st')}; iii) r0 = r0 ∪ {st'}; iv) f0 = f0 ∪ r0;
∀st = (q, y0, E0) ∈ r0 such as r0 ∈ f0, ∀(q, Σ, q') ∈ Δ such as (Σ = ↓Z ∧ Z ∉ Zy0) Do :
    i) ST = ST ∪ st' with st' = (q', y0, E0); ii) TR = TR ∪ {(st, Σ, st')}; iii) r0 = r0 ∪ {st'}; iv) f0 = f0 ∪ r0;

❷ While Ncou ≤ N Do :
A) /* To develop the inter-region transitions supported by the events ΣU and to create a new region */
∀st = (q, y, E) ∈ rr Do :
  1 ∀(q, ↑c, q') ∈ Δ such as (c ∈ E ∧ c = 0) Do :
      - If ∃ (((y, f(E), y') ∈ T) such as will f(E) = vrai) Then st' = (q', y'', E') ∧ (y'' = y') Else st' = (q', y'', E') ∧ (y'' = y)
        where the value of E' is distinguished from E only by the value of E which becomes 1 instead of 0;
      - TR = TR ∪ {(st, ↑E, st')};
      - If st' ∉ ST Then i) ST = ST ∪ {st'}; ii) to create the region r = {st'} ⊂ ST; iii) f0 = f0 ∪ r;
      - If Ncou = N-1 Then EM = EM ∪ {st}, EM = EM ∪ {rr}, RD = RD ∪ {r}; /* marking for the future intersection */

  2 ∀(q, ↓c, q') ∈ Δ such as (c ∈ E ∧ c = 1) Do :
      - If ∃ (((y, f(E), y') ∈ T) such as will f(E) = true) Then st' = (q', y'', E') ∧ (y'' = y') Else st' = (q', y'', E') ∧ (y'' = y)
        where the value of E' is distinguished from E only by the value of E which becomes 0 instead of 1;
      - TR = TR ∪ {(st, ↓E, st')};
      - If st' ∉ ST Then i) ST = ST ∪ {st'}; ii) to create the region r = {st'} ⊂ ST; iii) f0 = f0 ∪ r;
      - If Ncou = N-1 Then EM = EM ∪ {st}, EM = EM ∪ {rr}, RD = RD ∪ {r}; /* marking for the future intersection */

B) If ¬( ∃ (st, σ, st') ∈ TR) Then BLOC = BLOC ∪ {st} /* To memorize the blocking states */

C) /* To develop an unspecified region rC such as f0 = f0 ∪ rC */
  1 ∀(q, yC, EC) ∈ rC such as rC ∈ f0, ∀(q, Σ, q') ∈ Δ such as (Σ = ↑Z ∧ Z ∈ Zyc) Do :
      - If st' ∉ ST with st' = (q', yC, EC) Then i) ST = ST ∪ {st'}; ii) rC = rC ∪ {st'}; iii) f0 = f0 ∪ rC;
      - TR = TR ∪ {(st, Σ, st')};
  2 ∀(q, yC, EC) ∈ rC such as rC ∈ f0, ∀(q, Σ, q') ∈ Δ such as ( Σ = ↓Z ∧ Z ∉ Zyc) Do :
      - If st' ∉ ST with st' = (q', yC, EC) Then i) ST = ST ∪ {st'}; ii) rC = rC ∪ {st'}; iii) f0 = f0 ∪ rC;
      - TR = TR ∪ {(st, Σ, st')};

D) Ncou = Ncou + 1/* To increment the number of ΣU running */
```

Fig. 4. Development of the initial window f_0

```
Ncou = 0
❶ /* Recovery of the information calculated by the preceding intersection */
 A) /* Recovery of the states and the region marked */
 - ∀st = (q, y, E) ∈ rC such as {st, rC} ∈ EM Do : fC = fC ∪ rC;
B) /* Recovery of the already developed region */
 - ∀(st, ↑e, st')∈TR such as (st ∈rC and rC∈EM) and (st'=(q',y',E')∈r and r∈RD)
 and E' is distinguished from E only by the value E which becomes 1 instead of 0
 Do : fC = fC ∪ r;
 - ∀(st, ↓e, st')∈TR such as (st ∈rC and rC∈EM) and (st'=(q',y',E')∈r and r∈RD)
 and E' is distinguished from E only by the value E which becomes 0 instead of 1
 Do : fC = fC ∪ r;
C) EM = ∅ and RD = ∅ ; /* Initialization of sets EM and RD */
❷ While Ncou ≤ N Do :
 A) /* To develop the inter-region transitions and to create a new region */
  B) /* To memorize the blocking states */
 C) /* To develop an unspecified region rc such as fc = fc ∪ rC */
  D) Ncou = Ncou + 1; /* Incrementing of the number Σu running */
```

Fig. 5. Development of an unspecified window f_c

4.3. Treatment of deadlocks and generation of the automaton of partial controller

By analogy with the off-line synthesis, the principle of treatment of deadlocks consists in withdrawing the blocking states of $SYNC^N$. This states are identified by the set *BLOC*, as well as the transitions making it possible to reach these states. The removal of these blocking evolutions can possibly generate a controller for a number of evolutions lower than *N*. This step makes it possible to generate the automaton of partial controller OPT^N.

The reduction of the size of the automaton of controller OPT^N is significant compared to an off-line approach. In the case of a on-line synthesis, in worst case, the number of states proportional to $(|\Sigma_u|/2)^N$. For complexes systems, the profit memory capacity is then considerable. Concretely, as soon as the process generates an output, a new partial intersection/reduction is developed for a step moreover, and the automaton is updated.

Consequently, the architecture of establishment rests on the execution in parallel of two modules. The first module of on-line calculation of the controller carries out during the execution of the controller, a development for a step moreover. The second module of execution of the elaborate controller. Thus, the computing time of $SYNC^N$ and OPT^N influences the least possible the reactivity of the order. However, in spite of the optimisation of the size memory, the on-line synthesis does not guarantee the absence of deadlocks during operation. Indeed, since a partial intersection takes into account only one limited window of future evolutions of the system, no detectable deadlocks in the current window can occur at the time of the following partial intersections. By increasing size *N* of the window, these possible blocking states will be avoided or detected sufficiently early to be able to anticipate actions of correction. Consequently, the choice of *N* constitutes a compromise between the risk of deadlocks and the memory capacity necessary to the establishment of the elaborate controller.

The objective is now to establish a comparison between off-line and on-line synthesis for highlighting the profit in memory capacity.The table 2 recapitulates the sizes of the various automata for the pneumatic gripper.

GSS	*Process*	*Constraints*	*SUP*
11 states 13 trans.	288 states 2184 trans.	64 states 880 trans.	96 states 416 trans.

Table 2. Result of the various automata

For the simple example of pneumatic gripper, the step of generation of the order calculated off-line provides an automaton *SYNC* made up of 106

	Off-line synthesis	On-line synthesis	1st partial intersection	2nd partial intersection	3rd partial intersection	...
SYNC	52 states, 106 transitions	$SYNC^2$	6 states 6 transitions	6 states, 6 transitions	11 states, 10 transitions	...
OPT	30 states, 106 transitions	OPT^2	4 states, 3 transitions	5 states, 5 transitions	6 states, 6 transitions	...

Table 3. Automata generated off-line and on-line

transitions and 52 states grouped in 30 regions. The controller automaton *OPT* includes 30 states and 62 transitions. For a window of N=2 steps, the first partial intersections is given table 3.

We represented the optimal controller (OPT^2) in the figures 6.a and 6.b. In each state is associated the set with the orders to send (*ORD*) and to prohibit by *OPT* (*PROH*). These sets are used to inform the user about the orders, which are maintained, or not by the synthesis procedure. The first region of *SYNC* corresponds to state 0 of *OPT* (fig. 6.a) where no order is sent. The two states of the second region are aggregate in a single state1 where the order sent during the execution is *GO_DOWN*, no order is prohibited.

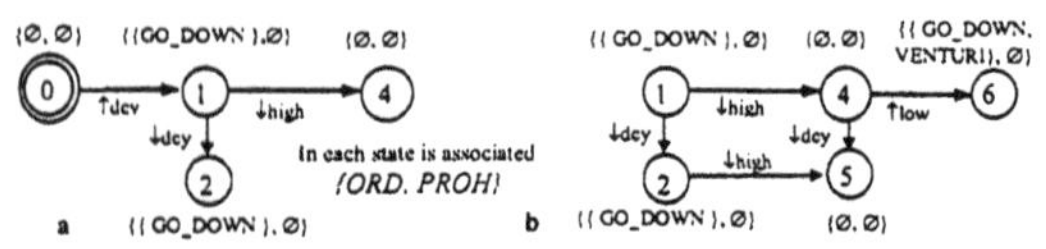

Fig. 6. Partial automata of controller (OPT^2)

The increase in the size of the window generates an increase in the number of states and transitions. Also a reduction in the number of intersection is a risk of reappearance of the problems involved in the off-line synthesis. A study on the choice of value *N* remains to be developed.

5. CONCLUSION

This paper present an approach for off-line and on-line synthesis of an optimal control implementation. This method must respect a number of safety and liveness constraints for a given Grafcet. The resulting controller can be viewed as a programing executing Grafcet in accordance with deterministic and reactive semantics.

Our current research proposes to enhance the practicability of the proposed formal framework. This, by adapting a dedicated assistance methodology for the modelling of the plant and an extension for the constraints. The user specifies the occurrence rules and the precedence relations. These rules are automatically translated into automata compatible with the supervisory theory.

For complex systems, the off-line synthesis approach requires a "huge" memory size for the implementation. We present the formalization of the on-line approach which overcomes this limitation. However, deadlocks are not guaranteed to be avoided in this case. Our future work must implement an heuristics to choose a pertinent value for the number of required evolutions. This, in the aim to avoid deadlocks and to guarantee a maximum "acceptable" size of the partial implementation.

6. REFERENCES

Balemi S., Hoffmann G.J., Gyugyi P., Wong H., Franklin G.F., "Supervisory control of has rapid thermal multiprocessor", IEEE Transactions one Automatic Control, vol. 38, n°7, p. 1040-1059, 1993.

Carré-Ménétrier V., Ndjab-Hagbebell C., Gellot F., Zaytoon J., "A case tool for the synthesis of optimal control implementation of Grafcet", IEEE International Conference on Systems, Man and Cybernetics, Tokyo-Japan, p.796-801, october 1999.

Chandra V., Kumar R., "A Discrete Event Systems Modeling Formalism Based one Event Occurrences Rules and Precedences", IEEE Transactions one Robotics and Automation, vol. 17, n°6, 2001.

Chung S.L., Lafortune S., Lin F., Limited lookahead policies in supervisory control of discrete event systems. IEEE Transactions on Automatic Control, vol. 32, Issue : 12, p.1921-1935, December 1992

IEC 60848 (International Electrotechnical Commission), Preparation of function charts for control systems. Publication 848, 2002.

Kumar R., "Supervisory Synthesis Techniques for Discrete Event Dynamical Systems", Thesis for pH D. degree, University of Texas, August 1991.

Philippot A., "Implantation d'une commande sûre dans un Automate Programmable Industriel à partir d'une spécificationexprimée en Grafcet", Memory of DEA, Reims, September 2002.

Philippot A., Tajer A., Gellot F., Carré-Ménétrier V., "Synthèse de la commande spécifiée en Grafcet : application à un préhenseur pneumatique", MSR' 03, Metz, France, p.61-75, 2003.

Philippot A., Tajer A., Gellot F., Carré-Ménétrier V., "Méthodologie de modélisation dans le cadre de la synthèse formelle des SED", submitted to CIFA'04, 2004.

Roussel J.M., "Analyse de Grafcet par Génération logique de l'Automate Equivalent", Thesis for pH D. Degree, ENS Cachan, 1994.

Wonham W. M., Ramadge P.J., "On the supremal controllable sublanguage of a given language", SIAM J Control Optimization, flight 25, n°3, p.637-659, 1987.

Copyright © IFAC Discrete Event Systems,
Reims, France, 2004

ELSEVIER
IFAC
PUBLICATIONS
www.elsevier.com/locate/ifac

MODULAR FINITE STATE MACHINES FOR LOGIC CONTROL

E. W. Endsley and D. M. Tilbury

Department of Mechanical Engineering
University of Michigan, Ann Arbor, MI 48109-2125
{ericend,tilbury}@umich.edu

Abstract: This paper presents a framework developed for logic control of manufacturing systems based on finite state machines. The framework is modular; FSMs are encapsulated into modules with well-defined communication between them. Events in the finite state machines are of two types, triggers and responses, allowing FSMs to react to and to force events to occur. A simple example is presented to illustrate the theory, and the application of the framework to two manufacturing system testbeds is described.

Keywords: Logic control, finite state machines.

1. INTRODUCTION AND BACKGROUND

Medium and high volume manufacturing has traditionally been accomplished by large-scale dedicated systems with thousands of inputs and outputs. Their logic controllers must handle sequences of operations, synchronization between components, error detection and recovery, as well as safety. Generally no model of the machining system is created for the purpose of control design. Additionally, the languages traditionally used to write the logic control do not lend themselves to analytic verification. As a result, the primary method for finding errors in the logic control is to run the controller on the actual system.

While improvements to the process of creating logic control would always have had economic benefits, changes in the state of manufacturing are making these improvements critical. Market pressures are driving the viable life of a product design and its manufacturing system down to the range of 3 to 5 years. In response to these changes, *reconfigurable manufacturing systems* (Koren et al., 1999) are being developed to allow manufacturers to keep up with consumer demands while still spreading the initial cost of the manufacturing system over a large number of parts. Current logic control programming practice is an impediment to the creation of reconfigurable manufacturing systems.

Both Finite State Machines (FSMs) and Petri nets have been proposed as a way to create and verify large logic control programs. FSMs are the basis of the supervisory control theory developed by Ramadge and Wonham (1989). While a few examples of using supervisory control on manufacturing systems exist in the literature (Brandin, 1996; Lauzon et al., 1995), there are several reasons why it is unlikely that its use will become widespread. Because a supervisory controller works by preventing events from occurring, additional constructs such as forced events need to be added to have the controllers *cause* an event to occur. The method requires detailed mathematical descriptions of the system and the desired behavior, but the creation of these descriptions may not be significantly easier or less error-prone than the direct creation of the control logic. Lastly, the computations required for synthesizing a controller for a large manufacturing system are likely to encounter complexity issues.

Statecharts, proposed by Harel (1987), introduce hierarchy to FSMs. Statecharts are good at describing the behavior of complex systems in a compact form. Like Petri nets they handle concurrency well. The complexity of the execution semantics makes verification of statecharts difficult, nonetheless some work has been done in this area (Damm et al., 1997; Rausch and Krogh, 1998). These works generally employ some limited or constrained semantics of statecharts.

2. MODULAR FINITE STATE MACHINES

This paper describes a new framework that we have developed for logic control called Modular Finite State Machines or MFSMs. MFSMs are a type of discrete event system; they react to and generate events. The framework is an extension of FSMs, with modularity and forced events, and allows for modular verification of the resulting logic controller (Endsley, 2004). Although the framework is related to many previously-defined frameworks based on FSMs, it has several important differences. In the classical Ramadge-Wonham supervisory control, the supervisor merely disables events, whereas in MFSMs the controller forces events to occur. This style of input-output FSMs (Mealy machines) are used in communicating FSMs (Brand and Zafiropulo, 1983), although there communication of events between subsystems is asynchronous whereas MFSM modules synchronize on the occurrance of events. Balemi (1994) also used input-output FSMs but only considered two modules (the plant and the controller); MFSMs allow for an arbitrary number of modules to be combined in a single system. In contrast to classical supervisory control, the design and verification of a controller using MFSMs does not require that a detailed model of the plant be available.

MFSMs are most similar to Statecharts in that there can be multiple events labeling a single transition, indicating that when an event occurs and a transition is taken, one or more events are forced to occur at the same time. The above-mentioned frameworks based on FSMs would break such transitions into multiple transitions, each with one event. Although the resulting FSM will generate the same language for a single FSM, the behavior is very different when multiple FSMs are combined. The key difference between MFSMs and Statecharts is that all communication of events between modules is specified using ports and filters; this precise specification allows for the modular verification of their global behavior.

2.1 Trigger / Response Finite State Machines

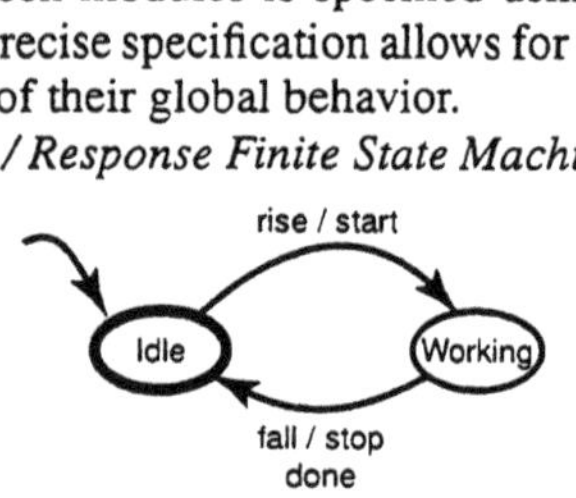

Fig. 1. A simple TR FSM.

The logic of how a particular MFSM behaves is defined by Trigger / Response Finite State Machines (TR FSMs). A TR FSM is always in one of a finite number of defined states. When a TR FSM witnesses an event, it reacts by changing state and/or generating other events as determined by its current state. The events that the TR FSM reacts to are called triggers and the events that it generates are called responses. On a transtion label, the trigger is the event before the slash and the responses are after the slash, see Fig. 1.

2.2 System of MFSMs

TR FSMs work well for defining small pieces of logic. To deal with the complexity found in large manufacturing systems, we structure the controller as a system of interacting TR FSMs, with each one controlling a portion of the system. To further aid in the creation of these large systems, we will employ several techniques from current general purpose programming practices. We will *encapsulate* the TR FSMs into **modules**. On a module, we will gather the events which communicate with other modules into sockets. Then we will define *interfaces* which specify the communication between modules. We will call these interfaces **filters**.

The primary purpose of a **module** is to hold a TR FSM and to indicate the source of trigger events and the destination of response events. Each event of the TR FSM is associated with a connection point called a socket. For example, the User module shown in Fig. 2 has two sockets. Socket 1 has rise and fall as triggers entering the module and no responses leaving the module; socket 2 has done as a trigger, and start and stop as responses. Events in the TR FSM are labeled with a name and socket number.

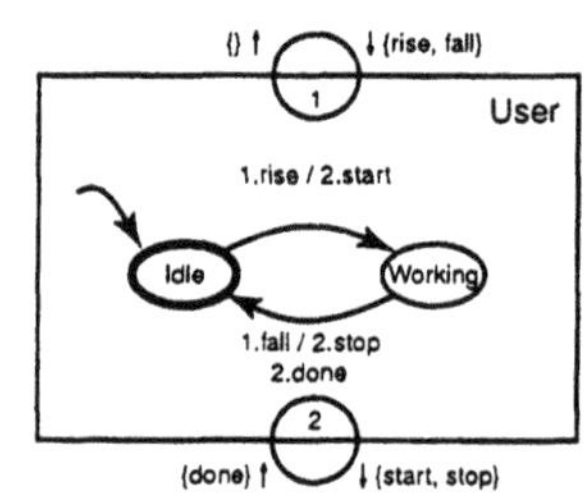

Fig. 2. A TR FSM encapsulated into the User module.

FSMs called **filters** are interfaces, or specifications, on the communication between modules. Filters play an important role in the design of MFSM systems, as they allow the designer to focus on the behavior of a particular module without having to be concerned with the details of other modules in the system. Filters encapsulate a TR FSM which has no responses, see Fig. 3. As filters are intended to be placed on the connection between two modules, they have two sockets, arbitrarily called a and b.

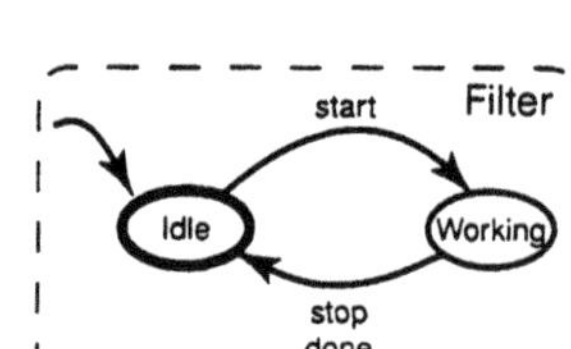

Fig. 3. A typical filter.

There are several steps in creating an **MFSM system**.

(1) Create the Structure, determining the number of modules and points of communication.
(2) Create the Filters specifying the communication between modules and with the plant.
(3) Create the Modules, defining the logical behavior of the system.
(4) Assemble the System, and verify.

This procedure possesses several characteristics of good programming practice. It is a top-down style of programming where the overall system architecture is designed first. A specification, in the form of the filters, is created for each module, and this specification is created at the beginning of the design.

2.3 Example of Designing a System

Before delving into the formal definitions of MFSMs, we will give an extended example illustrating how MFSMs can be used to construct a logic controller. The plant in this example consists of a push-button and light. The desired behavior of the plant is that the light will turn on when the button is depressed, and turn off after a timer expires or when the button is released, whichever comes first.

The system interacts with 3 external components: the button, the light, and a timer. We will have 2 modules, a User module and a Controller module. The User module interacts with the button, and the Controller module interacts with the light and the timer. The resulting skeleton system is shown in Fig. 4.

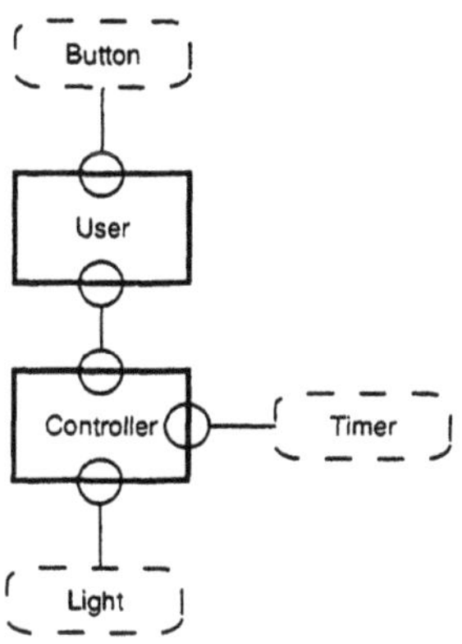

Fig. 4. The system structure.

We will begin by creating a filter on the communication between the User and the Controller module. This filter (shown in Fig. 3) has two states. From the Idle state the only thing that can happen is that the User module will tell the Controller to start. From the Working state the User can tell the Controller to stop or the Controller can tell the User module that it is done. The Button, Light and Timer filters represent communication with the environment and can be seen in Fig. 5.

To create the TR FSM for the User module, we consider that we want it to send start to the Controller module when the button is depressed, and to send stop to Controller if the button is released before the Controller sends the done event, see Fig. 2.

The way in which all of the elements of this system work together can be seen by tracing an event and the chain of responses that it generates through the system. The initial state of the system has all of the modules and filters in their initial state. From this state the only trigger which can enter the system is rise from the Button filter. This event causes Button to transition to the high state. It also causes the User module to transition to the Working state and generate the response start. This start passes into the Controller module through the filter Filter. In the process, it causes Filter to transition to the Working state. The Controller module reacts to start by transitioning to the Working state and generating on and start. The on event leaves the system through the Light filter causing it to transition to the on state. The start leaves the system through the Timer filter, causing it to transition to the running state.

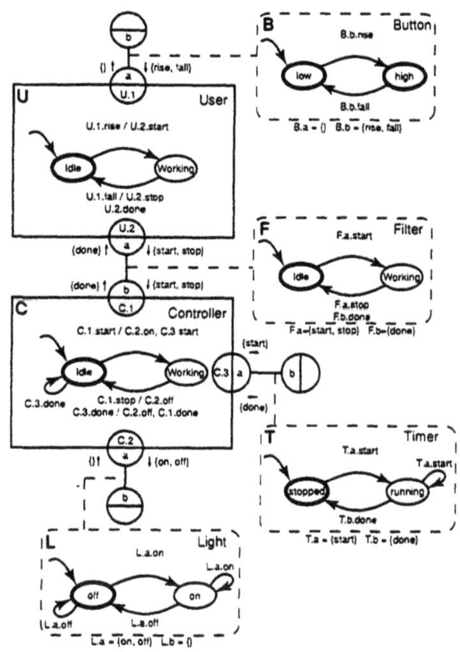

Fig. 5. The complete system.

3. MATHEMATICAL DEFINITIONS FOR MFSMS

Before defining the exact behavior of a MFSM system, we give formal definitions for a MFSM system and its components.

Definition 1. (Trigger / Response Transition). A trigger / response transition determines when and how a change of state occurs in a trigger / response FSM. The transition $\tau = (t_\tau, x_{s_\tau}, x_{d_\tau}, r_\tau)$ consists of a trigger t_τ, a source state x_{s_τ}, a destination state x_{d_τ}, and a sequence of responses produced when the transition occurs $r_\tau = [r_{\tau,1} \ldots r_{\tau,m_\tau}]$.

Definition 2. (Trigger / Response FSM). A trigger / response FSM observes trigger events and produces response events. The FSM F consists of a set of states X_F, a set of triggers T_F, a set of responses R_F, a set of transitions $\mathcal{T}_F$, an identification of the initial state x_{0_F}, an identification of the dump state x_{dp_F} and a set of marked states X_{M_F}. The FSM cannot have responses that trigger events within itself, $T_F \cap R_F = \emptyset$, and further, there must be no transitions out of the dump state. To ensure deterministic behavior of the FSM, there can be at most one transition for each combination of trigger and source state (t, x_s).

The dump state indicates that the TR FSM is in a terminal state, and that it is no longer able to process events. Transitions to the dump state are not usually included by design; they are used to identify errors that occurred in the creation of the composition of TR FSMs as defined in Definition 8.

Within a module or filter, events are bundled into defined connection points which we call sockets. We will commonly use the same letter to refer to both a module and the TR FSM within the module. We will use bold font for the modules (e.g. **M**) and italicized font for the TR FSMs (e.g. M). The same convention is used for filters and their TR FSMs.

Definition 3. (Socket). A socket bundles events of a module together for connection to another module. The socket S_{M_i} consists of a set of triggers $T_{\mathbf{M}_i}$ and a set of responses $R_{\mathbf{M}_i}$.

Definition 4. (Module). A module encapsulates a TR FSM for use within a system. The module **M** consists of a TR FSM and a set of sockets which can be connected to other modules. Every trigger in the FSM must be a trigger in exactly one of the sockets, and every trigger in a socket must be a trigger in the FSM. Every response in a socket must be a response in the FSM. It is permissable for the TR FSM to have internal responses which do not appear in a socket, but a response cannot appear in more than one socket.

Definition 5. (Filter). A filter encapsulates a TF FSM for specifying communication into and out of modules. The filter **F** consists of a TR FSM and two sockets, $S_{\mathbf{F}_a}$ and $S_{\mathbf{F}_b}$. Every trigger in the FSM must be a trigger in exactly one of the sockets, and every trigger in a socket must be a trigger in the FSM. The set of responses is empty.

Now that we have defined the components of a MFSM system, we can define how these components are connected to form a system. We first define pipes which are the explicit connection between modules and filters, and then proceed to define MFSM systems.

Definition 6. (Pipe). A pipe is a linking between objects of a system. The pipe $P = (S_{P_a}, S_{P_b}, \mathbf{F}_P)$ connects a socket on one module to a socket on another and identifies a filter for the connection.

There are 4 variations of pipes: connecting 2 modules and identifying a filter, connecting 2 modules without declaring a filter, connecting a filter with S_{P_a} but not connecting to another module, and connecting a filter with S_{P_b} but not connecting to another module.

Definition 7. (System of Modules). A system of modules $\mathbf{S} = (\mathcal{M}_\mathbf{S}, \mathcal{F}_\mathbf{S}, \mathcal{P}_\mathbf{S})$ consists of a set of modules $\mathcal{M}_\mathbf{S} = \{\mathbf{M}_1, \ldots, \mathbf{M}_{n_M}\}$, a set of filters $\mathcal{F}_\mathbf{S} = \{\mathbf{F}_1, \ldots, \mathbf{F}_{n_F}\}$, and a set of pipes $\mathcal{P}_\mathbf{S} = \{P_1, \ldots, P_{n_P}\}$ connecting those modules and filters.

To ensure that the system is well posed, we require that: (1) the pipes must connect to sockets of modules within the system; (2) the filters on the pipes must be in the system; (3) if a pipe connects two modules then the responses of one module must be a subset of the triggers of the other; (4) an event cannot be a trigger in more than one module, nor can it be a response in more than one module; (5) if an event is a trigger in one module and a response in another, then it is part of a pipe between those modules; (6) every filter is associated with exactly one pipe; (7) if a pipe has both a filter and a module at its a socket, then the responses of the module socket must be a subset of the triggers of the filter's a socket, and the triggers of the filter's b socket must be a subset of the triggers of the module socket, and similarly for the b socket.

The general concept for the behavior of a system of communicating modules of MFSMs is that responses from one module travel to another module, triggering transitions and possibly generating responses there. Due to space constraints a condensed version of the composition algorithm is given here. A more detailed and precise version can be found in (Endsley, 2004).

Definition 8. (Composition of TR FSMs). The composition of TR FSMs $F_1 \ldots F_n$, denoted $F_1 \| \ldots \| F_n$, is another TR FSM $F_{1\|\ldots\|n}$. Let each FSM $F_i = (X_{F_i}, T_{F_i}, R_{F_i}, \mathcal{T}_{F_i}, x_{0_{F_i}}, x_{dp_{F_i}}, X_{M_{F_i}})$. Then the composition is defined as $F_{1\|\ldots\|n}$ where the responses are the union of all responses, the triggers are the union of the triggers minus the responses, and the initial and dump states are the direct product of the initial and dump states of each F_i. The set of states $X_{F_{1\|\ldots\|n}}$ and the set of transitions $\mathcal{T}_{F_{1\|\ldots\|n}}$ are constructed by the composition algorithm on the following page.

While Definition 8 defines the composition of TR FSMs, we desire slightly different behavior when we compose a MFSM system consisting of modules and filters. We want it to be an error if the filters allow a trigger to occur, but a module does not. This can be accomplished by modifying Step 3 to also create a transition for each trigger t of the composition for which there exists a transition triggered by t from the current state of every filter that has that t as a trigger, regardless of the modules. Errors of this type will appear as disabled responses in Step 5.

A MFSM system can be composed into a single module by performing the composition of the TR FSMs contained within the modules and filters. The resulting single module also contains copies of the sockets which are on the perimeter of the system.

Definition 9. (System Composition). The module possessing the same behavior as a **S** is $\mathbf{M}_\mathbf{S} = (M_\mathbf{S}, \mathcal{S}_{M_\mathbf{S}})$ where $M_\mathbf{S}$ is the composition of the modules and filters of the system, and $\mathcal{S}_\mathbf{M}$ are copies of the sockets on the perimeter of the system.

Fig. 6 shows the results of combining the system seen in Fig. 5 into a single module. The presence of a

TR FSM Composition Algorithm

(1) Add the dump state and initial state to the set of states $X_{F_{1\|\ldots\|n}}$. Add the initial state to the queue of states to process Q_s.

(2) Remove x_s from the queue Q_s.

(3) Create a transition τ_n for each trigger t of the composition for which there exists a transition triggered by t from the current state x_{s_i} of every TR FSM that has that t as a trigger. The source of the transition is x_s, and the trigger is t. The response and the destination state are found with the following loop, which is initialized with an empty event queue Q_r and by setting the current event $\sigma = t$ and current state $x_c = (x_{c_1}, \ldots, x_{c_n}) = x_s$.

(4) If an **infinite response** is detected, set the current state x_c to the dump state $x_{dp_{F_{1\|\ldots\|n}}}$, clear the responses r_{τ_n} of the transition and skip to Step 8.

(5) If there exists a TR FSM F_i for which σ is a trigger, but there is not a transition triggered by σ from the current state x_{c_i} of that TR FSM, or the destination of such a transition is the dump state, then there is a **disabled response**. Set the current state x_c to the dump state, clear the responses of the transition and skip to Step 8.

(6) For each TR FSM F_i for which this event is a trigger, follow the transition for this event, updating the current state x_{c_i} of that TR FSM, and adding the responses of that transition to the event queue Q_r. If the queue and the response of the transition do not have the same events in the same order up to the length of the shorter of the two, there are **conflicting responses**. Set the current state x_c to the dump state $x_{dp_{F_{1\|\ldots\|n}}}$, clear the responses r_{τ_n} of the transition and skip to Step 8. If there is no conflict and the transition's response is longer, add the extra events onto the end of the queue.

(7) If the queue Q_r is not empty, remove the next event from the queue and set σ to it, add it to the response of the transition r_{τ_n} and return to Step 4.

(8) Set the destination $x_{d_{\tau_n}}$ of the new transition to the current state x_c. If the current state is not in the set of states $X_{F_{1\|\ldots\|n}}$, add it to the set of states and to the state queue Q_s.

(9) If there are more triggers to process for this state return to Step 3.

(10) If Q_s is not empty skip to Step 2.

transition to the dump state indicates that there is an error in the logic that should be corrected before the system is implemented.

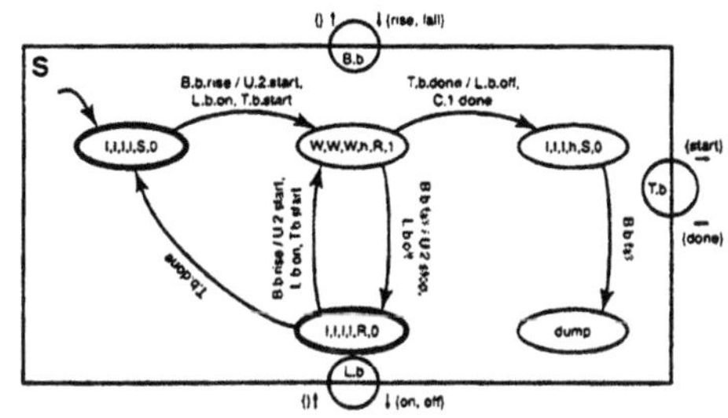

Fig. 6. The system combined into a single module.

4. IMPLEMENTATION EXAMPLES

A concern throughout the creation of MFSMs was that they be of practical use in creating logic control for manufacturing systems. To that end we developed a set of software tools (Endsley et al., 2000; Endsley, 2004) to create, verify and run MFSMs. We have also used those tools to create logic controllers for several manufacturing system testbeds. Fig. 7(a) shows the first testbed that we controlled using MFSMs. All control of the system is done through digital signals. The system has 14 bits of input and 15 bits of output; the machine could be in one of 5,038,848 states, not including the states of the parts.

(a) Linear testbed

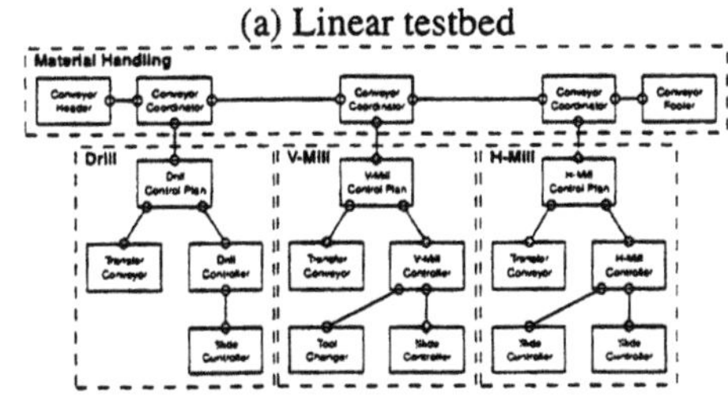

(b) MFSM Controller Structure

Fig. 7. The MSFM for the linear testbed.

The structure chosen for the MFSM controller, shown in Fig. 7(b), has 4 main subsystems: one for each station, and one for coordination. The controller for each station was created, verified and tested in isolation, demonstrating the localized design capabilities of MFSMs. Instead of a single module for coordination, a subsystem consisting of several modules was used. The resulting subsystem has the benefit that it could be adapted for a linear testbed with n stations by simply repeating one of the modules n times. The controller for this system consisted of 20 modules, but only 13 of these were unique. The transfer conveyor was used 3 times, the linear slide 4 times, and the conveyor coordinator 3 times. This demonstrates a fair amount of code reuse within this project. We have explored code reuse and reconfiguration within this testbed (Shah et al., 2002) and for a different, rotary testbed (Pollice et al., 2002). Currently we are using MFSMs to create cell level control for portions of the Reconfigurable Factory Testbed at the University of Michigan (Almeida and Tilbury, 2004).

5. CONCLUSIONS AND DISCUSSION

While MFSMs share characteristics with other frameworks, the particular combination of characteristics that MFSMs possess makes them unique. MFSMs are inherently modular and interactions between the modules are identified explicitly through pipes. The modules can be used as peers or in a multi-level hierarchy. There is no requirement that certain modules be subordinate to others. Modules can react to trigger events by generating response events, and a system of MFSMs will resolve this chain of responses before considering new input from the environment. The acceptable communication between modules can be explicitly defined using filters, allowing localized design, localized verification and code reuse. It is considered a design error when one module sends an event to a second module when the second module is in a state where it has no transition defined for that event.

Like any other framework, MFSMs have their strengths and weaknesses. Areas where MFSMs are strong include modularity, reconfigurability, verification, and execution. The modularity of MFSMs makes them suitable for creating logic control for large systems and enhances code reuse. The modular structure with filters also allows for the design of logic control that can be easily reconfigured when a manufacturing system is reconfigured. Because MFSMs are based on FSMs, verification tools can be easily applied to them. The modularity of MFSMs holds the key to enabling verification of the logic control of very large systems. And finally, as was found in the creation of our software tools, the semantics of MFSMs is directly executable. There is no need to resort to disabling or forcing events.

Areas where MFSMs are weak include concurrency and compactness. In creating the logic controllers for our testbeds, we learned that it is cumbersome to represent concurrency within a single module or filter. This is because a module or filter contains a single FSM, which does not represent concurrency well. A study on logic control methodologies (Lucas and Tilbury, 2003) showed that a logic controller written in MFSMs was 'larger' than controllers with the same functionality written in ladder diagram or Petri nets.

One of the primary reasons for creating MFSMs was to enable the verification of large logic controllers. While this paper does not present any results on verification, we have made progress in this area. With the help of Trigger / Response Languages, we have created a procedure for verifying the absence of conflicting and disabled responses within a MFSM system (Endsley, 2004). Significantly, this procedure is based on constructing the composition of a module with its surrounding filters, not constructing the composition of the entire system. While the state space of the composition of the module with its filters grows exponentially with the number of filters that the module communicates through, a good design should keep the number of connections to a single module small. Therefore, this composition does not fail due to state explosion, and the check can easily be applied to each module in the system. This procedure allows for the verification of those two properties in very large systems. We are hopeful that other procedures can be created for the modular verification of additional properties of MFSM systems.

Acknowledgements

This research was supported in part by the NSF-ERC for Reconfigurable Manufacturing Systems under grant EEC95-92125. The authors would also like to thank the anonymous reviewers for their helpful comments.

REFERENCES

E. Almeida and D. M. Tilbury. Automatic logic generation for reconfigurable cell-based manufacturing systems. In *Proceedings of the IFAC Workshop on Discrete Event Systems (WODES)*, Reims, September 2004.

S. Balemi. Input / output discrete event processes and communication delays. *Journal of Discrete Event Dynamic Systems: Theory and Applications*, 4:41 – 85, 1994.

D. Brand and P. Zafiropulo. On communcating finite-state machines. *Journal of the ACM*, 30(2):323–342, April 1983.

B. A. Brandin. Error-recovering supervisory control of automated manufacturing systems. *Integrated Computer-Aided Engineering*, 3(4):255 – 267, 1996.

W. Damm, B. Josko, H. Hungar, and A. Pnueli. A compositional real-time semantics of statemate designs. In *Proceedings of COMPOS*, pages 186 – 238, 1997.

E. W. Endsley. *Modular Finite State Machines for Logic Control: Theory, Verification and Applications to Reconfigurable Manufacturing Systems*. PhD thesis, University of Michigan, Ann Arbor, MI, May 2004.

E. W. Endsley, M. R. Lucas, and D. M. Tilbury. Software tools for verification of modular FSM based logic control for use in reconfigurable machine tools. In *Proceedings of the Japan–U.S.A. Symposium on Flexible Automation*, Ann Arbor, 2000.

D. Harel. Statecharts: A visual formalization for complex systems. *Science of Compter Programming*, 8(3):231 – 274, 1987.

Y. Koren, U. Heisel, F. Jovane, T. Moriwaki, G. Pritschow, G. Ulsoy, and H Van Brussel. Reconfigurable manufacturing systems. In *CIRP Annals*, volume 2, pages 6–12, November 1999. Keynote Paper.

S. C. Lauzon, A. K. L. Ma, J. K. Mills, and B. Benhabib. Implementing a discrete-event-system-based supervisory controller for a flexible manufacturing workcell. In *Proceedings of the 1995 IEEE International Conference on Robotics and Automation*, volume 2, pages 1429 – 1434. IEEE, May 1995.

M. R. Lucas and D. M. Tilbury. Comparing industrial logic design methods used in the automotive industry. In *Proceedings of the IEEE International Conference on Systems, Man, and Cybernetics*, pages 530 – 537, Washington, DC, October 2003.

S. L. Pollice, E. W. Endsley, and D. M. Tilbury. Logic control using modular FSMs: An evaluation of code re-use. In *Proceedings of the 2nd IFAC Conference on Mechatronic Systems*, 2002.

P. J. G. Ramadge and W. M. Wonham. The control of discrete event systems. *Proceedings of the IEEE*, 77(1):81 – 98, January 1989.

M. Rausch and B. H. Krogh. Symbolic verification of stateflow logic. In *Proceedings of the IFAC Workshop on Discrete Event Systems (WODES)*, Cagliari, 1998.

S. S. Shah, E. W. Endsley, M. R. Lucas, and D. M. Tilbury. Reconfigurable logic control using modular FSMs: Design, verification, implementation, and integrated error handling. In *Proceedings of the American Control Conference*, 2002.

ELSEVIER
IFAC
PUBLICATIONS
www.elsevier.com/locate/ifac

APPLYING AUTOMATED CONTROL SYNTHESIS METHODS TO CONDITION SYSTEMS REQUIRING STATE OBSERVERS

Yu Gong * **L. E. Holloway** **,[1]

* *VDC Solutions Inc., Cape Canaveral, FL*
** *Center for Manufacturing and*
Dept. of Electrical and Computer Engineering,
University of Kentucky
Lexington, KY, USA 40506

Abstract: Automated control synthesis methods have been presented in (Holloway *et al.*, 2000) for a class of fully observed plants modeled by condition systems, a class of discrete event systems. For plants which are not fully observed, an automated method to synthesize a state observer was presented in (Gong and Holloway, 2000). In this paper, we show how to apply automated control synthesis methods to plants which feed a state observer. In particular, we transform the original plant model into an intermediate form on which automated control synthesis methods are applied. We prove that the control thus generated is then *effective* at achieving high level control objectives when then applied to the original plant composed with a state observer. *Copyright © 2004 IFAC*

Keywords: Condition Systems, Observer, Control, Discrete Event Systems

1. INTRODUCTION

State feedback control of discrete event systems have been considered by many authors in the literature (Ramadge and Wonham, 1987) (Li and Wonham, 1988) (Yamalidou *et al.*, 1996). In particular, the closed-loop control of discrete event systems under partial state observations has been discussed by several authors in (Takai *et al.*, 1995) (Hanisch *et al.*, 1997). They synthesized a single feedback controller with the consideration of partial state information.

Giua considered the problem to create a Petri net state observer, which is used with a controller, to control the plant in closed-loop. He assumed event observations and either initial marking of the plant is unknown or the initial marking belongs to a "macromarking" (Giua, 1997). We consider condition systems, a form of Petri nets which interacts with other systems through condition signals (Holloway *et al.*, 2000). Same as Giua, we synthesize the controller and observer separately, and used them in feedback.

The main goal of this paper is to show that control synthesis methods previously shown in (Holloway *et al.*, 2000) can be used directly to synthesize control for plants which require an observer. This is done by creating an intermediate model, which we call an *x-plant*, which has input and output behavior equivalent in one sense to the input and output behavior of the composition of the real plant and the observer. It is then shown that the control that is synthesized to control the x-

[1] Email: holloway@engr.uky.edu. Supported by National Science Foundation Grant ECS-0115694, the Office of Naval Research Grant N000140110621, and the University of Kentucky Center for Manufacturing.

plant will then also be effective at control of the composition of the real plant and the observer.

In section 2, an introduction of condition systems is presented. In section 3, control concepts for condition systems are introduced. In section 4, the method of realizing feedback control by observer and controller is presented. Section 5 is an example and section 6 is a short discussion.

2. CONDITION SYSTEM

The systems that we consider interact with each other and with their outside environment through *conditions*. A condition can be considered as a signal that either has value "true", or "false". Let *AllC* be the universe of all conditions, such that for each condition c in *AllC*, there also exists a negated condition denoted $\neg c$, where $\neg(\neg c) = c$.

We represent a condition system G as a tuple $((P_G, T_G, A_G), M_G, \Phi_G())$, where (P_G, T_G, A_G) is a Petri net with place set P_G, transition set T_G, and arcs A_G, M_G is a set of possible net markings, and $\Phi_G()$ is a mapping $\Phi_G : (P_G \cup T_G) \longrightarrow 2^{AllC}$ of places and transitions to sets of conditions. For any place p, $\Phi_G(p)$ is called the set of *output conditions for* p, and for any transition t, $\Phi(G_t)$ is called the set of *enabling conditions for* t.

The next state mapping function $f_G : M_G \times 2^{AllC} \longrightarrow 2^{M_G}$ is defined from the net as follows: for any $m \in M_G$ and any set of true conditions $TrueC \subseteq AllC$, then $m' \in f_G(m, TrueC)$ if and only if there exists some transition set T such that

(1) T is *state enabled*, meaning $(\forall p \in P_G)$ $m(p) \geq |\{ \mid t \in T_G, p \text{ is input to } t\}|$
(2) T is *condition enabled*, meaning $(\forall t \in T)$ $\Phi_G(t) \subseteq TrueC$
(3) the next marking m' satisfies $\forall p \in P_G$: $m'(p) = m(p) + |\{t \in T \cap {}^{(t)}p\}| - |\{t \in T \cap p^{(t)}\}|$

In the above, we use the notation ${}^{(t)}p$ to represent the set of transitions with arcs leading to p, and $p^{(t)}$ to represent the set of transitions with arcs leading from p. Similarly, we use the notation ${}^{(p)}t$ to represent the set of places with arcs leading to t, and $t^{(p)}$ to represent the set of places with arcs leading from t.

We require that M_G be closed under $f_G(\cdot)$: if $m \in M_G$ and $m' \in f_G(m, TrueC)$ for some set of true conditions $TrueC$, then $m' \in M_G$.

The marking of the system drives conditions associated with marked places to true. In particular, for $m \in M_G$, define
$g_G(m) := \{c | \exists p \text{ s.t. } c \in \Phi_G(p) \text{ and } m(p) \geq 1\}$.

We define the output condition set (set of conditions which could be driven true by some marking) for a system G as $\Phi_{out}(G) := \{c \in \Phi_G(p) | p \in G\}$. Let $C^*_{out}(G) := \{c | c \in \Phi_{out}(G) \text{ or } \neg c \in \Phi_{out}(G)\}$ denote the set output conditions of G with their negations.

In this paper we will focus on condition systems that are *state graphs* and for which all markings in M_G are *safe*. A condition system is said to be a state-graph if each transition has exactly one input place and one output place (although it may have any number of condition inputs). The marking of a condition system is said to be *safe* if the marking of each individual place is always either zero or one.

A condition set sequence, called a C-sequence, is a finite length sequence of condition sets. The input/output behavior of a condition system can be described by C-sequences. Each C-sequence is of the form $(C_0 C_1 \ldots C_n)$ for some integer n and sets $C_i \subseteq AllC$ for all $0 \leq i \leq n$. Given C-sequences s_1 and s_2, let the expression $s_1 s_2$ indicate the concatenation of s_2 on the end of s_1. A set of C-sequences is called a language, and the language consisting of all C-sequences is denoted $\mathcal{L}$.

Definition 1. : (modified from (Holloway et al.*, 2000)) Given a condition system G and a marking m_0, define the language $L(G, m_0) \subset \mathcal{L}$ to be the set of condition set sequences such that $(C_0 C_1 C_2 \ldots C_n) \in L(G, m_0)$ if there exists some marking sequence $m_0 m_1 \ldots m_k$ and index mapping function $j(i)$ with $j(0) = 0$, $j(k) = n$ such that*

(1) Output conditions result only from the marking: $g_G(m_i) = C_{j(i)} \cap C^*_{out}(G)$,
(2) Markings evolve according to conditions: $m_{i+1} \in f_G(m_i, C_{j(i)})$ *for* $0 \leq i \leq k-1$.
(3) Sequencing is maintained: for any $0 \leq i < k$, $j(i+1) = j(i)$ *or* $j(i+1) = j(i) + 1$*;*

In this paper, we need to describe important characteristics of condition sequences without listing all details of all condition activity within the sequence. A convenient way to compare sequences is through a partial ordering "$\preceq$" which we refer to as the *descriptive ordering.* Further explanation of the ordering is given in (Holloway *et al.*, 2000). We define it again below:

Definition 2. : (Holloway et al.*, 2000) Define the* descriptive ordering $\preceq$ *over condition sequences such that*

(1) $(C_1 C'_1) \preceq (C_2)$ *if* $C_1 \subseteq C_2$ *and* $C'_1 \subseteq C_2$.
(2) $(C_1) \preceq (C_2 C'_2)$ *if* $C_1 \subseteq C_2$ *and* $C_1 \subseteq C'_2$.
(3) Given C-sequences s_1, s'_1, s_2, *and* s'_2 *such that* $s_1 \preceq s'_1$ *and* $s_2 \preceq s'_2$, *then* $s_1 s_2 \preceq s'_1 s'_2$

(4) If $s_1 \leq s_2$ and $s_2 \leq s_3$, then $s_1 \leq s_3$.

In brief, given s_1, s_2 as specifications of conditions which are known to be true sequentially over time, $s_1 \leq s_2$ means that s_2 contains at least as much specification of condition values as s_1, i.e. s_2 is at least as *descriptive* as s_1.

Example 1. : Given conditions a, b, c, define $s_1 = (\{a\}\{b\}\{c\})$, $s_2 = (\{a\}\{a, c\})$, and $s_3 = (\{a, b, c\})$. Then $(\emptyset) \leq s_1 \leq s_3$ and $(\emptyset) \leq s_2 \leq s_3$, but s_1 and s_2 are not comparable.

3. EFFECTIVENESS OF TASKBLOCKS

In (Holloway *et al.*, 2000), the control synthesis problem was considered for condition systems. In particular, control was broken into simple tasks, and the synthesized control logic was called a *taskblock*. A taskblock is associated with an *activation condition*, such that when the activation condition is true (via some higher level control command), the taskblock achieves a higher level control goal, such as driving the system to a desired goal state or keeping the taskblock in a set of allowable states.

Let G represent a plant model under consideration. Given some high level control goal for plant G, we will simply refer to the specific activation condition for that goal as $do \in AllC$, and the taskblock which is activated by it as $TB^*(do)$. The taskblock has as condition outputs signals to drive the plant model [2]. The taskblock also outputs certain status signals, including the condition $compl(do) \in \Phi_{out}(TB^*(do))$ which indicates task completion, and $idle(do) \in \Phi_{out}(TB^*(do))$, which indicates that the taskblock is not activated. In all subsequent discussion, we will assume a taskblock has an initial marking such that the signal $idle(do)$ is output true.

The high-level control goal associated with an activation condition do is associated with some condition $goal(do) \in \Phi_{out}(G)$ which is output from the component model. In particular, the high-level control goals we consider are that the taskblock must either drive the plant state such that $goal(do)$ becomes true, or must keep the plant state such that $goal(do)$ will remain true. Finally, let $C_{init}(do) \subseteq C^*_{out}(G)$ be a set of conditions output from the plant component that must be true for a taskblock to respond to its activation condition.

The following definition of *effective* formally describes the behavior of a taskblock when it is interacting with a system in its intended manner.

Definition 3. : (Holloway et al., *2000) Consider a condition do, let $\mathcal{G}$ denote a system consisting of a plant component G and some taskblock $TB(do)$, and let m_0 denote the initial marking such that $idle(do) \in g_{\mathcal{G}}(m_0)$. The condition do is* effective *for $\mathcal{G}$ under m_0 if each of the following statements are true:*

(1) Continued activation implies eventual completion: *For all $s \in L(\mathcal{G}, m_0)$ such that $(\emptyset\ (\{do\} \cup C_{init}(do))) \leq s$, for any set C_{ext} such that $do \in C_{ext}$ and $C_{ext} \cap (C^*_{out}(G) \cup \{\neg do\}) = \emptyset$, there exists s' such that $ss' \in L(\mathcal{G}, m_0)$, $(C_{ext}) \leq s'$, and*

$$(\{do\}\ \{do, compl(do)\}) \leq s'$$

(2) Completion implies earlier activation: *For all $s \in L(\mathcal{G}, m_0)$, if $(\emptyset\{compl(do)\}) \leq s$, then*

$$(\emptyset\ (\{do\} \cup C_{init}(do)\ )\ \emptyset) \leq s$$

(3) Completion implies achieved goal: *For any condition set string s and any condition set C such that $sC \in L(\mathcal{G}, m_0)$, if $\{compl(do)\} \subset C$, then*

$$\{compl(do), goal(do)\} \subseteq C$$

(4) Leaving completion implies earlier deactivation: *For all $s \in L(\mathcal{G}, m_0)$,*
if $(\emptyset\{compl(do)\}\{\neg compl(do)\}) \leq s$, then

$$(\emptyset\ \{\neg do\}\ \emptyset) \leq s$$

(5) Deactivation implies eventual return to idle: *For all $s \in L(\mathcal{G}, m_0)$, if $(\emptyset\{\neg do\}) \leq s$, then for any set C_{ext} such that $\neg do \in C_{ext}$ and $C_{ext} \cap (C^*_{out}(G) \cup \{do\}) = \emptyset$, there exists s' such that $ss' \in L(\mathcal{G}, m_0)$, $(C_{ext}) \leq s'$, and*

$$(\{\neg do\}\ \{\neg do, idle(do)\}) \leq s'$$

The first statement states that after do and $C_{init}(do)$ conditions are true, if do remains true, then there will eventually follow a completion condition $compl(do)$ from the taskblock, and completion is reached entirely through the interaction of the taskblocks and components in $\mathcal{G}$, and not from any other external conditions. The second statement of the definition states that if $compl(do)$ is true at the end of s, then at some prior time in s the conditions do and $C_{init}(do)$ were

[2] The paper (Holloway *et al.*, 2000) considered plants made up of multiple layers of condition systems, and the taskblock output activation conditions associated with inputs to a particular plant component could either feed into other taskblocks or to the plant through a *direct translator* which translated the activation signals directly into actuation inputs of the component. In this paper, we only consider a single layer, so to simplify discussion, we omit discussion of the direct translator. The asterix in the notation $TB^*(do)$ indicates the incorporation of the direct translator with the task block.

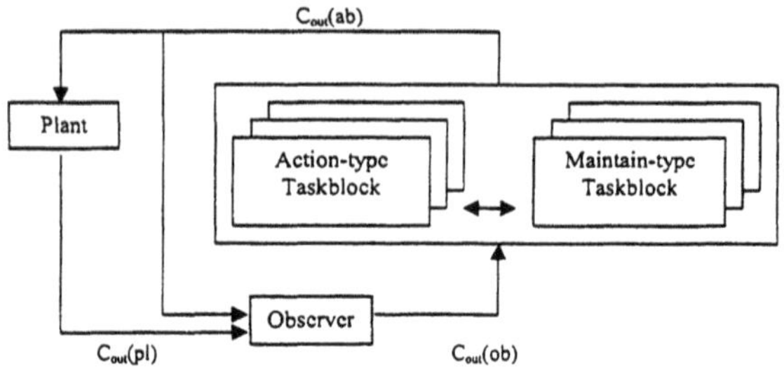

Fig. 1. Closed-loop System Structure.

true. The third statement in the definition states that whenever *do* and *compl(do)* are simultaneously true, then *goal(do)*, output from component $G_{compo}(do)$, must be true also. Statement 4 says that once a taskblock achieves completion, it will stay there until *do* becomes false. Finally, the last statement says that if *do* is false, then eventually *idle(do)* will become true.

The proposed closed-loop control of condition system plant using observer, and taskblocks is illustrated by figure 1.

Prompt Response Assumption (PRA): Consider a single-layer closed-loop system composed of a plant, an observer and taskblocks. The system satisfies the Prompt Response Assumption (PRA) if the following is true:

(1) If there are transitions enabled in the plant and at the same time there are enabled transitions in either the observer or the taskblocks, any enabled transitions in the observer and the taskblocks will fire first.
(2) If there are transitions enabled in the taskblocks and at the same time there are enabled transitions in the observer, the transitions in the observer will fire first.

4. CLOSED-LOOP CONTROL WITH OBSERVERS

The control synthesis procedure in (Holloway *et al.*, 2000) creates taskblocks that are proven to be effective for control of plants, as long as the plants satisfy the property that the marking of the net can be determined from the conditions which are true. We call this the *direct observability* requirement. In this paper we want to relax this requirement. In particular, we want to consider plants where there are some markings which map to the same output condition set, so the plant state is not directly observed. We will use a state observer cascaded to the output of the plant such that the observer will identify the state of the system uniquely. The goal of this paper is to show that we can use existing taskblock synthesis methods in order to create a taskblock that will

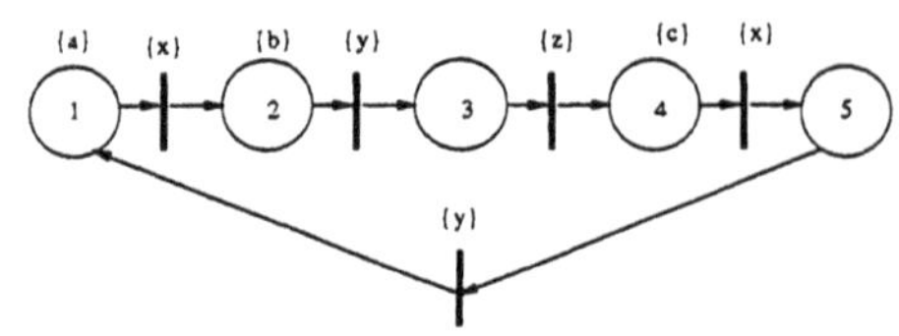

Fig. 2. The plant G_{plant}.

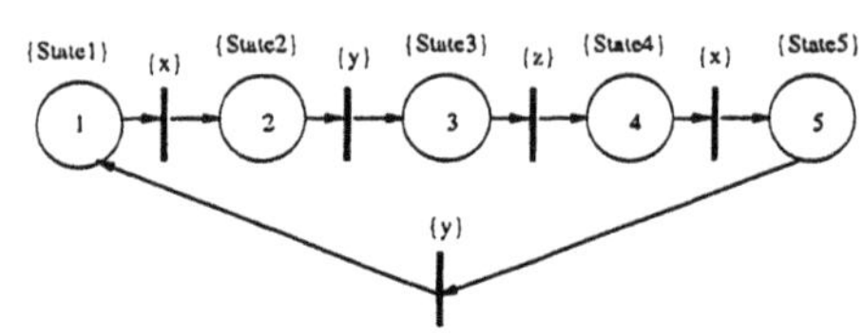

Fig. 3. The x-plant G_{xp} of G_{plant} in figure 2.

be effective for this cascade of plant and observer. This will be done by applying synthesis methods to an intermediate model, which we call the *x-plant*, G_{xp}.

Definition 4. : *Given a plant* G_{plant}, *define the* x-plant, *denoted* G_{xp}, *as a copy of the structure of* G_{plant} *but such that for any place* p_i *in* G_{xp}, $\Phi_{G_{xp}}(p_i) = \{State\text{-}p_i\}$, *where* $\{State\text{-}p_i\}$ *is the label uniquely identified to corresponding place* p_i *in* G_{plant}.

Given the plant in figure 2, the corresponding x-plant is shown in figure 3. Note that states 3 and 5 are equivalent in observation (i.e. no observation in this case), so the plant does not have direct observability. (However, it is *observable* in the sense of (Gong and Holloway, 2000), meaning that the state can be determined through knowledge of the past observations).

Note that the x-plant model has a unique output condition associated with each place, and identifying that place. We will create a taskblock to control the plant using these condition labels. This taskblock, $TB^*_{xp}(do)$, will be synthesized to be effective in closed loop with this fictitious x-plant. We will then show that this implies that $TB^*_{xp}(do)$ will be effective in closed loop for the composition of the plant and observer.

An observer synthesis method was presented in (Gong and Holloway, 2000). In this paper, we will not present the details of the method. However, we will rely on some properties proven in (Gong, 2003). Specifically, we will rely on properties of the composition of the plant and the observer.

Definition 5. : *Given a component plant* G_{plant} *and another system (the "observer")* G_{obv}, *define* $G_{cp} := G_{plant} \cup G_{obv}$. *We will refer to this composed plant as the* c-plant. *Given* G_{cp} *as a composed plant and its "observer", and given a known initial marking, we say that* G_{cp} *satisfies the*

Observer-Composed-Plant Property *(OCP Property) if the following are true:*

(1) for any non-transient state m of the composed system reachable from the initial marking, a place p_i is marked in G_{plant} if and only if there is a condition State-p_i such that State-$p_i \in g_{G_{cp}}(m)$.
(2) for each $s_{cp} \in L(G_{cp}, m_0^{cp})$, there exists a $s_{xp} \in L(G_{xp}, m_0^{xp})$, s.t.
  (a) $s_{xp} \leq s_{cp}$
  (b) $s_{xp}|_{AllC-C^*_{out}(G_{plant})} = s_{cp}|_{AllC-C^*_{out}(G_{plant})}$
(3) for each $s_{xp} \in L(G_{xp}, m_0^{xp})$, there exists a $s_{cp} \in L(G_{cp}, m_0^{cp})$, s.t.
  (a) $s_{xp} \leq s_{cp}$
  (b) $s_{xp}|_{AllC-C^*_{out}(G_{plant})} = s_{cp}|_{AllC-C^*_{out}(G_{plant})}$

It has been shown in (Gong, 2003) that the synthesized observers of (Gong and Holloway, 2000) satisfy the OCP property.

This finally brings us to our main result, that a taskblock synthesized for the x-plant will also be effective for the composition of the plant and the observer.

Theorem 1. : Consider a plant G_{plant}. Let G_{xp} be the corresponding x-plant (by definition 4), and G_{obv} be an observer for G_{plant} such that $G_{cp} := G_{plant} \cup G_{obv}$ satisfies the OCP property. Let m_0 denote the initial marking of all the systems such that for any place p_i in G_{plant} marked under m_0, the corresponding places p_i in G_{obv} and in G_{xp} are also marked. Consider any activation condition "do" and subsystem (taskblock) $TB^*_{xp}(do)$ such that "do" is effective for $G_{xp} \cup TB^*_{xp}(do)$ under m_0. Then, "do" is also effective for $G_{plant} \cup G_{obv} \cup TB^*_{xp}(do)$.

Proof: We prove the theorem by verifying each item in definition 3. Define $G_{TBx} := TB^*_{xp}(do) \cup G_{xp}$, and $G_{TBc} = TB^*_{xp}(do) \cup G_{cp}$.

(1) *Continued activation implies eventual completion:*
Assume $s_{xp} \in L(G_{TBx}, m_0)$ satisfies this statement, we should show there exists $s_{cp} \in L(G_{TBc}, m_0)$ that satisfies this statement too. From the assumption and definition of effective, we know that there exists s'_{xp} such that :
(a) $s_{xp}s'_{xp} \in L(G_{TBx}, m_0)$
(b) $(C_{ext}) \leq s'_{xp}$, $\forall C_{ext}$ s.t. $do \in C_{ext}$ and $C_{ext} \cap C^*_{out}(G) = \emptyset$,
(c) $s_\beta := (\{do\}\{compl(do)\}) \leq s'_{xp}$
For item (a), since $s_{xp}s'_{xp} \in L(G_{TBx}, m_0)$, then $s_{xp}s'_{xp} \in L(G_{xp}, m_0)$. By OCP property item 3, there exists $s^*_{cp} \in L(G_{cp}, m_0)$, s.t. $s_{xp}s'_{xp} \leq s^*_{cp}$ and $s_{xp}s'_{xp}|_{AllC-C^*_{out}(G_{plant})} = s^*_{cp}|_{AllC-C^*_{out}(G_{plant})}$. Divide s^*_{cp} into $s^*_{cp} = s_{cp}s'_{cp}$, s.t. $s_{xp} \leq s_{cp}$, $s'_{xp} \leq s'_{cp}$. Note that $s_{cp}s'_{cp}$ and $s_{xp}s'_{xp}$ are equivalent except for the conditions in C^*_{out}, which are not used as inputs or outputs of $TB^*_{xp}(do)$ anyway. Thus, $s_{xp}s'_{xp} \in L(TB^*_{xp}(do), m_0)$, implying that $s_{cp}s'_{cp} \in L(TB^*_{xp}(do), m_0)$, furthermore implying $s_{cp}s'_{cp} \in L(G_{TBc}, m_0)$.
For item (b), From assumption, $(C_{ext}) \leq s'_{xp}$, and since $s'_{xp} \leq s'_{cp}$, then $(C_{ext}) \leq s'_{cp}$.
For item (c), Also according to assumption, $s_\beta \leq s'_{xp}$, and since $s'_{xp} \leq s'_{cp}$, $s_\beta \leq s'_{cp}$.
We have thus shown that given any $s_{xp}s'_{xp} \in L(TB^*_{xp}(do)) \cup G_{xp}, m_0)$ that satisfies (a) (b) and (c), there exists a $s_{cp}s'_{cp} \in L(TB^*_{xp}(do)) \cup G_{cp}, m_0)$ that also satisfies similiar statements (a) (b) and (c) for G_{cp}.

(2) *Completion implies earlier activation:*
Consider any $s_{cp} \in L(G_{TBc}, m_0)$, s.t. $(\emptyset\{compl(do)\}) \leq s_{cp}$. By OCP property item 2, there exists an $s_{xp} \in L(G_{xp}, m_0)$, s.t. $s_{xp}|_{AllC-C^*_{out}(G_{plant})} = s_{cp}|_{AllC-C^*_{out}(G_{plant})}$, and $s_{xp} \leq s_{xp}$. Thus $(\emptyset\{compl(do)\}) \leq s_{xp}$. By effectiveness definition, then $(\emptyset(\{do\} \cup C_{init}(do))\emptyset) \leq s_{xp}$. $(\emptyset(\{do\} \cup C_{init}(do))\emptyset) \leq s_{xp} \leq s_{cp}$, therefore $(\emptyset(\{do\} \cup C_{init}(do))\emptyset) \leq s_{cp}$.

(3) *Completion implies achieved goal:*
Consider any s_{cp} and C_{cp}, such that $s_{cp}C_{cp} \in L(G_{TBc}, m_0)$ and $\{compl(do)\} \subset C_{cp}$. We want to show that $\{compl(do), goal(do)\} \subseteq C_{cp}$.
Assign $s^*_{cp} = s_{cp}C_{cp}$. Since $s^*_{cp} \in L(G_{TBc}, m_0)$, and $L(G_{TBc}, m_0) \subseteq L(G_{cp}, m_0)$ (from item 4 of lemma 4 in (Holloway *et al.*, 2000)), then $s^*_{cp} \in L(G_{cp}, m_0)$. By OCP property item 2, there exists a $s^*_{xp} = s_{xp}C_{xp}$ s.t. $s_{xp}C_{xp} \in L(G_{xp}, m_0)$, and $s_{xp}C_{xp} \leq s_{cp}C_{cp}$, with $s_{xp} \leq s_{cp}$ and $C_{xp} \leq C_{cp}$. Also, $s_{xp}C_{xp}$ and $s_{cp}C_{cp}$ are equivalent when projected on $AllC - C^*_{out}(G_{plant})$. So, $s_{xp}C_{xp}$ only can differ by conditions in $C^*_{out}(G_{plant})$, which by construction are not inputs or outputs of TB_{xp}. Thus, since $s_{cp}C_{cp} \in L(TB^*_{xp}(do), m_0)$, then $s_{xp}C_{xp}$ is also, so $s_{cp}C_{cp} \in L(G_{TBx}, m_0)$.
We now show $\{compl(do)\} \subset C_{xp}$. Since, $s_{xp}C_{xp}$ and $s_{cp}C_{cp}$ are equivalent when projected on $AllC - C^*_{out}(G_{plant})$, then C_{xp} and C_{cp} must be similarly equivalent under projection. Since $\{compl(do)\} \notin C^*_{out}(G_{plant})$ and $\{compl(do)\} \in C_{cp}$, then this implies $\{compl(do)\} \in C_{xp}$.
From above two, we have $\{compl(do), goal(do)\} \subseteq C_{xp}$. Since $C_{xp} \leq C_{cp}$, then $\{compl(do), goal(do)\} \subseteq C_{cp}$.

(4) *Leaving completion implies earlier deactivation:*
Consider any $s_{cp} \in L(TB^*_{xp}(do)) \cup G_{cp}, m_0)$, such that $(\emptyset\{compl(do)\}\{\neg compl(do)\}\emptyset) \leq$

s_{cp}. By OCP property item 2, there exists an $s_{xp} \in L(G_{xp}, m_0)$, s.t. $s_{xp}|_{AllC-C^{*}_{out}(G_{plant})} = s_{cp}|_{AllC-C^{*}_{out}(G_{plant})}$, and $s_{xp} \leq s_{xp}$. This implies $(\emptyset\{compl(do)\}\{\neg compl(do)\}\emptyset) \leq s_{xp}$. By definition of effectiveness, then $(\emptyset\{\neg do\}\emptyset) \leq s_{xp}$. Since $s_{xp} \leq s_{xp}$, then $(\emptyset\{\neg do\}\emptyset) \leq s_{xp} \leq s_{cp}$, so $(\emptyset\{\neg do\}\emptyset) \leq s_{cp}$.

(5) *Deactivation implies eventual return to idle:* Following the same statement 1 proof procedure, we can verify this statement easily.

◇◇◇

5. A CLOSED-LOOP EXAMPLE

In the previous section, we discussed how to realize close-loop control using observers. To illustrate the procedure, we have a simple example. The plant is in figure 2.

The task of control is to drive the plant to p_5. First we create an observer in figure 4, for the plant model in figure 2. Then we construct an x-plant as in figure 3. The software *ActionMaker* in the Spectool software package ((Holloway *et al.*, 2000)) is then called using the x-plant of figure 3 to get the taskblock for shown in figure 5. Connecting the plant, the observer and the taskblock together as figure 1, the taskblock will drive the plant from any place to p_5 with the help of the observer.

6. DISCUSSION

This paper discusses the structure of closed-loop control using state observer. The key result is that we can create an intermediate plant model (the x-plant) which includes observer output labels within the states. Control synthesis methods that are effective on this x-plant will also be effective on the combination of the real plant and the observer. This means that existing control synthesis methods can be applied to the x-plant, resulting in a control that will work in conjunction with the ob

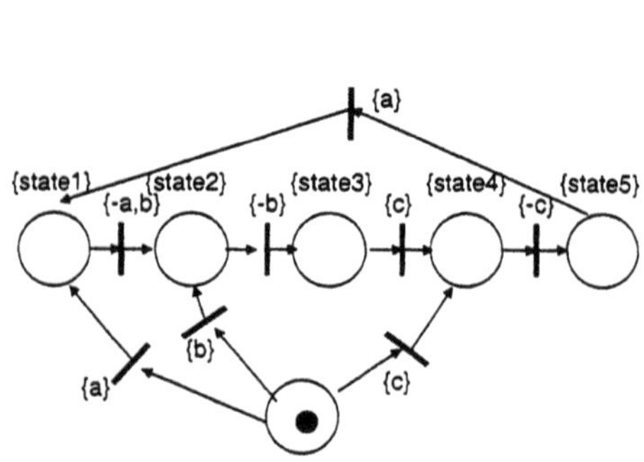

Fig. 4. The observer G_{obv} of G_{plant} in figure 2.

REFERENCES

Giua, Alessandro (1997). Petri net state estimators based on event observation. In: *In Proceedings of 1997 IEEE Conference on Decision and Contr ol.* San Diego.

Gong, Yu (2003). Observer Synthesis For Control Of A Class Of Condition Systems. PhD thesis. University of Kentucky, Lexington, Kentucky, USA.

Gong, Yu and L. E. Holloway (2000). State observer synthesis for a class of condition systems. In: *IEE International Workshop on Discrete Event Systems (WODES2000).* Ghent.

Hanisch, H. M., A. Luder and M. Rausch (1997). Controller synthesis for net condition/event systems with a solution for incomplete state observation. In: *European Journal of Control.* pp. 280–291.

Holloway, L. E., Xiaoyi Guan, Ranganathan Sundaravadivelu and Jeff Ashley Jr. (2000). Automated synthesis and composition of taskblocks for control of manufacturing systems. *IEEE Trans. On Systems, Man, and Cybernetics Part B.*

Li, Y. and W. M. Wonham (1988). Controllability and observability in the state feedback control of discrete-event systems. *Proc. 27th IEEE Conf. Decis. Contr.*

Ramadge, P. J. and W. M. Wonham (1987). Supervisory control of a class of discrete-event processes. *SIAM J. Contr. Optimization.*

Takai, S., T. Ushio and S. Kodama (1995). Static-state feedback control of discrete-event systems under partial observation. *IEEE Transactions on Automatic Control.*

Yamalidou, K., J. Moody, M. Lemmon and P. Antsaklis (1996). Feedback control of petri nets based on place invariants. *Automatica.*

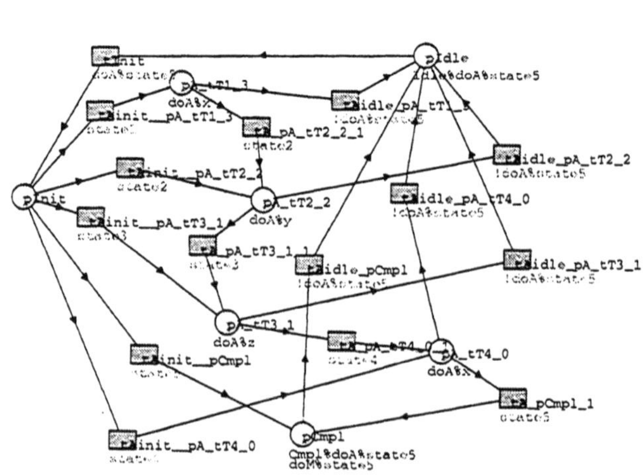

Fig. 5. The taskblock of G_{xp} in figure 3.

ELSEVIER
IFAC
PUBLICATIONS
www.elsevier.com/locate/ifac

A NOVEL HIERARCHICAL CONTROL ARCHITECTURE FOR A CLASS OF DISCRETE-EVENT SYSTEMS

Danjing Li[1], Eckart Mayer[1], Jörg Raisch[1, 2]

1. Systems and Control Theory Group, Max-Planck Institute, Dynamics of Complex Technical Systems, Sandtorstrasse 1, D-39106 Magdeburg, Germany
2. Lehrstuhl für Systemtheorie technischer Prozesse, Otto-von-Guericke Universität Magdeburg, Postfach 4120, D-39016 Magdeburg, Germany

Abstract: This paper proposes a novel hierarchical control architecture for a class of discrete-event systems. Under the proposed control scheme, a max-plus algebra model is introduced on the upper level to provide an optimal online plan. On the lower (implementation) level, min-plus algebra is used to solve cooperation problems between sub-plants as well as problems caused by unexpected events. A simple rail traffic example is given to show the effectiveness of the idea. *Copyright © 2004 IFAC*

Keywords: Discrete-event systems, max-plus algebra, min-plus algebra, rail traffic.

1. INTRODUCTION

In complex discrete-event systems, such as automated transportation systems, challenging issues include coordination of sub-plants and adequate handling of unexpected or uncertain events such as blockings, mechanical failure, etc. Under such circumstances, it is important that the coordination plan and its implementation strategy should adapt to the dynamic environment. In most current transportation systems, the control plan is predefined and a human operator is needed to handle unexpected situations in order to adjust the plan when unforeseen events occur. Inspired by the simplicity and the effectiveness of max-plus algebra models for discrete-event systems, a new planning and control scheme is proposed in this paper. Fig. 1 illustrates the control architecture discussed in the sequel.

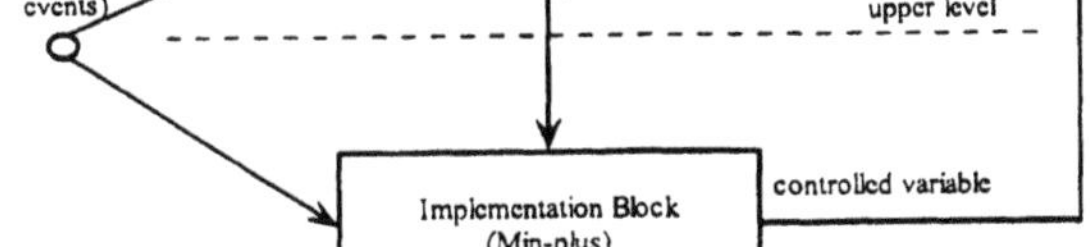

Fig. 1. Control structure

On the upper level, a max-plus algebra model is introduced to determine the time optimal plan. It also provides the event time specifications for the lower level. On the lower level, min-plus algebra not only coordinates the behaviour of sub-plants but also reduces energy consumption. In brief, the control system is a hierarchical structure: the upper level is a supervisory block, which produces the optimal plan for the lower level. The lower level is an implementation block based upon the specifications generated by the upper level. In this paper, a simple rail system is used to illustrate all concepts discussed. For this example, the plan generated by the upper control level specifies the sequence of trains and track segments where trains pass each other. The lower level generates velocity reference signals for the trains to implement the optimal plan.

This paper is structured as follows. Section 2 describes some basic ideas from max-plus algebra and min-plus algebra, which will be used in the sequel. Section 3 focuses on the supervisory block including methods for determining all feasible plans and, subsequently, for choosing the optimal plan in an online fashion. Section 4 explains how the specifications generated at the upper level can be implemented on the lower level with the help of min-plus algebra. In Section 5, a simple application example is given to illustrate the effectiveness of the proposed approach, followed by some conclusions in Section 6.

2. MAX-PLUS ALGEBRA & MIN-PLUS ALGEBRA

2.1 Max-plus algebra

In max-plus algebra (e.g. Baccelli, *et al.*, 1992), addition $\oplus$ is defined to be the maximum of two elements in conventional algebra, while multiplication $\otimes$ is defined to be conventional addition on the set $\mathbb{R}^* = \mathbb{R} \cup \{-\infty\}$:

$$a \oplus b := \max(a, b), \quad (1)$$

$$a \otimes b := a + b. \quad (2)$$

ε ($\varepsilon=-\infty$) is the additive identity in the max-plus algebra. e ($e = 0$) is the multiplicative identity. For a $l \times m$ matrix A over $\mathbb{R}^*$ and an $m \times n$ matrix B over $\mathbb{R}^*$, $A \otimes B$ is defined by

$$(A \otimes B)_{ij} = \oplus_{k=1}^{m} A_{ik} \otimes B_{kj} = \max_k (A_{ik} + B_{kj}). \quad (3)$$

2.2 Max-plus models

Max-plus algebra is a convenient modelling and analysis tool not only for cyclic systems but also for systems with non-cyclic behaviour. In this contribution, we will focus on the latter.

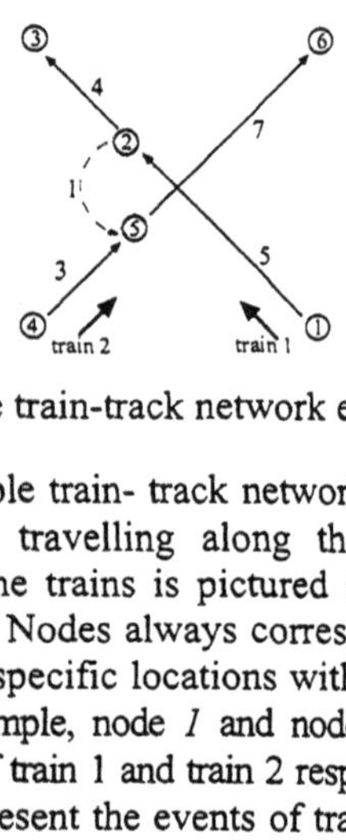

Fig. 2. A simple train-track network example

Consider a simple train- track network (Fig. 2), where two trains are travelling along their tracks. The movement of the trains is pictured as a graph with nodes and arcs. Nodes always correspond to specific trains crossing specific locations within the train-track system. For example, node *1* and node *4* represent the starting events of train 1 and train 2 respectively. Node *3* and node *6* represent the events of train 1 and train 2 arriving at their terminal stations. The arcs between the nodes represent the minimum time needed. For example, the event represented by node *2* may not happen earlier than 5 time units after the event time of node *1*. The time associated with an arc is also called weight of the arc. The light grey part of Fig. 2 represents the crossing area, which only one train can occupy at any instant of time in order to ensure safe crossing. Node *5* represents the event of train 2 entering the crossing area. Node *2* represents the event of train 1 leaving the crossing area. A "control arc" (dashed arc from node 2 to node *5* in Fig. 2) is introduced. This control arc reflects a safety requirement: the earliest time at which train 2 is allowed to enter the crossing area is 1 time unit after train 1 has left it.

Let x_i denote the event time for node i. Then, $\underline{x}_i$ is the earliest possible event time for node i, i.e. the lower bound of x_i. Train 1 and train 2 start their trips at time 0. The input to the system is a vector $u \in \mathbb{R}^{*2}$ representing the earliest starting times: $u=[0\ 0]^T$. The output of the system is defined by the earliest possible arrival times of the trains, $Y=[\underline{x}_3\ \underline{x}_6]^T$. It can be conveniently computed by

$$\underline{X} = \begin{bmatrix} \varepsilon & \varepsilon & \varepsilon & \varepsilon & \varepsilon & \varepsilon \\ 5 & \varepsilon & \varepsilon & \varepsilon & \varepsilon & \varepsilon \\ \varepsilon & 4 & \varepsilon & \varepsilon & \varepsilon & \varepsilon \\ \varepsilon & \varepsilon & \varepsilon & \varepsilon & \varepsilon & \varepsilon \\ \varepsilon & \varepsilon & \varepsilon & 3 & \varepsilon & \varepsilon \\ \varepsilon & \varepsilon & \varepsilon & \varepsilon & 7 & \varepsilon \end{bmatrix} \otimes \underline{X} \oplus \begin{bmatrix} \varepsilon & \varepsilon & \varepsilon & \varepsilon & \varepsilon & \varepsilon \\ \varepsilon & \varepsilon & \varepsilon & \varepsilon & \varepsilon & \varepsilon \\ \varepsilon & \varepsilon & \varepsilon & \varepsilon & \varepsilon & \varepsilon \\ \varepsilon & \varepsilon & \varepsilon & \varepsilon & \varepsilon & \varepsilon \\ \varepsilon & 1 & \varepsilon & \varepsilon & \varepsilon & \varepsilon \\ \varepsilon & \varepsilon & \varepsilon & \varepsilon & \varepsilon & \varepsilon \end{bmatrix} \otimes \underline{X} \oplus \begin{bmatrix} 0 & \varepsilon \\ \varepsilon & \varepsilon \\ \varepsilon & \varepsilon \\ \varepsilon & 0 \\ \varepsilon & \varepsilon \\ \varepsilon & \varepsilon \end{bmatrix} \otimes u, \quad (4)$$

$$Y = \begin{bmatrix} \varepsilon & \varepsilon & 0 & \varepsilon & \varepsilon & \varepsilon \\ \varepsilon & \varepsilon & \varepsilon & \varepsilon & \varepsilon & 0 \end{bmatrix} \otimes \underline{X}, \quad (5)$$

where $\underline{X} = [\underline{x}_1\ \underline{x}_2\ \underline{x}_3\ \underline{x}_4\ \underline{x}_5\ \underline{x}_6]^T$ is the earliest possible event time vector.

The general max-plus model in this contribution has the following form:

$$\underline{X} = A_{01} \otimes \underline{X} \oplus A_{02} \otimes \underline{X} \oplus B \otimes u, \quad (6)$$

$$Y = C \otimes \underline{X}, \quad (7)$$

where the elements of A_{01} correspond to "normal travelling arcs" representing train travelling times while the elements of A_{02} correspond to the control arcs. Of course, we can combine these two matrices into one matrix:

$$A_0 = A_{01} \oplus A_{02}. \quad (8)$$

Furthermore, an explicit max-plus model can be derived from (6), (7) and (8):

$$\underline{X} = A_0^* \otimes B \otimes u, \quad (9)$$

$$Y = C \otimes A_0^* \otimes B \otimes u, \quad (10)$$

$$A_0^* = I \oplus A_0 \oplus A_0^2 \oplus \cdots \oplus A_0^n \oplus A_0^{n+1} \oplus \cdots, \quad (11)$$

where I is identity matrix in max-plus algebra:

$$I = \begin{bmatrix} e & & \varepsilon \\ & \ddots & \\ \varepsilon & & e \end{bmatrix}. \quad (12)$$

A path is a sequence of arcs connecting a sequence of nodes. When the initial and the final node coincide, it is called a circuit (Baccelli, *et al.*, 1992). The length of a path or a circuit is equal to the number of arcs of which it is composed. For physically meaningful systems, the graph represented by A_0 does not contain circuits. For such noncyclic n by n matrices A_0, A_0^k, $k \geqslant n$, does not contribute to the sum (11), see e.g. (Baccelli, *et al.*, 1992).

With the system model (9) and (10), we can easily "simulate" the system for a given input to compute all event times $\underline{x}_i$. Subsequently, it will be assumed that the system input u contains lower bounds for the state vector X, i.e. $B=I$. The simulation results of the max-plus algebra model will be used both for determining the optimal plan and for generating the reference signals for train velocities.

2.3 Min-plus algebra

Min-plus algebra (Baccelli, *et al.*, 1992; Cuninghame-Green, 1979; Cuninghame-Green, 1991) is the system $\mathbb{R}_{min} = (\mathbb{R} \cup \{+\infty\}, \oplus', \otimes', \varepsilon', e)$. Here $\oplus'$ refers to minimisation, i.e.

$$a \oplus' b := \min(a, b), \tag{13}$$

For scalars, $\otimes'$ coincides with $\otimes$, i.e.

$$a \otimes' b = a \otimes b := a + b. \tag{14}$$

For matrices A, B, multiplication is defined by

$$(A \otimes' B)_{ij} = \oplus'^{m}_{k=1} A_{ik} \otimes' B_{kj} = \min_k (A_{ik} + B_{kj}). \tag{15}$$

The operation $\oplus'$ has the neutral element $\varepsilon' = +\infty$. The r-fold 'product' of a matrix X with itself will be written as:

$$X^{[r]} = \underbrace{X \otimes' X \otimes' \cdots \otimes' X}_{r \text{ times}}. \tag{16}$$

Similar to the $*$ operation in max-plus algebra, in min-plus algebra, the $*'$ operation can be defined:

$$A^{*'} := I' \oplus' A \oplus' \cdots \oplus' A^{[n]} \oplus' A^{[n+1]} \oplus' \cdots, \tag{17}$$

where I' is the identity matrix in min-plus algebra:

$$I' = \begin{bmatrix} e & & \varepsilon' \\ & \ddots & \\ \varepsilon' & & e \end{bmatrix}. \tag{18}$$

Note that the following relation holds for the $*$ and $*'$ operations of max-plus algebra and min-plus algebra:

$$(-A^T)^{*'} = -(A^*)^T. \tag{19}$$

where "-" is the conventional minus.

3. SUPERVISORY STRATEGY

As indicated in Fig. 1, to generate the optimal plan for the sequence and timing of the trains, the supervisory block for rail traffic systems does not only need information on the track topology, but also real time information on the current status of the trains. The optimal plan then specifies the future sequence of all train movements and their passing places. While determining the optimal plan is an online task, generation of all feasible plans can be done offline. In this contribution, we assume that the network is not too large, so that the complete set of feasible plans can be established. Once all feasible plans are known, it is possible to select the optimal plan online and send it to the implement block.

3.1 Feasible plans

A complex train-track network usually contains track segments which are used by several trains. For safety reasons only one train is allowed to occupy such a resource at any instant of time. For each resource, there are several possible train sequences, usually causing different combinatoric possibilities. Correspondingly, for the whole system consisting of a number of trains, there are several feasible plans. For example, while the control arc in Fig. 2 realises a special feasible plan, another feasible plan is to make train 2 pass the crossing area before train 1 enters it. Of course, to model this plan there is the need of adding nodes to the graph in Fig. 2. It is straightforward to find the time optimal plan within the set of feasible plans by simulating the max-plus model for each feasible plan. In the following, it is shown by use of an example, how the set of feasible plans and their max-plus models can be determined.

For a max-plus algebra model for train movements, all trains are assumed to move at full speed if possible. Such maximum speed assumption is only applied to find the time optimal plan. It is not necessary for the implementation block. Generally we know the distances between every two sequential nodes for each train on the corresponding tracks, under the maximum velocity assumption, we can determine the transition times for arcs connecting sequential nodes. In other words, for a given train track network, the travelling arcs and therefore the entries of matrix A_{01} are already determined.

On double-line track segments, forthcoming trains will not interfere. This is different for single-line track segments, which can only be occupied by one train at a time, so control arcs are necessary to ensure safe travelling for trains on these track segments. The structure of the control arcs carries the information about the sequence of trains. The weight of a control arc determines a possible safety time that should pass between the trains' movements. If two trains compete for the travel on one single-line track, there are two possibilities: either train 1 or train 2 can go first. If a train-track network includes several single-line track segments, the different combinations of those possibilities generate different plans. Of course, some of them could be infeasible, namely cause blocking.

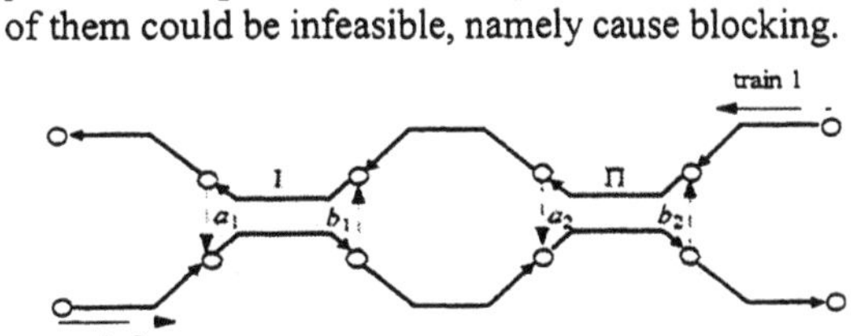

Fig. 3. A network with all possible control arcs

Fig. 3 shows the directed graph of a simple network with two single-line track segments and two trains, where train 1 moves from right to left and train 2 moves in opposite direction. For single-line segment I, the control arc labelled a_1, with $a_1 \geq 0$ represents the fact that train 2 starts to occupy this segment a_1 time units after train 1 has emptied it. The control arc labelled b_1 states the reverse sequence. Accordingly, only one of those two arcs can exist in any one plan. Instead of erasing a non-existent arc from a graph, we can also label it by ε. Hence, by definition, non-existing arcs have weight $-\infty$. Similarly, in segment II, the arcs labelled a_2 and b_2 represent the two possibilities. Table 1 shows all four combinations of the selection of arcs. Among these combinations, the case $a_1 \geq 0$, $b_2 \geq 0$ causes a conflict (blocking), thus this is an infeasible plan.

Table 1 Combinations of control arcs

combination	passing place of *train 1* and *train 2*
$a_1, a_2 >= 0$, $b_1 = b_2 = \varepsilon$	double-line track segment on the left of area I
$b_1, a_2 >= 0$, $a_1 = b_2 = \varepsilon$	double-line track segment between area I and area II
$a_1, b_2 >= 0$, $b_1 = a_2 = \varepsilon$	not feasible, see Fig. 4
$b_1, b_2 >= 0$, $a_1 = a_2 = \varepsilon$	double-line track segment on the right of area II

For large train-track network systems, we need a simple criterion to determine whether a plan is infeasible. Fig. 4 shows that the conflict of combing $a_1 \geq 0$ and $b_2 \geq 0$ comes with a circuit in the directed graph. A circuit with a positive weight causes a contradiction, because the events on the circuit would have to "happen some time before they actually happen". So the problem of determining infeasible plans can be solved by checking for the existence of circuits.

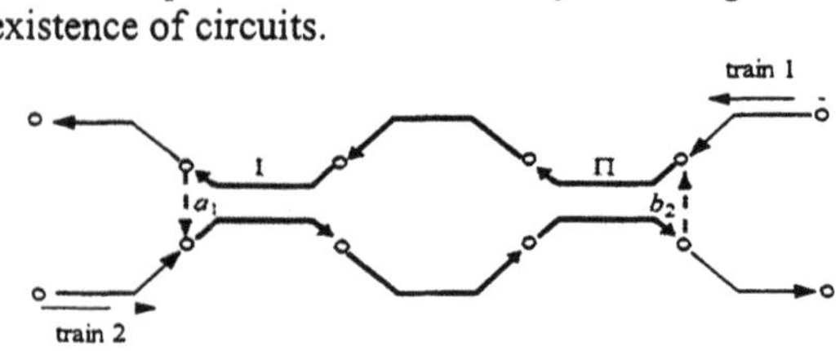

Fig. 4. Combination $a_1 \geq 0$, $b_2 \geq 0$ and the circuit

Within the A_0 matrix, the diagonal entry A_0 (i, i) represents an arc from node i to node i itself. If A_0 (i, i) is not ε, there is a circuit of length 1 containing node i. Suppose set $\alpha(i)$ is the set of circuits of length 2 containing node i. Considering the matrix product definition under the max-plus scheme, $A_0^2 = A_0 \otimes A_0$, A_0^2 (i, i) represents the largest sum of weights for any circuit of length 2 containing node i. Hence, if A_0^2 (i, i) is not ε, there is at least one circuit of length 2 containing node i. Similarly, if A_0^k (i, i) is not ε, there is at least one circuit of length k from node i to node i. In order to determine whether a plan is infeasible, we just need to calculate A_0^k, $k \leq n$, in the max-plus sense.

3.2 Online simulation

Based on the set of max-plus models for all feasible plans, the optimal plan with respect to a specific objective function can be easily determined by simulation. Max-plus simulation is based on the assumption that there is no unexpected event and that each train either waits for a synchronization condition to be met or otherwise moves at its maximum velocity. If any unexpected incident happens, it may affect some trains' travelling times either directly by blocking them or indirectly via synchronization with other trains. If this happens, the runtime event times do not match the event times calculated by a-priori model simulation any more. In order to do online plan optimisation, i.e. rescheduling at runtime, the state vector needs to be reinitialised based on the current status (time t_k) of the trains for online model simulation. The reinitialised state vector is denoted by $\underline{X}\_in$. In general, it contains three different kinds of elements:

1. For an event j which already happened the reinitialised state value is the actual event time, i.e.

$$\underline{x}\_in_j(t_k) = x_j. \tag{20}$$

2. For the next event of any train, the reinitialised state has to be calculated based on the current position of the respective train. Suppose at a time instant t_k, the distance a train has to move before it reaches the location associated with the next event (j), is $d_j(t_k)$. Under the assumption of maximum velocity v_{max}, the next reinitialised state $\underline{x}\_in_j(t_k)$ will be:

$$\underline{x}\_in_j(t_k) = \begin{cases} t_k + \dfrac{d_j(t_k)}{v_{\max}} & unblocked\ trains \\ t_{UE\_release} + \dfrac{d_j(t_k)}{v_{\max}} & blocked\ trains\,. \end{cases} \tag{21}$$

where $t_{UE\_release}$ is the estimated release time for the blocked train. If this time cannot be estimated beforehand, recalculation of the plan is based on the assumption of immediate release, i.e.

$$t_{UE\_release} = t_k\,. \tag{22}$$

3. For other events j in the future, the reinitialised states still remain undefined, i.e.

$$\underline{x}\_in_j(t_k) = \varepsilon. \tag{23}$$

Therefore, the reinitialised state at time t_k is:

$$\underline{x}\_in_j(t_k) = \begin{cases} x_j & if\ j\ is\ a\ past\ event \\ from\ (21) & if\ j\ is\ a\ next\ event\ for\ any\ train \\ \varepsilon & otherwise \end{cases} \tag{24}$$

Then, the updated vector of event times can be obtained for all feasible plans by simulation, i.e. by evaluating (9) with $u = \underline{X}\_in(t_k)$. (9) and (10) are rewritten as

$$\underline{X}(t_k) = A_0{}^* \otimes B \otimes \underline{X}\_in(t_k)\,, \tag{25}$$

$$Y(t_k) = C \otimes A_0{}^* \otimes B \otimes \underline{X}\_in(t_k)\,. \tag{26}$$

3.3. Online plan optimisation

During runtime, the progress of the trains is observed by the supervisory block. At regular time instants (e.g. in an equidistant discrete time scheme), simulation is done for all feasible plans, comparing the values of the objective function. The objective function is evaluated from the output vector Y. An example for the objective function would be the last trains' arrival time, i.e. we want to minimise max (Y_i).

Note that the number of feasible plans reduces as the trains progress through the network. As described in section 3.1, building the set of feasible plans is based on the different possible combinations of control arcs at the single-line track segments. During runtime, once a train enters a single-line segment, there is no need to consider the competing control arcs on this track segment because the choice has already been made.

4. IMPLEMENTATION

The output of the supervisory block generates the time optimal plan, i.e. the time specifications $\underline{X}$ for the events. Due to the maximum speed assumption under max-plus, $\underline{X}$ provides the earliest possible time for each event, and is therefore referred to as EPET (earliest possible event time) specifications. According to EPET, trains will always move at maximum speed, or otherwise stop to wait until the synchronization conditions are met. Clearly, this is undesirable from an energy saving point of view. An overall optimisation approach, minimising energy consumption for all trains in common, yields a large nonlinear problem and therefore is not realizable. Instead, a suboptimal more conservative approach is used here determining a velocity signal separately for each train. This is done as follows: first, as an upper bound for the event times, LNET (Latest necessary event time) is derived for each train. This is done in such a way that the trains' arrival time at the final destination according to EPET will be met. It is also ensured that LNET does not violate the EPET schedule of the other trains. In order to generate the LNET specifications, min-plus algebra, the dual system of max-plus algebra, is used.

Suppose Q is the event index set for all events corresponding to the arrival of trains at their final destinations and P_m is the index set for all events related to train m, then the LNET specifications $\overline{X}_m$ for train m can be calculated as

$$\overline{X}_m(t_k) = (-A_0{}^T)^{*'} \otimes' \underline{XR}_m(t_k), \quad (27)$$

where

$$(\underline{XR}_m)_i(t_k) = \begin{cases} \underline{x}_i(t_k), & i \in Q \text{ or } i \notin P_m \\ +\infty, & \text{otherwise} \end{cases}. \quad (28)$$

The min-plus algebra equation (27) represents "backward simulation". Using (19), it can be rewritten as

$$\overline{X}_m(t_k) = (-(A_0{}^*)^T) \otimes' \underline{XR}_m(t_k). \quad (29)$$

Now, for each train m, its EPET $\underline{X}$ and LNET $\overline{X}_m$ are available. Within this corridor, the velocity signal can be optimised locally for each train. If we neglect acceleration and braking effects, an energy optimal trajectory results from minimising

$$J_m = \int v_m{}^2 dt. \quad (30)$$

An example for an energy optimal trajectory is given in Fig. 5.

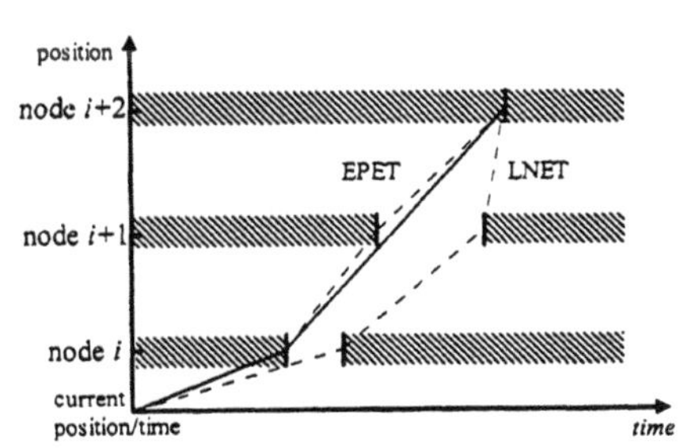

Fig. 5. Energy optimal trajectory (bold line)

5. RAIL TRAFFIC CASE STUDY

Various studies concerning the modelling and control of rail traffic system have been done during the past decades. For example, recently Schutter *et al.* extended the model predictive control framework to transfer coordination in railway systems (2002).

The effectiveness of the hierarchical control architecture proposed in the previous Sections is illustrated by a small rail traffic example (Fig. 6) involving 3 trains and 3 tracks. Initially, train 1, 2 and 3 are located at the end points of 3 tracks, i.e. A, B and C, respectively. The trains move along the tracks in the directions shown in Fig. 6. In the middle of both, track AO and track CO, double-line track segments are available, which make it possible that two trains pass each other between points N_1 and M_1 or N_2 and M_2, respectively. As in any real world application, unexpected events (e.g. blocking of the track, mechanic failure, etc.) may occur and have to be handled by the controller.

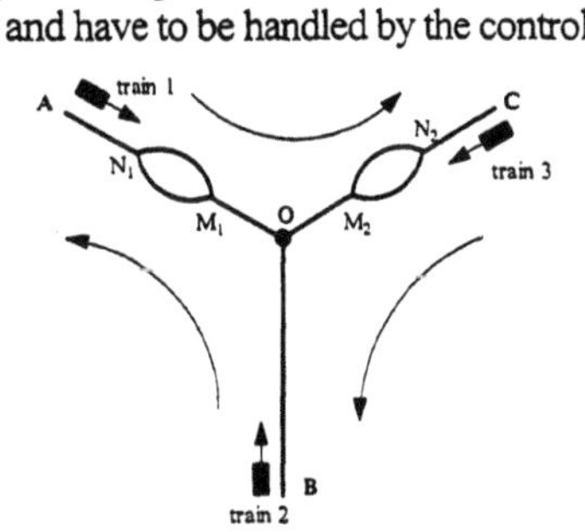

Fig. 6. A simple rail traffic network

5.1. Feasible plans

First, the set of feasible plans is computed offline. Single-line track segment AN_1 can either be occupied by train 1 or train 2. Nevertheless, the order is predetermined, as the track is already occupied by train 1 in the beginning. The same situation occurs for segments BO and CN_2. Therefore, although there are five single-line track segments, the order can only be chosen for segments OM_1 and OM_2. The proposed approach yields a total set of 3 feasible plans pictured in Fig. 7-Fig. 9. In plan 1, train 1 occupies track segment OM_1 just after train 2 empties it. Similarly, train 1 enters track segment OM_2 after train 3 leaves it. In plan 2 the sequence of trains is reversed in both segments. Plan 3 is similar to plan 1 except of the trains' order on OM_2. Fig. 10 shows the directed graph for plan 1 including the control arcs needed to enforce this plan.

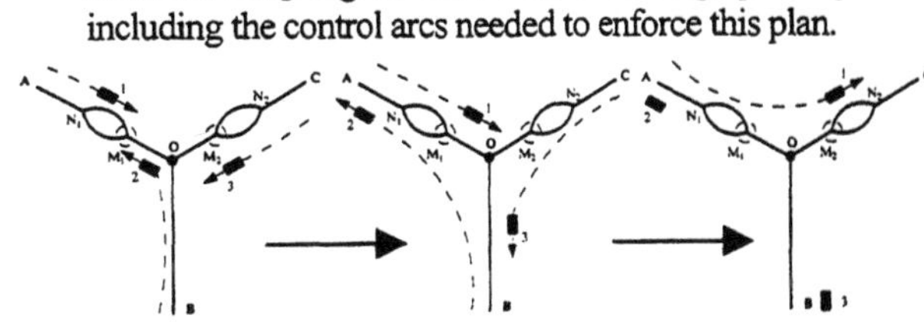

Fig. 7. Plan 1 (OM_1: train2 train1, OM_2: train3 train1)

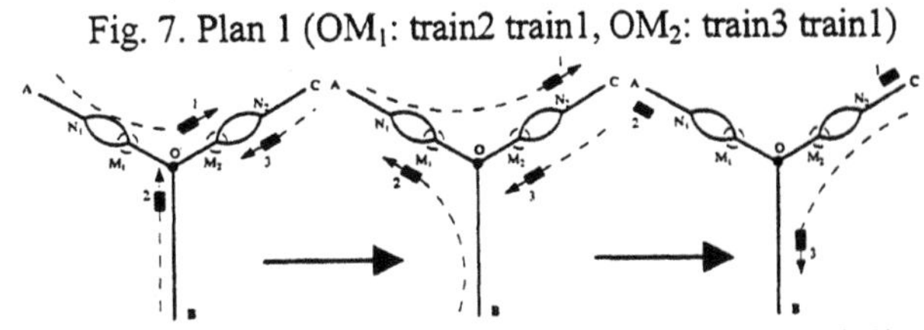

Fig. 8. Plan 2 (OM_1: train1 train2, OM_2: train1 train3)

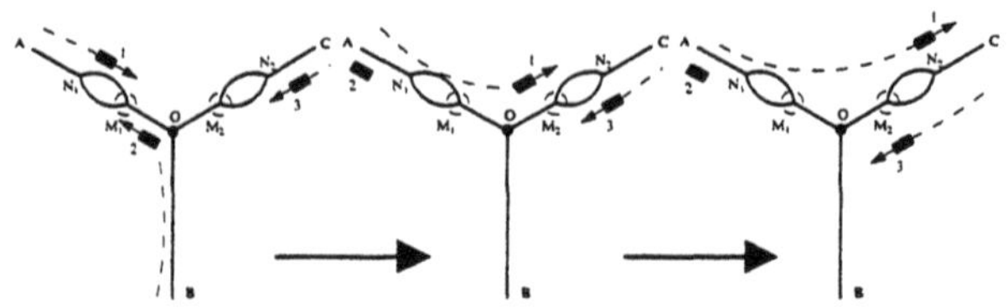

Fig. 9. Plan 3 (OM_1: train2 train1, OM_2: train1 train3)

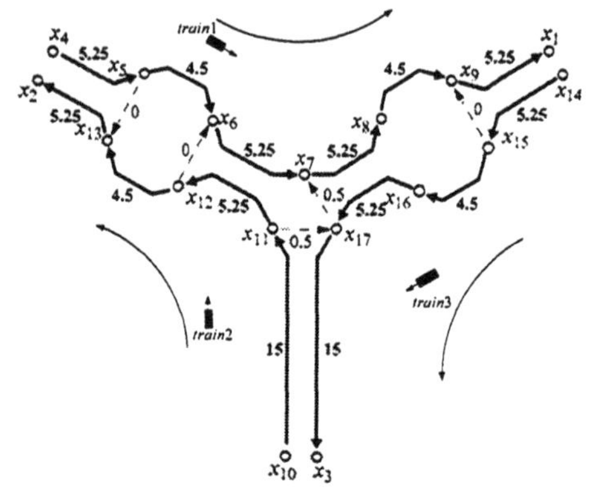

Fig. 10. Directed graph for plan 1

5.2. *Closed loop simulation results*

Plan 1 is the time optimal plan if there is no unexpected event. Fig. 11 shows the movements of the trains under the proposed hierarchical control scheme. In case of a disturbance (here train 2 is blocked on track BO), the supervisory level checks if it is necessary to change the plan. Such a situation is illustrated in Fig. 12. After train 2 has been blocked for some time, it is optimal to change from plan 1 to plan 2. This is determined in the supervisory block on the upper level and implemented by the implementation block on the lower level. This changes the operation sequence of trains: train 3 occupies track segment OM_2 after train 1 leaves it, train 2 enters track segment OM_1 after train 1 empties it. If the blocking is removed at time 12, this control policy will ensure that all trains arrive at final destinations at time 42.4. For comparison, Fig. 13 gives the result when there is no supervisory control. This implies that the original plan is maintained during the blocking and after it has been removed. Clearly it takes longer for the trains to finish their travel, as the last train arrives at 50.2. Note that in the case of blocking, the continuous adjustment of velocity signals leads to nonstraight trajectories in Fig. 12 and Fig. 13.

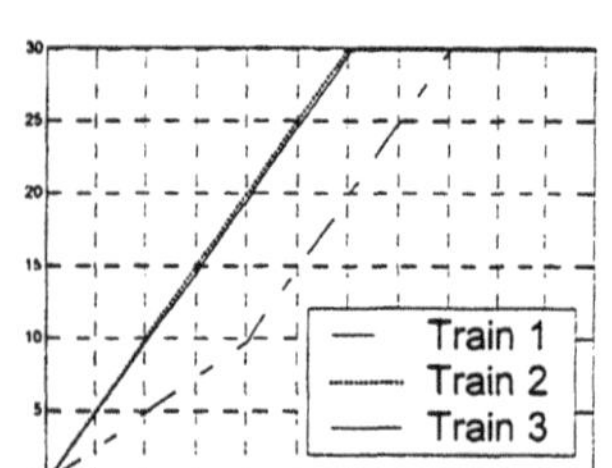

Fig. 11. Simulation result under normal situation

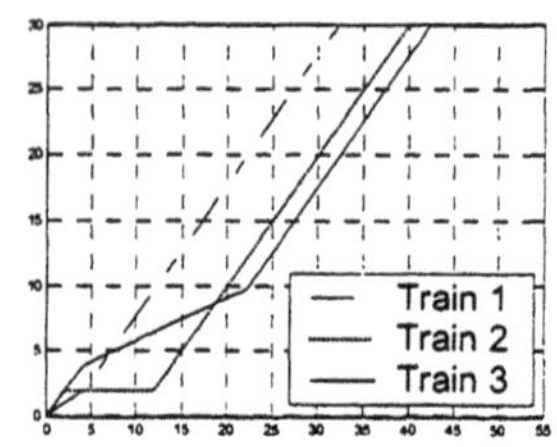

Fig. 12. Plan changing due to unexpected events

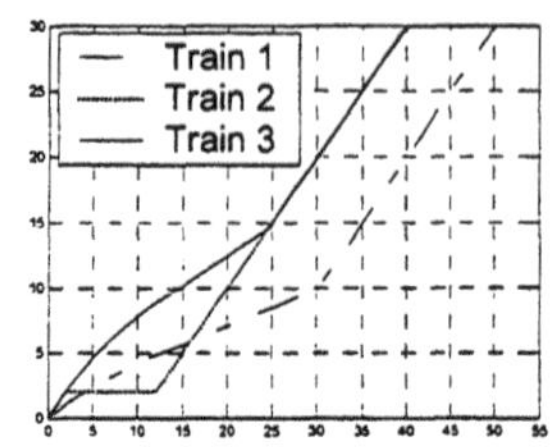

Fig. 13. Comparison study: No supervisory control

6. CONCLUSIONS

A novel hierarchical control architecture for a class of discrete-event systems has been proposed in this contribution. It has been applied to a rail traffic network. Based on a max-plus algebra model, the sequence of resource allocations is chosen such that minimum overall processing time is reached. An upper level supervisory block ensures the optimal sequence of train movements even under disruptive conditions. A lower level implementation block provides reference velocity signals for each train. By exploiting the remaining degrees of freedom, it reduces overall energy consumption. The implement policy is generated by use of the dual min-plus algebra model.

The method has been illustrated by means of a small rail traffic example. Simulation results show the effectiveness of the approach.

With respect to the extension of the method to large systems as well as for similar discrete-event systems in other fields, such as, flexible manufacturing or chemical process control, two aspects have not been treated in this contribution. The first is to find methods to cope with combinatoric explosion during the computation of the set of feasible plans for large systems; the second is the extension of the method to cyclically repeated processes where the advantages of the max-plus scheme can be further exploited.

REFERENCES

Baccelli, F., G.Cohen, G.J.Olsder, and J.P.Quadrat (1992). *Synchronization and Linearity*. John Wiley and Sons, New-York.

Cuninghame-Green, R.A. (1991). Minimax algebra and applications. *Fuzzy Sets and Systems*. **41**, pp.251-267.

Cuninghame-Green, R.A. (1979). *Minimax algebra*. Vol. 166 of Lecture Notes in Economics and Mathematical Systems, Springer-Verlag, Berlin.

Schutter, B.De, T. Van den Boom, and A. Hegyi (2002). A model predictive control approach for recovery from delays in railway systems. *Transportation Research Record*, no.1793, pp.15-20.

ELSEVIER
IFAC
PUBLICATIONS
www.elsevier.com/locate/ifac

STATE ESTIMATION AND CONTROL OF NONDETERMINISTIC λ-FREE LABELED PETRI NETS

Daniele Corona * Alessandro Giua * Carla Seatzu *

* *Dip. Ing. Elettrica ed Elettronica, U. di Cagliari, Italy*
Email: {daniele.corona,giua,seatzu}@diee.unica.it

Abstract: In this paper we present an original approach to estimate the marking of a labeled Petri net based on the observation of transition labels. In particular, we consider the case of nondeterministic transitions, i.e., transitions that share the same label and that can be simultaneously enabled. We also show how the estimate generated by the observer may be used to design a state feedback controller for forbidden marking specifications. More precisely, we discuss two different cases: the label-based feedback and the transition-based feedback, depending on the possibility of the controller to distinguish among transitions with the same label.

Keywords: Petri nets, λ-free Petri nets, marking estimation, control.

1. INTRODUCTION

In this paper we consider nondeterministic λ-free Petri nets where nondeterminism arises because two or more transitions with the same label may be simultaneously enabled from a reachable marking. We propose an approach to build a state observer that does not require the construction of the reachability graph, and thus works for both bounded and unbounded PN. Under some assumptions, we show that the set of markings that are consistent with the observation w — denoted $\mathcal{C}(w)$ — can be written as the solution of a linear system with a fixed structure that depends on some parameters that can be recursively computed. The main advantage of the proposed approach is that we need not exhaustively enumerate all consistent markings.
The characterization used to define the set of consistent markings — as well as the algorithm that is used to implement it — has been firstly proposed by the authors in (Corona *et al.*, 2003) as a conjecture. In this paper we present a slightly modified version of this algorithm (see Algorithm 6) for which a formal proof has been obtained: the complete proof is not reported here for lack of space but can be found in (Giua *et al.*, 2004).
Note that other authors (Meda *et al.*, 1998) have also discussed the problem of estimating the marking of a Petri net using a mix of transition firings and place observations. Finally, Zhang and Holloway (1995) used a Controlled Petri Net model for forbidden state avoidance under partial *event* observation with the assumption that the initial marking is known.
In this paper we also show how the proposed linear algebraic characterization of the set of consistent markings can be used to design a feedback controller for forbidden marking specifications. In particular, we assume that specifications are given in the form of Generalized Mutual Exclusion Constraints (GMEC) that limit the weighted sum of tokens in subsets of places.
We discuss two different control laws.
Label-based feedback. In this case the controller cannot distinguish among transitions that are undistinguishable by the observer (namely, transitions with the same label). Hence, if a transition labeled by the event e is disabled, then the control pattern must simultaneously disable all other transitions labeled by e.
Transition-based feedback. In this case we assume that the controller is not constrained by the observer observation mask, and may assign a different control pattern to different transitions even if they share the same label e.
It is important to observe that in the case of label-based feedback, Algorithm 6 can still be used to compute the set of markings in which the *closed-loop system* may be, given the actual observation. On the contrary, in the case of transition-based feedback the control pattern may reduce the nondeterminism of the net: when the nondeterministic event e is observed, only a subset of transitions labeled e may have fired, namely those transitions that are control enabled. This implies that Algorithm 6 needs to be slightly modified when used to compute the set of markings in which the *closed-loop system* may be, given the actual observation. This point is formalized in Algorithm 9.
An original example of state estimation and control, taken from the manufacturing domain, that motivates the interest for the particular considered model, is finally presented in Section 6.

2. BACKGROUND ON PETRI NETS

In this section we recall the formalism used in the paper. For more details on Petri nets we address to (Murata, 1989).

A *Place/Transition net* (P/T net) is a structure $N = (P, T, Pre, Post)$, where P is a set of m places; T is a set of n transitions; $Pre : P \times T \to \mathbb{N}$ and $Post : P \times T \to \mathbb{N}$ are the *pre*– and *post*– incidence functions that specify the arcs; $C = Post - Pre$ is the incidence matrix. The *preset* and *postset* of a node $X \in P \cup T$ are denoted $^\bullet X$ and $X^\bullet$ while $^\bullet X^\bullet = ^\bullet X \cup X^\bullet$.
A *marking* is a vector $M : P \to \mathbb{N}$ that assigns to each place of a P/T net a non–negative integer number of tokens, represented by black dots. We denote $M(p)$ the marking of place p. A *net system* $\langle N, M_0 \rangle$ is a net N with an initial marking M_0.
We write $M [\varsigma\rangle M'$ to denote that the firing of ς yields M'. We also denote $\boldsymbol{\sigma} : T \to \mathbb{N}$ the *firing vector* associated to a sequence ς.
The set of all markings reachable from M_0 defines the *reachability set* $R(N, M_0)$ of $\langle N, M_0 \rangle$.
A *labeling function* $L : T \to E$ assigns to each transition $t \in T$ a symbol from a given alphabet E. Note that the same label $e \in E$ may be associated to more than one transition while no transition may be labeled with the empty string ε. Using the notation of (Peterson, 1981) and (Gaubert and Giua, 1999) we say that this labeling function is λ*-free*.

Definition 1. A Petri net system $\langle N, M_0 \rangle$ with λ-free labeling function $L : T \to E$ is *deterministic* if for all markings $M \in R(N, M_0)$ and for any two transitions $t, t' \in T$:

$$t \neq t',\ L(t) = L(t'),\ M[t\rangle \implies \neg M[t'\rangle,$$

i.e., if two transitions are labeled with the same symbol they cannot simultaneously be enabled at M. ■

Determinism is a behavioral property but it is also possible to introduce a structural definition of determinism.

Definition 2. A Petri net N with λ-free labeling function $L : T \to E$ is *structurally deterministic* if for any two transitions $t, t' \in T$:

$$t \neq t' \implies L(t) \neq L(t'),$$

i.e., two different transitions cannot be labeled with the same symbol. ■

Note that if a Petri net N is structurally deterministic, then the net system $\langle N, M_0 \rangle$ is deterministic for all initial marking M_0.
In this paper we consider Petri nets that are not structurally deterministic. We say that a transition t is *nondeterministic* if its label is also associated to other transitions, otherwise a transition t is said to be *deterministic*. We also denote T^d the set of deterministic transitions and T^n the set of nondeterministic transitions. Clearly, $T = T^d \cup T^n$.
Analogously, we say that an event e is *nondeterministic* if there exists more than one transition t such that $L(t) = e$, otherwise we say that the event e is *deterministic*. Therefore, with no ambiguity on the notation, we may write $E = E^d \cup E^n$.
Note that the labeling function restricted to T^d is an isomorphism and thus, with no loss of generality we can assume $E^d = T^d$.
We denote as T_e the set of transitions labeled e:

$$T_e = \{t \in T \mid L(t) = e\}.$$

The restriction of the incidence matrix C to T_e (T^n) is denoted C_e (C^n) and the restriction of the firing vector $\boldsymbol{\sigma}$ to T_e is denoted $\boldsymbol{\sigma}_e$.
Finally, to each set of nondeterministic transitions T_e we associate the set $\mathcal{T}_e$ containing all possible subsets of transitions, apart from itself and the empty set, i.e.,

$$\mathcal{T}_e = \{\tau \subseteq T_e \mid \tau \neq \emptyset,\ \tau \neq T_e\} = 2^{T_e} \setminus \{\emptyset, T_e\}.$$

Clearly, $|\mathcal{T}_e| = 2^{n_e} - 2$ where n_e denotes the number of nondeterministic transitions labeled e.
We denote as w the word of events associated to the sequence ς, i.e., $w = L(\varsigma)$.

3. A LINEAR ALGEBRAIC CHARACTERIZATION OF THE SET OF CONSISTENT MARKINGS

In this paper we deal with the problem of estimating the marking of a net system $\langle N, M_0 \rangle$ whose marking cannot be directly observed.
After the word w has been observed, we define the set $\mathcal{C}(w)$ of w-consistent markings as the set of all markings in which the system may be given the observed behavior.

Definition 3. Given an observed word w, the set of *w-consistent markings* is $\mathcal{C}(w) = \{M \in \mathbb{N}^m \mid \exists$ a sequence of transitions $\varsigma : M_0[\varsigma\rangle M$ and $L(\varsigma) = w\}$. ■

In (Giua *et al.*, 2004) we provided a systematic and efficient procedure to estimate the set of markings that are consistent with an observed word under the following assumptions:

(A1) The structure of the net N is known.
(A2) The initial marking M_0 is known.
(A3) The label function is λ-free and labels associated to transition firings can be observed.
(A4) Nondeterministic transitions are *contact free*, i.e., for any two nondeterministic transitions t_i and t_j, it holds $^\bullet t_i^\bullet \cap {}^\bullet t_j^\bullet = \emptyset$ and $^\bullet t_i \cap t_i^\bullet = \emptyset$.

Clearly, $\mathcal{C}(\varepsilon) = M_0$ and $\mathcal{C}(w)$ is a singleton if for all e in w, T_e is a singleton. On the contrary, the degree of nondeterminism may increase as the cardinality of T_e increases.
In (Giua *et al.*, 2004) we have shown that, under the assumptions (A1) to (A4), a fixed number of constraints, not depending on the length of the observed word w, may be used to describe the set of w-consistent markings.
Let us first introduce the following notation.

Definition 4. Given a marking M and a transition $t \in T$, we define

$$z(M, t) = \min_{p \in ^\bullet t} \left\{ \left\lfloor \frac{M(p)}{Pre(p, t)} \right\rfloor \right\}$$

the enabling degree of transition t at M.
Given a set of transitions $\tau \subseteq T$, we also define

$$z(M, \tau) = \sum_{t \in \tau} z(M, t).$$

Finally, given a vector $\boldsymbol{\sigma} \in \mathbb{N}^n$, we denote as

$$\sigma(\tau) = \sum_{t \in \tau} \sigma(t).$$ ■

Note that if all transitions in τ are conflict free, then $z(\tau)$ represents the number of times transitions in τ may simultaneously fire at M.

Theorem 5. (Giua *et al.*, 2004) Let us consider a labeled Petri net system $\langle N, M_0 \rangle$ and let $L : T \to E$ be its labeling function. Let assumptions (A1) to (A4) be verified. Then, for all words $w \in E^*$ the set of w-consistent markings $\mathcal{C}(w)$ is equal to

$$\mathcal{C}(w) = \mathcal{M}(w) \stackrel{\text{def}}{=} \{M \in \mathbb{N}^m \mid \\ M = M_{b,w} + \sum_{e \in E^n} C_e \sigma_e; \quad \sigma_e \in \mathcal{S}_e(w)\} \qquad (1)$$

where

$$\mathcal{S}_e(w) \stackrel{\text{def}}{=} \{\sigma \in \mathbb{N}^{n_e} \mid (\forall \tau \in \mathcal{T}_e) \ \sigma(\tau) \le u_w(\tau), \\ \sigma(T_e) = u_w(T_e)\}, \qquad (2)$$

and the upper bounds $u_w(\tau)$ and $u_w(T_e)$, as well as the marking $M_{b,w}$, are computed using the recursive Algorithm 6.

Therefore, the number of constraints used to describe the set $\mathcal{S}_e(w)$ is equal to $2^{n_e} - 1$, regardless of the length of the observed word w.

Algorithm 6. (Upper bounds and basis marking computation)

1. Let $w = \varepsilon$ and $M_{b,w} = M_0$.
2. Let $u_w(\tau) = 0$ for all $e \in E^n$ and for all $\tau \in \mathcal{T}_e$.
3. Let $u_w(T_e) = 0$ for all $e \in E^n$.
4. Wait until an event e is observed.
5. Let $flag = 0$.
6. If $e \in E^d$, then
   let $t = L^{-1}(e)$,
   if ${}^\bullet t \cap ({}^\bullet T^{n\bullet}) = \emptyset$, then (Case A)
   $M_{b,we} = M_{b,w} + C(\cdot, t)$
   endif
   if $P_t \stackrel{\text{def}}{=} {}^\bullet t \cap (T^{n\bullet}) \neq \emptyset$, then (Case B)
   $\sigma_\alpha = \mathbf{0}$ (a vector of dimension $|T^n| \times 1$)
   $flag = 1$
   $T_{up} \stackrel{\text{def}}{=} T^n \cap^\bullet P_t$
   for all $\hat{t} \in T_{up}$, then
   $$\text{let } \sigma_\alpha(\hat{t}) = \max_{p \in P_t} \left\{0, \left\lceil \frac{Pre(p,t) - M_{b,w}(p)}{Post(p,\hat{t})} \right\rceil \right\}$$
   endfor
   for all τ : $\tau \cap T_{up} \neq \emptyset$, then
   $$u_{we}(\tau) = u_w(\tau) - \sum_{t \in \tau} \sigma_\alpha(t)$$
   endfor
   $M_{b,we} = M_{b,w} + C(\cdot, t) + C^n \sigma_\alpha$
   endif
   if ${}^\bullet t \cap ({}^\bullet T^n) \neq \emptyset$, then (Case C)
   if $flag = 0$, then
   $M_{b,we} = M_{b,w} + C(\cdot, t)$
   endif
   let $\mathcal{T}_r(t) = \{\hat{t} \in T^n \mid {}^\bullet t \cap^\bullet \hat{t} \neq \emptyset\}$
   for all $\hat{t} \in \mathcal{T}_r(t)$, then
   $u_{we}(\{\hat{t}\}) = \min\{u_w(\{\hat{t}\}), z(M_{b,we}, \hat{t})\}$
   for all $\tau \in \mathcal{T}_{L(\hat{t})}$: $\hat{t} \in \tau$ with $\tau \neq \{\hat{t}\}$, then
   $u_{we}(\tau) = \min\{u_w(\tau), u_{we}(\{\hat{t}\}) + u_w(\tau \setminus \{\hat{t}\})\}$
   endfor
   endfor
   endif
   else (Case D)
   for all $\tau \in \mathcal{T}_e$, then
   $u_{we}(\tau) = \min\{u_w(\tau) + 1, z(M_{b,w}, \tau)\}$
   endfor
   $u_{we}(T_e) = u_w(T_e) + 1$
   $M_{b,we} = M_{b,w}$
   endif
7. $w = we$

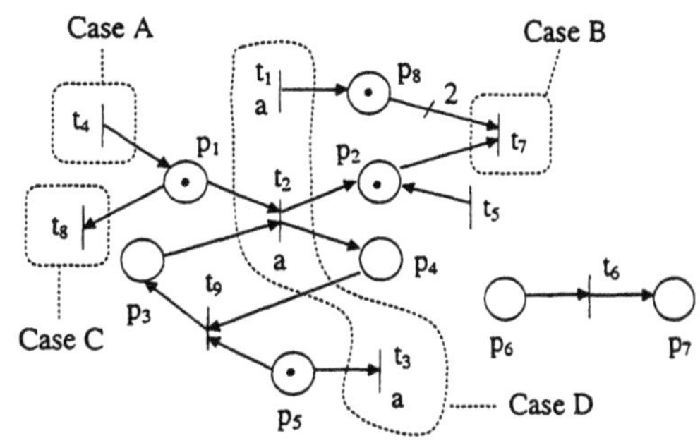

Fig. 1. The generic substructure of a more complex Petri net that satisfies the contact-free assumption.

8. Goto 4. ■

Let us first discuss the physical meaning of all the parameters characterizing the above set (1).
Let us preliminary observe that the firing of a nondeterministic transition t may be *detected* (or *reconstructed*) when a deterministic transition t_d is observed and the firing of t is strictly necessary to enable t_d. Therefore, using Algorithm 6, we define $M_{b,w}$ as the marking that we reach from the initial one by firing all the observed deterministic transitions, and all those nondeterministic transitions that have been detected. In the following we say that $M_{b,w}$ is the *basis marking* given the actual observation w.
Moreover, for each nondeterministic event e, the upper bound $u_w(T_e)$ denotes how many times the event e has been observed in w without being detected.
Finally, the upper bound $u_w(\tau)$ relative to a given subset $\tau \subset \mathcal{T}_e$, imposes a limit on the maximum number of times all transitions in τ may have fired, given the actual observation w, and taking into account that a certain number of nondeterministic transitions labeled e may have been detected.
Now, let us discuss in detail all cases in Algorithm 6. Consider the labeled Petri net in Figure 1 that represents the generic substructure of a more complex Petri net that satisfies the contact-free assumption (A4). Let us assume that in this subnet the only nondeterministic transitions are those labeled a. Let w be the actual observed word of events and let $M_{b,w}$ be the marking shown in Figure 1. Finally assume $|w|_a \geq 1$.
(Case A): *A deterministic transition t such that ${}^\bullet t \cap ({}^\bullet T^{n\bullet}) = \emptyset$ fires.* If t_4 fires we only update the basis marking taking into account that the deterministic transition t_4 has fired, but we deduce no information on the number of times the nondeterministic transitions have eventually fired.
(Case B): *A deterministic transition t such that ${}^\bullet t \cap (T^{n\bullet}) = P_t \neq \emptyset$ fires.* If t_7 fires, we know for sure that each place $p \in {}^\bullet t_7$ (namely, p_2 and p_8) contains a number of tokens that is greater or equal than $Pre(p, t_7)$. Now, given the basis marking $M_{b,w}$, if for some place $p \in {}^\bullet t_7$, $M_{b,w}(p) < Pre(p, t_7)$, we know for sure that the nondeterministic transition ${}^\bullet p$ has fired and we can also evaluate (see Algorithm 6) how many times it has fired. We consequently update the basis marking and the upper bounds relative to all subsets containing ${}^\bullet p$. In the case at hand, we can conclude that one of the previous observations of a was due to the firing of t_1. Therefore, the basis marking $M_{b,w}$ is updated to $M_{b,we} = M_{b,w} + C(\cdot, t_1) + C(\cdot, t_7)$.

(Case C): *A deterministic transition t such that* $^\bullet t \cap (^\bullet T^n) \neq \emptyset$ *fires.* If t_8 fires the upper bounds associated to subsets of nondeterministic transitions may decrease. In fact, if t_8 fires, we know for sure that if p is an input place of t_8, then it should contain a number of tokens that is greater or equal to $Pre(p, t_8)$. Therefore, if there is some nondeterministic transition exiting p, we know for sure that the maximum number of times it may have fired must ensure that in p there are at least $Pre(p, t_8)$ tokens. As an example, if in the actual case the upper bound associated to $\tau = \{t_2\}$ were 1, we reduce it to zero. Then, we update all the other $u_w(\tau)$'s relative to subsets τ containing t_2, as well as $u_w(T_a)$.

(Case D): *A nondeterministic event is observed.* If the nondeterministic event a is observed, we update the upper bounds $u_{wa}(\tau)$ relative to those subsets $\tau \in \mathcal{T}_a$ whose enabling degree at the current basis marking $M_{b,w}$ is greater than the bound $u_w(\tau)$. Furthermore, we always increment by one the value of the bound of T_a, i.e., $u_{wa}(T_a) = u_w(T_a) + 1$, that takes into account how many times the event a has been observed without being detected.

4. CONTROL USING THE SET OF CONSISTENT MARKINGS

The above linear algebraic characterization of the set of consistent markings can be efficiently used by a control agent to enforce a given specification on the plant behaviour. In particular, assume that the following hypothesis are verified.

(H1) The specification is given as a set of legal states of the form $\mathcal{L} = \{M \in \mathbb{N}^m \mid S^T \cdot M \leq \boldsymbol{k}\}$ where $S = [s_1 \cdots s_q]$ with $s_j \in \mathbb{Z}^n$ and $\boldsymbol{k} = [k_1 \cdots k_q]$ with $k_j \in \mathbb{Z}$. This kind of specifications, called *generalized mutual exclusion constraints* (GMEC) are denoted $(S, \boldsymbol{k})$.

(H2) The controller may disable transitions to prevent the plant from entering a forbidden marking, computing a marking dependent control pattern $f(t, M) : T \times \mathbb{N}^m \to \{0, 1\}$. If $f(t, M) = 0$ then t is disabled by the controller at M, while if $f(t, M) = 1$ it is enabled.

(H3) All transitions are controllable, i.e., they can be disabled by the controller.

It is well known that under the assumption that: the initial marking M_0 is legal, all transitions are controllable and the actual marking is known, an optimal (i.e., maximally permissive) control policy that enforces a given state specification is

$$f(t, M) = \begin{cases} 0 \text{ if } S^T \cdot M \leq \boldsymbol{k},\ M[t\rangle M', \\ \quad (\exists j)\ \boldsymbol{s}_j \cdot M' > k_j \\ 1 \text{ otherwise.} \end{cases}$$

When an observer is used in the control loop the actual marking M is not known and only a set of consistent markings $\mathcal{C} \subseteq \mathbb{N}^m$ is available to the controller. In this paper we discuss two different control laws, depending on the possibility of the controller to distinguish among transitions with the same label.

Label-based feedback (LBF). In this case the controller cannot distinguish among transitions that are undistinguishable by the observer (namely, transitions with the same label). Hence, if a transition labeled by event e is disabled, then the control pattern must simultaneously disable all other transitions labeled by e.

Transition-based feedback (TBF). In this case we assume that the controller is not constrained by the observer observation mask, and may assign a different control pattern to different transitions even if they share the same label e.

4.1 Label-based feedback

In this case the control law becomes a function $f(e, \mathcal{C}) : L \times 2^{\mathbb{N}^m} \to \{0, 1\}$ and can be given as follows.

Definition 7. (Optimal LBF for GMEC with observer) Given a GMEC $(S, \boldsymbol{k})$, a set of consistent markings $\mathcal{C} \subseteq \mathbb{N}^m$, and a λ-free labeling function $L : T \to E$, the firing of each transition labeled e should be prevented if and only if there exists at least one legal consistent marking M and a transition t labeled e, such that the firing of t from M leads to a forbidden marking, i.e.,

$$f(e, \mathcal{C}) = \begin{cases} 0 \text{ if } (\exists M,\ \exists t : L(t) = e) \\ \quad M \in \mathcal{C},\ S^T \cdot M \leq \boldsymbol{k}, \\ \quad M[t\rangle M',\ (\exists j)\ \boldsymbol{s}_j \cdot M' > k_j \\ 1 \text{ otherwise.} \end{cases}$$

■

The computation of the control law $f(e, \mathcal{C})$ may be carried out looking for the existence of a $t \in T_e$ and a $j \in \{1, \ldots, q\}$ such that the following constraint set admits a solution:

$$\begin{cases} M \in \mathcal{C} & (a) \\ S^T \cdot M \leq \boldsymbol{k} & (b) \\ M \geq Pre(\cdot, t) & (c) \\ M' = M + C(\cdot, t) & (d) \\ \boldsymbol{s}_j{}^T \cdot M' > k_j & (e) \\ M' \in \mathbb{N}^m & (f) \end{cases} \quad (3)$$

If system (3) admits a solution, there exists a consistent marking M – constraint(a) – that is legal – constraint (b) – from which the transition t may fire – constraint (c) – yielding a marking M' – constraint (d) – that is not legal – constraint (e) – because it holds $\boldsymbol{s}_j{}^T \cdot M' > k_j$. Note that, as a consequence of Equations (1) and (2), constraint (a) is linear with respect to M.

Let us finally observe that in the case of label-based feedback, Algorithm 6 can still be used to compute the set of markings in which the *closed-loop system* may be, given the current observation. In fact, if a nondeterministic event e is observed, this implies that a transition labeled e has fired (thus it was enabled by the controller). By assumption of LBF, this implies that all the other transitions labeled e were control enabled as well. Therefore, we may conclude that the controller does not influence the estimation procedure.

4.2 Transition-based feedback

When the controller may assign a different control pattern to nondeterminstic transitions with the same label the control pattern is a function $f(t, \mathcal{C}) : T \times 2^{\mathbb{N}^m} \to \{0, 1\}$ and can be given as follows.

Definition 8. (Optimal TBF for GMEC with observer) Given a GMEC $(S, \boldsymbol{k})$, a set of consistent markings $\mathcal{C} \subseteq \mathbb{N}^m$, and a λ-free labeling function $L : T \rightarrow E$, the firing of transition t should be prevented if and only if there exists at least one legal consistent marking M such that the firing of t from M leads to a forbidden marking, i.e.,

$$f(t, \mathcal{C}) = \begin{cases} 0 \text{ if } (\exists M)\ M \in \mathcal{C},\ S^T \cdot M \leq \boldsymbol{k}, \\ \quad M[t\rangle M',\ (\exists j)\ \boldsymbol{s}_j \cdot M' > k_j \\ 1 \text{ otherwise.} \end{cases}$$

■

The algorithm for the computation of the control law $f(t, \mathcal{C})$ is not given here for sake of brevity. It requires looking for the admissible solutions of a certain number of constraint sets analogous to Equation (3).

It is important to observe that the possibility of assigning a different control pattern to nondeterministic transitions with the same label may reduce the nondeterminism of the net. In fact, if a nondeterministic event e is observed and only a subset of transitions $T_e' \subset T_e$ was enabled by the controller, we know for sure that none of the transitions in $T_e \setminus T_e'$ has fired. This implies that Algorithm 6 needs to be slightly modified when used to compute the set of markings in which the *closed-loop system* may be, given the actual observation. This point is formalized in the following algorithm.

Algorithm 9. (Upper bound and basis marking computation under TBF)

When the controller implements a transition-based feedback, Case D of Algorithm 6 becomes:

```
let T'_e = {t ∈ T_e | f(t, C) = 1}
if T'_e = {t} then
  M_b,we = M_b,w + C(·, t)
else
    for all τ ∈ T_e
      if τ ∩ T'_e ≠ ∅ then
        u_we(τ) = min{u_w(τ) + 1, z(M_b,w, τ)}
      else u_we(τ) = u_w(τ)
      endif
    endfor
    u_we(T_e) = u_w(T_e) + 1
    M_b,we = M_b,w
endif
```

■

Therefore, in the case that there is only one transition t labeled e that is enabled by the controller, we update the basis marking and we do not increase the value of u_{we} (the event e is now deterministic). On the contrary, if there is more than one transition t labeled e that is enabled by the controller (i.e., $|T_e'| > 1$), we only update the upper bounds relative to the subsets of T_e containing transitions enabled by the controller, as well as the upper bound relative to T_e.

5. A FINAL EXAMPLE

Let us consider a job-shop system consisting of 13 machines $\mathcal{M}_1$, $\mathcal{M}_2$, ..., $\mathcal{M}_{13}$, and 3 jobs J_1, J_2, J_3. Each job requires three different operations: the first operation provides semi-finished products (we assume an infinite availability of raw materials); the second operation provides finished products;

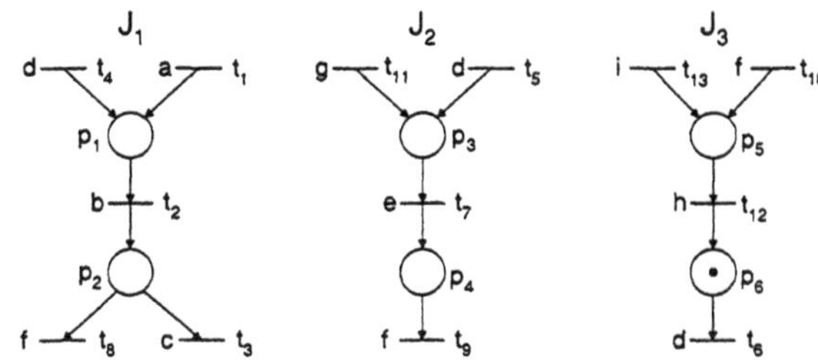

Fig. 2. The labeled Petri net system considered in Section 5.

finally, the third operation removes the finite product from the buffer. More precisely, the jobs must visit the machines in the following order:

$$J_1 = \{\mathcal{M}_1 \vee \mathcal{M}_4, \mathcal{M}_2, \mathcal{M}_3 \vee \mathcal{M}_8\},$$
$$J_2 = \{\mathcal{M}_5 \vee \mathcal{M}_{11}, \mathcal{M}_7, \mathcal{M}_9\},$$
$$J_3 = \{\mathcal{M}_{10} \vee \mathcal{M}_{13}, \mathcal{M}_{12}, \mathcal{M}_6\},$$

where $\vee$ denotes the logical *or*.

The operational state of each machine is detected by an appropriate symbol according to the following tables

$\mathcal{M}_1$	$\mathcal{M}_2$	$\mathcal{M}_3$	$\mathcal{M}_4, \mathcal{M}_5, \mathcal{M}_6$	$\mathcal{M}_7$
a	b	c	d	e

$\mathcal{M}_8, \mathcal{M}_9, \mathcal{M}_{10}$	$\mathcal{M}_{11}$	$\mathcal{M}_{12}$	$\mathcal{M}_{13}$
f	g	h	i

Thus, we are not able to distinguish among the operations of machines $\mathcal{M}_4$, $\mathcal{M}_5$ and $\mathcal{M}_6$ (labeled by the same symbol d), and the operations of machines $\mathcal{M}_8$, $\mathcal{M}_9$ and $\mathcal{M}_{10}$ (labeled by the same symbol f).

The Petri net model of the job-shop system is sketched in Figure 2. The firing of transition t_i model the operational process of machine $\mathcal{M}_i$, while its label is the symbol associated to machine $\mathcal{M}_i$, for $i = 1, \ldots, 13$. Finally, the marking of places p_1, p_3 and p_5 represents the semi-finished product; the marking of places p_2, p_4 and p_6 represents the finished product.

Assume that the initial marking is that one reported in Figure 2, namely $M_0 = [0\ 0\ 0\ 0\ 0\ 1]^T$.

Let us first assume that no controller is added to the system.

Initially, when no event is observed the basis marking is the initial marking and all the upper bounds are set to zero. As a new event is observed, the algorithm updates the basis marking and the upper bounds as listed in Table 1. Note that for simplicity of notation in Table 1 we omitted the dependence of the bounds on w and denoted as u_i, $u_{i,j}$, $u_{i,j,k}$ the upper bounds relative to the subsets of transitions $\{t_i\}$, $\{t_i, t_j\}$ and $\{t_i, t_j, t_k\}$, respectively.

Let us show how to use the table to compute the set $\mathcal{C}(d)$. It holds that $\mathcal{S}_d(d) = \{\boldsymbol{\sigma}_d = [\sigma_4\ \sigma_5\ \sigma_6]^T \in \mathbb{N}^3 \mid \sigma_4 \leq 1, \sigma_5 \leq 1, \sigma_6 \leq 1, \sigma_4 + \sigma_5 \leq 1, \sigma_4 + \sigma_6 \leq 1, \sigma_5 + \sigma_6 \leq 1, \sigma_4 + \sigma_5 + \sigma_6 = 1\}$, whose solutions are: $\boldsymbol{\sigma}^1 = [0\ 0\ 1]^T$, $\boldsymbol{\sigma}^2 = [0\ 1\ 0]^T$, $\boldsymbol{\sigma}^3 = [1\ 0\ 0]^T$. It also holds that $\mathcal{S}_f(d) = \{\mathbf{0}\}$. Thus, $\mathcal{C}(d) = \{M = M_{b,d} + C_d\boldsymbol{\sigma}_d^i + C_f\mathbf{0} \mid i = 1, 2, 3\} = \{[0\ 0\ 0\ 0\ 0\ 0]^T, [1\ 0\ 0\ 0\ 0\ 1]^T, [0\ 0\ 1\ 0\ 0\ 1]^T\}$.

Now, assume that a controller is used to enforce a given specification. Let assumptions (H1) to (H3) be verified.

w	*Case*	$M_{b,w}^T$	u_4	u_5	u_6	u_{45}	$u_{4,6}$	$u_{5,6}$	$u_{4,5,6}$	u_8	u_9	u_{10}	$u_{8,9}$	$u_{8,10}$	$u_{9,10}$	$u_{8,9,10}$
ε		[0 0 0 0 0 1]	0	0	0	0	0	0	0	0	0	0	0	0	0	0
d	*D*	[0 0 0 0 0 1]	1	1	1	1	1	1	1	0	0	0	0	0	0	0
dd	*D*	[0 0 0 0 0 1]	2	2	1	2	2	2	2	0	0	0	0	0	0	0
ddi	*A*	[0 0 0 0 1 1]	2	2	1	2	2	2	2	0	0	0	0	0	0	0
ddie	*B*	[0 0 0 1 1 1]	2	1	1	1	2	1	1	0	0	0	0	0	0	0
ddieg	*A*	[0 0 1 1 1 1]	2	1	1	1	2	1	1	0	0	0	0	0	0	0
ddiegb	*B*	[0 1 1 1 1 1]	1	1	1	0	1	1	0	0	0	0	0	0	0	0
ddiegbd	*D*	[0 1 1 1 1 1]	2	2	1	1	2	2	1	0	0	0	0	0	0	0
ddiegbdf	*D*	[0 1 1 1 1 1]	2	2	1	1	2	2	1	1	1	1	1	1	1	1
ddiegbdfa	*A*	[1 1 1 1 1 1]	2	2	1	1	2	2	1	1	1	1	1	1	1	1
ddiegbdfab	*B*	[0 2 1 1 1 1]	2	2	1	1	2	2	1	1	1	1	1	1	1	1
ddiegbdfabc	*C*	[0 1 1 1 1 1]	2	2	1	1	2	2	1	1	1	1	1	1	1	1

Table 1. The evolution of the net in Figure 2 when no control law is applied.

Let us first assume that all finite products are stored in a finite capacity buffer, and let 3 be its capacity. The set of legal markings is

$$\mathcal{L} = \{M \in \mathbb{N}^m \mid M(p_2) + M(p_4) + M(p_6) \leq 3\}.$$

The closed-loop behaviour of the system is obviously the same in the case of LBF and in the case of TBF because the set of transitions that may potentially violate the constraint is $J_t = \{t_2, t_7, t_{12}\} \subset T^d$, i.e., it only contains deterministic transitions. Thus the nondeterministic transitions are always enabled by the controller, because regardless of the actual observation, their firing may never lead to a violation of the constraint.

It is easy to verify that the previously considered word $w = ddiegbdfabc$ cannot be observed. More precisely, for all prefixes $w' \preceq ddiegb$ all transitions are enabled by the controller. On the contrary, after a further event d occurs all transitions in J_t are disabled by the controller, and the control patterns keep unaltered until the word $w = ddiegbdfa$ is observed, thus transition t_2 cannot fire (the event b cannot be observed).

Now, assume that all finite products are stored in an infinite capacity buffer, while semi-finished products relative to jobs J_1 and J_3 are stored in a finite capacity buffer, and let 2 be its capacity. The set of legal markings is

$$\mathcal{L} = \{M \in \mathbb{N}^m \mid M(p_1) + M(p_5) \leq 2\}.$$

In such a case the closed-loop behaviour of the system is no longer the same in the case of LBF and in the case of TBF because the set of transitions that may potentially violate the constraint is $J_t = \{t_1, t_4, t_{10}, t_{13}\}$, i.e., it also contains nondeterministic transitions.

A very simple example of this fact is given by the word $w = ddd$. In the case of TBF this sequence of events may occur, while in the case of LBF it cannot occur. In fact, when the first event d is observed all transitions are enabled (both in the case of TBF and LBF), thus a second event d may occur. At this point, the set $C(dd)$ contains the marking $M = [2\ 0\ 0\ 0\ 0\ 1]^T$ that enables a transition labeled d, namely t_4, whose firing would lead to a violation of the constraint. In the case of the LBF this implies that all transitions labeled d are disabled by the controller, thus a further event d cannot occur. On the contrary, in the case of the TBF, transition t_4 is disabled by the controller, but both transitions t_5 and t_6 may fire, thus a further event d may occur.

6. CONCLUSIONS

The contribution of this paper is twofold.

We first presented a marking estimation procedure that can be applied to λ-free labeled Petri nets (the formal proof of the correctness of this procedure can be found in (Giua *et al.*, 2004)). In particular, under the assumption that all nondeterministic transitions are contact-free, the set of markings consistent with an observed word can be described by a constraint set of linear inequalities that has a fixed structure that does not change as the length of the observed sequence increases.

Then, we shown how this marking estimation can be used by a control agent to enforce some specifications. In particular we discussed two different cases, depending on the possibility of the controller to assign a different control pattern to transitions that share the same label.

REFERENCES

Corona, D., A. Giua, J. Júlvez and C. Seatzu (2003). Observers for nondeterministic λ-free labeled petri nets. In: *Proc. 9th IEEE Int. Conf. on Emerging Technologies and Factory Automation*. pp. 307–314.

Gaubert, S. and A. Giua (1999). Petri net languages and infinite subsets of n^m. *J. Computer and System Sciences* **59**, 373–391.

Giua, A., D. Corona and C. Seatzu (2004). State estimation of λ-free labeled petri nets with contact-free nondeterministic transitions. Technical report. University of Cagliari, Italy. (also submitted to *Discrete Event Dynamic Systems)*.

Meda, M.E., A. Ramírez and A. Malo (1998). Identification in discrete event systems. In: *Proc. 1998 IEEE Int. Conf. on Systems, Man, and Cybernetics*. pp. 740–745.

Murata, T. (1989). Petri nets: properties, analysis and applications. *Proceedings of the IEEE* **77**(4), 541–580.

Peterson, J.L. (1981). *Petri net theory and the modeling of systems*. Prentice-Hall.

Zhang, L. and L.E. Holloway (1995). Forbidden state avoidance in controlled petri nets under partial observation. In: *Proc. 33rd Allerton Conference*. pp. 146–155.

ELSEVIER
IFAC
PUBLICATIONS
www.elsevier.com/locate/ifac

A MODEL OF COMPONENT CONSISTENCY IN DISTRIBUTED DIAGNOSIS

R. Su and W.M. Wonham

Dept. Electrical & Computer Engineering
University of Toronto
Toronto, ON M5S 3G4, Canada
surong,wonham@control.utoronto.ca

Abstract: In distributed diagnosis, each local diagnoser first generates a local estimate then communicates with other local diagnosers to reach suitably defined consistency. In this paper we propose two types of consistency and provide efficient computational procedures to achieve them. *Copyright © 2004 IFAC*

Keywords: discrete-event systems, distributed diagnosis, modeling and consistency

1. INTRODUCTION

So far there is a large body of research on fault diagnosis for discrete-event systems, e.g. finite-state automaton approaches (Sampath et al. (1998)), and Petri net approaches (Aghasaryaiu et al. (1997)). These can be roughly classified as follows: (1) centralized approaches like (Sampath et al. (1998)); (2) decentralized (Debouk et al. (2000)) (Su and Wonham (2000)); and (3) distributed (Su et al. (2002)) (Fabre et al. (2002)) (Benveniste et al. (2003)). It is well known that centralized approaches may lead to high space complexity in order to store the final diagnoser. Meanwhile, weak scalability and robustness are also features of centralized approaches. Decentralized approaches usually lead to relatively small local diagnosers and improved scalability and robustness. Nevertheless, designing a decentralized diagnoser still requires existence of a centralized target system model at the intermediate stage, which again forces a designer to face the space complexity issue. As for distributed approaches, they generally have the following features: (1) the system model is distributed; (2) diagnosis is locally generated; (3) constraints imposed by interaction among local components are satisfied by achieving a suitable consistency property. By introducing inter-component consistency, these approaches can avoid the potentially enormous intermediate system model. So space complexity in the design stage and during online diagnosis can be brought under control. To capture the interaction among local components, there are roughly two types of consistency in the literature: one is *global consistency*, which was more or less described in, e.g. (Baroni et al. (1999)) (Lamperti et al. (2000)) (Pencol et al. (2002)) (Fabre et al. (2002)) (Benveniste et al. (2003)); the other one is *local consistency*, as explained in (Su et al. (2002)). By using purely local computation and achieving appropriate inter-diagnoser consistency, we can greatly reduce space complexity of online diagnosis in many practical systems and substantially improve the scalability and robustness of diagnosis. This paper proposes a unified model to describe the different types of consistency and presents efficient computational algorithms to achieve them.

The paper is organized as follows. In Section 2 we present two types of consistency. In Section 3, we introduce the concept of supremal global support and propose an efficient computational procedure to achieve it. Then in Section 4, we introduce the concept of supremal local support and propose a computational procedure to achieve it. We introduce hierarchical computation in Section 5 and draw conclusions in Section 6.

2. GLOBAL AND LOCAL CONSISTENCIES

Let I be an index set. For each $i \in I$ let Σ_i be an event set. As usual (Wonham (2003a)), we use "$||$" to represent synchronous product. All natural projections in this paper obey the following rules of notation: given an index set J with a predefined collection of event sets $\{\Sigma_j | j \in J\}$, we define

$$\Sigma_J := \cup_{j \in J} \Sigma_j$$

Given two event sets Σ_i and Σ_j, we use $\mathrm{P}_{i,j}$ to mean the natural projection

$$\mathrm{P}_{i,j} : \Sigma_i^* \to (\Sigma_i \cap \Sigma_j)^*$$

and $\mathrm{P}_{J,i}$ to mean the natural projection

$$\mathrm{P}_{J,i} : \Sigma_J^* \to (\Sigma_J \cap \Sigma_i)^*$$

If we have another index set K with a predefined set $\{\Sigma_k | k \in K\}$, then we use $\mathrm{P}_{J,K}$ to mean the natural projection

$$\mathrm{P}_{J,K} : \Sigma_J^* \to (\Sigma_J \cap \Sigma_K)^*$$

Let $\mathcal{L} = \{\mathrm{L}_i \subseteq \Sigma_i^* | i \in I\}$ be a set of languages. Now we introduce two types of consistency.

Definition 2.1. $\{\mathrm{L}_i | i \in I\}$ is *globally consistent* if for each $i \in I$, $\mathrm{L}_i = \mathrm{P}_{I,i}(||_{j \in I} \mathrm{L}_j)$. □

Here $||_{j \in I} \mathrm{L}_j$ denotes synchronous product of all local components of I. It is well known that synchronous product is one way to combine local transitional behavior to form a global (or composite) transitional behavior. If we treat each local language L_i as a local estimate in diagnosis, then Def. 2.1 says that when a system reaches global consistency, knowing all other local estimates will not improve the current local estimate L_i.

Definition 2.2. $\{\mathrm{L}_i | i \in I\}$ is *locally consistent* if for each $i, j \in I$, $\mathrm{P}_{i,j}(\mathrm{L}_i) = \mathrm{P}_{j,i}(\mathrm{L}_j)$. □

Again, if we treat each local language L_i as a local estimate in diagnosis, then Def. 2.2 says that when a system reaches local consistency, knowing the local estimate of any local component that is a neighbor of component i will not improve the current local estimate L_i.

From above description we can see that the two types of consistency provide ways to determine whether or not all local components have reached an "agreement". On that basis we can decide when communication among local components should be terminated. The relation between global and local consistencies is described as follows.

Proposition 2.1. If $\mathcal{L}$ is globally consistent then it must be locally consistent. □

The reverse statement need not always hold (see (Su (2004))). It is interesting to know under what condition(s) global consistency is equivalent to local consistency. To this end, we need more technical preparation. Given $\{\Sigma_i | i \in I\}$ let $\mathrm{Gr} =< \mathrm{Ver}, \mathrm{Edg} >$ be a graph where $\mathrm{Ver} = I$ and $\mathrm{Edg} \subseteq \mathrm{Ver} \times \mathrm{Ver}$ such that

$$(\forall i, j \in \mathrm{Ver})\, (i, j) \in \mathrm{Edg} \iff i \neq j \,\&\, \Sigma_i \cap \Sigma_j \neq \varnothing$$

Clearly, if $(i, j) \in \mathrm{Edg}$ then so is (j, i). Thus Gr is an undirected graph. Figure 1 depicts such a graph.

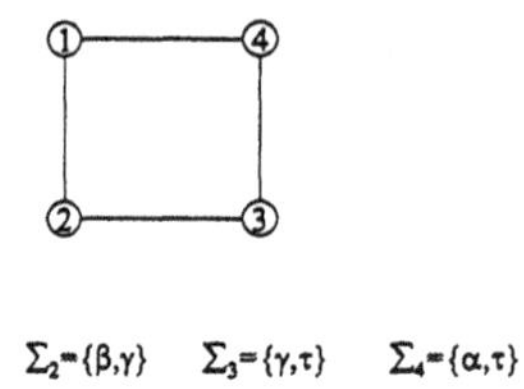

Fig. 1. Example of Graph Gr

Proposition 2.2. Given $\mathcal{L} = \{\mathrm{L}_i \subseteq \Sigma_i^* | i \in I\}$, if the corresponding graph Gr is a tree as commonly defined in graph theory (Diestel (2000)), then $\mathcal{L}$ is globally consistent if and only if it is locally consistent.

Proposition 2.1 indicates that local consistency is a necessary condition for global consistency, and Proposition 2.2 indicates that this necessary condition becomes sufficient if the graph Gr is a tree.

3. SUPREMAL GLOBAL SUPPORT

Definition 3.1. Given $\mathcal{L} = \{\mathrm{L}_i \subseteq \Sigma_i^* | i \in I\}$, a set $\{\mathrm{E}_i \subseteq \mathrm{L}_i | i \in I\}$ is a *global support* of $\mathcal{L}$ if it is globally consistent. Let $\Delta(\mathcal{L})$ be the set of all global supports of $\mathcal{L}$.

Clearly $\{\mathrm{E}_i = \varnothing | i \in I\} \in \Delta(\mathcal{L})$. We define a binary relation "$\leq$" among elements in $\prod_{i \in I} \mathrm{Pwr}(\Sigma_i^*)$, where $\mathrm{Pwr}(\Sigma_i^*)$ means the power set of Σ_i^*.

$$\{\mathrm{E}_i | i \in I\} \leq \{\tilde{\mathrm{E}}_i | i \in I\} \iff (\forall i \in I)\, \mathrm{E}_i \subseteq \tilde{\mathrm{E}}_i$$

$\mathcal{E} \in \Delta(\mathcal{L})$ is called the *supremal global support* of $\Delta(\mathcal{L})$, written $\mathrm{Sup}\Delta(\mathcal{L})$, if $(\forall \mathcal{E}' \in \Delta(\mathcal{L}))\, \mathcal{E}' \leq \mathcal{E}$.

Proposition 3.1. $\mathrm{Sup}\Delta(\mathcal{L}) = \{\mathrm{P}_{I,i}(||_{j \in I} \mathrm{L}_j) | i \in I\}$.

It is interesting in practice, e.g. for conservative diagnosis, to know how to compute such a supremal global support. Suppose $I = \{1, 2, \cdots, n\}$ and we want to compute $\mathrm{P}_{I,n}(||_{j \in I} \mathrm{L}_j)$. We first adopt an ordering on nodes in I which makes n the last node in the ordering. For simplicity, suppose the ordering is $[1, 2, \cdots, n]$. The ordering is an important factor for the space complexity of the following computational procedure. Due to limited space, we leave the ordering issue to a full version of this paper, where a heuristic ordering algorithm is given. Now we present our computational procedure, where the notation used below follows the notational rules described in Section 2.1.

Computational Procedure for Global Consistency: (CPGC)

(1) Initially set $T_0 := \Sigma_1$ and $W_0 := \Sigma_1^*$.
(2) For $k = 1, 2, \cdots, n-1$,
- $J_k := \{1, 2, \cdots, k\}$, $T_k := \Sigma_{J_k} \cap \Sigma_{I-J_k}$.
- $W_k := P_{T_{k-1}\cup\Sigma_k, T_k}(W_{k-1}||L_k)$

(3) $E_n := W_{n-1}||L_n$ □

Proposition 3.2. In CPGC above, $E_n = P_{I,n}(||_{j\in I}L_j)$.

To demonstrate CPGC, consider the following example. The system consists of 18 local components, namely $I = \{1, 2, \cdots, 17, 18\}$, whose transition structures are depicted in the left side of Figure 2. The right

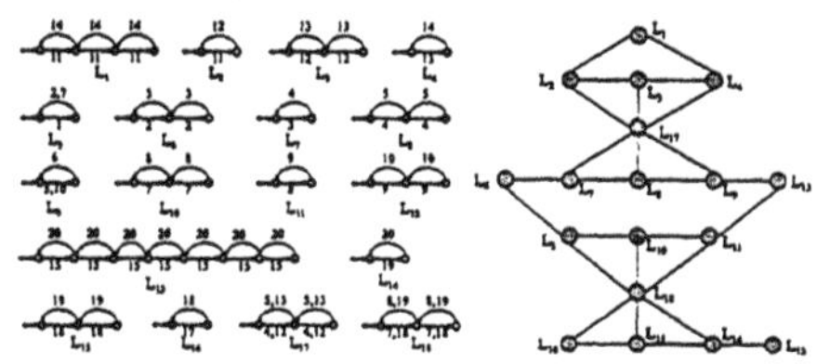

Fig. 2. Plant $\mathcal{L}$ (Left) and Graph Gr (Right)

picture shows the system's network structure. Suppose we want to compute $E_{18} = P_{I,18}(||_{k\in I}L_k))$. We select an arbitrary ordering, say $[1, 2, 3, \cdots, 17, 18]$. The computational result of CPGC is shown as follows. In Table 3.1, The entry $[a_1, a_2]$ means that the resultant finite-state automaton has a_1 states and a_2 transitions. From the table we see that our objective E_{18} is realized by an automaton with 6 states and 12 transitions. All results are obtained by using CTCT (Wonham (2003b)). Columns under the titles $W_{k-1}||L_k$ and W_k describe the size of all intermediate computational results before we obtain E_{18}. The maximum size of intermediate results is 216 states with 1242 transitions.

Table 3.1

k	L_k	$W_{k-1}\|\|L_k$	$W_k = P_{T_{k-1}\cup\Sigma_k, T_k}(W_{k-1}\|\|L_k)$
1	[4,6]	[4,6]	[4,6]
2	[2,2]	[8,13]	[5,8]
3	[3,4]	[15,30]	[15,30]
4	[2,2]	[6,9]	[3,4]
5	[2,3]	[6,17]	[6,17]
6	[3,4]	[18,60]	[3,10]
7	[2,2]	[6,20]	[3,10]
8	[3,4]	[9,33]	[9,33]
9	[2,3]	[18,78]	[9,42]
10	[3,4]	[27,135]	[27,135]
11	[2,2]	[54,279]	[54,279]
12	[3,4]	[162,756]	[27,108]
13	[8,14]	[216,1242]	[27,135]
14	[2,2]	[54,270]	[27,135]
15	[3,4]	[81,432]	[81,432]
16	[2,2]	[162,891]	[81,432]
17	[3,8]	[54,252]	[9,24]
18	[3,8]	[6,12]	-

The brute force computation completely fails in this example because the size of the result $||_{k\in I}L_k$ is too big to be handled by our software CTCT, which exhausts memory before it finishes the computation. By that time the computation has already reached 10^6 states and 8×10^6 transitions. If we compare $[216, 1242]$ with $[10^6, 8 \times 10^6]$, then clearly CPGC is much more efficient than the brute-force method. The correctness of E_{18} is guaranteed by Prop. 3.1. In order to obtain $\mathrm{Sup}\Delta(\mathcal{L})$, we need to repeat CPGC for each local component $i \in I$. At this point, we can see that CPGC may entail high time complexity if the system consists of many local components. We will see later that hierarchical computation may come to the rescue.

4. SUPREMAL LOCAL SUPPORT

4.1 Concept of Supremal Local Support

Definition 4.1. Given $\mathcal{L} = \{L_i \subseteq \Sigma_i^* | i \in I\}$, a set $\{E_i \subseteq L_i | i \in I\}$ is a *local support* of $\mathcal{L}$ if it is locally consistent. Let $\Gamma(\mathcal{L})$ be the set of all local supports.

Proposition 4.1. $(\exists \mathcal{E} \in \Gamma(\mathcal{L}))(\forall \mathcal{E}' \in \Gamma(\mathcal{L}))\, \mathcal{E}' \leq \mathcal{E}$.

We call $\mathcal{E}$ in Prop. 4.1 the *supremal local support* of $\Gamma(\mathcal{L})$, written $\mathrm{Sup}\Gamma(\mathcal{L})$. By Prop. 2.1 we can derive that $\mathrm{Sup}\Delta(\mathcal{L}) \leq \mathrm{Sup}\Gamma(\mathcal{L})$. In distributed diagnosis, $\mathrm{Sup}\Delta(\mathcal{L})$ leads to a tighter estimate than $\mathrm{Sup}\Gamma(\mathcal{L})$ does. But we have two reasons to be interested in $\mathrm{Sup}\Gamma(\mathcal{L})$. First, mathematically it imposes a much more difficult problem than $\mathrm{Sup}\Delta(\mathcal{L})$ does. Prop. 3.1 defines a concrete map from the initial set $\mathcal{L}$ to $\mathrm{Sup}\Delta(\mathcal{L})$. But we do not have a similar map between $\mathcal{L}$ and $\mathrm{Sup}\Gamma(\mathcal{L})$. Second, from a computational point of view, we will see that we can use a rather *scalable* computational procedure to compute $\mathrm{Sup}\Gamma(\mathcal{L})$, as long as the procedure finitely terminates. But to find a similar scalable procedure for $\mathrm{Sup}\Delta(\mathcal{L})$ is still an open question. In distributed diagnosis, by computing $\mathrm{Sup}\Gamma(\mathcal{L})$ we can find a tradeoff between quality of diagnosis and scalability and robustness of the diagnoser. Next, we introduce a computational procedure for the supremal local support.

4.2 Computing Supremal Local Support

In (Su et al. (2002)) the authors proposed a scalable computational algorithm that can be used here to achieve the supremal local support defined above. A similar algorithm appeared later in (Benveniste et al. (2003)) in the form of a Petri net computation. In this paper we revise our algorithm by imposing a binary relation on the nodes of the target network.

Given $\mathcal{L}$, let Gr = (Ver, Edg) be the associated graph. Suppose we pick one node, say node 1, as the *root* of Gr. As commonly used in graph theory (Diestel (2000)), we define the distance between the root node 1 and a node $v \in$ Ver, written $d(1, v)$, as the *length* of the shortest *path* between v and node 1. We assume that **the graph Gr is connected.** Let

$$\phi : \mathrm{Ver} \rightarrow \{1, \cdots, |\mathrm{Ver}|\}$$

be a one-to-one map such that

$$(\forall (i,j) \in \text{Edg})\, d(1,i) < d(1,j) \Rightarrow \phi(i) < \phi(j)$$

In general there may exist more than one choice for ϕ. Fix one choice of ϕ and define the following binary relation among vertices:

$$(\forall i,j \in \text{Ver})\, i \downarrow j \iff (i,j) \in \text{Edg} \,\&\, \phi(i) < \phi(j)$$

We call i a *father node* of j and the binary relation $\downarrow$ the *father-son* relation. The graph Gr equipped with $\downarrow$ can be treated as a directed graph, where $(i,j) \in \text{Edg}$ if and only if $i \downarrow j$. It is not difficult to see that the resulting directed graph is acyclic and reachable from the root. Due to the acyclic property, we can propose the following algorithm.

Computational Procedure for Local Consistency: (CPLC)

(1) (Initialization) $(\forall i \in I)\, \mathrm{E}_i^0 := \mathrm{L}_i$

(2) (Centripetal Propagation) When $n \in 2\mathbb{N}+1$, for each node $i \in V_j$ $(k \geq j \geq 0)$ compute

$$\mathrm{E}_i^n := \begin{cases} \mathrm{E}_i^{n-1} & \text{if } (\nexists r \in I)\, i \downarrow r \\ \mathrm{E}_i^{n-1} \| (\|_{r:i\downarrow r} \mathrm{P}_{r,i}(\mathrm{E}_r^n)) & \text{otherwise} \end{cases}$$

(3) (Centrifugal Propagation) When $n \in 2\mathbb{N}^+$, for each node $i \in V_j$ $(0 \leq j \leq k)$ compute

$$\mathrm{E}_i^n := \begin{cases} \mathrm{E}_i^{n-1} & \text{if } (\nexists r \in I)\, r \downarrow i \\ \mathrm{E}_i^{n-1} \| (\|_{r:r\downarrow i} \mathrm{P}_{r,i}(\mathrm{E}_r^n)) & \text{otherwise} \end{cases}$$

(4) **Termination:** $(\exists n \in \mathbb{N}^+)(\forall i \in I)\, \mathrm{E}_i^n = \mathrm{E}_i^{n+1}$ □

In step 2 before we compute E_i^n all those E_r^n with $i \downarrow r$ have been computed already. Similarly in step 3 before we compute E_i^n all those E_r^n with $r \downarrow i$ have been computed. Therefore CPLC is well defined. The main feature of CPLC is its *scalability*, in the sense that when we add a new node to the graph or remove one from the graph, only a few neighboring nodes need to update their local communication protocol, which is usually done by simply changing their list of father nodes and son nodes. As a comparison, in CPGC when we add a new node or remove one, we need to redesign the ordering among nodes. So CPGC is not scalable.

Figure 3 depicts how the CPLC procedure works in a simple example. In this example, the father-son

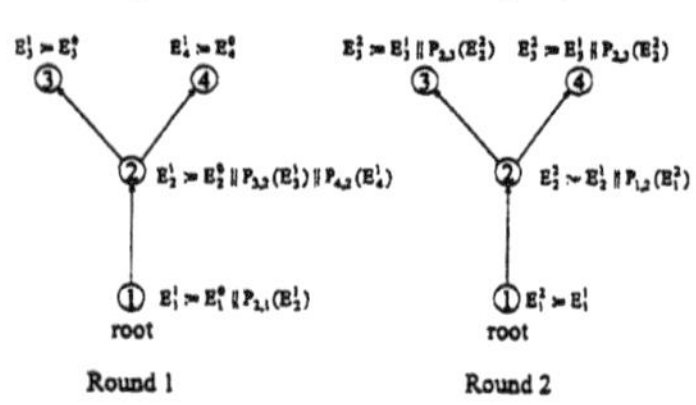

Fig. 3. 2-Round CPLC

relation is defined as follows: $1 \downarrow 2$, $2 \downarrow 3$, $2 \downarrow 4$. Let $\{V_0 = \{1\}, V_1 = \{2\}, V_2 = \{3,4\}\}$. During the computational process, in odd-number rounds we can see computation starts from nodes in V_2, then the node in V_1 and finally the node in V_0. In even-number rounds computation goes in the opposite direction, namely starts from the node in V_0, then the node in V_1 and finally nodes in V_2. So the computational process is well organized.

Proposition 4.2. If CPLC terminates at $n \in \mathbb{N}^+$, then we have that $\{\mathrm{E}_i^n | i \in I\} = \mathrm{Sup}\Gamma(\mathcal{L})$. □

So far we have shown in Prop. 4.2 that as long as CPLC finitely terminates, the result is the supremal local support $\mathrm{Sup}\Gamma(\mathcal{L})$. As shown in (Su (2004)), (1) if each L_i $(i \in I)$ is finite, CPLC terminates; (2) if Gr is a tree, then CPLC terminates at Round 2; (3) if Gr is a *simple net* with disjoint loops, by using the concept of *memory set*, CPLC terminates at Round 3. All these are positive results. But it turns out that (see Su (2004)) in some cases CPLC cannot terminate. The decidability of termination of CPLC is still unknown.

So far we have introduced concepts of supremal global support and supremal local support, and provided concrete computational procedures. Nevertheless, we have mentioned that CPLC may not always terminate and CPGC can only compute the supremal global support in a sequential way, which may result in high time complexity. Next we introduce hierarchical computation, which can partially reduce high time complexity of CPGC.

5. HIERARCHICAL COMPUTATION

5.1 Concept of Hierarchical Computation

First, we describe how to construct a two-level hierarchy. Given the index set I, let $\hat{I}$ be a second index set based on a partition $\{J_{\hat{k}} \neq \varnothing | \hat{k} \in \hat{I}\}$ of I. From now on we use $i, j, k, \cdots$ for elements of I, and $\hat{i}, \hat{j}, \hat{k}, \cdots$ for elements of $\hat{I}$. Similarly, we use $J, K, \cdots$ for subsets of I and $\hat{J}, \hat{K}, \cdots$ for subsets of $\hat{I}$. For each $\hat{k} \in \hat{I}$, recall that by the notational rules we have

$$\Sigma_{J_{\hat{k}}} := \cup_{i \in J_{\hat{k}}} \Sigma_i \text{ and } \Sigma_{I-J_{\hat{k}}} := \cup_{j \in I-J_{\hat{k}}} \Sigma_j$$

We define a new event set $\Sigma_{\hat{k}} \subseteq \Sigma_{J_{\hat{k}}}$ such that

$$\Sigma_{J_{\hat{k}}} \cap \Sigma_{I-J_{\hat{k}}} \subseteq \Sigma_{\hat{k}} \qquad (1)$$

namely, the new event set $\Sigma_{\hat{k}}$ should contain all events that are shared by $\{\mathrm{L}_i | i \in J_{\hat{k}}\}$ and $\{\mathrm{L}_j | j \in I - J_{\hat{k}}\}$. In fault diagnosis, besides those shared events, $\Sigma_{\hat{k}}$ usually contains only observable events which represent observation in the system.

As an illustration, let $I = \{1,2,3,4\}$ and suppose $\Sigma_1 = \{\alpha, \beta\}$, $\Sigma_2 = \{\beta, \gamma\}$, $\Sigma_3 = \{\sigma, \gamma, \delta\}$ and

$\Sigma_4 = \{\alpha, \delta\}$. Suppose we partition I into two subsets $J_a = \{1, 2\}$ and $J_b = \{3, 4\}$ with $\hat{I} = \{a, b\}$. Then

$$\Sigma_{J_a} = \{\alpha, \beta, \gamma\} \text{ and } \Sigma_{J_b} = \{\alpha, \gamma, \delta, \sigma\}$$

So $\Sigma_{J_a} \cap \Sigma_{I-J_a} = \Sigma_{J_a} \cap \Sigma_{J_b} = \{\alpha, \gamma\}$. To define a new set $\Sigma_a \subseteq \Sigma_{J_a}$ such that (1) holds, we can pick $\Sigma_a = \{\alpha, \gamma\}$ or $\Sigma_a = \{\alpha, \gamma, \beta\}$.

With the same notational rules for natural projections, for each $\hat{k} \in \hat{I}$ we define a new language on $\Sigma_{\hat{k}}^*$:

$$\mathrm{L}_{\hat{k}} := \mathrm{P}_{J_{\hat{k}},\hat{k}}(||_{i\in J_{\hat{k}}}\mathrm{L}_i) \qquad (2)$$

Let $\hat{\mathcal{L}} = \{\mathrm{L}_{\hat{k}} | \hat{k} \in \hat{I}\}$ be the *abstract model* of $\mathcal{L}$. Each element of $\hat{\mathcal{L}}$ is called an *abstract local module* of $\mathcal{L}$. The relation between $\mathcal{L}$ and $\hat{\mathcal{L}}$ can be summarized in the following commutative diagram (Figure 4), where

$$\mathbb{W} := \textstyle\prod_{i\in I} \mathrm{Pwr}(\Sigma_i^*) \text{ and } \hat{\mathbb{W}} := \prod_{\hat{k}\in\hat{I}} \mathrm{Pwr}(\Sigma_{\hat{k}}^*)$$

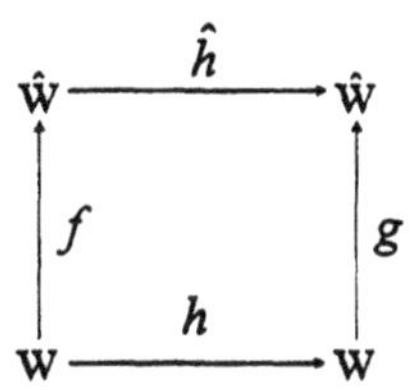

Fig. 4. Two-Level Hierarchy

and the maps in the diagram are defined as follows:

(1) $f(\{\mathrm{L}_i | i \in I\}) := \{\mathrm{P}_{J_{\hat{k}},\hat{k}}(||_{j\in J_{\hat{k}}}\mathrm{L}_j) | \hat{k} \in \hat{I}\}$
(2) $g(\{\mathrm{E}_i | i \in I\}) := \{\mathrm{P}_{I,\hat{k}}(||_{j\in I}\mathrm{E}_j) | \hat{k} \in \hat{I}\}$
(3) $h(\mathcal{L}) := \mathrm{Sup}\Delta(\mathcal{L})$
(4) $\hat{h}(\hat{\mathcal{L}}) := \mathrm{Sup}\Delta(\hat{\mathcal{L}})$

Nevertheless, it is more interesting in practice to know how to obtain the supremal global support $\{\mathrm{E}_i | i \in I\}$ of $\Delta(\mathcal{L})$ at the low level if the supremal global support $\{\mathrm{E}_{\hat{k}} | \hat{k} \in \hat{I}\}$ of $\Delta(\hat{\mathcal{L}})$ at high level is given. To this end we have the following result.

Proposition 5.1. Let f be defined as in Figure ??. Let $\mathcal{L} = \{\mathrm{L}_i | i \in I\} \in \mathbb{W}$, $\hat{\mathcal{L}} = \{\mathrm{L}_{\hat{k}} | \hat{k} \in \hat{I}\} \in \hat{\mathbb{W}}$ such that $f(\mathcal{L}) = \hat{\mathcal{L}}$. Suppose $\mathrm{Sup}\Delta(\mathcal{L}) = \{\mathrm{E}_i | i \in I\}$ and $\mathrm{Sup}\Delta(\hat{\mathcal{L}}) = \{\mathrm{E}_{\hat{k}} | \hat{k} \in \hat{I}\}$. Then we have that $(\forall \hat{k} \in \hat{I})(\forall i \in J_{\hat{k}})\, \mathrm{E}_i = \mathrm{P}_{J_{\hat{k}},i}((||_{j\in J_{\hat{k}}}\mathrm{L}_j)||\mathrm{E}_{\hat{k}})$.

By Prop. 3.1 we have $\mathrm{E}_i = \mathrm{P}_{I,i}(||_{j\in I}\mathrm{L}_i)$. Since computation is over I, if the size of I is large, then the time complexity may be a problem. Nevertheless, Proposition 5.1 tells us that if we can achieve the supremal global support $\mathrm{E}_{\hat{k}}$ at the high level, where $\hat{k} \in \hat{I}$ and $i \in J_{\hat{k}}$, then we only need the module $\{\mathrm{L}_j | j \in J_{\hat{k}}\}$ plus $\{\mathrm{E}_{\hat{k}}\}$ to achieve E_i. Notice that $J_{\hat{k}}$ is usually much smaller than I in a large scale system, and furthermore, concurrent computation can be applied on different local modules which are disjoint from each other. Both factors contribute to low time complexity of hierarchical computation. This is the main advantage gained from hierarchical computation. Due to the nature of abstraction, achieving the supremal global support in the abstract model at the high level is relatively easy. Prop. 5.1 has crucial application in multi-resolution diagnosis. It is not difficult to see that the current theory can be extended to a hierarchy with more levels such that, between every two adjacent levels we have a commutative diagram as Figure 4 and Proposition 5.1 holds. To demonstrate the hierarchical computational procedure, consider the following example.

5.2 Example of Hierarchical Computation

Suppose we have four languages $\mathrm{L}_1, \cdots, \mathrm{L}_4$ which are depicted in Figure 5 with the corresponding graph Gr. The index set $I = \{1, 2, 3, 4\}$. Suppose we partition

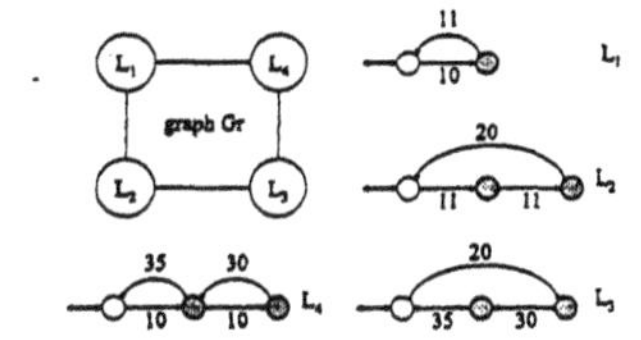

Fig. 5. Plant Model and Network Graph

I into two sets $J_a = \{1, 2\}$ and $J_b = \{3, 4\}$, where $\hat{I} = \{a, b\}$. Suppose we assign new event sets at the high level based on expression (1) as follows:

$$\Sigma_a := \{10, 20\} \text{ and } \Sigma_b := \{10, 20, 35\}$$

Then we can compute abstract local modules based on expression (2) as follows:

$$\mathrm{L}_a := \mathrm{P}_{J_a,a}(\mathrm{L}_1||\mathrm{L}_2) \text{ and } \mathrm{L}_b := \mathrm{P}_{J_b,b}(\mathrm{L}_3||\mathrm{L}_4)$$

where the results are shown in Figure 6. After that we

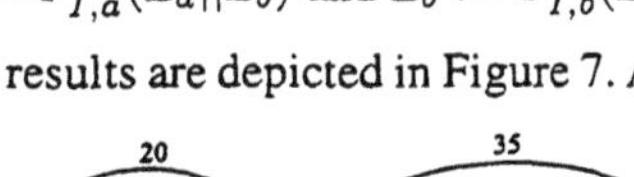

Fig. 6. Construction of $\hat{\mathcal{L}} = \{\mathrm{L}_a, \mathrm{L}_b\}$

can compute the supremal global support at the high level according to

$$\mathrm{E}_a := \mathrm{P}_{\hat{I},a}(\mathrm{L}_a||\mathrm{L}_b) \text{ and } \mathrm{E}_b := \mathrm{P}_{\hat{I},b}(\mathrm{L}_a||\mathrm{L}_b)$$

where the results are depicted in Figure 7. After we get

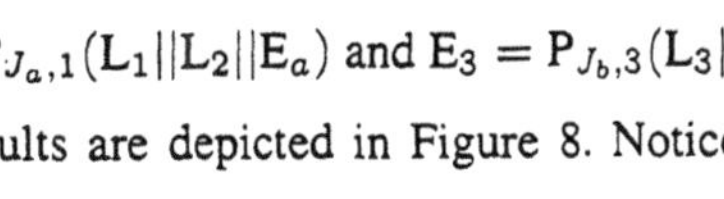

Fig. 7. $\mathrm{Sup}\Delta(\hat{\mathcal{L}}) = \{\mathrm{E}_a, \mathrm{E}_b\}$

E_a and E_b we can compute the supremal global support at the bottom level. Suppose we want to compute E_1 and E_3, which are shown as follows:

$$\mathrm{E}_1 = \mathrm{P}_{J_a,1}(\mathrm{L}_1||\mathrm{L}_2||\mathrm{E}_a) \text{ and } \mathrm{E}_3 = \mathrm{P}_{J_b,3}(\mathrm{L}_3||\mathrm{L}_4||\mathrm{E}_b)$$

The results are depicted in Figure 8. Notice that E_1

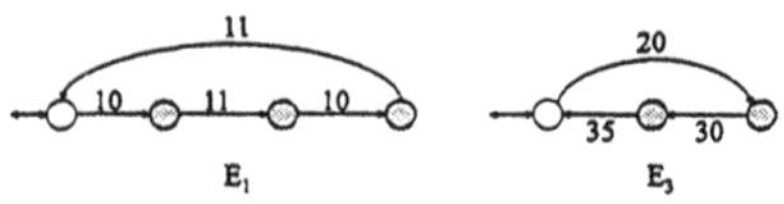

Fig. 8. Results of E_1 and E_3

and E_3 can be computed simultaneously because they belong to different modules. The results are verified to be isomorphic to the ones obtained by the brute-force method, namely

$$E_1 := P_{I,1}(L_1 || \cdots || L_4),\ E_3 := P_{I,3}(L_1 || \cdots || L_4)$$

The computation demonstrated above is realized by the software CTCT (Wonham (2003b)).

6. CONCLUSIONS

In this paper we introduced two concepts: the supremal global support and the supremal local support. In distributed diagnosis each type of supremal support can be used to model consistency among local diagnoses. Since global consistency is stronger than local consistency, applying the supremal global support in diagnosis usually results in a tighter estimate about fault status of a target system than applying the supremal local support does. Nevertheless, the computational procedure CPLC for achieving the supremal local support is much more scalable than CPGC for the supremal global support, as long as CPLC terminates. At this point, the supremal local support offers a tradeoff between diagnosis quality, in terms of the number of fault candidates contained in each local diagnosis, and scalability and robustness of the distributed diagnoser. To the best of our knowledge, CPLC first appeared in (Su et al. (2002)) then in (Benveniste et al. (2003)) a similar algorithm was given in the form of a Petri net computation. In this paper we improved it further by imposing the father-son relation $\downarrow$ among nodes, based on which we can tell in the worst case, how many rounds a user has to wait before termination, if indeed it terminates. Nevertheless, the decidability of the termination of CPLC is still unsolved. As for CPGC, from the example we can see that it is superior to the brute-force method as far as space complexity is concerned. Nevertheless, it is not scalable and the time complexity may be high. To partially overcome these disadvantages we introduced hierarchical computation, which by Prop. 5.1 is another application of divide-and-conquer. In online diagnosis, this may substantially reduce time complexity. Nevertheless, we admit that we have to pay the price of using extra memory to store the high-level abstract model. A rigorous space complexity analysis remains to be carried out. Our experience shows that, if we consider both static memory to store the hierarchical distributed model and dynamic memory to store the intermediate computational results, the space complexity of the hierarchical method is no worse, and in many cases much better, than the non-hierarchical method.

REFERENCES

A. Aghasaryaiu, E. Fabre, A. Benveniste, R. Boubour, and C. Jard. A Petri net approach to fault detection and diagnosis in distributed systems. In *Proc. 1997 IEEE Conference on Decision and Control (CDC'97)*, pages 720–725, December 1997.

P. Baroni, G. Lamperti, P. Pogliano, and M. Zanella. Diagnosis of large active systems. *Artificial Intelligence*, 110(1):135–183, May 1999.

A. Benveniste, S. Haar, E. Fabre, and C. Jard. Distributed monitoring of concurrent and asynchronous systems. In *Proc. ATPN-Workshop on Discrete Event Systems Control*, pages 97–142, Eindhoven, The Netherlands, June 2003.

R. Debouk, S. Lafortune, and D. Teneketzis. Coordinated decentralized protocols for failure diagnosis of discrete event systems. *Discrete Event Dynamic Systems: Theory and Applications*, 10(1/2):33–86, January 2000.

Reinhard Diestel. *Graph Theory*. Springer-Verlag, New York, 2nd edition, 2000.

E. Fabre, A. Benveniste, and C. Jard. Distributed diagnosis for large discrete event dynamic systems. In *Proc. 15th IFAC World Congress*, Barcelona, Spain, July 2002.

Gianfranco Lamperti, Marina Zanella, and Paolo Pogliano. Diagnosis of active systems by automata-based reasoning techniques. *Artificial Intelligence*, 12(3):217–237, May 2000.

Yannick Pencol, Marie-Odile Cordier, and Laurence Roz. Incremental decentralized diagnosis approach for the supervision of a telecommunication network. In *Proc. 2002 IEEE Conference on Decision and Control (CDC'02)*, pages 435–440, Las Vegas, USA, December 2002.

M. Sampath, S. Lafortune, and D. Teneketzis. Active diagnosis of discrete-event systems. *IEEE Transactions on Automatic Control*, 40:908–929, July 1998.

R. Su and W.M. Wonham. Decentralized fault diagnosis for discrete-event systems. In *Proc. 2000 CISS*, pages TP1:1–6, Princeton, New Jersey, March 2000.

R. Su, W.M. Wonham, J. Kurien, and X. Koutsoukos. Distributed diagnosis for qualitative systems. In *Proc. 6th International Workshop on Discrete Event Systems (WODES'02)*, pages 169–174, Zaragoza, Spain, October 2002.

Rong Su. *Distributed Diagnosis for Discrete-Event Systems*. PhD Thesis, ECE Dept., Univ. of Toronto, 2004.

W. M. Wonham. *Notes on Control of Discrete-Event Systems: ECE 1636F/1637S 2003-2004*. Systems Control Group, Dept. ECE, University of Toronto. URL: www.control.utoronto.ca/DES.html, 2003a.

W.M. Wonham. *Software Package for Supervisory Control*. Systems Control Group, Dept. of ECE, University of Toronto. URL: www.control.utoronto.ca/DES.html, 2003b.

ELSEVIER
IFAC
PUBLICATIONS
www.elsevier.com/locate/ifac

A FORMAL ANALYSIS OF A CAR PERIPHERY SUPERVISION SYSTEM [1]

Biniam Gebremichael * Tomas Krilavičius **
Yaroslav S. Usenko **

* *Nijmegen Institute for Computing and Information Sciences*
University of Nijmegen, P.O. Box 9010, 6500 GL, Nijmegen,
The Netherlands, `biniam at cs.kun.nl`
** *Formal Methods and Tools, University of Twente,*
P.O. Box 217, 7500 AE Enschede, The Netherlands,
`[t.krilavicius,usenko] at utwente.nl`

Abstract: This paper presents a formal model of the real-time service allocation unit for the Car Periphery Supervision (CPS) system—a case study proposed by Robert Bosch GmbH in the context of the EU IST project AMETIST. The CPS system is a hybrid system, which is modeled in terms of timed automata. It is done by splitting the values of nonlinear continuous variables into £nite set of regions and over-approximating the constraints on continuous variables into clock constraints. Safety properties of the timed model have been veri£ed using UPPAAL. This is a sufficient condition for validating the corresponding safety properties of the initial hybrid system. The difference in time scale between the CPS components have also been taken care of by over-approximating the timed model using the convex-hull over-approximation feature available in UPPAAL.

Keywords: hybrid systems, real-time systems, modeling, discretization, veri£cation, automotive

1. INTRODUCTION

A number of modeling and veri£cation tools for real-time and hybrid systems have been developed. For instance, the tools based on the theory of *timed automata*, Alur and Dill (1994), such as KRONOS by Bozga et al. (1998) and UPPAAL by Bengtsson et al. (1995); and the theory of *hybrid automata*, Henzinger (1996), such as HYTECH by Henzinger et al. (1997). Recent development in devising clever computational procedures have improved the ability of veri£cation tools to handle industrial-size problems automatically. Yet many essential problems, speci£cally in the area of hybrid systems, remain unsolvable with these techniques.

Two major problems are (1) the lack of adequate abstraction concepts for modeling large systems, Alur et al. (2000), (2) the exploding consumption of computing resources by the veri£cation algorithms, Burch et al. (1992).

The present paper addresses the £rst issue and provides evidence that correct abstraction and appropriate over-approximation techniques in modeling of large systems leads to veri£able models from which one can infer properties of the original model.

This is illustrated by the real-time service allocation case study for Car Periphery Supervision (CPS) system (cf. Kowalewski and Rittel (2003)) proposed by Robert Bosch GmbH in the context of the EU IST project AMETIST. The CPS system is a hybrid system that interacts with continuous environment via a discrete controller. Moreover, CPS safety properties

[1] This work was supported by the European Community Project IST-2001-35304 AMETIST `http://ametist.cs.utwente.nl`.

are parametrized with free variables, which should be determined in order to prove the safety of the system. The CPS model presented in this paper is a timed automata based abstraction of the system constructed manually by dividing the environment to a £nite set of regions.

Related Work Verifying whether a hybrid system H satis£es a property P can turn out to be undecidable for most cases. Appropriate abstraction can extract a £nite discrete system F from H by partitioning the state space of H into £nite number of regions. The survey by Alur et al. (2000) aims in £nding a class of hybrid systems which can be abstracted into F and whose veri£cation against a property P is decidable. The survey has shown that, proving that F satis£es P is equivalent or sufficient for proving that H satis£es P.

The approach used in the present paper is similar to the one of T. A. Henzinger and P. -H. Ho (1995), where, nonlinear continuous variables of the original hybrid system are over-approximated to de£ne a time-constrained automaton, and the approximated automaton will satisfy strictly fewer safety properties.

Another problem that leads to the state-space blowup in timed automata is the time scale difference between the different components of the system. This is often the case when an embedded system interacts with its environment (cf. Iversen et al. (2000)). One way to solve such problem is to use the convex-hull over-approximation method of Halbwachs et al. (1994) when verifying invariant properties.

Outline The organization of the paper is as follows: Section 2 presents an informal description of the CPS system. Next, in Section 3, the CPS system is formally modeled in a way that the desired properties are fully preserved and the model of the system is abstracted to allow veri£cation. Section 4 presents the properties and the veri£cation results. Finally Section 5 concludes the paper and lists some directions for future work.

An abstract of this paper appeared as Gebremichael et al. (2003). The complete UPPAAL model of the CPS system is available via the AMETIST project webpage.

2. CAR PERIPHERY SUPERVISION SYSTEM

Car Periphery Supervision (CPS) refers to the functionality and technology for obtaining information about the environment of a car. Applications like parking assistance, pre-crash detection and blind spot supervision depend on CPS for the basic operation and information sharing. Short Range Radar (SRR) sensors are mounted in-front of the car, and they scan the environment for nearby objects. The data collected by the sensors is sent to the computing part of CPS known as the Electronic Control Unit (ECU). The ECU processes the data and invoked applications based on the data. The structure of CPS and its environment is depicted in Fig. 1. A detailed description of the CPS system is given in Kowalewski and Rittel (2003), Moritz (2000) and Thiel et al. (2001).

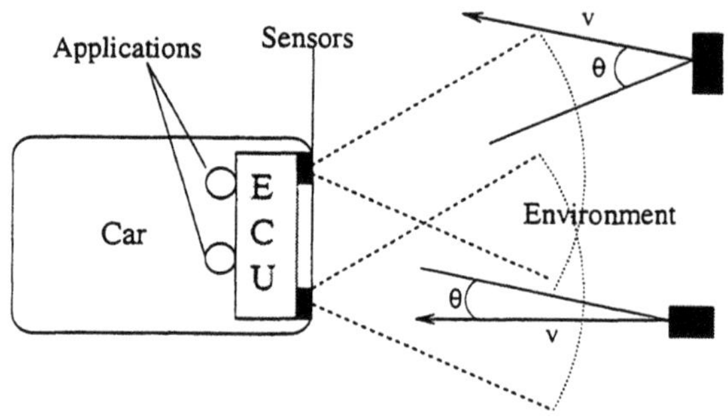

Fig. 1. The CPS system and its environment

The CPS system and its environment can be viewed as a system consisting of three parts, namely:

1. **Environment:** the environment of the CPS system is a dynamic system with moving and stationary objects in-front of the car. Object velocity and distance are important characteristics that determine the behavior of the CPS system in a continuous manner. There are several restrictive assumptions that apply to the environment in which the CPS system is supposed to operate. This is done to make the system more tractable.
2. **Sensors:** the Sensors are the interfaces through which the CPS is informed about the behavior of the environment. Sensors operate in discrete time, which can either be periodic or event-driven. The sensor component includes not only the equipment for sending and receiving radar signals but also a processor for basic data processing and control. Typically, the sensors that are situated in-front of the car return the distance to a nearest object (from their perspective) approaching the car.
3. **ECU:** the Electronic Control Unit is a board computer that performs a collection of tasks running on top of the OSEK operating system (`http://www.osek-vdx.org/`). These tasks are used to control the operation of the sensors, and to deliver accurate and on-time information about the environment to the applications.

3. FORMAL MODELING

Each of the three parts of the CPS system are operating in different modes. The environment is a dynamic and *continuously* changing system. The sensors operate on a *discrete* time scale, while ECU tasks are *real-time* tasks. An appropriate model for this system would be a hybrid automaton. However, proving correctness of a system using hybrid automata is difficult, if at all possible.

Another approach is to abstract from the unnecessary details of the model and transform it into a timed automata model, while preserving the desired properties. In this section, the CPS system is modeled in timed automata, and relevant properties are verified using UPPAAL. This model consists of six timed automata that are put in parallel. The structure of the system is shown in Fig. 2. The boxes represent the timed au-

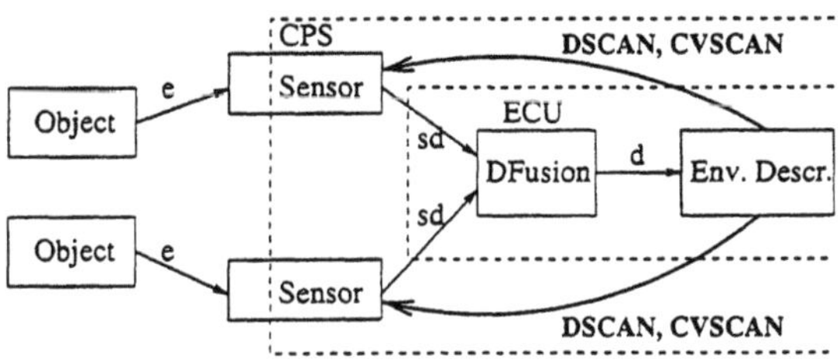

Fig. 2. Decomposition of CPS in timed automata.

tomata, the thin arrows represent the communication via shared variables, and the thick arrows represent the multiparty action synchronization.

3.1 CPS Environment

The environment is a collection of several objects moving with different velocities at different distances in-front of the car. An object i in the environment has the approaching velocity v_i and the angle θ_i relative to the direction of the car's movement (see Fig. 1).

Relative Velocity Let $\mathbb{T} = \mathbb{R}^+$ denote the time domain and let d_i denote the distance to an object i from the middle of the front of the car. Then the relative velocity of an object is defined as the function $\dot{d}_i : \mathbb{T} \rightarrow \mathbb{R}$ and it is given by the difference between the velocity of the object and the velocity of the car (v_{car}):

$$\dot{d}_i = v_i \cos(\theta_i) - v_{car}$$

The CPS system only tracks the objects that are close enough to the car. The distances to the remote objects and their velocities do not affect the behavior of the CPS system.

Regions As will be described in more detail in Section 3.2, the sensors in the CPS system scan for a nearest object, and return the distance to it. This can be given as:

$$d(t) = \min_{\forall i}(d_i(t))$$

The area in-front of the car is divided into twelve regions (see Kowalewski and Rittel (2003) for details). These regions are ordered in descending order and numbered from -1 to 9 (see Table 1).

Table 1. CPS regions

Region name	Region number (e)
FAR	-1
PreCV	0
RG_i	1...8
PreCrash	9

Assumptions The environment that can be handled by the limited capacity of the sensors and other components of the CPS is rather small. The following three criteria restrict the behavior of the environment that the CPS system can interact with.

(1) An object can approach the car with a relative velocity in the range from 13m/s to 56m/s.
(2) Only one object is present in the environment.
(3) The CPS system will be externally reinitialized when an object reaches the last region. In other words, the system terminates after the crash.

Environment behaviors outside the above assumptions have rare occurrence. Extra recovery treatments as in Kowalewski and Rittel (2003), which are not included in the current model, could be used to cope with such behaviors.

Timed Automaton of the Environment Figure 3 shows a timed automaton model of the CPS environment, which is based on the allowed relative velocity defined above. The automaton has four locations which correspond to the regions names in Table 1. Initially objects are far from the car (in location FAR and `e:=-1`). If an object comes closer to the car it will reach the location `PreCV`.

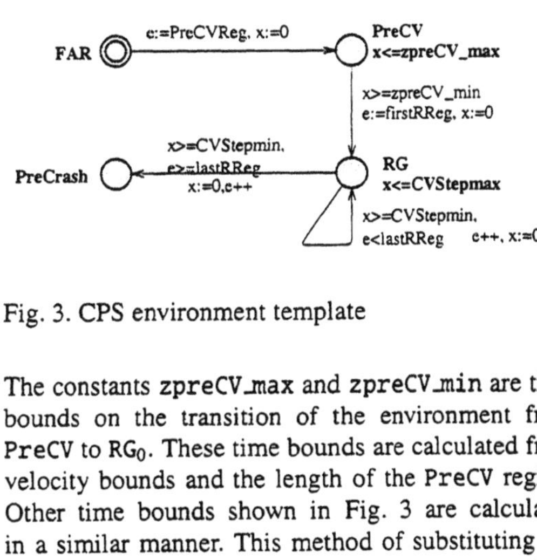

Fig. 3. CPS environment template

The constants `zpreCV_max` and `zpreCV_min` are time bounds on the transition of the environment from `PreCV` to RG_0. These time bounds are calculated from velocity bounds and the length of the `PreCV` region. Other time bounds shown in Fig. 3 are calculated in a similar manner. This method of substituting the maximal and minimal velocity constraints into clock constraints is done in accordance with the method of T. A. Henzinger and P. -H. Ho (1995).

An object has to cross all the regions up-to the `PreCrash` region. ($e \geq$ `lastRReg`) before it can go into `PreCrash`. As soon as the object is in `PreCrash`, an immediate action has to be taken by the ECU. The clock `x` is reset to measure the amount of time that the object spends in the `PreCrash` location before the

ECU does something about it. Section 4 lists important invariants that have to be satis£ed when an object reaches the PreCrash location.

3.2 Sensor

A sensor in the CPS system scans the environment for a nearest object in-front of the car, and returns the region value of the scanned object. A sensor has two modes of operation: DSCAN and CVSCAN, and one IDLE mode. These modes of operation are used as locations in the timed automaton of the sensor depicted in Fig. 4.

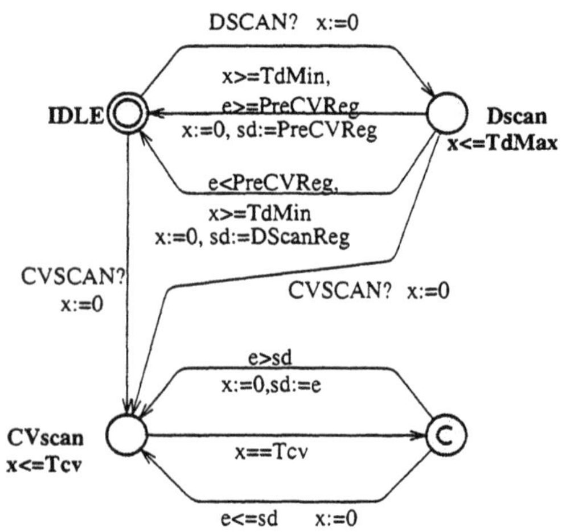

Fig. 4. Sensor Template

The sensor is initially in location IDLE and waits for a command from the ECU. If it receives a DSCAN command, it conducts a long range scan and, as soon as it £nds an object in PreCV, the sensor's distance reading sd is updated:

$$sd_i(t) = e_i(t),$$

where t is the time when the ith sensor scans the environment from its own perspective (e_i). For a CVSCAN command, the sensor produces a maximum of eight readings, one for each region RG_i. It scans the environment every Tcv time units, and sd_i is updated in a similar manner.

The CPS system usually has several sensors placed in-front of the car. In this paper only two sensors are considered. It is possible that the two sensors return different readings. This may be a result of the fact that they are situated at different positions, or the object may have an irregular form, or simply there are several objects in-front of the car. Thus, in general, there need to be no correlation between the readings of the two sensors (even though a visibility analysis shows that the difference is rather small).

3.3 Electronic Control Unit (ECU)

ECU stands for the collection of tasks running on a single processor. According to Kowalewski and Rittel (2003), the ECU executes more than two tasks. Most of these tasks are sequential and run in a predictable manner. Thus, they can easily be grouped into two tasks without affecting their behavior. These combined tasks are called DFusion (sensor fusion) and EnvDescription (environment description).

3.3.1. Sensor Fusion DFusion is a part of the ECU which receives individual distance values from all sensors and sends a "combined" single value to the remaining tasks. Combining several readings (sd_is) into one reading (d) can be done by the triangulation function as suggested by Kowalewski and Rittel (2003). The maximum function, instead of triangulation, is used here to compute the £nal value, since our model deals with one-dimensional value of the sd_is only. DFusion is a periodic task and in every jth step it computes $d(j)$ as follows:

$$d(j) = \max_{\forall i=\{1,2\}} (sd_i(j))$$

The variable $d(j)$ is a shared variable which is also readable by EnvDescription. Figure 5 shows the timed automaton of DFusion.

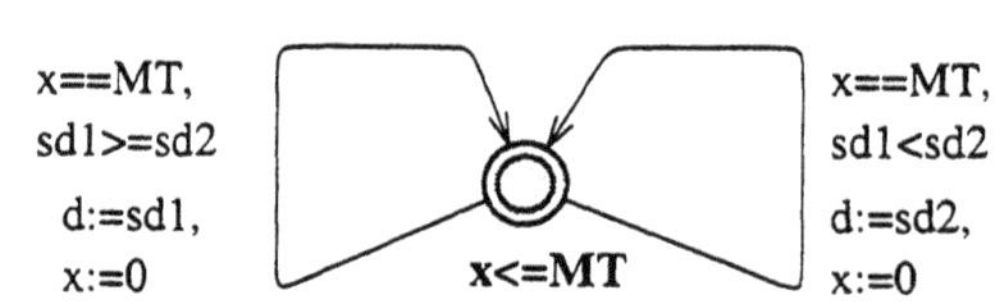

Fig. 5. Sensor Fusion

3.3.2. Environment Description EnvDescription is a part of ECU that receives environment data from DFusion and maintains accurate information about the environment. While doing so, EnvDescription controls the mode of operation of the sensors as well. The sensor-controlling part is de£ned in Kowalewski and Rittel (2003) as the "situation analysis" task. In the present paper it is combined with the EnvDescription task to avoid the state-space explosion.

The timed automaton of EnvDescription is shown in Fig. 6. Initially EnvDescription is in location IDLE, from which it periodically reads the value of d and broadcasts the DSCAN command to the sensors. If the value of d shows that an object is present in the PreCV region, then the automaton jumps to location (PreCV0), from which it sends the CVSCAN command. There is a time delay of CVComDL time units before this command happens. Once the sensors receive the CVSCAN command, the EnvDescription counts the number of RG_i regions by reading d computed by DFusion.

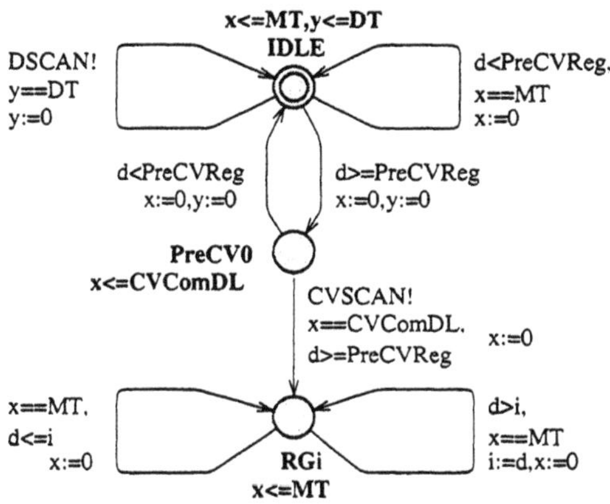

Fig. 6. Environment Description (ED)

4. VERIFICATION

4.1 Requirements

The primary goal of the CPS system is to provide accurate information about the environment of the car to the applications such as airbag in¤ation, parking assistance, pre-crash detection and others. The accuracy of the information provided by ECU is measured by the time delay between a change in the environment and the knowledge of the ECU about this change. The properties are stated below as parametrized temporal logic formulas, and they were veri£ed using UPPAAL on a workstation with a 512MHz CPU and 256Mb of RAM. The time required to verify these properties was drastically reduced using the convex-hull approximation feature of UPPAAL to less than a minute.

Property 1 (P1): EnvDescription has an accurate information about the object in the collision course. This property is modeled in the following way: the EnvDescription's information `ED.i` about the region of the object should not deviate too much from the environment's information `e1` about the same object. The goal is to £nd the maximal difference Q between these two values (the minimal Q for which the formula is satis£ed).

```
A[] (e1-ED.i <= Q and e2-ED.i <= Q)
```

Property 2 (P2): When an object reaches the pre-crash region (the environment automaton is in location `ENV1.PreCrash`), EnvDescription knows about this (`ED.i == lastRReg+1`) within a few time units (`ENV1.x > P`). Here `ENV1.x` represents the time after the environment automaton moved into location `ENV1.PreCrash`, and we are interested in £nding the minimal value of P.

```
A[] ((ENV1.PreCrash and ENV1.x > P)
     imply (ED.i == lastRReg+1))
```

Property 3 (P3): The ECU avoids false alarm. EnvDescription never reports advancement of an object toward the car before the object (the environment) actually does so.

```
A[]  (ED.i <= e1 or ED.i <= e2)
```

Property 4 (P4): The system is deadlock free.

```
A[] (not deadlock)
```

4.2 Results

The CPS model satis£es properties P1 for $Q \geq 3$, P2 for $P \geq 5$, P3 and P4. Note that Q can also be computed from P as $Q = \lceil P/CVStepmin \rceil$, since the difference in the number of regions can also be expressed as difference in time.

In the above model DFusion and EnvDescription run in arbitrary order, no prior scheduling is assumed. It is possible, however, to schedule the execution in such a way that, DFusion computes d and EnvDescription updates its counter immediately. Under such synchronization EnvDescription will update its information faster. Figure 7 shows the model of EnvDescription using this alternative schedule. The model of the new DFusion is not shown here, but it is similar to the one in Fig. 5 except that the new model sends a synchronization signal (`ND!`) to the EnvDescription as soon as a new value of d is computed. After the synchronization with DFusion, EnvDescription makes its decision without a delay (see the urgent locations in Fig. 7). For this setting the properties P1 for $Q \geq 2$ and P2 for $P \geq 3$ are satis£ed.

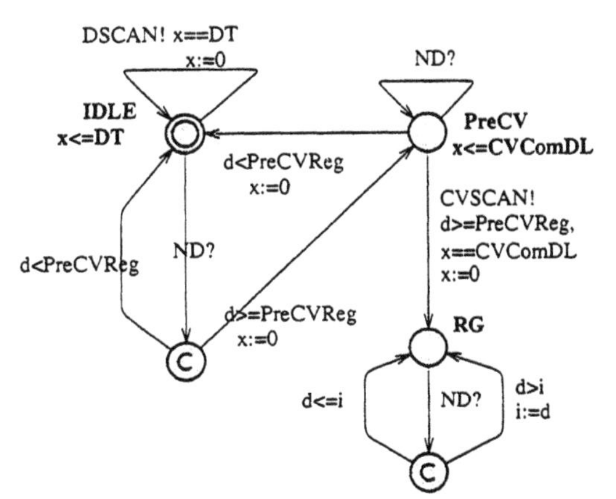

Fig. 7. Modi£ed Environment Description

In both cases P is equal to the time needed for an information to propagate from the environment to EnvDescription. That is the sum of the time spent by the sensor (Tcv), DFusion and EnvDescription.

$$P = Tcv + MT + MT$$

But in the scheduled model, both EnvDescription and DFusion need only one MT to update their information, and propagation time is reduced to $P = Tcv+MT$. These two cases show the worst and best cases for determining the value of P. In general, however, P is equal to $Tcv + MT$ plus the overhead associated with scheduling of the ECU tasks and the time spent by each task. When OSEK operating system, is used to schedule the ECU tasks, and if $OSEK_{SchTime}(T_i)$ is the time delay due to scheduling and running the tasks T_i, then the value of P is

$$P = Tcv + MT + OSEK_{SchTime}(T_i).$$

5. CONCLUSION AND FUTURE WORK

The car periphery supervision is a hybrid system. Verifying properties for hybrid systems is undecidable in general. However, the continuous variables of the model, the environment in this case, can be discretized to a £nite block of regions in order to make veri£cation of the properties possible.

The different time scale between the environment and CPS components have resulted in a state-space blow up. The convex-hull over-approximation technique of UPPAAL was used to verify the safety properties of the system. Another approach could be the exact acceleration method of Hendriks and Larsen (2002).

The assumptions made on the environment of CPS are too restrictive. Allowing only one object and having non decreasing e in the regions can be omitted by introducing a recovery mode in case when more than one object appears in the RG regions. As described in Kowalewski and Rittel (2003), the recovery mode is a third mode of sensor operation, which scans for follow-up objects when a nearest object disappears from the scene. This scenario may happen when one object in the collision course changes its trajectory and disappears; and later on, another object, which was close to the £rst one, enters a collision course. Adding recovery method to the model for such a scenario would be an interesting future work.

Acknowledgments

The authors would like to thank (in the reverse-alphabetic order) Frits Vaandrager, Stefan Kowalewski Holger Hermanns, Marko Auerswald and the members of the Sensor Group at Robert Bosch GmbH for their valuable comments and reviews.

REFERENCES

R. Alur and D.L. Dill. A theory of timed automata. *Theoretical Computer Science*, 126:183–235, 1994.

R. Alur, T.A. Henzinger, G. Lafferriere, and G. J. Pappas. Discrete abstractions of hybrid systems. In *Proceedings of the IEEE*, volume 88, 2000.

J. Bengtsson, K.G. Larsen, F. Larsson, P. Pettersson, and W. Yi. UPPAAL — a Tool Suite for Automatic Veri£cation of Real–Time Systems. In *Proc. of Workshop on Veri£cation and Control of Hybrid Systems III*, pages 232–243. Springer–Verlag, 1995.

M. Bozga, C. Daws, O. Maler, A. Olivero, S. Tripakis, and S. Yovine. Kronos: A model-checking tool for real-time systems. In A. J. Hu and M. Y. Vardi, editors, *Proc. 10th International Conference on Computer Aided Veri£cation, Vancouver, Canada*, volume 1427, pages 546–550. Springer-Verlag, 1998.

J.R. Burch, E.M. Clarke, K.L. McMillan, D.L. Dill, and L.J. Hwang. Symbolic model checking 10^{20} states and beyond. *I&C*, 98:142–170, June 1992.

B. Gebremichael, H. Hermanns, T. Krilavicius, and Y.S. Usenko. Hybrid modeling of a vehicle surveillance system with real-time data processing. In *Proceeding International Conference on Dynamical Systems Modeling and Stability Investigation*, page 419. 2003.

N. Halbwachs, Y. Proy, and P. Raymond. Veri£cation of linear hybrid systems by means of convex approximations. In *Static Analysis Symposium LICS*, pages 223–237. Springer Verlag, 1994.

M. Hendriks and K. G. Larsen. Exact acceleration of real-time model checking. In E. Asarin, O. Maler, and S. Yovine, editors, *Electronic Notes in Theoretical Computer Science*, volume 65. Elsevier Science Publishers, April 2002.

T.A. Henzinger. The theory of hybrid automata. In *Proceedings of the 11th Annual IEEE Symposium on Logic in Computer Science (LICS '96)*, pages 278–292, New Brunswick, New Jersey, 1996.

T.A. Henzinger, Pei-Hsin Ho, and Howard Wong-Toi. HYTECH: A model checker for hybrid systems. *International Journal on Software Tools for Technology Transfer*, 1:110–122, 1997.

T.K. Iversen, K.J. Kristoffersen, K.G. Larsen, M. Laursen, R.G. Madsen, S.K. Mortensen, P. Pettersson, and C.B. Thomasen. Model-checking real-time control programs — Verifying LEGO mindstorms systems using Uppaal. In *IEEE Euromicro Conference on Real-Time Systems*, pages 147–155, 2000.

S. Kowalewski and M. Rittel. IST Project AMETIST: real-time allocation for car periphery supervision. Technical report, Robert Bosch GmbH, 2003. Preliminary Description (Deliv. No. 3.1.3) see `http://ametist.cs.utwente.nl/`.

R. Moritz. Pre-crash sensing - functional. Technical report, Robert Bosch GmbH, 2000.

T.A. Henzinger and P. -H. Ho. Algorithmic analysis of nonlinear hybrid systems. In P. Wolper, editor, *Proceedings of the 7th International Conference On Computer Aided Veri£cation LICS*, volume 939, pages 225–238. Springer Verlag, 1995.

Steffan Thiel, Stefan Ferber, Thomas Fischer, Andreas Hein, and Michael Schlick. A case study in applying a product line approach for car periphery supervision system. Technical report, Robert Bosch GmbH, 2001.

ELSEVIER
IFAC
PUBLICATIONS
www.elsevier.com/locate/ifac

AN APPROACH TO ON-LINE BATCH SEQUENCING

Vladimir Jovan, Boštjan Hauptman

Department of Systems and Control
Jožef Stefan Institute, Jamova 39
1001 Ljubljana, Slovenia

Abstract: The paper presents an on-line algorithm that has been developed with the intention of predicting the adequate starting time for a single new batch in a system consisting of two chemical reactors sharing a common discharge vessel. The validation of the achieved results, derived from the algorithm testing, show that the proposed algorithm can significantly contribute to more stable production and can efficiently deal with the "no-wait" constraints and the non-uniform batch durations of the dedicated process.

Keywords: computer integrated manufacturing, production management, sequencing, production scheduling

1. INTRODUCTION

Sequencing and scheduling are forms of decision-making that play an essential role in manufacturing. Over the last three decades numerous approaches to the use of scheduling principles and methodologies in automatic control of industrial processes have been tried. Despite this, production scheduling often remains a manual process performed by experienced and skilled operators. In practice, there are many examples of industrial settings in which resource constraints exist, and there may be limits on the amount of time that a job can spend in a buffer between two consecutive operations. Often there is no intermediate buffer between two processing units and a blocking can occur. Sometimes the nature of a technological process itself demands that there should be no slack time between two subsequent operations and no-wait constraints should be taken into account. Until now, the literature has paid surprisingly little attention to flow shop production with blocking, no-wait constraints and uncertainty in the duration of a product cycle. Even less is known about how a scheduling algorithm should respond to deviations in production process, where the response should be on-line and should provide a feasible solution according to the actual situation on the factory floor. Mathematical based algorithms of such scheduling problems seem to be inappropriate for on-line usage (Raaymakers and Hoogeveen, 2000; Pinedo, 1995) due to NP-hardness. On the other hand, the expectations and needs of industrial partners are high, as the trend towards fully automated industrial production systems requires the efficient substitution of experienced operators.

The paper is organised into three sections: the first describes the batch process of primary and secondary gel washing, where a batch scheduling algorithm is required. The second section describes the concrete scheduling/control problem in the process under consideration and presents conclusions that should be taken into account during the design phase of a dedicated algorithm for automatic batch launching. The final section deals with the proposition for a batch-launching algorithm and its validation on the selected process.

2. DESCRIPTION OF THE TECHNOLOGICAL PROCESS

A production line for two-stage gel washing (Fig. 1) was chosen as a case study. The gel is washed first in one of four candle filters (Filter 1.2A and Filter 1.2B – Line A, Filter 1.2C and Filter 1.2D – Line B) on the primary washing and then additionally in one of three filters on the secondary washing. The procedure of washing is essentially of a batch nature and consists of a number of successive steps/operations (1-

charging with suspension, 2-idle, 3-formation of solid cake, 4-discharge of water, 5-cleaning of filter, 6-filling with water, 7-washing of cakes, 8-discharge of water, 9-drying of cakes, 10-decompression, 11-discharging) performed in pressurised candle filters.

There are some shared resources (dosing pumps, spraying pumps, compressed air) in the technological line that could affect the line's capacity, but a preliminary simulation of the line performance showed that the only shared resources which could really reduce the line's capacity were the discharge vessels 1.3 A and 1.3 B in first-stage washing (Jovan and Hauptman, 2003). Namely, the volume of these vessels is nearly the same as the volume of an individual filter, and only the content of one filter can be discharged at once into the vessel. The main problems regarding individual batch launching in the single filter are as follows:

- Due to technological requirements, a single batch that has been launched must be completed without any interruptions. This implies the "no-wait" constraints between all the batch steps.
- Because of the layout of the technological process, after being processed in a filter the suspension must be displaced into a common discharge vessel. This implies that the discharge vessel be empty at the beginning of the batch step "discharging", otherwise a "blocking" constraint occurs.
- The actual duration times of individual steps of a batch vary in practice and this uncertainty should be taken into account.

The actual batch duration varies to a certain extent, mainly due to the non-uniform quality of the input suspension and the current obturation of candle meshes, and for other indeterminable reasons. In practice, the typical batch duration time on a single filter is 70 +/- 7 mins. The batch steps that contribute the most to the dispersion of the whole batch duration are mostly "formation of solid cake" and "washing of cakes" (for example, see Figure 2).

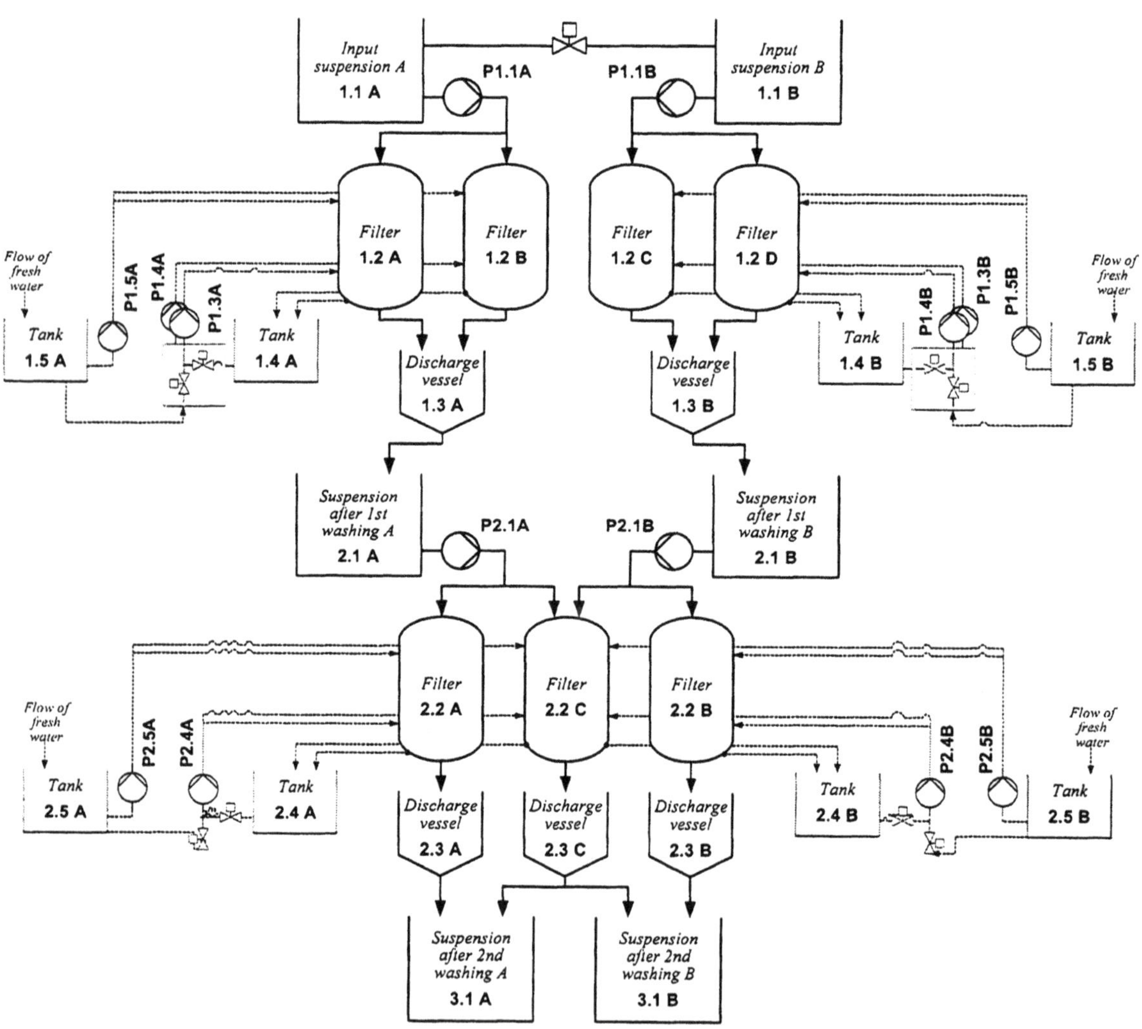

Fig. 1. Scheme of two-stage gel washing process

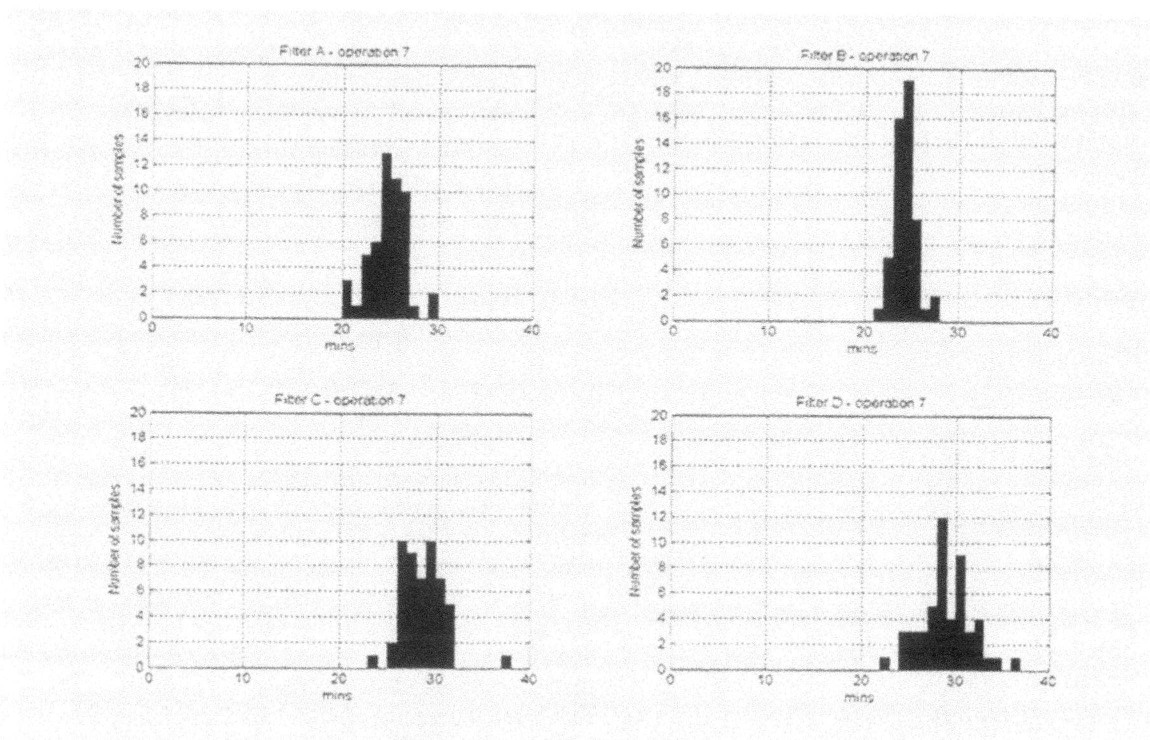

Fig. 2. Dispersion of the batch step "washing of cake" duration in the set of 50 samples for Filters A, B, C, D

The dispersion of actual processing times in an individual filter is the first reason for non-uniformity of material flow through the gel-washing sub-process. The second reason for production instability lies in the operator's decision to start an individual new batch. Once started, the batch has to be processed in a filter and then immediately discharged into the common discharge vessel. This means that too-early initiation of a new batch leads to violation of the "no-wait" constraint (waiting results in a degradation of output material), while the too-late start of a new batch results in a reduction of the prescribed capacity.

Both the uncertainty of the subsequent batch durations and manual launching of batches together contribute to the overall instability of production that can result in violation of the prescribed technological constraints, product quality degradation or can lead to the reduction of capacity of the gel washing process. For this reason a simple, reliable algorithm to initiate the batches automatically, has been developed.

3. A BATCH LAUNCHING ALGORITHM

The proposed algorithm for calculating the most appropriate starting time for a single new batch periodically monitors the current state of the process under observation. It is active when both two filters are empty and ready for filling, or when one of the filters is busy and the other one has been discharged and is now idle. When such a case occurs, a current status is assigned to each filter, which can be either "running" or "waiting" (that is, $F_A \rightarrow F_r, F_B \rightarrow F_W$ or $F_B \rightarrow F_r, F_A \rightarrow F_W$).

After passing through its *initialisation part*, the algorithm in the *prediction part* calculates the first approximation for the start of a new batch on an idle filter (time to start, t_{wtts}). The calculation takes into account the real duration of all steps that have already been completed and the average duration times for all steps that have not yet been processed. The average duration for all steps, that have not yet been processed on the F_r, are calculated as a moving average value from the last five actual duration times of a particular step. If calculated $t_{wtts} \leq 0$, the new batch on F_w begins immediately.

In the case that F_w is not eligible to start a new batch immediately (time to start, $t_{wtts} > 0$), the algorithm (in its *correction part*) periodically monitors the state of the F_r, and if a new batch step on this filter has been completed, then its average duration time is replaced with its actual duration and a new calculation for the start of a new batch on the idle filter is performed. Note that this calculation becomes more and more based on actual duration times (and so more realistic since it reflects the actual state of the observed process). The algorithm alternates between its prediction and correction parts until the condition for the start of a new batch $t_{wtts} \leq 0$ is fulfilled. The basic steps of the algorithm are given below.

Initialisation part

1. determine $\overline{T}_r, \overline{T}_w$, where $\overline{T} = \sum_{i=1}^{n} (\overline{t})_i$
2. determine $\overline{T}_{rDV}, \overline{T}_{wDV}$
3. determine t_r, k
4. determine $T_r' = \sum_{i=1}^{k} (t_r)_i + \sum_{i=k+1}^{n} (\overline{t}_r)_i$
5. determine $T_0 = 24*60/C_0$ (in mins)

Correction part

6. if new batch step on F_r has been completed, then $k:=k+1$, and recalculate T_r'
7. update t_r
8. calculate $T = \max\left\{ \begin{matrix} T_0 \\ T_r' \\ \overline{T}_w \\ \overline{T}_{rDV} + \overline{T}_{wDV} \end{matrix} \right\}$

Prediction part

9. determine $t_{wtts} = \frac{T}{2} + \left(T_r' - \overline{T}_w\right) + \frac{\overline{T}_{rDV} - \overline{T}_{wDV}}{2} - t_r$,

where:

F_r the filter where the batch is running

F_w the idle filter, waiting for a new batch launch

C_0 the desired number of batches/day

T_0 expected time interval for a single batch launch on an individual filter

T estimation of real time interval for a single batch launch on an individual filter

$\overline{T}_r$ average batch duration time on the active filter F_r

$\overline{T}_w$ average batch duration time on the idle filter F_w

$\overline{T}_{rDV}$ average processing duration time in the discharge vessel in the event that suspension comes from filter F_r

$\overline{T}_{wDV}$ average processing duration time in the discharge vessel in the event that suspension comes from filter F_w

t_r actual batch duration on F_r from the start to the moment of observation

k number of a batch steps on F_r completed until the time t_r

T_r' batch duration time on the active filter F_r, where the average duration times of all completed steps are replaced with the real duration times of these steps

t_{wtts} prediction of the remaining waiting time for a new batch launching on F_w

n number of a batch steps on a filter

$(t_r)_i$ actual duration of a completed batch step i on the filter F_r

$(\overline{t}_r)_i$ average duration of a batch step i on the F_r, calculated from the last five duration times of this step, $\overline{t} = \frac{1}{5}\sum_{j=1}^{5} t_j$.

The following example can be taken in order to explain the functioning of the algorithm: the desirable capacity is 18 batches per filter, the batch takes by average 70 minutes in the first filter and 60 minutes in the second, and the procedure in the discharge vessel lasts 30 minutes for the first filter and 20 minutes for the second. In order to achieve the desired capacity, batches must be initiated in the filter every 80 minutes (24×60 minutes per day for 18 batches per day). The actual batch initiation period in the same filter is defined as the maximum value from among the desired period, the duration of the batch in the first filter, the duration of the batch in the second filter, and the total duration of the assignment of the discharge vessel to the first and second filters. Because in the given example the desired batch initiation period is longer than the length of the batch in the first or second filters and is also longer than the total duration of the procedures in the discharge vessel for the batch from the first and second filters, the desired capacity can be achieved.

In each filter a batch will therefore be initiated every 80 minutes, and only coordination of the initiation between the two filters is further required. If the batches in the two filters were of the same length, and also the processing in the discharge vessels, the batches in the filters would be alternately initiated every 40 minutes. When the durations are not equal, it is necessary to adjust the half-period by the difference in the duration of the batches in the two filters, i.e. by 70–60 = 10 minutes, and by half of the difference in the durations of the procedures in the discharge vessel for the batches from the first and second filters, i.e. by (30–20)/2 = 5 minutes. If in this example a batch is initiated in the second filter 55 minutes after the initiation of the batch in the first filter, and a batch is initiated in the first filter 25 minutes after the batch is initiated in the second filter, the discharge vessel will be equally occupied (see Fig. 3). Given uniform occupancy of the discharge vessel, which is the sole critical joint resource, the likelihood of successful execution of the batch in the case of equal deviations in the duration of steps is equal for the both two filters.

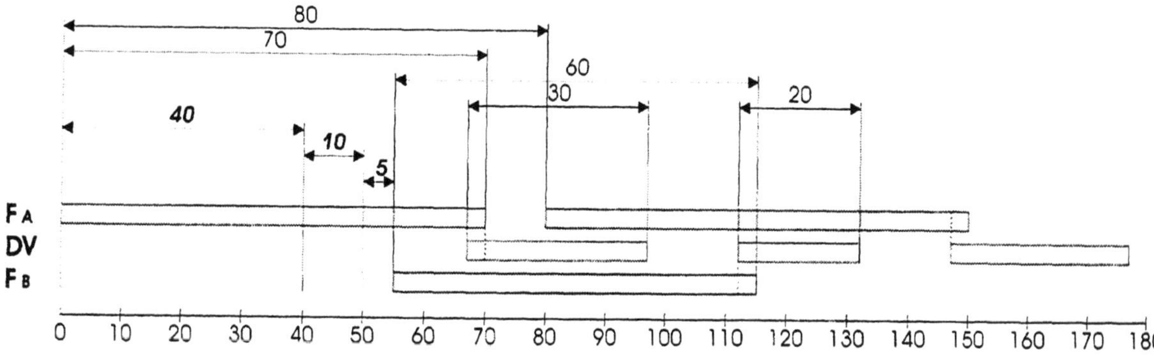

Fig. 3 – Principle of defining moment of batch initiation

4. TESTING OF AN ALGORITHM ON THE CASE-STUDY PROCESS

The algorithm has been implemented as a module of the existing process control system that is based on Siemens programmable logic controllers and uses the PCS7 programming environment.

The aim of the first test was to validate the functioning of the algorithm itself (Jovan and Hauptman, 2003a). The algorithm was installed on Line A of the gel washing sub-process in order to get a comparison among batch starting times performed by the algorithm and manual starting of the new batches performed by an operator on Line B. The current production was moderate so the reference capacity of the algorithm was set to 12 batches/day/filter. Figure 4 shows the results of the test (from 14:00 till 22:00).

From Figure 4 we notice that the occupancy of the shared discharge vessel (during the period from 14:00 till 22:00) is much more uniform on Line A, where automatic batch launching was performed, compared to Line B where the batches were launched manually. The histograms on Figure 5 show that the uniformity of batch launching is much higher in the case of automatic launching on Line A (the dispersion of the period of single batch start times is smaller).

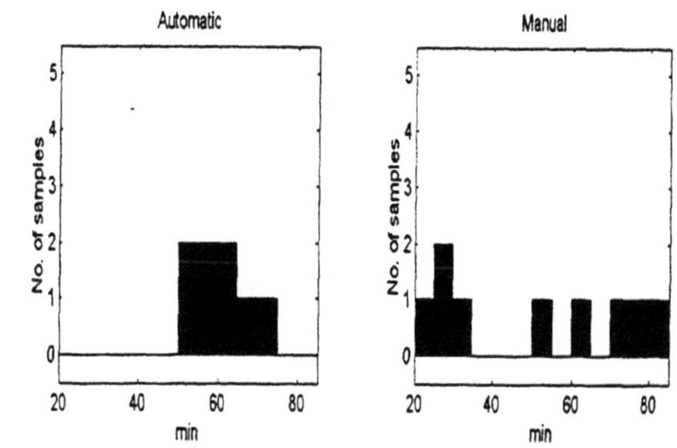

Fig. 5 Histograms of the dispersion of batch launching times for Line A (automatic) and Line B

The second test run of the algorithm was devoted to an investigation of its behaviour in the case where the production capacity of the gel washing sub-process was close to maximum. The automatic batch-launching algorithm was running on Line A and manual starts of the new batches were performed by an operator on Line B. As seen from Figure 6 the Filter 1.2A on Line A dictates the period of a single batch launching since it's batch durations are longer than those of Filter 1.2B. The algorithm tends to start a new batch on Filter 1.2A as soon the initial conditions for batch launching on Filter 1.2A are fulfilled and simultaneously adjusts the moment of starting a new batch on Filter 1.2B in such a way that the occupancy of the common discharge vessel tends to be uniform.

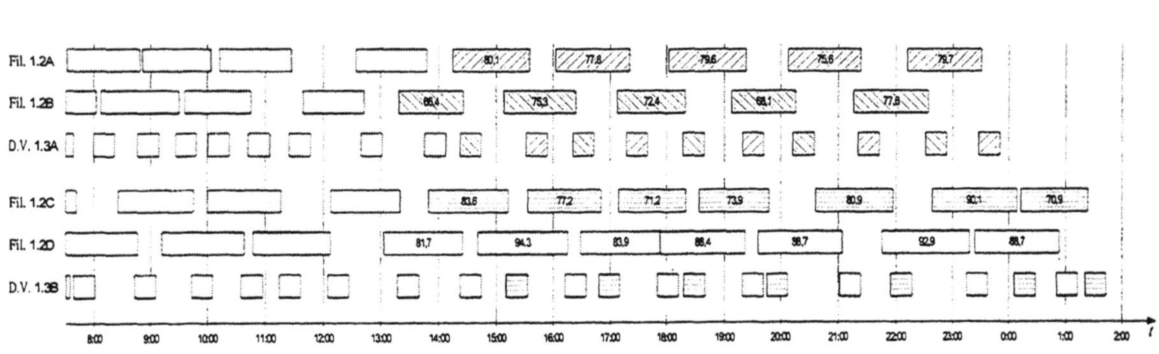

Fig. 4. Results of algorithm testing – experiment No. 1

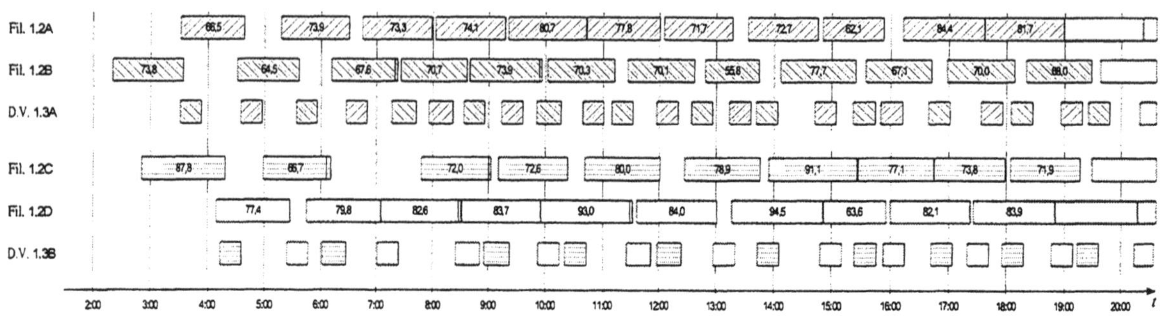

Fig. 6. Results of algorithm testing – experiment No. 2

The histograms in Figure 7 show that the dispersion of batch launching times is smaller in the case of automatic launching on Line A, in spite of the fact that the production rate on Line A was higher than on Line B. Two peaks appearing in the histograms on Figure 7 are the results of the changes in the reference production rate (from 18 batches/filter/day to 15 batches/filter/day) during the experiment.

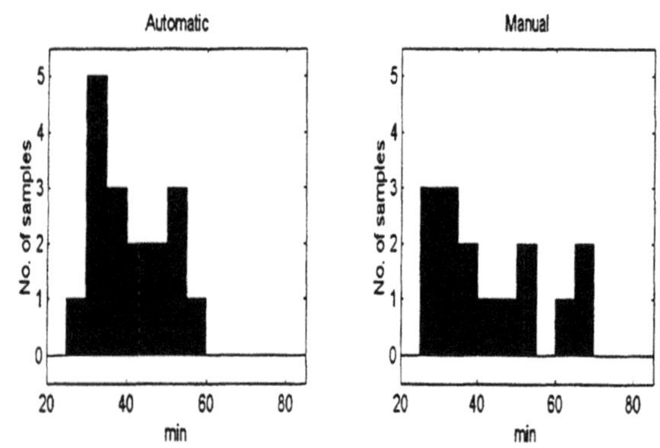

Fig. 7. Histograms of the dispersion of batch launching times for Line A (automatic) and Line B

5. CONCLUSIONS

The aim of the project was to define the dedicated algorithm that is capable to predict the most appropriate starting time for a single new batch in the two-stage gel washing process so that "no-wait" constraints imposed by the nature of the process are taken into account, "blocking" of any of batch operations due to shared resources must not occur and the material flow of the process is more uniform.

The proposed and tested algorithm continually adjusts predictions based on the current state, and correspondingly activates a new batch at the earliest possible starting time, simultaneously respecting all predefined constraints and the need for uniform and stable production. The algorithm was designed for the dedicated process; however, with minor modifications it is also applicable in other similar process environments (a system consisting of two reactors sharing a common resource, e.g. a reservoir, is quite common).

The validation of the achieved results, derived from the proposed batch algorithm testing of the real process show that the algorithm can significantly contribute to more stable production and can efficiently deal with the "no-wait" constraints and the non-uniform batch durations of the real case study process.

6. ACKNOWLEDGEMENTS

This work was supported in part by the 5th RTD Framework Programme under Contract G1RD-CT-2000-00318, "VIP-NET-Virtual Plant-Wide Management and Optimisation of Responsive Manufacturing Networks".

REFERENCES

Dilworth, J.B. (1996). Operations Management, second edition. *The McGraw-Hill Companies Inc.*, New York.

Jovan, V. and B. Hauptman, 2003, Functional specifications for a scheduling algorithm in a process oriented factory, *Workpackage No. 9's Deliverable No: D26 of the 5th RTD Framework Programme under Contract G1RD-CT-2000-00318, "VIP-NET - Virtual Plant-Wide Management and Optimisation of Responsive Manufacturing Networks"*

Jovan, V. and B. Hauptman, 2003a, Tested and validated scheduler for the case study, *Workpackage No. 9's Deliverable No: D27 of the 5th RTD Framework Programme under Contract G1RD-CT-2000-00318, "VIP-NET - Virtual Plant-Wide Management and Optimisation of Responsive Manufacturing Networks"*

Pinedo, M. (1995). *Scheduling: Theory, Algorithms, and Systems. Prentice Hall Inc.*, New Jersey (USA).

Raaymakers, W.H.M. and J.A. Hoogeveen (2000). Scheduling multipurpose batch process industries with no-wait restrictions by simulated annealing. *European Journal of Operational Research*, Nr. 126, pp. 131-15

ELSEVIER
IFAC
PUBLICATIONS
www.elsevier.com/locate/ifac

SEQUENCE ANALYSIS FOR TIME PETRI NETS

Valter Schastai, Evangivaldo A. Lima
Luis Allan Künzle

CPGEI/CEFET-PR, Av. 7 de setembro, 3165, 80230-901, Curitiba, Brasil

Abstract: In this paper is proposed a new methodology to analyse Time Petri nets. This new approach is based on the class graph and it allows temporal analysis of any valid sequence of transition firings. For that, it is proposed a procedure that find valid dates for all events over a transition firing sequence present in the graph. These valid dates are presented by means of an inequalities system, that is constructed following the sequence path in the class graph. Then, it is possible to calculate, for instance, the maximal time interval between two events in the sequence, as well as the latter date of a given event. *Copyright © 2004 IFAC*

Keywords: Time Petri Nets, Class Graph, Process, Concurrency, Time Analysis

1. INTRODUCTION

Since was introduced by Carl Adam Petri in firstly 60 (Petri, 1962), Petri Nets are used for modelling and analysis mainly concurrent and parallel systems. Addition, its graphic power and algebraic properties are used to describe dynamic behavior of systems. There exist a large research using different types of Petri Nets (Murata, 1989). When time constraints are considered, a important extension for ordinary Petri Net is Time Petri Nets (Ramchandani, 1974; Merlin, 1974; Sifakis, 1977).

To analyse Time Petri Nets two kinds of approaches must be considered: global approach analysis and behavior traces. The global approach analysis represent all reachable states and allow to extract properties such as deadlock, conflict, and boundedness. Analysis of behaviors traces are based on a specific scenario and avoid to represent the full state space. For instance, to calculate the largest temporal interval between two tasks and the latest date for starting a given task. In this paper we propose a new approach for temporal analysis of Time Petri Nets, combining global and behavior trace analysis. A procedure is presented to build valid dates for a sequence of transition firings using the class graph (Schastai and Künzle, 2002; Schastai, 2003).

For that, section 2 presents a formal definition for Time Petri Nets. In the section 3 two analysis methods for Time Petri Nets are discussed: the Class Graph proposed by Menasche and the Time Process Method proposed by Aura. In the section 4 we consider un accurate method to calculate valid dates for a sequence of transitions firings in a Time Petri Net using the class graph. An example of the proposed approach is showed in section 5. At last, the conclusion argues some limitations that still remain in the proposed method.

2. TIME PETRI NETS

Time Petri Nets, proposed by Merlin (Merlin, 1974), extends the conventional Petri Nets. Time Petri Nets are applied to describe systems in which time information is important. In a Time Petri Net each transition is associated with a time interval $(\theta_{min}, \theta_{max})$. θ_{min} represent the minimal time before firing a enabled transition, and θ_{max} represent the maximal time for firing a enabled transition. The transition firing does not take time and must be occurred in this interval.

Formally, Time Petri Nets is a tuple $NT = (P, T, Pre, Post, M_0, I)$, where P is a set of places, T is a set of transitions, Pre is an input application $Pre : (P \times T) \rightarrow \mathbf{N}$, $Post$ is an output application $Post : (T \times P) \rightarrow \mathbf{N}$, M_0 is the initial marking and $I : T \rightarrow (\mathbf{Q}^+ \times (\mathbf{Q}^+ \bigcup\{\infty\})) \mid I(t) = [a, b]$. I assigns a firing interval $I(t) = [a, b]$ for each transition t of the net. a is denominate static earliest firing time and b static latest firing time for each transition (Berthomieu and Menasche, 1982).

3. ANALYSIS METHODS FOR TIME PETRI NETS

There exists several techniques to analyse Time Petri Nets. The class graph (Menasche, 1982; Menasche, 1996) is the most important for a global analysis of modeled systems using Time Petri Nets. To analyse a specific scenario, (Aura, 1996; Aura and Lilius, 1997) proposes a interesting method, based on time processes. By means of the class graph is possible to describe the global behavior of safe nets. The analysis by means of processes is restricted for a valid sequence of events in the net. This method is also restricted for safe Petri nets. Other approaches allow temporal analysis of Time Petri Nets. In (Popova-Zeugmann, 1991) these nets are analyzed by the behavior "snapshots". In opposition to the class graph, time representation is not shown by intervals, but using a time sensitivity vector for each transition. The Popova method gives equivalents results to the ones of Menasche method.

3.1 Class Graph

The method presented in this section defines a finite space states representation of a Time Petri Net (Menasche, 1982). The class of states concept proposed by Menasche allows analysis by enumeration such that used in Conventional Petri Nets. The class graph is constructed observing the behavior in the time net.

A class of states is defined as the union of all possible firing dates for a given marking in the net. It is represented by a pair $C = (M, D)$ where M is the marking and D is a firing domain which describes all firing dates of the transitions enabled by marking M. D is defined as the solution of a system of inequalities in which each enabled transition at M has a time variable associated. For instance, let t_i and t_j two enabled transitions. Three kinds of inequalities can compose a domain D: $\alpha_i \leq t_i \leq \beta_i$, $\alpha_j \leq t_j \leq \beta_j$ and $t_i - t_j \leq \gamma_{ij}$ (Berthomieu and Menasche, 1982).

The conditions for firing transitions in a class follows the behavior of time model. For instance, firing t_k in $C = (M, D)$, leads a new class $C' = (M', D')$. M' is obtained as in conventional Petri Nets. The domain D is modified taking in account the possible firing dates for t_k. Transitions which remain enabled the time relations are keep. Also, new time intervals are added for transitions recently enabled in M' and the intervals of the transitions disabled by firing t_k are excluded from D.

For instance, in figure 1 is shown a Time Petri Net and its class graph. In the initial class C_0 both places p_2 and p_3 have one token each and all other places does not have tokens (marking [01100]). In this class transitions t_2 and t_3 are enabled. Both transitions can fire in the interval $[1, 2]$ (domain $1 \leq t_2 \leq 2$ and $1 \leq t_3 \leq 2$). Firing t_2 a new class C_1 is reached. C_1 is composed by marking [000110] and domain $0 \leq t_4 \leq 1$, $0 \leq t_3 \leq 1$. In the class C_1, t_3 remain enabled but its new firing interval is $[0, 1]$. As t_4 is a new enabled transition on C_1 class, its firing interval defined in the net model is added to class domain. All others classes are constructed following the procedure defined by Menasche method.

The analysis of Time Petri Nets for the methodology proposed by Menasche was implemented in a tool called PAREDE (Menasche, 1985). This tool run in MSDOS environment and has a frendly interface. It permits to construct the class graph of a time petri net. For bounded nets is possible to find some net properties such as liveness and reachability. In (Berthomieu *et al.*, 2003) is presented a tool to analyse Time Petri Nets – TINA. With TINA is also possible to construct the class graph of a Time Petri net using the method described above. TINA allows reachability analysis and verification of Linear Temporal Logic formulas using the class graph. As PAREDE, TINA is restricted to qualitative analysis of the net.

A great result from the Menasch's method is to identify all the classes of states and, for each class, all valid interval dates for each enabled transition. In this manner, all possible sequences are present in the paths of the class graph. However, the class graph is not intended to perform temporal analysis of transitions firings sequences. For instance, in the net of figure 1, to get class C_5 from initial class C_0, we have three paths in the class graph: $t_2t_3t_4$, $t_2t_4t_3$, and $t_3t_2t_4$. A temporal analysis of any of them gives a time duration of $[1, 4]$ time units. But, in the net, firing t_3 is concurrent to firing sequence t_2t_4. Then, we can get class C_5

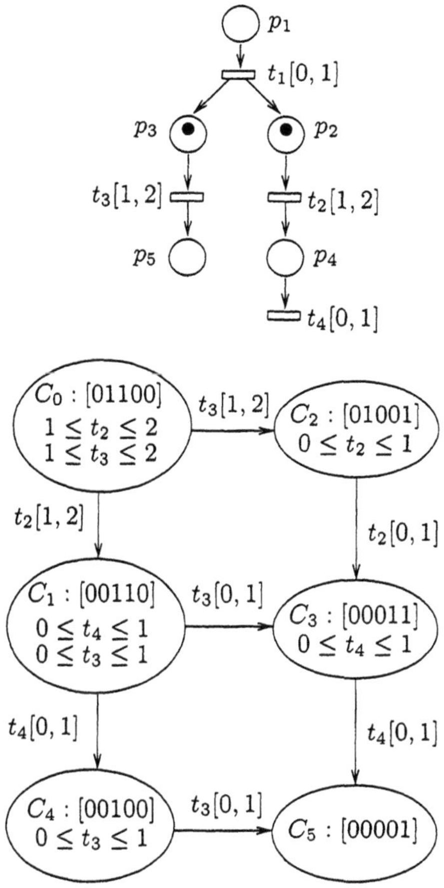

Fig. 1. A Time Petri Nets and its Class Graph

after $[1,3]$ time units from C_0. For this reason, we can not perform temporal analysis of transitions firings sequences using only the class graph, since it represents concurrence by interleaving. This implies a increase of imprecision when analyzing the time duration of a sequence of transitions firings, in which some transitions can be fired concurrently. So, the class graph it is not adjusted to do temporal analysis of net sequences.

3.2 Time Processes

The goal of the method proposed by Aura (Aura, 1996; Aura and Lilius, 1997) is to analyse time processes of a Time Petri Net, which are represented through a partial order semantics. In Conventional Petri Nets a partial order is established by processes (Reisig, 1985). A process is capable of representing a specific system behavior keeping the concurrence between events and causal dependence. In this way, the notion of time process is established, where occurrences dates are added to the process events.

A way used to represent a process is using Causal Nets (Engelfriet, 1991). In this kind of net it becomes possibie to describe causal histories of the system. In Aura method the association of a process (represented by a Causal Net) and a Petri Net is made by means of a homomorfism, which preserves the local structure of the net in the process. In the structure of a process the dates assigned to events must be checked according the time constraints presents in the Petri Net. The dates will be valid if, and only if, the behavior of net and the time limits assigned to the correspondent transitions are respected. So, in (Aura and Lilius, 1997), one algorithm to construct time processes is proposed. It allows to identify the set of valid interval dates for a process. The algorithm searches all occurrence dates of process events. This way, it analyse each event individually in relation to all transitions that are enabled. During the analysis, inequalities representing the interdependence among process events form a restrictions set. In fact, the restrictions are based in the causal relations among the events, as well as in the minimal and maximal limits of dates for occurrence of the same ones.

For example, figure 2 presents a Time Petri Net and its time process. The place b_1 represent the initial state of the process. The place b_1 is mapped from the place p_1 which is the initial making of the Petri Net. The first event in the process is e_1 and it represents t_1 firing in the Petri Net. Places and events in the process are assigned to places and transitions in the Petri Net. Using the algorithm proposed in (Aura and Lilius, 1997), all valid dates interval for events occurrence of a process are calculated. The valid time intervals are expressed by a inequalities set (table 1). Solving this inequalities set, it is possible to do a temporal analysis of a sequence of transitions firings, represented by a process, without increase of imprecision. The inequalities set correctly represents existing parallelism among enabled transitions. However, the analysis is restricted to a specific process. The method is not capable to find all valid sequences of a net.

Table 1. Constraints for the process of figure 2

Lower bound	Upper bound
$e_1 \ge 0 \wedge$	$e_1 \le 1 \wedge$
$e_2 - e_1 \ge 3 \wedge$	$e_2 - e_1 \le 5 \wedge$
$e_4 - e_1 \ge 1 \wedge$	$e_3 - e_2 \le 2 \wedge$
$e_3 - e_2 \ge 1 \wedge$	$e_4 - e_1 \le 2 \wedge$
$e_3 - e_1 \ge 1 \wedge$	$(e_5 - e_4 \le 1 \vee e_5 - e_2 \le 1) \wedge$
$e_5 - e_2 \ge 0 \wedge$	$(e_6 - e_3 \le 5 \vee e_6 - e_5 \le 5)$
$e_5 - e_1 \ge 0 \wedge$	
$e_5 - e_4 \ge 0 \wedge$	
$e_6 - e_3 \ge 1 \wedge$	
$e_6 - e_2 \ge 1 \wedge$	
$e_6 - e_1 \ge 1 \wedge$	
$e_6 - e_5 \ge 1 \wedge$	
$e_6 - e_4 \ge 1 \wedge$	

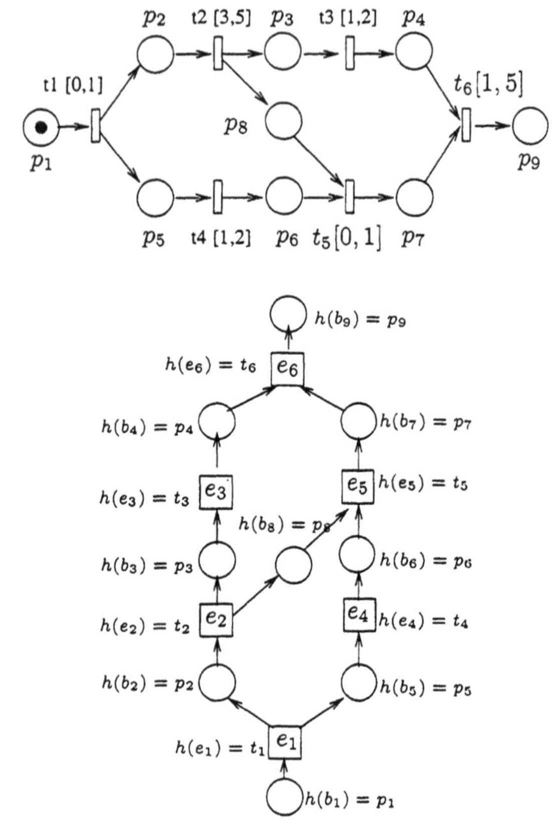

Fig. 2. Time Petri Nets and Time Process

4. PROPOSED APPROACH

The present work has a goal to propose a method for temporal analysis of any sequence of valid transitions firings. Using the behavior system representation based on the class graph, this procedure do not increase the imprecision of the time interval for firing transitions in any sequence of the net behavior.

In this proposed method a temporal analysis of a transitions firing sequence combines the structure of the class graph (section 3.1) and the set of inequalities of Aura's approach (section 3.2). In other words, the methodology proposed make, for each sequence to be analyzed in the graph, a inequalities system. The inequalities system describes all the possible firing dates of enabled transitions in a sequence (a path on the class graph, from the initial class to any other class).

Using the inequalities system constructed for a sequence, it is possible to calculate, for instance, the latest date and the earliest date for the occurrence of a determined event in a sequence or the greatest and least time distance between two events. Any program of mathematical optimization can be used to solve the inequalities system. An objective function will represent the analysis goal. The following application example will be used to illustrate our approach.

5. EXAMPLE

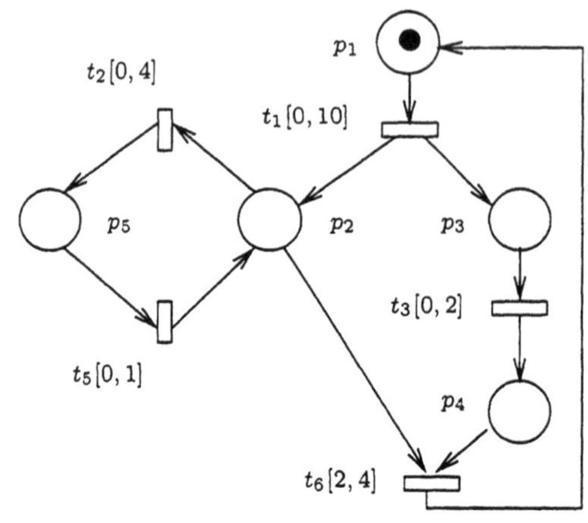

Fig. 3. Time Petri Net

Consider the Time Petri Net of figure 3. The class graph of the net is shown figure 4. Getting the transition sequence $s = t_1t_2t_3t_5t_6$ in the class graph we can observe that the latest valid firing date of transition t_6 is 18 time units. This date is obtained by adding the latest valid firing date of each transition in the sequence. However, if we observe the same sequence in the net of figure 3 the latest date for firing t_6 is 17 time units. It is clear there is a time imprecision between the sequence observed in the net and the same sequence observed in the class graph. Using the proposed approach, the increase of imprecision is eliminated.

To apply our proposed method, an inequalities system is constructed for the sequence transition to be analyzed. The path in the graph corresponds to the sequence to be analyzed, starting at the initial class C_0 and ending at the class reached by firing the last transition in the desired sequence. Based on the corresponding path of sequence $s = t_1t_2t_3t_5t_6$ in the class graph of figure 4, we can construct the inequalities system showed at table 2. The first inequality equations are obtained by analyzing the initial class C_0. For each transition in this class is assigned one inequality. These inequalities have the form $a_i \leq e_{t_i} \leq b_i$, where a_i is earliest firing date and b_i is the latest firing date of transition t_i in the respective class. For sequence s above the initial class is C_0 and in this class the inequality is $0 \leq e_{1t_1} \leq 10$.

Following the path, for each new class, it is necessary to add a new kind of inequality of the form: $a_i \leq e_{t_i} - e_{t_k} \leq b_i$, where t_i is a transition of the last visited class and t_k is a transition of the new class being analyzed. These new inequalities establishes the maximum time intervals for firing any transition of the new class. For example, leaving of initial class C_0 of sequence s, the next class to be visited is C_1 and the new inequalities are $0 \leq e_{1t_2} - e_{1t_1} \leq 4$ and $0 \leq e_{1t_3} - e_{1t_1} \leq 2$. So,

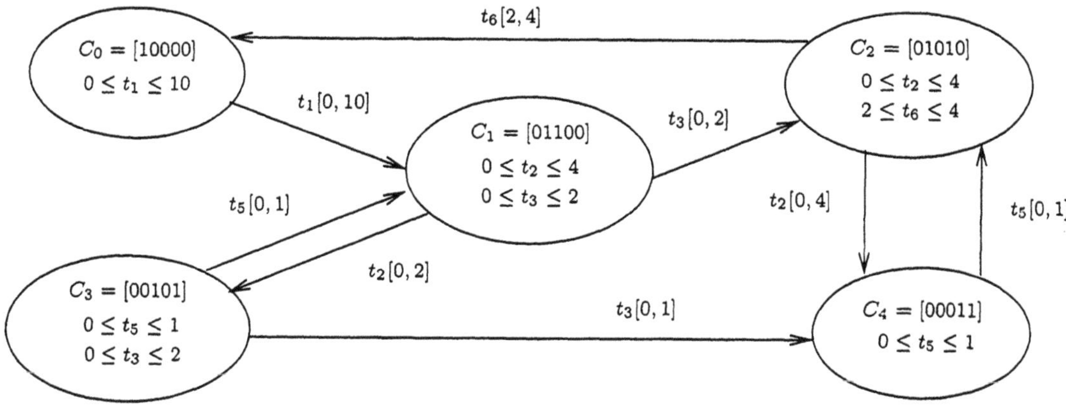

Fig. 4. Class Graph

Table 2. Constraints for the sequence: $s = t_1t_2t_3t_5t_6$

Interval	Temporal Constraints
I_1	$0 \le e_{1t_1} \le 10$
I_2	$0 \le e_{1t_2} - e_{1t_1} \le 4$
I_3	$0 \le e_{1t_3} - e_{1t_1} \le 2$
I_4	$0 \le e_{1t_5} - e_{1t_2} \le 1$
I_5	$0 \le e_{1t_3} - e_{1t_2} \le 2$
I_6	$0 \le e_{1t_5} - e_{1t_3} \le 1$
I_7	$0 \le e_{2t_2} - e_{1t_5} \le 4$
I_8	$2 \le e_{1t_6} - e_{1t_5} \le 4$
I_9	$0 \le e_{2t_1} - e_{1t_6} \le 10$

t_2 and t_3 are the new enabled transitions in the current class and t_1 is the fired transition of last visited class. The same procedure is used for next classes to be visited in the sequence.

The inequalities system build using our approach uses the same bounds used by Menasche's method to construct class graph. For this reason, it isn't necessary to consider bounds attached to the arcs in the class graph, once they have been obtained using the same inequalities. Menache's method calculates the bounds for a enabled transition based on inequalities of a current class. Our approach takes into account all inequalities for firing the multiset of transitions presents in the sequence, as shown table 2.

It is important to consider that a transition can be enabled several times during the execution of a transition sequence. In that case, to avoid errors during the construction of the inequalities system, each transition enabled twice or more times in the execution of a transition sequence, must be represented by different labels, each one corresponding to several firing of a same transition. Observing the sequence s in the class graph of figure 4, we can observe that t_2 is enabled in two classes (C_1 and C_2). In this way, t_2 is represented in the inequalities system by two labels: e_{1t_2} for the first enabled firing, and e_{2t_2} for the second one. In the same sequence the transition t_1 is also enabled twice and it is represented in the system by e_{1t_1} and e_{2t_1}. Solving this system, we can observe that the maximal firing date of t_6, in this sequence, is 17 time units. Moreover, other analysis can be done over the sequence using the same set of inequalities, such as the maximal time interval between firing two transitions. For instance, the maximal time interval between the firings of t_1 and t_2 is 2 time units.

6. CONCLUSION

In this paper it is proposed a method to perform temporal analysis of sequences of transition firings in a Time Petri net. The class graph is used to make a general representation of the system. As shown in the example, the increase of imprecision is avoided by analyzing an inequalities system constructed for the chosen sequence.

Existing restrictions found in the concepts used to construct the approach are also carried to it. As showed in (Menasche, 1996), the class graph is only applied to safe nets. If this restriction is not respected, the class graph may have an infinite number of classes.

REFERENCES

Aura, T. (1996). Time processes of time Petri nets. Research Report 38. Helsinki University of Technology, Department of Computer Science and Engineering, Digital Systems Laboratory.

Aura, T. and J. Lilius (1997). Time processes for time petri nets. In: *Lecture Notes in Computer Science: 18th International Conference on Application and Theory of Petri Nets, Toulouse, France, June 1997* (Azéma, P. and Balbo, G., Eds.). Vol. 1248. Springer Verlag. Berlin, Germany. pp. 136–155.

Berthomieu, B. and M. Menasche (1982). A state enumeration approach for analyzing time petri nets. In: *3rd European Workshop on Applications and Theory of Petri Nets*. Varenna, Italy.

Berthomieu, B., P. O. Ribet and F. Vernadat (2003). L'outil tina - construction d'espaces d'états abstraits pour les réseaux de petri et réseaux temporels. In: *Proceedings of MSR*. pp. 295–310.

Engelfriet, J. (1991). Branching processes of petri nets. *Acta Informatica* **28**(6), 575–591. NewsletterInfo: 38, 40.

Menasche, M. (1982). Analyse des réseaux de Petri temporisés et application aux systèmes distribués. PhD thesis. Université Paul Sabatier. Toulouse, France. Rapport LAAS n. 824.

Menasche, M. (1985). PAREDE: An automated tool for the analysis of time(d) petri nets.. In: *International Workshop on Timed Petri Nets, July 1-3*. IEEE Computer Society Press. Torino, Italy. pp. 162–169.

Menasche, M. (1996). Parâmetros temporais em redes de petri. Technical Report 8. Pontifícia Universidade Católica do Rio de Janeiro, Brasil.

Merlin, P. (1974). A Study of Recoverability of Computer Systems. PhD thesis. University of California IRVINE. available from Ann Arbor: Univ Microfilms, No. 75–11026.

Murata, T. (1989). Petri nets: Properties, analysis and applications.. In: *Proceedings of the IEEE*. number 4. pp. 541–580. NewsletterInfo: 33Published as Proceedings of the IEEE, volume 77, number 4.

Petri, C. A. (1962). Kommunikation mit Automaten. PhD thesis. Institut ffir Instrumentelle Mathematik. Bonn. English translation available as Communication with Automata, Tech. Rep. RADC-TR-65-377, Vol. 1, Suppl. 1, Applied Data Research, Princeton, NJ, 1966.

Popova-Zeugmann, L. (1991). On time petri nets. *EIK: Journal of Information Processing and Cybernetics EIK (formerly Elektronische Informationsverarbeitung und Kybernetik)* **27**(4), 227–244.

Ramchandani, C. (1974). Analysis of Asynchronous Concurrent Systems by Timed Petri Nets. PhD thesis. Cambridge, Mass.: MIT, Dept. Electrical Engineering. Project MAC TR-120.

Reisig, W. (1985). *Petri Nets: An Introduction*. Springer-Verlag New York, Inc.. Berlin, Germany.

Schastai, V. (2003). Métodos de Análise para Redes de Petri Temporais. PhD thesis. Brasil, Curitiba.: CEFET-PR, CPGEI Mestrado em Engenharia Elétrica e Informática Industrial.

Schastai, V. and L. A. Künzle (2002). Análise de redes de petri temporais combinando grafo de classes e processos. In: *Congresso Brasileiro de Automática 2002*. Natal-RN, Brasil.

Sifakis, J. (1977). Etude du comportement permanent des réseaux de petri temporiss. *Réseaux de Petri. Paris, 23–24 Mars 1977 AFCET. / Edité par l'Institut de Programmation de Paris* pp. 165–184.

ELSEVIER
IFAC
PUBLICATIONS
www.elsevier.com/locate/ifac

SENSOR ASSIGNAMENT FOR OBSERVABILITY IN INTERPRETED PETRI NETS

L. Aguirre-Salas*, O. Begovich and A. Ramirez-Treviño****

*CUCSUR. Univ. de Guadadalajara. Av. Indep. Nac. 151. Autlan, Jal., Mex. laguirre@cucsur.udg.mx
**CINVESTAV GDL. Lopez Mateos Sur 590. Guadalajara. Jal., Mex. [art,obegovi]@gdl.cinvestav.mx

Abstract: This paper addresses the minimal cost sensor selection problem for observability in Interpreted Petri Nets (IPN). A genetic algorithm to solve this problem is presented under the assumption that the sensor selection is made from a set of available state sensors which can be attached to several places of the net. The proposed procedure takes advantage of a characterization of the set of measurable places for event-detectability, reducing the computation time spent on testing this property. The algorithm is quite simple and helps to reduce the design time and effort of discrete event systems.

Keywords: Petri-nets, Observability, Sensor Systems, Genetic Algorithms.

I. INTRODUCTION

While designing a system, a common problem on engineering practice is to select the lowest cost sensor configuration such that the system state can be inferred from the information provided by the sensors, i.e. the system is observable. Several algorithms to solve this problem have been proposed for observable languages (Haji-Valizadeh and Loparo,1996), for fault detection (Bavichi and Chong, 1994) and for events sequence detection (Park and Chong, 1994). A common characteristic of most of these works is that the complete determination of the system events or the exact reconstruction of the system state is not addressed. On the other hand, using Petri nets, Aguirre-Salas, et al. (2002) have shown that the event-detectability and marking-detectability are sufficient conditions for observability. Taking advantage of this fact, in Aguirre-Salas (2003), a genetic algorithm has been presented, which uses a binary encoding of the candidate sensor configurations, where a locus of an individual represents a given place of the net. The algorithm assumes that an available sensor can be attached to a single place of the net. Although the algorithm is simple, most of the computation time is spent in finding individuals satisfying event-detectability, reducing its general performance. Moreover, the crossover and mutation operators do not guarantee that the resulting offsprings satisfy observability, so most of them are ignored and their parents are copied to the next population, reducing the average fittest of the new generation.

This paper addresses the minimal cost sensor selection problem for observability in IPN using a genetic algorithm. The sensor selection in made from a set of available state sensors which can be allocated to several places. It is assumed that the actuator signals are already attached to transitions. The proposed algorithm implements an integer encoding to represent candidate solutions, where a locus represents a transition of the net and its value represents the selected place that makes it detectable. In this case, the selection of a place for each transition is made by using a characterization of the set of measurable places that makes a net event-detectable. Thus, individuals satisfying event-detectability can be generated without the necessity of testing them for this property, increasing the algorithm performance. In this new approach, the crossover and mutation operators are slightly modified in order to guarantee that all the individuals of the population always satisfy observability. Thus, the resulting offsprings are fixed instead of rejected.

II. BASIC DEFINITIONS

A Petri net structure is a bipartite digraph represented by a 4-tuple $N = (P, T, I, O)$, where $P = \{p_1, p_2, ..., p_n\}$, $T = \{t_1, t_2, ..., t_m\}$, $I : P \times T \rightarrow Z^+ \cup \{0\}$ and $O : P \times T \rightarrow Z^+ \cup \{0\}$. The *incidence matrix* of a Petri net structure is $C = [c_{ij}]_{n\times m}$, where $c_{ij} = O(p_i,t_j) - I(p_i,t_j)$. Let $x,\ y \in T \cup P$, then ${}^\bullet(x) = \{y \mid \text{there is an arc from } y \text{ to } x\}$ and $(x)^\bullet = \{y \mid \text{there is an arc from } x \text{ to } y\}$ are the sets of *predecessors* and *successors* of the node x, respectively. Similarly, ${}^\bullet(x)^\bullet = (x)^\bullet \cup {}^\bullet(x)$.A *p-invariant* Y of a PN is a rational-valued solution of equation $Y^T C = 0$. The *support of a p-invariant* Y_i is the set $\|Y_i\| = \{p|Y_i(p) \neq 0\}$. The *marking* of a PN is a function $M : P \to \mathbb{N}$, that assigns to each place of N a non-negative number of tokens, depicted as black dots inside the places. The marking at the k-th instant is often represented by a vector $M_k = [M_k(p_1)\ M_k(p_2)\ \dots\ M_k(p_n)]^T$. A *PN system* (N, M_0) is a PN structure N with an initial marking M_0. A transition $t_j \in T$ of a given (N, M_0) is *enabled* at a marking M_k if $\forall p \in P$, $M_k(p) \geqslant I(p,t_j)$. The set of enabled transitions at a marking M_k is $E(M_k) = \{t|\forall p \in P, M_k(p) \geqslant I(p,t)\}$. If a transition $t \in E(M_k)$ is fired then a new marking $M_{k+1} = M_k + C(\bullet, t)$ is reached. This fact is denoted as $M_k \xrightarrow{t_j} M_{k+1}$. A *firing sequence* of a PN system (N, M_0) is a sequence $\sigma = t_i t_j \dots t_k$ such that $M_0 \xrightarrow{t_i} M_1 \xrightarrow{t_j} \dots \xrightarrow{t_k} M_k$. Observe that, a firing sequence is an enabled sequence at M_0. The set $\mathbf{R}(N, M_0) = \{M_k|M_0 \xrightarrow{\sigma} M_k\}$ is called the reachability set of (N, M_0). The firing language of (N, M_0) is $\mathcal{L}(N, M_0) = \{\sigma = t_i t_j \dots t_k\}|$ $M_0 \xrightarrow{t_i} M_1 \xrightarrow{t_j} \dots \xrightarrow{t_k} M_k\}$.

Definition 1: An Interpreted Petri Net (IPN) is a 4-tuple $Q = (N', \Sigma, \lambda, \varphi)$ where $N' = (N, M_0)$ is a PN system, $\Sigma = \{a_1, a_2, \dots, a_u\}$ is the input alphabet of the net, where α_i is an input symbol; $\lambda : T \to \Sigma \cup \{\varepsilon\}$ is a function that assigns an input symbol to each transition of the net with the following restriction: $\forall t_j, t_k \in T,\ j \neq k$, if $I(p_i, t_j) = I(p_i, t_k) \neq 0$, $\lambda(t_j) \neq \varepsilon$ and $\lambda(t_k) \neq \varepsilon$ then $\lambda(t_j) \neq \lambda(t_k)$. In this case, the symbol ε represents an internal system event; φ:$\mathbf{R}(N, M_0) \to [Z^+ \cup \{0\}]^r$ is a function that assigns an r-entry vector to every reachable marking of the net, where r is the number of outputs.

Remark 2: Hereafter it will be written (Q, M_0) instead of $Q = (N', \Sigma, \lambda, \varphi)$ to emphasize the fact that there exists an initial marking in an IPN. It is assumed that $\forall t_i, t_j$ if $t_i \neq t_j$ and $\lambda(t_i) = \lambda(t_j)$ then it holds that $C(\bullet, i) \neq C(\bullet, j)$. This is not a mayor constraint since if $C(\bullet, i) = C(\bullet, j)$ then either t_i or t_j is a redundant transition. Since in this work time is not considered, when two or more transitions fire simultaneously, it can be thought, without loss of generality, that their firing occur one after another separated by a small period of time. This work focuses on the case when the function φ is linear so it can be represented a $r \times n$ matrix, where r is the number of places representing the measurable states of the net. Each column of this matrix is an elementary or null vector and if the output symbol i is presented (turned on) every time $M(p_j) \geq 1$ then $\varphi(i,j) = 1$.

A transition $t \in T$ is said to be controllable, if $\lambda(t) \neq \varepsilon$, and uncontrollable, otherwise. The set of controllable transitions is $T_c = \{t|\lambda(t) \neq \varepsilon\}$. A controllable transition is represented as an unfilled bar, while an uncontrollable transition as a filled bar. Similarly, a place $p_j \in P$ is said to be measurable, if $\exists i$ such that $\varphi(i,j) = 1$; otherwise, p_i is called non-measurable. The set of measurable places is denoted by P_m. A measurable place is depicted as an unfilled circle, while a non-measurable place as a filled circle.

If $t \in E(M_k)$ and $t \notin T_c$ then it can be fired. On the other hand, if $t \in E(M_k)$ and $t \in T_c$ and $a = \lambda(t)$ is present then t must fire. In both cases, when t fires the new marking $M_{k+1} = M_k + C(\bullet, t)$ is reached and the output symbol $y_{k+1} = \varphi(M_{k+1})$ is observed. Let (Q, M_0) be an IPN and $\mathcal{L}(Q, M_0)$ be its firing language. The input language of (Q, M_0) is $\mathcal{L}_{in}(Q, M_0) = \{\omega = \lambda(t_i)\lambda(t_j)\dots\lambda(t_k)|\sigma = t_i t_j \dots t_k \in \mathcal{L}(Q, M_0)\}$. Each time an input word is given, different transition sequences can be enabled due to the existence of uncontrollable transitions; however, only one of them occurs generating a particular sequence of output symbols. Let (Q, M_0) be an IPN and ω be an input word of the net. The set of possible firing sequences generated by ω is $\Gamma_\omega = \{\sigma = t_i t_j \dots t_k \in \mathcal{L}(Q, M_0)|\lambda(\sigma) = \lambda(t_i)\lambda(t_j)\dots\lambda(t_k) = \omega\}$. The set of possible output words generated by ω is $Z_\omega = \{z = \varphi(M_0)\varphi(M_1)\dots\varphi(M_k)|M_0 \xrightarrow{t_i} M_1 \xrightarrow{t_j} \dots \xrightarrow{t_k} M_k \wedge t_i t_j \dots t_k \in \Gamma_\omega\}$. While modeling a DES in IPN terms, the knowledge of resources contained in the system is usually known (e.g. the number of available robots, machines, etc.). Let (Q, M_0) be an IPN and $M(p_j)$ a reachable marking of a place p_j. The set of equations $CML = \{\xi_i := \sum_{j=1}^{n} x_j^i M(p_j) = k_i | i \in \{1,2,\dots,s\},\ x_j^i \in \mathbb{Z}^+ \cup \{0\}\}$ is called conservative marking laws. A CML is well-defined, if $\forall p_j \notin P_m\ \exists \xi_i$ such that $x_j^i > 0$, i.e. all non-measurable places of the net are contained in at least one equation. The support of a conservative marking law ξ_i is the set $\|\xi_i\| = \{p_j \in P|x_j^i \neq 0\}$.

III. Observability

Definition 3: (Aguirre-Salas, et al., 2002) An IPN given by (Q, M_0) is *observable in k events* if $\forall \omega \in \mathcal{L}_{in}(Q, M_0)$, $\exists v$ with $|v| \leq k < \infty$ such that $\omega v \in \mathcal{L}_{in}(Q, M_0)$ and the information provided by the knowledge of ωv, the output word

$z \in Z_{\omega v}$, the net structure Q and the CML suffices to uniquely determine M_0 and the fired sequence $\sigma = t_i t_j \ldots t_r \in \Gamma_{\omega v}$.

Definition 4: (Aguirre-Salas, et al., 2002) An IPN given by (Q, M_0) is *event-detectable* if the firing of any transition $t \in T$ at a marking $M_k \in \mathbf{R}(Q, M_0)$ can be uniquely determined through the information provided by the input symbol $\alpha = \lambda(t)$ and the output signals $\varphi(M_k)$, $\varphi(M_k + C(\bullet, t))$.

Proposition 5: (Aguirre-Salas, et al., 2002) An IPN given by (Q, M_0) is event-detectable if and only if $\forall i \in [1, 2, \ldots, m]$, $\varphi C(\bullet, i) \neq 0$ and $\forall j \neq k \in [1, 2, \ldots, m]$ such that $\varphi C(\bullet, j) = \varphi C(\bullet, k)$, $\lambda(t_j) \neq \lambda(t_k)$, where $\varphi C(\bullet, k)$ is the column of φC corresponding to the transition t_k.

This result can be explained as follows: if φC has a null column, then the associated transition has no measurable input and output places, so its firing has no effect in the output y. Similarly, if there are two equal columns in φC, the firing of the corresponding transitions produce the same effect in the output y, so they can not be detected unless their input symbols are different. Observe that, this structural characterization does not indicate which places may be set as measurable in order to make event-detectable a net.

Proposition 6: An IPN given by (Q, M_0) is event-detectable if and only if 1) $\forall t_k \in T$, $\exists p_r \in P_m$ such that $p_r \in^\bullet (t_k)^\bullet \setminus^\bullet (t_k) \cap (t_k)^\bullet$ and 2) $\forall t_i, t_j \in T$ with $t_i \neq t_j$ and $\lambda(t_i) = \lambda(t_j)$, $\exists p_r \in P_m$, such that $p_r \in^\bullet (t_i) \setminus^\bullet (t_i) \cap^\bullet (t_j)$ or $p_r \in (t_i)^\bullet \setminus (t_i)^\bullet \cap (t_j)^\bullet$

Proof: (Necessary) Since (Q, M_0) is event-detectable, by Proposition 5 the columns in φC are no null and different from each other: let $\varphi C(\bullet, k)$ be the k-th column of φC corresponding to a transition, say t_k. Since there are no null columns in φC, $\exists r$ such that $\varphi C(r, k) \neq 0$, which corresponds to a measurable input or output place of t_k that is not in self-loop with t_k, i.e. $p_r \in^\bullet (t_k)^\bullet \setminus^\bullet (t_k) \cap (t_k)^\bullet$. Now, let $\varphi C(\bullet, i)$ and $\varphi C(\bullet, j)$ be two columns of φC, which correspond to two transitions, say t_i and t_j, respectively, such that $\lambda(t_i) = \lambda(t_j)$. Thus, by Proposition 5, $\varphi C(\bullet, i) \neq \varphi C(\bullet, j)$. It means that, they differ at least in one entry, say $\varphi C(r, i) \neq \varphi C(r, j)$. It is clear that, the rth row of φC corresponds to a non-common input or output measurable place of both transitions, say p_r, i.e. $p_r \in^\bullet (t_i) \setminus^\bullet (t_i) \cap^\bullet (t_j)$ or $p_r \in (t_i)^\bullet \setminus (t_i)^\bullet \cap (t_j)^\bullet$.

(Sufficiency) Let t_k be a transition such that $\exists p_r \in P_m$ and $p_r \in^\bullet (t_k)^\bullet \setminus^\bullet (t_k) \cap (t_k)^\bullet$. Thus, by Proposition 5, $\varphi C(r, k) \neq 0$ and, therefore, $\varphi C(\bullet, k)$ is a no null column.

Now, let t_i, t_j be two arbitrary different transitions $(i \neq j)$ such that $\lambda(t_i) = \lambda(t_j)$, and let p_r be a measurable place, such that $p_r \in^\bullet (t_i) \setminus^\bullet (t_i) \cap^\bullet (t_j)$ or $p_r \in (t_i)^\bullet \setminus (t_i)^\bullet \cap (t_j)^\bullet$. Thus, $\varphi C(r, i) \neq \varphi C(r, j)$ and, therefore, $\varphi C(\bullet, i)$ and $\varphi C(\bullet, j)$ are different from each other. Hence, (Q, M_0) is event-detectable. ■

In other works, a transition must have an input or output measurable place that is a non-common input or output place with other transitions.

Definition 7: (Aguirre-Salas, et al., 2002) An IPN given by (Q, M_0) is marking-detectable *in k steps* if and only if $\forall \omega \in \mathcal{L}_{in}(Q, M_0)$, such that $|\omega| \leq k < \infty$, the information provided by the knowledge of the input word ω, the output word $z \in Z_\omega$, the system structure and the CML suffices to uniquely determine $M_k \in \mathbf{R}(N, M_0)$, where $M_0 \xrightarrow{\sigma} M_k$ and $\sigma \in \Gamma_\omega$.

In a conservative IPN, the marking of each place has an upper and a lower bound, which determine the range of different values that it can reach. Let (Q, M_0) be a conservative IPN. The upper and lower marking bounds that a place p_j can reach are $M^{UB}(p_j) = \max(M(p_j) | M \in \mathbf{R}(Q, M_0))$ and $M^{LB}(p_j) = \min(M(p_j) | M \in \mathbf{R}(Q, M_0))$. In some classes of systems, it is possible to determine when a place reaches its upper or lower marking bounds. Thus, its making becomes known, and from this moment, having the knowledge of the subsequent firing sequence the marking of the place remains known. If the above is true for every place then the whole marking of the net becomes known, i.e. the net is marking-detectable.

Theorem 8: (Aguirre-Salas, et al., 2002) Let (Q, M_0) be a live, cyclic, and conservative IPN. (Q, M_0) is marking-detectable if i) it is event-detectable and ii) $\forall p \notin P_m$ $SD((Q, M_0), ^\bullet(p), (p)^\bullet) = M^{UB}(p) - M^{LB}(p)$, where $SD((Q, M_0), R, S)$ is the synchronic distance of R with respect to S (Silva, 1985).

Condition ii) of the previous theorem is called synchronic distance condition (SD condition).

Theorem 9: (Aguirre-Salas, et al., 2002) An IPN given by (Q, M_0) is observable in k steps if it is event-detectable and marking-detectable.

IV. Sensor selection

In this paper,the following points are assumed: 1)The actuator signals are already allocated to transitions. 2) The sensor selection is made from a pool of available state sensors S. Notice that there may be places for which there are not available sensors.

3) A sensor is allocated to a place by a function $\Xi : S \to P$, where $\Xi(s) = p$ if s can sense each reachable marking of p. The function Ξ is not necessarily an injective function, thus a sensor can be shared by two or more places. 4) The function $G : P \to S \cup \{\varnothing\}$ returns the sensor allocated to a given place, where the symbol $\varnothing$ is the null sensor. Thus, if $G(p) = s \in S$, it means that the sensor s is used to measure the marking of p (therefore, the marking of p is available at the output and p is a measurable place); on the other hand, if $G(p) = \varnothing$, it means that the place p has no allocated sensor to it (therefore, it is a non-measurable place). 5) The actuator signals are already assigned to the transitions of a non-measured IPN.

Definition 10: A non-measured IPN is a net (Q, M_0) such that $P_m = \{\}$. A sensor configuration $\varrho \in 2^S$ of an IPN given by (Q, M_0) is a set of sensors attached to the places of the net by an allocation function $G : P \to S \cup \{\varnothing\}$.

Definition 11: Let (Q, M_0) be a non-measured IPN, S be the set of available sensors for the places of (Q, M_0), $\varrho \in 2^S$ be a sensor configuration and $G : P \to S \cup \{\varnothing\}$ be an allocation function of the available sensors to the places of the net. The sensor configuration ϱ satisfies the observability property if the IPN resulting of assigning the sensors of ϱ to the places of (Q, M_0) by the function G is observable. The set of all sensor configurations of (Q, M_0) satisfying observability is denoted by $\hat{S}(Q, M_0)$.

Since the selection of a given sensor configuration is determined by its acquisition cost, the following cost function is defined. $w : S \to \mathbb{R}^+$, where $\mathbb{R}^+$ is the set of the positive real numbers including the zero. This function can be extended to a sensor configuration as follows: $W : 2^S \to \mathbb{R}^+$ where $W(\varrho) = \sum_{s_i \in \varrho} w(s_i)$. In this case, the problem of finding a minimal cost sensor configuration satisfying observability can be defined as follows.

Definition 12: Let (Q, M_0) be a non-measured IPN, S be the set of available sensors for (Q, M_0), and $W : 2^S \to \mathbb{R}^+$ be a sensor configuration cost function. The minimal sensor choice problem for observability is defined as to compute $\varrho^* \in \hat{S}(Q, M_0)$ such that $W(\varrho_i) = \min_{\varrho \in \hat{S}(Q, M_0)} W(\varrho)$.

From the above definition, an algorithm to find the minimal cost sensor configuration satisfying observability implies the computation of the set $\hat{S}(Q, M_0)$ (an exponential problem), evaluate each sensor configuration belonging to it and then finding out the one with the minimal cost (a polynomial problem). Thus, the minimal sensor choice problem is a NP problem. This paper proposes the following genetic algorithm for sensor choice taking advantage of the characterization stated in Proposition 6.

Each individual or candidate solution is an m-entry vector ϖ, where $m = |T|$ and $\varpi[i] = k \in \{j | p_j \in {}^\bullet(t_i) \cup (t_i)^\bullet$ and $G(p_i) \neq \varnothing\}$. In other works, the i-th locus of ϖ corresponds to the i-th transition and its value is the index of any input or output place of t_i that has an allocated sensor. Let ϖ be a candidate solution for a net (Q, M_0), then $\varrho = \{G(p_i) | i = \varpi[j]\ \forall j \in \{1, 2, ..., |T|\}\} \cup \{G(p_k) | p_k \in P_{SD}\}$ is called the sensor configuration corresponding to ϖ. Moreover, if ϱ satisfies event-detectability or observability then we will say that ϖ also satisfies event-detectability or observability, respectively. From Theorem 9, it is known that a live, conservative and cyclic IPN is observable if it is event-detectable and the SD condition for marking-detectability are fulfilled. Thus, given an PN system and the set P_{SD} of places that do not satisfy the SD condition of Theorem 8, an individual ϖ satisfying observability can be constructed as follows.

Algorithm 13: Candidate sensor configurations generator. Inputs: C - incidence matrix, V_C - control symbol allocation vector, G - sensor allocation function. Outputs: ϖ - a candidate solution.

1. For all $t_k \in Tx$: a) randomly, select a place $p_r \in {}^\bullet(t_k) \cup (t_k)^\bullet$ such that $G(p_r) = \varnothing$ and $\forall t_v \in T$ with $t_v \neq t_k$ and $V_C[v] = V_C[k]$, $p_r \in {}^\bullet(t_k) \setminus {}^\bullet(t_v) \cap {}^\bullet(t_k)$ or $p_r \in (t_k)^\bullet \setminus (t_v)^\bullet \cap (t_k)^\bullet$. b) set $\varpi[k] = r$

2. Return ϖ

Claim 14: Let (Q, M_0) be a non-measured IPN, S be the set of available sensors for the places of (Q, M_0), $G : P \to S \cup \{\varnothing\}$ be a function that allocates the available sensors to the places of the net and P_{SD} the set of places not satisfying the SD condition. An individual ϖ yield by algorithm 13 satisfies observability of (Q, M_0).

Proof: Let ϱ be the sensor configuration corresponding to the individual ϖ. The algorithm selects a set of places satisfying point 2 of Proposition 6, so their allocated sensors belong to ϱ. Thus, and since a place for each transition was selected, the resulting sensor configuration ϱ satisfies event-detectability. Since ϱ includes the allocated sensors of every place $p \in P_{SD}$ then $P_{SD} \subseteq P_m$ so the SD condition of theorem 8 is fulfilled. Thus, ϱ satisfies marking-detectability. Hence, by Theorem 9, ϱ satisfies observability. ■

Let ϱ be the corresponding sensor configuration of an individual ϖ. Since the objective is to compute a lowest initial cost sensor configuration, the sensor configuration cost function W defined above can be used as the fitness function. In order to simplify the computations, a r-entry vector V_W is used to represent the sensor costs, where $V_W[i] = w(s_i)$ if $s_i \in \varrho$ (i.e. $\varrho[i] = 1$) and $V_W[j] = 0$ if $s_j \notin \varrho$ (i.e. $\varrho[j] = 0$). A single point crossover binary operator with fitness proportional selection, roulette-wheel sampling is used with the following variation: if a resulting offspring does not satisfy event-detectability, then the value of some locus is affected in order to ensure that the individual satisfies event-detectability accordingly with the following algorithm.

Algorithm 15: ED correction. Inputs: C - incidence matrix, V_C - control symbol allocation vector, G - sensor allocation function, ϖ - a candidate solution. Outputs: ϖ - the fixed candidate solution.

- **1.** For all i, j such that $t_i \neq t_j$ and $V_C[i] = V_C[k]$: 1.1 Set $Lx = [^\bullet(t_i) \cap ^\bullet(t_j)] \cup [(t_i)^\bullet \cap (t_j)^\bullet]$. 1.2 If $p_r \in Lx$ and $p_v \in Lx$, where $r = \varpi[i]$ and $v = \varpi[j]$, then 1.2.1 Set k =random($\{i, j\}$). 1.2.2 Set q =random($\{h | p_h \in^\bullet (t_k) \cup (t_k)^\bullet \setminus Lx \wedge G(p_h) \neq \varnothing\}$). b.3) Set $\varpi[k] = q$.
- **2.** Return ϖ.

Notice that the algorithm verifies if point 2 of Proposition 6 is fulfilled for each pair of transitions. If not then another input or output place is randomly selected such that the proposition is fulfilled. So the sensor configuration ϱ corresponding to the resulting individual satisfies event-detectability and since the allocated sensors to every place belonging to P_{SD} is included, ϱ satisfies marking-detectability. Hence, ϱ satisfies observability. Therefore, the crossover operator preserves the observability satisfaction of the individuals of the population. The mutation consists of selecting a different place for each selected for mutation transition. Since this operator can yield an offspring that does not satisfy event-detectability after mutation, then algorithm 15 is called to ensure that event-detectability is satisfied.

Algorithm 16: Minimal cost sensor selection for observability. Inputs: C - incidence matrix of the non-measured net , V_C - control symbol allocation vector, S - set of available sensors, G - sensor allocation function, V_W - sensor cost vector, P_{SD} - set of places not satisfying the SD conditions. Outputs: The sensor configuration with the lowest cost $\varrho^*$and its cost w^*.

- **1.** Initialize the generation index: $ngen = 0$.
- **2.** Generate a random population Pop of r individuals with algorithm 13.
- **3.** Calculate the fitness W of each individual: $w_i = \varrho_i * V_W$.
- **4.** Set $nPop = \{\}$ and apply the crossover operator:
- **4.1** Select two individuals ϖ_i and ϖ_j with fitness proportional selection, roulette-wheel sampling.
- **4.2** With probability pc crossover the pair at a randomly chosen point to generate two new individuals $\hat{\varpi}_i$ and $\hat{\varpi}_j$.
- **4.2.1.** Apply the algorithm 15 to $\hat{\varpi}_i$
- **4.2.2** Include $\hat{\varpi}_i$ in the next population: $nPop = nPop \cup \{\hat{\varpi}_i\}$
- **4.2.3** Apply the algorithm 15 to $\hat{\varpi}_j$
- **4.2.4** Include $\hat{\varpi}_j$ in the next population: $nPop = nPop \cup \{\hat{\varpi}_j\}$
- **4.3** If no crossover takes place then $nPop = nPop \cup \{\hat{\varpi}_i\} \cup \{\hat{\varpi}_j\}$.
- **4.4** Go to step 4.1 until r individuals are generated.
- **4.5** Set $Pop = nPop$ and $nPop = \{\}$
- **5.** Apply the mutation operator:
- **5.1** For each individual $\varpi \in Pop$:
- **5.1.1** Set $\hat{\varpi} = \varpi$
- **5.1.2** For each locus $\hat{\varpi}[i]$ with probability pm: select a place $p_k \in^\bullet (t_i) \cup (t_i)^\bullet \setminus p_r$ where $r = \hat{\varpi}[i]$
- **5.1.3** Apply the algorithm 15 to $\hat{\varpi}$.
- **5.1.4** Include $\hat{\varpi}$ in the next population: $nPop = nPop \cup \{\hat{\varpi}\}$
- **5.2** Set $Pop = nPop$ and $ngen = ngen + 1$
- **6.** If $ngen < ngen_{max}$ then go to step 3.
- **7.** Transform each individual to a sensor configuration:
- **7.1** Set $\hat{S} = \{\}$ and For each $\varpi \in Pop$
- **7.1.1** Set $\varrho = \bar{0}$
- **7.1.2** For each locus $\hat{\varpi}[i]$: set $\varrho[G(\hat{\varpi}[i])] = 1$
- **7.1.3** For each $p_i \in P_{SD}$: set $\varrho[G(p_i)] = 1$
- **7.1.4** Set $\hat{S} = \hat{S} \cup \{\varrho\}$
- **8.** Calculate $\varrho^* \in \hat{S}$ such that

$$W(\varrho^*) = \min_{\varrho \in \hat{S}} W(\varrho)$$

- **9.** Output ϱ^* and $W(\varrho^*)$

V. EXAMPLE

Consider an automated manufacturing cell consisting of two incoming conveyors (IC1, IC2), two outgoing conveyors (OC1, OC2), five machines (M1 to M5), two AGV (AGV1, AGV2), a two-slot buffer (B) and four robots (R1 to R4) for loading and unloading parts. The system processes two different types of products (P1, P2) from two types of raw material (RM1, RM2). Arrows in figure 1 show the flow of each type of material through the system.

An IPN model of the system is shown in Figure 2. Notice that, in the IPN model all places are non-measurable. Moreover, all transitions are controllable and, by their physical meaning (which can be inferred from the model and the physical meaning of the places), it is possible to consider that $\forall t_i, t_j \in T$, $\lambda(t_i) \neq \lambda(t_j)$. Given the

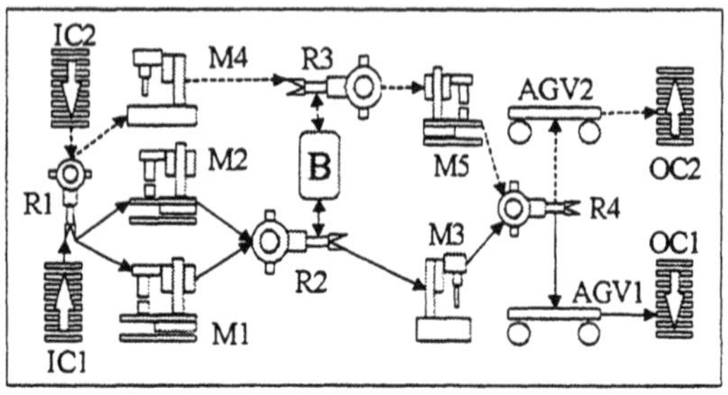

Fig. 1. Diagram of an automated manufacturing cell

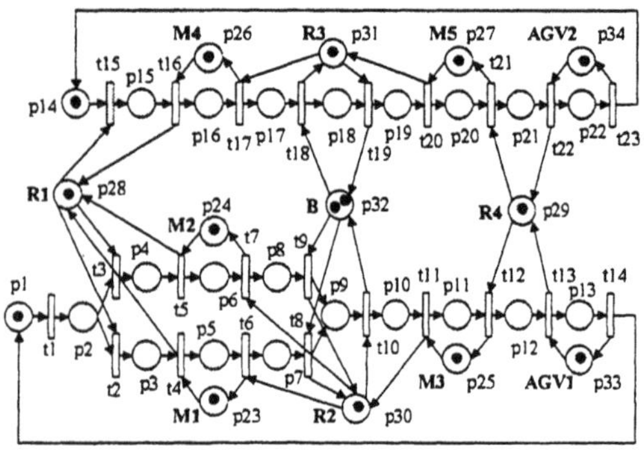

Fig. 2. An non-measured IPN model of the manufacturing cell.

initial marking of the net and using the classical methods of the PN literature, it is easy to prove that the net is conservative, live and has a cyclic behavior. Moreover, for every $p_i \in P - \{p_{32}\}$ it holds that $M^{LB}(p_i) = 0$, $M^{UB}(p_i) = 1$ and $SD((Q, M_0), (p_i)^\bullet, ^\bullet(p_i)) = 1$; and, $M^{LB}(p_{32}) = 0$, $M^{UB}(p_{32}) = 2$ and $SD((Q, M_0), (p_{32})^\bullet, ^\bullet(p_{32})) = 2$. It means that every place satisfies the SD conditions of Theorem 8; thus, P_{SD} is empty. Suppose that, there is an available sensor for each place of the net. In this case, $\forall p_i \in P$ $G(p_i) = s_i$. The initial sensor costs is given by the vector V_W = [30, 30, 40, 40, 50, 50, 40, 40, 40, 40, 50, 40, 50, 30, 40, 50, 40, 40, 40, 50, 40, 50, 50, 50, 50, 50, 50, 35, 35, 35, 35, 50, 50, 50]. The genetic algorithm parameters were set to the following values: population size $r = \min(2^n, 100)$, crossover probability $p_c = 0.75$, mutation probability $p_m = 0.001$ and $ngen_{\max} = 25$ (Chambers, 2000 and Mitchell, 1996). After running the algorithm several times the fittest sensor configuration found out is $\varrho^* = \{s_1, s_3, s_7, s_{12}, s_{14}, s_{15}, s_{28}, s_{29}, s_{30}, s_{31}, s_{32}\}$ and it has a cost of $W(\varrho^*) = 410$. As expected, this sensor configuration satisfies observability (by Propositions 5, 8 and Theorem 9).

VI. Conclusions

In this paper, the minimal sensor selection problem for observability in DES modeled by IPN was addressed. A genetic algorithm to solve this NP problem has presented based on a characterization of observable IPN presented in previous works plus a characterization of the set measurable places that make a net event-detectable. This characterization is used by the algorithm to select in a "smarted" way the measurable places for a net, given the set of available sensors. This leads to a improvement of the algorithm computational complexity with respect to the one presented in a previous work. The algorithm parameters were set to the values reported in most of the literature. It was run for a set of IPN models (one of them presented here) with randomly chosen sensor costs. As it is well known, the use of a genetic algorithm does not guarantee that the minimal is found. However, the algorithm proposed has an acceptable performance and in the worse case the algorithm returns a quite good solution. Moreover, the presented genetic algorithm has a better performance in time than the one presented in a previous work. Although, the algorithm has a pretty good performance, additional genetic operators can be included in order to avoid the "local minimum" effect that is present after a certain number of generations.

REFERENCES

Aguirre L., O. Begovich and A. Ramirez (2002). "Observability in Interpreted Petri Nets using Sequence Invariants." *Proc of the 41st IEEE Conference on Decision and Control*, Las Vegas, USA, pp. 3602-3607.

Aguirre-Salas L (2003). "Sensor selection for observability in Interpreted Petri Nets: a genetic approach." *Proc of the 42st IEEE Conference on Decision and Control*, Hawaii, USA, pp. 3760-3765.

Bavichi S. and E. Chong (1994). "Automated fault diagnosis using a discrete event systems framework." *Proc. of the IEEE Int. Symp. on Intelligent Control.* Columbus, Ohio.

Chambers L., editor (2001). *The practical handbook of Genetic Algorithms: Applications.* Florida, USA: Chapman&Hall/CRC Ed.

Haji-Valizadeh A and K.A. Loparo KA (1996). "Minimizing the cardinality of an events set for supervisors of Discrete-Event Dynamical Systems." In *IEEE TAC*; **Volume** 11, pp. 1579-1593.

Mitchell, M (1996). *An introduction to genetic algorithms.* Cambridge, Massachusetts, USA: The Mit Press Ed.

Park Y. and K. P. Chong (1994). "Sensor assignment for invertibility in interruptive timed discrete event systems." *Proc. of the IEEE International Symposium on Intelligent Control.* Columbus, Ohio, USA.

Silva M. (1985). *Petri nets on automatica and data processing.* Madrid, Spain: AC Ed.

Copyright © IFAC Discrete Event Systems,
Reims, France, 2004

ELSEVIER
IFAC
PUBLICATIONS
www.elsevier.com/locate/ifac

ALGORITHMIC SPEED CALCULATION FOR A TIMED CONTINUOUS PETRI NET

Calin Munteanu*, René David, Hassane Alla****

* *Politehnica University of Bucharest*[1]
** *Laboratoire d'Automatique de Grenoble (INPG-UJF-CNRS)*

Abstract: Continuous Petri nets with constant maximal speeds associated with transitions are concerned in this paper. Given an initial marking, the vector of instantaneous speeds can be obtained. This vector is constant for a time interval, and can change when the marking of a place becomes 0. An IB-state corresponds to a period of time when the speed vector is constant. Algorithmic calculation of the successive speed vectors is a difficult problem in the event of conflicts. This paper presents the bases of an algorithm which can be used if all the conflicts are solved by priorities. The more complex algorithm (available by internet) allowing conflict resolution by priorities for some conflicts and by sharing for other conflicts (in the same continuous Petri net) is mentioned. *Copyright © 2004 IFAC*

Keywords: Petri-nets, continuous, hybrid, timed, speed, algorithm.

1. INTRODUCTION

The reader is assumed to be familar with regular (i.e. discrete) Petri nets (PNs) defined by Petri (1962).

More recently, *continuous* PNs were defined (David and Alla, 1987). Autonomous continuous PNs have been shown to be a limit case of discrete PNs. In a continuous PN, the markings, arc weights and firing quantities are non-negative, as in a discrete PN, but are *not necessarily integers*. In a timed continuous PN, maximal speeds are associated with transitions. Other authors have proposed various results concerning these PNs (Demongodin, *et al.*, 2000; Recalde and Silva, 2000; Special issue, 2001).

The initial motivation leading to the concept of continuous PN was an approximate modeling of discrete systems with a large number of states (modeled by discrete PNs with large markings). It is clear that continuous PNs can also model continuous systems.

Hybrid PNs contain a discrete part and a continuous part, usually interacting (Le Bail, *et al.*, 1991; David and Alla, 2004). Given a timed hybrid PN (timings associated with discrete transitions and maximal speeds associated with continuous transitions), instantaneous behavior is analyzed in the following way: a stable marking of the discrete part is sought, then the instantaneous firing speeds of the continuous transitions are calculated. The semantics related to instantaneous firing speeds (i.e. local calculation given the marking and feeding flows of the input continuous places of a continuous transition) is relatively easy to define. However, automatic calculation for the whole continuous PN is difficult. Iterative algorithms presented in (David and Alla, 1987) and (Alla and David, 1998) do not work in all cases. Calculation by resolution of a linear

[1] This work was completed when the first author was working with the other authors in Grenoble.

programming problem (LPP) was used for some specific cases in (Balduzzi, *et al.*, 2000). The bases of a speed calculation method are presented in this paper. Many more details are given in (David and Alla, 2004) and a software is available at `http://sirphyco.lag.ensieg.inpg.fr/`.

The paper is organized as follows. Section 2 presents the concept of marking 0^+. In Section 3, the case where all the conflicts are solved by priorities is tackled and the concept of *passage* is illustrated. Resolution by sharing is explained in Section 4 and the software based on a general algorithm is briefly presented in Section 5. A comparison with the model and results in (Balduzzi, *et al.*, 2000) is given in Section 6 and Section 7 concludes the paper.

2. THE CONCEPT OF MARKING 0^+

Initially, this concept was introduced in order to increase the modeling power of hybrid PNs. However, *this concept appears naturally in a detailed analysis* of the behavior. This is illustrated in the sequel, the behavior of a continuous PN being a limit case of discrete (i.e. regular) PN. Normally, *token* and *mark* are synonymous. We use the word *mark* for the marking of the initial PN. Each mark is divided into k *tokens* and the new unit which is one kth of mark is called a *token*. This designation is consistent with the usual meanings since there is equivalence for $k = 1$.

Let us consider the timed continuous PN in Fig. 1a; the initial marking is (2, 0, 0) and the maximal speeds associated with T_1 and T_2 are $V_1 = 2$ and $V_2 = 3$. The discrete counterpart of this PN has the structure of Fig 1b, with $m_1 = 2$, $m_2 = 0$, $m_3 = 0$ (P_4 and P_5 specify the single server behavior: a single firing can occur at a time), and timings $d_1 = 1/2$ and $d_2 = 1/3$. Figure 1b is obtained from this discrete counterpart in the following way: each "moving" mark, i.e. involved in the *significant marking* (the *significant marking* is related to places whose markings are *not the same at any time*; e.g., in Fig. 1b, the marking m_4 of P_4 and m_5 of P_5 are not significant) is divided into k tokens, and the timings associated with the transitions are divided by k, i.e., $d_1 = 1/2k$ and $d_2 = 1/3k$. In this case, the token flow rates are $2k$ and $3k$.

The behavior of the continuous PN in Fig. 1a is the limit behavior of the discrete PN in Fig. 1b when k tends to infinity. Let us analyze the marking of P_2 in Fig. 1b (see also Fig. 1c). At time $t = 1/2k$, a token is deposited in P_2. This token enables T_2 which will be fired at $t = 1/2k + 1/3k = 5/6k$. The second firing of T_1 occurs at $t = 2/2k = 6/6k$. Hence, there is a token in P_2 between $t = 1/2k$ and $t = 5/6k$, P_2 is empty between $t = 5/6k$ and $t = 6/6k$, and so on. Figure 1c show that, between $t = 1/3k$ and $t = 1 + 1/3k$, a firing of T_2 occurs every $1/2k$, i.e. the firing rate is $2k$. If P_2 were never empty, the firing rate of T_2 would be $3k$, but in the present case, the firing rate of T_2 is slowed down by the firing rate of T_1.

Notation 1. The ratio α/k, where α is a finite positive number and $k \to \infty$, is denoted by 0^+.

According to the non-standard analysis (Robinson, 1966), 0^+ is a *non-standard number* infinitely close to 0. This non standard number has the property $0^+ + 0^+ = 0^+$.

The average marking of P_2 is $2/3k$ from $t = 1/3k$ to $t = 1 + 1/3k$. When k tends to infinity, it becomes $m_2(t) = 0^+$ for $t \in]0, 1]$ (according to Notation 1). During this time interval, the instantaneous[2] firing speed of T_2 is $v_2(t) = 2$. The firing speed of T_2 is slowed down by the firing speed of T_1: $v_1(t) = V_1 = 2$ and $v_2(t) = v_1(t) = 2$.

Thus, the variables related to the continuous timed PN in Fig. 1a, are expressed as follows *in standard numbers*. For[3] $t \in |0, 1|$: $v_1(t) = v_2(t) = 2$, $m_1(t) = 2 - 2t$, $m_2(t) = 0$, $m_3(t) = 2t$. For $t \in |1, \infty|$: $v_1(t) = v_2(t) = 0$, $m_1(t) = 0$, $m_2(t) = 0$, $m_3(t) = 2$.

Notation 2. When necessary to distinguish marking 0^+ from marking 0, notations $\tilde{m}_i$ for a place and $\tilde{\mathbf{m}}$ for a vector will be used.

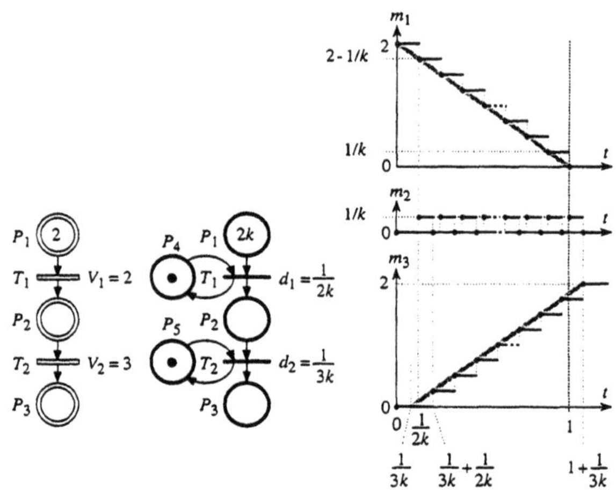

Fig.1 Illustration for marking 0^+.

---

[2] *Maximal firing speed* is noted with a capital V and *instantaneous firing speed* with a small v.

[3] Time interval $|a, b|$ means $]a, b[$ plus possibly a, possibly b, possibly both.

For example, according to Fig. 1c when $k \to \infty$, the marking of P_2 for the continuous PN in Fig. 1a is: $\tilde{m}_2(0) = 0$, $\tilde{m}_2(t) = 0^+$ for $t \in]0, 1]$, and $\tilde{m}_2(t) = 0$ for $t \in]1, \infty]$. For this PN, the significant marking is $\tilde{\mathbf{m}}(0) = (2, 0, 0)$ and $\tilde{\mathbf{m}}(1) = (0, 0^+, 2)$; as a matter of fact, one can see in Fig. 1c that $\tilde{m}_1(1)$ is exactly 0 whereas $\tilde{m}_2(1)$ is 0^+.

Let us now specify some vocabulary and notation. The sets of places and transitions are demoted by P and T. The marking is denoted by $\mathbf{m} = (m_1, m_2, m_3)$. The initial marking is $\mathbf{m}_0 = \mathbf{m}(0) = (2, 0, 0)$. Between $t = 0$ and $t = 1$, T_1 is *strongly enabled*, i.e. all its input places have a positive marking. Hence, it can be fired at its maximal speed: $v_1 = V_1 = 2$. Place P_2 is fed by the firing of T_1 at speed $v_1 = 2$. Since T_2 can be fired at a speed less than $V_2 = 3$, it is *weakly enabled.* An IB-state (invariant behavior state) is a period of time during which the instantaneous speed vector $\mathbf{v}$ is constant, i.e., the marking $\mathbf{m}$ evolves linearly. The *balance* of P_i, $B_i = I_i - O_i$ (I_i and O_i are the *feeding* and *drawing speeds*), corresponds to the time derivative of its marking, i.e., $\mathrm{d}m_i(t) / \mathrm{d}t = B_i(t)$.

For example, the behavior of the continuous PN in Fig. 1a corresponds to two successive IB-states, the first one from $t = 0$ to $t = 1$, and the second one $t = 1$ to $t = \infty$. During the first IB-state, $I_2 = v_1 = 2$, $O_2 = v_2 = 2$, $B_2 = I_2 - O_2 = 0$.

Property 1. (David and Alla, 2004) Let $|\,t_1, t_2\,|$ be a time interval corresponding to an IB-state. The marking of place P_i at the upper limit t_2 is $\tilde{m}_i(t_2) = 0^+$ if and only if: a) $\tilde{m}_i(t) = 0^+$ for $t \in |\,t_1, t_2\,|$, or b) $m_i(t)$ decreases to 0 AND $I_i(t) > 0$ for $t \in |\,t_1, t_2\,|$.

Consider for example the time interval $|\,0, 1\,|$ for the PN in Fig. 1a. According to Property 1a, since $\tilde{m}_2(t) = 0^+$ for the time interval, then $\tilde{m}_2(1) = 0^+$. Marking $m_1(t)$ decreases to 0 during the time interval, but $I_1(t) = 0$ during this time; hence $\tilde{m}_1(1) = 0$, according to Property 1b.

3. CONFLICT RESOLUTION BY PRIORITIES

Sections 3 and 4 are concerned with the calculation of the speed vector for *one IB-state*. The whole behavior will be obtained by a simple iteration of an algorithm related to an IB-state (Section 5). In this section, for any pair of transitions T_a and T_b involved in the same structural conflict, a priority rule $T_a < T_b$ (T_a takes priority over T_b) or $T_b < T_a$ is specified. The speed vector $\mathbf{v}_s$ related to IB-state s is obtained from the marking $\mathbf{m}_{s-1}$ at the beginning of this IB-state. The explanations and algorithm are illustrated with the example in Fig. 2. Four kinds of data are used. They correspond to sets of constraints denoted by C_1, C_2, C_3, and C_4. Sets C_1 and C_2 do not change whereas C_3 and C_4 change during the calculation related to an IB-state. Sets C_3 and C_4 change at each *passage.* The concept of *passage* will be presented just after the four sets of constraints.

1) *Speed limits.* According to model definition, an instantaneous speed must be positive and not greater than its specified maximal value: $C_1 = \{0 \le v_1 \le 2, 0 \le v_2 \le 2, \ldots, 0 \le v_6 \le 1\}$.
2) *Non-negative balances.* Since a marking cannot be negative, the balance of a place P_i whose marking is $m_i = 0$ at the beginning of an IB-state, must be non-negative. For our example, B_1, B_3, $B_4 \ge 0$, i.e.: $C_2 = \{v_3 + v_4 - v_1 - v_2 \ge 0, v_2 - v_4 \ge 0, v_4 - v_5 - v_6 \ge 0\}$.
3) *Transitions whose instantaneous speed is zero and surely firable transitions.* At some time in the calculation process, a set of surely firable transitions, denoted T_{SF}, is defined. All the transitions in this set are known to have a positive instantaneous speed. At the same time, all the other transitions have a zero speed. This set of zero speeds specifies C_3.
4) *Transitions whose instantaneous speed have already be calculated* at some time in the calculation process. This set of constraints is denoted by C_4.

The concept of *passage* is illustrated in the sequel with the example in Fig. 2; let $T_{SF}(h)$ and $C_3(h)$ denote the sets T_{SF} and C_3 at passage h.

Passage 1. Since T_3 is strongly enabled, $v_3 > 0$. Hence P_1 is fed and $v_1 > 0$ since $T_1 < T_2$. Right now, we do not know whether or not T_2 will be fired, or T_4, T_5, and T_6. Hence, $T_{SF}(1) = \{T_1, T_3\}$. Speeds v_1 and v_3 are then calculated: maximization of $J_1 = v_1 + v_3$, given the constraints C_1, C_2, and $C_3(1) = \{v_2 = 0, v_4 = 0, v_5 = 0, v_6 = 0\}$. Values $v_1 = 2$ and $v_3 = 3$ are obtained. Since $v_1 = V_1$ and $v_3 = V_3$, these speeds have been calculated once for all, and this provides a new set of constraints, denoted by $C_4(1)$, for the next passage: $C_4(1) = \{v_1 = 2, v_3 = 3\}$. From the speeds calculated up to now, the balance of place P_1, $B_1 = v_3 + v_4 - v_1 - v_2 = 1$ is positive. Hence, T_2 will be fired.

Passage 2. $B_1 > 0 \Rightarrow T_2$ fired $\Rightarrow P_3$ fed $\Rightarrow v_4 > 0 \Rightarrow P_4$ fed $\Rightarrow v_5 > 0$ since $T_5 < T_6$. Right now, we do not know whether or not T_6 will be fired. Hence, $T_{SF}(2) = \{T_1, T_2, T_3, T_4, T_5\}$.

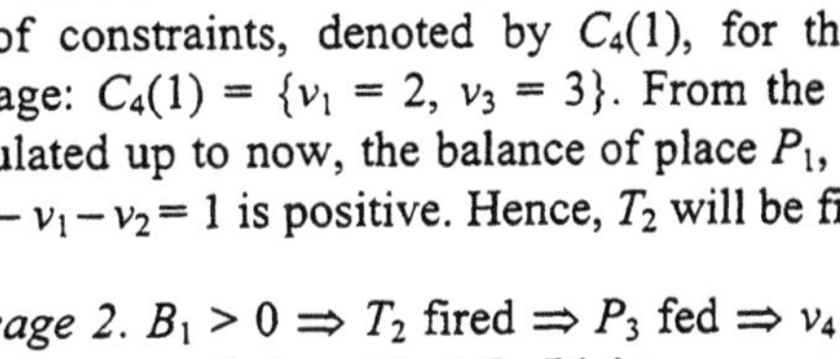

Fig. 2 Illustration for successive passages.

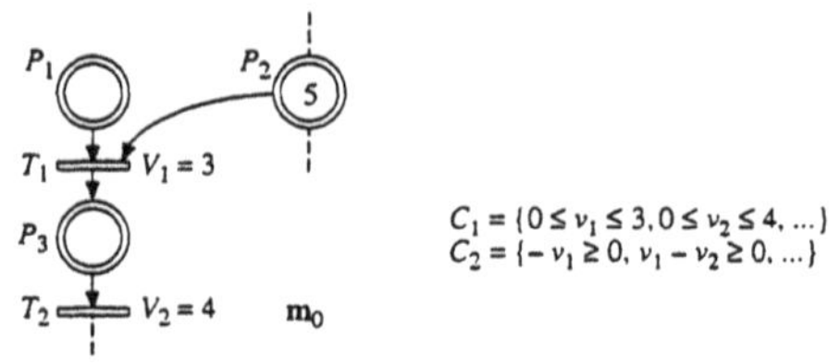

Fig. 3 Immediate results: $v_1 = v_2 = 0$.

Speeds v_2, v_4, and v_5 are then calculated: maximization of $J_2 = v_2 + v_4 + v_5$, given the constraints C_1, C_2, $C_3(2) = \{v_6 = 0\}$, and $C_4(1)$. Values $v_2 = 1.6$, $v_4 = 0.6$, $v_5 = 0.6$ are obtained. From the speeds obtained up to now, $B_4 = 0$. Hence, T_6 will not be fired (i.e. $v_6 = 0$). The speed calculation is finished: $\mathbf{v} = (2, 1.6, 3, 0.6, 0.6, 0)$ at the end of passage 2.

The *structure* of Algorithm 1 (used if all conflicts are solved by priorities) will be specified in the sequel. Sets C_1 and C_2 do not change during the speed calculation process; C_3 and C_4 change at each passage.

From C_1 and C_2, it can be immediately deduced that some instantaneous speeds must be 0. This is illustrated in Fig. 3. Since $m_1 = 0$, $B_1 \geq 0$ is required. Hence $B_1 = -v_1 \geq 0$ is in C_2. From $0 \leq v_1 \leq 3$ (in C_1) and $-v_1 \geq 0$ (in C_2), the value $v_1 = 0$ is obtained. Then, given $v_1 = 0$, $v_2 = 0$ is obtained from $0 \leq v_2 \leq 4$ and $v_1 - v_2 \geq 0$.

Structure of Algorithm 1 (resolution by priorities).

Step A. Initialization for IB-state s. Set up C_1, C_2. Find the T_j such that $v_j = 0$ is implied by C_1 and C_2, and place '$v_j = 0$' in $C_4(0)$. Let $h = 1$.

Step B. Passage h. Calculate $T_{SF}(h)$, $C_3(h)$, and J_h. Maximize J_h, given C_1, C_2, $C_3(h)$, and $C_4(h-1)$. Set up $C_4(h)$. **If** another passage is required **then** $h = h + 1$ and iterate *Step B*.

Step C. Ending. Calculate the various parameters related to IB-state s.

This structure is simple, but the complete algorithm is complex because it must take various difficulties into account. This new algorithm works in cases for which iterative algorithms presented in (Alla and David, 1998) were deficient, particularly for an example proposed by P. Caspi, from which the PN in Fig. 5b is inspired.

4. CONFLICT RESOLUTION BY SHARING

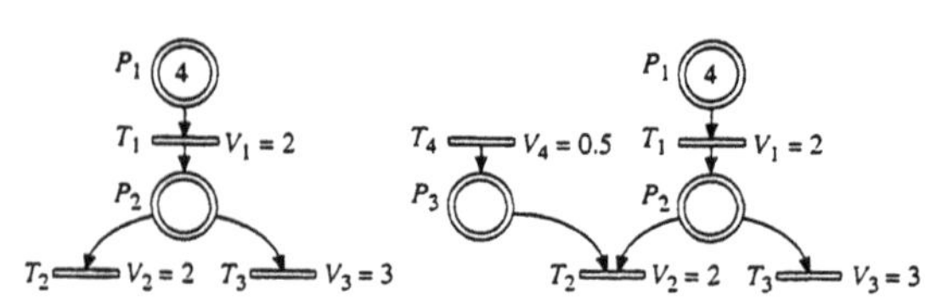

Fig. 4 Illustrations for sharings.

For the example in Fig. 4a, there is an actual conflict since $I_2 = v_1 = 2 < V_2 + V_3$. The conflict resolution, for this example, consists in finding values v_2 and v_3 such that:

$$v_2 + v_3 = 2. \qquad (1)$$

Priority $T_2 < T_3$ or priority $T_3 < T_2$ can be chosen. It is also possible to share the feeding flow I_2 between T_2 and T_3. A sharing rule is denoted by $[\alpha_2 T_2, \alpha_3 T_3]$, which means that the sharing must *aim at* $v_2 / \alpha_2 = v_3 / \alpha_3$. For example $[T_2, T_3]$ leads to $v_2 = 1$ and $v_3 = 1$.

If a sharing *aims at* some ratio between the speeds, it is possible that this ratio cannot be obtained because Equ. (1) must be satisfied. For the non-free choice PN in Fig. 4b, $v_2 \leq v_4 = 0.5$ because P_3 is empty. For the sharing $[T_2, T_3]$ the "part" allocated to each transition is 1 but T_2 cannot use completely its part since $v_2 \leq 0.5$. Hence, $v_2 = 0.5$ is obtained and $v_3 = 2 - v_2 = 1.5$ is obtained from (1).

An algorithm (Algorithm 2) mixing priorities and sharings has been obtained. If there is a single sharing together with priorities, there is no restriction about the structure of the PN, i.e. non-free choice and feedbacks are posible (for example, in Fig. 4b, v_4 could be a function of v_2 and v_3.

5. SOFTWARE

Algorithms 1 and 2 presented in the previous sections corresponds to the speed calculation for one IB-state. For the complete analysis behavior (i.e. speed vectors for all the successive IB-states), Algorithm 3 is used.

Algorithm 3 For all IB-states

Step 1. Initialization: timed continuous PN, initial marking $\mathbf{m}_0$, local conflict resolutions. Let $s = 1$.

Step 2. Setting the balances (literal expressions).

Step 3. Fire the transitions T_j such that $V_j = \infty$, which are enabled (discrete change). This allows to reach a stable marking. (A stable marking is such that evolution may be a continuous function of time but not a discrete change).

Step 4. Calculations for IB-state s using Algorithm 2. Let t_s be the time when this IB-state ends.

Step 5. **If** $t_s < \infty$ **then** let $\tilde{\mathbf{m}} = \tilde{\mathbf{m}}(t_s)$ and $s = s + 1$, and go to *Step 4* **else** END.

For the software available at `http://sirphyco.lag.ensieg.inpg.fr/`, the inputs (structure and initial marking of PN, maximal speeds, conflict resolution rules) may be given either in textual or graphic form. The results correspond to the evolution graph, i.e. marking $\mathbf{m}_{s-1}$, speed vector $\mathbf{v}$, and every place marking (function of time) for all successive IB-states.

6. EXAMPLE AND COMPARISON WITH THE MODEL AND RESULTS IN (Balduzzi, *et al.*, 2000)

In order to answer questions of reviewers, a comparative study of our results with the model and results in (Balduzzi, *et al.*, 2000) is required. In both cases, there is a "hierarchy" between the discrete transitions and the continuous ones: after a stable discrete marking has been obtained, the calculation of the speeds of continuous transitions can be performed (hence, in both cases, this calculation corresponds to a subjacent continuous PN).

6.1. The Models

Basically both models of hybrid PNs are similar: "First-order" in (Balduzzi, *et al.*, 2000) means that the speed vectors are piecewise constant and the markings evolve linearly. However, there are two differences.

1) In (Balduzzi, *et al.*, 2000), the instantaneous firing speed of transition T_j has a minimal value V_j' and a maximal values V_j (i.e. $V_j' \leq v_j \leq V_j$), whereas in our model there is only a maximal speed (i.e. $0 \leq v_j \leq V_j$). In the first case, it is possible that there is no solution, i.e. no speed vector can be found (for example if $V_1 = 2$ and $V_2' = 3$ in Fig. 1a). In the second case, there is always at least one solution ($\mathbf{v} = (0, 0, \ldots)$ in case of deadlock) and usually an infinity of solutions.

2) In (Balduzzi, *et al.*, 2000), a continuous transition is enabled if all its input discrete places have a sufficient marking: a continuous transition whose input continuous places are empty can be fired. With our notation, the model in (Balduzzi, *et al.*, 2000) is equivalent to having a marking 0^+ in all the empty places. In our model, as far as enabling is concerned, $\tilde{m}_i = 0^+$ and $\tilde{m}_i = 0$ must be distinguised.

6.2. Conflict Resolution

In (Balduzzi, *et al.*, 2000), global conflict resolution is suggested. There is no difficulty in adding a single criterion to a set of constraints in order to obtain the solution of a LPP. In case of *local conflict resolution* (which is consistent with the "philosophy" of Petri nets), the *resolution is restricted to free choice* PNs (priority or *fixed* ratios).

In our work, partially presented here and developped in (David and Alla, 2004), *local conflict resolution* is obtained *without restriction on the structure* if all the resolution rules are priorities between transitions or if there is at most one sharing between two transitions. Structural restrictions are only required if there are several resolutions by sharings in the PN.

6.3. Example

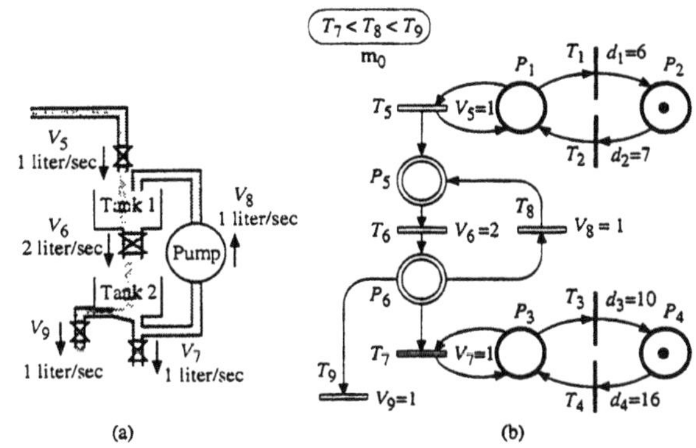

Fig. 5. (a) System to be modeled (represented when valve 5 open and valve 7 closed). (b) Timed hybrid PN model.

Let us consider the system in Fig. 5a. This system is made up of two tanks (numbers 1 and 2), four valves (numbers 5, 6, 7, and 9, corresponding to the index of speeds associated with them) which could be open or closed, and a pump. There is a permanent supply of tank 1, 1 liter/sec, when valve 5 is open. Liquid flows from tank 1 to tank 2 via valve 6 (2 liter/sec, assumed to be always open). Liquid in tank 2 can leave it by three ways, with their priorities obtained by gravity. The highest priority corresponds to valve 7 (1 liter/sec, if it is open). The second priority is the pipe leading to the pump (1 liter/sec). The third priority corresponds to valve 9 (1 liter/sec, assumed to be always open).

At initial time, it is assumed that *both tanks are empty* and both valves 5 and 7 are closed. Valve 5 will be open periodically from $t = 7$ to $t = 13$, then from $t = 20$ to $t = 26$ etc. Valve 7 will be open periodically from $t = 16$ to $t = 26$, then from $t = 42$ to $t = 52$ etc. Figure 5a shows the behavior when both tanks are empty while valve 5 is open and valve 7 is closed (between $t = 7$ and $t = 13$ for example). The behavior of this system can be modeled by the hybrid PN in Fig. 5b in which places P_5 and P_6 model the liquid quantities in tanks 1 and 2, and transitions T_5 to T_9 model the valves (same index) and the pump.

Result obtained by Algorithm 3. The behavior of the timed hybrid PN in Fig. 5a up to $t = 20$ is illustrated in Fig. 6. During the first IB-state, $v_5 = v_7 = 0$ since $m_1 = 0$ and $m_3 = 0$. Hence, since $v_5 = 0$, places P_5 and P_6 remain empty, i.e. $\tilde{m}_5 = 0$ and $\tilde{m}_6 = 0$. In Fig. 5a, the tanks and pipes under valve 5 are empty. At $t = 7$, P_1 becomes marked (valve 5 is open). From this time, $v_5 = 1$. Because of the priority $T_8 < T_9$ (T_7 is not enabled), $v_8 = 1$ and the speed vector is $\mathbf{v} = (v_5, v_6, v_7, v_8, v_9) = (1, 2, 0, 1, 1)$. This behavior is the sum of two flows: a flow whose speed is $v = 1$ from T_5 to T_9

and another flow whose speed is also $v = 1$ in the loop $P_5T_6P_6T_8P_5$. This is illustrated in Fig. 5a.
When valve 5 is closed, at $t = 13$, the flow from T_5 to T_9 disappears whereas the flow in the loop continues (due to priority $T_8 < T_9$). Then, the speed vector $\mathbf{v} = (0, 1, 0, 1, 0)$ is expected for the third IB-state. This result is effectively obtained with Algorithm 3. As a matter of fact, during the second IB-state, $\tilde{m}_5 = \tilde{m}_6 = 0^+$ because P_5 and P_6 are simultaneously fed and drained. It follows from Property 1a that $\tilde{m}_5(13) = \tilde{m}_6(13) = 0^+$, and this leads to the expected speed vector. The marking 0^+ in P_5 and P_6 corresponds to infinitely small values remaining in the places (illustrated by the liquid in the pipes in the loop in Fig. 5a). At $t = 16$, T_4 is fired and T_7 becomes enabled. This high priority transition empties the 0^+ marking in the loop (valve 7 open, the loop is emptied). The speed vector is then $\mathbf{v} = (0, 0, 0, 0, 0)$.

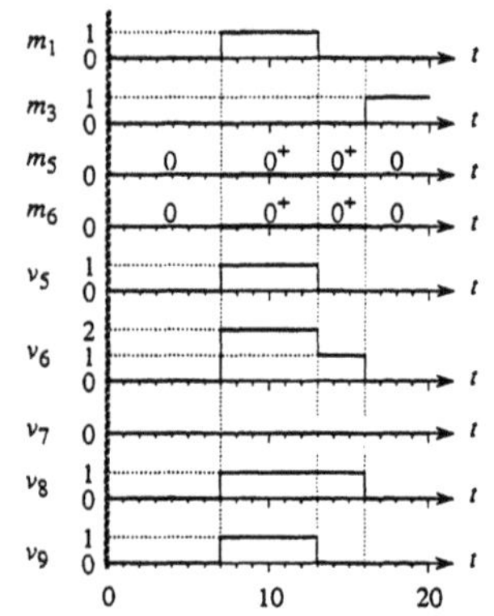

Fig. 6 Behavior of the timed hybrid PN in Fig. 5b up to t = 20, obtained by Algorithm 3.

Comparison with other algorithms. The results obtained by Algorithm 3 are summarized in Fig. 6. Let us emphasize that $(v_6(t), v_8(t)) = (0, 0)$ for $t \in]0, 7[$, $(2, 1)$ for $t \in]7, 13[$, $(1, 1)$ for $t \in]13, 16[$, $(0, 0)$ for $t \in]16, 20[$.

Algorithm in (Alla and David, 1998). This algorithm was not able to take the marking 0^+ into account. For the third IB-state (t = 13 to 16), the speed was calculated from $(m_5(13), m_6(13)) = (0, 0)$ and $v_6 = v_8 = 0$ was obtained instead of $v_6 = v_8 = 1$. (This was the problem pointed out by P. Caspi.)

Algorithm in (Balduzzi, *et al.*, 2000). This algorithm works as if a marking 0^+ were present in all the empty continuous places, *at any time* when a speed calculation is performed. This algorithm gives a correct value for the second and third IB-states, but it gives $v_6 = v_8 = 1$, instead of $v_6 = v_8 = 0$, for $t \in]0, 7[$ and $t \in]16, 20[$.

7. CONCLUSION

If there is no conflict, algorithmic calculation of the successive speed vectors is not difficult. If all the conflicts are solved by priorities, calculation of a single speed vector, corresponding to an IB-state, may require several passages, i.e. an iterated calculation. This is a difficult problem in the general case, the basis of which has been presented in the paper. In addition, resolution by sharing between transitions may be considered. The continuity between successive IB-states (Caspi's problem) described in this paper, has been solved thanks to conventional marking 0^+.

REFERENCES

Alla, H. and R. David (1998). A Modelling and Analysis Tool for Discrete Event Systems: Continuous Petri Net, *Performance Evaluation*, vol. **33**, pp. 175–199.

Balduzzi, F., A. Giua, and G. Menga (2000). First-Order Petri Nets: A Model for Optimization and Control, *IEEE Trans. on Robotics and Automation*, vol. **6,** N° 4.

David, R. and H. Alla (1987). Continuous Petri Nets, *8th European Workshop on Appli. and Theory of Petri Nets*, Saragosse (E), pp. 275-294

David, R. and H. Alla (2004?). *Discrete, Continuous, and Hybrid Petri Nets*, in preparation, to be published by Springer, Heidelberg.

Demongodin, I., M. Mostefaoui, and N. Sauer (2000). Steady State of Continuous Neutral Weighted Marked Graphs, in *Proc. Int. Conf. on Systems, Man and Cybernetics*, IEEE / SMC, USA, pp. 3186-3194.

Le Bail, J., H. Alla, and R. David (1991). Hybrid Petri Nets, *Proc. European Control Conference*, Grenoble (F).

Petri, C. A. (1962). *Communication with Automata*, Supplement 1 to Technical Report RADC-TR-65-337, N.Y., 1965. Translation by C.F. Greene of Kommunikation mit Automaten, PhD Dissertation, University of Bonn.

Recalde, L. and M. Silva (2000). PN Fluidification Revisited: Semantics and Steady State, *Proc. Int. Conf. on Automation of Mixed Processes: Hybrid Dynamic Systems* (ADPM 2000). Dortmund (D), pp.279-286.

Robinson, A. (1966). *Non Standard Analysis*, North-Holland, Amsterdam. Printed again: Princeton University Press, 1996.

Special Issue on Hybrid Petri Nets (2001). *Journal of Discrete Event Dynamic Systems: Theory and Applications*, vol. **11**, A. Di Febbraro, A. Giua, and G. Menga Eds.

ELSEVIER
IFAC
PUBLICATIONS
www.elsevier.com/locate/ifac

EARLY TERMINATION BY LOCAL STRING IN INCREMENTAL LANGUAGE CONTAINMENT TESTS

Arash Vahidi, Martin Fabian, Bengt Lennartson

The Control and Automation Lab. Department of Signals and Systems, Chalmers University of Technology, Sweden.
{ vahidi, fabian, bl } @s2.chalmers.se

Abstract: We investigate an incremental form of the two well-known language containment tests, verification of controllability and general language inclusion. An incremental algorithm is presented where in each verification cycle the results of previous rounds may be used for more efficient computation. Furthermore, we investigate reachability by local strings to allow early termination if it can be demonstrated that a test is bound to fail in the future. *Copyright © 2004 IFAC*

Keywords: Discrete Event Systems, Language Inclusion, Modular and Incremental Verification

1. INTRODUCTION

Two interesting topics in the class of *language containment* problems in formal methods are the *general language inclusion* test and the *controllability* test.

The language inclusion test verifies whether the behavioral properties defined as a *language* is included in some other such language. A special case of the language inclusion test is when the inclusion is allowed to be broken by a special set of events. The controllability test is such a test (Ramadge and Wonham, 1987).

As languages are sometimes infinite, they are often instead given as a set of small automata. The composition of such automata produces a single automaton which will generate the desired language. A *monolithic* approach to answer a language containment problem involves analyzing the automata in entirety by creating that single composite automaton. Due to the *combinatorial state explosion problem* however, this approach is very inefficient.

The *modular* and *incremental* algorithms presented in this work aim to avoid this difficulty by answering many smaller questions instead of a single but huge one. These algorithms also benefit from a decision criterion for *early termination* that exploits the *local behavior* of the involved automata. When this criterion holds, it can be proven that additional computation will no longer affect the outcome of the language containment test and the algorithm can terminate.

A demonstration of the performance of the introduced algorithm on a set of real-world and academical examples is given in the end of this paper, followed by a short discussion on the algorithms' strength and weakness and a brief comparison with some other approaches (Brandin *et al.*, 2000; Åkesson *et al.*, 2002).

2. PRELIMINARIES

We consider the deterministic finite automaton (Hopcroft *et al.*, 2001) $FA = \langle Q, \Sigma, \delta, q_i^i \rangle$ where the alphabet Σ consists of two disjoint controllable and uncontrollable sets of events $\Sigma = \Sigma_c \dot{\cup} \Sigma_u$. In addition, let $\delta(q, \sigma)!$ denote that δ is defined for the pair (q, σ).

Two or more automata can be *composed* by the $||$-operator to a single automaton by means of *Full Synchronous Composition* (FSC), as defined in Hoare's CSP (Hoare, 1985).

For the sake of readability, let superscript always distinguish automata in a composition. For example, $A = A^1|| \cdots ||A^n$ and $A^i = \langle Q^i, \Sigma^i, \delta^i, q_i \rangle$. Furthermore, for a set of automata indexes I, let $||(I)$ denote their composition (for example $X = ||(\{1,2\}) = X^1||X^2$), $\Sigma(I)$ be their union alphabet, and let $q_i(I)$ denote the initial state. In addition, $Q_r(I)$ is the set of *reachable* states in the composition $||(I)$.

For *online* full synchronous composition of several automata the *conjunctive transfer functions* are used. A conjunctive transfer function is created by modifying the transfer function of each automaton, δ^i to $\hat{\delta}^i$ by adding self-loops to the automaton for events outside its alphabet. This is due to the fact that an event not in alphabet of an automaton can not be blocked by that automaton:

$$\hat{\delta}^i(q^i, \sigma) = \begin{cases} \delta^i(q^i, \sigma) & \text{if } \delta^i(q^i, \sigma)! \\ q^i & \text{if } \sigma \notin \Sigma^i \\ \text{undefined} & \text{otherwise} \end{cases}$$

This is simple *event-compensation* and is used to modify a set of automata with possibly different alphabets to another set of automata with equal alphabets. After that, composition is a simple matter of computing the conjunction of transfer functions:

$$\delta = \{\langle\langle q^1, \ldots, q^n\rangle, \sigma, \langle \dot{q}^1, \ldots, \dot{q}^n\rangle\rangle \mid \bigwedge_{1 \le i \le n} \hat{\delta}^i(q^i, \sigma) = \dot{q}^i\}$$

From the definitions above, it should be clear that we allow automata to have different alphabet. To be able to work at a language level, inverse projection of the languages generated by each automaton (i.e., $P^{-1} : (\Sigma^i)^* \rightarrow \Sigma^*$) is applied. For the sake of readability, all references to such operations have been removed.

The conjunctive transfer functions can be used for *partitioned* reachability search. For example, $Q_r(I)$ is defined as **Reachable**$(I, \{q_i(I)\}, \Sigma(I))$, where the implementation of the algorithm *Reachable* is shown below:

Algorithm 1. Reachable
input : I, Q_0, Σ'
let $k := 0$
repeat
  $Q_{k+1} := Q_k \cup \{\langle \dot{q}^a, \dot{q}^b, \cdots \rangle \mid \exists \sigma \in \Sigma', \langle q^a, q^b, \cdots \rangle \in Q_k.\ (\forall i \in I)\ \hat{\delta}^i(q^i, \delta) = \dot{q}^i\}$
  $k := k + 1$
until $Q_k = Q_{k-1}$
return Q_k

3. LANGUAGE INCLUSION AND CONTROLLABILITY

In a language inclusion test, given two prefix-closed regular languages K and L, one desires to verify whether the first language *includes* the second language[1]:

$$L \subseteq K$$

If $K \neq \emptyset$, this can be written as:

$$K\Sigma \cap L \subseteq K \qquad (1)$$

Where Σ is chosen to be $\Sigma^K \cap \Sigma^L$. It is however sufficient to have $\Sigma = \Sigma^L$:

Theorem 1. Given two regular non-empty languages K and L

$$K\Sigma^L \cap L \subseteq K \ \Leftrightarrow\ L \subseteq K$$

PROOF.
If:
$K\Sigma^L \cap L \subseteq L$ and $L \subseteq K \ \Rightarrow K\Sigma^L \cap L \subseteq K$.
Only if:
$K\Sigma^L \cap L \subseteq K \ \Leftrightarrow\ K\Sigma^{L*} \cap L \subseteq K$
(see (Kumar et al., *1991))*
$K\Sigma^{L*} \cap L = [K\Sigma^{L*} \cup \Sigma^{L*}] \cap L \qquad (\epsilon \in K)$
$[K\Sigma^{L*} \cup \Sigma^{L*}] \cap L \subseteq K \ \Rightarrow\ L \subseteq K \qquad (L \subseteq \Sigma^{L*})$
□

Furthermore, (1) can be verified one event at a time.

Lemma 2. For a language K to include another language L with respect to an alphabet Σ, K must include L with respect to all events in Σ. The reverse also holds, that is

$$K\Sigma \cap L \subseteq K \ \Leftrightarrow\ (\forall_{\sigma \in \Sigma})\ K\{\sigma\} \cap L \subseteq K$$

3.1 A State-Based Approach

To disprove statements such as $K\Sigma' \cap L \subseteq K$, it is sufficient to find a counterexample in form of $s \in K \cap L$ and $st \in L - K \quad (t \in \Sigma')$.

Given that K and L are generated by automata X and Y respectively, a *state-based* approach is to find a suitable t and s as following: For all possible pairs of states $\langle q_x, q_y \rangle \in Q^X \times Q^Y$ such that some $t \in \Sigma'$ is enabled in q_y but not in q_x try to find the string $s \in (\Sigma^X \cup \Sigma^Y)^*$ such that $\delta^X(q_i^X, s) = q_x$ and $\delta^Y(q_i^Y, s) = q_y$.

[1] It is very important to note that the presented allows the automata to have different alphabets which are then implicitly modified to some global alphabet by means of inverse projections.

Let $\theta_{nc\Sigma'} \subseteq Q^X \times Q^Y \times \Sigma'$ denote the set of *incomplete transitions* associated with an event in Σ. An incomplete transition (q^1, q^2, σ) is a half transition such that $\delta^Y(q^1, \sigma)$ is enabled while $\delta^X(q^2, \sigma)$ is not.

All incomplete transitions are potential counterexamples to a language inclusion proof. The real counterexamples are those half-transitions where the states are *reachable* from the initial state. If the theoretical state space of the composition of X and Y is $Q = Q^X \times Q^Y$, let $Q_{nc\Sigma'} \subseteq Q$ be all the source-states of $\theta_{nc\Sigma'}$. Assume that Q_r is the set of states reachable from the initial state. Then if $Q_{nc\Sigma'} \cap Q_r$ is non-empty, the language inclusion have failed and each state in the intersection is a counterexample. This simple approach transforms the language inclusion problem into a *state reachability problem*.

In practice, X and Y are often represented as compositions of two disjoint sets of automata I and J, i.e. $X = ||(I)$ and $Y = ||(J)$. In such case, there are different approaches for solving the language containment problem. Three common approaches are illustrated in Figure 1. In a *monolithic* setup, the current problem is viewed as one single large problem. In this context, it would mean creating X and Y by means of composition in order to answer the language containment problem.

In *modular* form, one starts by solving a smaller sub-problem and only if that does not give an answer, a larger subproblem is chosen and so on. An *incremental* approach is like a modular one expect that it tries to reuse the previously computed results for better performance.

3.2 Modular Language Inclusion

A *monolithic* algorithm for language inclusion verification is often intractable due to its large space and time requirements. A *modular language inclusion* algorithm on the other hand, may be able verify Equation (1) efficiently. It is based on the following observations:

Observation 1. $\exists j \in J.\ L(Y^j) \subseteq L(X) \Rightarrow L(Y) \subseteq L(X)$

Observation 2. $\forall i \in I.\ L(Y) \subseteq L(X^i) \Rightarrow L(Y) \subseteq L(X)$

A corollary of Observation 1 is that

$$\exists\, \emptyset \subset J' \subseteq J.\ L\left(||(J')\right) \subseteq L(X) \Leftrightarrow L(Y) \subseteq L(X)$$

The combination of this and Observation 2 is the cornerstone of modular language inclusion:

Lemma 3. Let $X = ||(I)$, and $Y = ||(J)$, then

$$(\forall i \in I)\ (\exists J'.\ \emptyset \subset J' \subseteq J)\ L\left(||(J')\right) \subseteq L(X^i) \Leftrightarrow L(Y) \subseteq L(X)$$

Based on Lemma 3, a *modular language inclusion algorithm* works as follows. For every $i \in I$, a small J' is chosen and the language inclusion of $\{i\}$ against J' is verified. If this fails, the algorithm adds another automaton to J' and repeats the procedure above. If there are no more automata to add, then the test fails. Due to lack of space, a complete modular algorithm can not be presented in this paper. Such algorithms are however explained in great detail in (Brandin *et al.*, 2000) and (Åkesson *et al.*, 2002).

3.3 Incremental Language Inclusion

The modular algorithm aims to verify the emptiness of $Q_{nc\Sigma'}(\{i\}, J') \cap Q_r(\{i\} \cup J')$ for each $i \in I$ and some $J' \subset J$ [2]. To do this, $\Theta_{nc\Sigma'}(\{i\}, J')$, $Q_{nc\Sigma'}(\{i\}, J')$ and $Q_r(\{i\} \cup J')$ have been computed, which are abandoned if the test fails and a new automaton $j \in J - J'$ needs to be added to J'.

There is a significant amount of redundant work here as these sets are forgotten, and then recomputed. To remedy this, we present an incremental algorithm that will (to some extent) reuse the previous results for better efficiency. The idea behind incremental computation is probably best explained by Figure 1.c.

It is obvious that the new $\Theta_{nc\Sigma'}$ can be computed incrementally:

$$\Theta_{nc\Sigma'}(\{i\}, J' \cup \{j\}) = \{\langle q^i, q^{J'}, q^j, \sigma\rangle \mid \langle q^i, q^{J'}, \sigma\rangle \in \Theta_{nc\Sigma'}(\{i\}, J') \wedge \delta^j(q^j, \sigma)!\ \}$$

From which $Q_{nc\Sigma'}$ can be computed. Now, assume that the active alphabet of $Q_{nc\Sigma'}$, denoted Σ'', differs from Σ':

$$\Sigma'' = \{\sigma \in \Sigma' \mid \langle q^i, q^{J'}, \sigma\rangle \in \Theta_{nc\Sigma'}(\{i\}, J') \wedge \langle q^i, q^{J'}\rangle \in Q_r(\{i\} \cup J')\ \}$$

Then it has already been proved that incomplete states with respect to the events $\Sigma' - \Sigma''$ are not reachable. According to Lemma 2, the search can continue with respect to the smaller alphabet Σ''. Which is equal to the following modification to $\Theta_{nc\Sigma'}$:

$$\Theta_{nc\Sigma''}(\{i\}, J' \cup \{j\}) = \{\langle q^i, q^{J'}, q^j, \sigma\rangle \mid \langle q^i, q^{J'}, \sigma\rangle \in \Theta_{nc\Sigma'}(\{i\}, J') \wedge \delta^j(q^j, \sigma)! \wedge \langle q^i, q^{J'}\rangle \in Q_r(\{i\} \cup J')\ \}$$

Since both $\Theta_{nc\Sigma''}(\{i\}, J')$ and $Q_r(\{i\} \cup J')$ were computed in the last test, these values can be reused with no extra computation cost. This is the concept behind the incremental language inclusion algorithm:

[2] Note the slight change in the notation. Here sets of indices are used instead of automata.

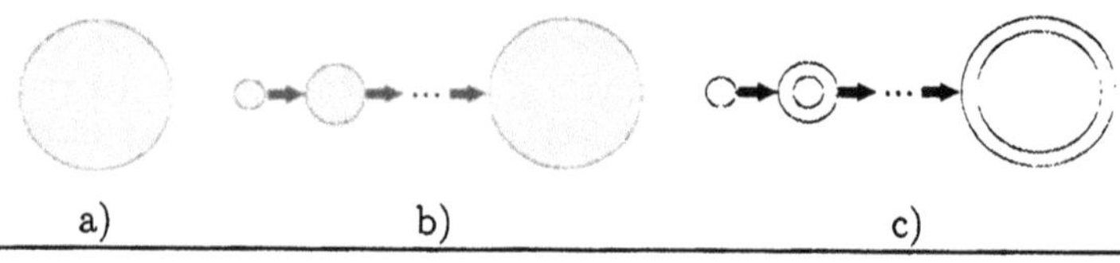

Fig. 1. Illustrating the difference between the (a) monolithic, (b) modular and (c) incremental approaches.

Algorithm 2. IncrementalLanguageInclusion
input : I, J
for *each* $i \in I$ **do**
  $\Sigma' := \Sigma^i \cap \Sigma(J), R := J, J' := \emptyset$
  $\Theta_{nc\Sigma'}(\{i\}, \emptyset) := \{\langle q, \sigma \rangle \in Q^i \times \Sigma' | \neg\delta^i(q, \sigma)!\}$
  while $R \neq \emptyset$ **do**
    choose *one automaton j from R*
    $J'_{old} := J'$
    $J' := J'_{old} \cup \{j\}$
    $\Theta_{nc\Sigma'}(\{i\}, J') := \{\langle q, q', \sigma \rangle \mid \langle q, \sigma \rangle \in \Theta_{nc\Sigma'}(\{i\}, J'_{old}) \wedge \delta^j(q^j, \sigma)!\}$
    if $Q_r(\{i\} \cup J') \times \Sigma' \cap \Theta_{nc\Sigma}(\{i\}, J') = \emptyset$ **then**
      - *test passed for i:*
      continue *with next automaton in I*
    end
    - *Update* Σ' *and* $\Theta_{nc\Sigma'}$
    $\Sigma' := \{\sigma \in \Sigma' \mid \langle q, \sigma \rangle \in \Theta_{nc\Sigma'}(\{i\}, J') \wedge q \in Q_r(\{i\} \cup J')\}$
    $\Theta_{nc\Sigma'}(\{i\}, J') := \Theta_{nc\Sigma'}(\{i\}, J') \cap Q \times \Sigma'$
  end
  return *test failed*
end
return *test passed*

3.4 Early Termination

Algorithm 2 is semi-incomplete in the sense that a negative answer is reported first after all automata in J have been added to J'. This is most often not intractable due to the exponential characteristic of the reachability test.

What is desired here is an *early termination criterion* that can guarantee the reachability of an incomplete transition without the need of building the complete J-automaton. If it can be proven that a corresponding state is reachable in the *global* state space, then there is no point trying to add more automata to make the trace (the string s) leading to this state to disappear.

Early termination is in fact used in the algorithm presented in (Åkesson *et al.*, 2002) (although this is not explicitly stated in the paper). Åkesson/Flordal/Fabian use a guided search algorithm to prove that a state in $Q_{nc\Sigma'}$ is globally reachable. Also (Brandin *et al.*, 2000) uses some type of early termination. While it is not described in the paper it seems that Brandin/Malik/Dietrich verify whether any of the strings leading into $Q_{nc\Sigma'}$ are accepted by all automata. Unfortunately, no implementation details are available about their approach.

As we intend to use Binary Decision Diagrams for better performance, a depth-first approach such as the guided search used in (Åkesson *et al.*, 2002) cannot be employed. Therefore, the *local string* early termination criterion presented below instead uses a breadth-first approach. It builds upon the observation that events not in the alphabet of an automaton are always accepted by that automaton. Consequently, strings of such events are also always accepted. This leads to the local reachability lemma:

Lemma 4. Let $X = \langle Q^X, \Sigma^X, \delta^X, q_i^X \rangle$ and $Y = \langle Q^Y, \Sigma^Y, \delta^Y, q_i^Y \rangle$. Then $(\Sigma^X - \Sigma^Y)^* \cap L(X) \subseteq L(Y||X)$.

As a corollary to Lemma 4, given a set of automata $M' \subset M$ and a state in $||(M')$ that is reachable from the initial state by a local string (i.e. a string in $[\Sigma(M') - \Sigma(M - M')]^*$), the projection of that state into the global state space of $||(M)$ is always reachable in $||(M)$. Using this method, it is sometimes possible to show that a half-transition is reachable in the global state space without actually constructing it (let $J' = M'$ and $J \cup I = M$). However, for early termination it is also needed to prove that such transition is also *enabled* in J (recall that this transition is disabled in I by definition).

There are various ways to show that a transition is enabled in J, with various degrees of incompleteness and computation cost. For example, one can easily verify that the considered event is enabled from the initial state [3] of $||(J - J')$. A better approximation is to check whether a transition is associated with an event that is local in J' with respect to J (note: not w.r.t. $I \cup J$), thus always enabled $J - J'$. This method seems to give a good balance between efficiency and accuracy and is used in our prototype implementation.

[3] As $||(J')$ has progressed by local strings, $||(J - J')$ is still in its initial state.

The local-reachable states can also be used in the incremental algorithm to make the reachability search itself incremental. The ideas of incremental local reachability builds upon the following observation:

Observation 3. Given a set of automata X and new automaton x, with $X' = X \cup \{x\}$, it is trivially shown that

$$\begin{aligned} & \mathit{Reachable}\,(X, \{q_i(X)\}, \Sigma_{local}(X)) \times \\ & \quad \mathit{Reachable}(\{x\}, \{q_i^x\}, \Sigma_{local}^x) \subseteq \\ & \mathit{Reachable}\,(X', \{\, q_i(X') \,\}, \Sigma_{local}(X')) \subseteq \\ & \quad \mathit{Reachable}\,(X', \{\, q_i(X') \,\}, \Sigma(X')) \end{aligned}$$

Where the local alphabets are with respect to some superset of X'.

3.5 Controllability

It is already known that controllability can be verified using a modular algorithm (Ramadge and Wonham, 1986). An interesting question is whether an incremental language inclusion algorithm such as Algorithm 2 can be used to verify controllability. The answer to this question is *yes, but with some modifications.*

If language containment algorithm is considering a strict subset of Σ^L, for example in a controllability test where $\Sigma' = \Sigma_u^K \cap \Sigma_u^L$, then Theorem 1 does not hold anymore. Lemma 3 gives therefore only a sufficient condition. The following adjustment to Lemma 3 makes the condition sufficient and necessary:

Lemma 5. Let $X = ||(I)$, and $Y = ||(J)$, then $L(X)\Sigma' \cap L(Y) \subseteq L(X)$ if and only if

$$\begin{aligned} & (\forall i \in I)\ (\exists I'.\ \{i\} \subseteq I' \subseteq I) \\ & \quad (\exists J' \subseteq J.\ \Sigma(J') \geq \Sigma^i \cap \Sigma(J)) \\ & \quad\quad L\,(||(I'))\,\Sigma' \cap L\,(||(J')) \subseteq L\,(||(I')) \end{aligned}$$

This has some minor effects on the implementation of the algorithm as $\Theta_{nc\Sigma'}$ must now be divided into two parts (one for I' and one for J'). Due to its complexity, the algorithm is not shown here.

4. INITIAL RESULTS

The DES tool Supremica currently includes efficient modular and monolithic algorithms for language inclusion and controllability verifications [4]. On top of this software, we have added four new algorithms: modular and incremental versions of the language inclusion and controllability test that all use a computation engine based on Binary Decision Diagrams (BDDs) (Bryant, 1986). Due to lack of good language inclusion benchmark models, only the controllability algorithms could be tested against a set of real-world and academical examples [5]:

- AGV: coordination of an Automatic Guided Vehicle from (Holloway and Krogh, 1990).
- C3: model of an alarm and central locking system in a car.
- AIP 1 and 2 are the models "AIP4, SYS4" and "AIP4, TOP TU1" of a highly automated manufacturing system (Leduc, 2002).

Table 1. Performance comparison for the controllability test.

	AGV	C3	AIP-1	AIP-2
C	N	Y	Y	Y
A	16	53	39	39
R	5×10^{10}	1.2×10^{26}	2.5×10^{25}	4.6×10^{22}
T_a	0.03	-	-	-
T_b	0.06	0.3	-	-
T_c	0.3	0.6	4.2	3.2
T_d	0.3	0.7	3.5	2.8

The results are presented in Table 1. The row C contains result of the controllability test. A is the number of automata in the systems and R is the (theoretical) number of states.

Tx is execution time in seconds where the letter x represents the used algorithm: a=monolithic, b=old modular, c=modular, d=incremental. Algorithms c and d were implemented by the authors while a and b use the implementation of (Åkesson *et al.*, 2002). A symbol "-" indicates execution failure (out of memory) [6].

As seen in Table 1, the new algorithms are not suited for problems that have an obvious or easy solution due to the large overhead introduced by the BDD engine. In the first two cases, the startup time of these algorithms dominates the computation time. For harder problems however, algorithm c and d are very successful while algorithm b fails due to the failure of its early termination procedure.

A disadvantage of the presented algorithms is that the termination criterion is incomplete. An enabled half-transition in $\Theta_{nc\Sigma'}$ may be reachable in the global states without having strings of local-events leading to it. Consequently, more automata than necessarily may be composed by these al-

[4] The original tool (at the moment without our modifications) can be found at http://www.supremica.org.

[5] We have deliberately avoided scalable benchmark examples such as dining philosophers or the transfer line as their inherited "shallow" structure can never stress a good modular or incremental algorithm.

[6] All tests were done on a low-end desktop PC (1.8 GHz P4 with 512 MB RAM of which only 64 MB was allotted).

gorithms. This is clearly seen in Table 2, where the modular algorithm of Supremica (Algorithm *b*) and the incremental algorithm (Algorithm *d*) are compared. In the table, *compositions* is the total number of compositions made while *depth* denotes the maximum number of automata in a composition.

Table 2. Comparing the compositions created by different algorithms.

	AGV	C3	AIP-1	AIP-2
	Algorithm *b*:			
Compositions	1	19	-	-
Depth	3	5	-	-
	Algorithm *d*:			
Compositions	10	50	105	99
Depth	11	8	18	16

The computation overhead introduced by these compositions is a clear disadvantage of our approach. The underlying BDD engine (based on the algorithm presented in (Vahidi *et al.*, 2004)) seems however to handle compositions with up to $25-30$ automata efficiently so this does not seem to be an immediate problem.

Beware that the efficiency of the presented algorithms is due to the combination of a good data structure (BDDs) and intelligent algorithms adopted to this data structure. A straight forward implementation with BDDs, or even a BDD version of the algorithm used in (Åkesson *et al.*, 2002) is bound to fail.

5. CONCLUSIONS AND FUTURE WORK

We have discussed algorithms for verification of language inclusion and controllability, which use an early termination procedure based on local-reachability. An efficient BDD-based implementation has been done on top of a freely available DES/SCT tool. Initial investigations indicate that this implementation successfully exploits the modularity of some very large real-world models.

Although the results are promising, there are places for improvements. For example, the automata selection heuristic must be studied more carefully. This heuristic seems to have significant impact on the running time of the algorithms. It also seems to affect the extra automata compositions discussed earlier.

Finally, it should be mentioned that although the focus in this work has been on verification, the presented algorithms can also be used for efficient synthesis of supervisor controllers (Åkesson *et al.*, 2002).

REFERENCES

Åkesson, Knut, Hugo Flordal and Martin Fabian (2002). Exploiting modularity for synthesis and verification of supervisors. In: *Proceedings of the 15th Triennial World Congress of the International Federation of Automatic Control.* Barcelona, Spain.

Brandin, B.A., R. Malik and P. Dietrich (2000). Incremental system verification and synthesis of minimally restrictive behaviors. In: *Proceedings of the 2000 American Control Conference.* Chicago, USA. pp. 4056–4061.

Bryant, Randal E. (1986). Graph-based algorithms for Boolean function manipulation. *IEEE Transactions on Computers* **35**(8), 667–691.

Hoare, C.A.R. (1985). *Communicating Sequential Processes.* Prentice-Hall International Series in Computer Science.

Holloway, L.E. and B.H. Krogh (1990). Synthesis of feedback control logic for a class of controlled Petri nets. *IEEE Transactions on Automatic Control* **35**(5), 514–523.

Hopcroft, J.E., R. Motwani and J.D. Ullman (2001). *Introduction to Automata Theory, Languages, and Computation.* Addison-Wesley.

Kumar, Ratnesh, V. K. Garg and S. I. Marcus (1991). On controllability and normality of discrete event dynamical systems. *Systems and Control Letters* **17**, 157–168.

Leduc, R.J. (2002). Hierarchical Interface Based Supervisory Control. PhD thesis. Department of Electrical and Computer Engineering, University of Toronto.

Ramadge, Peter J. and William Murray Wonham (1986). Modular supervisory control of discrete event systems. In: *Proceedings of 7th International Conf. Analysis and Optimization of Syst.* (A. Bensoussan and J. L. Lions, Eds.). Springer–Verlag. Antibes, France. pp. 202–214.

Ramadge, Peter J. and William Murray Wonham (1987). Supervisory control of a class of discrete event processes. *SIAM Journal of Control and Optimization* **25**, 206–230.

Vahidi, Arash, Bengt Lennartson and Martin Fabian (2004). Efficient supervisory synthesis of large systems. In: *Proceedings of the International Workshop on Discrete Event Systems (WODES'04).*

ELSEVIER
IFAC
PUBLICATIONS
www.elsevier.com/locate/ifac

TOWARDS OPTIMAL CTL* MODEL CHECKING OF TIME PETRI NETS

Hanifa Boucheneb *, Rachid Hadjidj *

* *Department of Computer Engineering,*
École Polytechnique de Montréal,
P.O. Box 6079, Station Centre-ville, Montréal, Québec
{hanifa.boucheneb,rachid.hadjidj}@polymtl.ca

Abstract: This paper considers the Time Petri Net model (*TPN* model) and proposes a contraction of its generally infinite state space. This contraction preserves all CTL^* properties of the model and produces finite graphs for bounded *TPN* models. When compared with other approaches ((Yoneda et al., 1998) ; (Berthomieu et al., 2003)), these graphs are smaller and much faster to compute. This paper shows also how to apply a fast computing bisimulation reduction rule to the obtained graphs so as to achieve or approach the optimal size with minor efforts. *Copyright © 2004 IFAC*

Keywords: Time Petri-nets; formal methods; real-time systems, verification; temporal logics; graph.

1. INTRODUCTION

Applying CTL^* model checking approach to the Time Petri net model (*TPN* model) needs contracting its generally infinite state space into some finite graph preserving CTL^* properties. Berthomieu and Diaz (Berthomieu et al., 1991) proposed a contraction of the *TPN* state space preserving *LTL* properties of the model but not necessarily CTL^* ones. Yoneda and Ryuba (Yoneda et al., 1998) developed another contraction called *Atomic State Class Graph* (*ASCG*). This contraction preserves CTL^* properties but has the disadvantage of yielding infinite graphs for some bounded[1] models. Recently, Berthomieu and Vernadat (Berthomieu et al., 2003) proposed an alternative construction of *ASCGs* producing smaller graphs, in shorter times. It also generates finite *ASCGs* for all bounded *TPN* models. This paper considers Berthomieu-Vernadat's approach and improves some of its steps to boost the computations and reduce the size of produced graphs. It also extends this approach by a very light extra phase to achieve further contractions. Berthomieu-Vernadat's approach proceeds in two main phases. In the first phase, the *TPN* state space is contracted into a graph preserving LTL properties, called *Strong State Class Graph (SSCG)*. In the second Phase, the *SSCG* is run through a refinement process to restore branching properties of the model. The improvements presented here focus mainly on the first phase of the process. The first modification consists in representing explicitly all states, including those reachable by time progression (hidden in Berthomieu-Vernadat's approach). In most cases, this explicit representation induces better contractions in both phases. The second modification consists in constructing a state class graph that does not necessarily preserve LTL properties of the model. The result is that the generated state class graphs are just a fraction of the size of their equivalent in the Berthomieu-Vernadat's approach. They are also computed in very shorter times. Using these graphs, it was possible to boost the second phase and obtain a noticeable gain

* Research supported by the NSERC grant num.238841-2001.
Corresponding author H. Boucheneb. Tel. +(540)-340-4711. Fax +(540)-340-4711.
[1] The model has a finite number of reachable markings.

in size too. For further contractions, an extra phase applies a fast computing bisimulation reduction rule. The results are compared to optimal sizes obtained using minimization under bisimulation [2]. Most of the time, these results are either optimal or very close to the optimal.

Section 2 of this paper is devoted to some definitions related to the *TPN* model and its different state space contractions. Section 3 presents the first phase of Berthomieu-Vernadat's approach. Section 4 develops the improvements to the first phase. Section 5 shows how to construct the *ASCG*. The bisimulation reduction rule is presented in section 6. Finally, section 7 reports some experimental results.

2. TIME PETRI NETS

2.1 Definition and behavior

A Time Petri Net (*TPN*) is a Petri Net augmented with time intervals associated to transitions. Formally, a *TPN* is a tuple $(P, T, Pre, Post, M_0, Is)$ where:

- P and T are finite sets of places and transitions such that $P \cap T = \emptyset$;
- Pre and $Post$ are the backward and the forward incidence functions $P \times T \longrightarrow \aleph$, $\aleph$ is the set of nonnegative integers;
- M_0 is the initial marking, $M_0 : P \longrightarrow \aleph$;
- $Is : T \rightarrow Q^+ \times (Q^+ \bigcup\{\infty\})$, Q^+ is the set of nonnegative rational numbers. Function Is associates with each transition t an interval $[tmin(t), tmax(t)]$ called the static firing interval of t. $tmin(t)$ and $tmax(t)$ are respectively the minimum and maximum firing delays.

Let M be a marking and t a transition. Transition t is enabled for the marking M iff all required tokens for firing t are present in M, i.e.: $\forall p \in P, M(p) \geq Pre(p,t)$. $En(M)$ denotes the set of all transitions enabled for the marking M.

The *TPN* state, called *clock state*, consists of a marking and a clock valuation function denoted V. This function associates with each enabled transition a clock. When a transition t becomes enabled, its clock is initialized to *zero*. The clock increases synchronously with time until the transition is fired or disabled by another firing. An enabled transition t can fire if the value of its clock is inside its static interval ($Is(t) = [tmin(t), tmax(t)]$). It must be fired immediately, without any additional delay, when its clock reaches its maximum firing delay ($tmax(t)$). Initially, the model is in its initial state (M_0, V_0), where M_0 is the initial marking and V_0 the null function [3]. The state of the model evolves either by time progressions (clocks increase with time) or by firing transitions. The firing of a transition is supposed to take no time, but leads to another marking (required tokens disappear while produced ones appear). Because of time density, the *TPN* model has in general infinite reachable states (i.e.: infinite state space), even if it is bounded. Its analysis by enumerative techniques must pass by state space contractions.

[2] This technique computes the coarsest bisimulation.

[3] $\forall t \in En(M_0), V_0(t) = 0$.

2.2 Concrete state spaces

The first abstraction of the *TPN* state space consists in hiding states reachable by time progressions. In the obtained graph, called the *concrete state space*, only states accessible via firing transitions are represented. This graph can be defined by a tuple $(\Sigma, \longrightarrow, \sigma_0)$ where:

- $\sigma_0 \in \Sigma$ is the initial state, i.e: $\sigma_0 = (M_0, V_0)$;
- Σ is the set of reachable states from σ_0 by firing sequences of transitions;
- $\longrightarrow \in (\Sigma \times T \times \Sigma)$ is the transition relation defined as follows:
  $((M,V), t_f, (M',V')) \in \longrightarrow$ iff:
  (1) $t_f \in En(M)$;
  (2) $\exists \theta \in \Re$,
  $(\theta \geq 0) \wedge (tmin(t_f) - V(t_f) \leq \theta) \wedge$
  $(\bigwedge_{t \in En(M)} \theta \leq tmax(t) - V(t))$;
  (3) $\forall p \in P$,
  $M'(p) = M(p) - Pre(p,t_f) + Post(p,t_f)$;
  (4) $\forall t' \in En(M')$,
  $V'(t') = 0$, if t' is newly enabled (by t_f);
  $V'(t') = V(t') + \theta$, if not;
  Condition $((\sigma, t_f, \sigma') \in \longrightarrow)$ is also denoted by $\sigma \longrightarrow_{t_f} \sigma'$.

2.3 State class spaces

State class spaces (or state class graphs) of a *TPN* model are defined as finite abstractions for the generally infinite concrete state spaces. Formally, a state class space can be defined as a structure $AS = (A, \Longrightarrow, \alpha_0)$ where:

- A is a cover [4] of the set Σ, such as all states of each element (*state class*) of A bear the same marking;
- α_0 is a state class of A containing σ_0;
- $\Longrightarrow \subseteq (A \times T \times A)$ is the successor relation satisfying at least the following condition denoted *EE*: $\forall(\alpha, t_f, \alpha') \in (A \times T \times A)$,
  $((\alpha, t_f, \alpha') \in \Longrightarrow) \Leftrightarrow (\exists \sigma \in \alpha, \exists \sigma' \in \alpha', \sigma \longrightarrow_{t_f} \sigma')$.
  The condition: $((\alpha, t_f, \alpha') \in \Longrightarrow)$ is also denoted by $\alpha \Longrightarrow_{t_f} \alpha'$.

The relation $\Longrightarrow$ may satisfy other conditions such as:

- *EA*: $(\alpha \Longrightarrow_{t_f} \alpha') \Rightarrow (\forall \sigma' \in \alpha', \exists \sigma \in \alpha, \sigma \longrightarrow_{t_f} \sigma')$.
- *AE*: $(\alpha \Longrightarrow_{t_f} \alpha') \Rightarrow (\forall \sigma \in \alpha, \exists \sigma' \in \alpha', \sigma \longrightarrow_{t_f} \sigma')$.

[4] A cover of Σ is a collection of sets whose union contains Σ.

A state class α is said to be *atomic* iff it satisfies AE i.e.:, $\forall t_f \in T, \forall \alpha' \in A$,
$(\alpha \Longrightarrow_{t_f} \alpha') \Rightarrow (\forall \sigma \in \alpha, \exists \sigma' \in \alpha', \sigma \longrightarrow_{t_f} \sigma')$.
A state class space is *atomic* iff all its classes are *atomic*. The following theorem, derived directly from results shown in (Berthomieu et al., 2003) and (Penczek et al., 2001), establishes a relation between conditions *EE*, *AE*, *EA* and the properties (*LTL* or CTL^*) of the model preserved in the state class space.

Theorem 1. Let $AS = (A, \Longrightarrow, \alpha_0)$ be a state class space of a *TPN* model. Then the following relations hold:

- If AS satisfies *EA* then it preserves LTL properties of the model.
- If AS satisfies AE then it preserves CTL^* properties of the model.

Notice that if a state class space is atomic, all states agglomerated in the same state class are bisimilar. Therefore, the atomic state class space is bisimilar to the concrete state space. Furthermore, in absence of silent transitions, bisimilarity implies preservation of CTL^* properties (Berthomieu et al., 2003).

To apply the CTL^* model checking technique, it is sufficient according to *Theorem 1* to build a state class space satisfying condition *AE*.
Berthomieu and Diaz (Berthomieu et al., 1991) proposed a contraction called *State Class Graph (SCG)* which captures LTL properties of the model but not necessarily CTL^* ones. Moreover, *SCGs* are not suitable, as a starting point for building state class spaces satisfying condition *AE*. The main reason is that the authors used an alternative definition of states based on intervals [5]. Two state classes reachable by firing two different sequences may have the same marking and the same firing domain, while one is atomic but not the other. Yoneda and Ryuba in (Yoneda et al., 1998) used the definition of *clock states* and developed a construction of another *state class space* satisfying both conditions *EA* and *AE*. Their approach consists of two phases. The first phase builds a state class space, called the *Strong State Class Graph* (*SSCG*), satisfying condition *EA*. The second phase translates the *SSCG* into another one, called the *Atomic State Class Graph (ASCG)*, satisfying both conditions *EA* and *AE*. Starting from an $SSCG$, the *ASCG* is obtained by repeatedly splitting classes not satisfying condition AE (i.e.: non atomic classes) until reaching an atomic graph. However, the proposed state class definition is some what complicated and the resulting state class space may be infinite, even if the model is bounded. Berthomieu and Vernadat (Berthomieu et al., 2003) brought recently some improvements to the state class definition and to the construction of both state class spaces ($SSCG$ and $ASCG$). They proposed to enforce condition *AE* only in the construction of the $ASCG$, since it is sufficient to preserve CTL^* properties of the *TPN* model. Their improvements led to smaller and finite state class spaces for all bounded *TPN* models.

[5] A state consists of a marking and a function which associates with each enabled transition, a firing interval.

The next section presents the approach of Berthomieu and Vernadat for constructing the $SSCG$. The improvements to this construction are developed in *section 4*.

3. STRONG STATE CLASS GRAPH

Consider a *TPN* model and its concrete state space $(\Sigma, \longrightarrow, \sigma_0)$. A strong state class agglomerates all states reachable by firing the same sequence of transitions. Let $\omega = (t_0 t_1 ... t_n)$ be a sequence of transitions firable starting from the initial state σ_0. The state class α corresponding to the firing sequence ω is the set:
$\{\sigma \in \Sigma | \exists \sigma_0, .., \sigma_n \in \Sigma, \sigma_0 \longrightarrow_{t_0} \sigma_1 .. \wedge \sigma_n \longrightarrow_{t_n} \sigma\}$.
This class is represented by a marking M and a formula F. F characterizes the clock domain of all states agglomerated in the class, where the clock of each enabled transition is represented by a variable with the same name.
The initial strong state class is the pair (M_0, F_0), where M_0 is the initial marking and
$F_0 = (\bigwedge_{t \in En(M_0)} t = 0)$.
Two strong state classes are said to be equal iff they share the same marking and their clock domains are equal (i.e.: their formulas are equivalent). To compare two strong state classes, each one is translated into its canonical form. The two classes are equal, if they have the same canonical form. The canonical form of a strong state class (M, F) is represented by the pair $(M, \mathbf{H})$, where $\mathbf{H}$ is a square matrix of order $|En(M) \cup \{o\}|$, o represents the time origin (the value 0), and $\forall (x, y) \in (En(M) \cup \{o\})^2$,
$\mathbf{H}(x, y) = Max((x - y)|F)$ [6].
The canonical form is computed using the shortest path Floyd-Warshall's algorithm, in $O(n^3)$, where n is the number of variables in F.
The *SSCG* of the *TPN* model is a state class space satisfying condition *EA*. It is defined as a structure $AS = (A, \Longrightarrow, \alpha_0)$ where:

- $\alpha_0 = \{\sigma_0\}$;
- A is the set of all reachable strong state classes;
- $\Longrightarrow \subseteq (A \times T \times A)$ is a transition relation defined by: $((M, F), t_f, (M', F')) \in \Longrightarrow$ iff:
  (1) There exists at least one state in the class (M, F) from which the transition t_f can fire;
  (2) The firing of t_f leads to the strong state class (M', F') which contains all states reachable from the states of (M, F) by t_f.

[6] $Max((x - y)|F)$ is the biggest value of $(x - y)$ in the domain represented by F.

The *SSCG* is obtained by applying the following firing rule to the initial *strong state class* and to each new class.
Let $\alpha = (M, F)$ be a strong state class and t_f a transition of T.

- t_f can occur from the class α iff,
  (1) t_f is enabled for the marking M and
  (2) the following formula (i) is consistent:
  $F \wedge (\theta \geq 0) \wedge (tmin(t_f) \leq t_f + \theta) \wedge (\bigwedge_{t \in En(M)} t + \theta \leq tmax(t))$.
- If t_f can occur from the class α, its firing leads to the *class* $\alpha' = (M', F')$ such that:
  (1) $\forall p \in P$,
  $M'(p) = M(p) - Pre(p, t_f) + Post(p, t_f)$
  (2) Starting from the formula (i) above, the clock domain F' is obtained by replacing each variable t with $(t - \theta)$[7], then eliminating by substitution t_f, θ, all variables associated with transitions conflicting with t_f for M, and finally, adding the constraint $t = 0$, for each newly enabled transition t in M'.

During the construction of the *SSCG*, each newly computed class is compared with the already computed ones. All equal *classes* are agglomerated in one node. With this technique, the finiteness of the $SSCG$ is not guaranteed for bounded TPN models with unbounded firing intervals. In (Berthomieu et al., 2003), the authors completed the construction of the $SSCG$ to handle bounded *TPN* models with unbounded firing intervals. They proposed to partition during the construction process each computed class with some unbounded transitions, into a set of subclasses, in the following way.
Let (M, F) be a strong state class such that some of its enabled transitions have unbounded static intervals. Let $En_{<\infty}(M) = \{t | t \in En(M) \wedge tmax(t) = \infty\}$. The class (M, F) is partitioned into the set: $\{(M, F_e) | (e = \emptyset \vee e \subseteq En_{=\infty}(M))\}$, where F_e is a consistent formula characterizing states of (M, F) in which all transitions of e have not yet reached their minimum delays, while those of $(En_{=\infty}(M) - e)$ have reached or over passed their minimum delays. F_e is computed in three steps:

(1) Initialize F_e with:
$[F \wedge (\bigwedge_{t \in e} t < tmin(t)) \wedge (\bigwedge_{t' \in En_{=\infty}(M) - e} t' \geq tmin(t'))]$;
(2) Eliminate, by substitution, all variables of:
$En_{=\infty}(M) - e$;
(3) Add the constraint:
$(\bigwedge_{t' \in En_{=\infty}(M) - e} tmin(t') \leq t' \leq \infty)$.

Note that the procedure described in this section generates strong state classes with convex clock domains. The convexity is a nice common feature to all approaches which participates in keeping computations simple.

[7] This substitution actualizes clock values of all enabled transitions

4. CONTRACTED STATE CLASS GRAPH

The *Contracted State Class Graph* ($CSCG$) can be seen as a contraction of the *SSCG* described in the previous section. The schema of computation is similar but requires relaxing the first class and each computed class, so as to explicitly represent states reachable by time progressions. An other difference resides in the way classes are agglomerated after they are relaxed. When a new class α is computed and relaxed, it is checked for inclusion with all previously computed classes instead of just equality. The details of agglomeration are given later on in this section. The relaxation operation is described in what follows.

4.1 Relaxation of classes

Definition 2. Let $\alpha = (M, F)$ be a strong state class. The relaxation of the class α, denoted by $Relax(\alpha)$ is the class $\alpha' = (M, F')$ obtained by adding all states reachable via time progressions from all states of α.

The following proposition shows how to compute the relaxed form of a strong state class.

Proposition 3. Let $\alpha = (M, F)$ be a strong state class. $Relax(\alpha) = (M, F')$ is computed in three steps:

(1) Initialize (F') with:
$F \wedge (\theta \geq 0) \wedge (\bigwedge_{t \in En(M)} t + \theta \leq tmax(t))$;
(2) Replace each variable t, by $t - \theta$;
(3) Eliminate θ by substitution.

PROOF. In step one, we add θ to represent a time progression and constrain it not to over pass the maximum allowed limit. The second step reveals all clock valuations reachable by time progressions.

During the construction of the *CSCG*, when a class α is computed, it is first relaxed before comparing it with classes already computed. The canonical form of the relaxed class is computed in a simple way, with no time overhead.

Proposition 4. Let $\alpha = (M, F)$ be a strong state class and $(M, \mathbf{H})$ its canonical form. The canonical form of its relaxation $\alpha' = Relax(\alpha)$ is computed in one step:

- For each t of $En(M)$, replace $\mathbf{H}(t, o)$ by $Min_{t' \in En(M)}(tmax(t') + \mathbf{H}(t, t'))$.

PROOF. The proof is based on *constraint graphs*. A *constraint graph* is an oriented and valuated graph, where nodes represent variables and arc weights are distances between variables. The *constraint graph* has one particular node, denoted o, representing the value 0. The arc (x, y, c) connecting x to y with weight c, corresponds to the constraint: $y - x \leq c$ (or $x - y \geq -c$). The rest of the proof is based on the two following known results about *constraint graphs*:

Let ϕ be a set of constraints representable by a *constraint graph*.

- ϕ is consistent iff the *constraint graph* of ϕ has no negative cycle.
- If the *constraint graph* of ϕ has no negative cycle, the weight of the shortest path in the graph, going from a node x to a node y, is equal to $Max((y - x)|\phi)$.

Consider the formula of F of α. This formula can be represented by a constraint graph. From the definition of $\mathbf{H}$, $\mathbf{H}(x, y)$ is the weight of the shortest path connecting node x to node y. Consider the *Proposition 3*. The constraint graph of the formula obtained after the second step can be computed using the constraint graph of F: first rename the node o by θ, then add a new node o, the arc $(\theta, o, 0)$, and finally, for each transition t in $En(M)$, add the arc $(o, t, tmax(t))$. The canonical form of the relaxation is deduced from the resulting constraint graphs by setting each $\mathbf{H}(x, y)$ to the weight of the shortest path connecting y to x in the modified graph.

4.2 *Contraction by inclusion*

In all other approaches, after computing a strong state class, it is checked for equality with previously computed classes. In the approach proposed here, the class is first relaxed then checked for inclusion. When a class is included in another one, it is eliminated, and all its input and output arcs are reconnected to the largest class. It is obvious that this contraction preserves the integrity of the model state space (no state is created nor destroyed) and also satisfies condition *EE*. The proof comes straight from the way the graph is constructed. Experimental results have shown that the *contracted strong state class graphs* are just a fraction of the size of their equivalent in the Berthomieu-Vernadat's approach, and computed in a fraction of the time. Furthermore, the gain increases significantly as the size of the model increases. This gain comes at the price of loosing condition *EA*, but fortunately condition *EE* is preserved. Condition *EE* is all what is needed for a state class graph to be a good candidate for constructing an $ASCG$.

5. ATOMIC STATE CLASS GRAPH

The $ASCG$ is computed starting from a state class space and refining it until it becomes atomic. Let $(\Sigma, \longrightarrow, \sigma_0)$ be the concrete state space of a *TPN* model and $AS = (A, \Longrightarrow, \alpha_0)$ a state class space. AS is atomic iff it satisfies condition *AE*, i.e.:
$\forall(\alpha, t_f, \alpha') \in\Longrightarrow, \forall\sigma \in \alpha, \exists\sigma' \in \alpha', (\sigma, t_f, \sigma') \in\longrightarrow$.
Let $Pred(\alpha', t_f)$ be the set of all states which can lead by firing t_f to some states in α'. To verify the atomicity of α for the arc (α, t_f, α'), it suffices to verify that the class α is equal or included in $Pred(\alpha', t_f)$. In case α is not atomic, it is partitioned into convex subclasses so as to isolate the predecessors of α' by t_f in α. This operation is called *splitting* (see Algorithm 1).

The following proposition shows how to compute the class $Pred(\alpha', t_f)$.

Let $(M', F') = \alpha'$ and $(M", F") = Pred(\alpha', t_f)$. We denote by $New(M', t_f)$ the set of all transitions newly enabled in M', i.e.: $New(M', t_f) = \{t \in En(M') | \exists p, M'(p) - Post(p, t_f) < Pre(p, t)\}$.

Proposition 5. The set $Pred(\alpha', t_f) = (M", F")$ is computed in five steps:

(1) $\forall p \in P$,
$M"(p) = M'(p) + Pre(p, t_f) - Post(p, t_f)$
(2) Initialize $F"$ to $(F' \wedge \bigwedge_{t \in New(M', t_f)} t = 0)$
(3) Eliminate by substitution all transitions in $New(M', t_f)$
(4) Add constraints: $(tmin(t_f) \leq t_f)$, $(\bigwedge_{t \in En(M")} 0 \leq t \leq tmax(t))$ and $\theta \geq 0$
(5) Replace each variable t by $t + \theta$, then eliminate θ by substitution.

PROOF. Since the firing of transition t_f sets the clock of each newly enabled transition to 0, step 2 extracts from α' the subset of states where the clocks of newly enabled transitions are equal to 0. Step 4 adds the firing constraints of t_f. Finally, step 5 goes back in time (each clock is decreased by θ time units).

Algorithm 1. **The splitting function**

```
Partition Function Splitting ( Class α, Formula F" )
{ // F" is the formula of Pred(α', t_f)
        Formula X := F; // F is the formula of α
        Partition Part : = ∅;
        for each constraint f of F" do {
            if (X ∧ f) is not consistent then
                return ∅; // No predecessor of α' by t_f in α
            if (X ∧ ¬f) is consistent then
                Part := Part ∪ {(M, X ∧ ¬f)};
            X := (X ∧ f);
        }
        Part := Part ∪ {(M, X)};
        return Part;
}
```

The refining procedure always terminates since the input graph and the number of splitting possibilities of each class are finite.

6. BISIMULATION REDUCTION RULE

The size of the *ASCG* obtained using the approach presented here, or all other approaches, is rarely optimal for several reasons. The convexity of classes is an important feature to keep calculations simple, however, it induces fragmentation of classes. The redundancy of states may induce a proliferation of classes too[8]. From these remarks, a bisimulation reduction rule can be applied to perform further agglomerations, but to the detriment of convexity. The optimal size of an *ASCG* could be obtained by minimizing it under bisimulation, but this technique is very time consuming. The proposed bisimulation reduction rule is very light weight. It consists in *agglomerating classes with identical successors for identical transitions*. It is obvious that two atomic classes with the same successors for the same transitions are bisimilar. This rule has to be applied repeatedly on the *ASCG* resulting from the second phase until it is no more applicable. Experimental results have shown that applying this rule produces the optimal graph most of the time. Another reduction rule consisting in agglomerating classes having states in common could be applied, since two atomic classes sharing states are bisimilar too. Nevertheless, this rule has been tested and found less interesting than the first one performance-wise.

7. IMPLEMENTATION RESULTS

Tests have been performed on several models including those used in (Berthomieu et al., 2003) (shown in Fig. 1). For the models given in Fig.1, Table 1 reports times and sizes (nodes/arcs) obtained using Berthomieu-Vernadat's approach (column 2) and the proposed approach (column 3). In columns 2 and 3, for each model, the first line indicates the size of the graph computed in the first phase (*SSCG*,(*CSCG* resp.)). Column 4 reports the results after applying the reduction rule given in *section 6*. Column 5 shows the optimal sizes obtained using minimization under bisimulation.

8. CONCLUSION

This paper considered the *TPN* model for *CTL** model checking. Knowing that the state space of this model is infinite, an approach is proposed to contract it as much as possible without loosing properties of interest. This approach improves the technique proposed in (Berthomieu et al., 2003) which proceeds in two phases. The first phase constructs a *strong state class graph (SSCG)*. The second phase refines this graph until *CTL** properties are restored. The resulting graph is called the *atomic state class graph (ASCG)*. The proposed approach constructs in the first phase, a *contracted state class graph (CSCG)*, much more compact than the *SSCG*, yet appropriate to be used by the second phase. The *CSCG* has also the advantage of boosting the second phase and reducing the size of the

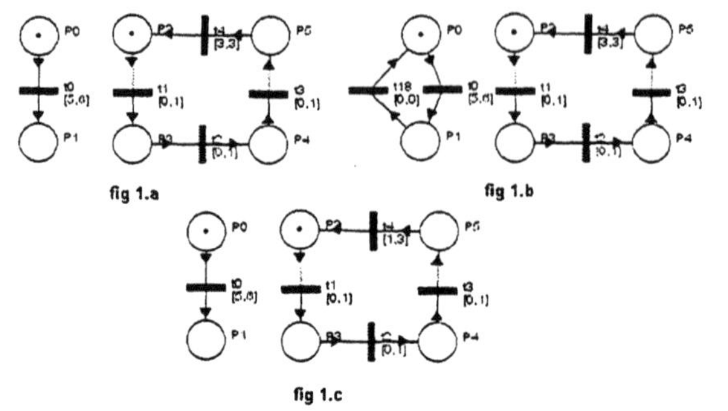

Fig. 1. TPN models used in experiments

ASCG. For further contractions, a third phase consisting in agglomerating classes with identical successors for identical transitions is added. Experimental results have shown that most of the time, obtained graphs are either optimal or very close to the optimal and computed in much smaller times.

As immediate perspective, we will study the possibility to generate more appropriate contractions for different classes of properties such as TCTL used in (Daws et al., 1996).

Table. 1 Experimental results

TPNs	Berthomieu	Proposed	Reduc.	Optimal
Fig. 1.a	21/29	12/17		
CPU (ms)	0	0		
ASCG	36/61	27/49	26/47	26/47
CPU (ms)	0	0	0	0
Fig. 1.b	60/93	15/25		
CPU (ms)	10	0		
ASCG	80/204	80/204	80/204	80/204
CPU (ms)	30	20	0	10
Fig. 1.c	39/63	12/17		
CPU (ms)	0	0		
ASCG	62/163	46/135	46/135	46/135
CPU (ms)	20	10	0	0
Fig.1.a+b	3862/8964	447/1532		
CPU (ms)	431	100		
ASCG	9349/36709	8806/35478	8599/34591	8599/34591
CPU (ms)	48350	12718	4035	211394
Fig.1.a+c	3725/10821	191 /590		
CPU (ms)	751	60		
ASCG	4634/22236	4076/19968	3898/19035	3898/19035
CPU (ms)	13710	5367	601	41780
Fig.1.b+c	9555/26002	426/1429		
CPU (ms)	2073	200		
ASCG	9868/45126	9556/44199	9366/43143	9359/43112
CPU (ms)	59726	15452	5047	350754

REFERENCES

Berthomieu, B. and Vernadat, F. (2003). State class constructions for branching analysis of Time Petri nests. *LNCS 2619*.

Berthomieu, B. and Diaz, M. (1991). Modeling and verification of time dependent systems using time Petri nets. *IEEE Transactions on Software Engineering*. **Vol. 17, no. 3.**

Daws,C. Olivero, A., Tripakis, S. and Yovine, S. (1996). The tool Kronos. In: *Hybrid Systems III. Verification and Control*. LNCS 1066, Springer-verlag. Ed, 1996.

Penczek,W. and Polrola,A. (2001). Abstraction and partial order reductions for checking branching properties of Time Petri nets, In: *Proc. of ICATPN*, LNCS 2075, pages 323-342.

Yoneda, T. and Ryuba, H. (1998). CTL Model Checking of Time Petri Nets Using Geometric Regions. *IEICE Trans. Inf. and Syst.*. **Vol. E99-D, no. 3.**

[8] In the *ASCG*, classes included in other classes may exist.

ELSEVIER
IFAC
PUBLICATIONS
www.elsevier.com/locate/ifac

BANDWIDTH REQUIREMENTS FOR EVENT-DRIVEN OBSERVATIONS OF CONTINUOUS-TIME VARIABLE

Marek Miśkowicz

AGH University of Science and Technology
Department of Electronics
al. Mickiewicza 30
30-059 Kraków, Poland
Email: miskow@agh.edu.pl

Abstract: The paper discusses results of the extension of the discrete-event approach to the continuous-time variables from the point of view of communication problems. The analytical formula for evaluation the mean and the maximum rate of event-based observation traffic is derived. For a time-monotonic signals the formula exactly estimates the mean observation rate if state changes are either equally or inequally spaced in the value domain. For signals that are not monotonic functions of the time the same formula evaluates the upper bound on the mean observation rate for the infinite resolution. *Copyright © 2004 IFAC*

Keywords: events, discrete-event dynamic system, discrete measurements, bandwidth allocation, signal levels, signals analysis, continuous variables, analytic approximations.

1. INTRODUCTION

Most discrete-event systems concern "man-made" products, like computer networks, complex software systems, production processes, etc. Many of quantities we deal with are "discrete", typically involving counting integer numbers. All activity in these systems is due to asynchronous occurrences of discrete events.

In real-time systems that interact with the physical environment and operate on continuous-time input signals the time-driven strategy is generally used. The arguments supported this approach stem either from control theory, or software architecture. The important reason is the existence of a well-established system theory for periodic discrete-time control systems. The analysis and design of such systems are simple because for linear time-invariant processes the closed-loop system becomes linear and periodic. The stroboscopic model that defines the behavior of the system at sampling instants can be described by the difference equations with constant coefficients (Aström and Wittenmark, 1997).

From the point of view of real-time software the periodic sampling naturally supports the time-triggered architecture (TTA). Thereby recommendations for software architecture harmonize with control problems, because time is indeed a critical issue for real-time systems. The concepts of time and speed play a major role in the process control area so the time-triggered strategy simplifies the relation between the software architecture and the basic model of a control system. Moreover, the explicit management of time at the basic level supports the evaluation of the temporal properties of software (De Paoli and Tisato, 1996).
However, there are important reasons to use the *event-driven* approach for sampling of continuous-time signals. The main motivation is the better resource utilization. Occupying the CPU for performing control calculations and sending data in a time-triggered fashion when nothing significant has happened in the process is an evident waste of resources. The problem is much more urgent if we

deal with *burst signals*, defined as a type of signals that do not change, much or at all, for most of time.
The proposal suggested above, because of its simplicity, is not new. Already in the late 1950s Ellis (1959) has noticed, that "*the most suitable sampling is by transmission of only significant data, as the new value obtained when the signal is changed by a given increment*". And as a result: "*it is not necessary to sample periodically, but only when quantized data change from one possible value to the next*". The similar approach can be found in hybrid systems, where the purely discrete-state models are constructed usually by building transitions based on finite partitions of the continuous state space.

But the better resource utilization is not the only reason to perform the event-based observations of the continuous-time physical environment. The event-triggered sampling is a possible alternative for an increasing number of applications where the assumption about uniform time-triggered sampling rate is no longer valid (Aström and Bernhardsson, 2002). The example is the networked control system, where the modern distributed architectures impose some constraints that make difficult to support the time-triggered paradigm. This is especially the case when control loops are closed over computer networks or buses, e.g., fieldbus area networks (FAN), or even the Internet (Arzén, 1999). Moreover, some of these networks use non-deterministic MAC protocols, e.g. Ethernet, LonTalk/EIA-709.1. It means that in fact such networked control system does *not* operate in the time-triggered control regime so the periodic sampling pattern loses its privileged position. Undoubtedly, a strong point of time-driven systems is the existence of a mature theory of control and scheduling. At the same time the field of event-based sampling is, as Astrom (2002) pointed out, "*still in its infancy*" and "*it would be very attractive to have a system theory similar to the one for periodic sampling*". The sampling theorem for event-based strategy parallel to the well-known Shannon criterion for periodic sampling has not yet been developed. Also a lack of clear guidelines for design a system with the event-based observations of physical magnitudes is an obstacle to develop and to recognize its properties.

The paper discusses results of the extension of the discrete-event approach to the continuous-time variables from the point of view of communication problems. In event-driven model the transition from the one state to another is caused by events so time is no longer a suitable independent variable in system modelling. Meanwhile, the communication requirements relate to temporal properties, e.g. measures that evaluate demands for required communication bandwidth.
The following problems should be considered in the event-based communication system: (1) How much communication bandwidth should be dedicated to send the event-based observations describing a given known continuous-time signal with a given accuracy? (2) If we know the available bandwidth, how to select a threshold that constitutes the triggering event occurrence?
We analyze the mean and the maximum rate of event-based observation traffic, i.e. the mean and the maximum number of state transitions of the controlled variable per a unit of time. It seems to be natural that both measures depend on the signal dynamics and the signal tracking resolution.

2. DISCRETE-STATE MODEL

Computers operate on time-continuous processes at time-discrete instants. Thus, the "knowledge" about physical world must be subjected to a discretization. The state of the environment is described by recording the values of significant variables at discrete-time instants. The information about the controlled variables at a particular point in time is captured by the notion of an *observation*.

2.1 Definition of the observation

The observation is an *atomic data structure*, consisting of the name of the variable, the point in real time when the observation was made (t_i) called a *timestamp*, and the observed value of the variable $x(t_i)$ (Kopetz, 1997). An observation should be transported in a single message from sensor node to the rest of the system because the message concept provides for the observation atomicity. The resolution of observations determines the rate of the *observation flow*, i.e. the amount of information that comes in a unit of time from the environment to the input of the system.

As a result of the observation a continuous-time signal is replaced by a sequence of numbers representing the signal at certain times. Let $Z = \{0,1,2...\}$ define the positive integers and let $\{t_i : i \in Z\}$ be a subset of the real numbers called the *observation instants*. Then, the discrete form of the signal is represented by a sequence:

$$X = \{x(t_i)\} : i \in Z\} \qquad (1)$$

Observations are performed by signal sampling usually.

2.2 Discrete-time vs. discrete-event model

A fundamental problem of conversion of the physical reality into the set of discrete-events is expressed by the following question. How the observation points have to be selected in such a way that the discrete approximation of a continuous-time signal is as accurate as possible on the one hand, and the number of observations is minimized on the other?
As it was mentioned, most applications in digital control and signal processing use the periodic observations, where observation instants are equally spaced in time such that:

$$X_{DT} = \{x(iT\} : i \in Z\} \qquad (2)$$

where T represents the *observation period*.

The periodic observation pattern is *not* optimal in terms of the observation rate.
Using the event-triggered strategy, signal is observed when the *significant event* occurs, i.e. the significant change of its state, which has to be handled by the computer system, is noted. If the signal is observed on a basis of its behavior in the value domain, i.e. only when measurements pass certain limits, then such a pattern is sometimes called the *magnitude-driven* algorithm. The magnitude-driven observation strategy is a particular case of the event-triggered one. However, in this paper we use both terms equivalently.

Let us consider the case, which is *dual* to the periodic observation pattern, namely, where the observations are equally spaced in the value domain. Thus, the following relationship occurs:

$$|x(t_i) - x(t_{i-1})| = E \qquad (3)$$

where $E > 0$, is called the *threshold*, or the *resolution* of the signal observations in the value domain; t_i, is the instant, and *i*, is the number of the current observation, respectively.

Let us specify the discrete-event state space, which consists of a set of the *observation levels*, i.e. the values, which trigger the observation, when the signal $x(t)$ passes them:

$$X_{DE} = \{x_k : k \in Z\} \qquad (4)$$

Let us assign each observation by the double index that represents either the observation number, or the observation level. Thus, for the *i*th observation consisting in that the signal $x(t)$ reaches the level x_k we can write:

$$x(t_i) = x_{k_i} \qquad (5)$$

where the double index $k_i \in \{1,...,N\}$ represents the *i*th observation taken at the instant t_i, and specifies the observation level x_k that has been passed.
For a continuous-time signal $x(t)$ we have

$$k_{i+1} - k_i \in \{-1,1\} \qquad (6)$$

because the observation occurs when the signal $x(t)$ passes one of the consecutive levels with respect to the most recent level. As it follows from the last relationship, the uniform event-driven observation can be reduced to information encoded in a single bit. For example it is used in one-bit AD converters, which also belong to the event-based applications. However, the relationship (6) is valid only if no sample might be lost. It eliminates the possibility of using only the sign one-bit data in messages sent over the networked control system with a MAC protocol that accept collisions, e.g. CSMA-based algorithms. It is because a loss or duplication of a single event observation results in a loss of state synchronization between the state of the observer and the state of the receiver.

2.3 Time-value ...ality

Either the dis... time or the discrete-event model defined above ...erived on a basis of time and value domain interdependence (Verissimo and Rodrigues, 2001). Thus, there is a duality between:

- *observing the value at a given time* – e.g. the temperature of the furnace amounts to 500°C at 10:00,
- *observing the time at which a given value occurs* – e.g. the temperature of the furnace reaches 500°C at 9:55.

However, the duality is not complete. Namely, in the time-triggered pattern exactly one signal value is assigned to each observation instant, whereas each observation level in the event-triggered algorithm might be crossed many times, if only the signal is not a monotonic function of the time.

3. MEAN OBSERVATION RATE ANALYSIS

Performing the event-driven observations we must consider the following problems: (1) How much communication bandwidth should be dedicated to send the event-based observations describing a given known continuous-time signal with a *given* accuracy? (2) If we *know* the available bandwidth, how to select a threshold that constitutes the triggering event occurrence?

3.1 Uniform event-driven observations

Let us consider the uniform event-driven observation strategy, where the signal *x(t)* is sampled when its value deviates from the value included in the most recent observation by an interval of confidence $\pm E$, called *threshold*, where $E > 0$ (Fig. 1).

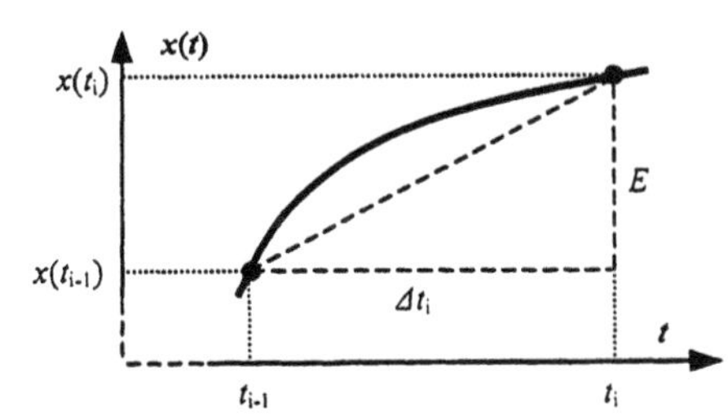

Fig. 1. The calculation of an observation interval in the uniform event-driven scheme.

Thus, there is a constant difference equal to *E* between signal values included in consecutive observations:

$$|x(t_i) - x(t_{i-1})| = E \qquad (7)$$

where $i = 1,2,...,n$.
By the definition the absolute value of the mean of the first time derivative $|\overline{\dot{x}_i}|$ of a continuous signal *x(t)* during the *i*th observation interval is given by the formula:

$$|\overline{\dot{x}_i}| = \frac{1}{t_i - t_{i-1}} \left| \int_{t_{i-1}}^{t_i} \dot{x}(t)dt \right| = \frac{|x(t_i) - x(t_{i-1})|}{t_i - t_{i-1}} \qquad (8)$$

Setting the equation (7) to the formula (8) we can find the length of the *i*th observation interval Δt_i in relation to the mean signal changes:

$$\Delta t_i = \frac{E}{\left|\dot{x}_i\right|} \quad (9)$$

where $\Delta t_i = t_i - t_{i-1}$.

Then, by the definition the mean length $\overline{\Delta t}$ of the observation interval calculated for a number of n intervals amounts to:

$$\overline{\Delta t} = \frac{1}{n}\sum_{i=1}^{n} \Delta t_i = \frac{d}{n} \quad (10)$$

where d represents the total length of all the observation intervals:

$$d = \sum_{i=1}^{n} \Delta t_i = t_n - t_0 \quad (11)$$

Thus, using the uniform event-driven pattern the mean observation rate λ, i.e. the number of observations in a time unit amounts to:

$$\lambda = \frac{1}{\overline{\Delta t}} = \frac{n}{\sum_{i=1}^{n} \Delta t_i} = \frac{n}{d} \quad (12)$$

Let us notice that by the definition, the absolute value of the mean of the first signal derivative in relation to the time during an interval (t_0, t_n) is given by the formula:

$$\left|\overline{\dot{x}(t)}\right| = \frac{1}{t_n - t_0}\left|\int_{t_0}^{t_n} \dot{x}(t)dt\right| \quad (13)$$

3.2 Observations of monotonic signals

If the signal derivative does not change its sign during the interval (t_0, t_n), then:

$$\forall i \in \{1,\ldots n\}: \left|\int_{t_{i-1}}^{t_i} \dot{x}(t)dt\right| = \int_{t_{i-1}}^{t_i} \left|\dot{x}(t)\right|dt \quad (14)$$

and therefore:

$$\left|\int_{t_0}^{t_n} \dot{x}(t)dt\right| = \left|\sum_{i=1}^{n} \int_{t_{i-1}}^{t_i} \dot{x}(t)dt\right| = \sum_{i=1}^{n} \left|\int_{t_{i-1}}^{t_i} \dot{x}(t)dt\right| \quad (15)$$

If we set the equation (15) to the formula (13) and if we additionally take into consideration (see Eq. (8)) that:

$$\sum_{i=1}^{n} \left|\overline{\dot{x}_i}\right| \Delta t_i = nE \quad (16)$$

we have:

$$\left|\overline{\dot{x}(t)}\right| = \frac{\int_{t_0}^{t_n} \left|\dot{x}(t)\right|dt}{t_n - t_0} = \frac{\sum_{i=1}^{n} \left|\overline{\dot{x}_i}\right| \Delta t_i}{d} = \frac{nE}{d} \quad (17)$$

Setting the equation (17) to the formula (12) we obtain the mean observation rate λ' for the continuous signal $x(t)$ that is a monotonic function of the time:

$$\lambda' = \frac{\left|\overline{\dot{x}(t)}\right|}{E} \quad (18)$$

The mean rate of the uniform event-driven observation flow is expressed by two factors:

- the absolute value of the mean of the first signal time-derivative, which is a measure of the observed signal $x(t)$,
- the resolution of the observation in the value domain.

On a basis of the formula (18) we can find also the maximum observation rate, which occurs during the fastest signal changes:

$$\lambda'_{\max} = \frac{1}{E}\left|\dot{x}(t)\right|_{\max} \quad (19)$$

or respectively, the minimum interarrival time between the consecutive observations:

$$\Delta t_{\min} = \frac{E}{\left|\dot{x}(t)\right|_{\max}} \quad (20)$$

The last relationship might be derived also directly from the equation (9).

Thus, the maximum observation rate is bounded implicitly by the threshold and the gradient of the signal value.

3.3 Non-uniform event-based observations

The validity of the formula (18) might be extended to the case when the event-driven observations are *inequally* spaced in the value domain, but the mean threshold $\overline{E}$ amounts to E.

The value domain is split into non-uniform ranges so there is not a constant difference between signal values at consecutive observation instants:

$$\left|x(t_i) - x(t_{i-1})\right| = E_i \quad (21)$$

where $i = 1,2,\ldots,n$. We assumed above that:

$$\overline{E} = \frac{1}{n}\sum_{i=1}^{n} E_i = E \quad (22)$$

thus (compare Eq. (9)):

$$\Delta t_i = \frac{E_i}{\left|\overline{\dot{x}_i}\right|} \quad (23)$$

Next, taking into account the fomulae (22) and (23) we get the relationship:

$$\sum_{i=1}^{n} \left|\overline{\dot{x}_i}\right| \Delta t_i = \sum_{i=1}^{n} E_i = nE \quad (24)$$

which is an identity with equation (16). As a result, we can directly derive the formula (18). Thereby the validity of the formula (18) is extended for the case of the non-uniform discrete-event state space of a continuous-time variable.

3.4 Observations of non-monotonic signals

As we assumed before, the formula (18) represents the mean rate of state transitions in discrete-event model for a variable that is monotonic function of the time. Many signals in control systems meet this assumption, e.g. first-order or higher order overdamped or critically damped step-responses.

Let us consider now the more general case, when the observed signal $x(t)$ is not a monotonic function of the time so its first time-derivative can change the sign during a given time interval (t_0, t_n) as it is shown on Figures 2a-2c.

In general the mean observation rate λ for non-monotonic signals is equal or less to that obtained by the formula (18) for monotonic signals, because:

$$\left| \int_{t_0}^{t_n} \dot{x}(t)dt \right| \le \int_{t_0}^{t_n} |\dot{x}(t)|dt = \sum_{i=1}^{n} \int_{t_{i-1}}^{t_i} |\dot{x}(t)|dt \qquad (22)$$

and depends on the number of local extema during the time interval (t_0, t_n).

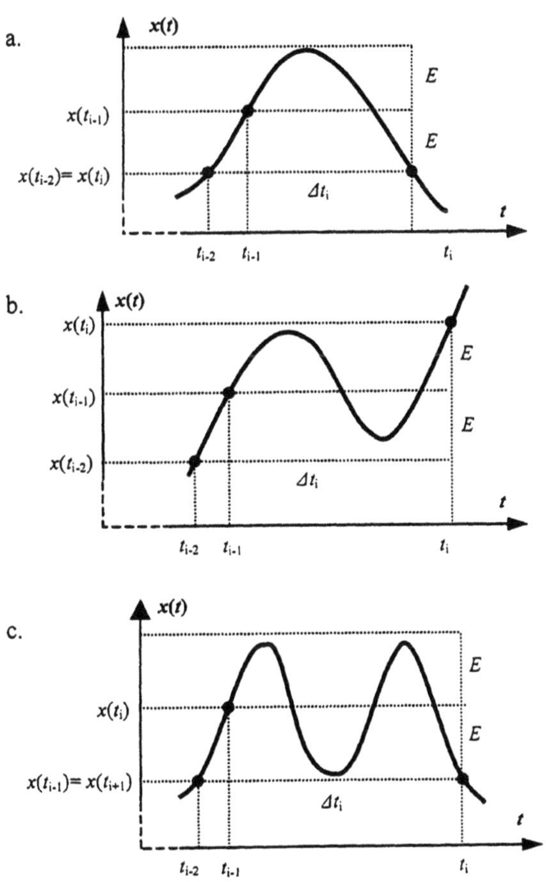

Fig. 2. Analysis of the influence of a number of local extrema for the mean observation rate.

Analyzing the influence of the number of local extrema during the ith observation interval (t_{i-1}, t_i) on the observation rate, we can notice that:

- if there is no local extremum during the ith observation interval, then (see Eqs (7) and (15)):

$$\left| \int_{t_{i-1}}^{t_i} \dot{x}(t)dt \right| = \int_{t_{i-1}}^{t_i} |\dot{x}(t)|dt = E \qquad (23)$$

- if at most one local extremum occurs during the ith observation interval, then (Fig. 2a):

$$\left| \int_{t_{i-1}}^{t_i} \dot{x}(t)dt \right| = E \le \int_{t_{i-1}}^{t_i} |\dot{x}(t)|dt < 3E \qquad (24)$$

- if there are at most two local extrema, then (Fig. 2b):

$$\left| \int_{t_{i-1}}^{t_i} \dot{x}(t)dt \right| = E \le \int_{t_{i-1}}^{t_i} |\dot{x}(t)|dt < 5E \qquad (25)$$

- if three local extrema at most occur, then (Fig. 2c)

$$\left| \int_{t_{i-1}}^{t_i} \dot{x}(t)dt \right| = E \le \int_{t_{i-1}}^{t_i} |\dot{x}(t)|dt < 7E \qquad (26)$$

Generalizing, if at most l_i local extrema during the ith observation interval occur, then:

$$\left| \int_{t_{i-1}}^{t_i} \dot{x}(t)dt \right| = E \le \int_{t_{i-1}}^{t_i} |\dot{x}(t)|dt < (2l_i + 1)E \qquad (27)$$

Transforming the right inequality we have:

$$\int_{t_{i-1}}^{t_i} |\dot{x}(t)|dt - 2l_i E < \left| \int_{t_{i-1}}^{t_i} \dot{x}(t)dt \right| \qquad (28)$$

Finally:

$$\int_{t_{i-1}}^{t_i} |\dot{x}(t)|dt - 2l_i E < \left| \int_{t_{i-1}}^{t_i} \dot{x}(t)dt \right| \le \int_{t_{i-1}}^{t_i} |\dot{x}(t)|dt \qquad (29)$$

Adding up the appropriate relationships for all $\{1, \ldots, n\}$ observation intervals we have:

$$\sum_{i=1}^{n} \int_{t_{i-1}}^{t_i} |\dot{x}(t)|dt - 2L_e E < \sum_{i=1}^{n} \left| \int_{t_{i-1}}^{t_i} \dot{x}(t)dt \right| \le \sum_{i=1}^{n} \int_{t_{i-1}}^{t_i} |\dot{x}(t)|dt \qquad (30)$$

where L_e is the total number of local extrema within the time interval (t_0, t_n).

Dividing the relationship (30) by the total length of all the observation intervals d and taking into account that:

$$\sum_{i=1}^{n} \left| \int_{t_{i-1}}^{t_i} \dot{x}(t)dt \right| = nE \qquad (31)$$

we have:

$$\frac{\overline{|\dot{x}(t)|}}{E} - 2\bar{l} < \lambda \le \frac{\overline{|\dot{x}(t)|}}{E} \qquad (32)$$

where $\bar{l} = L_e / d$, $\bar{l} \in R$ is the *local extrema density*, i.e. the average number of local extrema in a time unit, $\overline{|\dot{x}(t)|}$ is expressed by the formula (13). Finally:

$$\lambda' - 2\bar{l} < \lambda \le \lambda' \qquad (34)$$

As it follows from the formula (32), λ approaches λ', if a threshold E is small. Then:

$$\frac{1}{E}\overline{|\dot{x}(t)|} >> 2\bar{l} \qquad (35)$$

Finally if $E \to 0$, then $\lambda \to \lambda'$. Thus:

$$\lim_{E \to 0}(\lambda' - \lambda) = \inf_{E \to 0}(\lambda' - \lambda) = 0 \qquad (36)$$

Summing up, in general we can not estimate the mean observation rate accurately using the analytic approach if a signal is a non-monotonic function of the time. However, we can treat λ' as the approximation of the real value of λ, because it gives the *upper bound* on the mean observation rate. The last remark is important, because from the point of view of the bandwidth requirements λ' can be interpreted as the *worst-case* value of the mean observation rate for a known threshold E.

Thus, on a basis of best-effort strategy λ' is a useful measure to assign the communication resources for sending the event-based observation messages also if a signal is not a monotonic function of the time.

3.5 Heartbeat update

The inherent feature of the event-driven pattern is that the observation interval can theoretically go to infinity if changes of the monitored signal are less than the specified threshold. This feature is opposite to the reliability requirements, because it makes difficult to detect a possible sensor failure.

Therefore the event-driven observation strategy is usually equipped with the timeout mechanism. Namely, if during a specified time interval T_{max} the signal $x(t)$ does not change enough, the observation is performed in the time-triggered fashion, called *heartbeat observation*, e.g. see LonMark Interoperability Guidelines (2002) for control networks developed in LonWorks technology. In terms of node reliability it determines the *maximum latency* of error detection.

The heartbeat observation occurs especially during the time when the setpoint is reached and the system is in the equilibrium state. Thus, if the signal $x(t)$ does not change enough for a long time, the observation rate is constant and amounts to:

$$\lambda_{\min} = \frac{1}{T_{\max}} \tag{37}$$

The selection of the heartbeat observation period T_{max} is a trade-off between the dependability requirements and the available bandwidth.

3.6 Bandwidth requirements

Summing up, the current observation rate in the event-driven observation strategy for continuous-time signals is included within the range:

$$\lambda(t) = \left[\frac{1}{T_{\max}} ; \frac{\left|\dot{x}(t)\right|_{\max}}{E} \right] \tag{38}$$

4. CONCLUSIONS

The paper concerns the signal-oriented estimation of communication bandwidth requirements. We derived an easy analytical formula for evaluation the required communication resources to handle events that map state changes of a continuous-time variable. The derived formula expresses the balance between the resolution of the observation and the amount of the observation traffic. It has a significant consequence for the design practice. We can adjust the resolution of the observation to the available memory space or communication bandwidth.

For a time-monotonic signals the derived formula exactly estimates the mean observation rate if state changes are either equally or inequally spaced in the value domain.

In general we can not estimate the mean observation rate accurately using the analytic approach if a signal is a non-monotonic function of the time. However, on a basis of the derived formula we evaluated the upper bound on the mean observation rate for the infinite resolution.

Furthermore we calculated the minimum interarrival time for events. It is determined for a given resolution by the signal gradient with respect to time. The derived measures are useful for the best-effort strategy design, because we do not know *a priori* the interarrival times of events.

A great challenge for the further research is to adapt the event-based observation traffic for the real-time requirements.

REFERENCES

Arzén, K.E. (1999). A Simple Event-Based PID Controller. *Proceedings of IFAC World Congress*, vol.18, pp.423-8

Aström, K.J. and B. Bernhardsson (2002). Riemann and Lebesgue Sampling. *Proceedings of IEEE Conference on Decision and Control*, pp. 2011-2016, Las Vegas, Dec. 2002

Aström, K.J. and B. Wittenmark (1997). *Computer-Controlled Systems: Theory and Design*. Upper Saddle River, New Jersey.

De Paoli, F. and F. Tisato (1996). On the complementary nature of event-driven and time-driven models", *Control Engineering Practice*, vol. 4 (6), pp. 847-854.

Ellis, P.H. (1959). Extension of Phase Plane Analysis to Quantized Systems. *IRE Transactions on Automatic Control*, vol. AC-4 , pp. 43-59.

Kopetz H. (1997). *Real-Time Systems. Design Principles for Distributed Embedded Applications*. Kluwer Academic Publications.

Verissimo P. and Rodrigues L. (2001). *Distributed Systems for System Architects*. Kluwer Academic Publishers.

Layer 7 LonMark Interoperability Guidelines (2002). Version 3.2, LonMark Interoperability Association.

AUTHOR INDEX

Title/Year of publication	Editor(s)	ISBN
2002 continued		
Periodic Control Systems (W)	Bittanti & Colaneri	0 08 043682 X
Modeling and Control in Environmental Issues (W)	Sano, Nishioka & Tamura	0 08 043909 8
Computer Applications in Biotechnology (C)	Dochain & Perrier	0 08 043681 1
Time Delay Systems (W)	Gu, Abdallah & Niculescu	0 08 044004 5
Control Applications in Post-Harvest and Processing Technology (W)	Seo & Oshita	0 08 043557 2
Intelligent Assembly and Disassembly (W)	Kopacek, Pereira & Noe	0 08 043908 X
Adaptation and Learning in Control and Signal Processing (W)	Bittanti	0 08 043683 8
New Technologies for Computer Control (C)	Verbruggen, Chan & Vingerhoeds	0 08 043700 1
Internet Based Control Education (W)	Dormido & Morilla	0 08 043984 5
Intelligent Autonomous Vehicles (S)	Asama & Inoue	0 08 043899 7
2003		
Proceedings of the 15th IFAC World Congress 2002 (CD + 21 vols)	Camacho, Basanez & de la Puente	008 044184 X
Modeling and Control of Economic Systems (S)	Neck	0 08 043858 X
Mechatronic Systems (C)	Tomizuka	0 08 044197 1
Programmable Devices and Systems (W)	Srovnal & Vlcek	0 08 044130 0
Real Time Programming (W)	Colnaric, Adamski & Wegrzyn	0 08 044203 X
Lagrangian and Hamiltonian Methods in Nonlinear Control (W)	Astolfi, Gordillo & van der Schaft	0 08 044278 1
Intelligent Control Systems and Signal Processing (C)	Ruano, Ruano & Fleming	0 08 044088 6
Guidance and Control of Underwater Vehicles (W)	Roberts, Sutton & Allen	0 08 044202 1
Analysis and Design of Hybrid Systems (C)	Engell, Gueguen & Zaytoon	0 08 044094 0
Intelligent Manufacturing Systems (W)	Kadar, Monostori & Morel	0 08 044289 7
Control Applications of Optimization (W)	Gyurkovics & Bars	0 08 044074 6
Fieldbus Systems and Their Applications (C)	Dietrich, Neumann & Thomesse	0 08 044247 1
Intelligent Components and Instruments for Control Applications (S)	Almeida	0 08 044010 X
Modelling and Control in Biomedical Systems (S)	Feng & Carson	0 08 044159 9
2004		
Advances in Control Education (S)	Lindfors	0 08 043559 9
Robust Control Design (S)	Bittanti & Colaneri	0 08 044012 6
Fault Detection, Supervision and Safety of Technical Processes (S)	Staroswiecki & Wu	0 08 044011 8
Technology and International Stability (W)	Kopacek & Stapleton	0 08 044290 0
System Identification (SYSID 2003) (S)	Van den Hof, Wahlberg & Weiland	0 08 043709 5
Control Systems Design (C)	Kozak & Huba	0 08 044175 0
Robot Control (S)	Duleba & Sasiadek	0 08 044009 6
Time Delay Systems (W)	Garcia	0 08 044238 2
Control in Transportation Systems (S)	Tsugawa & Aoki	0 08 0440592
Manoeuvring and Control of Marine Craft (C)	Batlle & Blanke	0 08 044033 9
Power Plants and Power Systems Control (S)	Lee & Shin	0 08 044210 2
Automated Systems Based on Human Skill and Knowledge (S)	Stahre & Martensson	0 08 044291 9
Automatic Systems for Building the Infrastructure in Developing Countries (Knowledge and Technology Transfer) (W)	Dimirovski & Istefanopulos	0 08 044204 8
Intelligent Assembly and Disassembly (W)	Borangiu & Kopacek	0 08 044065 7
New Technologies for Automation of the Metallurgical Industry (W)	Wei Wang	0 08 044170 X
Advanced Control of Chemical Processes (S)	Allgöwer & Gao	008 044144 0

Customers wishing to obtain details of all available IFAC volumes, should contact their nearest Elsevier office or check the IFAC Publications website (www.elsevier.com/locate/ifac).